GEOTECHNICAL SPECIAL PUBLICATION NO. 90

GEO-ENGINEERING FOR UNDERGROUND FACILITIES

PROCEEDINGS OF THE THIRD NATIONAL CONFERENCE

SPONSORED BY
The Geo-Institute of the American Society of Civil Engineers

June 13–17, 1999
University of Illinois at Urbana–Champaign

EDITED BY
Gabriel Fernandez
Robert A. Bauer

1801 ALEXANDER BELL DRIVE
RESTON, VIRGINIA 20191–4400

Abstract: The papers collected in these proceedings summarize the practice of soil mechanics and engineering geology in the design and construction of underground facilities. The topics of the papers are related to the design of excavation support systems; the existing precedent in various geological environments; the development of recent excavation techniques (microtunneling) and the advance of ground modification efforts. The significance of these issues is illustrated by the description and evaluation of anticipated and actual behavior in several well-documented case histories. Other topics included in the papers relate to basic principles in the design of underground storage of pressurized fluids, with emphasis in the selection of an appropriate location and the treatment of the excavated walls in these openings; evaluation and treatment to control surface manifestations of underground excavation; evolving natural openings; use of specialized underground construction (slurry walls) and/or materials (geosynthetics) for geo-environmental remediation; and contractual aspects of underground construction including their interaction with the design.

Library of Congress Cataloging-in-Publication Data

Geo-engineering for underground facilities; proceedings of the 3rd conference: June 13-17, 1999 / sponsored by the Geo-Institute of the American Society of Civil Engineers, University of Illinois at Urbana-Champaign; edited by Gabriel Fernandez, Robert A. Bauer.
    p. cm. —(Geotechnical special publicaiton; no. 90)
    Includes bibliographical references and index.
    ISBN 0-7844-0434-8
    1. Engineering geology Congresses. 2. Underground construction Congresses. 3. Soil mechanics Congresses. I. Fernandez, Gabriel, 1949– . II. Bauer, Robert A., 1952– .
III. American Society of Civil Engineers. Geo-Institute. IV. Series.
TA703.5.042 1999
642.1′9—dc21   99-27427
                                                                                                             CIP

Any statements expressed in these materials are those of the individual authors and do not necessarily represent the views of ASCE, which takes no responsibility for any statement made herein. No reference made in this publication to any specific method, product, process or service constitutes or implies an endorsement, recommendation, or warranty thereof by ASCE. The materials are for general information only and do not represent a standard of ASCE, nor are they intended as a reference in purchase specifications, contracts, regulations, statutes, or any other legal document. ASCE makes no representation or warranty of any kind, whether express or implied, concerning the accuracy, completeness, suitability, or utility of any information, apparatus, product, or process discussed in this publication, and assumes no liability therefore. This information should not be used without first securing competent advice with respect to its suitability for any general or specific application. Anyone utilizing this information assumes all liability arising from such use, including but not limited to infringement of any patent or patents. *Photocopies:* Authorization to photocopy material for internal or personal use under circumstances not falling within the fair use provisions of the Copyright Act is granted by ASCE to libraries and other users registered with the Copyright Clearance Center (CCC) Transactional Reporting Service, provided that the base fee of $8.00 per article plus $.50 per page is paid directly to CCC, 222 Rosewood Drive, Danvers, MA 01923. The identification for ASCE Books is 0-7844-0434-8/99/ $8.00 + $.50 per page. Requests for special permission or bulk copying should be addressed to Permissions & Copyright Dept., ASCE.

Copyright © 1999 by the American Society of Civil Engineers, All Rights Reserved.
Library of Congress Catalog Card No: 99-27427   ISBN 0-7844-0434-8
Manufactured in the United States of America.

Geotechnical Special Publications

1. *Terzaghi Lectures*
2. *Geotechnical Aspects of Stiff and Hard Clays*
3. *Landslide Dams: Processes, Risk, and Mitigation*
4. *Tiebacks for Bulkheads*
5. *Settlement of Shallow Foundation on Cohesionless Soils: Design and Performance*
6. *Use of In Situ Tests in Geotechnical Engineering*
7. *Timber Bulkheads*
8. *Foundations for Transmission Line Towers*
9. *Foundations & Excavations in Decomposed Rock of the Piedmont Province*
10. *Engineering Aspects of Soil Erosion, Dispersive Clays and Loess*
11. *Dynamic Response of Pile Foundations–Experiment, Analysis and Observation*
12. *Soil Improvement: A Ten Year Update*
13. *Geotechnical Practice for Solid Waste Disposal '87*
14. *Geotechnical Aspects of Karst Terrains*
15. *Measured Performance Shallow Foundations*
16. *Special Topics in Foundations*
17. *Soil Properties Evaluation from Centrifugal Models*
18. *Geosynthetics for Soil Improvement*
19. *Mine Induced Subsidence: Effects on Engineered Structures*
20. *Earthquake Engineering & Soil Dynamics II*
21. *Hydraulic Fill Structures*
22. *Foundation Engineering*
23. *Predicted and Observed Axial Behavior of Piles*
24. *Resilient Moduli of Soils: Laboratory Conditions*
25. *Design and Performance of Earth Retaining Structures*
26. *Waste Containment Systems: Construction, Regulation, and Performance*
27. *Geotechnical Engineering Congress*
28. *Detection of and Construction at the Soil/Rock Interface*
29. *Recent Advances in Instrumentation, Data Acquisition and Testing in Soil Dynamics*
30. *Grouting, Soil Improvement and Geosynthetics*
31. *Stability and Performance of Slopes and Embankments II*
32. *Embankment of Dams–James L. Sherard Contributions*
33. *Excavation and Support for the Urban Infrastructure*
34. *Piles Under Dynamic Loads*
35. *Geotechnical Practice in Dam Rehabilitation*
36. *Fly Ash for Soil Improvement*
37. *Advances in Site Characterization: Data Acquisition, Data Management and Data Interpretation*
38. *Design and Performance of Deep Foundations: Piles and Piers in Soil and Soft Rock*
39. *Unsaturated Soils*
40. *Vertical and Horizontal Deformations of Foundations and Embankments*
41. *Predicted and Measured Behavior of Five Spread Footings on Sand*
42. *Serviceability of Earth Retaining Structures*
43. *Fracture Mechanics Applied to Geotechnical Engineering*
44. *Ground Failures Under Seismic Conditions*
45. *In Situ Deep Soil Improvement*
46. *Geoenvironment 2000*
47. *Geo-Environmental Issues Facing the Americas*

48	*Soil Suction Applications in Geotechnical Engineering*
49	*Soil Improvement for Earthquake Hazard Mitigation*
50	*Foundation Upgrading and Repair for Infrastructure Improvement*
51	*Performance of Deep Foundations Under Seismic Loading*
52	*Landslides Under Static and Dynamic Conditions–Analysis, Monitoring, and Mitigation*
53	*Landfill Closures–Environmental Protection and Land Recovery*
54	*Earthquake Design and Performance of Solid Waste Landfills*
55	*Earthquake-Induced Movements and Seismic Remediation of Existing Foundations and Abutments*
56	*Static and Dynamic Properties of Gravelly Soils*
57	*Verification of Geotechnical Grouting*
58	*Uncertainty in the Geologic Environment*
59	*Engineered Contaminated Soils and Interaction of Soil Geomembranes*
60	*Analysis and Design of Retaining Structures Against Earthquakes*
61	*Measuring and Modeling Time Dependent Soil Behavior*
62	*Case Histories of Geophysics Applied to Civil Engineering and Public Policy*
63	*Design with Residual Materials: Geotechnical and Construction Considerations*
64	*Observation and Modeling in Numerical Analysis and Model Tests in Dynamic Soil-Structure Interaction Problems*
65	*Dredging and Management of Dredged Material*
66	*Grouting: Compaction, Remediation and Testing*
67	*Spatial Analysis in Soil Dynamics and Earthquake Engineering*
68	*Unsaturated Soil Engineering Practice*
69	*Ground Improvement, Ground Reinforcement, Ground Treatment: Developments 1987-1997*
70	*Seismic Analysis and Design for Soil-Pile-Structure Interactions*
71	*In Situ Remediation of the Geoenvironment*
72	*Degradation of Natural Building Stone*
73	*Innovative Design and Construction for Foundations and Substructures Subject to Freezing and Frost*
74	*Guidelines of Engineering Practice for Braced and Tied-Back Excavations*
75	*Geotechnical Earthquake Engineering and Soil Dynamics III*
76	*Geosynthetics in Foundation Reinforcement and Erosion Control Systems*
77	*Stability of Natural Slopes in the Coastal Plain*
78	*Filtration and Drainage in Geotechnical/Geoenvironmental Engineering*
79	*Recycled Materials in Geotechnical Applications*
80	*Grouts and Grouting: A Potpourri of Projects*
81	*Soil Improvement for Big Digs*
82	*Risk-Based Corrective Action and Brownfields Restorations*
83	*Design and Construction of Earth Retaining Systems*
84	*Effects of Construction on Structures*
85	*Application of Geotechnical Principles in Pavement Engineering*
86	*Big Digs Around the World*
87	*Jacked Tunnel Design and Construction*
88	*Analysis, Design, Construction, and Testing of Deep Foundations*
89	*Recent Advances in the Characterization of Transportation Geo-Materials*
90	*Geo-Engineering for Underground Facilities*

PREFACE

Construction of the Chicago Subway in 1939 served as a proving ground for the emerging discipline of soil mechanics and underground construction today remains a fertile field for all disciplines of geotechnical and engineering geology practice.

In recent years, ground improvement techniques, including various forms of grouting, have been developed and expanded to deal with restrictive construction environment and difficult ground conditions. Ground improvement has also been applied to reduce the subsidence effects resulting from both natural and mined openings.

The papers collected in these proceedings summarize the practice of rock and soil mechanics and engineering geology in the design and construction of underground facilities. The topics of the papers are related to the design of excavation support systems, the existing precedent in various geological environments, the development of recent excavation techniques (microtunneling) and the advance of ground modification efforts. The significance of these issues is illustrated by the description and evaluation of anticipated and actual behavior in several well documented case histories. Other topics are related to basic principles in the design of underground storage of pressurized fluids, with emphasis in selection of an appropriate location and the treatment of the excavated walls in these openings.

Papers on evaluation and treatment to control subsidence are also included. The use of specialized underground construction (slurry walls) and/or materials (geosynthetics) for geo-environmental remediation is also treated in several papers. Finally, these proceedings also include papers which cover the contractual aspects of underground construction and management of risk.

The editors acknowledge the contribution of the organizing committee in the selection of the theme, main topics of the conference and papers to be presented. The members are:

Edward J. Cording, Co-Chair
Gabriel Fernandez, Co-Chair
University of Illinois at Urbana-Champaign

Kerry Allen
Dywidag-Systems International
Bernard Amadei
University of Colorado
Robert A. Bauer
Illinois State Geological Survey
Robert D. Bennett
USAE Waterways Experiment Station
Norma Cording
Urbana, IL

Amanda Ellioff
Engineering Management Consultants
Richard J. Finno
Northwestern University
John E. Garlanger
Ardaman & Associates, Inc.
Joseph Guertin
GZA Consultants
William H. Hansmire
Jacobs Associates

Youssef Hashash
University of Illinois at Urbana-Champaign
S. Scott Litke
ADSC: The International Association of Foundation Drilling
Russell K. McFarland
Nuclear Waste Technical Review Board
Harvey W. Parker
Harvey Parker & Associates
Jerry F. Parola
Case Foundation Co.
Michael Vitale
Montgomery Watson
W.D. (Toby) Wightman
Kiewit Construction Co.
Local Arrangements
James Onderdonk
Conferences and Institutes
University of Illinois at Urbana-Champaign

We thank all the peer reviewers who made valuable contributions. They are:

Lee Abramson	Vojtech Gall	Gholamreza Mesri
Thomas Anderson	John Garlanger	Gregory Miller
Mauricio Angulo	Michael Garlich	James Monsees
Roy Armstrong	Robert Gilbert	Levent Ozdemir
Robert Bauer	Safdar Gill	Jerry Parola
Barry Beck	Paul Hadala	John Piggott
R. David Bennett	William Hansmire	Eric Rehwoldt
Craig Benson	Youssef Hashash	David Rempe
Marco Boscardin	Michael Hendron	Mary Roth
John Bowders	Ronald Heuer	Lawrence Roth
Gary Brierley	Steven Hunt	Henry A. Russell
Ray Castelli	Mohammad Irshad	Kurt Staudtmeister
Alfonso Castro	Juan Carlos Jaramillo	Ray Sterling
Edward J. Cording	Jon Kaneshiro	Robert Thoms
H.W. Diamond	Stephen Klein	Erol Tutumluer
Prakash M. Donde	Gary Kuhns	Michael Vitale
Charles Dowding	Hugh Lacey	Stan Vitton
Amanda Elioff	Tracy Lundin	Stanley Walker
Augusto Espinosa-Silva	Gerry Marino	William Walton
Randy Essex	Gordon Mathesson	David E. Weatherby
Gabriel Fernandez	Ross Mcgillivrary	John Wolosick
Richard Finno		

Thank you also for all the hard work of the authors.

Gabriel Fernandez
University of Illinois at Urbana-Champaign

Robert A. Bauer
Illinois State Geological Survey

INTRODUCTION

The papers in this volume are focused on the conference topic "Geo Engineering for Underground Facilities" and can be separated into three categories: urban excavation and tunneling, underground conveyance and storage, and natural and mine-induced subsidence. Keynote presentations highlight the conference theme:

Integration of theory and practice in characterization, investigation of behavior, and use of fundamentals and precedent.

in the design and construction of underground facilities. Such a perspective is relevant in all areas of geo-engineering and thus appropriate for a national conference of the Geo Institute. Also of general interest is a wide range of geo-engineering topics presented in the short courses, workshops, exhibits, demonstrations, and special sessions of the conference.

Characterization

The ground in which we work cannot be specified, it must be explored. It is not fully described from borings or tests on samples. A geologic perspective, at the scale of the project, is key to the understanding of the character and variation of the ground. The geologic perspective should permeate all phases of the decision-making process, not only during exploration but throughout design and construction.

Investigation of behavior

Geo-engineering design does not proceed from a code or a classification system, useful as they may be, but from an understanding of the behavior of the ground and the structure within it — the mechanism of deformation, flow and failure. Behavior is influenced by the character and variability of the ground and by the processes, such as excavation, support, ground modification, and drainage, applied to it. These characteristics and processes occur at the scale of project and their effect on behavior must be observed at the project site.

Use of fundamentals and precedent

We build on our experience, using sound fundamentals of mechanics and careful observations of behavior to place the precedents in proper perspective. Understanding the observed behavior allows us to address the issues and manage the risks posed by new projects that extend the range of behavior beyond the precedents.

Challenges and opportunities

How will these principles be applied as we face new challenges and opportunities and employ new tools?

A major challenge is presented by increasing urbanization and the multiple demands placed on the use of urban space. Increasingly, the underground will be used for utilities, storage, transportation and communication systems, in order to free the surface for other activities. The need to replace, upgrade and expand aging infrastructure in major urban areas has resulted in vast underground projects of great complexity. Challenges reside in the design and execution of underground projects that must be constructed under increasingly tight constraints, in more difficult ground conditions, in close proximity to existing infrastructure, for a public increasingly less tolerant of even temporary disturbance to their environment.

Challenges and opportunities are also presented by the development of new tools. One such tool is numerical modeling, which is increasingly used to simulate ground behavior in geo-engineering problems. As the complexity and sophistication of these models grow, it becomes increasingly difficult yet increasingly important for the user to be able to understand the assumptions of the model and check the fundamental relationships controlling its behavior. The conference panel session on numerical methods is intended to produce a dialogue on these issues. As the simulations of numerical models become more realistic, it becomes more important to recognize that they are a simulation of behavior and not a replacement for the observation and measurement of actual behavior. Underground engineering is essentially an empirical endeavor and behavior must be observed and understood in order to be controlled in the design and construction process. Numerical models are a tool rather than an objective in this process.

Challenges and opportunities are provided by new developments in instrumentation for exploration, observation and monitoring. In the early 70's, instruments, for the first time, became capable of providing an adequate three-dimensional view of the pattern of ground movements during tunneling and excavation. Often, instruments failed because they were not suited to the underground environment and were not adequately field tested. To avoid unreliable electronic recording and transmission of data required using simple mechanical instruments that were difficult to gain access to read. Great benefits are being derived from more reliable, field-proven instruments. Some instruments can be calibrated and checked in situ. Micro-processors allow collection, processing, storage, and digital transmission of data within the instrument. On some projects, procedures have been developed to read instruments, collect the data in a central location and process them in a form that can be transmitted to remote terminals. Difficulties arise in understanding the significance of such data when the frequency of readings are not coordinated with construction events and the link to construction observations is missing. Instrumentation is a tool and not a replacement for the critical observations of a practiced eye.

Challenges and opportunities are presented by the interactions required to design and build underground. A tunnel is a structure in which great benefit is derived from the interaction between the structure and the surrounding ground, and, as Terzaghi noted in 1942, the designer who ignores this fact and does not take care of the ground must pay for it. In an analogous way, great benefits are derived from the interaction among the disciplines and parties engaged on an underground project. Without interaction, the contractor could be responsible for whatever initial support system and excavation method he wishes to use, based on his own assessment of ground conditions, in order to provide a safe working environment and maximize his profits. The designer would specify a final support system and lining that makes no assumptions as to what contractor placed, and would be fully capable of supporting the ground and functioning over the intended life of the program, no matter the condition in which it was left in after the initial support was placed. The above is a traditional approach and one of the simplest contractual arrangements. It is rarely used in its purest form for several reasons. On most projects, the way the ground is opened and supported is the most critical issue affecting the bid price, the progress of the job, the cost of the project, the impact on third parties, the installation of the final lining, and the nature of construction claims. Although the contractor is responsible for means and methods, many of the owner's interests are affected by the contractor's operations. The ground behavior itself is controlled by the interaction between the owner's site conditions and the contractor's means and methods, a condition which makes determination of differing site conditions difficult and often leads to disputes.

In the early 70's, such controversies caused the underground construction industry to become a pioneer in the development of contractual practices that would allow timely resolution of disputes without litigations. Recent efforts have been concentrated on contracting procedures that provide a team effort and more fully engage the expertise and energy of all the members of the team. A conference workshop and session on contracting practices are focused on procedures for managing and equitably assigning risk among the parties to the contracts.

The basic theme and main topics of this conference were conceived to promote the discussion, evaluation and analysis of the issues outlined above. Keynote speakers were chosen for their preeminent representation of an engineering philosophy that combined fundamentals, use of precedent, and construction knowledge to accomplish complex engineering endeavors. The papers and presentations reflect key aspects of the technical and contractual issues involved in underground projects.

Finally, we hope that this conference provides valuable opportunities for the exchange of knowledge among geotechnical engineers, engineering geologists, underground constructors, municipal engineers, users and regulators — all those who participate in geo-engineering for the underground.

        E. Cording
        G. Fernandez
        For the Organizing Committee
        Urbana, Illinois, 1999

ACKNOWLEDGMENTS

The editors thank Ms. Denise Ice for her assistance in setting up tracking systems and files for all papers, and papers reviews, and for handling most of the correspondence and communication in securing final papers and necessary attachments. We also thank Ms. Joyce Snider and Ms. Mary Ann Speck for their assistance with the collection of abstracts, acceptance letters and document generation.

Contents

Keynote Addresses

Six Decades of Subway Geo-Engineering: The Interplay of Theory and Practice 1
    Ralph B. Peck

The Art of the Possible 16
    Jack K. Lemley

Urban Underground Issues

A History of Deep Foundations in New York City: "Could We Build It Now?" 24
    Reuben Samuels

Major Sewer Tunnel Construction beneath Metro Tunnels in Singapore 29
    Richard Flanagan, Birger Schmidt and Tan Boon Tee

...And Back to Theory 41
    Augusto Espinosa-Silva

Tunnels through the History of Seattle 53
    Lee Abramson and Ralph Boirum

Geological Issues in Underground Conveyance and Storage

Geologic and Geotechnical Considerations for Pressure Tunnel Design 66
    Andrew H. Merritt

Groundwater Control for the Los Angeles Metro System beneath the
Santa Monica Mountains 82
    Roy F. Cook and Stuart Warren

Applications of Horizontal Sampling and Logging Technologies in Geotechnical
Site Investigations 93
    Samuel T. Ariaratnam, Erez N. Allouche, Kevin W. Biggar,
    Renato V. Clementino, and Peter K. Robertson

Slurry Walls for Control of Contaminant Migration: A Comparison of
United Kingdom and United States Practices 105
    Jeffrey C. Evans and Andrew R. Dawson

Ground Deformations Adjacent to a Soil–Bentonite Cutoff Wall 121
    George M. Filz, Diane Y. Baxter, David J. Bentler, and Richard R. Davidson

Hydraulic Conductivity Evaluation of Vertical Barrier Walls 140
    David E. Daniel and Hangseok Choi

Excavation Support Systems: Options

Practice and Performance of Excavation Support Systems in the Chicago Area 162
    Jerry F. Parola and Robert G. Lukas

Excavation Support Practices in the Pacific Northwest 176
   Tom A. Armour

High Capacity Micropiles: Basic Principles and Case Histories 188
   D.A. Bruce, M.E.C. Bruce, and R.P. Traylor

Suspension Wall Shoring System 200
   Chris J. Wolschlag, R. John Byrne, and David M. Cotton

Soil Nailing for the WHEDA Office Building in Madison, Wisconsin 212
   Eric W. Bahner and Joe Sirvinskis

Modeling Deformations around Excavations in Soft Clays 231
   Laurent X. Luccioni, Juan M. Pestana, and Demetrious C. Kousoftas

An Arch-Shaped Culvert Built by the Inflatable Form 243
   Karl H. Lewis, Carl Henderson, and Cathy Bażan-Arias

Site Characterization for Conveyance Tunnels

Identifying and Baselining Boulders for Underground Construction 255
   Steven W. Hunt and Mauricio Angulo

Tomography To Evaluate Site Conditions during Tunneling 271
   David M. Neil, Khamis Y. Haramy, David H. Hanson, and Jozef M. Descour

Characterization of Uncertainty in Underground Structures 282
   N.O. Nawari and R. Liang

Tomographic Imaging of Soil–Tunnel Interaction 294
   J. Carlos Santamarina and Americo L. Fernandez

Seismic Measurements To Investigate Disturbed Rock Zones 303
   B.A. Luke, K.H. Stokoe II, J.A. Bay, N.J. Lee, and P.P. Nelson

Mine-Induced Subsidence

Flooding-Induced Mine Subsidence at Restof, New York 315
   Leo L. Van Sambeek

Replacing a Bridge Adjacent to a Deep-Seated Sinkhole in Glacial Deposits 327
   Stanley E. Walker, Randall C. Divito, and Maureen S. Valentine

Measured Horizontal and Vertical Movements from a Sag Subsidence 339
   Gennaro G. Marino and James W. Mahar

Mine Shaft Stabilization Using Compaction Grouting 353
   Len Meier and Arthur G. Hoffmann

Time Domain Reflectometry (TDR) Monitoring of Overburden Deformation
To Verify Modeled Behavior 365
   Kevin M. O'Connor and John A. Siekmeier

Excavation Support Systems: Design

Excavations in San Francisco Bay Mud: Design for Deformation Control 377
   Demetrious C. Koutsoftas

Rationalizing the Practice of Strut Preloading for Braced Excavations.........................393
    S.J. Boone, H. Bidhendi, J. Westland, and M. Grabinsky

Design and Behavior of One-Tier Tieback Walls..405
    David E. Weatherby and Thomas C. Anderson

Toe Penetration Requirements for Tieback Walls...417
    C.G. Mueller, J.H. Long, and E.J. Cording

Evolution of Design: Los Angeles Metro Underground Structures.....................................429
    James E. Monsees and Amanda Elioff

Tunnel Design

Design of Pressure Tunnels..442
    Jeppe N. Eskilsson

Pressure Tunnels in Fractured Rock: Minimum Cover Criteria from the
Stability of Rock Wedges..459
    Tirso A. Alvarez, Jr., Edward J. Cording, and Gabriel G. Fernández

Stability Analyses of Underground Openings Using a Multiaxial Failure
Criterion...471
    Li Li, Michel Aubertin, and Richard Simon

Stress in Buried Pipes from Shear Distortions..483
    C.A. Davis

Determination of Vertical Loading on Underground Structures Based on an
Arching Evolution Concept..495
    Geraldo R. Iglesia, Herbert H. Einstein, and Robert V. Whitman

Analysis of Roof Truss for Underground Support..507
    Fusheng Zhu and Dae S. Young

Numerical Modeling for Underground Projects

Finite Element Analysis of Ground Response Due to Tunnel Excavation in Soils..........514
    Murad Y. Abu-Farsakh and Mehmet T. Tumay

Tunnel Support Design with Finite Element Analysis..526
    W.G. Louhenapessy

Numerical Analysis of the Powerhouse Cavern Setting for a Pumped Storage
Project..538
    Nasim Uddin

Transportation Tunnels

Design and Construction of the City Link Driven Tunnels...553
    Daniel N. Adams, Michael T. McRae, Bruce J. Hutchinson, and
    Stephen J. Porter

Development of Design and Construction Concepts for Jacked Tunnel Sections
of the Interstate-93/Interstate-90 Interchange: Central Artery/Tunnel Project,
Boston, Massachusetts..570
    Phillip M. Rice, Peter A. Mainville, Stephen Taylor, and Alan J. Powderham

Design and Construction of the Dallas Area Rapid Transit (DART) Starter Line
Tunnel Contract NC-1B in Dallas, Texas 582
    Gunars Richters, Charles Hood, and Tracy K. Lundin

Design and Performance of Large Tunnel Constructed in Saprolite 594
    William H. Hansmire, Lee W. Abramson, Glenn M. Boyce, and
    Clayton S. Mimura

Soft Ground Tunneling for Taipei Rapid Transit Systems 610
    Daniel H. Ju, Za-Chieh Moh, and Richard Hwang

Design and Performance of Tunnel Linings

Underground Concrete Construction 622
    Daniel P. O'Connor

Experience with Steel Seal Membranes for Liners in Pressure Shafts and Tunnels 634
    Alberto Marulanda and Ramiro Gutiérrez

Seepage Control Measures for an Underground Powerstation in a Semi-Arid
Region 647
    Jason E. Hedien

Seismic Performance of the Yerba Buena Island Tunnel 659
    Hubert K. Law and Ignatius P. Lam

Natural Sinkholes

Eagle Ridge Mall Sinkhole: Haines City, Florida 671
    James D. Hussin, Joseph A. Eduardo, and Dean A. Elliott

Geophysical Surveys of a Known Karst Feature: Oak Ridge Y-12 Plant,
Oak Ridge, Tennessee 684
    William E. Doll, Jonathan E. Nyquist, Philip J. Carpenter,
    Ronald D. Kaufmann, and Bradley J. Carr

Case Study: Site Characterization Methods in Karst 695
    J.R. Mackey, M.J.S. Roth, and J.E. Nyquist

Microtunneling

Geotechnical Engineering Aspects of Microtunneling in Soft Ground 706
    David C. Mathy, Robert A. Kahl, and Dru R. Nielson

Folsom East 2 Construction Proving Project: Field Evaluation of Alternative
Tunneling Methods in Cobbles and Boulders 720
    Kimberlie Staheli, David Bennett, Michael A. Maggi, Michael B. Watson,
    and Bruce J. Corwin

Jacking Loads Associated with Microtunneling 731
    David Bennett and Edward J. Cording

Tunneled Sewers in Houston, Texas 746
    Michael Hasen

Design and Planning of Urban Underground Construction Using Pipe Bursting
Techniques 756
    Samuel T. Ariaratnam, Jason S. Lueke, and Peter Strychowskyj

Columbia Slough Consolidation Conduit Tunnel 768
    Arlan H. Rippe, Michael Feroz, and Frank S. Buehler

Storage Cavern Design

Non-Halites and Fluids in Salt Formations: Effects on Cavern Storage
Operations 780
    Robert L. Thoms and Richard M. Gehle

Risk Assessment and Safety Criteria for Underground Caverns 797
    Massimo Guarascio and Carmela Di Bella

Experience in Underground Storage of Crude Oil in Salt 810
    James K. Linn and Jon Culbert

Design of a Deep Cavern Intersected by Weak Shale 821
    Bhaskar B. Thapa and Peter A. Dickson

Expected Performance of Thermal Stress Relief Elements for Ground
Support at the Yucca Mountain Repository 829
    Yiming Sun, Richard M. Nolting III, and John Cogan

Geosynthetics for Underground Applications

Waterproofing Tunnels with Geosynthetics in Europe 841
    Alberto M. Scuero, John A. Wilkes, and Gabriella Vaschetti

Control of Ground Behavior

Design and Performance of a Deep-Piled Circular Cofferdam 853
    Alan Powderham

Long-Term Behavior of Tunnels in Chicago Clay 866
    Nelson Kawamura and Edward J. Cording

Field Measurements during Construction of a Sewage Tunnel 879
    F.M. Abdrabbo, F.M. El-Nahhas, and H.A. Abd El-Lateef

Performance of Multiple Retention Systems during Cut and Cover Tunnel
Construction 888
    Ted D. Bushell, Daniel W. McCarthy, and Elaine McCluskey

Performance of Inclinometers in a Diaphragm Wall 900
    Donald D. Liou

Tunnel Rehabilitation and Grouting

Tunnel Rehabilitation 912
    Lee W. Abramson and Marco D. Boscardin

Rehabilitation of Tunnel Liners with Shotcrete 925
    Henry A. Russell

Rock Grouting: Contemporary Concepts in Materials, Methods, and
Verification 936
    D.A. Bruce, A. Naudts, and C. Gause

Polyurethane Grouting for Sealing Leakages in Tunnels 950
    Bert P. Kriekemans

Groundwater Control

Groundwater Control at Difficult Geological Interfaces 957
    Arthur B. Corwin, Thomas J. Tuozzolo, and Paul C. Schmall

Characteristics of Urethane-Grouted Sand for In Situ Seepage Control 969
    M.A. Gabr

Successful Urban Rock Tunneling in Spite of Encountering Gasoline, Cleaning
Solvents, and Methane Gas 983
    Raymond W. Henn and David Rogstad

Design Considerations Due to Groundwater Contamination: Minnesota Library
Access Center 995
    Stephen T. Jansen, Bruce D. Wagener, and Gordon J. Girtz

Ground Improvement

An Update on Deep Mixing Technology Worldwide 1007
    D.A. Bruce and M.E.C. Bruce

Jet Grouting in Cohesive Soils 1016
    Lawrence B. Gruner

Design and Construction of a Jet-Grouted Barrier Wall 1027
    D.M. White, M. Koelling, D.W. Ashcom, and J. Kurrus

Short Aggregate Piers Reinforce Soils Near Tunnels 1039
    Richard L. Handy, Nathaniel S. Fox, and Kord J. Wissman

Deep Mixing Technology for Deep Excavation 1048
    Ali Porbaha, Hideyuki Asada, and M. Javad Fatemi

Case Histories

Case Study of Support of Critical Utilities During Excavation 1060
    Julie Sharp, Ted Splitter, and John Sturman

Prediction of Settlement in Bangkok Ground Due to Tunnel Excavation 1070
    Tiew Wannipa and Teachavorasinskun Supot

A Field Observation and Prediction in Soft Ground Tunnel 1080
    Santiago F. Zapata, Jeffry Budiman, and Theodore Maynard

Adding Realism to the Planning of Rock Tunnel Boring Machine Projects 1091
    Christopher Laughton, Priscilla Nelson, and Yousof Abd Al-Jalil

Shoring for Leslie Station: Design Assessment and Construction Performance 1102
    J. Westland, S.J. Boone, P. Branco, D. MacDonald, and M. Meschino

True Behavior of High Pressure Tunnel for the Guavio Hydroelectric Project............ 1116
  Alberto Marulanda and Ramiro Gutiérrez

Geotechnical Lessons Learned for San Diego's South Bay Ocean Outfall.................. 1130
  Jon Y. Kaneshiro, Luciano Meiorin, Stephen J. Navin, and Svante Hjertberg

Contractual Issues

Positioning for Least Cost: Managing Risks in Underground Construction............... 1143
  Lawrence H. Roth

Professional Standards for the Geotechnical Engineer: A Legal Perspective............ 1155
  Karen P. Layng and Justin L. Weisberg

Removal of Subsurface Obstructions: An Alternative Contractual Approach............ 1164
  D.J. Mason III, Robert S.J. Berry, and David J. Hatem

Temporary Support of Buildings: Who Really Assumes the Responsibility?............. 1176
  Andrew J. Ciancia and Kevin J. O'Neill

Perspectives on Risk Assessment in Tunneling..1185
  Donald P. Richards

*Performance Monitoring of Tiebacks Using Instrumented Cables
  Ian F. Mountfort, W.F. Bawden, A.J. Hyett, M.W. Grabinsky, and P. Lausch

Indexes

Subject Index...1205

Author Index..1211

*Manuscript not available at time of publication.

xvii

SIX DECADES OF SUBWAY GEO-ENGINEERING
THE INTERPLAY OF THEORY AND PRACTICE

by

Ralph B. Peck, Hon. M. ASCE[1]

ABSTRACT

Highlights and reminiscences of advances in the art of tunneling for subway systems are recounted over a period of sixty years, from Chicago to BART, WMATA, Baltimore, and Los Angeles.

INTRODUCTION

This is not a technical treatise on subway tunneling. It is, rather, a series of only slightly technical reminiscences of one engineer's 6-decade love affair with subway tunnels.

It didn't start out that way. When I was a student at R.P.I. in the early 1930's I had every intention of being, like my father, a bridge engineer. My heroes were men like John and Washington Roebling, James B. Eads, Theodore Cooper, Lefferts Buck, Ralph Modjeski, and J.A.L. Waddell. Located, as R.P.I. was, only 150 miles upriver from New York City, we enjoyed lectures by D. B. Steinman on his bridge at Florianopolis in Brazil and the St. Johns Bridge over the Willamette River near Portland; by Shortridge Hardesty about vertical lift bridges; by Clarence Dunham about the Triborough Bridge then under construction; and by engineers

---
[1] Prof. Emeritus Foundation Engineering, Univ. of Illinois at Urbana-Champaign; Consulting Engineer, 1101 Warm Sands Dr. SE, Albuquerque NM 87123

of the Port of New York Authority about their pace-setting George Washington and Bayonne Bridges. All were marvels of precision, designed on the basis of mechanics and strength of materials. But I remember other lectures. Carlton Proctor, the junior partner of Moran and Proctor, talked about the procedures to restore to verticality one of the caissons for the new bridge across the Hudson at Poughkeepsie; I thought it unseemly that such a tilt should have been allowed to occur. And Ole Singstad talked about the Lincoln Tunnel under the Hudson, a project carrying on the Port Authority's expertise in subaqueous vehicular tunneling. But I found his lecture distressingly unscientific and lacking in engineering rigor. It was, instead, largely about leakage, grommets, and leaving enough muck in the tunnel to keep it from rising out of the river bed. I remember thinking that tunneling was hardly engineering, certainly not a profession in which I could see much of a calling.

All that, of course, was before I was introduced to soil mechanics or had even heard of Karl Terzaghi or Arthur Casagrande. Fortunately for me, Karl Terzaghi and I were both out of work at the same time - he because he had fled Austria as the Nazis were about to impress him into their service, and I because the American Bridge Company ran out of jobs in the recession of 1937 and I had come to Harvard to learn about the new soil mechanics in order to accept a job teaching it. Terzaghi accepted an offer from the City of Chicago to be a consultant on the new subway it was starting to build, and I became his representative on the job. I soon learned that there was much more to tunneling than calculations based on mechanics.

CHICAGO 1939-1943

For many years Chicago had set aside funds for a subway, funds strangely immune to appropriation for other purposes. So when depression threatened in 1939 and the federal government established the Public Works Administration to subsidize projects for relief of unemployment, it was an obvious step to supplement Chicago's traction fund with a federal grant. The grant came with a condition that the first contract be on the street in a matter of weeks. Ralph Burke, a Republican in a Democratic world, was appointed chief engineer. Built on the preliminary plans of the local engineering firm of Kelker and DeLeuw, and with Mr. Burke's background as a tunnel contractor and subsequently chief engineer of the Chicago Park District, the City's first contract, on North State Street, met the deadline.

Chicago was no stranger to fairly large tunnels in soft clays; the Sanitary District had an enviable reputation for its accomplishments, including large interceptor sewers, and the District was the major source of the designers and construction supervisors who entered Mr. Burke's organization. Burke was aware that the progress of the Sanitary District's tunneling, mostly beneath undeveloped property, could be followed by observing the settlement trough that appeared at the surface; he was also acutely aware that the first subway contract started at a large mined station beneath the busy intersection of North State Street and Chicago Avenue. So when he heard Terzaghi's lecture, arranged by the Raymond Concrete Pile Company's district manager, Al Cummings, and titled at Al's suggestion, "Dangers of Constructing Large Tunnels in Soft Clay under Large Cities", Burke outbid the State Street Property Owners Association for Terzaghi's services, took a personal interest in Terzaghi's approach, and threw his full weight behind Terzaghi's suggestions. This included extraordinary support for myself, as Terzaghi's representative on the job, and the crew of the "soils lab", as my small but select staff was called.

I am sure that Terzaghi already knew that whatever settlement occurred of the ground surface or of structures above or alongside the tunnels was the result of corresponding loss of ground into the excavation. Unless this could be demonstrated, however, the contractors could disclaim any of the detrimental consequences. The surface and building settlements were already being routinely measured. Terzaghi suggested that we observe and measure as best we could whatever movements were experienced by the clay around the tunnels, the movements of the lining itself, and the corresponding construction activities inside the tunnel. This we did, and correlated what we found with the unconfined compressive strength, our measure of the stiffness of the clay. The results, resisted at first by the contractors, soon proved their value and convincingly showed the way to improvement. This, without being distinguished by a name, was the observational method.

North State Street in those days was a depressed part of Chicago with few buildings of more than four stories.

State Street in the Loop area to the south was a different matter, the home of Marshall Field's, Carson Pirie Scott's, in short the shopping center of the city. But the soil tests showed clearly that the unconfined compressive strength of the clays along State Street in the Loop was only about half that of the clays at Chicago Avenue. Without any theory, it was obvious that a more conservative tunneling procedure would be required. Tunneling shields had not been used in Chicago, and the local contractors without shield-tunneling experience feared they would be supplanted by outsiders. Mr. Burke convened a board of consultants, of which Terzaghi as well as Col. Brinckerhoff and M. E. Chamberlain from the New York area, were members. Their unanimous recommendation was to use shield tunneling, particularly since the entire downtown stretch was to be one continuous station structure with a much larger cross section than that of the Chicago Avenue station. Thus, without any theory or calculation, a significant decision was reached on the basis of the settlement observations, the soil tests, and the measurements that showed the seats of lost ground in the liner-plate hand-dug tunnels.

I am sure you will have noticed that the crucial conclusions and decisions were reached on an essentially empirical basis. Those of you who are familiar with Terzaghi's papers on the liner-plate and shield tunnels of the Chicago subway will wonder what part the findings in those papers played on the job. Essentially none; the papers were written late in the game. Yet, Terzaghi suggested, and Mr. Burke approved, two test sections directed to the structural design of the tunnel linings. That Ralph Burke could persuade the City Council to appropriate funds (without the cooperation of the PWA) for two such sections, one each in liner-plate and shield construction, is no small indication of the confidence that Mr. Burke had in the merit of Terzaghi's work. Unfortunately, the considerable potential savings associated with the results of the test sections began to be realized only when tunneling resumed on the Congress Street extension after the interruption of World War II.

Indeed, those of you whose knowledge of the Chicago Subway Project was gained from the technical papers about the open-cut and tunnel measurements may have the impression that the subway organization as a whole enthusiastically embraced the findings and recommendations of Terzaghi and the soils group. Far from it. The structural section, under the guidance of a very able French-trained engineer, carried out highly sophisticated elastic analyses of even the most complex station structures, with a group of several score of analysts each operating the then most modern calculating machines. Many loading conditions, dependent on various realistic and not-so-realistic assumptions concerning lateral earth pressures, were taken into account, none of which considered the influence of the deformations of the surrounding clay on the distribution of earth pressure. As a result, the subway structures were so massive and so heavily reinforced that at some locations it was virtually impossible to place the concrete. The incongruity of the results was not lost on Mr. Burke, who noted that the newly designed replacements for several of the existing small freight tunnels that occupied the space beneath many of the streets in the Loop had heavily reinforced sections two feet thick, whereas the originals, already some 40 years old and in excellent condition, were completely unreinforced and only 10 inches thick.

The two test sections, one each for liner plate and for shield tunnels, were made circular in cross section, as flexible as practicable, with the capability of carrying the compressive ring stress without regard to capacity in bending. The underlying concept was that even the softest Chicago clay would be capable of redistributing the pressure around a circular tunnel in such a way as to result in an essentially equal all-around pressure, and thus there would be negligible moments in the lining. The redistribution would occur largely during and shortly after placement of the lining, so no significant bending stress would develop in any "final" lining that might be placed inside the lining that the structural engineers found so necessary to reinforce heavily.

That all was not comfortable in the subway design family is attested by a special issue of the Proceedings of the Western Society of Engineers devoted entirely to the initial Chicago Subway project. Details, including the various assumed loading conditions and corresponding structural analyses, are set out fully, but there is nothing whatsoever in the entire issue about the soil exploration or the work of Terzaghi or the soil laboratory. So the immediate contribution of soil mechanics to tunneling on the Chicago Subway was in the realm of relating construction procedures and soil properties to the effects of the tunneling on the adjacent and overlying facilities. In passing, it is notable that in a review of the condition of the subway structures after a half-century of service, one of the matters requiring attention was the leakage above the downtown station platforms. The leakage had been kept from being a nuisance by installing inconspicuous gutters, but it had taken its toll in corrosion of the mass of rebar resulting from the unrealistic loading conditions, which in turn derived from failing to consider the influence of deformation of the soil on the pressure acting on the structure.

The benefits to Chicago of the two test sections were not realized until after the war when construction was resumed west on Congress Street from Dearborn Street, where the shields had remained for several years. The Congress Street station, built in a deep open cut, marked the end of the shield tunneling; the line to the west reverted, not without difficulty, to liner-plate hand tunneling. But the reinforcement in the new section was nominal in contrast to that in the rest of the system.

The Chicago Subway project in the annals of geotechnical engineering assumed an importance far beyond its benefits at the time, largely because it demonstrated the enormous practical benefits that occur from even crude observations, crude at least in comparison to today's sophisticated instrumentation. Even today, it exemplifies the benefits that may be derived from simple but intelligently interpreted observations.

The Chicago Subway, incidentally, had a little bit of everything: hand-excavated liner-plate tunnels, shield tunnels (both under compressed air), open cut stations, crossovers in tunnel and in open cut, and even a prefabricated tube built to the south of the city, towed to the State Street crossing of the Chicago River and sunk into place. A great experience, mostly in soft clay.

BART 1963-1968

The Bay Area Rapid Transit system was the first major subway project after Chicago, an interval of some 20 years. Instead of only a few miles of downtown tunnels, BART was a regional urban and interurban transit facility largely at grade or on elevated structure, but including some subsurface portions east of the Bay and extensive tunneling in downtown San Francisco. Along Market Street the

GEO-ENGINEERING FOR UNDERGROUND FACILITIES   7

streetcar system, known as the Muni, was placed in tunnels above the rapid transit tunnels, and the then world's longest subaqueous sunken-tube tunnels were located in a trench beneath the Bay connecting San Francisco and Oakland/Berkeley. Unlike the Chicago Subway, which was essentially a downtown underground terminal system for the elevated network beyond, BART encompassed 75 miles of double-track line. Part was in the notorious Bay Mud, not unlike some of the Chicago clays, but the tunnels were driven through a wide variety of materials. And unlike in Chicago, earthquakes and aseismic design were a dominant consideration.

Because of its ductility, steel was selected as the material with which to line the running tunnels, and an early decision was made to let a separate contract to fabricate and furnish the segments to the contractors whenever they were to be used. Thus, the design of the segments became a matter of great economic and technical concern.

The designers were a combination of three widely known and experienced structural engineering firms; together they operated as Parsons Brinckerhoff-Tudor-Bechtel. One of the younger PB engineers was Tom Kuesel, a Yale product of Dimitri Krynine in soils and Hardy Cross in structures. He had studied the results of the two Chicago Subway "flexible" test sections and realized that they offered an attractive basis for design almost irrespective of the type of soil, a basis that would permit standardization of a fabricated circular segmental lining. Tom is reported to tell the story (I regret that I don't remember the incident, but my guest book says it was on April 14, 1964) that he came to Urbana for a one-day, crash course in circular flexible linings and why they would remain stable in even quite soft soils. Satisfied, he took the lead in designing the lining. So the Chicago Subway left a legacy to the Bay Area, and I became one of a group of consultants involved in the project. To be a consultant, I needed to obtain a Civil Engineer's license in California. It was purported to be heavy on seismic design, so I boned up on the California code and requested an oral exam, not without some trepidation. Fortunately (for me) the Corps of Engineers had appointed me to a Board of Consultants to investigate certain aspects of the landslide at Turnagain Heights that had occurred in the 1964 Anchorage Good-Friday Earthquake shortly before, and I had just returned from visiting the site. The examiners found my first-hand account exciting and spent the allotted time inquiring about Anchorage, so I may well owe my stimulating experience with BART to the Alaska Earthquake.

In general it was felt that the segmental steel lining would be inherently capable of accommodating most likely earthquake motions in soil, but rock was another story. The tunnels through the Berkeley Hills crossed the active Hayward Fault. Steel ribs and heavily reinforced concrete provided the primary support, but the tunnels were driven somewhat oversize in the vicinity of the fault so that, if necessary, the tracks could be realigned. Over the years movements have indeed occurred; so far, only slight cracking has been observed.

The Trans-Bay tube, at the time the world's longest sunken tunnel (the Chicago River crossing must have been one of the shortest!), required much detailed study. It profited, among others, from the expertise of Nate Newmark and George Housner in the budding art of aseismic design; I was concerned only with the stability of the trench slopes. Yet, it was a great satisfaction to learn that BART remained operational after the Loma Prieta quake even when the Bay Bridge suffered distress.

BART made early use of tunneling machines, generally with success in the slightly cohesive sandy soils encountered along much of Mission and Market Street. There were lessons to be learned, however. On Mission Street the cutter head turned without corresponding advance of the machine, whereupon a sinkhole developed uncomfortably close to a building. We learned that even a nearly closed face does not provide support under these circumstances. Nevertheless, under a variety of conditions, several types of machines proved themselves, and BART must be regarded as one of the pioneering projects for TBM's.

But I learned my biggest lesson from BART the hard way. As the tunnels approached the Ferry Building at the foot of Market Street, the upper pair of tunnels intended to serve the Muni system ended in the Embarcadero station, where the cars reversed direction and switched tracks; the lower tunnels descended beneath the Ferry Building and curved slightly to the south to connect to the Trans-Bay tubes. The line of the tunnels passed through a whole forest of timber piles. The piles no longer carried any structural loads except for piers that had remained in place when the buildings they supported had been removed. The two tunnels were being advanced by means of open-faced shields with compartments in which men could work. When a pile was encountered, the bay mud in which it was embedded was spaded away from around the pile and for about six inches above the shield's cutting edge, whereupon the pile was cut off by a chain saw or a burring tool. Thus, the piles were expected not to be in contact with the segmental steel lining after the shield advanced beyond the pile locations. The resident engineer was concerned that the piles would eventually descend onto the lining as a result of negative skin friction and possibly damage the lining. I did not share his concern, because the piles extended above the lining only a few feet and they carried no structural load.

On my next visit, the resident engineer could hardly wait to get me into the tunnels. There were major dimples in the lining where the piles were not located directly above a flange or stiffener, and it had been necessary to weld additional stiffeners, cut to fit the contours of the dimples where they occurred. It was an expensive operation (leading, of course, to an impressive claim), and I learned that, even in a fairly soft clay, substantial skin friction can develop in a short length of pile. Just to impress me, the resident engineer had painted all the repairs white. It wasn't a pretty sight.

WMATA 1996–

Starting in about 1966, shortly after BART, the Washington, D.C., rapid transit system has continued construction even to the present time. In contrast to BART, it carries only local urban traffic. Moreover, being a creation of Congress, it has consistently received the necessary appropriations to advance in an orderly fashion, and the design and construction organization has been able to maintain and develop its expertise. It has also maintained the continuity of its Board of Engineering Consultants, which has reviewed the work of the various designers as the designs have developed, and has kept abreast of construction. Successive partners of the geotechnical consultants, Mueser Rutledge Engineers, including Bill Mueser, Phil Rutledge, and Jim Gould, have been personally close to the field and laboratory investigations and have imparted an unusual uniformity and continuity to the geotechnical aspects of the work. The General Engineering Consultants, the DeLeuw, Cather organization, was represented in the earliest days of the project, among others, by Charles DeLeuw and Roy Anderson of Chicago Subway note.

Even in comparison to BART, the geology of the WMATA system is complex. The District of Columbia straddles the Fall Line, which separates the ancient Precambrian rocks and their weathering products from the younger Paleozoic rocks, and still younger Pleistocene and recent sediments are often present. As the system grew geographically, the advantage of a single, consistent geotechnical picture became evident. The idea of Geotechnical Design Summary Reports (GDSR) was nurtured and developed as the work progressed.

Among the noteworthy advances during the quarter-century of WMATA's history of construction and operation has been the evolution of means to combat groundwater. As I read again the minutes of the Board of Engineering Consultants over several decades of design, construction, and operation, I was struck by the persistence of the groundwater problems, some making themselves known early in construction, and some during operation. As successive segments were completed and put into operation, the focus broadened from construction difficulties to the annoyances and maintenance problems associated with leakage into the tunnels during operation. It became evident that concrete linings themselves were inadequate to prevent the leakage, that drain pipes installed to collect and direct water to invert or below-invert drains became clogged by mineral and bacterial deposits, and that measures to prevent the clogging were unsatisfactory even in the short term. Eventually it became apparent that the structural support, whether consisting of rock bolts and shotcrete or precast concrete segments, should be supplemented by an inner lining, usually of precast concrete segments, with a high-density polyethylene membrane between. The HDPE lining constituted a heat-welded unit that required great care during construction to avoid punctures and tears, especially where the continuity of running tunnels was interrupted by penetrations such as shafts or stations. The savings in maintenance costs, especially of electrical equipment, have more than justified the additional construction costs. The geotechnical requirements for adequate support remained undiminished, of course, but construction procedures and design had to satisfy the additional mandate for dry operating conditions.

The requirements for the design of the stations, conceived by architect Harry Weese, involved large vaulted precast concrete shells erected inside still larger openings, some comparable in size to major underground power plants but usually with little rock cover, and with the rock itself often appreciably affected by faulting, surface weathering and stress relaxation. Rock mechanics came into its own under the strong influence of Don Deere on the Board of Engineering Consultants. In this development the University of Illinois played a significant role in hands-on assessment of rock conditions during progress of construction, in instrumentation and interpretation of the data, and in rationalizing the findings during and after construction. A whole generation of graduate students made their contribution to the project under the direction of Dr. Ed Cording and themselves entered practice or became influential disciples and teachers of the observational method in applied rock mechanics.

Earth tunneling ran the range of possibilities, from the softest of clays to a wide variety of granular materials, and mining methods ranged from the use of a variety of shields to TBM's to NATM. Again, irrespective of the method, the struggle to anticipate and control groundwater conditions persisted. For all, WMATA was a great learning experience and, in the end, a great accomplishment. Not the least in the learning experience was the series of in-house reviews that methodically examined experience with the view of improving techniques, achieving economy, and enhancing safety.

Obviously, a project of such scope, variety of subsurface conditions, and long duration of construction cannot be discussed adequately in these reminiscences. It deserves a comprehensive history. Many innovations were put into practice and improved over the years - slurry walls in lieu of underpinning, chemical and compensation grouting, GDSR's in contractual procedures, . . . what a project!

BALTIMORE 1971-1981

Subway construction in Baltimore had its own challenges, but since design and construction occurred near mid-term of WMATA, and since Baltimore like the District of Columbia rests astride the Fall Line and has similar geology, WMATA experience had a strong influence on design and construction practice. In at least two soft-ground respects, however, Baltimore made significant contributions.

One of these was in the understanding and use of compensation grouting to reduce the effects of tunneling on adjacent or overlying structures. Much of the material within and above the tunnel alignment was slightly cohesive and granular, well suited to tunneling but not immune to loss of ground. Some years earlier, Ed Graff in California had developed a technique of injecting grout beneath oil tanks to compensate for settlement. His basic idea was to use a grout of stiff enough consistency that it would not fracture the soil and travel or escape uncontrollably, but instead would create relatively small balls that would expand near the injection points, compact the adjacent soil, and lift the overlying material. Successive injection through fairly closely spaced injection pipes permitted close control. It seemed likely that the procedure could be adapted to compensate for the loss of ground associated with shield tunneling.

To investigate this possibility, the specialty contractor, Hayward-Baker, conducted a carefully controlled test section with the collaboration of Prof. Cording. By means of inclinometers and settlement-measuring devices it was found that if the stiff grout was injected in the zone of loosening that occurs above and around the rear of the shield as the tailpiece void develops, movements that tend to be propagated beyond the usual loosened zone could be avoided. This understanding provided a rationale for deciding upon and controlling the process of compensation grouting.

At one location the alignment passed at considerable depth in a broad curve beneath several blocks of three- and four-story town houses and apartment buildings. The material within and above the tunnel alignment was well suited to tunneling but not immune to loss of ground that would lead to a settlement trough and corresponding building distortions. The movements were expected to be small and essentially harmless, but avoidance of unsightly cracking and at least some architectural damage could not be guaranteed. Underpinning, on the other hand, would undoubtedly be highly disruptive, even if the potential settlements should not occur. On the basis of the results of the test section, a decision was made to design and install an injection system for compensation grouting, but

GEO-ENGINEERING FOR UNDERGROUND FACILITIES   13

use it only to the extent indicated by settlement observations. In actuality, with the combination of careful tunneling and favorable ground, the injections proved to be unnecessary, but the expense of precautionary preparations was minor as compared to that which would have been entailed in underpinning.

Baltimore was also my personal introduction to the reality of contamination of pervious soils by hydrocarbons, more specifically to the results of leakage of gasoline from defective or abandoned buried tanks beneath automobile filling stations. Needless to say, the spectacle of flames burning in the working face of a tunnel heading made an impression that led not only to development of remedial measures, but also to considerable delay involving a new cycle of subsurface exploration, addressed to detection of contaminants, before work could continue. A foretaste of the problems with natural contaminants to come later in Los Angeles.

L. A. METRO 1983-

Technically, the Los Angeles Metro has been notable for two major advances: dealing with earthquakes including localized fault displacements; and tunneling and operating through pervious soils containing toxic and flammable gases such as methane and hydrogen sulfide. Non-technically, two principal factors have made difficult the realization of a viable underground system: the great spread of the populated area with no concentration along corridors that are obvious candidates for rapid transit lines, and numerous competing political entities within this great spread of territory. Hence, no "starter segment" could be expected to serve a significant part of the population and change its transportation habits enough to be financially viable. These "non-technical" aspects have had much to do with controversy over the project and seem likely to cause new construction to be discontinued at least temporarily. Even so, this account would be incomplete without mentioning some of the exciting technical aspects.

The stations for modern urban transit systems are often box-like structures several hundred meters long, sometimes contiguous and continuous with auxiliary structures including crossovers. Seismic waves moving in directions oblique to the structures have the potential to cause possibly severe racking motions. This possibility was addressed in depth, under the guidance of Jim Monsees, and represents a new chapter in aseismic design.

As in the BART system, the running tunnels must cross recently active faults. In contrast to the Berkeley Hills tunnels, however, at least one of the prospective crossings is through a thrust fault that displays a folded offset of several meters in comparatively young sediments. Obviously no man-made structure could resist major movements if they should be reactivated along such a fault, yet the likelihood that such a movement would occur during, say, a century of subway operation is small. The practical design was to provide the tunnels with somewhat oversize relatively flexible steel segmental linings that could accommodate movements that might reasonably be expected, but to leave for the future any remedial works required by greater movements, after the character of the actual events was known. Here, in spite of the merits of quantitative probabilistic assessments, the final conclusion fell back on the judgment of members of the engineering and seismotectonic community.

Much less problematic was the question of explosive and lethal gases. The area to be traversed by the subway includes several of the famous Los Angeles oil fields; producing wells still exist nearby. Flammable gas including methane occurs at a number of nearby basement locations and is piped away or burned. Minor explosions during drilling for foundations or tiebacks are by no means unknown. Moreover, the alignment originally proposed for one of the routes involved tunneling through the famous tar sands with their remarkable collection of fossils of unfortunate prehistoric mammals that ventured too close. As if the possibility of fire or explosion was not enough, the occurrence of H_2S, lethal in even very small concentrations, posed a threat not only during construction but in operation.

Obviously, the completed subway would need to include not only waterproofing where necessary, but gas-proofing as well, along with reliable detection systems and ventilation during operation. In constructing the running tunnels it proved feasible to erect segmental concrete linings in TBM's, backed up by gas-proof HDPE membranes held in place by an inner lining, as used for waterproofing on WMATA, but constructed to the more exacting requirements of gas leakage in comparison to water leakage. Considerable study had to be given to the longevity of the protection when subjected to chemical instead of aqueous environments.

The serious physical problems that I have described have been overcome, one by one, by sound but imaginative engineering. In this respect, the Los Angeles subway is, in my judgment, an outstanding success. Indeed, I have not mentioned the many other physical and environmental obstacles that were overcome by the engineers and contractors. Yet, for several reasons, the project appears to be, at least for the present, coming to a close. Some of the reasons include the inherently high cost of subsurface work in highly developed urban areas, the layout of the city which does not favor concentrated channels of transportation, the many political jurisdictions involved, and possibly the fondness of the Angelenos for their freeways and automobiles.

CODA

I have had a 60-year love affair with subway tunnels. The state of the art has changed radically, but the rate of change has not perceptibly decreased. Except with respect to seismicity, most of the changes have not been driven by advances in theory, but by observations based on experience. There is, I am confident, room for many more exciting careers in this most challenging aspect of geotechnics.

ACKNOWLEDGMENTS

With the exception of the Chicago Subway, for which all the engineering was done by the City's Department of Subways and Traction, the engineering for the projects was carried out by consulting firms engaged by the appropriate authorities. I have not attempted in this discussion to list them or name the principal players. They constitute an imposing array of talented and devoted individuals. The names I have mentioned are generally those, like myself, who acted as individual consultants.

The Art of the Possible

Jack K. Lemley[1]

Abstract

Up to the present time tunneling technology and geophysical needs have generally interacted to find the best possible solution to particular problems by making the necessary incremental improvements in technology and methods. Today, technology, practice, and theory demand that we in the tunnelling profession reassess this incremental, reactionary view of the "art of the possible." To move beyond the present and its problems, we must reexamine the role of geotechnology with regard to technology and TBMs, the political nature of the contractual environment, and the interaction of theory and practice. To advance, we must reassess and redefine our professional culture.

Introduction

Several years ago, I spoke on the Past, Present, and Future of Tunneling in an address subtitled, *Advancing the "Art of the Possible."* I concluded that we have always been practical folk who, perhaps, have not looked too far ahead. It was a simpler world. We did what we thought possible with the technology available.

With the approach of the year 2000, we can no longer rely on such short-term solutions. Under the stress of competition, our industry has undergone substantial consolidation into larger, more comprehensive companies while growing more global in seeking and prosecuting work. Technology has become very complex, a moving target. Our workforce is of necessity more educated, more sophisticated. Projects are larger, more complex, more demanding.

Professionally, our engineering/construction culture has reached an impasse. The old methods will no longer work. On large, complex projects, the chance for major failure is now too great. To maintain credibility and survive, we must invest the art of the possible with more predictable results.

---

[1] Chairman and CEO, American Ecology Corporation, 805 West Idaho, Suite 200, Boise, ID 83712

To ensure against major underground construction failures, we must educate all parties involved in public works. Project failures are often the result of poor contractual situations based on inadequate geotechnical data and coupled with poorly applied theory. Engineers, owners, contractors, and most importantly, public officials, must be made to realize that corners cannot be cut in geotechnical analysis. Unfortunately, the planning process is often driven by false economics and political expediency rather than on sufficient data incorporated by engineers who understand, first hand, what it means to work in subsurface space.

Before the tender process is begun, the design engineer must perform the needed geotechnical studies and use this data appropriately in the design process. The work must be done by competent, experienced people who are familiar with working underground. Without good data backed by good theory, major failures are inevitable. If this process is poorly done by the engineer in the planning stage, this initial failure can never really be overcome. It will only be exacerbated by reactive solutions sought after-the-fact during construction.

To update my assessment of our professional culture, I will consider our industry's past and present, then give my views on what I think should be done.
1. Background: The Art of the Possible
2. The Present and Its Problems: Where We Are Today
3. The Future: What Must Be Done

Background: The Art of the Possible

Up to the present time, tunneling technology and geophysical needs have generally interacted to produce the best possible solution to particular problems. In the process, we have made the necessary incremental improvements in technology and methods. This is the traditional approach.

We have essentially been reactive to geophysical conditions, usually successfully so. As practical people, with a keen eye to cost and schedule, we have generally used adequate data (with some notably unfortunate exceptions) to assess the situation, then used the best people, equipment, and technology available to get the job done in a timely, cost-effective manner. We were a little like the 18th century landscape architect known as "Capability" Brown. He acquired this nickname because, after inspecting the ground he was to develop, he would tell the owner what "capabilities" the terrain possessed as the work progressed. His clients had lots of money and time.

This "pay as you go" method was possible because older technology such as drill and blast was more adaptable to changing ground conditions. Progress was made by reacting to immediate problems. If the ground proved bad, alternate support, grouting, or other solutions could be applied so that some semblance of a schedule could be maintained. Even in the past, however, such thinking—usually applied by people with little practical knowledge of working underground—all too often lead to disaster.

Most of these disasters are pre-planned. Poor geotechnical work in support of the design process creates a negative situation for the project which can never really be overcome. If the data is bad or inadequate or, as often the case, misinterpreted, if the theory used to interpret the data is inappropriate or fallacious, then the design will be flawed. Consequently, the construction methods may be inappropriate with the bids based on this. The contractor will secure the wrong type of equipment. Cost and schedule inevitably suffer. The project can never recover from poor decisions made at this stage of the process.

The Present and Its Problems: Where We Are Today

At present, we have reached a point where the situation just described must be addressed if we are to preserve the creditability of our profession. Technology is changing at an increasing rate. The contractual environment is becoming more political, and, as a consequence, needed geotechnical work is sometimes slighted in favor of other cost/schedule pressures. Theory and practice have not merged to produce needed feedback as well as they should. These and other considerations must be addressed.

Technology and TBMs: Even more today, if the technology to be used is not well matched to the site's geophysical conditions, then the work can be seriously affected sometimes by orders of magnitude. From my consulting work, I can say that failure in underground work does not occur in small increments. On job after job, poor initial geotechnical investigation or interpretation applied by inexperienced and poorly qualified people combined with bad political and economic decisions do not produce petty problems. They produce major failures.

The key to a successful project does not lie strictly with a geologist, a tunnelling engineer, or a construction foreman. It lies with the full complement of organizations and personnel who must effectively use the information, materials, and technology at hand. A successful project can only be realized through broadening the experience base of our industry through all disciplines. Geotechnical work must blend seamlessly with mechanical, civil, electrical, and electronic engineering and related disciplines as well as with the integration of labor management skills into the process of project logistics. The project must always be considered as a whole, with each aspect, human and machine, fully appreciated and properly integrated into an overall plan which has been drawn based on adequately investigated physical conditions.

Today's TBM-driven tunneling culture relates more to a manufacturing process than to older, more flexible methods. TBM work is repetitive and cyclic. The technology and workforce must support its mechanical nature. Henry Ford would intuitively have understood the mechanical process by which the machine is guided, ground is cut, muck removed, liner segments or support installed, down-time managed, workers trained, and the other factors required for an orderly, efficient "assembly-line" to work. This demands that the manufacturing plant, the TBM, be specifically designed for the product to be produced.

If the geotechnical work has been adequately and accurately done and correctly analyzed, a TBM can be quite efficiently matched to ground conditions. If not, the project is almost preordained to fail. Failure of a magnitude that can break a job.

Such failures are of the compounding type. One thing leads to another. A TBM which has been launched is not an easily modified piece of equipment. If the machine does not fit the ground, it cannot be backed out and immediately replaced. TBMs are usually specifically matched with the expected ground. It takes time to determine what is wrong, redesign the machine, rebuild it, and carry on. This assumes changes can be made that will let the machine succeed. In addition, the same people who misjudged the situation in the first place will be trying to extricate themselves. Ground treatment is expensive and can only accomplish so much. Compounding this is that, rather than spending money for the most appropriate method based on additional investigation, the cheapest, quickest fix is usually tried. This is a recipe that compounds the disaster.

A mild example to this occurred on the Channel Tunnel project. Although the geotechnical work for the entire design process was generally well done, a five-kilometer portion of tunnel beginning at the Dover coast was slighted Early exploratory tunnels extending about a kilometer from the coast had found the same water-resistant quality of chalk marl that mid-channel test bores had encountered. The assumption was made that the Dover coastal area was the same.

Only when the open-mode TBMs began to encounter wet chalk, then very wet, blocky chalk, just beyond where the 1970s drive had ended, did we know we had a problem. As John King, the then UK Tunneling Director said, "I was responsible for progress and there wasn't any. I had no idea how long the wet ground would go on for."[2] It was solved by modifying the TBM's segment erection area and by grouting from the central service tunnel that had been driven ahead of the running tunnels. It would have had much less negative impact on cost and schedule if we had known of the rock conditions prior to beginning the tunnels.

The Political Contractual Environment: Project disasters usually start with poor engineering and site investigation in the design phase. If this work is not done properly, the project is predetermined to fail. The political nature of public works projects in particular squeezes cost and schedule considerations during the planning phase. The engineer, under pressure from the owner and public officials eager to state the earliest possible completion date, finds that in some cases he has less time than needed for site investigation and design. If the engineer is not well qualified for underground work and if he is susceptible to this political pressure, corners will be cut and the tender documents will not present an adequate picture on which bidding contractors can base their tenders.

---

[2] Cited in: Ty Byrd, The Making of the Channel Tunnel (a New Civil Engineer/Transmanche Link publication), Thomas Telford Ltd, 1994(?), p76.

Compounding this, there is a growing tendency for large, normally competent construction organizations to bid work which has an underground component they are only marginally able to undertake. They do this with the idea of hiring the needed expertise later. Frequently, the contractor makes assumptions based on insufficient data and experience, then selects means and methods, equipment, temporary support, and other items for the underground component of the project with the idea that this will be enough to get a foot in the door. Once the bid has been accepted, they feel, any needed corrections can be handled within their contract.

Another frequent scenario is that an owner, faced with developing a project that is heavily political in nature, will ally with a large, politically powerful engineering organization that has the clout to help with funding, public support, and proposal management. While this organization may be politically and economically adequate, it may lack the experience to design and manage the underground part of the work. In many cases, they do not hire competent, practical, experienced people for the underground work. The more of the work they control, the more profit can be made. And, after all, anyone can design a hole through the ground.

Under these and similar scenarios, once a project has been designed and tendered and the tender has been accepted, the project schedule should drive the work. The project develops inertia. If the underground portion has been poorly researched and designed, it will be overrun by the realities of the work. And recovery from such mistakes is extremely difficult for all parties.

Theory and Practice: Professor Robert Mair stated, in a recent edition of *T&T International*,[3] that theory and practice "are becoming far more interdependent. It is no longer acceptable to drive a tunnel and deal with the ground problems as they occur.... Geotechnical specialists are still all too often brought in at the point where things go wrong rather than at the start of the [site investigation and] design process." He sees cooperation between the "geotechnical sector" and the "high quality of practical geotechnical engineering being achieved by the tunnelling industry" as a "natural alliance."

As a classic example of this natural alliance, the Channel Tunnel's John Hester belongs to the elite group of experienced hands in the tunnelling industry. He understands geotechnical information and how to apply it under real conditions. He has done this on job after job. On the Channel Tunnel, he characterized himself as "one of the tunneling people, the guys at the sharp end of production." For the six British TBMs, he turned around a program that went from almost a year behind schedule to one that finished ahead of schedule, setting many records in the process.

---

[3] Robert Mair, cited in "Expert Vision," *Tunnels and Tunnelling International*, September 1998, 33.

We must continue to recognize and encourage men like John–hands-on professionals who came up through the ranks from TBM operators and supervisors to construction field executives. In a very real sense, they are the true innovators in our industry, the actual operators who access ground conditions, make things work, and bring this knowledge and organizational ability forward when they, in turn, become "management."

Professor Mair, in mentioning a "natural alliance" between theory and practice, feels there should be an "even closer relationship between tunnel builders and geotechnical engineers." (34) To be viable, tunnel construction, particularly with TBMs, must be as predictable as possible. Theory must be tested in practice, and practice must make good use of applicable theory.

Although we now have greatly improved technology and engineering techniques, we do not employ them as well as we should for reasons of poor communications, lack of training, and a tendency to manage in an older, hierarchical style. Dr. Kovari, Professor of Tunnelling at the Swiss Federal Institute of Technology, brought up this issue in a January 1998 *Tribune*[4] editorial. "We can assert that the theoretical foundations of modern tunnelling lag behind practice. It is, therefore, a matter of urgency to eliminate fallacious theories, in particular those concerning the behavior of the ground and its interaction with the support systems. The complexity of tunnelling problems and the rapid growth of projects worldwide have led to the emergence of pseudo-scientific theories. These must be identified as such. Only rational approaches guarantee safety and economy in underground construction."

As Dr. Kovari stated, much misinformation exists within the industry. Better analysis, better geophysical and technical development, is needed at all levels. Included in this must be client education. We must stress that, particularly under cost pressure, cutting back on geotechnical analysis is probably the worst mistake that can be made. The Owner always pays—one way or another. The ground itself is always saying: Pay me now or pay me later. We must find better, more creditable and verifiable methods of communicating this to clients, public and private. A stronger, mutually reinforcing alliance between theory and practice can help inject an air of reality here.

The Future: What Must Be Done

To correct the current, growing problems I have detailed, I suggest four specific action items. Others could be mentioned.

First: Ensure that the preliminary geotechnical engineering and site investigation are properly and completely done. This means allowing adequate time and money for this process. This also means educating those responsible for the project that if this is not done, massive failure can result.

---

[4] *Tribune*, January 1998, 5.

Second: Revise the pre-tender process to more realistically reflect the needs of the bid process and the contract. At present, the owner and his engineer often squander project time during the investigation and design phase. Since this eats into the actual time the contractor has to do the work, the engineer's schedule is frequently truncated to fit project politics. The contractor is forced to accommodate the work to an unrealistic timeframe, usually with no extra time for unexpected conditions.

The owner/engineer must allow adequate time, up front, for proper geotechnical analysis. If the expertise is not available in-house, the engineer must recognize his weakness and retain people with the expertise and knowledge to do the work properly. This information must be made available to the bidders. It must be complete and detailed, and it must include the Geologic Design Report that his design is based on. It must include the best analysis, based on the data, that the engineer can provide. It must be the truth, the whole truth, and nothing but the truth.

The owner/engineer must allocate risk according to the best information they have. Liability/responsibility must be spelled out and shared by the organization best able to manage the specific risk.

Third: The tender period itself must be long enough for the bidders to thoroughly determine what the conditions are that must be included in the tender. As a suggestion, after a bidder has been selected, the owner/engineer should agree on the performance standards of the contract, then sit down with the contractor and analyze the bid in detail to see if it can actually be done to specification and schedule. The object of this is to establish a performance baseline that is clearly understood by both parties. This baseline is also very useful if a Disputes Review Board is to be used.

Forth: The technology to be used must be integrated into a complete, consolidated, consistent package that the project team can live with. Technology is changing so rapidly over so many areas that even the best engineer can become lost in details. Technology has become a moving target.

Experience is imperative when attempting to integrate the geophysical setting into the design and then translating that information so that proper construction methods can be applied to achieve the design in an efficient, cost-effective manner. Unfortunately, there are not many in our industry who have the education, experience, management ability, and drive to do this. And there are fewer and fewer of them. Somehow we must develop the ability to find such people early in their careers and bring them forward so that their potential can be realized.

Conclusion

As the Greeks said, engineering is "the art of the possible." We have inherited this description, and it aptly fits our profession. As tunnelling becomes less expensive, safer, and more predictable, it is becoming a technology of choice. The world community is increasingly seeing underground construction as the economic and ecologically sound answer to competition for surface and air rights, particularly in dense urban environments.

The engineering and technology exist to do this. So does the desire to employ this old resource made new by the improved capabilities available to our profession. We must ensure that our professional culture is up to the challenge. To do this, we must ensure that the site investigation process is considered as a fundamental part of the planning for a project.

Tunnels connect. And as tunnels grow longer and more complex, they demand that we not only accept our current professional mission but move it forward. Technology, theory and practice must form a tighter, more mutually respectful alliance, and we need to find ways to increase feedback between them. We must also educate our clients and the public that the current nature of the bid process is often inadequate in terms of managing risk, cost and schedule, particularly for underground work. Together, we can make changes if we consolidate and focus our professional abilities to do so.

Engineering is the art of the possible, and what is possible is now very great indeed.

A HISTORY OF DEEP FOUNDATIONS IN NEW YORK CITY, "COULD WE BUILD IT NOW?"

Reuben Samuels

This presentation is a series of brief case histories of foundation projects in New York City relating to theory and practice as well as some "firsts" such as an early "top Down" project; using a seismograph to control blasting inside an operating powerhouse; the use of "soldier beams" at the perimeter of a blasted cut inside an operating subway station; and an early use (in a private project) of 4-position extensometers in the walls of a 70 foot deep rock cut adjoining a subway on one side and 14 story Park Avenue apartments on the other side.

Parsons Brinckerhoff, One Penn Plaza, New York, NY 10119-0061

1894 - 80th and 3rd - excavation of street car (original LRT's!); derbies stuffed with paper and rags provided the hard hats of the day; black powder and steam drills

1927 - Riverside Church; 120th and Broadway - large area rock foundation; timber sheeting to top of rock; stiff leg derrick(s)

26 GEO-ENGINEERING FOR UNDERGROUND FACILITIES

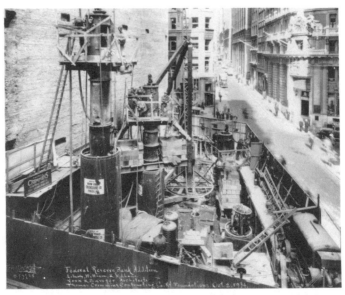

1934 - Federal Reserve Bank Addition, William St.; very tight site; vintage air caissons for new columns

1958 - 222 Broadway; another tight sight - subways on 2 sides and a major restaurant (with a bad foundation!) on the 32nd side; early example of "top down" construction

GEO-ENGINEERING FOR UNDERGROUND FACILITIES

1962 - deep foundation including blasting straddling operating subway line, i.e. 6th Ave.

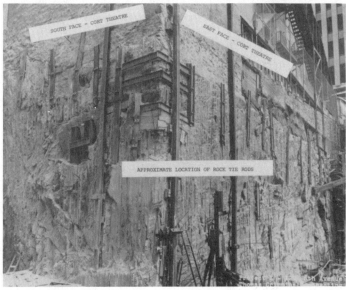

1971 - "Project Z", 47th - 48th & 6th - very large (800 ft x 200 ft) blasting job - reentrant rock corner with operating theater astride the corner

1971 - platform extension, 60th and 3rd - liner plates used as protection for blasting ove top of, between and under operating subway tunnels

1971 - platform extension, 60th and 3rd - view of blasted rock face; showing exposed li plate protection remaining cast in place original concrete liner

Major Sewer Tunnel Construction
beneath Metro Tunnels in Singapore

Richard Flanagan[1], MASCE; Birger Schmidt[2], FASCE;
and Tan Boon Tee[3]

Abstract

The Singapore Deep Tunnel Sewerage System (DTSS) project is being implemented by the Ministry of the Environment whereby a tunnel network will convey wastewater by gravity flow to a new treatment plant at the east end of the island. A feasible route through the Central Business District (CBD) could not be found because underground space at appropriate elevations were already occupied or reserved. The sewer tunnel system had to be detoured around the CBD. Even with the realignment, the DTSS needs to pass under a pair of soft ground tunnels for the currently under-construction mass rapid transit extension, the Northeast Line (NEL) rail line. Because a deeper elevation of the sewer tunnel would result in very substantial additional operations costs, it was desirable to place the sewer tunnel as close as possible beneath the NEL tunnels. The NEL would be operational at the time the sewer tunnel would be built, and the NEL tunnels would be extremely sensitive to ground movements.

Based on geotechnical analyses, risk assessments and consideration of geometric and schedule constraints, an early advanced construction of an underground chamber under the NEL has been adopted. This soft ground chamber would be constructed before

---

[1]Tunnels and Link Sewers Manager DTSS Project, Parsons Brinckerhoff International, CH2M/PB JV Pte. Ltd., 73 Tras Street, Singapore 079012
[2]Technical Director (Geotechnical Engineering), Parsons Brinckerhoff Quade & Douglas, Inc., Marathon Plaza, 303 Second Street, Suite 700 North, San Francisco, CA 94107
[3]Deputy Chief Engineer DTSS Project, Sewerage Department, Ministry of the Environment, 40 Scotts Road, Singapore 228231

the NEL tunnel construction and would permit the future DTSS Tunnel Boring Machine (TBM) to pass freely under the NEL tunnels at about a 2.3 m separation. This scheme was selected over other schemes because it removed as much as possible, the degree of interaction between the two projects and virtually all risk of undesirable displacements of the completed NEL tunnels.

Background

Singapore is one of the most densely populated Asian countries. Surface space is at a premium and there is a growing momentum to place more and more infrastructure in major tunnels. The DTSS project will help to preserve surface space for other uses. Metro (subway) lines and expressway tunnels have already been constructed under the CBD and the rest of the country. The newest Metro Rail Transit (MRT) system expansion is the 20-km-long NEL project. The (Singapore) Land Transport Authority (LTA) is managing this project. The running rail will be located primarily in twin soft ground tunnels, 5.8-m-diameter, excavated with Earth Pressure Balance, Tunnel Boring Machines (TBMs). The NEL project is currently under construction where tunnel excavation began in late 1998.

The DTSS project includes tunnel sections in soft ground, rock and mixed face ground conditions. All the DTSS tunnels will be excavated with TBMs that provide positive face control and utilize pre-cast concrete segment linings. A 7-m-diameter (excavated) DTSS soft ground tunnel crosses underneath twin NEL tunnels at Serangoon Road, which is NEL Contract C-705. At this right angle crossing, the two tunnel systems are separated by less than 2.5 m. The NEL tunnels will be supported by single pass linings (bolted, gasketed precast concrete segments). The rail will have direct-fixation system with only small allowable displacements and distortions.

The location of the crossing is at the intersection of Serangoon Road and the slip roads of the adjacent Pan Island Expressway (PIE). Traffic at this intersection is very heavy all day long with frequent major backups. There are also many buried utilities in the area including major communications, high voltage transmission and water main lines.

The LTA has very strict requirements and regulations for any excavations near MRT structures. Control of ground displacements induced by any DTSS tunnel construction on a MRT structure is very critical. MRT structure movements are limited to very small values so as to minimize risks to revenue service as well as operational factors and system components. For example, the tolerance available for adjusting the elevation of the direct fixation track is limited to 15 mm.

It is advantageous to keep the DTSS tunnel as high as possible. Deeper tunnels would be more expensive, and all the shafts would be deeper as well. Most importantly, a deeper tunnel would result in a deeper influent pumping station at the wastewater treatment plant, more expensive to construct, and resulting in significantly higher recurrent pumping costs. This DTSS crossing under the NEL tunnels sets the elevation of the entire DTSS tunnel system.

With cooperation between the Ministry of the Environment (ENV), the DTSS project managers (a joint venture of CH2M-Hill and Parsons Brinckerhoff), the LTA and the C-705 NEL Contractor, a solution was finally found that satisfied everyone with a minimum of risk exposure, reduced interface problems, and acceptable cost. In the following parts of this paper, we will discuss the options considered through the stages of development and the selected option in greater detail.

Ground Conditions

Ground types at the DTSS/NEL crossing consist of, in order of increasing depth; fill, fluvial deposits and the Old Alluvium (OA) Formation. The fill is comprised of loose sand and some building rubble to a depth of about 3 m. The fluvial deposits are very loose silty sands that terminate at a depth of 6 m. Both the NEL and DTSS tunnels will be driven through the OA. Groundwater is within 1.5 m of the ground surface.

The OA Formation is thought to represent a delta deposit, which was laid down in a slowly subsiding basin during the time of the Pleistocene interglacial periods. This formation for the most part is a lightly cemented silty sand (i.e., Granular OA) or silty clay (i.e. Cohesive OA). The former typically has 10 to 35 percent "fines" passing the #200 sieve. However, there are infrequent lenses of uniform uncemented sands.

The sand is quartz-feldspathic with feldspar weathering to varying depths. The weathering is indicated by a distinct color change from bluish to greenish gray when fresh, and to yellowish reddish and grayish brown when weathered.

The OA Formation is frequently divided into three units corresponding to varying degrees of density, weathering and degree of cementation. With increasing depth, weathering decreases while cementation increases. The Upper OA (UOA) is identified as those depths with Standard Penetration Test "N" Values less than 25. The Middle OA (MOA) and Lower OA (LOA) correspond to those sections with N values 25 to 80 and N values greater than 80, respectively. Both tunnel systems are located in the LOA. Geotechnical properties of the various strata are shown in Table 1.

Visual Observations in an Exploratory Shaft

Little was known about the behavior of the OA Formation as a tunneling medium as practically no tunnels to date have been driven through this material. Undisturbed samples are difficult to obtain although with very careful rock coring, core samples can be obtained.

Stratum	Modulus – E (MN/m^2)	Drained Cohesion (kN/m^2)	Drained Friction Angle (ϕ)
Fill	12.5	0	30^0
Fluvial deposits	15	0	30^0
Upper Old Alluvium	22.5	0	32^0
Middle Old Alluvium	56.0	10	35^0
Lower Old Alluvium	112.5	20	35^0

Table 1. Geotechnical Properties

Due to the lack of tunneling experience in the OA, the LTA geotechnical engineering unit in late 1997 implemented a shaft and short adit test trial in the OA. The experimental shaft was about 20 m deep and 4 m in diameter. A visit to this test section gave us very valuable information about the likely behavior of the OA as a tunneling medium.

The adit consisted of about a 2.5 m square opening supported by timber posts and some crown lagging. An examination of the crown suggested that the ground was nearly self-standing with very little load transferred to the lagging and posts. Groundwater inflows (in the adit) were negligible. The adit face was moist to the touch with no visible inflows. Some inflows (small) were pumped from the shaft bottom. The inflows came from leaks originating at the upper part of the shaft.

The adit face had been exposed for more than a day and was standing vertical with no slaking or raveling. A sample was attempted to be dislodged with a crow bar but it was very difficult to loosen any pieces. The adit had been excavated by using pneumatic spades. Indentations from the air spade excavation were still clearly visible after more than a day. Based on this test adit, the OA soil would behave better than might be expected from the borings and test data.

We expect that the OA will behave well during tunneling. However, as probable with any tunneling, there is always the potential for the unexpected. In the case of the OA, there have been experiences with infrequent clean uniform sand lenses with little or no cementation that could flow from an open face excavation.

The experimental adit showed that in the OA, a sequential excavation and support system with steel ribs and shotcrete support would be possible. The boring data from the area of the NEL/DTSS crossing were similar to that at the exploratory adit.

Crossing Studies and Alternatives

Based on the DTSS construction implementation schedule, the DTSS tunnel would be constructed after the NEL tunnel excavation, prior to revenue service. Thus, it was considered in the early stages, to construct the DTSS tunnel at a clear distance of about 2.5 m below the NEL tunnels. This was acceptable provided there would be non-damaging displacements of the NEL tunnels and that the direct fixation rails be installed after the DTSS excavation. Three-dimensional finite element analyses were made to estimate the displacements and impacts on the NEL tunnel linings. Displacements were also estimated based on the semi-empirical error-function method (Schmidt, 1969). With the apparent excellent soil conditions, the use of a TBM with positive face pressure control and careful attention to tunneling operations under the NEL tunnels, it was judged that the volumetric ground losses (expressed as a ratio of the unit face volume) could be held to under 1%. Results of these analyses indicated that the NEL tunnel would be subjected to displacements of about 30 mm and apparently acceptable stresses in the concrete lining.

It was considered that the DTSS tunnel excavations could have delays resulting in the latter being excavated after the NEL fixed rail was in-place. This later scenario was of concern for the various parties, as it could result in expensive extra work or rework by the NEL contractor. It soon became clear that reducing interface exposure between the two projects would be very desirable. Consequently, some other alternatives were considered as follows:

- Construct an inverted siphon
- Lower the sewer tunnel profile to gain additional vertical separation between the tunnel systems

Inverted Siphon:

The sewer tunnel would be depressed as a deep siphon, well below the NEL tunnels. The creation of such a siphon, on this particular project, would be very undesirable from an operational point of view, with risk of deposition and significant anaerobic activity.

Lower DTSS Tunnel Profile:

This consisted of dropping the alignment by about 4 m to gain one tunnel diameter clearance between the NEL and DTSS tunnels. For this alternative, however, the DTSS tunnel would still need to pass under the NEL and meet LTA restrictions. The major cost implication to ENV of this scheme is the long-term pumping cost at the wastewater treatment plant. The pumping cost was reviewed, and over a 100 year pumping operation, the cost implication is about $50 million in present value. Additional DTTS construction costs would also be incurred for deeper drop and access shafts.

Additional alternatives were considered to cover the possible condition that the DTSS tunnel would be excavated after rail installation, but prior to NEL revenue service. These additional alternatives are in the following paragraphs.

Ground Improvement by Permeation Grouting:

The soil strengthening would be done with chemical grouts. The use of chemical grouting in the OA is not technically advisable due to the low void ratio and high fines content, i.e.; the grout won't penetrate the soil mass properly.

Ground Reinforcement by Piles:

This alternative would not provide continuous support along the NEL and a sag would develop above the DTSS tunnel and impact the NEL tunnels.

Ground Reinforcement by Jet Grout Columns:

Jet grout column ground reinforcement is subject to the same problem with ground deformation developing above the DTSS tunnel as the preceding alternative. There is also no experience with jet grouting in the dense OA. Experience in similar dense soils to the OA has resulted in inconsistent column diameters and vertical continuity of the columns.

Cut & Cover Tunnel Box:

A structural concrete box would be supported by piles and would let the DTSS pass underneath while preventing ground deformations at the NEL tunnels. These structures would be highly effective but extremely costly. They also defeat the goal of the NEL tunnels to pass down Serangoon Road and through this intersection without impacting the traffic or neighborhood.

Slurrywall Underpinning:

The slurrywall underpinning is a more economical but slightly less effective variation of the cut & cover tunnel box alternative. It would consist of two rows of walls under each NEL tunnel. The slurrywalls would first be constructed and then the NEL TBMs would skim along the top of the walls. Later, the DTSS tunnel would "mine" through the walls (consisting of weak concrete). This alternative also impacts traffic but to a much less degree than the cut & cover tunnel box.

Ductile Cast Iron Lining:

This alternative consists of locally substituting ductile cast iron (DCI) tunnel lining segments for pre-cast concrete segments. This lining is thinner than precast concrete segments. Because it is ductile, it can withstand more ground deformation without structural distress than the brittle concrete. Its smaller thickness results in more clear internal tunnel space, thereby permitting more area for track adjustment. The NEL Contractor would need to have a factory modification to their TBMs (segment erector) as verified by talks with a TBM manufacturer. The NEL TBM tunneling would proceed normally until the TBMs reach the crossing. The tunneling would then be halted and the segment erectors re-configured to handle the DCI segments. Tunneling would then resume normally but with DCI segment ground support until the crossing area is cleared. The TBMs would again be halted, the segment erectors changed back to its original components and tunneling resumed with concrete segments.

Several other schemes were briefly considered including ground freezing, and underpinning by jacking pipes under the NEL tunnels from galleries on both sides. All of these latter schemes were considered too risky, too complicated or too expensive.

The slurrywall underpinning scheme was developed further as it was a desirable scheme. A three-dimensional finite element structural model of this underpinning was then used to analyze the effects of the DTSS excavation upon the NEL linings. Figure 1 illustrates the deformed mesh resulting from the DTSS excavation under the NEL.

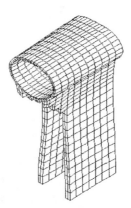

Figure 1. Deformed FEM Mesh of NEL Tunnel and Slurrywall Underpinning

36   GEO-ENGINEERING FOR UNDERGROUND FACILITIES

The three dimensional analyses of the slurrywall underpinning are summarized as :

- The NEL tunnel crown displaces vertically about 10 mm
- Maximum tensile stresses in the NEL lining are about 4.3 mPa
- Maximum compressive stresses in the NEL lining are about 16.3 mPa

The above results indicate that the induced stresses are small, lining distortion is low and distress or leakage at the tunnel segment joints unlikely.

Although the underpinning slurry wall was very feasible, construction would be very difficult due to site access and utilities. The cost of relocating some of these utilities would be very high. It was apparent that a buried 6.5 kV power line would require relocation, with a long relocation waiting period.

Construct a DTSS "Advance Tunnel":

This option consists of constructing a short length of an early DTSS tunnel **prior** to the NEL tunnel excavation. This tunnel would be a Sequentially Excavated Tunnel supported with shotcrete and steel ribs. The works would consist of a shaft to the side of the DTSS alignment and Serangoon Road, an access tunnel and about 43 m of an underground chamber or tunnel under the proposed NEL alignment (see Figures 2 and 3). This tunnel would be done with small excavation equipment. The future DTSS TBM would mine into this tunnel and the contractor would "walk" the TBM through this tunnel and resume mining at the other end. The length was sized enough such that ground movements and TBM excavation vibrations would not impact the operating NEL. The advantages of this solution include at least the following:

- Near-zero ground displacements on the finished NEL tunnels
- Insensitive to any changes in the DTSS implementation schedule
- No traffic diversions required
- No major utility relocations required

The Advance Tunnel could also provide useful data for the numerous mined connections between the main tunnels and the offset drop shafts most of which are located in the OA. The disadvantages of the Advance Tunnel are:

- There is limited experience in the use of shotcrete in Singapore
- This would be the first major sequential excavation and support tunnel project in the OA

GEO-ENGINEERING FOR UNDERGROUND FACILITIES

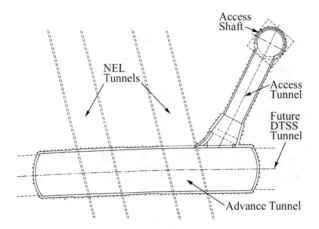

Figure 2. Plan View of Advance Tunnel Works

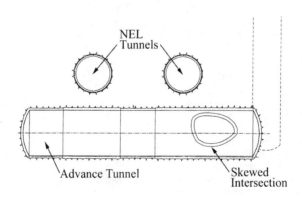

Figure 3. Section of the Advance Tunnel and NEL Tunnels

Advance Tunnel Implementation

Several implementation Advance Tunnel project delivery schemes were considered. These schemes were:

- Tender by the ENV as a Design/Build Tunnel Contract
- Tender by the ENV as a Tunnel Contract with a design prepared through ENV
- Construct as a Variation Order to the LTA C-705 Tunnel Contract with a design prepared through ENV
- Construct as a Design/Build Variation Order to the LTA C-705 Tunnel Contract

Each one of these contracting methods has advantages and disadvantages. In the end, all parties agreed that the best contracting procedure was the Design/Build Variation Order. The selected delivery method provided the fastest construction start and gave us confidence that the works would be completed before NEL tunnel construction.

A Design Brief was prepared along with a scope of work consisting of the following major components:

- Mobilization
- Shaft excavation and ground support
- Tunnel excavation and ground support
- Site work (fencing, grading, landscaping)
- Shaft access protection (secured shaft cover)
- Geotechnical instrumentation hardware and installation
- Extended operation of lighting, ventilation and pumping

Upon completion of the tunnel, an access ladder, lighting, ventilation duct and fan, pump, and equipment controls would be left in place for a limited time. These items would permit personnel access to the tunnel, particularly during the NEL excavation. The shaft would later be turned over to the DTSS tunnel contractor. This latter contractor would ultimately finish the works as a permanent DTSS access shaft.

Geotechnical instrumentation has been added so as to have monitoring of ground displacements of both the ground and structure movements in and around the advance tunnel during its construction, as well as during the NEL TBM excavations and long-term. Geotechnical instrumentation includes, but not limited to; piezometers, shallow and deep settlement points, inclinometers, optical surveys, and tape extensometers.

The major design requirements are as follows:

- Advance Tunnel clear inside diameter of 7.59 m
- Access tunnel and shaft to have specified geometry and certain minimum sizes
- Minimum design life of 5 years

- No particular finish
- Maximum groundwater inflow in shaft and tunnel walls not to exceed 50 litre/day/m^2
- Head walls of the Advance Tunnel not to contain any steel reinforcement

In addition, the contractor had to meet specific requirements, assessments and reporting regarding risks and the impact of the excavations on adjacent structures and utilities.

The Advance Tunnel works was awarded to the NEL C-705 Contractor, KSM-JV. KSM is a joint venture of Kumagai, Sembawang and Mitsui contractors. The tunnel and shaft excavation subcontractor is Zublin (Singapore). The contractor's principal design consultant is Geoconsult (Singapore).

Final Design of Advance Tunnel Works

KSM was permitted to select the actual shaft and tunnel ground support sizing and components, provided they meet certain required minimum geometry. In general, the final openings are much larger than required in order to accommodate construction equipment. A brief description of each component is presented in the following paragraphs.

Access Shaft:

The shaft has a total depth of about 31.5 m. The upper 15 m is a 6 m square sheet pile excavation. Below the sheet piles is a circular shaft with ground support by cast-in-place reinforced concrete rings. As the shaft approaches its bottom, mesh reinforced shotcrete will provide support and connection with the access tunnel.

Access Tunnel:

The 19 m long connecting Access Tunnel has a cross sectional area of 12.62 m^2. As it approaches the Advance Tunnel, it deepens to accommodate a future DTTS stepped maintenance ramp. This tunnel will be supported with steel lattice girders and mesh reinforced shotcrete.

Advance Tunnel:

Excavation will be in three stages; top heading, bench and invert. There are three major ground support classes. Each class includes full circular steel lattice girders and multiple layers of shotcrete (mesh reinforced). Final shotcrete thickness will vary from 350 to 450 mm. Excavation advance lengths are 800 to 1200 mm. The ground support system at the intersection with the access tunnel addressees three-dimensional load effects.

Because of the uncertainty of encountering undetected uncemented uniform sand lenses laden with groundwater, probing ahead of excavation faces will be done at all locations. As a contingency for ground with poor standup time, forepoling may be used. Dewatering is also planned as a backup contingency for running ground.

Prior to any excavation, approvals from various agencies were required. Due to the excellent cooperation by the various agencies, the approval was obtained relatively quickly. Excavation for the shaft began in December 1998. At the time this paper is presented, the works will be nearing completion.

References

Schmidt, B., "Settlements and Ground Movements Associated with Tunneling in Soil", thesis presented to the University of Illinois, at Urbana, Illinois, in 1969, in partial fulfillment of the requirements for the degree of Doctor of Philosophy.

... AND BACK TO THEORY
Augusto Espinosa-Silva[1]

Abstract

Bogotá soils conform a rather peculiar lacustrine deposit: the depth of quaternary soft silts and silty clays reaches 160 m at the subbasins and over 400 m at the center of the so called **sabana**; liquid limits are very high (LL > 100%), void ratios are also high ($1.0 < e < 6.0$) and consequently, compressibilities beyond 0.5 are common. Below a thin dessicated crust a **gibson** soil is encountered, meaning that strength and modulus expressions grow linearly with depth. Forced by urban pressures an increasing number of basement levels has been needed; in the short span of 30 years the city has passed from a maximum of 1 such level to 6 bringing about a correspondingly exponential growth of the magnitude of excavation technical and logistic problems. Besides this, conditions of the soil mass are not constant with time; instability is evident in progressive consolidation of a very soft deposit that is affected by hydrological unbalanced resources that cause dessication and even seismic events that unleash additional settlements to an already stable foundation. The experience of the author covers several excavations having 260 ft x 260 ft in plan and between 31 ft and 55 ft deep where he has acted as head of the consultant group and chieff advisor of the construction process. Several cases are presented to illustrate design considerations, agreement between anticipated and observed ground behavior, construction techniques, assesment and control of ground water, in-situ measurements by means of inclinometers, extensometers and piezometers. Finally an effort is made to go back to theory, strengthen its fundamentals and seek that expectations coincide with reality.

GENERAL BACKGROUND

Geologic Sketch

Bogotá city area corresponds to a late tertiary synclinal filled in by quaternary deposits mostly of a **lake bed origin** ([11], [12]).

In Cretaceous times the land was covered by a shallow ocean; this area slowly depressed to be filled by more than 16800 m of marine deposits, mainly shales, occasionally limestones and finally sandstones. Then the ocean slowly **retreated**; coal was deposited subsequently on an upgraded continental mass.

---

[1] Manager, AREAS LTDA., Carrera 20 No.84-14 of 304, Bogotá, COLOMBIA, S.A.

In early Tertiary the miocene orogeny folded, faulted and brought up previous deposits. Tranquil times made possible new depositions of continental claystones and occasional sandstones. New and persistent tectonic pulses have kept the orogeny active to present.

The whole area became finally a **synclinorium**. One of the valleys with cretaceous rocks as a basis, and partially filled with tertiary claystones, was flooded at the start of Pleistocene. In this lake, deposits of clays and sands were slowly accumulated at a rate of approximately 0.2 mm per year through all the Quaternary. This is the area on which Bogotá was built.

The depth of the deposit reaches 380 m (locally 780 m) or so in the open Sabana and to 160 m in the small **Bogotá Synclinal** located between the mountais of Suba and Usaquén, nearby towns absorved by continous city growth.

Bogotá Greenish-Gray Clay (BGC)

The upper part and more recent deposition within the Sabana formation, is what has been termed the Bogotá Greenish-Gray Clay (BGC). Consequently it refers to a lake-bed deposit of clays and silts with interspersed layers of peat, as remnants of swamps, and sand layers in the vicinity of ancient water-ways ([9], [10]). Defined depth of BGC goes to 60 m -the common reach of engineers and developers- and corresponds to the last 150000 years of deposition. The peculiarity of its characteristics lies in the volcanic origin of some of its clays and in the organic content which is evident in the Bogota synclinal that forms the sub-basin where most of present construction activity is taking place. Recent publications related with the detailed interpretation of more than 250 m of stratigraphic column confirmed a drastic change in mineralogical composition and clay content at 30 m depth: from a predominantly clayey (Kaolinitic and Illitic) material at the surface it passes to a mostly silty one; the ratio of feldspar/quartz content also changes abruptly. Through palinology it is also known that this change coincides with a **hiatus** in deposition, meaning that underlying soils are much older.

Some geotechnical properties of BGC are described below. Even though it is not enough to make BGC **unique**, in the terms of R.V. Whitman [19], it is sufficient to consider the deposit as exceptional.

. **Plasticity.-** Recent micro-zonation studies run by Ingeominas and the University of Andes gathered a vast sample of soils upon which it was possible to define the position of BGC in Casagrande's plasticity chart as (see **Figure 1**):
Ip = 0.772 (LL -15.2)

. **Compression and Recompression Indexes.-** Compression (see **Figure** 2) and recompression indexes were also investigated with the following results
Cc = 0.0089 (LL-14) more easily defined as
Cc = LL/100 and Cr = Cc/8

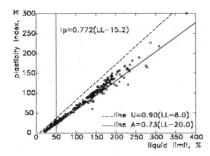

Figure 1. PLASTICITY CHART   Figure 2. COMPRESSION INDEX vs LIQUID LIMIT

- **Compressibility.** - It has become customary to define compressibility (C) as $Cc/(1+e_o)$. Following an old idea by I.Arango, **Figure 12** presents C versus the natural water content, w. The results are outstanding: BGC plots way out of the range previously defined. Based upon this, A. Espinosa [8] proposed four degrees of compressibility in the following way: **normal**, the area defined by Arango; **medium high**, up to 0.4 at w=100%; **very high**, up to 0.6 at the same water content; and **exceptional** for compressibilities greater than 0.8 where BGC is honorably accompanied by Mexico volcanic clay.

- **Void Ratio.** - Espinosa [7] presented several correlations of e v.s z and explained how void ratios were low near surface, where the deposit was preconsolidated by dessication; grew to a maximum denoting the start of normal consolidation and finally decreased slowly with depth in response to a growing effective geostatic stress. One curve corresponds to each zone as defined by the Microzonationg study as may be observed in **Figure 3**.

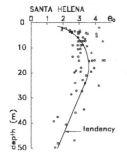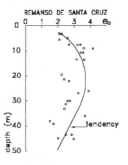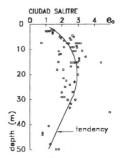

Figure 3. VOID RATIO

- **Strength.**- Strength, as measured by VST or CPT shows a growing tendency with depth below the surficial crust. Espinosa [6] has proposed that **Su=kz**. Several factors **k** have been defined for equivalent areas of Bogotá. In order to compare this with Mesri's [13] proposition related to Su/σ'_{vo} **Figure 4** has been deviced. It is easy to conclude that in fact Su/σ'_{vo} is meaningful for normally consolidated (young clays) where they show an increasing tendency of Su with depth compatible with the one expected in σ_{vo} effective. This happens typically in **gibson** type soils like BGC.

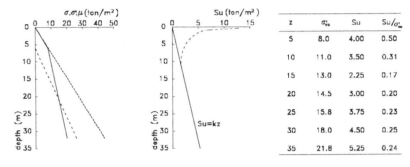

Figure 4. STRENGTH

EXCAVATIONS IN SOFT GROUND

Generalities

Bjerrum and Eide (1956) defined how to handle an excavation in soft cohesive ground as a foundation problem where, rather than loading, it is an unloading action which might cause failure.

Skempton (1951), few years earlier, had proposed a set of bearing capacity factors, Nc, to fit different geometries where the key parameter varied between 6.2 and 9.0 for circular or square foundations at a growing depth factor.

It was Peck (1943), who based on observations made at the Chicago subway, proposed the idea of apparent pressure diagrams to deal with calculations of the elements to support those excavations.

Peck's Contributions

In his state-of-the-art Lecture on the occasion of Mexico International Conference, Peck (1969) rounded up his conclusions on the problem of open and underground excavations in soft ground. Later on, Peck, Hanson and Thornburn [15] proposed an equivalence between the stability factor, Ns, and the progress of excavation; in their words if this factor is equal to or less than about 4, the behavior of the clay beside the cut is essentially elastic; on the other hand if it exceds 4 and approaches 6, the clay behaves in an increasingly inelastic way and the apparent pressure diagram is no longer truncated at the base. Base failure is imminent when γH/c_b exceds about 7. For larger Ns further investigation may be needed concerning possible base failure; for Ns as great as 10 or 12 heavy reinforcement of the cut may be needed and sophisticated techniques as top-down construction, prefounded piles or whatever is at hand. Terzaghi and Peck (1967) stated that the width of the trapezoid of pressures could vary between 0.4 γ H and 0.2 γ H for cohesive soils of increasing strength.

Factor of Safety

Bowles (1996) [3] presented an elaborate way to calculate the maximum depth of excavation. This is

$$D' = \frac{Su_{bc} \, Nc_b + \gamma H Nq - q_s}{\gamma - \frac{c_a}{r}} + F_b$$

It is based on Terzaghi's original idea expressed in his Theoretical Soil Mechanics (1943).

Espinosa (1998) [8] has pointed out the results of over-excavating, when the weight of soil removed is larger than the weight of the proposed structure. In some cases he has measured considerable heave even under the load of 12-story structures.

Miscelaneous Considerations

Peck (1969) when addressing the topic of berms recommended them, when used, to be substantial. Clough and Denby (1977) presented a procedure to calculate berms which results in rather large masses; otherwise their effect is completely irrelevant.

Xanthakos et.al. (1994) list a series of procedures, most of them of anonimous origin, in order to support excavations. They consider top-down construction, slurry walls with trousers in order to fix the base of the structure, prefounded columns to reach the base of excavation at an early date. All of those procedures have been tested in Bogotá.

Finite Element Method (FEM)

Clough and Tsui [4] presented a pioneering work on **tied-back walls** in clay; the study was conducted by means of the finite element method because of its ability to predict earth pressures, bending moments and soil and wall movements. While the cases described in the paper were hypothetical examples, the authors had the opportunity to callibrate the method comparing computed and observed behavior. This is how computational methods and facilities become **precedent** through the filter of reality.

FEM, besides being a desing tool which allows due consideration to soil-wall interaction, has helped greatly to understand how to procede once movements have started at an excavation. It is also a control tool of great efficacy. But it is not a game. Geotechnical engineers should not play computer games; this is a serious business. [22].

REVIEW OF PERTINENT CASES

Medium Ground

Four cases merit attention: all of them are rather tall anchored wall excavations that have been completed succesfully.

Case 1. XXI Century Project. This is a 26 m (80 ft) high excavation in terrace materials. The wall was constructed following top-down techniques, by shotcreting panels 4 m x 3m (13 ft x 10 ft) hold backwards by 60 ton cable tendons. Two features are worth mentioning; these are: first, inclination of the wall varied from vertical at the critical point to 80° with the horizontal as property line permitted; tiebacks at the steep side were active and gradually became passive when more inclined surfaces were possible. Secondly, a row of 10-story buildings, previously built, were to be preserved firmly in place even though they were founded on caissons at a mere 10 m depth. See **Figure 5**.

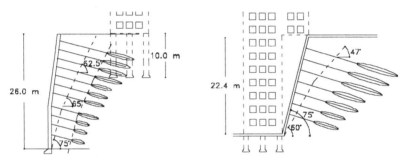

Figure 5. XXI CENTURY PROJECT   Figure 6. VIENTOLARGO PROJECT

Case 2. Vientolargo Project. This was a 23 m (76 ft) high excavation in badly weathered claystones and siltstones in a place where several land slides had taken place in previous excavations. The architect proposed a vertical wall but then he would have to go on caissons to found the interior portion of the building which was located just backwards of the main excavation. An alternate procedure was suggested: the main wall should be inclined to 75° to the horizontal and the loads originated at the small part of the building would travel through inclined columns, resting on the wall, to the row of caissons which were to receive the main building. Anchored wall techniques were used extensively. Buildings existing in back were not affected. See **Figure 6**.

Case 3. Colmena Project. This was a 16.5 m (55 ft) deep excavation having 55 m (180 ft) x 60 m (200 ft) rectangular shape. Four rows of cable tiebacks were provided at 2.5 m (8.0 ft) x 3.3 m (11 ft). The wall was shotcreted in a horizontal sequence, one row at a time. The cables were readily tensioned at rated capacity until it was evident that nearby houses were moving slightly backwards; since then tendons were tensioned in sequence at 30%, 60%, 100% of their rated capacity. Another problem encountered were pervasive water outcroppings at certain levels of the excavations that produced backward erosion; this was partly solved by injections and partly by pipe drainage projecting out of the excavation lines. See **Figure 7**.

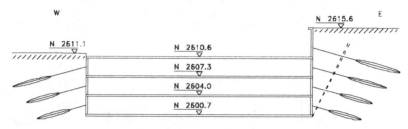

Figure 7. COLMENA PROJECT

Case 4. Hacienda Sta. Barbara Project. This was a huge commercial Mall surrounded by slurry walls 7.5 m (24.7 ft) to 13.5 m (44.5 ft) high. A distinct feature of this excavation solution was that the wall was supported by embeddment in a sand stratum at the base, and by means of one or two rows of anchors at the top. There was a sand stratum (1 m to 2 m thick) which was used to fix the anchors. Remaining soils were soft silts and clayey silts. See **Figure 8**.

Figure 8. HACIENDA SANTA BARBARA

Soft Ground

Five cases of excavations in soft to very soft ground deserve attention because a mixed and varied vademecun of solutions was necessary to solve the problems satisfactorily.

Case 5. Salitre Plaza. This was also a huge mall having 70000 m^2 (760000 ft^2), surrounded by a slurry wall 860 m (2 820 ft) long in plan and 12.0 m deep. When excavated at 8.6 m the wall was supported by a berm having 5.5 m wide at the top and 1:1 slope to the base (See **Figure 9**). Excavation proceeded by stages initially having 25 m x 25 m in plan dimensions. Foundation on 35 m deep piles was then completed promptly; columns were erected and the first floor slab was poured so that struts (metal pipes and wooden beams) could bear on the raker-diagonally-reinforced structure. Horizontal movements were kept to a minimum of less than 0.03 m, except in a zone where excavation had to procede reducing the berm; movements increased correspondingly to almost 0.13 m that were readily visible to the naked eye. In spite of this fact total movements were below 1% H.

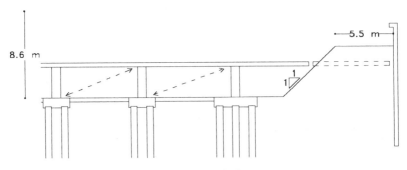

Figure 9. SALITRE PLAZA

Case 6. CEM Project. This particular office complex built on a lot 86 m (282 ft) x 86 m (282 ft) in plan was the first to be excavated to 10.6 m depth in this area newly developped on the way to El Dorado airport. A 15.0 m deep perimetral slurry wall was constructed from the floor of a first stage excavation made at elevation -3.0 m. It was provided with stilts or **trouser legs** [20] separated 8.5 m center to center and having 25.0 m long. Prefounded [20] columns within caissons were used to reach the pilecaps level of the first row of columns inside the excavation. It was believed that excavation of centers for elevators could procede safely. In fact, this was accomplished but at the cost of considerable distress and movement to the point that concrete facing of caissons was deformed to an ellipsoid. A row of 20 ton tiebacks was provided to help in restraining the upper part of the wall; even though it was proved afterwards that the elements sustained the load the movement allowed was of the same order of magnitude of the wall itself. Movement amounted to 0.12 m; something unheard of for promotors and developers. Construction proceeded from several centers toward the periphery where the caissons helped in obtaining an early support of the wall. Excavation was completed in a top-down sequence. In fact, the last m^3 of earth was moved out the day the structure was completed.

Case 7. Sheraton Project. Meant to serve for a Sheraton new Hotel in Bogotá, convention and office center, this project occupies a lot very similar to the one described in **Case 6,** above, and as a matter of fact the two are located only three blocks appart. The solution is similar to the one already mentioned, with one main difference: in this project it was decided to build a structural frame two or three bays width (See **Figure 10)**. The corresponding columns were erected within prefounded [21] caissons that reached the pilecaps level at an early date and permitted to pour the slab which was to serve as frame. Excavation proceeded in a top-down sequence while the structures were erected independently. Moderate heave was present even when a 12-story and 3-basement structure was finished on top. Considerable heave of 0.08 m to 0.10 m developped in a corner where excavation relief was not compensated by structural load. Movements were carefully controlled by inclinometers uniformily spaced along the wall, extensometers and complementary piezometers. A general FEM analysis was run for the project as a whole and then every stage of excavation was modeled for control purposes; a close agreement was attained between predicted and measured movements. **Figure 10** depicts how these movements progressed, depending on elevation, amount of excavation and structural bay. Maximum horizontal movement was registered in inclinometer No. 5 at elevation 12 m depth. It was clear the soil mobilized was at least twice the height of completed excavation.

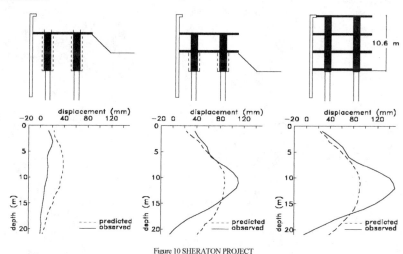

Figure 10 SHERATON PROJECT

Case 8. Maloka Center for the Advancement of Science. This peculiar project occupies a 60 m (180 ft) x 80 m (240ft) lot. The majority of the construction area is meant to be underground with a corresponding excavation of 8.6 m. Two problems have to be dealt with during the construction process; one was the stress relief ($\gamma z = 129$ kPa) that had to be counteracted by the strength of the clayey soil (Su Nc=132 kPa); and the other, that due to a thrifty investment policy the slurry wall -initially recommended- was supressed causing numerous localized slides one of which put in danger the domestic gas supply of western Bogotá. Good lessons were learnt now that a new stage of the project is being planned with much deeper excavations and meager counter-acting weights. This would conduce to a heavily over-compensated foundation and to the need of deep anchors to restore equilibrium.

Case 9, Riviera Project. This project was built in an area characterized by soft and exceptionally compressible soil deposits. Knowing this in advance every measure was taken to avoid failure. In fact, a surrounding wall 40 m (120ft) x 60 m (180ft) provided with trouser legs was built, corner beams were deviced to brace the wall and prefounded caissons were used to support a slab strap to bridge the lot across its middle line. Additionally, excavation was carefully planned in stages and top-down construction techniques were envisioned. All too good. Because then came pressures from the commercial department which was supposedly working faster than construction people could. The first stage of excavation was completed to the first underground level; cross bracing was built in concrete on 1st floor and 1st underground level. Excavation activities were accelerated and careful plans were broken because the geotechnical engineer was not corageous enough. Nature indeed was: soil plastification started, masses began to flow in search of deep seated balance, inclinometer movement was evident in records. When the make-shift structure accross the lot started to move out of plumb it was evident that general shear failure was a real menace. Fortunately in this case, the stand-up time was barely enough to allow for restoration of equilibrium. See **Figure 11**.

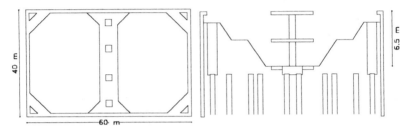

Figure 11. RIVIERA PROJECT

BACK TO THEORY

From Theory to Practice

If the profession has gone successfully from theory to practice under the expert guidance of the founding fathers it is necessary, 50 years after **Soil Mechanics in Engineering Practice** was published, that we continue going back to theory reviewing the fundamentals and seeking that expectations coincide with reality.

Dr. Laurits Bjerrum used to say in his lectures that it was necessary that the problem solved at the engineer's desk reflected reality. Our efforts cannot be founded, he said, on the terrible result of a problem correctly solved only that it was the wrong problem.

Geotechnical Indicators

It is recommended that a series of geotechnical indicators, based on simple but dependable calculations be part of the permanent arsenal at the engineer's work station. Some of those indicators come to the author's mind in the following way:

1. **The Factor N**. The factor N is omnipresent in the whole Soil Mechanics and Foundation Engineering. As stability factor it takes the value of 3.85 with Fellenius, and Taylor and goes to 5.52 for slopes in clay at any angle $\beta \leq 53°$ according to Taylor's circular failure analysis; for

vertical slopes and planar surfaces it becomes 4.0 in Therzaghi's Theoretical Soil Mechanics. With Peck, for braced cuts and slopes in soft cohesive soils, it has different connotations when valued 3.0, 4.0, 6.0, 7.0 or 8.0. Then as a Bearing Capacity Factor, appears as $\pi+2$ in Prandtl's smooth base analisis, goes to $1.5\pi+1$ in Terzagh's rough base, changes to 2π in Skempton's consideration of embedment which finally goes to 9.0 in the case of deep foundations. This should also be an omnipresent indicator in every day life of any geotechnical engineer.

2. **Compressibility C**. This parameter not only serves to evaluate the amount of consolidation taking place in any particular problem. It is proposed as a classification criteria and consequently as a very sensible indicator of the behavior of a mass of soil in dealing with the stability of soft clays (See **Figure 12**).

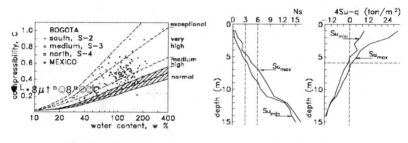

Figure 12. COMPRESSIBILITY   Figure 13. Ns and 4Su-q

3. **Ns and 4Su-q**. When dealing with excavations in soft ground it is very useful to know in advance how available strength is **invested** in holding the slope (Ns vs z) and how remainig passive force is used in supporting the base of the wall (4Su-q vs z). Illustrative examples for BGC are presented in **Figure 13**. When **Ns>6** and when **4Su-q→0** it is indicative that heavy supports are increasingly necessary and additional elements like stilts become mandatory.

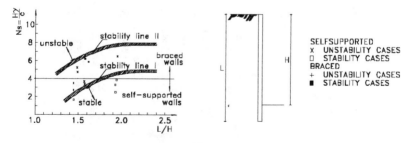

Figure 14. STABILITY CHART

4. **Stability Chart**. In **Figure 14** a stability chart is presented. It is meant to serve as an indicator for self-supported walls and braced walls; one stability line is proposed for each setting, as suggested by real cases, in the author's files, which are also plotted.

5. **Shear Wave Velocity** v_s. With the shear modulus, G, these have become the key parameters of **geotechnical earthquake engineering**. Particularly, v_s, is also a very valuable characterization element of soil deposits. It should also be used as an indicator with the restrictive connotation given here [1].

ACKNOWLEDGEMENT

From Ralph B. Peck we received the flag of Terzaghi that he himself helped to knit. The author had the honor to share in their legacy while being a student in Urbana and then passed it on to Gabriel Fernández, currently at the University of Illinois.

APPENDIX - REFERENCES

1. BORCHERDT, R.D. **New developments in estimating Site Effects on Ground Motion.** ATC 35-1, 1994.
2. BJERRUM, L. and EIDIE, O. **Stability of Strutted Excavations in Clay.** Géot. 6, 1956.
3. BOWLES, Joseph E. **Foundation Analysis and Design.** McGraw Hill, 5th. Ed 1996.
4. CLOUGH. G.W. and TSUI, Y. - **Performance of Tied-back Walls in Clay** ASCE JGD. Dec 1974
5. CLOUGH, G.W. and DENBY, G.M. **Stabilizing Berm Design for Temporary Walls in Clay.** ASCE, JGD, feb 1977.
6. ESPINOSA, Augusto. **Algo más sobre la Arcilla de Bogotá.** I Jornadas Geotécnicas SCI, 1976. (os)*
7. ESPINOSA, Augusto. **Geotecnia Sísmica de Bogotá.** SCI, SCG, VIII Jornadas Geotecnicas, II Foro sobre Geotecnia de la Sabana de Bogotá. Sept. 1995. (os)*
8. ESPINOSA, Augusto. **Indicadores Geotecnicos para una Problemática Renovada.** , VII Congreso Colombiano de Geotecnia, SCG, 1998. (os)*
9. ESPINOSA, A. y ARRIETA, A. **La Arcilla de Bogotá como Material de Ingeniería.** V Congreso Panamericano de Mecánica de Suelos e Ingeniería de Fundaciones. Buenos Aires, Argentina, 1975 (os)*
10. FERNANDEZ-DELGADO, G. and DUMAR, J. Engineering Properties of Bogotá Clay. Final Report to the Colombian Science Foundation, 1970. (os)*
11. HUBACH, Enrique. **Estratigrafía de la Sabana de Bogotá y Alrededores.** Boletín Geológico Vol. V No. 2. INGEOMINAS, 1957. (os)*
12. JULIVERT, Manuel. **Observaciones sobre el Cuaternario de la Sabana de Bogotá.** Boletín de Geología No. 7 UIS, 1961. (os)*
13. MESRI, Gholamreza. **New Design Procedure for Stability of Soft Clays.** Discussion of a paper by Ch, Ladd and R. Foot. ASCE, JGD, apr 1975.
14. PECK, R.B. **Deep Excavation and Tunneling in Soft Ground.** VII ICSMFI, México 1969.
15. PECK, R.B., HANSON, W.E. and THORNBURN, HT. **Foundation Engineering.** Wiley Int. 2nd Ed. 1974
16. SKEMPTON A.W. **The Bearing Capacity of Clays.** Proc British BHg. Rus. Conf., 1951.
17. TERZAGHI, Karl. **Theoretical Soil Mechanics.** John Wiley & Sons Inc. 1943.
18. TERZAGHI, K and PECK, R.B. **Foundation Engineering Practice.** 2nd. Ed. Wiley International. 1967. With MESRI, G. 3d. Edition, Wiley, 1996.
19. WHITMAN, R.V. **Are the Soil Depositions in Mexico City Unique?** Proc. of the ASCE Int. Conf. on the Mexico E.Q., 1986
20. XANTHAKOS., Petrus P. **Slurry Walls as Structural Systems.** 2nd Edition McGraw Hill 1994.
21. XANTHAKOS., Petrus P. et al. **Ground Control and Improvement.** Wiley Interscience 1994.
22. YOUSSEF,M. et. al. **Ground Movement Prediction for Deep Excavation in Soft Clay.** ASCE,JGD, june 1996.

* (os). Original in Spanish.

Tunnels Through the History of Seattle

Lee Abramson, M.ASCE[1] and Ralph Boirum, M.ASCE[2]

Abstract

The Denny Way/Lake Union CSO Control Project in Seattle, Washington consists of tunnels ranging in diameter between 42 in. (105 cm) and 14 ft 8 in. (4.5 m) and length between approximately 200 ft (61 m) and 6,200 ft 1,890 m). Seattle's history has had a dramatic effect on the subsurface conditions encountered on the project. The tunnels will cut through the heart of downtown Seattle and will encounter the original glacial and non-glacial period depositions, subsequent natural infills of valleys and shorelines with alluvium and colluvium, and man-made fill from regrading operations. Seattle has been the scene of massive cut regrading projects that have dramatically altered the topography of the city, including disposal of large hills, construction and reconstruction of sea walls and placement of shoreline fills along Elliott Bay and Lake Union. In addition to changing the topography, these industrial and infrastructure developments have sometimes contaminated the soil and water, impacting the procedures and costs of modern construction. This paper describes the multiple varieties of subsurface challenges encountered on this project related to tunnel and microtunnel construction and how these challenges were explored, evaluated, and are to be overcome.

---

[1]Director of Tunnels and Underground Services, Black & Veatch, 720 Third Avenue, Seattle, WA 98104
[2]Vice President, HWA GeoSciences, 19730 – 64th Avenue W., Suite 200, Lynnwood, WA 98036

Introduction

The Denny Way/Lake Union CSO Control Project is a joint undertaking of the City of Seattle and King County in Seattle, Washington. The City of Seattle is constructing two separate projects as part of the overall Denny Way/Lake Union CSO Control Project. The first project increases the capacity of the City's existing combined sewers along the east and south sides of Lake Union (Figure 1). This project eliminates two City combined sewer overflows through the installation of four flow-control structures and over two miles of pipe, ranging in size from 18-in. (45 cm) to 60-in. (150 cm) diameter. The second City of Seattle project eliminates an outfall into Lake Union, and links the expanded conveyance system with the storage constructed as part of King County's CSO control project. The County's Denny Way/Lake Union CSO Control Project consists of facilities that connect to the two City projects and accommodates flows. The net result of the combined projects is to reduce CSO overflow discharges in to Lake Union to no more than one untreated discharge per year. It will also reduce discharges at the Denny regulator station (into Puget Sound) to a minimum of 50 percent of the baseline CSO annual volume.

Project Description

King County's portion of the Denny Way/Lake Union CSO Control Project will include tunnels for the Mercer Street Tunnel, Lake Union Pipeline, South Lake Union Pipeline, Elliott West CSO Pipeline, Elliott West Effluent Pipeline, and Ancillary Pipe. The Mercer Street Tunnel will be a mile-plus-long conduit and storage tube extending beneath Mercer Street from near the south end of Lake Union to the shoreline of Puget Sound. It passes through the north side of Seattle Center in a commercial and residential neighborhood. At the eastern end of the alignment the Lake Union Pipeline and South Lake Union Pipeline will bring flow into the Mercer Tunnel from existing collection systems. Due to the depth of these pipelines and the presence of heavy traffic on surface streets, these pipelines will be constructed by tunneling with jacked or erected linings.

At the western end, effluent from the Mercer Tunnel, will flow into a new partial-treatment facility and then into an existing 108-in. (270 cm) ID pipeline (the Elliott Bay Interceptor, EBI) to the West Point Treatment plant or to a new marine outfall. Three new pipelines will be required leaving the partial-treatment facility. These include two 96-in. (240 cm) diameter pipelines and a 48 in. (120 cm) and 72-in. (180 cm) ID pipeline that will cross under eleven railroad tracks to a point near the shoreline of Puget Sound. Because the railroad tracks are very active and include two BNSF mainline tracks, these pipelines must be installed by trenchless techniques.

Design studies have determined that much of the alignment of the pipelines at the east and west end of the Denny Way / Lake Union CSO project consists of Areas that were filled during early development of the City of Seattle. Because the Seattle's history

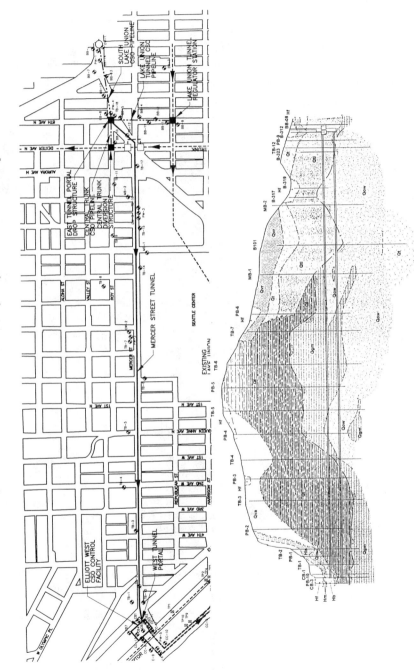

Figure 1 – Plan and Profile of the Denny Way/Lake Union CSO Project

has had a dramatic effect on the subsurface conditions to be encountered by these pipelines, and will greatly influence their construction.

Geotechnical explorations have disclosed several historic conditions that must be overcome during construction including the original glacial and non-glacial period depositions, subsequent infilling of valleys with alluvium and colluvium, excavation of large hills including the Denny Regrade and subsequent filling of ravines and shorelines, pile supported railroad trestles and wooden piers buried in uncontrolled debris fills, pile supported city streets now buried beneath regrade fills, sea walls and rip-rapped shorelines buried beneath fills along Elliott Bay and Lake Union, and previous industrial developments that have contaminated the soil and water.

This paper discusses the subsurface conditions anticipated for tunneling on the Denny Way/Lake Union CSO Project, and the design modifications resulting from these conditions.

Geology

Seattle is located in the central portion of the Puget Sound Lowland, an elongated topographic and structural depression bordered by the Cascade Mountains on the east and the Olympic Mountains on the west (Galster and Laprade, 1991). The Lowland is characterized by low-rolling relief and north-south trending hills and valleys with some deeply cut ravines. In general, the ground surface elevation is within 500 ft (152 m) of sea level.

Geologists have generally agreed that the Puget Sound area was subjected to four major glaciations during the Pleistocene Epoch. Ice for these glacial events originated in the coastal mountains and the Vancouver Range of British Columbia. The maximum southward advance of the ice was about halfway between Olympia and Centralia. Ice thickness in the Seattle area may have exceeded 1 mile.

The Pleistocene stratigraphic record in the central portion of the Puget Lowland is a complex sequence of glacially-derived and interglacial sediments. Erosion of certain deposits, as well as local deposition of sediments, further complicate the geologic setting. The oldest geologic unit encountered in the project area was Pre-Vashon glacial outwash and/or recessional soils deposited as the glacial ice retreated northward and eastward out of the Puget Sound Lowland prior to the last major glacial advance.

The glacial soils encountered in explorations for the Denny Way / Lake Union CSO project include hard, clays and silts from a large pro-glacial lake, which were subsequently overridden, outwash sands and gravels, basal lodgement till, and several ice contact and near-contact drift deposits. Local advances and retreats of glacial ice and location of the project site at the leeward end of a pre-glacial hill, has created a complex sequence of glacial soils. Except for the uppermost deposit of recessional outwash, these glacial soils are highly over-consolidated and are very dense or hard. The outwash sands and gravels are generally saturated, with permeabilities occasionally in excess of 0.1 cm/sec; however, the till, drifts, and clays generally have permeabilities less than 10^{-4} cm/sec. In general, the relatively dense glacial soils in the project area present predictable and mostly favorable tunneling conditions. Tunneling conditions in the fills at the eastern and western ends of the project are much less predictable and much less favorable for tunneling.

Tunneling Conditions at the Eastern End - South Lake Union

Since the turn of the century, the region of south Lake Union has been filled extensively. Records indicate the shoreline extended significantly further south, to about Mercer Street, and to about 9th Avenue on the west (Figure 2). The water level in Lake Union was typically about 10 ft (3 m) higher than today's level. Near the turn of the century, more than 20 sawmills ringed the shoreline of Lake Union, including one mill located at the present site of Valley Street between Westlake and Terry Avenues. This mill had a log slide that extended across the present Valley Street right-of-way into the lake. This ramp was apparently pile supported. Available records also indicate that Westlake Avenue consisted of a pile-supported plank road from the south side of Valley Street northward. It is likely that wood foundation piles for both Westlake Avenue and the log slide are still present along the proposed pipeline route. The fill in the area also includes substantial deposits of sawdust from the mills.

Tunneling Conditions at the Western End - Elliot Bay Shoreline

Three pipelines are to be installed by trenchless methods at the western end of the project near the shoreline of Elliott Bay. Two have 96-in. (240 cm) inside diameters, and the third will be 48 in. (120 cm) and 72 in. (150 cm) in diameter. They will be parallel to each other and approximately perpendicular to the shoreline, passing beneath eleven sets of railroad tracks and an existing 108-in. (270 cm) diameter sewer pipeline. The proposed pipeline alignments are shown on Figure 3.

Within the last 100 years, substantial fills have been placed to progressively move the shoreline of Elliott Bay westward from the base of the bluff along the uphill side of Elliott Avenue to its current location. Typically, this filling has pushed the shoreline out more than 600 ft (183 m). The railroad along the waterfront was initially built on pile supported trestles over the tide flats. Numerous piers and wharves and some city streets were also pile supported over the tidelands. As the tide flats were filled many of these structures were buried.

Figure 2 – Old South Lake Union Shoreline

Figure 3 – Plan of Elliott West Pipelines and Old Shoreline

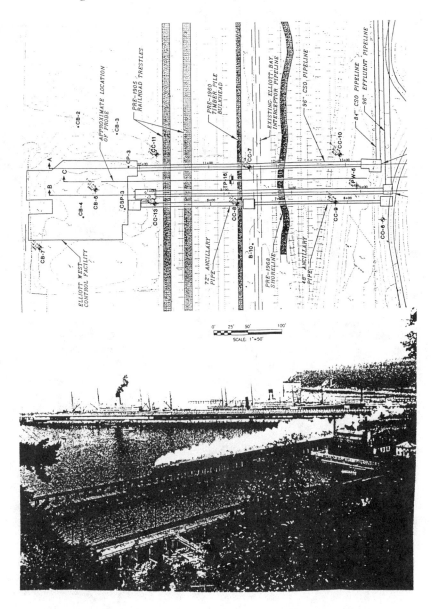

Explorations for the Denny Way / Lake Union CSO project indicate fill soils are present at the ground surface along all three tunnel alignments. Port of Seattle and Burlington Northern maps and historical aerial photos indicate at least three episodes of filling in this area. Near the turn of the century, the shoreline was east of the railroad tracks, near the eastern side of Elliott Avenue. At that time two sets of railroad tracks were located offshore on pile-supported trestles. Those piles are assumed to still be in place. At some time prior to 1960, and possibly in more than one episode of filling, the shoreline was extended to a point west of the existing BNSF tracks and east of the EBI pipeline. A timber pile bulkhead sea wall with loose riprap protection was reportedly constructed along that shoreline. Between 1960 and 1968 a second fill was placed extending the shoreline about 30 to 70 ft (9.1 to 21.3 m) to the west. Riprap shoreline armor was placed along the 1968 shoreline, and in 1968, the EBI pipeline was placed in a trench within this new fill. A profile illustrating the conditions along these pipeline alignments is shown on Figure 4.

According to Port of Seattle records, between 1968 and 1970, fill was dredged from near shore sediments and placed to extend the shoreline about 260 ft (79.3 m) further west to its present position. The dredged soils from this latest filling episode are considerably more uniform than the fills to the east. Dredge fill soils are predominantly clean sands with fine to coarse gravel and some zones of silt and clay, particularly to the south. Some cobbles were observed but boulders were not encountered within this unit during explorations. The dredged fill is generally loose to medium dense, becoming dense in some locations. SPT blow counts range from 2 to 60.

The older fills located east of the dredged fill consist predominantly of gray sandy silt to sandy clay, with large amounts of gravel, cobbles, boulders, and debris, including lumber, concrete, bricks, plastic, metal, wire, etc. These fills are generally loose to medium dense or soft to stiff and had SPT blow counts ranging from 2 to greater than 50 blows per foot.

While drilling, one of the borings had to be moved several times to the west after encountering refusal, apparently on the riprap shoreline armor. A zone of organics was encountered near the base of the older fill in Boring CC-7. A number of test pits were excavated in the fill materials along the railroad tracks, and all test pit sidewalls caved when groundwater was encountered during excavation.

Probings using horizontal directional drilling techniques confirmed the presence of riprap both east and west of the EBI pipeline. The locations of the old rip-rapped shorelines encountered by horizontal directional drilling are shown on Figure 4.

Past Tunneling Experience in the Seattle Area

The following section presents discussions of some of the relevant tunnel projects and the tunneling procedures used to complete tunnels through glacially overconsolidated

Figure 4 – Profiles of the Elliott West Pipelines

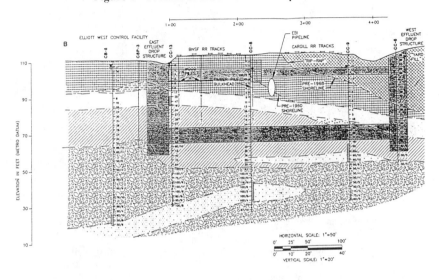

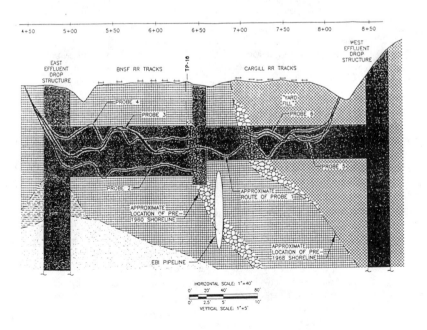

soils in the Seattle area. Tunneling methods utilized in the Seattle area have progressed in kind historically with the industry elsewhere in the U.S. and abroad. These tunnels are discussed in more detail below:

Lake Union Tunnel: The Lake Union Sewer Tunnel was constructed over a several year period in the late 1880's to 1890's. The tunnel extends from the south end of Lake Union from the intersection of Westlake and Republican Street, westward to 6th Avenue, and thence diagonally southwest to Denny Street and the current location of the outfall. The roughly 8-ft (2.4 m) wide by 8-ft (2.4 m) high tunnel had a total length of about 5,400 ft (1,646 m).

Project records mention the presence of "hardpan" (glacial till or glacial marine drift), lacustrine clays with saturated sand seams, and sands and gravel (outwash sand). As the tunnel progressed, thick seams of water bearing sand, boulders, and fractured clays were encountered. The fractured clays, coupled with the water flows led to face instability and caving conditions that necessitated the use of steel rail spiling to support the flat roof. One roof fall amounted to over 1,400 c.y. (1,077 c.m.) of material.

Groundwater inflows were reportedly so severe in this stretch of tunnel that the contractor resorted to drilling six 3-in. (7.5 cm) diameter pipes into the arch in order to tap the inflows. The pipes penetrated about 6 ft (1.8 m) of clay and then encountered a sand seam that yielded several hundred gallons per minute for several weeks.

The tunnel was accomplished by hand mining using shovels, picks and probably clay spades with the muck dumped into a rail mounted car and shoved by manpower out of the tunnel. Support of the 8-ft (2.4 m) by 8-ft (2.4 m) rectangular opening was provided by timber sets and cribbing. Advance rates were on the order of 10 ft (3 m) to 20 ft (6 m) per day.

Mt. Baker Ridge Tunnel: The Mt. Baker Ridge Tunnel was constructed in 1986. The tunnel is located about 2 mi. (3.2 km) southeast of downtown Seattle, adjacent to Lake Washington. The tunnel is about 1,360 ft (415 m) long, with an outside diameter of 84 ft (25.6 m) and a completed inside diameter of 63 ft (19.2 m). The tunnel was constructed by a stacked drift method which formed a flexible tunnel lining consisting of 24 individual 9.5-ft (2.9 m) diameter tunnels that were first constructed and backfilled with concrete to form a compression ring, followed by removal of the interior soils.

The tunnel was constructed primarily through glacially overridden, hard lacustrine silts and clays as well as minor amounts of dense outwash sand, and fill. The water table is 20 ft (6.1 m) to 40 ft (12.2 m) below the bottom of the tunnel. The more massively bedded clays and silts contained cobbles and boulders up to 10 ft (3 m) in diameter. Boulders from 3 ft (0.9 m) to 10 ft (3 m) were encountered at least 8 times in the approximately 6 mi. (9.7 km) of 9.5-ft (2.9 m) diameter tunnels.

The tunnels were constructed with an open faced, digger shield with support provided by an expanded, 5-piece concrete segment liner. Muck was removed with narrow gage locomotives and 2 cy cars.

Downtown Seattle Transit Project: The Downtown Seattle Transit Project travels beneath Pine Street and Third Avenue. The "L"-shaped alignment is about 1.3 mi. (2.1 km) long, and includes terminal stations at each end and three intermediate stations (Figure 5). The east-west leg of this project was constructed primarily by cut-and-cover methods, while the 5,000-ft (1,524 m) north-south leg of the project including a 285-ft (86.9 m) curve was constructed by bored tunnel methods. Tunneling began on May 29, 1987, from the south portal. Tunneling was completed on March 24, 1988, when the west shield holed through the west wall of the Westlake Station excavation.

The northern 3,200 ft (976 m) of the alignment consists of pre-Vashon glacial deposits of advance and recessional outwash, dense sands and gravels, glaciomarine drift (GMD), and glaciolacustrine hard silty clays and clayey silts. The glacial deposits overlie the Duwamish Formation, which comprises most of the southern 1,800 ft (549 m) of the alignment. The Duwamish Formation consists of older, interglacial, fluvial silty sands and lacustrine/estuarine silt, clay, and silty fine sand with local layers of volcanic ash, pumice, and hard peat.

Double-articulated digger shields with orange-peel breasting doors were supplied by the Robbins Company and were approximately 21.5 ft (6.6 m) long and had an outside diameter of 21.25 ft (6.5 m) with a 1-in. (2.5 cm)-thick overcut. In lieu of the specified one-pass grouted and gasketed segmental lining, the contractor elected to install reinforced concrete segments 6 in. (15 cm) thick, which were expanded out against the soils for initial support of the running tunnel section. Initial support in station areas consisted of steel ribs and timber lagging, which were easily removed when exposed during station excavation. Final support included a continuous double-welded PVC waterproofing membrane 0.06 in. (1.5 mm) thick and a cast-in-place unreinforced concrete lining 12 in. (30 cm) thick.

West Seattle Tunnel: The Alki Transfer/CSO Project was designed to convey sewage approximately three miles from the Alki Treatment Plant to the Elliott Bay Interceptor. The West Seattle Tunnel included construction of a 10,000-ft- (3,049 m) long, 13-ft-(4.0 m) diameter tunnel. Tunnel construction began in February 1995 and was completed in 51 weeks. It traversed glacial clays, silts, sands, gravels, and cobbles, all below the groundwater water table. The tunnel liner consisted of gasketed, concrete segments, with contact grouting used the fill the annular void between the excavated ground perimeter and precast liner segments.

Figure 5 – Map of Previous Tunneling in the Seattle Area

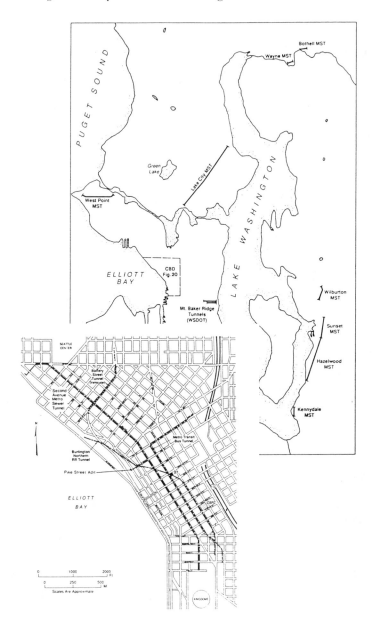

Conclusions

The Denny Way/Lake Union CSO Control Project in Seattle, Washington consists of tunnels ranging in diameter between 42 in. (105 cm) and 14 ft 8 in. (4.5 m), and in length between approximately 200 ft (61 m) and 6,200 ft (1,890 m). Seattle's history has had a dramatic effect on the subsurface conditions encountered on the project. The tunnels will encounter a variety of tunneling conditions including glacial deposits, infilled valleys, regraded hills, reconstructed sea walls and shoreline fills, and previous industrial and infrastructure developments that have contaminated the soil and water. Tunneling is expected to begin in Summer 1999.

Acknowledgements

Much of this paper is based on studies and evaluations carried out for the Denny Way/Lake Union CSO Project. The owner/agencies for this project are the King County Department of Natural Resources and the City of Seattle, Washington. Tunnel designs are being led by Black & Veatch, Rosewater Engineers, and Cosmopolitan Engineering Group. Geotechnical consultants on the project include Shannon & Wilson and HWA Geosciences. The authors gratefully acknowledge the contribution of their fellow team members and colleagues on the project including Judy Cochran, Calvin Locke, Vern Threlkeld, Val Wahbeh, Anne Symonds, Dan MacCarthy, Tom MacBriar, Ron Hines, Harvey Parker, Gary Brierley, Andre Marè, Jim Alexander and Red Robinson.

References

Black & Veatch (1998), "Geotechnical Baseline Report, Mercer Street Tunnel," Prepared for King County Department of Natural Resources, December.

Galster, R. W. and W. T. Laprade (1991), "Geology of Seattle, Washington, U.S.," Bulletin of the Association of Engineering Geologists, August.

HWA GeoSciences Inc., (1998), *Geotechnical Report, Denny Way / Lake Union CSO Project, South Lake Union Pipelines, Seattle, Washington*, consultant report dated November 20, 1998.

HWA GeoSciences Inc., (1998), *Denny Way/Lake Union CSO Project, South Lake Union Pipelines, Valley Connection Structure, Dewatering Evaluation, Seattle, Washington*, November 20,1998.

HWA GeoSciences Inc., (1998), *Geotechnical Report, Denny Way / Lake Union CSO Project, Elliott West Pipelines, Seattle, Washington,* November 6, 1998

Geologic and Geotechnical Considerations
for Pressure Tunnel Design

Andrew H. Merritt[1], M. ASCE

Abstract

The design of pressure tunnels for hydroelectric projects or other water conveyance systems requires the understanding of a number of geologic and geotechnical variables that include rock permeability, rock modulus, susceptibility of the rock to erosion or solution, and *in-situ* minimum rock stress. The definition of the groundwater table also is a key element for the design. Given the number of pressure tunnels constructed over the last 25 to 30 years that have not performed in a satisfactory manner, a review of the essential design requirements and geologic and geotechnical parameters is required and is given herein. Where the ground conditions have not been adequately addressed, excessive leakage of high pressure water has occurred during first filling and subsequent operation. The loss of water may be of such magnitude as to be financially unacceptable and the introduction of quantities of high pressure water into the rock mass may present hazards to adjacent facilities The costs involved in tunnel lining repair as a result of an inadequate definition of the rock conditions have been substantial in terms of lost energy generation, interrupted water supply to major centers of population, and for repairs done under difficult conditions. Because the cost of tunnel linings represents a major percentage of the total tunnel investment, the optimization of the lining design is an essential element of any water conveyance project.

Introduction

Site characterization for underground works presupposes a knowledge of the variables involved in rock excavation; principally the geological factors influencing rock support and groundwater inflow. Thus the procedures commonly used to determine

---

[1]President, Andrew H. Merritt, Inc., 7726 SW 36 Avenue, Gainesville FL 32608

anticipated underground conditions including: tunnel experience in similar rocks, regional and local groundwater flow systems, geologic mapping, exploratory drilling, core logging, water pressure testing to determine bedrock transmissivity, and rock classification may be suitable for tunnel excavation in general; however, this information may not be sufficient to define the geotechnical variables needed for pressure tunnel design.

When a review is made of water conveyance tunnels that have leaked unacceptably, it is found that the problems began with an inadequate or misleading site exploration program combined with the failure to make the appropriate observations during construction. This is not particularly surprising because there appears to be no systematic procedure used in the industry to define those variables necessary for pressure tunnel design. In many countries where the construction of hydroelectric projects has been decreasing, the profession has lost experienced design engineers and site geologists. Inexperienced project staff may not be adequately prepared to make the correct selection of various types of exploration methods or able to make the required observations and design decisions during tunneling.

Regarding the required observations during construction, other factors may be considered. Commonly the people who performed the original site work are not present during construction, thus there is limited continuity between these phases of the project. In addition, designers often are not involved in the construction - regardless of common perceptions related to design and construction practice. And finally, the construction managers and supervisors, of necessity, are more concerned with solving the construction-related problems and dealing with the Contractor than they are involved with the design. Thus the pertinent observations that could not be defined during the exploration program, no matter how detailed, often are not made during construction.

Moreover, considering the design-build trend of future projects, there may be insufficient time during the pre-construction phase to fully assess the ground conditions that later could lead to cost overruns or operational problems.

It is generally accepted in the tunneling profession that exploration and observation must continue throughout a project and therefore that the design is not finished until the project is finally terminated. This must be clearly understood for any water conveyance project and is especially true for pressure tunnels, as will be explained in more detail below.

Design Criteria - Statement of the Problem

General Comments   Simply stated, the challenge for the geologists and geotechnical engineers involved in water conveyance projects is to control the flow of water.

For that purpose a logical progression of events follows. The flow chart shown on Figure 1 defines the rock properties that must be determined to accomplish the design of a

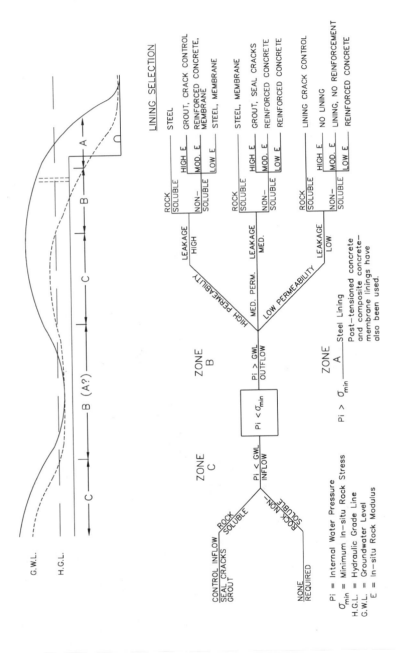

Figure 1   Pressure Tunnel Lining Selection: Geotechnical Considerations

pressure tunnel.

Definition of Groundwater Table   The hydraulic grade line is defined as the line of internal pressure governed by the level of the reservoir minus friction losses to which the water in the tunnel will rise. Where a stable groundwater table exists above this pressure surface, leakage from the tunnel will be minimal (Figure 1, Zone C). In this case, flow is towards the tunnel unless other project-related underground works function as permanent drains and draw down the original water table and permanently change the groundwater regime. During the initial exploration program the groundwater table needs to be defined.

Whereas some might argue that the term groundwater table is not appropriate to rock formations, the fact of the matter is that with an adequate number of borings, supplemented by the correct observations made during drilling, and followed by long term observations of water levels recorded in piezometers installed in the drill holes, a water table usually can be determined. Perched water levels do occur in rock; careful monitoring of water levels during drilling may assist in avoiding this possible confusion. It is recognized that a water table in some rocks may be difficult to define; the case of a massive granite or other igneous rocks with widely spaced and tight fractures having a low permeability would be one example.

Depending on the permeability of the rock mass and the depth of the tunnel, it should be expected that the water table will be drawn down partially or fully to tunnel invert elevation during construction. Experience has shown that the water table will recover, at least to the internal pressure line, following completion of the concrete lining and contact and/or consolidation grouting. Depending on the rock and tunnel lining permeability this recovery may take some time after tunnel filling. Thus, if the water table was originally high with respect to the hydraulic grade line (and leakage was therefore not considered to be a potential problem) these conditions can be expected to persist post tunnel filling. Some leakage during first filling is to be expected; however, it normally decreases with time.

Zone C generally does not require any lining for water control purposes and a concrete lining would be used only if the rock conditions require long term protection or the Owner or Engineer believe that it is good practice to line the tunnel regardless of the rock conditions. It may be decided that the exposed rock cannot withstand long term exposure to water and needs protection. This commonly occurs in the softer rock formations, in certain extrusive volcanic rocks containing minerals susceptible to swelling with subsequent deterioration, or for rock susceptible to solution and erosion. It is accepted that soft rock zones such as faults or shear zones and weathered seams can be protected with local concrete or shotcrete linings, as required.

Topographical conditions will cause the water table to rise and fall with respect to the internal pressure line (Figure 1) and eventually the water table will fall well below the internal pressure line at the downstream end of the tunnel (Zones A & B). A review of case histories of excessive tunnel leakage indicates that it is the downstream end of pressure tunnels where most of the problems have occurred. A review of the variables involved in Zone B readily explains this situation.

Rock Stress, Modulus, and Permeability   For the sections of tunnel located within Zones A and B the principal design parameters are governed by minimum *in-situ* rock stress, rock modulus, and rock permeability.

Zone A is defined as the section of tunnel where the minimum in-situ rock stress is less than the internal hydraulic pressure (not including surge which could add an additional 15% to 20 % to the static internal pressure). In this section of the tunnel, a steel lining is invariably used. It is noted that other alternatives to steel lining have been employed; i.e., post-tensioned cable reinforcement imbedded in the concrete lining or some form of "sandwich" lining consisting of a thin steel or plastic membrane inter-layered between two concrete linings or a shotcrete lining and concrete lining (Olive, 1975, Swanson, 1981). The post-tensioned and sandwich linings commonly are not used; the industry preference being for thicker steel linings. Generally the "sandwich" type linings are more common where the tunnel has not performed as planned and repair is necessary or in the situation where original design assumptions have not been validated during excavation and changes have had to be made during construction and the option of extending the thicker steel lining is not practicable.

The purpose of these linings is to take the full internal water pressure and avoid hydraulic jacking or fracturing of the rock that would release a variable quantity of high pressure water into the rock mass adjacent to an underground or surface powerhouse, water treatment plant, and adjacent high rock slopes. Hydraulic jacking has occurred in pressure tunnels and surge shafts on several projects. It is a serious situation to be avoided.

Zone B is that section of the tunnel where the groundwater table is below the hydraulic grade line and permanent leakage is theoretically possible. In this section of the tunnel, the minimum in-situ rock stress has been proved by field testing to be greater than the internal water pressure, thus hydraulic jacking, with its associated increase in rock permeability, is no longer a design consideration.

In Zone B of the water conveyance system, the tunnel lining may be either of reinforced or non-reinforced concrete depending on the in-situ rock modulus and the permeability (Fernandez, 1994). The erodibility or solubility of the rock mass also needs to be considered; however, these are somewhat special cases and will not be treated herein. It should be remembered that if the steel liner is relatively short, a high gradient will exist between the tunnel and, for example, an adjacent underground powerhouse and the erodibility of local soft rock zones must be considered.

In the most favorable conditions, a stiff rock mass of low permeability may need no lining or only a non-reinforced concrete or shotcrete lining. A high modulus rock mass of modest to high permeability may need only grouting behind a shotcrete lining to make the tunnel sufficiently water tight. The shotcrete lining, in this case, may serve only as a grout cap. If the rock is deformable under the internal operating pressure, a plain concrete lining may experience substantial longitudinal cracking and substantial quantities of water may leave the tunnel. This is of additional concern if the rock is pervious and if grouting cannot be relied upon to control leakage. The determination of the in-situ rock modulus of deformation is a design requirement under these conditions. Some years ago, a non-reinforced concrete lining in a low pressure (less than 8 bars) tunnel in Central America cracked excessively and about 4 m^3/s of water entering the intake did not reach the powerhouse at the downstream end (Deere, 1983).

Regarding the use of reinforcement in tunnel linings to control water leakage, some points can be considered. The first is that reinforcement should be be used only where required because its inclusion affects the quality of the concrete; segregation and honeycombing is common. Secondly, the reinforcement does not prevent cracking but distributes the cracks more evenly thereby reducing the amount of opening of the individual cracks. Thus the inclusion of reinforcement (perhaps 0.5 to 0.75%) mostly is to control the size of the cracks. Consolidation grouting of the rock mass surrounding the tunnel generally is combined with these reinforced linings.

<u>Unlined Tunnels</u>   There are many unlined water conveyance tunnels that have operated successfully for years without difficulties. Considering that the cost of a plain concrete lining may be around one-third of the total tunnel cost, the desirability of eliminating the lining is obvious. The writer has inspected unlined tunnels in hard rock after 25 to 30 years of operation and notes their success. Experiences of tunnel operators have been presented in the literature (Thomas and Whitham, 1964 and Riste and Lysne, 1987).

In selecting an unlined tunnel, one must accept the fact that there is a procedure that should be followed with this design option. The extent of the unlined sections can be determined only after excavation has been completed. The implication of such contract additions/modifications therefore has to be considered in the original contract documents. Following excavation the essential step is that the tunnel be mapped geologically in detail wherein all soft erodible seams are identified as well as all potentially unstable rock wedges that could lead to major rock falls. Once this has been done, the tunnel is inspected by the designers or appropriate parties and the final local rock treatment is defined. This local treatment may consist of short concrete lined sections, local rock anchoring, or pattern bolting with or without mesh-reinforced shotcrete. Anchored mesh-reinforced shotcrete or gunite has been proven to be effective in controlling erosion of soft or closely fractured rock. The rock bolts used in this procedure are fully grouted or resin encapsulated steel bars; i.e., rock dowels. The mapping, inspection, and support selection process is somewhat subjective and judgment is required; however, there is no alternative to its implementation.

Site Investigations

Borings   For long water conveyance tunnels that pass beneath mountainous terrain resulting in substantial cover over the tunnel and eventually terminating at a surface or underground generation or treatment plant, the cost of subsurface exploration can be substantial. All projects are faced with economic realities and the amount of exploration has to be controlled. Thus it is imperative that each boring be so located, drilled, and tested to satisfy basic design objectives. Boring logs should note the presence of special circumstances that could affect lining selection and design such as zones of water loss or gain, squeezing conditions (probably in fault zones), presence of gas, and abnormal water chemistry such as high pH (acidic) or high sulfate content.

There is a reasonable practical depth for borings for civil works of around 500 m. Thus where the rock cover exceeds 500 m one should not expect exploratory borings to go to tunnel grade. This fact is commonly accepted in the profession. Moreover there is no established boring frequency number and, for long deep tunnels, borings may be centered at hundreds of meters or kilometers. One should not apply a boring frequency criterion, which is based on subway or sewer projects in cities, to hydro or water conveyance tunnels in more remote locations. Clearly some judgment is necessary.

Water Table   Recalling the basic design criteria discussed above, the borings would be initially located following regional and site specific geologic mapping to determine the necessary design requirements. All borings drilled along the tunnel route need to record water losses, the position of the groundwater table, and to test for values of rock transmissivity in zones of suspected open fracturing. Piezometers should be left in all borings and read on a regular basis through to completion of the project.

At the end of the exploration program, there must be an understanding of the position of the water table; i.e., in simple terms whether or not it is above or below the system hydraulic grade line. If the water level fluctuates dramatically on a seasonal basis, the interest is in the lowest water level during the dry periods of the year. Under normal geologic conditions, the water table follows the topography; however, in karstic solution-prone limestone the water table is unpredictable and may be at or below major water courses and is expected to be below smaller tributaries. Water levels may likewise be unpredictable in highly pervious volcanic rock because of lava tubes and vastly different rock permeability between the different layers of lava and along intraflow materials.

In-situ Rock Stress   As noted above, steel linings are provided for those sections of tunnel where the internal static water pressure is greater than the minimum rock stress. In the past it was common to assume that the minimum rock stress was governed by the unit weight of rock (rock cover) above the tunnel. Case histories of Scandinavian tunnels are provided in the literature as are interesting formulae that attempt to accommodate the geometry of the topography. Experience has shown, however, that these empirical relationships are misleading and, in the writer's opinion, should not be used for design purposes. It is now understood that rock stress relief does occur, generally related to topographic and structural geological conditions, for which its prediction by such simplified relationships is not possible or reasonable. In short, the *in-situ* minimum rock stresses must be measured in the field.

There are a number of methods by which stress measurements are made such as hydraulic fracturing tests in deep boreholes and over-coring measurements in shallow borings drilled from exploration galleries. These techniques provide the full stress field; however, they are costly and are not necessary for a pressure tunnel. What is required is only the minimum *in-situ* stress with no need of knowing its direction. Thus, common practice is to perform hydraulic jacking tests. These tests are a variation of the water pressure pumping test or Lugeon test wherein water is pumped into the borehole slowly, in stages, until the desired pressure is reached or the pre-existing rock fractures are opened and the flow in the test increases notably. This test is believed to more closely simulate what would occur in a pressure tunnel and the results are taken as an indication of the minimum *in-situ* rock stress. It is understood that the desire is to open pre-existing fractures and not to create new fractures.

The borings wherein the tests are made are oriented to cross the fracture systems and especially those fractures that are expected to have been affected by stress relief. The tests are relatively simple to perform although the resulting graphs of pressure vs. flow may be difficult to interpret. This is especially true if the rock is pervious.

Hydraulic jacking tests essentially can be done at any depth in boreholes commonly drilled for civil works. The groundwater level has to be noted as the calculations consider effective stress. This approach is taken because it is recognized that the water table at the downstream end of a pressure tunnel will be permanently drained by adjacent access tunnels or other underground openings. Based on these data, the length of the steel lining can be estimated. However, there never will be enough borings and tests to accurately define the length of steel lining and additional tests must be made during construction to finalize the length of lining.

Anticipating that the tests done during construction may indicate more steel lining than shown on the construction drawings, provisions commonly are made in the bill of quantities for incremental lengths of steel lining.

In-situ Rock Modulus   Where the stiffness of the rock is insufficient to resist the internal water pressure, plain concrete (or even lightly reinforced concrete) linings have cracked and substantial longitudinal cracks have formed and opened to significant widths; several millimeters of opening have been recorded. For expected deformable rock, the tunnel designer will need a range of modulus values for parametric analyses in the early design stage that can be obtained during the exploration stage, perhaps by using one of the rock index methods. Eventually *in-situ* testing by plate jack methods probably will be required. Geophysical tests (Carrere, 1995) have been used on many projects; however, considerable judgment is required with this method. There is no reliable borehole method for measuring the required rock modulus values; borehole jacks sample too small a volume of rock to be reliable.

Rock Transmissivity   For rock tunnels, the common method of estimating rock permeability or transmissivity is by pumping-in tests performed in boreholes. The test is performed according to standard practice and the values expressed in Lugeon Units. The values obtained from such tests are approximate only and the gross permeability of the rock along the tunnel eventually needs to be defined in more detail during construction. Rock transmissivity measurements accomplished by pumping-out tests with observation wells are more representative tests but are not done for most tunnels, principally because of the greater cost. Eventually geologic mapping will define the areas of highest apparent permeability and pump-in tests can be made to obtain values for design purposes. Records of initial water inflow during tunneling are especially useful to highlighting areas of concern.

Concluding Remarks   At the completion of the exploration program a tunnel profile should be prepared similar to that shown on Figure 1. The general groundwater table is shown along with the hydraulic grade line. The probable zone of low minimum rock stress is defined as is the general rock permeability between the upstream end of the low stress zone and general area where the groundwater table rises above the hydraulic grade line. An evaluation has to be made on the likelihood of groundwater level rebound following tunneling and subsequent lining; i.e., will any of the construction practices or design elements result in permanent groundwater drawdown?

Once the field data have been analyzed, the design can be approximated; the length of the steel lining, amount of reinforcement, extent of concrete lining, and grouting details can be defined and shown on the contract drawings and reflected in the bill of quantities. The contract also needs to provide for probable changes in the type and extent of linings that will be defined during construction. Provisions for rock testing and adjustment in the final quantities should be defined in the contract and the testing process and possible additional materials subjected to competitive bidding.

Investigations During Construction

General Practice   It is accepted practice that the exploration for a water conveyance tunnel (or for any underground project) continues during excavation and therefore provisions have to be made in the project contract and budget for modifications to the bill of quantities and initial project cost. It should be understood that no amount of subsurface exploration can define the full range of probable rock conditions that will influence the design and performance of a pressured water conveyance tunnel. Therefore adjustments to the lining design are inevitable.

Assuming that not all members of the project geological/engineering design team have the same degree of experience, some level of common understanding needs to be achieved. The writer believes that this is best done if the original designers prepare a Design Memorandum wherein all the assumptions and calculations used for the design are included. With this document in hand and with some guidance, the field personnel have a basis for making the necessary observations of the geologic conditions along the tunnel to insure that the initial design assumptions remain valid.

The Design Memorandum is the essential document to insure continuity of the original exploration program, the project design phase, and final construction. It forms the guide for the final stage of the project; i.e., the required observations to be made during excavation and finalization of the lining design.

Geologic Observations   All tunnels are geologically mapped during construction noting the lithology and frequency, continuity, and characteristics of rock fractures and faults. The rock may be classified by the RQD, Q, and/or RMR methods. Consistency is desirable.

Observations of groundwater inflow are made noting initial inflows and rate of dissipation of flow. If possible, measurements of the initial water pressure should be made from time to time. This is done in specially drilled boreholes near the tunnel face using an inflatable packer and manometer. These data can provide a check on the level of the groundwater table used in the design. The piezometers installed in the original surface boreholes should be read regularly; the rate of decrease of the water table as the tunnel passes beneath the instrument may be useful in predicting the future rate of rise of the water table when the lining is completed and before the tunnel is filled.

The downstream end of a pressure tunnel that contains the surge tank and penstock shaft are often located in topographic ridges to accommodate the surge tank as an underground structure. These ridges are commonly narrow and the rock therein may be stress relieved. This phenomenon has been observed on many projects and the degree of stress relief confirmed by *in-situ* measurements. In these cases the minimum rock stress often bears no relation to rock cover. During excavation in these ridges, the open nature of the rock fractures may be apparent and attention should be paid to this situation during the geologic mapping of the area. Hydraulic jacking of the rock has occurred in surge tanks and the upper portion of penstock shafts on several projects.

In-situ Stress Measurements   It is now common practice to perform simple hydraulic jacking tests at the upstream end of the design limit of the steel lining to verify that the minimum *in-situ* rock stress is above the static operating pressure of the tunnel. It is accepted that, during the exploration period, not enough tests could have been done to fix the limit of the steel lining. In the writer's opinion, these final tests should be performed on all projects.

One should anticipate performing at least 20 tests total in a number of borings drilled in varying orientations to insure testing the relief joints. There should be a 20% safety factor applied to the data. Tests should continue until there is some consistency to the amount of pressure causing the rock fractures to open.

Rock Permeability   There are situations where the permeability of the rock along the tunnel should be determined. For example, the rock upstream of the beginning of the steel liner may appear not to require any concrete lining and the only consideration is the permeability of the fractures. In another situation at the same relative location, the rock may appear to be deformable and, if the tunnel is lined, the concrete may be expected to crack and water loss would likewise be of concern if the rock is pervious. In both these situations, Lugeon tests are advisable. Any Lugeon values above 10 would be of concern and would require grouting. If the rock is deformable and not considered groutable, induced hydraulic gradients may lead to erosion of the surrounding rock. Heavy reinforcement, sandwich-type linings or other techniques may be considered.

Rock Modulus   Softer sedimentary or volcanic rocks or otherwise deformable closely fractured zones may require specialized linings including reinforcement. In these cases a knowledge of the rock modulus would be useful and *in-situ* testing is required.

Two basic approaches to determining rock modulus are common. The first, and better of the two, is to perform plate jack tests using a plate diameter of at least 0.80 m. The second method, and less direct, is by geophysical testing adopting either the so-called "Petite Sismique" (Scarabee) method or a calculation of dynamic modulus using compressional or shear wave velocities. Both methods use field established relationships between dynamic and static moduli. The seismic methods have to be used with caution because there is considerable scatter in graphs of the static vs. dynamic test values. It seems that the best approach is to classify the rock along the area of tunnel in question using geophysical methods and then test the areas of concern with the plate jack method. For soft rocks, having an *in-situ* modulus less than about 1 GPa, rock creep is likely and the plate jack test loads should be maintained until deformations have ceased. In this case, the rock modulus may decrease with time.

Contract Matters

There have been a number of water conveyance tunnels that have not performed as anticipated mainly related to hydraulic jacking of the surge tank or tunnels in the vicinity of an underground powerhouse or to a large amount of leakage elsewhere along the tunnel. Reviewing the history of these projects, it is apparent that detailed observations were not made during excavation and that *in-situ* testing before the lining was constructed would have been beneficial. Where an inadequate design was provided, the cost to the project to correct the deficiency, either for the direct cost of repairs, lost generation of energy over a 1 to 2-year period during the repairs, or added interest payments over the same period, has been substantial. The reaction of the project's Owner in these cases was not complementary to the Engineer. Although the Contractor has nothing to do with the design and normally should be held blameless, there have been situations where the Owner has sought some compensation from the Contractor perhaps because of a perceived more favorable financial situation when compared with the Engineer.

One possible remedy for this situation is for the Engineer to clearly inform the Owner of the variables inherent in underground works and, in the case of water conveyance tunnels, to present the matters that have to be addressed during construction and the testing that could be involved. Because the Owner controls the funds for both the investigation program as well as the Engineer's personnel on site during construction, an understanding should be achievable because the satisfactory performance of the tunnel is clearly in the best interest of the Owner.

Thus, at some time during the excavation of the tunnel and well before the decisions on final lining are made, the Engineer needs to inspect the tunnel, assess the conditions encountered, verify that the assumptions made in the Design Memorandum are still valid and make those *in-situ* tests judged necessary.

Because, inevitably, some design changes will result, causing increases in time and material costs, flexibility needs to be built into the construction contract. Such flexibility may include incremental additions to the length of steel lining, additional reinforcement for the concrete lining, or additional grouting - all of which should be bid competitively. Provisions to excavate test galleries, provide equipment for hydrojacking tests or probe holes ahead of the tunnel face should be included.

The Engineer should present this concept to the Owner early in the project and the Contractor needs to understand that certain demands will be made on his services that are not directly related to the construction process.

Monitoring and First Filling

Water Levels and Flows   Once the tunnel has been excavated, concrete lined, and grouted, monitoring consists only of reading piezometers initially installed during the exploration period or placed specifically for particular geologic or topographic reasons. It has been found, as expected, that the groundwater levels begin to rise as drainage towards the tunnel ceases following lining. Whether or not the water table returns to its former level depends on the rock permeability and rainfall during the period prior to first filling.

There may be locations along the tunnel in areas of substantial rock cover that large groundwater flows have continued throughout construction and drainage into the tunnel was allowed to continue principally for construction purposes. These inflows should be measured prior to filling the tunnel in the event that future water balance calculations are necessary to assess total leakage.

Tunnel Survey   It is assumed that a geologic record has been made of the entire tunnel and that the records of piezometer water levels have been recorded throughout construction. The geologic records are important for both unlined and lined tunnels.

A survey of the concrete lined sections should be made and cracks in the concrete should be mapped. Transverse shrinkage cracks and construction joints should be noted as well as any possible longitudinal shrinkage cracks. The latter are generally an indication of a high cement content in the concrete mix. This survey is to record the initial condition of the lining because a different crack pattern may be superimposed on the initial pattern when the tunnel is pressurized. This information may be useful in the event that tunnel leakage does occur and the tunnel has to be unwatered for future inspection. Crack patterns are a reflection of the response of the liner/rock interaction.

First Filling   Tunnel filling should be a slow deliberate process wherein the water pressure is allowed to rise at a water head rate of about 10 m/hour. This rate would be for favorable conditions where the water level behind the liner has already started to rebound naturally. If no external water pressure exists, consideration should be given to a slower rate of filling. Filling rates of 2 m/hr have been recommended by Benson (1988). Caution should be exercised to avoid rapid tunnel filling especially if the reservoir water is cold and the liner only has been recently completed. Too rapid filling also has resulted in excessive concrete cracking along the tunnel arch where the contact grouting was omitted or poorly done.

Filling rates for unlined tunnels can be about the same as for lined tunnels; however, the unwatering rate (for occasional inspection) should be slower depending on the nature of the rock in the tunnel. If the rock mass is normally jointed, an unwatering rate consistent with the filling rate is reasonable. Where the rock contains faults or is otherwise relatively impervious, the unwatering can create large head differentials due to its inability to readily drain; collapses may occur unless these zones have been well supported and drained. A normal shotcrete treatment placed over low permeability rock may be especially susceptible to collapse if the unwatering is rapid.

Unwatering of lined tunnels is not required as part of the regular inspection of hydroelectric projects or water conveyance systems. It would be done only if there is evidence of excessive leakage, indication of head loss, rising pressure of nearby piezometers placed specifically for monitoring purposes, or evidence of ground loss around the tunnel. In the absence of the above, neither a concrete lined tunnel nor an unlined tunnel should be subjected to the differential heads created by unwatering. Arbitrary inspection requirements are not justified. Unlined tunnels often are inspected within 6 months of first filling, repaired as necessary, and then not emptied unless required because of particular concerns..

In the event that an inspection is desired and it is impracticable to empty the tunnel due to system demands or the concern of pressure changes in the rock around the tunnel, then the use of ROV technology (remotely operated vehicles) is a solution. The advance in these systems has extended the ability to safely inspect tunnels, at present to distances up to 10 km, (Aquatic Sciences Inc. personal communication) and it is certain that the use of this technology will increase in the coming years.

Leakage   The data shown on Figure 2 represent the performance of 12 tunnels on first filling. Leakage rates of about 5 l/s/km or less may be considered acceptable for most projects. Many tunnels actually gain water in areas of high water tables and persistent aquifers. Several of the tunnels shown had obvious excessive water loss and had to be emptied for major repairs. Such repairs are not discussed herein. The flow chart shown on Figure 1 is provided in an attempt to avoid these adverse conditions.

Conclusions

The successful operation of pressure tunnels requires that the geologic and geotechnical variables inherent in any rock mass have been recognized and considered in the design of the tunnel lining or that unlined tunnels have been locally supported where required. The project team involved in the investigations is obligated to have a knowledge of these variables such that the required information is extracted from the exploratory borings and field geologic mapping and that required rock testing is done.

The contract documents need to include the probability that rock testing will be required during construction and that lengths of steel lining, amount of reinforcement, and extent of grouting may increase depending on encountered rock conditions. Accordingly, the construction staff must be aware of the initial design assumptions such that the rock

GEO-ENGINEERING FOR UNDERGROUND FACILITIES

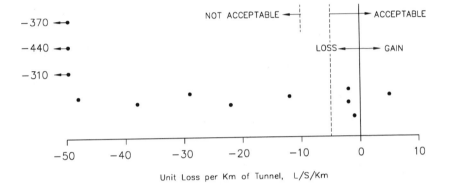

Figure 2   Tunnel Water Gain-Loss on First Filling - 12 Projects

conditions are evaluated during excavation and further testing and modifications made as necessary.

Unsatisfactory performance of pressurized water conveyance tunnels may result in severe financial consequences and perhaps influence the safety of adjacent structures. The cost to repair such tunnel linings may be substantial and likely involves engineering time in studying the problem, possible ROV inspection costs, eventual unwatering and removing the facility from operation, and the direct costs related to additional lining, repair to existing lining, and probable re-grouting.

An organized approach to identifying the rock conditions affecting lining design such as given on Figure 1 should reduce the likelihood of unsatisfactory pressure tunnel performance.

References

Benson, R. P., (1988). "Design of Unlined and Lined Pressure Tunnels", International Symposium on Tunnelling for Water Resources and Power Projects", New Delhi, India.

Carrere, A., (1995). " Application of the SCARABEE method to Re-evaluation of Concrete Dams Safety", Coyne et Bellier, GEODIA Geophysique, internal document, pp 1-25.

Deere, D. U., (1983). "Unique Geotechnical Problems at Some Hydroelectric Projects," VII Pan American Conference on Soil Mechanics and Foundation Engineering, Vancouver, Canada, pp 865-888.

Fernandez, G., (1994). "Behavior of Pressure Tunnels and Guidelines for Liner Design," ASCE Journal of Geotechnical Engineering, Vol 120, No. 10, pp 1768-1789.

Olive, R. T., (1975). "The Prestressing of Gordon Power Tunnel Lining by Interface Grouting," Institute of Civil Engineers, Australia, pp 145-154.

Riste, O. and Lysne, D. K., (1987). " Preliminary Survey of Unlined Tunnels Exposed to High Velocities," Conference of Underground Power Plants, Oslo, pp 878-891.

Swanson, D.T., (1981). "Prestressed Concrete Pressure Tunnels," RETC Proceedings, Vol 1, pp 519-532..

Thomas, H. H. and Whitham, L. S., (1964). "Tunnels for Hydroelectric Power in Tasmania," ASCE Power Division, October, pp 1-28.

Groundwater Control for the Los Angeles Metro System beneath the Santa Monica Mountains

Roy F. Cook[1] MASCE and Stuart Warren[2]

Abstract

The paper presents the measures adopted during construction of the Los Angeles Metro System beneath the Santa Monica Mountains to control groundwater inflows to the tunnels. Groundwater inflow control is required to maintain the water table within the mountains at pre construction levels and to protect the natural environment within the canyons of the mountains.

Groundwater movement pathways have been identified based on the geologic materials and structures within the mountains, tunnel inflow records, water level changes in monitoring wells and spring flow data. Since the primary porosity for the bedrock is very low, the water-bearing characteristics of the rocks, particularly in the plutonic and volcanic rocks, are largely controlled by secondary permeability where the predominant mode of water flow is through interconnected fissures and discontinuities. Groundwater control measures for the tunnels have included a program of face and formation grouting, and the design of a final tunnel lining in certain reaches to reduce water inflows through it.

Introduction

As a result of the underground construction for the Los Angeles Metro System beneath the Santa Monica Mountains, measures have been implemented to avoid impacting the natural environment within canyons in the mountains. Drawdown of the groundwater from the construction activities and from the tunnels long-term could result in changes to canyon springs and creeks. This could subsequently affect the vegetation within the canyons. In response to these concerns, groundwater inflows are being limited in specific tunnel reaches that might impact natural surface water features.

---

[1] Senior Engineering Manager, Parsons Brinckerhoff Quade & Douglas, Los Angeles, California
[2] Senior Engineering Geologist, Hatch Mott MacDonald, Pleasanton, California

Project Description

Twin running tunnels with nominal finished diameters of 5.4 m have been driven approximately 4,000 m between Hollywood in the Los Angeles Basin and Universal City in the San Fernando Valley. The tunnel alignment passes through the relatively steep terrain of the eastern portion of the Santa Monica Mountains. The mountains rise steeply about 200 m above the surrounding plain of the Los Angeles Basin. Maximum tunnel depth is about 300 m below the crest of the mountains. Figure 1 provides a profile of the tunnel through the mountains.

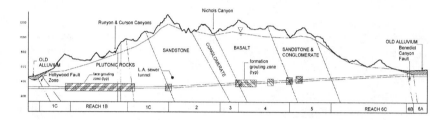

Figure 1 Profile of Tunnel through the Santa Monica Mountains

Geotechnical investigations for the tunnels included more than 70 borings to characterize the rock mass along the alignment including such fault structures as the Hollywood and Benedict Canyon Faults. Groundwater levels along the alignment were monitored at 16 locations.

Observations taken during construction, and spanning a period of relatively dry weather followed by El Nino storms in early 1998, demonstrated that the rock mass forming the mountains has a relatively low storativity. Therefore, drawdown of the groundwater could potentially impact the long-term hydrologic conditions unless measures were taken to limit drainage into the tunnels.

Three major groundwater movement pathways were identified during tunnel construction. They are fractured zones in the Late Cretaceous granodiorite, contacts between sandstone lenses and basalt within the Middle Topanga Formation, and the contact zones between the Middle Topanga Formation basalt and adjacent strata. Measures have been adopted within these areas to reduce tunnel inflows. These have included face and formation grouting to provide a rock mass annulus about each tunnel of reduced permeability and the design of the final tunnel lining to resist the water pressures and limit inflows.

The tunnel contractor initially supported the tunnels with mesh and steel ribs, and then installed a final reinforced cast-in-place concrete lining having a 0.3-m nominal thickness. Cross passages have also been excavated at intervals between the running tunnels. Initial ground support for cross passages in rock consist of grouted dowels and fiber-reinforced shotcrete placed immediately after excavation with a final lining consisting of plain shotcrete.

Geology
The mountains are a complex structural geologic unit formed by a faulted anticline and comprise a core of metamorphic and plutonic basement rocks partially covered with sedimentary and volcanic strata. Along the alignment, the tunnels pass through six highly distinctive bedrock units ranging in age from Cretaceous to Upper Miocene, as indicated in Table 1.

The Hollywood fault zone is the dominant tectonic feature and marks the south side of the mountains. It is inclined to the north at 60 to 70 degrees and displays vertical and left lateral displacement. The fault juxtaposes Quaternary-age alluvium of the Los Angeles Basin to the south against the Cretaceous age granitic rocks within the mountain mass. The width of the fault gouge and crushed rock may be up to 150 m (normal horizontal distance) and the fault zone forms an effective groundwater barrier, trapping water on the north side in the mountains. Several other faults and shear zones and many structural discontinuities are also present within the mountains.

APPROX LENGTH	GEOLOGIC UNITS	ROCK TYPE
1,550 m	Plutonic Rock	Mainly granodiorite
335 m	Chico Formation and Simi Conglomerate	Conglomerate & sandstone
210 m	Las Virgenes Sandstone and Lower Topanga	Sandstone and conglomerate sandstone
540 m	Middle Topanga	Basalt, basalt breccia and sandstone
750 m	Lower section of Upper Topanga	Sandstone, partly conglomerate
910 m	Upper section of Upper Topanga	Interbedded sandstone and siltstone/shale

Table 1 Summary of Tunnel Reach Geology

The tunnel reaches identified that could impact groundwater resources of the area as follows:

Granodiorite: The undifferentiated plutonic rocks (termed granodiorite) are generally massive, weakly foliated, granodiorite, quartz diorite, and quartz monzonite. Dikes of basalt, felsite, and aplite are present ranging from a few centimeters to several tens of meters thick and occur at irregular intervals.

The rock is differentially weathered and fractured. In general, the greater the overburden the less the rock is weathered. Nearly decomposed and intensely sheared and brecciated, granitic rocks are present close to the Hollywood fault zone. The degree of weathering within the tunnel envelope decreases northward as overburden increases and transitions to fresh rock by approximately the middle of the reach.

Discontinuities consist of joints, shears, and foliation. Joint spacing in the plutonic rocks varies from moderately close to very close with the highest percentage of joints spaced between 50 mm and 200 mm apart. Most of the joints appear in two principal sets, two widespread secondary sets with a random set. Dip angles vary from horizontal to vertical. Surfaces are mostly planar and are slightly rough. Discontinuities are typically partially infilled with calcite or clay with sand.

Middle Topanga Formation: The Middle Topanga Formation consists of a thick sequence of extrusive and possibly minor intrusive volcanic basalt breccia and basalt flows that are nearly massive and lack any apparent flow structure. The breccia dominates and is composed of angular basalt fragments up to several centimeters across in an altered (chlorite) volcanic fine-grained matrix. The basalts are jointed and sheared. Occasional and widely spaced sandstone lenses up to 25 m thick lie conformably within the volcanics and dip northeast, similar to the regional structural trend.

At tunnel depth, the rock is generally unweathered, but some minerals appear to be hydrothermally altered to chlorite and smectites. Many joints, shears and vesicles are infilled with zeolite, calcite, or chlorite minerals. The rock has a relatively low strength (average intact compressive strength of 20 MPa) atypical of most hard intact basalts.

The volcanic rocks contain at least two systematic sets of joints superimposed with one random set. Continuous joints are typically widely to moderately spaced (2 m to 200 mm apart). Locally, impersistent joints are closely to very closely spaced (200 mm to 10 mm apart). Surfaces are planar, often polished (smooth), and clay coated or filled.

Contact between Middle Topanga and Adjacent Strata (Lower Topanga and Upper Topanga): The sandstone rocks at the contacts with the intrusive basalt of the Middle Topanga Formation consist of 10 m to 25 m of intact rock with very persistent, moderately spaced (20 mm to 600 mm) open joints.

Hydrologic Conditions
The canyons of the Santa Monica Mountains are deeply incised into the terrain and contain surface water occurring mainly as springs and intermittent streams. To maintain flows, the perennial and seasonal springs depend on precipitation, domestic irrigation and a groundwater table close to the surface.

The springs are hydraulically connected with the groundwater at tunnel depth in only two strata units, the interlayered Middle Topanga basalts and sandstones, and the granodiorite. Fault zones and contacts with other more impermeable rock units form groundwater barriers that create the surface springs. Groundwater level measurements obtained during the geotechnical investigations prior to the start of tunneling indicated that the generalized piezometric surface tends to follow the ground surface.

Springs in Basalt: The basalt contains or controls the perennial and seasonal springs in the Nichols Canyon, Oakshire and Passmore Catchments. These springs issue from the Middle Topanga Basalt/basalt breccia, the sandstones within the basalts and from contact zones between the Upper or Lower Topanga sandstones with the Middle Topanga basalts.

Springs in Granodiorite: Perennial and seasonal springs in the Runyon Canyon and Curson Canyon Catchments issue from the Granodiorite. The southern part of Runyon Canyon Park does not have any seasonal or perennial springs but its vegetation suffers distress when the water table is lowered.

Rock Mass Conditions

A summary of the rock mass characteristics is given in Table 2 based on Rock Mass Quality 'Q' (Barton et al, 1974) observed by field inspection during tunneling. .

Rock Type	Q Values	
	AR Tunnel	AL Tunnel
Lower Topanga - Sandstone	0.5 - 10	0.5 - 5
Middle Topanga - Basalt	0.01 - 10	0.1 - 5
Upper Topanga - Sandstone	0.5 - 10	1 - 50
Granodiorite	0.1 – 40 primarily 1 – 10	0.1 - 40 primarily 1 – 10

Table 2 Rock Mass Characterization

As part of the geotechnical exploratory program for the tunnels, the rock mass hydraulic conditions were investigated. Packer tests were used to estimate rock permeabilities. Mean permeabilities were calculated using a geometric mean with the single lowest measured permeability value within each rock type excluded. It should be noted that the location of the packer tests do not necessarily represent the highest permeability zones encountered during tunnel construction. A summary of the results from these investigations is given in Table 3. Hydrostatic heads up to 2.8 MPa occur along the tunnel alignment at discrete hydrogeological units bounded by numerous faults acting as aquitards.

	Granodiorite	Basalt	Sandstones
Rock Mass Permeability (cm per sec)	$5 \times 10^{-8} - 2 \times 10^{-4}$ mean 1×10^{-5} Fissure 9×10^{-4} Matrix 2×10^{-4}	$8 \times 10^{-7} - 3 \times 10^{-5}$ mean 7×10^{-6}	$3 \times 10^{-6} - 3 \times 10^{-7}$

Table 3 Rock Mass Permeabilities

Model Studies
Models of the groundwater system within the Santa Monica Mountains have been developed to estimate the effects of tunnel excavation on the groundwater regime within

the mountains. A water balance model (Geoscience, 1998) of the conditions for the area was developed that incorporated inflow terms such as precipitation, recharge and supplemented surface flow, and outflow terms such as evapotranspiration, surface runoff, subsurface outflow, and tunnel inflow. This model has indicated that the return of the water table would occur provided long-term inflows to the tunnels were reduced to below 40 percent of the levels observed in the granodiorite and the basalt at the completion of tunneling.

Two-dimensional numerical models were developed to estimate steady state groundwater conditions and assess grouting and lining requirements. These included a simple closed form approach (see Fig 2) to model the steady state groundwater flow conditions developing about an unlined circular opening at depth (Freeze and Cherry, 1979).

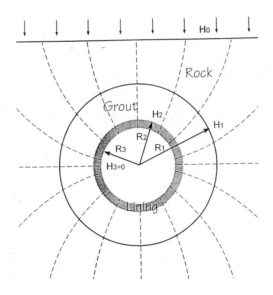

Figure 2 Model of Hydraulic System about Tunnel

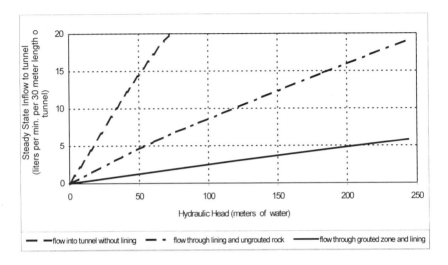

Figure 3 Estimated Steady State Tunnel Inflow vs. Water Head

The average rainfall for the area is approximately 300 mm per year. A groundwater recharge equivalent to 125 mm per year is assumed for the model with the remaining being lost to surface runoff and evapotranspiration. Using a simplified water balance model, an estimate was made of the permeability requirements needed in the lining and the grout in order to limit tunnels inflows into the tunnel to levels that would maintain steady state conditions. The approach does not provide an assessment of how long recharge would take to reach the pre construction conditions but rather an estimate of the permeabilities needed at the lining and in the grout zone to maintain steady state conditions that represent original conditions.

Impacts from Construction of Tunnels
During construction, the main groundwater migration pathways into the tunnel were through fractured fault zones in the granodiorite, contact surfaces between different formations and contact surfaces between different materials within the formation (e.g. contacts between sandstone lenses and basalt in the Middle Topanga Formation. Maximum inflow through the face was approximately 1,000 liters per minute and occurred in the granodiorite. Generally, groundwater flows diminished to one-third or less of their initial levels within 24 hours.

Some changes in surface water flow systems were observed during tunnel construction. Springs in the 2,000-m wide monitored corridor of the basalts had decreased flow rates or temporarily dried up, coincident with a three-year period of low precipitation. Anecdotal information indicates that this was an abnormal situation and that lowering of the water table due to tunnel excavation was a possible contributing factor.

The condition of some springs and creeks in the Nichols Canyon Catchment that depend on the water table in the basalt did deteriorate following excavation of the tunnels and prior to the lining of the tunnels. For example, records for Nichols Canyon Creek indicated a significant flow decline after the first tunnel was excavated through sandstone lenses within the Middle Topanga Formation. Groundwater levels in observation wells in the basalt indicated declines of more than 40 m and some springs fed by the basalt decreased in flow rates and seepage lengths.

The springs at the heads of Runyon and Curson Canyons located on the granodiorite were not apparently affected by tunnel construction. They have generally maintained a pattern similar to that observed prior to the start of construction taking into account the rainfall patterns.

The springs that issue at the faulted northern boundary of the granodiorite are thought to emanate primarily from water perched in the weathered sandstones immediately north of the granodiorite. Therefore, they may not be impacted by changes in the water table elevation. As an example, one spring in Runyon Canyon was dry prior to the start of tunnel construction in September 1995 and remained dry until larger than normal precipitation associated with the El Nino weather patterns in early 1998. The spring lies between two discreet subzones of faulting within a fault zone of the granodiorite. It appears to be fed from a localized area and is influenced by relatively small changes in water table elevation.

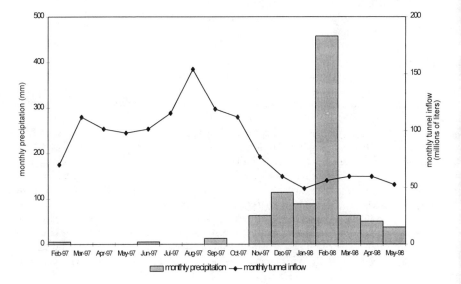

Figure 4 Precipitation and Tunnel Inflow Records

On the basis of records of springs further away from the tunnel alignment showing smaller declines in flow rates, the width of the influence was estimated to be 1,800 m. A precise correlation between the status of creeks and springs and the groundwater inflow to the tunnels cannot be made due to the influences of seasonal fluctuations and human interference.

Groundwater Control in Tunnels

The original approach for controlling groundwater in the tunnels was the installation of a concrete lining with weepholes at regular intervals along the length of each tunnel. The weepholes prevented the build up of groundwater pressures on the outside of the lining. This approach could however, lead to changes in the groundwater system within the mountains. As a result of this, the tunnel design was modified and measures were adopted to prevent inflows to the tunnels beneath areas that could affect the springs. Thus, the groundwater table in the mountains could return to pre construction levels.

The objectives of the ground water control measures adopted were to ensure that the tunnel inflows do not exceed 5 liters per minute per 30 m length of tunnel once the water table has recovered to its pre construction levels and to maintain it at levels that reflect the seasonal variations in precipitation levels. To achieve this objective, additional measures were implemented during construction. These included:

- Formation and face grouting in areas of the basalts, granodiorite and sandstones to reduce the rock permeability within an annulus around tunnels.
- A final lining in the granodiorite designed for the hydrostatic water pressures and reduced concrete permeability.

Grouting Program: The contract specification contained a provision that face grouting was required should steady-state groundwater inflows through the face exceed 700 liters per minute or a sustained water inflow through a tunnel probe hole of 30 m length exceeds 175 liters per minute for two hours. As a result of the increased concerns regarding groundwater inflows once construction started, the Authority also authorized tunnel face grouting when flows exceeded 7 liters per minute per 30-m long probe hole. The tunnel face grouting program for groundwater control consisted of the injection of cement-based grout through holes drilled ahead of the tunnel boring machine from drilling positions within the tunnel heading. Grouting was carried out primarily using microfine cement grouts except during major takes where Type III Portland cement was used. Often, Type III cement grout would meet with refusal and a switch to microfine cement would result in further significant grout takes in the hole. A refusal criterion of zero flow at a maximum pressure of 0.35 MPa above the hydrostatic head was used to control grout volumes. As a result of grouting ahead of the face, face inflows were generally less than 35 liters per minute when excavating through the grouted zones of the basalts and the granodiorites.

To further reduce inflows during tunneling, a supplemental formation grouting program was carried out in the areas of fractured basalt and sandstones where inflows were observed. The purpose of this grouting was to fill the fissures and discontinuities within the annulus with cement grouts and more fully form a grout annulus about each tunnel begun by the face grouting program. A total of approximately 2,000 holes mainly 3 m in length were drilled and Type III Portland cement and microfine grouts pumped into the formation. Generally, microfine was pumped first and then forced into the smaller fractures with Type III cement except in areas identified as very fractured where Type III cement was used initially. Maximum grout pressures were 1.4 MPa with a refusal criterion taken as zero flow at maximum pressure. With this approach, steady inflows of groundwater were reduced from approximately 75 liters per minute to less than approximately 30 liters per minute per 30-m lengths of tunnel through the basalt and sandstones.

In the basalt, the supplemental formation grouting program was largely effective without a lining in place, reducing water inflows by approximately 60 percent from the pre-grouted intermediate flow levels. To achieve the inflow criteria, additional supplemental formation grouting is planned with the lining in place. The purpose of the grouting is to develop an annulus, 3 to 6 m's wide about each tunnel. With the permeability of the annulus reduced to less than 2×10^{-7} cm per second, the original design for the final cast-in-place concrete lining can be maintained and the tunnel flows reduced to acceptable levels.

	Length of Grout Holes (m)	Microfine Cement (kg)	Type III Cement (kg)
Face Grouting	1,573	42,133	42,491
Formation Grouting	4,883	68,750	134,625

Table 4   Summary of Face and Formation Grouting Program

Since the hydraulic head will not have fully recovered at the completion of the supplemental grouting program, its effectiveness in reducing permeability within the rock annulus will be evaluated on the basis of the hydraulic head measured by piezometers located in the rock beyond the grouted zone and the measured inflow into the tunnel within discreet reaches. The basis for the evaluation is given in Figure 3.

Final Lining Redesign: Trials of supplementary formation grouting in the granodiorite prior to the installation of the lining were not effective. Observations indicated that the clay material plugged joints during grouting and infill inhibited grout flows. Therefore, the grouting program was discontinued and measures adopted to reduce inflows through the lining in the granodiorite. The lining redesign included the elimination of weepholes, the specification of a higher strength (4.2 MPa – 28 day strength) in sections beneath the highest hydraulic head (400 m) and the specification of more stringent shrinkage requirements for the concrete (less than 0.004) thereby making the concrete more impermeable.

Conclusion

A series of measures were undertaken to minimize the effects of tunneling on the lowering of the water table within the Santa Monica Mountains and its potential impact on surface hydrology and the natural environment within the canyons. The groundwater control measures implemented can be broadly categorized into those related to i). Reducing the permeability of the rock mass in the vicinity of the tunnels, and ii). Improving the capability of the cast-in-place concrete tunnel lining to limit leakage.

Although the water table has not yet reestablished itself to pre construction levels, a substantial recovery has occurred following the reduction in the rock mass permeability around the tunnel as a result of grouting in the basalt, installation of a lining through the granodiorite and groundwater recharge from the 1997-98 El Nino rainfall.

In the basalt, the clean and relatively tight nature of the joints allowed the use of cement and microfine grouts to be used prior to the installation of the final lining to reduce rock mass permeability within an approximately 3 m wide annulus about the tunnels. The grouting program has reduced inflows to the tunnel and the permeability of the grouted annulus has been estimated to be about 10^{-6} cm per second. Further grouting with microfine and chemical grouts is planned to reduce the permeability further.

The shear zones in the granodiorite with extensive sand, clay and gouge infill material inhibited the effectiveness of the grouting program. Therefore, an alternative approach has been adopted with the final cast-in-place tunnel lining designed to resist water pressures. This has required in some reaches the use of a low shrinkage concrete with a 28-day strength of 4.2 MPa.

References
Barton, N., et al., 1974, "Engineering Classification of Rock Masses for the Design of Tunnel Support,", Rock Mechanics, Vol. 6, No. 4, pp. 189-236.

Freeze R.A. and J.A. Cherry, 1979, Groundwater, Prentice Hall.

Geoscience Support Services Incorporated, 1998, Geohydrological Study of Impacts of C0311 Redline Tunnel though the Santa Monica Mountains, prepared for the Los Angeles County Metropolitan Transportation Authority.

Acknowledgements
The Authors wish to acknowledge the Los Angeles County Metropolitan Transportation Authority for allowing us to present the information contained in this paper.

APPLICATIONS OF HORIZONTAL SAMPLING AND LOGGING TECHNOLOGIES IN GEOTECHNICAL SITE INVESTIGATIONS

Samuel T. Ariaratnam,[1] AM. ASCE, Erez N. Allouche,[2] Kevin W. Biggar,[3] Renato V. Clementino,[4] and Peter K. Robertson[5]

Abstract

Horizontal Directional Drilling enables the drilling of horizontal boreholes 50 mm (2") to 1200 mm (48") in diameter beneath surface/subsurface obstacles, over extended horizontal distances. Recently, this technology has expanded its application beyond the oil, gas, and utilities industries to include the sampling and logging of soils in geotechnical site investigations. This paper provides a short introduction to horizontal directional drilling. Horizontal sampling and logging tools including a prototype multiple-port soil sampler and a horizontal directional pre-bored cone penetration device (HD-PB-CPT), currently under development as part of an academia/industry collaborative research program, are described. Overall, horizontal directional drilling technology has valuable applications in site investigation, offering a more efficient and cost-effective method for collecting geotechnical and geo-environmental data.

Introduction

Traditionally, ground sampling and monitoring of suspected contaminated soils and ground water have been done using vertical drilling technology; however, vertical drilling presents some technical limitations, as the drilling rig

---

[1] Asst. Prof., Dept. of Civ. and Envir. Engrg, Univ. of Alberta, Edmonton, Canada T6G 2G7
[2] Grad Res Asst., Dept. of Civ. and Envir. Engrg, Univ. of Alberta, Edmonton, Canada T6G 2G7
[3] Assoc. Prof., Dept. of Civ. and Envir. Engrg, Univ. of Alberta, Edmonton, Canada T6G 2G7
[4] Dept. of Structures and Foundations, Rio de Janeiro State University – UERJ, Rio de Janeiro, RJ Brazil
[5] Prof., Dept. of Civ. and Envir. Engrg, Univ. of Alberta, Edmonton, Canada T6G 2G7

must be located generally above the location of interest and determination of the lateral extent of contamination requires multiple boreholes. This may make the collection of soil samples below surface/subsurface structures prohibitive. In addition, vertical drilling is associated with the risk of penetrating impermeable layers, potentially leading to the cross contamination of deeper aquifers. For geotechnical investigations, horizontal sampling and logging can provide additional information, supplementing that obtained from vertical drilling. Geotechnical studies of tunneling and micro-tunneling projects and slope stability assessments are but two potential applications of these techniques.

Horizontal Sampling

Horizontal drilling technology provides the ability to recover undisturbed, high quality, samples from areas that cannot be reached using vertical drilling technology, such as beneath structures. Multiple target points, at different depths, distances, and directions can be collected without a need to reset the drilling rig. Additionally, this highly automated, remote access drilling technique offers an elevated safety level to field personnel since exposure to contaminants may be dramatically reduced. Finally, the risk of penetrating impermeable layers, potentially increasing the extent of contamination is significantly reduced. Monitoring wells installed in vertical holes only indicate contamination that has reached the ground water and migrated laterally from the source to the well. In this instance, considerable contaminant leakage into the soil above the ground water table may have already occurred. Furthermore, determining the lateral extent of a plume requires the drilling of a large number of vertical boreholes. Recent developments in soil sampling using horizontal boring methods provide a means to overcome these limitations.

Existing Methods of Horizontal Sampling

The earliest sampler to be used in conjunction with horizontal directional drilling rigs for the purpose of characterizing contaminated sites was developed in the early 1990's (Karlsson 1993). The sampler is capable of collecting a 1.5 m long, 55 mm diameter sample from soft to medium soils. Due to the large size of the sampler (nearly 6 m long, 121 mm in diameter and 270 kg in weight) it requires a relatively large drilling rig. The sampler operates on a principal similar to a Shelby Tube core sampler. Initially, a pre-determined length of bore is drilled. The drill string must then be withdrawn and the cutting head removed before the sampler can be attached. The sampler is then advanced into a bore to the target area while the load on the outer tube is kept constant with an applied hydraulic pressure. At a pre-determined location, an inner tube is accelerated into the formation by a hydraulic pressure to a calculated punch release force. The sample is then drawn back into the outer tube while pressure is maintained to prevent drilling media from contaminating the sample. This process is repeated for each sample. The repetitive and tedious nature of the sampling process, and the need for a large drilling rig, tends to make the utilization of this sampler rather costly.

In the mid-1990's, a second generation core sampler, significantly smaller and lighter than its predecessor, was developed capable of being used with medium and small sized drill rigs typically employed in the utility installation industry (The Charles Machines Works 1997). During the sampling process, the drilling rig is located a distance away from the zone of interest and the drilling head is navigated, below ground, to a distance of 0.3 to 0.6 m from the target area. The drill string is then retracted, the cutting head removed, and a soil sampler is connected to the end of the drill string. The sampler is pushed through the bore and then continues to be pushed through the soil until the target area is reached. The drill string is retracted approximately 0.45 m, and the sampler tube is automatically locked in an open position. The sampler is then pushed forward 0.3 to 0.6 m, filling the tube with soil. The sampler and drill string are then withdrawn from the bore. The sampling tube is removed and replaced with the drilling head, and the process is repeated. While eliminating the need for a large drilling unit, this sampler suffered from the same deficiency of its predecessor, the need to withdraw the entire drill string twice for each sample collected.

An attempt to solve this problem was done by SINTEF, a company located in Trondheim, Norway (Tokle 1997). The system developed contains a collapsible drilling head which can be withdrawn through the drill string to the surface where it is replaced with a special core-barrel with a diamond coring bit which is pumped down the drill string. Therefore, instead of removing the entire drill string, only the drill bit needs to be extracted to collect a new sample. The coring-barrel is then pulled back to the surface using a trip-wire and the sample is extracted. Core samples up to 3 m long may be collected during a single sampling operation. This method is limited to drilling and sampling in rock formations, and requires specialty equipment. Therefore, it cannot be adopted for use in conjunction with commercially available directional drilling equipment. Subsequently, the SINTEF method is used primarily for site investigations along the proposed alignment of tunnels or for collecting samples from concrete dams.

Future Directions in Horizontal Sampling

Currently, there are no known devices on the market capable of collecting multiple samples during a single operation, as the drill string must make two trips for every soil sample taken. Each sample involves drilling to the target area, retrieving the drill rod, installing the sampler, and then pushing the sample back to the target area. This is a time consuming process that may be quite cost prohibitive given that the cost of operating small or medium sized drilling rigs may range from $4,000 to $8,000 per day (Wilson and Losonsky 1995).

Research was initiated to develop a device capable of collecting multiple soil samples during a single pass along the wellbore. Emphasis was placed on universality to ensure that the device can be operated by typical horizontal directional drilling rigs currently available in North America, sophisticated controls, and sample quality. One such device, currently under development at the University of Alberta, is presented in the following section.

Multiple-Port Soil Sampler

The multiple-port soil sampler, developed through a joint industry and university research & development collaboration, is designed to be pulled back through a continuous borehole while extracting six samples. Drilling proceeds horizontally to the desired depth using regular drilling procedures. Once the target depth has been reached, dry drilling is done horizontally. Finally, the drill resurfaces at an exit location using normal (wet) drilling procedures. The drill head is then removed and the multiple-port soil sampler placed on the end of the drill string. This assembly is subsequently pulled back through the borehole, stopping at pre-specified locations to extract soil samples from the borehole wall. The above sampling process differs conceptually from the horizontal sampling methods previously described, as the samples are taken during the pull back stage as opposed to being ahead of the drilling head. The design of the multiple-port soil sampler includes a 19 cm diameter, 193 cm long outer casing in which the sampling mechanism is housed. Six-19 mm diameter, 18 cm long sampling tubes, placed on a rotating carousel, are used to collect individual samples. The sampling tubes are arranged so that they may be rotated into a position where they can be pushed out of the outer casing at an angle of 39^0 and into the soil. A 2200N linear actuator is used to push and withdraw the sample tubes, while a 100N motor connected to a gear mechanism rotates the carousel. Another linear actuator powers a cylinder which moves a distance of 50 mm along the inside of the sampler body, covering and uncovering the sampling window, effectively preventing soil particles and water from entering its interior. The "nerve center" of the device is a CPU board mounted at the rear end. Information regarding the position and status of the various components is monitored and transmitted to the surface via a control cable. The sampler is operated using a laptop computer with special-purpose software, as the operating scheme is hard coded into the CPU board. A schematic diagram of the multiple-port sampler is shown in Figure 1.

The ability to extract several samples in one sampling pass can reduce sampling time by up to 70%, thus resulting in significant productivity improvement (Allouche et al. 1998). Additionally, the reduction in effort to decontaminate sampling equipment results in improved safety levels and reduced costs.

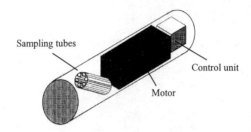

Figure 1. Schematic of the multiple-port soil sampler

Testing Procedures

The testing program was divided into two phases. Phase I involved full-scale laboratory testing, while Phase II consisted of a field test. The laboratory tests were designed to evaluate the performance of the multiple-port soil sampler and identify functionality problems. Additionally, these tests were intended to identify the technical limitations of the sampler.

Full-Scale Laboratory Tests

A special loading frame to house the sampler was constructed from 25 mm angle steel bars, 1220 mm long, 300 mm high, and 300 mm wide. Full-scale tests were performed on three type of soils; Athabasca Clay (M.C.=18%), Athabasca Clay (M.C.=21%) and Dense Sand. A mold was positioned to accommodate extraction of two soil samples as each test was repeated twice as to achieve statistically meaningful data.

Test results revealed that the sample recovery rates during full-scale testing ranged between 95-100% for non-cohesive soils and between 60-80% for cohesive soils. The sampler functioned properly when used in non-cohesive soil; however, it failed to withdraw the sampling tube due to inadequate design of the pulling mechanism. As a result, the pull-back mechanism was re-designed and the carousel re-built prior to initiation of the field tests.

Field Tests

Once modifications to the sampler were made, field tests were conducted at the University of Alberta field laboratory. The soil profile at the site is essentially Lake Edmonton Clay to about 4 to 7 m in depth, overlying glacial till. The upper 4 m is predominantly clay and below this depth the profile starts to become predominantly silty-sand (Clementino et al. 1998). A 3.5 m x 3.5 m x 3.75 m deep pit was excavated along the proposed alignment and filled with sand, in order to create a mixed face condition test section. The first test was conducted above the groundwater table. A 46 m long, 100 mm diameter, pilot bore was drilled along the profile shown in Figure 2. Upon arriving at the exit location south-west end of the bore, the drilling head was replaced with a 200 mm reamer behind which the sampling device was connected. The entire assembly was then pulled back towards the rig, with soil samples taken at predetermined locations along the path. During the sampling process the operator, sheltered inside the back of a truck, continuously monitored the operation via a laptop computer. The sampling process lasted less than 30 minutes, during which all systems appeared to perform satisfactory with no apparent difficulties. Inspection upon retrieval of the sampler from the borehole revealed that no drilling fluids, formation water or soil had entered the sampler cavity. The samples collected were not visibly distorted, and 90-100% recovery rate was achieved in both clay and sand.

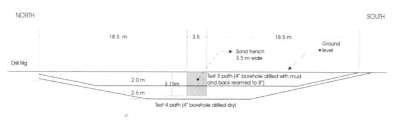

Figure 2. Field test alignment for the multiple-port soil sampler

The following day, a second test was performed along the same alignment at a depth of 4.6 m. The drilling of the pilot bore and the attachment of the reamer and sampler proceeded as described previously. The collection of the first sample took place as planned; however, during the collection of the second sample, difficulty arose as the carousel failed to index into its next position. The sampler was subsequently retrieved and inspected. The host formation was a silty-sand aquifer with a high hydrostatic pressure. It appeared that during the sampling operation, as the sampling port opened, some sand infiltrated into the sampler, jamming the rotation mechanism. Modifications to correct this problem are currently underway.

Summary of Test Results

The laboratory and field tests demonstrated that the main limiting factors for the use of the multiple-port soil sampler in geotechnical site investigation are small sample size and possible infiltration of formation fluids in the case of a high hydrostatic pressure. The infiltration of drilling fluids can be eliminated by adding a second mechanism to seal the gap between the sampling tube and wall of the sampling port. Also, the rotating mechanism is to be modified to make it more robust in case of sand infiltration. In terms of sample size, collecting two soil samples at the same location may provide a solution to acquiring the necessary amount of soil for testing. An alternative may be to build a carousel that will carry four 32 mm (instead of six 19 mm) sampling tubes, thus increasing sample volume by about 250%.

Horizontal Logging

One of the biggest concerns about searching for Dense Non-Aqueous Phase Liquids (DNAPLs) by means of vertical wells is the risk of aggravating the contamination by creating a vertical channel through an impervious layer holding the DNAPL. Horizontal wells, drilled by a horizontal directional drilling technique, offer the possibility of accessing DNAPL pools without creating a vertical conduit, thus minimizing the risk of spreading the contamination through aquifers.

With the introduction of mechanized tunnel boring, a need for soil data along the alignment of the tunnel has arisen (Broere and van Tol 1998). It is important for tunneling designers to know in advance the location and characteristic of sand pockets along the tunnel alignment.

In this context, horizontal site characterization is extremely desirable. A probe with the ability to obtain geotechnical parameters in a horizontal path and at the same time delineate the lateral extent of DNAPL pools, in a cost-effective manner, can greatly improve, not only remediation, but tunneling designs as well.

Horizontal CPT for tunneling design has been achieved in the past; however, potential problem with buckling of the rods and the need for opening a large pit to perform the test renders this option unattractive (ConeTech 1989).

The unique characteristic of HDD drilling a borehole with an entry and an exit point initiated the idea of developing a logging device based on CPT technology where the penetrometer is pulled back through the borehole rather than pushed ahead of the drill stem. This probe, called a Horizontal Directional

Pre-Bored Cone Penetration Test (HD-PB-CPT), is under development at the University of Alberta.

Despite any disturbance that the drilling process may introduce into the soil, the probe is capable of producing a large enough displacement into the soil to achieve a limit pressure in the natural soil (Ferreira 1992), similar in principal to a Pre-Bored Pressuremeter Test (PBPT).

Probe Development

Due to the geometry of the problem, pulling a CPT through a pre-bored hole created by horizontal directional drilling can seem to be analogous to a pre-bored pressuremeter test. Both tests are carried out in a pre-opened hole, therefore, stress release and soil disturbance may be of the same order of magnitude for both kinds of tests.

One main concern is the degree of disturbance that the drilling procedure may induce on the soil. The HD-PB-CPT must generate sufficient strain so that limit pressure (p_l) is achieved, capturing the natural soil response. If disturbance during drilling is small, a cavity expansion by the probe of 10 or 15% may be adequate to capture the natural soil response (Carter et al. 1986; Ferreira 1992). However, if the degree of disturbance is high, a large cavity expansion is required (Ferreira and Robertson 1992). This is likely the case for PBPT because the pre-drilling procedure always causes soil unloading (Palmer 1972). An annulus of perturbed soil is formed and large deformations during the subsequent pressuremeter test are required until the natural soil response can be recorded. This will also likely be the case for the HD-PB-CPT, where the perturbed soil zone is not only due to load stress release but as well due to cutting and reaming.

The rigidity index (I_r) of the soil, introduced by Vesic (1972), has a direct relationship with the amount of disturbance caused by pressuremeter installation as shown by Equation 1 (Ferreira 1992). The value of the rigidity index of most clay soils lies within a limited range from about 100 to 500 for soft and stiff clays, respectively.

$$\frac{R_{PE}}{R_0} = \sqrt{\varepsilon \cdot I_r (2+\varepsilon)} \qquad [1]$$

where;
R_{PE} = radius of the plastic-elastic transition (disturbed zone)
R_0 = initial radius (borehole radius)
ε = cavity strain
$I_r = G/S_u$ → G = shear modulus; S_u = undrained shear strength

Assuming that the disturbed zone is equal to the radius of the plastic-elastic transition (R_{PE}), creating a disturbed annulus 20 times larger than the borehole radius (R_0) will overcome the soil disturbance due to boring and achieve the limit pressure of the natural soil for stiff clays ($I_r = 500$). Using Equation 1, it is possible to calculate a cavity strain of 34% to satisfy this condition. On the other hand, for soft clays ($I_r = 100$) a disturbed annulus of 10 times the borehole radius seems to be enough to achieve the natural soil response and surpass boring disturbance, in this case, the cavity strain to be induced into the soil from Equation 1 is 41%.

A critical point in the design of the HD-PB-CPT is to define the correct probe diameter that can produce a displacement large enough to reach the limit pressure and also to ensure that the drill rig will have enough capacity to pull the probe back through the borehole. In order to address this issue and check the above numbers, pilot field tests were performed in Lake Edmonton clay soil at the University of Alberta field laboratory (see Clementino et al. 1998 for additional details). The results of the pilot tests have shown that, at least for Lake Edmonton clay, a cavity strain of 37% was enough to achieve the limit pressure of the natural soil. This result supports the theoretical cavity strains calculated by Equation 1.

With the above cavity strain results and concerns about the drilling rig capability to pull back the probe through soils like dense sands, it was decided to build the first HD-PB-CPT prototype with a 15.24 cm (6") diameter. This will induce 50% strain to a 10.16 cm (4") in diameter borehole drill by a horizontal directional drill rig. It was expected that this level of strain would produce sufficient soil displacement to achieve the limit pressure in the natural soil for various types of soils. This was also determined to be within the drill rig pulling capability for most soil types, depending on the depth of the test.

To measure the borehole diameter and correct the cone-soil contact area during the test, a caliper may be placed ahead of the probe and behind the reamer, as shown schematically in Figure 3.

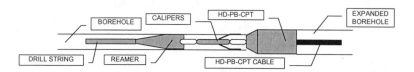

Figure 3. Schematic view of the HD-PB-CPT

Due to economical and time constraints, the first prototype was built without the caliper, instead a 10.16 cm (4") diameter bullet-shaped device was placed in the front section of the probe to ensure a constant borehole diameter.

Field Tests and Results

Several tests were performed at the University of Alberta field laboratory. As mentioned previously, the soil is mainly Lake Edmonton clay; therefore, to create a well defined contrast between two soil zones, a sandpit was dug intersecting the middle of the test path as shown in Figure 4.

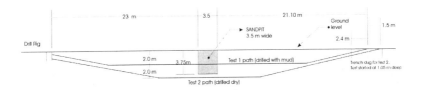

Figure 4. Field test alignment for the HD-PB-CPT

Test 1 was drilled using drilling mud. This resulted in a friction sleeve reading of zero. To increase the friction between the soil and the probe, a second borehole was drilled without using drilling fluid (dry). Figure 5 illustrates the results of this second test.

Compared with a standard vertical CPT test performed in the same clay, the HD-PB-CPT results for cone and friction resistance was lower. Nevertheless, it revealed a consistent result as cone and friction resistance were more or less constant through out the clay, as expected, and dropped significantly through the sandpit because of stress relief due to the excavation. Additionally, the friction ratio indicated a significant increase in the sandpit zone. Although a perfect match with the standard CPT was not obtained during the first test, it was easy to identify the correct location of the sandpit by examining the test profile.

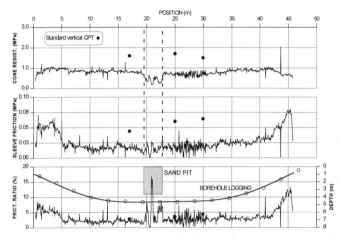

Figure 5. HD-PB-CPT test number 2

Conclusions

Horizontal sampling and logging possess the potential for major breakthroughs in geo-environmental, geo-construction, and mineral exploration. The idea of a borehole with an entrance and exit location, as well as one that stretches over an extensive horizontal distance, opens new, exciting opportunities. It is now feasible to collect multiple samples below buildings and other structures and to remotely sample environmentally sensitive or hazardous areas. Plumes can now be delineated using a few horizontal boreholes with greater accuracy than provided by numerous vertical wells. New horizontal sampling (i.e. multiple-port soil sampler) and logging (i.e. horizontal directional pre-bored cone penetration device) tools are presented which overcome the main limitations of current site characterization methodologies, namely a tedious and expensive process. Each device was successfully tested both in the laboratory and in the field. It is envisioned that over the next ten years, a new generation of horizontal soil sampling and logging tools including multiple-port soil samplers, ground water samplers, penetration tools, and borehole geophysical logging devices, will be utilized in increasing frequency for geo-environmental and geo-technical applications as engineers and owners become more familiar with the technology and its inherent advantages.

Acknowledgments

The authors would like to recognize Mobile Augers and Research Ltd., ConeTec Investigation Ltd., and SubTerra Corporation for their valuable assistance. The fourth author would also like to acknowledge the Brazilian agency CAPES, as well as Rio de Janeiro State University - UERJ, for their financial support. The financial support from the National Research Council is also acknowledged.

References

Allouche, E.N., S.T. Ariaratnam, and K.W. Biggar. (1998). Environmental remediation using horizontal directional drilling: applications and modeling. *Practice Periodical of Hazardous, Toxic, and Radioactive Waste Management*, ASCE, 2(3), pp. 93-99.

Broere,W. and A.F. van Tol. (1998). Horizontal cone penetration testing. *Proceedings of the International Conference on Site Characterization*, ISC'98. April 19-22, Atlanta, Georgia, pp. 989-994.

Carter, J.P., J.R. Booker, and S.K. Yeung. (1986). Cavity expansion in cohesive frictional soils. *Geotechnique 36*, No. 3, pp 349-358.

Clementino, R.V., P.K. Robertson, K.W. Biggar, and G. Cyre. (1998). Horizontal directional pre-bored cone penetration test (HD-PB-CPT). *Proceedings of the 51st Canadian Geotechnical Conference*, Edmonton, Alberta, pp. 815-819.

ConeTech (1989). Presentation and interpretation of CPT data: lutose creek cover. *Technical Report prepared for the Canadian National Railways*, Calgary, AB.

Ferreira, R.S. (1992). Interpretation of pressuremeter tests using a curve fitting technique. *Ph.D. thesis*, University of Alberta, Edmonton, AB.

Ferreira, R.S. and P.K. Robertson. (1992). Undrained pressuremeter interpretation based on loading data. *Proceedings of the 45th Canadian Geotechnical Conference*, Toronto, Ontario, pp. 22-1-22-10.

Karlsson, H., and R. Bitto. (1990). New horizontal wellbore system for monitor of remedial wells. *Proceedings of Contaminated Groundwater Control & Well Types*, Washington, D.C., pp. 357-362.

Palmer, A.C. (1972). Undrained plane strain expansion of a cylindrical cavity: a simple interpretation of the pressuremeter test. *Geotechnique 22*, No. 3, pp. 451-457.

The Charles Machine Works Inc. (1997). Company brochure. Perry, Oklahoma.

Tokle, V. (1997). Tunneling need not be a gamble. *No-Dig International*, 8(9), pp. 17-19.

Vesic, A.S. (1972). Expansion of cavities in infinite soil mass. *Soil Mechanics and Foundations Division – ASCE 98*, pp. 265-289.

Wilson, D.D., and G. Losonsky. (1995). Horizontal environmental well cost evaluation methods. *Energy and Environmental Expo*, PD-Volume 65, Drilling Technology, ASME, Houston, Texas, pp. 43-50.

SLURRY WALLS FOR CONTROL OF CONTAMINANT MIGRATION A COMPARISON OF UK AND US PRACTICES

By Jeffrey C. Evans[1], Member, ASCE and Andrew R. Dawson[2]

Abstract: In both the UK and the US, low permeability, vertical barriers are widely used in environmental control systems designed and constructed for the control of contaminant migration from contaminated sites. In the US, soil-bentonite slurry wall technology is the predominant technology of choice for the construction of these low permeability barriers for environmental applications. In contrast, cement-bentonite slurry wall technology is the most widely employed technique in the UK. The paper presents an examination, comparison, and assessment of each of these practices. Explanations for the reasons of national preference of one technology over the other are explored. It is believed the leading reason for the continued use of the selected technique is the confidence and experience base built up over the years in the respective countries. Also explored is the question of whether or not it is reasonable for increased use of soil-bentonite technology in the UK and cement-bentonite technology in the US. It is concluded that there are no major barriers for the adoption, where appropriate, of both soil-bentonite and cement-bentonite technologies by both the UK and the US. For the UK to adopt soil-bentonite technology, it would be necessary to revise the draft national specification that suggests minimum unconfined shear strength be specified. For the US, the inclusion of slag in the traditional cement-bentonite mixtures would be required for the adoption of this technology.

INTRODUCTION

In both the UK and the US, low permeability, vertical barriers are widely used in environmental control systems designed and constructed for the control of contaminant migration from contaminated sites. In the US, soil-bentonite slurry wall (SB) technology is the predominant technology of choice for the construction of these low permeability barriers for environmental applications. In contrast, cement-bentonite (CB) slurry wall technology is the most widely employed technique in the UK. The paper presents an

---

[1] Professor, Department of Civil Engineering, Bucknell University, Lewisburg, PA 17837, USA
[2] Senior Lecturer, School of Civil Engineering, University of Nottingham, Nottingham, NG7 2RD, UK

examination, comparison, and assessment of each of these practices. Explanations for the reasons of national preference of one technology over the other are explored. Conclusions are reached regarding the basis for the national preferences and to the question of whether or not it is reasonable for increased use of SB technology in the UK and CB technology in the US.

HISTORICAL CONTEXT

The precise historical origins of cut-off walls constructed using the slurry method of excavation are not clear. In a recent literature review (Jefferis 1997) reports the first field trials of a slurry trench cut-off began in 1945 under the direction of the US Army Corps of Engineers as a cut-off for seepage control through and beneath a levee along the Mississippi River. In another review (Ressi di Cervia 1991), it is reported that the US Navy built slurry trenches in the mid-1940s in California. Another (Nash 1974) assigns the year of 1950 to a slurry trench excavated with a trenching machine and used in California as a barrier to salt water entering a fresh water zone. Using current terminology, this would certainly be considered an environmental application. All of these early US projects were backfilled with a soil material; that is, they were SB projects using current terminology. The subsequent introduction of the hydraulic excavator and long stick was a key factor in reducing the cost of the technique and no doubt contributed to increased usage. Ressi di Cerva points out that it was in Italy in the late 1940's that patents were first acquired on slurry wall construction. While the earliest uses of slurry walls in the US dates back to the 1940's, most were installed since the mid-1970's and SB has accounted for the overwhelming majority of these installations (Ryan 1987). The first CB in the US is reported to be in 1973 at a dam site in Georgia (Ryan 1987).

The early history of the use of CB in Europe, and specifically the UK, is uncertain. Research into the early history of European slurry walls (Jefferis 1997) shows that by the early to mid-1970's, the use of CB walls was well established. Interestingly, as in the US, SB walls were used early on in the UK. A SB wall for control of ground water in a gravel pit is reported to have been installed in 1963 and two-phase walls using plastic concrete had been used to repair dam cores (Jefferis 1997). A turning point in the development of CB technology in the UK came in 1974 with the onset of research under the direction of Stephan Jefferis into improved mixtures incorporating ground granulated blast furnace slag. This research demonstrated the positive effects of blast furnace slag on CB mixtures in terms of reduced hydraulic conductivity and reduce bleed (Jefferis 1981). At about the same time, a CB wall without slag was constructed (Prentice 1974). The first CB wall with slag was constructed in 1975 as a cut-off beneath a dam (Jefferis 1997). Finally, the first UK use of CB for the control of leachate at a landfill is reported as being in 1983 (Jefferis 1983).

Currently, slurry cut-off walls are employed in waste and pollution management in a number of ways including:
1. as barriers around existing older active landfills to permit environmentally safe expansion,
2. as barriers around existing landfills where gas or contaminated ground water migration is of concern,

3. as components in contaminated land remediation schemes to enhance performance of alternative principle means (such as pump-and-treat, funnel-and-gate, or dig and dump) and
4. as a principal component in contaminated land remediation based on containment.

Recent examples of the use of CB slurry walls for waste and pollution management in the UK include the Millennium Dome project, London and the Pride Park Project, Derby (Barker, et al. 1997). Similarly, recent uses of SB slurry walls on Superfund sites in the US are the McColl in southern California (which is now redeveloped as a golf course) and the Allied Signal, Ironton, Ohio project (in conjunction with a pump-and-treat system).

GENERAL DESCRIPTION OF METHODS

A comparison of various elements of US and UK practices is summarized on Table 1 to supplement the discussion below. The entries are to represent typical practices and outcomes.

Table 1 Comparison of US and UK Practices and Outcomes

ELEMENT	US PRACTICE	UK PRACTICE
Barrier composition	Soil-bentonite (SB)	Slag-cement-bentonite (CB)
Hydraulic conductivity	$< 1 \times 10^{-7}$ cm/s	$< 1 \times 10^{-7}$ cm/s after 90 days
Solids content (M_s/M_T)	~ 70%	~20%
Unconfined compressive strength	~0	> 100 kPa @ 28 days
Strain to failure	plastic	brittle
Time dependency	Consolidation: rapidly (within a few days)	Initial set: within one day Complete hydration reactions: 90 days or more
Construction Stages	Two phase	One phase
Excavation Equipment	Backhoe, clamshell	Backhoe, clamshell
Depth (typical)	20 m	15 m
Width (typical)	0.75m	0.6m
Length (typical)	> 1 km	< 1 km
Working space needed	Large for slurry plant & backfill mixing	Small for slurry plant & excavation spoil disposal
Material assessment	Hydraulic conductivity, compatibility during design	Hydraulic conductivity, strength, strain at failure

US Practice

Soil-bentonite slurry trench cut-off walls are constructed in two phases: excavation and backfill. European terminology differentiates the type of wall not by their materials (SB and CB) but rather their methods (two-phase method and one phase method. The first phase (excavation) is performed using conventional excavating equipment. A narrow trench (0.5 to 1.5m) wide is excavated, usually with a backhoe, and trench stability is maintained with the aid of a slurry composed of bentonite (about 5%) and water (about 95%).

Once the desired depth has been reached, the slurry in the trench is then displaced by a soil-bentonite backfill (the second phase). The backfill is a mixture of soil, bentonite-water slurry, and occasionally dry bentonite. This backfill is made from either material excavated from the trench, imported materials or some combination of the two. The hydraulic conductivity of soil-bentonite, as measured in the laboratory, is typically between 1×10^{-7} cm/s and 1×10^{-8} cm/s. While not typically specified or measured, other characteristics of SB backfill include no unconfined compressive strength, high plasticity with very high strains at failure (greater than 20% ore even strain hardening), and a high solids content (typical about 70%).

UK Practice

Cement-bentonite slurry trench cut-off walls are constructed in a single phase. Excavating equipment is similar. Excavation is as shown on Fig. 1, except that the slurry is the CB that will form the permanent backfill rather than temporary bentonite-water slurry. The slurry, however, is left to cure (harden) in the trench and forms the permanent cut-off. In US practice, the bentonite-water slurry is primarily a construction expedient. CB slurry is commonly made of water (80%) and cementitious material (20%). In the UK, the cementitious material is ordinary Portland cement (OPC) with approximately 60-80% granulated ground blast furnace slag replacement. Example mix proportions are 35 kg/m^3 bentonite, 120 kg/m^3 ground granulated blast furnace slag, 30 kg/m^3 ordinary Portland cement and 934 l water (Barker, et al. 1997). The hydraulic conductivity of CB, as measured in the laboratory on cured samples, is typically less than 1×10^{-7} cm/s. Other properties of CB material include a moderate unconfined compressive strength in the range of 100 to 1000 kPa (BRE 1994), a low to moderate strain at failure (typically less than 2% in an unconfined compression test), and a low solids content (about 20% solids and 80% water). Note that membranes have, in the past, been frequently used in the wall.

It should be noted that CB walls are occasionally constructed in the US. The primary difference is that the cementitious material is OPC (without slag replacement) and the resulting hydraulic conductivity is higher (1×10^{-5} cm/s to 1×10^{-6} cm/s). The UK practice of using slag as a partial cement replacement was adopted following studies by Jefferis (1997) which demonstrated the lower conductivity that could be achieved. It should also be noted that a information regarding strength, compatibility, flow and transport conditions is provided by Manassero, et al. (1995) for CB walls formulated in a similar way to UK practice.

DESIGN

US Practice

In the US, the two principal performance requirements are low hydraulic conductivity and permanence. Permanence is that property of the slurry wall defining its ability to maintain the as-built properties in the presence of contaminants. Permanence is often synonymously expressed as compatibility with the specific environment, or durability. A maximum hydraulic conductivity of 1×10^{-7} cm/s has long been the standard as this is readily achievable with appropriate materials and construction technique. More recently, it has been shown that, under typical gradients, contaminant transport is predominantly diffusion controlled for hydraulic conductivity values below this threshold (Evans 1997a). While the hydraulic conductivity of SB is very stress dependent (Evans 1996), designers are free to specify the effective consolidating pressures deemed appropriate for these tests. Guidance to designers presented by the US Army Corps of Engineers identifies the need for compatibility testing and cautions designers to allow about 6 months during the design phase to conduct these studies (USACOE 1996). This recent specification reflects US design practice where site specific compatibility testing is undertaken for the vast majority of projects. Since SB has no unconfined compressive strength and is an extremely plastic material, there is not generally a performance requirement for stress-strain behavior. Finally, although the SB wall is used as a barrier for contaminant migration, design studies rarely explicitly consider the expected time and rate of contaminant transport through the barrier. The implicit assumption is that if the hydraulic conductivity is sufficiently low ($<1 \times 10^{-7}$ cm/s), the contaminant transport rate is sufficient for adequate performance throughout the design life of the barrier.

UK Practice

The performance requirements for CB are typically low hydraulic conductivity, stress-strain behavior and durability (Privett, et al. 1996). Similar to US practice, a maximum hydraulic conductivity of 1×10^{-7} cm/s is usually specified. While there may be some influence of stress on the measured value of hydraulic conductivity, the time of testing is substantially more important. Designers normally choose a curing time of either 28 or 90 days. Since CB walls include cementitious materials, a strain at failure (usually a minimum of 5%) had been historically specified to avoid a barrier that is too brittle. Current design practice typically omits a strain at failure criteria or specifies a value of 2 or 3%. For a strain at failure criterion to be meaningful an effective confining pressure must also be specified since the strain at failure is stress dependent (the higher the confining pressure the higher the strain at failure). The trade-off for mix designers is that altering the mix for increased flexibility tends to increase hydraulic conductivity. A minimum strength of 50 kPa is included in the draft national specification (Doe and Jefferis 1996a). Lastly, there is neither current requirement nor practice in the UK for documenting durability. Current guidance from the Construction Industry Research and Information Association (CIRIA) (Privett et al. 1996) using information from Jefferis, (1993) points to surviving examples of ancient construction materials as evidence of long-term internal stability of slurry walls. While a national Draft Specification identifies durability as a required property of the barrier (Doe and Jefferis 1996a), testing is left to the designer and no specific guidance is provided. In the accompanying guidance document (Doe and Jefferis 1996b), it is noted that there is "...almost no design data..." and many of the factors associated with compatibility are discussed. These recent publications reflect UK design practice where site specific compatibility testing is not undertaken for the vast majority of projects. As with the US design practice, although the CB wall is used as a barrier for contaminant migration, design studies rarely explicitly consider the expected breakthrough behavior of the barrier.

Assessment of Design Practices

In US and UK practices, the primary differences for the design of slurry walls as barriers for pollution migration relate to the UK need for a minimum unconfined compressive strength (to assure adequate set of the CB material) and the US practice of conducting site specific compatibility/durability testing. As noted above, a minimum strength of 50 kPa is included in the draft national specification (Doe and Jefferis 1996a). This minimum strength is included to provide some assurance that the slurry hardens rather than for any specific need for a minimum shear strength. For the vast majority of sites, there is little geotechnical rationale for a minimum unconfined compressive strength. Since the critical equilibrium conditions in the ground occur during construction when the trench is filled with a fluid, the post construction condition is not critical except perhaps on slopes where high shear stresses may remain. Further, material strength is composed of a stress-dependent component as well as a stress independent component and a material with a zero unconfined compressive strength may be very strong indeed (e.g. sands and gravels).

Compatibility testing is generally performed in the US and not generally performed in the UK. Two major reasons for this fundamental difference are suggested:
1) Early (1980's) research in the US on clay liners and soil-bentonite backfill indicated the potential for dramatic increases in hydraulic conductivity under certain contaminant chemistry conditions. Given the complexity of the ground water chemistry at most sites, this leads logically to the need to test each slurry wall design against each sites specific ground water.
2) "Design" of the actual slurry wall mixtures is usually left to the specialist contractors and the project designers provide a performance specification. By the time a performance specification is received by the specialty contractor for bid (perhaps only two to three weeks before work is to commence), there is usually insufficient time in the project schedule to permit a long-term durability-testing program (such as the 6 months suggested by the USACOE).

An assessment of the two practices must also consider that CB is a manufactured product in the sense of the use of a mixing plant and a list of ingredients. As such, it is arguably a more homogeneous and controllable product than SB. This may engender greater confidence in the reliable performance of CB (although there is no hard evidence to support CB's increased reliability over SB).

CONSTRUCTION: MATERIALS AND METHODS

Slurry preparation

Similar techniques for slurry preparation are used for both SB and CB. Contractors from both countries stress the importance of proper bentonite hydration for the satisfactory performance of the barrier. In the UK, bentonite-water slurry is mixed in a high-speed colloidal shear mixer, although an additional 4-8 hours of hydration time is required. In the US colloidal shear mixers are used without additional hydration time. If low shear mixers are used, hydration ponds or tanks are employed to permit adequate hydration time. Essentially, low quality bentonite and/or low quality mixing requires longer hydration time.

The cement and slag are added to hydrated slurry in a separate mixer in the UK. In the US, cement is typically added to hydrated bentonite-water slurry in the same mixer used to mix the bentonite water slurry.

Excavation

Similar techniques are used for both SB and CB barrier wall excavation. Typically, backhoes are used for excavations to depths of 15m or so and crane mounted clamshells are employed for greater depths. Since the slurry for a SB project is composed of bentonite and water and does not set, excavations can be stopped at any depth and continued later (the next day for example). CB slurry hardens and thus if the excavation to the trench base in not be completed within the working day the mix must be altered with retarders to delay the hardening. The life of fluid slurry can be extended to about 24 hours. If excavation is not completed in a timely enough fashion, the excavated slurry would need to be excavated and wasted, resulting in considerable cost to the contractor. While this occurs occasionally, it is relatively uncommon.

Excavated materials are often reused as backfill in SB projects and therefore may either be trucked to a remote mixing area or dumped directly along the trench for reworking into backfill. Materials excavated in UK practice are typically wasted or used as site fill.

Backfill

Properly prepared backfill is placed in the trench in a manner to avoid entrapment of slurry or segregation of coarse fraction. The material typically appears and behaves as high slump concrete (100mm-150mm slump is typical).

Since the CB is left in the trench to harden, no displacement of construction slurry is necessary.

PERFORMANCE

For the purpose of discussion, performance related issues are classified as short term and long term issues. Short-term issues are those relating to the as-built properties of the barrier whereas long-term issues relate to durability and contaminant transport through the barrier.

Short Term Performance

US Practice

For the two-phase construction methods needed for SB, field quality control measures are essential to insure the homogeneity of the constructed barrier. While much has been written on this topic (Spooner 1984, Evans 1991, Evans 1993), the focus here is relative to the comparison of UK and US practices. The most commonly asked question from those unfamiliar with the SB technique is: How does one insure that the SB backfill completely displaces the excavation slurry? While it is not possible to send inspectors into the trench, 1) prohibiting placement of backfill directly into the slurry and 2) a comprehensive program of soundings of the backfill and excavation surfaces ensures the displacement. In this way, the backfill can be expected to displace the slurry and form a homogeneous barrier. The most common defect in SB walls in inadequate key as a result of either poor design or construction quality control. Sources of construction defects in SB barriers also include improperly mixed backfill, entrapped sediments at the bottom of the trench, and entrapped slurry pockets in the backfill (LaGrega, et al. 1994).

Documentation of material properties is routinely done using laboratory testing of field mixed samples taken just prior to their introduction into the trench. Mixed samples of SB would be tested in the field for slump and sampled for grain size, moisture content and permeability tests. In some projects, a field laboratory is established for this testing. In situ testing is rarely routinely undertaken. In case of dispute, in situ testing (with limited success) and/or in situ sample and laboratory testing may be conducted.

UK Practice

For single phase construction needed for CB, the continuity of the barrier is more easily insured. The physical passing of the excavation bucket through the trench section demonstrates the trench continuity and the slurry hardens in place. Samples are taken from various depths of the trench to check that any settling of coarse fraction from the slurry prior to slurry set does not result in a material at the bottom of the trench that is substantially more permeable than the remainder of the trench. It is implicitly assumed that the trench remains stable while the slurry hardens and any further sedimentation after sampling does not result in a material of greater permeability.

Documentation of material properties is routinely done using laboratory testing of field mixed samples. Mixed samples of CB are obtained from various locations within the trench, allowed to set in the field and then cured and tested in the laboratory in strength and permeability tests. For example, on a recent project 3 km long and 7 m deep, 76 consolidated-undrained strength tests at 14 days, 81 permeability tests at 28 days, 71 strain at failure tests at 90 days and 77 unconfined compression tests at 28 days were conducted (Barker et al. 1997). Like the US practice, in situ testing is rarely routinely undertaken but in case of dispute, in situ testing and/or in situ sample and laboratory testing may be conducted. While in situ testing may initially appear to be a more reliable way of assessing the hydraulic conductivity of the barrier, penetration of the completed barrier is required and in situ testing programs are not without complications of their own (Tedd, et al. 1995a, Tedd, et al. 1995b).

Long Term Performance: Field Measures

For the most part, long term measures of barrier wall performance are not undertaken. While there may be site ground water monitoring programs, these are not usually specifically designed to detect defects or degradation in performance of the barrier wall. In both the US and the UK the implicit philosophy appears similar: based on measurements and observations of selected barrier properties during design and construction, the long-term performance is assumed adequate.

RESEARCH TO DATE

For both SB and CB, research efforts to date have revealed a number of significant findings applicable to the use of slurry walls for the control of contaminant migration. Assessment of these findings can help define areas in which a high degree of confidence exists as well as areas regarding further research.

UK Research

Early research into the formulation of CB mixtures with slag was carried out under the direction of Jefferis and included studies of strength, strain, and hydraulic conductivity (Jefferis 1997). The influence of slag replacement upon the hydraulic conductivity was recognized and typical results are shown on Fig. 1. Without slag, rather higher values of hydraulic conductivity result, similar to those obtained in the US under similar mix designs. To test the universality of this relationship developed for UK slurry walls, 3 series of samples were prepared at Bucknell University using US sourced materials (Trietley 1996) and tested after one year of curing (Veracco and Smith, 1997). The averages of the replicate tests for each of 3 slag replacements are shown as solid square symbols on Fig. 1. These data support the expectation that UK style CB mixtures using slag are technically feasible in the US as well.

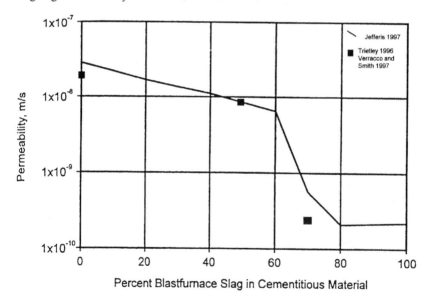

Fig. 1 Influence of Slag on Hydraulic Conductivity

Although compatibility of CB mixtures with contaminants is not routinely tested at each site, research into this issue has been undertaken. As contractors often "design" the mixtures, they have considerable corporate experience as well. For example, one contractor (Barker et al. 1997) suggests UK CB mixtures are likely to remain durable if the following limits in ground water are taken:

- pH > 4.5
- Sulfides < 5 mg/l
- Sulfate < 6000 mg/l
- Ammonium < 100 mg/l
- Phenols < 10 mg/l
- PAH's < 10% (in slurry)

Jefferis (1992) sheds light on the expected shape of the hydraulic conductivity versus time relationship for CB materials as shown on Fig. 2. The y-axis presents the ratio between the current hydraulic conductivity and the original hydraulic conductivity as a function of time on the x-axis. Time is normalized as the time for full chemical reaction of the entire sample. Two significant conclusions can be derived from this work. First, for situations where the permeability increases, there is no way to predict what the final hydraulic conductivity will be until the reaction has gone to completion. Second, the reaction volume needed may be many times greater than the pore volume displacement of 3 generally recognized in the US as suitable for compatibility testing. This is because the mechanisms that cause permeability increases in CB (chemical reactions such as sulfate attack) are different from those causing permeability increases in SB (decreasing the diffuse ion layer causing bentonite shrinkage resulting in larger pore sizes in the backfill).

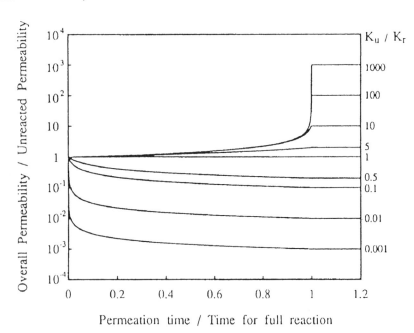

Fig. 2. Permeability-Time Curves for CB Mixtures (from Jefferis 1992)

BRE has evaluated immersion tests for compatibility (Garvin, et al. 1994 and Garvin and Hayles 1998). BRE have observed CB degradation by some chemicals. They also note that observed degradation under unconfined conditions was not observed when the sample was confined as tested in long-term permeability tests. These findings indicate the conservative nature of immersions tests to evaluate CB compatibility. They have also conducted long-term laboratory compatibility tests (Tedd, et al., 1998) which show some physical deterioration with a very aggressive sulfate permeant.

Leaching tests (CEN, 1996) on the slag and cement used as principle ingredients for UK slurry walls have shown that nickel, cadmium and lead can be leached from the unreacted materials (Hill, 1999). This includes both the slag and the cement. No leaching test data is available on the reacted CB product.

US Research

The permeability of SB backfill mixtures has been widely studied (Evans 1985, Evans 1994) and more recently, the transmission of contaminants by diffusion has been shown to be more important than by advection (water flow) when the permeability of the slurry is as low as 10^{-7} cm/s (Evans, et al. 1997a). At present the relationship between contaminant transport rates including advection, diffusion and sorption is not known for slurry walls and contaminants present in ground water. Research has also shown substantial degradation in permeability for soil-bentonite barrier mixtures permeated with non-aqueous phase liquid (Evans, et al. 1985).].

Research has also begun to evaluate possibility of the slurry material behaving as an 'Active' containment medium by which the contaminants are taken into the structure of the slurry material (especially by being adsorbed onto minerals within the mixture). It has been shown for US style soil-bentonite slurry walls that the adsorptive capacity of the barrier may be significantly enhanced by the addition of specific adsorbents. At present barriers walls are not explicitly designed to account for the potentially active nature of the barrier. Lack of detailed understanding of the long-term chemical interaction with slurry walls and their capacity to remove contaminants from ground water is the principal reason for this.

SOCIO-POLITICAL AND TECHNOLOGICAL CONSIDERATIONS

In addition to the technical issues described above, the development of different technologies for similar purposes is no doubt shaped by the social circumstances within which the development takes place (Mackenzie, 1993). The following section identifies these circumstances and their impact on the development of slurry wall technology.

Local Custom and Practice

As described above, since the development of slurry wall technique in the 1940's until 1973, all slurry walls in the US were SB. This amounts to nearly 20 years experience by contractors, owners, and regulators. Further, since many of these projects were for control of ground water inflow for excavations, the "proof of concept" was repeated many times over as those involved witnessed dry excavations (including the first author in the late 1970's). Thus, contractors built up experience and expertise in bidding and constructing SB walls, designers became confident in their ability to specify and design SB walls, purchasers became confident in spending their money for SB walls, and regulators became confident in the ability of SB barriers to resist ground water flow.

Similarly, as described in the Historical Context section earlier in this paper, early walls in the UK were two phase walls using plastic concrete backfill and since the early 1970's, CB was the technique of choice. Since CB is also widely used in Europe, and the two leading UK slurry wall contractors are based on the continent (Keller and Bachy-Soletanche), it is reasonable to expect continental influence on the early stages of slurry wall development in the UK.

Applications

In both the US and UK, early uses of slurry walls were for control of uncontaminated ground water. In the UK, cut-off walls (using two-stage plastic concrete) through dam cores dominated early uses and such applications led designers to prefer a material relatively strong to avoid the potential for hydraulic fracturing. Thus, research begun in the early 1970's and continuing into the 1980's focused on improving cementitious barriers. This led to the use of slag in the mixtures and hydraulic conductivity values in the range of 1×10^{-7} cm/s and lower. This trend regarding their usage remained until the 1990's when they were employed more widely than in the past for control of contaminant migration. In the US, heightened environmental awareness due to events such as Love Canal led to a nationwide emphasis on the clean-up of contaminated sites which includes the Superfund program (LaGrega, et al., 1994). As a result, vertical barriers that could readily achieve a hydraulic conductivity of 1×10^{-7} cm/s and lower were commonly employed. As the CB technology in the US did not include slag in the mixtures, the resulting hydraulic conductivity was 10 to 100 times higher than that achievable with CB. Hence, research efforts focused on SB data regarding their compatibility was developed in the early 1980's. This formed a scientific framework for it's continued use.

Legal Framework

In the UK, CB walls as pollution barriers are often installed as a result of property owners desire to sell a Brownfield site. They are also used for ground water protection. At this stage, the local authority requires demonstration of adequate environmental protection from any damage resulting from site contamination. Similar environmental protection is required from site developers wanting local authority approval for their development plans.

In the US, environmental authorities are an integral part in the process of selecting remedies for contaminated sites. Further, no imminent plans for site development are necessary to trigger the need for site remediation. As a result of different legislative authorities, remediation of contaminated sites is more aggressively pursued in the US than in the UK at present. Further, the lack of redevelopment pressure in the US allows for more time (and thus more testing) prior to the commencement of work.

Other Factors

There are a number of additional factors that may account for the early divergence in choice of technology. These factors are also frequently mentioned when consideration is given to the present basis for continuing the current practices.

Depth

It is argued that, due to geological factors, the vast majority of walls in the UK are constructed to depths of 15 m or less. For example, the deepest wall done by in the UK one of the leading UK contractors is 32 m. Since CB hardens, it is necessary to complete excavation to grade prior to the slurry hardening to avoid redigging hardened slurry and the associated costs. Hence, depth does not generally impact the use of CB as a suitable technology.

Many walls in the US exceed 20 m in depth and thus, with SB, the ability to stop at any depth at the end of a working day offers an advantage over CB. This is particularly true when considering backhoes are used to a depth of up to 32 m.

Length

It is argued that most slurry walls in the UK are relatively short compared to the US practice. While no data on this is available, it is reasonable to conclude this is correct. Thus, SB requiring a backfill slope within the excavated trench leads to longer section of slurry trench being worked simultaneously. For example, a 20m deep wall with a backfill having a slope of 10h:1v beneath the slurry surface, requires a minimum 200 m of open trench (plus that open for the excavation activities which adds another 10 m or so). So, in fact, CB technology lends itself to projects with relatively short wall lengths.

Space

SB technology requires a considerable space for the mixing of backfill. This can be either along the trench or in a remote batch area. It is easy to conclude, for the reasons just given, that space in the area of most slurry wall projects in the UK is at a premium and CB can be constructed using less. In contrast, many US sites offer adequate room for SB backfill mixing.

Cost

A UK author writes "Cement-bentonite walls are constructed in a one phase operation and are therefore cheaper to install" (Potter, 1995). Similarly, a US author writes, "The principle disadvantages of CB are typically higher cost and higher permeability" (Ryan, 1987). It is clear that these two assessments of cost reflect the experience of the contractors in local custom and practice. Closer examination of cost, however, leads to the conclusion that SB may have cost advantages on larger projects (long and deep) where as CB might have cost advantages on smaller projects (shorter and shallower). Hence if anecdotal data regarding the length and depth of projects in the US and UK discussed above are reliable, then both statements beginning this section can be accepted. Further, even in the US where the cost of SB is generally lower, CB may have cost advantages for sites with limited access and/or trench stability problems.

Bentonite Quality

US slurry walls are constructed using processed "Wyoming" sodium bentonite. In contrast, UK walls are constructed using calcium bentonite (often from Greece) that has been processed into sodium bentonite. As far as the writers know, there have been no studies to compare the difference in performance between these two products when used in slurry wall applications.

SUMMARY AND CONCLUSIONS

Presented in Table 1 was a summary of the specific differences and similarities of each of the two techniques. Advantages and disadvantages of each are summarized on Table 2.

It is concluded that CB technology incorporating slag into the slurry mixtures can be readily implemented in the US. The ingredients (OPC and slag) and the equipment (excavators and slurry mixers) are readily available. There is even some experience as CB walls (without slag) are, on occasion, constructed in the US. The biggest obstacle is the lack of data in the published literature readily available to US designers, consultants, contractors and regulators. This paper begins to rectify that deficiency. In addition, it would be necessary to confirm that the leaching of the cured CB material does not leach cadmium, nickel or lead even though these are present in both the cement and the slag ingredients

It is also concluded that SB technology can be used in the UK. Adoption of SB would require a revision of the draft specification that suggests to designers that a minimum unconfined shear strength be specified.

Finally, it is concluded that both compatibility and transport consideration be incorporated into the design of SB and CB cutoff walls. Since these techniques are used as long-term barriers to contaminant transport, site specific investigation of their long term durability and transport properties is necessary, regardless of the regulations which may currently be in force in either the US or UK.

Table 2 Advantages/Disadvantages of US and UK Practices

TYPE	ADVANTAGES	DISADVANTAGES
UK-style CB	<ul><li>Slurry quality readily controlled</li><li>Hydraulic conductivity readily obtained ($< 1 \times 10^{-7}$ cm/s)</li><li>Minimal space requirements</li><li>Design independent of site geology (not independent of contamination)</li><li>Better trench stability</li></ul>	<ul><li>Excavated soil wasted</li><li>Properties change with time (hydration)</li><li>Limited concerns of shrinkage and cracking above water table</li><li></li></ul>
US-style SB	<ul><li>Process not unduly sensitive to excavation rates</li><li>Reuse of excavated material common</li><li>High solids content likely to enhance long-term reliability</li><li>Economic on longer/deeper walls</li></ul>	<ul><li>Hydraulic conductivity sensitive to soil gradation</li><li>Requires local availability of suitable soil</li><li>Large construction area needed for backfill mixing</li><li>Short projects difficult due to long backfill slope</li></ul>

ACKNOWLEDGEMENTS

The authors appreciate the time the following individuals spent discussing the details of US and UK slurry wall practices. These individuals, along with their affiliations, include: Ian Martin and Robert Harris, Environment Agency; Paul Tedd and Stephen Garvin, Building Research Establishment; Stephan Jefferis, Golder Associates;

Kelvin Potter, ICI; Lucy Philip, Leeds University; Robert Essler, Keller; Peter Barker, Bachy-Soletanche; Peter Braithewaite, Arup Consultants; and Steven Day, Geo-Solutions, Inc. The authors gratefully acknowledge the support of EPSRC provided by their Visiting Fellowship Research Grant for Dr. Evans.

REFERENCES

Building Research Establishment (1994). "Slurry trench cut-off walls to contain contamination," BRE Digest 395, July.

Barker, P., Esnault, A. and Braithewaite, P. (1997). "Containment Barrier at Pride Park, Derby, England," *Proceedings of the 1997 International Containment Technology Conference*, St. Petersburg, FL.

CEN (1996) "Compliance testing for leaching of granular materials," Draft European Standard, PrEN 12457 LEACHING. CEN TC 292. July.

Doe, G. and Jefferis, S. (1996a) "Draft Specification for Slurry Trench Cut-off Walls (as Barriers to Pollution Migration)," commissioned by Ground Board of the Institution of Civil Engineers in conjunction with Construction Industry Research and Information Association (CIRIA) and Building Research Establishment (BRE), dated 13 April, 43 pp.

Doe, G. and Jefferis, S. (1996b) "Draft Notes for Guidance for Slurry Trench Cut-off Walls (as Barriers to Pollution Migration)," commissioned by Ground Board of the Institution of Civil Engineers in conjunction with Construction Industry Research and Information Association (CIRIA) and Building Research Establishment (BRE), dated 13 April, 43 pp.

Evans, J.C., Fang, H. Y., and Kugelman, I. J., (1985). "Containment of Hazardous Materials with Soil-Bentonite Slurry Walls," *Proceedings of the 6th National Conference on the Management of Uncontrolled Hazardous Waste Sites*, Wash., D. C., Nov., pp. 249-252.

Evans, J C. (1991). "Geotechnics of Hazardous Waste Control Systems," Chapter 20 in *Foundation Engineering Handbook, 2nd ed.*, Ed. H. Y. Fang, Von Nostrand Reinhold Company, New York, NY.

Evans, J. C. (1993) "Vertical Cutoff Walls," Chapter 17 in *Geotechnical Practice for Waste Disposal*, Ed. D. E. Daniel, Chapman and Hall.

Evans, J.C. (1994). "Hydraulic Conductivity of Vertical Cut-off Walls," Hydraulic Conductivity and Waste Contaminant Transport in Soils, ASTM STP 1142, D. E. Daniel and S. J. Trautwein, Eds. American Society for Testing and Materials, Philadelphia, pp. 79-94.

Evans, J.C., Adams, T.L. and Prince, M.J. (1997a)"Metals Attenuation in Minerally Enhanced Slurry Walls," *Proceedings of the 1997 International Containment Technology Conference*, St. Petersburg, FL.

Garvin S L, Tedd P and Paul V (1994). "Research on the performance of cement-bentonite containment barriers in the United Kingdom," Second Int. Symp. on Environmental Contamination in Central and Eastern Europe. September, Budapest.

Garvin, S.L. and Hayles, C.S. (1998). The chemical compatibility of cement-bentonite cut-off wall material," paper submitted for publication in Construction & Building Materials Journal.

Hill, A. (1999) "Progress Report: Environmental Assessment of Alternative Materials for Road Construction," University of Nottingham, Notttingham, UK, February.

Jefferis, S.A. (1981) "Bentonite-cement slurries for hydraulic cut-offs", *Tenth International Conference*, International Society for Soil Mechanics and Foundation Engineering, Stockholm, Vol. 1, pp. 425-440.

Jefferis, S.A. (1993). "In-ground barriers," In: *Contaminated Land – Problems and Solutions*, ed. T. Cairney, Blackie A+P, London. Pp. 111-140.

Jefferis, S.A. (1997). "The Origins of the Slurry Trench Cut-off and a Review of Cement-Bentonite Cut-off Walls in the UK," Proceedings of the International Containment Technology Conference, St. Petersburg, FL, pp. 52-70.

LaGrega M.L., Buckingham, P.L., and Evans, J.C., *Hazardous Waste Management*, McGraw-Hill Book Company, New York, NY, 1994.

Mackenzie, D. (1990*)* *Inventing Accuracy: A Historical Sociology of Nuclear Missile Guidance*, MIT Press, Cambridge, MA.

Manassero, M., Fratolocchi, E., Pasqualini, E., Spanna, C., and Verga, F. (1995) "Containment with Vertical Cutoff Walls", *Proceedings, Geoenvironment 2000*, ASCE, Vol. 2, pp. 1142-1172.

Nash, K.L. (1974). "Stability of Trenches Filled with Fluids," *J. Construction Div.*, ASCE, CO4, pp. 533-542.

Philip, L. (1998) Personal communication

Potter, K.J. (1995). "Owner's Perspective in Selection of Vertical Barrier Materials," paper prepared for Containment Technology Workshop, Baltimore MD, 29 August cited in: Assessment *of Barrier Containment Technologies*, ed. R.R. Rumer and J.K. Mitchell, (http://em-52.doe.gov/ifd/scfa/textindx.pdf)

Prentice, J.E. (1974). "Pollution Control in Alluvial Terrain," *Ground Engineering*, Vol. 7, No. 5, pp. 26-28.

Privett, K.D., Matthews, S.C. and Hedges, R.A. (1996)."Barriers, liners and cover systems for containment and control of land contamination," CIRIA Special Publication 124, 278pp.

Ressi di Cerva, A.L. (1992). "History of Slurry Wall Construction," Slurry Walls: Design, Construction and Quality Control, ASTM STP 1129, D.B Paul, R.R. Davidson, and NJ Cavalli, Eds., American Society for Testing and Materials, Philadelphia, pp. 3-15.

Ryan, C.R. (1987) "Vertical Barriers for Pollution Control," *Geotechnical Practice for Waste Disposal '87*, ASCE Geotechnical Special Publication No. 13, pp. 182-204.

Spooner, P., et al. (1984) *Slurry Trench Construction for Pollution Migration Control*, EPA-540/2-84-001, US Environmental Protection Agency, Cincinnati, OH.

Tedd, P., Butcher, A. P., and Powell, J. J. M., (1995a). "Assessment of the piezocone to measure the in-situ properties of cement-bentonite slurry trench cut-off walls," *Geoenvironmental Engineering, Contaminated ground: fate of pollutants and remediation*, Yong, R. M. and Thomas (eds.), H. R., pp. 48-55.

Tedd, P., Quarterman, R. S. T., and Holton, I. R., (1995b). "Development of an instrument to measure in-situ Permeability of slurry trench cut-off walls," *Proceedings of the 4th International symposium on Field Measurements in Geomechanics*, Bergamo, Italy, April 10-12, pp. 441-446.

Trietlley, A.C. (1996), "The Effects of Slag Addition on the Permeability of Cement-Bentonite," Honors thesis, Bucknell University, Department of Civil Engineering.

US Army Corps of Engineers (1998) "Soil-Bentonite Slurry Trench for HTRW Projects," Guide Specification for Construction, CEGS-02260.

Veracco, T. and Smith, B. (1997) "The Effects of Slag Addition and Curing on the Permeability of Cement-Bentonite," research report, Bucknell University, Department of Civil Engineering.

Ground Deformations Adjacent to a Soil-Bentonite Cutoff Wall

George M. Filz[1], Diane Y. Baxter[2], David J. Bentler[3], and Richard R. Davidson[4]

Abstract

A 30 m deep soil-bentonite cutoff wall was constructed around a contaminated site located in Mountain View, California. The cutoff wall alignment passed within 6 m of a building that housed a semiconductor manufacturing operation. Construction of the cutoff wall was successful in most respects, but lateral and vertical movements of the ground adjacent to the cutoff wall were so large that the building was rendered unusable for the manufacturing operation it contained, and a lawsuit resulted. This paper provides descriptions of the site, the cutoff wall construction, and ground deformations that were caused by the cutoff wall and by ground water extraction. Hydraulic conductivity test procedures and results from the QC/QA program are also presented.

Introduction

The Raytheon Company produces semiconductors at facilities located in Mountain View, California, about 30 miles southeast of San Francisco. Because volatile organic compounds (VOCs) had been detected in the ground at the site, remedial measures consisting of a perimeter soil-bentonite cutoff wall and ground water extraction were implemented. The cutoff wall served to contain contaminated ground water at the site and to divert off-site ground water flow around the Raytheon property. The cutoff wall was not keyed into an impermeable layer, and ground water was extracted from within the area enclosed by the cutoff wall to

---

[1] Associate Professor, Virginia Tech, Blacksburg, VA 24061-0105
[2] Graduate Research Assistant, Virginia Tech, Blacksburg, VA 24061-0105
[3] Assistant Professor, University of Kentucky, Lexington, KY 40506-0281
[4] Principal, URS Greiner Woodward Clyde, Level 6, 486-494 Pacific Highway, St. Leonards, NSW 2065, Australia

prevent downward migration of contaminants. Extracted ground water was treated to remove contaminants from the site.

The cutoff wall construction was successful in many respects. For example, the work was completed on schedule, without change orders, and within the project budget at a cost consistent with costs for similar projects. Specifications for the bentonite-water slurry, trench excavation, soil-bentonite backfill, dust, and vibrations were all satisfied, as documented by an extensive quality control and quality assurance (QC/QA) program.

In one important way, however, the work produced an undesired consequence. The cutoff wall alignment passed within 6 m of a building used for semiconductor manufacturing. Construction of the cutoff wall caused the building to settle about 10 cm, producing cracks in the walls and the floors and disrupting manufacturing operations. A large lawsuit resulted. Smaller settlements were measured at other buildings.

Prior to construction, the designers performed calculations using the finite element method to estimate the amount of settlement that the cutoff wall construction would induce in nearby buildings. The observed settlements exceeded those calculated during design, in some cases by a factor of about five.

The authors became involved with the project during post-failure evaluations of the ground movements at the Raytheon site and in subsequent research studies concerning the mechanical and hydraulic performance of soil-bentonite cutoff walls. To the authors' knowledge, established and verified methods for numerically modeling all phases of soil-bentonite cutoff wall construction and subsequent backfill consolidation were not available at the time the design calculations for this project were made. Three of the authors are currently involved in a research project that includes development of procedures for making reliable predictions of ground movements adjacent to soil-bentonite cutoff walls.

The ground movements at the Raytheon site were well documented through use of inclinometers and surface settlement points. As described below, data from this instrumentation suggest that settlements of the buildings near the cutoff wall resulted from consolidation of the soil-bentonite backfill. According to this hypothesis, the ground near the cutoff wall moved laterally towards the cutoff wall during backfill consolidation, and this lateral ground movement produced settlement of the ground surface.

A principal purpose of this paper is to present ground movement data that was collected at the Raytheon site. Necessary background information is provided, including site conditions and cutoff wall construction procedures. An extensive QC/QA program, with emphasis on hydraulic conductivity testing of the soil-bentonite backfill, was an integral part of the cutoff wall construction. While the hydraulic conductivity testing is not critical for understanding the ground movements that occurred, the results are interesting and are included here to provide a more complete description of the Raytheon cutoff wall case history.

Numerical analyses of ground deformation are not included in this paper. Such analyses are being performed by three of the authors as part of an ongoing research study. Results of the numerical analyses will be prepared for publication elsewhere.

Site Conditions

The following description of the conditions at the Raytheon site, which is shown in Figure 1, is based on the Interim Remedial Measures Report (Golder 1987) and other descriptions of the project (Burgess et al. 1988, Burke and Achhorner 1988).

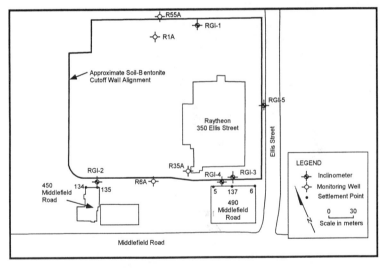

Figure 1. Site Plan (after Earth 1989)

In 1982, the Raytheon Company began an extensive investigation to determine the types, concentrations, and extent of contamination. Over 200 borings were drilled, over 100 ground water monitoring wells were installed, and over 2,000 samples (soil, water, and gas) were tested for organic chemicals.

Subsurface conditions at the site are complex, consisting of interbedded layers of gravel, sand, silt, and clay. These heterogeneous conditions were produced by repeated fluctuations of the sea level during the last ice age. At times of low sea levels, sands and gravels transported from surrounding higher areas were deposited at the site. During periods of high sea levels, the site was submerged and silts and clays were deposited. These repeated sea level fluctuations produced a subsurface profile consisting of interbedded coarse- and fine-grained strata of variable thickness and lateral extent. A typical subsurface profile at the site is shown in Figure 2. The sand and gravel deposits are typically medium dense to dense. The fine-grained deposits are typically stiff in consistency.

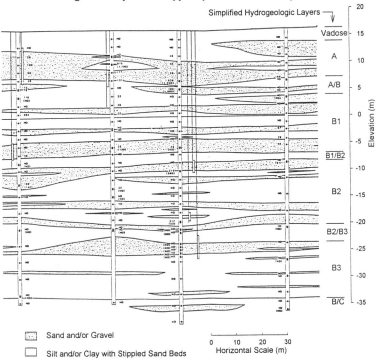

Figure 2. Typical Subsurface Profile (After Golder 1987)

The complex site conditions were reduced to a simpler conceptual framework for use in interpreting hydrogeologic conditions at the site. The near-surface aquifers, which consist primarily of sand and gravel deposits, are referred to as the A, B1, B2, B3, and C aquifers. The results of slug tests and well pumping tests indicate that the hydraulic conductivity of the aquifers is about 10^{-2} to 10^{-4} cm/s. The aquitards separating these aquifers are designated the A/B, B1/B2, B2/B3, and B/C aquitards. These aquitards consist of clay and silt deposits with interbedded sand and gravel lenses. The results of laboratory tests indicate that the hydraulic conductivity of the fine-grained portions of the aquitards is in the range of 10^{-6} to 10^{-9} cm/s. The arrangement and average depths of the aquifers and aquitards along the southern leg of the cutoff wall are shown in Figure 3. By comparing Figures 2 and 3, it can be seen that the hydrogeologic model in Figure 3 is a vast simplification of the actual subsurface conditions at this site. In the project documentation, it is recognized that the A and B aquifers are interconnected by "leaky" A and B aquitards. The B/C aquitard seems to be more intact than the overlying aquitards.

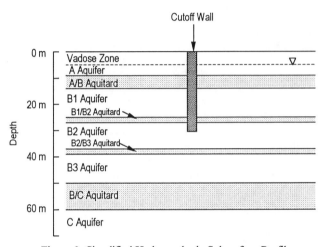

Figure 3. Simplified Hydrogeologic Subsurface Profile

The ground water table is generally encountered at a depth of about 5 to 6 meters below the ground surface. Natural ground water flow is generally to the north. Measurements of piezometric levels indicate that there is an upward gradient from the B3 aquifer to the A aquifer and a downward gradient from the B3 aquifer to the C aquifer. The primary sources of the municipal ground water supply for the city of Mountain View are the C aquifer and deeper aquifers.

Contaminants in the soil and ground water consisted of VOCs, primarily trichloroethylene and 1,2-dichloroethylene. VOC concentrations ranged up to 37 ppm in the soil and 200 mg/l in the ground water. Concentrations were highest in the vadose zone and decreased with depth. Significant contaminant concentrations were not detected below the B2 aquifer. Hydrogeologic studies indicated that contaminants were being carried off the site by ground water flow. It was also found that VOCs from adjacent upgradient property were being carried onto the site by ground water flow.

Cutoff Wall Construction

The following description of the cutoff wall construction is based on information provided by Burgess et al. (1988) and Burke and Achhorner (1988).

The cutoff wall was constructed during the period from June through October, 1987. An important aspect of the project was that semiconductor manufacturing, which occurred in the buildings at 350 Ellis Street and 490 Middlefield Road (see Figure 1), had to remain in operation throughout construction. This manufacturing operation employed over one thousand people and was in continuous operation around the clock, seven days per week.

As shown in Figure 4, the trench was excavated under a bentonite-water slurry using a backhoe with a 15 m reach and two cranes equipped with clamshell grab buckets to complete the excavation to depth 30 m. The trench was 0.9 m wide. The minimum unit weight of the slurry in the trench was specified to be 11.0 kN/m^3 in order to reduce the potential for ground movements during excavation. According to QC test results, the average unit weight of the slurry in the trench was 11.6 kN/m^3.

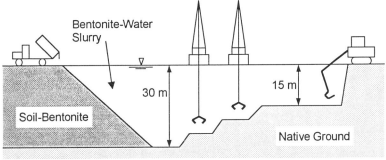

Figure 4. Soil-Bentonite Cutoff Wall Construction (After Burke and Achhorner 1988)

The excavated soils were loaded into trucks for transportation to an on-site mixing area. Only excavated soil with less than 0.5 ppm VOCs could be used in the backfill. The mixing operation began by blending excavated soil with slurry using wide track bulldozers to produce a homogenous material with the desired consistency. This material was then pushed into a pit where dry bentonite was added and mixed with a backhoe to achieve the minimum specified bentonite concentration of two percent in the backfill. As described below, all of the hydraulic conductivity tests that were performed as part of the QC/QA program resulted in hydraulic conductivity values below the specified minimum value of 10^{-7} cm/sec.

Thorough field mixing of the soil-bentonite was undoubtedly important in achieving this result.

The soil-bentonite backfill mixture was loaded into trucks, transported back to the cutoff wall, and end dumped into the trench at the point where the backfill slope reached the top of the trench. Because the backfill is very soft, the backfill slope was shallow, and the length of the slurry-supported portion of the trench was at least 320 m at the top of the trench. The specifications required that the separation distance between the backfill and the toe of the excavation at the bottom of the trench be at least 15 m. As indicated by inclinometer measurements discussed below, the relatively long slurry-supported portion of the trench permitted movement of the trench walls towards the trench centerline prior to backfill placement.

An extensive QC/QA program was completed during construction. Tests and observations were required for the slurry, the backfill, and the trench geometry and continuity. Of particular interest to the project team were determinations of hydraulic conductivity. Laboratory tests of hydraulic conductivity were made on grab samples of the backfill mixture by the contractor, the design engineer, and an independent consultant retained by the owner. The test results from all three groups were quite consistent, as shown in Figure 5. The geometric mean hydraulic conductivity was about 2×10^{-8} cm/s, and all laboratory test results were below the maximum specified hydraulic conductivity of 1×10^{-7} cm/s.

Performing hydraulic conductivity tests on soil-bentonite can be challenging due to its high water content and very soft consistency. Typically, some form of consolidation is required prior to performing the tests. In addition, the time required to run the tests can be lengthy because the hydraulic conductivity of soil-bentonite is low. In order to run hydraulic conductivity tests in a manner that was timely for QC purposes, the contractor consolidated the soil-bentonite in a standard consolidometer and then permeated in the same consolidometer. A significant detail of the procedure used for QC/QA hydraulic conductivity testing on this project is that the specimens were consolidated to pressures of 170 kPa or more before testing. This pressure was selected based on an effective vertical pressure calculated at the mid-depth of the trench by assuming that no arching occurs, i.e., that no shear stresses develop on the trench sidewalls during consolidation of the backfill. Evans et al. (1995) showed that arching does in fact occur in soil-bentonite cutoff walls, and the resulting consolidation pressures are much smaller than would be calculated using the assumption of no arching. Procedures for calculating more realistic consolidation pressures in soil-bentonite backfilled trenches are provided by Evans et al. (1995) and Filz (1996). Published hydraulic conductivity test results (Adams et al. 1997; *Barrier* 1995) show that hydraulic conductivity can decrease by one-half to one order of magnitude in response to an increase in consolidation pressure of one order of magnitude for a soil-bentonite mixture. These results indicate that the hydraulic conductivity of the cutoff wall at the Raytheon site is probably higher than the hydraulic conductivity determined by the QC/QA test results for the project. Studies have shown that hydraulic conductivity tests performed on soil-bentonite mixtures can be run effectively at low pressures in a standard consolidometer and in the API filter press (Adams et al. 1997; Heslin et al. 1997).

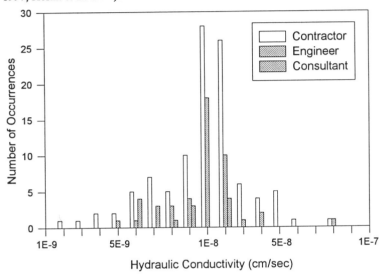

Figure 5. Hydraulic Conductivity Tests on Soil-Bentonite (After Burgess et al. 1988)

There are several important issues relating to consolidation pressures for hydraulic conductivity tests performed on soil-bentonite. First, consolidation pressures should always be reported with hydraulic conductivity values. Second, if backpressure saturation is applied, it should be implemented without overconsolidating the specimen. Third, the construction specifications for soil-bentonite cutoff wall projects should state the consolidation pressure range to be used for QC/QA testing, and this pressure range should be calculated based on consideration of the arching that is expected to occur in the trench backfill. Underlying all of these issues is the concern that if hydraulic conductivity tests on soil-bentonite are run at consolidation pressures that are higher than *in situ* pressures in the cutoff wall, the hydraulic conductivity test results will be unconservative.

In addition to laboratory tests, the project specifications for the Raytheon cutoff wall required that ten *in situ* hydraulic conductivity tests be performed in the cutoff wall. The contractor made six unsuccessful attempts to perform these tests.

The first three attempts were made using pushpoint piezometers, and the other three attempts were made using standpipes installed through hollow stem augers. According to Burke and Achhorner (1988), these attempts failed or gave erroneous results because of the unconsolidated nature of the backfill. After these unsuccessful attempts, the requirement that the contractor perform *in situ* hydraulic conductivity tests was abandoned.

Deformation of the Ground Adjacent to the Cutoff Wall

Movements of the ground adjacent to the cutoff wall were monitored using surface settlement points and slope inclinometer casings installed in bore holes (Burgess et al. 1988). Some of the slope inclinometer casings were disturbed during construction or were installed after cutoff wall construction was completed. Two of the inclinometer casings, however, seem to provide reliable data that includes the cutoff wall construction period. These are designated RGI-2, which is located 3.7 m from the centerline of the cutoff wall near the building at 450 Middlefield Road, and RGI-3, which is located 8.5 m from the centerline of the cutoff wall near the building at 350 Ellis Street. Figure 6 shows that the top of inclinometer casing RGI-2 moved about 1½ cm towards the trench during trench excavation. No significant movement of RGI-2 occurred during the backfill placement process. RGI-2 moved an additional 3½ cm towards the trench during the 7½ month period following backfill placement. It is hypothesized that this movement of RGI-2 towards the cutoff wall was due to consolidation of the soil-bentonite backfill. Figure 6 shows an abrupt displacement at depth 26 m, which is near the bottom of the cutoff wall. This abrupt displacement developed between 1 and 3 months after backfill placement.

Figure 7 shows the movements of inclinometer casing RGI-3. Even though RGI-3 exhibited some tendency for small erratic movements, the data in Figure 7 indicates that the ground movement was towards the trench during trench excavation, back away from the trench during backfill placement, and towards the cutoff wall after backfill placement. Six and one-half months after backfill placement, the maximum lateral movement was about 3 cm. The movements at RGI-3 are generally smaller than those at RGI-2, which is consistent with RGI-3 being farther from the trench than RGI-2. It can also be seen in Figures 6 and 7 that the movements at RGI-3 are distributed over most of the inclinometer casing length, in contrast to the abrupt displacement at depth 26 m in RGI-2. Both inclinometers show very little movement below the trench bottom.

Another perspective on the lateral movements of the ground adjacent to the cutoff wall can be seen in Figure 8, which shows the incremental lateral movement that occurred one month and six months after backfill placement. The figure illustrates that both inclinometers moved towards the trench and also that the movement decreased with increasing distance from the trench. It is hypothesized that

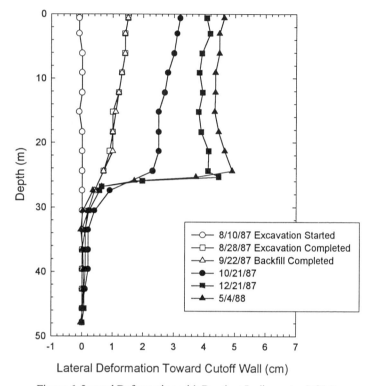

Figure 6. Lateral Deformation with Depth at Inclinometer RGI-2

the thickness of the cutoff wall decreased during backfill consolidation, and the trench sidewalls moved toward the trench centerline. This process would produce the lateral movement of the adjacent ground exhibited by inclinometer casings RGI-2 and RGI-3.

Time histories of lateral and vertical ground movements in the vicinity of the 450 Middlefield Road building are shown in Figure 9. The maximum lateral movement of RGI-2, which occurred at the top of RGI-2, is plotted versus time. Also shown in Figure 9 are the time histories of excavation, backfill placement, settlement at survey points 134 and 135, ground water level inside the area contained by the cutoff wall, and ground water level outside the area contained by the cutoff wall. Survey points 134 and 135 are located on the north side of the building at 450 Middlefield Road, about 11.2 m from the trench centerline. The ground water levels shown in Figure 9 were obtained from piezometers located near the southern leg of the cutoff wall. The piezometer locations are shown in Figure 1. It

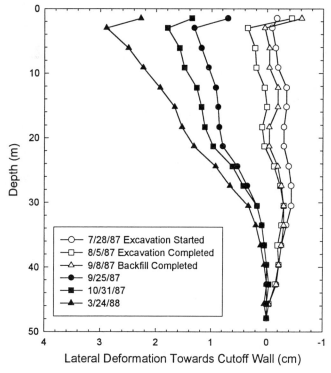

Figure 7. Lateral Deformation with Depth at Inclinometer RGI-3

can be seen in Figure 9 that completion of the cutoff wall, in combination with ground water extraction from within the area contained by the cutoff wall, produced a drop in ground water level inside the cutoff wall and mounding of ground water outside the cutoff wall. Ground water mounding outside the southern leg of the cutoff wall is logical because the natural ground water flow direction is to the north at this site. The difference between ground water levels across the cutoff wall demonstrates that the cutoff wall serves to inhibit lateral ground water flow. Nine months after construction of the cutoff wall, survey points 134 and 135 had both settled about 1½ cm. After the same time period, the top of RGI-2, which is located 3.7 m from the trench centerline, had moved about 6½ cm towards the cutoff wall. Concerns about ground settlement and potential damage to the buildings near the cutoff wall caused the project team to reduce the amount of ground water extraction from within the area contained by the cutoff wall. The effects of reducing ground water extraction can be seen in Figure 9 by the trend towards

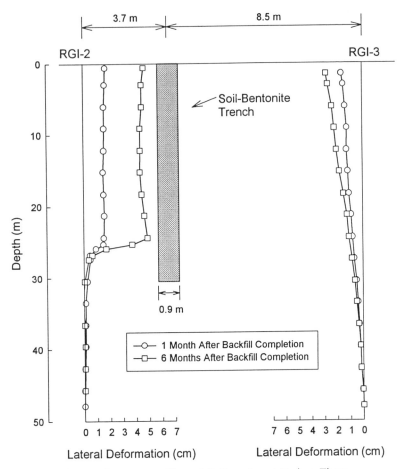

Figure 8. Increments of Lateral Deformation at Various Times After Backfill Placement at RGI-2 and RGI-3

equalization of ground water levels across the cutoff wall during late 1987 and into 1988.

Figure 10 shows the ground water levels obtained from piezometers located near the northern leg of the cutoff wall. The piezometer locations are shown in Figure 1. Inside the cutoff wall, the piezometers show a drop in ground water level that occurred at the time the cutoff wall was completed. After pumping inside the cutoff wall was reduced, the ground water level inside the wall rose.

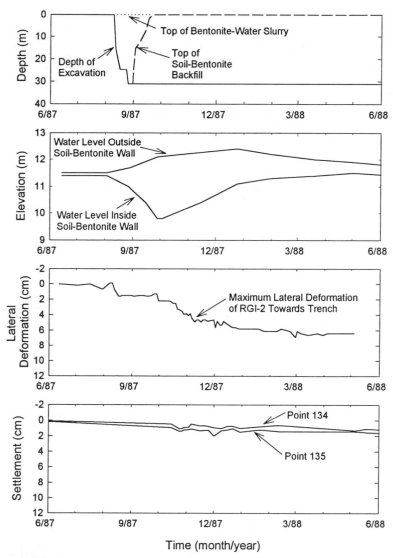

Figure 9. Timelines for Excavation Sequence, Ground Water Levels, and Deformations in the Vicinity of RGI-2

Outside the cutoff wall, the piezometers indicate a continual decrease in ground water levels. This may be due to a shadowing effect that the cutoff wall has on the natural ground water flow, which is to the north.

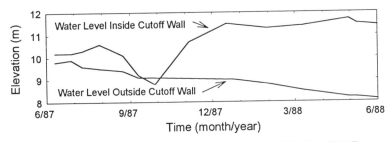

Figure 10. Ground Water Levels at the North Leg of the Cutoff Wall

Figure 11 shows data similar to that in Figure 9, except the information is for a location near the building at 490 Middlefield Road. In Figure 11, the maximum lateral ground deformation at RGI-3, which occurred near the top of RGI-3, is plotted. The settlement data is from survey points 5, 6, and 137, which are located on the north side of the building at 490 Middlefield Road, about 6 m from the cutoff wall. The ground water levels shown in Figure 11, which are the same as those shown in Figure 9, are from piezometers located near the southern leg of the cutoff wall. Nine months after construction of the cutoff wall, the range of settlement of the survey points was 2 to 10 cm. After the same time period, the top of RGI-3, which is located 8.5 m from the cutoff wall centerline, moved about 3 cm towards the cutoff wall. These movement magnitudes are different than those shown in Figure 9 for instrumentation near the building at 450 Middlefield Road. These differences may be due to differences in subsurface conditions and/or cutoff wall construction procedures at the two buildings, but it is also noted that the instrumentation is located at different distances from the cutoff wall. Survey points 134 and 135 are located 11.2 m from the cutoff wall and settled 1½ cm, whereas survey points 5, 6, and 137 are located 6 m from the cutoff wall and settled between 2 and 10 cm. RGI-2 is located about 3.7 m from the cutoff wall and moved laterally about 6½ cm, whereas RGI-3 is located about 8.5 m from the cutoff wall and moved laterally about 2 cm. As expected, there is a trend of decreased deformation with increasing distance from the trench.

It can be seen in Figure 11 that the 490 Middlefield Road building experienced substantial settlement even though ground water levels beneath the building, i.e., outside the cutoff wall, increased. Thus, consolidation of the ground under the 490 Middlefield Road building due to ground water withdrawal inside the contained area cannot account for the settlements that were observed. The inclinometer and settlement point data suggest, as described above, that the most likely cause of the settlement was trench excavation and post-construction consolidation

GEO-ENGINEERING FOR UNDERGROUND FACILITIES 135

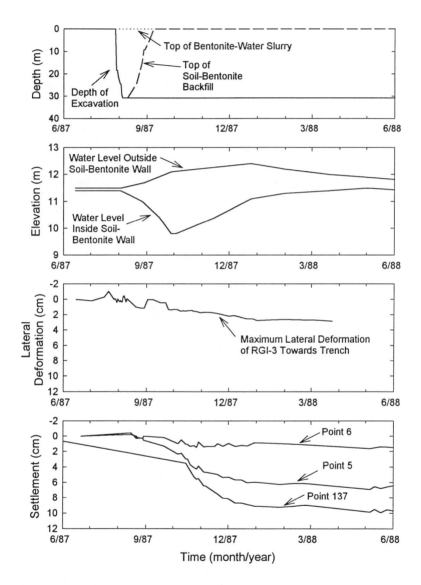

Figure 11. Timelines for Excavation Sequence, Ground Water Levels, and Deformation in the Vicinity of RGI-3

of the soil-bentonite backfill. Ground water extraction from within the contained area could have contributed to consolidation of the soil-bentonite and, by this mechanism, induced additional settlement of the 490 Middlefield Road building.

Settlement of the building at 490 Middlefield Road caused cracking of the building walls and, more importantly, disruption to semiconductor manufacturing operations taking place inside the building. Certain pieces of equipment in the building became so far out of level that they could not perform their functions properly. A large lawsuit was initiated to recover the costs of building repair and manufacturing down time.

The instrumentation data collected at the Raytheon site illustrate the type of ground movements that can take place adjacent to a soil-bentonite cutoff wall. A schematic diagram of these movements is provided in Figure 12. During trench excavation, the trench walls move in laterally towards the trench centerline, and the adjacent ground surface moves down. During trench backfilling, the trench walls may move slightly away from the trench centerline. During subsequent consolidation of the trench backfill, the trench walls again move towards the trench centerline and the adjacent ground surface experiences additional settlement. Ground water extraction within an area contained by a soil-bentonite cutoff wall would be expected to increase the effective stresses within the soil-bentonite backfill, thereby causing additional backfill consolidation, inward movement of the trench walls, and adjacent ground settlement. Ground water extraction could also cause compression of the native ground in the area where the ground water extraction occurs.

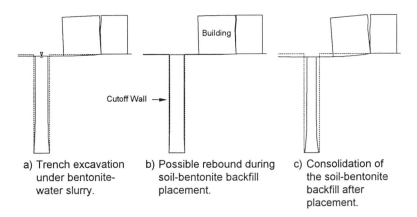

Figure 12. Movements of Adjacent Ground During Soil-Bentonite Construction and Consolidation

Conclusions

Observations of ground movements adjacent to the Raytheon soil-bentonite cutoff wall clearly show that large vertical and lateral ground movements can occur near soil-bentonite cutoff walls. Vertical movements of up to 10 cm and lateral movements of up to 6 cm were recorded. Movement magnitudes generally decreased with increasing distance from the cutoff wall. A building located 6 m from the cutoff wall was so severely affected by ground movements that the building was rendered unusable for the semiconductor manufacturing operation that had been taking place there. The observed patterns of ground movement indicate that post-construction consolidation of the soil-bentonite caused most of the ground movements.

Consolidation of soil-bentonite occurs due to dissipation of excess pore pressures that exist after placement. Ground water extraction can also cause additional consolidation of soil-bentonite by increasing the effective stresses within the soil-bentonite. Ground water extraction can also cause settlement of the native ground where the extraction occurs.

In retrospect, it would have been preferable to use a cutoff wall material containing cement in combination with a panelized construction technique to minimize adjacent ground deformations in areas where the cutoff wall passed close to existing buildings.

The finite element calculations made by the designers prior to construction substantially underestimated those observed during construction. To the authors' knowledge, established and verified methods for numerically modeling all phases of construction and backfill consolidation were not available at the time the calculations for this project were made. Research studies are currently underway at Virginia Tech to model the construction processes and develop procedures that can be used to estimate the magnitudes of expected ground movements as a function of distance from soil-bentonite cutoff walls. The intent is to use the data from the Raytheon cutoff wall to calibrate the finite element analysis procedures that will be employed to make these estimates.

The presence of a perimeter soil-bentonite cutoff wall can alter local ground water flow. Mounding of ground water can occur up-gradient of an enclosure, and decreases in ground water levels can occur down-gradient of an enclosure.

Laboratory tests of hydraulic conductivity should be performed using consolidation pressures that reflect the *in situ* pressures in soil-bentonite cutoff walls. Because arching can be expected to occur in the cutoff wall, these pressures are less than those computed assuming geostatic conditions. Performing hydraulic conductivity tests assuming geostatic conditions can lead to unconservative results. Hydraulic conductivity tests can be performed effectively at low pressures in both consolidation permeameters and the API filter press.

It is difficult to perform *in situ* measurements of the hydraulic conductivity of soil-bentonite cutoff wall backfill material. It would be very useful to develop reliable procedures for such measurements.

Acknowledgements

Clark Morrison contributed to the studies that formed the basis for this paper, and his work is gratefully acknowledged. Financial support for preparation of this paper was provided by Woodward Clyde Consultants and the National Science Foundation (Grant No. CMS-9502448). The findings expressed in this paper are those of the authors and do not necessarily reflect the views of Woodward Clyde Consultants or the National Science Foundation.

References

Adams, T., Baxter, D., Boyer, R., Britton, J., Henry, L., Heslin, G., and Filz, G. (1997). "The mechanical and hydraulic behavior of soil-bentonite cutoff walls." *Charles E. Via, Jr. Department of Civil Engineering Report*, Virginia Polytechnic Institute and State University, Blacksburg, Va.

Barrier Containment Technologies for Environmental Remedial Applications. (1995). John Wiley & Sons, Inc., R. R. Rumer and M. E. Ryan, eds., New York.

Burgess, A. S., Leonard, M. S., and Laird, G. S. (1988). "Design and construction of a soil bentonite slurry wall around an operating facility superfund site." *Proc., 2nd Int. Conf. on Case Histories in Geotech. Engrg.*, S. Prakash, ed., 81-88.

Burke, G. K. and Achhorner, F. N. (1988). "Construction and quality assessment of the in situ containment of contaminated groundwater." *Proc., 5th National Conf. on Hazardous Wastes and Hazardous Materials*, The Hazardous Materials Control Research Institute, 6-10.

Earth Sciences Associates. (1989). "Geotechnical investigation and evaluation of remedial measures for repair of the building at 490 Middlefield Road." Palo Alto, Ca.

Evans, J. C., Costa, M. J. and Cooley B. (1995). "The state-of-stress in soil bentonite slurry trench cutoff walls." *Geoenvironment 2000, Geotech. Spec. Publ. No. 46*, Y. B. Acar, and D. E. Daniel, eds., ASCE, New York, 1173-1191.

Filz, G. M. (1996). "Consolidation stresses in soil-bentonite backfilled trenches." *Proc., 2nd Int. Congress on Environmental Geotechnics,* M. Kamon, ed., Balkema, Rotterdam, 497-502.

Golder Associates. (1987). "Raytheon Semiconductor Division, Mountain View California." Rep., Interim remedial measures, Vol. I and II, Redmond, Wash.

Heslin, G. M., Filz, G. M., and Baxter, D. Y. (1997). "An improved method for interpreting API filter press hydraulic conductivity test results." *Proc., Int. Containment Technology,* 71-77.

HYDRAULIC CONDUCTIVITY EVALUATION OF VERTICAL BARRIER WALLS

David E. Daniel[1] and Hangseok Choi[2]

ABSTRACT

Vertical barrier walls are routinely used to contain contaminated groundwater and vapors around old landfills and remediation sites. Three methods are available for evaluating the hydraulic conductivity of vertical barriers: (1) laboratory tests on reconstituted samples; (2) laboratory tests on "undisturbed" samples; and (3) in situ tests. Test procedures and relative advantages of each method are summarized. Typical test results are provided for three projects. Different methods for measuring hydraulic conductivity can produce hydraulic conductivity's that vary by orders of magnitude, and not always in the same direction on different projects. Numerical analyses were performed for in situ slug tests, and comparative results are presented to clarify the significance of critical variables for single-well borehole tests on vertical barrier walls.

INTRODUCTION

Vertical barriers are frequently employed at contaminated sites to restrict the lateral spreading of contaminants or gases, or to limit the inward migration of fresh groundwater into an area from which contaminated groundwater is being pumped. Vertical barriers are usually constructed in one of three ways: (1) a trench is excavated using a viscous slurry to prevent the trench from collapsing, and the trench is backfilled with a relatively impermeable material (e.g., soil-bentonite, cement-bentonite, or soil-cement-bentonite); (2) soils are mixed in situ or permeated with additives such as grout to form a barrier (e.g., deep soil mixing, jet grouting, permeation grouting, and grouting of hydrofractured openings); or (3) relatively impermeable materials are inserted into the ground (e.g., steel or plastic sheet piling and geomembrane panels). Vertical cutoff walls are discussed in a variety of papers in the literature, but comprehensive overviews may be found in

---

[1] Prof. of Civil & Env. Eng., Univ. of Illinois, Urbana, IL 61801
[2] Graduate Research Assistant, Univ. of Illinois, Urbana, IL 61801

D'Appolonia (1980), Spooner et al. (1984), Xanthakos (1994), Evans (1994), Rumer and Ryan (1995), and Rumer and Mitchell (1995).

One of the most important aspects of creating a vertical barrier is verification of low hydraulic conductivity. In the U.S., most vertical barrier walls that have been constructed to date have been formed from bentonite-amended materials (sometimes with Portland cement or other additives, as well) constructed with slurry trenching or deep mixing. Such barriers are typically 0.5 to 1 m wide, with depths ranging from a few meters up to several tens of meters. The design maximum hydraulic conductivity varies, depending on the application, but is typically either 1×10^{-6} cm/s or 1×10^{-7} cm/s.

Three methods are available for evaluating hydraulic conductivity: (1) laboratory tests on reconstituted samples; (2) laboratory tests on "undisturbed" samples; and (3) in situ tests. Although tests on reconstituted samples (sometimes called "wet test") are the industry standard method for verifying hydraulic conductivity, there is interest in developing better methods of hydraulic conductivity evaluation. The purpose of this paper is to review methods of hydraulic conductivity testing, present comparative data showing differences in test results, and provide insight into factors that can affect in situ tests performed with a single well ("slug") test.

METHODS OF HYDRAULIC CONDUCTIVITY TESTING

The primary methods of hydraulic conductivity testing are discussed in the succeeding sections. The relative advantages and disadvantages are summarized in Table 1.

Laboratory Tests on Reconstituted Samples

By far the most commonly used type of hydraulic conductivity test for verifying hydraulic conductivity of vertical barriers is the laboratory test on a reconstituted sample of the barrier material. For soil-bentonite backfilled slurry walls, the sample is usually a "grab" sample removed from the backfill mixing area just prior to placement in the slurry-filled trench. For barrier walls constructed in situ (e.g., deep soil mixed walls), the sample is removed from the actual barrier, near the surface and before the backfill has had time to set up or to cure. Some laboratories have developed special sampling devices to enable a relatively undisturbed sample to be taken from beneath the surface of the barrier, e.g., in cement-bentonite slurries.

Table 1. Advantages and Disadvantages of Hydraulic Conductivity Testing Methods for Vertical Barriers.

Category	Method	Advantages	Disadvantages
Laboratory Tests on Reconstituted Samples	Test in Fluid Loss Apparatus	1. Inexpensive 2. Minimal Sample Handling Issues	1. Void Ratio of Test Material Highly Variable 2. No Control over Stresses or Saturation 3. Reconstituted Sample May Not Be Representative of Barrier Material Constructed in Field
	Consolidation-Cell Test	1. Compressive Stress Controlled 2. Easy to Form a Test Specimen by Consolidating Test Material from a Loose or Nearly Fluid State 3. Convenient for Permeation with Contaminated Liquids	1. Equipment Relatively Complex and Expensive 2. No Control over Saturation 3. Reconstituted Sample May Not Be Representative of Barrier Material Constructed in Field
	Flexible-Wall Permeameter	1. Flexible-Wall Permeameter Is Industry Standard Method of Testing Low Permeability Materials 2. Full Control Over Stresses 3. Sample Completely Saturated	1. Difficult to Form a Test Specimen for a Very Soft Test Material 2. Equipment Relatively Complex and Expensive 3. Reconstituted Sample May Not Be Representative of Barrier Material Constructed in Field

Table 1. Advantages and Disadvantages of Hydraulic Conductivity Testing Methods for Vertical Barriers (cont.).

Category	Method	Advantages	Disadvantages
Laboratory Tests on "Undisturbed Samples"	Test in Sampling Tube	1. Low Cost	1. Potential Sidewall Leakage 2. Lack of Control over Stresses 3. Lack of Control over Saturation 4. No Flexibility over Direction of Fluid Flow (Vert. or Horiz.) 5. Sample Almost Certain to Be Disturbed to Some Extent
Laboratory Tests on "Undisturbed Samples"	Flexible-Wall Permeameter	1. Flexible-Wall Permeameter Is Industry Standard Method of Testing Low Permeability Soil 2. Full Control Over Stresses 3. Sample Completely Saturated 4. No Restriction on Size of Sample 5. Can Trim Sample to Permeate in Any Direction	1. Soft Sample May Be Difficult to Handle 2. Difficult to Test at Very Low Effective Stress 3. Sample Almost Certain to Be Disturbed to Some Extent
In Situ Tests	Pizeocone	1. Additional Information Besides Hydraulic Conductivity Collected 2. In Situ Barrier Material Is Tested	1. Permeated Volume Very Small 2. Experience Very Limited
	Single Well ("Slug") Test	1. Large Volume of Material Tested 2. In Situ Barrier Material Is Permeated	1. Borehole May Be Smeared 2. Proximity of Well Screen Material to Edge of Barrier Unknown 3. Methods for Calculating Hydraulic Conductivity Not

There are three ways to measure the hydraulic conductivity of reconstituted barrier materials: (1) fabricate a cylindrical test specimen of the barrier material and then place the test specimen in a permeameter, e.g., a flexible-wall permeameter conforming to ASTM D5084; (2) place the backfill in a rigid-wall permeameter, apply a confining stress (e.g., in a consolidation-cell permeameter), and then measure hydraulic conductivity after the backfill has consolidated; and (3) place the backfill in a filter-press device, apply an air pressure, and measure the apparent hydraulic conductivity. Daniel (1994) and Evans (1994) provide background information.

The simplest and least expensive method for measuring hydraulic conductivity is the API fluid loss test. The fluid loss test (e.g., ASTM D5891) is routinely performed on bentonitic slurries and provides an excellent indicator of the quality of bentonite as a gelling agent. Figure 1 shows a schematic diagram of the fluid loss testing equipment adapted to hydraulic conductivity testing of soil-bentonite backfill. The soil-bentonite backfill is placed in the device, an air pressure is applied, and a "filter cake" forms near the interface with the underlying filter sand. After some consolidation has occurred, the rate of flow through the material is measured and hydraulic conductivity is calculated (Heslin et al., 1997). It is best to use a layer of sand above the test material to provide confinement (Barvenik and Ayres, 1987; and Heslin et al., 1997).

One of the problems with the fluid loss test is that the void ratio is lower near the bottom of the material. Although the test method appears relatively crude, it is used fairly frequently and experience suggests that the method does seem to provide hydraulic conductivity's in the same general range as other testing methods (Barvenik and Ayres, 1987; Heslin et al., 1997).

Another method of laboratory hydraulic conductivity testing involves placing the barrier material in a consolidation-cell permeameter, tamping or "rodding" the material to remove air voids, consolidating the material, and then permeating it. This provides a very convenient method for forming a consolidated test specimen in the same rigid-wall cell that will be used for permeation.

An additional method of testing reconstituted samples is to form a test specimen in a cylinder (e.g., by tamping the material with a rod into a compaction mold), transferring the test specimen to a flexible-wall permeameter, and the measuring the hydraulic conductivity in the flexible-wall cell per ASTM D5084. This method is probably the best of the three test methods on reconstituted samples in the sense that the flexible-wall method provides the greatest degree of control and industry acceptance. The problem, however, is that many barrier materials are extremely soft and very difficult to shape into a cylindrical specimen that will stand under its own weight. Backfills with cement can be cured prior to transfer to the flexible-wall permeameter, although it is best to transfer the test specimen to the permeameter and subject the specimen to the effective confining stress that simulates in situ conditions as quickly as possible.

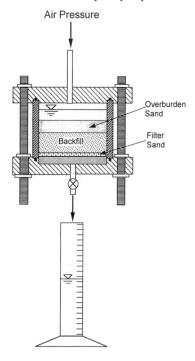

Figure 1. Schematic Diagram of Fluid Loss Apparatus for Hydraulic Conductivity Testing.

The fundamental disadvantage of all tests involving reconstituted samples is that the material being tested may be quite different from the in situ material. "Rodding" the backfill material into a mold may or may not produce a material with similar pore size distribution and hydraulic conductivity as the actual in situ backfill material. Despite this problem, however, laboratory tests on reconstituted samples (so-called "wet" tests) are today the industry standard method for evaluating hydraulic conductivity of vertical barriers in the U.S. because the other methods (discussed below) have major problems and uncertainties associated with them, and because tests on reconstituted samples are far faster and simpler than the alternative procedures.

Laboratory Tests on "Undisturbed" Samples

Laboratory tests on "undisturbed" samples of material recovered from vertical barriers are performed in essentially the same manner as tests on undisturbed samples recovered from soil strata. Although hydraulic conductivity tests can be performed on the sample contained inside the thin-walled sampling tube (Table 1), such tests are rare. Rather, the normal procedure is to extrude a sample from the sampling tube, trim a cylindrical test specimen, and permeate the specimen in a flexible-wall permeameter.

The process of measuring hydraulic conductivity of "undisturbed" samples is fraught with potential difficulty. First, there is a high probability that the barrier material will be damaged to some extent by the sampling process. This is particularly true for backfills containing a small amount of cement, which have a fragile structure and which are easily cracked (e.g., Yang et al., 1993). Because the hydraulic conductivity of barrier materials is dominated by the characteristics of the few largest pores or occasional cracks, the material is highly sensitive to disturbance. Further, the test specimen must be handled very carefully when it is set up in the permeameter; very soft samples (e.g., soil-bentonite backfill in slurry trenches) can be extremely difficult to handle. Time must be allowed for consolidation of the backfill to occur before the material is sampled, which creates highly undesirable delay between construction and hydraulic conductivity verification. In addition, the material must be consolidated in the permeameter to a pre-determined effective stress, but experience indicates that because of arching, the actual in situ vertical effective stress may be far lower than the value computed from geostatic conditions (Evans et al., 1995). If the specimen is consolidated to the geostatic vertical effective stress or higher, the measured hydraulic conductivity will be too low. In addition, if the constructed barrier wall is not perfectly vertical, if the borehole used to obtain a relatively undisturbed sample is not vertical, or if the borehole is drilled off center from the barrier wall, the "undisturbed" sample may be taken from outside the vertical barrier. Finally, if the vertical barrier does contain an occasional defect ("window"), the probability of sampling from that particular zone is very small. For example, assume that windows 0.5 m by 0.5 m occur in a 20-m-deep vertical barrier at a spacing of 50 m between windows. The "windows" occupy an area that is only 0.025% of the total area, yielding a probability of encountering the window in any single sample of approximately one in 4,000. Thus, even if a relatively undisturbed sample is obtained from within the boundaries and consolidated to the proper vertical effective stress, there is no assurance that the more permeable zones within the barrier have been sampled or tested.

Data are available from three projects in which hydraulic conductivity's were measured with tests on reconstituted and undisturbed samples. Project 1 is described by Barvenik and Ayres (1987) and involved a soil-bentonite backfill in a slurry trench. Project 2 is described by Yang et al. (1993) and involved a soil-bentonite-cement mixture in a deep soil mixed wall. Information for Project 3 was obtained from data submitted to a state regulatory agency for a soil-cement-bentonite backfill in a slurry trench. The data are compared in Table 2 and indicate that hydraulic conductivity's measured on reconstituted samples are consistently lower than values measured on "undisturbed" samples. The ratio of conductivity's measured on reconstituted samples, compared to "undisturbed" samples, is remarkably consistent and in the range of 0.1 to 0.2. Yang et al. (1993) present information for Project 2 indicating a possibility of severe disturbance of the soil-cement-bentonite barrier wall material, which suggests that the conductivity's measured on "undisturbed" samples may have been too large.

Table 2. Comparative Data on Hydraulic Conductivity's (k's) Measured on Reconstituted and Undisturbed Samples (Average Is Geometric Mean).

Project	Average k (cm/s) Reconstituted Samples	Average k (cm/s) "Undisturbed" Samples	Ratio of k's for Reconstituted/ Undisturbed
1	1×10^{-7}	5×10^{-7}	0.2
2	1×10^{-6}	7×10^{-6}	0.14
3	1×10^{-7}	8×10^{-7}	0.13

In Situ Hydraulic Conductivity Tests

In situ hydraulic conductivity tests on vertical barriers, like all in situ hydraulic conductivity tests, offer the opportunity to permeate the actual backfill material with minimal disturbance. Two types of in situ tests have been performed. Manassero (1994) describes the use of the piezocone test for evaluating the hydraulic conductivity of a cement-bentonite barrier. The piezocone provides interesting opportunities for evaluating hydraulic conductivity but permeates a small volume of material and may, upon insertion into the barrier, create sufficient disturbance to alter hydraulic conductivity.

The more commonly used test in the U.S. (although still rarely used) is a single-well, falling- or rising-head test, commonly termed "slug test". The idea is to employ aquifer slug-test procedures for a well installed in a vertical barrier, interpreting the data with the solution of Bouwer and Rice (1976). Teeter and Clemence (1986) and Yang et al. (1993) describe the basic concept, and present flow net solutions to take into account the boundaries of the vertical barrier. A more fundamental limitation, however, to the Bouwer and Rice (1976) solution is that the tested material is assumed to be incompressible, and clearly certain vertical barrier materials (particularly soil-bentonite backfill) are highly compressible.

Another significant problem with the slug test is the unknown separation between the well screen and the edge of the barrier. As suggested in Figure 2, if the well is not perfectly centered in the barrier, if the barrier is not perfectly vertical, or if the borehole is not perfectly vertical, the screen will not be centered in the barrier wall. With normal construction tolerances, it is possible that the screened section will not even lie within the barrier wall but, instead, may intersect the adjacent aquifer. If the slug test is used, extraordinary care must be taken to construct as close to a perfectly vertical barrier as possible, drill the well in the center of the vertical barrier, and to drill an essentially vertical well. Instruments such as a slope indicator can be used to verify the verticality of the cased well after installation.

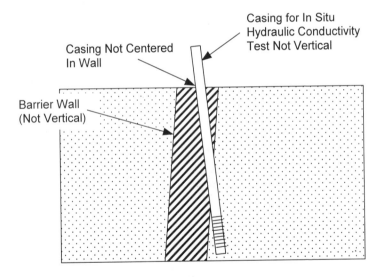

Figure 2. Potential Problems with Centering a Screened Well in a Vertical Barrier.

Other potential problems with slug tests include the possibility that vibrations from the drilling operation will fracture the barrier material. Drilling may also cause a smearing of material within the hole. With falling head tests, there is a risk of hydraulic fracturing, such as reported by Bjerrum et al. (1972) because of the low vertical stresses created when the backfill arches during consolidation (Evans, 1995). Another practical limitation is that the slug test must be performed after the wall has had time to consolidate or cure, which can take months. Hydraulic conductivity results normally will not be obtained until after the construction is finished, which severely limits the value of the test for construction quality control.

Laboratory Vs. In Situ Test Results

In situ test results are compared with laboratory test result for Projects 2 and 3 mentioned earlier. The differences in relative values of hydraulic conductivity's measured on the two projects are striking. On Project 2 (a deep soil mixed wall), laboratory hydraulic conductivity tests on "undisturbed" samples yielded the highest hydraulic conductivity's while the slug tests yielded the lowest hydraulic conductivity's. In situ hydraulic conductivity test tend to yield higher hydraulic conductivity's than laboratory tests for most soils, but just the opposite was observed here. Some of the hypotheses surrounding the differences are described by Yang et al. (1993). One possible explanation is that the "undisturbed" samples were in fact badly disturbed as a result of sampling the very weakly cemented, sandy material. The physical condition of the samples, which contained cracks, is consistent with this explanation. The reconstituted samples may not have been prepared in the laboratory in a manner that replicates conditions in the field, which would not be surprising because rodding a soil-bentonite-cement mixture into a mold in the laboratory may not necessarily yield a material that has similar hydraulic properties as those of a material formed in the field with deep soil mixing equipment. In any case, it is possible and perhaps probable that the in situ tests results are the most accurate hydraulic conductivity's on this project. On the other hand, it was argued by others involved in this project that the drilling operation for the in situ tests smeared the walls of the borehole, producing artificially low hydraulic conductivity's. The causes for the differences in hydraulic conductivity's were never isolated.

Table 3. Comparative Data on Hydraulic Conductivity's (k's) Measured on Reconstituted Samples, Undisturbed Samples, and In Situ Slug Tests.

Project	Avg. k (cm/s) Reconstituted Samples	Avg. k (cm/s) "Undisturbed" Samples	Avg. k (cm/s) In Situ Slug Tests	k (slug test)/ k (reconst.)	k (slug test)/ k (undist.)
2	1×10^{-6}	7×10^{-6}	3×10^{-7}	0.3	0.04
3	1×10^{-7}	8×10^{-7}	2×10^{-6}	20	2.5

On Project 3 (a soil-bentonite-cement backfill in a slurry trench), the slug tests yielded a hydraulic conductivity that averaged about 20 times higher than results from laboratory tests on reconstituted samples, and about 2.5 times higher than results from tests on "undisturbed" samples. There was major disagreement between the various parties involved in this project about which hydraulic conductivity values were correct, and, again, insufficient data were available to isolate the causes for the differences. The hydraulic conductivity's from the slug tests were calculated using the Bouwer and Rice (1976) procedure, and later were corrected based on numerical simulations. Even so, significant uncertainties existed concerning effects of backfill compressibility, proximity of the edge of the screened well with the edge of the vertical barrier wall, and other factors. It was these uncertainties that motivated the authors to explore issues surrounding interpretation of the results of slug tests in more detail.

INTERPRETATION OF RESULTS OF SLUG TESTS

The procedure usually used for interpreting the results of slug tests is described by Bouwer and Rice (1976) and is commonly referred to as the Bouwer-Rice (B-R) method. Although other techniques may be used to analyze the results of single-well "slug" tests, the B-R method is the only one known by the authors to have been described in the literature for analysis of vertical barriers (Teeter and Clemence, 1986; and Yang et al., 1993). In addition, the B-R method is the most popular method for analysis of slug test data (Hyder an Butler, 1995).

The limitations in the application of the B-R method (and to other methods of aquifer analysis applied to vertical barriers) are the following. First, the specific storage (compressibility) of the formation is assumed to be negligible. However, many vertical barrier materials (particularly soil-bentonite backfill placed in a slurry trench) are highly compressible. Because the slug test involves a varying head, the barrier material tends to compress or expand, depending on the change in water pressure relative to the long-term equilibrium pressure established by the water table position. The effect of consolidation may be evaluated as follows. During a slug test the total stress in the material adjacent to the well remains approximately constant. In a rising head test (where the water level in a well is lowered relative to the water table position), the water pressure has been reduced relative to the long-term equilibrium value, which causes an increase in effective stress and a tendency for the backfill to consolidate and to compress. Consolidation water will partly flow into the surrounding formation and partly into the well. Consolidation water flowing into the well is causes the water level in the well to rise faster than it would have if the material had been incompressible. The accelerated rate of rise of the water level in the well caused by expulsion of consolidation water from the backfill in turn causes the calculated hydraulic conductivity to be too large. The calculated hydraulic conductivity is also too large in a falling head test for the same reason, although in a falling head test the water level is raised above the long-term equilibrium value, the backfill tends to swell, and the swelling tends to withdraw water from the well (again causing an overestimation of k). Although the impact of backfill compressibility is clear in principle, it is not known whether the compressibility effect would have a significant impact upon the calculated hydraulic conductivity. Similarly, it is not known whether enough time would pass for consolidation to have a significant effect during a typical slug test on a vertical barrier wall backfill.

A second critical assumption in the B-R solution is that the aquifer is semi-infinite. A vertical barrier has boundaries a short distance from the well. Teeter and Clemence (1986) and Yang et al. (1993) attempted to correct for this problem with flow net solutions, but such solutions only apply to steady flow and not to transient flow in an unsteady medium.

A third assumption is that the changes in the water table during a slug test are so small that the water table can be treated as a constant head boundary at the outer radius of influence of the slug test. Over large horizontal distances in an aquifer, this assumption may be reasonable, but over distances of a fraction of a meter in a vertical barrier, the changes in the water table can be significant. Because vertical barrier materials are more fine textured than aquifer materials, flow above the water table may be more significant in a vertical barrier than in an aquifer.

Because of these and other issues, applicability of the B-R solution to slug tests performed on vertical barriers is questionable. Hyder and Butler concluded that the B-R solution provided estimated hydraulic conductivity's in aquifers that were within 30% of their true values. However, they note that in less-permeable, clay-rich formations, the B-R method may overpredict hydraulic conductivity by more than 100%.

NUMERICAL MODEL

To evaluate some of the variables that affect interpretation of data from slug tests, the authors developed an explicit, three-dimensional finite difference solution was developed to solve the basic ground-water flow equation (Freeze and Cherry, 1979):

$$k_x \frac{\partial^2 h}{\partial x^2} + k_y \frac{\partial^2 h}{\partial y^2} + k_z \frac{\partial^2 h}{\partial z^2} = S_s \frac{\partial^2 h}{\partial t^2} \qquad (1)$$

where k_x, k_y, and k_z are the hydraulic conductivities in the x, y, and z directions, S_s is the specific storage, h is total hydraulic head, and t is time. The specific storage is defined as follows:

$$S_s = \rho g \alpha + n \beta \qquad (2)$$

where ρ is the bulk mass density of the soil, g is the acceleration due to gravity, α is the compressibility of the soil (= volumetric strain of the soil divided by the change in effective stress, often called coefficient of volume compressibility or m_v in soil mechanics literature), n is porosity, and β is the compressibility of water. The compressibility of water is often considered to be negligible to that of soil for compressible, clayey soils such as often used in vertical barrier walls.

A three-dimensional, explicit finite difference solution was developed to solve Eq. 1. Initially, a fully implicit scheme was used, but the memory requirements substantially exceeded those of an explicit scheme, and for this reason the explicit solution was found to be more convenient. The solution was written in Fortran using block-centered node points. The full solution will be described in the junior author's dissertation, which is still in development. Extensive calibrations were developed for problems involving radial flow and flow in cylindrical coordinates to ensure that the model was robust and reliable, and provided results that were identical to various analytic solutions, including the B-R method under the assumed conditions employed in the B-R solution.

In this paper, results are presented for one well analyzed on Project 3 mentioned earlier. The well casing and screen were installed to a depth of 5.2 m in a 200-mm-diameter borehole that was drilled into the 600-mm-wide vertical barrier wall. The screen at the bottom of the well had a length of 0.55 m and a radius (filter pack) of 36 mm. The hydraulic conductivity calculated by the engineering consultant on the project was 2×10^{-6} cm/s. The required hydraulic conductivity was $\leq 1 \times 10^{-6}$ cm/s. Therefore, the hydraulic conductivity measured from this slug test failed in the original analysis performed by the consultant. The failure of this and other tests led to questions about the validity of the data interpretation and of the B-R method itself as applied to slug tests on vertical barrier walls.

Typical Values of Specific Storage

In order to analyze a vertical barrier, one of the essential pieces of information that is needed is the specific storage (S_s) of the vertical barrier material. The higher the compressibility of the material (α), the higher the value of S_s, as shown in Eq. 2. Very little data have been published on the compressibility of vertical barrier materials. Khoury et al. (1992) presented data on volumetric strains for specified ranges in effective stress, from which specific storage values could be calculated. The S_s for a backfill with 1.18% bentonite was 2.83×10^{-4} cm^{-1}, while that of backfill with 1.65% bentonite was 3.6×10^{-4} cm^{-1}.

Data obtained from Project 3 included information on the change in void ratio that occurred when about 35 kPa of effective confining stress was applied to "undisturbed" samples of backfill to consolidate the samples prior to hydraulic conductivity testing. Data from approximately 25 samples yielded an average S_s of 4.8×10^{-4} cm^{-1}. Therefore, data from these two sources indicate that soil-bentonite backfill likely has a specific storage in the range of 3 to 5×10^{-4} cm^{-1}. A value of $S_s = 5 \times 10^{-4}$ cm^{-1} was assumed for analysis of the well from Project 3.

Effect of Specific Storage on Hydraulic Conductivity

One of the most significant issues concerning use of aquifer methods for analysis of slug test data on vertical barriers is the effect of the compressibility of the barrier material. To isolate this effect, an aquifer was first analyzed with the numerical model. Analysis of an aquifer case eliminated geometrical issues from the analysis. In the B-R test, a plot (herein referred to as the B-R style plot) is made of the log of the ratio of head at time t, $H(t)$, divided by initial head (H_0), versus time. The plot is linear for the assumed incompressible material. If a plot of the actual data is non linear, a linear fit is made to the data. As shown in Fig. 3, as compressibility increases, the B-R style plot becomes increasingly nonlinear.

When one tries to fit a linear curve to a non-linear data set, the fit must be made at some point in the data set. Where the fit is made impacts the calculated hydraulic conductivity. For example, Fig. 4 shows a family of linear curves for different hydraulic conductivity's for the same aquifer case used to generate Fig. 3, but ignoring the compressibility of the material. Figure 5 shows the ratio of the hydraulic conductivity that would have been determined from the B-R method (k_{B-R}) to the actual hydraulic conductivity used in the numerical simulation (k_{real}), versus the fitting or reading point. For example, if the reading point is 0.5, this means that the fit was made at $H(t)/H_0 = 0.5$. Clearly the error in the B-R method becomes more and more significant when the fit is made based on early data, when the head has not changed much. The tendency would normally be to fit to the relatively early data (often $H(t)/H_0$ in the range of 0.6 to 0.8) because these early data are more reliable (large-time data can be subject to water table fluctuations caused by tidal changes or shifts in atmospheric pressure), and because when testing compressible materials in general one tends to try to keep the head changes as small as possible to minimize compressibility effects. For a typical case in which the fitting is done at $H(t)/H_0$ of say 0.6, the B-R method yields a hydraulic conductivity that is too high by a factor of about 1.5 to 2.

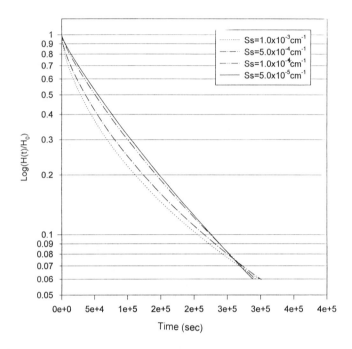

Figure 3. Results of Numerical Analysis of the Effect of the Specific Storage on Bouwer-Rice Style Data Plot for Aquifers of Varying Specific Storage.

Geometry of Vertical Barrier Wall and Location of Well

Another factor considered was the fact that the well under consideration for Project 3 was not vertical and was determined from slope indicator measurements performed inside the well casing to be located 17 cm off center at the center of the screened section. The edge of the well screen, rather than being located 264 mm (11 inches) from the edge of the barrier, was theoretically located 94 mm (4 inches) from the edge of the barrier due to lack of verticality in the borehole into which the casing was inserted. These distances assume that the vertical barrier was absolutely vertical, but some deviation from verticality of the barrier itself is to be expected. The actual separation between the edge of the well and the edge of the vertical barrier may have been more or less than 94 mm.

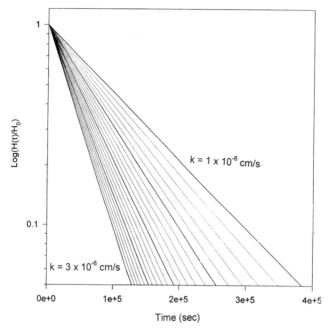

Figure 4. Family of Theoretical Curves for Bouwer-Rice Style Plot for Aquifers of Negligible Compressibility and Different Hydraulic Conductivity.

Figure 6 shows theoretical and measured plots of $H(t)/H_0$ versus log of time. Theoretical curves were generated assuming $S_s = 5 \times 10^{-4}$ cm^{-1} (actual average value based on laboratory data) and a range of hydraulic conductivity. The measured data match theoretical curves in the range of hydraulic conductivity of $1.5 - 2 \times 10^{-6}$ cm/s.

It is known from slope indicator data that the well screen was not centered in the vertical barrier wall. Analyses were performed assuming that the well was off-centered by the measured amount, and results are shown in Fig. 7. The hydraulic conductivity was again in the range of 1.5 to 2×10^{-6} cm/s, indicating that the fact that the well was off-centered had practically no impact on the water level in the well or upon the calculated hydraulic conductivity.

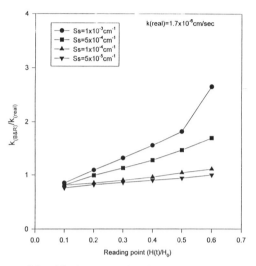

Figure 5. Effect of Specific Storage and Fitting Point on Hydraulic Conductivity.

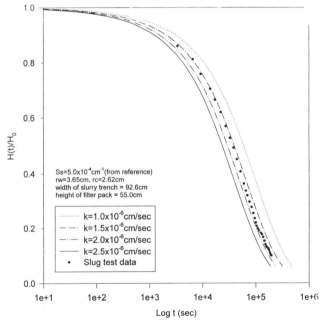

Figure 6. Theoretical and Measured Response of Slug Test for Well Assumed to be Centered in Barrier.

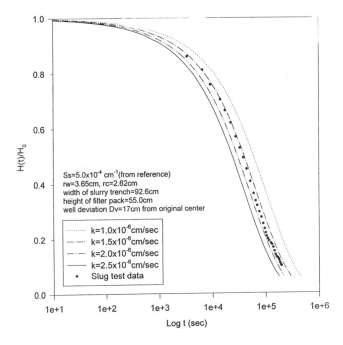

Figure 7. Theoretical and Measured Response of Slug Test for Well Assumed to be Off Center in Barrier.

The results of analyses just presented provide insights into the B-R method. The assumption in the B-R method that the aquifer is negligibly compressible can be a significant problem when applying the B-R method to analysis of slug tests in vertical barriers. Failure to account for compressibility of the barrier material leads to calculation of k that is too high by a factor of two or more. The problems are two-fold: (1) the fact that the material is compressible causes the B-R style data plots to be non-linear, and a B-R linear curve must be fit through some point in the measured data – a different hydraulic conductivity results depending on the point of fitting; and (2) the B-R method tends to yield a hydraulic conductivity that is too high when the material being tested is compressible. On the positive side, the error in the B-R method is offset slightly by other compensating errors. For instance, note from Fig. 5 that for materials of relatively low compressibility, and fits of data made at very large times, the B-R method yields a hydraulic conductivity that is slightly too low. This is probably because of the other simplifying assumptions in the B-R method. The initial estimate of k made by the consulting engineer on Project 3 ($k = 2 \times 10^{-6}$ cm/s), which ignored compressibility effects and the proximity of the well to the wall of the barrier, differed little from the estimate of 1.5 to 2 x 10^{-6} cm/s based on finite difference analysis for an off-centered well in compressible material. The errors in the B-R method tend to offset.

The effect of finite geometry is far less important than compressibility. The theoretical impact of an off-center well seems of far less concern than the practical concern that the actual distance between the well screen and wall of the barrier can only be known if: (1) the exact horizontal offset of the well casing due to lack of verticality of the borehole is known, and (2) the precise location of the walls of the barrier is known. While (1) can be determined through slope indicator measurements, (2) cannot. Perhaps this problem of unknown separation between the wall of the barrier and the screen is the most fundamental difficulty in the interpretation of slug test data. There is a risk that the well screen may actually lie outside the limits of the barrier wall, or lie so close to the edge of the barrier that the calculated hydraulic conductivity would be much too large.

Finally, space limitations did not permit a detailed description of the details of the numerical model, but the following information is provided. The node space was determined by a series of trials using different spacings. The spacing in the horizontal direction was much more critical than in the vertical direction because most of the flow to the well is horizontal flow. A node spacing of slightly over 10 mm was selected. The number of node points used in the analyses varied depending on problem being analyzed but was typically in the hundreds. The boundary conditions were as follows. Along the vertical outer boundaries where the barrier wall met the surrounding formation, the ground water table was assumed to be constant and, thus, the head was assumed to be constant. The base of the barrier wall was assumed to be keyed into a relatively impermeable formation and, therefore, this boundary was a no-flow boundary. The upper, horizontal boundary was similarly a no-flow boundary. The boundary condition along the interface between the non-screened section of the well and the vertical barrier was a no-flow boundary. The only difficulty boundary condition to describe was at the interface between the well screen and vertical barrier wall material, and the condition was difficult because the head in the well varied. A constant-head boundary condition at the well screen/barrier interface was assumed for any one time step. The heads at new time $t+\Delta t$ were calculated, and the fluxes (horizontal and vertical) at nodes adjacent to the well screen were determined for the time step. The fluxes were summed to determine the total amount of flow into or out of the well over the time step. The change in water level in the standpipe was determined based on the quantity of flow and cross-sectional area of the standpipe. For the next time step, the new "constant" head based on the revised water level in the standpipe was used for the well screen/barrier interface.

CONCLUSIONS

Three types of hydraulic conductivity tests can be performed on vertical barriers: (1) laboratory tests on reconstituted samples; (2) laboratory tests on "undisturbed samples", and (3) in situ tests such as slug tests. While in principle slug tests might appear to be the most attractive, and tests on undisturbed samples to be the second most attractive, in reality both types of tests are fraught with potential problems. For the slug test, a suitable methodology for analyzing the test results (taking into account the compressible nature of the barrier material and boundaries of the vertical barrier) has yet to be developed. Numerical analysis appears to be the only hope for reasonably accurate analysis. Numerical results presented herein demonstrate the effect of compressibility is far more important than that of geometry. However, practical problems also plague the slug test: the separation between the well screen and edge of the vertical barrier is unknown, there is a risk that the well screen may actually lie outside the limits of the vertical barrier, the barrier material may be smeared or fractured by the drilling process, and the test must be performed weeks or months after construction (after consolidation or curing has occurred) thus greatly reducing the value of the test for construction quality control.

Laboratory hydraulic conductivity tests on undisturbed samples seem like a good idea in principle, but practical problems are significant. The material cannot be sampled until weeks or months after construction, again making the value of the test for construction quality control very limited. Sampling disturbance may be inevitable and may be particularly severe for backfills containing cement, which produces a fragile material that can crack during sampling. The sample must be consolidated to the effective stress expected in the vertical barrier wall at a particular depth, but the proper stress is likely to be far lower than geostatic stress and to be difficult to determine accurately. Finally, the probability that the sampling process will lead to detection of occasional small defects ("windows") in the barrier is very low because such windows make up a small percentage of the volume of the wall.

Although tests on reconstituted samples may appear to be the least desirable type of test, they are by far the most practical type of test. Results can be obtained quickly enough to be of use for construction quality control purposes. The tests are reasonably simple and inexpensive, and there is extensive experience with these types of tests. The key limitation of these tests is that the prepared sample may not have the same pore size distribution as the in situ material and, therefore, the sample that is tested may not be representative of the actual barrier material.

Clearly, all three types of hydraulic conductivity tests discussed in this paper have limitations. The state of the art for hydraulic conductivity testing of vertical barriers cannot be described as well developed. Further research is needed to develop better ways of testing vertical barrier materials for hydraulic conductivity. The most important needs appear to be development of techniques to permit accurate analysis of the results of slug tests in vertical barriers and development of geophysical or other techniques that will enable precise determination of the location of a borehole relative to the vertical walls of a barrier.

REFERENCES

Barvenik, M.J., and J.E. Ayres (1987), "Construction Quality Control and Post-Construction Performance Verification for the Gilson Road Hazardous Waste Site Cutoff Wall," U.S. Environmental Protection Agency, EPA/600/2-87/065, Cincinnati, OH.

Bjerrum, L., Nash, J.K.T.L., Kennard, R.M., and R.E. Gibson (1972), "Hydraulic Fracturing in Field Permeability Testing," *Geotechnique*, 22(2): 319-332.

Bouwer, H., and R.C. Rice (1976), "A Slug Test for Determining Hydraulic Conductivity of Unconfined Aquifers with Completely or Partially Penetrating Wells," *Water Resources Research*, 12(3): 423-428.

D'Appolonia, D.J. (1980), "Soil-Bentonite Slurry Trench Cutoffs," *Journal of the Geotechnical Engineering Division, ASCE*, 106(4): 399-417.

Daniel, D.E. (1994), "State-of-the-Art: Laboratory Hydraulic Conductivity Tests for Saturated Soils," *Hydraulic Conductivity and Waste Contaminant Transport in Soil, ASTM STP 1142*, D.E. Daniel and S.J. Trautwein (Eds.), American Society for Testing and Materials, Philadelphia, 30-78.

Evans, J.C. (1994), "Hydraulic Conductivity of Vertical Cutoff Walls," *Hydraulic Conductivity and Waste Contaminant Transport in Soil, ASTM STP 1142*, D.E. Daniel and S.J. Trautwein (Eds.), American Society for Testing and Materials, Philadelphia, 79-94.

Evans, J.C., Costa, M.J., and B. Cooley (1995), "The State-of-Stress in Soil-Bentonite Slurry Trench Cutoff Walls," *Geoenvironment 2000*, Y.B. Acar and D.E. Daniel (Eds.), American Society of Civil Engineers, New York, 2: 1173-1191.

Freeze, R.A., and J.A. Cherry (1979), *Groundwater*, Prentice-Hall, Englewood Cliffs, NJ.

Heslin, G.M., Filtz, G.M., Baxter, D.Y., and Davidson, R.R. (1997), "An Improved Method for Interpreting API Filter Press Hydraulic Conductivity

Test Results," *Proceedings, International Containment Technology Conference*, St. Petersburg, FL, pp. 71-78.

Hyder, Z., and J.J. Butler (1995), "Slug Tests in Unconfined Formations: An Assessment of the Bouwer and Rice Technique, *Ground Water*, 33(1):16-22.

Khoury, M.A., Fayad, P.H., and R.S. Ladd (1992), "Design, Construction and Performance of a Soil-Bentonite Cutoff Wall Constructed in Two Stages," *Slurry Walls: Design, Construction and Quality Control, ASTM STP1129*, D.B. Paul, R.R. Davidson, and N.J. Cavalli (Eds.), American Society for Testing and Materials, Philadelphia, 289-308.

Manassero, M. (1994), "Hydraulic Conductivity Assessment of a Slurry Wall Using Piezocone Test," *Journal of Geotechnical Engineering*, 120(10): 1725-1746.

Rumer, R.R., and Mitchell, J.K. (1995), *Assessment of Barrier Containment Technologies*, National Technical Information Service, #PB96-180583, Springfield, VA, 435 p.

Rumer, R.R., and Ryan, M.E. (1995), *Barrier Containment Technologies for Environmental Remediation Applications*, John Wiley & Sons, New York, 170 p.

Spooner, P., Wetzel, R.S., Spooner, C.E., Furman, C.A., Tokarski, E.F., and G.E. Hunt (1984), *Slurry Trench Construction for Pollution Migration Control*, U.S. Environmental Protection Agency, EPA-540/2-84-001, Cincinnati, Ohio.

Teeter, R.M., and S.P. Clemence (1986), "In-Place Permeability Measurement of Slurry Trench Cutoff Walls," *Proceedings*, Use of In Situ Tests in Geotechnical Engineering, American Society of Civil Engineers, New York, pp. 1049-1061.

Xanthakos, P.P. (1994), *Slurry Walls as Structural Systems*, Second Edition, McGraw Hill, New York.

Yang, D.S., Luscher, U., Kimoto, I., and S. Takeshima (1993), "SMW Wall for Seepage Control in Levee Construction," *Proceedings: Third International Conference on Case Histories in Geotechnical Engineering*, Univ. of Missouri-Rolla, Vol. 1, pp. 487-492.

PRACTICE AND PERFORMANCE OF EXCAVATION
SUPPORT SYSTEMS IN THE CHICAGO AREA

Jerry F. Parola[1] and Robert G. Lukas[2] Members, ASCE

Abstract

The practice and performance of excavation support systems are dependent upon the subsoil profile, site geometry and construction considerations. The geology environment as well as the urban construction considerations in Chicago influence the selection of earth retention systems with the final selection as determined by the cost-performance relationship. Several case histories are provided to exemplify the practice and performance of excavation support systems in the Chicago area.

Introduction

Earth retention construction in urban environments creates a multitude of design considerations and an array of possible earth retention systems. In current practice, earth retention systems for both temporary and permanent conditions have become quite diversified and include ground modifications, soil structure and structural systems. Ground modification systems include numerous ground strengthening techniques such as grouting, jet grouting or freezing. Soil structure systems include soil nailing and reinforced earth methods.

---

1. Vice President, Case Foundation Company, 1325 West Lake Street, Roselle, Illinois 60172.
2. Consultant, Ground Engineering Consultants, Inc., 350 Pfingsten Road, Suite 106, Northbrook, Illinois 60062.

Structural systems include soldier piles and lagging, steel sheeting, drilled shaft walls, and reinforced slurry walls. The performance of a structural wall is dependent upon lateral support systems, which include cantilever, cross-lot or rakers, tie-backs and top-down construction.

The selection of an earth retention system is dependent upon subsurface information, site and construction considerations, performance and cost. The soil and water conditions dictate the proposed loads and constructability of systems. Site considerations include existing utilities and substructure, existing and/or abandoned foundations, adjacent structures with imposed loads, contaminants, and site dimensions. Construction considerations include the feasibility of installing a type retention in the geo-environment.

Selection of a retention system requires consideration of performance, namely possible movements as well as basal stability and possible impact on adjacent utilities and structures. In recent years, there have been extensive efforts to predict movements during the excavation sequence; however, other movements related to the earth retention systems are often ignored or not given proper consideration. Related movements are caused by obstruction removal and improper placement of backfill prior to retention wall installation, vibrations, loss or heave of ground during deep foundation installation, berm stability, workmanship, improper backfill between the retention system and new wall, and temporary support removal. Also, considerations should be given to the effect of sheeting extraction after completion of the basement. Clayey soils adhere to the sheeting, therefore voids may create substantial off-site movements. For this reason, earth retention systems are left-in-place in the Chicago urban environment.

Ultimately, the best cost-performance relationship for a particular project will determine the earth retention selection. The scope of this paper discusses the practice and performance of excavation support systems in the Chicago area. Several case histories illustrate the current practice in the Chicago area as dictated by the geology profile and urban environment. A history of the excavation support system is also presented.

Typical Soil Profile

Prior to development, the downtown portion of Chicago was once a low lying land that was elevated above Lake Michigan by only 1 to 3 m. There were numerous wetlands and low lying areas with poor drainage. The soil surface consisted of a thin layer of sand which frequently grades to a sandy silt with increasing depth. This deposit increases in thickness in an easterly direction towards Lake Michigan. Below the silt and sand is a low strength clayey soil deposit commonly referred to as "Chicago Blue Clay". After 1855, the grade over the entire Chicagoland area was raised so as to provide for better drainage and sewers. The fill that was placed to raise the grade consisted of a mixture of all types of materials that were available from either excavations or from demolition of buildings such as the wreckage from the 1871 Chicago Fire. Geology and Chicago subsoils are described in detail by Peck, et al (1954). A typical soil profile from a boring in the downtown area of Chicago is shown in Figure 1.

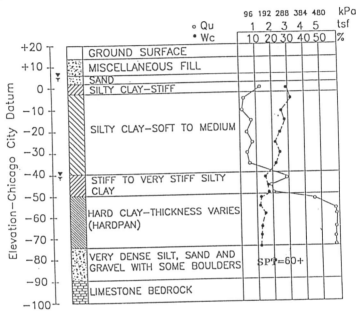

Figure 1. Generalized Soil Profile

At the present time, the grade is approximately +14 Chicago City Datum (CCD). Elevation 0 CCD corresponds to low water datum of Lake Michigan. However, lake elevations fluctuate about 1.5 m from high to low levels. The fill and the underlying sand and sandy silt extends to about elevation +2 CCD whereupon a silty clay soil deposit is encountered to approximate elevation –40 CCD. The clayey soils above approximate elevation –2 CCD are in a stiff to very stiff condition as a result of past partial desiccation. Below –2 CCD, the unconfined compressive strengths are typically on the order 29 to 86 kPa (0.3 to 0.9 tsf) with the lowest strength soils in the range of –2 to –25 CCD. Vane shear tests performed in these deposits have indicated a shear strength on the order of 24 to 29 kPa (500 to 600 psf) above approximate elevation –20 CCD, and 10 to 43 Kpa (200 to 900 psf) below this elevation. The natural water contents are typically on the order of 20% to 25% although the water contents near the upper portion of this deposit are frequently in the range of 25% to 30%. These soils are considered to be slightly overconsolidated. Ground water is typically present at about +3 to +6 CCD.

Below elevation –40 CCD, the soil strength increases to a very stiff condition and becoming hard at about elevation –60 to –65 CCD. This hard layer is classified as "hardpan" and many of the foundations from the high rise buildings are supported on drilled piers extended to this layer. These structures frequently have basements extending from 1 to 4 levels deep which penetrate through the upper fill deposits, the water bearing sand and extend into the soft to medium strength clays.

History of Excavation Support Systems

During the period from 1870 to 1910, almost all of the mid-rise buildings in Chicago were constructed on shallow spread footing foundations. Most of these structures had at least one level of basement which placed the footings at about elevation 0 to +3 CCD which is in the stiff crust just above the soft clay. Some buildings had two or more levels of basement which placed the footings within the soft clay. One level basement excavation in near proximity of existing buildings generally did not produce a significant amount of movement because the upper 4.6 m of soil profile consisted of fill and sand which only deformed a slight amount when properly braced. Two level basements extend into the underlying softer clays and settlements plus lateral deformations can be on the order of 50 mm or more adjacent to the braced cut. Movements of this order of magnitude were frequently accepted during the early years of construction in Chicago because the buildings adjacent to the cut were frequently experiencing an equal or greater amount of settlement due to consolidation of the softer clays under the weight of the structure.

Measurements of settlement at varying distances from basement excavations for 1, 2 and 3 or more basements during the period of 1920 to 1940 were summarized by Peck (1969) as shown in Figure 2. The bracing systems at that time usually consisted of a sheetpile wall or soldier piles and lagging that extended to a slightly greater depth than the base of the excavation and were supported by inclined rakers. The data shown in Figure 2 provided the initial rational basis for estimating ground movements adjacent to braced cuts in Chicago. Movement measurements as a result of excavation were extended by Goldberg, et al (1976) to include different soil types and types of bracing system used. The data used to prepare these prediction tables was obtained from many different sites in the U.S.A.

It is interesting to note that beginning around 1900, methods were used to construct deep basements of 3 to 4 levels deep without causing significant off site movements. The initial step was to install deep foundations such as hand dug piers which were dug from ground surface and concreted to the future basement elevation. Next, trenches were hand dug from grade to the lowest basement level around the perimeter of the structure as well as cross trenches at locations where columns would be situated. The trenches were braced with wood lagging as the excavation extended in depth. When the proper depth was reached, reinforcing steel and concrete was placed within the perimeter trenches thereby constructing the perimeter walls before the interior was excavated. As the interior was excavated, concrete struts were placed in the cross trenches to brace the walls as well as to form support for the future floor slabs. In effect, this was an earlier version of the top down method of construction.

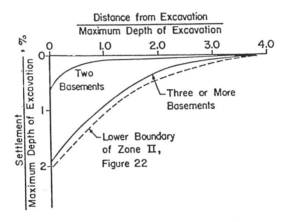

Figure 2. Summary of Settlements due to Basement Excavations in Downtown Chicago (Peck, 1969)

Subsequent to approximately 1915, many of the mid to high-rise buildings were supported upon deep foundations consisting of hand dug piers or piles. These structures experienced very little settlement under the loads imposed by the structure. When basement excavations were constructed adjacent to these facilities, settlement and lateral deformations induced by the excavation frequently resulted in some movement of either the foundation system or floor slab resulting in localized distress. Frequently, this has resulted in litigation. In many cases, the general contractor obtained bids from the bracing contractors and used the cheapest type of retention system regardless of the potential consequences for off-site movement. The bracing of the cut was considered to be the contractor's responsibility. No calculations were made by the design team as to the minimum sizes or length of the sheeting or the spacing of the braces and the potential for off site movement. Instead, the design of the bracing system was partially based on the contractor's experience.

Because of the litigation that occurred over the last 2 to 3 decades, it became apparent that all of the below grade excavations need to be carefully analyzed for the potential of off site movement which in turn is a function of the type and design of the various members of the bracing system. Clough (1986) developed a procedure for predicting movements of braced cuts in clayey soils. This procedure takes into account the factor of safety against base heave in the low strength clay plus the stiffness of the sheeting and bracing system to predict the off site movements as shown in Figure 3. If the movements for the initial design of the bracing system are predicted to be greater than can be tolerated using this procedure, a stiffer bracing system or underpinning prior to excavation will be required.

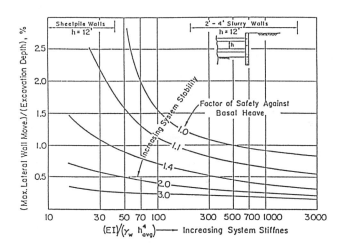

Figure 3. Theoretical Relationship Between Maximum Lateral Wall Movement, Factor of Safety Against Basal, Heave and System Stiffness. (Clough, 1986)

At the current time, most braced cuts in Chicago are design/build with the aid of a design team and/or the contractor's organization. While it is not possible to stop off site movements, the use of the proper type of wall and bracing system can be selected to either minimize the movements or plans can be developed for underpinning of adjacent structures to supplement the bracing system to minimize off site movements and damage.

Design Procedures

A thorough design for an earth retention system for deep excavation in Chicago would consist essentially of three main components. These are:

- Estimation of the lateral pressures that will develop and must be resisted by the bracing system. The procedure recommended by Peck (1969) has been successfully used for the major projects in the Chicagoland area. Measurement of strut loads using strain gauges has confirmed that the strut loads are frequently less than predicted by the procedures recommended by Peck. There are cases where the strut loads are higher than predicted but on average, the strut loads agree reasonably well with the values suggested by Peck.

- Structural design of the earth retention system. This design has to take into account the geometry of the site plus the presence of adjacent structures and utilities. All of these factors affect the size, depth, and the spacing of the lateral support system.

- Prediction of off-site movements. After steps 1 and 2 have been completed, it is necessary to estimate the off site movements using either or both the predictions by Peck (1969) or Clough (1986). If the predicted off site movements are too large, then a stiffer bracing system must be used. This is usually accomplished with an additional row of walers and braces. However, it may also be necessary to change systems such as using a slurry wall versus a steel sheeting or soldier pile system. If the economics do not justify the stiffer bracing system then underpinning of the adjacent facilities may be necessary.

The importance of this last item is demonstrated by a project located at the south side of downtown Chicago. The proposed basement excavation extended over an entire city block. While still in the planning stages, different bracing systems were considered including soldier piles and lagging, steel sheetpiling and a slurry wall. Two and three levels of basements were also under consideration. The soldier pile and lagging system was immediately apparent as being too flexible and would result in significant off site movements. In order to evaluate the steel sheetpile bracing system and the slurry wall system, calculations of off site movement were made using the procedure recommended by Clough (1986) and assuming two levels of bracing with inclined rakers. The loads for the inclined rakers were transferred across the site by grade beams which were supported by drilled pier foundations installed prior to construction of the retention system. The predicted lateral movement for a conventional steel sheet piling system for a two level basement was 45mm and for a three level basement 64mm. The vertical movements were predicted to be 70 mm and 104 mm for the two and three level basements, respectively. A slurry wall would reduce the lateral and vertical movements to 50 mm for a three level basement. Because of the large off site movements predicted for the three level basement using steel sheeting, it was decided to abandon the three level basement in favor of the two level basement. The cost for the slurry wall was significantly higher than for the steel sheetpile system so the sheetpile system was selected even though the predicted lateral movements at the edge of the excavation would be on the order of 45 mm. On three sides of the site, streets were present and this amount of movement was not considered significant enough to cause damage to utilities. On the fourth side, shallow spread footing foundations were present supporting an elevated transit system. The amount of predicted movement could affect the shallow foundations so elevation readings were taken on the columns of the elevated transit system to determine if and when these foundations needed to be underpinned.

Six inclinometers were installed around the perimeter of the site immediately behind the sheeting. The maximum lateral deformations ranged from 63 to 89 mm with an average of approximately 76 mm. These movements are greater than predicted. However, not all the braces were installed promptly and the portion of the basement slab adjacent to this sheeting was not constructed as quickly as planned. These deviations resulted in additional movement. After the rakers were removed, 5mm to 15 mm of lateral deflection occurred as a result of the compression of the lightly compacted granular backfill that was placed between the sheeting and the completed basement wall. The majority of the lateral deformations occurred in the range of elevation –8 to –18 CCD which is the zone of the softest clay at the site.

At the inclinometer station that recorded the most movement, a localized excavation preceded the installation of a corner brace and the exposed height of sheeting was sufficient to cause the additional movement. This required one of the columns of the elevated train structure to be jacked back to its original position and then shimmed. Cracks also appeared in the asphalt pavement adjacent to one corner of this site but utilities were not damaged. The street was re-paved after construction.

Two or Less Basement Levels

Typical earth retention systems for two or less basement levels within the 9 m (30 ft.) depth range include the standard structure system of steel sheeting and soldier piles and lagging. The performance of the structural wall is dependent on the system stiffness and the type of lateral support which may include rakers, cross-lot braces or tie-backs in areas where the upper sand layer is distinct. The following case histories represent typical examples of support systems for the two or less basement levels.

Case History 1 This project required a temporary retention system to protect the perimeter of city streets with utilities for one city block approximately 79 m x 67 m and existing active retail building 33 m x 19 m on Michigan Avenue within the city block. The existing four story brick building is supported on concrete spread and wall footings founded on top of the sand layer. The performance requirements included less than 25 mm movement of the existing structure and less than 50 mm at the perimeter streets.

Sheeting piling with one level of tie-backs in the upper sand layer was used to support the excavation depth, H, 5.5 m to 6.1 m (18 ft. to 20 ft.). The soil profile at this site consisted of of an upper water bearing sand layer from 2.4 m to 4.6 m below grade overlying a thin layer of stiff clay underlain by soft silty clay. The pressure grouted tie-backs in the sand layer were designed for 445 kN (100 kips) except at the west perimeter street where an existing 18 m brick sewer located within the sand layer only 9 m from the perimeter wall limited the bond zone. Inconsistent grout pressure along the west perimeter as result of variability of the sand layer required regroutable and additional tie-backs. The concern for existing building movement during sheeting installation as well as the excavation sequence required underpinning prior to the foundation construction. Underpinning was accomplished by jacking 102 mm diameter grout filled pipe to a design capacity of 267 kN (60 kips) and maximum spacing of 1.8 m. The pipe pier was jacked and locked-off to a bracket anchored to the existing foundation wall.

The movement data obtained by means of surface settlement and lateral points, soil and utilities settlement indicators, and inclinometers indicated the performance criteria were satisfied. The settlement, δ_V, was 13 mm to 37 mm (δ_V/H = 0.2 to 0.6%) and inclinometer data indicated 13 mm to 33 mm (δ_H/H = 0.2 to 0.5%) maximum lateral movement, δ_H, along the perimeter. The upper limits of movements occurred along the west perimeter as a result of the building construction crane pad at the west side. The building settlement data indicated negligible movement.

Case History 2   Sheet piling and two cross-lot brace levels were installed to temporarily support the excavation depth (H) of 6.1 m to 6.7 m (20 to 22 ft.). The L-shape basement approximately 55 m x 30 m is bounded by two streets, an alley, and an existing one to four story brick building supported on wall footings. With the L-shape basement width of approximately 18 m, cross-lot pipe bracing and corner braces were used. The bottom brace level was removed after the mat slab was poured against the sheeting. The sheeting was used as an outside form with slip form construction of the concrete wall. The top wale level was installed on the outside sheeting face. Underpinning piers as used in Case History 1 were installed to protect building movements during sheeting installation and excavation.

Cross-lot bracing systems eliminate the effect of berm creep; therefore, the performance of the support system is expected to be improved. Inclinometer data showed maximum lateral displacements less than 19 mm (δ_H/H = 0.3%) at street locations and less than 13 mm (δ_H/H = 0.2%) at building location. The building settlements were negligible except at one location where 16 mm was recorded.

Case History 3   Sheet piling and two brace levels were installed to temporarily support the excavation depth (H) of 6.7 m to 7.3 m (22 ft. to 24 ft.) for the perimeter of streets and utilities of a city block approximately 70 m x 70 m. Even though this project is within a few blocks of case history 1 and 2, the sand layer has limited thickness therefore eliminating the upper level tie-back option. With the site dimensions, two raker brace levels were used and supported by a reinforced concrete grade beams between the caisson caps. The reaction grade beams were designed to transfer raker reactions from opposite sides of the excavation. Sloped berms with 5 m minimum horizontal top width were used to temporarily support the perimeter sheet piling during installation of the reaction grade beams. Berms were maintained during installation of upper rakers. After completion of the upper level bracing, the lower brace level was installed in a construction sequence similar to the upper level. All rakers were jacked and prestressed.

The movement observations corresponded to the anticipated and projected performance. The maximum observed surface settlement, δ_V, varied from 25 m to 83 mm (δ_V/H = 0.4 to 1.2%) around the site perimeter. The maximum lateral movement, δ_H, from inclinometer data varied from 51 mm to 76 mm (δ_H/H = 0.7 to 1.0%) The effect of the berm creep, basal stability and workmanship are certainly influencing factors for the raker bracing system. The following case history will highlight the effect of berms on behavior.

Case History 4   At a project site on the west side of downtown Chicago, soldier piles and a top waler were installed as part of a retention system (Gill and Lukas, 1990). Earth berms were left in place until the interior of the site could be excavated to grade, pile caps formed on previously driven piles, and rakers installed to pile caps. During excavation to a depth of 7.3 m to 8.5 m (24 ft. to 28 ft.) within the interior, the ground shifted and a number of the piles were displaced. This required digging around the piles to remove the upper portions that deflected and splicing on new sections. The repair resulted in many months of delay. In addition, there was a concrete strike so the berms remained in place for a period of 7 months which was much longer than originally intended. The typical profile of the site with the berm is shown in Figure 4. A slope stability analysis indicates that the factor of safety against a deep seated movement was on the order of 1.2. This is considered satisfactory for temporary stability, however, creep may occur.

In spite of the wide berm and the adequate factor of safety against a deep seated failure, movements of the soldier piles occurred as shown in Figure 5. Near the edges of the site where there was end restraint, the lateral deflections over a period of 7 months were on the order of 25 mm. However, at the center of the site, the deflections were as high as 230 mm (δ_H/H = 2.7%).

An attempt was made to restrain the movement by installing temporary rakers near the center of the wall. Bearing pads were embedded in soft clays to provide a reaction for the inclined rakers. The lateral movement slowed after the temporary rakers were installed but did not stop.

172   GEO-ENGINEERING FOR UNDERGROUND FACILITIES

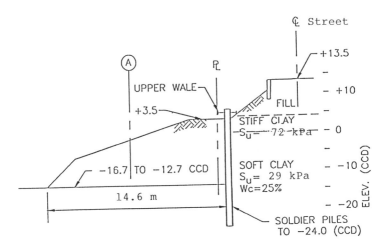

Figure 4. Typical Profile

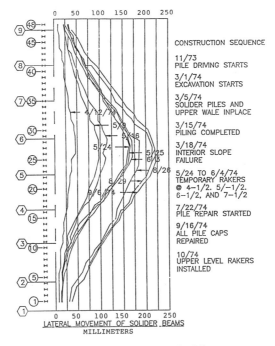

Figure 5. Lateral Movements of Soldier Piles

This data indicates that berms are not very effective in restraining the soil mass adjacent to braced cuts where the base of the berm extends into the soft clay deposit. For these reasons, berms should only be considered to be effective on a short term basis.

Three or More Basement Levels

Typical earth retention systems for three or more basement levels include a structural system with increased stiffness. Slurry walls are predominately used for performance and economics of the deeper support system. The slurry wall stiffness is typically 10 times or greater than the sheet pile wall. Cost effective support systems for the deep basements may be the top-down construction which includes time savings of structure completion.

Case History 5   Slurry walls with top-down construction were installed to support both temporary and permanent loadings for three basement levels and excavation depth of 10.7 m (35 ft). The L-shaped basement with dimensions of 72 m and 34 m is bounded by streets on 3 sides and existing buildings supported by footings on the west side.

The construction procedure included pre-trenching in segments around the basement perimeter to remove slurry wall obstructions. Cantilever soldier piles and lagging were used to protect the streets during pre-trenching and guide wall installation. The slurry wall 610 mm (24 in.) thick and 17.7 m deep was installed in alternating panels with a notched joint segment. Keyways (76 mm) with bent-out bars were used at the slurry wall and basement floor connections. Upon completion of the slurry wall, caissons and below grade building columns were installed, and excavation with installation of the upper level basement slab completed. Cap beams on the top of slurry wall and slab at ground level were then installed. The remaining basement installation was completed by top-down method.

The movement monitoring system included six perimeter inclinometers, settlement points at the existing buildings, and horizontal points at grade and roof structure of the three existing buildings. The inclinometer data indicated a range of horizontal movement within the excavation depth from 7.6 mm to 17.8 mm ($\delta_H/H = 0.1$ to 0.2%) with larger movements occurring at grade. Slight lateral movements of 2 mm to 5 mm were recorded at 16.8 m depth. The vertical and horizontal movements of the existing building were negligible. The performance supports the top-down method for this deep basement excavation.

Case History 6   A recent design/build project of four basement levels with an excavation depth of 15 m (49 ft.) will start construction in the Spring of 1999. The basement perimeter will enclose a city block approximately 183 m x 61 m. Slurry walls 762 mm (30 in.) thick with top-down construction will be utilized, similar to Case History 5. The differences are that caissons supporting perimeter wall columns will be installed prior to the slurry wall construction. The slurry wall panels will butt against and connect to the caissons. Also, the core structure with shear walls for the high rise requires a bottom-top construction sequence. The construction of the core structure installation will be accomplished with a circular sheeting cofferdam 25 m in diameter and supported with steel ring beams. The core structure must be completed to grade to accommodate the top-down construction of the adjoining basements. Partial basement and superstructure construction occurring simultaneously will create time savings for the structure completion schedule.

<u>Case History 7</u> This case history exemplifies the multitude of design and construction considerations for an earth retention system in an urban environment. The three level basement approximately 67 m x 70 m is bounded by an existing building with equivalent three level basement depth, an existing building with one basement level depth, and lower street levels beneath and adjacent to upper supported streets. The upper level streets are supported by a truss and column system on caissons. The upper level street is 10.7 m above lower street grade with 6.7 m clearance below truss system.

Slurry walls 610 mm (24 in.) thick were used to support the 9.1 m (30 ft.) deep excavation with limited overhead clearance on one side. The slurry wall construction was created with different panel joints. Permanent wide flange beams were used under the upper street level because the temporary joint sections could not be readily extracted in a cost effective manner. Conventional temporary joint sections and caisson shafts were used on other walls.

The slurry walls were supported with upper level tie-backs and lower level corner braces and rakers. The upper level tie-back system varied as a result of the site conditions. Deadmen sheeting with tie-rods were used as an anchorage on the north side due to interference of existing utilities and structures. Pressure grouted tie-backs in the sand layer were used on the east side, and hardpan tie-backs were used on the south side adjacent to an existing parking garage basement. Lower level rakers were installed against grade beam reactions. The existing column and caisson support for the upper street were supported laterally within the basement area until the completion of the basement.

This hybrid support system as a result of site and construction considerations performed within expected and tolerable movement limits. The maximum surface settlement was approximately 45 mm. ($\delta_V/H = 0.50\%$)

Conclusion

- The design of any retention system should include a prediction of offsite movements to determine if adjustments in the bracing or structural members are needed to limit movements. This prediction will also make the owner/developer aware of the magnitude of movements before authorizing the work.

- In Chicago, the low strength clay deposit in the depth range of 5 m to 12 m below grade dictates the type of retention system that is needed to minimize lateral movements. For two basements or less, conventional braced sheeting or soldier piles and lagging are usually sufficient. For three basements or more, slurry walls with top down construction may be required.

- The observational method as described by Peck (1969) and the procedure of Clough (1986) provide reasonable methods for predicting ground movements for excavation support systems in the Chicago area. With the movement monitored by inclinometer and settlement readings, the assumed behavior of the support system can be verified as well as the need for remedial shoring as required during the construction operation.

- With the numerous types of earth retention systems available today, the best cost-performance relationship for a particular project will decide the earth retention selection. Each project is unique and should be evaluated for the proper selection of the earth retention system.

References

Clough, G.W. (1986) "Effects of Excavation Induced Movements in Clays on Adjacent Structures", Chicago Geotechnical Lecture Series, Illinois Section ASCE, Chicago, Illinois, pp. 1-22.

Gill, S.A. and Lukas, R.G. (1990) "Ground Movement Adjacent to Braced Cuts" Conference on Design and Performance of earth Retaining Structures, ASCE, Cornell University, pp. 471-488.

Goldberg, D.T. et al (1976) "Lateral Support Systems and Underpinning, Volume 1. Design and Construction", FHWA-RD-75-128, pp. 1-28.

Peck, R.B and Reed, W.C. (1954) "Engineering Properties of Chicago Subsoils", University of Illinois Engineering Experiment Station Bulletin No. 423, Urbana.

Peck, R.B. (1969) "Deep Excavations and Tunneling in Soft Ground", Proceedings of the 7th International Conference on Soil Mechanics and Foundation Engineering, State of the Art Volume, Mexico, pp. 225-296.

Excavation Support Practices in the Pacific Northwest

Tom A. Armour, P.E., M. ASCE[1]

Abstract

Temporary and permanent excavation support techniques are constantly evolving in the United States. It is clear from observing excavation support practice across the country that there remain considerable regional variations in the contract procurement, design, and construction techniques. Many of these differences reflect the logical response to local conditions-both geological and economic-but certain other variations now appear to be due to a relative inertia in research and development at one extreme and, regrettably, a lack of understanding of fundamentals at the other extreme. This paper reviews aspects of the procurement, design, and construction of excavation support systems typically used in the Pacific Northwest. Attention is focused on local geology, history of use, and current contract procurement practices. Three case histories are presented to further illustrate the innovative excavation support techniques used in the Pacific Northwest.

---

[1] President, DBM Contractors, Inc., 1220 S. 356th Street, Federal Way, WA. 98003

Introduction

In the United States there has been steady growth in the use of innovative excavation support systems in the years since their introduction in the early 1960's. This expansion has to a large extent been driven by specialty geotechnical contractors and engineers who both imported foreign developments and conducted their own basic researches, in locally prevailing conditions. With such data and experience, these engineers have had the confidence to promote "alternative solutions" for difficult excavation support problems to receptive owners, designers, and general building contractors. In turn, being engaged in the details of design and performance monitoring has given these parties the closed exposure necessary for the real transfer and understanding of the technological concepts, and this knowledge has generated positive feedback for the contractors.

In the United States, most earth support projects are designed and constructed with reference to State Department of Transportation recommendations, while the Federal Highway Administration remains a focal point for ongoing research.

Against this generally positive background, it is clear from observing excavation support practice across the country that there remain considerable regional variations in contract procurement, design, and construction techniques. Many of these differences reflect the response to local geological and economic conditions. However, certain other variations appear to be due to a growth in knowledge from research and development at one extreme and, the lack thereof at the other extreme.

In the Pacific Northwest, such disparities are not so prevalent in the field of excavation support. Open minded owners, experienced design engineers and specialty geotechnical contractors, and quality construction equipment assist greatly in the development of innovative solutions, positive results, and technological growth. This teamwork often results in the most technically and economically feasible excavation support projects.

It is the purpose of this paper to first review the evolution of excavation support systems in the Pacific Northwest, and then to illustrate the innovative techniques being designed and constructed today.

Puget Sound Geology

Very little bedrock is typically exposed at the surface in the Northwest. Bedrock is exposed in isolated outcrops throughout the region. Most of the Pacific Northwest is underlain by sediments deposited during the Quaternary Period, popularly called the "ice ages," when the Puget Sound lowland was repeatedly invaded from the north by glaciers. Although the exact number of ice advances is not definitely known, studies of the glacial drift suggest that it has happened at least 4 or 5 times.

The typical undisturbed stratigraphy in the central Puget Sound area consists (from the top down) of a thin soil layer overlying relatively impermeable very compact glacial till (hard pan), permeable sands with advance outwash sands and gravels, and/or nearly impermeable clays and silts. However, in many areas, the stratigraphy can be more complicated.

Throughout much of the area, the glacial till, approaching a thickness of up to 10 m, forms a relatively strong and resistant cap that covers much of the highlands. Although till is in many places impermeable to ground water, fractures and gulling in the till surface allow percolation into the lower sedimentary layers.

In the typical sand deposits, the upper part may be dry, even in winter, whereas ground water flows rapidly through its basal zone, where the water is perched on underlying clays and silts. Permeability within the sand unit varies laterally and vertically, and ground water piping can occur along weak zones.

Any of these Pleistocene units may overlie knobs and fault blocks of impermeable Tertiary bedrock.

Excavation Support History

In the early 1970's with the growth of major metropolitan area such as Seattle, Vancouver, B.C., and Portland, many of today's more conventional "top-down" excavation support techniques were designed and installed. Anchored soldier pile and lagging earth retention methods were typically used for major downtown excavations. Due to high groundwater issues, anchored slurry walls and secant/tangent pile walls were also used for excavation support for the realignment of Interstate 90 in Mercer Island, Washington.

Due to the compact nature of the glacial till, driven pile support systems are generally not feasible in the Northwest. Drilled pile and anchor systems typically offered the most economically feasible "top-down" excavation support system. Over the years, through monitoring and instrumentation of the many excavation support systems, technological advancements and cost saving benefits were realized.

In the early 1980's, the European technology soil nailing was first developed (in the United States) in the Northwest. Puget Sound geologic conditions provided an excavation ideally suited for soil nailing and shotcreting.

Soil nailing is steadily gaining popularity and acceptance because it can be used with conventional equipment, it reduces excavation time, it allows construction related activities to proceed in restricted space, and it can produce significant cost savings over conventional anchored excavation support techniques.

Ground freezing was first introduced in the Northwest in the late 1980's, but is just recently being developed and implemented. With the variable groundwater conditions and dense cobbly and bouldery ground conditions, ground freezing offers economical and technical merits for excavation support projects.

Almost as important as the technology itself is the procurement process the owner uses for the design and construction of excavation support systems. In the 1970's and 1980's, permanent and temporary excavation support systems in the Northwest were typically owner designed and specialty contractor built. More often than not, specialty contractors value engineered the owner's design with a more efficient and less costly earth retention system.

Today more and more owners are realizing the benefits from design/build contract procurement methods. Typically, in the Northwest, all temporary excavation support walls are contractor designed and approximately 50% of permanent shoring walls are contractor designed and built.

Design and construction of "top down" earth retention systems remains a very sophisticated construction technique. The work of designing, fabricating and installing drilled and grouted excavation support systems requires the professional experience of a specialty geotechnical engineer and contractor. Technical specifications and directions cannot replace professional experience. This experience, if given the opportunity, will benefit the owner with more cost effective solutions to his/her excavation support project.

Three design/build excavation support projects, typical of the Pacific Northwest, are presented below:

KOMO TV Building, Seattle, Washington

This unique design-build project began from the need to supply a temporary earth retention system for a local TV station's new building and below grade parking structure. The building was designed to reach about 15 m below the adjacent street and existing building grades. The original recommendations for support of the excavation included traditional soldier pile and tie-back support in order to limit wall and adjacent structure movement. Wall movement was required to be less than 13 mm and smaller at the structures. The results of the preliminary review by the design-build team (DBM Contractors, Inc. and Golder Associates) indicated the site conditions allowed for a certain amount of innovation that could save significant construction time and dollars. The innovations presented to the owner (Fisher Broadcasting) and the general contractor (Sellen Construction) included top-down soil nailed wall construction to support nearby heavily loaded monorail piers, an operating broadcast building, high voltage power vaults and city bus lanes.

The project was unique in that this was a completely new application for soil nail ground reinforcement where closely-spaced (1.5 m on center), grouted steel bars were inserted next to and under structural piers, building footings and sensitive utilities. The 2 m diameter piers supported an active City of Seattle Monorail transit system. The bent supported rail runs overhead, angling toward the excavation. The train imparts forces that transfer, according to 1961 as-built drawings, to four 7.5 m deep piles that were as close as 2.5 m from the top-of-wall and "belled" out to about 3 m wide at the base. It was determined during construction that two of these supports were actually shallow spread footings, prompting a wall redesign. Monorail loads for design were significant, with lateral loads of 205 kN, vertical loads of 2200 kN and moments of 2300 kN-m. This loading required the thickest and heaviest reinforcement on the soil nail project.

It is also the first time, to the designer's knowledge, that soil nails and shotcrete facing were used for building support of an "inside" wall corner that jutted into the excavation. The existing KOMO TV building and associated high voltage vaults are extremely sensitive issues with regard to construction disturbances. The existing building footing loads are as high as 125 kN/m imparting high surcharge pressures on the nails and wall face.

Most of the supported soil materials originated from glacial till although some fractured lacustrine clay deposits were encountered just a meter thick near the bottom of the excavation. Some ground water seepage was removed in sumps at the base of the cut.

Utilities near the excavation required special consideration because they were within just one meter of the top of the wall. Special steel and shotcrete cantilever sections extended up to 2.5 m above the top of wall and were built by back forming and then applying the shotcrete application on the "splash board" forms. Several of the utilities were in place under a high traffic City of Seattle bus transit lane. The City of Seattle street use department (CTRANS) placed this project under careful scrutiny, requiring minimal wall movements and restricting encroachment distances between nails and underground utilities. The designer was able to obtain variances for instances where the actual underground utility elevation could be observed and the nails placed accordingly to miss the utility.

One of the largest challenges during construction came when the soil nail drill crew discovered that two of the four large diameter piles were "missing" and in their place spread footings has been formed during the construction of the Monorail supports. These spread footings imparted larger pressures higher on the wall than planned and a more heavily reinforced section was required for redesign and construction within a weeks time.

The City of Seattle required every building wall segment and monorail pier to be monitored. Wall and pier deflections were measured with borehole inclinometers and optical survey. The measured deflections were well below the required 13 mm deflection limit measured at the top of the soil nail wall.

The final product, a temporary soil nailed earth retaining wall and shotcrete facing covered over 4000 m^2 of excavation face, using 1,200 soil nails for soil reinforcing. The owner and contractor consider this project a success largely owed to the 20% cost savings over conventional anchored soldier pile and lagging systems and the cooperative and innovative attitudes of the design-build construction team.

Photograph 1. View of wall under construction at monorail piers

Photograph 2. View of completed walls at Komo Building

King Street Center, Seattle, Washington

The King Street Center project is located in the Pioneer Square District of Seattle, This area historically was tidal lands for the Puget Sound. Turn of the century photos of the area show elevated railroad tracks supported on timber pilings. The entire area was eventually filled by washing down surrounding slopes, raising the site elevation 3 m to 8 m above sea level.

The site itself has been used for multiple buildings in its history, and as such, contains substantial construction debris, brick walls, timber piles and concrete/granite footings from the last hundred years.

DBM Contractors, Inc. was contracted with by Lease Crutcher Lewis to provide the design and construction of a temporary excavation support system to facilitate the construction of King Street Center. Soils information recommended a cutoff wall along the North alignment as well as the North 75 feet of the East and West alignment.

DBM's initial proposal consisted of sheet piling along the cutoff alignment and anchored soldier pile and timber lagging walls along the remaining alignment. The City of Seattle expressed concern over potential damage to turn of the century bell and spigot water lines that were still in operation along a portion of the sheet pile wall.

DBM proposed an alternative of secant piles in place of the sheet piles, and additionally performed a test installation of sheet piles on site away from the utilities of concern. Vibrations induced by the test sheet pile installation were monitored. All parties involved worked together in developing a cost effective excavation support system within the constraints noted. The majority of the sheet pile cutoff wall alignment was allowed to proceed with vibration monitoring required. It was decided DBM would pre-drill locations if excessive vibrations occurred during sheet pile installation.

Major obstacles in the construction of the sheet pile wall including buried granite footings, timber piles, and surficial construction debris, caused the wall to have to be realigned during construction.

Another portion of the cutoff wall alignment was simply too close to the "rickety" water lines for sheet pile installation. In this area, secant piles were installed.

The end result was 700 m^2 of anchored soldier pile and lagging wall, 750 m^2 of anchored sheet pile wall and 200 m^2 of anchored secant pile wall.

The soils for the project were primarily characterized as silty, loose, fine to medium sand. The water table was typically within the top 1.5 m.

Soldier piles extended a total depth of 10 m and the secant piles were drilled a total depth of 15 m. Every hole encountered timber pile obstructions causing a very time consuming drilling operation. Dewatering of the site was not initially allowed due to the surrounding historical buildings and risk of settlement.

It was understood that a certain degree of dewatering would be incidental to the excavating but multiple monitoring wells were installed in proximity to the surrounding structures. As the excavation commenced, it became immediately apparent that the mode of excavation was crucial to the success of the excavation support system. Water was boiling through the silts and sands in the excavation. As a result, excavation had to occur some distance from the shoring wall to allow drainage to occur. The excavation could then slowly proceed to the wall alignment. Some limited dewatering was then allowed, providing the area of influence did not encroach on the surrounding structures. The end result of the cutoff wall, revised excavation system, and the dewatering wells, was a much more stable site that ultimately became nearly dry to begin construction of the grade beams and base slab.

DBM was able to complete the project on schedule due to their early recognition of the site problems and the overall team atmosphere from all involved parties.

Photograph 3. View of drilling obstructions along west wall

Photograph 4. View of completed walls at King Street Center

SAFECO Campus, Redmond, Washington

During the design development of the new SAFECO Campus Project, in Redmond, Washington, design/build engineers and specialty geotechnical contractors evaluated several value added ideas that would bring substantial cost and schedule savings using top-down soil nail wall construction methods for the permanent building walls during the excavation.

SAFECO Insurance Company will be moving its entire operations group to their existing facility in Redmond, Washington. Included in the upgrade project are two below grade parking garages, cafeteria building, main office building, data center and one other office building. Design/build temporary top-down soil nailed excavation support systems were proposed for the below grade permanent building walls at Garages F and G, main office building and the cafeteria. Permanent design/build soil nailed earth retention systems were proposed for the data center building which was designed as a base isolated structure.

In general, subsurface conditions consist of glacial till with perched water conditions throughout the site. Utilities near the excavation required special consideration due to their location near the top of the walls and back of the proposed soil nail layout.

The main design and construction constraints, in addition to designing around several existing utilities, included the close proximity of the existing Region Building to the West Wall of Garage F, where positive underpinning with high capacity micropiles was required prior to soil nail wall construction. Soil nails would then be splayed around the micropiles to avoid interference's.

The walls at Garages F and G, main office building and the cafeteria were designed as permanent top-down reinforced shotcrete, (250 mm to 500 mm thick) soil nail walls with temporary soil nails. Heights varied up to 8 m in height. Each one of the sites were relatively small in plan and required construction sequencing between the soil nail and excavation operations.

The data center building contains all of SAFECO's computer operations and was required to be designed to maintain operations throughout a high magnitude earthquake. The designers called for a building supported on a base isolation system surrounded by a permanent soil nailed earth retention system with a maximum height of 7 m.

Construction of the soil nailed excavation support system is expected to be completed by August 1999. The final product will include fifteen soil nailed walls and reinforced shotcrete facing covering over 8,000 m^2 of excavation face, using 3,100 soil nails for soil reinforcing.

Similar projects of this nature and magnitude using permanent top-down soil nail construction techniques were also recently completed for Microsoft on their Augusta and Troon campuses.

Acknowledgements

The author wishes to thank the following parties for the opportunity to work together on the three projects discussed in this paper:

KOMO TV Project
Owner: Fisher Properties
Contractor: Sellen Construction
Engineer: Shannon & Wilson
DBM Engineer: Golder & Associates

King Street Center
Owner: Wright Runstead and Company
Contractor: Lease Crutcher Lewis
Engineer: Hart Crowser, KPFF
DBM Engineer: Civil Tech

SAFECO Campus
Owner: SAFECO
Contractor: Baugh Enterprises
Engineer: Entranco, KPFF
DBM Engineer: Hart Crowser

High Capacity Micropiles -
Basic Principles and Case Histories

D. A. Bruce[1], M. ASCE
M. E. C. Bruce[2], M. ASCE
R. P. Traylor[3], M. ASCE

Abstract

Small diameter, drilled, grouted-in-place micropiles have been installed throughout the world since 1952. A recent international study, funded by the U.S. Government, and summarized herein, provides new insight into design, construction and testing practices used in various countries. Two recent case histories, reflective of contemporary U.S. practice, illustrate the major innovations in theory and construction.

Background

Between 1993 and 1996, the Federal Highway Administration (FHWA) funded the single most significant and comprehensive review of global micropile practice so far conducted. This effort also underlined the desire of the FHWA to contribute to a contemporary French national research project's five-year effort named FOREVER (Fondations REnforcées VERticalement) and designed to conduct a variety of integrated experimental programs relating to micropiles. The FHWA study featured the formation of an International Advisory Panel comprising specialists from North America and Europe.

Not only did this group ensure that a comprehensive review of practice was conducted, but also they were able to resolve a number of fundamental issues regarding various aspects of the classification, design, construction and performance of micropiles.

---

[1]ECO Geosystems, L.P., Pittsburgh, PA, USA
[2]geotechnica, S.A., Pittsburgh, PA, USA
[3]Structural Preservation Systems, Baltimore, MD, USA

These issue had been a source of confusion and misunderstanding and therefore restricted the use of micropiles in certain engineering circles.

Scope

Micropiles are, generically, small-diameter, bored, grouted-in-place piles incorporating steel reinforcement. They have been used throughout the world since 1952 for various purposes, and this has spawned a profusion of national and local names, including pali radice, micropali (Italian), pieux racines, pieux aiguilles, minipieux, micropieus (French), minipile, micropile, pin pile, root pile, needle pile (English), Verpresspfähle and Wurzelpfähle (German) and Estaca Raiz (Portuguese). All, however, refer to the "special type of small diameter bored pile" as discussed by Koreck (1978).

Such a pile can sustain axial and/or lateral loads, and may be considered as either one component in a composite soil/pile mass or as a small-diameter substitute for a conventional pile, depending on the design concept. Inherent in their genesis and application is the precept that micropiles are installed with methods that cause minimal disturbance to structure, soil and environment. This therefore excluded other related techniques from the FHWA study such as those that employ percussive or explosive energy (driven elements), ultra-high flushing and/or grouting pressures (jet piles) or large diameter drilling techniques that can easily cause lateral soil decompression (auger cast piles).

Characteristics and Definitions

Micropiles are a small-diameter subset of cast-in-place replacement piles. With such conventional cast-in-place replacement piles, most, and occasionally all, the load is resisted by concrete as opposed to steel. Small cross-sectional area is therefore synonymous with low structural capacity. Micropiles, however, are distinguished by not having followed this pattern: innovative and vigorous drilling and grouting methods like those developed in related geotechnical practices such as ground anchoring, permit high grout/ground bond values to be generated along the micropile periphery. To exploit this potential benefit, high capacity steel elements, occupying up to 50 percent of the hole volume, can be used as the principal (or sole) load bearing element, with the surrounding grout serving only to transfer, by friction, the applied load between the soil and the steel. End-bearing is not relied upon, and in any event, is relatively insignificant given the pile geometries involved. Early micropile diameters were around 100mm, but with the development of more powerful drilling equipment, diameters of up to 300mm are now considered practical. Thus, micropiles are capable of sustaining surprisingly high loads (compression loads of over 5000 kN have been recorded), or conversely, can resist lower loads with minimal movement.

The development of highly specialized drilling equipment and methods also allows micropiles to be drilled through virtually every ground condition, natural and artificial, with minimal vibration, disturbance and noise, and at any angle below horizontal. Micropiles are therefore used widely for underpinning existing structures, and the equipment can be further adapted to operate in locations with low headroom and severely restricted access.

All of these observations of its traditionally recognized characteristics therefore lead to a fuller definition of a micropile: a small-diameter (less than 300mm, although in France the limit is set as 250mm), replacement, drilled pile composed of placed or injected grout, and having some form of steel reinforcement to resist a high proportion of the design load. This load is mainly (and initially) accepted by the steel and transferred via the grout to the surrounding rock or soil, by high values of interfacial friction with minimal end bearing component, as is the case for ground anchors and soil nails. They are constructed by the type of equipment used for ground anchor and grouting projects, although micropiles often must be installed in low headroom and/or difficult access locations. They must be capable of causing minimal damage to structure or foundation material during installation and must be environmentally responsive. The majority of micropiles are between 100 and 250mm in diameter, 20 to 30 m in length, and 300 to 1000 kN in compressive or tensile service load, although far greater depths and much higher loads are not uncommon in the United States. It is also the case that many short piles of 75-mm diameter or less are installed routinely for residential underpinning. These typically have service loads less than 150 kN, but are often of the displacement category discussed above.

Classification of Micropiles

It has been common to find micropiles sub-classified according to diameter, some constructional process, or by the nature of the reinforcement. However, given the definition of a micropile provided above, the FHWA team concluded that a new, rigorous classification be adopted based on two criteria:

- The philosophy of behavior, and
- The method of grouting.

The first criterion dictates the basis of the overall design concepts, and the second is the principal determinant of grout/ground bond capacity.

Classification Based on Philosophy of Behavior. Micropiles are usually designed to transfer structural loads to more competent or stable strata. They therefore act as substitutes or alternatives for other conventional pile systems (Figure 1). For axially loaded piles, the pile/ground interaction is in the form of side shear and so is restricted to that zone of ground immediately surrounding the pile. For micropiles used as in situ reinforcements for slope stabilization, research by Pearlman et al. (1992) suggests that pile/ground interaction occurs only relatively close to the slide plane, although above this level, the pile group may also provide a certain degree of continuity to the pile/ground composite structure. In both cases,

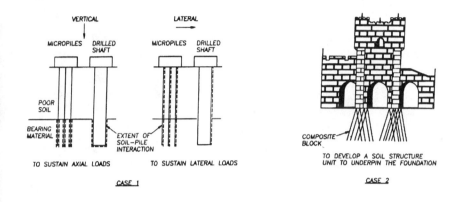

Figure 1. Classification based on design philosophy. (Bruce et al., 1995)

however, the pile (principally the reinforcement) resists directly the applied loads. This is equally true for cases when individual piles or groups of piles are used. In this context, a group is defined as a tight collection of piles, each of which is subjected to direct loading. Depending on prevailing codes relating to pile group design, individual pile design capacity may have to be reduced in conformity with conventional "reduction ratio" concepts. These concepts were typically developed for driven piles, and so this restriction is almost never enforced for micropiles, given their mode of construction which tends to improve, not damage, the soil mass between piles.

When axially-loaded piles of this type are designed to transfer their load only within a remote founding stratum, pile head movements will occur during loading, in proportion to the length and composition of the pile shaft between structure and the founding stratum, and the load. Piles of this type can be preloaded (Bruce et al. 1990) to ensure that the overlying structure can be supported without further movements occurring. Equally, if suitably competent ground conditions exist all the way down from below the structure, then the pile can be fully bonded to the soil over its entire length and so movements under equivalent loads will be smaller than in the previous case. These directly loaded piles, whether for axial or lateral loading conditions, are referred to as CASE 1 elements. They comprise virtually all North American applications to date, and at least 90 percent of all known international applications.

On the other hand, one may distinguish the small group of CASE 2 structures. Dr. Lizzi introduced the concept of micropiling when he patented the "root pile" (palo radice) in 1952. The name alone evokes the concept of support and stabilization by an interlocking, three-dimensional network of reticulated piles similar to the root network of a tree. This concept involves the creation of laterally confined soil/pile composite structure that can provide underpinning, stabilization or earth retention, also as illustrated in Figure 1. In this case, the piles are not heavily reinforced since they are not individually and directly loaded: rather, they circumscribe a zone of reinforced, composite, confined material that offers resistance with minimal movement. The piles are fully bonded over their entire length and so for this case to work, the soil over its entire profile must have some reasonable degree of competence. Lizzi's research (1982) has shown that a positive "network effect" is achieved in terms of load/movement performance, such is the effectiveness and efficiency of the reticulated pile/soil interaction in the composite mass.

It is clear, therefore, that the basis of design for a CASE 2 network is radically different from a CASE 1 pile (or group of piles). This is addressed in Volume 2 of the FHWA study. Notwithstanding this difference, however, there will be occasions where there are applications transitional between these designs (although this attractive possibility is currently, conservatively, ignored for pile groups), while a CASE 2 slope stability structure may have to consider direct pile loading conditions (in bending or shear) across well defined slip planes. By recognizing these two basic design philosophies, even those transitional cases can be designed with appropriate engineering clarity and precision.

Classification Based on Method of Grouting. The successive steps in constructing micropiles are, simply:

- Drill;
- Place reinforcement; and
- Place and typically pressurize grout (usually involving extraction of temporary steel drill casing).

There is no question that the *drilling* method and technique will affect the magnitude of the grout/ground bond which can be mobilized, while the act of placing the *reinforcement* cannot be expected to influence this bond development. Generally, however, international practice both in micropiles and ground anchors confirms that the method of *grouting* is generally the most sensitive construction control over grout/ground bond development. The following classification of micropile type, **based primarily on the type and pressure of the grouting** is therefore adopted. It is shown schematically in Figure 2.

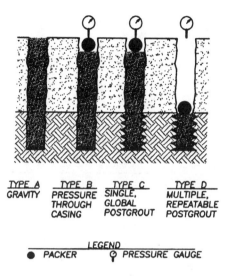

Figure 2. Classification based on construction (grouting) method (Bruce et al., 1995)

- *Type A:* Grout is placed in the pile under gravity head only. Since the grout column is not pressurized, sand-cement "mortars", as well as neat cement grouts, may be used. The pile drill hole may have an underreamed base (largely to aid performance in tension), but this is now very rare and not encountered in any other micropile type.

- *Type B:* Neat cement grout is injected into the drilled hole as the temporary steel drill casing or auger is withdrawn. Pressures are typically in the range of 0.3 to 1 MPa, and are limited by the ability of the soil to maintain a grout tight "seal" around the casing during its withdrawal, and the need to avoid hydrofracture pressures and/or excessive grout consumptions, particularly in upper, permeable, fills.

- *Type C:* Neat cement grout is placed in the hole as for Type A. Between 15 and 25 minutes later, and so before hardening of this primary grout, similar grout is injected, once, via a preplaced sleeved grout pipe at a pressure of at least 1 MPa. This type of pile, referred to in France as IGU (Injection Globale et Unitaire), seems to be common practice only in that country to date.

- *Type D:* Neat cement grout is placed in the hole as for Type A. Some hours later, when this primary grout has hardened, similar grout is injected via a preplaced sleeved grout pipe. In this case, however, a packer is used inside the sleeved pipe so that specific horizons can be treated, several times if necessary, at pressures of 2 to

8 MPa. This is referred to in France as IRS (Injection Répétitive et Sélective), and is common practice worldwide.

Table 1 summarizes the relationships between the classification and application.

Application	Structural Support	In Situ Earth Reinforcement			
Sub-applications	Underpinning of existing foundations New foundations Seismic retrofitting	Slope stabilization and Excavation support	Soil strengthening	Settlement reduction	Structural stability
Design concept	CASE 1	CASE 1 and CASE 2 with transitions	CASE 2 with minor CASE 1	CASE 2	CASE 2
Construction type	Type A (bond zones in rock or stiff clays) Type B and D in soil (Type C only in France)	Type A (CASE 1 and 2) and Type B (CASE 1) in soil	Type A and B in soil	Type A in soil	Type A in soil
Estimate of on application	Probably 95% of total world applications	0 to 5%	Less than 1%	None known to date	Less than 1%

Table 1. Relationship between micropile application, design concept, and construction type.

Combined Classification. Micropiles can therefore be allocated classification numbers denoting the philosophy of behavior (CASE 1 or CASE 2), which relates fundamentally to the design approach, and a letter denoting the method of grouting (Type A, B, C, or D), which reflects the major constructional control over capacity. For example, a repeatedly post-grouted micropile used for direct structural underpinning is referred to as Type 1D, whereas a gravity grouted micropile used as part of a stabilizing network is Type 2A.

Case Histories

Brief details from two recent case histories are provided to illustrate typical current U.S. practice in the use of CASE 1 and CASE 2 micropiles respectively.

Old Post Office and Courthouse Building, San Juan, Puerto Rico (Zelenko et al., 1998)

Introduction. This structure was built in 1914 and added to in 1940. It was founded primarily on Raymond step taper piles and timber piles, and is underlain by potentially liquefiable fine, silty sand. Design requirements were for 217 high capacity CASE 1B micropiles, each of service load 533kN in compression, 356kN in tension and 44kN in lateral capacity (at a maximum allowable deflection of 13mm). This is an increasingly common application of high capacity micropiles. These piles were to be installed both through existing pile caps, and in other locations where new pile caps would be later created.

Geology and Site. Under 2.4 - 3m of variable fill, the 8m thick zone of potentially liquefiable sands overlaid weathered limestone. Access for drilling equipment was difficult, given the nature of the overhead and underground obstructions (in the fill, and including the old piles and pile caps).

Design Details. To satisfy the Owner's performance requirements, each pile comprised a 244-mm o.d. steel casing socketed 0.9m into the limestone. Below this, a 200-mm diameter hole was drilled a minimum of 4.6m deep to accommodate a 60-mm diameter reinforcing bar. In addition, a 3.1-m length of 178-mm diameter steel casing was placed in the upper part of the pile to satisfy lateral load deflection characteristics.

Construction. Diesel hydraulic track rigs were used with water flush to rotary drill both overburden and rock. Pressure grouting through the drill head to a maximum pressure of 0.7 MPa was conducted with a retarded low water cement ratio Type I-II grout, produced in a high speed, high shear mixer.

Preproduction Load Tests. Load tests were conducted on a limited number of test piles prior to the production work proceeding. These tests were conducted in accordance with the relevant ASTM standards, enhanced with extra load cycles, to the criteria shown in Table 2. Results are shown in Table 3.

Final Remarks. Following the successful testing program, the remaining piles were installed during a three month period, despite severe construction impediments, including severe water rationing and ongoing building demolition. The overall structural remediation is now complete.

Compression Test	1. Movement less than theoretical pile compression + 3.8mm + January 20, 19991% pile diameter. 2. Permanent movement after test load less than 13mm. 3. 150% of design load (801 kN) to reach limestone. 4. Maximum test load of 1068 kN to be achieved.
Tension Test	1. Permanent movement after test load less than 13mm. 2. No creep at test load. 3. Maximum test load of 712 kN to be achieved.
Lateral Test	1. Maximum head deflection less than 13mm at design load. 2. Design load is 44 kN, test load 88 kN.

Table 2. Acceptance Criteria for test piles, Puerto Rico.

Compression (Pile 212)	1. 15.2mm 2. 5.41mm 3. Acceptable 4. Achieved	Tension (Pile)	1. 5.87mm 2. Achieved 3. Achieved	Lateral (Pile)	1. 11.4mm 2. Achieved
Conclusion	Pass		Pass		Pass

Table 3. Summary of results obtained on test piles, as judged against Table 2 criteria.

Underpinning of Machine Foundation, E. Pennsylvania (Cadden et al., 1998)

Introduction. Heavy machinery used for plastic injection molding showed signs of unacceptable movements due to its weight, the high and repeated dynamic forces exerted during its operation and the subsoil conditions. Due to overriding scheduling restrictions, the problem had to be solved with the equipment in situ, to ensure angular distortions of less than 0.13mm in 3m, and no differential settlement between the two halves of the machine.

Geology and Site. Figure 3 shows the generalized conditions under the machinery. The soil proved to be medium stiff silts and clays, and the rock was shale. The operational facility comprised a large steel framed structure providing access via roll top doors at each end. Space around the equipment was severely restricted for drilling access.

Design Details. The relatively thin and weak reinforced concrete slab was deemed unable to sustain the loads imposed by high capacity CASE 1 micropiles, and so an array of CASE 2 piles was designed (Figure 4). Each pile had a nominal service load of 200 to 400 kN, and comprised a 25-m diameter reinforcing bar in a 76-mm hole, drilled 3m into bedrock.

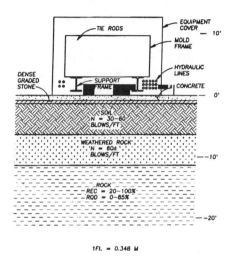

Figure 3. Simplified geologic layers beneath machinery, E. Pennsylvania (Cadden et al., 1998)

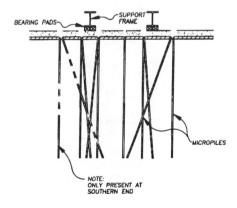

Figure 4. Micropile installation cross section, E. Pennsylvania (Cadden et al., 1998)

Construction. Special modular rotary drilling rigs were used to ensure appropriate access and hole inclination. The pile grout was designed to have low slump to facilitate clean-up in the building, while a high early strength mix was used to grout the pile reinforcement to the slab to enhance load transfer.

Performance. Following the underpinning, a uniform movement of around 0.2mm occurred, but without distortion or rocking during the transient loading applied by the machine. Within one week, no further movements were recorded, and the facility continues to operate without movement problems.

The Future

In the United States, as in the case worldwide, new geotechnical and structural challenges for both static and seismic retrofit are fostering the continuing growth of micropile technology. In particular, the demands of seismic engineering continue to provide new impetus to the study and understanding of pile performance in general, and pile networks especially.

Aided by the classification breakthrough made by FHWA, researchers in the United States, France, and Japan are poised to close the gap that still exists between the level of analytical understanding, and the excellence of the construction, testing, and performance knowledge. One consequence will be a rapid growth in the application of CASE 2 structures, optimally and rigorously designed to ensure efficient and economic solutions especially for seismic applications.

The relative ease of global information retrieval and exchange systems, coupled with the momentum established by micropile researchers in the late 1990s will ensure that developments in micropile technology will continue apace, and provide a fitting reflection of the foresight of their progenitor, Fernando Lizzi.

References

Bruce, D.A., S.L., Pearlman, and J.H. Clark. (1990). "Foundation Rehabilitatin of the Pocomoke River Bridge, MD, Using High Capacity Preloaded PinPiles." Proc. 7th Annual International Bridge Conference, Pittsburgh, PA June 18-20, Paper IBC-90-42, 9 pp.

Bruce, D.A., A. F. DiMillio, and I. Juran. (1995). "Introduction to Micropiles: An International Perspective". In *Foundation Upgrading for Infrastructure Improvement.* American Society of Civil Engineers, Proceedings of the ASCE Convention, San Diego, CA, Oct. 23-26, Geotechnical Specialty Publication No. 50, pp. 1-26.

Cadden, A.W., D.A. Bruce, and R.P. Traylor. (1998). "Micropile Underpinning of a Machine Foundation," *Concrete International*, Vol. 20, No. 1, January, pp. 53-57.

Federal Highway Administration (1996). "Drilled and Grouted Micropiles: State-of-Practice Review", 4 Vol. FHWA-RD-96-016, FHWA-RD-96-017, FHWA-RD-96-018, FHWA-RD-96-019, Co-Principal Investigators: D. A. Bruce and I. Juran.

Koreck, H.W. (1978). "Small-Diameter Bored Injection Piles." *Ground Engineering*, Vol. 11, No. 4 pp. 14-29.

Lizzi, F (1982). *Static Restoration of Monuments*. Sagep Publisher. Genoa, Italy, 146 p.

Pearlman, S.L., B.D. Campell, and J.L. Withiam. (1992). "Slope Stabilization Using In Situ Earth Reinforcements". Proceedings of Specialty Conference, American Society of Civil Engineers, Berkeley, CA, June 29-July 1, Geotechnical Specialty Publication No. 31. pp. 1333-1348.

Zelenko, B.H., D.A. Bruce, D.A. Schoenwolf, and R.P. Traylor. (1998). "Micropile Application for Seismic Retrofit Preserves Historic Structure in Old San Juan, Puerto Rico." American Society of Civil Engineers, Annual Conference, Boston, MA, October 18-22, Geotechnical Specialty Publication No. 80, pp. 43-62.

Suspension Wall Shoring System

Chris J. Wolschlag[1], P.E., Ph.D., A. M. ASCE,
R. John Byrne[2], P.E., Ph.D., and David M. Cotton[3], P.E.

Abstract

An excavation shoring system referred to as a "Suspension Wall" (patent pending) has been developed to meet three challenges of below-grade construction in congested urban environments: (1) clearance restrictions, either overhead or adjacent to the property, that preclude the installation of vertical soldier piles; (2) deep utility conflicts that prohibit the installation of subhorizontal soil nail or ground anchor elements; and (3) adjacent surface facilities or structures that are sensitive to the lateral movements induced by conventional shoring wall construction. The suspension wall shoring system utilizes steeply-inclined pretensioned soil nails and subvertical compression soil nails to suspend a shotcrete facing that is prestressed against the soil to produce near at-rest horizontal earth pressures and restrain lateral movement. The suspension wall shoring system was proven successful on a project in downtown Portland, Oregon.

Project Background

Rouse Management Inc., a nationwide developer and real estate manager, contracted Golder Associates Inc. (GAI) in January 1998 to provide geotechnical engineering and temporary shoring wall design services and for a new retail center located in downtown Portland, Oregon. The development consists of one level of below-grade retail shops and a multi-level above-grade retail center. The project site comprises Block 50, bounded by S.W. Morrison Street on the North, S.W. Yamhill Street on the South, S.W. 4th Avenue on the West, and S.W. 3rd Avenue on the East. The existing grades vary from elevation 11.5 m (38 ft) in the northeast corner, to elevation 13.4 m (44 ft) in the southwest corner.

---

[1] Associate, Golder Associates Inc., 4104 148th Avenue NE, Redmond, WA, 98052
[2] Principal, Golder Associates Inc., 4104 148th Avenue NE, Redmond, WA, 98052
[3] Principal, Golder Associates Inc., 4104 148th Avenue NE, Redmond, WA, 98052

The excavation base was at elevation 7.3 m (24 ft), resulting in excavation depths ranging from 4.3 to 6.1 m (14 to 20 ft). The north and south sides of the site were immediately adjacent to a critical metropolitan light-rail system.

A site plan is reproduced in Figure 1. The site is approximately 70-m by 70-m (230-ft by 230-ft) in plan dimension. The below-grade excavation extends beyond the property lines into the street on all sides of the site. Along the east side of the site, the permanent below-grade wall is located in the middle of the nearest traffic lane, 7.9 m (26 ft) beyond the property line. Along the north and south sides of the site, the permanent below-grade walls are located within 0.3 m (1 ft) of the curb line, 1.5 m (5 ft) from the nearest track rail, and 4.9 m (16 ft) beyond the property lines. Temporary shoring walls were required along the north, south, and east sides of the site.

Subsurface Conditions

In general, the subsurface conditions at the project site are comprised of three distinct soil layers. The surface soil layer is characterized as fill, reaching depths of 0.6 to 3.7 m (2 to 12 ft), and is comprised of loose to medium dense silt and sand with intermixed gravel, boulders, and rubble. Underlying the fill layer, a firm silt, with varying amounts of fine to medium sand, extends to depths of 4.0 to 7.6 m (13 to 25 ft). These two upper units are underlain by dense to very dense gravels and cobbles in a matrix of silt and sand. The top of the dense gravels varies from elevation 5.8 m (19 ft) in the southwest corner of the site, to elevation 7.6 m (25 ft) in the north end of the site, and to elevation 8.2 m (27 ft) in the southeast corner of the site. Therefore, the less competent soil units controlled the lateral shoring wall loads. For the purposes of design, the upper soil units were combined into one design soil unit. The shoring design parameters for the upper unit, including anchor adhesion for the lower gravels, are summarized below:

- At rest horizontal earth pressure coefficient, $K_o = 0.45$
- Total soil unit weight, $\gamma = 19.6$ kN/m^3 (125 pcf)
- Soil friction angle, $\phi = 34°$
- Soil cohesion, $c = 0$ kPa (0 psf)
- Design adhesion in upper soils (silts & sands), $A_{DU} = 29$ kN/m (2 k/ft)
- Design adhesion in lower soils (gravels), $A_{DL} = 73$ kN/m (5 k/ft)

The Willamette River, located several blocks north of the site, controls the groundwater level at the project site. During the summer months when the shoring walls were built, the water level consistently occurred at about elevation 3 m (10 ft), or 4.3 m (14 ft) below the base of the excavation. Therefore, groundwater was not a design issue for the shoring wall system.

Shoring Wall Issues and Design Constraints

The east wall along S.W. 3rd Avenue is 4.6 to 5.5 m (15 to 18 ft) high (Figure 2). The main design constraints for the east wall consisted of a shallow gas line, and a 0.76-m (2.5-ft) diameter high-pressure water line located roughly 5.5 m (18 ft) behind the wall and 3 m (10 ft) below grade. Because the adjacent roadway was not considered sensitive to movement, a conventional soil nail wall was constructed along the east side of the site, with slightly steepened row 1 nails that missed the water line combined with strut nails to provide vertical support. In firm silts and compact sands, this type of shoring system would be expected to experience lateral movements on the order of 0.2 to 0.4% of the wall height, or 13 to 25 mm (0.5 to 1 in).

The north wall, along S.W. Morrison Avenue, is 4.3- to 4.6-m (14- to 15-ft) high. The south wall, along S.W. Yamhill Avenue, is 5.5- to 6.1-m (18- to 20-ft) high (Figure 3). Both shoring walls are located within 1.5 m (5 ft) of the nearest train rail, and within 0.6 m (2 ft) of the track slab. In fact, the light-rail train car rear-view mirrors extended into the excavation when the trains passed the site. The light-rail system is owned and operated by the Tri-County Metropolitan Authority (Tri-Met). Tri-Met specified that the construction-induced lateral and vertical movement of the light-rail system must be essentially zero. Therefore, soil nailed walls or conventionally-designed anchored soldier pile walls would not be adequate. It became evident that the selected shoring system would have to prestress the adjacent ground to at-rest levels of horizontal stress in order to limit wall movements. The design uniform apparent earth pressure was selected at 5.5H kPa (35H psf), plus additional surcharge loading.

The overhead high-voltage DC power lines, located roughly 6.1 m (20 ft) above each track, were within about an 2.4 m (8 ft) horizontal distance from the shoring walls. Therefore, vertical soldier pile drill rig leads would fall within about 1.5 m (5 ft) of the power lines. The distance between the power lines and drill rig leads would require the power to be shut down during the installation of soldier piles. A train passed by the site about every seven minutes during the day. Tri-Met would only agree to shut the power down for a relatively short period of time from 2 am to 6 am. Therefore, although piles could be installed during the 4-hour power-off window, the slow progress would adversely impact the construction schedule and project budget. Another method of ground support was needed that did not require a shut down of the light-rail power lines.

In addition to the above constraints, a 0.9-m (3-ft) diameter combined sewer was located behind each of the north and south walls. Each sewer was generally 4.3- to 4.6-m (14- to 15-ft) deep and located roughly 4.0 to 4.3 m (13 to 14 ft) behind the shoring walls. The location and size of the sewer made the installation of traditional sub-horizontal ground anchors impossible. The anchor declinations would have to be at least 45° to provide the minimum 0.9-m (3-ft) clearance specified by the City of Portland.

Suspension Wall System

Based on the three constraints imposed on the north and south shoring walls, a new type of shoring wall, referred to as a suspension wall system was developed. The shoring design constraints are summarized as follows:

- Limit wall movements to essentially zero.
- Avoid overhead power obstruction.
- Avoid deep combined sewer obstruction.

Steeply inclined, prestressed tension anchors were utilized to provide the high lateral prestress and to avoid the deep combined sewers. Subvertical compression anchors were utilized to resist the vertical component of load from the tension anchors, while maintaining more than 2.7 m (9 ft) of horizontal clearance from the overhead power lines. A shotcrete facing, constructed in staged vertical lifts, was suspended by the group of tension and compression anchors, and provided the means to transfer the resultant horizontal force to the retained soil.

A typical cross-section is presented in Figure 3. Each shoring wall contained three 1.5-m (5-ft) high lifts of structural shotcrete. For areas where the wall height exceeded 4.6 m (15 ft), the upper portion of the wall was sloped back and flashcoated. The first lift contained both tension and compression anchors, while the second and third lifts contained only tension anchors. Because of the differing declinations, the tension and compression anchors were staggered along the wall.

All tension anchors were installed at a 45° declination from horizontal except for the row 1 anchors of the north wall which were installed at a 60° declination. All compression anchors were installed at a 75° declination. Typical wall facing details are illustrated in Figure 4. Note that the tension and compression anchors were located vertically so that their lines of action coincided within the wall facing, so as to not impose significant rotation on the first lift of shotcrete during prestressing.

The tension and compression anchors may also be referred to as soil nails mainly because the density of the drilled elements is more in line with a soil nail system. Except for the fact that the elements are pretensioned, and include a sheathed no-load zone, the details of the element are very much like that of soil nails. However, the tension elements function more like anchors than soil nails. Both terms may be used to refer to the tension and compression elements of the suspension wall system.

Suspension Wall Construction

Because the relative mix of silt and sand in the soils (and therefore stand-up time) was thought to be unpredictable and quite variable, the shotcrete facing was shown on the plans as a two-application system. The intended construction sequence for areas of poor face stability is described as follows. After all of the anchor installations for the lift were complete, the shoring wall face was cut to the neat-line in roughly 6.1-m (20-ft) wide alternating slots. The first application of shotcrete consisted of a 50 mm (2-in) thick flashcoat, and would commence as soon as the geocomposite drain board and mesh were placed against the soil, and the compression nail cushion, nut, and plate were installed (Figure 4). A small bearing plate and nut were wet-set and attached to the tension nail to secure the flashcoat. After a period of at least four hours, the intermediate slots were cut and flashcoated as well. The following day, the second and final shotcrete application would take place. The main mat of deformed reinforcing bars would be placed at the center of the full wall, the first nut would be sleeved, and the remaining 100 to 150 mm (4 to 6 in) of shotcrete would be applied. The final bearing plate and nut were wet-set at an angle appropriate for stressing (Figure 4).

Installation of the tension and compression nails in lift 1 of the north wall is illustrated on Figure 5. The light-rail track is shown on the left half of the figure. Two Klemm 806 duplex air-rotary drill rigs are seen in the right half of the figure. The construction fence shown in the middle of the figure separates the drill rigs and track slab. Note the power lines in the upper left part of the figure.

The slot-cutting and flashcoating procedures described above and shown on the plans are illustrated on Figure 6 for the first lift of the north wall. Figure 7 illustrates the wall preparation for the second and final application of shotcrete. The cut has been extended down 0.6 m (2 ft) for the vertical steel lap and the reinforcing steel has been placed. Note that unlike any of the other north or south wall shotcrete lifts, lift 1 of the north wall was battered to miss the deep combined storm sewer. The logistics of the utilities were different enough for the north and south walls that a battered lift 1 was necessary.

Initially, the slot-cutting and flashcoating procedures were considered mandatory by Tri-Met for the first lift along the north and south walls, because of the risk of movement associated with loss of ground if caving conditions were encountered. However, after about 75 percent of the north wall lift 1 and 50 percent of the south wall lift 1 were constructed, GAI and Tri-Met were convinced that because the ground was mostly a silt, the stand-up time was excellent and slot-cutting and flashcoating were not necessary. Figure 8 illustrates the wall preparation prior to shotcrete placement for north wall lift 1 after the flashcoat process was discontinued.

A view of the south wall during preparation of the final lift for shotcreting is illustrated in Figure 9, while Figure 10 contains a view of the south wall after completion. Note the trolley in Figure 9 and the light-rail train in Figure 10.

The combination of suspension and conventional soil nail walls at this project were bid by the specialty shoring subcontractor at approximately $540/m^2 ($50/ft^2). This bid included the entire wall construction, except for excavation, control, and other standard general contractor scope items. Extra work due to unanticipated utilities, obstructions, and delays resulted in a final shoring subcontractor price for the 1,115 m^2 (12,000 ft^2) wall of roughly $540/m^2 ($63/ft^2).

Performance and Conclusions

Along the north and south walls, where the suspension wall system was installed, the measured lateral wall movements and vertical track slab movements were less than the 1.5-mm (0.06-in) accuracy of the optical survey. Therefore, the suspension wall system performed as expected with essentially no wall movements.

Along the east soil nail wall, both lateral and vertical wall movements ranged from 5 to 20 mm (0.2 to 0.8 in), with an average of about 13 mm (0.5 in). This magnitude of wall movement was anticipated in the design of a conventional soil nail wall for the firm silts and loose to compact sands at the site. Therefore, the east wall demonstrated that the lack of wall movement observed along the north and south walls can in fact be attributed to the use of the suspension shoring wall system.

Acknowledgments

The authors wish to thank David Zastrow of Rouse Management Inc. for the opportunity to work on the Pioneer Place Block 50 project. We also wish to thank Barry Reynolds, Al Bergman, Brett Harrison, and Ted Gay of Howard S. Wright Construction Company, and Ed Chappelle of Condon-Johnson Associates.

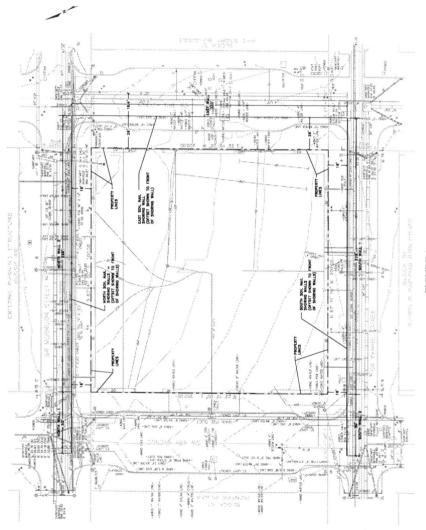

FIGURE 1
Site Plan

GEO-ENGINEERING FOR UNDERGROUND FACILITIES

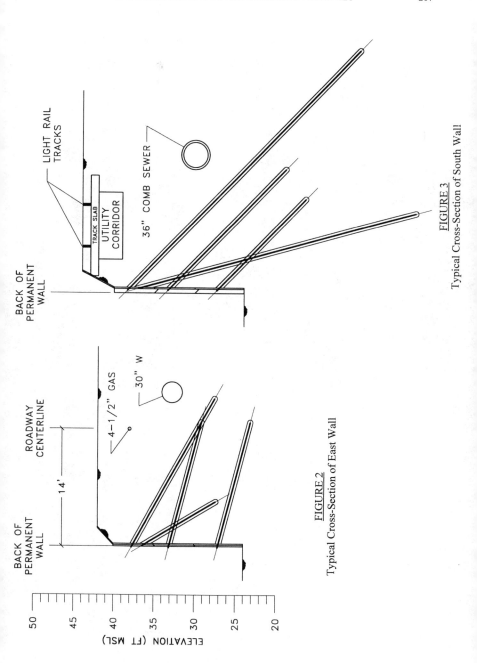

FIGURE 2
Typical Cross-Section of East Wall

FIGURE 3
Typical Cross-Section of South Wall

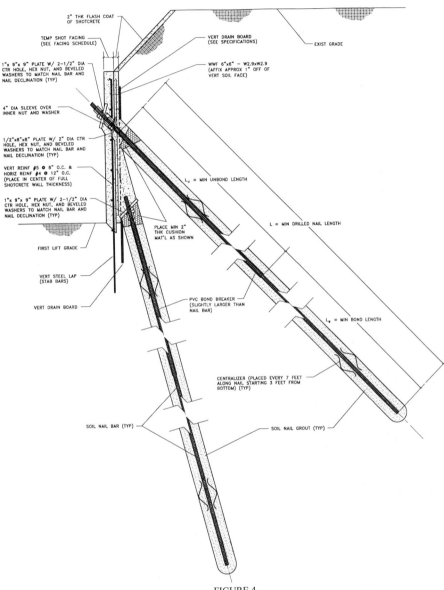

FIGURE 4
Typical North and South Wall Facing Details

FIGURE 5
Drilling Row 1 and Row A Soil Nails Along North Wall

FIGURE 6
Slot-Cut and Flashcoat Procedures for North Wall Lift #1

FIGURE 7
Preparation for the Final Temporary Facing for North Wall Lift #1

FIGURE 8
Close-up View of Wall Preparation Prior to Shotcrete Placement

FIGURE 9
View of South Wall During Lift #3 Preparation

FIGURE 10
View of Completed South Wall

SOIL NAILING FOR THE WHEDA OFFICE BUILDING IN MADISON, WISCONSIN

By Eric W. Bahner[1] M.ASCE, and Joe Sirvinskis[2]

ABSTRACT: Underground construction for large urban developments is often driven by a number critical issues including space needs, site limitations, and functionality requirements specific to the principal occupant. In Madison, Wisconsin, above ground space limitations associated with a downtown located on an isthmus between 2 lakes, and local ordinances prohibiting the construction of buildings which rise above the tops of the columns of the state capitol has resulted in the inclusion of substantial plans for underground parking on a number of the major downtown developments. For its new office building, the Wisconsin Housing and Economic Development Authority (WHEDA) developed plans which included 5 levels of underground parking. The required excavation for the parking structure was completed within 1.5 to 3 meters of 2 major city streets, within 1 to 1.5 meters of the local historical landmark St. Raphael Cathedral, and extended to a depth of 15.2 meters below street level. The engineer-contractor team of Woodward-Clyde and Edward E. Gillen Company proposed a soil nail retention system to retain the bulk of this excavation. Where the excavation was to be completed in close proximity to St. Raphael Cathedral, a pile and lagging system with multiple tieback levels was used to retain the cut and eliminated the need for underpinning. Optical survey monitoring during construction of the soil nail wall indicated no substantial ground movement, consistent with observations made by Denby and others that soil nail walls constructed in self-supporting soils typically exhibit negligible movement. Lateral movements associated with the tieback wall were within the envelope of lateral movement predicted using 1-dimensional finite element modeling. The temporary retention system was completed ahead of schedule and within budget, and provides an excellent example of contractor-engineer team work and soil nailing practice in the upper midwest.

By E.W. BAHNER[1] Geotechnical Engineering Manager, URS Greiner Woodward Clyde, Milwaukee, WI, USA, Email: ewbahneØ@wcc.com, Ph.414.513.0577. AND Joe Sirvinskis[2] Estimator/Project Manager, Edward E. Gillen Company, Milwaukee, WI, USA, Email: inform@gillenco.com, Ph. 414.769.3120

INTRODUCTION

The city of Madison has consistently been ranked as one of the most livable cities in the United States, and is the capitol of the great state of Wisconsin. The location of the city's downtown is unique, in that it is situated on an isthmus which rises as much as 20 to 30 meters above adjacent lakes Menona and Mendota. Local ordinances prohibit the construction of new buildings taller than the elevation of the base of capital dome. These local ordinances and significant above ground space limitations have resulted in the inclusion of substantial underground parking plans for major downtown developments. To address the earth retention needs for this project, the specialty contractor Edward E. Gillen Company of Milwaukee, and geotechnical engineer Woodward Clyde successfully designed and constructed soil nail walls, and adjacent to St. Raphael Cathedral, a tiedback soldier pile and lagging wall to safely and cost effectively retain the excavation.

PROJECT DESCRIPTION

In 1995, the state of Wisconsin moved forward with plans to construct a new office building for the Wisconsin Housing and Economic Development Authority (WHEDA). WHEDA officials along with the architect-engineer team of Potter Lawson Architects and Arnold & O'Sheridan both of Madison evaluated 2 sites within 2 blocks of the State capitol, and ultimately chose the one block site located at the southwest corner of West Washington Avenue and Fairchild Street. The site location is shown in Figure 1.

The site is bounded by major Madison arterials West Washington Avenue and Fairchild Street on the north and east sides of the site, respectively. The Madison Business College bounds the west side of the site. St. Raphael Cathedral and its surface parking lot occupies the entire south side of the site. Refer to Figure 2. The Madison Business College is a single story concrete and masonry structure. St. Raphael Cathedral, a city historical landmark, was built in the mid 1800's, and is constructed of locally mined sandstone. The building features ornate stained glass windows, carved stone fixtures, ornate ceilings and a large steeple. Elevations across the site range from Elevation 71(ft), Madison City Datum (MCD) on the south to Elevation 52 MCD on the northwest side, resulting in about 4.5 meters of relief across the site.

The proposed development consisted of an 8 story office building and five levels of underground parking. The base grades for the underground parking structure were expected to range from Elevation 18 to 13.5 MCD, some 12 to 16 meters below existing grade. The structure was to occupy the entire block, requiring an excavation

214　　　GEO-ENGINEERING FOR UNDERGROUND FACILITIES

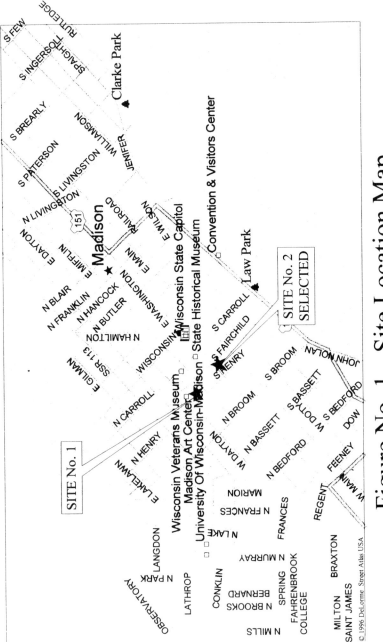

Figure No. 1 - Site Location Map

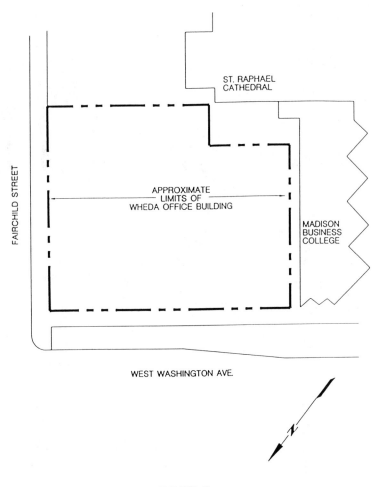

FIGURE 2
SITE PLAN

meters of and to a depth of 12 meters below the foundations of the adjacent cathedral. The excavation would be located approximately 4.5 meters east of the existing Business College buildings, and extend 9 meters below the foundations of that structure.

SUBSURFACE CONDITIONS

The Madison capitol square is constructed on glacial drumlin and end moraine deposits consisting of dense silty sands with varying amounts of cobbles and boulders. Some of the boulders can be as large a 1.5 to 2 meters in diameter. Two to three meters of miscellaneous fill is typically present immediately below street level. The groundwater table rises from the adjacent lake levels to depths of 9 to 18 meters below existing grades, but can be deeper in areas due to pumping from municipal water supply wells.

RETENTION SYSTEM DESIGN

Based on our experience, the ground below Madison's capitol square is well-suited to soil nailing, and with good control of the excavation face, ground movement due to excavation is generally very small or negligible. This is consistent with experiences shared by Denby from his involvement in the design and construction of soil nail walls in very similar ground conditions in downtown Seattle. (Ref. 1). Of more concern was the safety of the excavation through the shallow fill, and that portion of the excavation immediately adjacent to St. Raphael Cathedral. In the conceptual design phase, sloped open-cuts protected with short dowels, visqueen and wire mesh or chain link fence was established as an economic approach to excavation support through the fill. However, safe excavation adjacent to the cathedral was a main focus of the design effort.

From experience with similar historic structures in the city, it was unlikely that the cathedral was constructed on a continuous spread footing. Continuous spread footings typically have some structural capacity to span between individual support points, and underpinning can generally be safely undertaken in such situations. However, the foundation walls of many of the older structures in Madison as well as other Wisconsin cities are typically constructed of field stone or rubble, and are generally not mortared in place. Such foundations tend to be brittle and very sensitive to excavation and changes in support. Consequently, cracking or wall failure associated with changes in support during underpinning was a significant concern. Similarly, control of structural distress associated with the development of lateral strain resulting from horizontal deflection of an adjacent retention system was also a major concern.

The soil nail walls were designed using a Braced Cut approach as described by Nicholson (Ref. 2), and by Bahner (Ref. 3). The design was checked using the computer program SNAIL, a limit equilibrium computer program developed by the California Department of Transportation, Division of New Technology, Material and Research Office of Geotechnical Engineering (Ref.4). The nails for the project consisted of 4 in. diameter drilled holes with grouted in place mild steel bars. The nails were installed on a 5 ft vertical by 6 ft horizontal grid pattern as the excavation was advanced. The nail length was determined based on the expected and later confirmed frictional resistance offered by the soil, and optical lateral loads from the soil and adjacent surcharges loads. In general, the total nail lengths were on the order of 0.6 to 0.8 times the total wall height. A typical soil nail wall cross-section is shown in Figure 3. A typical shotcrete facing detail is shown in Figure 3A. Photographs of completed portions of the soil nail walls are shown on the following page.

The excavation adjacent to St. Raphael Cathedral was retained with a tieback soldier pile and lagging wall. The design of the wall was completed using the soil-structure interaction program RIDO. The RIDO analysis provided an assessment of the likely lateral deflections, bending moments and shear forces associated with the excavation sequence. RIDO translated as "curtain" in French (Robert Fages, 1993), uses the one dimensional finite element method in which retaining wall structures are represented as a series of bar elements that are free to displace laterally as well as rotate. The soil is modeled as a linear elastic-plastic material having a modulus of subgrade reaction that is limited by the fully active and passive Rankine states. The program uses the linear elastic-plastic soil model to adjust soil pressures as the wall deflects based upon the structural behavior of the wall and support system using earth pressure theory. Loads in the tieback anchors were also provided by the RIDO analysis.

For the expected ground conditions and building surcharge loads, a soldier pile and lagging wall with HP14 x 73 beams spaced at 2.5 meter centers, and 4 rows of tiebacks at 2 meter(±) centers was expected to undergo 1/2 to 3/4 of an inch of lateral movement. The tieback loads were expected range from 512 kN in the top anchor row to 454 kN in the bottom row. The tiebacks were not pressure grouted, but installed using gravity grouting procedures, similar to the soil nail installation. As a result, the total tieback lengths ranged from 20 meters in the top row to 11 meters in the bottom row. Figure 4 provides a typical soldier pile wall cross section. Assuming satisfactory control of the excavation face during installation of the lagging, and minimal caving of the boreholes during tieback installation, settlement of the adjacent cathedral was expected to be 1/4 in. to negligible. Graphical output from the RIDO model is shown in Figure 5. Based on this analysis, it was

Photograph No. 1: Completed open cut & soil nail walls

Photograph No. 2: Completed soil nail walls & tied back pile & lagging wall.

GEO-ENGINEERING FOR UNDERGROUND FACILITIES 219

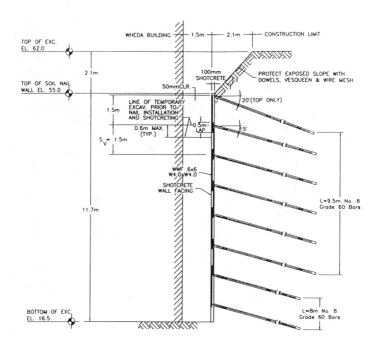

FIGURE 3
TYPICAL X-SECTION: SOIL NAIL WALL
SCALE: METERS

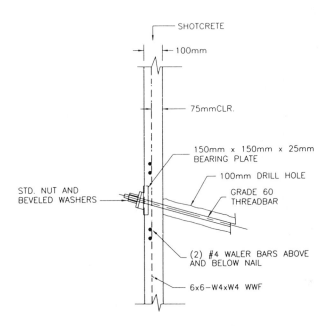

FIGURE 3a
SHOTCRETE FACE DETAIL

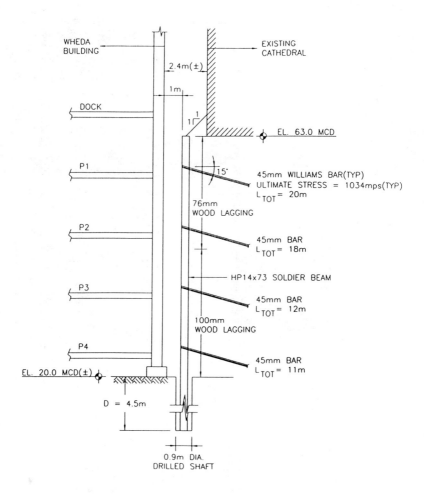

FIGURE 4
CROSS-SECTION: TIED BACK SOLDIER PILE WALL
SCALE: METERS

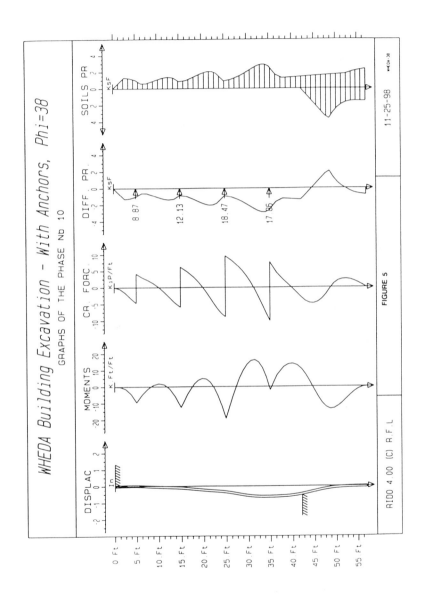

determined that underpinning of the Cathedral could be eliminated, at a significant cost savings.

CONSTRUCTION PROCEDURES

The soil nail walls were installed in 5 ft vertical lifts to excavation base grade. This process is illustrated in Figure 6. To prevent overbreak of the excavated face which can add to the cost of shotcreting, each excessive lift was rough cut with a slight batter, and fine-trimmed to vertical after nail installation. The reinforcing bars were installed in the 4 in. diameter drill holes, and gravity grouted by the tremie method. The adequacy of the nail lengths for the soil nail walls were confirmed using verification tests at the onset of construction, and with proof tests as necessary during production. Upon fine trimming the cut, the wire mesh, waler bars and shotcrete were applied to exposed excavation face.

The soldier piles for the pile and lagging wall were installed in one meter diameter drilled shafts terminated 4.5 meters below base of the excavation. The soldier piles were set in the shafts and anchored in place with concrete. Low strength lean concrete was used to backfill the balance of the shafts above excavation base grade. The tiebacks were installed using methods similar to those used to install the soil nails. Each tieback was proof-tested in accordance with Post Tensioning Institute guidelines. None of the tiebacks failed proof testing.

CONSTRUCTION MONITORING

The construction monitoring program consisted of daily observation of wall construction by Gillen Company and Woodward-Clyde engineers, and geotechnical instrumentation measurements. Excavation performance was monitored primarily using optical survey methods. An inclinometer was also installed behind the front flange of one of the soldier beams to check the behavior of the pile and lagging wall. The monitoring program was specified by Woodward Clyde, and carried out by engineers hired by the owner's construction manager.

The survey points were used primarily to check settlement of the adjacent buildings, street curbs and utilities. Vertical movement of these points would likely be the result of lateral wall movement or ground loss during excavation. The monitoring points consisted of masonry nails set into the mortar joints of the existing buildings, and chisel marks established along the curbs of the adjacent streets. The points were measured to the nearest 1.5 mm (0.005 ft). The inclinometer casing was supplied by Slope Indicator Company, and was grouted in a lightweight metal casing tack welded to the back of the front flange of a soldier beam. Optical survey and inclinometer measurements were taken and distributed on a weekly to biweekly basis

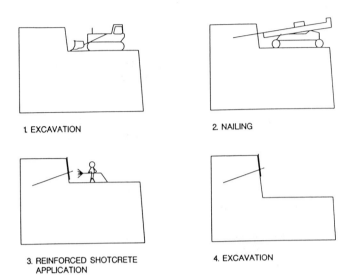

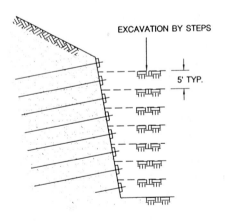

FIGURE 6
SOIL NAIL WALL
CONSTRUCTION SEQUENCE

to begin. Later when it was determined that the observed ground movements were very small to negligible, the frequency of the readings was reduced to bi-weekly to monthly.

WALL PERFORMANCE

Wall construction began in February 1996 with the installation of the soldier beams, and verification testing of the nails, and generally proceeded without incident into July 1996. Instrumentation monitoring was terminated in November 1996. Optical survey monitoring indicated deviations in the base readings about 1.5 to 3 mm (0.005 to .01 ft) at monitoring points located on St. Raphael Cathedral. These deviations were attributed to the distance of the points from the survey instrument and ongoing construction activity, and are considered within the accuracy of the instrument. Survey measurements at points located along the adjacent streets indicated no measurable movement. A summary of the measurements at selected optical survey points is provided in Table 1. The location of the survey points is provided in Figure 7. Inclinometer measurements were typically less than 1/4 in., and generally within the envelope of wall movement predicted during design. Refer to Figure 8. As shown in Figure 8, approximately 1/10 in. of lateral deflection *away* from the excavation was noted. This movement may have occurred during tieback stressing, and might be attributed to crushing failure of the low strength lean concrete backfill placed in the shafts during soldier beam installation.

Visual observations of the adjacent streets and buildings generally revealed no apparent signs of damage that could be associated with excavation induced ground movement. However some minor cracking was observed in the bituminous pavement of the Cathedral parking lot at a distance of approximately 4.5 meters south of the south face of the excavation. It was not clear if the cracks observed were the result of routine pavement distress or lateral deflection of the soil nail wall.

In mid-June 1996, the mettle of the retention system was severely tested when a summer thunderstorm dumped 10 to 18 in. of rain on the Madison area over a 48 hour period. In the days following the storm, over 100,000 gallons of water was pumped from the excavation. Post-storm observations indicated no major damage to the retention systems, adjacent buildings or utilities. However, some minor damage to the shotcrete face was noted along the south wall of the excavation where washouts at excavation base grade caused some sagging and horizontal cracking of the shotcrete. Further damage was avoided by backfilling the areas where washouts had occurred. Instrumentation readings completed shortly after the storm indicated no unusual wall movements.

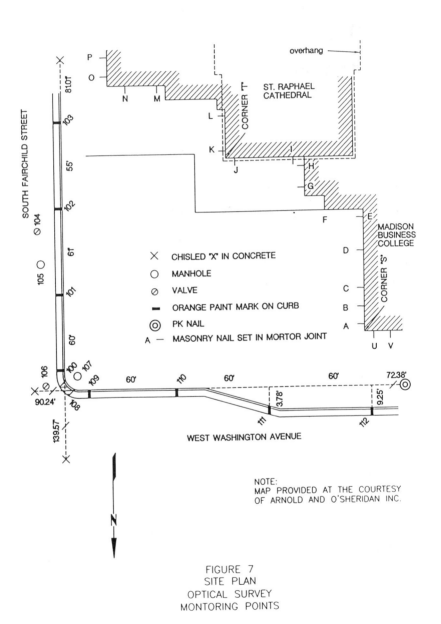

FIGURE 7
SITE PLAN
OPTICAL SURVEY
MONTORING POINTS

GEO-ENGINEERING FOR UNDERGROUND FACILITIES 227

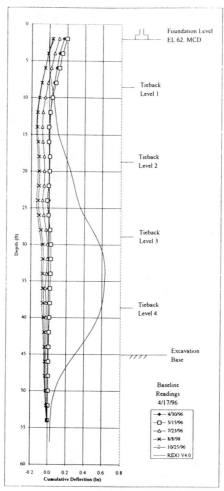

Figure 8
Selected Inclinometer Measurements
Soldier Pile & Lagging Wall

CONCLUSIONS

From our work on the project, and our experience with the design and construction of soil nail walls in similar ground conditions, we conclude the following:

- Soil nailing offers a very cost-effective and adaptable retention system for dense tills, especially in the presence of cobbles and larger boulders.

- In self-supporting soils, ground movements associated with soil nail walls is often negligible or very small, consistent with Denby's observations during construction of the first two soil nail walls in Seattle

- Self-supporting soils such as those in the Madison area could lend themselves to underpinning using soil nailing. However, underpinning using soil nailing carries more risk. This risk must be managed by developing an installation system to control ground loss during excavation, and undertaking a comprehensive program of geotechincal instrumentation to evaluate wall performance, and provide reliable information concerning the performance of the underpinned structures during the time retention system is constructed and in use.

The retention systems constructed for this project provide an excellent example of team work between specialty contractor and designer, and soil nailing practice in the upper midwest.

ACKNOWLEDGMENTS

The authors wish to acknowledge the following parties involved on this project: Potter Lawson Architects and Arnold & O'Sheridan, Inc. of Madison, the project architect and structural engineer. J.P. Cullen the General Contractor, and the efforts of Mr. Gene Sheedy, Project Superintendent for the Gillen Company.

REFERENCES

1. Denby, Gordon M. Two Soil Nailing Case Histories, First Use in Seattle (note: Author is attempting to find information about this reference)

2. Nicholson, Peter J., "Insitu Ground Reinforcement Techniques presented at the International Conference on Deep Foundations, Beijing, China, September 1986 (Meeting reprint).

3. Bahner, Eric W., "Soil Nailing in Madison Wisconsin" Proceedings - Woodward Clyde Share Holders Conference, Phoenix, Arizona, October 1994.

4. California Department of Transportation - Division of New Technology, Material and Research Office of Geotechnical Engineering, SNAIL Program Version 2.06, A Users Manual, February 10, 1993.

TABLE 1
SUMMARY OF MOVEMENT @ SELECTED OPTICAL SURVEY MONITORING POINTS

POINT	LOCATION	2/29	3/18	4/1	5/3	5/15	5/30	6/17	7/11	8/27	10/3	11/1
G	Cathedral[1]	112.805	+0.005	+0.005	(3)	(3)	+0.055	-0.015	+0.035	--	--	--
K	Cathedral[1]	114.190	+0.005	-0.005	(3)	(3)	+0.040	-0.020	0.000	--	--	--
L	Cathedral[1]	113.620	+0.010	0.000	+0.05	+0.065	+0.060	-0.065	+0.060	+0.050	+0.050	--
M	Cathedral[1]	115.295	-0.015	-0.055	-0.03	-0.015	-0.020	-0.03	+0.045	+0.080	+0.030	+0.045
N	Cathedral[1]	115.205	+0.010	+0.005	+0.025	+0.045	+0.025	+0.035	+0.015	-0.015	-0.015	-0.030
100	Curb[2]	98.12	0.000	0.000	0.000	0.000	0.000	0.000	0.000	0.000	0.000	0.000
101	Curb[2]	101.45	0.000	0.000	0.000	0.000	0.000	0.000	0.000	0.000	0.000	0.000
102	Curb[2]	104.97	0.000	0.000	0.000	0.000	0.000	0.000	0.000	0.000	0.000	0.000
103	Curb[2]	107.13	0.000	0.000	0.000	0.000	0.000	0.000	0.000	0.000	0.000	0.000
104	Valve	103.08	0.000	0.000	0.000	0.000	0.000	0.000	0.000	0.000	0.000	0.000
105	Manhole	102.53	0.000	0.000	0.000	0.000	0.000	0.000	0.000	0.000	0.000	0.000
106	Valve	98.060	0.000	0.000	0.000	0.000	0.000	0.000	0.000	0.000	0.000	0.000
107	Manhole	96.20	0.000	0.000	0.000	0.000	0.000	0.000	0.000	0.000	0.000	0.000
108	Curb[2]	95.02	0.000	0.000	0.000	0.000	0.000	0.000	0.000	0.000	0.000	0.000
109	Curb[2]	94.55	0.000	0.000	0.000	0.000	0.000	0.000	0.000	0.000	0.000	0.000
110	Curb[2]	92.44	0.000	0.000	0.000	0.000	0.000	0.000	0.000	0.000	0.000	0.000
111	Curb[2]	90.80	0.000	0.000	0.000	0.000	0.000	0.000	0.000	0.000	0.000	0.000
112	Curb[2]	89.50	0.000	0.000	0.000	0.000	0.000	0.000	0.000	0.000	0.000	0.000

*READINGS IN FEET

Notes:
1. Cathedral Monitoring points were masonry nails set in mortar joints
2. Points on curb were marked with a chiseled X.
3. Point was inaccessible during construction activities.

MODELING DEFORMATIONS AROUND EXCAVATIONS IN SOFT CLAYS

Laurent X. Luccioni[1], Student Member, ASCE
Juan M. Pestana[2], Member, ASCE
and Demetrious C. Kousoftas[3], Member, ASCE

Abstract

This paper presents a numerical formulation to accurately describe soil displacements around ground openings in soft soil deposits. The proposed technique uses a new simplified soil model, referred to as Bear-Clay, that captures the key aspects of lightly overconsolidated clay behavior and an efficient implementation into both non-linear finite element and finite difference computer codes. Instrumentation data collected during the construction of San Francisco's largest rail project in 20 years, the MUNI Metro Turnback (MMT), are used to illustrate the capabilities of a proposed numerical. This approach enables practicing engineers with a tool to perform parametric studies and to produce realistic estimates of ground deformation. These studies provide valuable information to achieve an optimum solution between construction cost and associated risks related to underground projects in congested urban areas.

Introduction

Civil engineers working on deep excavations in congested urban environment, are particularly concerned with soil movements and their effects on adjacent structures (e.g., Peck 1969, Burland and Worth 1974, Clough et al. 1974, O'Rourke 1981). Currently, the State-of-the-Practice relies heavily on empirical or

---
[1] Ph.D. Candidate, Depart. of Civil and Environmental Engrg., University of California, Berkeley
[2] Assistant Professor, Depart. of Civil and Environmental Engrg., University of California, Berkeley
[3] Principal, Dames and Moore, 221 Main Street Ste. 600, San Francisco, CA 94105

semi-empirical methods to estimate ground movements (e.g., Peck 1969, O'Rourke 1981). Continuum mechanics theory, on the other hand, coupled with numerical methods such as finite elements or finite differences have the potential to provide designers with timely information about soil movement and the response of nearby structures to these movements (e.g., Clough et al. 1974, Hashash and Whittle, 1996). Depending on the level of uncertainties associated with subsurface characterization, construction sequences, and other parallel activities, numerical modeling may be used as a predictive tool to provide insight or to verify performance through methodical comparison with high quality instrumentation data. The usefulness of these methods for the analysis of excavations is determined primarily by the following parameters: a) "accurate" modeling of the (typically complex) construction sequence (e.g., Clough and Davidson, 1977); and b) extensive characterization of the mechanical behavior of the soil material (e.g., Finno and Harahap 1991, Whittle et al. 1993), and c) stable, accurate, and efficient numerical implementation (Borja 1991)

This paper presents a numerical procedure and its application to the calculation of deformation around a deep excavation in soft clay. The proposed numerical procedure consists of a new soil model, referred to as Bear-Clay, and an efficient and accurate implementation into numerical codes. The proposed method is unique in the sense that the constitutive model and numerical implementations were developed in parallel in the search for an optimum procedure. Bear-Clay describes the key elements of lightly overconsolidated clay behavior, namely: a) anisotropic stress-strain-strength properties and b) small strain nonlinearity in undrained shear. The model uses a relatively small number of parameters which can be obtained from standard laboratory tests. This new soil model has been implemented into user friendly nonlinear finite element (GEOFEAP, Bray et al., 1995) and finite difference (FLAC, Itasca, 1993) computer codes, using newly developed accurate, efficient, and robust algorithms (Luccioni and Pestana 1999a). Validation of the numerical procedure capabilities is achieved through comparison with a well documented case history of a cut-and-cover section in San Francisco Bay Mud, which includes: a) extensive field instrumentation, b) complete characterization of the soil profile and soil parameters and c) well documented history of the construction sequence and relevant events during project execution.

Constitutive laws

The Bear-Clay model was developed to provide practicing engineers with a new numerical tool that takes advantage of greater understanding of soil behavior. The formulation is based on the incrementally linearized theory of rate independent infinitesimal elasto-plasticity (e.g. Prevost, 1978). The model assumes that the 1-D compression of normally consolidated specimens can be represented through the Limiting Compression Curve, LCC, characterized by a linear relationship between void ratio, e and mean effective stress, σ' when plotted in a log e-log σ' space (Pestana and Whittle, 1994). The small strain non-linearity in shear is described by a Perfectly Hysteretic formulation with isotropic, tangential moduli related to the most recent stress reversal (Jamiolkowski et al., 1994). The model assumes a constant Poisson's ratio chosen to match the measured unloading from normally consolidated states to moderately overconsolidated states OCR=3-4. Critical state conditions are represented by an isotropic function of the form proposed by Matsuoka and Nakai (1974), while a new single yield surface is used to describe the plastic behavior of soil specimens consolidated along a radial stress path (Pestana and Luccioni, 1998, 1999). The model introduces new expressions for the flow rule describing plastic strains and new forms of the density and kinematic hardening for the yield surface.

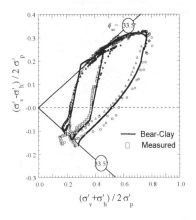 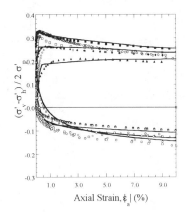

Figure 1. Comparison of Bear-Clay prediction and measured data for K_0 consolidated undrained triaxial tests

The model uses seven input parameters, four of which can be directly measured from laboratory tests while the others must be determined by parametric studies, matching model predictions with measured undrained triaxial compression and extension stress-strain and effective stress paths. Parameter $ρ_c$, describing the slope of the VCL in a log e-log p space, is obtained from one dimensional consolidation tests. The input parameter, m, is obtained by matching the effective stress path and peak undrained strengths in triaxial compression. The critical state friction angle, $Φ_{cs}$, determines the failure criteria at about 10% strain, above which the small strain theory assumption start breaking down. The very small strain stiffness parameter, G_b, is estimated from laboratory or shear wave velocity measurement, and the Poisson's ratio is chosen to describe the 1-D unloading stress path from the normally consolidated state to equivalent OCR of 4. Parameter w_s is estimated through parametric study to match stress-strain curve in undrained triaxial extension test with capabilities to accurately measured strains in the range of 10^{-3}% to 10^{-1}%. Input parameter Ψ is obtained from a small parametric study to match the stress-strain response in undrained extension triaxial test for axial strain greater than 0.1%. In general, triaxial compression and extension tests at OCR equal to 1 are used to calibrate the model, whereas simulations for other OCRs are first class predictions. Figure 1 shows comparison between the effective stress paths and shear stress-strain response predicted and measured in undrained triaxial compression and extension tests on initially Ko consolidated specimens of Boston Blue Clay for OCR= 1, 2 and 4. The tests on normally consolidated specimens (i.e., OCR=1) were used to determine the parameters of the model, thus good agreement is expected. The model gives good predictions of the undrained strength in compression and extension. The predicted strength in extension is achieved at larger strains, where as testing results are typically unreliable for strains larger than 5-10% due to necking of the sample. The predicted stress-strain curves are in good agreement with measured data at strains less than 10% well in the range of the practical application at hand. In summary, the model is able to capture the soil response under different stress paths with several OCRs. A complete description of the constitutive laws, model parameters, and predictive capability of Bear-Clay can be found in Luccioni and Pestana 1998, 1999a.

Numerical platform

The new model has been implemented into an explicit finite difference computer code, FLAC, and in a finite element computer code, GeoFEAP (Bray et al., 1995) that can perform coupled flow and deformation analyses. The results presented here pertain only to those obtained with FLAC, although preliminary analyses with GeoFEAP yielded similar results. FLAC is a user friendly numerical platform which has been used successfully in a wide variety of projects (Itasca, 1993) and features an embedded language that allows implementation of user defined constitutive laws similar to those of other computer codes (e.g., ABAQUS^{TM}). In FLAC the full dynamic equations are solved for each time step using an explicit scheme. Time step sizes are chosen such that the calculation "wave speed" always keeps ahead of the physical wave speed. This approach is best suited for very non-linear behavior and has the advantage of being able to track instability by taking into account the kinematic energy. On the other hand, to reach a quasi-static solution, some artificial damping has to be introduced. The main disadvantage of the explicit method seems to be the small time step, which means that large numbers of steps must be taken with significant impact in the solution time. The code incorporates a mixed discretization in order to avoid 'mesh locking' when modeling near incompressible material such as clays in undrained condition.

Numerical implementation

It has long been recognized that global performance of numerical techniques such as finite element and finite difference is very sensitive to local numerical implementation of constitutive laws (Ortiz and Popov, 1985, Simo et al., 1997). In this context a new fully explicit automatic substepping scheme with error control stress integration algorithm was developed following the original work of Sloan (1987). This explicit algorithm falls under the category of return mapping algorithm with standard operator split procedure. It does not require neither determination of initial yield nor any form of stress adjustment to prevent drift from the yield surface. Full details of discrete equations and algorithm performance is given in Luccioni and Pestana (1999b).

Application to the MUNI turnback project

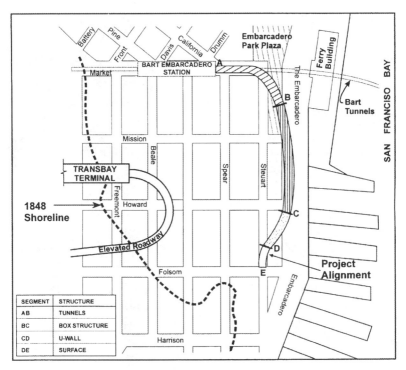

Figure 2. MUNI Turnback project location

Project background

The MUNI Turnaround project, located in downtown San Francisco, includes 280 m of twin tunnels connected to 434 m of cut-and-cover box tunnel structures and about 134 m of open U-structure where the tunnel rises to ground surface (cf., figure 2). The twin tunnels below lower Market street are 6 m in diameter with 12 m center to center lateral separation. The depth of cover over the MUNI tunnels ranges from 6 to 8 m. The vertical separation between the MUNI tunnel invert and the top of the BART tunnels ranges from 1.5 m at the Spear street shaft to 4 m at Justin Herman Park. The cut and cover structure was constructed from Justin Herman plaza under the Embarcadero roadway. It is separated from the bored tunnels by a ventilation shaft. The cut and cover excavation is approximately 33 m deep and 18 m wide.

Description of the site

The geological profile of the project is typical of downtown San Francisco, and generally consists of rubble fill underlain by Young Bay Mud deposits (see figure 3). The stratigraphy along the MUNI cut and cover section consists of the following major strata:

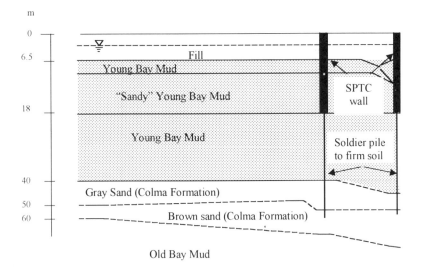

Figure 3. Representative cross section

- At the surface there is a layer of loose to medium dense fill, approximately 5 to 7 m thick in the cut and cover area, and 8 to 13 m thick in the tunnel section.
- Underlying the fill, there is a clay strata mainly normally consolidated, commonly referred to as Young Bay Mud, with a thickness varying from 22 to 40 m. Some sand lenses can be found in the upper layers of the Young Bay Mud.
- Below the Young Bay mud, there are various sand layers 5 to 10 m thick. The upper 3 to 5 m of this deposit is considered to be an alluvial or marine beach deposit, while the lower 3.5 to 5 m, brown in color, is very dense, and is considered part of the Colma Formation.
- Below the Colma Formation is the Old Bay Mud, a very stiff clay deposit of variable thickness.

The composition of the fill is highly non-homogenous, with the predominant component being sands and random-sized gravel and cobbles. Hence, engineering properties such as moisture content and dry density were found to vary widely, especially in the upper portion of the fill. The Young Bay Mud is a medium stiff clay with a variable amount of sand, shells and organic matter. It has a medium to high plasticity, with the plasticity index (PI) varying between 20 to 40, with an average total unit weight of 1,682 Kg/m^3. The upper portion of the Young Bay Mud, to a depth of approximately 17 m, appears to be normally consolidated. At depths greater than 17 m, the Young Bay Mud appears to be overconsolidated, with preconsolidation pressure increasing with depth. The engineering properties of the Colma formation are variable, with formation densities ranging from dense to very dense. The material has a total unit weight of 2,083 Kg/m^3 with a drained shear strength described by an average friction angle of 35 degrees and zero cohesion. The Old Bay Mud is a very stiff clay with undrained shear strength between 1 to 2 Kg/cm^2 and a total unit weight of 1,842 Kg/m^3. From consolidation tests, the compression ratio, ρ_c, was found to be about 0.212. Overconsolidation ratio is estimated to range from 1.5 to 2.0. Summary of the input parameters used for the analysis is given in table 1.

Table 1. Summary of material models and parameters

	Young-Bay Mud	Fill	Alluvium	Old Bay Mud
Constitutive law	Bear-Clay	Drucker-Prager	Drucker-Prager	Linear Elastic
Total Unit Weight (Kg/m^3)	1,681	1,921	2,082	1,841

	Young-Bay Mud	Fill	Alluvium	Old Bay Mud
Stiffness K or G_b (MPa)	$G_b = 625$	$K = 35.6$	$K = 86.2$	$K = 105.4$
Poisson's Ratio, ν	0.4	0.3	0.3	0.4
Small Strain Nonlinearity, w_s	6.0	N/A	N/A	N/A
Critical State Friction Angle, Φcs	34	32	35	N/A
Compression, ρ_c	0.212	N/A	N/A	N/A
Yield Surface, m	1.0	N/A	N/A	N/A
Evolving Anisotropy, Ψ	15.0	N/A	N/A	N/A

Instrumentation

The MUNI Turnback project is located in a crowded urban area, surrounded by buildings founded on relatively shallow wooden piles in Young Bay Mud. Hence the main challenge was to control the deformation around the excavation in order to prevent excessive differential settlements resulting in damage for nearby buildings and the existing BART tunnels. As a result, the project area was heavily instrumented to follow the evolution of deformation as the excavation proceeded (i.e., observational approach). Inclinometers were installed in the Soldier Pile Tremis Concrete (SPTC) wall and away from the SPTC wall; an extensive array of settlement markers was deployed perpendicularly to the excavation wall to measure settlements away from the excavation.

Comparison of measured and calculated displacements

Numerical analysis of the excavation does not involve any calibrations using field measurement results. Model parameters needed for the numerical analysis, were derived from laboratory or in-situ testing procedures conducted by Dames and Moore during the investigation phase of the project. Instrument records used in this analysis consist of inclinometer records. Inclinometer No. 20, located 1 m from the SPTC wall was chosen for comparison between measured and calculated horizontal displacements.

Figure 4 presents comparisons between measured horizontal displacement and calculated deformations using the Bear-Clay model. Four different recordings of the same inclinometer No. 20 representing measurements taken after excavation for bracing level A, B, C, and base slab were performed, are compared with the calculated deformation field. The shapes of measured and calculated horizontal displacement are very similar. The relative error between measured and calculated

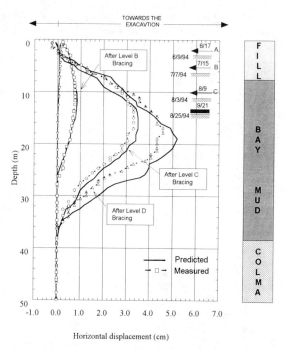

Figure 4. Measured vs. calculated displacement at 1 m behind the wall

is in the order of 10-15%, with a systematic bias toward larger calculated displacement. This number is quite reasonable considering the level of uncertainties in the overall procedure.

Conclusions

This paper has presented a numerical procedure and its application to the calculation of deformation around a deep excavation in soft soils. The numerical procedure uses a new soil model, referred to as Bear-Clay, and an efficient and accurate numerical implementation into an explicit finite difference code, FLAC. Bear-Clay requires only seven parameters to describe key attributes of the measured behavior of K_0 slightly overconsolidated clays as small strain non-linearity and anisotropic stress-strain-strength. Validation of model capabilities is achieved through comparison with the performance of a well-documented case history of the MUNI Turnback project involving a deep cut-and-cover section in San Francisco Bay Mud. Predicted ground deformation is in very good agreement with measured response from inclinometer and settlement markers. The proposed study shows that the new relatively simple soil model and a robust implementation technique can produce results of similar quality to those of more sophisticated soil models. It also provides a practical tool to satisfy the demands associated with the prediction of impacts of ground openings on existing nearby structures and other facilities.

Acknowledgment

Supports for this research was provided by the National Science Foundation (NSF) through grant No. CMS 9612136. This support is gratefully acknowledged. The authors would also like to thank Bechtel Corporation for a visit of the new MUNI underground system, as well as providing us with valuable insight into the project.

References

Borja R.I., (1991) "CAM-CLAY Plasticity, Part II: Implicit integration of Constitutive equations based on a nonlinear stress predictor", Computer Methods In Applied Mechanics and Engineering, 78, 49-72

Bray, J.D., Espinoza, R.D., Soga, K. and Taylor, R.L.(1995) "GeoFEAP: Geotechnical Finite Element Analysis Program, Volume I, Theory", Technical Report UCB/GT/95-09, Department of Civil and Environmental Engineering, Berkeley, CA.

Burland, J.B. and Wroth C.P. (1974) "Allowable and differential settlement of structure including damage and soil-structure Interaction" Proc. Int. Conf. on Settlement of Structures, 661-654.

Clough G.W. and Davidson (1977) "Effect of construction on Geotechnical Performance" Proc. 9th Int. Conf. on Soil Mech. & Found. Engr. Tokyo 15-53

Clough G.W. and Hansen L.A. (1974) "Clay anisotropy and braced wall behavior", ASCE, Journal of Geotechnical Engineering, 107, 337-345

Finno, R.J. and Harahap I.S. (1991) "Finite element analysis of HDR-4 excavation", ASCE, Journal of Geotechnical Engineering, 117, 1590-1609.

Hashash, Y.M.A. and Whittle, A.J. (1996) "Ground movement prediction for deep excavation in soft clays" ASCE, Journal of Geotechnical Engineering, 122, 474-486.

Itasca Consulting Group, Inc., (1993) "FLAC Version 3.2-Fast Lagrangian Analysis of Continua- User's Manual"

Jamiolkowski, M, Lancellota, R. and LoPresti, DCF (1994) "Remarks on the stiffness at small strains of six italian clays," Sapporo Conference, Vol. I, 95-114.

Luccioni, L. and Pestana, J.M. (1998) " BEAR-CLAY: A New Simplified Model for Lightly Overconsolidated Clays," University of California, Berkeley, Report UCB/GT (in press).

Luccioni, L.X. and Pestana, J.M. (1999a) "A Simplified Constitutive Model for Lightly Consolidated Clays", to be submitted Geotechnique.

Luccioni, L.X. and Pestana, J.M. (1999b) "Part II: Explicit Implementation of a Nonlinear Elasto-Plastic Model for Clays", to be submitted.

Matsuoka, H. and Nakai, T. (1974) "Stress-deformation and strength characteristics under three different principal stresses", Proc. of the Jap. Soc. of Civil Engrg. 232, 59-70.

O'Rouke T.D. (1981) "Ground Movement Caused by Braced Excavations", Journal of Geotechnical Engineering, ASCE, 107 1159-1177

Ortiz, M and Popov, E.P. (1985) "Accuracy and Stability of Integration Algorithms for Elastoplastic Constitutive Equations", International Journal for Numerical Methods in Engineering, 21, 1561-1576

Peck, R.B. (1969) "Deep Excavation and Tunneling in Soft Ground" Proc. Seventh Int. Conf. on Soil Mechanics and Foundation Engineering, 225-290.

Pestana, J.M. (1994) "A unified constitutive model for clays and sands", ScD. Thesis, MIT, Cambridge, MA.

Pestana, J.M. and Whittle, A.J. (1994) "Model prediction of anisotropic clay behavior due to consolidation stress history," Proc. 8th Intl. Conf. on Computer Methods and Advances in Geomechanics, Siriwardane & Zaman, Eds., Vol. 2, 1527-1532.

Pestana J.M. and Luccioni, L.X. (1998) Description of Drained and Undrained Behavior of Soft Marine Clay Deposits, Proc.12th Engineering Mechanics Conference, Murakami and Luco (Eds.), La Jolla, California, 1013-1016

Pestana J.M. and Luccioni, L.X. (1999) "Modeling the Drained Behavior of Lightly Overconsolidated Clays", to be submitted ASCE, Journal of Geotechnical and Geoenvironmental Engineering.

Prevost J.H. (1978) "Plasticity Theory for Soil Stress-Strain Behavior" ASCE, J. of Mech. Div., 104(EM5), 1177-1194

Simo, J.C. and Hughes, T.J.R. (1997) Computational Inelasticity, Interdisciplinary Applied Mathematics (IAM) Springer 1997

Sloan, S.W. (1987) "Substepping Schemes for the Numerical Integration of Elastoplastic Stress-Strain Relations", International Journal for Numerical Methods In Engineering, 24, 893-911

Whittle, A.J., Hashash, Y.M.A., and Whitman, R.V. (1993) "Analysis of Deep Excavation in Boston", Journal of Geotechnical Engineering, ASCE, 69-90

AN ARCH-SHAPED CULVERT BUILT BY THE INFLATABLE FORM

Karl H. Lewis[1], Carl Henderson[2], Cathy Bazán-Arias[3]

Abstract

Significant aging and deterioration of the U.S. infrastructure will present a difficult and costly problem. The inflatable form process is one that can facilitate the rebuilding of the infrastructure; however, it has not been embraced by industry. To demonstrate the adequacy of the above process, it was used to construct a large diameter arch-shaped culvert in Lorain, Ohio. The culvert was instrumented with concrete and soil pressure cells, vibrating wire strain gages, and observation bolts; and the field results were compared with those of a finite element model. Comparisons between field measurements and the finite element results were generally good but there were some discrepancies. It was concluded that construction, using the inflatable form process for culverts, can be safe, time and cost effective provided proper planning, management and workmanship are employed.

Introduction

Due to significant aging and deterioration of the U.S. infrastructure, the rebuilding of public facilities such as highways, bridges, sewers, etc. will present a difficult and costly problem. Procedures which limit the cost and time of construction of these facilities include the inflatable form process using shotcrete. Presently, two firms (Concepts in Concrete and HP Domes) utilize some type of inflatable form to construct arch-shaped culvert/tunnels.

---

[1] Associate Professor of Civil Engineering, University of Pittsburgh, Pittsburgh, PA 15261
[2] Project Engineer/Laboratory Manager, Ph.D., American Geotechnical and Environmental Services, Inc., McMurray, PA 15317
[3] Ph.D. Candidate, University of Pittsburgh, Pittsburgh, PA 15261

Current data indicates that an arch-shaped shotcrete culvert can be built at a 20% to 40% reduction of the cost and time needed for the construction of its competitor, the conventional box-shaped concrete culvert. Although this technology has been widely used to construct domes at sites in the United States and Canada, it has seldom been used to construct large diameter culverts/tunnels. Moreover, in the few cases where it has been used, the behavior of the constructed culvert was never analyzed or evaluated. As a result, questions and concerns about its structural integrity and long term performance and durability have limited its use, and contractors continue to use the traditional box-shaped versions for culverts.

Thus, a project was undertaken to develop a method of analysis and to study the long term behavior of an arch-shaped culvert/tunnel built via the inflatable form process. The Ohio Department of Transportation planned to widen State Road 113 in Elyria, Lorain County, Ohio, and was willing to consider the use of an arch culvert founded on a mat foundation in place of the proposed three-sided box culvert founded on piles. Nor-Con Inc. of Norwalk, Ohio, was the contractor selected for the job, and HP Domes Co. was responsible for advising the contractor on construction. Lewis outlined a plan for the instrumentation and long term evaluation of this culvert which was built by the inflatable form process between June and September, 1993 (Lewis, 1991).

The properties of the shotcrete placed at the site were determined by conventional strength and durability tests published by the American Society of Testing Materials (ASTM, 1993). The finite element code ANSYS was used to generate a discrete, two dimensional model for the project (ANSYS, 1995). Field behavior is being monitored via the taking of measurements and readings from strain gages and observation bolts installed on the arch and base slab of the culvert, and from earth pressure cells in the surrounding soil system. The finite element results were compared to these measurements.

Culvert and Embankment/Roadway Construction

The original site was a two lane highway built over a bridge under which a stream flowed from north to south. First the bridge, roadway, and surrounding area were cleared, poor soil was removed, and the site was prepared for the construction of the culvert and the embankment that would accommodate the widening and shifting of Route 113. On completion of the above work, the stream was temporarily diverted and the base slab was constructed by conventional means. Then, the inflatable form (Figure 1) was placed over the base slab, and the arch was constructed on the form. Finally, the form was deflated and removed, the culvert was covered with fill, and the roadway was constructed over the fill as shown in Figure 2.

Figure 1 - The Inflatable Form

The basic structure of the highway culvert consists of a 6.4 m (21 ft) wide by 30.5 m (100 ft) long base slab of 38.1 cm (15 in) thickness, an arch covering 2/3rds of a circle with a radius of approximately 3.2 m (10.4 ft) and a length of 30.5 m (100 ft), and wing walls. The arch has an average thickness of 20.3 cm (8 in), whereas the thickness of the wing walls (curved "shells" cantilevered from the ends of the arch) equals 17.8 cm (7 in). Base reinforcement consists of two layers of rebar with transverse and longitudinal unbonded tendons for post-tensioning, whereas arch reinforcement consists of two layers of heavy wire mesh with transverse rebar and longitudinal unbonded tendons for post-tensioning. Reinforcement of the wing walls consists of one layer of rebar.

Material Properties of Shotcrete Used For Culvert

Laboratory tests were conducted to determine the strength and durability characteristics of the shotcrete placed at the Elyria, Ohio culvert. These characteristics were determined under normal curing conditions, and after accelerated freeze-thaw cycles. Shotcrete for testing was obtained by gunning wet mix shotcrete from the Elyria, Ohio culvert shotcrete mix, into prepared master panels. Individual samples were later cut or cored from the slabs formed in the master panels. Of particular concern was the effect that freeze-thaw cycles (40, 80, 300 cycles) would have on bond strength between shotcrete layers. The shotcrete mix design was as follows:

- 390 kg/m^3 (658 lbs/yd^3) of cement.
- 148 kg/m^3 (250 lbs/yd^3) of water.
- 60 kg/m^3 (100 lbs/yd^3) of Class "F" fly ash.
- 1,231 kg/m^3 (2,075 lbs/yd^3) of sand.
- 415 kg/m^3 (700 lbs/yd^3) of #8 limestone.
- 1 kg/m^3 (3 oz./CWT) of water reducer.
- 0.7 kg/m^3 (2 oz./CWT) of retarder.
- 0.4 kg/m^3 (1 oz./CWT) of air entraining agent.

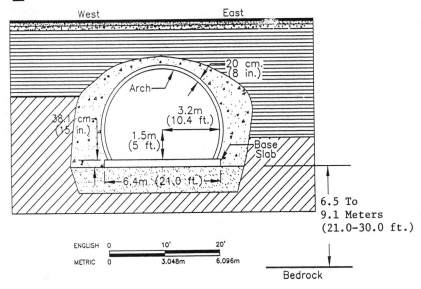

Figure 2 - Culvert Built via the Inflatable Form Process

Test results indicated that strength was directly related to sample quality. Compressive strength ranged from 29 kPa (4,300 psi) to 50 kPa (7,200 psi), and flexural strength ranged from 4 kPa (583 psi) to 5.4 kPa (782 psi). Rapid freeze-thaw cycling did not significantly affect the compressive and flexural strength of good quality samples. However, freeze-thaw cycling was found to significantly reduce splitting tensile strength. Splitting tensile samples were cut so as to intersect the interface between two shotcrete layers (Henderson, 1998).

Instrumentation of Soil-Culvert System

In locating the instruments, extreme care was taken to gather relevant data and avoid damage due to construction. In addition, to prevent tampering with instruments and minimize environmental effects, protection was installed for some of the instruments (Henderson, 1998). Figures 3 through 6 show the locations of the instruments (excluding four concrete pressure cells) installed at the site. The breakdown of the instruments is given below. The four concrete pressure cells, installed at the interface of the arch and base slab, are not shown because they have not functioned properly or rendered realistic results.

- 13 Soil Pressure Cells (A-J,O-Q) installed in the granular backfill and clay fill. Each soil pressure cell was surrounded by a 3-inch thick sand lens.
- 9 Vibrating Wire Strain Gages (SA-SI) installed on the arch to measure strains.
- 7 Observation Bolts (1-5,A,B) installed on the arch and base slab to measure the relative displacements of the culvert.

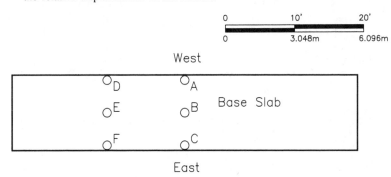

Figure 3 - Pressure Cells Underneath Base Slab

Finite Element Analysis

The Ohio soil-culvert system was studied in two dimensions using the finite element program ANSYS (ANSYS, 1995). This program allows the modeling of the relevant properties of the system, including geometric irregularities, nonhomogeneity

248 GEO-ENGINEERING FOR UNDERGROUND FACILITIES

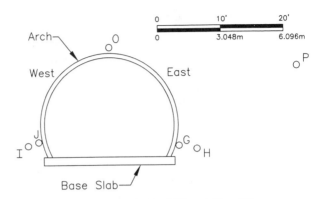

Figure 4 - Pressure Cells in Granular Backfill and Clay Fill

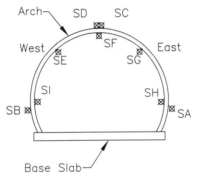

Figure 5 - Location and Designation of Strain Gages

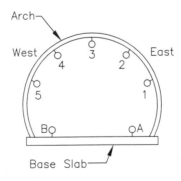

Figure 6 - Location and Designation of Observation Bolts

and material nonlinearity. In this study, isoparametric elements represent the soil and backfill, beam elements represent the culvert, and interface elements account for the relative displacements between the culvert and its surrounding backfill. The model, schematically shown in Figure 7, was verified against a theoretical elastic solution (Katona, 1976). The average difference between the finite element and theoretical results was 7% for the radial pressures, 3% for the radial displacement at the crown, and 4% for both the moment and thrust at the crown. To model field construction, an incremental solution simulating eight stages of construction was adopted.

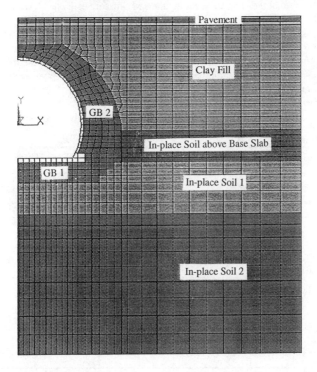

Figure 7 - Ohio Soil-Culvert Finite Element Model (East Side)

Table 1 presents the linear material properties implemented in the finite element model. Nonlinear properties were accounted for; however, since the system remained within the linear elastic range, the stress, strain and displacement results for the linear and nonlinear runs were identical. Tables 2 and 3 show the comparison of the relative field displacements and the finite element results. Figures 8 and 9 present select comparisons of measured stresses and strains and the finite element results.

As anticipated, the culvert is deflecting vertically at the crown, and bulging at the sides. The relative displacements, some stresses and strains (Soil Pressure Cell O in Figure 8 and Strain Gage SA in Figure 9, for example) compare favorably with field measurements. In other cases, the model yields significantly higher values than those measured in the field for the stress cells, and lower values for strain gages (Cell H in Figure 8 and Strain Gage SG in Figure 9, respectively). Among the reasons for the discrepancies are stress releases in the field (due to deformation of soil around the pressure cells) and concrete creep that are not considered in the finite element model. It is also likely that the sand surrounding the soil pressure cells may have intruded the granular backfill and may not have fully transferred pressure to the cells. In addition, assumptions such as two dimensional behavior, finite media, possible variations in material properties in areas taken to be uniform, and estimates of actual geometry, made to generate the analytical model may also have contributed to the difference in results. Lastly, some field equipment may not have functioned properly and may have yielded some unreliable data.

Table 1 - Material Properties for the Ohio Culvert Model

Material Description	Elastic Modulus E (kPa)	Unit Weight γ (kN/m^3)	Poisson's Ratio ν	Compressive Strength f' (kPa)	Density ρ (kg/m^3)
Arch	2.7E+07	22	0.15	36000	2241
Base Slab	2.6E+07	24	0.15	27560	2401
Granular Backfill under Base Slab (GB 1)	2.4E+04	19	0.3		1921
Granular Backfill around Arch (GB 2)	2.3E+04	19	0.3		1921
In-place Soil above Base Slab	5.1E+03	20	0.35		2081
In-place Soil 1 below Base Slab	8.3E+03	20	0.35		2081
In-place Soil 2 below Base Slab	1.8E+04	20	0.35		2081
Compacted Clay Fill	2.4E+04	21	0.35		2126
Aggregate Subbase (Dense Sand)	4.6E+04	21	0.35		2161
Concrete Base (for pavement)	2.6E+07	24	0.15	27560	2401
Asphalt Surface	2.4E+06	24	0.35		2401

Table 2 - Comparison of Incremental Analytical Results and Stages of Construction Field Data: Displacement of Bolts on the East Side

	Field		Model	
	U_x (cm)	U_y (cm)	U_x (cm)	U_y (cm)
Bolt No.	August 7, 1993			
1	0.22	-0.21	0.28	-0.40
2	0.17	-0.39	0.09	-0.55
3	0.05	-0.60	0.00	-0.73
Bolt No.	August 13, 1993			
1	0.24	-0.25	0.28	-0.43
2	0.17	-0.44	0.09	-0.58
3	0.08	-0.62	0.00	-0.79
Bolt No.	August 24, 1993			
1	0.30	-0.30	0.31	-0.43
2	0.23	-0.52	0.09	-0.58
3	0.08	-0.71	0.00	-0.79
Bolt No.	September 13,1993			
1	0.35	-0.36	0.31	-0.46
2	0.23	-0.61	0.09	-0.61
3	0.07	-0.87	0.00	-0.85

Table 3 - Comparison of Incremental Analytical Results and Stages of Construction Field Data: Displacement of Bolts on the West Side

	Field		Model	
	U_x (cm)	U_y (cm)	U_x (cm)	U_y (cm)
Bolt No.	August 7, 1993			
3	0.05	-0.60	0.00	-0.61
4	-0.06	-0.48	-0.06	-0.49
5	-0.22	-0.30	-0.21	-0.37
Bolt No.	August 13, 1993			
3	0.08	-0.62	0.00	-0.64
4	-0.10	-0.55	-0.06	-0.52
5	-0.25	-0.37	-0.22	-0.40
Bolt No.	August 24, 1993			
3	0.08	-0.71	0.00	-0.67
4	-0.09	-0.61	-0.07	-0.55
5	-0.27	-0.38	-0.24	-0.40
Bolt No.	September 13,1993			
3	0.07	-0.87	0.00	-0.70
4	-0.06	-0.73	-0.07	-0.55
5	-0.30	-0.46	-0.25	-0.40

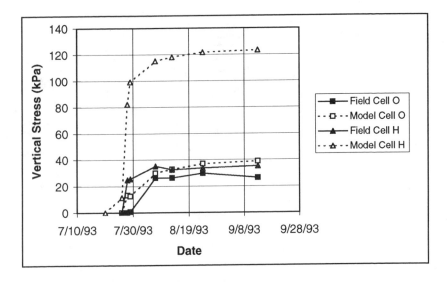

Figure 8 - Soil Pressures Above the Crown (Cell O) and on the Side (Cell H) of the Culvert

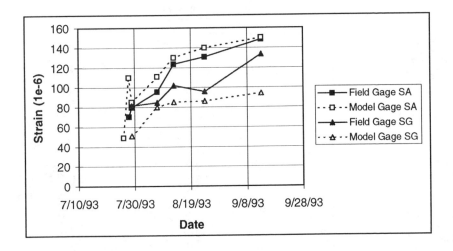

Figure 9 - Strain Measurements (SA & SG) on the Side of the Arch

Conclusions and Recommendations

After careful evaluation of the data generated from the case study described in the preceding paragraphs, the following conclusions have been made:

1. Construction using the inflatable form process can be a safe, time and cost effective way of building arch-shaped shotcrete culverts/tunnels. However, good planning, management, and workmanship are necessary to alleviate possible problems of construction. Potential problems including bulging in the inflatable form, setting of the wire mesh reinforcement, leaving voids and hollow spots around the wire mesh, delamination between layers, and wasting of shotcrete, may be encountered if proper care is not taken.

2. The mix design used for the project provided a strong, durable, and very workable (ease of placing and finishing) shotcrete product. In addition, very little rebound was observed when placing this mix on the inflatable form. However, the care in placement and workmanship has a great influence on the quality of the shotcrete structure (independent of the shotcrete mix).

3. The instrumentation has provided a better understanding and confirmation of the behavior of large, arch-shaped, shotcrete culverts/tunnels.

4. Some field measurements compared well with values obtained utilizing the finite element model. Others, however, showed significant disparities which are attributed in part to stress releases in the soil surrounding pressure cells, concrete creep, assumptions made regarding the geometry and material properties to generate the analytical model, and possible malfunctioning of some instruments.

5. Visual inspections of the structure indicate no major cracking or deterioration of the interior surface of the culvert. As of July 5, 1996, readings from the instruments indicate that the culvert is structurally sound and trends from the data imply that it should remain so. Further monitoring of the culvert is advised to evaluate its long term performance.

6. To avoid the above mentioned problems related to the use of wire mesh, and for overall superior shotcrete performance, it is proposed that steel fibers be used for secondary reinforcement in lieu of wire mesh.

Acknowledgements

We thank the Ben Franklin Technology Center of Pittsburgh, PA for funding the research, and the Department of Civil and Environmental Engineering at the University of Pittsburgh for its general assistance. We are grateful to the Ohio Department of Transportation for giving us the opportunity to study an arch-shaped culvert built via the inflatable form process.

References

Annual Book of ASTM Standards, Section 4, Vol 0402, Concrete and Aggregates, American Society for Testing and Materials, Philadelphia, PA, 1993.

ANSYS User's Manual, Volumes I - V, ANSYS Revision 5.2, SAS IP, August 31, 1995.

Henderson, C.F., "Determining the Reliability of an Arch-Shaped, Steel Fiber Reinforced Shotcrete (SFRS) Culvert, Built via the Inflatable Form Process", Ph.D. Dissertation, University of Pittsburgh, April 1998.

Katona, M.G., Smith, J.M., Odello, R.J. and Allgood, J.R., "CANDE: A Modern Approach for the Structural Design and Analysis of Buried Culverts," Civil Engineering Laboratory, Naval Construction Battalion Center, October 1976.

Lewis, K.H., "Proposal for Instrumentation and Monitoring of Elyria, Ohio Construction Site", Proposal, University of Pittsburgh, 1991.

Identifying and Baselining Boulders for Underground Construction

Steven W. Hunt[1] and Mauricio Angulo[2]

Abstract

This paper focuses on identification of boulders during subsurface exploration and on baselining of boulder quantities for construction of underground facilities. Comments are provided on the sensitivity and reliability of conventional and other subsurface exploration methods in identifying boulders based on a literature study and survey sent to exploration drillers. Geologic assessment and documentation of drilling observations are two keys to boulder identification. Boulder quantities may be baselined by utilizing statistical relationships between boulder indications in boreholes and quantities encountered in excavations, experience databases, and probabilistic methods of data evaluation.

Introduction

Boulders are often a cause of significant excavation difficulty for many shafts and soft-ground tunnels. They are a major concern when micro-tunneling. They can result in delays, lost ground problems, face instability and tunnel alignment difficulty (Cording et al., 1989; Tarkoy, 1994). A 1984 study of many tunnels found that boulders and other obstructions accounted for approximately 12 percent of tunneling problems and are a common cause of differing site condition claims (US National Committee on Tunneling Technology, 1984). Schmidt (1974) listed obstructions (including boulders) as one of five high priority parameters to be assessed in a subsurface investigation for a tunnel project.

To appropriately allocate risks and minimize unnecessary claims for extra work, various tunneling committees and support organizations have concluded that pertinent subsurface parameters such as boulder quantities should be thoroughly investigated during design and baseline quantities included in contract documents for use by bidders. The culmination of this effort is an American Society of Civil Engineers guideline document entitled *Geotechnical Baseline Reports for Underground Construction*

---

[1] P.E., M. ASCE, Senior Geotechnical Engineer, Harza Engineering Company, Milwaukee, WI 53202
[2] Ph.D., M. ASCE, Geotechnical Engineer, Harza Engineering Company, Chicago IL, 60606-6392

(Technical Committee on Geotechnical Reports of the Underground Technology Research Center, 1997). The guideline explains how baselining may be performed.

The focus of this paper is on baselining boulder quantities. Three primary components to boulder baselining are presented:
1. Researching available geologic and excavation case history information.
2. Properly monitoring and documenting boulder signs during subsurface exploration.
3. Estimating boulder quantities based on empirical correlations with boring data, statistical databases, and probabilistic methods.

A boulder is defined as a stone or rock fragment that has a diameter or width of at least 300 mm (12 inches). Boulder properties of most interest to underground professionals are considered to be:

- *Frequency* – volumetric density, occurrence per foot of tunnel (probably the most important property to baselined since it directly affects excavation effort and cost).
- *Distribution* – random (scattered) or geologically concentrated (lags or nests).
- *Size* – approximate diameter or width, length and depth (many tunnel boring machines [TBMs] can not process boulders that are larger than approximately 20 to 30 percent of the excavated diameter).
- *Shape* – spherical, cubic, slaby, irregular, and angularity or roundness of corners (affects breaking by TBM cutters and disks and passage through mucking system).
- *Composition* – rock mineralogy, compressive strength, lithology, degree of weathering (affects ease of breaking by TBM cutters and the rate of cutter wear or the effort needed for manual drilling and splitting).
- *Matrix soil composition* – density, strength, grain-size distribution, permeability (affects ability to push partially contacted boulders aside, to hold them rigid for fracturing by disc cutters, to dig around them with excavating tools or to control soil and groundwater inflows at the heading when exposing them for drilling and splitting).

Geologic Setting

The distribution and properties of boulders within a soil unit should be assessed in conjunction with the geology of the formations involved. Baselining of boulders based on soil type, e.g. silty clay or gravelly sand, may not result in reasonable correlations. Instead, boulder occurrence should be assessed for individual geologic soil units with consideration of common geologic characteristics and anomalies. The importance of using a geologic framework when assessing boulder occurrence (and other parameters that affect tunneling) has been explained by Essex (1993), Gould (1995), Heuer (1978), Legget (1979), Terzaghi et al. (1996) and others.

Typical characteristics of boulders and likelihood of occurrence within various glacial units and morphological features are discussed by Flint (1971). If available, local geologic papers should be studied to better understand the regional structure of geologic features and typical characteristics of the units and formations that are likely to be encountered. The previously listed boulder properties can be better understood and explained in reports, and baselined if subsurface exploration data is evaluated in a geologic context.

Subsurface Exploration for Identifying Boulders

Subsurface exploration is the most important component of baselining boulders for a specific project location or alignment. It is the source of site-specific data on geologic conditions and of indications that boulders are present. Despite its importance, cobble and boulder presence may not be addressed and quantities may often be underestimated within many geotechnical reports (Gould, 1995 and Cording et al., 1989). Poor exploration practices often fail to identify boulders where they are abundant. Tarkoy (1992) cites poor subsurface exploration and reporting practices as a primary cause of unanticipated boulder conditions and related tunneling problems and claims.

Hunt and Fradkin (1991) describe a Milwaukee project where drilling was not monitored in the field by an engineer or geologist. As a result, the logs of 16 borings that were drilled into bedrock through moderately bouldery till failed to report any boulders. In addition to missing scattered boulders in an upper till, an underlying bouldery lodgment till was misinterpreted as weathered bedrock. Excavation and ground support problems at two shafts resulted in differing site condition claims and extra costs over $500,000.

Tarkoy (1994) cites three case histories where borings did not properly identify boulders, bedrock or both. Wallis (1993) describes a Seattle microtunneling project where boulders were not anticipated by the geotechnical report, but where frequent cobbles and small boulders were encountered. Abbott (1995) describes another Seattle project where no boulders were reported on boring logs, but where frequent boulders up to 0.6 m in size stopped and severely damaged a 1.17 m Isecki Unclemole. Stoll (1976) describes a Michigan project where 17 borings failed to indicate any boulders but where numerous boulders were encountered during tunneling. Many more examples can be cited.

A study of these projects and other references (Gould, 1995; Osterberg, 1978) suggests the following reasons why many subsurface investigations and geotechnical reports fail to adequately predict boulder occurrence:

1. Lack of geologic assessment in planning, evaluating and reporting of subsurface exploration data resulting in a poor understanding of the geologic setting, character and uncertainties.

2. Lack of a phased exploration program that progressively reduces key uncertainties that remain from the previous phase.
3. Lack of focused drilling instructions and monitoring of drilling by a geotechnical engineer or geologist (or one with proper training and experience).
4. Inadequate documentation and reporting of drilling observations and historical information.
5. Over reliance on drilling refusal or extremely high Standard Penetration blow counts as the primary indicators of cobbles and boulders.
6. Insufficient exploration budget (including boring spacing) and lack of redundancy in exploration methods causing over-reliance on only conventional drilling methods that are often chosen for speed and cost of drilling rather than sensitivity to cobbles and boulders.

Conventional Drilling and Sampling Methods

Drilling and sampling methods that are selected affect the relative volume of ground sampled, the sensitivity of drilling tools to boulder presence and the ability to recover boulder samples. Very little guidance was found in the literature on selecting drilling and sampling methods that help improve cobble and boulder identification. To supplement this deficiency, a survey on boulder drilling and identification was sent to 15 subsurface exploration drillers and drilling managers in Wisconsin, Michigan and Illinois. Five responses were received. The results are summarized below:

- 4 of 5 said that a 205 mm (8-inch) hollow-stem auger is usually more sensitive to cobble and boulder presence than 75-100 mm (3-4 inch) rotary wash drilling.
- 4 of 5 said that smaller, lower torque drill rigs are generally more sensitive to cobbles and boulders than bigger, higher torque drill rigs.
- 4 of 5 said that larger 150-200 mm (6-8 inch) wash rotary drilling would be more suitable for identifying cobbles and boulders than similar size auger drilling.
- 5 of 5 said that larger diameter and stiffer rotary wash drill rods help increase sensitivity to cobbles and boulders by minimizing deflections and bending.
- 5 of 5 said that cobbles and boulders could still be detected (despite less sensitivity) when using a high torque rig with augers equipped with tungsten carbide fingers.
- 3 of 5 said that augers often wrap around some cobbles and boulders thereby decreasing sensitivity.
- 5 of 5 said that they could distinguish between gravel and cobbles based on the nature of "chatter" and degree of rig "jumping."
- 4 of 5 said that they could distinguish between cobbles and boulders based on the nature of "chatter" and degree of rig "jumping."
- 3 of 5 said that very high (refusal or near refusal) Standard Penetration Test blow counts are not a reliable indicator of cobbles and boulders unless drill rig chatter also suggests the presence of rocks larger than gravel.

- 4 of 5 said that drilling with a rock-coring barrel is not a better method than auger or rotary wash drilling for identifying cobbles and boulders.
- 3 of 5 said that boulder samples could be obtained by use of rock coring, but warned that if the cobble or boulder moves the bit may be destroyed or the core barrel may become stuck.
- 4 of 5 said that boulder thickness could be better determined by auger or rotary wash drilling than by rock coring.
- 3 of 5 said that documenting drilling rates (e.g. mm/min) was not practical and not a good indicator of relative drilling resistance.

Boring Spacing

An appropriate spacing between borings for boulder identification is dependent on many factors including nature and complexity of the site geology, availability of previous borings and subsurface experience, the borehole diameter, use of supplementary exploration methods (such as probe holes or seismic reflection/refraction), tunnel depth, anticipated tunneling methods and subsurface exploration budget. A National Research Council subcommittee on geotechnical site investigations found that an average boring spacing of approximately 80 m (260 feet) was completed for 84 non-mountainous tunnel projects studied (US National Committee on Tunneling Technology, 1984). To further reduce tunneling risks and provide better overall exploration results they recommended an average of 1.5 linear feet of borehole per route foot of tunnel alignment. This converts to a much closer average spacing of 20 m (66 feet) assuming average borehole depths of 30 m (100 feet) are appropriate. Terzaghi et al. (1996) recommend a minimum spacing of 30 to 60 m for subway [tunnel] projects. Essex (1993) recommends a minimum of one borehole per shaft and a maximum spacing from 46 m to 61 m (150 to 200 feet) for micro-tunnel and pipe jacking projects. Klein (1996) made essentially the same recommendation, but with the maximum spacing at 91 m (300 feet).

Horn and Ciancia (1989) provide a detailed evaluation of the effect of boring spacing variations from 3 m to 30 m (10 to 100 feet) on the number of boulders encountered within a well documented, 91 m (300 foot) long bouldery stretch of the Red Hook Intercepting Sewer in New York. Although they found that more borings hit boulders as the spacing decreased, the percentage of borings hitting boulders did not increase appreciably. An average spacing of 30.5 m (100 feet) was determined to be reasonable for that project. They also concluded that borings should be spaced closer in the vicinity of geologic features that may contain boulder concentrations.

Stoll (1976) found from a study of randomly distributed 305 mm (1 foot) diameter boulders projected onto a plane, that about five times as many 102 mm (4-inch) diameter holes as 914 mm (3-foot) diameter holes are required to achieve the same degree of certainty on boulder concentrations. This study demonstrates that borehole diameter should also be considered when determining the spacing of borings and probes.

Large Diameter Auger Drilled Shafts

Large diameter, 760 mm to 1.22 m (2.5 to 4 feet), auger drilled shafts appear to be a very effective method of boulder exploration. The unit volume explored for a 914 mm (3 feet) diameter drilled shaft is over 80 times greater than for a 102 mm (4 inch) diameter borehole. Cobbles and boulders up to approximately 305 mm (1 foot) can be removed and examined using a 914 mm (3 feet) diameter single helix or bucket auger such as commonly used for drilled shaft construction (Stoll, 1976). In addition, temporary casings can be set to allow ground conditions to be examined and sampled at the shaft base or within windows cut within the casing (Brierley et al., 1991; Smirnoff and Lundin 1985). Selected large diameter holes could be left accessible through a bidding period to allow contractors and their geotechnical advisors to directly examine ground conditions. Alternatively, the holes can be converted to wells and utilized for full scale pump tests. Multiple uses should help improve cost-effectiveness of large diameter drilled shafts in an exploration program.

Stoll (1976) demonstrated the value of large diameter holes. On one southeastern Michigan project, 17 convention boreholes failed to identify one boulder. However, fifty percent of the drilling logs from 150 dewatering wells having a 914 mm (3 foot) diameter indicated boulder presence. Stoll found from a study of randomly distributed 315 mm (1 foot) diameter boulders projected onto a planar surface, that when a "boulder concentration" ranges from 1 to 16 percent, 914 mm (3 foot) diameter holes encounter boulders from three to five times more often than do 102 mm (4-inch) diameter holes.

Hammer and Rotosonic (Percussive) Drilling

References indicate that percussive hammer and rotosonic drilling methods are effective at penetrating through bouldery soil at faster and more economical rates than by conventional drilling methods. Brierley et al. (1991) describe exploration in bouldery ground on a Tempe, Arizona project using a hammer drill. The method involved driving of a double-walled casing with a diesel pile hammer that cuts and fractures the soil encountered. The material encountered was removed by air forced down the casing annulus and up the inner casing. The resulting disturbed sample material and measured penetration rates were analyzed by a geologist to estimate the nature of cobbles, boulders and matrix soil encountered. Hammer drilling allows boulder presence and composition to be determined, but may not provide adequately reliable information on boulder size and frequency.

Rotosonic drilling involves advancement of a casing with a drill head having tungsten carbide embeds. The drill head is vibrated at a high frequency and progressively rotated downward. As the casing cuts through the ground, an inner core barrel advances around the cut material to obtain continuous core samples from 75 to 228 mm (3 to 9-inches) in diameter and 0.3 to 6.1 m (1 to 20 feet) in length (Davis and Oothoudt, 1997). Case histories show that rotosonic drilling can obtain core samples of cobbles and boulders at relatively high production rates. Boulder composition and size (chord length cored) can be determined. In addition to its ability to core boulders, rotosonic drilling is well suited for piezometer installation and environmental sampling.

Exposed Cuts, Test Pits, Shafts and Pilot Tunnels

Available boulder information from large excavations in the vicinity of a site should be gathered and reviewed to help plan future subsurface explorations and to provide data for boulder baselining. Relatively large excavations (e.g. shafts, test pits, cofferdams, pilot tunnels and roadway cuts) or eroded bluffs that expose geologic units of interest provide a good opportunity to determine cobble and boulder characteristics within relatively large volumes of soil. If available, case histories of previous cofferdam, shaft and tunnel excavations that documented cobble and boulder occurrence within the same or a similar geologic units are also a valuable source of information.

Geophysical Methods

Geophysical methods may in some cases successfully complement conventional drilling and sampling, but are generally ineffective for identifying and quantifying boulders. Geophysical methods including active seismics (surface and borehole cross-hole reflection/refraction), ground penetrating radar, magnetometer surveys, resistivity and others have been performed along tunnel alignments with limited success. Past discussions on use of geophysical methods in tunneling are presented in Alsup (1974) and Schmidt (1974). De Pasquale and Pinelli (1998), Hindle (1995), and Miller (1996) present more recent discussions of geophysical use on tunnel and directional drilling projects.

In general, geophysical methods have the highest resolution at shallow depths (less than 3 to 10 m) and where significant contrast in properties exists between boulders and the soil matrix. In addition, signal penetration is limited to shallow depths for some methods. For example, ground penetrating radar penetration is generally limited to 2 to 3 m depths in clay and saturated granular strata. Most tunnels are deeper than 3 m and are overlain by some clay or saturated strata. Furthermore, boulders at the tunnel zone may be embedded in dense soils that provide poor contrast in reflective properties.

Surface application of seismic reflection/refraction is perhaps the most useful of the geophysical methods, particularly when the tunnel zone is within 10 m of bedrock. Although boulders are unlikely to be identified, this method can be effective at identifying variations between boreholes in the level of bedrock and possibly of a dense bouldery till surface. Knowing the proximity of the tunnel zone to bedrock is an important factor in baselining boulders because bouldery ground is often more likely within approximately 3 m of bedrock.

Monitoring, Documenting and Reporting of Drilling Observations and Data

Monitoring, documenting and reporting of drilling observations and data are critically important aspects of identifying cobble and boulder presence, (Brierley, 1996;

Gould, 1995; Legget, 1979; Neyer, 1985; Osterberg, 1989). Despite its importance, many subsurface exploration programs for tunneling projects are performed without field monitoring by an experienced geologist or geotechnical engineer. Drillers should not be relied upon to document their observations of cobble and boulder presence. They are busy operating the equipment and trying to be profitably efficient at advancing the borehole. They are generally reluctant to spend extra time carefully describing depths and indications of cobbles and boulders. However, most drillers are willing to orally report their cobble and boulder observations and assessments of relative drilling resistance. With good cooperation between the driller and a monitoring professional, more accurate and thorough information can be recorded. In addition, a monitoring professional should be able to make timely decisions on drilling program changes such as sample type and interval in response to the conditions encountered.

Boring Logs

Final boring logs included in geotechnical reports often fail to adequately report information on drilling methods and observations such as chatter, bouncing, relative drilling resistance and estimated boulder and boulder zone thickness (Gould, 1995). Even if drilling information and observations are fully documented on field logs, its value is diminished unless it is transferred to the final logs. Field logs are seldom conveniently available to bidders and their consultants. Therefore, the final boring logs should contain detailed information on drilling behavior and boulder observations (Neyer, 1985).

Limited Milwaukee-Chicago Area Boulder Experience in Tunnels

Records from five tunneling projects in the Milwaukee and Chicago metropolitan areas were examined to assess boulder quantities encountered and boulder indications on boring logs. Pertinent data from these projects is summarized in Table 1. These tunnels were mostly mined within Wisconsinan age Oak Creek/Wadsworth till and associated outwash and ice-margin soils that are more than 3 m (10 feet) above bedrock.

Without other guidance, the boring log boulder data for these projects seems to suggest that only a small number of boulders might be expected during tunneling. The percentage of borings hitting boulders ranged from 0 to 35 percent. The percentage of total length of boulders drilled as compared to lengths bored within potentially bouldery till, outwash or ice-margin soils varied from 0 to 3.5 percent.

The reported number of boulders encountered during tunneling on the project segments studied varied from as few as 5 to as many as 282. Boulder sizes generally ranged from 0.3 to 1.5 m (1 to 5 feet) diameter. Shapes mostly ranged from spherical to rounded corner cubes. Boulder compositions were not well documented, but are estimated to have been approximately two-thirds erratics (non-native igneous boulders) and one-third native dolomite boulders. The average number of boulders ranged from 0.2 to 3.4 and averaged 1.9 per 30 m (100 feet) of tunnel. The percentage of estimated boulder volume relative to total excavated volume ranged from 0.01 percent to 0.72 percent (boulder volume was estimated as $0.7 D^3$ which is approximately the average

Table 1 – Limited Milwaukee and Chicago Area Boulder Experience

Case No.	1	2	3	4	5
Case Name	Milwaukee, Interplant Solids Pipeline	Milwaukee, South Pennsylvania Ave. MIS	Milwaukee, Ramsey Avenue Relief Sewer	Milwaukee, CT-7 Collector, Bruce St.	Elgin IL, Northeast Interceptor
Tunnel Length, m (ft)	2295 (7529)	2498 (8195)	941 (3088)	656 (2151)	1313 (4308)
Excavated Diameter, m (ft)	2.29 (7.5)	1.40 (4.6)	1.00 (3.3)	3.54 (11.6)	1.98 (6.5)
No. of Borings	16	44	8	8	17
Avg. Boring Spacing, m (ft)	143.3 (470)	131.4 (431)	125.9 (413)	82.0 (269)	77.1 (253)
% Boulder Length Drilled in Borings	2.9%	3.5%	0.4%	0.0%	2.4%
No. of Boulders Hit By Borings	7	25	1	0	8
% Borings Hitting Boulders	25%	25%	12.5%	0%	35%
No. of Boulders In Tunnel	232	282	7	5	112
Avg. Boulders per 30m (100 ft)	3.1	3.4	0.3	0.2	2.6
Avg. Till Boulders per 30m (100 ft)	7.1	9.0	0.3	0.2	2.6
Max. Boulders per 30m (100 ft)	49.0	27.1	7.0	1.0	32.3
Estimated Boulder Volume, m^3 (yd^3)	44.3 (58.0)	27.4 (35.9)	0.6 (0.8)	0.6 (0.8)	19.3 (25.2)
Avg. % Boulders By Volume	0.47%	0.72%	0.08%	0.01%	0.35%
Avg. % Boulders By Volume Mined in Till/Outwash/ Ice-Margin	1.08%	1.82%	0.08%	0.01%	0.35%
Max % Boulders By Volume, 60m (200') Segment +/-	10.52%	5.11%	1.21%	0.03%	1.12%

volume of a sphere with diameter D and a cube with width D). At four of the projects (Cases 1, 2, 3 and 5) 10 to 60 m (30 to 200 foot) long tunnel segments encountered boulder concentrations (nests and lag zones) that accounted for much of the total boulder quantities. Maximum boulder volume concentrations by volume within 60 m (200 foot) long segments ranged from 0.03 to 10.52 percent.

The limited number of Milwaukee-Chicago area tunneling projects evaluated provide some useful information for baselining boulders, particularly in the study area. A comparison of boulder volume mined to boulder length drilled in borings is shown in Figure 1. The normalized percent boulder length drilled in borings (x-axis) was computed as the percent boulder length drilled in borings times the average boring spacing in feet divided by 30.5 m (100 feet). The percentages of both the boulder length drilled in borings and volumes of boulders to volumes of tunnel excavated are limited to the potentially bouldery soils encountered, i.e. till, outwash and ice-margin units. Although based on only five cases, Figure 1 shows that the data points fall reasonably close to a linear trend line. This relationship could be used as a guideline for estimating the boulder volume likely to be encountered when tunneling through potentially bouldery glacial soil.

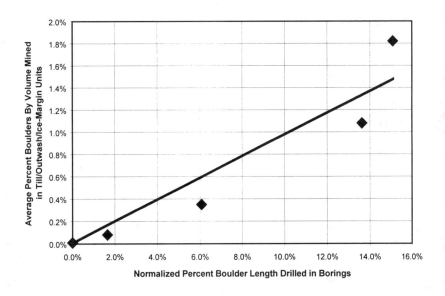

Figure 1. Boulder Volume from Normalized Length Drilled in Borings

Establishing A Boulder Baseline

Guidelines for baselining quantities to be presented in a geotechnical baseline report and contract documents are thoroughly discussed in an ASCE publication entitled *Geotechnical Reports for Underground Construction,* (Technical Committee on Geotechnical Reports of the Underground Technology Research Center, 1997). Much thought should be given on what boulder sizes are important, how quantities will be measured, how payments will be made and upon the conservatism used for the quantities listed. For example, on micro-tunnel and small tunnel projects all boulder sizes are likely to be important. On larger tunnels and those with more face access, only large boulders may need to be baselined.

Use of Conventional Borings and Empirical Relationships

Percent boulder volume and a boulder quantity may be estimated using boulder fraction data interpreted from conventional soil borings and probes. Boulder fraction data (boulder length divided by length drilled in a unit) from borings should be geologically evaluated to assess boulder distribution for specific units. For example, one data set may be for a unit with randomly distributed boulders and another data set for concentrated boulders within a lag zone unit. Distinctions might also be made for different segments of alignment, particularly if most of the borings that encountered boulders are located in a segment.

After estimating average boulder fractions for the units and segments selected, percent boulder volume may be estimated using a relationship such as that shown in Figure 1. Boulder volumes may be estimated for estimated soil volumes to be excavated/mined within each of the soil types and segments involved.

Even if a relationship between boulder fraction in borings to percent boulder volume is not locally available, boulder data from excavation projects within local geologically similar soil units may be used as limits for baselining boulders. Table 1 shows that the minimum average boulder volume within the till evaluated could be assumed as approximately 0.05 percent. The maximum boulder volume averaged over a 60-m (200-foot) segment within this unit could be assumed as approximately 11 percent. Additional data for common soil units or concentration zones could be collected from adequately documented excavations in a region. For example, observations made from a 12.2 m (40 foot) square shaft that was excavated into rock in southern Milwaukee (Hunt and Fradkin, 1991) indicate that bouldery till within approximately 2 m of bedrock had an average cobble and boulder concentration of approximately 70 percent by volume at that site.

Probabilistic Methods for Analysis of Boring Data

Probabilistic approaches may also be used for baselining boulder quantities where sufficient high quality data exists. Given that conventional borings at a typical spacing actually sample only a very small portion of the soil volume to be excavated and due to inherent geologic variability, data uncertainties will exist for the frequency, distribution, size, shape, composition and other characteristics of boulders in a soil stratum. Probability tools may be useful in analyzing uncertainties in a systematic manner to complement more traditional geotechnical evaluation methods rather than to replace them.

Stoll (1976) evaluated the sensitivity of conventional and larger diameter boreholes in indicating boulder presence. Stoll assumed a random concentration of 315 mm (1 foot) diameter boulders projected onto a plane. The diameter of the boring and the concentration of boulders are the only variables that controlled the likelihood of detecting boulders. A chart was presented to determine the number of borings required for a 95% certainty of encountering at least one boulder for different boulder concentrations. The chart is of limited practical use since it is related to the non-encounter of only one boulder size. Although the paper presents a valid approach for statistical analyses, it appears to be somewhat limited for practical application. An improvement over this method would be to incorporate a variable boulder concentration (dependent on the site geology) and to incorporate boulder size variability.

Tang and Quek (1986) developed statistical models of boulder size and boulder fraction. The first model uses intercepted boulder chord length data to estimate the distribution of boulder diameters for the entire boulder population. Bayesian analysis is used to update the mean boulder size. This method is very thorough from a mathematical perspective but may not be practical due to the difficulty in measuring chord lengths during drilling and where sufficient data is lacking. In addition, the method assumes that all the boulders are spheres.

The second probabilistic model presented by Tang and Quek (1986) relates the fraction of boulder volume encountered at boring locations within a bouldery stratum to boulder volume within the stratum. This parameter may be easier to use in practice since the required data may be directly obtained from boring logs. A key assumption with the method is that the boulders are randomly distributed. Based on two Singapore sites studied, Tang and Quek found that a lognormal cumulative distribution function correlated well with the data. They hypothesized that Bayesian updating methods as described in Ang and Tang, 1975 could be used to incorporate data from additional sites. If the parameters of the distribution function can be updated or verified for a site, then boulder fraction data may be used to estimate boulder volume within a selected volume of the same soil mass.

An effort was made to evaluate the data from the five Milwaukee/Chicago area cases that were previously discussed. A simple method based on the percentage of borings that encountered boulders proved to be ineffective. The data in Table 1 shows that the percentage of borings encountering boulders is not directly related to either the number of boulders or the boulder volume encountered. Such a simple method fails to account for borehole diameter, boring spacing, boring depth, occurrence within specific geologic units and the excavated tunnel volume.

Attempts were also made to develop cumulative distribution functions for correlating boulder fraction encountered in borings to boulder volumes encountered during tunneling at the five Milwaukee/Chicago area sites. Insufficient data was available to establish useable relationships. Our results suggest that this probabilistic approach may not be applicable for typical tunnel projects unless a database correlation between boulder fraction in borings and percentage boulder volume within the soil mass can be developed for soil units within a region. Even if an applicable cumulative distribution function can be established for a soil unit, this method would only help provide baseline estimates for the more randomly distributed boulders. Boulder estimates for concentrations and anomalies would have to be independently assessed.

Conclusions

The primary conclusions of this paper are as follows:

1. Boulder quantities may be baselined by combining local geologic knowledge and boulder experience with the results of a phased subsurface investigation program.
2. Boulder baselining should start with a desk study of available information on the geologic setting of the site and of local boulder experience within similar geologic units. This information should be utilized to plan an initial subsurface exploration program. Subsequent exploration phases should reduce geologic uncertainties by more intensively exploring the ground, reducing data gaps and providing redundancy.
3. Conventional subsurface exploration should provide useful boulder data if drillers are instructed to carefully report indications of boulders and if a properly qualified geotechnical professional is assigned to monitor drilling in the field and to document all pertinent observations. Final boring logs should include field observations.
4. Redundant or supplementary exploration methods should be considered for one or more phases of exploration. These methods might include use of auger borings where rotary wash borings were previously used. Larger diameter borings, more sensitive (less powerful) drill rigs, percussive methods such as rotosonic drilling, test pits, 0.9 to 1.2 m diameter auger drilled shafts and geophysical methods such as seismic refraction/reflection may also provide useful supplementary or redundant information.
5. The normal criteria recommended in the literature for determination of boring spacing on tunnel projects should be suitable for boulder detection. An average borehole spacing ranging from 30 to 90 m (100 to 300 feet) is typical for soft-ground tunnel projects. For randomly distributed boulders, the average percentage of borings hitting boulders is generally not sensitive to borehole spacing within this range. Closer

borehole spacing (e.g. 30 m or less) guided by geologic interpretation may be selectively utilized within specific areas to help minimize the risk of missing large boulder nests or lag deposits.

6. Normalized percent length of boulders intercepted in borings within potential bouldery soil appears to have a linear relationship to boulder volume as a percentage of excavated volume. Where local experience is lacking, the relationship shown in Figure 1 could be utilized to estimate an order of magnitude for boulder volume. Local experience should be accumulated to verify or modify this relationship.

7. A local experience database should be developed to document ranges in boulder quantities, sizes, and composition for common local soil units and for geologic features such as nests within filled gullies or ice-margin deposits, lag zones between geologic units of different age, and soil-rock interface zones. This database could be used as a guide when evaluating subsurface exploration results. Local boulder experience on projects should be published to help a local database to grow.

8. Probabilistic methods may be useful where there is sufficient understanding of the site geology and where a large enough population of high quality data exists to perform a Bayesian analysis and to establish a probability density function. Where sufficient quantity and quality of data is not available, probabilistic methods are not likely to be useful.

References

Abbott, D.G. (1995). "Another Perspective on "Teamwork ...." *Trenchless Technology.* July 1995, 6-7.

Ang, A. H.-S., and Tang, W. H. (1975). *Probability Concepts in Engineering Planning and Design, Vol. 1 - Basic Principles*, John Wiley & Sons, New York, NY.

Alsup S. (1974). "Recommended Borehole Investigation for Soft Ground" *Subsurface Exploration for Underground Excavation and Heavy Construction.* American Society of Civil Engineers, New York, 117-127.

Brierley, G.S. (1996). "Microtunneling! Schmicrotunneling!" *Trenchless Technology.* September 1996, 8.

Brierley G.S., Howard, A.L. and Romley R.E. (1991). "Subsurface Exploration Utilizing Large Diameter Borings for the Price Road Drain Tunnel." *Proceedings, 1991 Rapid Excavation and Tunneling Conference.* Society for Mining Metallurgy and Exploration. Littleton CO. Chapter 1, 3-15.

Cording, E.J., Brierley, G.S., Mahar J.W., and Boscardin M.D. (1989). "Controlling Ground Movements During Tunneling." *The Art and Science of Geotechnical Engineering.* Editors: Cording, E.J., Hall W.J., Haltiwanger J.D., Hendron, A.J. Jr., and Mesri G. Prentice Hall. New Jersey, 478-482.

Davis R. and Oothoudt T. (1997). "The Use of Rotosonic Drilling in Environmental Investigations," *Soil and Groundwater Cleanup,* May 1997, 34-36.

De Pasquale, G. and Pinelli G. (1998). "No-Dig Application Planning Using Dedicated Radar Techniques." *No-Dig International*, Mining Journal LTD, London, February 1998, I-12 – I-14.

Essex, R.J. (1993). "Subsurface Exploration Considerations for Microtunneling/Pipe Jacking Projects." *Proceedings of Trenchless Technology: An Advanced Technical Seminar.* Trenchless Technology Center, Louisiana Tech University, Ruston, LA. 276-287.

Flint, R.F. (1971). *Glacial and Quaternary Geology.* Wiley. New York. 147-197.

Gould, J.P. (1995). "Geotechnology in Dispute Resolution." *Journal of Geotechnical Engineering.* ASCE. New York. Vol. 121, No. 7. July 1995, 523-534.

Heuer. R.E. (1978). "Site Characterization for Underground Design and Construction." *Site Characterization & Exploration.* ASCE. New York, 39-55.

Hindle D.J. (1995). "Geotechnical Appraisal." *World Tunneling.* London. Nov. 1995, 371-373.

Horn H. M. and Ciancia A.J. (1989). "Geotechnical Problems Posed by the Red Hook Tunnel." *The Art and Science of Geotechnical Engineering.* Editors: Cording, E.J., Hall W.J., Haltiwanger J.D., Hendron, A.J. Jr., and Mesri G. Prentice Hall. New Jersey, 367-385.

Hunt, S.W. and Fradkin, S.B. (1991). "Costly Environmental and Geotechnical DSC Claims Resulting from Exploration Program and Reporting Inadequacies." *Proceedings, 34th Annual Meeting of Association of Engineering Geologists.* Association of Engineering Geologists, Greensburg, PA, 127-136.

Klein S.J., Nagle G.S., Raines, G,L. (1996). "Important Geotechnical Considerations in Microtunneling." *No-Dig Engineering.* Vol. 3. No. 4. July/August (1996), 9-12.

Legget R.F. (1979). "Geology and Geotechnical Engineering." *Journal of the Geotechnical Engineering Division.* ASCE. New York. Vol. 105, No. GT3. March 1979, 342-391.

Miller R.J. (1996). "Hazard Recognition in Trenchless Technology." *No-Dig Engineering.* Vol.3. No. 6. November/December 1996, 13-15.

Neyer, J.C. (1985). "Geotechnical Investigation For Tunnels in Glacial Soils." *Proceedings, 1985 Rapid Excavation and Tunneling Conference.* Society for Mining Metallurgy and Exploration. Littleton CO. Chapter 1, 3-15.

Osterberg, J.O. (1978). "Failures in Exploration Programs." *Site Characterization & Exploration.* ASCE. New York, 3-9.

Osterberg, J.O. (1989). "Necessary Redundancy in Geotechnical Engineering." *Journal of Geotechnical Engineering.* ASCE. New York. Vol. 115, No. 11. November 1989, 1513-1531.

Schmidt, B. (1974). "Exploration for Soft Ground Tunnels - A New Approach." *Subsurface Exploration for Underground Excavation and Heavy Construction.* American Society of Civil Engineers, New York, 84-96.

Smirnoff T.P. and Lundin T.K. (1985). " Design of Initial and Final Support of Pressure Tunnels in the Phoenix "SGC". *Proceedings, 1985 Rapid Excavation and Tunneling Conference.* . Society for Mining Metallurgy and Exploration. Littleton CO. Chapter 26, 428-438.

Stoll, U.W. (1976). "Probability That A Soil Boring Will Encounter Boulders." *Conference on Better Contracting for Underground Construction.* Michigan Section of American Society of Civil Engineers. Detroit. 34-48.

Tang, W. and Quek S.T., (1986). "Statistical Model of Boulder Size and Fraction." *Journal of Geotechnical Engineering.* ASCE. New York. Vol. 112, No. 1. January 1986, 79-90.

Tarkoy, P.J., (1992). "The Achilles Heel of Trenchless Technology: An Editorial Comment." *Trenchless Technology.* September/October 1992, 25, 41.

Tarkoy P.J., (1994) "Case Histories in Trenchless Excavation." *No-Dig Engineering.* Vol. 1. No. 1, 17-21.

Technical Committee on Geotechnical Reports of the Underground Technology Research Center. (1997). *Geotechnical Baseline Reports for Underground Construction.* Randall J. Essex, Editor. ASCE. New York.

Terzaghi, K., Peck, R.B. and Mesri, G., (1996). "Article 12 - Program for Subsurface Exploration," *Soil Mechanics in Engineering Practice,* Wiley, New York, 3rd Ed., 55-58.

U.S. National Committee on Tunneling Technology, (1984). Geotechnical *Site Investigations for Underground Projects.* National Academy Press. Washington D.C. Vol. 1

Wallis, S. (1996). "Shepard Subway." North American Tunneling (World Tunneling). London. Vol. 9. No. 7, N17-N23.

Tomography to Evaluate Site Conditions during Tunneling

David M. Neil[1], Khamis Y. Haramy[2], David H. Hanson[3], and Jozef M. Descour[4]

Abstract

A near-real-time three-dimensional tomographic imaging system called Rock3D^{TM} is used extensively in the hard rock and coal mining industries to characterize the rock mass structural setting and to identify changing stress conditions caused by mining. Rock3D^{TM} has been used successfully to map and characterize contact zones, large joints or structures, shear zones, faults, karst cavities, and to determine the changing three-dimensional stress regime around an opening and within the rock mass. NSA Engineering, the developers of this technology, has adapted this mining tool to civil tunneling applications to image actual geologic site conditions along a proposed tunnel or alignment prior to and during tunnel construction. These data can be used on a day-to-day basis to look ahead of the TBM, or to keep permanent records after completion.

Introduction

The demand for underground tunnel and shaft structures for the civil and mining industries has risen sharply in the last decade and will likely double in total excavated volume in the next 10 years. Tunnels and shafts are becoming larger in diameter, lengthier, and chosen over open cut for underground structures. This is especially true in civil infrastructure projects that are undertaken in metropolitan areas. This demand has spurred an unprecedented growth in technologies that identify the myriad of challenges presented by changing ground conditions. One of the major challenges is the need by all parties of a tunneling venture to know the detailed nature of the ground environment that will affect the driveage, support requirements, and long-term maintenance of the finished structure.

---

[1] President & CEO
[2] Sr. Project Manager-Rock Mechanics
[3] Manager, Core Technologies
[4] Sr. Geophysicist
NSA Engineering, Inc.
15000 West Sixth Avenue, Suite 100
Denver, Colorado 80401

Geological and geotechnical engineers have struggled for years with the difficult task of developing a comprehensive interpretation of geologic conditions influencing the selection of a tunnel alignment or construction method from an often-limited empirical database. A number of geophysical methods have been applied in an effort to expand the knowledge of subsurface conditions in advance of tunneling operations. Seismic refraction and reflection and ground-penetrating radar are the most commonly used. One alternative to time-consuming and expensive drilling programs or to shutting down operations for standard geophysical investigations to characterize excavation alignments is the use of Rock3D™ tomographic imaging. Although tomographic techniques are not new, expanded capabilities and near-real time imaging have recently been adapted to applications in geologic/geotechnical environments.

NSA Engineering, Inc., has developed an advanced geophysical mapping technology that now provides a near-real time, multidimensional tomographic image. The software system is called Rock3D™. When combined with one of several suites of data acquisition and storage hardware, an extremely versatile package is available that is capable of processing both refraction and reflection data. The signal sources may range from artificial seismo-acoustic data produced by drilling or tunneling machines to ground-probing radar signals. Modes of operation may be velocity based (S-wave or P-wave) or attenuation based. Processing and evaluation of the data sets can be fully accomplished in a matter of hours. The Rock3D™ interactive companion graphics software provides an operator-designed display map for the tomographic image within the coordinate system used for the data survey.

Rock3D™—A Tomographic Imaging System

Advances in computer technologies and data processing have allowed geophysical techniques to become more popular in the geotechnical field. The Rock3D™ software is the heart of a complete system capable of imaging the interior structure of a rock/ground mass several hundred feet ahead of a TBM. The principle behind Rock3D™ is that seismic energy travels through different material types with different attenuation and velocity levels. Seismic waves will travel faster and with less attenuation through competent or highly stressed rock than through broken rock or voids (Nur, 1987, Shea-Albin, et al., 1991, Yu, 1992).

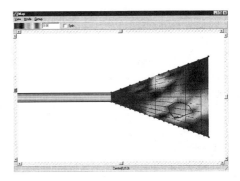

Figure 1. A schematic representation of a tomographic cross-hole survey performed in front of a tunnel face. Two boreholes were used for seismic sources and receivers required to construct the tomographic image.

Geologic anomalies such as voids, structural anomalies such as faults, or composition anomalies such as hard inclusions can all be imaged when surveyed using the proper techniques and instrumentation.

Tomography is a generic term that technically means to "draw a slice or section" of a target material. This is accomplished, in general terms, by passing a signal from a source through a target and on to a receiving sensor. The signal captured by the sensor is processed using one or more computerized iterative algorithms. The output of these processes can be used to produce an image that represents the interior of the target material by virtue of the influence of that material on the passage of the original signal. Much of the pioneering work for this technique occurred in the medical field, where CAT (Computer Assisted Tomography) scans have become commonplace. Activity in adapting tomographic techniques to the more restrictive geometries encountered in geophysical applications has accelerated since Devaney (1984) introduced geophysical diffraction tomography (Figure 1).

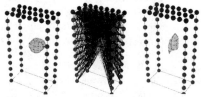

Figure 2. Forward modeling is used to evaluate potential effectiveness of various source/receiver arrays before field installation. Here an embedded boulder is modeled. The resulting inverse model can be used to evaluate the ray coverage and resolution of the reconstructed image.

Data sets are collected as full waveform signals that are captured with a sampling rate and duration appropriate for the signal source, earth materials, and physical scale of the study. Source frequencies must be considered in this process. If the source is artificial, some discretion is often available to the operator in source selection. If the source is natural, or not subject to selection, accommodations must be made in the receiver array design and the data sampling, but proper array design is not so straightforward. NSA Engineering has developed and uses forward modeling software, which predicts the performance of a particular source/receiver configuration and permits comparison of the projected layout with the results of a reconstructed model of that same layout. Where the reconstructed image matches well with the image created from the forward model, the quality of the results can be predicted. This approach allows development of a good preliminary concept of the optimum field installation for the particular goal of the tomographic survey (Figure 2).

Once the sensor array geometry has been established, raw data are collected by the data acquisition system on an operator-established time interval for a continuous source, or on a trigger device basis for non-continuous sources. The data collection may be done on any appropriate data acquisition hardware, but the stored file format may require use of a custom file read program for the Rock3D™ system. The raw data waveforms are read into the Rock3D™ program, where an additional file containing a partially processed version of the data is created. From these two files, lists of processing options are exercised, depending upon the data type and intended use of the output images.

The basic properties of the data used for creating tomographic images are the velocity of the wave through the medium and the relative amplitude attenuation of the signals caused by the medium. For either case, a simultaneous iterative reconstruction process (SIRT) is used to converge the solution set to a range of values that is within operator-specified limits. In this process, an initially assumed value is iteratively improved until an acceptably small change is observed in the tomogram from one iteration to the next. Typically, three or four cycles of this process provide sufficient resolution and accuracy for most geophysical purposes.

Options available with Rock3D^{TM} for observing data character and analyzing details include graphical presentations of the signal waveform, manual first-pick adjustments, first-pick alignments, point plots of the first arrival picks vs. distance, comparison of distances to signal amplitudes, signal frequency distributions, etc. Interactive spreadsheet displays are included for ray parameters, automated source movements, data set delineation, CAD display design, site and tomogram geometry parameters, etc. Generally, provisions are made for manipulation of some data characteristics so that the user can accommodate varying sensitivity between sensors, multiple arrivals at a sensor, noise filtering, and other such impediments to resolving data sets into useful tomograms.

Recent Rock3D^{TM} Field Applications

Tomographic images are presented in a format that includes the actual mine plan and prior identified structural features. The output can also be combined with other available rock mechanics data to better understand any trends that may cause mining difficulties. Such capabilities within the system software provide proactive information that can potentially increase operational efficiency and improve mining safety.

The most important criterion for success of any tomography system is that the targeted anomalies must be of some minimum level of contrast when compared to the surrounding rock mass or soil. This contrast can be a difference in the physical properties or stress condition of the media through which the source signal travels. Some obvious conditions that result in this contrast include hard inclusions in soil or void spaces in rock material, or stress focal points that may develop as a result of structure geometry. Tunneling operations are of particular interest in application of this technology since all the above conditions may exist within a single proposed tunnel alignment.

NSA Engineering, Inc., has completed numerous projects using offset vertical seismic profiling (VSP) and cross-borehole tomography surveys. Three recent case studies are presented here including the tomographic images from which the analyses

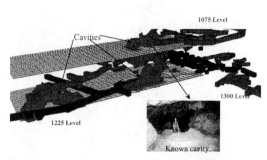

Figure 3. Potential voids as determined from tomographic inversion are shown by the 3-D purple contours. Also shown here is a photograph taken of one of the voids delineated by seismic tomography.

Case Study #1: At a gold mine in the United States, calcite-lined cavities varying in size and shape were encountered, impacting miners' health and underground mine safety during development and production. Without knowledge of the location of these voids, consistent structural integrity could not be dependably established during development in the vicinity. Production above a void could cause a floor failure that could claim both equipment and personnel when the strength of the rock above the void is exceeded. In addition, the gasses within the voids could pose an immediate health risk if directly encountered, or a long-term risk if ventilation air should short circuit through the karstic voids and push the gasses into the fresh-air supply. Rock3D™ was used to evaluate the viability of locating the extent of known cavities and to map other unknown cavities ahead of mining.

Tomographic imaging of a large underground area, approximately 300 m in the N-S direction and 120 m in the E-W direction was conducted using both production blasting and a sledgehammer as the seismic sources. Velocity tomography was used to analyze the mine structure. A 3-D image of the mine was created, which included the locations of drifts and headings, geophone positions, and the locations of known voids. Figure 3 illustrates the mine layout, sensor location, and ray paths from one source used as part of the tomographic reconstruction. The cavities that were imaged tomographically agreed well with known cavities at the mine. The images also indicated areas of highly stressed rock surrounding the cavities. The project demonstrated that Rock3D™ could successfully image the location and extent of cavities in the mine with acceptable accuracy, as well as the presence of high stress zones.

Case Study #2: Another study was completed in an Australian copper mine to delineate stope boundaries, evaluate ore stock quantities, determine stress distribution, and identify structures within stope areas. The ore was extracted from 12 different stopes simultaneously; the stopes exceeded 100 m in height and ranged in volume from 70,000 to 500,000 tonnes. The completed stopes were generally backfilled with a mixture of mill tailings, mullock, and fly ash to form a fill material capable of standing unsupported over the full vertical stope height. The backfill material was generally less dense than the host rock and could be mapped using tomographic imaging. Because of ongoing expansion at the mine and with production expected to triple, a monitoring system capable of accurately detecting subtle changes in the behavior of the rock mass was deemed beneficial. Although NSA's tomographic imaging technology had never been applied specifically for this purpose, based on results, it was shown that differences in the travel time provides an indication of the internal structure or content of a rock mass. Figure 4 shows a tomographic slice through the monitored area. Structural anomalies or other variations of material content within a particular area were identified and located based on analysis of relative changes or differences that occurred in the physical properties of the area covered.

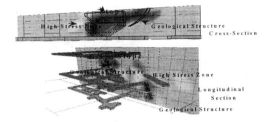

Figure 4. Tomographic imaging of an Australian copper mine. Areas of relatively higher stress are shown in red and lower competency material, such as rubble in a stope, is depicted in purple.

The Rock3D™ system was deployed in an array that had been determined through forward models. Tomographic images were produced that depicted evidence of the four parameters under study. Because this study was completed on a relatively large scale, definition of these features was limited by the geometry and distances over which the survey was completed.

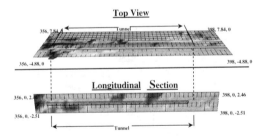

Figure 5. Two sections along the tunnel alignment. Reds are regions of higher stress or more competent material, and purples the lower stress or less competent rocks.

Case Study #3: Rock3D™ was successfully used in a micro-tunneling application, to perform a cross-borehole survey to assess the presence or absence of ancient stone walls and foundations in the alignment of a proposed 1-m civil tunnel in Ireland. The concern was that the micro-tunneling machine, boring at a depth of 4 to 5 meters below the street level, might encounter old stone walls and foundations that could stop the progress of the micro-tunnel machine and possibly require a costly recovery operation.

Two candidate sites were selected, and a total of 11 boreholes were drilled between 7 and 10 m deep; each borehole was cased with 100-mm-diameter perforated pipe. A buffalo gun was used as the seismic source, and hydrophones were used to receive the seismic signals. Figures 5 and 6 show a top view plane cut through the 3-D tomogram (5A), and a 2-D longitudinal section (5B), and a 3-D tomogram with selected low and elevated velocities that were developed from the cross hole data sets (6A and B). Subsequent driveage of the micro-tunnel confirmed the accuracy of the images. The elevated velocity contours refer to velocities of approximately 2,500 m/s.

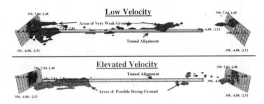

Figure 6. This view of the same tunnel as in Figure 5 shows the use of 3-D contours to delineate areas of stronger and weaker materials. This type of presentation allows a 3-D visualization of the materials in advance of the tunnel.

Application of Rock3D^{TM} for Tomographic Imaging Ahead of a TBM

Techniques that attempt to image ground conditions in advance of a tunnel driveage have been reported in the literature for some time. Several published and unpublished reports have shown that, in some cases, these techniques provide a good indication of discontinuities present ahead of the tunnel face (Sattel, et al., 1992, 1994, Hasegawa, et al., 1992, Inazaki, et al., 1996). Most of these techniques derive information from reflected seismic energy. Data are processed in a similar manner to reflection seismic surveys and vertical seismic profiles or VSP (Inazaki, et al., 1996). Thus, the positions of reflecting boundaries are determined by an average or interval velocity found from the curvature and dip of reflection events on seismic trace gathers. Standard processing includes wavefield separation, deconvolution, and migration. Sattel, et al., (1994) use a technique termed diffraction stack migration to obtain images. However, none of these techniques presents a true tomographic image of the velocity structure of the material. The following discussion presents methods to obtain such a tomographic image of the geologic boundaries and velocity structure present ahead of a TBM using the Rock3D^{TM} processing package.

The most straightforward technique involved in tomographic imaging ahead of a TBM requires the use of boreholes. Using drill holes placed along the tunnel alignment prior to excavation, a tomographic cross-borehole survey may be performed well in advance of the face (Figure 1). Often, the same holes that are drilled to obtain core to assess the geologic material physical properties and strata analyses may be used for the source/receiver arrays. This technique has been applied to many situations with reliable results (Ivansson, 1987). Unfortunately, the expense and inconvenience of drilling holes in front of an active TBM sometimes makes this technique impractical.

Velocity tomography is based on the concept that the arrival time of a direct wave from a source at a receiver is determined by the integral of the slowness of the material over the path traveled. Reflection tomography is based on a similar concept, except that the reflector location is constrained to lie on a three-dimensional surface defined by the source location, receiver location, two-way travel time, and seismic velocity. In a constant velocity medium, this surface is an ellipsoid. Deviation from a constant velocity will introduce varying degrees of distortion to the ellipsoid. Thus, the estimation of reflector location and seismic velocity may be formulated as a general inverse problem, similar to direct arrival velocity and attenuation tomography. This type of generalized inversion has been performed on surface reflection seismic data by Bishop, et al. (1985) and Bording, et al. (1987). One of the motivations for these works was the fact that the influence of lateral velocity changes cannot be adequately represented by the velocity field obtained from normal-moveout (NMO) analysis. Unfortunately, the inversion algorithms used were computationally expensive. With the increase in computing power now available and development of the proprietary Rock3D^{TM} algorithm, this problem may now be revisited.

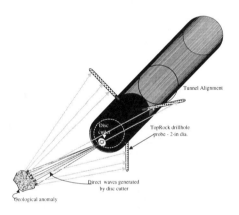

Figure 7 illustrates the configuration designed for use in tunnels being driven in hard-rock conditions. The approach considered uses a series of strong seismic signals generated by individual cutters during tunnel boring operations. The signals generated by the selected cutter or cutters are monitored over predetermined time windows for each cutter path, as shown in Figure 8. The cutter position represents a localized source at the tunnel face. This source signal is monitored separately to provide a reference for subsequent processing (Rector and Hardage, 1992, Rector and Marion, 1991, Descour and Ozdemir, 1991).

An array of seismic receivers detects waves reflected from structural/ geological anomalies in the surrounding rock mass, along with

Figure 7. Schematic illustrating the system envisioned for imaging in advance of a TBM using seismic signals reflected off geologic targets. The cutters would be used as the source, and forward-looking sensors arrayed around the tunnel circumference would be used to detect the seismic energy.

direct signals from the source. Receivers may be mounted on the grippers of the TBM, or, as shown in Figure 7, they may be inserted into three or more drill holes extending radially from the tunnel walls. The holes should be drilled as close to the face as possible. Exact location of the holes depends upon TBM design. These receivers are the multisensor TopRock™ probes, and consist of six or more unidirectional sensors mounted in a single unit. Since the receivers are unidirectional, by pointing the axis of the receivers parallel to the tunnel axis, reflections from the head of the tunnel face are preferentially detected. Seismic waves reflected by geologic anomalies ahead of the TBM carry the source signature (Figure 8). The Rock3D™ reflection tomography software is designed to identify reflected signals using correlations with the source signature and to generate 3-D images of the structural anomalies without stopping the tunneling operation.

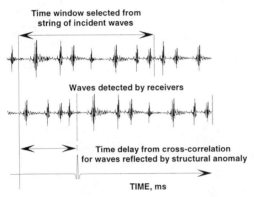

Figure 8. Using a correlation technique, the cutter signal may be used as the energy source in forward-looking reflection tomography. Selected time windows are recorded and processed to image the reflector structure ahead of the TBM.

Conceptually, the reflectors are considered to be a dense set of point diffractors. Thus, every receiver will, theoretically, receive a signal from each point on the reflecting horizon, because each point on the horizon can be thought of as a point diffractor that generates secondary waves. A migrated image of the geologic target is produced because of the constructive and destructive interference among signals detected at adjacent traces. The Rock3D™ software implements this algorithm that has been referred to as a holistic migration (Robinson, 1998) to determine locations for the reflectors. Given this location of the geologic target and the two-way travel time, a back propagation algorithm solves for the seismic velocity in the rock between the source, reflector, and receiver. The reflector position and velocity structure are iteratively adjusted to minimize residual travel times. Thus, the velocity structure between source-reflector-receiver and the position for each diffraction event on a seismic trace is determined for each geologic target, and a true tomographic image of conditions ahead of the face is constructed.

Summary

With the development of Rock3D™, a tool to provide mine operators and tunnel contractors with a rapid means of assessing ground conditions has become available. Numerous applications have shown the efficacy of this data acquisition and processing package in constructing tomographic images of velocity and attenuation properties that are related to geologic targets and changing stress conditions. It has been shown that accurate images of targets such as voids and geologic boundaries may be easily developed by analyzing the first arrival signals from seismic sources.

The restricted access imposed by tunneling operations requires that more of the seismic waveform be used to extract information on unknown ground conditions. Analysis of reflected and refracted energy provides at least partial compensation for inadequate spatial coverage of sources and receivers. We have presented a technique using the Rock3D™ tomographic imaging software whereby these secondary arrivals may be inverted for both the target location and the media velocity structure. This approach differs significantly from other methods in use in that a true tomographic image of the area in the immediate vicinity of the TBM is constructed.

References

Bishop, T. N., K.P. Bube, R. T. Cutler, R. T. Langan, P. L. Love, J. R. Resnick, R. T. Shuey, D. A. Spindler, and H. W. Wyld, 1985. Tomographic determination of velocity and depth in laterally varying media. Geophysics, v. 50, no. 6, pp. 903-923.

Bording, R. Phillip, A. Gersztenkorn, L. R. Lines, J. A. Scales, S. Treitel, 1987. Applications of seismic travel-time tomography. Geophys. J. R. Astr. Soc., v. 90, pp. 285-303.

Descour, J. M. and L. Ozdemir, 1991. Directed S-waves for remote sensing in front of a mechanical excavator, Proceedings International Symposium on Mine Mechanization and Automation, V. 1, pp. 2-57 to 2-66.

Devaney, A. J., 1984. Geophysical diffraction tomography, IEEE Trans. on Geoscience and Remote Sensing, v. GE-22, no. 1, pp. 3-13.

Hasegawa, M., M. Usui, and K. Gotoh, 1992. An example of prediction of geological structure in fracture zone, 29th IGC Abstracts, v. 3, p. 862.

Inazaki, T., T. Kurahashi, Research Group on HSP, 1996: Imaging and Characterizing Fractures Ahead of Tunnel Face Using In-Tunnel HSP Method, Proceedings of the Symposium on the Application of Geophysics to Engineering and Environmental Problems, Environmental and Engineering Geophysical Society, pp. 597-604.

Ivansson, S., 1987. Crosshole transmission tomography. In Seismic Tomography With Applications in Global Seismology and Exploration Geophysics, (G. Nolet ed.), D. Reidel Publishing Co., pp. 159-188.

Nur, A., 1987. Seismic rock properties for reservoir descriptions and monitoring. In Seismic Tomography With Applications in Global Seismology and Exploration Geophysics, (G. Nolet ed.), D. Reidel Publishing Co., pp. 203-237.

Rector, J. W. III, and R. A. Hardage, 1992. Radiation pattern and seismic waves generated by a working roller-cone drill bit. Geophysics, v. 57, no. 10, pp. 1319-1333.

Rector, J. W. III, and B. P. Marion, 1991. The use of drill-bit energy as a downhole seismic source. Geophysics, v. 56, no. 5, pp. 628-634.

Robinson, E. A., 1998. Holistic Migration. The Leading Edge, March 1998, pp. 313-320.

Sattel, G., B. K. Sander, F. Amberg, and T. Kashiwa, T., 1994. Tunnel Seismic Prediction, TSP® - some case histories, Amberg Measuring Technique Technical Article 18.20, 16 pp.

Sattel, G., P. Frey, P., and R. Amberg, R., 1992. Prediction ahead of the tunnel face by seismic methods – pilot project in Centovalli Tunnel, Locarno, Switzerland, First Break, v. 10, pp. 19-25.

Shea-Albin, V. R., D. R. Hanson, and R. E. Gerlick, 1991. Elastic wave velocity and attenuation as used to define phases of loading and failure in coal. USBM Report of Investigation 9355, 43 pp.

Yu, G., 1992. Elastic properties of coals. Ph.D. Thesis, MacQuarie Univ., Sydney, Australia, 133 pp.

Characterization of Uncertainty in Underground Structures

N. O. Nawari[1] and R. Liang[2]

Abstract

In the design of underground facilities, inherent uncertainties arise in the reduction of actual site geological-hydrogeological conditions to a representative analytical model, and in the determination of physical properties of the subsurface formations relevant to the site and the proposed underground structure.

Gravity, tectonic, weathering and erosion brought about by the environment are factors that contribute eventually to the instability of underground structures. Such factors are generally difficult to quantify with present probabilistic approaches. Moreover, the associated rules and data in our current modeling are viewed as non-fuzzy. But, our preliminary investigations with geological and engineering geological maps, boreholes, field tests, sampling, laboratory work, up to the analysis and design, are all connected with vague, ill-defined, undeterministic and to a great extent non-statistical data.

In this paper a new procedure for characterizing uncertainty in the design of underground structures will be presented using fuzzy-safety techniques. This will contribute to the reduction of the difficulties mentioned above in quantifying the imprecise geological and environmental data. The new technique includes basically two approaches: (i) fuzzy synthetic analysis and (ii) computation with uncertain parameters utilizing the concept of fuzzy variables and preference functions. The application of this method in the practice is illustrated by numerical examples.

---

[1] Assistant Professor, Department of Civil Engineering, University of Akron, Akron, OH 44325-3905
[2] Professor of Civil Engineering, Department of Civil Engineering., University of Akron, Akron, OH 44325-3905

Introduction:

Uncertainties may be associated with physical phenomena that are inherently variable or with prediction and estimation of real state of nature performed under conditions of incomplete or inadequate information. From this perspective, uncertainty is associated with the inherent variability of the physical process and/or with the imperfection in the modeling of the physical process.

In underground structures, site conditions vary spatially and with time. Moreover, field scale and long-term performance is not known for many underground systems. Before construction the geological and hydro-geological conditions are largely unknown. During construction, parameters, which affect excavation and support, are partially known. Also in case of known geological and hydro-geological conditions, the advances rate of excavation vary due to effects of labor and equipment performance, construction material properties and unforeseen construction events. Moreover, the impact of external factors, such as tectonic and climate changes is difficult to predict. Thus, in underground structures one has to deal with variety of uncertainties, probably more so than other fields of civil engineering.

A simple case study showing the uncertainties in the analysis and design of an underground structure can be seen in tunnel structure through water-bearing rock formations (Wittke,1990). During the construction stage in such conditions, the water table will be lowered and may extend to the level of the cross-section. However, depending upon the permeability of the rock mass and the draining action, the lowered water table may remain above the tunnel. The following uncertainties arise in conjunction with the water pressure: (a) water pressure does not have to be applied, (b) water pressure must be applied partially, (c) water pressure may develop in tunnel sections through areas of rock mass which were dry during tunnel driving, if water can flow along the tunnel via an annular gap or zone around the tunnel's circumference loosened by driving. Other uncertainties may include the degree of grouting to reduce permeability of rock mass, changes in head due to future tunnel leaks or construction of other future underground drains.

A good example illustrating the uncertainties and their consequences in underground structures is the failure of the Krieberg tunnel (Germany) (Duddeck, 1990). This tunnel was constructed for the German federal railway for a high-speed rail line. The Krieberg tunnel passes through quaternary deposits in the northern area consisting of loess, silt with sand lenses overlaying clay basin. The tunnel was carefully designed so that extensive failure is not possible. The intensive instrumentation and in situ measurement of the surface settlement and interior deformations were an integral part of the design model. As the excavation of the adit's vault advanced after finishing the sidewalls, failure of the vault in the least fractured zone (the sandstone area) started only a few days later. The vault collapsed continuously over 55 m. The back analysis indicated that the local geological conditions and the interpretation of the clay-water-sand lenses interaction and their influence on the stability of tunnel were not considered in the design model.

Uncertainties in Underground Structures:

The Uncertainty space in underground structures is significantly large. There are difficulties in obtaining precise information as to how the ground will perform and to anticipate everything that is going to take place in an underground project.

To characterize uncertainties in ground properties, we need to combine, in addition to the actual test data, knowledge about the quality of the data, and most importantly engineering judgement and experience. Then, an appropriate mathematical formulation of uncertainty capturing this information may be developed.

Sources of uncertainties in underground systems are numerous. However, we can summarize the basic elements of uncertainties as follows:
(a) Geological Formation: variations in ground type, continuity of formations, material properties of rock, soil, mixed ground, joints, fissures, folds, and faults. Difficulties in determining sampling distribution and size, accessibility to specific location, and sample disturbance.
(b) Difficulties in predicting the behavior of ground because of test limitations, for instance the interaction of ground material, external load, geometry, and construction procedure.
(c) Excitation: externally applied loads, tectonic forces, climate impacts such as rainfall, and temperature fluctuations. Except traffic loads, other excitations are difficult to specify.
(d) Performance over time: is the prediction model good enough to define the behavior of the structure during and after construction? Stress conditions, loosening and movement, squeeze or creep, volumetric stresses including swelling and slacking.

At present, most of the techniques employed for uncertainty characterization in underground problems are of a stochastic nature (e.g. Vanmarcke, 1977; Einstein/Baecher, 1982; Einstein, 1991; Einstein et.al 1996). Probability theory was viewed as the unique methodology to handle uncertainty. This has led to improper applications and overestimation of the capabilities of the probabilistic theory itself. There are many uncertainties, which have no place within the probability theoretical framework of modeling. Uncertainties involving description, linguistic terms, judgmental opinions, as well as those based on very scarce information have not been incorporated satisfactory in probability theory (Zadeh, 1965; Kosko, 1992). In this respect, uncertainty is a multidimensional concept. This multidimensional nature of uncertainty is obscured when uncertainty is conceived solely in terms of probability theory.

Theoretical Background

Fuzzy logic theory is attractive to geotechnical engineering for the same reasons that the probabilistic concept has been found lacking (Nawari/Hartman 1996, 1998). Subjective information, judgmental opinions, scant evidence, tentative facts, suspicious beliefs, ill defined boundary, and disjoint data can be modified and manipulated using fuzzy logic theory in a manner which reflects the degree of imprecision in the original information and data. Fuzzy variables behave as a family of elastic constraints having a well defined semantic meaning and possessing a natural ability to make all necessary perspective adjustments. Thus, they can be adapted to any variation of cognitive perspectives in geotechnical system.

The essential theoretical backbone for the fuzzy-uncertainty models includes fuzzy variables, fuzzy relations and the extension principle. Explanation and definitions of these concepts can be found, for instance, in Dubois/Prade, 1980 and Bandemer /Gottwald 1995.

Characterization Models

A. Fuzzy Variables:

The geological uncertainties mentioned in the pervious sections can be described using the concept of fuzzy variable. To illustrate this, consider the following examples:

(i)- In this example consider the hydraulic conductivity of a sandstone formation. The test results showed an average value of: $k = 10^{-2}$ m/s. This formation is normally described as permeable layer. Now, we can assign the fuzzy variable
A = "permeable": $\lambda_A(k) = \{(1/10^{-2}); (0.7/10^{-3}); (0.2/10^{-6}); (0.0/10^{-9})\}$
The assignment of the support values of the fuzzy variable defined above represents the engineer appraisal. It is specified here for the purpose of illustration.

Depending upon the structure and composition of the sandstone formation, different locations can be described by the appropriate value of the fuzzy variable (see table 1).

Table (1): Fuzzy variable: hydraulic conductivity k of sandstone formation

Symbol	Value	$k=10^{-6}$ m/s	$k=10^{-9}$ m/s	$k=10^{-3}$ m/s	$K=10^{-2}$ m/s
k_1	Permeable	0.2	0	0.7	1
$(k_1)^2$	Very permeable	0.04	0	0.49	1
$(k_1)^{1/2}$	More or less permeable	0.45	0	0.84	1
$((k_1)^2)^C$	Not very permeable	0.96	1	0.51	0
$k_1 \cap ((k_1)^2)^C$	Permeable but not very permeable	0.2	0	0.51	0

These variables can be then incorporated in the fuzzy synthetic analysis as will be shown below in (B).

(ii)- Volume changes in the underground formation, have led to serious damages in the past which provided the impetus for a serious consideration in every underground design (Wittke, 1981). A simple parameter used to define the swelling potential of sedimentary formation is the water adsorption index. A high value of this index indicates a high swelling potential. Since we have, in this definition, uncertainty associated with the descriptive term, the measured value and its representation for the whole formation, it is more convenient to consider it as a fuzzy variable. For instance, let: A = "High swelling potential"; X = Water Adsorption Index (%). Then, we can establish the fuzzy variable (see figure 1): $\lambda_A(x) = 1-e^{-kx^2}$ $k>1.0$

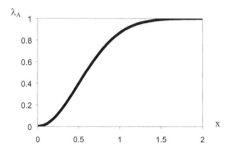

Figure 1. The fuzzy variable: "High swelling potential"

The selection of the above continuos function for the fuzzy variable is based upon engineering judgement (e.g. soils or rocks having water adsorption index greater than 2 is generally considered as very expansive and less than 1 and greater than 0.5 is regarded as moderately expansive).

(iii)- Similar characterizations can be defined for other underground design parameters, such as discontinuity spacing, size, and orientation, angles of strike and dip, degree of weathering, and so forth.

B. Fuzzy Qualitative Analysis

In broader sense, this model is a synthetical analysis, which can be interpreted as a qualitative modeling scheme by which the system behavior is qualitatively explained by using fuzzy variables, judgmental opinions and causal relations.

The behavior of underground structures can be delineated by values of its variables (forces, deformations, pressures, material properties, ...etc.). Such a description sometimes fails to provide much insight into how the system parameters change and to what extend these changes affect the system function.

The fuzzy synthetic model will help to provide a qualitative description of the deep mechanism, capable of representing incomplete knowledge of the structure and behavior of the system. The characterization model is based upon qualitative reasoning, and approximate reasoning (Zadeh, 1992; Dubois/Prade 1988). This simply refers to a collection of fuzzy IF-THEN rules in which the conditions and consequents involve fuzzy variables. For example, if X and Y are input fuzzy variables and Z is the output fuzzy variable, the relation among X, Y, and Z may be expresses as:

IF X is A_1 AND Y is B_1, THEN Z is C_1
IF X is A_2 AND Y is B_2, THEN Z is C_2
.... (1)

IF X is A_n AND Y is B_n, THEN Z is C_n

Where A_i, B_i and C_i, i=1,...n, are values of the fuzzy variables.

Given the dependence of Z on X and Y in the form of equation (1), we can employ the compositional rule to compute the value of Z given the values of X and Y. A very simple example is illustrated below:

Given:
IF the degree of weathering is high AND discontinuity spacing is large, THEN rock strength is low
IF the degree of weathering is very low AND discontinuity spacing is small, THEN rock strength is very high
Required: what is the rock strength IF the degree of weathering is medium AND discontinuity spacing is more or less small?
Applying the principles of qualitative reasoning the answer to this conditon could be:
IF the degree of weathering is medium AND discontinuity spacing is more or less small THEN rock strength is high.

Of course, much more such schemata can be written down and studied. Because of this wealth of possibilities, no more details will be addressed here. However, a more practical application of this model can be seen in the assessment of the risk potential of underground structures. These systems are commonly characterised by the interaction between many varied interrelated state parameters and attributes. The significant factors are generally the geological, hydrogeological and environmental causes. These parameters can not be considered in the classical engineering stability analysis. For instance, in the case of the stability analysis of caverns, the causes or actions which affects the safety of the underground opening can be described using fuzzy variables and relations depicted in table 2.

Table (2): Fuzzy Variables and Relations for the risk analysis of a cavern:

i	Causes Complex	Action (W_i)	Consequence (K_i)
1	Geology and Hydrogeology (e. g. joints pattern, unknown material properties, Tectonic, and Groundwater)	significant	very significant
2	Climate and environmental (Temperature, Gases, rainfall, snow, and frost)	medium	very significant
3	Collapse computation model	small	Significant

C. Computation with uncertain parameters:

Multi-attribute engineering design starts with the specification of a set of functional requirements. These represent the functional space consisting of the minimum number of mutually independent and realizable elements, which describe the design properties of the structure to be designed. The elements of the set of solutions satisfying the functional requirements are called the fuzzy design variables. They represent the physical space in which the structure will be realized. Within the limit state context this can be described by the following: let

x = a vector of algebraically independent design input variables in the mathematical model $\aleph$ (loads, geometric parameters, material properties, environmental conditions, ...)
$\aleph$ = a mathematical model of a quasi-static physical system which can be defined by a vector of mapping $\mathbf{f}$ of arbitrary large dimensions from the model input space X to a model performance space Z consisting of a vector output. Then, formally the system model $\aleph$ can be defined as a trinomial set $(\mathbf{f}, X, Z)$, where X and Z are taken to be the universe of discourse of possible values of vector x and z. The safety against a given collapse mode can be defined by specifying an inequality constraint:

$$R_s \geq R_a \quad \text{or} \quad z = \frac{R_s}{R_a} \geq 1 \quad (2)$$

where R_s = Resistance forces; R_a = External excitations (actions); z = Grade of safety. Note that, z is referred to as a grade of safety and not the conventional safety factor involved in the ASD. Moreover, uncertainties are included in R_s and R_a through the fuzzy variables.

In the fuzzy failure event, there is no one-limit state surface to provide a crisp portioning of strict dilapidated and survival sets. Instead a family of limit state surfaces will be introduced to reflect the real structural environment:
(a)- Safety state (I): absolute safe
(b)- Safety state (II): safe
(c)- Safety state (III): more or less safe (slightly damaged)
(d)- Failure state (I): partial collapse (require maintenance)
(e)- Failure state (II): absolute collapse

To see the explanation of these definitions see figure 2. The functions in figure 2 can be described in different manners, depending upon the type of structure, its importance, level of risk, ...etc. This model will be illustrated with the application example below.

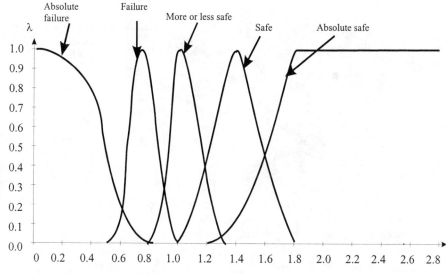

Figure 2. Failure and Safety limit states

Application Example

The stability investigation of a jointed rock slope is to be performed. The geometry and the soil properties are shown in figure 3. In this problem all dimensions, material properties and loading are fuzzy design variables.
The stability of the slope is defined with the following relation:

$$z = \frac{cL + [W \cos\omega - U + V \sin\omega + T \cos v]\tan\varphi}{W \sin\omega + V \cos\omega - T \sin v} \geq 1.0 \qquad (3)$$

Where,
C = cohesion of the joint; L = length of the joint
W = total weight of the slope; φ = Frictional angle of the joint
T = Anchor force; U, V = pore water pressure
θ = Inclination of the slope from the horizontal;
v = Angle between the anchor and the discontinuity;
ω = Inclination of the discontinuity from horizontal

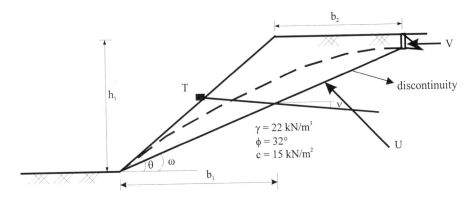

Figure 3: Stability of a Rock Slope

The model (C) will now be implemented. The vector of fuzzy design variable is

$$X = \left\{\tilde{h_1}, \tilde{b_1}, \tilde{b_2}, \tilde{\gamma}, \tilde{c}, \tilde{\varphi}, \tilde{T}, \tilde{\theta}, \tilde{\omega}, \tilde{v}\right\} \tag{4}$$

Table (3) summarizes the definition of the fuzzy design variables. The differences between constants and fuzzy variables will be shown with the sign ~ above the symbol. The constant vector is given by equation (5):

$$X = \{h_1 \ b_1 \ b_2 \ \gamma \ c \ \varphi \ T \ \theta \ \omega \ v\} \tag{5}$$
$$= \{10.0m \ 10.0m \ 5.0m \ 220.0kN/m^3 \ 15.0kN/m \ 32.0° \ 100.0kN \ 45.0° \ 30.0° \ 20.0°\}$$

In Table (3) there are linear and non-Linear functions, defined as follows

$$\lambda_x = \begin{cases} L(\xi) = L((a_1 - x)/u) & x \le a_1, \ u > 0 \\ R(\xi) = L((x - a_2)/v) & a_1 \le x_1 \le a_2 \\ 0 & x < a_1 - u, \ x > a_2 - v \end{cases} \tag{6}$$

In case of non-linear functions, the reference functions L(ξ) and R(ξ) are given by the following relations:

$$L(\xi) = \sqrt{1 + \frac{(x - a_1)}{u}} \quad (18) \qquad\qquad R(\xi) = 1 - \left(\frac{x - a_2}{v}\right)^2 \tag{7}$$

An example of these functions is described graphically in Fig.4. The values of the parameters u and v can be determined using different techniques. For instance if a limited statistical laboratory data is available then u and v can be taken as the standard deviation. If information about the maximum and the minimum plausible values is available from experience or database, then u or v can be assumed to equal the difference between those values. More elaborate methods, such as neural network and genetic algorithm are also available (Lee et.al., 1993; Kosko 1992).

GEO-ENGINEERING FOR UNDERGROUND FACILITIES 291

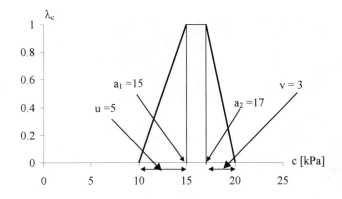

Figure 4: The fuzzy variable $\tilde{c}$ - Linear Function

For the design computations, the influence of the variance of the variables in the safety grade will be now investigated. The following two analysis cases are considered:
(a) All design variables are fuzzy variables according to Table (3).
(b) Gamma is constant (22.0 kN/m²); all other design variables are fuzzy variables,

Table (3): Definition of the fuzzy Variables

Variable	Function parameters			
	a_1	a_2	u	v
$\tilde{h_1}$ [m]	9	9	1.5	1
$\tilde{b_1}$ [m]	10	10	1.5	0.5
$\tilde{b_2}$ [m]	5	5.5	0.5	0.5
$\tilde{\gamma}$ [kN/m³]	22	23	2	4
$\tilde{c}$ [kN/m²]	15	17	5	3
$\tilde{\varphi}$ [°]	30	32	3	3
$\tilde{T}$ [kN/m]	100	110	10	15
$\tilde{\theta}$ [°]	45	47	3	3
$\tilde{\omega}$ [°]	30	30	3	4
$\tilde{\nu}$ [°]	20	21	1	1

The results of the computations are depicted in Fig.4. This figure shows the analysis cases (a) and (b), which describe the fuzzy safety grade. This diagram depicts that the influence of the fuzziness of the weight of the slope in the safety grade is appreciable. Now, for the assessment of the safety state, we need to recognize that the safety against failure is no longer a crisp set. From Fig. 4, it is clear that the domain less than 1 ($z < 1.0$, failure domain) is smaller than the domain > 1.0 (safe domain). One can in this situation assess the stability as more or less safe (Safety State III, Fig. 3).

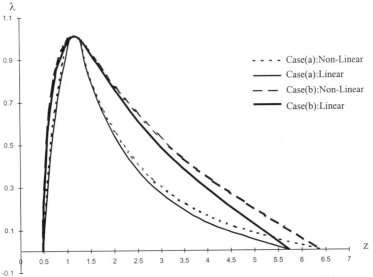

Figure 4 : Safety grade for the analysis cases (a) and (b)

The stability investigation according to the probability method (Ang/Tang, 1984; LRFD AASHTO, 1998) was performed to compare results with the proposed procedure. The resistance factors are 1.6 for cohesion, 1.25 for friction angle and 1.3 for the anchor force. The results indicate that the grade of safety is: $z = 0.79$, which represents an unsafe structure. However, when the variance of all design variables is taken into account (not only c, φ and T, but also the geometry of the slope, joints and the rock unit weight), Fig. 4 shows evidently that this slope is never in an absolute failure state.

Conclusions

In the design of underground facilities, inherent uncertainties arise in the reduction of actual site conditions to a representative analytical model, and in the determination of physical properties of the subsurface formations. Not all these uncertainties can be considered adequately within the probabilistic safety concept.

The fuzzy characterization models are based upon the inclusion and handling of the uncertainty in underground systems through a combination of objective and mostly explicitly expressed data and from information that is eminently subjective. This characterization acts as elastic constraints over universe of discourses and can be adapted to variations of cognitive perspective in underground problems. Thus, the safety evaluation and economy of the structure will be enhanced.

Reference:

AASHTO, 1998, LRFD Reference manual: Load and Resistance factor Design for Highway Bridge substructures. American Association for Transportation and Highway Officials, Washington, D.C.

Ang, H-S., and Tang, W.H. (1984). Probability Concepts in Engineering Planning and Design. V.2, John Wiley and Sons, New York.

Bandemer, H. and Gottwald, S. (1995). Fuzzy Sets, Fuzzy Logic, Fuzzy Methods with Applications. John Wiley and Sons, New York

Dubois, D. and Prade, H. (1980). Fuzzy Sets and Systems. Academic Press, London.

Dubois, D. and Prade, H. (1988). On Fuzzy Syllogisms. Comput. Intell. 14, pp.171-179.

Duddeck, H. (1990). Tunneling in soft ground and sedimentary rock for high-speed double-track railway lines in Germany. Tunneling and Underground Space Technology 5, 257-263

Einstein, H.H. and Baecher, G.B.(1982). Probabilistic and statistical methods in engineering geology. Rock Mechanics, Supp.12, pp.47-61.

Einstein, H. (1991). Reliability in Rock Engineering. Proceeding of Geotechnical Engineering Congress 1991. Geotechnical Special Publication No. 27, ASCE, 608-633

Einstein, H., Halabe, V. B., Dudt, J., Descoeudres F. (1996). Geological Uncertainties in Tunneling. Proceeding of Uncertainty'96. Geotechnical Special Publication No. 58, ASCE, pp.239-253

Lee, M., and H. Takagi (1993). Integrating design stages of fuzzy systems using genetic algorithms. IEEE Trans. Paper 0-7803-06t4-7/93.

Kosko B. (1992). Neural networks and fuzzy systems. Prentice Hall-Englewood Cliffs.

Nahmias, S. (1978). Fuzzy variables. Fuzzy sets and systems I, pp.97-110.

Nawari, O. and Hartman, R. (1997). Determination of the Characteristic values with respect to the new European Codes in Civil Engineering using Fuzzy Modeling. (in German) Journal of "Die Bautechnik" 74, Heft 4, pp.227-232, Berlin, April.

Nawari, O. and Hartman, R. (1998). Fuzzy Logic Concept in the Limit States of Geotechnical Structures. The XI Danube-European Conference, 25-29 May, , Porec, Croatia, pp.855-862.

Vanmarcke, E.(1977). Probabilistic modelling of soil profiles. ASCE Journal of Geotechncial Engineering Division, Vol.103, NO. GT11, pp.1227-1246.

Wittke, W. (1981). Some Aspects of the design and construction of tunnel in swelling rock. Proc. Int. Symp. On Week Rock, Tokyo.

Wittke, W. (1990). Rock Mechanics-Theory and Applications. Springer-Verlag, Berlin.

Zadeh, L. A.(1965). Fuzzy Sets. Inform. And Control, Vol. 8, pp. 338-353.

Zadeh, L. A.(1992). Knowledge representation in fuzzy logic. In Yager and Zadeh, eds., An Introduction to fuzzy Logic Applications in Intelligent Systems, pp. 1-25, Boston, Kluwer Academic

TOMOGRAPHIC IMAGING OF SOIL-TUNNEL INTERACTION
J. Carlos Santamarina[1] and Americo L. Fernandez[1]

Abstract
Recent innovations in field testing devices, developments in signal processing and effective algorithms for inverse problem solving support the implementation of tomographic imaging as a reliable tool for site characterization. One important application is the imaging of the state of stress in a soil mass. This paper presents an experimental tomographic study of the state of stress around a tunnel. Inversions based on the parametric characterization of the stress field and pixel based solutions are attempted. Results show the potential of tomographic imaging in the characterization of geotechnical systems and in the monitoring of subsurface processes.

Introduction
Elastic wave tomography consists on inverting the distribution of wave propagation velocity, V, from travel time measurements, t. For a travel distance, L, the travel time is:

$$t = L/V \qquad (1)$$

In terms of slowness, $s = 1/V$, equation (1) becomes:

$$t = L \cdot s \qquad (2)$$

This expression presumes wave propagation through an infinite homogeneous, isotropic, elastic medium. When the wave-front travels through different materials, Equation 2 is written in integral form:

$$t = \int_L s \, dL \qquad (3)$$

---

[1] Professor and Graduate Student, School of Civil and Environmental Engineering, Georgia Institute of Technology, Atlanta, GA 30332-0355

where L is the ray path from the source to the receiver. The propagation velocity for elastic waves depends on parameters like the shear modulus, G, constrained modulus, M, and soil density, ρ:

$$Vs = \sqrt{\frac{G}{\rho}} \quad \text{and} \quad Vp = \sqrt{\frac{M}{\rho}} \tag{4}$$

In soils and fractured media, the stiffness G and M depend on the state of stress. The P-wave velocity is proportional to the effective stress in the direction of wave propagation, σ'_p:

$$V_p \propto \sigma'^{\alpha}_p \quad \text{(unsaturated media)} \tag{5}$$

The S-wave velocity is proportional to the effective stress in the direction of wave propagation, σ'_p and in the direction of the particle motion, σ'_m:

$$V_s \propto \sigma'^{\alpha}_p \sigma'^{\beta}_m \tag{6}$$

Cascante and Santamarina (1996) show that the velocity-stress correlation can be written in terms of the mean stress, σ'_{mean}, and the deviatoric stress, σ'_{dev} on the polarization plane:

$$Vs = A \cdot \left(\frac{\sigma'_p + \sigma'_s}{2}\right)^m \cdot \left(\frac{\sigma'_p - \sigma'_s}{2}\right)^n \tag{7}$$

where A, m and n are experimentally obtained constants (e.g., triaxial cell with bender elements or resonant column). Laboratory tests suggest that n is very small and it can be disregarded ($n \approx 0$), while m takes values around 0.18 to 0.3.

Applying concepts discussed above, the state of stress can be determined from travel time measurements, by combining Equations (7) and (3):

$$t = A^{-1} \cdot \int_L \left(\frac{\sigma'_p + \sigma'_s}{2}\right)^{-m} \cdot dL \tag{8}$$

The purpose of tomographic inversion is to obtain the spatial distribution of stress given crosshole travel time measurements (see earlier numerical and experimental studies in Santamarina et al. 1994 and Santamarina & Potts 1994)

Experimental Study
The feasibility of this technique is evaluated in the laboratory with a scale model of a tunnel. A 28.3 lt. (1 ft^3) true triaxial box is filled with a clean, fine, semi-angular, poorly graded silica sand (D_{50}= 0.32 mm, coefficient of uniformity, Cu=2.12, and coefficient of curvature, Cc = 0.98). The tunnel is modeled with a thin teflon tube (4 mm ID), covered with a very flexible latex sleeve filled with deaired water. This

setup permits controlling the pressure within the tunnel, P_{int}; the far field boundary stresses are imposed with the bladders on the walls of the true triaxial cell (Fig.1).

The "tunnel" is buried in the center of the triaxial box. The initial external diameter of the pressurized tunnel is ~10mm, however this value changes depending on the variations of the internal pressure and the far field boundary pressure. Two vertical arrays of 8 bender elements, separated L_{cross}=190 mm, are installed at both sides of the tunnel to simulate a crosshole test. The vertical separation between bender elements is 25 mm (Fig. 2). Note that the diameter of the tunnel is significantly smaller than the crosshole distance, $L_{cross}/d \approx 19$.

Shear waves are generated with the vertically oriented bender elements by applying a 5 Volt step function with a signal generator. Crosshole tomographic data are gathered with the horizontally polarized S-waves. The signals captured by the receiving bender elements are registered with a multi-channel digital storage oscilloscope (sampling rate of 500 kHz). Stacks of 32 signals are used to improve the signal-to-noise ratio. The frequency band measured in the tests ranges between 6 kHz and 12 kHz. Stored signals are filtered with a 20 kHz low-pass filter to remove high frequency noise. The travel time for each ray is measured with a 0.002ms precision.

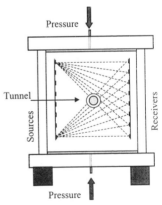

Fig 1. Triaxial box   Fig 2. Tomography inside the box

The test sequence includes the following steps: (1) The vertical confining pressure is raised to 205 kPa (applied by rubber diaphragms at the top and bottom of the box) while zero lateral strain conditions are maintained on the other four sides of the box. (2) The pressure inside the tunnel is raised at the same time in order to keep the internal volume constant inside the tunnel. The first tomographic test is performed at this point. (3) The pressure in the membrane is raised to $P_{int} = 350$ kPa and the vertical confining pressure is kept constant at 205 kPa. The second tomographic study is conducted at this stage. A typical set of signals and average velocity shadows are presented in Figs. 3 and 4

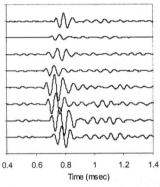

Fig. 3 A tomographic shot
(Source No.4; $P_{int} = 205$ kPa)

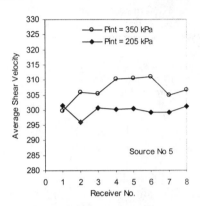

Fig. 4 Typical Velocity Shadows
($P_{int} = 205$ kPa)

Pixel Based Inversion

A straight ray path condition is assumed for this analysis to reduce non-linear effects and consequent convergence difficulties. Therefore, travel length is determined from the geometry of the test setup. The area between "boreholes" is divided into a grid of square pixels, where the velocity of every pixel is to be determined. Equation 2 becomes:

$$[t] = [L].[s] \quad (9)$$

where [t] is a vector of travel times measured for each ray (known), [L] is the matrix of travel lengths in every pixel (model assumption) and [s] is the vector of pixel slowness (unknowns). The inverse problem consists of computing [s].

Fuzzy logic algorithms are used to interpret patterns in the arrival times, and to define the velocity shadows for each test in order to produce an image of the tunnel. (Details of inversion algorithms are presented in Santamarina & Fratta, 1998). Afterwards, the velocity distribution in the box is computed by matrix-based inversion, using a least square solution with regularization (RLSS), using the fuzzy logic solution as initial guess. The RLSS expression is,

$$[s] = [s_o] + ([L]^T.[L] + \lambda.[R].[R]^T)^{-1}.[L]^T . ([t] - [L].[s_o]) \quad (10)$$

Where λ is a nonnegative weighting factor and R is the smoothing regularization matrix (Santamarina and Fratta, 1998). The optimal value of λ is obtained by plotting the minimum and maximum values of velocity inverted for each pixel together with the residual. The solution in Equation 10 allows the addition of a-priori information about the solution in the model, such as the regularization matrix, R, and the initial guess of slowness, s_o (from fuzzy logic). Regularization corrects ill-conditioning.

Once the velocity distribution is determined, the mean stress field can be computed from Equation 6. Results are shown in Fig. 6. The difference velocity tomogram is presented in Fig. 7

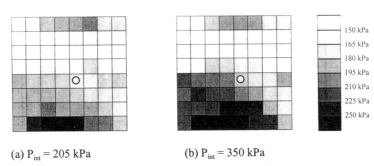

(a) P_{int} = 205 kPa   (b) P_{int} = 350 kPa

Fig. 6 Tomographic images of the stress field for different pressure conditions inside the tunnel

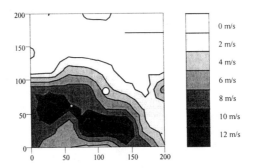

Fig. 7 Difference tomogram. This image is obtained by subtracting the velocity values obtained for the case P_{int} = 350 kPa and the case P_{int} = 205 kPa

These images show the prevalent effect of a region of higher stiffness in the lower left corner. Furthermore, the stiffness in this region experiences the greatest increase after the tunnel is pressurized. A careful study of the signals shows that this is not a mathematical artifact or an image ghost.

Parametric-Based Inversion
The pixel-based solution imposes a large number of unknowns, weakening invertibility. The alternative approach explored herein is to presume a function of the stress field that depends on few parameters, such as the far field stresses, and the internal pressure of the inclusion.

The selected function has two components: (1) A first-order Fourier series approximation to solve for the stress field around a circular cavity in an biaxially loaded homogeneous medium (Fig. 8) and (2) an axisymmetric stress field induced by the internal pressure (Fig. 9). Without the explicit intent to presume elasticity, a Kirsch-type solution is selected in agreement with these criteria (Poulos & Davis, 1974 and Goodman, 1989. Note that cavity expansion solutions provide similar trends for the first component, e.g., Borden and Yan, 1989):

$$\sigma_r = P_{int}\left(\frac{a^2}{R^2}\right) + \left(\frac{p1+p2}{2}\right)\left(1-\frac{a^2}{R^2}\right) + \left(\frac{p1-p2}{2}\right)\left(1-\frac{4a^2}{R^2}+\frac{3a^4}{R^4}\right)\cos 2\theta \quad (11)$$

$$\sigma_\theta = -P_{int}\left(\frac{a^2}{R^2}\right) + \left(\frac{p1+p2}{2}\right)\left(1+\frac{a^2}{R^2}\right) - \left(\frac{p1-p2}{2}\right)\left(1+\frac{3a^4}{R^4}\right)\cos 2\theta \quad (12)$$

$$\tau_{r\theta} = -\left(\frac{p1-p2}{2}\right)\left(1+\frac{2a^2}{R^2}-\frac{3a^4}{R^4}\right)\sin 2\theta \quad (13)$$

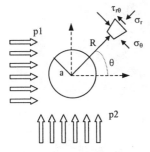
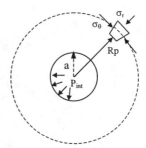

Fig. 8 Stresses around a hollow cavity in a biaxial field

Fig. 9 Stress due to P_{int}

The parametric inversion is performed to back-calculate the pressure inside the tunnel, P_{int}. This inversion is based on the minimization of the L-2 norm between the measured travel times, t^{meas}, and the predicted travel times obtained by estimating the velocity distribution inside the box, t^{pred}, successively selecting different values of P_{int},

$$\text{L-2 norm} = \sum_i \left(t_i^{pred} - t_i^{meas} \right)^2 \qquad (14)$$

The acting state of stress for the analytical model is computed for the pressure conditions in the tunnel of 205 and 350 kPa. Results are shown in Figs. 10 and 11.

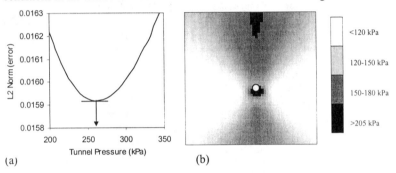

Fig. 10 Parametric inversion for the applied tunnel pressure P_{int}
(Acting P_{int}=205 kPa)
(a) Inverted tunnel pressure $P_{int} \approx 260$ kPa
(b) Radial stress around the tunnel (thresholded image)

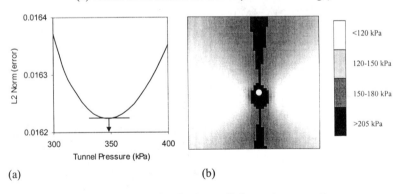

Fig. 11 Parametric inversion for the applied tunnel pressure P_{int}
(Acting P_{int}=350 kPa)
(a) Inverted tunnel pressure $P_{int} \approx 349$ kPa
(b) Radial stress around the tunnel (thresholded image)

The inverted internal pressure in the first case, is 260 kPa instead of the applied value of 205 kPa. As for the second case, the inverted pressure is very close to the applied pressure of 350 kPa. It is assumed that the diameter of the tunnel increased due to the pressure increment, therefore a 15mm O.D. is used for this inversion. In fact, there is a trade-off between diameter and pressure, so that the same minimum error can be obtained with a combination of small diameter and high pressure, or large diameter with low pressure. Hence, one of the parameters should be defined accurately before attempting an inversion for the other one.

It is important to highlight that the pixel-based images in Fig. 6 display the stress in the polarization plane. However, images in Fig. 10 and 11 are the computed radial stress around the tunnel for the inverted value of P_{int} (Equation 11).

Final Comments

Measured travel times and the travel times predicted with inverted parameters are plotted in Fig. 12, for both pixel-based and parametric inversions. In both cases, travel times are adequately predicted. Clearly, error minimization is not a sufficient criterion in obtaining valuable results.

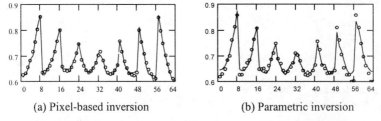

(a) Pixel-based inversion   (b) Parametric inversion

Fig.12 Difference between travel times measured (ball-points) and travel times predicted (lines)

Data gathering and inverse problem solving leading to tomographic images of the state of stress are challenging tasks. The setup is critical to reduce model errors that are magnified during inversion. Considering the typical shear wave velocity in sands, the dimensions of the box and the resolution of the acquisition systems used in this study, errors as small as 1-2 mm in the geometry definition affect the inversion of velocities.

The effect of anisotropy is not considered in this case. A relatively simple approach is to elliptically stretch the pixel geometry to take into consideration anisotropy in shear wave velocity. This alternative would correct for the velocity anisotropy in the background, but would negatively impact the inversion of induced stresses. A more cumbersome approach is to attach three unknowns to each pixel: maximum velocity, minimum velocity and principal direction. However, this approach severely increases the number of unknowns and ill-conditioning.

Acknowledgement
This research was supported in part by Petroleos de Venezuela. Julio Valdes helped in the development of the laboratory instrumentation.

Further Information
Feedback, errata and further references related to this publication will be posted in the authors web sites.

References
Borden, R. H. and Yan, S. L. (1989) "Axial capacity of pressure-injected piles: A cavity expansion model" *Foundation Engineering: Current Principles and Practices*, ASCE GSP 22: 1505-1519.

Cascante, G. and J.C. Santamarina (1996) " Interparticle Contact Behavior and Wave Propagation" *ASCE Journal of Geotechnical Engineering*, 122(10): 831-839.

Goodman, Richard E. (1989) "Introduction to Rock Mechanics" *John Wiley & Sons*, 2nd Edition.

Poulos, H.G. and Davis, E.H. (1974) "Elastic Solutions for Soil and Rock Mechanics" *John Wiley and Sons*.

Santamarina, J.C. and Fratta, D. (1998) "Introduction to Discrete Signals and Inverse Problems in Civil Engineering" *ASCE Press*.

Santamarina, J.C., Graham, J., MacDougall, C., and Roy, V. (1994) "Tomographic Imaging Changes in Effective Stress in Granular Media (Simulation Study)" *Transportation Research Record, No. 1415*, pp. 95-99.

Santamarina, J.C., and B. Potts (1994) "On the Imaging of Stress Changes in Particulate media: An Experimental Study" *Canadian Geotechnical Journal*, Vol. 31, No.2, pp. 215-222

Seismic Measurements to Investigate Disturbed Rock Zones

B. A. Luke,[1] Associate Member, ASCE, K. H. Stokoe, II,[2] Member, ASCE,
J. A. Bay,[3] Associate Member, ASCE, N. J. Lee,[4]
and P. P. Nelson,[5] Associate Member, ASCE

Abstract

Seismic crosshole and SASW measurements were made in underground openings in salt and tuff to evaluate the extent of the disturbed zone resulting from excavation. Results are summarized and then interpreted to investigate the effects of opening geometry, age, and mining method on the extent of the disturbed zone.

Introduction

The mechanical and hydrologic properties of rock surrounding an underground excavation will be impacted by the presence of the excavated opening (Borns and Stormont 1988). The extent of this disturbed rock zone is highly variable and depends on many factors including the virgin stress state, constitutive properties of the intact rock and rock joints, jointing patterns, opening geometry, age of the excavation, artificial supports used, and mining method. Knowledge of the extent of the disturbed area can shed light on these variables, and is useful in understanding the performance and longevity of the structure.

The velocity of seismic waves transmitted through a rock mass can be related to its stiffness. Near the surface of an excavation, reduced confining pressures will cause fractures to open and, in general, the rock will expand toward the cavity. This will result in reduced seismic wave transmission velocities. We can investigate both the magnitude of the reduction in velocity and its variation with distance away from the excavation surface to infer the extent of the disturbed rock zone.

---

[1] Assistant Professor, University of Nevada, Las Vegas, Las Vegas, Nevada 89154-4015
[2] Cockrell Family Regents Chair in Engineering No. 9, University of Texas at Austin, Austin, Texas
[3] Assistant Professor, Utah State University, Logan, Utah
[4] Research Associate, Texas Department of Transportation, Austin, Texas
[5] Program Director, National Science Foundation, Arlington, Virginia

Over an eight-year period, the authors used seismic crosshole and surface wave measurements to investigate disturbed zones around underground openings in salt and in a welded tuff. In this paper, we discuss the utility of these methods to investigate the disturbed rock zone, and the implications of our findings with respect to the effects of age, mining method, and opening geometry on the integrity of the mined opening.

Two U. S. Department of Energy sites were visited as part of this study. The first was the Waste Isolation Pilot Plant (WIPP), a facility in New Mexico developed to investigate the disposal of defense-generated transuranic waste in a bedded salt formation. Several visits were made to the WIPP facility by researchers from the University of Texas at Austin over several years, which allowed the monitoring of changes to rock mass stiffness over time. The second site was the Exploratory Studies Facility of the Yucca Mountain Project (YMP), a site in Nevada which is being developed to investigate disposal of high-level radioactive waste from civilian sources in an ashfall tuff. Researchers from the University of Nevada, Las Vegas visited the YMP facility to study effects of excavation methods on rock mass stiffness. In the case of geologic disposal of radioactive waste, the extent of the disturbed rock zone must be known in order to predict the permeability of the rock mass to fluids, as well as the long-term stability of the excavated openings (e.g., Olsson et al. 1995).

Crosshole and Spectral-Analysis-of-Surface-Waves (SASW) measurements were made at various locations at WIPP and Yucca Mountain, as summarized in Fig. 1.

Crosshole and SASW Measurements in Salt

The underground portion of the WIPP facility near Carlsbad, New Mexico, is located at a depth of about 655 m in the Salado Formation, which consists of uniformly bedded evaporites composed primarily of halite (rock salt), with layers of argillaceous halite, polyhalite and anhydrite interspersed (Matalucci and Munson 1988). The repository horizon is situated roughly at the lower third-point in the 600-m-thick formation. Most of the underground openings were created using a roadheader machine.

SASW and crosshole measurements were made on walls of rectangular drifts at Site 1, Site 2a, Site 2b, and Room D. Measurements were also made in two cylindrical cavities. In conjunction with the Small-Scale Seal Performance Tests (SSSPT), measurements were made in a 0.91-m diameter horizontal drilled shaft located in Room D, and in Room Q, a 2.8-m diameter drift with a circular cross-section, excavated by full-face tunnel boring machine. Site 1 was visited and tested repeatedly during 1988, 1990, and 1994. Site 2a was visited in 1988 only. Site 2b was visited twice, in 1990 and 1994. Rooms D and Q were visited in 1994 only.

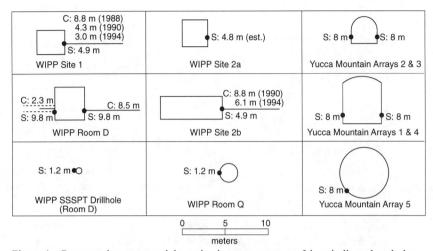

Figure 1. Cross sections summarizing seismic measurements. Lines indicate boreholes used for crosshole testing, and dots represent locations of SASW arrays. Values indicate maximum depths of crosshole (C) measurements and maximum spacings for SASW (S) arrays. Dashed lines in WIPP Room D delineate a large-diameter drilled shaft.

Crosshole Testing Procedure

The crosshole tests were performed using paired sub-horizontal drillholes. With one exception, piezoceramic transducers in a system know as Bebop (Roblee 1990) were used as both source and receiver. The paired devices are wedged in the drillholes, approximately opposite one another, and both trigger and response signals are monitored. The system can be used to measure both compression (P) and shear (S) waves by rotating the source and receiver in their respective drillholes. Time records are averaged over 5 to 25 tests at each location. Sample P- and S-wave records are shown in Fig. 2. The seismic wave velocity is determined by dividing the travel path length by the travel time. The travel time is determined by comparing the initiation of the source energy with the time of the first arrival of the wave at the receiver. The system delay time, which is predetermined in the laboratory, must also be factored out. For the S-wave, which is more challenging to identify since its arrival is preceded by the P-wave, a butterfly pattern is created by repeating measurements with the source pointed in opposing directions. This reverses the polarity of the shear wave between the two sets of measurements.

In the case of the 0.91-m-diameter SSSPT shaft, crosshole measurements were made using a small-diameter horizontal drillhole which angled toward the back of the large-diameter hole, and the wall of the large-diameter borehole. The source was placed

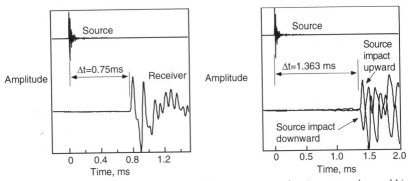

Figure 2. Sample time records from crosshole measurements in a) compression and b) shear

in the small-diameter hole, and signals were received on the wall of the large-diameter hole, using the same transducers that were used for SASW testing.

SASW Testing Procedure

The surface wave measurements employed the SASW method, for which the rationale and procedures have been presented in detail elsewhere (Stokoe et al. 1994). This method involves the frequency-domain measurement of motion perpendicular to a free surface resulting from a seismic disturbance applied on the same surface, a short distance away. At a sufficient distance from the source, the ground motion will be dominated by the Rayleigh-type surface wave. The surface wave comprises combined shear and compressional motion, propagates along cylindrical wavefronts from a point source on a free surface, and exhibits dispersion in layered systems, meaning that its velocity will vary as a function of wavelength. Rayleigh wave dispersion can be related to changes in material stiffness with depth. Traditional interpretation of an SASW measurement invokes the assumptions that the fundamental mode of the Rayleigh wave is dominant, and that the site is composed of planar layers. A schematic of the process involved in an SASW measurement is shown in Fig. 3.

To record near-surface behavior on rock, piezoelectric accelerometers with frequency response as high as 50-100 kHz were used. The accelerometers were coupled to the rock with magnets and either ferrous nails in drilled holes or, for the earlier tests, ferrous washers, which had been epoxied in place. To resolve the stiffness profile over an adequate range, data were collected using receiver spacings ranging from 0.15 to 9.8 m. Seismic energy was applied using hammers of various sizes. The sensor output was recorded and viewed using a dynamic signal analyzer.

WIPP Site 1

Site 1 was occupied during all three visits (Lopez 1989; Stacks 1991; Madianos 1991; Stokoe et al. 1996). Both SASW and crosshole testing were conducted there (Fig. 1). Boreholes used for crosshole testing were 8.8 m deep, 4.5 m apart, and 0.1 m in diameter. The initial crosshole measurements were made to the full depth of the

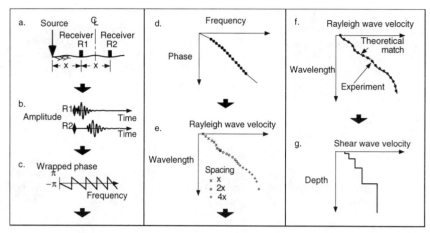

Figure 3. Schematic of a typical SASW measurement: a) source energy is applied at the ground surface; b) the resulting ground motion at the receivers is detected and digitized; c) the time signals are converted to the frequency domain and the phase difference between the two receivers is determined; d) the phase data are unwrapped and masked; e) the dispersion curve is generated from unwrapped phase data at several different receiver spacings; and f) the experimental dispersion curve is matched with a theoretical curve to yield g) the shear wave velocity profile for the site. After Luke and Stokoe (1998).

boreholes. Over time, however, the holes pinched down in the vertical dimension, so the crosshole testing equipment could not be inserted beyond a depths of 4.3 and 3.0 m during the second and third visits, respectively.

The rock face had not been scaled between tests, so the same accelerometer positions could be reoccupied for each SASW test. The maximum receiver spacing used in the SASW measurements was 4.9 m. Results of the measurements are summarized in Fig. 4. Taken together, the crosshole measurements do not indicate a consistent pattern of change in velocity over time; however, they do indicate a leveling off to a constant

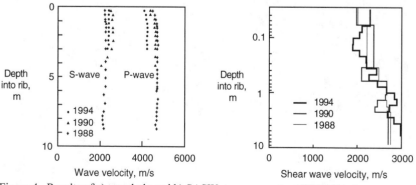

Figure 4. Results of a) crosshole and b) SASW measurements at WIPP Site 1

velocity at a depth of about 1.2 to 1.8 m. This depth does not seem to change significantly over time. The SASW measurements support the conclusion that the disturbed zone extends about 1.5 m. The most recent SASW measurements indicate a lower-velocity zone at shallow depths (0.06-0.3 m) than seen in earlier measurements. Note that the SASW measurements give much higher resolution near the surface than do the crosshole. On the other hand, the SASW measurements may have greater uncertainty at depth, because near-surface fractures inhibit transmission of surface wave energy at depth. This effect can be seen by inspecting the degree of scatter present in the Rayleigh wave dispersion curves.

WIPP Site 2a

Site 2a, located in Room 3, was occupied for testing in 1988 only. Only SASW measurements were conducted there. Results of measurements are summarized in Fig. 5. Again, the reduced-velocity strata appear to extend to a depth of about 1.2 m.

WIPP Site 2b

Site 2a was inaccessible in 1990 due to safety reasons. An alternate Site 2b was selected about 100 m away in Alcove L-2. This site was reoccupied in 1994. Both SASW and crosshole measurements were conducted at this site during both visits. However, the rock surface had been scaled between visits, and wire mesh installed, so the identical SASW testing positions could not be reoccupied. The amount of rock removed in the scaling process was not recorded. Boreholes used for crosshole testing were 4.5 m apart and 0.1 m in diameter. During 1990, crosshole measurements were made to a depth of 8.8 m; however, due to vertical pinching of the holes, testing was limited to a depth of 6.1 m in 1994. The maximum spacing used for SASW measurements was 4.9 m. Results of the measurements are summarized in Fig. 6.

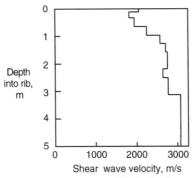

Figure 5. Results of SASW measurements at WIPP Site 2a

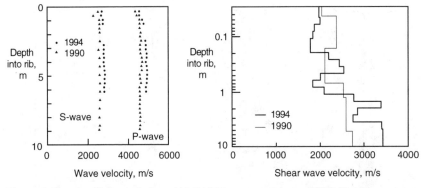

Figure 6. Results of a) crosshole and b) SASW measurements at WIPP Site 2b

The crosshole velocities are systematically higher at the later time, which can not be logically explained. The crosshole measurements indicate a depth of disturbed zone of about 1.2 m, and suggest that the depth may be increasing over time. The more recent SASW measurements support the indication of disturbed rock zone depth of approximately 1.2 m. Despite the surface scaling that went on between site visits, the SASW measurements indicate a decrease in stiffness near the surface over time.

WIPP Room D

All testing in Room D was conducted in 1994. Five SASW arrays were laid out on the walls of the room, three on the east side and two on the west, with maximum receiver spacing of 9.8 m. Crosshole measurements were made on the east wall to a depth of 8.5 m, using subparallel drillholes 3.4 m apart. On the west wall, crosshole measurements used a steeply angled drillhole and the wall of the horizontal drilled shaft, as described previously. This netted compression and shear wave velocity measurements to a depth of 2.3 m from the drift surface. Results are summarized in Fig. 7. The SASW profiles shown represent arithmetic averages of data from the multiple arrays from each side of the room. Differences between adjacent arrays were small. Differences in velocity between the average profiles for either side of the drift were on the order of 10%.

The crosshole data indicate a disturbed zone of approximately 2.5 m deep. The trend for increasing shear wave velocity with depth is more gradual in the SASW measurements. Both crosshole and SASW measurements indicate that shear wave velocities are lower on the west side of the drift, which forms one wall of a broad pillar within the WIPP facility, and which is perforated by three large-diameter drilled shafts (recall Fig. 1). The east side of the room forms an outer boundary of the excavated facility. For comparison, the average shear wave velocity profile from testing on the east rib is also plotted with the crosshole data. For depths greater than a few tenths of a meter, velocities determined from SASW testing were higher than those determined from crosshole measurements.

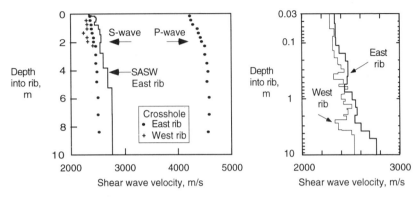

Figure 7. Results of a) crosshole and b) SASW measurements at WIPP Room D.

WIPP SSSPT Drilled Shaft and Room Q

SASW measurements were also conducted inside one of three parallel large-diameter (0.91 m) horizontal shafts cut into the west rib of Room D. These shafts had been sealed with concrete shortly after excavation, at a depth of 3.0 m. Thus, the maximum possible receiver spacing for SASW measurements was 1.2 m. Measurements were conducted along the crown, along the south springline, and circumferentially; although only the springline measurements are discussed here.

Based in part on the data collected at WIPP, recent research to investigate the propagation of surface waves in cylindrical openings has provided insight important to this study. Kalinski and others (1998) found that the speed of waves propagating along the axis of a cylindrical cavity varied with wavelength: waves with short wavelengths propagated as Rayleigh waves, but at longer wavelengths, the wave observed on the wall of the cylinder becomes a shear wave. Shear waves are pure body-type waves, which have different propagation and attenuation characteristics than surface waves. In a homogeneous medium, the velocity of a shear wave will be 3 - 15% higher than the velocity of the Rayleigh wave, depending on the Poisson's ratio of the material. The transition from Rayleigh to shear wave is gradual, and is complete at a wavelength approximately equal to three times the cylinder radius. It has also been found that more stringent limits for data masking are required in order to avoid near-field effects (Kalinski 1998).

The shear wave velocity profile resulting from SASW testing in the SSSPT drillhole is provided in Fig. 8. A model appropriate for surface waves travelling along a cylindrical surface was used to develop the shear wave velocity profile (Kalinski et al. 1998). The simple solution indicates a shallow disturbed zone, about 0.03 m deep. Beyond this, the velocity is quite uniform.

GEO-ENGINEERING FOR UNDERGROUND FACILITIES   311

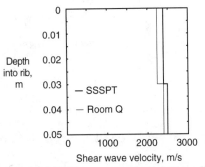

Figure 8. Results of axial SASW measurements in 0.91- and 2.8-m diameter circular openings (SSSPT and Room Q, respectively); after Kalinski et al. (1998).

An SASW array was laid out along the south springline in the entrance to Room Q, a cylindrical opening 2.8 m in diameter, which is sealed with a bulkhead at a depth of 3.3 m. In the space available, the maximum possible receiver spacing was 1.2 m. Results are summarized in Fig. 8, and are remarkably similar to those for the smaller diameter horizontal shaft. The depth of disturbed material is the same, and shear wave velocities are within 5% of one another. Note that these velocities are higher than the near-surface velocities measured in the rectangular openings.

SASW Measurements in Welded Tuff

The second site to be studied was the Exploratory Studies Facility (ESF) of the Yucca Mountain Project. The ESF is situated in welded tuff, a brittle rock of pyroclastic origin. The facility is built around an 8-km long, 7.6-m diameter tunnel which was excavated by full-face tunnel boring machine. Testing alcoves were excavated using drill-and-blast methods in some areas and roadheader in others. SASW measurements were conducted at five locations to study the effects of excavation method on the nature and extent of the disturbed zone.

Procedures used for SASW testing at Yucca Mountain were virtually the same as those used at WIPP. Receiver spacings ranged from 0.25 to 8 m. The measurements were made in or near the Ghost Dance Fault Alcove (Number 6), before it was extended to intercept the fault. The alcove is located 3.7 km along the main drift from the North entrance. This area is in the crystal-poor middle non-lithophysal zone of the Topopah Spring Tuff rock unit. Overburden thickness is about 230 m. All measurements were made within a 50-m radius of one another. Excavation methods and array numbers correlate as follows: Arrays 1 and 4 – roadheader; Arrays 2 and 3 – drill-and-blast; and Array 5 – full-face tunnel boring machine (Luke and Tomlinson 1997). Refer to Fig. 1 for opening geometry.

Shear wave velocity profiles from the SASW testing at Yucca Mountain are shown in Fig. 9. Depths of disturbed zones range from about 4.5 to 7 m. In general, the rock mined by mechanical methods shows higher shear wave velocities and a thinner disturbed rock zone than the rock mined by drill-and blast. The rock mined by roadheader appears to have slightly higher shear wave velocity than that mined by full face tunnel boring machine. However, one must consider the size of the data set and the natural variability of the rock when considering the significance of this observation. Opening geometry may also play a role. Although the opening mined by TBM is circular, which is an inherently stable shape, its diameter is significantly larger than the effective diameter of the portion of Alcove 6 where the SASW measurements were made. In a jointed rock mass, larger opening sizes will necessarily cause greater perturbations to the in situ stress state.

Implications for Rock Mass Disturbance

The seismic testing was successful in evaluating zones of decreased compression- and shear-wave velocities. This decrease in velocity can be attributed to several factors. All openings are affected by stress redistribution, however, other factors will be different for different sites. One must also keep in mind the natural variability of the rock mass, which will contribute to variations in results from one test location to another. Time-dependent changes are at least in part attributable to scaling of the excavation surface, which is considered routine maintenance, and causes different surfaces to be tested at different times.

At WIPP, the depths of disturbed rock zones were surprisingly consistent: on the order of 1 to 2 m for walls of rectangular openings, and much smaller for cylindrical cavities. It appears that seismic velocities will be higher and the disturbed rock zone will be shallower in a cylindrical opening, which is intrinsically less disruptive to the virgin

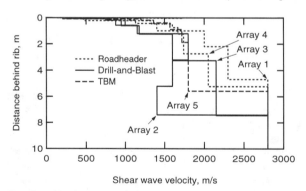

Figure 9. Results of SASW measurements in the Ghost Dance Fault alcove at Yucca Mountain

stress trajectories than is a rectangular opening. However, the opening size must also be taken into consideration. The depth of the disturbed rock zone was considerably higher in the tuffs of Yucca Mountain than in the salts of WIPP, due to the drastically different mechanical properties of the rock.

Effects of mining methods were readily apparent in the SASW results. On the other hand, results of both SASW and crosshole measurements were inconclusive regarding the effects of age of the openings. A longer monitoring period may be needed to sort out apparent inconsistencies.

Much work remains to be done to help quantify the extent and nature of disturbed rock zones surrounding excavations. Another feature which can shed light on the issue is the presence of scatter in the surface wave measurements. A rock with open fissures will cause the waves to reflect, refract, and attenuate, resulting in a dispersion curve with a great deal more scatter than would be apparent in an intact rock, or even a fractured rock under moderate confining pressures. More work is needed to quantify these effects.

The transition from Rayleigh to shear waves seen in cylindrical cavities at longer wavelengths should also be significant in excavated openings of other shapes. These effects could have artificially elevated interpreted shear wave velocities at longer wavelengths, and should be further investigated.

Conclusion

Disturbed zones are caused and affected by many factors including stress redistribution, opening geometry and age, and excavation method. Seismic measurements conducted rapidly at numerous sites around an underground facility can be used to establish the nature of the disturbed zone, and its variability. This information can then be correlated to cause. A dual approach combining convenient, non-intrusive SASW measurements with limited intrusive crosshole testing is recommended.

Acknowledgements

This work involved the efforts of many students at the Universities of Texas and Nevada who helped collect and interpret the data. Funding was provided by Sandia National Laboratories and the U. S. Bureau of Mines Generic Mineral Technology Center on Mine Systems Design and Ground Control. The authors gratefully acknowledge this help and support. We also thank Bjorn Sundquist for his help in preparing the manuscript.

References

Borns, D. J. and J. C. Stormont, 1988. *An Interim Report on Excavation Effect Studies at the Waste Isolation Pilot Plant: The Delineation of the Disturbed Rock Zone.* SAND87-1375, Sandia National Laboratories, Albuquerque, NM; 30 pp.

Kalinski, M. E., 1998. *Determination of In Situ Vs and Gmax Using Surface Wave Measurements in Cased and Uncased Boreholes.* Ph.D. Dissertation, Department of Civil Engineering, University of Texas at Austin, 407 pp.

Kalinski, M.E., K. H. Stokoe II, J. M. Roesset and D. S. Cheng, 1998. "Measurements and Modeling of Surface Waves in Drilled Shafts in Rock." *Non-Destructive and Automated Testing for Soil and Rock Properties,* ASTM STP 1350, W. A. Marr and C. E. Fairhurst, Eds., American Society for Testing and Materials.

Luke, B. A. and K. H. Stokoe, II, 1998. "Application of the SASW Method for Underwater Sites." *Journal of Geotechnical and Geoenvironmental Engineering* 124(6):523-531.

Lopez, M., 1989. *Investigation of Variations in Rock Stiffness by the SASW (Spectral-Analysis-of-Surface-Waves) Method Following Excavation by Blasting and Mining.* Master's Thesis, Department of Civil Engineering, The University of Texas at Austin, 250 pp.

Luke, B. A. and K. L. Tomlinson, 1997. *Assessment of Excavation Effects in the Northern Ghost Dance Fault Alcove of the Exploratory Studies Facility, Yucca Mountain.* Report of a study conducted under agreement with Sandia National Laboratories, Contract number AS-0320, 16 Jan. University of Nevada, Las Vegas; 94 pp.

Madianos, M., 1991. *Field and Laboratory Investigation of Rock Masses Using Surface Wave Seismic Testing (SASW).* Master's thesis, Department of Civil Engineering, The University of Texas at Austin; 296 pp.

Matalucci, R. and D. Munson, 1988. "Planning, developing and organizing in situ tests for the Waste Isolation Pilot Plant (WIPP)," *The Mechanical Behavior of Salt II:* Proceedings of the Second Conference, Hanover, Germany, pp. 329-360.

Olsson, O., K. Slimane, and N. Davies, 1995. "ZEDEX – an in-situ study of the importance of the Excavation Disturbed Zone to repository performance," *Proceedings,* Sixth Annual International Conference on High Level Radioactive Waste Management, 30 April-5 May, 1995, Las Vegas, Nev. ASCE, New York, pp. 203-205.

Roblee, C., 1990. *Development and Evaluation of Tomographic Seismic Imaging Techniques for Characterization of Geotechnical Sites.* Ph.D. dissertation, Department of Civil Engineering, The University of Texas at Austin; 680 pp.

Stacks, R. E., 1991. *Borehole Seismic Studies of Rock Stiffness Variations in Mined Environments.* Master's thesis, Department of Civil Engineering, The University of Texas at Austin; 231 pp.

Stokoe, K. H., II, N-K. J. Lee, B. A. Luke, J. A. Bay, and M. E. Kalinski, 1996. *Seismic Investigation of Disturbed Rock Zones around Excavations at the Waste Isolation Pilot Plant.* Geotechnical Engineering Report GR96-4, prepared for Sandia National Laboratories. University of Texas at Austin; 84 pp.

Stokoe, K. H., II, S. G. Wright, J. A. Bay, and J. M. Roësset (1994). Characterization of geotechnical sites by SASW method. *Geophysical Characterization of Sites,* ed. R. D. Woods. Oxford and IBH, New Delhi, India; 15-25.

Flooding-Induced Mine Subsidence at Retsof, New York

Leo L. Van Sambeek, Member[1]

Abstract

During 1994-95, the Retsof underground salt mine near Geneseo, New York, experienced an unstoppable inflow of fresh water, which flooded the mine. During the 21 months of inflow, two types of surface subsidence were experienced in addition to normal mining-induced subsidence. The first type was sinkholes, up to 20-m deep caused by piping of valley sediments through fractured hard rocks (limestones and shales) over the mine. The second was subsidence, significantly faster and deeper than mining-induced subsidence, measured over a 4-sq-km area around the inflow area. Surface subsidence rates increased tenfold to a hundredfold during the inflow compared to mining-induced subsidence rates measured before the inflow began. Dissolution of unmined salt by the inflowing fresh water explains the subsidence behavior along primary underground flow paths from the inflow to the higher elevations of the mine. While the sinkholes formed over just a few days or weeks, the dissolution-induced subsidence occurred throughout the 21 months of active inflow but slowed abruptly after the mine was filled. Current subsidence rates over all areas of the mine are now at or less than corresponding preinflow rates.

Introduction

On March 12, 1994, a seismic event was recorded in western New York. On the same day, a subsidence feature was highlighted in the snow over the southeastern part of the Akzo Nobel Salt Company's Retsof salt mine near Geneseo, New York. The highlights, which closely mirrored the footprint of the underground mine, took the form of concentric

---

[1] Manager-Mine & Field Services, RESPEC, P.O. Box 725, Rapid City, South Dakota 57709 (leo.van.sambeek@respec.com)

tension cracking and compression ridges in snow-covered fields adjacent to state Highway Route 20A. Damage to a bridge on Route 20A was noted as the asphalt compressed laterally to form "speed bumps" on both ends of the bridge. At the time of the seismic event, no workers were underground in the mine, but a high methane concentration in the exhaust ventilation indicated something had occurred underground. Inspection of the mine by mine-rescue teams working their way from the shaft found an undisturbed mine until a roof fall was encountered in a yield-pillar mining panel known as 2-Yard-South. Inspection of the panel by mine personnel and mining consultants revealed that a 10-m-thick mass of shale roof rock had fallen into the panel and that water was flowing on the floor from the eastern access to the panel.

The water entering 2-Yard-South was generally believed to be flowing into the mine through broken or fractured shales from the overlying limestone rocks (Onondaga, Cobleskill, and Bertie Formations). Sometime during April 1994, water may have also begun to flow into the mine within another mining panel known as 11-Yard-West, which was also mined as a yield-pillar panel. Although never directly observed, indications for inflow into this panel were an increase in the rate of water level rise in the mine and changes in surface-subsidence patterns.

The inflow rate into 2-Yard-South was first measured at more than $0.3\text{-m}^3/\text{s}$ (5,000 gallons per minute (gpm)). Later, particularly after inflow into 11-Yard-West was indicated, the inflow rate is estimated to have been more than $1\text{-m}^3/\text{s}$, but less than $1.3\text{-m}^3/\text{s}$, based on the volume of the mine being filled as the underground shoreline advanced up dip. The time required to fill the mine (approximately 21 months) substantiates a steady inflow rate approaching $1.3\text{-m}^3/\text{s}$ (20,000 gpm).

The first sinkhole developed directly above the 2-Yard-South panel a few days after the start of the inflow into that panel. The sinkhole began as a small depression but expanded to 100-m-diameter and deepened to more than 6 m over the next several weeks. On May 25, 1994, a second sinkhole suddenly developed directly above the 11-Yard-West mining panel. By early June 1994, the sinkhole over 11-Yard-West was about 200 m in diameter and 20-m deep. Based on June 8, 1994, measurements, the 2-Yard-South sinkhole volume was $100,000\text{-m}^3$ and the 11-Yard-West sinkhole volume was $260,000\text{-m}^3$. The size of these sinkholes has not visibly increased since June 1994.

The inflow into the Retsof Salt Mine posed several immediate and long-term problems and concerns. Example problems or concerns and their outcome were:

- Subsidence of the Route 20A bridge made it necessary to close the road to traffic. (The bridge and damaged highway were rebuilt and reopened in 1997.)
- Media reports proclaimed that the entire valley area over the mine was susceptible to collapse similar to the sinkholes that formed over the yield-pillar panels. (While general subsidence occurred over a 4-sq-km area, neither additional sinkholes nor catastrophic subsidence was experienced elsewhere within the valley.)

- Structural damage could occur to buildings over the mine and adjacent areas. (One farm homestead was severely damaged. Three other houses that were vacated were subsequently reoccupied.)
- Several water wells, both north and south of the mine, experienced substantial drawdown. (Recovery in these wells is occurring but full recovery will take several years. Yeager et al. (1998) presents summary data and analysis of the hydrological impacts for the valley.)
- Damage could occur to underground and surface utilities: sewers, water lines, telephone and electrical lines, and buried cables. (Vulnerable utilities were moved to less susceptible areas or reinstalled to include subsidence absorbing components.)
- Two large sinkholes developed. (The sinkholes stabilized soon after their creation and subsidence around them virtually stopped after the mine filled with water. They now function as two ponds with seasonal recharge from Beard's Creek, which intersects them.)
- Cropland within the Genesee River Valley will be more susceptible to flooding because of inverted drainage patterns in areas of significant elevation change.
- Local creeks experienced undercutting because of the increased gradients and flow velocities. (Streambed damage-control measures installed by the mining company mitigated this effect. Mussetter Engineering Inc. (1996) presents the anticipated effects of subsidence on local creeks and the Genesee River.)

The recent subsidence over the Retsof Mine can be considered for three distinct areas shown in Figure 1. Area 1 (inflow area) encompasses a 0.2-sq-km surface area directly over 2-Yard-South and 11-Yard-West, which is the immediate area of the sinkholes. Area 2 (significant dissolution area) is over that part of the mine up to about 3,000 m north of the first area and covers about 4-sq-km. Area 3 includes the surface over the rest of the mine, about 22-sq-km.

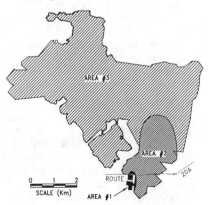

Figure 1. Retsof Mine Layout by Area

In Area 1, surface subsidence (and damage) manifested itself early on. Two sinkholes developed over the panels where water entered the mine. Apparently, the rock layers above the mine fractured because of unusual geologic conditions in combination with mining. These fractures allowed fresh water from the overlying water-bearing glacial valley till to flow downward and through limestone and shale formations. As the water passed through the shales, and later as it flowed through the mine, salt was dissolved. The removal of salt caused a loss of support for the shales and other rock above the mine. Eventually, where dissolution was severe, the shales and limestone layers above the shales broke and collapsed into the void created by dissolution. When the limestone layer broke, it allowed the glacial till to slump downward, creating depressions on the surface (sinkholes). The sinkholes and damage to the highway, the Route 20A bridge, and a nearby farm are localized consequences of where water entered the mine and are not the result of general subsidence over the mine.

Area 2 is defined as where fresh water dissolved the salt left to support the rock above the mine (overburden). This area of significant dissolution is outside the immediate inflow area and extends northward for about 3 km. Because underground support was removed, the overburden moved downward in response. For example, before the inflow, the surface over the southeastern area of the mine was subsiding at a rate of about 0.15 m per year as a combined effect of salt removal by mining and creep shortening of pillars. In the same area approximately 5 months later, the surface subsided at measured rates of 1 to nearly 15 m per year because the pillars were being dissolved. A large amount of salt was dissolved from this area; therefore, subsidence was larger and sooner than would have been expected for a dry mine situation. On the other hand, because a large amount of salt was dissolved in this area, less salt was dissolved in the more distant areas of the mine.

Area 3 was filled with brine that became saturated by dissolving salt from Areas 1 and 2 before it flowed up gradient. Flooding with saturated brine did not produce deleterious effects, and subsidence rates over Area 3 were reduced when the brine pressure provided beneficial confinement of salt pillars. Future subsidence over this large area of the mine will still approximate the magnitude expected for a dry mine scenario but will occur at a slower rate. Consequently, centuries may pass before ultimate subsidence is achieved.

Figure 2 shows a contour plot of the measured subsidence over Areas 1 and 2, as described above. The maximum subsidence is about 5.5 m (just north of the inflow areas).

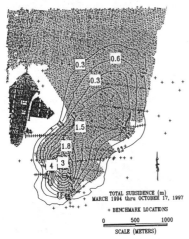

Figure 2. Measured Subsidence

Most of Area 2 experienced from 0.5 to 3-m of subsidence during the 24-month period. The magnitude and pattern for the subsidence can be explained by the dissolution of salt within the mine as it filled. Van Sambeek (1996; 1997) presents complete sets of subsidence plots for the period from March 1994 to March 1997.

Background

Surface subsidence is a natural downward movement of the earth's surface in response to the removal of fluids or solids from underground. At the Retsof Mine, the surface subsides because solid salt was removed. The speed at which the surface subsides depends on the amount of material removed, the thickness of overburden (depth), and types of rock involved. At Retsof, the amount of material removed by mining was about 70 percent of the salt in a mining horizon about 3.81-m tall. The depth is about 335 m, and the type of rocks involved include salt (within and also above and below the mining horizon), shales, limestones, and glacial till. Dissolution of salt from pillars and the roof and floor by the inflowing fresh water effectively increased the percentage extraction of salt local to the inflow area.

The term surface subsidence can be resolved into four specific quantities related to vertical and horizontal displacements of the surface:

Subsidence	The actual vertical (downward) displacement of the surface.
Slope	The gradient produced by different amounts of subsidence. Normally, the slope is largest near the mining perimeter and tends toward zero both away from the mine and toward the center of the mine. Severe slope can disrupt the usefulness of a structure by causing tilting.
Curvature	The rate of change in the slope. For example, because of curvature, a rigid building foundation must either bend to conform to the new ground shape or separate from the ground and behave like a cantilevered beam.
Horizontal Strain	Compression or extension of the ground by differentials in horizontal displacements. Just as points move downward, surface points move horizontally toward the center of the mine. Because the magnitude and direction of horizontal displacements vary, strains are induced in the ground. Generally, compressive horizontal strains occur inside the perimeter of the mine and tensile strains occur outside the perimeter. Compressive strains can cause buckling and bowing, while tensile strains can cause separations and cracking.

Normally, the greater the subsidence, the greater the potential for damage; however, even large subsidence does not always cause damage. The values of allowable deformation are determined according to experience in different countries, types of structures, and the type of damage. Generally, smaller allowable quantities apply to critical structures (e.g., pipelines, power plants, and dams). Greater allowances are made for more flexible or more easily repaired structures, such as small buildings, highways, and power lines.

Even without the inflow, the surface over the Retsof Mine would have subsided more than 2.4 m over a broad area at the center of the mine tapering down to virtually zero about 300 m outside the perimeter (or edge) of the mine. This subsidence process started 110 years ago when the first salt was mined and would normally not have been complete for centuries after mining finished. Because subsidence was occurring within a time frame measured in centuries, neither was it readily detectable, nor was it apparently damaging surface properties. During the early stages of the inflow, anticipated conditions were that the inflow might reduce the time required for complete subsidence to decades, depending on salt dissolution patterns and the potential for pillar punching because of wetted shale softening within the roof and floor (Van Sambeek, 1994).

Subsidence Measurements

Before the inflow, surface subsidence was monitored on a single line of benchmarks installed in May 1990 over the active mining area of the Retsof Mine. This line coincidentally ended up being situated near the inflow. Four annual elevation surveys were performed before the inflow and provide the basis for understanding the "normal" subsidence situation, which applies to the mine within the glacial valley where mining was active throughout the 4-year measurement period (i.e., the southeastern part of the mine). The surface subsidence over the western part of the mine, which is not under the glacial valley and had been mined earlier, would have had different subsidence characteristics.

After the inflow began, Akzo Nobel Salt greatly expanded the area where surface movements were monitored. Within just a few days, mine engineers had established new subsidence survey lines over the area that would eventually be the sinkhole area. Eventually 35 subsidence survey lines with 605 individual survey points (benchmarks) were monitored (most on a weekly basis). Subsidence lines were concentrated over the mine, but some extended several kilometers beyond the mine boundary. Measured subsidence information was reported by Akzo Nobel on a monthly basis (now quarterly) to state and town agencies (and placed in publicly accessible repositories).

Subsidence-rate plots highlight the dynamic nature of the subsidence both spatially and temporally. Examples of the subsidence-rate contours are shown in Figure 3 (August 1994) and Figure 4 (August 1995). The greatest subsidence rates occurred in August 1994. By December 1994, the rates had decreased by more than one-half from their peaks and by the

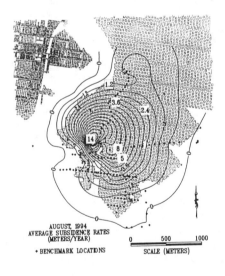
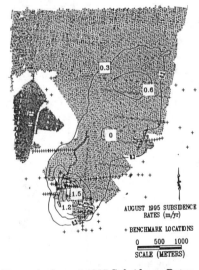

Figure 3. August 1994 Subsidence Rates   Figure 4. August 1995 Subsidence Rates

time the mine filled in December 1995, the rates had diminished to nominal preinflow magnitudes. A total subsidence of approximately 4.2 m and the sudden decrease in rates thereafter is believed indicative of nearly complete dissolution of the Retsof salt bed with simultaneous closure of the mine openings. The greatest cumulative subsidence concentrated around the inflow areas and the southeastern portion of the mine where most dissolution took place. Beyond about 3 km north of the inflow area, the postinflow cumulative subsidence is 0.3 m or less.

Figure 5 illustrates stabilization by plotting the subsidence rate for four benchmarks (Pole C, Pole 1, Pole 2, and Pole 3) near the Route 20A bridge and immediately north of the sinkholes. Two episodes of stabilization are observable. The first occurred in response to an increasing area of contact between the roof and floor shales following complete dissolution of the Retsof salt bed in the vicinity of the inflows. The second episode occurred during late December 1995 and early January 1996 when the subsidence slowed from a steady 1 m/yr to about 0.15 m/yr. This stabilization is attributed to 1) the cessation of dissolution and 2) the structural support provided by the increased brine head when the mine filled. The four benchmarks in Figure 5 reflect underground closure, not only directly below, but also within a zone of influence (an angle of draw). The stabilization illustrated was typical for the mine-wide area, not just the area near the bridge.

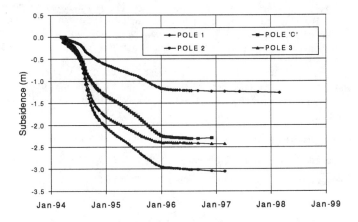

Figure 5. Subsidence Near Bridge on Route 20A

Surface subsidence, particularly around the perimeter of a mine, also causes horizontal displacements. Variations in horizontal displacements cause what are called horizontal strains, which can either compress the surface or cause it to stretch, depending on the location above the mine. Surface-strain measurements were initiated over the Retsof Mine in December 1994. Nine separate measurement lines were established on roads crossing the mine perimeter. These lines were generally about 1,000 m long, and where possible, the midpoint of the line was roughly centered over the edge of mining. The surface-strain lines consisted of markers spaced about 30 m apart and recessed into the road pavement or shoulders. Surveyors taped the distance between adjacent markers to measure any change in the interval lengths that reflected differential displacements, which could then be resolved into surface strains. The repeatability of the measurement was about 1.5 millimeters, so strains as small as 50 microstrain could be detected.

The surface strain line on Route 20A at the east side of the mine (see Figure 1) was nearest the most active subsidence and would, therefore, be the most likely of all the lines to measure horizontal strains. Early-on strain measurements on the Route 20A line weakly demonstrated the theoretical horizontal strain pattern expected at the mine boundary (i.e., a sinusoidal distribution of compressive strains over the mined area and tensile strains over adjacent solid ground). Scatter in later data either obscured the expected strain distribution, or horizontal strain development essentially stopped after the first few measurements. None of the other strain lines showed consistent or resolvable strains.

Dissolution-Enhanced Subsidence

Surface subsidence magnitudes are ultimately directly proportional to the void space created underground. Before the inflow, the void space produced by mining was fairly evenly distributed over the 22-sq-km area with an areal extraction ratio of about 70 percent and a mining height of about 3.8 m. Fresh water inflow changed the void distribution in some areas by dissolving salt. Both the depth of the sinkholes and localized subsidence in excess of mining height suggest the role dissolution had in causing subsidence. The effects of dissolution are not as readily apparent away from the immediate inflow area.

Water entering the mine was basically fresh water that dissolved unmined salt until it became saturated. At the temperature of the mine 14°C (58°F), fresh water dissolves about 0.16 volume of salt per volume of water. Therefore, about 35×10^6 kg-salt per day would be dissolved by a 1.2 m^3/s freshwater inflow. Salt dissolution could have occurred from:

- Vertical pillar (rib) surfaces, which were initially about 3.8 m tall.
- Roof surfaces, where unmined salt was generally 0.6 to 1.2 m thick.
- Floor surfaces, where unmined salt was generally 0.3 to 0.6 m thick.
- Interbeds of salt within the overlying and underlying shales.

From solution-mining experience, dissolution rates from surfaces, such as those listed above, have been empirically defined (e.g., Cummins, 1973). Of the three surfaces applicable in submerged mined rooms, the most efficient dissolution occurs on an overhead horizontal surface (roof) because unsaturated brine "floats" on more saturated brine and remains in contact with the salt. The second most efficient dissolution is from vertical surfaces where density-driven convection cells circulate less saturated brine against the vertical surface. The least efficient dissolution is from the floor because there is no buoyancy or density-driven mechanism to displace the most saturated (hence, most dense) brine from the horizontal floor surface. Where a turbulent flow situation exists, dissolution rate does not depend as strongly on surface orientation.

The inflowing water dissolved salt from the pillars while that part of the mine was being filled. Dissolution from pillar ribs reduced the area of the pillar and thereby increased the effective areal extraction ratio. An increase in overall extraction ratio causes faster subsidence rates and larger ultimate subsidence over the area of dissolution. In the extreme, dissolution may have removed pillars completely. Water also dissolved salt from the roof and floor; the pillar height could have increased from the nominal 3.8 m (as mined) to as much as 4.8 to 5.5m, which is representative of the Retsof salt bed thickness.

Within the stratigraphy at Retsof, there is an abundance of salt besides that in the mining horizon. At least 15 m of salt occurs in major and minor beds from the mining floor upward. From the mining floor downward, there are several major salt beds. Each of these beds of salt was potentially exposed to fresh water and could have been dissolved (at least locally). If these salt beds were dissolved, additional void volume would have been created that must eventually close and contribute to surface subsidence.

Analysis Results

Subsidence over the Akzo Mine has been intensely monitored since shortly after the inflow started. A numerical modeling analysis was performed based on information from 16 of the survey lines (approximately 290 individual benchmarks), which are located over the portion of the mine where the most active subsidence took place. Measurements covering the first 20 months of inflow (May 1994 through December 1995) were used to estimate the underground mine closures. Subsidence after December 1995 was insignificant in terms of this analysis.

The numerical analysis related the monthly incremental subsidence from each benchmark to the amount of underground closure required to cause such subsidence. Because subsidence occurs over a surface area wider than the underground area, a simultaneous solution was necessary for many discrete areas of closure (active dissolution), causing a superposed subsidence at numerous benchmark locations on surface for each of the 20 months. The analysis thus considered both spatial and temporal correlation between closure (dissolution) and surface subsidence.

About 600 salt cells with side lengths of 75 to 150 m were used to represent the unmined salt subject to potential dissolution. The required monthly underground closure for each cell was simultaneously calculated to match the monthly subsidence increment measured on the surface. A least-sum-of-squared error approach was used in the analysis, and individual cell closures were constrained to be positive (i.e., closure could not be reversed) and cumulatively no more than 5.5 m (the thickness of the salt seam).

Numerical analysis results indicate that significant portions of the mine must have undergone substantial closure. Closures up to 5.5 m are calculated which, if true, requires all the salt in the Retsof bed to have been dissolved and the roof and floor shales to have contacted each other. In adjacent areas with calculated cell closures less than 5.5 m, the roof shale probably subsided until being supported by remnant pillars remaining after selective dissolution from tops of those pillars. Once the shale subsided into the freshwater layer, dissolution in that area stopped because the remnant salt pillars were then fully submerged in saturated brine.

The dissolution scenario suggested by the numerical analysis involved several stages:

1. Initially, water entering the mine flowed down dip (southward) until the mine workings were submerged up to the region of 2-Yard-South. After 2-Yard-South became submerged, buoyant freshwater flow flowed along the roof and dissolved salt from the roof and tops of pillars. As water exited from the submerged portion of the mine, it would have dispersed among the pillars and toward the shoreline, which existed because of the 1 percent dip of the mine.

2. As salt was dissolved from the submerged roof, surrounding pillars were made effectively taller and even more salt was exposed and could be dissolved from pillars. A solution mining phenomena termed "helicopter wings" likely developed at the top of pillars, and dissolution was able to cut across entire pillars. This combined roof salt removal and pillar topping would have removed all the roof salt (from 0.5-to 1.2-m thick over 100 percent of the area involved) and allowed subsidence of the roof onto the remaining portion of the pillars. The volumetric subsidence (closure) rate would have matched the volumetric dissolution rate.

3. After the roof salt was dissolved, the roof would have subsided even further as more salt was dissolved from the pillars. Dissolution of the pillar would have stopped after the insoluble roof shale subsided enough to displace the replenishable freshwater layer. Then, new fresh water would have been diverted along another flow path leading to a higher elevation. The as-mined roof elevations and buoyancy forces dictated the original flow path; after the roof began to subside, however, fresh water would have flowed along the highest elevations and up the steepest gradient available. Shifting of the flow paths because roof shales were lowered caused the dissolution front to spread from primarily northwest of the inflow area across the mine expanse to the east and even to the south during the late stage of mine filling.

4. Wherever salt was exposed to the fresh water along a flow path, the salt was dissolved and an even more favorable gradient developed. Eventually, dissolution attacked the pillars (pillar topping) and caused subsidence of the roof, which initiated another change in the flow path and perpetuation of the process.

5. Near the inflow area, flows would have been turbulent and buoyancy forces could not stratify a freshwater layer. Because of turbulence, dissolution could occur on the roof, pillar ribs, and floor equally well. Moreover, flows became increasingly turbulent as flow channels were squeezed tighter by a subsiding roof.

After cumulative mine closures were calculated, horizontal strains were calculated based on the mine closure as of December 1995. Figure 6 shows contours of calculated horizontal strains over the southeastern portion of the mine. Potentially damaging strains are concentrated near the sinkhole area. Significant is the relatively small horizontal strains associated with areas that sustained relatively large vertical movement.

Expectations and Conclusions

Now that the Retsof Salt Mine has filled with brine, conditions of increased stability have been attained in several regards. First, because fresh water is no longer flowing into the mine, dissolution of support rock has stopped. Second, the hydrostatic pressure from the brine is providing some structural support to pillars. The worst is over with respect to rapid, damaging subsidence over the general mine area. Subsidence will, however, continue until the voids created by mining and dissolution have totally closed.

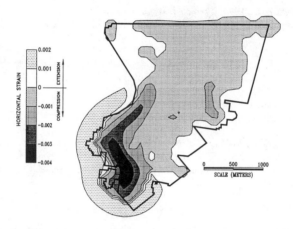

Figure 6. Calculated Horizontal Strains Based on Underground Closure

The rise in water level at the Retsof Mine was monitored using a borehole, which was drilled from the surface and penetrated the mine workings near the southern end of the mine. Since the mine filled, the height of water in the borehole is approximately 230 m. The water produced confining pressure of about 2.5 MPa on pillars at the southern end of the mine. The confining pressure becomes gradually less at locations up dip from the southern end of the mine. For pillars that suffered minimal or no dissolution effects, the closure rate (and the corresponding subsidence rate over them) will be less than the preinflow rates because of the structural support provided by the confining pressure.

Subsidence during the 24 months since the mine filled has been almost indistinguishable from normal surveying errors and seasonal movements of the ground (soil wetting from snowmelts, frost heaves, etc.). Even over areas where substantial subsidence had occurred, the subsidence rates are now a few centimeters per year (and apparently still slowing). No sinkholes have developed other than the two original sinkholes. New sinkholes, or expansion of the existing sinkholes, is not expected because water is no longer flowing into the mine.

References

Cummins, A. B. (ed.), 1973. *SME Mining Engineering Handbook*, Society of Mining Engineers of the A.I.M.E., New York, NY.

Mussetter Engineering Inc., 1996. *Evaluation of Potential Effects of Subsidence on the Genesee River and Tributaries*, prepared by Mussetter Engineering Inc., Fort Collins, CO, for Akzo Nobel Salt Inc., Clark Summit, PA.

Van Sambeek, L. L., 1994. *Predicted Ground Settlement Over the AKZO Nobel Retsof Mine*, RSI-0525, prepared by RE/SPEC Inc., Rapid City, SD, for Akzo Nobel Salt Inc., Dobbs Ferry, NY.

Van Sambeek, L. L., 1996. *Retsof Mine Surface Subsidence: Postflooding Update*, RSI-0699, prepared by RE/SPEC Inc., Rapid City, SD, for Akzo Nobel Salt Inc., Dobbs Ferry, NY.

Van Sambeek, L. L., 1997. *Retsof Mine Surface Subsidence: 1997 Update*, RSI-0918, prepared by RE/SPEC Inc., Rapid City, SD, for Akzo Nobel Salt Inc., Retsof, NY.

Yaeger, R. M., T. S. Miller, and W. M. Kappel, 1998. *Simulation of the Effects of Mine Collapse on Ground-Water Flow and Land Subsidence in a Regional Aquifer System in the Genesee Valley, Livingston County, New York*, USGS Professional Paper, U.S. Geological Survey, Ithaca, NY.

Acknowledgements

The author expresses his gratitude to Akzo Nobel Salt Inc. for assistance in data collection and permission to publish this paper. The author is forever indebted to Dr. Hamish D. Miller, a fellow consultant on this and other mining projects. Dr. Miller's untimely death robbed the mining- and civil-engineering professions of a critical thinker and unabashed friend. We had planned to write this paper jointly; so in his absence, I dedicate this paper to his memory.

REPLACING A BRIDGE ADJACENT TO A DEEP-SEATED SINKHOLE IN GLACIAL DEPOSITS

Stanley E. Walker, P.E., M. ASCE[1], Randall C. Divito, Assoc. M. ASCE[2], Maureen S. Valentine, P.E., M. ASCE[3]

Abstract

Between March 1994 and January 1996, the U.S. Route 20A bridge over Beards Creek in western New York settled more than 4.5 m (15 ft) and moved approximately 1.5 m (5 ft) upstream toward the nearer of two large sinkholes. The sinkholes, centered on the creek 120 m and 400 m (400 and 1,300 ft) upstream of the bridge, developed during the catastrophic inundation of an underlying, 2600 ha (6,500-ac), active salt mine. At the bridge, the mine's overburden consists of approximately 210 m (700 ft) of shale and limestone and over 150 m (500 ft) of glacial and alluvial deposits.

The mine owner assumed responsibility for designing and constructing a replacement bridge. The authors' firm was engaged to provide geotechnical engineering assistance to the designer in planning and overseeing the construction of the new bridge. Construction began in December 1996 and the bridge was completed and opened to traffic in mid-October 1997.

The bridge replacement was completed in a "fast-track" design-and-build sequence. In response to the placement of more than 27000t (30,000 tons) of earthfill, steel and concrete within 90m (300 ft.) of the sinkhole, the subgrade settled as much as 20 cm (8 in.), moved laterally as much as 25 mm.(1 in.), well within the predicted response. Closely monitored instrumentation indicated no signs of foundation instability.

BACKGROUND

In 1936 a 30-m (100-ft) long, three-span, pile-supported, reinforced concrete bridge was constructed to carry NY Route 39 (later also designated U.S. Route 20A) over Beards Creek. This waterway is a tributary of the Genesee River in the hamlet of

---

[1] V. Pres., Haley & Aldrich, Inc., 189 N. Water St., Rochester, NY 14604
 Ph:(716) 327-5533, Fax:(716) 232-6768, e-mail:SEW@HaleyAldrich.com
[2] Staff Engr., Haley & Aldrich, Inc., [3] Sr.Engr., Haley & Aldrich, Inc.

Cuylerville near the town of Geneseo, Livingston County in western New York. This bridge replaced the second of two earlier, flood-destroyed bridges at this same location.

Geologic Setting

The bridge was located near the western edge of a 3 km (2-mi) wide, flat-bottomed valley through which the Genesee River meanders northward toward Lake Ontario. The bridge location is underlain by about 165 m (540 ft) of alluvial and glacial soil deposits filling a glacially scoured bedrock valley. The post-glacial alluvium, consisting of irregularly stratified silt, sand, and gravel is about 10 m (35 ft) thick. The alluvium, in turn, is underlain by at least 90 m (300 ft) of glacial lakebed sediments including two 30-m (100-ft) thick soft silt/clay sequences (NYSDOT, 1994). A mixture of coarse-grained sand and gravel outwash and broadly graded glacial till extends from the lakebed deposits to the bedrock surface. The uppermost bedrock unit is the Onondaga Limestone, a prolific regional aquifer, which is underlain by other limestone, shale, and evaporite units of Devonian age, including the salt-rich Salina Group.

An approximately 3.7 m (12-ft) thick, southward dipping zone of nearly pure halite at a depth of approximately 370 m (1,200 ft) has been mined by room-and-pillar techniques since the 1880s from works 8 km (5 mi) to the north. By the 1990s the mine's works had been extended beneath and about 0.8 km (0.5 mi) south of the bridge, bringing the mine's total area to about 2600 ha (6,500 ac) (Moran, 1995).

Mine Inundation and Piezometric Responses

In the early weeks of March 1994, fresh groundwater from the Onondaga aquifer began seeping into the mine cavity immediately south of the bridge. The seepage increased rapidly to an estimated rate in the range of 1.0 to 1.2 m^3/s (15,000 to 20,000 gpm). Despite efforts to stop it, the inflow continued until early 1996, when the entire mine was inundated. The inflow of fresh water immediately triggered dissolution of the mine's supporting pillars. The resulting fracturing and subsidence of the overlying rock apparently allowed even greater flows of fresh water from the limestone aquifer into the mine.

The mine, although more than 150 m (500 ft) below sea level, had been operated in a dry condition and therefore had an atmospheric piezometric head. The limestone aquifer, 180 to 210 m (600 to700 ft) above the mine cavity had an historic piezometric level of about El.165m (540 ft) (NGVD) above sea level. By August 1994, the piezometric head in the aquifer, near the bridge, had dropped below El. 70 m (230 ft) and to El. 45 m (150 ft) in December 1995 when the mine became completely flooded. Within the next six months, the piezometric level in the aquifer rose to El. 110 m (360 ft) about equivalent to the brine level in the flooded mine cavity (Akzo,1994-1996).

Concurrently, piezometric levels in the lower valley-fill sediments experienced a similar, but somewhat subdued, cycle of lowering and rebound. At a well, screened in glaciofluvial deposits at El. 54 m (177 ft), about 425 m (1,400 ft) southwest of the bridge, the water level dropped to El 84 m (275 ft) by late-January 1995 and rebounded to El. 108 m (354 ft) by the end of June 1996. In a well screened at El. 98 m (322 ft) in the lower of the two thick silty clay lacustrine deposits, located about 1370 m (4,500 ft) southeast of the bridge, the water level dropped to El. 149 m (487 ft), but showed no perceptible rebound through June 1996, following the flooding of the mine. It appears that the upper silty clay deposit (El. 125 m to 160 m (415 to 520 ft)) experienced little or no change in piezometric levels during the flooding of the mine.

Although the piezometric changes experienced by the deep soil and rock formations were dramatic, those experienced by the more compressible thick lacustrine silty clay deposits were, at most, very modest. Based on a review of the detailed subsidence data gathered during the 27 months following the start of inflow to the mine and calculations based on the results of several consolidation tests conducted on samples of the lacustrine strata (Gowan et al, 1996) (NYSDOT, 1981), it appears that less than 0.3 m (1 ft) of the observed surface subsidence is attributable to consolidation of the thick lacustrine clay deposits.

Surface Subsidence and Strain
As part of the mine owner's on-going operations, surface subsidence over the mine was routinely monitored. Soon after the inflow started, the monitoring program was expanded and intensified to include bi-weekly surveying of more than 600 points within and surrounding the mine area.

By late March, a 40-ft deep, approximately 260-ft diameter sinkhole, centered on the creek about 400 ft upstream of the bridge, developed. By late-May a second, 65-ft deep 600-foot diameter sinkhole, also centered on the creek, developed about 1,300 feet upstream of the bridge. Figure 1 shows the locations of the sinkholes relative to the Route 20A bridge and indicates the subsidence measured between March 1994 and March 1996. Subsidence has been most significant in the area just south of the Route 20A bridge.

Within an approximately 170 m (550-ft) radius of the smaller sinkhole, numerous concentric tension cracks, as much as 15 cm (6 in) wide with vertical offsets of up to 0.6 m (2 ft), developed, indicating as much as 1.5 m (5 ft) of concentric lateral movement and a similar amount of subsidence of the near-surface soil surrounding the sinkhole. The extent of these observed features are also shown on Figure 1.

The surveys showed that the ground surface began to subside almost immediately and continued to settle, but the rate diminished markedly after the mine was completely inundated (Akzo, 1994-7). Figure 2 is a plot of settlement versus time at the bridge's east abutment. Between March 1994 and January 1996, the existing U.S. Route 20A bridge settled more than 4.5 m (15 ft) and moved approximately 1.5 m (5 ft) upstream toward the nearer of the two large sinkholes.

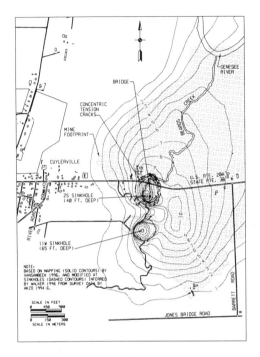

Figure 1. Site Map Showing Sinkholes and Subsidence Contours

The rapid development of the sinkholes is believed by the authors to have resulted from the loss of large volumes of soil from immediately above the bedrock surface. This soil loss was most likely the result of "piping" or internal erosion caused by the sudden discharge of groundwater from the lower soil deposits and the limestone aquifer associated with the flooding of the mine. Figure 3 presents the authors' hypothesis of the sinkhole development, based on their review of the site topographic and subsurface data and the application of sinkhole development concepts developed by Hendron, Fernandez, and Lenzini (1979).

PLANNING AND DESIGN

The mine owner assumed responsibility for designing and constructing a replacement for the damaged bridge. The authors' firm was engaged to assist the designer in planning and overseeing the construction of a new bridge meeting current service and design standards. The new bridge was to be constructed and opened to traffic as soon as practicable.

Feasibility Assessment

The design team's first task was to assess the feasibility of replacing the 30 m (100-ft) long, three-span bridge to its initial elevation within its existing right-of-way. Although other sites for the new bridge were considered, it was concluded that because the mine roof would be the most fully in contact with the mine floor in the area of greatest subsidence (in the areas adjacent to the sinkholes), the potential for additional long-term settlement would be less at the existing bridge site than in the other undermined and partially collapsed areas surrounding the sinkholes. Rebuilding the bridge at its original location would not only minimize the potential for greater post-construction settlement, but reduce the time and cost associated with acquiring additional right-of-way and constructing a new roadway.

As a part of the feasibility study, new design flood levels were computed for the recently altered terrain (Matzat, 1996) and a preliminary estimate of the potential long-term surface subsidence at the bridge was made.

Based on the assessment of the available data and projected settlements, it was concluded that a single-span bridge with adjustable bearings could be safely and economically built at the existing bridge site.

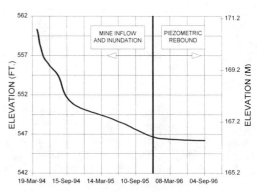

Figure 2. Measured Settlement versus Date East Abutment

Design Objectives

The principal objective of the design was to re-establish, as soon as practicable, a level of highway service to the affected communities, including emergency access during floods equivalent to that provided by the previous bridge.

Other significant objectives of the design included: 1) providing a 50-year service life, 2) complying with present-day bridge and roadway configuration criteria, 3) avoiding any expansion of the right-of-way onto an existing historic park between the western approach and the nearby sinkhole, and 4) minimizing the acquisition of additional right-of-way.

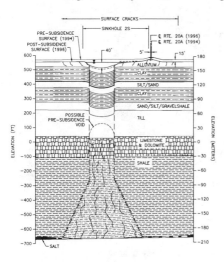

Figure 3. Conjectured Sinkhole Development

Geotechnical Analyses

Long-term settlement and foundation stability were the principal geotechnical concerns associated with the design of the new bridge (Walker, 1996).

The mechanism causing the on-going sinkhole-induced subsidence has not been determined.

However, it is believed to be a combination of plastic deformation of the deep, underlying lacustrine clay toward the more lightly loaded sinkhole and the related concentric shear failures in the shallower alluvial deposits. However, the regression analysis of the extensive time-series subsidence monitoring data for the bridge and several nearby points provides a rational basis for projecting the long-term effect of the on-going sitewide plastic deformation.

Due to the strained and cracked condition of the near-surface alluvial soils it was impractical to construct an analytical model for predicting the ground's short-term response to the new embankment and bridge loads. Therefore, it was concluded that the most practical means of dealing with short-term settlements would be to monitor subgrade settlements as the embankments were raised and to adjust the construction schedule to accommodate their demonstrated rates and magnitudes.

To refine the estimate of the long-term magnitude of the on-going, collapse-induced subsidence, a predictive analog, as presented in Figure 4, was developed by regression analysis of the bridge settlement readings following the inundation of the mine (Walker, 1996). This analog indicated a logarithmic reduction in the settlement rate and an estimated 20 cm (8 in) of collapse-induced general subsidence over the 50-year design life of the new bridge. The potential for short-term compression of the near-surface soils and long-term consolidation of the thick underlying clay strata due to the addition of the required embankment fills and the bridge itself was assessed, indicating a total of as much as 60 cm (24 in) of combined short- and long-term settlement over the next 50 years.

A review of the results of numerous consolidation tests conducted on samples of the local lacustrine clays (NYSDOT, 1981) (Gowan et al, 1996) showed the upper clay layer to be modestly compressible and to be slightly over-consolidated. However, rough calculations showed that the new embankment and structure loads would cause only minor stress increases and little consolidation-related settlement in the underlying, thick clay layers.

Based on a general assessment by Gowan, et

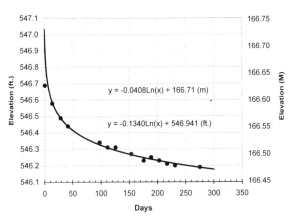

Figure 4. Regression Analysis of Post-Inundation Subsidence

al (1996) and the foregoing considerations, the authors concluded that the land beneath the new bridge would likely undergo between 20 and 60 cm (8 and 24 in) of settlement during construction and the subsequent 50 years, allocated to approximately 20 cm (8 in) of rapid elastic compression as the fills were placed, 20 cm (8 in) of long-term plastic deformation associated with the on-going adjustment of the upper soil deposits to the presence of the nearby sinkhole, and 0 to 20 cm (0 to 8 in) of long-term consolidation of the underlying thick clay layers.

The need to raise the roadway 5.5 to 6 m (18 to 20 ft) above its sunken level, and even higher above the creekbed and the nearby sinkhole, posed the potential for embankment instability. The areas of principal concern were the abutments and the flanks of the embankments at the creek edge, especially on the south side of the approach embankments, toward the sinkhole. To provide stability against shear failure of the abutments toward the creek, it was necessary to refill the creekbed to its original level with stone fill and to leave the old abutment foundations in place in addition to founding the new abutments on tightly spaced vertical friction piles.

The new embankment side slopes could be flattened sufficiently to assure lateral stability on the north side of the roadway. However, the need to avoid the park land and the sinkhole on the south posed a challenge to assuring short- and long-term stability of the embankments. Rotational and block stability analyses indicated that a combination of deeply driven, anchored sheet piling and additional stone filling upstream in the creekbed and at the base of the vertical sheet piling could provide the stability needed to support the raised embankments.

Design Features

As depicted in Figure 5, the new bridge consists of a 38-m (125-ft) long single-span, steel girder superstructure with an integral reinforced concrete deck seated on pile-supported concrete stub abutments.

To accommodate the anticipated 20 to 40 cm (8 to 16 in) of post-construction settlement and related potential distortion resulting from the installation of the more than 27000 t (30,000 tons) of earthfill, concrete, and steel, the bridge was designed with repositionable, flexible expansion bearings at both ends of the girders. Additionally, extra clearance between the girders and the abutment backwalls, and flexible, strip-sealed expansion joints at both ends of the deck were provided to allow the abutments to move as much as 5 cm (2 in) longitudinally.

CONSTRUCTION

The design and construction were advanced in a tandem ("fast-track") sequence to allow the early placement of the lower portions of the heavy earthfill and to promote (and get a measure of) anticipated settlement and pore-pressure responses and to see if the new loads would cause measurable lateral movements in the subgrade soil. Staging of the

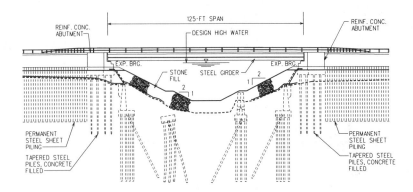

Figure 5. Configuration of Replacement Bridge

Reprinted from: Beck, Barry F. (ed.), Sinkholes - Proceedings of the seventh multi disciplinary conference, Harrisburg, USA, 10.14 April 1999. 1999. c.480 pp., EUR 100/US$115/GEP71.00. Please order from: A.A. Balkema, Old Post Road, Brookfield, Vermont 05036 (telephone: 802-276-3162; telefax: 802-276-3837; e-mail: info@ashgate.com).

design and construction provided a logical ("critical path") sequence in which the details of subsequent elements of the bridge and its foundations could be refined and adjusted, if necessary, to accommodate unexpected ground responses.

Response Monitoring

To minimize the uncertainty associated with the possible responses of the severely disturbed subgrade soils and yet, allow the earthwork to advance as rapidly as would be prudent, a network of settlement platforms, inclinometers, and piezometers was installed in critical locations before the earthwork was started. The eight settlement platforms were set and secured on the existing subgrade along the roadway centerline with one seated in the creekbed. The four inclinometers and vibrating-wire piezometers were installed in pairs. Two pairs were set immediately creekward of the proposed abutments at the centerline of the bridge and the other two about 3 m (10 ft) south of the abutment wingwalls, toward the sinkhole. The inclinometers extended to 18 m (60 ft) below the initial subgrade. The piezometers were set about 10 m (30 ft) below subgrade, where the maximum shear stresses from the new embankments and abutments were expected to develop.

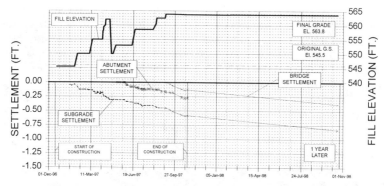

Figure 6. Ground Response to Construction of Replacement Bridge

These instruments were calibrated before the initiation of the earthfilling and were read frequently throughout the construction of the approach embankments and the bridge until its opening in October 1997. The data from these readings, along with close surveys of the positions of several critical points on the abutments as they were raised, were carefully evaluated and used to control the progress of the construction.

Construction Progress and Ground Responses

The earthwork was started in late December with the placement of crusher-run aggregate fill to raise the northern flanks of the new embankments to the approximate level (El 167 m (548 ft)) of the remnant road grade. In early February 1997, 2 to 3 m (7 to 10 ft) of medium stone fill was placed in the sunken creekbed to stabilize the severely ruptured subgrade and to re-establish the creek's channel. This filling helped to advance the compression of the subgrade and provided a working level well above normal spring run-off creek levels. As depicted and noted on Figure 6, the initial stabilization filling was followed by installation of the embankment-retaining sheet piling along the southern side of the roadway and the raising the roadway embankments to El. 169 m (555 ft), about 2.5 m (8 ft) above the grade of the sunken roadway. An additional 2 m (7 ft) of temporary surcharge fill was placed at the abutment locations.

In the meantime, design of the abutments and superstructure was completed and the girders were ordered. In early May, after settlements of about 10 cm (4 in), the surcharge was removed and the abutment foundation piles were driven and the pile caps cast. By early June, after about 11 cm (4.5 in) of total subgrade settlement, the abutment backwalls and bridge seats (girder-supporting pedestals) were cast. The embankments were then raised to El. 170 m (558.5 ft), about 1.6 m (5.3 ft) below the final roadway grade. The elevations of and distance between the two abutments were closely monitored. The abutments were found to have settled about 2.5 cm (1 in) with no significant lateral (convergent) movement detected by the control surveys or inclinometer measurement. On 10 July, the four girders were installed. In early August, the embankments were raised to within 0.5 m (1.5 ft) of final roadway grade. With the abutments having settled about 5 cm (2 in) since their completion, on 28 August, the bridge's concrete deck was cast and finished. By 6 October, the embankments had been brought to final grade and the roadway was paved. (Walker and Divito, 1997).

As shown on Figure 6, the old ground beneath the bridge had settled 18 to 20 cm (7 to 8 in) under the weight of the rebuilt roadway and new bridge. However, due to the construction sequencing, the finished deck and approaches had undergone only 5 cm (2 in) of essentially uniform and imperceptible settlement. At no time did the closely monitored instrumentation indicate any excessive pore pressure or instability in the underlying soil.

On 20 October 1997, ten months from the start of its construction, the bridge was opened to the public.

One Year Later

In late October 1998, the authors revisited the bridge site to observe its condition. The established settlement monitoring points were resurveyed. The survey indicated that the approach embankments and the bridge had undergone approximately 7.5 cm (3 in) of additional general settlement with a slight .25 to 1.0 cm (0.1 to 0.4-in) cant downward to the south, toward the sinkhole. It was observed that the flexible bearings on both ends of all four girders were uniformly strained about 1.2 cm (0.5 in), indicating that the abutments had converged about 2.5 cm (1 in) since the original setting of the girders.

Figure 7 presents a comparison of the actual settlements during and through the first year following construction to those projected prior to the start of construction.

Follow-up readings of the inclinometers have indicated lateral ground movements of approximately 0.6 cm (0.25 in) toward the creek and to the south, toward the sinkhole since completion of the construction and 2.5 cm (1 in), respectively, since the start of construction (NYSDOT, 1998).

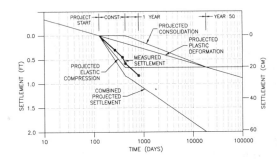

Figure 7. Comparison of Actual vs. Projected Settlements

PERFORMANCE SUMMARY

During the 10-month-long reconstruction of the Beards Creek bridge, when more than 27000 t (30,000 tons) of earthfill, steel, and concrete were placed within 90 m (300 ft) of the sinkhole, the ground beneath it settled as much as 20 cm (8 in), moved laterally as much as 25 mm (1 in), but demonstrated no detectable signs of foundation instability. During the year following its completion, the bridge and its approach embankments have undergone an additional 7.5 cm (3 in) of general settlement with some tilt toward the nearby sinkhole and some modest (in the order of 12 mm (0.5 in)) lateral shifting of the abutments toward the creek. Movements of the nature and magnitudes observed were anticipated in the design and are well within the projected values.

U.S. Route 20A is very heavily traveled, for a rural highway, and the new bridge and reconstructed roadway are performing exceptionally well, considering the site's recent history.

CONCLUSIONS

Prediction of the responses of severely disturbed ground to new loadings associated with the repair or replacement of facilities damaged by such disturbances presents a significant challenge to design professionals.

An understanding of the course and destiny of the damaging ground disturbance, the geologic setting, and the fundamentals of soil mechanics are essential for predicting the ground's responses to the new loads and stresses associated with the stabilization and re-use of the disturbed ground. Documented time-series observations of the ground's failure, even if just during the failure's aftermath, are crucial to establishing a rational basis for predicting future responses of the disturbed ground.

The failure and replacement of the Beards Creek bridge, has provided a useful set of observations, projections, and measured responses to the imposition of significant new earthfill and structure loads on a severely disturbed site.

APPENDIX

References

Akzo Nobel Salt Inc., 1994-1996. "Land Subsidence and Well Water Level Monitoring Data, March 1994 through September 1996", Unpublished.

Gowan, S. W., VanSambeek, L. L., and Brekken, G. A., 1996. "Anticipated Conditions for the Route 20A Highway and Bridge Reconstruction at Cuylerville, New York", prepared by RE/SPEC, Inc., Rapid City, SD and Alpha Geoscience, Albany, NY.

Hendron, A. J., Jr., Fernandez, G., and Lenzini, P., May 1979. "Study of Sinkhole Formation Mechanisms at Carey Salt Brinefield, Hutchinson, Kansas". Solution Mining Research Institute, Inc. Flossmoor, Illinois.

Matthews, S. L., 1996. "Structure Study Report for Replacement of US Route 20A and State Route 39 Bridge Over Beards Creek (BIN 1016090)", prepared by LaBella Associates, P.C., Rochester, NY.

Matzat, J. W., 1996. "Hydraulic Report for US Route 20A and State Route 39 Bridge Replacement Over Beards Creek", prepared by LaBella Associates, P.C., Rochester, NY.

Moran, R. P., Scovazzo, V. A., Strieb, D. L., 1995. "Impact Analysis, Retsof Mine, Livingston County, New York", prepared by John T. Boyd Company, Pittsburgh, PA.

New York State Department of Transportation, 1981. Geotechnical Laboratory Test Results, I-390 Highway Crossing of Genesee River Valley, Livingston County, New York, Albany, NY.

New York State Department of Transportation, 1994. Logs of Test Borings made at Beards Creek Bridge, March-April, 1994, Albany, NY.

New York State Department of Transportation, 1998. Summary of Field Instrumentation Readings, October 1997-September 1998, BIN 1016090, Albany, NY.

Walker, S. E., 1996. "Geotechnical Evaluation - Replacement of US Route 20A and State Route 39 Bridge Over Beards Creek", prepared by Haley & Aldrich of New York, Rochester, NY.

Walker, S. E. and Divito, R. C., 1997. "Geotechnical Monitoring Summary Report, Replacement of U.S. Route 20A and State Route 39 Bridge over Beards Creek BIN 1016090, Town of Leicester, Livingston County, New York", prepared by Haley & Aldrich of New York, Rochester, NY.

Walker, S. E. and Matzat, J. M., 1997. "Planning the Replacement of the Beards Creek Bridge", *The Engineering Geology and Hydrogeology of Karst Terranes, Beck & Stephenson (eds)*, pp 373-380. Balkema, Rotterdam. ISBN 90 5410 867 3.

Walker, S. E. and Matzat, S. L., 1999 (in publication). "Replacing the Beards Creek Bridge", *Proc. 7th Multidisciplinary Conf. on Sinkholes and the Engineering and Environmental Impacts of Karst^{tm}*, April 1999, Harrisburg, PA.

Measured Horizontal And Vertical Movements From a Sag Subsidence

Gennaro G. Marino, Ph.D., P.E., [1]
and James W. Mahar, Ph.D., P.G.[2]

Abstract

This paper presents data collected on sag subsidence which resulted over an abandoned room and pillar coal mine in east central Illinois. The mine is approximately 40 meters [130 ft] deep and generally has extraction ratios ranging from 60 to 75%. Mine subsidence in the study area occurred over a period of 15 years. During this subsidence investigation mine collapse was expected to progress into unsubsided areas because of the past subsidence history. A ground surface monitoring program was set up over an area of possible future subsidence. Both vertical and horizontal displacements were measured. Subsidence eventually occurred in this area and both horizontal and vertical measurements were obtained. This paper contains the displacement data as well as characteristics of the other sags on the site. To authors' knowledge this is the most complete record of lateral and vertical displacement information where measurements were taken before and after unplanned sag subsidence.

Introduction

The information contained in this paper was part of a joint study performed by the Civil Engineering Department at the University of Illinois in cooperation with the U. S. Bureau of Mines in Minneapolis, Minnesota. The study included investigation

---

[1]President, Marino Engineering Associates, Inc., 907 W. Fairview Ave., Urbana, IL 61801

[2]President, Geotechnical Consultants, Inc., 4 College Park Court, Savoy, IL 61874

of the mine stability and surface subsidence effects over an abandoned room and pillar coal mine near Danville, Illinois. The scope of work performed over a seven year investigation period included: determining the subsidence profile characteristics, the subsidence induced damage; the subsurface conditions, and the rock mechanics properties of the coal measures, as well as evaluating the ground movements, the associated damage, and the mine stability conditions. This investigation covered 6.9 hectares [17 acres] and included mine subsidence that took place from months to about 20 years after mining.

The focus of information presented in this paper is related to measurements of displacements on the ground surface. This information is useful because subsidence over abandoned coal mines occurs unexpectedly and a complete history of displacement is not available. This case data was obtained because the ground surface was instrumented prior to the subsidence. Another important aspect of the measurements is the occurrence of overlapping mine subsidence sags above abandoned room and pillar workings. This overlap which occurs at different times has caused extensive damage to structures and utilities.

Site Description

The study area is located in the town of Hegeler, in east-central Illinois. The site lies a few kilometers south of Danville and about 8 kilometers [5 miles] west of the Illinois-Indiana border. The town of Hegeler can be characterized as a light industrial and agricultural area with a population of about 1,600. The area is extensively underlain by room and pillar coal workings that were operated between 1870 and 1974.

The study area as shown in Figure 1, is bounded on the south side by Spelter Avenue. Subsidence affected the area north of Spelter Avenue and includes a radio station, the three associated radio towers, and residential homes (most of which have now been abandoned and demolished because of subsidence damage). Farmlands lie to the north and east of the radio station property. The site topographically is flat to gently rolling.

Geologic Conditions

The exploratory work consisted of drilling six boreholes (B-1 to B-6) at the project site to depths from about 46 to 49 m [150 to 160 ft] (see Figure 1). The holes were drilled about 2.4 to 6.7 m [8 to 22 ft] below the mine floor (below the Herrin No. 6 Coal).

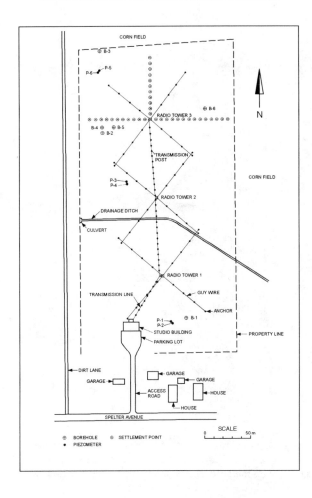

Figure 1. Site Plan of Hegeler Study Site

A general geologic column at the site is given in Figure 2. The soil at the site essentially consists of 9 to 14 m [30 to 47 ft] of glacial till that lies directly on bedrock. Coal measures at the site are in the Illinois Basin. The mined-out No. 6 coal has a rock cover of 26.5 to 32.3 m [87 to 106 ft] and is 39.9 to 41.2 m [131 to 135 ft] below the ground surface. The rock above the mined-out seam mainly consists of shale.

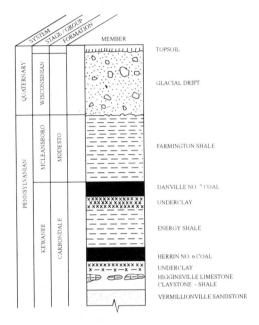

Figure 2. Geologic Column In The Hegeler, Illinois Area

Mining Conditions

The V-Day Coal mine underlies most of the town of Hegeler (which is just south of Danville) and for an extensive area north of the site. The Herrin (No. 6) Coal was worked from the mine between 1946 and 1967. This mine was the last operating underground mine in the area and was closed in 1974.

The 1.8 to 2.1 m [6 to 7-ft] thick coal seam was mined using a modified room and pillar system (Hunt, 1980) with mine openings oriented north-south and east. In the study area the V-Day mine generally has rectangular pillars ranging in width from 3.0 to 7.6 m [10 to 25 ft], rooms are 6.1 to 13.7 m [20 to 45 ft] wide, and cross-cuts are typically 3.0 to 7.6 m [10 to 25 ft] wide (see Figure 3). The cross cuts are generally located every 25.9 to 48.8 m [85 to 160 ft]. The pillar height to width ratio ranges from 0.34 to 0.4 and the panel width to panel depth ratio varied from 1.9 to 2.8. The percent extraction in the mine is typically 66 to 74 percent (average 70%).

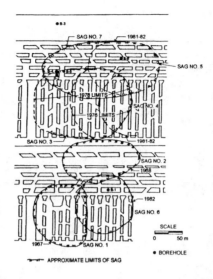

Figure 3. Subsidence Over The Mined-Out Areas

History of Subsidence

Figure 1 shows a plan view of the study site. Sag subsidence was first reported at the site in July, 1967 and has extended across essentially the entire site. A plan of the approximate sag limits observed at the surface with their respective dates of occurrence are given in Figure 3. The sags have been numbered in their chronological order of occurrence. The time at which each sag developed was determined either by surface damage observations or based on available measurement data. Mine subsidence occurred over both panels and entries.

The first sag (Sag No. 1), which occurred in July, 1967 developed in the southwest section of the study site (see Figure 3). Sag No. 1 occurred with little warning. Ground movements cracked and buckled asphalt pavement, destroyed three homes, severely damaged a radio station building, and ruptured service utilities to these structures. In addition, some of the guy wires to the nearest radio tower for the station were tensioned. Most of the major displacement occurred over a period of two days.

Sag No. 2 developed north of Sag No. 1 about one year later (see Figure 3). The center of the sag coincides approximately with the foundation of Radio Tower 2. Most of the settlement appeared to occur rapidly and after six weeks the tower had subsided 0.84 m [2.75 ft]. The guy wires supporting the tower loosened and had to be retightened several times to prevent damage to the tower.

Sag No. 3, the third area that subsided on the site, formed in November-December, 1976 just north of Sag No. 2 (see Figure 3). Sag No. 3 almost caused a radio tower to fall when the foundation settled significantly and tilted about 17 percent. In July, 1978, an adjacent area of mine collapsed and extended Sag No. 3 to the west. This mine subsidence damaged two towers by straining guy wires. Surface, subsurface, and mining conditions related to Sag Nos. 1, 2, and 3 were discussed in detail by Marino et al (1982).

The subsided area north and east of Sag No. 3 appears to have occurred in three stages and is indicated by Sag Nos. 4, 5, and 7 in Figure 3. The sag numbers 4, 5 and 7 were assigned based on the appropriate order of subsidence on the site. Sag No. 7 was marked by excessive settlement of Tower No. 3 on July 22, 1982. The settlement affected the transmission from the tower and required retensioning of loosened guy wires.

Sag No. 6 (Figure 3) developed on June 16, 1982 about one month before Sag No. 7. This subsidence occurred in the southern section of the study site and damaged two homes (see Figure 1 and 3). Sags 4 through 7 are described in detail by Marino and Devine (1985).

Subsidence Characteristics

General

The subsidence limits and profile characteristics were assessed by determining the difference between the post- and pre-subsidence elevations. Pre-subsidence elevations for Sag Nos. 1 to 3 were estimated by linear extrapolation between points located outside the main area of downward movement. For Sag No. 1, the limits of the subsidence were determined from a map of perimeter tension cracks and compression ridges prepared by staff of the Illinois State Geological Survey (1967), accounts of damage, and differential settlement of the radio station building. Ground surface profiles prior to subsidence were estimated by interpolating linearly across the access road and the parking lot to points estimated outside the edge of the depression.

For Sag Nos. 2 and 3 in the field north of the station building, a topographic map was prepared from elevation measurements on a 6.1 to 7.6 m [20 to 25 ft] grid in order to establish postsubsidence profiles. The presubsidence profiles were determined by linearly interpolating between presubsidence elevation points from outside the area of major downward movement.

For Sag Nos. 4, 5 and 7 a subsidence contour map was prepared based on before and after topographic surveys made in the field north of the radio station building. Also for Sag Nos. 5 and 7 settlement points were surveyed before and eight times after the main movements. The settlement points were installed on a 6.1 m [20 ft] spacing and extend outward from Tower 3 in the north, east and west directions (see Figure 1). To the south, surface ground shots were taken adjacent to transmission poles leading to Tower 3 (see Figure 1).

Table 1 summarizes the measured subsidence profile characteristics for the sags at the site. The maximum subsidence ranged from 0.6 to 1.1 m [1.8 ft to 3.5 ft] over sag areas 76 to 183 m [250 to 600 ft] wide. The maximum subsidence slope and curvatures are severe and generally increased with maximum subsidence. Subsidence profile characteristics could not be determined for Sag Nos. 4 and 6.

Table 1. Summary of Measured Sag Characteristics

Sag Number	1	2	3	5	7
Maximum Diameter (m)	106.7	135.6	173.7	182.9	182.9
Minimum Diameter (m)	93.0	71.6	125.0	121.9	121.9
Maximum Settlement (m)	1.07	0.91	0.91	0.8	0.55
Maximum Slope	0.055	0.038	0.034	0.044	0.024
Maximum Curvature (m^{-1})					
Compression	3.9×10^{-3}	2.8×10^{-3}	2.1×10^{-3}	3.2×10^{-3}	1.6×10^{-3}
Tension	3.9×10^{-3}	2.3×10^{-3}	2.1×10^{-3}	3.0×10^{-3}	1.3×10^{-3}
Extraction Ratio (%)	66	74	71	65	60
Seam Height (m)	1.95	1.95	1.86	1.86	1.86
Modified Subsidence Factor*	0.83	0.64	0.70	0.66	0.49

* Modified Subsidence Factor: $SF' = S_{max}/eH$ Where S_{max} = Maximum Subsidence; e = Extraction Ratio; and H = Seam Height

Vertical Displacement for Sag No. 7

Sag No. 7 is an oval shaped depression about 183 m [600 ft] long and 122 m [400 ft] wide, and about 0.55 m [1.8 ft] deep. The subsidence encompassed by Sag Nos. 5 and 7 resulted in an elliptical shaped depression about 274 m [900 ft] long and 122 m [400 ft] wide (see Figure 3). Most of the settlement points installed along both the E-W and N-S lines were affected by Sag No.7. Figure 4 shows the vertical displacement of the settlement monuments 4 and 496 days after Sag No. 7 was first noticed. There is about a 0.15 m [0.5 ft] high spot (hump) in the east-west subsidence profile just east of Tower 3 (see Figure 4). The topographic "hump" was caused by superposition of Sags 5 and 7.

346　　　　　　　GEO-ENGINEERING FOR UNDERGROUND FACILITIES

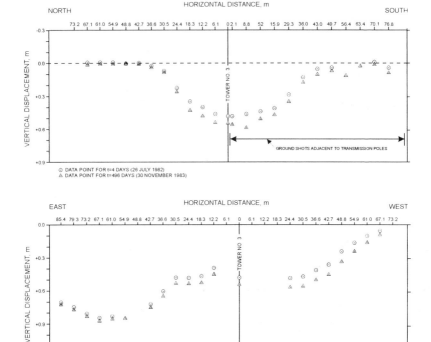

Figure 4. North-South and East-West Settlement Profiles of Monument Points at Tower 3 at 4 and 496 Days After Sag No. 7 Was First Noticed.

Time displacement data show that between the interval of 4 to 496 days a maximum of 0.12 m [0.4 ft] of subsidence occurred and was concentrated mostly near the center of the sag. The displacements after 4 days generally decrease to zero from the center to the sag limits. Compared with other sags in Illinois the residual movement of the Sag No.7 subsidence is relatively small.

Lateral Displacement

To monitor lateral displacement induced by subsidence at the ground surface, the horizontal distances between the existing transmission poles were measured periodically. Pole locations are shown in Figure 1. The horizontal distances between transmission poles were measured with a calibrated tape extensometer. The extensometer measurements for poles north of Tower 2 (Pole Nos. 22 to 34) were taken before and 4, 8, and 473 days after the main movement in Sag No. 7.

Two sets of lateral displacement readings (i.e., the upper and lower measurements shown in Figure 5) were made to assess the sag induced curvature effects on the measured changes in contraction or extension between the poles. However, it was found that the maximum strain correction was only 0.06% at the point of maximum curvature and therefore only the lower readings were used to determine the lateral ground displacements at the base of the poles.

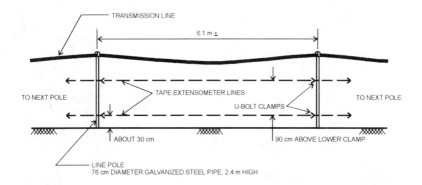

Figure 5. Tape Extensometer Measurements Between Transmission Posts

Figure 6 shows a plan view of the subsidence contours with the transmission pole locations and the geometric relationship between the pole displacements and directions to the change in the lateral distance between the poles. The orientation of the pole displacement was taken to be in the direction of the dip of the sag at the pole location (see Figure 6). Pole displacements relative to Pole No. 24 are plotted in Figure 7. From level survey data the zero subsidence for Sag No. 7 was approximately at Pole No. 24 (see Figure 4). As shown in Figure 7, the maximum lateral displacement is 0.25 m [0.82 ft], which decreased with increasing time to 0.23 m [0.75 ft]. Noting that the maximum associated subsidence is 0.53 m [1.75 ft], the ratio of the lateral to vertical displacement equals 0.47.

Figure 7 also shows the strain profiles, calculated in 6.1 m [20 ft] intervals, assessed from the two pole displacement curves at 4 and 473 days after the initial movements. About 40% of the profile was in compression (or 60% of the profile was in the tension zone). The maximum tensile strain measured is 1.9% but occurred close to the mid-profile zero strain areas. The maximum compressive strain, however, could not be determined because of insufficient data in the region of maximum subsidence. Based on the available data the maximum compressive strain could have been on the order of 1 to 2.2 %. The 4 and 473 day strain profiles indicate a slight strain relaxation with time.

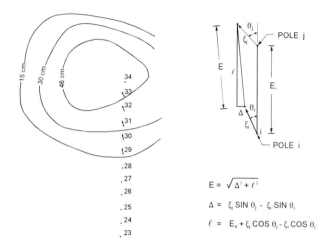

$$E = \sqrt{\Delta^2 + \ell^2}$$
$$\Delta = \xi_j \sin\theta_j - \xi_i \sin\theta_i$$
$$\ell = E_o + \xi_j \cos\theta_j - \xi_i \cos\theta_i$$

Figure 6. Plan and Geometry of Transmission Pole Displacements for Sag No. 7

For longwall mining, where subsidence is planned, empirical lateral displacement and strain relationships have been developed. Table 2 shows ratios of maximum lateral displacement to maximum subsidence for longwall mining in various coal mining regions. Also in modeling subsidence movements on the ground surface from high extraction mining of flat lying coal beds Kratzsch (1983) reports using a ratio of 0.45. The maximum lateral to vertical displacement ratio measured at the site study of 0.47 is on the high end of the reported range (see Table 2).

Maximum lateral displacement values are used in profile precalculation methods. Important applications of profile precalculations are in determining the subsidence potential of an undermined, proposed construction site and in designing subsidence resistant foundations (Marino, 1991 and 1998). Precalculation methods predict lateral displacements by distributing the maximum displacement in proportion to the slope profile (Brauner, 1973, Peng and Geng, 1982, and Marino, 1998). The lateral strain can then be calculated for any interval along the predicted profile. Because the change in slope is proportionally equal to the change in lateral displacement over some interval, the curvature is empirically proportional to the ground strain. Empirical relationships have been developed between the curvature and horizontal strain (e.g., NCB, 1975, and Bauer and Hunt, 1981).

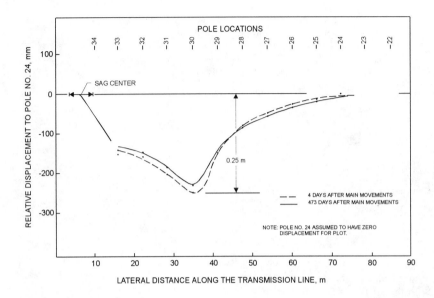
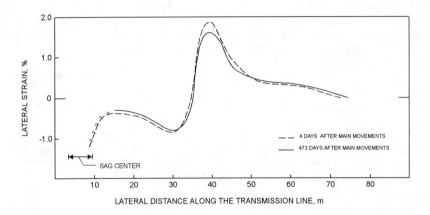

Figure 7. Lateral Displacements and Strains Along the Transmission Line for Sag No. 7

Table 2. Maximum Lateral to Vertical Displacement Ratios, V_{max}/S_{max}, from Longwall Mining

Location	V_{max}/S_{max}	Reference
U.S. Appalachian Field	0.3	Peng and Geng, 1982
Germany	0.35 - 0.50	Brauner, 1973; Muller, 1973; and Pflaging, 1978
USSR	0.30 - 0.35	Brauner, 1973
France	0.4	Brauner, 1973
Great Britain	0.04 - 0.32 (G. B. Ave: 0.22 all 0.15 limestone cases 0.24 others)	Breeds, 1976

In the field, the maximum calculated strain depends on the measurement interval, i.e., the maximum strain increases with a smaller measurement interval. Although the smaller the interval the larger the effects of measurement inaccuracies. Horizontal measurement intervals of 0.05 and 0.1 times the mine depth have been used (NCB, 1975, Bauer and Hunt, 1981). Until this paper, however, no measured profiles of lateral displacement or strain have been reported over abandoned mines because of the unpredictable character of sag subsidence. Bauer and Hunt (1981) determined the maximum ground and "tar and chip" road extension strain occurring over a distance of 0.1 times the mine depth for a number of abandoned mine cases. The results are shown on Figure 8 where maximum tensile strains are plotted against the associated maximum tensile curvatures for both longwall and room and pillar subsidences. A large strain variation exists for curvatures greater than a 0.61×10^{-4} m^{-1} [2×10^{-4} ft^{-1}].

To compare the measured maximum tensile strain of 1.9% and the corresponding tensile curvature of 1.14×10^{-4} m^{-1} [3.73×10^{-4} ft^{-1}] at the study site to other case data the study results have been plotted in Figure 8. As can be seen the study site data indicate strain greater than the data measured using ground surface cracks.

Summary

An extensive study of the subsidence history and related conditions of an abandoned mined-out area just south of Danville, Illinois was conducted. The mine is about 40 m [130 ft] deep. The entire study area progressively subsided over 15 years with 7 separate subsidence sags developing on the ground surface. This paper summarizes the subsidence measurements including horizontal ground surface displacements where the ratio of maximum horizontal displacement to maximum vertical displacement was found to be 0.47. The measured maximum settlement of the sag subsidence ranged from 0.55 m to 1.07 m [1.8 to 3.5 ft].

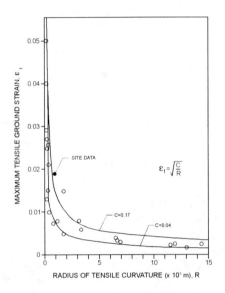

Figure 8. Maximum Tensile Ground Strain Compared With the Radius of Tensile Curvature in Illinois (After Bauer and Hunt, 1981)

REFERENCES

Bauer, R., and Hunt, S. R., 1981, Profile Strain and Time Characteristics of Subsidence from Coal Mining in Illinois, Proc. Workshop on Surface Subsidence due to Underground Mining, Morgantown, WV, pp. 207-218.

Brauner, G., 1973, Subsidence Due to Underground Mining (in two parts), 1. Theory and Practices in Predicting Surface Deformation, BuMines IC8571, 56 pp.

Breeds, C.D., 1976, A Study of Mining Subsidence Effects on Surface Structures with Special Reference to Geological Factors, Ph.D. Thesis, University of Nottingham, October, 279 pp.

Hunt, S.R., Surface Subsidence Due to Coal Mining in Illinois, Ph.D. Thesis, University of Illinois at Urbana-Champaign, 1980, 129 pp.

ISGS, 1967, Subsidence at Hegeler, Illinois State Geological Survey, Internal Field Report, 9 pp.

Kratzsch, H., 1983, Mining Subsidence Engineering, Springer-Verlag, New York, pp 543.

Muller, W.: Die Berechnng der Bewegungen und Spannungen des Geirges vom Abbau bis zur Tagesoberflache nach der Methode der endlichen Elemente, Diss. Berlin1973.

Marino, G.G., Mahar, J.W., Dobbels, D.J., and Kiesling, D.R., 1982, Mine Subsidence and Related Structural Damage, Report to USBM, Minneapolis, MN, 160

Marino, G.G., and Devine, A., 1985, Mine Subsidence and Structural Damage, Hegeler, Illinois, from July 1981 to February 1985, U.S.B.M. Report, 45 pp.

Marino, G. G., 1991, Foundation Design in a Subsidence Prone Area in Indiana, Proceedings of the 34th Annual Meeting of the Association of Engineering Geologists, "Environmental and Geotechnical Challenges for the Decade", Chicago IL, 11 pp.

Marino, G.G., 1998, The Siting of a Prison Complex above an Abandoned Underground Coal Mine, ASCE Geotechnical Journal, October, 1998, Vol. 124 No. 10.

National Coal Board (NCB), 1975, Subsidence Engineer's Handbook, United Kingdom National Coal Board Production Dept., 2nd Edition, London, 1975, 111 pp.

Peng, S.S., and Geng, D.Y., 1982, The Appalachian Field: General Characteristics of Surface Subsidence and Monitoring Methods, Surface Mining Environment Monitoring and Reclamation Handbook, L.V.A. Sendlein, H. Yazicigil, and C.L. Carlson, Editors, Elsevier Science Publishing Co., Inc., NY, pp 627-645.

Pflaging, K.: Aktualisierung der mathematischen Anatze fur Vorausberechnungsverfahren fur Bodenbewegungen, Prodeeings, "Lagerstattenerfassung, Bodenbewegungen, Bergschaden, Ingeniervermessung", Clausthal 1978, Vol. 8, 187/223.

Mine Shaft Stabilization Using Compaction Grouting

Len Meier[1] and Arthur G. Hoffmann, P.E.[2], Member

Abstract

A home owner in Des Moines, Iowa experienced cracking basement walls, sticking doors, basement slab settlements, cracking stone fascia and settlement of her front yard and sidewalk for the second time in three years. The U.S. Department of the Interior, Office of Surface Mining, Reclamation and Enforcement investigated the claims and initiated a subsurface investigation. Using literature review, drilling and geophysical investigations, it was determined that the house was located on top of an abandoned underground coal mine shaft. The loss of soil into the shaft was resulting in subsidence and progressively worsening structural damage to the house. A remedial alternatives analysis was performed and compaction grouting of the mine shaft and surrounding loose soils was found to be the most viable and cost effective alternative. A five week compaction grouting program was successfully performed both outside and within the basement. Approximately 153 cubic meters (200 cubic yards) of grout were injected, the ground was stabilized and the shaft was filled. Subsequently, portions of the basement walls and floor slab were replaced and the house was leveled.

The Initial Complaint and OSMRE Response

In August 1997, a homeowner in Des Moines, Iowa, reported to the U.S. Department of the Interior, Office of Surface Mining, Reclamation and Enforcement (OSMRE) Mid-Continent Regional Coordination Center in Alton, Illinois, that basement walls on the north and east sides of her residence had recently cracked, doors of the home did not close properly, the basement slab had settled approximately 2.5cm (1 inch), and that the exterior stone fascia on the north wall had cracked (Figure 1) and displaced approximately 2.5cm (1 inch) horizontally. She also reported that the yard and the sidewalk leading to her front door had recently settled. OSMRE investigated the claim and collected historic information about the

---

[1] Project Manager, OSMRE, MCRCC, Alton Federal Building, 501 Belle Street, Alton, IL 62002.

[2] Geotechnical Section Manager, Gannett Fleming, Inc., Three Foster Plaza, Pittsburgh, PA 15220.

house. It was learned that in 1994, the owner hired a contractor to repair the basement floor slab, which had cracked. The slab was removed and a 1.2m (4 foot) deep, 3.4m (10 foot) wide circular depression was uncovered near the northeast corner of the house.

The contractor was recalled by OSMRE in August of 1997 to investigate the basement conditions. Backhoe trenches excavated to a depth of approximately 3.7m (12 feet) on the north and east sides of the house revealed firm subsurface soils on the east side of the house and soft, easily probed soils on the north side. Small diameter holes drilled through the previously replaced basement slab indicated that the underlying material had settled approximately 25cm (10 inches). The recent basement wall cracks were located among cracks which had been repaired more than 18 years earlier. Repairs of the older cracks were still competent in 1994 when plywood paneling was removed from the concrete walls by the homeowner. In July, 1997, the homeowner observed renewed settlement of the floor slab and fresh cracks in the wall.

Figure 1
Cracked Exterior Stone Fascia
North Exterior Wall

Investigations

Preliminary Site Reconnaissance: OSMRE retained Gannett Fleming, Inc. of Pittsburgh, Pennsylvania to provide structural and geotechnical investigation and evaluation services for the project. Gannett Fleming conducted a site reconnaissance and a preliminary literature review of subsurface conditions. The Gannett Fleming site reconnaissance confirmed the visual observations made by OSMRE and identified the potential for further structural damage due to continuing movements, with the potential to produce direct injury to the inhabitants. A review of published literature indicated that coal mining had occurred approximately 46m (150 feet) below the site in the Cliffland coal seam and approximately 55m (180 feet) below the site in the Black Oak coal seam. Available underground mine mapping indicated a vertical mine entrance in the immediate vicinity of the property.

The preliminary conclusion, based upon the reconnaissance, was that unstable earth movements were the cause of the structural distress and settlement and that the source of the movements was likely related to subsidence of the abandoned mines or to soil movement into the vertical mine entry. A geotechnical investigation program was initiated, with the goal of identifying the relationship between the surface deformations and the abandoned deep mine workings and/or shafts.

Geotechnical Investigations: The geotechnical investigations included a literature review and interview process as well as structural monitoring. Additionally, both geophysical and conventional subsurface boring investigations were performed.

GEO-ENGINEERING FOR UNDERGROUND FACILITIES 355

Literature Review: The historical sequence of development of the neighborhood, the extent of undermining in both coal seams and the potential for existence of large, unknown subsurface structures were investigated using aerial photographs, mine and geologic mapping, Sanborn fire insurance mapping, and interviews with private and public utilities. This review revealed that widespread mining activities ceased in approximately 1935. The mining activities predated development of the neighborhood. Additionally, it was determined that no large, unmarked subsurface structures, such as tanks or cisterns, were operated by utility owners in the area.

Structural Monitoring: The residence was evaluated for structural integrity. Exposed structural members were inspected for excessive deformations and deflections and loss of support for joists and the center main floor support beam.

Deformations and deflections were identified in the basement foundation walls at the east end of the residence. In addition to visual observations, four Avonguard crack gauges were installed and monitored. Three were installed in the basement and one was installed outside the house on the stone fascia. During the four month investigation and design period, no significant and reliable movements of the interior gauges was observed. The exterior gauge revealed 10mm of both vertical and horizontal movement.

Geophysical Investigations: Due to the potential presence of an unsafe, subsurface, vertical mine shaft, geophysical investigations were performed in advance of borings. Both seismic refraction profiling and microgravity mapping surveys were performed by Enviroscan, Inc. of Lancaster, Pennsylvania. The purpose of the program was to detect and delineate subsurface soil disturbances and/or mass deficiencies that might be associated with the suspected mine shaft. It was assumed that the shaft was either open (i.e., a void) or loosely backfilled. The seismic refraction profiling was performed, primarily, to determine the density of subsurface strata for use in interpreting the gravity data, and, secondarily, to detect and delineate any loose backfill or disturbed soils associated with the shaft. The microgravity mapping was performed to directly detect and delineate mass deficiencies, such as the shaft.

A seismograph and a weight-drop source were used to record seismic travel times at 3m (10 foot) intervals along 13 linear arrays of geophones across the site. Data analyses resulted in the identification of three distinct subsurface soil layers. No loose backfill or disturbed soils were directly identified.

Figure 2 - Site and Boring Location Plan

Gravity readings were collected at stations located on approximately 4.6m (15 foot) centers across the site (including the basement of the house) using a gravimeter. The gravity data displayed a distinct "bulls-eye" low (i.e., an inferred mass deficiency) in the front lawn adjacent to the north corner of the house. The mass deficiency was interpreted as being consistent with the presence of a loosely filled vertical mine shaft. The location identified correlated well with the observed subsidence of the front yard.

<u>Boring Program</u>: A total of six borings were performed (Figure 2). Initially, three borings were performed on the north side of the house, in the front yard, by J&R Drilling Services of Grimes, Iowa. Boring B-1 was advanced at a distance of 15m (50 feet) from the estimated center of the suspected shaft location. The purpose of this boring was to collect background subsurface data, such as soil stratigraphy and characteristics, depth to bedrock and coal seams, and any indications of deep mine subsidence observable in bedrock. Boring B-2 was drilled at the center of the suspected shaft. Boring B-3 was drilled between B-2 and the northeast corner of the house. All three borings encountered approximately 15m (50 feet) of soft silts, clayey silts and silty sands overlying shale bedrock. The soils exhibited unconfined compressive strengths of between 120 to 240 kPa (1.25 and 2.5 tsf) and Standard Penetration Test results ranging between six and ten blows per foot. Static water levels were observed at the top of bedrock. The base of the upper coal seam was encountered at a depth of 47.2m (155 feet) below grade. The coal was intact. No evidence of deep mine subsidence was observed.

Figure 3 - Angle Borings

Based upon these data, it was determined that there was no shaft at the suspected location. Subsequently, two angled borings, B-4 and B-5, were drilled beneath the house, from the front yard, by Maxim Technologies, Inc. of St. Paul, Minnesota (Figure 3). Each of the angled borings encountered wood and very soft soils or voids beneath the northeast corner of the house. It was estimated, based on the visual and physical evidence and the geometry of the borings, that the drill strings had encountered a wood cribbed, approximately 1.5m (5 foot) diameter, vertical air shaft which was partially filled with loose soils. The shaft was located 1.8m (6 feet) west and 1.8m (6 feet) south of the northeast corner of the house, within the footprint of the house.

In order to verify the presence of the shaft, and to determine the depth and composition of the materials within the shaft, one additional boring, B-6, was advanced from within the basement. The interior drilling work was performed by

Freddie Van's drilling company of Pittsburg, Kansas using a customized hydraulic horizontal boring machine which was turned on its side (Figure 4). Very soft soil, apparent voids of up to 3m (10 feet) in height and loose rubble were encountered in the shaft. The boring confirmed the presence of a shaft that extended to the depth of the Cliffland coal seam.

Investigation Findings and Conclusions

Based upon the investigation of site utilities, observations of settlements made on site, the deformation characteristics of the structure, the observations and repairs made in 1994, and the boring data, the following findings and conclusions were developed:

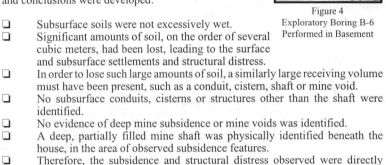

Figure 4
Exploratory Boring B-6
Performed in Basement

- ❏ Subsurface soils were not excessively wet.
- ❏ Significant amounts of soil, on the order of several cubic meters, had been lost, leading to the surface and subsurface settlements and structural distress.
- ❏ In order to lose such large amounts of soil, a similarly large receiving volume must have been present, such as a conduit, cistern, shaft or mine void.
- ❏ No subsurface conduits, cisterns or structures other than the shaft were identified.
- ❏ No evidence of deep mine subsidence or mine voids was identified.
- ❏ A deep, partially filled mine shaft was physically identified beneath the house, in the area of observed subsidence features.
- ❏ Therefore, the subsidence and structural distress observed were directly related to the movement of subsurface soils into the mine shaft. This was the accepted mechanism of failure.

The settlements and structural distress was characterized as a chronic problem. The evidence of crack repairs dating back 18 or more years, the settlement of the basement floor slab twice in three years and the observation of continuous movements of the exterior walls, which were supported independently from the interior walls, supported this conclusion. There was no evidence that the failure mechanism had stabilized or would do so in the future. Unless the ongoing failure was halted, the structural stability of the house would continue to deteriorate. The condition of the structure at the time of the investigation was adequate. However, it was determined that the structure would become unsafe if settlement was allowed to continue. The presence of identified but unmapped voids also presented the possibility of sudden, catastrophic collapse of the affected portion of the foundation if stabilization was not accomplished. OSMRE determined that the abandoned underground mine conditions that were the source of the ongoing soil loss required remediation in order to halt the degradation of the structure.

Remedial Alternatives

The goal of the remedial program was to eliminate the effects of the mine shaft on the structure. Densification of the subsurface soils and filling of the shaft using compaction grouting techniques as well as support of the structure on deep foundations and relocation of the house, on the same lot, away from the shaft, were the remedial alternatives considered. The advantages and disadvantages of each alternative were evaluated.

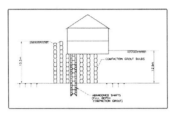

Alternate 1
Compaction Grouting

Compaction grouting would address all of the failure mechanisms and provide a long term solution by halting soil migration into the shaft and by densifying loose soils between bedrock and the surface. The technique is flexible and field directed and adjusted to suit site conditions. The resident would not be displaced during construction. The technique is routine for specialty contractors and has been successfully applied in the past by OSMRE. Work would be performed inside the house, which would be an inconvenience due to fugitive dust and noise. Potential existed for damage to the structure due to grout injection and drilling activities.

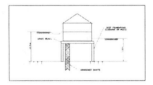

Alternate 2
Deep Foundations

This would be a long term solution with few unknowns relative to support of the structure. Most of the work could be performed outside without displacing the resident. The work is expensive and would require excavation and structural support during construction. Although the structure itself would be supported, the mechanism of failure would not be halted and it is possible that the basement slab would have to be replaced again in the future. Future subsidence would be likely, possibly affecting utilities and neighboring structures.

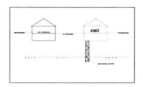

Alternate 3
Relocation of Residence

Relocation of the structure would be routine construction work. The effects of the failure mechanism would be mitigated for the structure itself, however, utilities and neighboring structures would remain at risk. In order to relocate the house, a waiver from the local 10m (30 foot) property line setback would be required. Winter weather would increase costs due to earthwork and concrete work requirements and the resident would likely have to be temporarily displaced during construction.

Cost Benefit Analyses

It was decided that support of the house on deep foundations (Alternate 2) would not be carried forward. This alternative would not mitigate the underlying failure mechanism and future settlements could be expected, with the potential to adversely affect the yard, utilities, and the neighbor's house. This alternative was not considered to be an effective solution.

Cost estimates for the remaining alternatives were developed. The estimate for compaction grouting (Alternate 1) was based upon historical costs for compaction grouting projects designed by Gannett for, and constructed by private contractors for OSMRE. The estimate assumed that approximately 40 grout injection holes would be advanced at 1.2m (4 foot) centers from within the basement and in the yard in a grid pattern centered on the shaft. Each hole would extend to bedrock and would be grouted, in stages, from bottom to top, with approximately 0.37 cubic meters of grout injected per linear meter (4 cubic feet per linear foot) of hole. The cost estimate would include 7% of the total bid cost for mobilization, 5% for demobilization and 5% for restoration.

The total estimated cost for compaction grouting was approximately $126,500 including contingencies.

The relocation of the house on the existing property (Alternate 3) was estimated using published construction cost data adjusted for working in Des Moines, in winter, on a developed property. The total estimated cost for structure relocation was approximately $137,000 including contingencies.

Based upon the favorable cost estimate, the anticipated effectiveness of the treatment and the success of previous residential compaction grouting projects for mine subsidence remediation, the compaction grouting alternative was carried through final design.

Compaction Grouting Program

A total of 48 injection holes for compaction grouting were scheduled. The borings were carried to bedrock at a depth of approximately 12.8m (42 feet) below the basement floor. The contractor was given the option to advance a 5cm (2 inch) steel grout casing to bedrock using either air, water or direct drive techniques.

Each of the possible methods of advancing grout casing presented potential risks. Exploratory drilling with water experienced frequent loss of return water, without evidence of surface discharge. Because investigators had already determined that there were no sewer pipes in the area, it was determined that the drilling water was finding its way into the shaft. Large volumes of drilling water flowing into the shaft presented the potential of carrying away additional soil from under the foundations. Drilling had shown communication between several of the exploratory holes, presumably along soil fractures or sand lenses. This raised concerns that advancing the grout casing with water, under the house, might cause further migration of soil into the shaft, which could lead to destabilization of the structure.

Percussion methods of advancing casing posed potential for different problems. Exploratory drilling had demonstrated that mine cribbing or railroad ties were located in unknown positions under the house. The existence of mine cribbing or ties presented two problems. The ties were suspected to be adjacent to voids and not tightly held in place. Driving casing through the ties raised the possibility of their displacement and binding on the casing. Percussion drilling also raised a concern about dislodging the mine cribbing, potentially initiating a rapid collapse of the shaft and catastrophic loss of soil into the shaft. In addition, discussions with the homeowner indicated that there was as much as 15 cubic meters (20 cubic yards) of cast-in-place concrete and the earlier broken basement slab buried under the existing floor. These conditions raised concerns that vibratory energy imparted by percussion methods would cause densification of the loose soil beneath the house and acceleration of foundation settlements.

Use of air raised the possibility of fracturing and further weakening the soils. The use of rotary auger methods was economically and practically infeasible due to the slow progress of drilling the amount of waste cuttings that could be expected, and the limited headroom within the basement.

During the pre-bid meeting with prospective bidders and the subsequent preconstruction meeting with the successful contractor, the advantages and disadvantages of each potential drilling method were discussed with the bidder, but the contractor was allowed to select the method he would use.

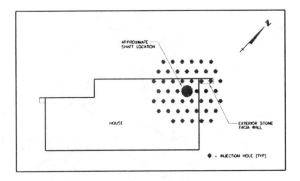

Figure 5 - Compaction Grouting Program

The injection holes were located on the project plans and the order of drilling and grouting was specified in order to assure the stability of the structure and the effectiveness of the treatment (Figure 5). The holes were sequenced in a staggered fashion, both within and outside the house, in order to allow for pressure dissipation and grout set. The holes immediately adjacent to the basement walls were grouted first. This allowed for the densification of the soils beneath the footings and served to stabilize the structure, which was the primary concern. The remaining holes were grouted in a radial fashion extending out from the shaft location. The shaft was grouted last, to its full 45.7m (150 foot) depth.

Grouting was designed to be performed in stages. After full depth installation, the grout casing was withdrawn 30cm (12 inches) and the grout was injected. Grouting continued from this discharge point until pre-determined termination criteria (described below) were achieved. The grout casing was then withdrawn from 0.3 to 0.9m (1 to 3 feet) and the grouting process was then repeated. The initial plan was to withdraw and grout until the discharge point of the pipe was within 0.9m (3 feet) of the outside ground surface or 0.3m (1 foot) of the basement slab.

Damage of existing subsurface utilities and the house due to the grouting pressures was a concern, so the contractor was restricted from grouting within 0.9m (3 feet) of any utility line. Also, the contractor was required to monitor the structure for vertical displacements during grouting. The concern was that the grouting pressures would result in heave of the ground and/or house.

The grouting operations for any stage of grouting in any hole were halted when one of the following termination criteria were met:

- Grout flow ceased at a predetermined maximum threshold pressure, or
- Ground and/or structure movements were detected, or
- Maximum grout quantity versus maximum back pressure versus depth criteria were met, or
- A predetermined maximum grout quantity was injected.

In order to effectively compact the existing weak, loose subgrade soils, the grout was designed to have a 2.5cm (1 inch) maximum slump. Although mix design was the responsibility of the contractor, the mix was required to be composed of at least one part cement and six parts aggregate with potable water as required to achieve the appropriate slump. Mineral fillers, such as fly ash, were allowed as a replacement for aggregate. In addition to the slump and mix requirements, the grout was required to provide a 28 day compressive strength of 28 kg/cm^2 (400 psi).

Construction

Five contractors submitted bids for the project. The successful contractor, Structural Preservation Systems, Inc. of Baltimore, Maryland, submitted a bid of $97,526, which was approximately $29,000 below the estimated cost.

The drilling and grouting program was performed, substantially, as per the project plans and specifications. For drilling in the basement, the contractor used a TEI electric rock drill. A Davey Kent hydraulic drill rig was used for the exterior holes (Figure 6). SPS used air as the drilling fluid for the 389 linear meters (1,275 linear feet) of drilling performed. During the drilling of an injection hole 2m (6 feet) from the basement wall, soil was blown into the basement through a crack in the wall and the wall shifted both vertically and horizontally. This incident revealed that the air used for drilling was fracturing the weak, loose subsurface soils. Injection of air while advancing grout casing into the water and mud filled mine shaft was problematic. Large volumes (1 cubic meter ±) of mud and water were discharged at the top of the hole in the basement. Containment and clean up of this backwash material was difficult.

Staged grouting procedures initially called for grout injection to within 0.3m (1 foot) of the basement slab. In the early stages of grouting it was possible to grout to this elevation. Later in the process, as the soils became denser, very small structure displacements began to occur while the grout discharge point was relatively deep. At the completion of the program, grouting was terminated between 2.4 to 3 meters (8 to 10 feet) below the basement slab because the compaction of the ground was causing structure movements.

Figure 6 - Compaction Grouting Operations

A total of approximately 153 cubic meters (200 cubic yards) of grout were injected. The quantity of grout was estimated at 260 cubic meters (340 cubic yards) prior to construction. The drilling and grouting work began on January 22, 1998 and was completed on February 28, 1998 with only six days of downtime.

Grouting under the basement floor resulted in unanticipated cracks and minor heaving of the floor slab. Demolition and replacement of the floor slab in 1993 left large pieces of concrete rubble, masses of cast in place concrete of various sizes, and sand backfill under the slab. These items combined with unmapped voids under the slab made monitoring the floor for lift more difficult than was anticipated during design. Floor heave sometimes occurred 3 to 4.5m (10 or 15 feet) away from the grout injection location, most likely because of the large pieces of rubble under the floor. Several cracks formed in the floor due to this heave before grouting could be terminated.

Upon completion of compaction grouting work, OSMRE contracted with Rex Wolfe Contractors to replace portions of the basement slab and portions of the north and east basement walls. The project, was completed on June 29, 1998. Project construction, including compaction grouting, wall replacement and slab restoration, was completed for less than the original engineer's estimate.

Conclusions

Through the use of visual observations and angled borings, the suspected mine shaft location was identified beneath the house. In this case, the microgravity techniques did not accurately identify the actual shaft location. Based upon the surface and subsurface data and observations, as well as literature reviews and interviews, the mechanism of failure was identified. The following goals for a remediation program were identified:

- Densify and strengthen the weak, loose soils supporting the building,
- Arrest soil migration into the mine shaft, and
- Fill the mine shaft.

Compaction grouting was identified as being suitable for the site subsurface conditions. Ultimately, it was selected as the most cost effective remedial alternative which could achieve the remediation goals. Compaction grouting, even inside the house, was an efficient operation which created acceptable levels of disturbance to the house and the subsurface soils. Drilling techniques were a source of concern with respect to both aggravation of the existing structural distress and potential for damage to the contents of the house. Drilling with air caused some undesirable disturbance to the subsurface soils but, for all but the final injection hole, allowed for a relatively clean drilling operation, both inside and outside the house. It also reduced the risk of more substantial settlement of the house during placement and set-up of the compaction grout columns. Bottom to top, staged grouting with low slump cement grout was effective at densifying the subsurface soils, arresting soil migration and filling the shaft.

When grouting under a structure that has already undergone previous foundation repairs or where very-near-surface voids may have already formed, floor monitoring for heave and settlement should be increased to avoid later replacement of the floor slab.

The flexibility of the compaction grouting process allowed OSMRE to compensate for unforseen subsurface conditions and complete project construction within the engineer's cost estimate.

Acknowledgments

The authors would like to thank all of the contractors involved in this project. They all worked diligently and with care in the basement of the house to protect personal belongings, limit disturbance to the structure and minimize mess. They would especially acknowledge and thank the homeowner for her patience and cooperation throughout the entire disturbing process of our work.

TDR Monitoring of Overburden Deformation to Verify Modeled Behavior

Kevin M. O'Connor[1], Member
John A. Siekmeier[2], Member

Abstract

This paper presents a comprehensive procedure for sampling, monitoring, modeling, and numerical simulation of rock mass behavior. Motivated by research into high extraction mining within bedded strata, this procedure has proven to be effective in a variety of mining and geotechnical applications. It begins by systematically estimating a rock mass rating for every stratum identified in drill-core logs and computing an empirical bending stiffness for each stratum. Then significant horizontal discontinuities are identified where adjacent strata have a large contrast in stiffness, and displacements along these discontinuities are monitored by placing a coaxial cable in each drill hole and anchoring it to the rock with expansive grout. As mining occurs and shear stresses are redistributed, displacements along the rock mass discontinuities cause deformation of the embedded cables which is monitored using time domain reflectometry (TDR). Measurements have verified modeled behavior of horizontally bedded sedimentary rock along the centerline of a coal mine longwall panel. The locations and magnitudes of simulated displacements are consistent with those measured using TDR. The importance of incorporating variations in rock stiffness and plastic sliding along "stiffness discontinuities" into any numerical model that is intended to accurately simulate rock mass behavior has been confirmed by experience with TDR monitoring on many projects.

---

[1]President, GeoTDR, Inc., 297 Pinewood Dr., Apple Valley, MN 55124

[2]Senior Engineer, Office of Materials and Road Research, Minnesota Dept. of Transportation, 1400 Gervais Avenue, Maplewood, MN 55109-2044.

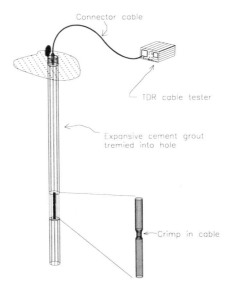

Figure 1.-Schematic of TDR monitoring cable installation.

Introduction

The emphasis in this approach is identification and explicit modeling of displacements along discontinuities as an integral component of rock mass response to changes in shear stress. The approach is outlined in this paper using an example in which we modeled rock mass response to high extraction underground coal mining. Verification was done by comparison between measured and predicted surface subsidence and measured and predicted subsurface displacements. The paper closes with a brief discussion of experience with TDR measurements in other projects which have demonstrated repeatedly that plastic slip along "stiffness discontinuities" is a predominant component of rock deformation.

Measurement of Rock Mass Displacements using Time Domain Reflectometry

Rock mass displacements within the strata overlying several mines have been monitored using TDR technology (O'Connor and Dowding, 1999). Basically, a solid-aluminum coaxial cable is crimped at equally-spaced locations then grouted into a borehole drilled from the surface (Figure 1). A TDR cable tester is connected to the cable and sends voltage pulses down the cable. At every location where the cable is crimped or rock movement has caused cable deformation, a reflection is sent back to the tester which displays and records a TDR waveform.

Reflections from the crimps provide distance reference markers in a TDR waveform so that deformation due to rock movement can be located accurately. The shape and magnitude of a reflection at each location where deformation is occurring correlates with the type and magnitude of cable damage at that location. Based on laboratory calibrations, it is possible to distinguish shear deformation from tensile deformation and to quantify shear displacement (Dowding et al, 1988).

A particular strength of TDR technology is that it allows remote, continuous monitoring of rock movement. When monitoring subsurface deformation in response to active mining, it is important to make measurements as frequently as possible over an extended period of time. This is important not only to accurately quantify changes in the rate of movement at multiple depths but also to acquire as much information as possible before the magnitude of movement exceeds the measurement range. Commercially-available TDR units (Tektronix, 1989) can be interfaced with an off-the-shelf datalogger and modem (Campbell Scientific, 1991) that is accessed via telephone or telemetry as shown schematically in figure 2. This type of system was used to remotely collect the data presented in this paper.

Modeling Methodology

The modeling approach can be illustrated by outlining a study (Siekmeier and O'Connor, 1993, 1994a) which was conducted to define the probable mechanics of rock mass deformation and surface subsidence in response to longwall coal mining. The approach was patterned after guidelines outlined by Starfield and Cundall (1988) who encouraged researchers to remember that the focus of numerical analysis should be on gaining an understanding of the mechanisms that characterize the system and not on obtaining unique numerical values for specific parameters. Hart and Cundall (1992) go on to explain that this goal is accomplished by using the computer model as a laboratory to perform experiments on the system. This can result in an improved understanding of mechanisms, knowledge of parameter dependence, and a means by which to check theories or hypotheses. Ultimately, the new knowledge may lead to new theories or simple conceptual models that can then be used in design.

The commercially-available Universal Distinct Element Code (UDEC; Itasca, 1993) was used to perform desktop experiments and identify parameters that significantly influence model behavior. The code allows internal deformation of discrete blocks and also allows plastic sliding and separations along discontinuities. UDEC is a particular type of discrete-element method that uses deformable contacts and an explicit time-stepping solution of the equations of motion (Cundall and Strack, 1979). Output in the form of displacements, velocities, and stress distributions were

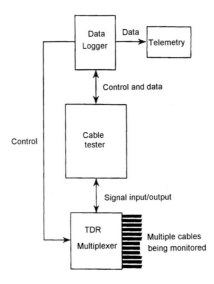

Figure 2.-Schematic of system used for remote monitoring.

compiled into movies that made it possible to visualize the influence of various parameters on model response.

For purposes of our study, simulated surface and subsurface displacements were compared with field measurements. In particular, simulated subsurface shear displacements along horizontal discontinuities were compared with TDR measurements and simulated surface subsidence profiles were compared with survey measurements.

Simulation of Horizontal Discontinuities

The distinct-element model is composed of deformable blocks separated by discontinuities. Since it was hypothesized that caving and bulking over the longwall panel are affected by the location and character of discontinuities within the rock mass, a modified Rock Mass Rating (RMR) system (Bieniawski, 1989; Golder Associates, 1989) was used to systematically define the location of horizontal discontinuities in the model. A commercially-available computer spreadsheet program was used to calculate a RMR for each lithologic bed based on drill core logs and engineering property tests (Siekmeier and O'Connor, 1994b; Siekmeier et al, 1992). The RMR was then used to calculate a deformation modulus and bending stiffness for each lithologic bed. Large contrasts in the bending stiffness between adjacent lithologic beds have been shown to correlate with horizontal shear displacements measured using TDR in studies of high extraction mines (O'Connor et al, 1995). Thus the locations of these large contrasts in bending stiffness were used to define horizontal discontinuities in the distinct element model.

The RMR includes six basic parameters: Intact rock strength, R_{IRS}, drill core quality, R_{RQD}, spacing of discontinuities, R_{JS}, condition of discontinuities, R_{JC}, groundwater condition, R_{JW}, and orientation of discontinuities, R_{JO}. The sum of these parameters is adjusted using factors which account for blasting damage, A_B, *in situ* stress or stress change, A_S, and major faults or fractures, A_P,

$$RMR = [R_{IRS} + R_{RQD} + R_{JS} + R_{JC} + R_{JW} + R_{JO}] * [A_B * A_S * A_P]. \quad (1)$$

Values for these parameters are selected based on information from drill core logs, other site data, and engineering judgement. The need for subjective engineering judgement and interpretation is due to the fact that data may have been collected in a form that is not compatible with the tables used to select the parameter values.

The *in situ* deformation modulus, E, of each lithologic bed is estimated with the empirical relationship proposed by Serafim and Pereira (1983) for RMR < 50,

$$E = 10^{[(RMR-10)/40]} * 1000 \quad <MPa>. \quad (2)$$

Figure 3 shows the correlation between this estimated RMR modulus and the dynamic modulus determined by sonic velocity logging. Note that there is an order-of-magnitude difference in the scale, but the locations and relative size of spikes in the profiles are similar.

The moment of inertia, I, per unit width for each bed was determined assuming that its thickness, t, was constant

$$I = \frac{1 * t^3}{12} \quad <m^4/m>. \quad (3)$$

The product of Equation (2) and Equation (3), E∗I, is the bending stiffness per unit width of each lithologic bed and has units of <MPa-m^4/m>. This stiffness is graphically displayed as a histogram (Figure 3) and used to identify locations where there is a large contrast in stiffness between adjacent beds. The thicknesses of the lithologic beds are indicated in figure 3 by the distance between horizontal lines in the bending stiffness histogram. Note that the bending stiffness is heavily dependent on the bed thickness via

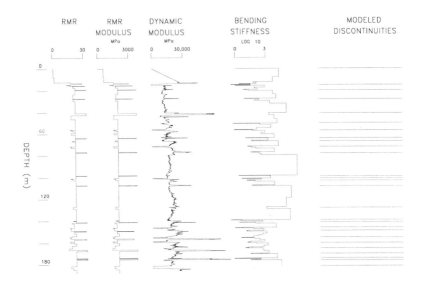

Figure 3.-Horizontal discontinuities defined by a large contrast in stiffness between adjacent beds.

the moment of inertia so that thick beds dominate the graphical stiffness histogram. The logarithmic scale is used to allow the location of thin beds of low stiffness to be visible. Adjacent lithologic beds with similar bending stiffnesses were grouped together to form geostructural elements, which are defined in Figure 3 by the modeled discontinuities.

Numerical Modeling

The model geometry used in the study summarized in this paper consisted of twenty-six deformable blocks stacked as layer-cake to simulate a horizontally bedded rock mass 900 m long by 200 m deep. The deformable block thicknesses were defined by the modeled discontinuities shown on the right side of figure 3. Vertical discontinuities were not included in the model so each block was numerically discretized using finite-difference triangles to permit internal elastic deformation. A stress boundary condition was used along the vertical sides of the model. The horizontal stress was varied from 0.5 to 2 to 4 times the vertical stress during different computer simulations.

The material properties listed in Table 1 were judged to be reasonable approx-

Table 1.- Properties assigned to intact material.

	Density (kg/m³)	Bulk modulus (MPa)	Shear modulus (MPa)	Mohr- Coulomb Failure		
				Friction angle (degrees)	Cohesion (MPa)	Tensile strength (MPa)
Soil	1,750	107	23	30	0	0
Rock	2,500	1,000	600	30	6	2

Table 2.-Properties assigned to horizontal discontinuities.

Normal stiffness (MPa/m)	Shear stiffness (MPa/m)	Coulomb Slip Criterion	
		Friction angle (degrees)	Cohesion (MPa)
15,000	1,500	10	0

imations of the intact properties and consistent with values of stiffness estimated in the spreadsheet analysis (Figure 3). Yielding of the intact blocks was prescribed using a Mohr-Coulomb criterion. The constitutive law that defined shear behavior along discontinuities assumed a linear elastic relationship between applied shear stress and displacement until the stress exceeded the Coulomb slip criterion listed in Table 2. When this criterion was exceeded, plastic slip occurred with kinematic restraints provided by adjacent blocks and the boundary conditions.

The model size and complexity was limited by the PC's processor speed rather than available memory since run times for models using all available memory were impractical. To simulate longwall mining, 1-meter-long blocks were deleted at mine level in a step-wise fashion. Each meter of excavation required 250 to 1000 calculation cycles at 0.3 to 0.5 Hz to distribute the effects throughout the simulated rock mass.

Comparison of Simulated and Measured Displacements

Figure 4 shows TDR waveforms acquired and the horizontal discontinuities defined in figure 3. The waveforms were recorded as the longwall face approached and advanced past the borehole. The regularly spaced spikes identified with asterisks are associated with crimps made in the cable prior to placement in the borehole and are used as distance reference markers. The reflection spikes which increased in magnitude indicate where cable deformation occurred due to rock movement. Large-scale curvature of the TDR waveform which is evident on May

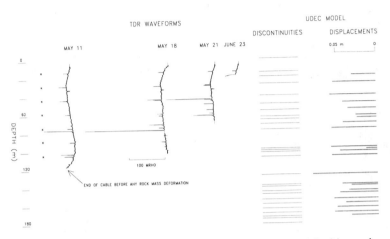

Figure 4.-Relationship between TDR waveforms, modeled discontinuities, and model displacements. Asterisks indicate TDR reflections from reference crimps made in cable prior to installation in borehole.

11 and May 18 was due to electrical losses along the coaxial cable and not associated with rock mass deformation.

The TDR reflection magnitude is proportional to the magnitude of cable deformation caused by shearing along the horizontal discontinuities (O'Connor and Dowding, 1999). These shear displacements occurred where relatively thick shale, sandstone, siltstone, and claystone beds were adjacent to thin limestone, shale, and coal beds. The UDEC model behavior is consistent with the location and magnitude of cable deformation due to shear along these discontinuities. For example, at a depth of 77 m the TDR reflection on May 18 indicates a shear deformation of 52 mm while the ultimate UDEC shear displacement was 56 mm. At a depth of 48 m, the TDR reflection on May 21 indicates a shear deformation of 32 mm while the ultimate UDEC shear displacement at this depth was 58 mm. These results confirm behavior observed with TDR measurements at other sites—that subsurface displacements are not only controlled by the presence of rock mass discontinuities but also by the relative stiffness of strata in contact along those discontinuities.

Figure 5 compares the longitudinal surface subsidence profiles of models without and with horizontal discontinuities (profiles A and B, respectively). It is apparent that incorporating horizontal discontinuities has a major effect on the magnitude and shape of the profile. The surface profile (A) generated by the model without discontinuities could be more closely matched to the actual surface

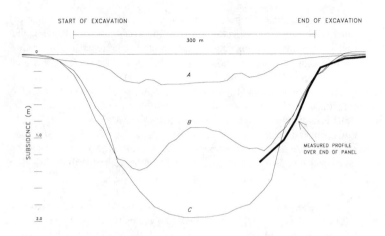

Figure 5. Comparison of modeled subsidence profiles and measured subsidence profile; *A*, incremental excavation without horizontal discontinuities; *B*, incremental excavation with horizontal discontinuities; *C*, instantaneous excavation of entire block.

measurements if the stiffness of intact block material was reduced. However, this type of curve fitting would only allow the model to accurately predict subsidence at a specific site or in areas with very similar geologic characteristics. No knowledge would be gained about the mechanisms responsible for subsidence. Incorporation of horizontal discontinuities into a distinct-element model provides a more systematic and physically justifiable approach to approximate the stiffness of the rock mass.

Figure 5 also shows the importance of simulating mining with step-wise excavation. When the entire block of coal was removed in one step (profile *C*), the model behaved as an elastic continuum and estimated 2 m of subsidence for a 2 m extracted thickness. The actual maximum measured subsidence was only 1.3 m. Step-wise excavation allowed for redistribution of shear stresses through the model and separation along horizontal discontinuities near the mined layer. This simulated caving of the mine roof and resulted in a surface profile that more closely matched the measured profile over the end of the panel. It should be noted that the obvious hump in the center of profile *B* is a numerical artifact which was eliminated when the simulated horizontal stress was increased from 0.5 to 4.0 times the vertical stress.

Confirmation from Experience at Other Sites

The mechanisms suggested by these measurements and numerical experiments indicate that subsurface processes involved with subsidence include progressive delamination, slip, and separation along horizontal discontinuities between lithologic beds that previously formed a massive laminated linear arch. Kneisley and Haramy (1992), O'Connor and Dowding (1992), Hazen and Sargand (1988), and Holt and Mikula (1984) have also suggested a sequence of main roof caving that is strongly affected by the presence of horizontal discontinuities and thick, stiff beds within the overburden over coal mines. Nieto et al. (1983) also found this behavior occurs over brine cavities. A valid model of strata behavior must (a) incorporate explicit discontinuities, (b) incorporate the relative stiffness of strata in contact along these discontinuities, and © allow for redistribution of shear stress along these discontinuities as mining progresses.

TDR cables have been used to monitor rock mass behavior at active coal mines (O'Connor et al, 1995), abandoned coal mines (O'Connor and Murphy, 1997), potash mines (O'Connor and Zimmerly, 1991), and Frasch sulphur mines (O'Connor and Norland, 1995). In all cases, shear displacement along horizontal discontinuities was a predominant component of rock deformation. Dowding and Huang (1994) found that the cumulative sum of all subsurface shear displacements measured with TDR approximated the total horizontal displacement measured by surveying on the surface above a longwall panel. Furthermore, recent experience with monitoring slope movement in both stiff soils and softer soils has demonstrated that displacements tend to occur along contacts between strata with large contrast in stiffness. Whether the contrast is a consequence of fracture density or intact stiffness, these contact planes are weak links as shear stresses are redistributed.

References

Bieniawski, Z. T. *Engineering Rock Mass Classification.* John Wiley and Sons, New York, NY, 1989, 251 p.

Campbell Scientific, Inc. (Logan, UT), Time Domain Reflectometry for Measurement of Rock Mass Deformation. Product brochure, July, 1991, 2pp.

Cundall, P. A. and O. D. L. Strack. A Discrete Element Numerical Model for Granular Assemblies. *Geotechnique*, 29, 1979, pp. 47-65.

Dowding, C. H. and F.-C. Huang. Telemetric Monitoring for Early Detection of Rock Movement With Time Domain Reflectometry. *J. Geot. Eng.*, Am. Soc. Civ. Eng., v. 120, No. 8, 1994, pp. 1413-1427.

Dowding, C.H., M.B. Su, and K.M. O'Connor. Principle of Time Domain Reflectometry Applied to Measurement of Rock Mass Deformation. *Int. Journal of Rock Mech., Mining Sci, and Geomechanical Abst.*, Vol. 25, No. 5, 1988, pp 287-297.

GEO-ENGINEERING FOR UNDERGROUND FACILITIES 375

Golder Associates. Geotechnical Study of Rock Mass Classification Parameters for Sydney Coal Field, Nova Scotia. Report to CANMET, DSS Contract No. 09SQ.23440-8-9002, Ottawa, Ontario, Aug., 1989, 81pp.

Hart, R. D. and P. A. Cundall. Microcomputer Programs for Explicit Numerical Analysis in Geotechnical Engineering. Presented at the International Seminar on Numerical Methods in Geomechanics, Moscow, Mar., 1992.

Hazen, G. A. and S. M. Sargand. Methods for assessing Effects of Longwall Mining on Surface Subsidence. *Mining Engineering*, Vol. 40, No. 6, 1988, pp. 451-454.

Holt, G. E. and P. A. Mikula. Evaluation of Subsidence Prediction Methods for Coal Mining in Eastern Australia. Australian Coal Industry Research Laboratories, Published Report 84-15, 1984, 168p.

Itasca Consulting Group (Minneapolis, MN). UDEC Version 2.0. Vol. 1 and 2, June, 1993.

Kneisley, R. O. and K. Y. Haramy. Large-Scale Strata Response to Longwall Mining: A Case Study. BuMines RI 9427, Washington, D. C., 1992, 25p.

Nieto, A. S., D. Stump, and D. G. Russel. A Mechanism for Sinkhole Development Above Brine Cavities in the Windsor-Detroit Area. Proceedings, Sixth International Symposium on Salt, Vol. 1, 1983, pp. 351-367.

O'Connor, K.M. and C.H. Dowding. Distinct Element Modeling and Analysis of Mining-Induced Subsidence. *Rock Mechanics and Rock Engineering*, Vol. 25, 1992, pp 1-24.

O'Connor, K.M. and C.H. Dowding. *GeoMeasurements by Pulsing TDR Cables and Probes*. CRC Press LLC, Boca Raton, 1999, 402 p.

O'Connor, K.M. and E.W. Murphy. TDR Monitoring as a Component of Subsidence Risk Assessment Over Abandoned Mines. *Int. Journal of Rock Mechanics & Mining Science*. Vol. 34, Nos. 3-4, Paper 230. 1997.

O'Connor, K.M. and M.R. Norland. Monitoring Subsidence Mechanisms Using Time Domain Reflectometry. Proceedings, Joseph F. Poland Symposium on Land Subsidence, Sacramento, October, 1995, pp. 427-434.

O'Connor, K.M., J.A. Siekmeier, and L.R. Powell. Using a Computer Spreadsheet to Characterize Rock Masses Prior to Subsidence Prediction and Numerical Analysis. U. S. Bureau of Mines RI 9581, 1995, 69pp.

O'Connor, K.M. and T. Zimmerly. Application of Time Domain Reflectometry to Ground Control. Paper in Proceedings of the 10th International Conference on Ground Control in Mining (Morgantown, June, 1991). WV Univ., 1991, pp. 115-121.

Serafim, J. L. and J. P. Pereira. Considerations of the Geomechanics Classification of Bieniawski. Proceedings, Int.. Sym. Eng. Geol. Underground Constr., A.A. Balkema, Boston, 1983, pp. 33-34

Siekmeier, J.A. and K.M. O'Connor. Modeling Overburden Response to Longwall Mining. Proceedings, First Canadian Sym. on Numerical Modeling App. in Mining and Geomechanics, March, 1993, pp. 110-118.

Siekmeier, J.A. and K.M. O'Connor. Explicit Modeling of Rock Mass

Discontinuities and Mining Subsidence. Proceedings, Eighth Int. Conf. on Comp. Meth. and Adv. in Geomechanics, May, 1994a, pp. 1895-1900.

Siekmeier, J.A. and K.M. O'Connor. Description of Laminated Overburden Using Rock Mass Classification. Proceedings, Fifth Conf. on Ground Control for Midwestern Coal Mines, June, 1994b, pp. 1-13.

Siekmeier, J.A., K.M. O'Connor, and L.R. Powell. Rock Mass Classification Applied to Subsidence over High Extraction Coal Mines. Proceedings, 3rd Workshop on Subsidence Due to Underground Mining, Morgantown, June, 1992, pp. 317-325.

Starfield, A. M. and P. A. Cundall. Towards a Methodology for Rock Mechanics Modeling. *Int. J. Rock Mech. Min. Sci. and Geomech. Abstr.*, 25 (3), 1988, pp. 99-106.

Tektronix, Inc. (Redmond, OR), 1502B Metallic Time Domain Reflectometer Operator Manual, 1989.

EXCAVATIONS IN SAN FRANCISCO BAY MUD: DESIGN FOR DEFORMATION CONTROL

Demetrious C. Koutsoftas[1]

Abstract

The design of the shoring to support deep excavations in soft clays is highly empirical and involves arbitrary simplifying assumptions. The end result is that the design does not deal with ground deformations in a satisfactory manner. A new approach for the design of excavations and shoring has been made possible by the recent development of a methodology described by Clough and O'Rourke (1990), which relates deformations to system stiffness and base stability. Key elements of the design, such as wall stiffness (EI), embedment of the shoring below excavation subgrade, and spacing of horizontal supports can be specified to control deformations within tolerable limits. Once these elements have been specified, the completion of the shoring design becomes a relatively simple problem of structural analysis. More importantly, this approach leaves little room for manipulation of the design by prospective bidders, which means that owners can realistically expect to get what they paid for. This paper presents three case histories to illustrate the application of this method and identifies key factors that need to be considered in the overall design of the shoring and excavations and the impacts of other construction activities on the overall deformations of the system.

Introduction

The design of the shoring system to support deep excavations in soft clays is one of the most difficult and controversial aspects of underground construction. The difficulties stem from the highly empirical nature of the design process and the rather arbitrary simplifying assumptions required to reduce a highly indeterminate soil-structure

---

[1]Principal, Dames & Moore, 221 Main Street, Suite 600, San Francisco, CA. 94105

interaction problem into a simple structural system for analysis and design purposes. The design process is complicated further, by the common contractual process that often delegates the responsibility for the design of the shoring to the contractor, on the premise that the shoring is a temporary structure. Given that the cost of the shoring can be a substantial portion of the overall cost of many projects, it is almost an irresistible temptation for the bidders to find ways to minimize the cost of the shoring, which frequently involve overoptimistic design assumptions and manipulations of the design process to simply satisfy minimum safety criteria without regard to the real factors that control ground deformations. While this practice has generally yielded satisfactory results, issues arising from the design and construction of the excavation support system are among the most frequent sources of difficulties and disputes during construction that often lead to litigation between owners, contractors, and geotechnical engineers. Often, the key reason for such disputes is because project specifications are not sufficiently clear regarding the design and performance requirements. This creates an opportunity for creative contractors to underbid the job and then claim extra costs and delays because what the contractor assumed in his bid may be far from what the designer and owner really expected. The thesis of this paper is that the geotechnical engineer together with the owner should take primary responsibility for the design of key elements of the shoring and should specify these elements to ascertain that the expected performance criteria can be met. This approach also "levels the playing field" and makes it possible for conscientious contractors to be successful and not be penalized because they might have included in their bids realistic costs for the design and construction of the shoring to satisfy the required performance criteria.

The methodology proposed by Clough and O'Rourke (1990) for estimating excavation deformations, together with the requirements to satisfy base stability, are the key to the development of the necessary design criteria, which can be the basis for the development of clear requirements to be included in the shoring specifications. The paper presents three case histories of excavations constructed in the San Francisco Bay Area during the past 10 years, which were designed by the owner's design team specifically to control deformations within tolerable limits, and protect adjacent structures from the impacts of ground deformations. The owner, through the geotechnical engineer, assumed primary responsibility for the development of design criteria for deformation control, and for specifying key elements of the shoring system consistent with the design criteria.

Basic Design Approach

Clough and O'Rourke (1990) and Clough et al. (1989) presented a semi-empirical method for estimating excavation deformations in soft clays. The basic premise of the method is that the maximum lateral deformations caused by excavation are dependent on the system stiffness, S, and the factor of safety against basal heave. The system stiffness is determined from the wall stiffness and spacing of the horizontal supports as

$$S = \frac{EI}{\gamma_w h^4} \quad \text{...(1)}$$

where:
- E is the modulus of elasticity of the wall;
- I is the moment of inertia of a unit length of the wall;
- γ_w is the unit weight of water (included for the sole purpose of making the system stiffness a dimensionless parameter); and
- h is the average spacing between supports

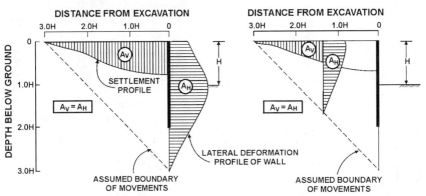

Figure 1: Geometric Relationships: Settlement and Lateral Deformation Profiles

A simple (semi-empirical) design chart allows estimation of maximum lateral deformations as a percentage of the depth of the excavation, once the system stiffness has been selected and the factor of safety against basal heave has been estimated. The factor of safety calculated as proposed by Clough et al. (1989) is not a true factor of safety, because it ignores the embedment of the wall, soil anisotropy, and other factors. It should be viewed as simply an "index" factor of safety. Clough et al. (1989) also provide guidelines for determining the approximate profile of lateral deformations, the distribution of lateral deformations with distance from the excavation, and the profile of normalized settlements behind the excavation. Clough and O'Rourke (1990)· also presented guidelines for estimating settlement profiles behind excavations. Using the procedures and guidelines included in the above .references, the shape of the lateral deformations of the wall can be approximately estimated. The next step is to estimate the settlement profile behind the wall consistent with the estimated deflection profile of the wall. Simple geometric relations can be used as shown on Fig. 1 to estimate settlements and lateral deformations behind the wall on the assumption that the areas under the lateral deformation and settlement profiles must be equal. The assumption about equality of the areas under the lateral deformation and settlement profiles is based on the premise that (in the absence of groundwater lowering due to dewatering), the deformations caused by excavations in soft clays take place under undrained conditions.

For design purposes, once the excavation stability has been evaluated, maximum lateral deformations and settlements can be estimated as a function of system stiffness, which in turn can be used to design the shoring for the selected level of tolerable deformations.

For each of the case histories, key aspects of the subsurface conditions are first described, followed by base stability considerations and presentation of key elements of the design. Each case concludes with presentation of the most salient aspects of the performance of the excavations.

Case History No. 1: EBMUD Deep Storage Basin

As part of the Wet Weather Improvements program of the East Bay Municipal Utility District (EBMUD), a deep storage basin was constructed at the site of a treatment plant in Oakland, California. The basin was 81 m by 72 m in plan and 13 m deep. It was located within 9 m of an existing effluent channel that had to remain operational at all times, and was also within 10.7 m to 15 m from various other facilities including the existing energy building and a number of above ground storage tanks. There was a serious concern about the impacts of the basin excavation on the adjacent structures.

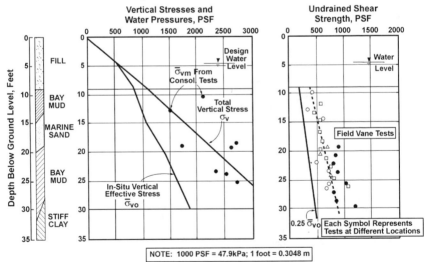

Figure 2: Subsurface Conditions: Storage Basin Site

The subsurface conditions are illustrated on Fig. 2. At the surface is a layer of very loose sandy fill, 2.4 m thick underlain by a soft layer of Bay Mud that extended to a depth of 9 m. Within portions of the site, a very loose, but thin, marine sand layer was sandwiched between the Bay Mud. Below the Bay Mud is a layer of very stiff clay that extends to depths greater than 35 m below ground surface. Results of consolidation tests indicate that the Bay Mud was preconsolidated, apparently due to construction dewatering from previous projects at this site. The undrained strength of the Bay Mud measured from field vane shear tests increases linearly with depth. Comparison of strengths estimated following the SHANSEP (Ladd and Foott, 1974) method with the measured vane shear strengths confirms that the mud was preconsolidated. The stiff clay below the Bay Mud had strengths on the order of 100 kpa (2,000 psf), and increasing with depth.

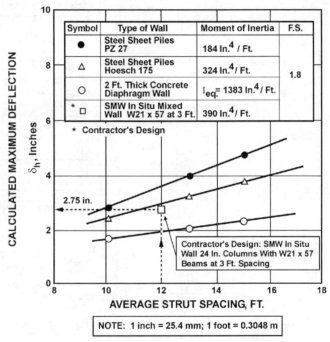

Figure 3: Shoring Design Criteria for Deformation Control

The specifications for the shoring required that the maximum lateral deformations measured with inclinometers installed behind the shoring should not exceed 70 mm (2¾ in.). The specifications included requirements for a minimum EI of the wall in combination with a maximum strut spacing of 3 m (10 feet), but allowed other alternatives to be considered. A range of options were included in the geotechnical report as combinations of wall stiffnesses, expressed in terms of the EI per lineal meter of the wall, and average strut spacings, for which estimates of deformations were provided. The results included in the geotechnical report are illustrated on Fig. 3, as plots of maximum lateral deformations versus strut spacing for three different types of shoring walls. The contractor elected to use an in-situ mixed soil-cement wall 0.6 m thick, with steel soldier piles consisting of W21x57, spaced 0.91 m (3 feet) apart; and average strut spacing of 3.66 m (12 feet, average of 11 feet, 11 feet, and 14 feet). The estimated maximum lateral deformation for the contractor's system met the specified 70 mm limit.

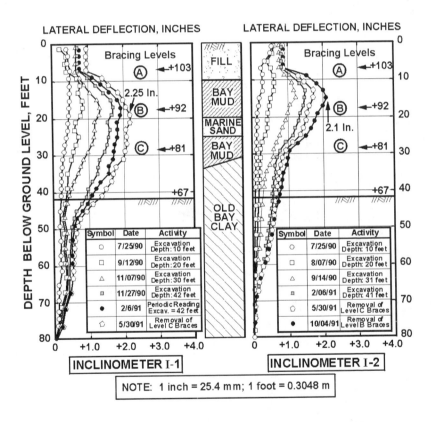

Figure 4: Lateral Deformation Profiles at Various Excavation Stages: Storage Basin Site

This case, which was designed in 1988, was probably the first, or at least one of the first, excavations to be designed specifically for deformation control. It was also the first major excavation in the United States to utilize in-situ deep soil mixing to construct the shoring wall.

Fig. 4 shows the lateral deformations measured with two inclinometers, located 1 m to 1.5 m behind the shoring wall, at various construction stages, and Fig. 5 plots maximum lateral deformations versus excavation depths. At the final excavation depth, the maximum lateral deflections ranged from 41 mm (1.6 in.) to 46 mm (1.8 in.). It should be noted that at each of the three strut levels, there were some incremental deformations, after strut installation, presumably due to compression of the struts (that were as much as 56 m long). Additional deformations developed during strut removal for the construction of the basin. The final deformations were 53 mm (2.1 in.) and 57 mm (2.25 in.), respectively. The lower deformations measured at end of excavation than the estimated values suggest that the "index" factor of safety against basal heave was probably higher than the estimated value of 1.8. However, the final deformations were in reasonably good agreement with the estimated values.

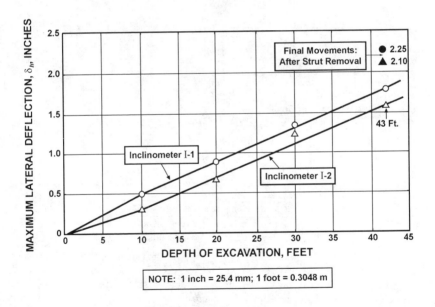

Figure 5: Maximum Lateral Deformations Versus Excavation Depth

Case History No. 2: MUNI Metro Turnback Project

The San Francisco MUNI Metro Turnback project involved construction of deep excavations and tunneling along the Embarcadero waterfront at the foot of Market Street. The excavations were 16 m (52 feet) wide and varied in depth from 11 m (36 feet) to 13.1 m (43 feet). The excavations had to be carried out in an area that is characterized by loose fills up to 6 m (20 feet) deep, underlain by a layer of soft Bay Mud up to 30 m (100 feet) thick. The stress history and strength characteristics of the Bay Mud are shown on Fig. 6. The excavations pass within 10 m of a block of old, settlement-sensitive buildings, and obviously there was considerable concern with the potential ground movements caused by the excavations and their impacts on the adjacent buildings. The designer and owner decided that the shoring system had to be designed by the Engineer of Record to control ground deformations within tolerable limits.

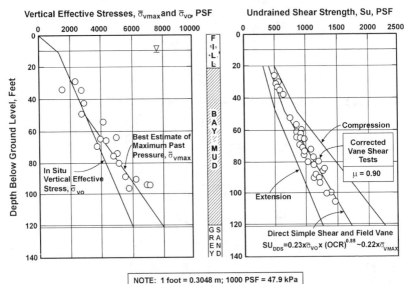

Figure 6: Stress History and Strength Characterization of the Bay Mud at the MUNI Site

The shoring system consisted of a soldier pile and tremie concrete (SPTC) wall and three levels of struts. The wall was 0.9 m thick concrete with steel soldier piles consisting of W36x211 sections. The concrete wall extended to a depth of 20 m (65 feet) below ground surface, penetrating only deep enough to intercept the critical failure surface (for basal heave). The primary soldier piles were spaced 3.66 m (12 feet) apart and extended below the Bay Mud layer either into a very dense sand layer, or into rock. The secondary soldier piles, (also W36x211) which were placed between the primary piles, extended only 6 m below the concrete wall.

The struts and walers were also designed by the engineer. Only minor details of the connections were left to the contractor. The specifications included a comprehensive list of requirements governing the excavation process and included threshold values of deformations which would trigger implementation of contingency plans to control deformations within tolerable limits. A thoroughly instrumented test section was constructed to verify the design before the excavations in front of the settlement-sensitive buildings were undertaken.

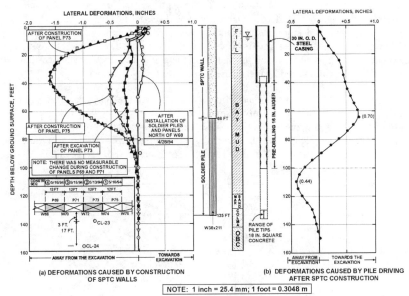

Figure 7: Lateral Deformations Caused by SPTC Wall Construction and Pile Driving

In addition to excavation settlements, there was considerable concern that the construction of the SPTC walls could cause substantial additional deformations that could include up to 25 mm of lateral deformations and settlements. Also, installation of the foundation piles (which had to be performed from existing ground level using 15 m long followers) through the soft Bay Mud could cause significant heave and subsequent consolidation settlements. Removal of the followers could cause collapse of the holes with additional settlements. To control soil heave and settlements during pile driving, the specifications required predrilling through the fill and Bay Mud strata down to within 3 m of the bottom of the Bay Mud, and that holes left after withdrawal of the followers be backfilled with cement grout or gravel. Steel casings were specified through the fill to minimize settlements due to cave-in of the predrilled holes. Also, the specifications required that the SPTC wall adjacent to the buildings had to be completed before pile driving would be allowed to proceed.

Lateral deformations and settlements were estimated following the Clough and O'Rourke (1990) method. For the 11 m deep excavations, the "index" factor of safety was 1.45 and the system stiffness 1500. The maximum lateral deformations were estimated to be 64 mm with a corresponding maximum settlement of 58 mm. These estimates included the effects of slurry wall construction and the excavation and bracing process. It was assumed that net settlements due to pile driving would be negligible.

Fig. 7 shows lateral deformations during construction of the SPTC wall and during pile driving, measured with inclinometer, CL-23, located 1 m behind the SPTC wall. It is evident that large lateral deformations developed during construction of the SPTC wall. The actual deformations were in the opposite direction of the assumed deformations. It is worth noting that incremental deflections developed due to construction of each of several panels close to the monitoring location.

In spite of all the precautions taken, lateral deformations due to pile driving were significant, although they were only a small fraction of the deflections that could have developed without the precautions described earlier. The net effect of the lateral deformations due to pile driving shown on Fig. 7 (as reflected by the net area under the lateral deflection diagram) was that the associated settlements were very small.

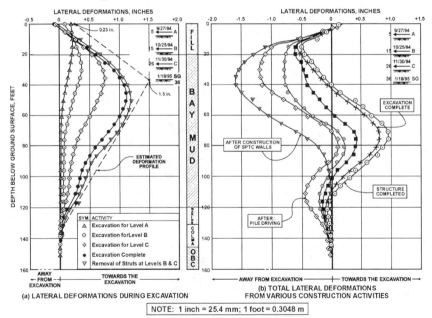

Figure 8: Lateral Deformations Caused by Excavation and Combined Effects of Construction Activities (inclinometer CL-23)

Fig. 8 presents lateral deformations measured during various stages of excavation, and the combined deflections resulting from SPTC wall construction, pile driving, and excavation. The estimated lateral deformations caused by the excavation process alone are also shown on Fig. 8a for comparison.

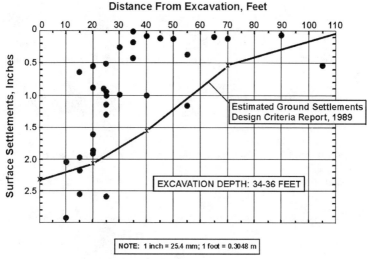

Figure 9: Measured and Estimated Settlements at End of Construction: MUNI Site

Fig. 9 shows the settlements measured behind the SPTC wall in the area of the 11 m deep excavation, compared with the envelope of settlements estimated during design.

This case illustrates that attention to all sources of deformations is essential. The deformations caused by excavation were predicted reasonably well, as were the settlements. The pile driving procedures were successful in controlling deformations due to pile driving to relatively small values. Although the lateral deformations were significant, the net effect was that the settlements due to pile driving were very small. The deformations due to construction of the SPTC walls were in the opposite direction than anticipated and were surprisingly large. Similar outward movements during concreting of slurry trenches were reported by Poh and Wong (1998) for diaphragm walls constructed in soft clays, similar to the Bay Mud, for a project in Singapore. When the deformations from all construction activities are combined, the actual shapes of the deformation profiles are completely different from the profiles of deformations caused by excavation alone. These are the ground deformations that need to be considered in evaluating potential impacts on adjacent structures.

Case History No. 3: Islais Creek Transport/Storage Project

The Islais Creek transport/storage project involved over 2 km of deep underground box sewers constructed in an area where extensive deposits of soft soils are present. Contract D, which is one of the five contracts that comprise the project, involved approximately 457 m of underground box sewers, 11 m to 12 m deep and 11 m wide. The subsurface conditions varied substantially along the project alignment. For design purposes, the alignment was divided into three segments: Army Street segment, Indiana Street segment, and the Islais Creek channel segment, according to the subsurface conditions. The shoring was designed accordingly to reflect the variations in subsurface conditions.

At the surface is a layer of fill, 4.5 to 6 m thick, underlain by a layer of Bay Mud of variable thickness. Along the Army Street segment of the alignment, the Bay Mud varied in thickness from 6 m to 12 m, and was medium stiff to stiff, and slightly overconsolidated. Along the Islais Creek channel, the Bay Mud was as much as 30 m thick and was much softer. Along Indiana Street, the thickness of the mud increased from 12 m at the north end to 25 m at the south end. Fig. 10 illustrates the variations in undrained strength versus depth from field vane shear tests made adjacent to the creek and near the north end of the alignment.

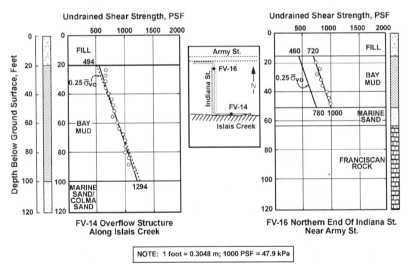

Figure 10: Subsurface Conditions and Undrained Shear Strength of Bay Mud - Islais Creek Site

This project actually represents three case histories because of the major differences in ground conditions along the three segments of the alignment and the different designs developed to deal with these conditions. In the interest of brevity, this paper deals only with the performance of the segment along Army Street to illustrate the design approach.

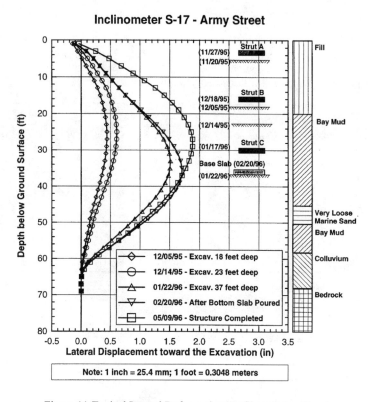

Figure 11:Typical Lateral Deformation Profiles: Army Street Segment at Islais Creek Site

The specifications required a wall stiffness (EI) of 134,219 kN-m^2/m (99,000 k-ft.^{2}/ft), a strut spacing of 3 m, and that the wall should penetrate into the stiff colluvium or rock strata below the Bay Mud. The design corresponds to a system stiffness of 159. The contractor's actual shoring system consisted of a 0.9 m thick soil-cement wall with W30x108 steel soldier piles spaced 1.22 m (4 feet) apart; and three levels of support with average spacing of 4.27 m (for the 11.6 m deep excavation), corresponding to a system stiffness of 94. For the calculated "index" factor of safety of 1.7, the estimated maximum lateral deformations of the proposed system were 0.5% of the depth of the excavation, or 58 mm. Fig. 11 shows typical lateral deformation profiles measured, at a distance of 1 m behind the shoring wall, at various times during construction. Fig. 12 plots maximum lateral deformations versus excavation depths. The calculated maximum deformation value of 58 mm compares favorably to the measured maximum deformations that ranged between 38 and 56 mm.

It is worth noting that additional deformations developed during removal of the struts that altered the shape of the lateral deformation profile (Fig. 11), although the maximum value of deflections did not change. Also, some small creep deformations developed between the end of the excavation and construction of the base slab.

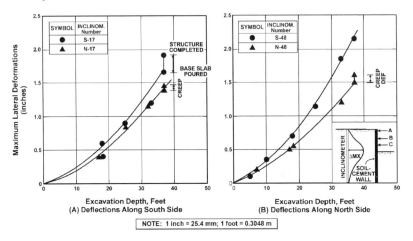

Figure 12: Maximum Lateral Deformations Versus Excavation Depth: Islais Creek Site

Conclusions

The three case histories presented in this paper demonstrate that the method described by Clough and O'Rourke (1990) can be used successfully to design excavations and shoring in soft clays to control deformations within tolerable limits. The estimated deformations were in reasonable agreement with the results of the measurements made during construction. Equally important, the process of designing excavations for deformation control forces the designer and contractor to consider all possible sources of deformations, not just excavation. Clear specification requirements eliminate much of the disputes between the contractor on the one side and the owner and designer on the other side. This approach levels the "playing field" and allows a competitive bid process that provides the specified product at a fair price.

The MUNI case illustrates that sources of deformation other than excavation and shoring can contribute significant additional deformations. They include pile driving, construction of diaphragm walls, and installation of soldier piles. In addition to excavation deformations, significant lateral deformations and settlements were observed at the deep storage basin site (Case History No. 1), which were caused by the pile driving operations, although in the interest of brevity they were not included in this paper. Also, very significant lateral deformations were observed at the Islais Creek site caused by jet-grouting (as part of work for the two other segments of the project performed concurrent with Case History No. 3). Large lateral deformations were also observed on several other projects in the Bay Area where jet-grouting was performed. Deformations caused by jet-grouting in some cases may overshadow all other sources of deformations. Other significant sources of deformations include construction vibrations, dewatering, construction of drilled piers, and perhaps other activities, although no specific information was included in the paper about these other sources of deformations. Other important factors that need to be considered, as demonstrated by the three case histories, include deformations due to creep (while the excavations remain open) as well as deformations that develop during removal of the struts to construct the underground structures.

Acknowledgements

The author would like to acknowledge professors Thomas D. O'Rourke of Cornell University and Wayne G. Clough of Georgia Institute of Technology for their contributions in the applications of the methodology described in this paper. Professor Clough was generous enough to make available early drafts of the Clough et al. (1989) paper that made application of the method possible for the EBMUD and MUNI projects. Professor O'Rourke was special consultant to Dames & Moore on all three projects and was instrumental in helping the author streamline the application of this method; and for his sound advice and unfailing support throughout these projects. The author also benefited from the advice of professors Ralph B. Peck and G.W. Clough, who were members of the board of consultants for the MUNI project. Their support was critical to the owner's acceptance of this approach. The author also wishes to thank project owners, EBMUD, MUNI, and the Department of Public Works of the City of San Francisco for their support and cooperation. Many other colleagues and professional organizations contributed to the success of these projects. I only wish it could be possible to mention all of them by name. Finally, Professor Charles C. Ladd is acknowledged for his thorough review of the paper and his helpful comments.

References

Bjerrum, L. (1972), "Embankments on Soft Ground," ASCE Specialty Conference on Earth and Earth Supported Structures, Vol. 2, pp. 1-54.

Clough, G.W. and O'Rourke, T.D. (1990), "Construction Induced Movements of In-Situ Walls," ASCE Geotechnical Special Publication No. 25 – Design and Performance of Earth Retaining Structures, pp. 439-470.

Clough, G.W. Smith, E.M. and Sweeney, B.P. (1989), "Movement Control of Excavation Support Systems by Iterative Design," Proc. ASCE, Foundation Engineering Congress: Current Principles and Practices, Vol. 2, pp. 869-884.

Ladd, C.C. and Foott, R. (1974), "New Design Procedure for Stability of Soft Clays," ASCE, Journal of the Geotechnical Engineering Division, Vol. 100, No. GT7, pp. 863-786.

Poh, T.Y. and Wong, I.H. (1998), "Effects of Construction of Diaphragm Wall Panels on Adjacent Ground: Field Trial," Proc. ASCE Journal of Geotechnical and Environmental Engineering, Vol. 124 No. 8, pp. 745-756.

Rationalizing the Practice of Strut Preloading for Braced Excavations

S.J. Boone[1], H. Bidhendi[2], J. Westland[1], and M. Grabinsky[2]

Abstract

Preloading of struts has been considered good practice when control of ground deformations is of concern but little quantitative field data has been published demonstrating the effects of this practice. This paper presents a comparison of numerical models and detailed field data regarding the benefits of strut preloading for braced excavations. Conceptual mechanisms based on both numerical and measured behaviour are provided to explain how strut preloading enhances general control of ground deformation during excavation shoring.

Introduction

Where excavation-induced movements are a concern, it has been considered "good practice" over the past 20 to 30 years to "preload" the lateral supports of excavation walls. Tensioning of tie-backs prior to further excavation is common because the elastic deformation of these restraints can be relatively large for economically-sized steel tendons (e.g. Clough and Tsui 1974). For braced excavations the struts are often larger than necessary for pure compression loading because of the need to limit self-weight bending and beam-column effects arising from span length. As a result, elastic compression of stiff struts is typically much less than the elastic elongation of anchors and is a relatively small proportion of shoring-induced ground movements. Discussions on preloading of braced excavations have typically focused on the need to remove the "slack" from the connections between the struts, wales, and wall system, though some have also suggested that there are beneficial effects within the soil mass behind the wall.

Although the concept of strut preloading for deformation control is intuitively attractive, there has been little published field data that quantitatively demonstrates its beneficial effects.

---

[1] Golder Associates Ltd., 2180 Meadowvale Blvd., Mississauga, ON, L5N 5S3

[2] Dept. of Civil Engineering, University of Toronto, Toronto, ON

The mechanisms of deformation surrounding an excavation are complex and involve: shoring wall stiffness; brace stiffness; connection details; soil stiffness; base stability; workmanship; and earth pressure redistribution in a sequential construction process. Most published work has focused on the results of parametric studies illustrating that support preloading has some benefit but these have not delved specifically into the mechanisms involved in movement reduction. This paper uses published and new results of numerical modelling in comparison with field data to evaluate and illustrate strut preloading and its effects within the soil mass. If the effects of preloading can be quantified, it should be feasible to optimize the use of materials and labour required for wall construction, and labour required for strut preloading in future shoring designs where ground movements are to be controlled.

Background and Conceptual Models

As early as 1961, the beneficial effects of strut preloading were recognised in published research. Rowe and Briggs (1961) constructed a model excavation support system that consisted of steel sheeting, "non-yielding" struts, wales, and dry sand fill. The wall in the model utilized by Rowe and Briggs (1961) fully penetrated the sand below the excavation; however, the wall and model extended only 10% to 30% of the wall height below the excavation bottom. By removing the fill from in front of the instrumented wall, they were able to determine loads at support levels and earth pressures between supports. Their results indicated that higher prestress values and a greater number of struts within the excavation significantly reduced surface settlements behind the wall.

Tomlinson (1963, and later editions) implies that strut preloading is part of standard construction methods by stating that "...the usual procedure is to use the struts to apply pressure to the ground by wedging or jacking rather than to allow earth pressure to come on to the supports." R.B. Peck (1969) noted that "...settlements adjacent to an open cut can be reduced only by decreasing the lateral movements of the earth supports and the rise of the bottom," and "...prestressing of struts and tiebacks significantly reduces the [ground] movements." Peck et. al. (1973) note that "a common and desirable precaution" for reducing movements of shoring systems is to prestress the struts as soon as they are installed. They further suggest that prestress loads of between 40% and 70% of the anticipated maximum strut load are "customary."

Clough and Tsui (1974) completed parametric analyses of tied-back excavations supporting normally consolidated clay. A two-dimensional, non-linear elastic, finite-element model was utilized to evaluate soil/structure interaction. They concluded that prestressing the anchors reduces deformations of the wall and soil, and that the effect of prestressing was especially prominent at the top of the wall. Preloading that was based on the trapezoidal apparent earth pressure diagrams (AEPD) suggested by Peck (1969) was shown to be more effective in controlling movements than use of triangular at-rest earth pressures - therefore suggesting that "...the concept of restoring the initial soil pressures...does not work." Wall deformation profiles from the study also suggested that, although prestressing of struts or tie-backs could not *eliminate* movement at the base of the excavation, preloading had some effect on both maximum total and deep movements. From the study results, prestressing of tie-backs to about 100% of the design AEPD (where $\sigma_{apparent} = 0.4\gamma H$) could be expected to result in movements of about 50% of those without prestressing, although these

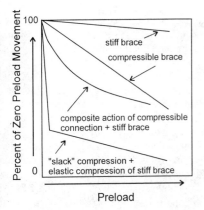

Fig. 1. Conceptual relationship between preload and relative movement for different strut and connection conditions.

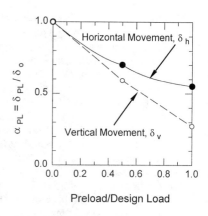

Fig. 2. Effect of preloading in soft to stiff clay, expressed as % of AEPD design load. (modified from Mana and Clough 1981).

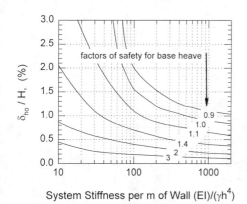

Fig. 3. Relationships between wall stiffness, factor of safety, and lateral movement (modified from Clough, et al. 1989).

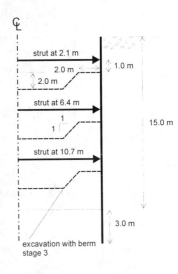

Fig. 4. Geometry and excavation stages used for numerical modelling.

deformations included some movement of anchorages (even though these were placed outside of the "active" earth zone). Of the more important conclusions, their work indicated that the proportional influence of prestressing was greater than the effects of either wall or tie-back stiffness for their particular conditions.

Palmer and Kenney (1972) and O'Rourke (1981) focused on the effective stiffness of struts where reductions in stiffness arise because of imperfect connections between struts and wales. Goldberg, et al. (1976) note that "Preloading of bracing is done for the purpose of removing elastic compression in the brace and the slack that may exist in the support wall, between sheeting and wales, in connection members, and between the soil and wall;" NAVFAC (1982) echoes this view of preloading. These concepts are schematically illustrated in Fig. 1. The implication of this conceptual figure is that if a stiff brace is used and if there is good workmanship when constructing the connection, then there is little benefit in preloading. Recent Canadian publications (CNBC 1990 and CFEM 1992) go further than Peck et al. (1973) and Goldberg et al. (1976) and suggest that "wedging and Jacking" should be carried out to prevent horizontal movements and that preload levels of 100% may be appropriate under some circumstances.

Recognising the importance of designing shoring systems to control ground movements, Mana and Clough (1981) completed non-linear finite element modelling of braced excavations in soft to stiff clays. Graphs were provided to assess the effects of: wall stiffness and strut spacing; strut stiffness; depth to an underlying firm layer; excavation width; strut preload; and the ratio of elastic modulus to undrained shear strength (modulus multiplier). As Fig. 2 shows, the maximum movement for a preloaded braced excavation (sheet-pile wall), δ_{PL}, could be estimated by multiplying the maximum movement for a non-preloaded braced excavation, δ_o, by a preloading factor, α_{PL}, related to the AEPD design pressure (where $\sigma_{apparent}$ = 0.65γH). Although the study numerically quantified a benefit of preloading, the mechanisms responsible for the effect and its relationship to wall stiffness were not discussed in detail. The analytical work for non-preloaded excavation support systems was extended by Clough et al. (1989) and supplemented with case history data where they illustrated the direct links between the factor of safety for base stability, relative wall stiffness per unit length of the excavation, $EI/\gamma h^4$, and maximum lateral movement, δ_{ho}, relative to excavation depth, H (see Fig. 3).

Numerical Modelling

FLAC (Itasca 1996), an explicit-finite-difference based program, was used to model the performance of a hypothetical excavation supported by soldier-piles and lagging. Soil behaviour was modeled as a Mohr-Coulomb elastic-perfectly-plastic constitutive material (ϕ' = 35°, E = 60 MPa, K_o = 1-sinϕ'). The wall and struts were considered linear-elastic structural elements, and cohesive-frictional elements that allowed slippage were used at the soil-wall interface to account for the discontinuity imposed by the spacing of the soldier piles and the cuts for the lagging in between the piles. The excavation stages were based on the typical construction practice in Toronto, Canada, where a temporary central trench is cut to allow equipment access, leaving berms at the sides, struts are then installed, and the remainder of the earth is removed to the next stage. Figure 4 shows the sequence of excavation and location of struts.

Following the work of Mana and Clough (1981) and Clough et al. (1989), the approximate relative non-dimensional stiffness (S_r) of a particular vertical section of a soldier-pile and lagging wall can be estimated using:

$$S_r = EI/(\gamma h^4 s) \tag{1}$$

where s is the horizontal spacing between piles, and h is the average vertical spacing between supports. Different levels of preloading, expressed as a percentages of the AEPD strut loads (where $\sigma_{apparent} = 0.65 K_a \gamma H$), and magnitudes of relative wall stiffness (S_r) typical of those utilized in competent soils were evaluated in the parametric study. To examine the changes induced within the soil mass, contours of maximum shear strain are considered a useful indicator.

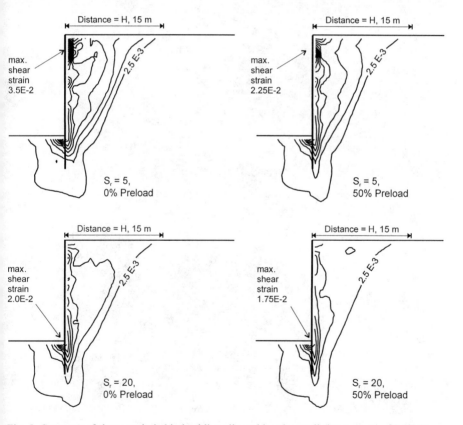

Fig. 5. Contours of shear strain behind soldier-pile and lagging wall (increments of 2.5E-3).

Figure 5 illustrates that preloading the struts to 50% of the AEPD load reduces the maximum shear strain close to the wall as well as limiting the intesity of shear deformations within the soil mass (see Fig. 6 for example). It can also be inferred from Fig. 5 that, for relatively flexible walls, localized zones of excessive straining in the soil at the strut levels are induced during excavation; however preloading significantly reduces the maximum shear strain. When S_r increases the intensity of localized zones of straining decreases (Fig. 5) and the predominant effect of another mechanism emerges. Preloading affects stiffer walls mainly through the reduction in the extent of failed zones by applying a more uniform confining stress along the height of the wall.

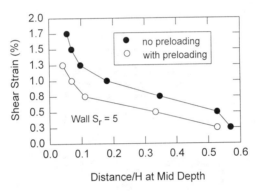

Fig. 6. Intesity of shear strain within soil mass behind the wall at excavation mid-depth.

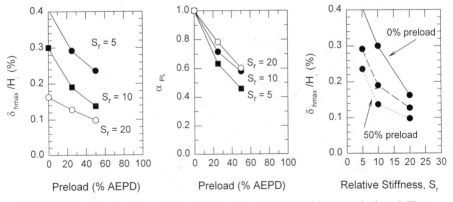

Fig. 7. Effect of preloading on lateral deformation as estimated by numerical modelling.

In Fig. 7 the maximum lateral movements (δ_{hmax}) are plotted against percentage of preload. The beneficial effect of preloading on movement control is obvious. For all the walls in the analyzed range, there are reductions of 40% or more in δ_{hmax} compared to the zero preload movements for any given wall stiffness when the struts are preloaded to 50% of the APED-based strut loads. Relative wall stiffness primarily increases the effectiveness of the structural system in reducing absolute deformations and, when preloading is carried out, stiffer walls more evenly distribute the applied loads.

For all of the analyzed no-preload cases the middle strut picks up more load compared to the top and bottom struts as the excavation progressed. The numerical modelling also indicates for this case that while the maximum load in the bottom strut did not change significantly, the preloading induces interaction between the top strut and middle strut, and the bottom strut has the lowest axial load among struts.

A Case History

A braced excavation, over 650 m long, from 9 m deep at the west end to nearly 20 m deep through the middle and east end, was made for construction of a new section of subway. Because of topography and structural design, the excavation width coincidentally nearly matched the excavation depth throughout the project. The dewatered excavation was made through glacial till and highly overconsolidated glaciolacustrine sand, silt, and clay deposits. Properties of the various strata are provided in Table 1 and the general stratigraphy has been presented by Busbridge et al. (1998). Over 50 buildings were in the immediate vicinity of the project. Analysis and discussion of the building responses to the excavation has been presented by Boone et al. (1999).

During design empirical relationships were used for estimating the potential performance of various excavation support systems. Predicted ground movements were then used to asses the response of surrounding structures in an iterative, two-step risk-evaluation process (e.g. Boone et al. 1998 and 1999) to select the best cost/risk balance for the project. Based on these analyses, it was considered that if a number of minimum criteria were included in the contract, adequate control of ground movements could be achieved with soldier-piles and lagging; preloading struts to at least 50% of their design load was one of these criteria.

The final shoring design included soldier-piles placed in pre-bored holes with 75 mm thick wood lagging between the piles. Soldier-piles were installed 3 m apart and were supported by deck beams and two to three pre-loaded pipe struts at each pile (i.e. there were no wales). Each strut was connected to the piles by wide plate-steel flanges (see Fig. 8) with the end area of the plate-flange equivalent to that of the strut. Preloading of struts was accomplished by inserting a flat-jack into a notch cut within the flange, jacking in the preload, welding the connection and removing the jack. Vertical spacing between struts ranged from 2.4 to 5.8 m.

Table 1. Generalized geotechnical properties of deposits.

Parameter	West End	Middle and East Ends
Total Unit Weight, γ (kN/m^3)	Fill - 18 Sand/Silt - 22 Clay/Till - 20 to 22	Fill - 18 Sand/Silt - 22 Clay/Till - 22
Effective Internal Friction Angle - Granular Soils, ϕ', (degrees)	Fill - 25 Sand/Silt - 38	Fill - 25 Sand/Silt - 36
Design Undrained Shear Strength - Cohesive Soils, S_u (kPa)	150 in upper part of clay deposit, 400 in lower	400
Approx. Average SPT "N" Value per borehole (blows/0.3 m)	20 to 60	40 to 70
Approx. Calculated Base Stability Factor of Safety	4 to 6 Depth = 9 to 13 m	4 to 6 Depth = 13 to 20 m

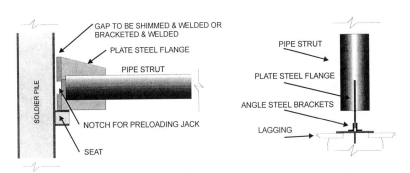

Fig. 8. Schematic detail of strut-to-pile connection: elevation view on left; plan on right.

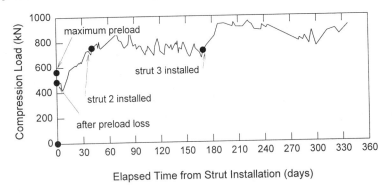

Fig. 9. Preload loss for an example strut where the preload loss was about 14%.

To monitor the contractor's performance an instrumentation program was undertaken on the owner's behalf. Instrumentation relevant to this paper included inclinometers installed in the ground behind the shoring system and vibrating wire strain gauges installed in arrays where each strut and deck beam in a vertical section was instrumented (settlement data are provided in Boone et al. 1999). During construction the strain gauges were read immediately before preloading to provide a "zero" reading while the strut sat on its supports. Readings were subsequently taken at full jack load and following jack removal. These strain gauge readings demonstrated that part of the preload was being lost (see Fig. 9). It was determined that the welds at the connection between the strut and pile were only being partially completed prior to removing the jack. The combination of weld quality, weld area, and remaining gap between the pile and strut plate-flange, and redistribution of stresses within the plate flange allowed compression (and thus load loss) to take place once the jack was removed. After recognizing this issue, the contractor made efforts to ensure adequate load transfer. In general, since the excavation was competed starting at the east and west ends and progressed toward the middle of the project, preloading effectiveness was at its best near the middle of the excavation. After preloading and preload losses, it was observed that all strut loads increased, illustrating that the combination of earth and temperature loads exceeded the realised preload (Fig. 9). Final strut loads compared well with the AEPD used for design (Fig. 10).

Figure 11 illustrates the maximum lateral movements (discounting premature strut removal and over-excavation that occurred in a few instances) as related to system stiffness. The average realised preloading for each inclinometer location, expressed as a percentage of the design load derived from the apparent earth pressure diagram is also shown for each data point. Elastic compression of the struts has been estimated to be in the range of 10% of the total movement at maximum loads. For the typical construction conditions lateral and vertical movements associated with the excavation were primarily influenced by wall stiffness and preloading effectiveness. Maximum lateral movements and the percentage of these that occurred below the base of the excavation at each inclinometer location is plotted in Fig. 12. Figure 12 indicates that in the middle of the project, where preloading was most effective, the movement below the excavation base was less than in other areas even though all areas had similar factors of safety as calculated based solely on shear strength, excavation width, and depth. It is expected that the process of averaging preloading effectiveness for particular instrumented locations and variations in measurements contribute to the observed scatter in data.

Discussion of Results

Although the AEPD-based strut loads have been used to compare relative preloads (in this and prior studies), AEPDs are simply a design tool used to arrive at a particular wall stiffness and strut size, are subject to considerable judgement, and implicitly include temperature effects (Goldberg et al. 1976). Nevertheless, the trends illustrated by Figs. 2, 3, 6, 11, and 12 are all consistent. For the range of wall stiffness investigated these figures also suggest that, whatever the basis for determining preloads, the beneficial effect of improving both settlement and horizontal movement performance of braced excavations in cohesive and granular soils is supported by both numerical and field data.

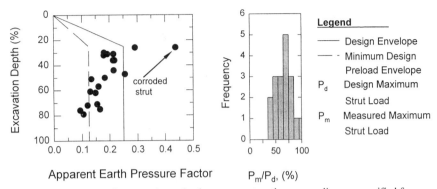

Fig. 10. Comparison of measured strut loads to apparent earth pressure diagram specified for final design of soldier-pile and lagging excavation support system.

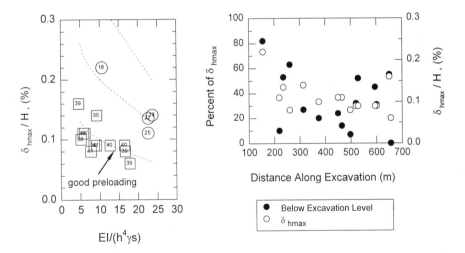

Fig. 11. Lateral movement relative to stiffness and preloading. Numbers indicate % of AEPD-based strut design load. Approximate contours shown for 0%, 20% & 40% preload.

Fig. 12. Percent of total lateral deformation below excavation and δ_{hmax} at inclinometer locations.

This study indicates that preloading excavation braces reduces deformation by:

1) inducing elastic compression deformations immediately behind the wall system prior to further excavation;
2) increasing confining stresses prior to stress redistribution (from excavation) and therefore improving the deformation response of the soils; and
3) contributing to the base stability and limiting associated deformations below the base of the excavation.

The field data also illustrate another important issue: in this case study, following preload losses and full welding, the compression of "slack" at the wall-strut connection was insignificant by virtue of the direct connection of the strut to the pile. Elastic compression of the strut was likely less than 10% of the minimum recorded movements. If elimination of connection slack or compression were meaningful factors in the effectiveness of preloading, it could be expected that the data in Fig. 11 would not exhibit the improvements seen between the poor preloading (about 20%) and good preloading (about 40%) data points rather than the observed near-linear and proportional improvements in deformation control.

Figure 11, illustrating the lateral movements and the range of case study S_t and realized preloading values, further suggests that preloading has a greater effect in practice than could be anticipated based on numerical models (when compared to Fig. 7). With soldier-pile and lagging walls, the lagging typically does not experience the full earth pressure exerted on the shoring system (Peck 1969, Goldberg et al. 1976). Use of flexible lagging takes advantage of arching between soldier-piles where sufficient deformation can occur and loads are shed to the piles. It is also likely that pre-development of such "arching" by preloading the struts, and thus preloading the horizontal stresses on the piles, is partly responsible for the improved response of the field system when compared to the sheet-pile laboratory model of Rowe and Briggs (1961) or finite element models. Such two-dimensional representations of wall systems fail to account for the composite action of soldier-pile and lagging systems that are comprised of "flexible" wood members and "stiff" steel members. Further research on this important three-dimensional mechanism of deformation is currently underway.

Conclusions

It is clear that preloading of struts in braced excavations improves the ground movement control beyond the immediate effects of reducing "slack" in the shoring system or elastic compression of the braces. Ground deformations associated with soldier-pile and lagging walls arise because of deformation of the wall system (piles), the soils between the piles and below the excavation, as well as other important workmanship considerations. Preloading imposes horizontal stresses within the soil mass prior to the stress-redistribution that occurs during subsequent excavation. These changes in the stress-regime within the soil induced by preloading also limit ground deformations below the base of the excavation. Preloading is especially effective for soldier-pile and lagging walls and further work is needed to examine the three-dimensional effects of pre-developing "arching" between piles. This work demonstrates through comparison of numerical models, laboratory models, and field evidence that significant improvements in deformation control for braced excavations can be achieved by the combined effects of pile/wall stiffness and strut preloading.

References

Boone, S.J., Westland, J., and Nusink, R. (1999). Comparative Evaluation of Building Responses to an Adjacent Braced Excavation. *Canadian Geotechnical Journal* (in press).

Boone, S.J., Garrod, B., and Branco, P. (1998). Building and Utility Damage Assessments, Risk and Construction Settlement Control. Proc. 24th World Tunnelling Congress, Sao Paulo, Brazil. A. Negro and A. Ferriera, eds., A.A. Balkema, Rotterdam, 243 - 248.

Busbridge, J.R., Westland, J., Boone, S.J. and Pennington, B.N. (1998). Underground Construction for the Extension of the Toronto Subway. ASCE Geotechnical Special Publication No. 86, 273 - 296.

CFEM (1992). Canadian Foundation Engineering Manual. Canadian Geotechnical Society, Bitech Publishers, Richmond, B.C.

Clough, G.W. and Tsui, Y. (1974). Performance of Tied-Back Walls in Clay. Journal of the Geotechnical Engineering Division, ASCE, 100(12), 1259 - 1273.

Clough, W.G. and O'Rourke, T.D. (1990). Construction Induced Movements of Insitu Walls. Design and Performance of Earth Retaining Structures, Geotechnical Special Publication No. 25, ASCE, P.C. Lambe and L.A. Hansen, eds., 439 - 470.

Clough, W.G., Smith, E.M., and Sweeney, B.P. (1989). Movement Control of Excavation Support Systems by Iterative Design. ASCE Geotechnical Special Publication No. 22, F.H Kulhawy, ed., ASCE, 869 - 884.

CNBC (1990). Canadian National Building Code, Supplement to the National Building Code Commentary L, Foundations, Temporary Excavations. National Research Council of Canada, 229 - 236.

Goldberg, D. T., Jawarski, W. E., and Gordon, M. D. (1976). Lateral Support Systems and Underpinning, Volume II. U.S. Department of Transportation.

Itasca (1996). FLAC User's Manual. Itasca Consulting Group Inc. Minneapolis, MN.

Mana, A.I. and Clough, G.W. (1981). Prediction of Movements for Braced Cuts in Clay. Journal of the Geotechnical Division, ASCE, 107(6), 756 - 777.

NAVFAC DM-7.1 (1982). Soil Mechanics, Design Manual 7.1. United States Department of the Navy, Alexandria, VA.

O'Rourke, T. D. (1981). Ground Movements Caused by Braced Excavations. Journal of the Geotechnical Division, ASCE, 107(9), 1159 - 1177.

Palmer, J.H.L. and Kenney, T.C. (1972). Analytical Study of a Braced Excavation in Weak Clay. Canadian Geotechnical Journal, Vol. 9, 145 - 164.

Peck, R. B. 1969. Deep Excavations and Tunnelling in Soft Ground: State-of-the Art Report. Proc. 7th Int. Conf. on Soil Mech. and Found. Engrng., Mexico, 225-290.

Peck, R.B., Hanson, W.E., and Thornburn, T.H. (1973). Foundation Engineering, Second Edition. John Wiley & Sons, New York.

Rowe, P.W. and Briggs, A. (1961). Measurements on Model Strutted Sheet Pile Excavations. Proc. 5th Int. Conf. on Soil Mech. and Found. Engnrg., Vol. 2, 473 - 478.

Tomlinson, M.J. (1963). Foundation Design and Construction, First Edition. Pitman Publishing Ltd., London.

Design and Behavior of One-Tier Tiedback Walls

David E. Weatherby[1] and Thomas C. Anderson[2] Members, ASCE

Abstract

Research on an instrumented, 7.62 m high, soldier beam wall and model wall tests show that apparent earth pressure diagrams predict the support load and the bending moments in walls supported by one-row of tiebacks. Measurements show that the bending moments are small below the bottom of the excavation, and assuming a hinge in the soldier beam at the bottom of the excavation gives good bending moment predictions. Measurements show that the support load and bending moments for the final construction stage are the maximum load and moments. Therefore, it is unnecessary to design for intermediate construction stages.

The research shows; that the soldier beam distributes the tieback load to the ground, that the earth pressures were concentrated at the tieback location, and that trapezoidal apparent earth pressure diagrams describe the earth pressure behavior of the walls better than triangular or rectangular pressure diagrams.

Background

Apparent earth pressure diagrams or envelopes were developed to give the magnitude and distribution of earth pressures on braced soldier beam or sheet piling excavation support systems. Apparent earth pressure envelopes are rectangular or trapezoidal in shape. Common apparent earth pressure diagrams are described by Terzaghi, et al. (1996) and Schnabel (1982).

---

[1] Research & Development Manager, Schnabel Foundation Company, 45240 Business Court, Sterling, VA 20166
[2] Midwest Regional Manager, Schnabel Foundation Company, 210 Cleveland Street, Cary, IL 60013

There is general agreement that apparent earth pressure diagrams are appropriate for the design of soldier beam walls supported by multiple rows of tiebacks. There is no agreement as to whether these diagrams apply to walls supported by one row of tiebacks. For example, AASHTO's *Standard Specification for Highway Bridges* (1996) recommends that walls supported by one row of tiebacks be designed to resist Rankine triangular earth pressures, and that bending moments and tieback loads should be determined for different construction stages.

A triangular earth pressure diagram is appropriate for anchored bulkheads with a yielding support. Bulkheads are backfilled after the sheet piling has been driven and the deadman anchors are loaded as the wall moves outward in response to raising the backfill. Tiedback walls are built differently. They are built from the top down, and the tieback is fully loaded when the excavation is at the tieback level. Rowe (1952) demonstrated that bulkheads supported by one level of fixed supports (pre-loaded tiebacks) did not rotate outward sufficiently far enough to reduce the earth pressures to an active triangular distribution. Rowe's tests showed that the earth pressures were higher than active at the support and lower than active between the support and the bottom of the excavation. Terzaghi, et al. (1996) stated that the earth pressures will have a resultant near the center of the wall when the top of the wall is restrained.

Full-Scale Wall Research

A 7.62 m high, instrumented, tiedback wall was constructed as part of a research program (Weatherby et al., 1998). One objective of the research was to determine the appropriate earth pressure diagram for the design of driven and drilled-in soldier beam walls supported by one row of tiebacks.

The tiedback, H-beam and wood lagging wall was constructed at Texas A&M's National Science Foundation designated site for Geotechnical Experimentation. Vibrating wire strain gages were located at 0.3 m intervals along the flanges of the soldier beams. Inclinometer casings were installed between the flanges and behind the wall. Vibrating wire load cells were installed on the tiebacks, and optical settlement points were located on the soldier beams and the ground surface behind the wall.

Subsurface Conditions. The wall was built in an alluvial sand deposit. Average standard penetration resistances are shown in Figure 1, a cross-sectional view showing the soil profile at the center of the wall. An average total unit weight of 1842 kg/m^3, and an angle of internal friction of 32° was representative of the deposit.

Wall Design and Construction. Soldier beams and tiebacks were designed to support the 25H Trapezoidal apparent earth pressures shown in Figure 2. A soldier beam spacing of 2.44 m, 76 mm thick wood lagging, and a 1.52 m toe penetration was selected for the test section. Bending moments and the tieback load were determined assuming a hinge at subgrade. The tieback was located 2.74 m from the top of the beam. The maximum design bending moment was 150.2 kN-m, and it occurred at the tieback location. The horizontal component of the tieback load was 347.4 kN. Pressure-injected tiebacks with a 4.57 m unbonded length, and a 7.32 m bond length supported the wall. The tiebacks were installed at 30 degrees from the horizontal and their design load was 400 kN. Calculations were the same for driven and the drilled-in soldier beams.

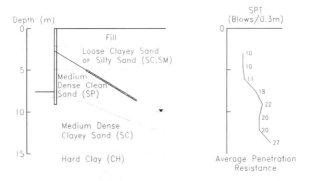

Figure 1. Wall Cross-section and In Situ Test Results

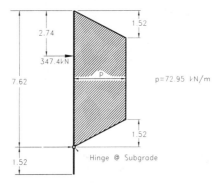

Figure 2. 25H Trapezoidal Apparent Earth Pressure Diagram (Design Diagram)

Soldier beams were HP 250×85 beams with a 76×76×6 angle welded to each flange (moment of inertia = 173.9 × 10^6 mm^4). The angles protected the strain gauges during soldier beam installation. Two beams were driven (SB15 and SB16), and two beams were drilled-in. The drilled-in beams (SB13 and SB14) were installed in cased holes. Structural concrete was used to backfill the toe of SB13, and lean-mix backfill was used around the toe of SB14.

Tiebacks were tested to 534 kN in accordance with PTI Recommendations (1996). The average lock-off load for the four tiebacks was 301 kN, 75 percent of the design load.

Observed Performance. The average tieback load was 322 kN upon completion of construction, a 6.9 percent increase from the lock-off load. Eighty-four days later, the average tieback load stabilized at 330 kN, 82 percent of the design load. This indicates that the tieback design load determined from the apparent earth pressure diagram in Figure 2 was greater than the load required to support the wall.

Figure 3 shows the predicted and measured soldier beam bending moments. The maximum design moment was 150 kN-m and was located at the tieback. Average measured bending moments at the tieback were; 119 kN-m after stressing the tiebacks, 151 kN-m upon completion of construction, and 156 kN-m at the end of the observation period. The bending moment at the tieback location increased as the tieback load increased.

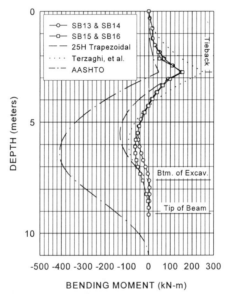

Figure 3. Predicted and Measured Soldier Beam Moments

The maximum design bending moment in the span below the tieback was 132 kN-m. Bending moments below the tieback were small after the tieback was stressed and increased as the excavation deepened. Upon completion of the wall, the average measured moments between the tieback and the bottom of the excavation was 57 kN-m, and it increased to 65 kN-m at the end of the observation period. Maximum bending moments in the driven beams between the tieback and the bottom of the excavation were 23 percent larger than the moments in the drilled-in beams. Drilled-in soldier beams may have behaved as a partial composite section. Bending moments between the tieback and the bottom of the excavation depended upon the earth pressure. Small moments were observed at the bottom of the excavation and along the toe of the soldier beams. The drilled-in beams did not act as a composite section in the toe.

Average settlement was 16 mm for the drilled-in beams, and 11 mm for the driven beams. All four beams settled sufficiently to fully mobilize skin friction along the beam toes. Driven beams apparently developed higher skin friction than the drilled-in beams. The soldier beam backfilled with structural concrete continued to settle after completion of the excavation. The two driven soldier beams and the beam backfilled with lean-mix backfill did not settle significantly during the monitoring period.

The horizontal movement at the top of the wall, soldier beam settlements, and ground surface settlements are consistent with those reported by Peck (1969), Goldberg, et al. (1976), and Clough and O'Rourke (1990). Figure 4 shows the inclinometer profiles for the drilled-in beams and Inclinometer E5, which was located between the beams and 0.61 m behind the lagging. Figure 5 shows similar profiles for the driven beams and Inclinometer E6. Lateral movements of the driven beams was less than the drilled-in beams. Differences in lateral movements are related to differences in vertical settlements of the beams; the greater the settlement the greater the lateral movement. Above the bottom of the excavation, the inclinometers between the beams (E5 and E6) moved out about 5 mm more than the adjacent soldier beams. This difference reflects the bowing of the lagging boards between the soldier beams.

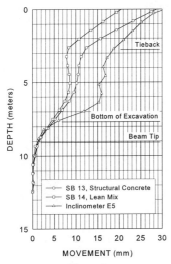

Figure 4.
Lateral Movement of Drilled-in Soldier Beams and Inclinometer E5

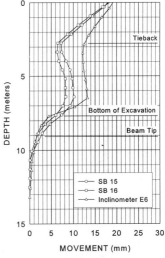

Figure 5.
Lateral Movement of Driven Soldier Beams and Inclinometer E6

Below the bottom of the excavation, the lateral movements were small and the movement patterns indicated that the lateral load-carrying capacity of the beams was adequate. Similar behavior was observed for the beams backfilled with structural concrete and lean-mix. In the toe, the drilled-in soldier beams moved the same as the inclinometer between the beams, while the driven soldier beams moved out slightly more than the inclinometer between the beams. Driven beams were 254 mm wide and the drilled shafts for the drilled-in beams were 610 mm in diameter. The increased area gave the drilled-in soldier beams have a greater lateral load-carrying capacity.

Measured Bending Moments Compared with Predicted Moments. Predicted bending moment diagrams for the wall assuming Terzaghi, et al. (1996) and the 25H Trapezoidal apparent earth pressure diagrams, and American Association of State and Highway Transportation Officials' (AASHTO, 1996) triangular pressure diagram are shown in Figure 3. Figure 2 shows the design earth pressure diagram (25H Trapezoid), and Figure 6 shows the Terzaghi, et al. and AASHTO design diagrams. Average measured bending moments for the drilled-in and driven soldier beams are shown in Figure 3.

The AASHTO triangular pressure diagram predicts large bending moments between the tieback and the bottom of the excavation and low moments at the tieback location, and it requires an 3.38 m embedment depth to resist the large load applied to lower portion of the wall. The actual embedment depth was 1.52 m. The AASHTO maximum bending moment in the span between the tieback and the bottom of the excavation is six times the measured moment, and three times the moments calculated from the apparent earth pressure diagrams. Below the bottom of the excavation the measured moments were small, and less than 10 percent of those predicted from the AASHTO design diagram. Small bending moments and the deformed shape of the beams below the bottom of the excavation are consistent with the mobilization of small lateral toe resistances.

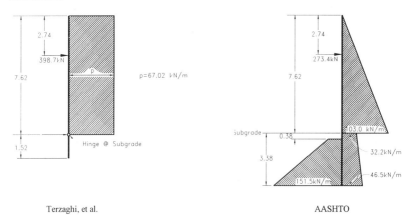

Figure 6. **Earth Pressure Diagrams for the Design of One-Tier Walls**

These results are consistent with Rowe's (1952) anchored bulkhead experiments with fixed supports, and Terzaghi, et al. (1996) observations regarding the position of the earth pressure resultant for walls restrained at the top. The measurements support the practice of using apparent earth pressure diagrams and the assumption of a subgrade hinge for the design of walls supported by one row of tiebacks. A subgrade hinge would not be valid in weak ground.

Model-Scale Wall Research Results

Two model soldier beam and lagging walls supported by one row of tiebacks were constructed in a medium-dense, fine to medium sand at the University of Illinois' large-scale soil model test facility (Mueller et al., 1998). The walls were scaled to be one-quarter the size of the full-scale wall. The walls were 1.90 m high and had a 0.38 m embedment depth. Soldier beams were 0.61 m on center. Each soldier beam was instrumented with temperature compensated electrical resistance strain gauges. Soldier beams for Test Wall 1 had a relative soil/wall stiffness consistent with the beams installed in the full-scale wall. Test Wall 2 was constructed using more flexible beams. The tieback test load for both walls was 7507 kN, but the load upon completion of excavation was 3843 N for Wall 1 and 5600 N for Wall 2. Differences in tieback load resulted from construction details for the model tiebacks.

Figure 7 shows computed earth pressures and the design earth pressure for the full-scale wall and computed earth pressures for the model walls. Lateral earth pressures were interpreted from measured bending strains. The procedure consisted of curve fitting, with appropriate boundary conditions, to obtain a twice differentiable function that could be used to compute the earth pressures. Figure 7 shows that the earth pressure distributions for the full-scale and the model walls were similar, and that the mobilized pressures conform to the trapezoidal design pressure diagram with a hinge at subgrade. Earth pressures for Wall 1 were less than the those observed for Wall 2 because the tieback load for Wall 1 was less than the load for Wall 2. A larger lateral resistance was mobilized along the embedded portion of Wall 1 than Wall 2. Since the tieback load was smaller in Wall 1, more load was transferred to the toe.

Non-symmetrical Trapezoidal Apparent Earth Pressure Diagram

Figures 5 and 7 suggests that a trapezoidal apparent earth pressure diagram should be used to design soldier beams supported by one row of tiebacks. The research also shows that the soldier beams distributed the tieback load to the ground. If the soldier beam distributes the tieback load to the ground, then the shape of the earth pressure diagram will depend upon the location of the tieback. Several different trapezoidal diagrams were investigated. The total load for each diagram was the same as that given by the design diagram (Figure 2). The apparent earth pressure diagram in Figure 8 predicted reasonable moments and tieback load, and considered soil-structure interaction aspects

412 GEO-ENGINEERING FOR UNDERGROUND FACILITIES

of wall behavior observed during the research.

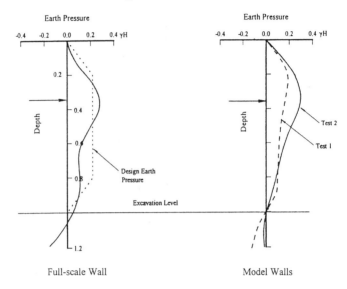

Figure 7. Comparison of Lateral Earth Pressures on the Full-scale and Model Walls

Earth pressures in Figure 8 increase to a maximum at a depth equal to two-thirds the distance to the tieback, and remain constant for a distance equal to one-third the height of the wall. Below that depth, the pressure decrease to zero at the bottom of the excavation. A hinge is assumed at subgrade or the bottom of the excavation. A similar diagram has been developed for walls supported by multiple rows of tiebacks (Weatherby, 1998 and Weatherby, et al., 1998).

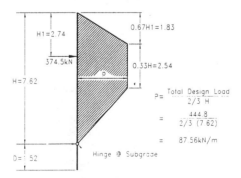

Figure 8. Non-symmetrical Trapezoidal Apparent Earth Pressure Diagram

Figure 9 shows the bending moments for the design pressure diagram (Figure 2), the Terzaghi et al. pressure diagram (Figure 6), the pressure diagram in Figure 8, and the measured soldier beam moments in the full-scale wall. Table 1 shows the total load from each apparent earth pressure diagram, the horizontal component of the tieback load, and the bending moment at the tieback. The results show that the two trapezoidal apparent earth pressure diagrams predicted the bending moment at the tieback better than the rectangular diagram. Bending moment at the tieback from the Terzaghi et al. diagram is about 65 percent greater than the design or measured moment, while the tieback load is only 15 percent higher than the design tieback load. Total load from the Terzaghi et al. diagram is 15 percent larger than the load from the other two diagrams, and that accounts for the larger tieback load. The large predicted moment from the Terzaghi et al. diagram results primarily from the rectangular shape of the diagram not the modest increase in total load. Results shown in Figure 9 and calculations summarized in Table 1 further support the use of trapezoidal earth pressure diagrams for the design of tiedback walls supported by one row of tiebacks.

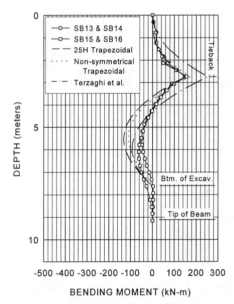

Figure 9. Comparison of Measured and Predicted Soldier Beam Bending Moments

Table 1. Comparison of Aspects of Three Apparent Earth Pressure Diagrams

Diagram	Total Load (kN)	Tieback Load (kN)	Moment at Tieback (kN)
Design	444.8	347.4	150.3
Non-symmetrical Trapezoidal	444.8	374.5	170.8
Terzaghi et al.	510.4	398.7	252.0

Staged Analysis

The practice of checking bending moments for the "cantilever stage" excavation to allow the tiebacks to be installed, developed when one-tiered walls were designed to support active triangular earth pressures. Apparent earth pressure diagrams are envelopes that included the effects of each stage of construction. Consequently, it should be unnecessary to check intermediate construction stages, when the walls are designed to support an appropriate trapezoidal apparent earth pressure diagram for the final excavation stage. For example, apparent earth pressure bending moments over the upper portion of the wall are much larger than moments predicted from a staged analysis using active earth pressures. The moments are larger because the apparent earth pressures are 2 to 3 times greater than Rankine triangular earth pressures over the upper portion of the wall.

Figure 10 shows the measured bending moments for the three construction stages for the full-scale wall, and the predicted moments for the final construction stage. The predicted moment diagram for the final construction excavation stage brackets the measured moments for the three construction stages. These results show that apparent earth pressure diagrams reasonably predict the bending moments in the wall, and that it is unnecessary to design tieback, H-beam walls for intermediate construction stages.

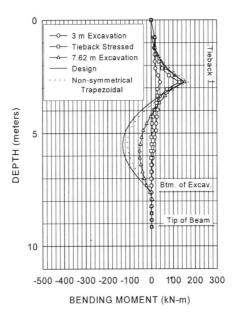

Figure 10. Measured Bending Moments for Each Construction Stage Compared to Moments Predicted from Apparent Earth Pressure Diagrams

Observations and Conclusions

The following observations and conclusions regarding the design and behavior of one-tier, flexible, tiedback walls in competent ground are made:

1. The full-scale wall and model walls, designed to resist the 25H Trapezoidal apparent earth pressures, behaved satisfactorily.
2. Driven and drilled-in soldier beams in the full-scale wall performed satisfactorily. Driven beams settled less and moved laterally less than the drilled-in beams. Drilled-in beams appeared to develop larger lateral toe resistances.
3. Tiebacks in the full scale wall were locked off at a load equal to 75 percent of the design load. After construction, the load quickly stabilized at a value equal to 82 percent of the design load.
4. Trapezoidal apparent earth pressure diagrams (25H or Non-symmetrical Trapezoidal diagrams) describe the behavior of wall supported by one row of tiebacks better than triangular or rectangular pressure diagrams.
5. Bending moments in and lateral movements of the soldier beam toes are small. A 1.52 m embedment depth was adequate to develop the required lateral resistance.
6. Assuming a hinge at subgrade gives a conservative prediction of bending moments when a trapezoidal apparent earth pressure diagram is use.
7. Bending moments at the tieback were a response to the applied tieback load.
8. Bending moments below the tieback are a result of applied earth pressures.
9. The Non-symmetrical Trapezoidal apparent earth pressure diagram in Figure 8 can be used to design walls supported by one row of tiebacks. The total design load for the diagram must be specified.

Acknowledgments

This work was done as part of a research project funded by the Federal Highway Administration and Schnabel Foundation Company. The work was performed at Texas A&M under the direction of Professor Jean-Louis Briaud, at the University of Illinois under the direction of Professors James H. Long and Edward J. Cording, and at Schnabel Foundation Company. Texas A&M graduate students M. Chung, N.-K. Kim, and W. F. Powers III collected the data for the full-scale wall research reported in this paper. C. G. Mueller, University of Illinois graduate student, conducted the model tests and performed the model wall analyses presented in this paper. The development of the non-symmetrical trapezoidal apparent earth pressure diagram was guided by Professor Cording.

References

AASHTO (1996). *The Standard Specification for Highway Bridges*, 16th Edition, Washington, DC.

Clough, G.W. and O'Rourke, T.D. (1990). "Construction Induced Movement of In-situ Walls," *Proceedings ASCE Specialty Conference Design and Performance of Earth Retaining Structures*, Cornell University, Ithaca, New York.

Goldberg, D.T., Jaworski, W.E., and Gordon, M.D. (1976). *Lateral Support Systems and Underpinning*, Report No. FHWA-RD-75-128, Vol. 1, Federal Highway Administration, Washington, DC.

Mueller, C.G., Long, J.H., Weatherby, D.E., Cording, E.J., Powers III, W.F., and Briaud, J-L. (1998). *Summary Report of Research on Permanent Ground Anchor Walls*, "Volume III: Model-scale Wall Tests and Ground Anchor Tests," Report No. FHWA-RD-98-067, Federal Highway Administration, McLean, Virginia.

Peck, R.B. (1969). "Deep Excavations and Tunneling in Soft Ground, State of Art Report," *7th ICSMFE*, Mexico City, pp. 225-290.

PTI (1996). *Recommendations for Prestressed Rock and Soil Anchors*, 3rd Edition, Post-Tensioning Institute, Phoenix, Arizona.

Rowe, P.W. (1952). "Anchored Sheetpile Walls," *Proceedings Institute of Civil Engineers*, London, England, Vol. 1(1), pp. 27-70.

Schnabel, H.J. (1982). *Tiebacks in Foundation Engineering and Construction*, McGraw-Hill Book Company, New York.

Terzaghi, K., Peck, R.B., and Mesri, G. (1996). *Soil Mechanics in Engineering Practice*, 3rd Edition, John Wiley & Sons, New York.

Weatherby, D.E. (1997). *Design Manual for Permanent Ground Anchor Walls*, Report No. FHWA-RD-97-130, FHWA, McLean, VA.

Weatherby, D.E., Chung, M., Kim, N-K., and Briaud, J-L., (1998). *Summary Report of Research on Permanent Ground Anchor Walls*, "Volume II: Full-scale Wall Tests and a Soil-Structure Interaction Model," Report No. FHWA-RD-98-066, FHWA, McLean, VA.

Toe Penetration Requirements for Tieback Walls

C.G. Mueller[1], J.H. Long[2], and E.J. Cording[2]

Abstract

Instrumented performance of four model tieback walls and available full-scale measurements have been used to examine tieback wall behavior. Specifically, conclusions have been developed with respect to the distribution of lateral and axial forces, and requirements for design of the embedded length of wall (toe). The test measurements generally support use of a design earth pressure envelope for walls with multiple levels of tiebacks that is consistent with experience for internally braced structures. A modified earth pressure envelope provides improved interpretations of required tieback forces and toe penetration depths for walls with a single level of support. Model test and full-scale measurements indicate that significant axial forces due to downdrag can develop during tieback wall construction. Design axial and lateral forces must be carefully evaluated with respect to tolerable levels of wall deformation.

Introduction

Tieback walls have been used in earth support applications for over 30 years. While a considerable base of knowledge exists with respect to the deformations of these structures, comparatively few measurements of the lateral and axial forces that develop on the walls have been made. As a consequence, design procedures for tieback walls are based principally on experience with anchored bulkheads and internally braced walls. A wide range of tieback forces and toe penetration depths can be computed, depending upon the magnitude and distribution of lateral earth pressure assumed for design. Thus, a fundamental investigation of the behavior of tieback walls has important implications for design of these structures.

---

[1]Associate Member, Geotechnology, Inc., 2258 Grissom, St. Louis, MO 63146.

[2]Member, University of Illinois at Urbana-Champaign.

Instrumented performance of four model-scale tieback walls are used to examine the distribution of lateral and vertical forces that develop during prestressing and excavation. Field measurements from a companion study (Chung and Briaud, 1994) supplement the model test observations. Significant conclusions concerning appropriate earth pressures upon which to base design and toe penetration requirements are developed. Current design practice is reviewed to provide a practical framework for presenting results and extending them to practice.

Current Design Practice

Lateral earth pressures for design of tieback walls are generally based on either classical earth pressure theories or practical experience with internally braced walls or anchored bulkheads. Classical earth pressure theories can be used to evaluate upper and lower bounds of design lateral thrust, and conceptually, provide a basis for selecting lateral earth pressures with consideration to acceptable levels of ground deformation. General purpose limit equilibrium methods of analysis can be used to evaluate the minimum force required for stability, and provide a rational approach for evaluating the effects of irregular backslope geometries, surcharge loadings, seepage, and layered soil profiles. The minimum force can then be factored and distributed to the wall based on experience with anchored bulkheads and internally braced walls. More typically, lateral earth pressure envelopes derived from strut load measurements are used to evaluate required tieback forces and toe penetration depths. These apparent earth pressure envelopes are either uniform or trapezoidal in shape, and provide resultant thrusts that are typically intermediate to active and at-rest states of stress. The principal disadvantage of current approaches for selecting design lateral earth pressures is that a rationale for incorporating the effects of wall stiffness and tieback locations does not currently exist.

For walls with a single level of support, the distribution of lateral forces to the tiebacks and toe can be evaluated from force and moment equilibrium assuming either free- or fixed-earth support conditions. Simplifying assumptions are required to determine the distribution of forces for walls with multiple levels of tiebacks. In the tributary area method (Goldberg et. al., 1976), tieback forces are assumed to correspond to the portion of the earth pressure diagram bounded by the half-distances between adjacent levels of support. The required toe reaction is computed as one-half of the earth pressure diagram between the lowest level of tiebacks and design grade. In the hinge method (Williams and Waite, 1993), the top and intermediate spans of the wall are analyzed as simply supported beams. The bottom span can then be evaluated with a single level of support, assuming either free-earth or fixed-earth support conditions. Smaller toe reactions are calculated for a uniform pressure distribution compared with a triangular distribution of pressure providing the same resultant thrust. Fixed-earth support conditions result in greater depths of toe penetration compared with a free-earth support analysis, but yield a lower tie force.

The distribution of axial forces for tieback wall design has received comparatively little attention in the literature. As a minimum, walls must be designed with sufficient axial capacity to accommodate the vertical component of force introduced by tieback prestressing. It is conceivable, however, to develop additional vertical load due to relative downward movement of the ground with respect to the wall. Although tieback walls have generally performed satisfactorily, poor performance associated with wall settlement has been observed (Shannon and Strazer, 1970).

Model Testing Program

Four large-scale model tieback walls were constructed in the Illinois Large-Scale Model Test Facility maintained by the Department of Civil Engineering at the University of Illinois. Detailed discussion of the test facility is given by Long et. al. (1998). Brief descriptions of the model walls, test variables, instrumentation, and construction procedures are given herein.

Description of the Walls. The model-scale walls were designed as soldier pile and lagging structures, and were approximately 1.9 m in height, with 0.4 m of pile toe penetration below grade. Nine soldier piles were used in each test and were spaced at approximately 0.6 m intervals across the width of the test chamber. Model soldier piles consisted of steel tubing with scaled bending stiffnesses consistent with full-scale walls ranging in height from about 7 to 9 m and supported by one to two levels of tiebacks. Steel plate (3 mm thick) was used to simulate the 76 to 102 mm thick hardwood boards used to retain soil between soldier piles in full-scale applications. Small diameter (6 and 9.5 mm) steel bars, with scaled axial stiffnesses consistent with post-tensioned prototype tiebacks, were used to model the anchors. The bars were connected to a reaction frame inside the test chamber, so that wall deformations associated with mass movements behind the anchors, or load redistribution along the anchorage, were not modeled.

The soil used for the model tieback studies consisted of a medium-grained, uniform sand (SP) with a fines content of less than 1 percent. Dry sand was used in the model tests to simplify material handling and deposition, and to decrease uncertainties that moisture might introduce to behavior of the material. Sand was placed inside the test chamber by pluviation through air. Mass rate of flow and height of fall were used to control sample density. Based on cone penetration soundings and direct measurements, the relative density of the soil was estimated between 45 and 55 percent. A medium dense deposit was used in the model tests to provide a material with shear strength, stress-strain, and volume change characteristics consistent with a dense sand for the range of stresses normally encountered in practice.

Test Variables. Variables of the model-scale testing program included the number of levels of tiebacks, relative soil/wall stiffness, axial pile capacity, and mode of failure. The walls were constructed with either one or two levels of tiebacks. Relative soil/wall stiffness was varied approximately one order of magnitude, so that a full range of behavior could be studied. Axial pile tip capacities were varied so that the significance of pile settlements for vertical load transfer and lateral wall movements could be assessed. In one test, pile dimensions provided an axial pile tip capacity that was significantly less than the vertical component of tieback force. In the remaining tests, piles were fitted with an enlarged bearing plate to increase available axial capacity. Failure modes at the end of construction included unloading of tiebacks and over-excavation of the embedded length of wall.

Instrumentation. Test measurements during construction of the model-scale walls included lateral and vertical wall and ground movements, bending and axial pile strains, and tieback loads. Wall and ground movements were measured using dial gages and multiple position extensometers formed by connecting DC-DC LVDT's in series. Bending and axial pile strains were measured using bonded resistance type strain gages installed on the front and back flanges of the soldier piles, and compensated for temperature variation and drift with dummy gages installed adjacent to the active gages. The repeatability of the strain gages was approximately ± 1 $\mu\epsilon$, compared with maximum measured bending and axial pile strains on the order of $500\pm$ $\mu\epsilon$ and $30\pm$ $\mu\epsilon$, respectively. Tieback loads were estimated using one pair of bonded resistance strain gages placed near the head of the anchor.

Construction. Sand was placed inside the test chamber by air pluviation from concrete buckets hoisted into position with a radio-controlled crane. Once sand had been placed inside the test chamber to the pile tip elevation, soldier piles and tiebacks were installed. Tiebacks were temporarily connected to the soldier piles, but were disconnected before complete filling of the test chamber. Geotextile fabric was placed between the soldier piles during filling to simulate "stand-up" time and facilitate placement of lagging without excessive ground loss.

Excavation was performed in 76 mm increments. With the geotextile fabric exposed by excavation, lagging was slipped over machine screws installed through the soldier piles and secured with coupler nuts. Voids which developed between the lagging and geotextile fabric were backfilled with a sand/glue mixture. All electronic instrumentation was read immediately prior to and following installation of a single lagging board. Dial gages at the ground surface were read after completion of a complete row of lagging.

Excavation continued in the manner described above until the tiebacks at a support level were exposed. The tiebacks were connected to a load control device and wale used to transfer and distribute tieback forces to adjacent soldier piles. Each tieback was loaded to approximately 120 percent of design to simulate proof and performance tests used in the field. The loads were then decreased and locked-off at 75 percent of the design load. Tieback loads were calculated using a trapezoidal pressure distribution with an intensity of 25H as suggested by Schnabel (1982) for a wide range of soil types including generally stiff clays and medium to dense sands. All electronic instrumentation and dial gages were recorded with loads in the tiebacks at 120 and 75 percent of design. After completing excavation to design grade, pile toes were exposed and tieback loads reduced to study the behavior of the walls at large deformations.

Lateral Earth Pressures

Lateral earth pressure interpretations for the model tests were developed using measured bending strains and tieback forces, and assumed boundary conditions at the top of the piles. Detailed discussion of the procedure is given by Mueller (1999), but in general, the approach consisted of developing "best-fit" curves to the test measurements using Hermetian polynomials, then twice differentiating the "best-fit" functions to develop interpretations of shear and lateral earth pressure. Earth pressure interpretations developed from the model test measurements are discussed in this section. Lateral earth pressures have been normalized with respect to vertical ground stress, γH, where γ represents the unit weight of the soil and H is the depth of the excavation at design grade. Lateral wall movements, expressed as a percentage of the maximum depth of excavation, are also shown.

Thrust Development. The most significant changes in lateral earth pressure in the model tests developed during tieback prestressing (Fig. 1). Tieback prestressing pulled the soldier piles back into the soil, resulting in development of a zone of

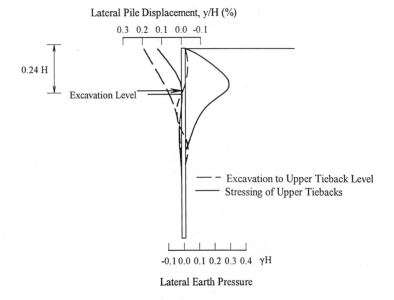

Fig. 1 - Effect of Tieback Prestressing on Lateral Earth Pressures, Model Test 4

compression adjacent to the wall to support the applied prestress loads. An approximately symmetric pressure "bulb" developed at the tieback level, the shape of which, was a function of relative soil/wall stiffness. Stiff walls tended to develop a wide and flat pressure "bulb" compared with wall which were more flexible.

Small changes in lateral earth pressures were observed in the model tests during excavation below a tieback (Fig. 2). Excavation below the tiebacks resulted in lateral bulging of the walls below the support. Lateral bulging developed with the walls essentially fixed against displacement at the tieback level, and additional resistance to movement developed along the fully embedded length of the wall. Lateral bulging resulted in small decreases in pressure in the unsupported span of the walls, and corresponding increases in pressure near the base of the walls and in the vicinity of the tiebacks.

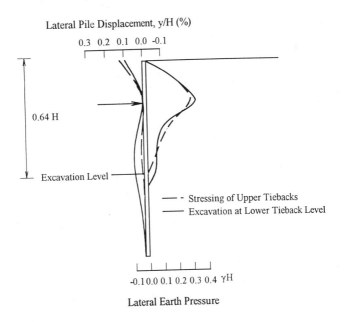

Fig. 2 - Changes in Earth Pressures During Excavation Below a Tieback

Stressing of a lower level of tiebacks resulted in a similar wall deformation response to that which had been observed during stressing of the upper tiebacks, and development of a passive pressure "bulb" in the vicinity of the support. During excavation below the lower level of tiebacks, lateral bulging developed in the span between the tiebacks and design grade. Wall deformations also included a significant component of outward rotation and translation, which apparently resulted from pile settlements associated with development of downdrag and corresponding mobilization of end-bearing and side resistance. This behavior is consistent with previous model studies performed by Hanna and Matallana (1970), who observed that wall settlement geometrically contributes to outward translation without change in length of the tiebacks.

The magnitude of model wall rotation and translation was sufficient to develop a triangular distribution of pressure on the wall, based on Terzaghi's (1936) rigid retaining wall experiments in sand, and Rowe's (1952) model-scale studies of anchored bulkhead behavior. Despite the development of significant rotation and translation in the model tieback studies, the lateral earth pressures retained the characteristic pressure distribution locked-in during tieback prestressing. Thus, development of rotation and translation in tieback wall construction is not necessarily consistent with a triangular distribution of pressure, as is sometimes assumed in design. In fact, the model tests observations generally support use of earth pressure envelopes for walls with multiple levels of support that are consistent with experience with internally braced walls (Fig. 3).

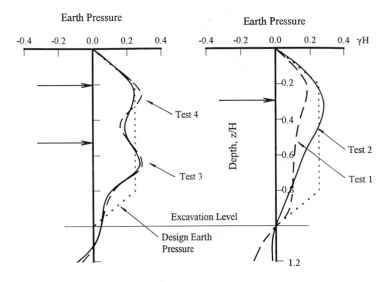

Fig. 3 - Lateral Earth Pressures with Excavation at Design Grade

Significance of Lateral Toe Resistance. Tieback prestressing in the model tests had a significant effect on the magnitude of lateral pressures observed near the base of the walls. Earth pressures near the base of the walls were significantly less than the Coulomb earth pressure computed for the full height of the wall and distributed hydrostatically. This observation is generally consistent with that of Rowe (1952) for flexible bulkheads prior to anchor yield. As a consequence of the small pressures near the base of the walls, mobilized lateral toe resistance was significantly smaller than the computed required reaction calculated using either the tributary area or hinge methods. In fact, mobilized lateral toe resistance in the model tests was typically less than 10 percent of the total design lateral thrust. The model test observations were in basic agreement with the interpretations of lateral toe resistance for the full-scale walls constructed as part of a companion study (Chung and Briaud, 1994). For walls supported by multiple levels of tiebacks, a balanced trapezoidal distribution of earth pressure provides a rationale basis for evaluating tieback forces and estimating bending moments, but may overestimate required toe penetration for lateral forces, particularly for excavations in generally stiff ground. As discussed in a subsequent section, however, toe penetration for walls with multiple levels of support may become more important to satisfy axial force considerations. For walls supported by a single level of tiebacks, a balanced non-symmetrical trapezoid provides more reasonable interpretations of required lateral toe reactions (Fig. 4).

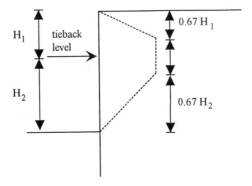

Fig. 4 - Balanced Non-Symmetric Trapezoid for Design of Tieback Walls with a Single Level of Support (Cording et. al., 1998)

Axial Loads

Maximum axial loads in the model tests in relation to the progress of excavation and tieback prestressing activities are summarized in Fig. 5. Several significant observations are made, including that maximum axial loads generally increased throughout excavation and that comparatively small relative downward movements of the ground with respect to the wall were required to mobilize full downdrag.

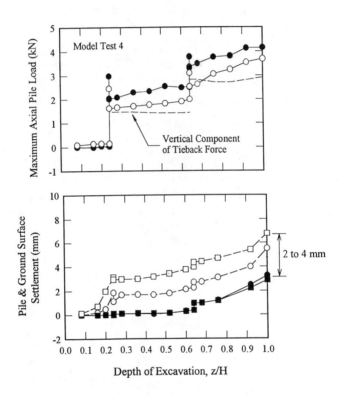

Fig. 5 - Maximum Axial Pile Loads in Model Test 4

From a practical perspective, it is necessary to consider only the maximum axial load that manifests itself in the pile toe, since pile toe penetration is a variable in design and the maximum depth of cut is fixed by project requirements. Fig. 6 summarizes maximum axial loads, Q, normalized with respect to the vertical component of tieback force, T, observed in the toes of the model and full-scale walls. In the model tests, significant axial load was supported by the piles above design subgrade. Maximum axial loads observed in the pile toe ranged from 1.0 to 1.2 times the vertical component of tieback force. The full-scale test section supported by a single level of tiebacks generally supports the model test observations. For the full-scale test section supported by two levels of tiebacks, however, significant pile settlements developed and maximum observed axial loads in the pile toe were about 50 to 60 percent of the vertical component of tieback force. The consequence of the large pile settlements, however, was correspondingly large lateral movements of the wall ($\approx$ 0.45% H), compared with the test section supported by a single level of tiebacks ($\approx$ 0.24% H). Thus, while it may be possible to design pile toes for axial loads less than the vertical load introduced by tieback prestressing, the consequences of wall settlement on adjacent ground movements must be carefully considered.

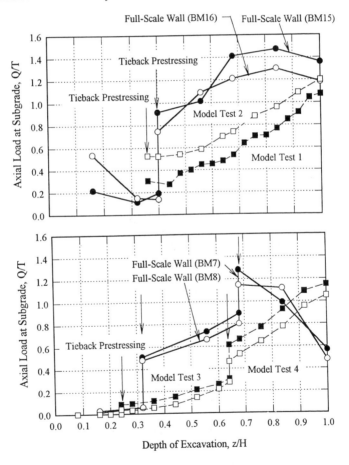

Fig. 6 - Maximum Axial Loads in Pile Toes in Model and Full-Scale Walls

Summary and Conclusions

Model test and full-scale field measurements have been used to examine tieback wall behavior and develop conclusions concerning toe penetration requirements for design. The significant findings of the study can be summarized as follows:

1. The available measurements generally support use of a balanced symmetric trapezoidal distribution of earth pressure for design of tieback walls with multiple levels of support. The calculated toe reaction obtained using a balanced symmetric trapezoid is generally conservative, however, for excavations in stiff clays and dense sands. For excavations with an intervening soft layer, toe penetration can be useful in controlling certain components of wall deformation. The decision in design to include toe penetration for lateral loads should be based on full assessment of the effect on wall and corresponding ground deformations.

2. For tieback walls with a single level of support, a modified earth pressure distribution (balanced, non-symmetric trapezoid) provides improved estimates of required toe penetration for consideration of lateral forces.

3. Axial forces associated with downdrag can be significant during tieback wall construction, and comparatively small relative movements of the ground with respect to the wall are required to mobilize full downdrag. Appropriate design axial forces must be evaluated with careful consideration to acceptable levels of wall and ground movements. Model and full-scale test observations have shown that significant translation and rotation of a wall can develop due to settlement. Where ground control is critical to performance, toes should be designed as a minimum to support the vertical component of tieback force associated with prestressing.

Acknowledgments

Funding for the study described herein was provided by the Federal Highway Administration and Schnabel Foundation Company as part of a research project titled "Permanent Tieback Walls", FHWA Contract DTFH61-89-C-00038. Funding for the model test facility was provided under a series of grants from the U.S. Army Research Office to the University of Illinois Advanced Construction and Technology Center, DAA1 03-87-K-0006, DAA1 03-85-G-0186, and DAA1 03-86-G-0188. The contributions of the FHWA, Schnabel Foundation Company, and the U.S. Army Research Office are gratefully acknowledged.

References

Chung, M. and Briaud, J. (1994) "Behavior of a Full-Scale Tieback Wall in Sand", Geotechnical Engineering, Department of Civil Engineering, Texas A&M University, College Station, Texas, 249 p.

Cording, E.J., Long, J.H., Mueller, C.G., and Ghahreman, B. (1998) "Excavation, Ground Movements, and Building Damage", Presentation to New York Metropolitan Section, ASCE, by E.J. Cording.

Flaate, K.S. (1966) "Stresses and Movements in Connection with Braced Cuts in Sand and Clay", Ph.D. Thesis, University of Illinois, Urbana-Champaign, 264 p.

Goldberg, D.T., Jaworski, W.E., and Gordon, M.D. (1976) Lateral Support Systems and Underpinning, Vol. I, II, and III, Federal Highway Administration, FHWA-RD-75-103, April.

Hanna, T.H. and Matallana (1970) "The Behavior of Tied-Back Retaining Walls", Canadian Geotechnical Journal, Vol. 7, pp. 372-396.

Long, J.H., Cording, E.J., and Mueller, C.G. (1998) "Analysis and Behavior of Tieback Walls", Report to Schnabel Foundation Co., FHWA Permanent Tieback Wall Study.

Mueller, C.G. (1999) "Load and Deformation Response of Tieback Walls", Ph.D. Thesis, University of Illinois.

Schnabel, H. (1982) Tiebacks in Foundation Engineering and Construction Practice, McGraw-Hill Book Company, New York, New York, 170 pp.

Rowe, P.W. (1952) "Anchored Sheet Pile Walls", Proceedings of the Institution of Civil Engineers, Part 1, pp. 27-70.

Shannon, W.L. and Strazer, R.J. (1970) "Tied-Back Excavation Wall for Seattle First National Bank", ASCE, Civil Engineering, March, pp. 62-64.

Williams, B.P. and Waite, D. (1993) The Design and Construction of Sheet-Pile Cofferdams, CIRIA Publication 95, London, Thomas Telford Publications, 198 pp.

EVOLUTION OF DESIGN - LA METRO UNDERGROUND STRUCTURES

James E. Monsees, Ph.D, P.E, Fellow, ASCE[1]
Amanda Elioff, P.E. Member, ASCE[2]

ABSTRACT

The Los Angeles Metro Rail project has provided unique tunneling challenges over the past 16 years. The project is near completion - for the foreseeable future - and this paper presents a review of the major design and construction issues, solutions, and the authors' vision for advancement in future design and construction. Included in the discussion are tunneling in the presence of hazardous gasses, tunneling machine specifications, innovations in segmental linings, and design for seismic loads in light of a general "ratcheting up" of predicted earthquake movements that happen after major earthquakes.

INTRODUCTION

The Los Angeles Metro Rail project has had a series of ups and downs. Currently the plan is to finish the original 17.6 miles of subway and postpone extensions to an uncertain future date. We are not going to dwell on that, however. Rather we will document the achievements and discuss briefly some of the major advances that have been made on underground design and construction as a result of this project and then alert the profession to one area that we believe requires significantly more research and development for future projects to be constructed in similar environments

The major underground challenges faced on the project over its 15 year (phased) design to be discussed herein are:

---

[1] Technical Director for Underground Engineering, Sr. Vice President, Parsons Brinckerhoff, Inc., Orange 505 S. Main Street, Orange, CA 92868
[2] Professional Associate, Parsons Brinckerhoff, Inc., 444 S. Flower St., Los Angeles, CA 90071

- Subsurface Gases
- Tunneling Machines
- Tunnel Linings
- Seismic Design

SUBSURFACE GASES

When we started work on the project, it was known that Los Angeles had a history of oil production (parts of the basin had produced locally more crude oil per acre than any spot on earth) and that the crude seeping to the surface at the La Brea Tar Pits, just west of downtown and near the initial subway alignment, was a manifestation of underground petroleum. Thus, we fully anticipated encountering methane gas in the tunnels.

A 1971 gas explosion in the relatively close Sylmar tunnel caused 17 deaths, and as a result, California has the most rigorous tunnel safety code in the nation. Once the extent of the methane was determined, we were confident that the tunnels could be safely constructed using these restrictions. The contractor had to provide the right equipment, provide sufficient ventilation, perform required training of personnel, follow the established procedures and conduct the specified gas measurements.

These actions were taken and the tunnels constructed without a major incident. However, constructing the tunnels was only half the challenge: the subway also must both operate and transport people safely in the methane environment.

To meet this latter challenge we elected to encase all underground structures (tunnels, stations and appendages) in a continuous high density polyethylene (HDPE) membrane that was sandwiched between the initial support and final lining. We argued that this would also waterproof the structures, thereby performing two roles. Being inert, the 60 to 100 mil HDPE can be joined only by welding and its proper installation required great effort by the contractors and great diligence by the construction management staff. With continuing improvements in product (e.g., going to a white-black layered HDPE; the better to see holes, etc.), in installation techniques, and in inspection, contractors are able to obtain (nearly) a gas-tight (and, therefore, water-tight) tunnel. As a tertiary measure, automatic gas sensors have been installed to continuously monitor the underground environments. Should a leak be detected, alarms are sounded and emergency fans are activated.

With all of these steps the first tunnels were successfully built and put into service and they have been operating successfully since 1993. However, the latest section of subway to be designed, the East Side Extension, presented a new challenge - along with the methane, the ground also contains hydrogen sulfide (H_2S). This extension was to have added 4 miles to the system running from Downtown to East Los Angeles (Figure 1). Previously, a Mid-City alignment design was delayed and later re-aligned to avoid high concentrations of H_2S in unsaturated ground (Elioff et.al, 1995). While the effect on humans of H_2S vary to some extent, it can be extremely dangerous to health at a concentration in air of 40 parts per million (ppm). Thus worker exposure is limited by law to a weighted average of only 10 ppm for an eight hour work day and at a concentration of 15 ppm all workers must be evacuated. Above about 100 ppm the concentration may be fatal.

These factors all played major roles in our design decisions on the next two challenges, the tunnel methods and the tunnel linings. On the East Side Extension, concentrations of dissolved H_2S in the groundwater exceeded 100 mg/l in some areas, which could produce well over 100 ppm in air if allowed to collect in unventilated areas. Encountering H_2S in the tunnels or in tunnel construction may never be realized however: both the Mid City and East Side Extensions have been postponed indefinitely as local legislation was passed in November 1998 prohibiting use of county sale tax revenues on underground transit projects.

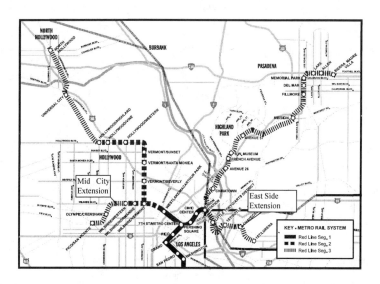

Figure 1. System Map Los Angeles Metro

TUNNELING MACHINES

For the most part, the early subway tunnels in Los Angeles were driven under public right of way and contractors were able to construct them using open face tunneling machines and "two-pass" construction with initial supports of ribs and boards or unbolted precast segments, the HDPE membrane, and a final lining of cast-in-place concrete. Contrary to sensationalized press reports, the two-pass approach provided some of the least expensive subway tunnels in the world.

For the East Side Extension, the design criteria changed dramatically:

- The alignment's zone of influence passes under some 250 buildings and other structures, some dating to the 1920's, 30's and 40's.
- The ground contains H_2S.
- Although more than 90% of the L.A. tunnels have been constructed trouble free, highly publicized and unacceptable exceptions have occurred (Monsees, 1998). A comparison with other tunnels around the world, undertaken by the Los Angeles Metropolitan Transit Authority (MTA) showed L.A. Metro tunnels to be equal or below in cost and equal or better in quality (Eisenstein et. al., 1997). However, due to an infamous sinkhole and excessive settlement at a few other sites, the MTA and its outside panel of consultants decided that open face shield construction would no longer be acceptable for settlement control.

Therefore, the designers were to be required to go "high tech", i.e., specify closed, pressure-face machines for all tunnel drives on the East Side Extension. More specifically, on the drives from Union Station to Little Tokyo Station, the specifications as they now stand, require use of a slurry-face machine, and on the drives from Little Tokyo to First/Lorena they require either an Earth Pressure Balance Machine (EPBM) or a slurry face machine.

The rationale for requiring the slurry machine between Union and Little Tokyo Stations springs from the hydrogen sulfide in both the ground and groundwater. Professionals can (and do) have different opinions regarding how much H_2S will be encountered and how much might be released into the tunnel. However, all agree that, in comparison to our experience with methane, the toxic property of H_2S meant that only extremely small amounts could be allowed into the work environment.

With a slurry machine, all water and muck from the face is trapped in a closed system until it arrives at the separation plant. Thus, risk to workers in the confined space of the tunnel is limited to the relatively unlikely event of an uncontrolled spill from the machine face or rupture in the slurry line running from the machine to the surface. At the separation plant on the surface, the exposure to workers can be controlled by treating (neutralizing) the H_2S in the muck stream, by high volume ventilation, and/or by keeping the whole plant isolated in its own mini-environment and treating or filtering any effluent from the plant.

On the remaining runs, (beyond Little Tokyo Station) the design required use of a pressure-face machine: either an EPBM or slurry face machine. With the use of pressure-face machines, which have not been used or developed in the U.S. as much as in either Japan or Europe, the contractor would be expected to have much better control of the face, and better support of the ground behind the machines than has been the case in the typical U.S. soft ground tunnel. Thus, the contractor would be expected to limit settlements to less than 2.5 cm for a 5.5-m tunnel. For added assurance that settlements of the more sensitive or proximate buildings would be controlled, a compaction grouting program using pre-installed pipes was designed.

It is acknowledged that U.S. use of pressure-face machines lags that in Japan and Europe by a decade or more. It is further acknowledged that requiring such machines be used would mean that U.S. contractors might have to associate with contractors from Japan or Europe to bring this technology and the specified experience to the U.S. Forcing such associations is exactly what we intended to achieve (Monsees, 1998).

TUNNEL LININGS

The use of pressure-face machines necessitates ground support with bolted, gasketed precast segments with immediate back-fill grouting after the ring is assembled. Costs of this "one-pass" system are typically higher than for the two-pass system, as was verified by previously bid LA metro work. Thus alternatives to the HDPE and final CIP liner were highly desirable.

With the primary concern of H_2S intrusion into the tunnels, a proven gasket design was sought to ensure a seal against H_2S containing water, even under cyclic, seismic distortions. We designed a system providing several levels of redundancy and tested full scale gasketed segments at the University of Illinois. The redundant system included: 1) A double gasket design combined with convex shaped longitudinal joints, (Figure 2), 2) an oversized tunnel (152mm on the radius) such that HDPE and a CIP lining could be added at a later date should leaks occur, and 3) gas detection and ventilation systems similar to the rest of the system.

The testing program provided data to convince the designers and the Authority that the tunnels would not leak significantly, even when subjected to cyclical loads induced by the design earthquake. The seismic design criteria is discussed further in the next section. The segment testing program included eight elements; a brief discussion of the program and results follows:

Relationships between gasket leakage, gas pressure and straight gasket offset: These tests showed no leakage of gaskets (in metal test frames) at gas pressures exceeding those that would be found surrounding the tunnels. Leakage did begin when gaskets the (nominal

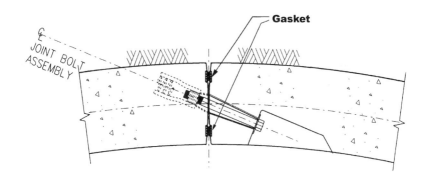

Figure 2 Radial Joint Detail

width 33mm) were offset about 5.5cm., which is greater than construction tolerances allowed in the contract documents.

Static compression and relaxation tests of three gasket materials: EPDM, (ethylene-propylene-Diene Terpolymer) and two compounds of Neoprene rubber (as supplied by gasket manufacturers) were compressed to determine the extent of relaxation under load with time. Gasket relaxation properties are used to evaluate long term performance under the anticipated water pressures. The testing at the University of Illinois showed EPDM material to have the greatest relaxation with time, and as projected, somewhat greater relaxation than the 40% at 100 years required by specifications. Tests by gasket manufacturers report more favorable results, and considering the additional data and conservative loads used for pressure testing, it is believed that specifications for long term relaxation can be met by EPDM.

Structural capacity of convex to convex joints: Tests were performed both in direct axial loading and with a slight rotation of segments to simulate earthquake induced distortion. In both cases, load at failure (due to splitting) was approximately one-half of that for the total theoretical section of segment, but still several times the design load, indicating adequate factors of safety in structural capacity.

Cyclic loading: Precast segments were rotated under working soil loads to simulate cycles of ovaling or racking due to an earthquake. The segments were subjected to 120 cycles of deformation while under an axial load of 27.3T and again at an axial load of 57T. No damage to segments due to cycling was observed.

Leakage testing after cycling: After the cyclic testing, water pressure was reintroduced between the gaskets to test leakage. Initially, no leakage occurred up to a water pressure of about 3.4 bar, when water leaked through bolt pockets. Bolt pockets were sealed using a polyurethane sealant, and tests were repeated.

Actual field water pressures are estimated to be a maximum of 2 bar, but higher test pressures are used to account for gasket relaxation. After bolt pockets were sealed more than 3.4 bar was reached before minor leakage occurred at gasket corners. Cyclic testing did not reduce pressures at leakage, indicating good performance after an earthquake.

Gas leakage subsequent to cycling was tested through portions of the gaskets. These tests found minor leakage at high pressures and very low leakage values at anticipated field conditions. The gas leakage tests after segment cycling were performed using a dilute methane/air gas mixture. Hydrogen sulfide gas leakage is expected to be lower due to the larger H_2S molecule size.

T-Joint test to account for offsets between segment rings: The T-Joint represents the staggered joint between segment rings. Axial load was applied to radial joint and the assembly was loaded cyclically to induce an horizontal deformation to the joint, simulating tunnel "ovaling" during an earthquake. Water leakage tests through the joint after cycling were then performed. No leakage occurred up to about 3.4 bar pressure before cycling and about 2.5 bar after. Some damage to the outside edge of one segment occurred during cycling and leakage occurred at this location.

Repair schemes: Short length model segments were cast with grout access grooves in front of the outer gasket (Figure 3). This would facilitate grouting as a repair option in the event of water or gas leakage. Two types of swelling urethane grouts were used for the trials. Results showed that such a detail can be added to segment design allowing a secondary method for sealing leakage. Similar details using reduced groove dimensions were also effective.

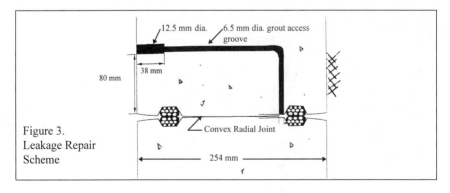

Figure 3. Leakage Repair Scheme

Gasket materials: The three gasket materials (described above) were exposed to water containing contaminates found along the East Side alignment. The EPDM sample appeared superior to others with respect to gas permeability and durability upon exposure to the contaminants.

Results of these tests gave the Authority the assurance needed to allow tunnel design and construction to proceed without installing a second cast-in-place lining (which was to have been fiber reinforced) and an HDPE membrane. Eliminating the inner liner would reduced the tunnel cost by an estimated $20M.

SEISMIC DESIGN

When the Red Line project began in 1982/83, an attempt was made to use nuclear power plant seismic criteria for design of the underground structures. Immediately, J. Merrit and J. Monsees recognized that this was not appropriate for underground structures. Instead, they developed a criteria in which the seismic input was primarily a racking effect as caused by the earth movements. Thus, the structures were designed for a seismic distortion that was superimposed on the usual design loads. This approach, sometimes modified for local conditions, has been shown to be far more appropriate than the "pseudo static" loads often used above ground and has been used around the world.

The seismic design criteria is also based on two levels: 1) The Operating Design Earthquake (ODE) under which the structures are expected to receive no more than architectural damage and to go immediately back into service; and 2) The Maximum Design Earthquake (MDE) under which the structures are expected to sustain structural damage (crack, form plastic hinges) but not collapse, i.e., not pose a threat to life. We can report that the 1994 Northridge quake subjected the structures to the ODE accelerations and the structures behaved exactly as planned – in fact we are not sure any new cracks were formed.

New seismic design criteria was developed for recent Red Line structures, considering new seismic data available for the Los Angeles Region. The use of probabilistic rather than deterministic methods of earthquake prediction and the recognition of buried or blind faults and the earthquake motions they produce. Specifically, the discovery of the buried thrust fault in North Ridge and the Coyote Pass escarpment in East Los Angeles.

These new studies (1996-97) evaluated ground motions at the studied sites and ground deformations at the Coyote Pass Escarpment tunnel crossings. The resulting revised design criteria for the East Side Extension, which was under design at the time, provided design ground accelerations about 50 percent higher than the existing (1984) criteria, and racking deformations up to approximately two times the 1984 criteria. Load factors for the contribution from the Operating Design Earthquake were reduced to 1.0 for consistency with trends in current practice. These changes were concurred with by experts advising the Authority on seismic issues.

Structural design was aided using a Soil-structure Interaction computer model. During the course of the program, the model was modified for changes in ground motion inputs and material properties of the modeled structure. Station designs in progress at the time of the studies were subsequently revised to reflect the new distortions.

In summary, this effort provided updated design criteria for the East Side Extension in line with the latest knowledge and practice in the earthquake engineering field. Incorporated in the criteria and subsequent design of the East Side stations were new information on ground motions and near-field effects, recently discovered causative faults, computer simulations, laboratory research on the behavior of reinforced concrete structures under cyclic loads, and appropriate load factors considering site- specific studies.

All of these advances have the net effect of raising the levels of the design earthquakes. This, in turn requires corresponding changes in reinforcing amount and details: For this discussion, we consider the impact of the assumption that a design level of 1.0g is a possibility. We also make the simple approximation that free-field motion will scale upwards approximately as the acceleration. The results of these assumptions are shown on Figures 4a and b, and they are quite interesting.

Previously the excursions beyond the assumed maximum displacement of 36mm (Figure 4a) are seen to be two to three. Now, however, with the higher ground motions the excursions (Figure 4b) are seen to be greater than 10. Referring to Figure 5, it is readily seen that a typical reinforced concrete structure (structural wall shown in Figure 5) would degrade in capacity to below 20% of its initial capacity if subjected to this higher input.

From this simple evaluation, we draw three conclusions that appear to require more investigation:

1. Usual designs for reinforced concrete do not provide sufficient reinforcement to provide the repeated plastic behavior predicted.

2. These much greater displacements coupled with the structure's natural behavior to shed excess "load" to adjacent sections indicate that more plastic hinges will be formed and that the initial plastic hinges will probably be "smeared" out, i. e., made into wider plastic zones.

3. Current methods of analyses will have to be revisited to ascertain that the behavior hypothesized is reasonable and, if so, to provide appropriate, codified means of responding to the challenge.

For the Metro design, the objective was to convert the behavior from that represented in Figure 5, structural wall (typically reinforced, to that represented in the fully confined condition. With sufficient confinement, i.e., with large numbers of ties or by jacketing a column following work pioneered by Caltrans (California State Highway Department), this ductile behavior can be obtained for at least 8 or 10 cycles (or more). For the walls and slabs in a Metro station, however, we have to obtain this behavior by greatly increasing the number of ties in the zones which must behave elastically.

The. Metro structures in North Hollywood were actually under construction when the new criteria was developed. Degenkolb Engineers provided an in-progress design that greatly increased the number of ties in the station walls. Simply stated, ties were required as follows:

$$s < 6 d_b$$
$$s < d/4$$
$$s < 152 \text{ mm},$$

where s is the tie spacing and d is the depth of section and d_b is the main reinforcing bar diameter. The tightest of the three limits is to be used.

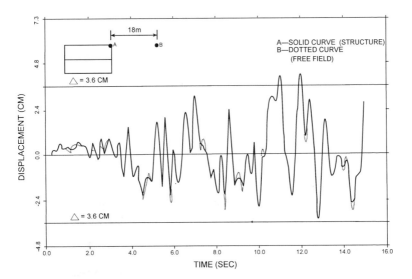

Figure 4a. Comparison of Structure and Free-Field Motion

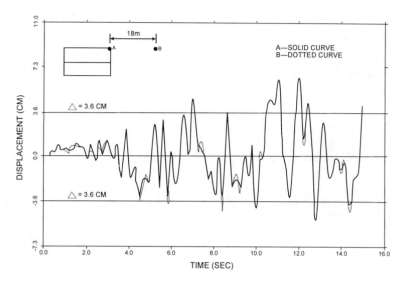

Figures 4b. Comparison of Structure and Free-Field Motion
(Ground motions increased 50%)

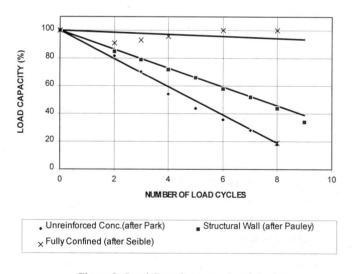

Figure 5. Load Capacity versus Load Cycles

As stated, these new requirements greatly increase the number and complexity of ties. To develop the maximum portion of the cross section, the ties must act directly, that is, each main bar must be tied by a direct tensile tie. All this means that the amount of the tie steel increases greatly as does the labor required to place all the bars. In addition, it is necessary to increase greatly the vibration effort to get the concrete consolidated around the tighter spaces formed by the additional ties.

These tie spacings apparently are based upon rules of thumb developed by Pauley (1992). The net impact of these recommendations is to nearly double the amount of tie reinforcement in the structure. (Note that each main bar must be cross-tied directly rather than relying on rectangular loop ties). It must be remembered that these structures are walls and slabs, not individual columns. Thus, the increase in steel represents a very large increase in cost. Approximately 10,000 to 14,000 square meters of 1.2 to 1.8 m thick wall and slab are involved per station. With the postulated "smearing out" of the hinges, we may be approaching the requirement that all or the majority of these walls and slabs be reinforced with closely spaced ties in their entirety instead of near discreet joints only.

Certainly, every indication is that these modifications are in the right direction. What appears to be needed is sufficient research to conclusively prove that this does provide enough (or is it too much?) reinforcement to result in the extremely ductile and reproducible hysteritic elastic behavior required to meet the new set of demands, imposed by the new, more critical design earthquake levels.

CLOSURE

Although new work on the LA Metro is stopped for now, the project has built some of the most cost effective tunnels in the world. In addition, the project has introduced a number of innovations and advances to the North American tunneling community. These include:

- Use of HDPE membranes to keep gas (and therefore water) out of tunnels, stations, and other structures.
- Specifications for pressure face tunnel boring machines for metro tunnels
- Development of a one-pass precast tunnel lining segment system with double gaskets and a pre-installed method for leak mitigation.
- Development of distortion-based seismic design criteria for underground structures.

In the area of seismic design, we see the need for innovation in designing and building underground stations which are large structures of slabs and walls. New methods, materials and approaches are needed to make these structures continue to remain stable under high levels and large numbers of earthquake induced distortion cycles.

REFERENCES

Dowrick, David J., (1987) Earthquake Resistant Design, Second Ed., John Wiley and Sons, New York

Eisenstein, Z, et. al, (1995), "Report on Tunneling Feasibility and Performance," for Los Angeles County Metropolitan Transportation Authority.

Elioff, M. A., et. al. (1995), *"Geotechnical Investigations and Design Alternatives for Tunneling in the Presence of Hydrogen Sulfide Gas - Los Angeles Metro,"* Rapid Excavation and Tunneling Conference.

Monsees, James E., (1998), *"Deja-Vu on the Los Angeles Metro - Facing Old (But New) Challenges All Over Again,"* Proceedings, ASCE Annual Convention, Boston.

Park, R. and M.J.N Priestly, (1975), Reinforced Concrete Structures, John Wiley and Sons, New York

Paulay, T. and M.J.N. Priestly (1992), Seismic Design of Reinforced Concrete and Masonry Buildings, John Wiley and Sons, New York.

Seible, F., and M.J.N. Priestly (1993), *"Retrofit of Rectangular Flexural Columns with Composite Fiber Jackets,"* Proceedings Second Annual Seismic Research Workshop, Sacramento, CA.

Design of Pressure Tunnels
Jeppe N. Eskilsson[1]
Member A.S.C.E.

Historical Briefs

Technology for the design and construction of tunnels continues to improve as new and more economic solutions are brought into practice. Innovation takes place incrementally, often by necessity, sometimes by ingenuity and sometimes based on not-so-successful experiences. Pressurized water conveyance tunnels which serve purposes including supplying entire regions with water for life sustenance, agricultural use, conveyance and storage of wastewater, and hydroelectric power have been subject to all of these types of developments. The design of pressure tunnels imposes specific conditions on the nature of the geologic characteristics and on the designer's appreciation of these characteristics relative to facility design criteria. This article will discuss the fundamental considerations and approaches for the design of pressure tunnels from both historical and technical perspectives, based on the Author's work with design of this type of facility during the 1980s and 1990s.

The scale of some of the wide range of facilities that include pressure tunnels is impressive, as measured in kilometers (miles) of length, diameter, megawatt, millions of m^3 of water per year, thousands of square kilometers of irrigated land, size of population supplied with water, or in terms of construction cost. Hundreds of kilometers (miles) of water supply tunnels and numerous shafts have been built deep in rock to supply New York City, Boston, Denver, San Francisco, Los Angeles, Phoenix, and other major metropolitan areas in the U.S. with drinking water. Some of these facilities continue uninterrupted operation as sole water transmission mains after more than half a century of service following construction. A portion of a schematic profile of a typical urban water supply tunnel system is shown in Figure 1. Another important category of pressure tunnels and shafts are components of numerous extensive hydropower generation facilities around the world. A schematic profile of a common tunnel and shaft system for power generation is shown in Figure 2.

James Sanborn, a geologist for the New York Board of Water Supply, reports in the "Berkey Volume" by the Geological Society of America (see Sanborn (1950)) that

---

[1] Senior Associate, Jacobs Associates, 500 Sansome Street, San Francisco, CA 94111.

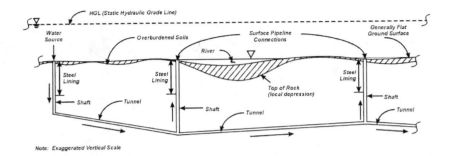

Figure 1.
Schematic Water Supply Tunnel System
Pressure Tunnels and Shafts

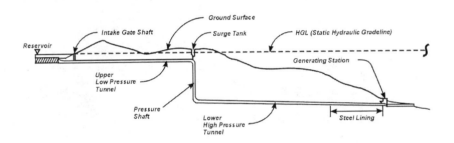

Figure 2.
Schematic Hydroelectric Power Facility
Pressure Tunnels and Shafts

analytical aspects of design of pressure tunnels were developed as early as 1906 for the ongoing design of the 160 kilometer (100 mile) long Catskill Aqueduct by Thomas H. Wiggin, the design engineer for the New York City Board of Water Supply. Mr. Wiggin was assisted by Dr. Charles P. Berkey, professor of geology at Columbia University, who played an important role in the development of sound pressure tunnel design approaches in the United States during the first half of the 20th Century.

A quote from the design of the Catskill Aqueduct is significant both in technical content and in the context of the era when it was applied:

> "The theory in the design of a pressure tunnel in rock is that the water column, reckoned to the hydraulic gradient, is balanced by the dead weight of rock and earth. Sound rock is the chief reliance for successful use in this structure, and a concrete lining of rich mix and substantial thickness is used to reduce leakage, to support the rock, and to provide a smooth waterway. High carrying capacity, without

excessive progressive reduction in the same, is accomplished by the smooth concrete lining."

Important refinements of this basic design approach have been developed over the last several decades. Their application in the current state of practice of design of pressure tunnels is briefly summarized below.

Fundamental Design Considerations for Pressure Tunnels

A number of fundamental design criteria and other important considerations must be defined, or identified, early during the planning phase of design of a facility as these will affect the definition of the necessary facility components, facility geometry and feasibility of various design approaches. The following discussion outlines the most important considerations, which generally apply for design of all types of pressure tunnels. These fundamental considerations include:

- Serviceability Requirements
- Prevailing Geological Conditions
- Environmental Considerations
- Essential Constructability Considerations
- Definition of Horizontal Alignment
- Definition of Vertical Alignment

Serviceability Requirements

The most fundamental effort of engineering for any new facility is to generally define the facility. This initial design effort may be called Project Definition or Conceptual Design and may result in the identification of a pressure tunnel as one of the key facility components. The general scope of this early planning work may, depending on the type of facility, include identification/definition of:

- Points of origin and termination of the transmission facility.
- Points of intermediate connections to existing, or planned local transmission, storage or treatment facilities.
- Essential operational transmission requirements (e.g., transmission rate, capacities, and flow rates).
- Static Hydraulic Gradeline (HGL)
- Conceptual Hydraulic Analysis and preliminary determination of conduit size, mode of operation, and transient pressure.
- Evaluation of alternative conceptual designs, e.g., pipelines, and tunnel and shaft alternatives.
- Identification of Preferred Alternative.

Prevailing Geological Conditions

The design of a pressure tunnel requires a progression of geologic/geotechnical studies and evaluations. The work requires a thorough study of regional characteristics with increasing levels of attention to detail along prospective horizontal alignment "corridors." It would be difficult to attempt to make an all-inclusive list of geotechnical factors to be considered, investigated and evaluated. Only a general list of fundamental geotechnical parameters for the design of unlined and hydraulically lined pressure tunnels is provided here with brief comments. These parameters are:

- In Situ Stress
- Groundwater Regime

- In Situ Permeability
- Modulus of Deformation
- Topographic Characteristics

The nature of each of these parameters affect design of unlined and concrete lined pressure tunnels; i.e., tunnels dependent on the host rock mass to contain and confine pressurized water within the conduit. Brief comments are provided below for each of these parameters. Tunnel design alternatives continuously provided with impermeable linings, e.g., pressure pipe, do not impose strict requirements on these parameters.

In Situ Stress - In situ stress must consequently exceed the water pressure in all directions along the entire length of all unlined or hydraulically lined conduits of a pressure tunnel facility. The state of in situ stress along the alignment of a specific facility may over geologic time have been affected by one or several significant events. The nature, and variation, of in situ stress conditions in the specific region need to be understood by the designer from the very beginning of the design of a pressure tunnel, because of its significance for feasibility. In situ testing to investigate the site-specific in situ stress regime (critical stress) is required as design proceeds, to confirm feasibility of the design approach.

In Situ Permeability - Leakage from a pressure tunnel constructed without an impermeable membrane (e.g., steel lining) results from flow of pressurized water through imperfections in the rock mass, consisting of joints, shears, shear zones, faults, faultzones, and contact surfaces. Hydraulic conductivity (permeability) is a function of interconnected, open discontinuities (aperture), and the extent, type and durability of any infillings. In situ permeability varies throughout a rock mass depending on the frequency of occurrence of high permeability discontinuities. In situ permeability can be altered by hydraulic jacking due to too high water pressures, from a leaking pressure tunnel, or over time due to dissolution or erosion from water leaking through discontinuities.

Modulus of Deformation - Modulus of deformation or rock mass stiffness, is generally important along the entire length of concrete and steel linings which have been designed to depend structurally on interaction with the rock mass. The magnitude of modulus of deformation of the host rock mass is of less importance for the design of unlined tunnels, or design of nominally reinforced concrete linings to be applied locally along the alignment of a pressure tunnel. However, modulus is important for design of heavily reinforced concrete linings designed to control leakage, and also where internal pressure conditions, combined with load sharing with the rock mass, governs the design of steel linings.

Geologic factors affecting modulus of deformation include the properties of the intact rock material, the attitude of the bedding, presence of dense jointing or shears, fault gouge, and weathering/alteration. As with all other geotechnical characteristics, modulus generally varies significantly along a tunnel. Consequently, the parameter is generally handled most efficiently in design analysis by the development and use of ranges of "order of magnitude" values to identify critical ranges triggering requirements for special design features. Extensive areas of low modulus may require a specific lining type and can affect the siting, or design, of special facility components, such as shafts and intersections. Any adverse effects of construction methods employed require additional consideration when establishing this design parameter.

Groundwater Regime - The natural level of the phreatic surface, which may vary seasonally, is significant for the design of pressure tunnels. Its location relative to topographic contours depends on in situ permeability characteristics and seasonal recharge from precipitation. The location of the phreatic surface relative to the HGL of an unlined or concrete lined tunnel affects leakage from the tunnel, and may affect the feasibility and economics of a specific design approach.

Topographic Characteristics - Topographic characteristics affect numerous aspects of the design of pressure tunnels, including in situ stress and consequently horizontal and vertical alignments of unlined and concrete lined tunnels, requirements for shafts and other types of special structures, depths of shafts, design pressure, types of lining systems required locally and globally along the tunnel, as well as numerous constructability factors.

Geotechnical Characterization

Adequate geotechnical characterization is essential for successful design of all tunnels. Furthermore, it is inherent that thorough characterization based on limited surface level information, and a limited number of borings, will only reveal a portion of the varying geological/geotechnical characteristics that will be encountered along the alignment of a deep tunnel in rock. The basic approach for development of a reasonable geotechnical exploration program for a specific project, is to adequately explore and characterize known, or suspected, geological features along a "continuum" and to characterize the "continuum", i.e., the prevailing conditions, by having an adequate database for generalized conclusions. The resulting qualitative and quantitative descriptions of conditions and parameters require use of "ranges" rather than "exact numbers" to reflect the variability of conditions present.

Geotechnical characterization for a long tunnel generally requires three efforts -- Preliminary, Detailed and Supplementary Characterization. It is essential that adequate input by a qualified engineering geologist is provided throughout all phases of characterization and that a qualified designer actively participates in all planning of characterization work, reviews all significant data and in the interpretation of this data.

Preliminary Characterization – The objective of this phase is to establish feasibility of the design approach and to define Prevailing Geological Conditions based on thorough evaluation of readily available information. The Preliminary Report resulting from the work should include a site geologic plan and one or several geologic sections (profiles) through the pressure tunnel alignment, and any special structures, or areas, of concern to the extent possible. Following this phase of characterization the designer proceeds to evaluate and conceptually define the facility, including the "essentially final" horizontal and vertical alignments and "final" siting of special structures, e.g., shafts.

Detailed Characterization – Detailed characterization involves application of all relevant geologic/geotechnical approaches, methods and tools necessary with the objective of developing a thorough, detailed understanding of the geology along the pressure tunnel alignment. Three categories of work in site characterization are discussed below in more detail -- Borings, Laboratory Testing and Downhole Testing.

Borings – Deep borings for tunnels in rock are expensive. Consequently, a limited number of borings are executed for deep tunnels. It is imperative to obtain the maximum possible, accurate, and relevant information from each, often widely spaced, deep boring. The emphasis applies to lithologic information, information about rock mass discontinuities, essential rock material information and drilling information. It is therefore suggested that all borings in rock are continuously core drilled, and that qualified geologists be continuously present during drilling. All borings should extend below the anticipated tunnel invert.

Laboratory Testing – Limited unit weight or density tests, compressive strength (occasionally with Young Modulus and Poisson's Ratio data) and pointload data is required as a minimum. A basic approach is to obtain a minimum three to four tests of laboratory strength and unit weight information for each lithologic material in each geological "region" along the tunnel. Relatively inexpensive pointload tests are often performed in the field in each interval of ten feet of core during drilling. Mineralogic assessments (including thin sections) are required to investigate presence of minerals affecting durability of tunnel lining materials and for constructability evaluations.

Downhole Testing – For design of pressure tunnels, data on rock mass permeability and in situ stress is essential. It is therefore recommended that core borings be water pressure tested (packer testing) along their entire length below the weathered zone of rock. In selected borings, some of the test intervals should be subjected to Hydraulic Jacking Tests to assess in situ stress conditions (see Doe, et. al. (1987)).

Supplementary Characterization - Unanticipated locations with geologic characteristics of particular importance may be identified as characterization proceeds. This must be recognized in the Characterization Plan by allowing for necessary adjustments in the exploration program and by allowing for an adequate contingency for contingent borings locations.

Environmental Considerations

This aspect of pressure tunnel design would be largely identical for any type of tunnel, except with respect to the acceptable risk of leakage. The issues included in this category include numerous "non-technical" considerations important for implementation of any major project, including work to satisfy regulatory issues related to environmental impact, and required mitigation, permitting, rights-of-way considerations, project procurement, community outreach and public relations. The nature of these aspects may affect the outcome of constructability evaluations, feasibility of horizontal alignment alternatives, and type of tunnel lining system(s) required. The concepts of "acceptable risk," "acceptable leakage" and "acceptable intrusion" vary, depending on a variety of considerations such as remoteness of the site as opposed to an urban/suburban setting. Environmental considerations may affect horizontal and vertical alignments, and even the implementability of a specific facility.

Constructability Considerations

The design of all tunnels requires thorough consideration of factors affecting construction and consequently the cost of a facility. Most of these important considerations are similar for all types of tunnels, but several are of particular importance for design of pressure tunnels. These particular considerations include tunnel construction accessibility, construction shaft locations, and tunnel depth.

Construction accessibility affects construction costs and time required for construction, particularly in terrain of steep topographical relief, e.g., for water transmission tunnels in mountainous terrain with few alternative sites for construction access, staging and development of tunnel headings. These considerations will often affect both horizontal and vertical alignments of a water transmission tunnel, including locations of portals and construction shafts. Identification of suitable locations for construction shafts may be particularly challenging in urban/suburban areas. It is desirable to find suitable sites allowing tunnel headings with maximum lengths of 5 to 7 miles within a relatively narrow alignment corridor, which is often pre-defined by requirements for connections to local transmission pipelines at the surface. It is desirable to identify a construction shaft site with a size of 15,000 m^2 to 25,000 m^2 (3 to 5 acres), but tunnels have been successfully constructed from sites no larger than 2,000 m^2 (0.5 acres) in urban areas.

Definition of Horizontal Alignment

Once the design considerations discussed above have been incorporated, "mature" horizontal alignments can be identified. The optimal horizontal alignment must be selected prior to initiation of detailed geotechnical characterization and significant efforts of design. An intermediate step involving localized, preliminary geotechnical exploration may be required to investigate localized, potentially significant geotechnical issues affecting feasibility or selection of an optimal horizontal alignment.

The establishment of the optimal horizontal alignment may be an iterative effort also involving evaluations of alternative vertical alignments and lining system requirements. Optimization of alignments, e.g., utilization of topography to optimally align a pressure tunnel under and parallel to a mountain ridge, may offer significant savings in requirements for impermeable lining and, consequently, overall construction costs.

Definition of Vertical Alignment

The definition of the vertical alignment (profile) required for a pressure tunnel involves a complex series of evaluations. A fully developed vertical alignment satisfies an envelope of minimum requirements resulting from evaluations of all Fundamental Design Considerations discussed above, and Pressure Tunnel Design Principles discussed in the following section of this article.

Pressure Tunnel Design Principles

The type of lining system applied depends on numerous factors including:
- Geotechnical characteristics
- Serviceability requirements
- Hydraulics
- Lining material cost and availability
- Limitations on suitable maximum flow velocities
- Excavated tunnel diameter
- Overall construction cost and schedule
- Regulatory requirements

There are three basic types of pressure tunnel lining systems: Unlined and Hydraulically lined tunnels, Impermeable lining, and Semi-permeable lining. Design of each of the three different types of lining systems is briefly summarized below.

Design of Unlined and Hydraulically Lined Pressure Tunnels

Hundreds of miles of hydraulically lined pressure tunnels continue to provide satisfactory service as important water supply transmission mains, for electric power generation, and for storage and transport of waste water after decades of use. A modern, hydraulically lined pressure tunnel features a cast-in-place concrete lining. The objective of this type of lining is to provide a smooth, durable flow surface, to protect and support the rock, to facilitate operational requirements of reliability and low maintenance, and when required, provide for safe access for inspection and maintenance work. For design purposes, an unreinforced concrete lining does not serve to control leakage, i.e., the basic design criterion for this type of tunnel is identical to that of an unlined tunnel. The main lining component confining and containing pressurized water within the flow conduit is the surrounding rock mass itself. Consequently the characteristics of the rock mass, both in terms of quality and quantity, are critical for successful design.

All hydraulically lined and unlined pressure tunnels leak to some extent. The owner of such a facility need to be advised of this characteristic and must come to terms with the relative concepts of "acceptable" and "unacceptable" leakage. High costs of lost water, or lost generating capacity, may make a particular leakage rate either acceptable or unacceptable depending on the geographical location or other reasons. Even relatively small quantities of leakage concentrated to a specific location may be unacceptable if operation of the tunnel causes identifiable damage to the environment or to adjacent facilities. There are well tested approaches to mitigate these risks by applying observational methods during construction, and by installation of appropriate special lining treatment or occasionally, by providing drainage components in areas observed to be of concern. This approach will be discussed in more detail under "Design Efforts During Construction" below.

Water supply tunnels should generally be provided with a hydraulic lining to increase reliability of operation, and to facilitate inspection and maintenance work. Water supply tunnels are community lifeline-type facilities, which require the highest level of reliability while offering few opportunities for maintenance. It is not uncommon that these facilities operate continuously for 50 or even 100 years without opportunities for any below ground maintenance work due to absence of costly, redundant transmission mains. Unlined tunnels used for hydroelectric power generation require rock and sand traps for the protection of mechanical equipment, even if these tunnels are excavated by a Tunnel Boring Machine. Regulatory considerations (e.g., FERC) often demand frequent inspection of power tunnels. Seasonal operation of agricultural water transmission tunnels allow for safe and efficient maintenance of unlined tunnel facilities during annual shut-downs.

The prevailing rock mass permeability must be low for consideration of a hydraulically lined or unlined tunnel. In qualitative terms this means that the rock mass exhibits only infrequent occurrences of significantly permeable discontinuities along the length of the tunnel. In quantitative terms this may be expressed by stating that standard packer pressure tests in borings should generally show hydraulic conductivity smaller than 5×10^{-8} m/sec with infrequent, localized occurrence of data showing conductivity in the range up to 5×10^{-6} m/sec. Infrequently occurring zones of higher permeability along the excavated tunnel to be defined following excavation may be treated as "special lining treatment areas;" i.e., provided with supplementary lining components, including consolidation grouting, heavy reinforcement or substitution with steel lining, as appropriate.

Cover Requirements - In situ stress must generally exceed internal water pressure in all directions surrounding a pressure tunnel which depends on the rock mass to control leakage. Where gravity controls minimum in situ stress, minimum depth of overburden rock and soil materials is critical for determining required depth of the tunnel. Topographic characteristics along a zone parallel to the horizontal alignment consequently affect depth requirements. The general case of required minimum depth is decribed by the so-called Norwegian Criterion for Confinement and was originally published by Bergh-Christensen et al. (1971).

In regards to topographic considerations, and concern for side cover requirements, it can be concluded that no serious concerns apply for alignments parallel to valley sides sloping up to 10° to 15°. At a sideslope of 15°, a maximum possible error of minimum depth on the order of seven percent is created. This margin is covered by the factor of safety, even considering additional uncertainties related to overburden soil depth, and unit weight of geological materials along the tunnel. These uncertainties make $F.S.$ a "qualified" or "apparent" factor of safety, which must be evaluated based on the level of thoroughness and actual conditions revealed from the geotechnical characterization effort.

The critical stress is not always governed by gravity and numerous exceptions to this characteristic (e.g. stress relief) have led to the failure of facilities. Furthermore, the approach of using a "rule-of-thumb" is generally appropriate only for determining preliminary vertical alignments, due to the important basic assumptions of favorable in situ stress and in situ permeability conditions. With the development of the relatively simple Hydraulic Jacking Test (see Geotechnical Characterization below), the critical stress across discontinuities can be readily investigated both during the pre-construction design phase and during excavation for the facilities to validate assumptions relating to in situ conditions. A thorough geotechnical characterization program also results in adequate evaluation of the prevailing rock mass permeability. This parameter must be evaluated in detail along every foot of the tunnel following excavation as discussed under "Design Efforts During Construction" below.

The fundamental requirements for determining the minimum depth of pressure tunnels were well understood by workers at the turn of the century (see Sanborn(1950)). Numerous other workers have presented minimum depth requirements over the years. Dr. Karl Terzaghi provided the following comments (as quoted in Spencer et al. (1964):

"...the construction of an unlined pressure tunnel through rock always involves a risk even in the event that the computations indicate a conservative factor of safety with respect to the opening-up of joints under internal pressure..."

Dr. Terzaghi (1963), further states that, "The most unfavorable condition develops if the water enters a continuous joint or set of joints oriented more or less parallel to the ground surface. Let:

H_C = depth of rock above the tunnel, measured at right angles to the ground surface

H = Hydrostatic head or water in the tunnel

γ_w = Unit Weight of Water

γ_R = Unit Weight of Rock

The rock above the joint will rise, unless H_C = is equal to or greater than $H'_C = \dfrac{H \gamma_w}{\gamma_R}$ if $\gamma_w = 1$ and $\gamma_R = 2.6$:

$$H'_C = 0.385 H$$

This fundamental relationship has been known for many decades. Therefore it has become common practice to specify a water-tight lining in all those parts of pressure tunnels, above which the depth of the overburden is smaller than $H/2$. This rule-of-thumb corresponds to a minimum factor of safety of about 1.3..."

The rule-of-thumb of requiring a minimum rock cover of: $D = 0.5H$ has a sound analytical basis for application in generally flat terrain, as a special case of the Norwegian Cover Criterion where $\beta = 0$ and $F.S. = 1.2$ to 1.3, and the criterion used by the early New York Water Tunnel designers (see Sanborn (1950)) without an expressed factor of safety. Sound judgement, preferably based on extensive experience, must be applied, along with a thorough understanding of the critical assumptions related to in situ stress and in situ permeability, in developing minimum depth of unlined and hydraulically lined pressure tunnels based on the "0.5 Cover Criterion."

For facilities in generally flat terrain with locally significant depth of overburden soils, the Author developed the "0.5 Cover Criterion" in more detail for efficient application in design of recent pressure tunnel projects (see Eskilsson et al. (1997). The resulting format of the "0.5 Cover Criterion" is:

$$CR = 0.5 = \frac{D}{H} = \frac{TOR' - INV}{HGL - INV} = \frac{\left[TOR + (TOG - TOR)\frac{\gamma_{OB}}{\gamma_R} - INV\right]}{HGL - INV}$$

CR = Cover Ratio, = $0.5 = \frac{F.S.\gamma_w}{\gamma_R}$
    $F.S.=1.2$ to 1.3
D = Minimum Depth of Rock (actually measured to tunnel invert, m or ft.)
H = Static Internal Head in tunnel (m or ft) = HGL - INV
TOR' = Modified Rock Surface Elevation (m or ft) =
    $= TOR + (TOG - TOR)\frac{\gamma_{OB}}{\gamma_R}$

TOG = Top-of-Ground Elevation (m or ft)
TOR = Top-of-rock Elevation (m or ft)
INV = Invert Elevation (m or ft)
γ_{OB} = Unit Weight of Overburden Soils
γ_R = Unit Weight of Rock
HGL = Static Hydraulic Gradeline Elevation (m or ft)

As discussed earlier in this article, several essential constructability considerations, and occasionally localized special lining requirements, are required to be accounted for in determining the vertical alignment for a specific facility, in addition to minimum depth requirements resulting from application of the minimum cover criterion. Following the selection of a preliminary vertical alignment, detailed geological/geotechnical characterization may proceed. This process is discussed later in this article.

Lining Design

The design of an unreinforced concrete lining for a hydraulically lined pressure tunnel is relatively straightforward. It is common to phase concrete materials with a compressive strength in the range of 25 to 35 MPa (4000 psi to 5000 psi) in all types of tunnels. Cement and aggregates must be durable materials and construction quality requirements must be high. The concrete lining must be contact grouted in the crown and overbreak areas. Pressure grouting is performed to seal shrinkage cracks as a final operation of lining construction. The minimum thickness of a concrete lining is on the order of 12 in. for constructability reasons and considering the need for high quality construction. It is recommended that in areas where cracks in excess of 3.8×10^{-3} m (0.15 in.) in width are anticipated (areas of lower modulus) due to initial, internal pressure loading, minimum hoop reinforcing and nominal longitudinal steel ("temperature steel") required by the ACI code is incorporated.

Design of Semi-permeable Concrete Linings

Pressure tunnels are generally not systematically lined with a semi-permeable lining. However, localized areas of pressure tunnels, designed to be hydraulically lined with unreinforced concrete lining, or as unlined tunnels, may require special lining treatment to control leakage. Such treatment may consist of heavy reinforcement of the concrete.

A fundamental requirement of a semi-permeable lining designed to control leakage is that its permeability is much smaller than the rock mass in which it is constructed. If this requirement is not satisfied, internal pressure and porewater pressure in the rock outside the lining will be equal, leading to the conclusion that the reinforced concrete lining does not carry any load due to the internal pressure. The tunnel lining then behaves as an unreinforced concrete lining, which may lead to failure to perform as intended.

Leakage through a reinforced concrete lining occurs primarily through longitudinal cracks, theoretically well distributed around the circumference of the lining. The leakage rate through a crack is theoretically proportional to the cube of the width of the crack. Design of these linings is consequently based on control of the width of cracks. Cracking of unreinforced concrete occurs at very low strain levels on the order of 10^{-4}, which correspond to very low stress in the hoop reinforcing, on the order of 20 MPa (3 ksi). The allowable crack width of 2.5 x 10^{-4}m (0.01 in.) for applied design results in stress levels in the steel reinforcement on the order of 100 MPa (10 ksi to 15 ksi). Recommended longitudinal reinforcing steel corresponds to reinforcing percentages on the order of 0.4 to 0.5 of concrete lining crossectional area. The structural theory behind the required analysis is discussed in Hendron et al. (1987) and USACOE (1997). The latter reference includes MathCAD applications for expediency of analysis. Readers are encouraged to review these references for further details of the required analyses.

Due to significant sensitivity to modulus of deformation of rock for both disturbed and undisturbed rock mass, modulus of "elasticity" for cracked concrete, assumptions of uniform cracking, and uniform quality of construction, caution is recommended in applying semi-permeable linings. It should also be noted that this approach only applies to control of leakage in localized sections of tunnels, identified during construction, and not for reliable control of porewater pressure in the ground. The vertical alignment for this type of tunnel should consequently fulfill the requirements for the otherwise hydraulically lined or unlined tunnel.

Design of Impermeable Steel Linings

Several types of impermeable linings have been applied, either experimentally or as a standard practice design approach. Concrete-encased, steel cylinder lining is currently the most common type, due to its cost and constructability advantages.

Only fine-grained steels with a medium range yield strength are recommended for tunnel and shaft steel linings. A practical limitation on design is reached at a steel plate thickness of 1 ½ in. when a requirement for post-weld heat treatment according to governing welding standards (ASME Boiler and Pressure Vessel Code and AWS D1.1 Structural Welding Code) becomes mandatory. This requirement would necessitate access to the outside of the steel lining field joint, an impractical requirement. The common joint between lining pipe segments is a full penetration butt-weld constructed from the inside of the lining and provided with an external backing ring. The common quality control requirement for this type of joint is by ultrasonic testing.

Design of steel cylinder linings require analysis for internal pressure applied during facility operation, external pressure following future rapid unwatering of the facility, and minimum steel plate thickness required for handling and installation of steel lining sections. The thickness requirement for design will be governed by the largest thickness resulting from analysis for each of these criteria. Many considerations are involved with each of these criteria as summarized briefly below.

Internal Pressure – Analysis is required for normal operating pressure, including static pressure corresponding to the static HGL, and normal transient pressures. Additionally, analysis for exceptional dynamic pressures, including water hammer pressures, are required. A common guideline is to allow 50 percent of yield stress to be used as allowable stress based on a design pressure corresponding to static HGL, with an increase to 75 percent for design pressure including normal transient pressure, and an increase to 100 percent of yield stress for infrequent emergency-type total pressures. Guidelines for the development of design pressures, allowable steel stress, and analytical requirements are provided by various national agencies, manuals and handbooks (see Stutsman (1993), and AWWA (1989).

In rock materials with low modulus, or significant uncertainties or variations of modulus, it may be appropriate to disregard load sharing with rock. Analysis is consequently similar to what is required for design of pipelines and only involves use of the "thin cylinder" equation.

Under adequate cover of competent rock, and where realistic ranges of relatively uniform and sufficiently high magnitudes of modulus of deformation prevail, consideration of load sharing to account for interaction between steel lining and rock mass is prudent due to potential savings of steel plate thickness. The interaction equation commonly used for load sharing analysis was presented in its current form by Vaughn (1956). The reader is encouraged to review the original articles as space limitations prevent presentation and discussion in this article.

In the interaction equation quoted above, the gap between the outside surface of the steel lining and the backfill concrete plays a significant role. This gap consists in reality of a complex function of steel lining diameter changes resulting from thermal contraction of the steel lining, shrinkage of the backfill concrete after placement, and permanent deformation of backfill concrete and rock due to radial loads. For load sharing analysis, it is common practice to assume that the gap is only a function of thermal contraction of the steel lining based on the maximum temperature differential anticipated (temperature at time of placement of the backfill concrete and the coldest temperature anticipated for transported water).

Selection of appropriate magnitudes of modulus of deformation for the various concrete and rock materials requires sound judgement based on experience and the state of practice. For preliminary analysis it is recommended to use a low modulus for the cracked backfill concrete layer on the order of 7,000 MPa (1×10^6 psi), or to use the same magnitude selected for the disturbed zone of rock whichever is lower. For the disturbed zone of rock, the issue of selection of layer thickness is also presented. The extent of this zone depends on the undisturbed rock mass characteristics and method of excavation. It has been determined that the thickness of this zone is only on the order of 1 ft. to 2 ft. in tunnels excavated in massive rock by TBM, and on the order of 0.5 – 1.3 m (2 ft to 4 ft.) for tunnels excavated by conventional drilling and blasting (see Nishida (1982). Earlier authors recommended a significantly larger extent of this zone (see Brekke et. al. (1987)). The degree of reduction of undistrubed rock modulus to assign appropriate values to the disturbed zone is again a judgement call. A suggested range of reduction would be from 50 percent to 75 percent for this type of analysis. Numerous publications provide magnitudes of measured modulus of deformation for undisturbed rock based on in situ testing.

It is recommended that short sections of reinforced concrete lining be provided as structural transitions at the ends of sections of steel lining. These transitions are provided to reduce adverse impacts of otherwise very abrupt changes of lining stiffness, resulting in significant cracking of an unreinforced concrete lining. The transitions should have a relatively high percentage of reinforcing steel in the immediate vicinity of the end of the steel lining, gradually reducing to nominal reinforcing over a distance of one tunnel diameter.

External Pressure – On unwatering of pressure tunnels the pressure on the lining reverses from internal pressure due to operation to external pressure from groundwater loading, resulting in a tendency for steel linings to buckle. Currently applied buckling analyses for a plain concrete embedded steel cylinder were developed primarily by Amstutz (1970), and Jacobsen (1974). Comparisons of both approaches were made by Moore (1990), and in Stutsman (1993), resulting in recommendations to generally apply the analytical approach presented by Jacobsen. The theoretical analysis required for design is not repeated in this article due to its complexity and space limitations. The reader is encouraged to study the articles referenced above, and to review the now available MathCAD applications presented in USACOE (1997), to expedite analysis.

The magnitude of the gap between backfill concrete and steel cylinder, discussed above under load sharing analysis for internal pressure, is also very significant for the analysis of buckling. Prudent judgement is required in selection of the appropriate magnitude of this gap to be applied in buckling analysis. This gap should be increased from the basic temperature gap used for load sharing analysis, to account for permanent deformation of the ground at a future unwatering event.

Where particularly large external heads govern steel plate thickness, and where required steel plate thickness would exceed 0.03m (1-1/2 in.), external stiffening is generally selected for design. The thickness of plate for a steel lining to be stiffened externally is governed by the requirements for internal pressure. Analysis of externally stiffened steel cylinders is commonly made using the approach presented by Windenburg, et al. (1934). The reader is encouraged to study the original article for this type of analysis.

A steel lining provided with an internal, cast-in-place concrete lining to resist buckling is first mentioned in articles regarding an Austrian hydroelectric plant designed in 1940. The war resulted in the scarcity of steel plate for standard steel linings, which led to the use of a very thin, highly stressed steel lining with an internal concrete structure to resist buckling. Variations of this approach are still applied in the Alps region in Europe. A recent application of steel plate pipe, provided with an internal concrete cylinder to resist buckling, was developed for a water supply tunnel in Southern California with involvement of the Author. This 3.6m (12-ft) I. D. pipe lining is in certain areas designed for 390m (1300 ft) of external head. This lining, essentially designed as a modified Reinforced Concrete Cylinder Pipe (RCCP), includes an external steel cylinder with an internal reinforced concrete core produced in a pipe casting yard. These heavy, and consequently short, lining sections will be transported into the tunnel and welded at the joints prior to placing backfill concrete between the outside steel cylinder surface and the rock surface. As for the alternative of externally stiffened steel cylinders, the excavated tunnel diameter required for this design is larger than the diameter required for a plain steel cylinder lining.

Minimum Steel Lining Thickness – A minimum steel lining wall thickness must be provided to allow practical handling of lining sections during manufacturing, transportation and installation.. The analysis required for minimum thickness is described in Brekke et al. (1987).

Protective Linings - Steel linings for tunnels are commonly lined internally with a cement mortar lining applied in situ following the completion of lining installation and backfill concrete as a means of corrosion protection. Mortar lining applied by centrifugal action is not practical for steep tunnel grades and shafts, which are also commonly provided with epoxy coating. High velocity penstocks, and appurtenances are increasingly coated internally with epoxy corrosion protection.

Design Efforts During Construction

The standard practice of requiring the designer to provide Design Services During Construction, i.e., to maintain a continuous, adequate level of insight into construction activities to the extent that the design intent is indeed executed, is of particular importance for successful design of pressure tunnels. The important assumptions made during the pre-construction design phase must be verified during construction to establish actual final lining requirements and to select appropriate final lining components. The required documentation and design efforts consist of mapping of tunnel surfaces, monitoring of groundwater inflow, evaluation of information and final lining determination. Due to space limitations only mapping and groundwater monitoring is briefly discussed below.

Tunnel Mapping

The common "cut-cylinder type" map described in "Standard Guide for Geotechnical Mapping of Large Underground Openings in Rock, "ASTM D 4879-89, provides an adequate basic guideline for mapping of tunnels. Some adjustments of emphasis of the mapping efforts, are required to record geologic information of particular importance for observation of pressure tunnels during excavation.

The information required specifically for the purpose of pressure tunnel lining determination includes major geologic rock structures, e.g., faults, shears, intrusions; occurrence and characteristics of prevailing rock mass discontinuities, e.g., spacing, aperture, infillings, continuity; qualitative strength (e.g., hardness, "soundness," degree of weathering) of the rock; water occurrence, qualitatively and quantitatively; and primary support requirements. These categories of information relate to ground permeability characteristics and factors indicating issues related to the modulus of deformation. In some geolgical settings, geologists need to be alert to indicators of low in situ stress, e.g. open joints. Information should be recorded to scale by qualified engineering geologists. Construction contracts must include provisions to facilitate mapping crews during excavation.

Monitoring of Groundwater

Groundwater occurrence during and following tunnel excavation is the single most important indicator of localized permeability and potential for future leakage along a pressure tunnel alignment. Groundwater occurrence characteristics vary along the tunnel alignment and with time elapsed following excavation.

The initially occurring rate of inflow from a specific set of discontinuities is the most significant indicator of ground permeability at the location. It is thus important to continuously attempt to obtain an "initial inflow rate," preferably quantitatively, within hours of excavation continuously along the entire alignment of a pressure tunnel. Inflow rates commonly diminish over time due to the depletion of stored water in portions of interconnected discontinuity features, while recharge of others is maintained. The change of inflow rate will need to be observed until stabilization of inflow occurs. Monitoring of changes in the groundwater level from the surface may also be important for a number of reasons including detection of effects on the environment and final lining requirements.

References

American Water Works Association, "Steel Pipe - A Guide for Design and Installation," Manual M11, AWWA, Denver, 1989.

Amstutz, E., "Buckling of Pressure Shafts and Tunnel Linings," Water Power, November 1970, pp. 391-399.

Bergh-Christensen, J. and Dannevig, N.T., "Engineering Geological Considerations Concerning the Unlined Pressure Shaft at the Mauranger Power Project," Unpublished Report, GEOTEAM A/S, Oslo, 1971.

Brekke, T.L. and Ripley, B.D., "Design Guidelines for Pressure Tunnels and Shafts," Electric Power Research Institute, Report AP-5273, June 1987.

Doe, T.W. and Korbin, G.E., "A Comparison of Hydraulic Fracturing and Hydraulic Jacking Stress Measurements," Proceedings 28th U.S. Symposium, Tucson, 1987.

Eskilsson, J.N., "Design of the MetroWest Pressure Tunnel," Proceedings of the Rapid Excavation and Tunneling Conference, AIME/ASCE, Las Vegas, 1997, pp. 25-43.

Hendron, A.J., Fernandez, G., Lenzini, P.A. and Hendron, M.A., "Design of Pressure Tunnels," in "The Art and Science of Geotechnical Engineering," Prentice Hall, Englewood Cliffs, 1989, pp. 161-192.

Jacobsen, S., "Buckling of Circular Rings and Cylindrical Tubes Under External Pressure," Water Power and Dam Construction, December 1974, pp. 400-407.

Nishida, T., Matsumaru, Y., Miyanaga, Y. and Hori, M., "Rock Mechanical Viewpoint on Excavation of Pressure Tunnel by Tunnel Boring Machine," International Symposium on Rock Mechanics, Achen, May 1982, pp. 815-826.

Sanborn, J.F., "Engineering Geology in the Design and Construction of Tunnels," Engineering Geology (Berkey) Volume, The Geological Society of America, November 1950, New York, pp. 45-81.

Spencer, R., Laverty, B. and Barber, D., "Unlined Tunnels of the Southern California Edison Company, ASCE Power Journal, Paper No. 4087, No. 3, October 1964, New York, pp. 105-132.

Stutsman, R.D., editor, "Steel Penstocks," ASCE Manuals and Reports on Engineering Practice, No. 79, New York, 1993.

Terzaghi, K., "Stability of Steep Slopes on Hard Unweathered Rock," Geotechnique, The Institution of Civil Engineers, London, September 1963, pp. 251-271.

U.S. Army Corps. of Engineers, "Tunnels and Shafts in Rock," EM 1110-2-2901, Washington, D.C., May 1997.

Vaughn, E.W., "Steel Linings for Pressure Shafts in Solid Rock," Proc. ASCE Journal of the Power Division, Paper No. 949, April 1956, pp. 949-1 to 949-39.

Windenburg, D.F. and Trilling, C., "Collapse by Instability of Thin Cylindrical Shells Under External Pressure," Transactions of ASME, April 1934.

PRESSURE TUNNELS IN FRACTURED ROCK: MINIMUM COVER CRITERIA FROM THE STABILITY OF ROCK WEDGES

Tirso A. Alvarez[1], Edward J. Cording[2] & Gabriel G. Fernández[3]. Members ASCE

ABSTRACT. The applicability of existing rock cover criteria for siting unlined pressure tunnels has been reviewed in this paper and limitations placed based on site geology and fracturing. Hydrojacking modes of failure revealed by numerical models are presented. A new cover criterion has been developed that predicts hydrojacking of steeply-dipping rock joints as the mass moves along shallow-dipping joints, including the role of geology.

INTRODUCTION

Selecting the location of pressure tunnels within the rock mass is one of the key decisions for successful hydroelectric power development and other water conveyance schemes. To prevent hydrojacking it is typically required that the pressure inside the tunnel remain below the minimum in situ total normal stress in rock joints near the tunnel ($\sigma_{n\,min}$). The requirement is accepted as a conservative design postulate, since at most only individual joints near the tunnel would open when this condition is met. The key unknown is the minimum in situ total normal stress in rock joints in the mass ($\sigma_{n\,min}$).

Current practice in pressure tunnels is to first select the location of the tunnel within the rock mass using one of the rock cover criteria, typically the Norwegian criterion (Bergh-Christensen & Dannevig, 1971). *In situ* stress measurements are performed after the tunnel has been excavated, most commonly hydrojacking and hydrofracturing tests, to verify that the state of stress is indeed adequate (or inadequate). In some instances the stress measurement tests are conducted from the surface before the excavation of the tunnels. The in situ stress measurement program is therefore used to define lining requirements (length of steel and reinforced concrete liner), after the key decision of where to locate the tunnel has been made.

---

[1] Partner, Geoconsult, Inc., P.O. Box 362040. San Juan, PR 00936-2040.;
[2] Professor of Civil Engineering, University of Illinois at Urbana-Champaign. Urbana, IL 61801
[3] Research Eng., Dept. of Civil Engineering, University of Illinois at Urbana-Champaign. Urbana, IL 61801

If the in situ stress conditions prove adequate near the estimated location of the upstream end of the steel liner, when the minimum total normal stress across existing discontinuities ($\sigma_{n\ min}$) is larger that the static internal water pressure (p_i), then the procedure works fine and may be cost effective. On the other hand, if it turns out that adequate in situ stresses are not present near the intended upstream end of the steel liner, the required length of steel and/or other measures can be substantially larger than anticipated. Major unforeseen costs arise not only from the cost of the additional lining but from the lost revenue due to delays in putting into service the generating stations or water supply schemes (Hendron et al, 1987).

A need exists for more reliable initial guidelines for the location of pressure tunnels so that the post-excavation confirmation of in-situ stresses does not result in substantial changes in expected location of liners.

ROCK COVER CRITERIA

The existing rock cover criteria attempt to estimate the minimum in situ stresses in the rock mass using only the depth of the tunnel and inclination of the ground. The estimated minimum stress is then compared with the water pressure and the required depth of rock cover is computed. Broch (1984) presented a review of the development of rock cover criteria in Norway.

Following the 1970 hydrojacking failure of the Åskora pressure tunnel in Norway, Bergh-Christensen and Dannevig (1971) proposed the Norwegian rock cover criterion:

$$\frac{L_r}{h_i} > \frac{\gamma_w}{\gamma_r} \frac{1}{\cos\beta} \qquad (1)$$

where h_i is the internal piezometric head, β is the inclination of the valley side from the horizontal, L_r is the minimum distance ($\perp$) between the tunnel or shaft and the ground surface, and γ_r and γ_w are the unit weights of rock and water, respectively. Equation (1) was derived from the limit equilibrium of a slice of rock mass perpendicular to the ground surface, which is equivalent to lifting a plane parallel to the valley wall.

Brekke and Ripley (1986) summarized the experience with unlined tunnels in a chart of L_r/h_i versus β. A significant portion of the database was composed of Norwegian cases. They concluded, "... for preliminary layout in terms of minimum requirements, it appears that the Norwegian and Snowy Mountain criteria are very useful tools. Where leakage has occurred in tunnels and shafts that satisfy the criterion, factors other than inadequate confinement were sometimes the cause".

After reviewing the state of the art and past experiences, some reported in his paper, Deere (1983) had concluded: "Conventional rules-of-thumb have been found wanting. They work well when rock conditions and in situ stresses are favorable; where the contrary is true, the performance of the pressure tunnel can be unsatisfactory".

To reconcile the opposing conclusions expressed by Deere (1983) and Brekke & Ripley (1986), the pressure tunnel database of Brekke & Ripley (1986) has been updated with six additional projects from South and Central America and the United States, presented in Table 1. The results were re-plotted using geology as a discriminating factor and uncovered trends that place limits on the applicability of the Norwegian and similar cover criteria based on geologic setting.

In four of the new cases added, hydrojacking failures took place. The cover ratio was computed at the deepest point that suffered hydrojacking (largest L_r/h_i). In some instances cover ratios are reported at the end of the steel liner instead of the more significant location previously described, because of lack of additional information.

In the other two projects reported in Table 1, post-excavation in-situ stress measurements were performed which indicated that the original design would have been inadequate due to lower than expected minimum normal stresses in the joints. Two cover ratios were computed, one for the original location of the end of steel liner using the static head in the tunnel. The other cover ratio reported was computed at the same location but assuming that the internal piezometric head (h_i) in the tunnel equals the equivalent water head corresponding to the average of the measured minimum normal stress ($\sigma_{n\,min}$). This is close to the head at which hydrojacking is onset.

The successful performance of the Norwegian criterion in Norwegian pressure tunnel cases was evident in the cover ratio vs. slope angle chart presented by Broch (1984). Indeed, the conclusions of Brekke & Ripley (1986) apply to pressure tunnels in massive metamorphic and plutonic rocks, common in Norway, as shown in Figure 1a). Only the pressure tunnel at Bjerka met the criterion and had hydrojacking failure due to the presence of adversely oriented joint sets. Another exception occurred in the Balsam Meadow project in California (Brekke & Korbin, 1990), where persistent vertical joints caused very small minimum normal stresses, which would have hydrojacked if the Norwegian rule had been used exclusively.

Figure 1b) shows the cover ratio versus slope angle plot for pressure tunnels and shafts in sedimentary and volcanic extrusive formations. As observed from the figure the prediction of cover ratio using the Norwegian rule is not good. In fact, no clear relation between the slope angle and the cover ratio is observed and hydrojacking failure is much more common than satisfactory performance. It is evident that in these formations the conclusion reached by Deere (1983) is supported by this figure and that of Brekke & Ripley (1986) can not be sustained.

A common feature of pressure tunnels in sedimentary and extrusive igneous rocks is the presence of near-vertical joint sets, which control the state of stress. In general, the minor principal stresses are near perpendicular to these vertical sets (horizontal) independently of slope angle. Thus, the effect of the horizontal to vertical stress ratio (k_o) and shear strength of critical shallow joint sets is in many instances more important than the slope angle as demonstrated by the results of a numerical study on the effect of rock joints presented in the next section.

Table 1. Pressure tunnel cases added to the database of Brekke & Ripley (1986). (λ & α are the dips of steep & shallow joints, respectively)

#	Project	Country	Geologic Description	λ/α [°]	β [°]	L_r [m]	h_i [m]	L_r/h_i [-]	Location cover ratio	Reference
50a	Rocky Mountain 1985	Georgia USA	Bedded sediments w/ slight dip into slope, w/ vertical joints	90/ -10	25	112	238	0.47	Original end of steel liner	Swiger (1991)
50b				90/ -10	25	112	210	0.53	$h_i = \sigma_{n\,min}/\gamma_w$ measured	
57a	Deere case 1: 1970s	Central America	Volcanic, blocky lava, altered, soil like joints w/ interbedded tuff and few layers of hard lava.	?	4	71	58	1.22	600 m U/S end cont. cracks in plain conc.	Deere (1983)
57b	Deere case 1: 1970s			?	10	40	58	0.69	U/S end of steel liner	
58	Deere case 2: 1970s	Central America	Volcanic, tuff & tuff breccia w/ hard ignimbrite & lava.	?	7	80	80	1.00	U/S end of steel liner	Deere (1983)
59a	Mesitas: 1984	Colombia	Sedimentary: Sandstone / Siltstone with shale sequences. Vertical joints	90/ -15	55	225	224	1.00	U/S end of steel liner	Fernández (1984)
59b				90/ -15	55	197	195	1.00	Surge shaft EL 2380	
60	WSSC Bi-Cnty 1987	Maryland USA	Pre-cambrian quartz mica GNEISS, sheared (Wissahickon)	60/ -60	1	50	125	0.40	Zone 4-5: severe crack liner	Cording Fernández (1988)
61A	Balsam Meadow 1986	California USA	Dinkey Creek Pluton: Massive quartz diorite. Open vertical joints	90/ 0	35	245	430	0.57	Orig. End steel liner	Brekke & Korbin (1990)
61B				90/ 0	35	245	260	0.94	Using $h_i = \sigma_{n\,min}/\gamma_w$ measured	

COUPLED HYDROMECHANICAL BEHAVIOR OF PRESSURE TUNNELS IN FRACTURED ROCK

Alvarez (1997) conducted numerical analyses using the Discrete Finite Element Method (DFEM), to understand the hydromechanical rock-joint interactions leading to hydrojacking. Filling of a pressure tunnel was simulated until hydrojacking occurred and the modes of failure were identified.

The DFEM (Ghaboussi, 1988; Barbosa, 1990) permits the large deformation and finite strain analysis of an ensemble of deformable bodies (intact rock blocks) which interact through deformable, frictional contacts (rock joints) and incorporates the effects of water pressures in the joints. The deformable bodies are modeled using Finite Elements. Water flow occurs through rock joints and their coupled hydromechanical behavior was modeled in accordance with the findings by Alvarez, Cording & Mikhail (1995).

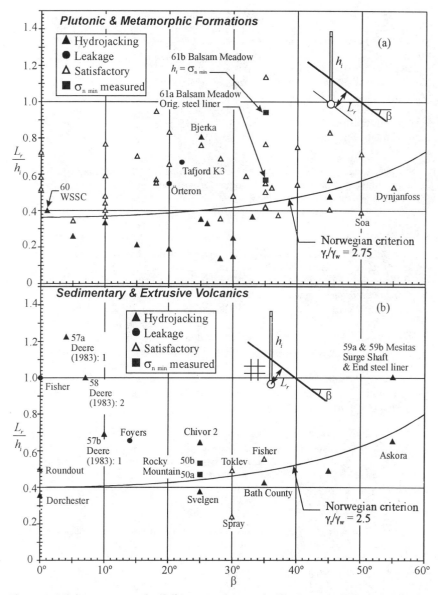

Figure 1. Minimum cover ratio (L_r/h_i) versus slope angle (β). Pressure Tunnels and Shafts in (a) Plutonic & metamorphic rocks and (b) Sedimentary and extrusive Volcanics. Database : Brekke & Ripley (1986) and Table 1.

The mechanics of hydraulic jacking of a fractured rock mass can be summarized as follows. As pressure inside the tunnel increases, flow from the tunnel into the mass occurs when the internal piezometric head in the tunnel becomes higher than the background head ($h_i > h_o$). Further increases in head inside the tunnel cause several notable events to take place, namely (a) the mounding of the water table, which intersects the ground surface forming surface springs when the internal head is $h_i = h_{i\ SS}$; (b) the initiation of failure as *local* hydrojacking first opens rock joints near the tunnel ($h_i = h_{i\ LJ}$) and finally (c) the coalescence of open joint segments and formation of a hydraulic connection to a lower head outlet, which causes large increases in flow rate and thus causes failure by *global* hydrojacking ($h_i = h_{i\ GJ}$).

The decisive effect of rock mass discontinuities on the hydrojacking mode of failure is illustrated in Figure 2. The figure summarizes the distribution of open rock joints after global hydrojacking had taken place in six numerical cases analyzed by Alvarez (1997). The thickest lines indicate open joints and the thinnest lines indicate closed joints.

The effects of slope inclination are in many instances overshadowed by the larger effect of rock joint orientation, initial state of stress and shear strength of shallow dipping shear zones. The following discussion sheds light into the behavior of pressure tunnels in fractured ground. In all cases decompression of the rock mass led to global hydrojacking as explained by Fernández & Alvarez (1994).

PRESSURE TUNNELS IN FRACTURED HORIZONTAL GROUND

Fractured rock masses with near-vertical ($\lambda > 80°$) and sub-horizontal ($\alpha < 10°$) joint sets admit the existence of a full range of in situ stress conditions; from a columnar condition in which the horizontal stresses are zero and the vertical loads are carried by rock columns ($k_o = 0$), to a condition where the horizontal stresses are only limited by the intact strength of the rock blocks ($k_o > 3$).

The local and global hydrojacking heads are intimately related to the mode of failure. For $k_o < 1$, hydrojacking failure takes place by opening vertical joints, therefore $h_{i\ LJ}$ and $h_{i\ GJ}$ depend on the magnitude of the stress ratio. When $k_o > 1$, horizontal planes are lifted and the hydrojacking heads are independent of the magnitude of the stress ratio and depend on the vertical overburden stress. In cases where k_o varies with depth, the mode of failure changes accordingly, always opening joints with minimum normal stress.

The traditional vertical cover criteria is unconservative in horizontal ground with vertical and horizontal joints when the total stress ratio at tunnel elevation, k_o, is less than 0.8. For $k_o > 0.8$ these criteria become adequate and conservative when $k_o > 1$. The following relationship predicted well the occurrence of local hydrojacking in the numerical models studied by Alvarez (1997):

$$\frac{L_r}{h_{iLJ}} = \frac{\gamma_w}{\gamma_r} \frac{1}{k_o} \; ; \; k_o \leq 1 \qquad (2)$$

This simple equation explains the difficulties associated with the prediction of minimum cover using the Norwegian rule when the ground surface is near horizontal ($\beta < 20°$) and vertical joints are present in the rock mass. The horizontal to vertical stress ratio must be known or estimated *a priori*, either from pre-excavation stress measurements or experience.

PRESSURE TUNNELS NEAR INCLINED VALLEY WALLS

In fractured rock masses, the state of stress existing in the slope before pressurization of the tunnel is controlled by the tectonic history, the slope angle and the attitude and strength of the joint sets. Rock masses which experienced tectonic horizontal stresses before erosion of the valley have higher in situ horizontal stresses because the shear strength in the shallow dipping joints was mobilized during unloading and relief caused by erosion, thus locking-in a portion of these stresses. On the other hand, masses which only experienced gravitational loading before erosion of the valley, do not mobilize a substantial portion of the shear strength of the shallow-dipping joints.

The upper bound of total normal stresses across steep rock joints is limited by the shear strength of shallow dipping joints and can be estimated by performing limit equilibrium analyses using kinematically admissible wedges of rock bounded by the discontinuity sets. Where kinematically admissible wedges do not form, such as when the joints are perpendicular and parallel to the ground surface, the in situ stresses could be higher.

Joint orientation and initial state of stress controlled the pattern of open joints during hydrojacking, as shown in Figure 2. In the cases with vertical and horizontal joints, global hydrojacking involved opening vertical joints, with a significant reduction of the initial total normal stress in these joints as the slope deformed toward the valley during pressurization of the tunnel.

Hydrojacking of the mass was preceded by movement of a wedge along the horizontal joints, which together with the deformability of the medium permitted the opening of vertical joints through which water escaped to the ground surface. The DFEM analyses showed that the water pressure in the tunnel at which global hydrojacking occurred was closely related to the pressure at which the wedge slid toward the valley.

MINIMUM COVER CRITERIA FROM THE STABILITY OF ROCK WEDGES

The observation reported in the previous section was used to formulate a criterion for failure with the premise that global hydrojacking occurs *when steeply dipping joints open as movement of the wedge begins along shallow dipping joints.*

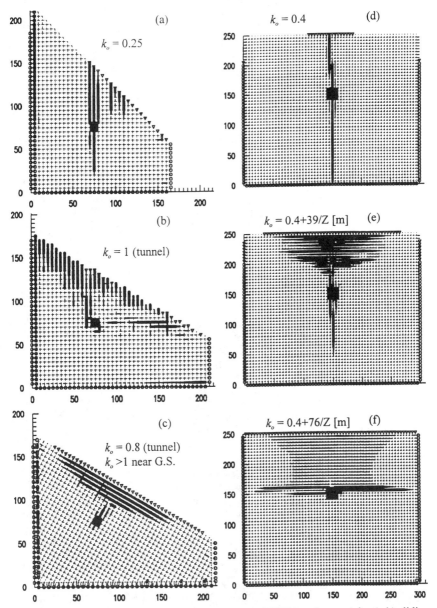

Figure 2. Global hydrojacking modes of failure from DFEM analyses. a) k_o <1; b) sliding ;c) lifting planes ||; d) V joints k_o <1; e) V & H joints k_o <1 & >1; f) H joints open (k_o >1)

The movement of the wedge occurs because of increases in uplift water pressures along shallow dipping joints, which reduce their shear strength, and increases in water pressures along steep joints, which push the wedge toward the valley.

The Illinois Criterion was derived from the limit equilibrium of the wedge of rock illustrated in Figure 3 (Alvarez, 1997) and is given as:

$$\frac{L_r}{h_i} = \frac{\gamma_w}{\gamma_r} M ;$$
$$M = \frac{\cos\beta [A - K + C]}{D} \qquad (3)$$

where

$$A = (B_1 - B_2)\left(\frac{\tan\beta}{\tan\lambda} - \frac{\tan\alpha}{\tan(\phi_2 - \alpha)}\right); K = \frac{\tan\alpha}{\tan\lambda}\left(B_1 + B_2 \frac{\tan\beta}{\tan(\phi_2 - \alpha)}\right)$$

$$C = \left(B_2 + B_1 \frac{\tan\beta}{\tan(\phi_2 - \alpha)}\right) \text{ and } D = \left(1 - \frac{\tan\alpha}{\tan\lambda}\right)$$

The cover coefficient, M, of equation (3) contains information regarding the site geology, through the dip angles of steep (λ) and shallow (α) joint sets and the friction angle of the shallow joints (ϕ_2). In addition, the coefficient includes the valley wall inclination (β) and indirectly, information on the initial in-situ stresses through the water force ratios (B_1 and B_2).

For rock masses with low initial stresses, the excess water pressure due to filling of the tunnel varies approximately linearly with distance. Therefore $B_1 \sim B_2 \sim 1$. On the other hand, in tight rock masses the water pressure distribution varies logarithmically with distance from the tunnel, hence $B_1 < 1$ and $B_2 < 1$.

If the joint sets are orthogonal and joint set ❶, the lifting plane, is parallel to the ground surface ($\lambda = \beta$, $\alpha = \beta - 90°$), equation (3) reduces to:

$$\frac{L_r}{h_i} = \frac{\gamma_w}{\gamma_r} M ; M = B_1 \cos\beta (1 - \tan\beta \tan(\phi_2 - \beta)) \qquad (4)$$

If it is further assumed that the joint set perpendicular to the valley wall ❷ is frictionless ($\phi_2 = 0°$) and that the pressure distribution is linear ($B_1 = 1$), the Norwegian cover criterion is recovered as a special case of the Illinois Criterion:

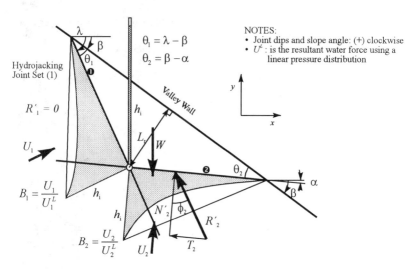

Figure 3. Limit equilibrium. Hydrojacking of steep joints and sliding along shallow joints

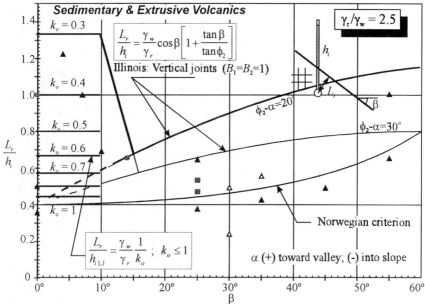

Figure 4. Minimum cover ratio (L_r/h_i) versus slope angle (β) & horizontal to vertical stress ratio (k_o). Norwegian and Illinois criteria. Database for pressure tunnels in sedimentary and extrusive volcanics: Brekke & Ripley (1986) and Table 1.

$$\frac{L_r}{h_i} = \frac{\gamma_w}{\gamma_r} M \; ; \; M = \frac{B_1}{\cos\beta} = \frac{1}{\cos\beta} \Rightarrow \frac{L_r}{h_i} = \frac{\gamma_w}{\gamma_r} \frac{1}{\cos\beta} \tag{5}$$

For rock formations with vertical joints ($\lambda = 90°$), equation (3) reduces to:

$$\frac{L_r}{h_i} = \frac{\gamma_w}{\gamma_r} M \; ; \; M = \cos\beta \left[(B_2 - B_1) \frac{\tan\alpha}{\tan(\phi_2 - \alpha)} + B_2 + B_1 \frac{\tan\beta}{\tan(\phi_2 - \alpha)} \right] \tag{6}$$

and for the special case of vertical joints and linear water pressure distribution ($B_1 = B_2 = 1$), the Illinois Criterion reduces to:

$$\frac{L_r}{h_i} = \frac{\gamma_w}{\gamma_r} \cos\beta \left[1 + \frac{\tan\beta}{\tan(\phi_2 - \alpha)} \right] \tag{7}$$

The simplified form of the Illinois Criterion for masses with vertical joints, equation (7), is plotted in Figure 4 for a range of friction angles (ϕ_2-α =20° & 30°). The dip of the shallow joints (α) is positive toward the valley and negative into the slope. The database of pressure tunnels in sedimentary and extrusive volcanic formations is also presented in the figure. In addition, equation (2) has been incorporated to explain hydrojacking in tunnels in near horizontal ground.

In near horizontal ground ($\beta<20°$), sliding along shallow-dipping joints is not a critical mode of failure. Instead the horizontal to vertical stress ratio dominates, as illustrated in Figure 4. The Illinois Criterion is a reasonable envelope of the hydrojacking failure cases shown in Figure 4 for slopes with inclinations greater than 20°.

The cover criteria presented in this section can be used to select the depth of the tunnel for specific geologic conditions present at a given site. This added flexibility is a key contribution for preliminary and feasibility level studies where information from mapped geology and structural features can be incorporated into an initial conservative estimate of the required cover. In situ stress measurements must be performed for final lining selection.

CONCLUSIONS

The Norwegian rock cover criterion is applicable for siting pressure tunnels in massive rock, such as plutonic and metamorphic formations, where minimum in situ stress is strongly correlated to slope inclination and hydrojacking occurs by lifting of planes parallel to the ground surface. The Norwegian criterion is not a reliable predictor of hydrojacking failure in fractured rock masses, such as sedimentary and extrusive volcanic formations, where other factors such as stress history and joint orientation become important.

Equations have been presented to predict rock cover to prevent hydrojacking of steeply-dipping joints when the rock mass slides along shallow dipping joints. The

Illinois criterion incorporates the effects of slope inclination, joint attitude and shear strength. For near horizontal ground, the horizontal to vertical stress ratio controls.

ACKNOWLEDGEMENTS

This research was conducted at the University of Illinois under the auspices of the National Science Foundation (NSF), Grant MSS-921652. The Stanley D. Wilson and Ralph B. Peck fellowships helped support the first author during the initial years of his stay at Illinois. Finally, the first author thanks Mr. Alan R. Crumley and Dr. Nelson Kawamura of Geoconsult, Inc. for their review of this paper.

REFERENCES

Alvarez, T. A., 1997. *A study of the coupled hydromechanical behavior of jointed rock masses around pressure tunnels.* Ph.D. dissertation. Dept. of Civil Engineering. University of Illinois at Urbana-Champaign.

Alvarez, T. A., Cording, E.J. & Mikhail, R. 1995. Hydromechanical behavior of rock joints. A re-interpretation of published experiments. *Proc. 35th U.S. Symposium on Rock Mechanics.* Daemen & Schultz, Eds., 665-671.

Barbosa, R., 1990. *Discrete element models for granular materials and rock masses.* Ph.D. dissertation. Dept. of Civil Engineering. University of Illinois at Urbana.

Brekke, T.L. and Ripley, B., 1986, *Design strategies for pressure tunnels and shafts*, Report EPRI contract No. RP-1747-17.

Brekke, T.L. & Korbin, G.E., 1990. Recent Sierra Nevada Hydroelectric Projects Geological Engineering Lessons. *Proc. H. Bolton Seed Memorial Symposium.* Univ. of California, Berkeley. Duncan (Ed.), Vol. 2, 43-66, BiTech Publishers.

Broch, E., 1984, Development of unlined pressure tunnels and shafts in Norway, *Underground Space*, Vol. 8, pp 177-184.

Cording, E.J. & Fernández, G., 1988. *Walk-through inspection report. WSSC Bi-County tunnel.* Washington Suburban Sanitary Commission (WSSC), Maryland.

Deere, D. U., 1983, Unique geotechnical problems at some hydroelectric projects, *7th Pan Am. Conf. Soil Mech. Found. Eng.* Vancouver, 865-888.

Fernández, G., 1984. Consulting report to Empresa de Energía Eléctrica de Bogotá (EEEB) regarding the Mesitas Tunnel. Colombia. November 6, 1984. (In Spanish)

Fernández, G. and Alvarez, T.A., 1994. Seepage-Induced Effective Stresses and Water Pressures around Pressure Tunnels, *Journal of Geotechnical Engineering*, Vol. 120, No. 1, 108-128 ASCE.

Ghaboussi, J., 1988. Fully deformable discrete element analysis using a finite element approach. *Int. J. Computers and Geotechnics.* 5: 175-195

Hendron, A.J., Fernández, G., Lenzini, P. and Hendron, M.A., 1987, Design of pressure tunnels, *The Art and Science of Geotechnical Engineering at the Dawn of the Twenty First Century*, (Cording, et al., Eds.), 161-192, Prentice Hall.

Swiger, W.F., 1991. Recent problems with pressure tunnels. *1991 Stanley D. Wilson Memorial Lecture.* University of Washington, Seattle. 45 pp.

Stability Analyses of Underground Openings Using a Multiaxial Failure Criterion

Li Li[1], Michel Aubertin[2], and Richard Simon[3]

Abstract

The induced stress state around underground excavations in rock is three-dimensional. Accordingly, stability analyses should be based on a multiaxial failure criterion. Recently, the authors have introduced a general failure surface that relies on specific fracture mechanisms for tensile and/or compressive stress states. In this paper, this new criterion is presented. It is then shown how it can be applied to evaluate the stability of underground openings. As a specific example, the case of boreholes in a high stress field in homogeneous and isotropic rocks is analyzed to estimate the major principal stress.

Key words: rock, brittle, failure, borehole, breakouts, stability.

Introduction

Natural stress fields and induced stress states around openings in rock masses are three-dimensional. To adequately evaluate the stability of man made excavations, a multiaxial failure criterion is generally required by modern methods of analysis.

Over the past few decades, a number of investigations have been performed to study rock failure and the associated physical mechanics. Through these, it became apparent that the processes involved were numerous and complex (e.g. Gramberg 1989; Kenemy and Cook 1991). For engineering applications, many have considered that the failure conditions could be defined in a simplified manner.

---

[1] Postdoctoral Researcher, [2] Professor, [3] Assistant Professor, Department of Civil, Geological and Mining Engineering - École Polytechnique de Montréal, C.P. 6079, Succursale Centre-ville, Montréal, QC, H3C 3A7, Canada, phone: 1 514 340 4711 extension 3279, fax: 1 514 340 4477, email: Li.Li@mail.polymtl.ca .

Accordingly, various expressions have been proposed using linear (Coulomb type) formulations or non linear functions that make use of two or three principal stress components (see reviews by Lade 1993; Andreev 1995). Although slightly more complex to use, it is generally recognized that 3D non linear expressions are more representative of rock behavior (Song and Haimson 1997), and better suited for numerical calculations used for stability analyses.

Existing criteria for intact rock all aim at describing the shape of the failure surface as given by the peak strength of samples loaded under different testing conditions, such as uniaxial traction, diametral compression, uniaxial compression, conventional triaxial compression, reduced triaxial extension, etc. As these criteria are relying on more or less similar experimental results, most resemble each other, and it is often difficult to properly differentiate between existing formulations. The criterion presented in the following, named $MSDP_u$, offers some distinctive features when compared to most other criteria.

To explicitly define these particular differences, one can first recall briefly the physical mechanisms associated to progressive damage which may eventually lead to material failure. Such processes have been studied extensively in recent years (e.g. Ashby and Hallam 1986; Atkinson 1987; Wittaker et al. 1992; Germanovitch et al. 1996). These studies have shown that failure of rocks and rock-like materials essentially results from the initiation, propagation and coalescence of microcracks, and that these phenomena are controlled by the local stress state. Depending on the loading conditions, cracks may be opened or closed, and in the latter case, frictional shearing along crack surfaces must occur to allow crack extension. Hence, failure depends on the frictional component that exists as soon as some of the propagating cracks are closed. However, contrary to some existing theories, it can be expected that the apparent friction angle ϕ_a on shearing planes (or ϕ_b+i in a Patton type shear strength model—see Ladanyi and Archambault 1970) progressively decreases as normal stress increases. During conventional triaxial compression tests on low porosity rocks, such decrease in the apparent friction angle (ϕ_a) as confining stress increases stops when shearing surface geometry is no more influenced by normal stress. This aspect differentiates criteria with a constant ϕ_a (Coulomb type), those with continuously decreasing ϕ_a (Griffith type), and those, like the one proposed below, that include a minimum value of ϕ_a (which should be close to ϕ_r or ϕ_b, the residual or basic friction angle).

On the other hand, experimental results on various materials and fracture mechanics considerations suggest that a small confining pressure in a multiaxial test may increase the absolute value of the axial tensile strength, while a larger pressure eventually decreases this strength by inducing additional tensile stresses near crack tips (Aubertin and Simon 1997, 1998). This distinction is used to differentiate the $MSDP_u$ criterion from those in which tensile strength is not affected by the presence of a small compressive stress (Griffith type). Another feature that separates $MSDP_u$

from some other formulations is material strength under biaxial compression, which has been shown to be larger than the uniaxial compressive strength C_0 (Lade 1993). It can also be expected that biaxial and triaxial tensile loads will induce failure at a smaller absolute value of the minimum principal stress than for the unidimensional case (e.g. Brady 1970). Hence, it should be considered incorrect (and non conservative) to assume that tensile failure only occurs when the minor principal stress reaches the uniaxial tensile strength T_0 of the rock.

Finally, it is well known that the failure surface in a Cartesian stress space is not fully symmetric around the hydrostatic axis, the peak strength being lower in extension than in compression (Lade 1993). The effect of the Lode angle θ should then be included in a 3D criterion to adequately define the influence of the loading geometry on the failure surface formulation.

In the following section, a failure criterion bearing the above mentioned characteristics is proposed. It can be used to evaluate stability of underground openings. As a specific application, the case of boreholes is used to illustrate the utilization of this new formulation in the simple case of homogeneous and isotropic rocks with little or no fractures.

$MSDP_u$ Criterion

A multiaxial failure criterion, named MSDP, was recently developed and applied successfully to brittle rocks and to rock-like materials (Aubertin and Simon 1996, 1998). The MSDP criterion differs from other existing criteria in various ways that have been presented briefly in the previous section. One advantage of the criterion is that the four material parameters required for its application can be obtained by simple and independent tests, namely: uniaxial compression tests, uniaxial tension (or equivalent) tests, and tilt or shear tests on plane surfaces; one can also use biaxial compression, conventional triaxial compression, and reduced triaxial extension tests to obtain and/or confirm the parameters value.

The MSDP criterion is somewhat phenomenological, but nevertheless captures the main features of rock failure. The good correspondence between MSDP and various experimental results has been shown in previous publications. Nevertheless, despite its relative simplicity and physical significance, the MSDP criterion has a few minor drawbacks, which are mainly due to its mathematical formulation (and not to its descriptive capabilities). These have recently been reviewed by Aubertin et al. (1998). To overcome these limitations, an updated formulation, named $MSDP_u$ (for unified formulation of MSDP), is proposed. The $MSDP_u$ criterion is expressed as follows:

$$\sqrt{J_2} - F_0 F_\pi = 0 \tag{1a}$$

or written in explicit form

$$\sqrt{J_2} - b\left\{\frac{\alpha^2\left(I_1^2 - 2a_1 I_1\right) + a_2^2}{b^2 + (1-b^2)\sin^2(45° - 1.5\theta)}\right\}^{1/2} = 0 \qquad (1b)$$

where F_0 gives the shape of surface in the $\sqrt{J_2}$ - I_1 plane; and F_π is the controlling function in the octahedral (π) plane perpendicular to the hydrostatic axis. In the equations, the material parameters are defined as follows:

$$\alpha = \frac{2\sin\phi}{\sqrt{3}(3-\sin\phi)} \qquad (2)$$

$$a_1 = \left(\frac{\sigma_c - \sigma_t}{2}\right) - \left(\frac{\sigma_c^2 - (\sigma_t/b)^2}{6\alpha^2(\sigma_c + \sigma_t)}\right) \qquad (3)$$

$$a_2 = \left\{\left(\frac{\sigma_c + \sigma_t/b^2}{3(\sigma_c + \sigma_t)} - \alpha^2\right)\sigma_c\sigma_t\right\}^{1/2} \qquad (4)$$

where σ_c (= C_0) and σ_t (= $|T_0|$) are the uniaxial strength in compression and tension respectively; the value of ϕ in Equation (2) is equal to ϕ_r or ϕ_b; parameter b is used to define the shape in the π plane (usually b $\cong$ 0.75); I_1 is the first invariant of the stress tensor σ_{ij}; J_2 is the second invariant of the deviatoric stress tensor S_{ij}; θ is the Lode angle defined in the octahedral (π) plane (-30° $\leq \theta \leq$ 30°).

Figure 1 shows the representation of the MSDP$_u$ criterion in the $\sqrt{J_2}$ - I_1 plane for CTC (θ = 30°) and RTE (θ = -30°) loading conditions.

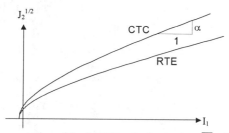

Figure 1. Representation of the MSDP$_u$ criterion in the $\sqrt{J_2}$ - I_1 plane; CTC: conventional triaxial compression; RTE: reduced triaxial extension.

The validity of the MSDP$_u$ criterion has been recently assessed for different types of rock (Aubertin et al. 1998). For instance, Figure 2 shows the criterion with results on sandstone samples submitted to conventional triaxial compression tests. Figure 3 shows a comparison between the criterion and experimental results obtained from hollow cylinders tests on a limestone. These results illustrate the good agreement between MSDP$_u$ and experimental data.

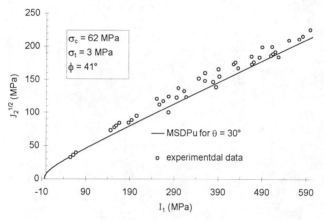

Figure 2. Application of the MSDP$_u$ criterion to sandstone under CTC (data from Schwartz 1964).

Application to Underground Openings

The MSDP$_u$ criterion is particularly well suited to evaluate the stability of underground openings. However, in the case of large scale excavations such as those created for civil and mining engineering projects, the material parameters introduced into the criterion have to correspond to the rock mass in situ conditions. Because there actually exists a large uncertainty on such parameters (for any criterion), and on how to go from laboratory scale to in situ properties (Aubertin et al. 1999), the authors have chosen to illustrate this first application of the criterion to boreholes in homogeneous and isotropic rocks where laboratory properties can be used directly for the material around actual openings. Here, vertical boreholes are considered with one of the principal natural stress (σ_v) being parallel to the borehole axis. This allows the use of simple analytical solutions to establish the elastic stress field (e.g. Hiramatshu and Oka 1962). In the following applications, the size of the borehole breakout, as defined by angle Θ_b (see Figure 4), is used for validation of the criterion. Over the years, breakouts have frequently been used to estimate components of the in situ stress field (e.g. Zoback et al. 1985; Zoback 1992). In many cases however, past interpretations have been limited to the Coulomb or other simplified criteria, which may not be adequate for such applications (e.g. Maury 1987; Song and Haimson 1997).

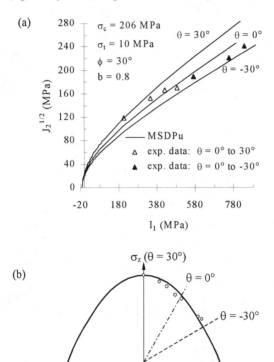

Figure 3. The MSDP$_u$ criterion in the $\sqrt{J_2}$ - I_1 plane (a) and π plane (b) applied to limestone under different loading conditions with various Lode angles θ (data from Hoskins 1969).

The value of σ_v is taken as the overburden stress, and the value of the minimum horizontal principal stress σ_h is known following measurements, such as hydraulic fracturing tests for instance (e.g. Zoback et al. 1985). The critical angle Θ_b is then used to calculate the value of the largest horizontal stress component σ_H.

Failures around boreholes occurs when the value of the safety factor becomes critical. In this application, the safety factor is defined simply as:

$$SF = \frac{F_0 F_\pi}{\sqrt{J_2}} = \frac{\text{Capacity (allowable stress calculated from MSDP}_u)}{\text{Demand (induced deviatoric stress)}} \quad (5)$$

The breakout then appears at B' and B (see Figure 4) where FS equals unity.

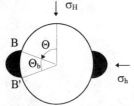

Figure 4. Definition of borehole breakout characteristics.

The value of angle Θ_b can be obtained by observations in the borehole. The principal stress state at point B and B' then becomes:

$$\begin{aligned}\sigma_1 &= \sigma_H + \sigma_h - 2(\sigma_H - \sigma_h)\cos 2\Theta \\ \sigma_2 &= \sigma_v - 2\nu(\sigma_H - \sigma_h)\cos 2\Theta \\ \sigma_3 &= 0 \end{aligned} \quad (6)$$

where σ_1, σ_2 and σ_3 correspond to the tangential, axial and radial stress respectively; σ_H and σ_h are the maximum and minimum horizontal in situ stresses respectively; σ_v is the vertical principal in situ stress; ν is the Poisson's ratio; Θ is the angle of borehole breakout between point B and the maximum horizontal principal stress direction σ_H. The corresponding Lode angle θ is then given by:

$$\theta = \tan^{-1} \frac{\sigma_1 - 2\sigma_2}{\sqrt{3}\sigma_1} \quad (7)$$

while the induced invariant stresses are given by:

$$I_1 = \sigma_1 + \sigma_2 \quad (8)$$

$$\sqrt{J_2} = \frac{1}{\sqrt{3}}(\sigma_1^2 - \sigma_1\sigma_2 + \sigma_2^2)^{1/2} \quad (9)$$

Using Equation (5) with SF = 1, one obtains an equation that describes the relationship between the maximum horizontal stress σ_H and the angle of borehole breakout Θ_b (= 180° - 2Θ):

$$\sqrt{J_2}\,[\sigma_H, \Theta] = F_0[\sigma_H, \Theta]\,F_\pi[\sigma_H, \Theta] \quad (10)$$

Using Θ_b for such calculation is justified by the fact that once the borehole breakout initiates, the depth may increase with time but the width usually remains unchanged (Zheng et al. 1989). So, by solving the above equation, the magnitude of σ_H can be deduced if Θ_b is known.

The $MSDP_u$ criterion is used with laboratory test results published by Song and Haimson (1997) on Berea sandstone blocks with long holes. The measured angle of the borehole breakouts Θ_b was about 31° for applied far-field stresses of σ_H = 80 MPa, σ_h = 20 MPa, σ_v = 30 MPa; with applied far-field stresses of σ_H = 90 MPa, σ_h = 30 MPa, σ_v = 40 MPa, the value of Θ_b became about 35°. Using these known values of Θ_b, σ_h and σ_v, the value of σ_H can be back-calculated. The material parameters σ_c, σ_t, ϕ and b are based on test results also published by Song and Haimson (1997), as shown in Figure 5.

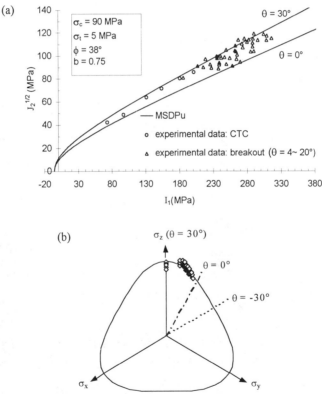

Figure 5. The $MSDP_u$ criterion in the $\sqrt{J_2}$ - I_1 plane (a) and π plane (b) applied to Berea sandstone under different loading conditions with various Lode angle θ (data from Song and Haimson 1997).

The value of σ_H can be obtained by solving Equation (10). Because the latter can only be solved in an iterative manner, a graphical approach is rather used. To do so, a plausible range of σ_H is first assumed (from σ_{Hmin} to σ_{Hmax}). Changing σ_H modifies the Lode angle, so each value corresponds to a different curve in the $\sqrt{J_2}$ - I_1 plane. Figure 6(a) shows a schematic representation of the MSDP$_u$ criterion at Lode angles corresponding to σ_{Hmin} and σ_{Hmax}; also shown are the corresponding values of $I_1(\sigma_{Hmin})$ and $I_1(\sigma_{Hmax})$. From the straight line that joints the two points, two new points of intersection (F and G) are obtained on the corresponding failure curves; these correspond to $I_1(\sigma_{H1})$ and $I_1(\sigma_{H2})$. From these, σ_{H1} and σ_{H2} can be deduced from Equation (8), each corresponding to a Lode angle that can be calculated from Equations (6) and (7) (i.e. $\theta(\sigma_{H1})$ and $\theta(\sigma_{H2})$). The Lode angle θ corresponding to the actual value of σ_H can then be obtained by interpolation (see Figure 6b):

$$\frac{\theta(\sigma_H)-\theta(\sigma_{Hmin})}{\theta(\sigma_{Hmax})-\theta(\sigma_{Hmin})} = \frac{\theta(\sigma_H)-\theta(\sigma_{H1})}{\theta(\sigma_{H2})-\theta(\sigma_{H1})} \tag{11}$$

and the third curve of MSDP$_u$ can be drawn in the $\sqrt{J_2}$ - I_1 plane. Thus, one obtains the point of intersection at $I_1(\sigma_H)$. The value of σ_H is finally deduced from Equations (6) and (8). Figure 6(b) illustrates how to converge on point H and the actual σ_H using θ values deduced from Equation (8) and from the application of the MSDP$_u$ criterion with Equations (6), (7), and (10).

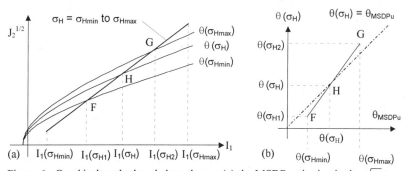

Figure 6. Graphical method to deduce the σ_H; (a) the MSDP$_u$ criterion in the $\sqrt{J_2}$ - I_1 plane with various Lode angles; (b) interpolation to find point H where $\theta(\sigma_H)$ is given from Equations (7) and (8) while θ_{MSDPu} is obtained from Equations (6), (7) and (10).

For tests performed on Berea sandstone, the value of σ_H back-calculated by the above procedure is 75 MPa for $\Theta_b = 31°$ and 80 MPa for $\Theta_b = 35°$, which compare fairly well with the values provided by Song and Haimson (1997). Equally good results were also obtained with tests performed on Westerly granite, with back-calculated values of σ_H within 10% of those given by the same authors.

Discussion

The above presentation shows that the $MSDP_u$ criterion can be used to evaluate failure around underground openings. As shown above, analytical solutions can be used to estimate the stress field and corresponding safety factor from elastic solution for tunnels, shafts and other circular excavations. An example of such calculations is shown in Figure 7 which presents contours of SF values for different stress conditions. These contours compare well with actual observed fracture geometry by Rawlings et al. (1993) for similar stress fields. However, it must be mentioned that elastic analyses can not provide the full extent of the failed zone because material properties beyond peak strength are no longer elastic, and the observed softening causes stress redistribution. This usually increases the size of the failed zone when compared to an elastic analysis. Nevertheless, results such as those shown in Figure 7 can be used to evaluate, at least in a preliminary manner, the stability of openings and the locations of the critical points at its boundary.

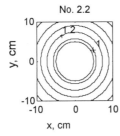 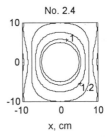

Figure 7. Calculation of SF using the $MSDP_u$ criterion and elastic solution; No. 2.2, corresponding to Test 2.2 of Rawlings et al. (1993), $\sigma_v = 2.65$ MPa, $\sigma_h = \sigma_H = 1.8$ MPa, with a hole parallel to σ_v; No.2.4, corresponding to Test 2.4 of Rawlings et al. (1993), $\sigma_v = 2.2$ MPa, $\sigma_h = \sigma_H = 1.32$ MPa, with a hole of 60° deviation from the σ_v in the σ_v, σ_H plane.

Of course, for large scale excavations in rock masses, the material properties must be adjusted to the in situ conditions. This aspect, and other influence factors such as time effects and anisotropy, are not discussed here because of space limitations, but will be addressed in upcoming publications (e.g. Aubertin et al. 1999; Li. et al. 1999).

Conclusion

Some of the main characteristics of rock failure are briefly presented, and these are described using the MSDP$_u$ criterion. This three-dimensional formulation, which is based on a mathematical modification of the MSDP function previously proposed by the authors, is used to represent the failure surface of rocks submitted to different stress fields. This multiaxial criterion can be used to analyze stability of underground openings. As a specific application, the MSDP$_u$ criterion is utilized to estimate the major principal stress from borehole breakout size.

Acknowledgments

Part of this work has been financed by grants from NSERC (OGP 0089749) and IRSST, whose contributions are gratefully acknowledged. The authors also thank Professor Haimson who provided additional information on testing results. The first author thanks IRSST for a post-doctoral research grant.

References

Andreev, G.E. (1995). "*Brittle Failure of Rock Materials - Test results and Constitutive Models.*" Balkema, Rotterdam.
Ashby, M.F., and Hallam, S.D. (1986). "The failure of brittle solids containing small cracks under compressive stress state." *Acta Metall.*, 34, 497-510.
Atkinson, B.K. (1987). "*Fracture Mechanics of Rock.*" Academic Press, London.
Aubertin, M., Li, L., Simon, R., and Khalfi, S. (1998). "Formulation and application of a short term strength criterion for isotropic rocks." Submitted to the *Canadian Geotechnical Journal*.
Aubertin, M., Li, L., Simon, R., and Khalfi, S. (1999). "A unified representation of the damage and failure criteria for rocks and rock masses." Submitted to the *9th ISRM Congress, Paris, 1999*.
Aubertin, M., and Simon R. (1996). "A multiaxial failure criterion that combines two quadric surfaces." *Tools and Techniques in Rock Mechanics*, M. Aubertin, F. Hassani and H. Mitri, eds., Balkema, 1729-1736.
Aubertin, M., and Simon, R. (1997). "A damage initiation criterion for low porosity rocks." *Int. J. Rock Mech. & Min. Sci.*, 34(3-4), Paper No. 017, CD-ROM, Elsevier.
Aubertin, M., and Simon, R. (1998). "Un critère de rupture multiaxial pour matériaux fragiles. *Can. J. of Civil Engng.*, 25(2), 277-290.
Brady, B.T. (1970). "Effect of the intermediate principal stress on the fracture of brittle rock." *Rock Mechanics—Theory and Practice: Proc. 11th Symp. Rock Mech.*, W.H. Somerton, ed., Am. Int. Min. Metall. and Petroleum Engnrs., Inc., 267-279.
Germanovitch, L.N., Carter, B.J., Ingraffea, A.R., Dyskin, A.V., and Lee, K.K. (1996). "Mechanics of 3D crack growth under compressive loads." *Tools and Techniques in Rock Mechanics*, M. Aubertin, F. Hassani and H. Mitri, eds., Balkema, 1151-1160.
Gramberg, J. (1989). "*A Non-Conventional View on Rock Mechanics and Fracture*

Mechanics." Balkema, Rotterdam.
Hiramatshu, Y., and Oka, Y. (1962). "Stress around a shaft or level excavated in ground with a three-dimensional stress state." *Mem. Fac. of Engng. Kyoto Univ.*, 24, 56-76.
Hoskins, E.R. (1969). "The failure of thick-walled hollow cylinders of isotropic rock." *Int. J. Rock Mech. Min. Sci.*, 6, 99-125.
Kemeny, J.M., and Cook, N.G.W. (1991). "Micromechanics of deformation in rocks." *Toughening Mechanisms in Quasi-Brittle Materials*, S.P. Shah, ed., Kluwer Academic Press, 155-188.
Ladanyi, B., and Archambault, G. (1970). "Simulation of shear behavior of a jointed rock mass." *Proc. 11th U.S. Symp. Rock Mech.*, 105-125.
Lade, P.V. (1993). "Rock strength criteria - The theories and evidence." *Comprehensive Rock Engineering - Principles, Practice and Projects*, J.A. Hudson, ed., Pergamon Press, 1, 255-284.
Li, L., Aubertin, M., and Simon, R. (1999). "A multiaxial failure criterion with time and size effects for intact rock." Submitted to the *37th US Rock Mech. Symp.*, Vaill, 1999.
Maury, V. (1987). "Observations, Researches and recent results about failure mechanisms around single galleries." *Proc. 6th Int. Cong. on Rock Mech.*, G. Herget and S. Vongpaisal, eds., Montreal, 2, 1119-1128.
Rawlings, C.G., Barton, N.R., Bandis, S.C., Addis, M.A., and Gutierrez, M.S. (1993). "Laboratory and numerical discontinuum modeling of wellbore stability." *J. Pet. Technol.*, 45, 1086-1092.
Schwartz, A.E. (1964). "Failure of rock in the triaxial test." *Proc. 6th U.S. Symp. Rock Mech.*, 109-151.
Song, I., and Haimson, B.C. (1997). "Polyaxial strength criteria and their use in estimating in situ stress magnitudes from borehole breakout dimensions." *Int. J. Rock Mech. & Min. Sci.*, 34(3-4), paper No. 116, CD-ROM, Elsvier.
Wittaker, B.N., Singh, R.N., and Sun, G. (1992). *Rock Fracture Mechanics - Principle, Design, and Applications.*" Elsevier, Amsterdam.
Zheng, Z., Kemeny, J., and Cook, N.G.W. (1989). "Analysis of borehole breakouts." *J. Geophys. Res.*, 94(B6), 7171-7182.
Zoback, M.D., Moos, D., Mastin, L., and Anderson, R.N. (1985). "Well bore breakouts and in situ stress." *J. Geophys. Res.*, 90(B7), 5523-5530.
Zoback, M.L. (1992). "First- and second-order patterns of stress in the lithospheric stress patterns." *J. Geophys. Res.*, 97(B8), 11703-11728.

STRESS IN BURIED PIPES FROM SHEAR DISTORTIONS
C. A. Davis[1], Member ASCE

Abstract

An analytical elastic solution is presented to evaluate the effects of an applied shear stress on buried pipes to better understand their performance during earthquakes. The solutions identify the relative importance of physical variables involved in the shearing problem. The problem was investigated for three conditions at the soil-pipe interface: (1) no slip, (2) complete slip, and (3) friction. Compared to an applied normal stress, the addition of a shear stress increases the magnitude of stress and displacement in a buried pipe. The location of the maximum stress is shifted by as much as $45°$. An applied shear stress causes slippage to occur around significant portions of the pipe which leads to considerable soil yielding. Analytical results provide reasonable bounding solutions to the actual field conditions. The solutions were compared to case studies of flexible corrugated pipes shaken by the 1994 Northridge earthquake and were shown to adequately assess the observed pipe performance.

Introduction

Shear distortions from strong earthquake shaking can damage large and small diameter buried pipes. The 1971 San Fernando, 1994 Northridge, and 1995 Kobe earthquakes have revealed the vulnerability of buried concrete, steel, plastic, and fiberglass pipelines to transverse shear distortions (Subcommittee on Water and Sewage Systems, 1973; Davis and Bardet, 1998; Davis and Bardet, 1999; Tohda et al., 1996). Even so, underground pipes are rarely designed for transverse shearing forces (Moser, 1990).

The effect of transverse shear distortions on the response of buried pipes is not well documented. There have been limited field or laboratory tests and very few analytical studies to investigate the effects of transverse shear stresses. Luco and De Barros (1994), Lee and Karl (1992), and Wong et al. (1985) investigated the diffraction of elastic plane SV shear waves by an unlined cylindrical cavity near a free surface. Datta et al. (1984)

---

[1]Geotechnical Engineer, Los Angeles Department of Water and Power, Los Angeles.

investigated a similar case for a buried pipe that remained completely bonded to the elastic medium. All of these studies neglected pre-earthquake stress conditions and present results only for very large pipes. Davis (1999) performed a parametric study of underground cavities subjected to elastic plane SV waves and showed that the dynamic response of pipes up to 10 m diameter can be evaluated using a simplified static solution.

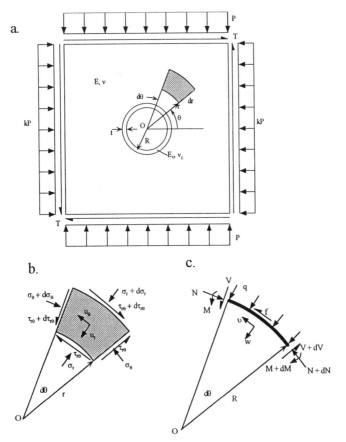

Figure 1. Elastic model: (a) load conditions and geometry; (b) Element of elastic medium; (c) Shell element.

An elastic-static analytical solution is presented herein to investigate the soil-pipe interaction for the combined effects of applied vertical, horizontal, and shearing stresses. This report presents preliminary results of a work in progress (Davis, 1999). The objective of this investigation is to determine how the magnitude and location of maximum pipe stress changes with an applied shear stress, and the effects of slippage along the cylinder-medium interface. The analytical solution is applied to case studies of flexible pipes shaken by the 1994 Northridge earthquake.

Model

Historically, evaluation of the interaction between buried pipes and the surrounding soil medium has been limited to the case of applied vertical and horizontal loads. Marston developed a limit equilibrium analysis method (Moser, 1990); however, this method is not easily adapted to the evaluation of shear distortions. Closed form elastic solutions (e.g. Burns and Richard, 1964; Höeg, 1968) can be easily modified to evaluate the effects of an applied shearing stress and will be used in this investigation.

Figure 1 shows the two-dimensional model used for this analysis, which is similar to that used by Burns and Richard (1964) and Höeg (1968). The full space is made of a massless, elastic, isotropic, homogeneous medium and is uniform except for the circular cylindrical section that is removed to form a cavity. The cavity is lined with an elastic, isotropic, and homogeneous cylindrical shell of thickness t and mid-surface radius R. The elastic soil medium and pipe shell are defined by the r - θ coordinate system with the origin located at the cylinder center axis O. The medium properties are determined by the elastic modulus E and Poisson's ratio v and the cylinder properties are determined by the elastic modulus E_c and Poisson's ratio v_c.

Figure 1a shows the cylinder subjected to symmetric loading consisting of a uniform vertical pressure P and a uniform horizontal pressure kP. The cylinder is also subjected to an anti-symmetric shear stress T applied along the same boundaries as P and kP. For the mathematical model, the loading boundaries are assumed to be far from the cylinder-medium interface (i.e. a distance approximately $2R$ from the interface (Höeg, 1968)).

Solution to the Problem

The stress and displacements in the cylinder and medium can be determined directly from the results of Burns and Richard (1964) or Höeg (1968) by using an axes transformation. Using Höeg's (1968) solution for the no slip and complete slip boundary conditions at the interface $r = R$ shown in Table 1 gives the medium radial stress σ_r, tangential stress σ_θ, and shear stress $\tau_{r\theta}$, equations:

$$\sigma_r = \frac{P}{2}(1+k)\left[1 - A_o\left(\frac{R}{r}\right)^2\right]$$

$$-\left\{\frac{P}{2}(1-k)\left[1 - 3A_2\left(\frac{R}{r}\right)^4 - 4B_2\left(\frac{R}{r}\right)^2\right] - T\left[3C_2\left(\frac{R}{r}\right)^4 + 4D_2\left(\frac{R}{r}\right)^2\right]\right\}\cos 2\theta \qquad (1a)$$

$$-\left\{T\left[1 - 3A_2\left(\frac{R}{r}\right)^4 - 4B_2\left(\frac{R}{r}\right)^2\right] - \frac{P}{2}(1-k)\left[3C_2\left(\frac{R}{r}\right)^4 + 4D_2\left(\frac{R}{r}\right)^2\right]\right\}\sin 2\theta$$

$$\sigma_\theta = \frac{P}{2}(1+k)\left[1+A_0\left(\frac{R}{r}\right)^2\right] + \left\{\frac{P}{2}(1-k)\left[1-3A_2\left(\frac{R}{r}\right)^4\right] - 3TC_2\left(\frac{R}{r}\right)^4\right\}\cos 2\theta$$

$$+ \left\{T\left[1-3A_2\left(\frac{R}{r}\right)^4\right] - 3\frac{P}{2}(1-k)C_2\left(\frac{R}{r}\right)^4\right\}\sin 2\theta \qquad (1b)$$

$$\tau_{r\theta} = \left\{\frac{P}{2}(1-k)\left[1+3A_2\left(\frac{R}{r}\right)^4 + 2B_2\left(\frac{R}{r}\right)^2\right] + T\left[3C_2\left(\frac{R}{r}\right)^4 + 2D_2\left(\frac{R}{r}\right)^2\right]\right\}\sin 2\theta$$

$$- \left\{T\left[1+3A_2\left(\frac{R}{r}\right)^4 + 2B_2\left(\frac{R}{r}\right)^2\right] + \frac{P}{2}(1-k)\left[3C_2\left(\frac{R}{r}\right)^4 + 2D_2\left(\frac{R}{r}\right)^2\right]\right\} \qquad (1c)$$

Table 1. Boundary conditions at the interface $r = R$ used for elastic models.

Boundary condition	Radial stress	Shear stress	Radial displacement	Tangential displacement
No slip	$\sigma_r = q(\theta)$	$\tau_{r\theta} = f(\theta)$	$u_r = w(\theta)$	$u_\theta = \upsilon(\theta)$
Complete slip	$\sigma_r = q(\theta)$	$\tau_{r\theta} = 0$ $f(\theta) = 0$	$u_r = w(\theta)$	
Frictional	$\sigma_r = q(\theta)$	$\tau_{r\theta} = \pm\mu\sigma_r$ $f(\theta) = \pm\mu\sigma_r$	$u_r = w(\theta)$	

The uniform coefficient A_o in Eq. 1 is applicable to all boundary conditions:

$$A_o = \frac{(1-2\nu)(C-1)}{(1-2\nu)C+1} \qquad (2)$$

The variable coefficients for the no slip boundary condition are:

$$A_2 = \frac{F(1-2\nu)(1-C) - \frac{1}{2}C(1-2\nu)^2 + 2}{F[3-2\nu+C(1-2\nu)] + C\left(\frac{5}{2} - 8\nu + 6\nu^2\right) + 6 - 8\nu} \qquad (3a)$$

$$B_2 = \frac{F[1+C(1-2\nu)] - \frac{1}{2}C(1-2\nu) - 2}{F[3-2\nu+C(1-2\nu)] + C\left(\frac{5}{2} - 8\nu + 6\nu^2\right) + 6 - 8\nu} \qquad (3b)$$

and for the complete slippage interface condition:

$$A_2 = -\frac{2F+1-2v}{2F+5-6v}, \quad B_2 = \frac{2F-1}{2F+5-6v} \tag{4}$$

The coefficients $C_2 = D_2 = 0$ apply to both boundary conditions. The coefficients in Eqns. 3 and 4 are described in terms of the relative stiffness of the medium to the cylinder by use of the compressibility ratio C and flexibility ratio F:

$$C = \frac{ER(1-v_c^2)}{E_c A(1+v)(1-v)}, \quad F = \frac{ER^3(1-v_c^2)}{6E_c I(1+v)} \tag{5}$$

where A and I are the cross-sectional area and moment of inertia of the pipe wall.

Effects of Applied Shear Stress T

Figure 2 shows how σ_r varies with flexural stiffness at $\theta = 45°$ for given values of T/P using $t/R = 0.075$. Figure 2, and all other results in this report, are presented for values of $k = 0.4$, $v = 0.35$, and $v_c = 0.3$.

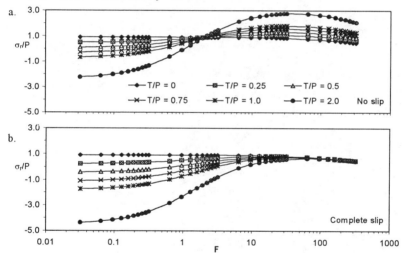

Figure 2. Variation of normalized radial stress σ_r/P, at the cylinder – medium interface, with flexural stiffness F for normalized applied shear stress T/P: (a) no slip boundary condition; (b) complete slip boundary condition.

Figures 3 and 4 show the variation of σ_r, σ_θ, and $\tau_{r\theta}$ with angle θ for $F=11$, $C=0.045$ and different levels of T/P for the no slip and complete slip interface conditions, respectively. The maximum interface stress location shifts as a function of the magnitude of T. Although not shown, the maximum displacements, hoop force, and moments are shifted similarly. The shift is approximately 40° at $T/P = 2.0$ and reaches a maximum of 45° at $T/P \approx 10$ to 20, indicating the cylinder is responding mainly to shear deformation when $T \geq 10P$. Figures 2 to 4 show that σ_r and σ_θ may develop tensile stress (i.e. negative

σ_r and σ_θ) along significant portions of the pipe, depending on the boundary conditions, magnitude of T, and the relative stiffness of the pipe and soil.

Evaluation of Slippage at the Soil-Pipe Interpace

Figure 5 shows how the friction ratio $f_r = \tau_{r\theta}/\sigma_r$ varies with the angle θ over a range of T using $t/R = 0.05$ with the no slip boundary condition. Defining μ as the coefficient of friction between the medium and cylinder and assuming no adhesive forces, in the regions where $|f_r| > |\mu|$ the medium will slip across the cylinder surface. The stress distribution shown in Fig. 5 indicates that the boundary conditions in Table 1 are not uniformly applied around the entire cylinder perimeter. The regions where $|f_r| > |\mu|$ decreases with increasing k.

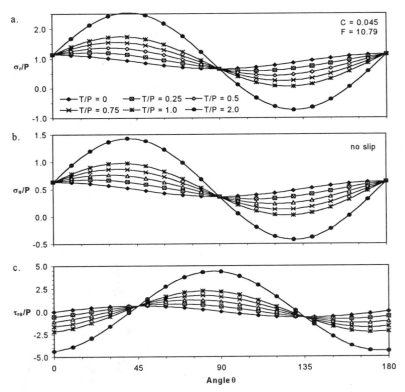

Figure 3. Variation of stress at the cylinder – medium interface with angle θ and normalized applied shear stress T/P for the no slip boundary condition: (a) normalized radial stress σ_r/P; (b) normalized tangential stress σ_θ/P; (c) normalized shear stress $\tau_{r\theta}/P$.

GEO-ENGINEERING FOR UNDERGROUND FACILITIES 489

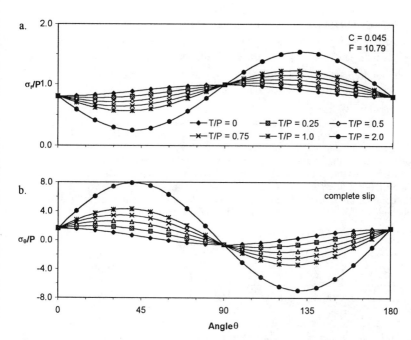

Figure 4. Variation of stress at the cylinder – medium interface, with angle θ an normalized applied shear stress T/P for the complete slip boundary condition; (a normalized radial stress σ_r/P; (b) normalized tangential stress σ_θ/P.

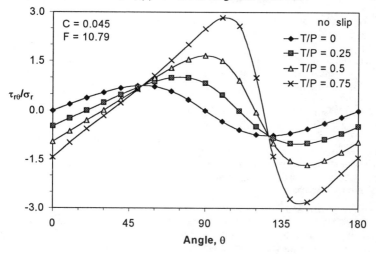

Figure 5. Variation of friction ratio f_r, at the cylinder – medium interface with angle θ.

O'Rourke (1996) reported the coefficient of friction range $0.2 \leq \mu \leq 0.8$. Figure 5 shows that the shear loading T greatly increases the interface shear stress, which can lead to the slippage condition $|f_r| > |\mu|$ around significant portions of a pipe. Therefore, a complete understanding of buried pipe behavior can not be assessed without an investigation of the interface slippage and resulting frictional shear stresses.

Höeg (1968) recognized and described how partial interface slippage develops. However, a thorough literature review did not uncover any information describing how a frictional interface shear stress influences pipe performance. As a result, an investigation was carried out as part of this study to assess the effects of the frictional interface shear stress $\tau_{r\theta} = \pm\mu\sigma_r$ that exists once slippage has initiated. The elastic coefficients were solved using the same procedure outlined by Burns and Richard (1964) and Höeg (1968) with the frictional interface boundary conditions shown in Table 1. The resulting medium stress equations are represented by Eq. 1. The uniform coefficient A_0 is defined by Eq. 2. The variable coefficients for the frictional interface condition are:

$$A_2 = -\frac{[2F + 5 - 6v\,][2F + 1 - 2v] + \mu^2[2(2-v) - F\,][2(2-3v) + F]}{[2F + 5 - 6v]^2 - \mu^2[F + 2(2-3v)]^2} \quad (6a)$$

$$B_2 = \frac{[2F + 5 - 6v\,][2F - 1] + \mu^2[2 - F\,][2(2-3v) + F]}{[2F + 5 - 6v]^2 - \mu^2[F + 2(2-3v)]^2} \quad (6b)$$

$$C_2 = -\frac{12\mu[F + 2(1-v)](1-v)}{[2F + 5 - 6v]^2 - \mu^2[F + 2(2-3v)]^2} \quad (6c)$$

$$D_2 = \frac{6\mu[F + 1](1-v)}{[2F + 5 - 6v]^2 - \mu^2[F + 2(2-3v)]^2} \quad (6d)$$

The coefficients in Eq. 6 were solved using $+\mu$, however, the alternate condition of $-\mu$ must also be satisfied. To evaluate the effects of the interface frictional shear stress, the range of accuracy of the solution must first be established, then compared with results of the complete slip solution. Figure 6a compares the radial stress frictional solution using $+\mu$ and $-\mu$. For purposes of this comparison an arbitrary accuracy of 5% was selected. As seen in Fig. 6a, the frictional solution accuracy increases with increasing F and varies with the magnitude of T. The frictional solution is inaccurate for $F<7$ for all values of T. Figure 6b compares the frictional and complete slip results for σ_r. Figure 6 shows that for very flexible pipes (i.e. $F>100$) the shear stress resulting from friction at the interface may be neglected. In addition, within the accuracy of the frictional solution for $7 \leq F \leq 100$, the complete slip solution provides similar results. It is therefore concluded that the frictional shear stresses developed at the soil-pipe interface during slippage does not significantly effect the analysis results, and where slippage occurs the equilibrium can be adequately defined by $\mu=0$.

Höeg (1968) showed how slippage induces soil yielding. Interface slippage and yielding reduces the amount of stress transmitted to the cylinder. Thus, the true behavior of pipes is complicated by the fact that the actual interface conditions are a combination of slip and no slip. The degree to which boundary condition has the greatest influence is dependent upon the magnitude of T and P, relative stiffness F and C, and lateral stress coefficient k. Describing in terms of the no slip solution, the presence of interface slippage reduces the radial and shear stresses in the vicinity of the springline ($\theta=0$). Radial displacements increase at the crown and decrease at springline. As a result, slippage and soil yielding effectively dampen and reduce the variable stress and displacement components (i.e. the sin 2θ and cos 2θ terms). This effect can be accounted for by modifying the no slip boundary solution to allow for damping of the variable terms and has been idealized for the design of flexible pipes by allowing k→1 and C→1 (i.e. assume soil and pipe stiffness are equal, Moser, 1990).

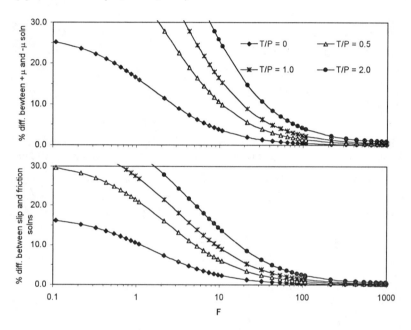

Figure 6. Comparison of results for the frictional interface solution; (a) Percent difference in solution between +μ and -μ as a function of flexibility ratio F; (b) Percent difference between frictional solution (+μ) and no slip solution.

Case Study

The January 17, 1994, Northridge Earthquake subjected four corrugated metal pipes located at the Lower San Fernando Dam (LSFD) to large shear distortions resulting from strong near-source seismic shaking. Table 2 summarizes the properties and performance of the four pipes. Two 2.4 m diameter pipes (Pipes 1 and 4) sustained damage while two 1.5 m diameter pipes (Pipes 2 and 3) withstood the shaking without any damage. Figure 7 shows the damaged sections to Pipes 1 and 4. As seen in Fig. 7a, Pipe 1 sustained complete lateral collapse while Fig. 7b shows that Pipe 4 deformed up to 14 cm. Pipe 4 also suffered some axial damage. The construction history and seismic performance of Pipes 1 through 4 has been described in detail in Davis and Bardet (1996, 1998, and 1999) and will not be repeated here.

Ground motions were well recorded in the LSFD vicinity during the Northridge earthquake. Davis and Bardet (1999) evaluated the strong ground motions and pipe orientations and determined that the horizontal peak particle ground velocity (pgv) oriented transverse to the pipe longitudinal axis is the most important ground motion parameter influencing buried pipe performance. Table 2 presents the range of pgv the four pipes experienced during the 1994 earthquake. The transverse ground shear strain γ can be estimated from:

$$\gamma = \frac{pgv}{C_s} \tag{8}$$

where C_s is the average shear wave velocity within the top 30 m of the ground surface. The applied vertical pressure is $P=\rho gH$ and the normalized applied shear stress is evaluated from:

$$\frac{T}{P} = \frac{pgvC_s}{Hg} \tag{9}$$

Where ρ is the soil density, g is the acceleration of gravity, and H is the depth of pipe. The maximum hoop force N_{max} can be determined from T and P and then compared to the critical hoop force N_{cr} necessary to cause failure. The critical hoop force to cause buckling (Moore, 1989) can be evaluated with units of force per unit length using:

$$N_{cr} = 0.66(E_c I)^{1/3} \left(\frac{E}{1-v^2}\right)^{2/3} \tag{10}$$

from which the factor of safety FS against buckling is determined:

$$FS = N_{cr}/N \tag{11}$$

All four pipes are buried to a depth of at least $2R$ and can therefore be evaluated using the elastic solutions presented herein. Table 2 shows the results for the LSFD pipes using no slip and complete slip interface solutions. Scattering of seismic waves off the LSFD pipes is not significant and will not be considered in this evaluation (Davis, 1999). Calculations were performed for Table 2 without modifying the solution of Eqn. 1 using values of E, E_c, v, v_c, k, and ρg of 12.4 Mpa, 200 Gpa, 0.35, 0.3, 0.4, and 19.6 kN/m³, respectively. Table 2 also shows the variable units. The results show that this method of analysis adequately bounds the observed pipe behavior. Assuming no slippage, all four pipes were subjected to stresses near or above the stability limit. Table 2 indicates all pipes have an adequate FS against failure using the complete slip solution. Davis and Bardet (1998; 1999) showed that soil stiffness reduction and pore pressure buildup during seismic shaking reduces E and N_{cr}. The analysis presented herein also indicates that a reduction in E decreases F and increases N_{max}. Thus, modulus reduction from seismic shaking can have a compounding effect on pipe stability, and may be a primary factor in the performance of Pipe 1. The combination of axial and lateral stress in Pipe 4 would reduce the buckling resistance below that of Eq. 10 and may help explain the damage.

A modified analysis was also performed to evaluate the effect of the stress redistribution that results in flexible pipes from interface slippage and soil yielding prior to the earthquake. An effective lateral stress ratio of $k_{eff} = 1$ was used. The results using the complete slip solution predicted buckling failure of all four pipes during the earthquake.

This analysis shows that the relation between pipe stiffness and interface slippage before and during the earthquake may be a significant parameter for understanding the seismic response of buried pipes.

This case study shows that the practical evaluation of buried pipes in the presence of shearing stress is very complicated. Nevertheless, the solutions presented herein provide reasonable bounding values for assessing the observed seismic performance of Pipes 1 through 4.

Table 2. Characteristics of corrugated metal pipes at the Lower San Fernando Dam.

Pipe	R	t	A	I	C	F	H	pgv (cm/s)		C_s	T/P	FS	
	(m)	(mm)	(x10⁻³ m²/m)	(x10⁻⁸ m⁴/m)			(m)˙	min	max	(m/s)		no slip	full slip
1	1.22	4.27	4.515	9.384	0.037	128	11.9	59	97	500	6.5-11	0.94-0.6	7.7-7.3
2	0.76	3.51	3.692	7.430	0.029	41	2.4	103	103	450	17	0.96	12
3	0.76	3.51	3.692	7.430	0.029	41	5.5	43	130	500	4-12	1.7-0.65	8.6-6.4
4	1.22	4.27	4.515	9.384	0.037	128	2.1	86	86	350	13	1.1	15

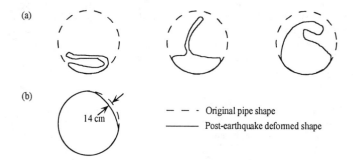

– – – Original pipe shape
——— Post-earthquake deformed shape

Figure 7. Seismic induced deformations of corrugated metal pipes at the Lower San Fernando Dam during the 1994 Northridge Earthquake: (a) Pipe 1; (b) Pipe 4.

Conclusion

The performance of buried pipes subjected to shear distortions has been investigated herein using an analytical elastic closed form solution. The investigation showed that slippage at the soil-pipe interface is an important parameter in understanding the pipe performance. A mathematical solution derived to evaluate the effects of a frictional interface shear stress after slippage has initiated showed that this shear stress does not significantly influence pipe performance and can be neglected. Shear distortions cause tensile tangential stresses to result in the soil medium around the pipe for all pipe stiffness' of interest. Soil yielding resulting from the tensile stresses can significantly effect pipe performance. An evaluation of four corrugated metal pipes that were subjected to large shear distortions during the 1994 Northridge earthquake showed that the complete slippage and no slippage elastic solutions can be used to adequately evaluate the range of forces that a buried pipe may experience.

Acknowledgements

Support from the Los Angeles Department of Water and Power and Prof. J.P. Bardet from the University of Southern California is greatly acknowledged. Prof. Bardet provided valuable comments on the frictional interface solution.

References

Burns, J. Q., and R. M. Richard, 1964, "Attenuation of Stress for Buried Cylinders," *Proc. Symp. Soil Struct. Interaction*, Univ. of Arizona, Tucson, AZ, pp. 378-392.
Datta, S. K., A. V. Shah, and K. C. Wong, 1984, "Dynamic Stress and Displacements in Buried Pipe," *Jour. of Engr. Mech.*, ASCE, Vol. 110, EM10, pp. 1451-1466.
Davis, C. A., 1999, "Study of Near Source Earthquake Effects on Flexible Buried Pipes," Draft thesis to be presented to the University of Southern California in partial fulfillment of the requirements for the degree of Doctor of Philosophy.
Davis, C.A., and J.P. Bardet, 1996b, "Performance of Four Corrugated Metal Pipes during the 1994 Northridge Earthquake" *Proc. 6th Japan-US Workshop on Earthquake Res. Dsgn. Lifeline Fac. and Countermeasures against Soil Liq.*, Tokyo, June, pp. 77-93.
Davis, C.A., and J.P. Bardet, 1998, "Seismic Analysis of Large Diameter Flexible Underground Pipes" *Jour. Geot.& Geoenv. Eng. Div.*, ASCE, 124, 10, pp. 1005-1015.
Davis, C. A., and J.P. Bardet, 1999, "Case History and Analysis of Responses of Buried Corrugated Metal Pipes to Earthquakes," Sub. *Jour. Geo. & Geoenv. Eng. Div.* ASCE.
Höeg, K., 1968, "Stresses Against Underground Structural Cylinders," *Journal of the Soil Mechanics and Foundations Division*, ASCE, Vol. 94, SM4, pp. 833-858.
Lee, V. W., and J. Karl, 1992, "Diffraction of SV Waves by Underground, Circular, Cylindrical Cavities," *Soil Dynamics and Earthquake Engineering*, Vol. 11, pp. 445-456.
Luco, J.E. and F. C. P. De Barros, 1994, "Dynamic Displacements and Stresses in the Vicinity of a Cylindrical Cavity Embedded in a Half-Space," *Earthquake Engineering and Structural Dynamics*, Vol. 23, pp. 321-340.
Moore, I. D., 1989, "Elastic Buckling of Buried Flexible Tubes – A review of Theory and Experiment," *Jour. Geot. Eng. Div.*, ASCE, Vol. 115, GT3, pp. 340-358.
Moser, A.P., 1990, "Buried Pipe Design," *McGraw Hill*, New York.
O'Rourke, T. D., 1996, "Lessons Learned for Lifeline Engineering from Major Urban Earthquakes," *11th World Conf. Earthquake Engineering*, June 23-28, Acapulco, Mexico.
Subcommittee on Water and Sewage Systems, 1973, "Earthquake Damage to Water and Sewerage Facilities," U.S. Dept. of Commerce, *NOAA*, Vol. II, pp. 75-193.
Tohda, J., H. Yoshimura, and L. Li, 1996, "Characteristic Features of Damage to the Public Sewerage Systems in the Hanshin Area," *Soils and Foundations Special issue*, *Japanese Geotechnical Society*, pp. 335-347.
Wong, K.C., A. H. Shah, and S. K. Datta, 1985, "Diffraction of Elastic Waves in a Half-Space. II. Analytical and Numerical Solutions," *Bull. Seism. Soc. Am.*, 75, pp. 69-92.

Determination of Vertical Loading on Underground
Structures Based on an Arching Evolution Concept

by Geraldo R. Iglesia[1], Herbert H. Einstein[2], and Robert V. Whitman[3]

Abstract

By combining experimental data from centrifuge trapdoor tests with some theories on load redistribution due to arching, a novel approach for determining the vertical loading on underground structures in granular soils has been developed. This approach eventually leads to the main result of a ground reaction curve, which is a plot of support loading versus displacement, for each type of underground conditions considered. The methodology presented here is validated against available field data found in the literature.

Introduction

For designers of tunnels and other underground structures, the *ground reaction curve* (GRC) is quite an invaluable tool. The GRC has been called several other names, such as *characteristic curve* (e.g., Einstein and Schwartz, 1979), *ground convergence curve* (e.g., Wong and Kaiser, 1991), or *ground response curve* (e.g., Brown et al., 1983), but they all characterize the variation of support pressure as the surrounding ground tends to displace or converge toward the underground opening.

The emergence of the GRC concept as a design aid has long been preceded by the recognition of the arching phenomenon in soils. When an inclusion is present within a soil mass, arching occurs due to the difference in stiffness between the inclusion and the soil. A recent study (Iglesia et al., 1999) involving centrifuge experiments with the classic trapdoor setup has provided fresh insight into the soil arching phenomenon.

---

[1]Consulting Engineer, 11580 Kirby Place, San Diego, CA 92126
[2]Professor of Civil and Environmental Engineering, Massachusetts Institute of Technology, 77 Massachusetts Avenue, Cambridge, MA 02139
[3]Professor *Emeritus* of Civil and Environmental Engineering, Massachusetts Institute of Technology, 77 Massachusetts Avenue, Cambridge, MA 02139

Centrifuge trapdoor tests with granular media show that an actual arch forms as the vertical displacement at the top of an underground structure is simulated by the downward movement of the trapdoor (active arching mode). As the trapdoor is gradually lowered, the arch evolves from an initially curved shape to a triangular one, before ultimately collapsing with the appearance of a prismatic sliding mass bounded by two vertical shear planes emanating from the sides of the trapdoor. This evolution of the arch is illustrated in Figure 1.

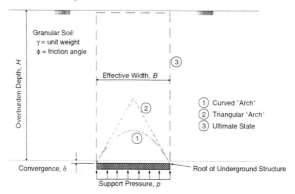

Figure 1. Arching Evolution

The evolution of the arch depicted above provides the basis for estimating the vertical loading on an underground structure. Indeed, a methodology has been proposed (Iglesia et al., 1999) not only for determining the vertical loading on the structure, but also for generating the ground reaction curve for the overlying soil. This paper presents the proposed scheme and then evaluates its applicability to actual cases by comparing generated GRC's against field data documented in the literature.

Proposed Methodology

In the proposed scheme, the GRC is constructed in the dimensionless space of normalized loading (p^*) vs. normalized displacement (δ^*). The support pressure (p) is normalized with respect to the geostatic (free-field) stress (p_o) at the crown level to obtain the normalized loading ($p^* = p/p_o$), while the converging displacement (δ) is normalized with respect to the effective width (B) of the structure to arrive at the normalized displacement ($\delta^* = \delta/B$). (Refer to Figure 1 for the basic notation.) A generalized GRC such as shown in Figure 2 consists of four main segments -- the initial arching phase, the maximum arching (minimum loading) condition, the loading recovery stage, and the ultimate state -- which will now be discussed.

Initial Arching. -- Starting from a geostatic condition, the initial convergence of the soil toward the underground opening causes a rather abrupt drop in the load on the crown of the structure. The loading on the structure decreases rather abruptly because only a relatively small portion of the overlying material tends to follow the initial convergence toward the underground opening; i.e., the "arch" starts to form. Inasmuch as the reduced loading is mainly attributed to (active) arching, the rate of initial stress decrease (slope of the approximately straight line in Figure 2) in the normalized loading vs. normalized displacement plot has been called the *modulus of arching*, M_A. Based on the centrifuge trapdoor experiments with granular media, the modulus of arching has a value of about 125 (Iglesia et al., 1999). In other words, at small relative displacements, for each incremental convergence of 0.1% of the effective width of the underground structure, the change in vertical stress on the structure is about one-eighth (⅛) of the geostatic (free-field) vertical stress.

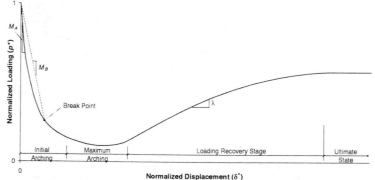

Figure 2. Generalized Ground Reaction Curve

Break Point. -- As the underground opening converges toward a state of maximum arching (minimum loading), the GRC starts to deviate from its initially linear trend and tends to curve around a bend. The point on this bend at which the curvature approaches a relative maximum is referred to as the *break point* (see Figure 2).

A parameter called the *relative arching ratio*, A_R, is defined as the ratio of the vertical stress reduction at a particular stage of initial convergence to the maximum vertical stress reduction that could occur due to active arching. Assuming that the minimum vertical stress (p_m) on the structure can be determined, this parameter can be expressed in equation form as:

$$A_R = \frac{p_o - p}{p_o - p_m} = \frac{1 - (p/p_o)}{1 - (p_m/p_o)} \qquad (1)$$

where p is the vertical stress on the structure at a particular stage of convergence, and p_o is the geostatic (free-field) vertical stress at the level of the top of the structure. A_R goes from 0 to 1 as the vertical stress drops from the geostatic to the minimum value.

The relative arching ratio at the break point appears to be a function of the overburden depth aspect ratio, which is the ratio of the overburden depth (H) to the width of the structure (B). This function has been empirically derived from the centrifuge experiments (Iglesia et al., 1999) as:

$$A_{RB} = 1 - 0.41 e^{-0.54(H/B)} \qquad (2)$$

Moreover, at the break point, the *secant modulus of active arching*, M_B, which is the slope of the line from the geostatic intercept ($p/p_o = 1$ at $\delta = 0$) to the break point (as in Figure 2), has been deduced from the centrifuge data to be about 63.

Maximum Arching. -- The vertical loading on the underground structure approaches a minimum value when a state of maximum arching is reached. This corresponds to a condition in which a physical arch, somewhat parabolic in shape, forms just above the underground structure. This tends to occur when the relative displacement between the underground structure and the surrounding soil is about 2 to 6 percent of the effective width of the structure.

Based on salient aspects of the arching theories by Engesser (1882), Bierbaumer (1913), and Evans (1983), it is possible to quantify the minimum loading on an underground structure. As a starting point, the shape of the stable physical arch is presumed to be a parabola, following the suggestion by Engesser (1882). Let θ be the inclination, with respect to the horizontal, of the tangent line through each end of the parabolic arch spanning the width B of the structure (Figure 3). While Engesser equates θ to the friction angle ϕ of the soil, it is assumed here that θ is the complementary angle of ϕ, i.e., $\theta = 90° - \phi$, or that each end of the parabola makes an angle ϕ with the vertical. This assumption corresponds to the orientation prescribed by Bierbaumer and Evans, although their "arch" is triangular in shape. Applying concepts of plasticity theory, Evans (1983) proposes that a "slip line" develops at each edge of the settling trapdoor at a dilation angle v with respect to the (vertical) direction of straining; this dilation angle v approaches the friction angle ϕ, corresponding to the associated flow rule. Observations of a relatively tall arch (at the stages of relatively small displacement) in various trapdoor experiments [e.g., Evans (1983), Stone (1988), and Iglesia (1991)] tend to support the assumed (parabolic) arch configuration.

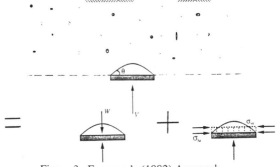

Figure 3. Engesser's (1882) Approach

As the roof of the underground structure displaces with respect to the adjacent substratum, the material below the arch tends to separate from the rest of the soil. Consequently, the pre-existing vertical load at the top of the arch gets partially transferred to the sides. This load redistribution also brings about an increase in the lateral stress across the bottom of the arch. As illustrated in Figure 3, the effective vertical load V acting on the structure consists, then, of the weight W of the soil below the arch and the contribution of the vertical stress σ_{vr} induced by the intensified lateral stress σ_{hr} just above the structure.

The weight W per unit length of structure can be computed from the cross-sectional area of the material below the arch multiplied by the specific weight of the soil. For a parabolic arch with an inclination angle θ at the sides with respect to the horizontal, this results in:

$$W = \frac{\gamma B^2 \tan\theta}{6} = \frac{\gamma B^2 \tan(90°-\phi)}{6} = \frac{\gamma B^2 \cot\phi}{6} \qquad (3)$$

Following Engesser's (1882) approach, the effect of the additional vertical stress σ_{vr} brought about by the transfer of soil pressure to the sides can be quantified by considering an imaginary structural arch, of thickness dh, whose shape is defined by the same parabola as above, and uniformly loaded over its entire span, shown in Figure 4. The horizontal normal stress σ_{hr} at the bottom of the arch (Figure 3) is then the lateral thrust reaction dF_h divided by dh (Figure 4). The underlying assumption is that the stresses at the ends of the arch are the same in the imaginary structure as in the actual case.

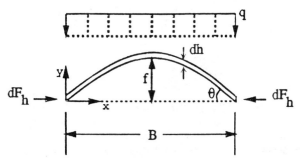

Figure 4. Imaginary Structural Arch in Engesser's (1882) Analysis

Assuming that σ_{hr} is constant across the bottom of the arch and that the arch itself represents a failure surface within the soil mass, the vertical normal stress σ_{vr} can be written as:

$$\sigma_{vr} = K_E \sigma_{hr} \qquad (4)$$

where

$$K_E = \frac{\cos^2\phi}{1 + \sin^2\phi} \qquad (5)$$

The foregoing expression can be derived by considering a Mohr diagram, such as shown in Figure 5. That the arch is a failure surface implies that the state of stress on that surface corresponds to the point on the Mohr circle tangent to the failure envelope, i.e., the point of maximum obliquity.

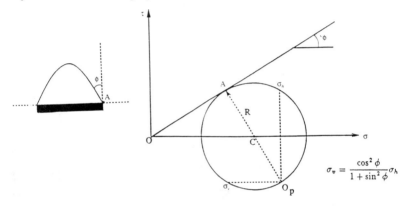

Figure 5. Mohr Diagram Assuming Failure Stress State at Edge of Structure

The equivalent uniform loading on the imaginary arch has been intuitively expressed by Engesser (1882) as

$$q = dh \cdot (\gamma - \frac{\sigma_{vr}}{H}) \qquad (6)$$

in which γ is the specific weight of the granular medium and H is the overburden depth. In essence, the vertical stress redistributed to the sides of the structure is approximated as the geostatic stress (γH) minus the vertical normal stress (σ_{vr}) at the level of the top of the structure (see Figure 3); the redistributed vertical stress is then divided by the overburden depth (H) to obtain a reasonable estimate of an equivalent unit weight of the imaginary structural arch.

For a parabolic arch of span B forming an angle θ with the horizontal at the ends (Figure 4), the lateral thrust dF_h due to the uniform loading q can be shown to be [see, e.g., Leontovich (1959)]

$$dF_h = \frac{qB^2}{8f} = \frac{qB}{2\tan\theta} = \frac{qB}{2\cot\phi} \qquad (7)$$

where $f = B\tan\theta/4$ is the rise of the arch. From Equations 6 and 7, the lateral stress σ_{hr} is, thus,

$$\sigma_{hr} = \frac{dF_h}{dh} = \frac{B}{2\cot\phi}(\gamma - \frac{\sigma_{vr}}{H}) \qquad (8)$$

Substituting σ_{hr} from Equation 8 into Equation 4 and simplifying, one arrives at

$$\sigma_{vr} = \frac{HB\gamma K_E}{2H\cot\phi + BK_E} \qquad (9)$$

The effective load V acting on the underground structure is the sum of the force from this vertical stress σ_{vr} and the weight W of the volume of soil underneath the arch (Equation 3). Hence,

$$V = \gamma B^2[\frac{HK_E}{2H\cot\phi + BK_E} + \frac{\cot\phi}{6}] \qquad (10)$$

In terms of average stress, the expression becomes

$$p_m = \gamma B[\frac{HK_E}{2H\cot\phi + BK_E} + \frac{\cot\phi}{6}] \qquad (11)$$

Normalizing this with respect to the geostatic stress $p_o = \gamma H$, one obtains

$$\frac{p_m}{p_o} = \frac{B}{H}[\frac{K_E}{2\cot\phi + \frac{B}{H}K_E} + \frac{\cot\phi}{6}] \qquad (12)$$

A similar approach can be applied to a triangular arch configuration, to try to model the evolving deformation pattern. The resulting expression leads to normalized loading values that are only slightly higher than those obtained from Equation 12 (Iglesia et al., 1999). This might provide a clue as to why the load tends to stay at a low level with continued displacement as the curved arch gradually transforms into a triangular shape.

Loading Recovery Stage. -- This stage refers to the transition from the maximum arching (minimum loading) condition to the ultimate state. Idealizing this transition to be approximately linear, the rate of increase in normalized loading for each increment of normalized displacement (slope of the GRC loading recovery line as in Figure 2) has been called the *load recovery index*, λ.

Based on the centrifuge trapdoor experiments, the load recovery index increases with increasing structure-to-grain ratio, B/D_{50} (where D_{50} is the average grain size), and tends to decrease with greater H/B ratios. The following empirically derived relation may be used to estimate the load recovery index (Iglesia et al., 1999):

$$\lambda = [2.5 + 5.7\log(\frac{B}{10D_{50}})]e^{-0.65(H/B)} \qquad (13)$$

Ultimate State. -- As the surrounding soil continually converges toward the underground opening, the arch will eventually collapse, and a prismatic sliding mass, bounded by a pair of vertical shear planes emanating from the sides of the inclusion, will emerge. This classic case has been the subject of numerous theoretical models, which are generally based on the silo theory originated by Janssen (1895). The derivation involves an infinitesimal layer of soil subjected to its own self-weight, forces above and below the layer, and lateral normal and shearing stresses. The resulting differential equation is then solved, and appropriate boundary conditions are imposed. For cohesionless soils with no surcharge, the formula for the reduced loading has the form

$$\frac{p}{p_o} = \frac{B}{2HK\tan\phi}[1 - e^{-2K\tan\phi(H/B)}] \qquad (14)$$

The main source of disagreement among the various theories lies in the value of K. The active lateral earth pressure coefficient K_a [$= (1-\sin\phi)/(1+\sin\phi)$] is commonly used, although this inherently assumes zero shear stress along the vertical planes. Terzaghi (1943) treats K as an empirical constant and tends to favor a value of unity for practical applications. Perhaps the most theoretically sound suggestion comes from Krynine (1945), who has noted that, if the state of stress along the vertical planes falls on the failure envelope in a Mohr diagram, the value of K should be $\cos^2\phi/(1+\sin^2\phi)$. Krynine's recommended expression for the K coefficient has been endorsed by many investigators and is also proposed here.

Note that Krynine's K has the same magnitude as the K_E given by Equation 5. The latter, however, is a ratio of the vertical stress to the horizontal stress, unlike the former. Moreover, K_E is based on an inclined shearing surface, while Krynine assumes a vertical failure plane. Thus, even though their Mohr circle construction may look similar, their respective poles (or origins of planes) are not at the same location. [Recall that the pole (O_P) is a point on the Mohr circle with the following property: a line through O_P and any point A of the Mohr circle will be parallel to the plane on which the stresses given by point A act.] Since $K_E < 1$ for typical values of ϕ, the implication is: at the maximum arching condition, the horizontal stress σ_h is greater than the vertical stress σ_v; however, at the ultimate state, $\sigma_v > \sigma_h$.

Procedure for Constructing GRC. -- To recapitulate, the following are the proposed steps in creating a ground reaction curve:

1. From the geostatic intercept ($p/p_o = 1$ at $\delta = 0$), draw the initial straight line portion of the GRC by assuming a value of 125 for the modulus of arching.

2. Calculate the minimum normalized loading on the structure using Equation 12, based on soil properties, structural dimensions, and overburden depth. Using this calculated minimum value as the ordinate, sketch a flat bottom of the GRC for the following range of normalized displacements: $0.03 < \delta_m^* < 0.05$. (This corresponds to the middle half of the range of δ_m^* values observed in the centrifuge trapdoor tests.)

3. Based on the known H/B ratio and Equation 2, estimate the relative arching ratio, A_{RB}, at the break point. Compute the corresponding break point load level from Equation 1, then establish the break point using a secant modulus of active arching of about 63. Connect the initial straight line (from Step 1) to the flat bottom portion (from Step 2) with a smooth arc through the break point.

4. Determine the ultimate normalized loading on the inclusion using Equation 14 and Krynine's lateral pressure coefficient, $K = K_E$ (in magnitude). Then construct a horizontal line corresponding to this ultimate value at the tail end of the plot.

5. Estimate the load recovery index for the current H/B and B/D_{50} ratios, based on Equation 13. Using this load recovery index, draw a line from the bottom portion of the plot (obtained in Step 2) to the ultimate load level, maintaining continuity in slopes.

Comparison with Data from Case Histories

A couple of documented case histories in which GRC-type data were obtained are available in the literature. The case histories considered here are: the Kennedale Tunnel (Wong and Kaiser, 1991) and the Washington D.C. Metro (Hansmire and Cording, 1985). Both these projects involve tunnels with circular cross-sections. Since the methodology proposed here is based on experimentation and theoretical formulation with a flat-top structure, simplifying assumptions are made when the proposed scheme is applied to circular tunnels.

For load calculations, NAVFAC (1986) recommends using the diameter of the circular tunnel for the width B of the structure. While this seems appropriate for the ultimate state, it may not be for the minimum loading condition. When the initial arch forms above the tunnel, it tends to be localized over a narrower span. Thus, intuitively, for calculating the minimum normalized loading, the effective width to use in Equation 12 is assumed to be the radius of the tunnel. Also, the effective width for normalizing displacements is assumed to be the radius of the tunnel. When estimating the load recovery index (Equation 13), however, as a conservative simplification, the diameter of the tunnel is used as the width parameter.

Furthermore, in the comparisons presented below, the overburden depth H is defined to be the vertical distance between the ground surface and the crown of the tunnel. This is consistent with NAVFAC (1986) and seems to make sense when one tries to apply the equations derived for a flat-top structure to circular tunnels. This differs, however, from the available data from the two case histories considered here, in which the overburden depth is measured from the ground surface to the tunnel axis or springline. Thus, it should be noted that the data have been modified accordingly to conform with the convention adopted in this paper.

Kennedale Tunnel (Wong and Kaiser, 1991) -- This tunnel was constructed in Edmonton, Alberta, Canada to convey stormwater. It was driven by a Lovat tunnel boring machine through predominantly cohesionless soils with friction angles ranging from about 30° to 35°. The excavated diameter of the tunnel was 3.2 m over a total length of 1,670 m, with an average overburden cover to the tunnel roof of 12.1 m.

An extensive field instrumentation program was devised to monitor the relationship between the vertical settlements and the actual support pressures induced by tunnelling. Twenty-six centerline surface settlement points and five multi-point extensometers were installed to detect the vertical settlement profiles along the tunnel axis. In addition, the crown pressure was measured by calibrated segmental laggings in 16 sections along the tunnel length.

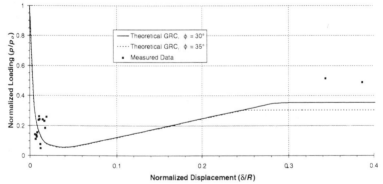

Figure 6. Comparison with Data from Kennedale Tunnel

The recorded lagging pressures (normalized by the geostatic stress at the crown) vs. surface settlements (normalized by the tunnel radius R) are shown in Figure 6. Generated GRC's based on the method proposed in this paper corresponding to friction angles of 30° and 35° (assuming an average grain size, D_{50}, on the order of 2.5 mm) are superposed in Figure 6. It is remarkable that the minimum loading calculated using Equation 12 matches the recorded data quite well. Ultimate load predictions are about 30 to 50% too low; however, it has been reported that, at the two locations where high lagging pressures were recorded, roof collapse was initiated because poor ground control was exercised. (Tunnelling with poor ground control could loosen the surrounding soil, leading to a lower effective friction angle and potentially higher loads on the underground structure.) Also, the comparison is limited since the GRC δ values refer to displacements at the tunnel crown, while the reported values are the observed surface displacements.

Washington D.C. Metro (Hansmire and Cording, 1985) -- Extensive field monitoring programs were undertaken during the construction of the Washington D.C. rapid transit system, the Metro. Two 6.4-m-OD tunnels were advanced with a shielded tunnel boring machine primarily through sand and gravel with drained shear strengths corresponding to a friction angle of about 33°. Inasmuch as field measurements indicated that the behavior of the second tunnel was influenced by the first, only the ground responses of the first tunnel are reexamined here. The depth of overburden to the crown of the first tunnel was about 11.4 m.

Inclinometers, multi-point extensometers, and surface-settlement gages were used to measure the three-dimensional pattern of soil displacement near the advancing tunnel. The pressure on the lining was determined from steel ribs instrumented with vibrating-wire strain gages. Based on the measurements, a ground reaction curve was deduced, as shown in Figure 7.

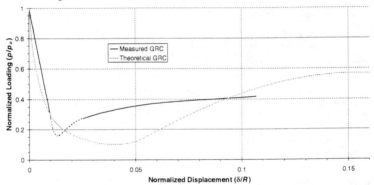

Figure 7. Comparison with Washington D. C. Metro (First Tunnel) Data

Using the steps outlined in this paper, a "theoretical" GRC was constructed as in Figure 7 (assuming an average grain size, D_{50}, on the order of 5 mm). Although the minimum loading on the tunnel has only been inferred and not directly measured, the minimum loading prediction based on Equation 12 may be considered to be reasonably satisfactory. The predicted ultimate loading is about 35% higher than measured, and the "theoretical" GRC does not exactly resemble the measured GRC. However, as a design supplement or for preliminary analysis, the derived GRC may not be a bad representation of the ground-lining interaction at the crown of the Washington D.C. Metro tunnel.

Summary and Conclusions

A simple but potentially effective method for determining the vertical loading on underground structures in granular soils has been presented in this paper. More significantly, a procedure for constructing a ground reaction curve for the crown of an inclusion, for virtually the entire spectrum of convergence of an underground opening, has been laid out. Comparison of GRC's generated using the proposed scheme with data from documented case histories indicates that the minimum load levels are satisfactorily predicted. The proposed scheme offers a relatively simple and cost-effective way to come up with reasonable GRC's as a supplementary alternative to existing methods or for preliminary design of underground facilities.

References

Bierbaumer, A. (1913). Die Dimensionerung des Tunnelmauerwerks, Engelmann, Leipzig.

Brown, E. T., Bray, J. W., Ladanyi, B., and Hoek, E. (1983). "Ground Response Curves for Rock Tunnels," Journal of Geotechnical Engineering, vol. 109, no. 1, pp. 15-39.

Einstein, H. H. and Schwartz, C. W. (1979). "Simplified Analysis for Tunnel Supports," Journal of the Geotechnical Engineering Division, American Society of Civil Engineers, vol. 105, no. GT4, pp. 499-518.

Engesser, Fr. (1882). "Ueber den Erdduck gegen innere Stützwande (Tunnelwande)," Deutsche Bauzeitung, No. 16, pp. 91-93.

Evans, C. H. (1983). An Examination of Arching in Granular Soils, S.M. Thesis, Department of Civil Engineering, Massachusetts Institute of Technology.

Hansmire, W. H. and Cording, E. J. (1985). "Soil Tunnel Test Section: Case History Summary," Journal of Geotechnical Engineering, vol. 111, no. 11, pp. 1301-1320.

Iglesia, G. R. (1991). Trapdoor Experiments on the Centrifuge: A Study of Arching in Geomaterials and Similitude in Geotechnical Models, Ph.D. Thesis, Department of Civil Engineering, Massachusetts Institute of Technology.

Iglesia, G. R., Einstein, H. H., and Whitman, R. V. (1999). "Investigation of Soil Arching with Centrifuge Tests," Journal of Geotechnical and Geo-Environmental Engineering, American Society of Civil Engineers (submitted for publication).

Janssen, H. A. (1895). "Versuche über Getreidedruck in Silozellen," Z. d. Vereins deutscher Ingenieure, vol. 39, p. 1045 (partial English translation in Proceedings of the Institute of Civil Engineers, London, 1896, p. 553).

Krynine, D. P. (1945). Discussion of "Stability and Stiffness of Cellular Cofferdams," by Karl Terzaghi, Transactions, American Society of Civil Engineers, vol. 110, pp. 1175-1178.

Leontovich, V. (1959). Frames and Arches: Condensed Solutions for Structural Analysis, McGraw-Hill Book Company, Inc.

Naval Facilities Engineering Command (NAVFAC, 1986). Design Manual DM7.01 -- Soil Mechanics.

Stone, K. J. L. (1988). Modelling of Rupture Development in Soils, Ph.D. Dissertation, Wolfson College, Cambridge University.

Terzaghi, K. (1943). Theoretical Soil Mechanics, John Wiley and Sons, Inc.

Wong, R. C. K. and Kaiser, P. K. (1991), "Performance Assessment of Tunnels in Cohesionless Soils," Journal of Geotechnical Engineering, American Society of Civil Engineers, vol. 117, no. 12, pp. 1880-1901.

Analysis of Roof Truss for Underground Support

Fusheng Zhu, Dae S. Young[1]

Abstract

In the design of excavation supports, the rock arch may offer a better model of the voussior roof than the beam theory. The truss system provides an effective method of controlling roof in especially weak formations where the conventional supports simply do not work well. In this paper, a set of closed-form solutions for roof truss system design has been proposed based on the rock arching theory. The application and field observations made in a coal mine showed that it is simple but practical for the preliminary design of truss systems.

Introduction

In the case where the roof beams are fractured and unstable, a roof truss system can supply supplemental support. After the White truss (White, 1970) was introduced in the late 1960s, the angle-bolt truss (Mangelsdorf, 1980) received broader acceptance in the early 1980s. Because the members of the truss are tensioned as installed, an uplifting force to the immediate roof can be provided by the truss, and it can remove some of the initial deformation in the lower strata. Generally, the truss system provides an effective method of controlling the roof where the conventional supports simply do not work, or may cause some additional problems. Unlike separated angle bolts, which also provide reinforcement across potential shear planes along the abutments, the roof truss provides reinforcement to the roof from all of its participating elements as an integrated unit, and this makes it work as an effective supporting system.

---

[1]Department of Mining Engineering, Michigan Technological University, Houghton, Michigan, U.S.A.

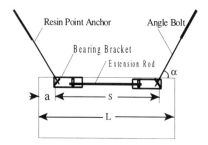

Figure 1. The Angle-bolt Truss

The truss system usually consists of ordinary point-anchored bolts, a connection bar, a turnbuckle to give the proper tension to the bar, adjustable member, and /or bearing blocks. Figure 1 shows the typical hardware and geometry of the angle-bolt truss.

The roof trusses have been used in excavations of both mining and many civil projects, and some new devices have been developed. Despite significant technological advances associated with the hardware of bolts and trusses, the truss system was rarely designed by using a rational methodology. In this paper, a set of closed-form solutions has been derived for roof truss to prevent the cracking and shearing failure of roof beams. The solution is simple and practical to use in the preliminary design and/or check of parameters of a proposed truss system. A case study is given to show its practical applications.

Determination of the Parameters of Roof Truss

The results of photoelastic model studies of laminated roof show that the tensile stress in the mid-span of the roof is reduced, and the neutral axis of the strata moves upward because of effect of a truss system (Neal, Townsend, and Johnson, 1983). The roof truss not only has the advantages of a bolting system, but also creates a pair of active upward forces against the roof beam. These forces support the roof against roof deflection. As a result, the tensile stress due to the deflection is decreased and the mid-span is completely in compression as prestress in the tension members of the truss system increases. Suppose that the tensile stress in the connection element can be adjusted to make the bending stresses at both the center and the end of the roof be exactly zero, an equation in respect to the bolt hole span (S) can be given as:

$$A_1 S^2 + B_1 S + C_1 = 0 \tag{1}$$

where $A_1 = 4.5\cos\alpha$
  $B_1 = -6(L\cos\alpha + h\sin\alpha)$
  $C_1 = 1.5L^2\cos\alpha + 4Lh\sin\alpha$
  h = height of an immediate roof beam
  L = span of the entry
  α = angle of inclination of the bolt holes

The equation (1) is equivalent to the equilibrium state of the entire system consisting of the roof beam and the roof truss. Therefore, it can be applied to calculate the important parameter; distance from hole to hole. It is important to notice that the typical angles of the inclined chord from the horizontal direction should be $38°$ to $60°$, and the anchorage point must be over the rib to maintain the support system stability. This must be observed in careful installation (Peng, 1986). The lengths and the angle of inclination of the bolts then can be determined by taking account of these two factors.

Installation of the roof trusses normally creates a compressive force vertically downwards near the rib edge of a pillar. This compressive force would increase the rock strength at the roof intersection to prevent a possible cutter failure of the roof. In that case, the roof truss should provide sufficient supplemental thrust to keep the shearing resistance at the abutment greater than the required abutment vertical reaction. This gives the required minimum tension in the extension rod for the single truss as:

$$T_1 = WL^3/[12(L-S)^2\cos\alpha + 16(L-S)h\sin\alpha]\sin\alpha \tag{2}$$

where W = weight of the roof beam and overburden.

Two multiple trusses may be used in wide rooms or entries. Multiple trusses can be installed within the other or with one overlapping the other, as shown in Figure 2. Assume that the bending stresses at both the center and the end of the roof are exactly zero again, the distance S can be obtained as

$$A_2 S^2 + B_2 S + C_2 = 0 \tag{3}$$

where $A_2 = 4.5\cos\alpha$
  $B_2 = -6(L\cos\alpha + h\sin\alpha)$
  $C_2 = 1.5L^2\cos\alpha - 6aL\sin\alpha + 18a^2\cos\alpha + 2Lh\sin\alpha + 12ah\sin\alpha$
  a = distance from side wall to bolt hole

For multiple trusses with one in the other as shown in Figure 2a, the distance S_2 is:

$$S_2 = L - 2a \quad (4)$$

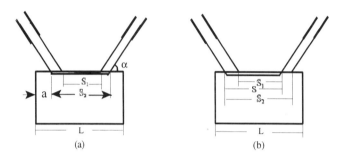

(a)Truss with one in the other; (b) Truss with one overlapping the other

Figure 2. The Multiple Roof Truss

For the truss with one overlapping the other, as shown in Figure 2b, the distance S from the geometry is

$$S = 0.5 (S_1 + S_2) \quad (5)$$

The tension in the horizontal chord can be obtained by

$$T_2 = WL^3/\{[12(L - S)^2 + 4a^2]\cos\alpha + 8[(L - S) + 2a]h\sin\alpha\}\sin\alpha \quad (6)$$

The possibility of shear failure of the roof beam at the abutments is assessable by comparing the maximum frictional resistance and the abutment shear force. This gives the roof safety factor against shearing as

$$F_s = \{[3\cos\alpha(L^2 - S^2) + 6hS\sin\alpha]T_1\sin\alpha / LW - L^2\}\tan\phi/Bh + F_o \quad (7)$$

where B is truss spacing along the longitudinal direction, and F_o is the factor of safety against the abutment shear failure for an unsupported roof as given by Wright (1973):

$$F_o = L^2 \tan\phi/(3.16hL - 1.76h^2) \tag{8}$$

Note that the F_o should be changed when the truss spacing along the longitudinal direction is different from the unit of thickness.

Similarly, the roof safety factor against shearing failure for the multiple truss is

$$F_s = \{[3\cos\alpha(L^2 - S^2) + 4a(L - a)] + 6(L + S - 2a)hS\sin\alpha]T_2\sin\alpha / LW - L^2\}\tan\phi/Bh + F_o \tag{9}$$

The buckling of the roof beam is another important factor to be checked in design (Duvall, 1976), which is currently being reviewed by the authors for improvements. A reasonable selection of the stiffness of the truss may retain the roof deflection sufficiently small to assure the roof safety. The design of angle-bolt truss installation based on the interaction between truss and roof deflection has been discussed in Mangelsdorf (1985).

Case Study

Single truss support of long wall panel entries

The seam thickness of No. 4 coal seams at Hong Miao Coal Mine is about 2.24 m with a dip of 2°-3°. Geological investigation showed that the immediate roof consists of 0.15 m shale, 0.4 m siltstone and 1.2 m laminated sandstone. The steeply dipping joints with 85°/88° (dip direction/dip) are in the direction that the entries were driven, and the critical joints are those which are near to and dipping toward the abutments.

The rock bolts of 1.8 m in length and 0.9 m resin grouted were used originally to support the roofs, but cutter roof problems appeared frequently. The roof truss was selected to prevent the cutter roof failure.

The entries are 5.2 m wide. The unit weights of the siltstone and the sandstone are 0.0268 MN/m³ and 0.0224 MN/m³, respectively. The internal friction angle of the roof is 28°. Substituting these parameters into Eq. 1, the distance from hole to hole can be calculated as $S' = 2.6$ m. The length of the bolts, 2.0 m long, and their inclination angle of 50°, can be determined based on the stable anchorage and the experience of bolting of the mine. Substituting $S = 2.6$ m into Eq. 2, the minimum tension in the tensile member is $T_1 = 0.078$ MN, but the safety factor against shear failure by using Eq. 7 showed an unacceptable value. In order to obtain the sufficient value F_s, a series of calculations were made by changing the tension T_1 and the truss spacing B along the longitudinal direction. Finally, tension $T_1 = 0.174$ MN, and $B = 1.2$ m yielded $F_s = 3$, which is acceptable. These parameters have been proved in the mining practice, and adopted as basic design criteria for the entries of new longwall panels. Roof failure has not occurred since the roof truss system was applied.

The typical displacements versus a time plot of the roof and the convergence versus time curve of the wall are given in Figure 3. The movement of the roof and the walls occurred after installation of the support system at $t = 0$, but the deformation of the rock surrounding the entries was limited in the expected range. The roof moved down about 52 mm after installation of the roof truss system. The convergence of the walls was about 63 mm, but the entries were maintained in a reasonably stable condition.

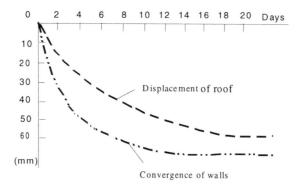

Figure 3. Displacements Versus Time Plots for Roof and Wall

Conclusion

The truss system provides an effective method of controlling the roof where conventional supports simply do not work, or may cause some additional problems. As well known, there are some rock bolt design guidelines based on rock mass classifications, but the truss systems are rarely designed using a rational methodology, including an empirical design method. The proposed closed-form solutions about the hole-hole distance and the required minimum tension in the extension rod provide a possibility to select and check the parameters of a truss system in a preliminary design. The determination of the length and diameter of the extension rod are based on analytical results, rather than the experience of others, so that it is possible to reduce the waste of the supporting materials. The proposed solutions are based on the failure modes of a roof beam; cracking and shearing, and they are simple and practical. The typical displacements versus a time plot of the roof and the convergence versus time curve of the wall in the case study demonstrates that the stabilization of surrounding rock of the openings definitely occurred after the roof truss installation, and bolted roof failure has not occurred in that mine after the use of the roof truss system. The proposed closed-form solutions have been applied successfully to the preliminary designs of the roof trusses of the panel entries and the face openings in several longwall coal mines in China.

References

1. White, C. C. Roof support of underground mines and openings, *Patent No 3,505,824*, U.S. Patent Office, Washington D. C., 1970.
2. Mangelsdorf, C. P. Current trends in roof truss hardware, *Proc. 2nd Conf. on Ground Control in Mining*, S. S. Peng (Ed.), West Virginia Univ., pp.108-112, 1980.
3. Neal, G. M., J. M. Townsend, and L. P. Johnson. Optimizing roof truss installation with body-loaded photo elastic models, *Mining Engineering*, **30**(6), pp.660-666, 1983.
4. Peng, S. S. Coal Mine Ground Control (Second Edition), John Wiley & Sons, 1986.
5. Wright, F. D. Roof control through beam action and arching. *SME Mining Engineering Handbook*, Amer. Inst. Min. Met. Engr., New York, 1, pp.13.80-96, 1973.
6. Mangelsdorf, C. P. Design of a roof truss bolting plan for Bear Mine, *Proc. 2nd Conf. on Ground Control in Mining*, S. S. Peng (Ed.), West Virginia Univ., pp.11-17, 1985.
7. Duvall, W.I. General principles of underground opening design in competent rock, *Proc. 7th U.S. Sympo. on Rock Mechanics*, University of Utah, S.L.C., Utah, pp. 73A1, 1976.

Finite Element Analysis of Ground Response due to Tunnel Excavation in Soils

Murad Y. Abu-Farsakh[1], Associate Member, ASCE, and Mehmet T. Tumay[2], Fellow, ASCE

Abstract

Excavation of tunnels in soft ground is usually associated with ground deformations around and ahead of the tunnel boring machine. Therefore, the prediction of these deformations before excavation is very important for engineering design. In this paper, a finite element computational model is used to predict the ground deformations caused by the Earth Pressure Balance (EPB) shield tunneling in cohesive soils. This study presents the effect of different soil parameters, such as the coefficient of earth pressure at rest, k_o, overconsolidation ratio, OCR, the effect of applied heaving pressure and the depth of the soil over the tunnel crown, H, on the predicted ground deformations. Results of this work show a wide variation of soil deformations can occur due to changes in soil parameters, heave pressure and cover depth.

Introduction

Construction of tunnels in soft ground is usually associated with ground deformation. Careful consideration must be given to the magnitude and distribution of these deformations and their effect on pre-existing adjacent structures and utilities. Therefore, there is a need to predict the ground deformations before the tunnel construction. Excavation of tunnels in soft ground was first carried out using compressed air shields followed by the slurry shields in which fluid was used to stabilize the cutting face instead of air. In 1974 the Japanese introduced the earth Pressure Balance (EPB) shield, in which the cutting face is supported by the excavated soil itself. The pressure is applied to the EPB tunnel face to

---
[1] Research Associate, Louisiana Transportation Research Center, LA.
[2] Professor and Associate Dean for Research, Louisiana Transportation Research Center, LA.

counterbalance the existing underground overburden and hydrostatic pressure. The EPB shields were used in USA for the first time in 1981 to excavate the N-2 tunnel in San Francisco. It was shown that the ground deformation associated with the EPB shield tunneling is lower than those caused by the other conventional methods (Clough and Leca, 1993).

Recently, the predictions of ground deformations and stress patterns during tunneling are carried out using numerical models based on finite element analysis. The finite element method is capable of simulating and analyzing the tunnel excavation under different soil conditions, for different tunnel geometries, different construction procedures, for both the short and long term conditions. Due to simplicity and cost effectiveness, the two-dimensional plane strain or axi-symmetric numerical analysis is adopted (e.g. Ng et al. 1986, Finno and Clough 1985, Rowe and Lee 1992).

In this study a simplified two-dimensional computational model that is based on the plane strain "transverse-longitudinal" sections (Abu-Farsakh and Voyiadjis, 1999) has been used. This model is capable of simulating the continuous advancement of the shield and incorporating the three-dimensional deformation of the soil around and ahead of the shield face. Elasto-plastic nonlinear governing equations (Voyiadjis and Abu-Farsakh, 1997) that is based on the concept of theory of mixtures for inelastic porous media for finite deformations are used to describe the time-dependent deformation of the saturated cohesive soils. The validity of his model was verified elsewhere (Abu-Farsakh and Voyiadjis, 1999) by predicting the soil deformations for the N-2 tunnel in San Francisco, California. Herein, this model is used to study the effect of different soil parameters, such as the coefficient of earth pressure at rest, k_o, overconsolidation ratio, OCR, heaving pressure and the depth of the soil above the tunnel crown, H, on the resulted soil deformations.

Numerical Model

The tunneling process is simulated using a finite element numerical model that is based on the combination of the "longitudinal-transverse" plane strain sections developed by Abu-Farsakh and Voyiadjis (1999). Analysis of the longitudinal section is used to simulate the continuous advance of the shield and to provide the needed information in the transverse section analysis such as the initial surface heave/settlement and the distribution of excess pore pressure around the tunnel opening. The effect of the three-dimensional deformations is then incorporated into the transverse section analysis to predict the short and long term deformations and stress patterns around the tunnel opening. The transverse section analysis is simulated using the following stages: (1) Simulate the initial heave/or settlement by applying/or unloading an incremental radial pressure until the excess pore pressure at the springline reaches the value obtained from the longitudinal analysis. (2) Repeat stage (1) with a new mesh using different tunnel opening such that the actual tunnel opening is reached at the end of this stage. (3) Incrementally unload the pressure around the tunnel opening until the tail gap is closed. (4) Activate the soil-lining interaction and apply the weight of the erected lining incrementally. (5) Simulate the time-dependent deformation (consolidation) by allowing the excess pore pressure to dissipate with time.

Figure 1 presents the finite element mesh in the transverse section analysis that is used to analyze the excavation of a 5m diameter tunnel in clayey soils for the cover/diameter ratio, H/D = 2.5. The 8-noded isoparametric elements, Q8P4, are used in the finite element mesh. The modified Cam clay model is used to describe the nonlinear elasto-plastic behavior of the clayey soil. The soil parameters of the clayey soil are: the slope of the isotropic compression line, λ = 0.326, the slope of the isotropic unloading line, κ = 0.043, the slope of the critical state line, M = 1.2, Γ = 2.72, void ratio, v = 0.3, total unit weight, γ_t = 20 kN/m^3, and the coefficient of hydraulic conductivity, k_h = 5 k_v = 10^{-8} m/s.

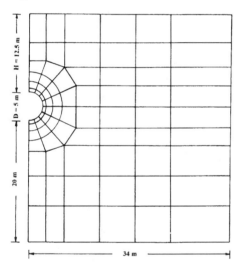

Figure 1. Finite element mesh for the transverse section analysis.

To incorporate the three-dimensional effect into the transverse section analysis, the magnitude and distribution of the heave/settlement pressure is needed to simulate the initial heave/settlement. Previous study (Abu-Farsakh and Voyiadjis 1999) shows that the distribution of the heaving pressure around the tunnel opening is elliptical with crown : springline : invert ratios depend on the coefficient of earth pressure at rest, k_o. The magnitude of the radial heave pressure is controlled by the excess pore pressure value at the springline. In this study, the applied heave pressure (the pressure the EPB machine exerts on the shield face), P_{heave}, in the longitudinal section analysis is kept at 75 kPa.

Analysis and Results

Analysis was carried out for different cover/diameter ratios (H/D = 1.5, 2.5, 3.5, 4.5) with coefficient of earth pressure at rest, k_o, ranges between 0.4 to 1.0, and overconsolidation ratios, OCR, varies from 1 to 10.

Effect of the coefficient of earth pressure at rest, k_o

The coefficient of earth pressure at rest, k_o, represent the ratio of the horizontal and vertical in situ stresses. The effect of k_o on the vertical and lateral deformations was studied for normally consolidated clay (OCR=1). As k_o increases, the lateral confining stress increase and hence result in increasing the vertical deformations while decreasing the soil deformations in the lateral direction. The predicted centerline surface settlements as a function of the coefficient of earth pressure at rest, k_o, for different cover/diameter ratios (H/D) are presented in Figures 2 and 3, without and with heaving pressure respectively. As the coefficient of earth pressure, k_o, increases, the centerline surface settlement increases. The effect of k_o decreases for higher values. As example, for H/D=2.5 and no heaving pressure, an increase in k_o from 0.4 to 0.5 resulted in increasing the centerline surface settlement by 60%. On the other hand, an increase in k_o from 0.8 to 1.0 resulted in increasing the centerline surface settlement by 6%.

The distribution of surface settlements for different k_o (H/D=2.5) are shown in Figures 4 and 5 for both cases with and without initial heave, respectively. In general, as k_o increases, the distribution of the surface settlement tend to increase at the region closer to the tunnel centerline and to decrease at further distances. Figures 6 and 7 present the distribution of lateral displacements at 12m from tunnel centerline for different k_o (H/D=2.5) predicted at maximum heave and after consolidation respectively. An increase in coefficient of earth pressure at rest, k_o, resulted in decreasing the lateral displacement drastically, and hence decreasing the extend of the plastic zone laterally around the tunnel opening. As a general trend, the increase in k_o reduces the displacement in the lateral direction while increasing the displacement in the vertical direction in a limited region around the tunnel. The correct evaluation of the in situ coefficient of earth pressure at rest, k_o, is an important key factor in predicting the effect of tunnel excavation on adjacent structures and utilities.

Effect of overconsolidation ratio, OCR

Analysis was carried out for different overconsolidation ratios (OCR = 1, 2, 5, 10) to study the effect of the OCR on the soil deformations around the tunnel. The OCR is defined as the ratio of the preconsolidation pressure to the present effective overburden pressure. For overconsolidated clays, the coefficient of earth pressure at rest, $k_{o(OC)}$, is related to the OCR and has to be adjusted accordingly. In this study the following relation is used to approximate the $k_{o(OC)}$ (Das 1985):

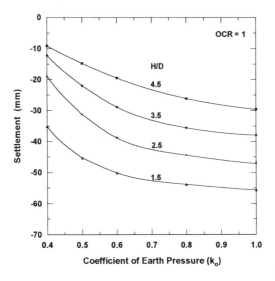

Figure 2. Variation of centerline surface settlement with k_o (no heave).

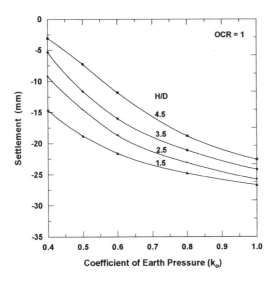

Figure 3. Variation of centerline surface settlement with k_o (with heave).

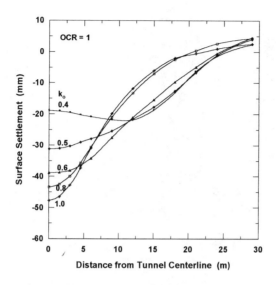

Figure 4. Distribution of surface settlement with k_o for H/D = 2.5 (no heave).

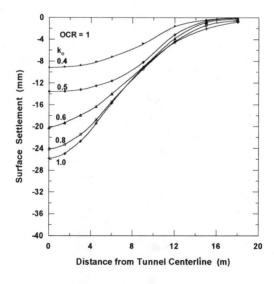

Figure 5. Distribution of surface settlement with k_o for H/D = 2.5 (heave).

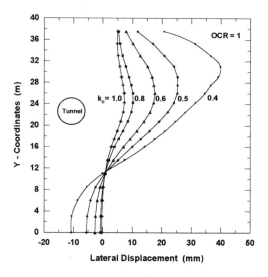

Figure 6. Variation of lateral displacements at maximum heave with k_o for H/D = 2.5 (at 12 m from tunnel centerline).

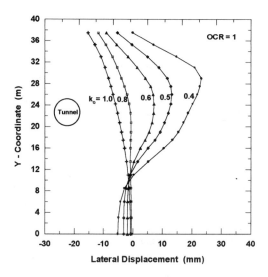

Figure 7. Variation of final lateral displacements with k_o for H/D = 2.5 (at 12 m from tunnel centerline).

$$k_{o(OC)} = k_{o(NC)}\sqrt{OCR} \qquad (1)$$

Figures 8 and 9 presents the variation of the centerline surface settlements with OCR for the heave and no heave cases respectively. The increase in OCR of the soil decreases the associated vertical settlements up to a certain limit (OCR<5). Beyond this limit, the effect of OCR is considered minimal. The effect of OCR depends on the amount of stress change and the stress path followed during tunnel excavation and construction. The stress state in the overconsolidated soils lies inside the yield surface, causing the soil to behave elastically until the stress state touches the yield surface. As a result, this reduces the plastic behavior of the soil and decreases the extent of the plastic zone around the tunnel.

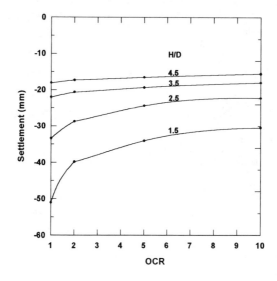

Figure 8. Variation of centerline surface settlement with OCR (no heave).

Effect of heaving pressure

The Earth Pressure Balance (EPB) shield can be operated such that the rate of excavation is less than the rate of advancing the machine, by adjusting the applied earth pressure at the shield face, thus forcing the soil away from the shield face causing small initial heave. Imposing an initial surface heave will reduce the final surface settlement. Figure 10 presents the variation of the centerline surface settlement with the increase in the heave pressure, P_{heave} (the simulated pressure the EPB machine exerts on the face) at $k_o=0.5$. Increasing the applied heave pressure tends to increase the initial surface heave, and hence reduce the final surface

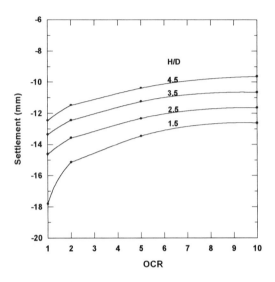

Figure 9. Variation of centerline surface settlement with OCR (with heave).

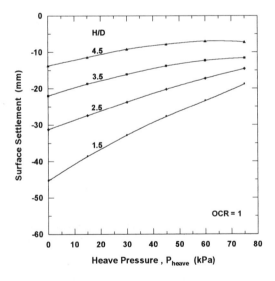

Figure 10. Variation of centerline surface settlement with P_{heave}.

settlement. However, imposing large heave can cause damage to surface and subsurface structures and utilities, as well as, excessive settlements.

Excess pore pressures

Contours of excess pore pressures at $k_o=0.8$ for H/D=2.5 and OCR=1 are presented in Figures 11(a) and 11(b) at the maximum heave and after closure of the tail gap, respectively. The excess pore pressure after heaving varies from about 20 kPa near the tunnel crown and invert to about 75 kPa near the tunnel springline. After the gap closure, this excess pore pressure reduces to about 60 kPa near the springline, 10 kPa near the crown, and -10 kPa near the invert. The dissipation of this excess pore pressure is responsible for the long-term deformation (consolidation) of the soil.

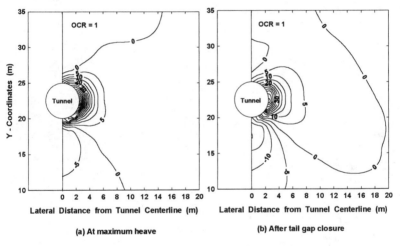

(a) At maximum heave (b) After tail gap closure

Figure 11. Contours of excess pore pressure (kPa) for $k_o=0.5$ and H/D=2.5.

Figures 12(a) and 12(b) compare the contours of excess pore pressures after the closure of tail gap for H/D=2.5 and OCR=1 with no heave pressure for $k_o = 0.5$ and 0.8 respectively. Even though there is no applied heave pressure, positive excess pore pressures (up to 45 kPa for for $k_o = 0.5$ and 35 kPa for $k_o = 0.8$) tend to develop near the springline causing long-term deformations.

Comparison between the contours of excess pore pressures after the closure of tail gap for H/D=2.5 are presented in Figures 13(a) and 13(b) for OCR=1 and 5 respectively. Contours of positive excess pore pressures for overconsolidated soil extend more in the lateral direction and less in the vertical direction than the normally consolidated soil.

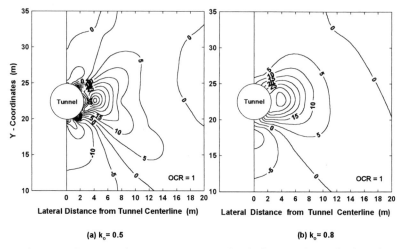

Figure 12. Contours of excess pore pressure (kPa) after gap closure (no heave).

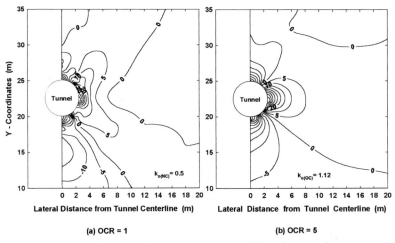

Figure 13. Contours of excess pore pressure (kPa) after gap closure.

Conclusions

A finite element analysis has been carried out to study the effect of the coefficient of earth pressure at rest, k_o, overconsolidation ratio, OCR, applied heaving pressure and the cover depth, H, on the ground deformations around the tunnel. Based on this parametric study, the following conclusions can be obtained:

- As the coefficient of earth pressure at rest, k_o, increases, the lateral confining stress increases thus reducing the lateral movement of soils ahead of the shield face. As a result, the soil deformation in the lateral direction is decreased, while the vertical settlement is increased in the region above the tunnel.

- An increase in OCR of the soil will decrease the vertical settlements associated with tunneling up to a certain limit (OCR<5), beyond that the effect is considered minimal. This is because the soil tends to behave more elastically as the OCR increases.

- The increase in the applied heave pressure tends to decrease the final surface settlement by imposing an initial heave.

- Increasing the depth/diameter ratio (H/D) reduces the surface settlements due to the increase in overburden pressure above the tunnel machine.

- Positive excess pore pressures tend to develop near the springline while negative pore pressures tend to develop near the tunnel crown and invert.

It should be noted that this study presents a general view of the effect of different parameters on the soil response due to shield tunneling. Application to real-world problems will require more than picking numbers off the graphs. Experience and engineering judgment on the specific site conditions and construction procedures, which are beyond the scope of this paper, will have to be exercised.

References

Abu-Farsakh, M. Y., and Voyiadjis G. Z., 1999, "Computational Model for the Simulation of the Shield Tunneling Process in Cohesive Soils," International Journal for Numerical and Analytical Methods in Geomechanics, vol. 23, pp. 23-44.

Clough, G. W., and Leca, E., 1993, "EPB Shield Tunneling in Mixed Face Conditions," Journal of Geotechnical Engineering, ASCE, vol. 119, No. 10, pp. 1640-1656.

Das, B. M., 1985, Principles of Geotechnical Engineering, PWS Publishers, Boston, Massachusetts, USA.

Finno, R., J., and Clough, G. W., 1985, "Evaluation of Soil Response to EPB Shield Tunneling," Journal of Geotechnical Engineering, ASCE, vol. 111, No. 2, pp. 155-173.

Ng, M. C., Lo, K. Y., and Rowe, R. K., 1986, "Analysis of Field Performance – the Thunder Bay Tunnel," Canadian Geotechnical Journal, vol. 23, pp. 30-50.

Rowe, R. K., and Lee, K. M., 1992, "Subsidence Owing to Tunneling. II. Evaluation of a Prediction Technique," Canadian Geotechnical Journal, vol. 29, pp. 941-954.

Voyiadjis, G. Z., and Abu-Farsakh, M. Y., 1997, "Coupled Theory of Mixtures for Clayey Soils," Computers and Geotechnics, vol. 20, No. ¾, pp. 195-222.

Tunnel Support Design with Finite Element Analysis

W.G. Louhenapessy [1]

Abstract

One of the major shortcomings of the Q classification system is that it does not take into account the orientation of rock joints with respect to the exposed surface of the tunnel excavation. This paper proposes a rational methodology by taking into consideration the influence of the orientation of joint sets. Using the finite element method and the multilaminate model for jointed rock masses (Zienkiewicz & Pande, 1977), a set of design charts (rose diagrams) useful to the practising engineers for circular tunnels are presented.

Introduction

Rock mass classification systems, such as the Q classification system (Barton *et al.*, 1974) and RMR system (Bieniawski, 1990), as well as others are efforts to classify rock mass properties and the rock condition as a single number. Advanced numerical methods have been frequently adopted but are not suitable for routine analysis especially when unforeseen conditions are encountered on-site during construction.

Stability in rock tunnels

Collapse of the tunnel roof, sides or face takes place when the stresses which are imposed on the rock mass due to excavation exceed its strength. Thus, the factors which influence the collapse of a tunnel are, strength of the

---

[1] Department of Civil Engineering, University of Wales, Swansea, Singleton Park, Swansea, SA2 8PP, UK

jointed rock mass and factors affecting stresses imposed due to excavation.

The strength of jointed rock masses is affected by the strength of the intact rock, the presence of joints, mechanical properties of rock joints and also the presence of water. The properties or characteristic of rock joint are: dip and orientation, spacing of parallel joint set, number of joint sets and surface roughness. Some factors affecting stresses imposed by excavation on the rock mass are insitu stress ratio and the depth of excavation.

Constitutive models and Fundamental equations

The Mohr-Coulomb constitutive model for rock joints has been adopted as.

$$\tau_j = C_j + \sigma_n \tan\phi \qquad (1)$$

where ϕ and C_j are *friction angle* and *cohesion* respectively for the joint, τ_j is the shear stress on the joint plane and σ_n is the normal stress on the joint plane. The failure criterion for intact rock used is the Mohr-Coulomb criterion as follows,

$$\tau_i = \sigma_{ni} \tan\phi_o + C_o \qquad (2)$$

where σ_n is the normal stress on the failure plane and ϕ_o and C_o are material constants for intact rock.

A general framework for constitutive models for jointed rock masses

Multilaminate framework for developing a constitutive model for jointed rock masses has been discussed in detail in various publication (Zienkiewicz & Pande, 1977; Pande & Williams, 1990). The elastic constants of a jointed rock mass and rock joint can be determined from large scale in-situ experiments or large scale triaxial tests. However, these experiments are expensive and time consuming. An alternative approach is to derive the elasticity matrix of the jointed rock mass from the constitutive properties of its constituents. The philosophy here is to treat jointed rock mass as a composite material with intact rock and rock joints as its constituents. It is to find the normal stiffness and shear stiffness of the rock joint. The elasticity matrix of rock mass, $\mathbf{D}_e^{rm}$, is

$$\mathbf{D}_e^{rm} = \left[\sum_{i=1}^{n} \mathbf{T}\, \mathbf{C}_l^j\, \mathbf{T}^T + \left[\mathbf{D}_e^i\right]^{-1} \right]^{-1} \qquad (3)$$

where "n" is number of joint sets, $\mathbf{T}$ is a transformation matrix, $\mathbf{C}$ is the compliance matrix contain joint stiffness data and $\mathbf{D}_e^i$ is the conventional elasticity matrix of the intact rock (Pande & Williams, 1990).

Methodology of computation of pressure on tunnel supports

The methodology proposed here for the computation of pressure on tunnel supports is based on a practical approach which is commonly used in many areas of engineering design. Since the behaviour of a jointed rock mass is highly non-linear, considerable computational effort is needed in solving the complex rock-structure interaction problem. This is not practical and not economic from a numerical modelling point of view. An alternative is to apply the so called 'stress path method' in which an estimate is made of the stress path experienced at a few typical points in the structure (Pande & Williams, 1990). The stability of these points in the rock mass is considered based on the adopted failure criterion and the support pressure is computed, if required, in such a way that the rock mass is prevented from collapse.

For example, consider a point such as A on the roof of the tunnel (Figure 1) excavated at a certain depth in a jointed rock mass. Before excavation this point experiences geostatic stresses. The stresses at this point after excavation which may be in stages can be computed assuming jointed rock mass as an anisotropic multilaminate material having the elasticity matrix given by equation 3. The deviation of stress from the geostatic condition is readily obtained and gives the stress path to which a rock mass will be subjected at point A. This stress path at a point is imposed on rock mass and computation made to judge if failure in any of the following modes is possible:

a. failure of intact rock: The strength parameters of the intact rock are examined and the failure function is checked.

b. failure of joint sets: The strength parameters are examined and failure in shear or tension is checked. A numerical algorithm which determines the place of failure onset, ie. intact or jointed rock is presented elsewhere (Louhenapessy, 1998).

If failure is observed in any of the modes, a pressure (p) normal to the periphery of tunnel is computed which would prevent the failure of the rock mass at that point. The above procedure is repeated at a number of points on the periphery of the tunnel and simple engineering calculations are made to determine the spacing of passive rock bolts of a given diameter.

Numerical Examples

In this section, analysis of a 12.8 m diameter circular tunnel excavated at various depths in the jointed rock having one or two sets of joints is presented for the illustration of the methodology of computing support pressures. Here, a two-dimensional idealisation is adopted, though extension to a three-dimensional situation is straight forward. The notation for describing the fabric of the the rock joints is shown in the inset. Figure 2(a) depicts the geometry

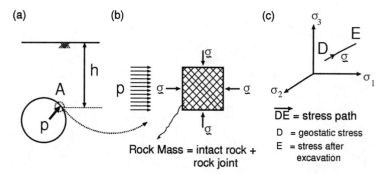

Figure 1 : Pressure applied normal to the tunnel periphery at point A to prevent failure of joint rock. (a) geometry, (b) application of normal support pressure to prevent failure under the stress path experienced by point A, (c) stress path experienced by point A in principal stress space

of the problem. The material parameters assumed for illustration are shown in Tables 1.

Support pressure has been computed for the rock mass having one set of joints at various orientations as well as for two sets of joints. The results for one set and two sets of joints are presented here. Some of the results for multiple sets are presented elsewhere (Louhenapessy & Pande, 1997; Louhenapessy & Pande, 1998). For the calculation of stress paths, elastic finite element an&.ysis is undertaken. Figure 2(b) shows the typical finite element (FE) mesh used for the analysis which consists of 736 nodes and 224 eight-noded *isoparametric* elements. In view of the approximate nature of the method of calculation, the density of the mesh is not crucial and it is assumed that the mesh shown in Figure 2(b) gives accurate stress paths for practical purposes.

Eight points have been chosen on the circumference of the tunnel for studying the requirement of support pressure. Three cases of in-situ stress corresponding to $K_o = 0.333$, 1.00 and 2.00 have been studied. It is noted that cohesion for joints is adopted as zero and the friction angle is varied between $10°$ to $50°$.

Rose diagrams of support pressure

Rose diagrams are useful tools for presenting results of parametric studies of tunnel support pressure analysis. Here the support pressure required at a point on the periphery of the tunnel is plotted as a radial line, the length of which represents the support pressure. Such diagrams are shown in Figures 3

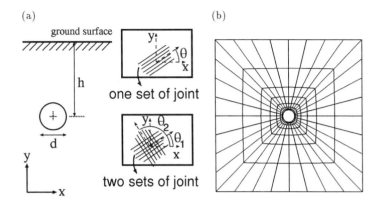

Figure 2 : (a) Tunnel Geometry and fabric of rock joint (inset). (b) Finite Element mesh

to 6. The support pressure has been normalised with reference to geostatic stress at the centre of the tunnel before excavation (γh). Figures 4(a) and (b) are for **sd** tunnel for the case when the joints are inclined at $\theta = 0°$ and $45°$ respectively. Figure 5 and 6 are for two sets of joint. For reason of symmetry, it is sufficient to look at only half side of rose diagram (Figures 4 to 6).

Normalised Support Pressure is calculated from the following equation,

$$N_p = \frac{P}{\gamma h} \qquad (4)$$

where N_p is the normalised support pressure, P is the tunnel support pressure obtained from finite element analysis, γ is the unit weight of rock and h is the depth of tunnel. From these rose diagrams, support pressure can be obtained based on the depth of tunnel, joint friction angle ϕ, in situ stress ratio, K_o, and orientation of the joints, θ. The requirement for support pressure varies from point to point on the periphery of the tunnel. Obviously, engineering judgement has to be used and provision should be made for maximum required support pressure in any section. It should be noted thats in most cases failure takes place due to sliding on joints but there are also situations in which joint open or intact rock fails, the latter case arising at very deep tunnel (vd).

Comparison with the Q classification system

Here we examine two specific cases of the support pressure requirements and compare them with those obtained from the Q classification system. The first case is that of a shallow tunnel, **sd** (CASE I), with K_o being 0.333, 1.0 and 2.0, while the second case is that of very deep tunnel, **vd**, (CASE II). The tunnels are in sandstone with one set of joint and intact rock having compresive strength (σ_c) of 100 MPa.

The rock mass quality, **Q** has been proposed based on 6 parameters (Barton et al., 1974). The values of **Q** and P_{roof} are defined by,

$$Q = \frac{RQD}{J_n} \cdot \frac{J_r}{J_a} \cdot \frac{J_w}{SRF} \quad \text{and} \quad \mathbf{P}_{roof} = \frac{2}{3} J_n^{\frac{1}{2}} J_r^{-1} Q^{-\frac{1}{3}}$$

where RQD is the Rock Quality Designation, J_n is the joint set number, J_r is the joint roughness number, J_a is the joint alteration number, J_w is the joint water reduction factor, SRF is the stress reduction factor and P_{roof} is permanent roof support pressure.

The following data been assumed for comparison (Deere, 1968; Oberti et al., 1986; Hoek et al., 1995; Natau et al., 1995; Palmstrom, 1995; Louhenapessy & Pande, 1998):
- RQD=72 %, J_n= 2 (one joint set), J_r=1.5, J_a=1.0 and J_w=1.0 (dry),
- σ_1 is the maximum principal stresses (from FE analysis).
- *for* **sd** *tunnel:* $\sigma_1 \approx 1.09$ MPa. $\frac{\sigma_c}{\sigma_1} \approx 91.5$ (medium stress, $SRF = 1.0$)
- *for* **vd** *tunnel:* $\sigma_1 \approx 26.67$ MPa. $\frac{\sigma_c}{\sigma_1} \approx 3.75$ (high stress, $SRF = 7.0$)

Based on the above parameters:
for CASE I, **Q** = 54.00 and P_{roof} = 16.6 kPa and,
for CASE II, **Q** = 7.71 and P_{roof} = 31.8 kPa.

Assuming, 25 mm (diameter) steel bolts, their spacing is obtained as 2950 mm c/c. and 2100 mm c/c respectively. Tables 2 and 3 show the comparison of spacing of rock bolts for cases I and II respectively computed on the basis of the proposed methodology. It is obvious that whilst the Q system gives a single spacing for each case, the spacing based on the theory of this paper varies depending primarily on the orientation of joints. The Q system does not always give a safe spacing.

Discussion

In general rose diagram give more extensive information i.e. the zone and extent of area to be rock bolted is indicated. Moreover, they provide a more rational and practical solution as compared to that proposed by any classification system.

It may be noted that rock bolt normal to the periphery of the tunnel may not be effective in certain situations i.e. the orientation of the joints may be such that no amount of normal pressure would prevent joint failure or the spacing of rock bolts may be too small for adoption in practice, in such a case a combination of shotcrete lining and rock bolts of at inclination maybe required.

It is shown that, as expected, the use of Q classification system leads to an over-conservative design in some cases whilst leading to unsafe design in others.

Conclusion

Analsysis and design of tunnel support system is a complex problem of rock sturcture analysis. In this paper a rational but practical method of computing support pressure has been suggested. It is based on the 'stress path' method of analysis. The stress path at a number of point on the periphery of the tunnel is computed using an elastic finite element method.

A multilaminate theory is used to compute the support pressure which would prevent the collapse of the rock mass. The methodology is explained by a set of rose diagram. It is proposed that the engineers should develop similar diagram for the tunnel based on actual laboratory / field data. These design charts can be readily read for any situation during construction.

The methods of excavation of support pressure based on a classification system lack rationale and should be used with caution.

REFERENCES

Barton, N., Lien, R., & Lunde, J. 1974. Engineering classification of rock masses for the design of tunnel support. *Rock Mech.*, **6**, 189–236.

Bieniawski, Z.T. 1990. *Engineering rock mass classification.* 1 edn. Chichester: Wiley.

Deere, D.U. 1968. Geological consideration. *Pages 1–20 of:* Stagg, K.G., & Zienkiewicz, O.C. (eds), *Rock Mechanics in Engineering Practice.* London: John Wiley & Sons.

Hoek, E, Kaiser.P.K., & Bawden, W.F. 1995. *Support of Underground Excavations in Hard Rock.* 1 edn. Rotterdam: A.A.Balkema.

Louhenapessy, W, & Pande, G. N. 1997. *A Rational Finite Element Analysis (FEA) Based Procedure for The Analysis of Pressure on Tunnel Supports.* Internal report no: CR/964/97. Department of Civil Engineering: Univ. of Wales, Swansea.

Louhenapessy, W.G. 1998. A Rational Finite Element Analysis Based Procedure for The Analysis of Pressure on Tunnel Supports. *Pages 236-243 of: Proc. Canadian Soc. Mech. Engineering FORUM 1998*, vol. 2. Ryerson Polytechnic University, Toronto.

Louhenapessy, W, & Pande, G. N. 1998. On a rational method of analysis and design of tunnel supports based on the finite element technique. *J. of Rock Mech. & Tunnelling Tech.*, **4**(2), 97–124.

Natau, O, Buhler, M, Keller, S, & Mutschler, T. 1995. Large scale triaxial test in combination with a FEM analysis for the determination of the properties of a transversal isotropic rock mass. *Pages 635–643 of: Fuji, T (ed), 8th International Congress on Rock Mechanics*, vol. 2. ISRM, Tokyo.

Oberti, G, Bavestrello, F, Rossi, P, & Flamigni, F. 1986. Rock Mechanics Investigation, Design and Construction of the Ridracoli Dam. *Rock Mechanics & Rock Engineering*, **19**, 113–142.

Palmstrom, A. 1995. RMi - a system for characterizing rock mass strength for use in rock engineering. *J. of Rock Mech. & Tunnelling Tech.*, **1**(2), 69–108.

Pande, G.N., & Williams, J.R. 1990. *Numerical Methods in Rock Mechanics*. Chichester: John Willey. 327p.

Zienkiewicz, O.C., & Pande, G.N. 1977. Time dependent multi-laminate model of rocks - a numerical study of deformation and failure of rock masses. *Int. J. Numerical and Analytical Meth. in Geomech.*, **1**(1), 219–247.

Table 1 : Properties for tunnel in jointed rock

Intact rock	$E = 7 \times 10^7$ kPa Rocktype: SANDSTONE $\nu = 0.3$ $C_i = 28870$ kPa $\phi_i = 30°$ $\gamma = 24.5$ kN/m^3	Joint rock	$C_n = 1 \times 10^{-7}$ kPa^{-1} $C_s = 2 \times 10^{-7}$ kPa^{-1} $Cohesion = 0$ $\phi = 10°, 20°, 30°, 40°, 50°$ rock joint spacing = 1 m
$K_o = 0.333$, 1.0 and 2.0		Tunnel type **sd** : depth, **h**= 80 m Tunnel type **vd** : depth, **h** = 1600 m	

Table 2 : Roof Bolt Spacing (Diameter = 25 mm)

Orientation θ	insitu stress ratio K_o	Q system (mm)	Circular tunnel : CASE I (sd) ; depth , h = 80 m				
			Proposed Method (mm) joint friction angle, ϕ				
			$10°$	$20°$	$30°$	$40°$	$50°$
$0°$	0.333	2950	2150	2150	2150	2150	2150
	1.0	2950	NO*	NO*	NO*	NO*	NO*
	2.0	2950	NO*	NO*	NO*	NO*	NO*
$45°$	0.333	2950	970	960	930	850	NO*
	1.0	2950	220	220	220	210	NO*
	2.0	2950	X**	X**	X**	X**	NO*
$90°$	0.333	2950	NO*	NO*	NO*	NO*	NO*
	1.0	2950	NO*	NO*	NO*	NO*	NO*
	2.0	2950	NO*	NO*	NO*	NO*	NO*

NO* = no support required   X** = spacing too small, (not practical)

Table 3 : Roof Bolt Spacing (Diameter = 25 mm)

Orientation θ	insitu stress ratio K_o	Q system (mm)	Circular tunnel : CASE II (vd) ; depth , h = 1600 m				
			Proposed Method (mm) joint friction angle, ϕ				
			$10°$	$20°$	$30°$	$40°$	$50°$
$0°$	0.333	2150	440	440	440	440	440
	1.0	2150	280	280	280	280	280
	2.0	2150	X**	X**	X**	X**	X**
$45°$	0.333	2150	X**	X**	X**	X**	X**
	1.0	2150	X**	X**	X**	X**	X**
	2.0	2150	X**	X**	X**	X**	X**
$90°$	0.333	2150	NO*	NO*	NO*	NO*	NO*
	1.0	2150	NO*	NO*	NO*	NO*	NO*
	2.0	2150	X**	X**	X**	X**	X**

NO* = no support required   X** = spacing too small, (not practical)

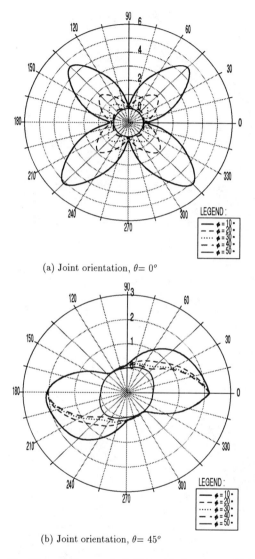

(a) Joint orientation, $\theta = 0°$

(b) Joint orientation, $\theta = 45°$

Figure 3 : Normalised Support Pressure for "sd" tunnels, for one set of joint, $K_o = 0.333$

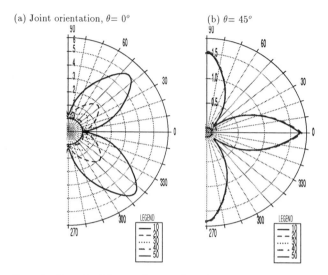

Figure 4 : Normalised Support Pressure for "sd" tunnels, for one set of joint, $K_o = 1.0$

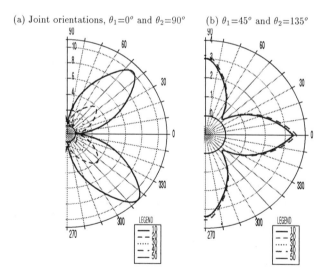

Figure 5 : Normalized Support Pressure for "sd" tunnels, for two sets of joint, $K_o = 1.0$

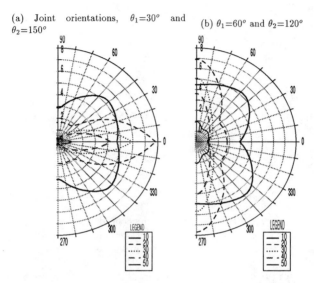

Figure 6 : Normalised Support Pressure for "sd" tunnels, for two sets of joint, $K_o = 1.0$

Numerical Analysis of the Powerhouse Cavern setting for a Pumped Storage Project

Nasim Uddin, Ph.D., P.E., M. ASCE[1]

Abstract: The existence of a limestone mine located some 2200 ft (671 m) below the ground surface and having a volume of 338 million cubic feet, provides a unique opportunity for the development of a high head pumped storage power facility. An extensive geotechnical program of drilling, testing, and grouting showed the presence of high pressure brine water in a Oriskany sandstone zone and at the depth of proposed powerhouse excavations. The data also indicated that ideal rock conditions for the power house excavation exists between the mine and the high-pressure brine water. It was immediately realized that it might be preferable or even necessary, to raise the powerhouse to avoid any hydraulic connection with the brine. The question is, what setting for the bottom of the powerhouse excavation is sufficient to meet the requirements. This paper presents the studies that were carried out to establish a powerhouse setting that would provide both a structurally stable excavation under the loading conditions imposed by the high-pressure brine, and eliminate the possibility of significant brine inflow to the excavated caverns.

Introduction

The existence of a limestone mine, located some 2200 feet (671 m) below the ground surface and having a volume of 338 million cubic feet (9.6 million cubic meter), provides a unique opportunity for the development of a high head pumped storage power facility (a schematic sketch is shown in Fig. 1). In the early engineering phases of the project, state-of-the-art single stage reversible pump-turbines (SSRPT) were selected as the optimum hydro-mechanical equipment for this development. With this type of equipment, the bottom of the powerhouse must be located some 265 feet (81 m) below the bottom of the mine (lower reservoir) to provide the submergence necessary to prevent cavitation of the runner during pumping operations. A deep borehole was drilled into the rock strata below the mine where the powerhouse would be located, and a brine water zone under high

---
[1]Assistant Professor, University of Evansville, 1800 Lincoln Ave., IN 47722

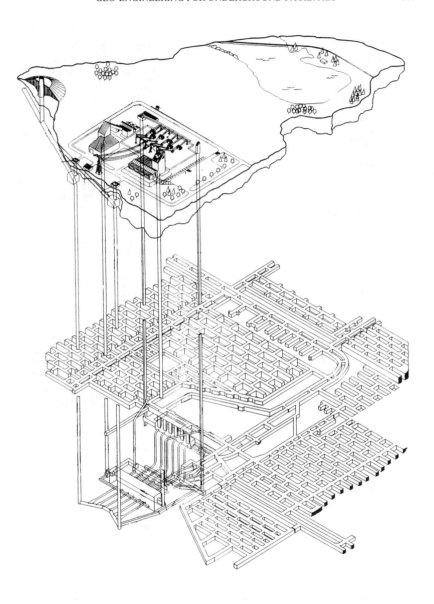

FIGURE 1 Underground Pump Storage Power facilities

pressure was encountered at the depth of proposed powerhouse excavations. This would place much of the lower part of the powerhouse well below the Oriskany stratum and into the high-pressure brine bearing Helderberg and Bass Island strata. To deal with this problem, two primary alternative solutions have been identified: either the area in which the power facilities are to be constructed must be maintained free of water by some combination of grouting and draingae; or some other type of pump-turbine equipment must be selected which would not require such deep submergence below minimum tailwater level at which the effects of the high-pressure brine would not endanger the facilities.

An extensive geotechnical program was designed to investigate all of the rock strata between the level of the mine and the original powerhouse level, and gather all of the geotechnical information necessary for final design of the selected hydro-mechanical equipment. The data indicated that ideal rock conditions for the powerhouse excavations exist between the mine and the high-pressure brine water. The high-pressure brine water is isolated to a zone at and below the Oriskany sandstone. The top of the Oriskany sandstone is 200 feet (61 m) below the mine and is a clearly defined marker that was identified in all of the 20 borings that were carried to this depth. In the course of the investigations from the mine tests were made to assess the feasibility of the use of grout to cut off the high pressure brines. Although not completely conclusive, these tests indicated that, with the appropriate techniques, and using a combination of cementitious and chemical grout, it would be possible to construct a grout curtain which would be capable of cutting off a large percentage of the brine inflow to the propose dexcavation for the power facilities. But the grout curtain requirements would add a major cost to the project, estimated to be of the order of 200 million dollars and would significantly increase the construction schedule. However, if the powerhouse is raised sufficiently using other type of hydro-equipment the need for a major grouting system is eliminated. On the basis of all of the foregoing consideration, it was decided to raise the powerhouse sufficiently to completely avoid the brine water problem. But what setting for the bottom of the powerhouse excavation is sufficient to meet this requirement?

This paper describes the studies that were carried out to establish a powerhouse setting that would provide both a structurally stable excavation under the loading conditions imposed by the high-pressure brine, and eliminate the possibility of significant brine inflow to the excavated caverns. (Detailed design for the project is currently completed and bidding for the project will start soon.)

Geotechnical Conditions (ref. 1)
- The geology, hydrogeology, and anticipated behavior due to excavation have been established from observations in the mine, and from exploratory drilling undertaken from the surface in 1991, and from the mine in the fall of 1992.
- The rock at or below mine level, at least to the top of Oriskany, is essentially unfractured, with no evidence of faulting or steeply dipping joint sets.

- Brine water pressure observed in the rock strata at or below the Oriskany ranges up to 1250 psig (8.62 MPa) as measured at mine floor level. There is however, no evidence of upward migration of this water into the mine.
- In situ stress measurements indicated a wide range of major principal stress direction and magnitude. In the analysis, the major principal stress taken as perpendicular to the powerhouse with a magnitude of the ratio of horizontal stress to vertical stress ranging from 1.0 to 2.3, representing full range of stress conditions likely in the powerhouse area.

Approach
In considering possible alternative settings for the power facilities to avoid the high-pressure brine, the first question that arises is, "How high is high enough?" To answer this question, two basic criteria which must be met have been identified as follows: (1)The proposed excavations must be structurally stable under the particular loading conditions imposed by the high-pressure brine water (and, of course, under the loading conditions normally associated with an underground excavation in rock), (2) There should be no possibility of significant brine flow into the excavated caverns. In order to confirm the criteria governing the safe powerhouse a number of possible failure modes are considered first as portrayed in the figure 2. Then numerical analyses aimed at clearly establishing the rock stress and fluid flow conditions in the rock strata between the bottom of the powerhouse and the top to the Oriskany were carried out. These analyses comprised:
- A finite element method (FEM) analysis of the zone of rock surrounding the power facility caverns, to determine rock stresses, strains, and displacements, both to assess the potential for hydrofracture of the zone below the powerhouse and to confirm the structural stability of this zone at the selected setting.
- A parallel FEM analysis of the existing mine to confirm the parameters adopted for the study of the power facility caverns.
- Wedge and beam analyses of the rock below the powerhouse floor as a conservative overall check on the FEM results, and
- Seepage analysis of potential matrix and fracture flow in the strata below the bottom of the powerhouse and the Oriskany.

Numerical Analysis:
The methodology adopted, input data used, and results obtained from these analyses are presented in the following sections.

Finite Element Analysis
General: The purpose of this analysis was to establish the required powerhouse setting above the high-pressure brine in the rock below the Oriskany sandstone. The analysis included modeling the major underground openings, and no attempt was made in this study to investigate stresses and/or strain around openings which are not relevant to the setting of the powerhouse. Hydrofracture can occur when the fluid pressure exceeds the minimum in situ rock stress. In intact rock, the fluid pressure must also overcome the tensile strength of the rock; in addition, in order to allow initiation of the fracture, a discontinuity or stress concentration must be

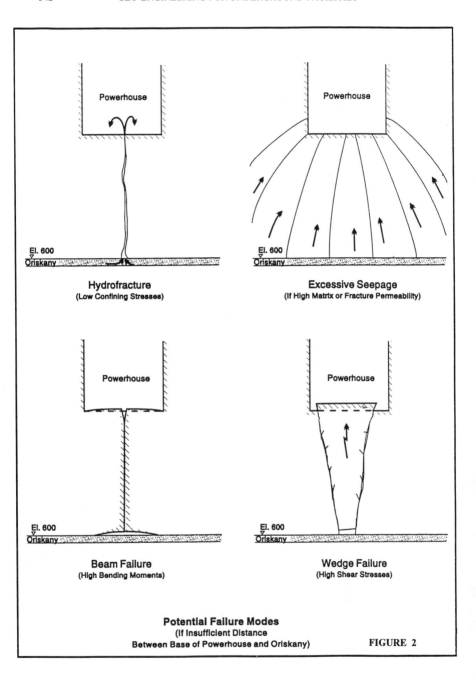

FIGURE 2. Potential Failure Modes (If Insufficient Distance Between Base of Powerhouse and Oriskany)

present. In appraising the potential for the occurrence of hydrofracture in the zone of rock between the bottom of the power facilities and the Oriskany stratum, it was (conservatively) assumed that there would be an existing fracture present. So that the rock could be ascribed zero tensile strength and hence, given the necessary fluid pressure relative to the in situ rock stress, hydrofracture could occur. The susceptibility of the rock to hydrofracture therefore, can be expressed as the ratio of the minimum in situ rock stress to the fluid pressure, and this ratio can be considered as a measure of the "factor of safety" against hydrofracture. Thus, if *in situ stress/fluid pressure* is greater then unity, hydrofracture will not occur. The principal concern relates to the potential for the occurrence of a vertical hydrofracture between the Oriskany and the bottom of the power facilities. Therefore, the focus of the analysis was on the determination of the magnitude and distribution of horizontal stresses in this region, particularly on the identification of the minimum stress levels (which would provide the lowest factor of safety against hydrofracture).

Numerical Model: In order to assess the state of stress and displacement in the rock mass between the bottom of the excavation for the power facilities and the top of the Oriskany stratum, two-dimensional finite element modeling (FEM) was employed. The study was undertaken using a University of California Berkeley finite element program adapted for the analysis of cavities in rock. The program allows the effect of the pore water pressure to be modeled, and can handle both plane stress and plan strain elements. Since the powerhouse, transformer, and draft tube manifold excavations are over 500 feet (152 m) long, a two-dimensional analysis is appropriate for determining the stress distribution around the principal openings. However, the penstock and draft tube tunnels are subparallel with the plane of the section and only 13 and 19 feet (4 and 5.8 m) in diameter respectively. So an approximation of the effect of these openings was made by adjusting the thickness of the elements at the openings, using a plane stress condition in these elements. The mesh adopted for the two-dimensional finite element model of the power facilities is shown in Figure 3. It comprises nine node quadrilateral elements which allow linear stress change across the element and provides more accurate results without having to use small elements. The overall mesh contains approximately 10,000 nodes and 2,500 elements, and extends 500 feet (152 m) above and below the power facility openings, and 500 feet (152 m) to each side, to ensure that the model boundaries do not affect the stresses around the modeled excavations. The analysis generally assumes plane strain within the section except for the area of the bus gallery and draft tube tunnels where plane stress elements are used. The penstocks were not included in the model because they are not in the plane of the analyzed section and are relatively small (nine feet in diameter).

In order to provide a measure of validation both of the model and the criteria adopted in the analysis of the proposed excavations for the power facilities, an analysis was made of the existing mine openings, which are known to be stable. The purpose of the mine analysis was to compare the stress levels predicted in the powerhouse with the observed condition in the mine. If similar finite element analysis is performed on the mine and powerhouse openings, it can be postulated as

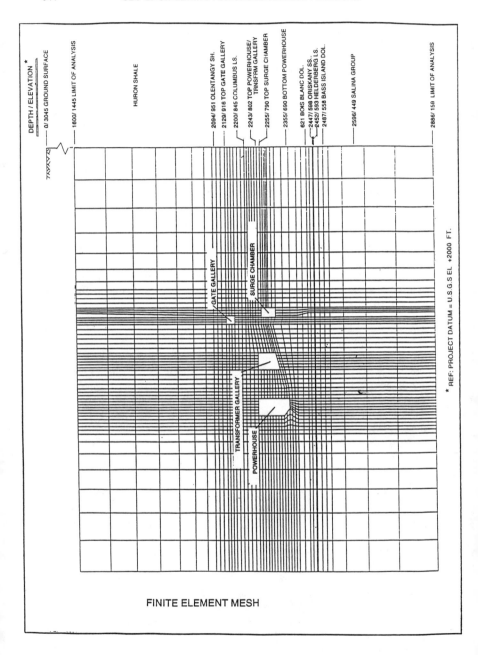

FIGURE 3

a limit condition that the highest stresses calculated in the mine are at the point of failure. As the mine is stable it therefore must have a factor of safety of one or greater. If this failure criteria is used in the powerhouse analysis, then provided this analysis shows a factor of safety greater than one, the powerhouse must then have a factor of safety greater than the mine which has been in a stable condition for up to 50 years. The mine openings were modeled with the same rock properties and in situ stress conditions. Mine openings 32 feet (9.8 m) wide, and 17 feet (5.2 m) and 46 feet (14 m) high, at 75-foot (23-m) centers were modeled. The entries between rooms were approximately modeled by adjustment of the model thickness in the same manner as the penstock and draft tube tunnels for the powerhouse analysis. The analysis results (not shown here for the brevity) indicate that the mine is generally stable. Since the mine is stable, either the adopted failure criteria is too conservative or that the high in situ stress is not present at mine levels. The use of both the high in situ stress case and the same failure criteria in the analysis of the powerhouse is, therefore, conservative.

Material Properties: Following table shows the material properties selected based on in-situ and laboratory testing:

Stratum	FEMA INPUT		POST-PROCESSOR INPUT			Mohr-Coulomb
	Modulus of Elasticity psi (Mpa)	Poisson's Ratio	Intact Rock Uni. Compres. Strength Psi (MPa)	Hoek & Brown Failure Parameters		
				M	S	ϕ
Olentangy Shale	2.17×10^6 (14.9×10^3)	0.24	9600 (66)	3.42	0.0353	37
Delaware Limestone	2.75×10^6 (19.0×10^3)	0.24	11600 (80)	12.6	0.2358	50
Columbus Limestone	2.75×10^6 (19.0×10^3)	0.24	12300 (85)	12.6	0.2358	50
Bois Blanc Limestone	2.75×10^6 (19.0×10^3)	0.24	13050 (90)	12.6	0.2358	50
Oriskany Sandstone	2.17×10^6 (14.9×10^3)	0.24	13050 (90)	6.6	0.0776	47
Helderberg Limestone	2.17×10^6 (14.9×10^3)	0.24	10150 (70)	8.8	0.0776	47
Bass Islands Dolomite	2.17×10^6 (14.9×10^3)	0.24	8700 (60)	8.8	0.0776	47

Powerhouse Setting: Three alternative powerhouse settings were modeled; 35 feet (10.7 m), 60 feet (18.3 m), and 90 feet (27.5 m) above the top of the Oriskany which was assumed to be at El 600 feet (183 m). [In the area investigated, the highest observed level of the Oriskany was El 600 feet (183 m), ranging down to El 593 feet (180.7 m)].

In-Situ Stress Conditions: Because of the uncertainties attached to the in situ stress conditions measured during the 1992 field investigations, as discussed in Ref. 1, the following range of conditions was modeled: (1)Vertical Stress (σ_v): The vertical stress in the rock prior to the introduction of the mine or power facility caverns was assumed to be γ (rock density) x h (depth below ground level), (2) Horizontal Stress (σ_H): In order to encompass the range of possible powerhouse settings and in situ horizontal stress conditions following cases are considered: Below Oriscany effective horizontal stress $\sigma'_H = 1.6\ \sigma'_v$ and above Oriscany two ranges considered - for case 1, $\sigma'_H = 2.3\ \sigma'_v$ and for case 2, $\sigma'_H = 1.0\ \sigma'_v$ (3) Stress Direction: It was assumed that the major horizontal principal stresses are orientated East-West, parallel to the proposed longitudinal axis of the powerhouse, and that the minor horizontal principal stresses are orientated North-South, parallel to the plane of the powerhouse cross section, and (4) Hydrogeology: In the Oriskany formation and below, the fluid pressures were modeled as a pore pressure load. Assumed pressure at the top of the Oriskany was conservatively set at 1400 psi (9.7 MPa). [The effective stress = total stress - 1400 psi (9.7 MPa)].

Sequence of Analysis: Openings were modeled using the approach prescribed by Wittke (Ref. 3) which can be summarized as follows: (1) Set up finite element mesh, including meshing of openings to be excavated. (2) Generate the initial stress state in the rock using bulk forces from the rock's dead weight and any tectonic forces (The deformations caused by the initial forces occur before the excavation is created). (3) Adjust the material properties of the elements to be excavated so that the stiffness and unit weight of these elements is set to approximately zero. The intact rock elements maintain the same loads and stiffness, and the analysis is repeated. The deformations due to the excavation are obtained by subtracting the deformation due to the initial state of stress from those obtained from the post-excavation analysis.

Results of Finite Element Analysis
At the lower ratio of horizontal to vertical stress considered in the analysis (Case 2, $\sigma'_H = 1.0\ \sigma'_v$), the major principal effective stresses, as shown in Figure 4, are generally horizontal below the powerhouse. The stresses increase from a far field value of 2700 psi (18.6 MPa), compressive to approximately 5600 psi (38.6 MPa), compressive in the region between the base of the powerhouse opening and the Oriskany stratum. As can also be seen in Figure 4, the minor principal effective stress in this region is generally vertical, decreasing from 2700 psi (18.6 MPa) in the far field to zero at the invert of the powerhouse opening, and 2500 psi (17.2 MPa) at 10 feet above the Oriskany. The variation of minor effective principal stress in the region between the powerhouse and the Oriskany stratum for invert settings of El 690, 660, and 635 feet (El 210, 201 and 193.5 m) are summarized in Figure 5.

As the major effective principal stress exceeds 1400 psi (9.6 MPa) (the fluid pressure at the Oriskany) throughout the region below the base of the powerhouse, the only zones in which hydrofracture might occur are those in which the minor effective principal stress falls below 1400 psi (9.6 MPa). As can be seen in Figure 5, this zone only occurs in a limited region immediately below the base of the powerhouse: the minor effective principal stress levels below this zone, extending

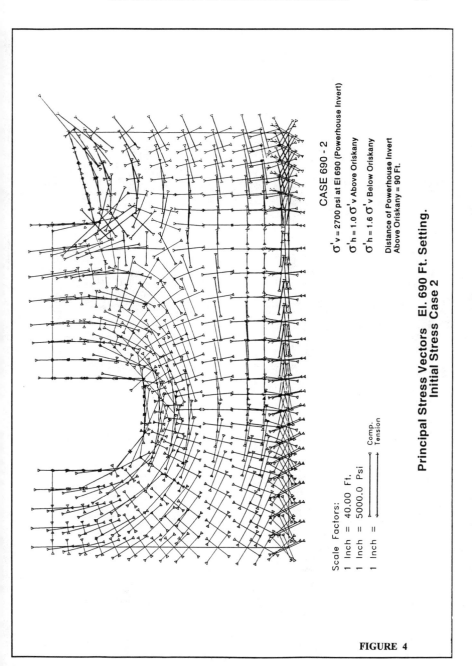

FIGURE 4

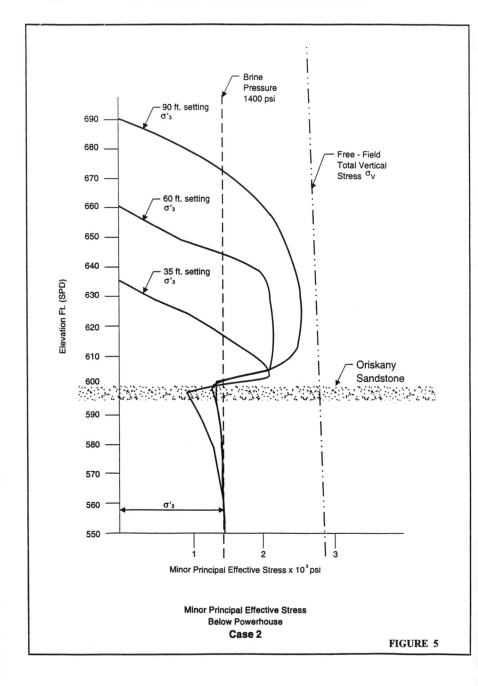

FIGURE 5. Minor Principal Effective Stress Below Powerhouse, Case 2

down to the Oriskany, are significantly above 1400 psi (9.6 MPa). The variation in minor effective principal stress between the base of the powerhouse and the Oriskany is plotted for the three settings in Figure 5. The corresponding thickness of rock with minor principal stress above 1400 psi (9.6 MPa) between the base of the powerhouse and the Oriskany stratum is summarized in following Table for the three settings considered.

Thickness of Rock Below Powerhouse not Susceptible to Hydrofracture

Powerhouse Setting	Thickness of Rock Below Powerhouse where: $\sigma_3/P > 1$
90 feet (27.4 m)	72 feet (22 m) (80%)
60 feet (18.3 m)	43 feet (13 m) (72%)
35 feet (10.7 m)	17 feet (5 m) (49%)

It can be seen from Figure 5 and the above Table that the percentage of the thickness of rock below the powerhouse susceptible to hydrofracture increases as the powerhouse setting is lowered. In all cases, the zone of rock where the minimum principal effective stress is less than brine pressure is immediately adjacent to the excavations. Therefore, in none of the cases analyzed, can hydrofracture actually occur.

It is recognized that although the rock between the bottom of the powerhouse and the Oriskany stratum is generally of high quality, uniform and essentially unfractured, variation in composition and microstructure will occur locally. Prudency suggests that to ensure safety against hydrofracture, a minimum thickness of rock of 60 feet (18.3 m) with a ratio of effective stress to fluid pressure greater than one should be established. In addition, the requirements for the selected hydromechanical equipment are also satisfied with the El 690 feet (El 210.3 m) setting of the powerhouse.

Displacements: Displacement changes have been computed for vertical (Y) and horizontal (X) directions for powerhouse settings of 90, 60 and 35 feet (27.4, 18.3 and 10.7 m) above the Oriskany for both Case 1 ($\sigma'_H = 2.3\ \sigma'_v$) and Case 2 ($\sigma'_H = 1.0\ \sigma'_v$) stress ratios. Below the power facilities openings a maximum horizontal (X) displacement in the order of 0.7 inches (18 mm) and maximum vertical displacement (y) in the order of 0.1 inch (2.54 mm) was obtained. These displacements are within the expected range for this size of opening. The displacements are generally dependent on the magnitude of the in situ horizontal stress and are independent of the setting of the powerhouse above the Oriskany.

Strain Change: The change in horizontal strain in the rock below the powerhouse is generally compressive indicating a tendency to close vertical joints. Tensile strain changes due to unloading occur in the walls and floor of the powerhouse, but are limited to about 45 feet (13.7 m) from the cavern. It should be again noted that the positive strain change, although tensile in direction, is not associated with tensile stress. Vertical strain changes in the floor are also due to the Poissons ratio effect caused by the increase in horizontal stress under the opening.

Hoek and Brown Strength Factors: Hoek and Brown strength factors (factors of safety) have been calculated in accordance with the criteria set out (ref 4), and for brevity not presented in the paper. Even at the lowest setting, a zone of rock with a strength factor greater than 1.0 is present beneath the openings. At the proposed setting of 90 feet (27.4 m) above the Oriskany, a 35-foot-thick (10.7 m) zone of rock with a strength factor greater than 2.0 at the higher stress ratio, and 60 feet (18.3 m) thick zone of rock with a strength factor greater than 2.0 at the lower stress ratio is indicated. Here again, some very shallow failure zones (factor <1) can be seen to occur around the walls and roofs of the openings. The design of the caverns will take this into account with shapes and rock support measures designed to ensure stability at locally over-stressed zone. These conditions will not affect the ability of the rock below the powerhouse excavation to prevent the upward flow of brine.

Wedge Analysis:
FERC Board recommended that analysis be performed to assess the stability of a rock wedge formed by two inclined joints in the base of the powerhouse and a horizontal joint at the top of the Oriskany sandstone. This analysis is very conservative, and represents an extreme condition. The analysis of a failure wedge below the powerhouse was performed using the limit equilibrium method. Although no joints exist which would provide the necessary failure wedge, a range of potential joint was postulated to allow the analysis proceed. The wedge analysis (Fig. 6) showed that, even under the very conservative assumptions made with regards not only to the presence of joints, but also with regard to the friction angle ($\phi = 35°$), wedge sliding could not occur unless the power house opening was lowered to within about 40 feet (12.2 m) (El 642) of the top of Oriskany. At the proposed powerhouse inverse setting of El 690 feet, a minimum factor of safety against wedge failure of 1.8 was found. This can be compared with Hoek & Brown strength factor of safety in excess of 2.0 found in the finite elements of analysis of the region between the bottom of the powerhouse and the Oriskany.

Beam Analysis:
As the further check on the safety of the proposed setting, the rock below the powerhouse was analyzed as a beam, to determine the minimum depth to prevent failure. The analysis showed that in limiting case, a beam thickness of 45 feet (13.7 m) would be sufficient to resist the loads imposed by the brine fluids at the Oriskany without exceeding the conservatively assumed allowable bending and shear stress levels in the rock.

Seepage Analysis:
To determine the potential seepage into the power facilities from the high-pressure brine and below the Oriskany, a flow net analysis, assuming a homogeneous unfractured rock, was performed. Based on a permeability of 1×10^{-8} cm/sec, the calculated inflow to the powerhouse would be less than 4 ft^3 per hour (0.5 gpm), a

GEO-ENGINEERING FOR UNDERGROUND FACILITIES

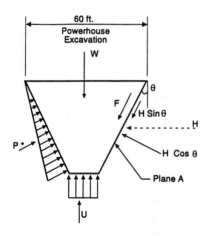

For Equilibrium of Wedge $\quad 2F \cos \theta + W = U + 2P \sin \theta$

$$F = \frac{U - W + 2P \sin \theta}{2 \cos \theta}$$

when W = Weight of wedge
U = Uplift
P = Total Hydrostatic Force at the Joint
F = Shear Force

For Equilibrium on Plane A $\quad \tan \phi_m = \dfrac{F + H \sin \theta}{H \cos \theta}$

H = Horizontal Rock Load
ϕ_m = Shearing Resistance Angle Mobilized

Factor of Safety Against
Movement of Wedge $\quad = \dfrac{\tan \phi_f}{\tan \phi_m}$

ϕ_f = Angle of Shearing Resistance at Failure

Force Diagram For Wedge Stability

* Fluid Pressure Diagram on Right Side of Wedge
Omitted for Clarity.

FIGURE 6

negligible amount. A further analysis of possible fracture flow was performed to evaluate the potential flow along an assumed discontinuity between the powerhouse and the top of the Oriskany. Based on a paper by W. J. Baker "Flow in Fissured Formation" (ref. 4) the calculated results indicate that if a fractures do exist—and geotechnical investigations have indicated that any discontinuities that are present are widely spaced and often calcite filled – the poential of inflow to the powerhouse would be less than 25 ft^3 per hour (3 gpm). However, if such a discontinuity were to be encountered, it would be possible to seal against a flow of this magnitude by grouting.

Conclusions

The analysis have conclusively shown that, from both the structural and seepage standpoint, a setting of the powerhouse which will provide 90 feet (27.4 m) between the bottom of the powerhouse excavation and the top of Oriskany stratum will be safe and stable with no risk of failure caused by the presence of fluids in and below the Oriskany sandstone. Excavation methods will be required to endure that the rock below the bottom of the excavation is maintained in sound condition with a minimum of disturbance by the excavation.

References

1. Acres international Corporation, "S. Hydroelectric Pumped Storage Project, Report on Investigations from the Mine, Fall 1992." 2 vols., Report prepared for S. Energy Storage Inc., March 1993.
2. Hoek, E. and E.T. Brown, "The Hoek-Brown Failure Criteria – an 1988 update" Rock Engineering for Underground Excavation, 15th Canadian Rock Mechanics Symposium, Toronto, 1988.
3. Wittke, *Rock Mechanics Theory and Application with Case Histories*, Springer and Verley, New York, NY, 1990, p. 1075.
4. Baker, W. J. "Flow in Fissured Formation." Proceedings, 4th World Petroleum Congress, Rome, 1955.

Acknowledgments

A number of Acres International Corporation personnel participated in this project including the author and would like to acknowledge their contributions: Staff Engineer Dan Curtis, Project Engineer Nigel J. Bond, and Project Manager D. C. Willet.

Design and Construction of the City Link Driven Tunnels
Daniel N. Adams[1], Michael T. McRae[2],
Bruce J. Hutchison[3] and Stephen J. Porter[4]

Abstract

The Melbourne City Link project includes two three-lane highway tunnels, 1.6 and 3.4 km long, which are both approximately 16 m wide and 9 m high. The tunnels were excavated through interbedded, moderately to highly jointed siltstone and sandstone of the Silurian Age Melbourne Mudstone formation over the major parts of their lengths. A section of one of the tunnels extends through mixed face ground conditions consisting of basalt overlying a very stiff silt deposit. This paper describes the design of both the primary support and the final lining for the tunnels and documents the design modifications that were made as a result of the monitoring performed during construction.

Introduction

Melbourne, Australia's City Link project involves construction or upgrading of approximately 22 km of twin three- or four-lane freeways, which will connect three major freeways to create a much needed bypass of the Central Business District (CBD). The project itself was broken into two geographical sections – the Western Link and Southern Link. The Western Link consists of elevated structures and widened existing above grade freeways. The Southern link consists of the Domain and Burnley Tunnels, and widening of the Southeastern Freeway. The focus of this paper is on the Domain and Burnley tunnel portion of the Southern Link.

The Burnley Tunnel is 3.4 km in length, of which 2.9 km is driven tunnel and 0.5 km is cut and cover. The Domain Tunnel is 1.6 km long, with 0.7 km of driven tunnel and 0.9 km of cut and cover. Both tunnels are sized to accommodate three lanes of traffic, 500 mm shoulders, and one walkway. Adding space for ventilation, including an emergency smoke duct, signage and lighting, yields a finished tunnel cross section

---

[1] Associate, Jacobs Associates, 500 Sansome St., Suite 700, San Francisco, CA 94111
[2] Associate, Jacobs Associates, 880 Apollo Street, Suite 331, El Segundo, CA 90245-4701
[3] Senior Geotechnical Engineering, ABM-Savage River Mine, Box 659, Burnie Tasmania, Australia
[4] Associate, Hyder Consulting Pty Ltd, 601 Pacific Hwy, St. Leonards, NSW 2065, Australia

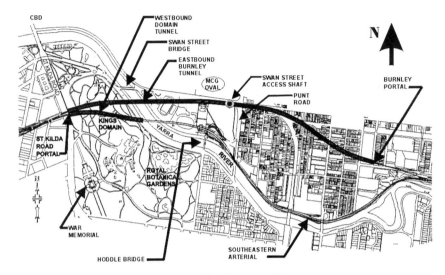

Figure 1. Site Location Plan

approximately 13.8 m wide by 8 m high. Both driven tunnels begin at the eastern side of St. Kilda Road, east of the King's Domain and the Royal Botanical Gardens. The Burnley Tunnel passes beneath the Yarra River, to the low point at the Swan Street shaft, where it begins a 5.2 percent grade up beneath the inner suburb of Richmond, daylighting at the Burnley Street interchange. The Domain Tunnel, extends beneath the King's Domain to the south bank of the Yarra River, where it crosses the river in cut and cover. Figure 1 shows the alignment for each of the tunnels.

The project is being delivered under a Build-Own-Operate-Transfer ("BOOT") arrangement with the Victorian State Government. Transurban won by tender the "Concession" from the Victorian Government to build, own and operate the Melbourne City Link (MCL) as a toll road for a period of 34 years. After this period, ownership will return to the State. Transurban contracted the Transfield Obayashi Joint Venture (TOJV) to design and construct the Project. TOJV engaged the Hyder/CMP Joint Venture (a joint venture of the consulting engineers Hyder Consulting and CMPS&F) to undertake the design of the Southern Link component of MCL which includes the tunnels. Jacobs Associates of San Francisco has provided specialist expertise in tunnel design. Geotechnical investigations have been undertaken by Golder Associates.

Geology and Geotechnical Model

Geologic Setting

The MCL tunnels extend primarily through a marine sedimentary formation of Silurian age, referred to commonly as Melbourne Mudstone or the Silurian formation.

This formation (called Silurian for the remainder of the paper) consists of interbedded siltstone and sandstone with siltstone comprising approximately 80 percent of the rock mass. The siltstone beds are typically 100 to 800 mm thick. The 10 to 20 percent of the rock mass consisting of sandstone typically occurs in thin laminations ranging from 20 to 100 mm thick, though the sandstone layers can be up to 2 m thick. The unconfined compressive strength of the intact siltstone typically varies from 60 to 120 MPa where it is fresh to slightly weathered, to less than 5 MPa where the siltstone is highly to completely weathered. The unconfined compressive strength of the sandstone is generally in the range of 100 to 200 MPa. The Silurian is relatively durable and is not highly susceptible to slaking, softening or swelling upon excavation and exposure.

The majority of the Burnley Tunnel extends through moderately weathered to fresh rock, and 60 percent of the tunnel extends through slightly weathered to fresh rock. However, extremely to highly weathered rock was encountered in each of the portal and cut and cover sections of the Burnley tunnel. The Domain tunnel is much shallower than the Burnley tunnel and as such was driven predominantly through moderately weathered rock.

At the eastern portal of the Burnley Tunnel, the tunnel passes through a buried valley and encounters mixed face conditions consisting of basalt flows overlying silts, clays, and colluvium. The Swan Street Shaft, located approximately halfway along the Burnley Tunnel, encountered similar conditions, including a major sand and gravel aquifer below the Basalt. A geological profile along the Burnley Tunnel is shown on Figure 2.

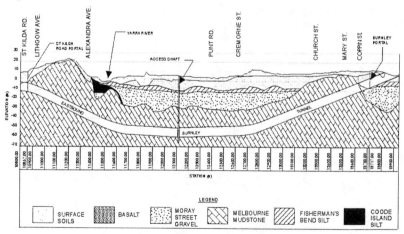

Figure 2. Profile Through Burnley Tunnel

Geologic Structure

The Silurian has been folded into a series of synclines and anticlines, with the fold axes dipping gently to the north and south. The tunnel alignments are near perpendicular to the fold axes and, therefore, for the most part cut across the strike of the beds. During excavation, two major faults were encountered in the project area, one beneath the Kings Domain and the Yarra River, and another beneath Richmond (toward the eastern end of the Burnley Tunnel). A major fault below the Richmond area was encountered over a 500 m length of the Burnley Tunnel. The fault zones are characterized by zones of intensely sheared and crushed material.

Occasional bedding parallel faults and shears were also encountered during the excavation of the tunnels. These features can typically be characterized as seams of sheared, crushed, and altered rock, typically on the order of 20 mm thick. Several thrust type faults, dipping 20 to 40 degrees to the northwest, were encountered during the excavation of the tunnels. Significant sheared and crushed zones, comprised of rock fragments in a matrix of high plasticity clay up to 1 m in width, were associated with these thrust faults. Steep reverse faults, dipping predominantly to the northeast and southwest were also encountered in the tunnel excavations. These faults are characterized by sheared and crushed zones up to 500 mm thick.

Four dominant joint sets dissect the rock mass, one parallel to bedding, and three orthogonal joint sets. Typical joint spacing ranges from 80 to 3000 mm. In the siltstone, these joints are generally tightly closed. In the sandstone the joints are often slightly open (less than 2 mm aperture).

Geology at the Portals

At the western ends of the two tunnels the Silurian is covered by mixed Tertiary aged soils. The soil profile includes layers of sandy clays, clayey sands, silty sands, and sensitive silts. The portion of the tunnels between the western portals of the cut and cover section and the east side of St. Kilda Road was constructed using a top-down sequence. During the drilling of the piers, monitoring was performed to evaluate the properties of the siltstone overlying the crown of the driven tunnels at the portals. During this program, an intrusive body was identified that crossed over the Burnley Tunnel portal about 0.5 m above the tunnel crown. The dyke material is extremely weathered and consists primarily of clay.

The initial 100 m of the Burnley Tunnel adjacent to the eastern portal was driven through mixed ground, consisting of high strength basalt over very stiff clay of the Fishermens Bend Silt formation (FBS) and colluvium (see Figure 2). The basalt ranges from blocky to very blocky and seamy and the degree of weathering within the basalt typically varies from slightly to moderately weathered. However, extremely weathered zones altered to clay were observed within the basalt.

Hydrogeologic Profile

Lamb, Coulsell and Hutchison (1998) have described the evolution of the hydrogeologic regime of the Burnley and Domain Tunnels. The main aquifers within the project area consist of confined alluvial sand and/or gravel known as the Moray Street Gravels (MSG) and Holocene Alluvium (HA) aquifers, and the confined/unconfined fractured Silurian and basalt rock aquifers. These units are shown in Figure 2. At the start of construction, groundwater levels ranged from 2 to 3 m below the ground surface along the majority of the tunnel alignment. At the western end of the tunnels, the groundwater level beneath the high ground of the King's Domain was up to 20 m below the ground surface prior to the commencement of tunneling.

As reported above, the tunnels were driven, for the most part, through the Silurian. The Silurian is a fractured rock aquifer, connected to the overlying alluvial aquifers. Excavation of the tunnels allowed temporary groundwater inflows which substantially de-pressurized portions of the various aquifers. This de-pressurization provided the potential to accelerate on-going secondary consolidation rates of the Holocene aged Coode Island Silt (CIS) and the older Quaternary Fishermens Bend Silt (FBS) deposits, in areas where these materials directly overly the rock aquifers.

Initial Rock Support Design Criteria, Philosophy and Approach

General

The objectives of the initial primary support, and the governing criteria relevant to the design of it, were as follows:

- ensure the stability of the ground around the tunnels, and thus prevent rockfalls;
- limit overbreak which must be filled with concrete; and
- limit deformation of the ground so as to avoid encroachment into the space reserved for the final lining.

The design methods used for the primary support, are described in detail by McRae, et al. (1997). In summary, the design process involved an assimilation of several methods, including:

1. A review of empirical design methods, including a review of specific precedent experience in the Melbourne area and in similar geologic conditions around the world.

2. An assessment of the possible extent of regions around the tunnels where stresses may be low. Within these zones, potential blocks and wedges of rock that are not restrained against displacement under gravity, and may consequently fall, unless restrained by support, were estimated. Rock reinforcement was designed to support the weight of loosened rock, which might lie within such envelopes around the tunnels.

3. Numerical analyses were performed using the computer program FLAC (Itasca, 1995) to confirm basic design assumptions, identify important behavioral mechanisms, estimate the loads in the tunnel support elements, and to estimate deformations within the rock mass around the tunnels in the various anticipated geological conditions.

4. Analyses of structural support elements, such as steel sets and wall beams, were performed using the computer programs SPACE GASS (ITS, 1998) and STRAND 6 (G+D Computing, 1993) and appropriate closed form calculation methods.

Typical Geological Conditions

Utilizing the above methods, the initial support design work was designed to suit a number of Typical Geological Conditions (TGC's) expected to be encountered during tunneling. The TGC's were developed based on site specific geologic properties including:

- unconfined compressive strength (UCS) of the intact rock
- orientation of the bedding in relation to the tunnel axis
- spacing of fractures or joints
- condition along bedding planes and fractures

Eleven different rock support arrangements were designed to provide support for the anticipated TGC's. A typical support arrangement is shown in Figure 3 for high strength mudstone (UCS > 25 MPa). Generally, in the stronger rock, support consisted of 4.5 m long, 32 mm diameter resin-grouted rock bolts, which were tensioned to 80 kN, and installed in a pattern sequence. Patterns ranged from 2 x 2 m, to 1 x 1 m, depending

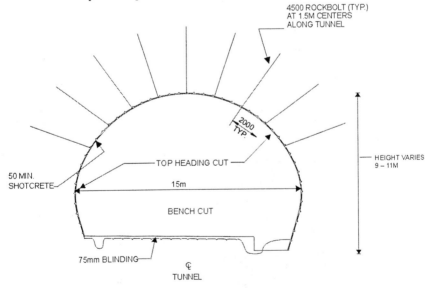

Figure 3. Typical Initial Support In Strong Rock

on observed structure and conditions. Generally for moderate strength rock (5MPa<UCS<25MPa), support consisted of a combination of the 4.5 m long and 6 m long, 32 mm diameter bolts, at a variety of patterns depending on observed conditions. In the weak rock, (UCS<5 MPa), the design support ranged from 6 m long bolts, and 7.5 and 9 m long cable bolts to steel sets (250UC89). In addition to the rock support designs for the TGC's along the tunnel drives, additional support arrangements were developed for support of special ground conditions, consisting of faults and dykes.

The initial support design called for the use of both plain and fiber reinforced shotcrete applied, for most TGC's, after the installation of the rock bolts. For strong rock, in favorable structure, a 50-mm layer of plain shotcrete was required to be applied within 30 m of the excavation face. In moderate strength rock or rock with unfavorable structure, a 50-mm initial layer of fiber-reinforced shotcrete was required to be applied shortly after rock bolt installation and tensioning. For the weaker siltstone, and in blocky ground, fiber reinforced shotcrete was applied immediately after excavation, prior to bolting or placement of steel sets. Following bolt/set installation in the weaker rock conditions, a second layer of shotcrete was required.

Initial Support for Portals

Adjacent to the St. Kilda Road portal, the primary support was designed to support the full weight of the soil overburden (15m of soil) which extended to within 500 mm of the excavated profile of the tunnels. In order to limit surface ground movements, and provide face stability during the start of tunneling, steel pipe spiling, supported on 250UC89 steel sets was utilized. The pipe spiles consist of 114 mm diameter, 12-m long pipes, spaced at 250 mm from shoulder to shoulder around the arch of the tunnels. The pipe spiles had holes drilled through their section at approximately 300-mm intervals and were filled with cement grout subsequent to installation.

Adjacent to the Burnley Portal, the entire tunnel cross section is within the basalt, however, almost immediately into the tunnel, as the tunnel slopes downward, the invert encounters a silty clay (FBS, see Figure 2). Approximately 50 m into the tunnel, conditions at the face were mixed, with the lower half the excavation in clay and the upper half in basalt. The tunnel then extended through a zone of colluvial soil before encountering the Silurian. The initial support design for this section of tunnel, known as the Burnley Decline, focused on reinforcing the basalt such that it could arch over the tunnel opening, limit the bearing stresses on the FBS immediately adjacent to the tunnel opening, and thereby control the convergence of the ground into the tunnel opening. The design of the primary support was developed using a combination of closed form analyses and the program FLAC to investigate the stress distribution around the tunnel opening.

Initial support of the Burnley Decline included the use of both rock bolts and steel sets, to support the tunnel excavation. The philosophy behind the design involved using the rock bolts to reinforce the rock mass and enhance its ability to arch over the tunnel opening and to utilize the steel sets so that a closed ring could be established following the excavation of the bench. In areas where the basalt was extremely weathered, 36 mm bar spiling was required.

Temporary and Permanent Groundwater Control

The design recognized that both long term and construction term drawdown of the aquifers had to be addressed. Modeling the hydrogeologic regime was carried out using the computer program AFPM, a pseudo three-dimensional groundwater flow model. Initially, the model was calibrated based on pump tests and groundwater levels established in a series of piezometers spread over the area from Richmond to South Melbourne. Throughout the project the model was updated seven times, to take into account hydrogeologic data that was being systematically collected. The analysis indicated that during construction up to 80 l/s could flow into the tunnel and that drawdown of the alluvial aquifers would reach into the South Melbourne area. The water within the rock and alluvial aquifers is saline in nature, and thus the effects of drawdown on potable wells, due to tunnel inflows, were not an issue. Concerns were however focused on the potential effects of increased consolidation of overlying compressible deposits due to aquifer depressurization.

Based on the considerations related to consolidation of the compressible deposits, the design specified a maximum drawdown criteria of 1 m for the alluvial aquifers under South Melbourne and the Yarra River upstream of Hoddle Bridge (see Figure 1). The area along the Yarra River, between South Melbourne and Hoddle Bridge, is public parkland and roadways overlying silt deposits. Settlement in the majority of these areas was considered to be quite tolerable, so the risk level was assessed as being low. As this area was in the region of largest expected drawdown, and effective widespread recharge was not considered practical, it was decided to allow the drawdown to occur and to target specific assets, such as the Swan Street Bridge and the Royal Botanical Gardens, with recharge wells.

As the tunnel design was nearing finalization, construction at the centrally located Swan Street Shaft was underway. At this time in the project, groundwater control had included a variety of options, including pre-excavation grouting, to limit flows into the tunnel. On the basis of grouting trials conducted during the sinking of the Swan Street Shaft and initial tunnel drives, and consideration of the effect on construction programming, TOJV determined that temporary drawdown would be limited by placement of temporary recharge wells at the ground surface in critical areas and that the recharge effort would be combined with an extensive monitoring program. The decision was also made to limit pre-excavation grouting efforts to those areas where large inflows interfered with construction activities. Long-term groundwater control was to be provided by the construction of a sealed (tanked) tunnel within the length of tunnels where it was determined that the rock mass around the tunnels was directly connected to the aquifers which could depressurize the settlement-prone soils. This decision resulted in the design of drained tunnel sections at each end of the Burnley Tunnel and the south end of the Domain Tunnel, and a tanked tunnel section through the majority of each Tunnel's length.

Final Lining Design

General

The general design criteria for the permanent lining of the tunnels included:

- assuring the stability of the ground around the tunnel for the 100 year design life of the project;
- controlling groundwater inflows to the levels required for environmental and tunnel drainage purposes and protecting the road surface from water seepage; and
- providing a suitable finish to the internal surface of the tunnel for tunnel fit-out, ventilation efficiency, and aesthetic purposes.

Drained tunnel sections utilized a waterproofing membrane and geotextile fleece (for protection of the membrane) over the tunnel arch and drainage provisions under the pavement section consisting of a permeable sub-base and suitable drainage collection systems. Undrained (tanked) tunnel sections have a waterproofing membrane and geotextile fleece around the complete perimeter of the tunnel. The invert of the tanked tunnel consists of a structural arch of unreinforced concrete. Drained and tanked tunnel lining cross sections were subsequently designed to have an identical excavation profile above the tunnel pavement level. This allowed for excavation to proceed, as the design was completed, to invert level regardless of eventual locations chosen for drained or tanked configurations.

The City Link Project generally adopted use of the AustRoads Code for Roadworks and Bridgeworks. Relevant provisions of this code, including load factors, phi factors, material properties and material behavior, were applied to the lining design. Where these concrete codes did not address matters required in the design, reference was made to American Concrete Institute (ACI) and Committee European du Beton (CEB) Codes.

Design Loads Used for Final Lining

Maximum (or upper bound) loads that can feasibly be exerted on the final lining are limited to maximum loads which the primary support is capable of supporting, plus maximum feasible groundwater pressures. Load cases used as a basis for design of the final lining were derived from the most severe primary support loading cases incorporating rock bolts and steel sets respectively, plus various anticipated long term groundwater pressures.

Loads which derive from the water pressure conditions in the rock mass were applied to the lining in a radial direction. It was determined that peak water pressures at the crown of the tunnel would, in the case of drained sections, be limited to 100 kPa (equivalent to 10m of water head) due to drainage through the rock into the drainage layer under the invert. At the invert of the drained tunnel, it was assumed that a negligible groundwater pressure would exist due to the presence of the drainage system beneath the tunnel pavement. The hydrostatic load cases used for the tanked tunnel included three different hydrostatic pressures, and were combined with the rock load cases as necessary to determine the most severe loading for design purposes.

Methods of Analysis and Design

The preliminary design for the final lining was performed using manual calculation techniques and simple computer models which could quickly analyze a range of load cases, material properties, and alternative tunnel cross sections. Detailed design utilized the two-dimensional, non-linear, finite element programs STRAND6 and ABAQUS (HKS, 1997). The rock mass was modeled as a continuum with properties appropriate for the rock mass. Rock loads were modeled by using both applied nodal forces and by simulating actual wedges in the rock mass which applied the appropriate self-weight loads. The concrete lining was modeled using continuum elements with a non-linear stress-strain relationship in compression. A sufficient number of concrete elements were used to accurately characterize the stress variation across the thickness of the lining. "Gap" elements, separating the lining from the rock mass, were used to represent the gap resulting from the inclusion of the drainage fleece. The gap elements allow the transfer of compression forces only. The model accounts for cracking in the concrete by using the appropriate tensile strength for the elements simulating the concrete lining. The effect of shrinkage on the structural behavior of the invert was included in the analyses by incorporating gap elements in the model. In the tanked tunnel, the presence of a membrane below the invert was expected to minimize restraint of the invert during curing and hence minimize shrinkage cracking.

The key output of the numerical analyses which was used to evaluate the required lining thickness and concrete strength were the stress distributions in the elements modeling the concrete lining. In addition to the FE work, a static analysis using approximate closed form calculation methods was performed to provide an independent procedure to verify the design.

Results of Analysis

Based on an evaluation of the potential stress distributions in the lining elements for the range of anticipated loading conditions, it was determined that the concrete used for the tunnel arch must have a compressive strength of 50 MPa and a minimum thickness of 450 mm.

The results of the numerical analyses also indicate that heave of up to 80 mm should be expected to occur down the centerline of the tunnel floor as the groundwater pressures increase around the tanked tunnel section. The analysis predicted that approximately 1/4 of this heave will take place after sealing the temporary drainage systems, as the shrinkage gaps close with the applied load. The remainder of the heave will occur slowly as the water table returns to pre-construction levels and the full hydrostatic load develops. It was specified that the invert be cast with a 75 mm downward pre-camber in the deeper sections, and that placement of the tunnel pavement asphalt overlay be delayed to near completion of fit-out of the tunnel. This was to allow

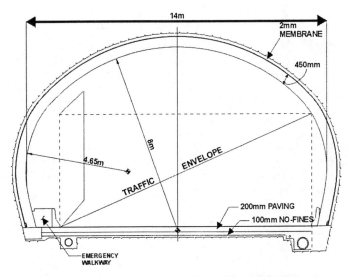

Figure 4: Typical Drained Tunnel Cross-Section

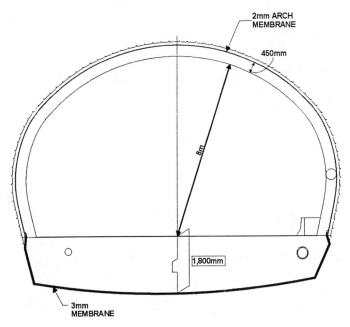

Figure 5: Typical Tanked Tunnel Section

the majority of such deflection to occur prior to paving. Typical tunnel design cross sections showing the final lining are included as Figures 4 and 5.

Burnley Decline - Final Lining Design

The Burnley Decline was analyzed as a special case because the soil materials in the sidewalls of this section of tunnel exhibit stiffness significantly lower than in the other areas of the tunnel. It was recognized that the reduced lateral restraint available from these materials could result in higher bending moments developing in sections of the tunnel lining. A beam-spring model was initially used to analyze the behavior of the lining in the Burnley Decline. Spring stiffness were derived from the moduli of deformation for the various geological units encountered along this section of tunnel. These analyses predicted relatively high bending moments and relatively large deformations at the crown, haunch and invert. These results indicated that the lining would require significant reinforcing steel and a design for a reinforced final lining was developed for this section of tunnel. However, during initial tunneling, the FBS encountered was significantly stiffer than assumed during design and plate load tests were performed to determine a revised modulus of deformation. With this revised information, non-linear finite element analyses were performed which showed that an unreinforced lining would suffice in the mixed ground conditions.

Construction Sequence and Approach

Construction on the City Link Driven tunnels commenced in late July 1996 with the excavation of the central Swan Street Shaft. In September of 1996 excavation commenced in the basalt at the Burnley Decline, followed by start up at the west portals. By January of 1997, tunnel excavation was proceeding at 5 headings. The excavation was performed by mainly by Mitsui S300 road-headers using a full-face top heading and bench excavation sequence. Typically, shortly after each excavation cycle, rockbolts were installed, followed by at least 50 mm of shotcrete. In blocky ground and within sheared rock and dyke materials, steel set support was used. Installation of the steel sets involved application of a layer of shotcrete following excavation, then standing of a set, installation of blocking, and shotcrete encasement of the set. Excavation lengths in advance of installed support ranged from 1.0 m within the crushed or fault zones, to greater than 2 m where rock exhibited high strength and favorable structure.

Approximately 400 m behind the top heading drive, the bench was excavated using rock breakers, drill and blast, and hydraulic excavators. Road headers were used to trim after the top heading hole-through. The bench was taken out in ½ tunnel widths to allow for continuous traffic to the face, and to accommodate invert membrane and concrete operations. Once the bench had progressed in from each portal about 200 m, the invert was prepared by placement of no-fines concrete in the drained section, and by placement of blinding concrete followed by fleece and membrane within the tanked section. Once enough of the leading half invert concrete was in, it was used as an access way into the tunnel, such that the other half width of bench could be constructed. In this way, the invert was advanced down the tunnel in two halves, with a ramp in the middle. After the invert had been placed full width, final survey measurements were made of the crown to check for tights and mark rough areas of overbreak. A smoothing layer of unreinforced shotcrete was then placed over the initial support shotcrete, and the arch fleece and membrane were then installed. The membrane was placed by heat-welding to roundels. All seams were double welded and pressure tested. Circumferential water stops were required at approximately 12 m centers, such that one waterstop would be placed within each invert and arch pour.

The concrete arch was placed shortly behind the arch membrane. A 12.5 m long traveling form was used to place the concrete. High early strength unreinforced concrete with a minimum strength of 50 MPa was used for the construction of the final lining. The concrete was batched on-site (at three locations); super-plasticizer was added in the tunnel at a one of two quality control stations, where slump and mix type were also checked, and cylinders were taken. Placement was done bottom up, on alternating sides of the form, in accordance with the form's self-supporting design. Transverse construction joints on the concrete arch were treated with a wax based bond breaker, in an attempt to concentrate concrete shrinkage cracks at the joints. Concrete placement was monitored through ports in the form walls and via the use of sight pipes set along the crown-centerline of the formwork. The sight pipes were cut a minimum of 25 mm greater than the minimum concrete thickness required at the section; and cement paste showing up in them indicated that the minimum thickness of arch was cast. These pipes were later used as contact grout pipes.

Instrumentation Results & Confirmation of Design Assumptions

The following instrumentation was utilized during construction to confirm design assumptions, and monitor the work in progress:

- Surface Extensometers
- Surface Level Monitoring Points
- Inclinometer Monitoring at the Burnley Decline
- Detailed Tunnel Arrays consisting of:

  Multi-Point Rod Extensometers

  Rockbolt Loadcells

  5 Point Convergence arrays
- Individual Multi-Point Rod Extensometers
- 3 and 5 point Tunnel Convergence Arrays

Layout of Instrumentation

Surface extensometers, and surface level monitoring points were installed directly above the tunnel centerline in the areas of shallow cover near the east and west portals, and in the area of low cover along Alexandra Ave. The surface level monitoring points were set up along the centerline on grids running perpendicular to the tunnel axis out to widths of about 2 to 3 tunnel diameters. The extensometers were located above the tunnel centerline. The instruments were surveyed regularly, as often as daily as excavation proceeded beneath the points, to weekly as the face advanced away from these areas, followed by monthly, and subsequently on an as requested basis.

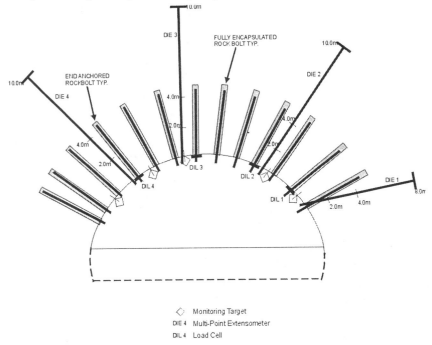

Figure 6. General Arrangement of a Detailed Instrumentation Array

Four detailed instrumentation arrays were installed at various locations within the tunnels to monitor both excavation and support performance in various ground conditions and TGC's. Each array had up to four multi-point rod-extensometers, four rockbolt load cells, and a 5-point convergence monitoring array. A typical instrumentation array is shown in cross section in Figure 6, above.

Survey tunnel convergence arrays were installed regularly along the tunnel as either 3-point or 5-point arrays (with targets attached to either rock bolt heads or to steel sets) to monitor tunnel convergence over time. The majority of these arrays were initially installed in the Burnley Tunnel through the mixed face conditions. In this area the arrays were 6 m apart, increasing to around 100 m apart as the tunnel face advanced away from the mixed face conditions. Similarly spaced convergence arrays were also installed in both the Burnley and Domain Tunnels during the pipe canopy installation near the portal area. Subsequent arrays were again gradually spaced out as the heading progressed.

Individual extensometers were also installed in several underground locations to monitor displacements at specific areas of concern, such as the brow at the base of the Swan Street shaft and within the pillar between the Burnley and Domain tunnels at the St. Kilda Portal.

Two inclinometers were installed on either side of the tunnel at the Burnley Decline, approximately 1.5 m from the walls of the tunnel excavation.

Measured and Predicted Performance

St. Kilda Road Portal

Surface settlements ranged from 3 to 28 mm close to the St. Kilda road portal above the pipe spiling, with the majority of the measured settlements being within the range of 4 to 8 mm. Subsurface displacements recorded by the extensometers at St. Kilda Road indicated that the rock mass over the Domain Tunnel displaced between 11 and 18 mm, with most of the recorded movement occurring as the top heading excavation was excavated directly beneath the instrument. Extensometers above the Burnley Tunnel at St. Kilda Road showed that only 3 mm of movement occurred above the third pipe canopy. The extensometer installed horizontally through the rock pillar between the tunnels at the St. Kilda Road portal showed a total displacement between the two tunnels of 5 mm.

Domain Tunnel

A detailed instrumentation array (Array No. 3) was installed in a section of the Domain Tunnel within high strength siltstone with unfavorable structure. This array indicated negligible convergence and little or no additional load on the rockbolts above the tensioning load. Five convergence arrays were installed in a variety of moderately strong siltstone and dykes encountered in the Domain Tunnel. These arrays indicated up to 5 to 9 mm of convergence.

The FLAC analyses performed for the Domain Tunnel predicted convergences of less than 2 mm except where the effects of unfavorable rock structure or clay coated joints were simulated in the models. In moderate to high strength siltstone with unfavorable structure and zones of altered rock, the predicted convergences ranged from 9 to 12 mm.

Burnley Tunnel

Detailed Instrumentation Array No. 4 was installed in a section of the Burnley Tunnel within moderate strength siltstone with unfavorable structure. This array recorded negligible convergence and no increase in rockbolt load above the tensioning load. An adjacent convergence array indicated up to 5 mm of convergence in this location. Detailed Instrumentation Array No. 2 was installed in a section of the Burnley Tunnel within high strength siltstone with unfavorable structure. This array recorded up to 8 mm of convergence and no increase in rockbolt load above the tensioning load.

Of the 19 convergence arrays installed within the rest of the Burnley drive, all but one recorded no more than 5 mm of settlement. At one steel set array, located within highly fractured blocky ground, the survey indicated 13 mm of convergence.

The FLAC analyses performed for the Burnley Tunnel predicted 10 to 17 mm of convergence in strong siltstone with unfavorable structure.

Burnley Decline

Surface installed extensometers at the Burnley Decline indicated movement within the ground ranged from 2 mm to 30 mm, however, most of this movement showed up as compression in the clay layer above the Basalt (due to drought conditions during the work), and not displacement of the rock. The convergence arrays installed on the steel sets within the Burnley Decline showed up to 4 mm of convergence. The inclinometers installed adjacent to the Burnley Decline showed lateral squeezing of between 4 and 7 mm, concentrated at the tunnel shoulder and invert levels.

The FLAC analyses performed for the Burnley Decline predicted 50 to 100 mm of convergence at the crown and 10 mm of lateral convergence at the tunnel walls. The excessive convergence predicted at the crown is probably the result of underestimating the stiffness of the silt underlying the basalt as discussed above.

Modification of Initial Design

As excavation proceeded, and the ground behavior was clearly established, the information obtained from the instrumentation program was used as the basis for reducing the ground support requirements. Of particular significance was the elimination of the 7.5 and 9 m long cable bolts, and about 40 percent of the steel set "legs" (posts).

As the instrumentation showed that significantly less convergence was occurring than the 50-mm allowed for in the design, the arch form was modified to reduce the total amount of concrete placed.

Tunnel Lining Convergence

At the time of writing this paper (February, 1999), monitoring of the tunnel lining indicated that the crown/arch movements are negligible but that the invert has undergone up to 45 mm of heave along the centerline. Monitoring of the groundwater recovery indicates that the water level has reached a level approximately 25 m above the tunnel crown, which is approximately 50 percent of the total recovery.

Conclusions

The design of the City Link tunnels included a multi-staged approach to develop practical and economical designs for both the primary support and permanent linings. The design and construction procedures used on this project clearly illustrate the value of instrumentation as an indispensable component of the observational approach, as it allows the behavior of the ground and support to be clearly established. Instrumentation and monitoring during construction was successfully used to both confirm the adequacy of the design and to refine the structural requirements to optimize the design as construction proceeded.

The numerical methods utilized during the design provided a useful tool for the design of both the primary support and the final lining. These methods allowed the analyses of complex problems, and, in a number of cases, resulted in more economical designs than would have been derived from more traditional closed form or empirical design methods. The deformations predicted by the models were in some cases quite close to the measured deformations and in other cases were quite divergent, primarily due to the difficulty of accurately characterizing the properties of the ground.

REFERENCES

G+D Computing Pty, Ltd., 1993, STRAND6, Version 6.1, Ultimo, NSW, Australia

Hibbitt, Karlsson & Sorensen, Inc., 1997, ABAQUS/Standard V5.7, Pawtucket, RI, USA.

Hutchison, B.J. and Lamb, I.A., 1999, Construction Management of Hydrogelogical Aspects of Melbourne's City Link Project, 10th Australian Tunnelling Conference, Melbourne Australia, March 1999.

Itasca Consulting Group, 1995, FLAC Fast Lagrangian Analysis of Continua, Version 3.3, Itasca Consulting Group, Minneapolis, MN.

Integrated Technical Software, SPACE GASS, Version 8, December, 1998; www.spacegss.com

Lamb, I.A., Coulsell, J.B., and Hutchison, B.J., Melbourne's City Link Tunnels – Part 1: Engineering Geology in an Urban Environment, Proceedings of the 8th Congress of International Association of Engineering Geologists, Vancouver, Canada, 1998.

McRae, M.T., Wilson, C., Porter, S.J. Hutchison, B..J Primary Support Design for Large Span Tunnels In Weak Rock: The Melbourne City Link Tunnels, *Proceedings of the Rapid Excavation and Tunneling Conference,* Las Vegas, Nevada, RETC 1997.

Development of Design and Construction Concepts for
Jacked Tunnel Sections of I-93/I-90 Interchange
Central Artery/Tunnel Project, Boston, MA

Philip M. Rice[1], M.ASCE, Peter A. Mainville[2], Stephen Taylor[3], M.ASCE, and
Alan J. Powderham[4]

Abstract

The Central Artery/Tunnel Project, now under construction in Boston, is the largest infrastructure project in the United States. A major element of the project is the new South Bay Interchange for Interstates I-90 and I-93, and includes the extension of the existing I-90 under the tracks leading into the South Station Railway Terminal. Three tunnel sections will be constructed under the South Station approach track system. Several construction methods were considered for the tunnels, including cut and cover techniques involving a sequence of track closures and shifts; mined tunnels, multiple drifts excavated by microtunneling, and jacking methods. The selected approach consists of constructing each tunnel section as a series of full cross-section reinforced concrete boxes and jacking them into place. Ground freezing will be used to stabilize the soft soils in the tunnel face prior to jacking, and during the jacking operation, the frozen ground will be excavated by roadheaders, working in a compartmentalized shield in front of the lead tunnel box section. The features of the various construction methods considered are discussed, as well as the decision process through which the construction approach finally adopted was selected.

Introduction

The $10.8B Central Artery/Tunnel Project is the largest infrastructure project in the United States. The project will replace the aging elevated Central Artery (Interstate I-93) in the center of Boston, MA with a modern underground expressway, provide a

---

[1]Senior Professional Associate, Parsons Brinckerhoff, Boston, MA; [2]Supervising Construction Engineer, Parsons Brinckerhoff, Boston, MA; [3]Principal Engineer, Hatch Mott MacDonald, Boston, MA; [4]Divisional Managing Director, Mott MacDonald, Croydon, England.

new South Bay Interchange between I-93 and the Massachusetts Turnpike (Interstate I-90) and extend I-90 from downtown Boston to the City's Logan International Airport. Proceeding eastward from the I-93/I-90 Interchange, the I-90 extension will pass under the extensive network of tracks leading into the South Station Railway Terminal. Amtrak and Massachusetts Bay Transportation Authority (MBTA) commuter rail use South Station for hundreds of train movements a day. Maintaining normal train operations into and out of this regional transportation hub, therefore, was the critical requirement to be addressed in developing construction methods for the new sections of highway.

The challenge of minimizing construction impacts on the Railroad operations was complicated by limited site area adjacent to the tracks for staging the construction, and the difficulties posed by the soil profile, which, in addition to major layers of soft soils, contains several active and abandoned utilities, and numerous major obstructions from two centuries of harbor, industrial and railroad-related activities in the area.

This paper describes the development of design and construction concepts for the I-90 highway tunnel sections passing under the South Station tracks. The evolution of the concepts for building these highway sections is discussed. The key elements of the alternatives considered are presented, as well as the basis for selecting the construction approach that was finally adopted, which consists of constructing each tunnel section as a series of full cross-section concrete boxes and jacking them into place from large open thrust pits excavated immediately west of the tracks. The adopted alternative represents the largest tunnel jacking project, in terms of the number of tunnels, the dimensions of the individual tunnels, and the complexity of the site conditions, ever undertaken in the United States.

The owner of the CA/T Project is the Massachusetts Turnpike Authority (MTA), which took over this role from the Massachusetts Highway Department (MHD) in 1997. Major funding for the project is provided by the Federal Highway Administration (FHWA). The MTA's Management Consultant for design and construction of the project is the joint venture of Bechtel/Parsons Brinckerhoff (B/PB). Design of the South Bay Interchange, including the jacked tunnel sections, was performed by the Maguire Group and Frederick R. Harris (M/H) joint venture as the Section Design Consultant (SDC). Hatch Mott MacDonald (HMM) assisted the SDC in the development of the specialized tunnel jacking design.

Site and Subsurface Conditions

A general view of the tunnel site is shown in Figure 1. The site is bounded on the north side by the MTA CA/T Project headquarters building and the MBTA South Station Rail and Bus Terminal facilities. The US Postal Service's Boston central distribution facility is situated at the northeast corner. The Fort Point Channel forms the eastern and southern boundaries of the site, and the west side abuts the traveling lanes and associated exit and ramps of the existing I-93/I-90 Interchange. The site area for the construction and installation of the tunnels beneath the railway tracks has approximate dimensions of 300 m by 200 m, and is outlined in black in Figure 1.

The site area has a varied history of development. Reclamation progressed primarily between the 18th century and the early part of the 20th century, and the remnants of successive wharf retaining structures remain buried there. The current wharf line was constructed circa 1905 as part of the South Station development. The deepest layer of fill material is related generally to wharf construction in the first half of the 19th century, and is fairly coarse-grained with a medium dense consistency. The upper fill material in the area occupied by the present track network and immediately behind the present sea wall line was associated with the railroad system constructed during the second half of the 19th century. This material is quite variable and is typically finer-grained and less dense than the lower fill material. The upper fill also has numerous large obstructions such as cobbles, boulders, and fragments of concrete, steel, wood, bricks and granite blocks, as well as an abandoned depressed trackway. This trackway is a major reinforced concrete U-shaped structure, supported on a mass of timber piles, which ran from the ground surface at the southern end of the site, and gradually deepened as it progressed towards South Station. Groundwater levels at the site are typically 2 to 3 m below existing grade.

Figure 1. General View of Tunnel Site Looking North

Underlying the fill are extensive, continuous deposits of organic materials, with a variable horizon. In the vicinity of the planned tunnel jacking operations, these deposits are in the range of 3 to 6 m thick, and consist largely of organic silt with fine sand and some peat. The deposits also contain obstructions such as stone ballast, construction debris and other materials that were dropped into the water at the time the site was an active harbor.

Below the organic deposits are local lenses of relatively dense sand and inorganic silt of alluvial origin, generally less than 1.5 m thick, but ranging up to 3 m thick in places. The thickest soil deposit at the tunnel alignments is marine clay, also known locally as the Boston Blue Clay, consisting of clay and silt with occasional partings and seams of fine sand and silt. The deposit is stronger and less compressible at the top, over a thickness of approximately 4.5 m. The lower section of the marine clay is considerably softer. Figure 2 shows the typical soil profile at the site.

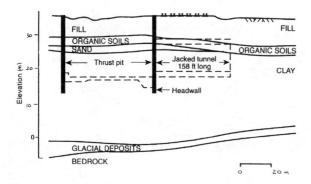

Figure 2. Typical Soil Profile at Tunnel Site (Ref. 2)

Preliminary Design Concept

The preliminary design and construction staging for the section of the I-90 roadway that would pass under the tracks leading to South Station were developed by B/PB. A primary design criterion was the requirement to maintain all existing MBTA/Amtrak railroad service during construction. Much of this I-90 construction would be adjacent to and below eight active railroad tracks. With the South Bay Interchange located adjacent to the west side of the tracks, the space available for construction staging was a major issue, and therefore controlled the possible construction sequencing at the site and the preliminary tunnel design. In total, there were nine highway tunnel sections beneath the rail tracks, as shown in Figure 3. These tunnels were referred to (from north to south) as Ramp D, Ramp F, Ramp DD, I-90 WB, Ramp L (North), I-90 EB (West), I-90 EB (East), Ramp L (South) and Ramp CD.

Construction sequencing constraints dictated a need for two types of tunnel construction, cut and cover, and mined. Cut and cover construction was selected for tunnel sections where there was a possibility of staged rail track relocations, permitting the use of conventional deep braced excavations with dewatering for control of groundwater inflows. This method was applied to all but the two northernmost tunnels at the site. For these tunnels, there were five stages of track relocations. For each relocation stage, the signals and communication systems and the planned overhead electrification had to be moved and then re-established. Additionally, railroad safety and maintenance of rail service schedules were primary concerns for the tunnel designers and construction planners. This greatly complicated the construction sequencing issues.

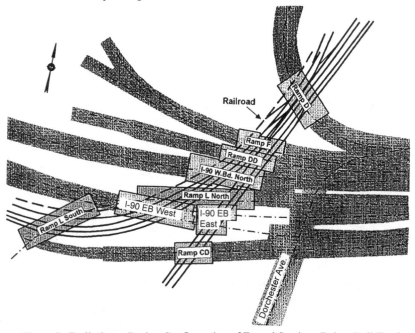

Figure 3. Preliminary Design Configuration of Tunnel Sections Below Rail Tracks

Cut and cover construction, however, could not be used for the two tunnels closest to South Station, Ramp D and Ramp F. This was due to the close proximity of the station platforms, which precluded the ability to relocate the tracks sufficiently, and the extensive track switching and turnouts in the area. Consequently, these tunnels had to be constructed by mining. Each mined tunnel was designed to be approximately 120 feet in length so it could accommodate all of the overlying tracks.

Two mined tunnel schemes employing underpinning techniques (Schemes A and B) were incorporated into the preliminary design drawings for Ramps D and F. Scheme A, as shown in Figure 4, specified the use of soldier piles [refer to (1) in Figure 4] and 610-mm diameter pipes (1a) to be installed amongst the tracks above the area of the mined tunnel. Parallel on each side of the track complex, a sheet pile wall would be constructed as a bulkhead for the approach of the staged cut and cover tunnels. Cementitious grout curtains (2) would be installed to seal off water inflows transverse to the tracks. The granular fill material above the roof of the mined tunnel area would be stabilized by chemical grouting (2a). After the new cut and cover tunnels were constructed up to the sheet piling on either side of the tracks, parallel 1.8-m diameter steel pipes (3) would be jacked laterally to form a deck approximately 9 m below the Railroad. These jacked pipes were to span the width of the trackage and extended far enough to be supported by the cut and cover tunnel sections that were previously built adjacent to each side of the mined tunnels.

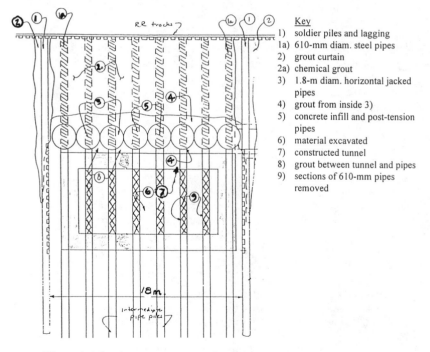

Figure 4. Mined Tunnel Scheme A Developed During Preliminary Design

Upon completion of the lateral jacking of the 1.8-m diameter pipes, the soil above would be grouted from inside the pipes (4) to prevent soil and water intrusion during the mining operation. Subsequent to the grouting process, post-tensioning ducts and cables would be placed in the pipes and then concreted and post-tensioned (5). As shown in Figure 4, the vertical 610-mm pipe piles in the path of the jacked pipe would be cut off during the jacking operation and used for intermediate supports. These supports are required to reduce the vertical deflection of the jacked pipes while mining. After cutting away the steel sheet piling bulkhead, and as the mining progresses, wood lagging boards would be installed between the support of excavation soldier piles. After excavating within this pre-formed support, the rigid framed concrete tunnel sections (7) would be constructed in the mined tunnel area (6) and connected to the previously constructed adjacent cut and cover tunnels. Grouting (8) would be used to ensure permanent transfer of the pipe load to the newly built tunnel roof. Removal of the vertical 610-mm diameter pipe sections (9) within the vehicular passageway could then take place.

Scheme B is similar in design to Scheme A, but with a primary difference being that the 610-mm diameter pipe piles are eliminated and replaced by several large transfer girders.

Both of the two mined tunnel schemes and the relocation of the railroad for the building of the cut and cover tunnel sections were included in the preliminary design drawing package given to the bidding SDCs.

Development of Jacked Tunnel Alternative

On the CA/T Project, Section Design Consultants typically complete the preliminary design presented to them by B/PB to final design, including completed bid drawings and special specifications to supplement the standard CA/T specifications. The proposal process allows prospective SDC teams to consider alternatives or variations to the preliminary design. It was during this process that the SDC team of M/H, together with its subconsultant, HMM, proposed the use of full size tunnel jacking as an alternative for the construction of the tunnels beneath the rail tracks.

Full size tunnel jacking is a technique for constructing monolithic concrete rectangular tunnel sections, which are of sufficient size to carry multi-lane highways, under operating railway tracks. The technique is generally applied in soft ground for relatively short tunnel sections which, without the need to maintain the overlying rail track, would usually be constructed by cut and cover methods. Tunnel jacking evolved from pipe jacking, retaining many of the same features. Pipe jacking was introduced on the Northern Pacific Railroad in the late 1890s as a method for installing culverts without severe disruption to the overlying rail service (Ref. 1).

Technological advancement from pipe jacking to tunnel jacking was achieved in the 1960s, when circular pipe jacked sections were found to be either too small or inefficient for their intended purpose, and they were replaced by rectangular sections. In the 30 or so years since, the technique has developed significantly, with applications to sections ranging from 3 m square to 25 m wide x 12 m high and larger (Ref. 4). The method has been used in many parts of the world, including Europe (particularly in the UK and Germany), Australia, India, South Africa, Canada, and Japan. Tunnel jacking, however, has not been used to any significant extent in the USA. Figure 5 shows the key features of tunnel jacking.

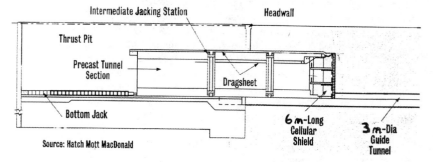

Figure 5. Key Features of Tunnel Jacking

Tunnel jacking was considered initially as an alternative for the mined sections of tunnel described above for Ramps D and F. As the concept was considered in greater detail, it developed into a viable alternative for all of the tunnel sections to be constructed under the rail tracks at the site. The decision to make the significant change from the preliminary design to the use of full size tunnel jacking as an alternate to both the mined sections of tunnel and the cut and cover sections of the tunnel was reached through tremendous teamwork among the MHD, FHWA, B/PB and the SDC, together with the MBTA and Amtrak. The final decision to proceed with the alternate design was based on detailed comparative evaluations of constructability, cost, schedule, and risk associated with both the preliminary design and the alternative under review. This process was subjected to extensive scrutiny from eminent engineers within the project review team. Recognizing the need to fully relate design to construction methods, HMM were assisted by Trafalgar House Technology, now Kvaerner Technology, who are specialist tunnel jacking contractors.

In evaluating the alternative, several key aspects relating to the preliminary design were carefully reviewed. One disadvantage of the preliminary concept was the need to realign the tracks over five stages to create space for each length of tunnel construction. More importantly from the Railroad Operator's perspective, it would be necessary to move and reconnect signaling and safety control cabling and equipment at each stage. Each change in the track configuration would have affected one of the 180 separate interlock controlling switches to South Station's platforms, and furthermore, adjusting the computer programming of the rail system's electronic monitoring and control systems would have required software modification, implementation and operational checking.

Other features of the preliminary design included the requirement for excavations over 21 m deep to be constructed close to and between operational tracks, posing both safety and construction vehicle access issues. The areas that could be made available for construction of the intermediate tunnel sections were very restricted, and would inevitably have required a large number of adjacent rail possessions to ensure operational safety. The configuration of the tracks in their various alignments, which would have minimal space between them, would have resulted in heavily skewed tunnel structural elements, complicating the construction joint detailing, reinforcement and waterproofing. Additionally, the sequential unloading and reloading of the subgrade as each section of tunnel was excavated, constructed and then backfilled would potentially have created undesirable complex differential ground movements.

Tunnel jacking offered reductions of the risks associated with the aforementioned issues. A major benefit would be elimination of the track relocations and associated work required for cut and cover tunnels as described previously. Other advantages of the alternative identified at the concept stage included the flexibility to shift the deep excavations adjacent to the railroad farther away from the tracks by relatively small increases to the length of jacked tunnel. Excavation support walls could be moved farther away from tracks, giving the contractor more space to work without fouling the tracks. The need to perform construction work in an "island" bounded by railroad tracks during an intermediate cut and cover stage would be eliminated. Total volume of excavated spoil would be minimized, a particular environmental benefit considering the urban setting of the site, since much of the spoil from the fill stratum was expected to be contaminated.

Many of the subsurface obstructions present in the fill could be avoided by jacking beneath them. Tunnel jacking would also reduce the interface between the construction contractor and the track relocation contractor. For the preliminary design concept, five mobilization and demobilization phases of the track relocation were contemplated during the construction work, all of them on the critical path. The jacking alternative initially reduced this to two mobilization/demobilizations, only one of which would be on the critical path. During final design development, efforts were made successfully to further reduce the track relocation.

Other advantages offered by the tunnel jacking alternative included elimination of the sequential unloading/heave and then backfilling/recompacting of the soil for small lengths of tunnel, followed by the same process for adjacent sections of tunnel construction. This reduced the potential for excessive differential ground movements, and consequently for rail track movements, and tunnel serviceability issues. The alternative also avoided the need to construct tunnel sections with multiple construction joints at heavily skewed angles, compounded by difficult access and little room for laydown. A final notable advantage gained was a significant decrease in potential conflict between the tunnel construction and the proposed construction activities associated with the overhead electrification of the tracks for Amtrak's high speed rail project, which was scheduled to proceed concurrently with the CA/T Project work.

B/PB studied the staged railroad/mined tunnel and the jacked tunnel methods for risks of possible claims by the contractor, compatibility with railroad operations, construction and railroad safety, possible over-run of schedules, potential for costs being exceeded during design and construction, and the interaction of construction with the proposed Amtrak Electrification Project, which would be progressing concurrently with CA/T construction in the South Station area. It was determined that the jacked tunnel method offered the CA/T Project significant savings in track relocations, would minimize railroad safety issues, and would allow better control of the construction costs for this section of I-90.

After this decision was reached, the overall layout of the interchange continued to evolve to accommodate strategic changes to the highway system. This resulted in the number and layout of the jacked tunnels shown in Figure 3 changing significantly, as a result of the deletion of Ramps CD and DD, the relocation of the Ramp L alignment into the Fort Point Channel, and the combination of Ramp F and I-90 WB, and I-90 EB (West) and I-90 EB (East), into single tunnels. The rationale, however, behind the decision to make the change from the preliminary design and construction concepts to the alternative full size tunnel jacking methods, remained valid for all tunnels and at each stage of the global evolution of the I-90/I-93 Interchange design.

The final design development was a complex, extensive and innovative process that "pushed the envelope" of engineering practice on a number of technical fronts including concrete technology, ground treatment techniques and soil/structure interaction. (Refs. 3, 6). Some of the key features of the final design include the use of "waterproof" concrete as an alternative to an external waterproofing membrane; stabilization of the soft material in the tunnel face and beyond by using a combination of chemical grouting, horizontal jet grouting, soil nailing and dewatering; and the provision of guide/pilot tunnels at the base of the final tunnels.

The final layout of the interchange with three jacked tunnels is shown in Figure 6. The tunnels have similar transverse dimensions (25m wide x 12m high) and have lengths that vary from 50m to 100m. A Baseline Design was established and, due to the highly specialized nature of the jacked tunnel construction, the CA/T Project established detailed pre-qualification requirements for contractors wishing to bid on the work (Ref. 5). All main contractors bidding for this section of the CA/T Project were required to be either pre-qualified themselves for the tunnel jacking, or to include on their team a pre-qualified tunnel jacking specialist.

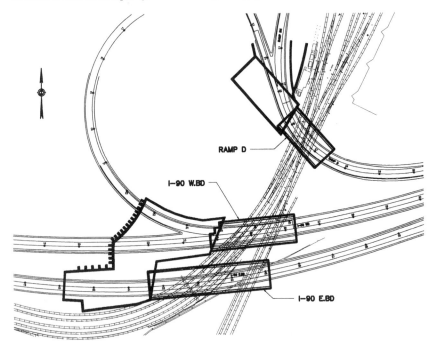

Figure 6. Final Layout of the I-90/I-93 Interchange with Three Jacked Tunnels

Value Engineering Changes Leading to Current Design

The contract which includes a portion of the I-93/I-90 Interchange and the I-90 jacked tunnel sections, designated Contract C09A4, was awarded by the MHD to the Slattery/Interbeton/JF White/Perini Joint Venture (SIWP) with a value of approximately $397M. Notice to Proceed was given in late January, 1997. Shortly after starting work on C09A4, SIWP submitted a value engineering change proposal (VECP) that proposed several significant changes to the bid design. These changes included the substitution of ground freezing for the dewatering, grouting and soil nailing techniques specified in the Baseline Design. This modification offered the advantages of enabling complete treatment of the soil mass prior to the start of tunneling, improved face stability and encapsulation of obstructions, and lower risk of ground losses as a result of "windows" in the ground treatment.

Another significant change offered by the VECP was the planned use of longer lead tunnel units, which, combined with the addition of an anti-drag system below the base of the tunnel box, is expected to provide increased directional stability during jacking. This modification has led to the elimination of the pair of guide tunnels included in the Baseline Design ahead of each tunnel to maintain tunnel alignment.

The VECP was accepted by MHD and B/PB after an extensive review, and the work is currently proceeding with the changes described above. The first tunnel section to be built, Ramp D, is scheduled to be jacked in mid-1999. Completion of the I-90 tunnels is planned for the end of 2001. Overall completion of the CA/T Project is scheduled for 2004.

Acknowledgments

The authors express their gratitude to the Massachusetts Turnpike Authority's senior management staff for the CA/T Project, and the senior project management group of Bechtel/Parsons Brinckerhoff, with special thanks to Tony Lancellotti and Jim Roop, for their support of the publication of this paper.

References

1. Drennon, C.B., "Pipe Jacking: State of the Art", Journal of the Construction Division, Proceedings of the American Society of Civil Engineers, Vol. 105, No. C03, September 1979.
2. Lambrechts, J.R., Winsor, D.S., and Roy, P.A., "Central Artery Project Innovations for Excavation and Tunnel Support," Proceedings, Central Pennsylvania Section, ASCE, Seminar on Excellence in Geotechnical Engineering, Harrisburg, PA, October, 1997.
3. Smirnoff, T.P., Taylor, S., Das, P.K. and Valenti, R., "World Class Jacked Tunnels: Boston Central Artery/Tunnel Project," Proceedings, 13th FIP Congress on Challenges for Concrete in the Next Millennium, Amsterdam, May, 1998.
4. Taylor, S., "An Introduction to Tunnel Jacking," Proceedings, Rapid Transit Conference, American Public Transit Association, June 1997.
5. Winsor, D. and Taylor, S., "The Development of a Baseline Design for Jacking Highway Tunnels Beneath Mainline Tracks at Boston's South Station," ASCE Geotechnical Special Publication Number 83, 1998.
6. "Digging it in Boston", Engineering News Record, August 19, 1996, pp. 24-28.

Design and Construction
of the DART Starter Line Tunnel Contract NC-1B
Dallas, Texas

Gunars Richters[1], M. ASCE, Charles Hood[2], M. ASCE,
and Tracy K. Lundin[3], M. ASCE

ABSTRACT

The Dallas Area Rapid Transit (DART) light rail starter system consisted of 32km light rail alignment to be constructed above grade, at grade and below grade. During early project planning, a 5.2km section was identified as being constructed below grade concurrently with the reconstruction of US75. This section included an underground station, two 600m sections to be constructed as mined tunnel, and 4km constructed as open cut.

In the spring of 1991 it became apparent that the approach of constructing the 5.2km section concurrently with the reconstruction of US75 was schedule prohibitive. Therefore DART requested the Principal Section Designer (PSD), a team led by Huitt-Zollars, Inc. with Sverdrup Civil Inc. as the primary subconsultant, to prepare a study to evaluate construction options for this 5.2km section. This study indicated that constructing the entire 5.2km alignment as a bored tunnel was the only option that met DART's schedule for completion of the light rail starter system.

In May 1991, the PSD team was given six months to prepare Construction Contract Documents for the underground CityPlace Station and the 5.2km alignment as a bored tunnel. To meet the aggressive schedule a decision was made to bid the work with the design of the tunnel 100% complete and the underground CityPlace Station at a 30% complete stage. This approach enabled the fast track design to successfully meet the projects schedule constraints. Design and construction of Contract NC-1B was completed on time and within budget.

INTRODUCTION

DART had developed a master systems plan, which included the light rail system shown in Figure 1. To execute design and construction of the 32km light rail system, DART retained a PSD team led by Huitt-Zollars, which acted as Program Manager in March 1990. The Huitt-Zollars Team included Sverdrup Civil, Inc., Hellmuth Obata & Kassabaum, Inc., Barton-Ashman Associates, Inc., Arrendondo, Brunz & Associates, Inc., LS Transit Systems, Inc., and Southwestern Laboratories, Inc. DART also retained

---
[1] Geotechnical Consultant, Sverdrup Civil, Inc., Boston, MA.
[2] Senior Structural Engineer, Sverdrup Civil, Inc., Dallas, TX.
[3] Tunnel Principal, Sverdrup Civil, Inc., Boston, MA.

the services of a number of Section Designers (SD) who were to design various portions of the 32km system. The SD for mined portions of the work was the HDR Team that included tunnel consultants Lachel & Associates and Dr. G. Sauer Corp. The SD for pen cut portions of work was a team led by HNTB under contract to TXDOT.

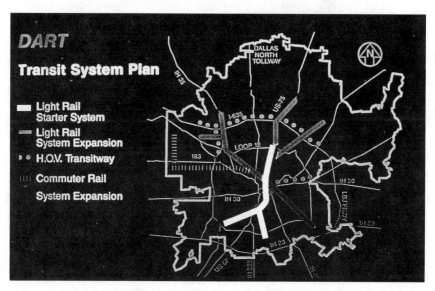

Figure 1: DART Transit System Plan

During early project planning, the 5.2km section was identified as being constructed below grade concurrently with the reconstruction of the rebuilt frontage roads of US75, also known as the North Central Expressway. This section included an underground station, two 600m sections to be constructed as mined tunnel and 4km constructed as open cut. The two tunnel segments were located at each end of the 5.2km section and were required to avoid interferences with the Woodall Rogers Freeway on the south end, and Mockingbird Lane at the north end of the alignment.

On the south end of the alignment, the tunnel was located below the elevated interchange between the Woodall Rogers Freeway and US75 and was aligned to avoid drilled shaft supported piers and dipped below US75. The north portal was located about 9m north of Mockingbird Lane and had about 8m of cover between the tunnel crown and the pavement. The HDR Team designed the two 600m tunnels as being either conventionally mined or mined using the New Austrian Tunneling Method (NATM). The remaining 4km long middle portion of the alignment was designed as conventional open cut construction.

The key to this approach was that the large open cut portion of the contract was to be constructed concurrently with US75 reconstruction. However, by the spring of 1991 the schedule for the US75 work was slipping and concurrent construction was no longer deemed possible. DART's concern was that the 5.2km section would not be completed during the same time frame as the entire 32km starter system (Note: as of January 1999 reconstruction of US 75 continues). Therefore, in March 1991, DART requested the PSD to prepare a study to evaluate construction options for this 5.2km section which could satisfy their primary constraint which was schedule. Additional constraints included minimizing project costs and keeping the alignment on or under public and/or right-of-way controlled by DART.

The study of construction alternatives included consideration of a bored or mined tunnel and an elevated structure which were both compared to the original open cut alternative. The study concluded that a 5.2km bored tunnel alternative:

- was the only alternative that could be constructed in a time frame consistent with completion of the overall 32km light rail starter system,
- would greatly reduce utility coordination and relocation,
- would minimize the impact to US75 reconstruction whenever it were to occur and,
- would reduce traffic impacts caused by US75 reconstruction by having the DART starter system available to commuters.

The study estimated the cost for design and construction of the bored tunnel alternative to be $194 million.

As a result of the study, DART requested the PSD to prepare construction contract documents using a bored tunnel for the 5.2km starter segment. To meet an aggressive six month schedule for this task a decision was made to bid the work with the design of the tunnel 100% complete and the underground CityPlace Station at a 30% complete stage. The PSD worked closely with the SD to incorporate previous design efforts into Contract NC-1B. Bids were opened on December 17, 1991 and a construction Notice To Proceed (NTP) was issued to the S.A. Healy Company on February 29, 1992.

SITE GEOLOGY

At the time Contract NC-1B was being designed, several other underground projects were being executed near the project site and in the same geologic media as would be encountered on the DART project. Nearby projects included the Superconducting Super Collider (SSC) project in Waxahachie, the Cole Park Detention Vault and two drainage tunnels feeding into the detention vault in Dallas, and the reconstruction of US75 which was planned as being depressed below grade. The SSC and Cole Park projects are described in PB/MK, 1992, and Freeland and Nordell, 1993, respectively. Geotechnical investigations for these projects showed that, in general, both the regional geology and mechanical properties of individual geologic units were consistent. During tunnel design, the PSD took advantage of all investigations for underground elements, which were under construction. All pertinent information collected during design was made available to bidders.

On a regional scale, the thick sequence of sedimentary rocks under Dallas dips southeastward toward the Gulf of Mexico. The units decrease in age from Paleozoic inland to Quaternary at the Coast. The outcroppings in the Dallas area belong to the Upper Cretaceous Gulf Series except for the thin deposits of alluvial deposits. The Gulf Series includes the Taylor, Austin, Eagle Ford, and Woodbine Groups.

In the vicinity of the project site there were at least seven different geotechnical investigations performed for either DART or the US75 improvements as described by Wallis, 1992. The initial borings for US75 improvement were taken in 1967 and the DART studies were started in 1987. Once the decision was made to construct Contract NC-1B as a bored tunnel additional geotechnical investigations were performed including an investigation of the characteristics of bentonite seams and discontinuities at CityPlace Station.

At the project site, investigations revealed that facilities would encounter recent alluvial soils, Austin Chalk, and possibly the Eagle Ford Shale. The recent alluvial soils were primarily unconsolidated materials consisting of fills overlying terrace deposits that were primarily dark gray to tan, calcareous clay, silt, and sand. The basal portion consisted of stratified water bearing clay, sands, and gravels.

Most of the length of the tunnel would be excavated in light to medium gray Austin Chalk (microgranular calcite) which is found directly below the soil profile. Geologists subdivide the Austin Chalk based on fossil zones and characteristics of surface exposures. The lithic characteristics that distinguish subdivisions are variations of bed thickness, concentration of fossil materials, and marly zones containing bentonite developed from volcanic ash falls. The physical characteristics of Austin Chalk are quite uniform. The vertical profile of the tunnel was sited about 3m above the Eagle Ford Shale which underlies the chalk. The Eagle Ford Group is a marine shale consisting of a dark gray to black, calcareous to noncalcareous montmorillonite shale with high shrink/swell properties (PB/MK, 1992). Properties of these materials are presented in Table 1.

Normally groundwater was at a depth of about 7.5m. It appeared that, except in the terrace materials at the south portal, groundwater recharge was from ground surface and bentonite seams prevented groundwater from migrating deeper. This was observed in open excavations where water seeped out on top of the bentonite seams. However at the south portal, as much as 6m of groundwater was above the Austin Chalk in the lower terrace sands and gravels.

Typically in the Austin Chalk, two types of faults can be encountered; simple and major. A simple fault is characterized as a series of single, very tight faults similar to joints with displacements of up to about 1.5m (PB/MK, 1992). The spacing between these faults generally ranges from 30m to 100m as mapped in the US75 drainage tunnels. One major fault was encountered in the DART tunnels and was located just south of the CityPlace

Station. This fault was characteristic of major faults in the chalk and was associated with a shear zone and slickensides. As expected, this fault was permeable and yielded water when encountered in the tunnels.

Table 1
General Stratum Description

Stratum	Classification	Total Unit Weight (kg/m3)	Friction (deg.)	Cohesion (kPa)	Description
Soil Strata					
Man made Fill					Erratic mixture of earth & debris
Upper Terrace	CL	2.08	15	143	Deposits with occasional silt, sand, gravel and lime deposits
	CH	2.05	17	72	
Lower Terrace	Non-cohesive	2.1	35	0	Sand coarser with depth, gravel pockets in sand
Austin Chalk	Residual Clay	1.9	5	86	Limey CL to CH w/ occasional lime
Unweathered Austin Chalk					
Water Content, %		Range		2 – 20	Austin Chalk Limestone fresh bluish gray
		Mean		11	
Unit Dry Weight, kg/m3		Mean		2.1	
		Std. Deviation		0.13	
Unconfined Compression, MPa		Mean		17.5	
		Std. Deviation		4.0	
Weathered Austin Chalk					
Water Content, %		Range		2 – 19	Austin Chalk Limestone severely to highly weathered.
		Mean		13	
Unit Dry Weight, kg/m3		Mean		1.9	
		Std. Deviation		0.10	
Unconfined Compression, MPa		Mean		6.3	
		Std. Deviation		2.8	
Eagle Ford Shale					
Water Content, %		Range		13 – 17	Dark bluish gray-black shale.
		Mean		15	
Unit Dry Weight, kg/m3		Mean		1.9	
Unconfined Compression, MPa		Range		1.8 – 3.3	
		Mean		2.3	

TUNNEL ALIGNMENT

Initially the plan was to locate the light rail alignment under the surface frontage roads on both sides of the reconstructed US75 right-of-way. The Texas Department of Transportation (TXDOT) reconstruction of US75 sited the roadway grade approximately 6m below grade. To provide stormwater drainage for the depressed roadway, a 2.5m (in the south) and 5.5m diameter (in the north) TXDOT drainage tunnels were constructed under the US75 right of way to deliver flows to the Cole Park Detention Vault. Therefore the initial DART plan for Contract NC-1B was to site the light rail alignment with the rail grade in the frontage road roughly matching the US75 grade and above the TXDOT drainage tunnels.

Once the decision was made to construct the rail line as a bored tunnel, the vertical alignment had to be deepened to a point below the TXDOT drainage tunnels which drove CityPlace Station deeper. At the south end, positioning the final horizontal and vertical alignment came down to clearances of fractions of a meter to avoid interference with the Woodall-Rogers Freeway/US75 elevated interchange piers. In addition, the tunnel grade beneath the interchange also needed to rise to meet the adjacent contract's south U-Wall alignment which was a boat section formed by secant piles and a base slab. Subsurface conditions at this location were not favorable because of water bearing sands overlying the Austin Chalk. The south portal location was selected to avoid ramps and piers; however, this forced the portal to be located in mixed face conditions.

North of the southern portal, geotechnical investigations revealed the presence of a prominent bentonite marker bed up to 0.6m thick which varied in depth from 18m to 30m with the Eagle Ford Shale located at a depth of about 40m. Previous experience in the area, which was confirmed by project geotechnical testing, was that bentonite marker beds and the Eagle Ford Shale were to

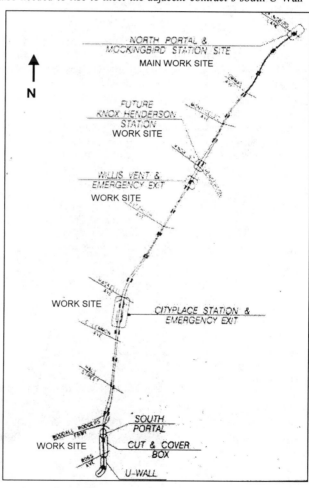

Figure 2: DART Contract NC-1B Plan

be avoided if possible to minimize adverse impacts to the project. To avoid these features, and to provide adequate clearance to the TXDOT drainage tunnel, the light rail bore was sited between the bentonite marker bed and the Eagle Ford Shale.

With the entire 5.2km project alignment located underground, new project requirements were established which included a minimum of two emergency egress exits and five connections to the surface for ventilation purposes. Several of these features were incorporated into the design of CityPlace Station which had an invert elevation 36m deep, 3m above the Eagle Ford Shale, and avoided the drainage tunnel.

Once the alignment was determined, the locations of work sites and construction staging areas were re-evaluated. The revised plan was for the tunnel to be mined using a Tunnel Boring Machine (TBM), therefore muck removal and construction material staging and transportation became more focused on a single work site as compared to a more widely distributed construction zone which was possible assuming open cut construction. Therefore the main construction work site and staging area was relocated from the south end of the contract to the north portal which offered with more flexible transportation access and a larger staging area. Since the north portal area was the planned location of Mockingbird Station (Contract NC-2) the design of the station had to be advanced to define the station configuration to ensure compatibility with Contract NC-1B activities. Additional work sites were provided at the south portal, at City Place Station, and at Willis Avenue as shown in Figure 2.

TUNNEL DESIGN

The tunnel and CityPlace Station were located primarily in the bluish gray unweathered Austin Chalk which is an ideal tunneling media. Except for the presence of bentonite seams the chalk is homogenous with self-healing properties, has sufficient strength to be self-supporting, and is soft enough for high production excavation. The exposed Austin Chalk in the air blast suppression chamber excavation at CityPlace Station is shown in Figure 3.

Figure 3: Air Blast Chamber Near CityPlace Station with the Ventilation Tunnel on the Right (Note the Exposed Austin Chalk)

Bentonite seams were of concern with respect to fallout of rock blocks or slabs because of their expansive properties, and the fact that they represented a plane of weakness in the rock mass. When located sufficiently high above the crown of the excavation bentonite seams played a positive role by acting as an aquaclude by intercepting groundwater seepage and preventing it from penetrating the excavations.

To address these concerns initial support for the running tunnel consisted of patterned rock dowels. The design also incorporated 100mm of fiber reinforced shotcrete because exposed Austin Chalk weathers rapidly and can begin to ravel. See Figure 4. In addition, a detail was developed to over-excavate bentonite seams encountered and replace the bentonite with concrete.

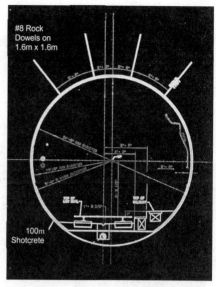

Figure 4: Running Tunnel Section

Subsurface investigations and construction of the nearby TXDOT drainage tunnels indicated that groundwater would be encountered when tunneling through the fault zones of the unweathered Austin Chalk. To control groundwater flows encountered in localized areas during construction, a detail using filter fabric was provided to direct seepage to a drain in the tunnel invert. An impervious membrane that was later covered with a 230mm thick cast-in-place concrete final lining retained the filter fabric.

During construction, groundwater was encountered in both running tunnels flowing from a fault located just south of CityPlace Station. The TXDOT drainage tunnel excavation also encountered groundwater from the same fault. This fault, and other locations where seepage was encountered, were controlled using either the waterproofing detail or by using panning to intercept flows and direct them to the tunnel drain.

SOUTH PORTAL

The south portal of the tunnel daylights at the deep U-Wall boat section of the adjacent contract. The Austin Chalk in this area is relatively deep and is overlain by water bearing sand and silty sand. As a transition between the boat section and the mined tunnel, an open cut section approximately 84m long consisting of slurry walls with tie-backs formed the ground support system that provided for the start of the mixed face mined tunnel. The slurry wall was founded in Austin Chalk to achieve a groundwater cut-off. Contract Documents permitted either NATM or standard tunneling methods using a roadheader with steel ribs and lagging. The water bearing sand in the crown was to be grouted at contractors option.

The elevated interchange between the Woodall-Rogers Freeway and US75 was supported on drilled piers and is shown in the background of Figure 5. The pier tips were within 1.5m of the mixed face tunnel springline. To complicate matters, a bentonite seam was located about 2.5m below the drilled piers and 1.5m below the tunnel springline.

Figure 5: South Portal with Woodall-Rogers Freeway and US 75 Interchange

To start mining the mixed face tunnel, the contractor first grouted the water bearing sand from the ground surface and installed spiling over the tunnel crown to stabilize the water bearing sands and to reinforce the weathered rock and soil/rock interface before breaking through the slurry wall.

Spiling was also used to reinforce the rock when tunneling adjacent to the piers. A roadheader was used to excavate both soil and rock. Steel ribs with timber lagging in the crown and shotcrete in rock were used for tunnel initial support. Tunnel final support was designed as a 230mm thick unreinforced concrete lining. During construction, the final liner thickness was decreased in some areas to accommodate alignment deviations and to maintain clearances for the dynamic envelope of the train.

CITYPLACE STATION

The CityPlace Station configuration, shown in Figures 6 and 7, provides for station access to the CityPlace Tower to the east. The station was sited north of the CityPlace Tower centerline, underneath and parallel to US75. This location provided direct access from potential development on the southwest side of the station and for future station entrances north of Haskell Avenue on both sides of US75.

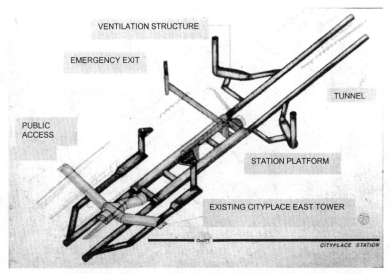

Figure 6: CityPlace Station Complex in Three-Dimensions

The overall platform configuration utilizes the "binocular" station design. This approach was chosen to limit the clear spans required and to minimize the volume of excavation. The platform width of 4.7m and length of 122m was established to accommodate possible future use of four car trains. Each track will be served by a side platform between the two tunnels. The distance between tunnel centerlines is 28m

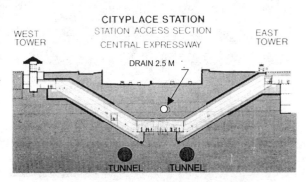

Figure 7: CityPlace Access Relationship to Rehabilitated US 75

and access between the two platforms was provided by 6m wide passageways at each end and at the midpoint of the platforms. A mezzanine level was located below the TXDOT diameter drainage tunnel, which served the depressed US75. To expedite the design, Dr. G. Sauer Corporation was retained as a subcontractor to the PSD Team to perform finite element analysis on critical station cross sections and to aid in the preparation of Bid

Documents. Analysis efforts also assumed the use of NATM construction techniques. Figure 8 shows typical station construction activities.

CityPlace Station is only part of the underground facilities at the track level. Ancillary electrical and mechanical equipment at track level provides services such as fire protection, water, sanitary and storm sewers, ventilation, emergency power, backup lighting, and emergency egress systems. Since the construction contract was bid at the 30% design completion stage, bidders were given sufficient structural detail to estimate quantities and price the work. However, due to a lack of definition regarding system design, a $16 million bid allowance was established for mechanical, electrical and plumbing work. DART retained the HDR Team including Lachel and Associates and Dr. G. Sauer Corporation to develop final drawings in nine months, and to deliver the design in 138 calendar days after NTP which was required to maintain the construction schedule.

Figure 8: CityPlace Station Installation of Groundwater Control at Cross Passage

In early 1999, the CityPlace Station finish construction contract was awarded. Since January 1997, the light rail trains have been transporting passengers to and from Mockingbird Station through the CityPlace Station without stopping.

NORTH PORTAL

The north portal was within 9m of Mockingbird Lane, a major east-west artery. The tunnel crown was less than 8m from the pavement surface and included about 4.5m of fill and terrace materials and 3m of weathered Austin Chalk. Borings indicated that joints in the area dipped vertically with a strike of 10 degrees off the tunnel alignment. Therefore the portal was reinforced with three rows of tensioned rock bolts with the middle row skewed to intercept the shear planes.

After award of the contract, the contractor proposed a change which was to rough grade a large portion of the adjacent NC-2 Contract which would allow better TBM access to the portal. As a result, the PSD designed a soil nailed retained wall to support the approach to the tunnel portal. The soil nail wall was used due to right-of-way restrictions and to fit the wall within the finished NC-2 Mockingbird Station configuration.

The contractor elected to construct both the southbound and the northbound starter tunnels simultaneously. However, during construction the unsupported rock between tunnel bores became over stressed which generated cracks at tunnel springline. The Construction manager, Stone & Webster, directed the contractor to install rock bolts through the cracks in a crisscross pattern, which stopped further crack propagation.

The contractor elected to use a conveyor belt to transport muck from the TBM to the stockpile at the portal. After TBM operations were fine-tuned, the contractor was able to establish production records with this construction technique.

CONCLUSIONS

1. NC-1B contract proved that Austin Chalk TBM-bored tunnels could be economical, less disruptive and faster than construction of elevated structures or cut-and-cover tunnels.

2. The DART NC-1B contract was constructed within budget and on schedule using a combination of 100% and 30% complete Bid Documents. This success can be attributed to cooperation of all the parties involved.

3. The Austin Chalk proved to be an ideal tunneling medium with less groundwater infiltration than expected.

ACKNOWLEDGEMENTS

The authors gratefully acknowledge the support of the Dallas Area Rapid Transit Authority and some of the key persons who made the project a success; Mr. Richard A Brown, Assistant Vice President, Facilities Engineering; Messrs. Ronnie Smith and Keith Smith, NC-1B Design Project Managers and Mr. Bo Cung, NC-1B Construction Manager.

REFERENCES

(1) PB/MK Team, "CCU A-650, N25 to N40 Tunnel Basic Construction Geotechnical Design Summary Report, Addendum No. 1", for the Superconducting Super Collider Laboratory, 27 April 1992.

(2) Freeland, P.E., and L.A. Nordell, "Cole Park Detention Vault Project" Rapid Excavation and Tunneling Conference Proceedings, Boston, 1993.

(3) Wallis, Shani "Dallas Area Rapid Transit", World Tunnelling, May 1992.

Design And Performance of Large Tunnel
Constructed in Saprolite

William H. Hansmire[1], F. ASCE
Lee W. Abramson[2], M. ASCE
Glenn M. Boyce[3], M. ASCE
Clayton S. Mimura[4], M. ASCE

Abstract

Large highway tunnels were constructed as part of the Interstate Route H-3 in Honolulu, Hawaii. At one portal, these highway tunnels were started in saprolite, an extremely to highly weathered basalt that has the character of hard soil. Excavated size was about 15 m wide and 12 m high (49 ft by 39 ft). Finished width after concreting is about 13 m (43 ft). Exploration, engineering and design, construction, and performance during tunnel excavation are summarized. History of the tunnel is given with respect to contemporary projects in the United States and to significant precedents in Hawaii.

Introduction

Interstate Route H-3 is a 4-lane highway that traverses the topographic features of Haiku Valley, the Koolau Range of mountains, and North Halawa Valley. The tunnel is about 15 km (9 miles) from the urban center of Honolulu, Hawaii. The highway is a major transportation link across the island, and is the third and most ambitious to be constructed. A pair of tunnels, commonly referred to as the Trans-Koolau Tunnels and each about 1.6 km (1 mile) long, were constructed through the mountains. At the Halawa Portals, the closest to Honolulu, about 60 m (200 ft) were tunneled in saprolite.

Extensive exploration and ground characterization took place during design. Special techniques not common practice at that time in Hawaii were used to recover continuous samples of saprolite that permitted detailed geological characterization and laboratory testing. Pilot drifts in the top headings of the tunnels were excavated to permit

---
[1] Principal, Jacobs Associates, 500 Sansome Street, San Francisco, CA 94111
[2] Black and Veatch, 720 3rd Avenue, Seattle, WA 98104
[3] Senior Professional Associate, Parsons Brinckerhoff Quade & Douglas, 303 Second Street, Suite 700 North, San Francisco, CA 94107
[4] President, Geolabs Hawaii, 2006 Kalihi Street, Honolulu, HI 96819

in-situ testing and detailed geologic mapping. See Abramson and Hansmire (1989b) and Boyce and Abramson (1991a,b,c) for more details. See Fig. 1 for the geologic profile and ground characterization summary, and Fig. 2 for the typical tunnel cross section.

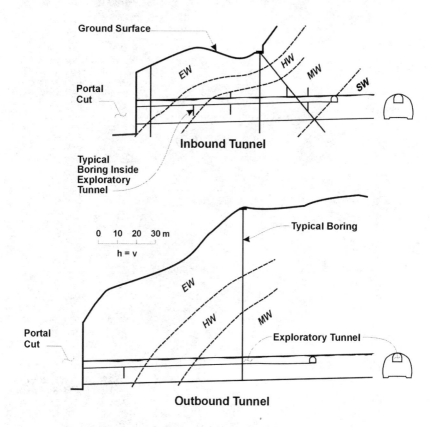

Figure 1. Geologic Profiles and Zones of Weathering
Inbound and Outbound Tunnels at North Halawa Portal
Trans-Koolau Tunnel

Exploration started in 1988, exploratory tunneling started in 1989, and contracts for the highway tunnel construction were awarded starting in late 1990. The tunnels were satisfactorily completed by 1994 and the entire highway was put into service finally in 1997. Construction history of the whole tunnel is given in Hansmire et al (1993).

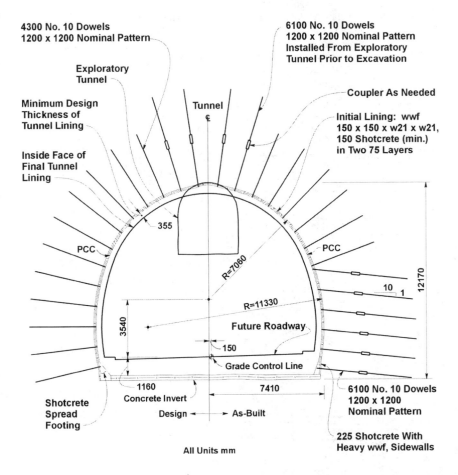

Figure 2. Typical Cross Section for Tunnel in Saprolite

Historical Perspectives and Precedents

Many large tunnels were designed and constructed with the Interstate Highway System. The Trans-Koolau Tunnel was the last to be completed. (The Trans-Koolau Tunnel was dedicated December 12, 1997 as the Tetsuo Harano Memorial Tunnel. Harano was a Hawaii Department of Transportation employee for over 50 years and was Chief Engineer for many of those years.) The whole H-3 highway project with the tunnel as a key element was a landmark of delay. Started in the late 1960's, it was not opened until 1997, about 30 years later. In the end the public has found the highway both useful and beautiful. Professionally the Interstate Route H-3 project including the tunnel was recognized by receiving the 1998 Outstanding Civil Engineering Achievement Award by the American Society of Civil Engineers (ASCE, 1998).

Historical precedents and how they affected this project are reviewed in the following paragraphs. The design process was certainly affected by the dramatic and unfortunate experience much earlier with a major failure during construction of the nearby Wilson Tunnel. Also a highway tunnel in a similar geologic setting and about 4 km (2.5 miles) from the Trans-Koolau tunnel, the Wilson Tunnel experienced a catastrophic collapse in 1954 during construction. The first bore was eventually completed a few years later by soft ground tunneling using hand-mining techniques. The tunnel drive had started in rock, but toward the end of the drive weathering increased and highly weathered rock to eventually residual soil conditions were present. Driving a single large tunnel heading, fast raveling ground, and un-timely ground support were major issues in the failure. A second bore was completed without incident with appropriate construction procedures.

Peck (1981) gives the definitive summary of the Wilson Tunnel and the many years of delay in publication reflects deference to sensitivities of the families of workers who lost their lives in the collapse. The everyday impression of people who remembered the incident decades later was that the tunnels were somewhat mysterious and dangerous. This was a real public concern that the Trans-Koolau tunnel planners and project engineers had to deal with. Ralph Peck's experience on the Wilson Tunnel made him the natural choice as the key consultant for design of the new tunnels.

Tunnel design and construction precedents elsewhere in the United States are relevant. In another mountainous situation, the first bore of what is loosely termed the Eisenhower Tunnel (originally the Straight Creek Tunnel) in the Colorado Rocky Mountains experienced substantial difficulties in construction. It was completed in the 1970's after great publicity and controversy. The tunnel in poor ground conditions was completed with a multiple–drift scheme. See Hopper et al (1972), whose paper greatly underplays the drama associated with that project. The second bore was designed from the start with multiple drifts in the poor ground and pioneered risk-sharing contracting practices of Disputes Review Boards and Escrow Bid Documents (see McOllough, 1981). A. A. Mathews was a consultant to the contractor for the first Eisenhower tunnel and was involved in the second tunnel design. With this background, Al Matthews also became a consultant to the Trans-Koolau tunnel design when it restarted in the 1980's.

In parallel to design and construction of highway tunnels, many advances were taking place with cavern and tunnel construction on the Washington, D. C. Metro. Cording (1984) documents the industry in a time of both technological and attitudinal change regarding design for large underground structures. The transit stations that are described in Ed Cording's paper are different from highways tunnels in many significant ways. Both require, however, design and construction of initial and final ground support of large span underground openings in rock, and sometimes very poor rock or soil like conditions. Later in the 1990's (see Heflin et al, 1991), shotcrete and dowel supported underground structures with membrane linings for waterproofing had gained full acceptance in the United States. The Washington, D. C. Metro was the arena in the United States where ground *reinforcement* by rock bolts, dowels, and shotcrete replaced ground *support* by structural steel as the first choice of many designers and constructors. It was also where risk-sharing and disputes resolution contracting practices were implemented on a large scale and culminated in setting industry guidelines (ASCE, 1989). A whole generation of engineers gained tunnel design and construction experience on the many projects in Washington, D. C. The H-3 project benefited from that experience.

Early designs on the Trans-Koolau tunnel reflected their time in history, as documented in Dodds and King (1972) and Mandel (1974). Project records also show a design in the 1970's for an earlier H-3 highway alignment that had large invert drifts in the tunnel section (for what was called the Red Hill Tunnel) that were clearly a reaction to the Wilson Tunnel experience. Dodds and King focused on instrumentation in a pilot tunnel to learn about ground and ground reinforcement behavior as the key to establishing a tunnel design. Of note in their paper is the concept of a 'rib in rock' ground support scheme where bundled steel reinforcing is shotcreted in place to form ground support. This concept had come from Europe and this thinking was definitely a precursor of modern day use of lattice girders and shotcrete in the United States and world-wide for initial tunnel linings in soil and rock. Dodds and King were trying to change things, but were stopped by project delays.

The use of pilot bores for large tunnels also went through substantial development during this time. Abramson and Hansmire (1989a) summarize the rationale for what was actually done in Hawaii. An Exploratory Tunnel was constructed to serve also as a drain below the whole tunnel. In the saprolite conditions however, pilot tunnels as top heading excavations for the future large tunnels were constructed as a part of an in-depth exploration and in-situ testing program (see Figs. 1 and 2).

In retrospect, the Trans-Koolau tunnel benefited from the long delay from its inception in the 1960's to its completion in the late 1990's. During this delay, which spanned a generation, the experience and lessons learned on other tunnels, from technology and materials improvements, to changes in design practices, were usefully employed on this project.

Exploration and Site Conditions
    Site geology is weathered basalt that originated as lava that form the Hawaiian Islands. Where this volcanic rock is highly to extremely weathered, it is known as saprolite. Saprolite acts like a clayey silt material when disturbed. It commonly loses much of its strength when drilled or excavated. This sensitivity to disturbance was a key factor in past problems in tunnel construction. For this reason, great attention was given to characterizing the saprolite. See Boyce and Abramson (1991a,b,c). Three exploration

programs were undertaken to retrieve core sample for testing. Within the saprolite, the standard practice had been to use split spoon sampling techniques. For the H-3 project, foam and polymers were introduced to the drilling fluids for the first time to retrieve intact core samples for testing. Degrees of weathering were defined for exploration and geologic mapping purposes. The definitions were based on the physical disintegration due to chemical alteration of the minerals in the rock. Terms and abbreviations used to describe weathering shown in Fig. 1 were:

<u>Extremely Weathered (EW)</u> - Original minerals of the rock almost entirely altered to secondary minerals, even though the original fabric may be intact. Also termed saprolite.

<u>Highly Weathered</u> (HW) - The rock is weakened to such an extent that a 50 mm (2 in.) diameter core can be broken readily by hand across the rock fabric. Also termed saprolite.

<u>Moderately Weathered</u> (MW) - Rock is discolored and noticeably weakened, but a 50 mm (2 in.) diameter core cannot usually be broken by hand across the rock fabric.

<u>Slightly Weathered</u> (SW) - Rock is slightly discolored, but not noticeably lower in strength than fresh rock.

<u>Unweathered</u> (UW) - Rock shows no discoloration, loss of strength, or any other effect of weathering. Also termed fresh rock.

Analysis and Design – Initial Tunnel Lining

Numerical analyses were a key part of design to evaluate constructibility and adequacy of the initial tunnel ground support. Boundary element analyses (BELP) (Hoek and Brown, 1980) provided a basis for evaluating length and density of ground reinforcement to maintain mass strength of the ground about the tunnel. The intent was to achieve a design that had an adequate theoretical basis, yet be practical to construct. It was possible to demonstrate by analysis and judgment that the tunnels could be built in a 'top-down' sequence, rather than in a 'bottom-up' stacked drift sequence which had been used in the past for other large tunnels constructed in what was perceived to be poor ground. The following assumptions were made during the design:

- Initial ground support: permanent. Resin grouted.
- Final lining: cast-in-place concrete. Some long-term structural capacity.
- Ground water: drained.
- Tunnel separation: far enough apart so neither tunnel affects the other.
- Excavation: mechanical.
- Rock behavior: poorer than in exploratory tunnel because of size effects.
- Stiffness of saprolite: 140 MPa (20,000 psi).

A range of ground strength parameters were used in analysis. The most representative and the basis for design are summarized in Table 1. Typical analysis for each stage of construction is shown in Fig. 3.

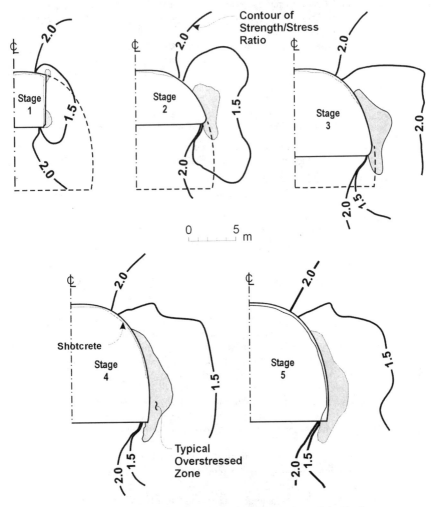

Figure 3. Design Analysis: Stresses About Tunnel At Each Stage of Excavation

Contours of stress to strength ratios are shown and are analogous to a factor of safety. Stress concentrations and yielding of the ground are clearly exhibited in the lower sidewalls of the tunnel opening. It was on the basis of these analyses that top down construction was considered feasible and the length of ground reinforcement of 4.3 m (14 ft), slightly more than one fourth of the tunnel diameter, was established.

Analysis Case	Mohr-Coulomb Failure Criteria		Hoek and Brown Failure Criteria			K_0	Support Pressure, kPa (psi)
	c, kPa (psf)	Ø, Degrees	m	s	σ_c, kPa (psf)		
Original	335 (7000)	10	0.7	1	865 (18100)	0.4	0
Revised	215 (4500)	10	0.7	1	865 (18100)	0.4	40 (6)

Unit weight of saprolite: 1680 kg/m³ (105 pcf).
Tunnel depths: Inbound 18 m (60 ft), outbound 30 m (100 ft)

Table 1 Ground Strength and Analysis Parameters

The permanent ground support for the tunnels in saprolite is shown in Fig. 2 and summarized as follows:

- Previously excavated exploratory tunnel: becomes center top heading of new tunnel and had four, 6 m (20 ft) dowels at a 1.2 m (4 ft) longitudinal and transverse spacing and 150 mm (6 in.) thick reinforced shotcrete, installed before main tunnel excavation.
- Side slashes of heading and benches: 4.3 m (14 ft), No. 10 (32 mm) dowels at a 1.2 m (4 ft) longitudinal and transverse spacing and 150 mm (6 in.) reinforced shotcrete. Installation before next advance.
- Bench: reinforced shotcrete and 4.3 m (14 ft) dowels installed in bench side before making the next advance.
- Shotcrete lining footings: shotcreted at the base of the bottom side slashes.
- Supplementary support of dowels and shotcrete as needed in localized areas.

As construction problems developed on the project (see below), a general concern developed for the stability of the tunnels. Strength characteristics of the saprolite and the tunnel analyses were reexamined. Considerable data had been used to establish the strength parameters and judgement had been used in selecting suitable parameters for analysis and a design basis. Pressuremeter tests in the exploratory tunnel indicated that the cohesion of the saprolite material was approximately 215 kPa (4500 psf), significantly less than the 335 kPa (7000 psf) used in the original design. It was thought that perhaps the value of 335 kPa was too high due to the fact that it was based too heavily on laboratory testing. The pressuremeter tests performed in the field may more accurately represent the strength characteristics of the saprolite, and thus be more appropriate for design. The re-calculation using the BELP program with the revised saprolite strengths agreed with the field observations and showed that the dowels needed to be lengthened.

The results of BELP indicated an overstressed zone as shown in Fig. 4, which was approximately 4 m (13 ft) out in the radial direction. This overstressed zone would not be

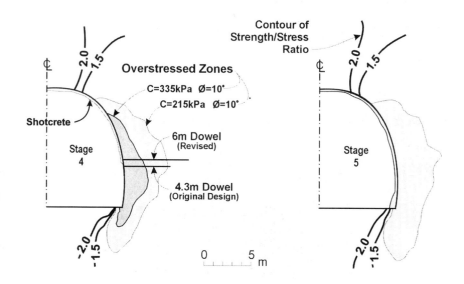

Figure 4. Updated Design Analysis: Stresses for Fully Excavated Condition

significantly reinforced by the 4.3 m (14 ft) dowels. The remedial design included lengthening the dowels to 6 m (20 ft) and adding shotcrete, thereby increasing the confining pressure within the tunnel. Revisions were made to the support and tunnel construction sequence for the outbound tunnel as follows:

- All new dowels in saprolite: cement grouted.
- Sidewall dowels: increased to 6 m (20 ft). No. 10 bars (32 mm) cement grouted in a 100 mm (4 in.) diameter hole. Dowels added to portions of the bench already excavated.
- Additional shotcrete: 75 mm (3 in.) installed for the first 7 m (24 ft).
- Shotcrete footing: increased to 1.2 m (4 ft).
- Excavation and sequence limitations revised: 1.2 m (4 ft) increments for full height and width of bench, 1.2 m (4 ft) limit per 8 hours for initial increments,

to allow shotcrete to set, for first 12 increments, which was about 15 m (48 feet) beyond the area of shotcrete cracks. Saprolite invert treatment was required after 2.4 m (8 ft) advance of bench.

Final Tunnel Lining

Design of the final cast-in-place concrete tunnel lining relied upon the substantial initial ground support that was installed. The initial support was assumed to prevent significant loading of the concrete final lining. As finally analyzed, a vertical load of about one-fourth the tunnel width (equivalent to a height of saprolite of 4.3 m (14 ft)) was input to structural analysis. Structural capacity for this loading would also accommodate contact grouting pressures. The resulting design was 355 mm (14 in.) minimum thickness concrete reinforced with No. 8 bars (25 mm diameter) at 300 mm (12 in.) spacing (150 mm (6 in.) in the crown). For just the tunnel as a circular shape with the final lining, the flexibility ratio is about 50. Actual flexibility is less with the concrete ceiling and roadway. Overall, the goal was achieved of a fairly flexible lining of minimal thickness with capacity to deform and take load if necessary. There has been no indication that the final lining is taking significant load.

Construction Experience and Performance of Tunnels

The story of how the tunnels performed is complicated by several activities proceeding in parallel and construction problems. This narrative simplifies the history by first summarizing key events for the portal excavations, next summarizing the inbound tunneling, and finally presenting the outbound tunneling.

Portal Excavations and Initial Tunneling

Portal excavations had a direct influence on the tunnel design and construction. Late in the summer of 1991 the scene was one of simultaneous portal cut excavation in many stages and top heading excavation of the outbound tunnel. Portal cuts were high (50 m or about 165 ft) and there had always been concern for stable slopes. As the project developed, one outright slope failure and on-going concern for the remaining slopes strongly affected getting the tunnels in saprolite completed.

The terrain at the portals is mountainous and naturally steep. In the vicinity of the tunnel portals, weathered slopes in rock stand vertically. In saprolite, vertical slopes were known to be stable for months or years. The topography, however, is a continual scene of mass wasting where oversteepened slopes fail (usually in times of wet weather) as shallow slides, and occasionally, with a deep seated failure surface. One deep seated example of a major slope failure occurred in a cut on the access road to the tunnel. The scarp of this landslide can be seen in the background on the cover of the January 1995 issue of *Civil Engineering*. At the tunnel portals, there was no geologic evidence that a deep seated failure was credible. Ironically, the project had started out assuming that some type of major excavation support by tiebacks would be necessary for the portal cuts. An in-depth and long-term tieback testing program was implemented two years before portal construction and a prototype permanent tieback design was developed. During design it was judged that only reinforcement of the cut faces by 3 to 6 m long (10 to 20 ft) pattern dowels with welded wire fabric and shotcrete was required. The design of the long tiebacks was set aside.

As the first (outbound) tunnel was started and the cuts continued to be widened and lowered away from the outbound tunnel portal, tension cracks in ridges adjacent to the tunnel were noticed. Inclinometer instrumentation indicated the portal cuts had displaced toward the excavation on the order of 25 mm (1 in.) or more. Notable cracks in the outbound tunnel shotcrete lining also developed. The first 6 m (20 ft) of the tunnel was pre-supported with horizontal spiling in the upper 120°. Distinct circumferential cracks as wide as 25 mm (1 in.) developed just beyond the spiling. The cracking was considered to result from the natural horizontal stretching of the mountainside that was occurring with excavation. Where the lining horizontal reinforcing dramatically changed, at the end of the spiling, cracking was concentrated in a zone about equal to one-half tunnel diameter. At the time it was not considered to be indicative of a problem with tunnel stability. Despite concerns for tension cracks developing, both the cracks in the ridge and the tunnel lining were recognized as a result of portal excavation, not necessarily tunnel advance, which was by that time well beyond any influence at the portal. Convergence was being measured that would be of concern if the tunnel were in hard rock, but not so much of a concern if the tunnel were considered to be in hard soil.

Thus at the end of August 1991, the project was dealing with its first major tunnel heading, some disconcerting cracks in the tunnel lining, and cracks in adjacent ridges of the portal cuts. Then on August 31, 1991, portal cut face 'Q' failed (cut faces for the complex portal geometry were given alphabetical references, starting with 'A'). Cut face Q is in the approach cut to the inbound tunnel and not directly associated with the then current outbound tunneling. This failure brought all work to an immediate halt while engineering solutions to the problems were worked through. One group on the project felt a major slope failure was possible at the portal of the outbound tunnel. See Abramson et al (1993) for more discussion on the portal excavation and this failure.

Definite measures were taken to ensure stability of the portal cuts and tunnels. The geological engineering debate on stability of the outbound portal cut, and what safety factor might exist for the present condition, was set aside. Definite measures were taken in order that the project could proceed. Cut faces were reinforced with longer dowels. Long tiebacks were installed over the outbound tunnel to further ensure portal stability. The design was readily adopted from the prototype tieback testing that had been conducted. Ground support in the tunnels was increased as addressed in detail elsewhere in this paper. Tunneling was switched to the inbound tunnel while the remedial measures were implemented.

Inbound Tunel

Inbound tunnel construction was started cautiously. Considering the past problems with the slopes and the uncertainty of how the partially excavated outbound tunnel would eventually perform, no chances were taken. Inbound tunneling was very orderly in that the top heading was advanced completely through the saprolite and beyond for more than 60 m (200 ft) before any bench excavation was started. Top heading excavation proceeded in 0.6 m (2 ft) increments for 23 m (76 ft) of tunnel, then increased to 1.2 m (4 ft) increments.

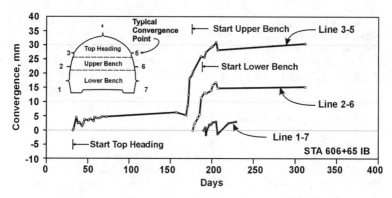

Figure 5. Convergence Measurements in Inbound Tunnel

In this situation, inbound tunnel instrumentation was given special emphasis and reliable measurements were obtained. Seven lines of convergence point instrumentation were installed near the portal in the typical array indicated in the tunnel cross section inset in Fig. 5. Convergence is the shortening between the two measuring points on the tunnel walls. In all cases, the difference between the convergence lines, which are geometric chords across the tunnel, and true diametric positions was ignored. In the end, only the horizontal lines in the top heading and upper bench were useful.

Convergence slowed distinctly with time and with distance of tunnel excavation from the reference points. Fig. 5 is characteristic of the measurements. Movements for each stage of excavation are shown. These are relative movements from the time of instrument installation, and not total displacements. Some movement had to have taken place during the Exploratory Tunnel excavation, which had been about 2 years earlier. Movement was also missed because convergence points could not be installed until some top heading or bench excavation took place, which in general was about 6 to 9 m (20 to 30 ft) or about 40 to 60 % of the tunnel diameter. Despite these limitations, a useful and very orderly pattern of movement was recorded for all of the instrument lines in this tunnel.

Only small inward movement of about 5 mm (0.2 in.) was recorded for the top heading excavation. It is believed by the authors that, for the top heading especially, most movement occurred before the convergence instrumentation could be installed. Much more notable from the practical perspective was that no significant movement was occurring with time. About 20 mm (0.8 in.) additional movement occurred with upper bench excavation. Another 5 mm (0.2 in.) occurred with excavation of the lower bench. As with the top heading essentially no movement occurred with time for the bench excavations. Final readings were made almost one year after excavation and the maximum total convergence recorded remained about 30 mm (1.2 in.).

As noted above, the convergence measurements by their nature cannot measure total movement from the pre-excavation conditions. If only half of the convergence were recorded, true total convergence would be about 60 mm (2.4 in.), which is about 0.4% of the tunnel diameter. This convergence compares to about the same amount of theoretical elastic displacement for excavation of an unlined tunnel in a uniform state of stress equivalent to the overburden load and a material stiffness of the saprolite. The conclusion was that the tunnel, particularly with the ground reinforcement, was behaving within the elastic range of deformation of the saprolite.

Outbound Tunnel
Compared to the inbound tunnel, this tunnel exhibited much more deformation and had a much more complex history of construction. It was started first and finished last. The top heading had advanced over 60 m (200 ft) before tunneling stopped and the remedial works were implemented as described previously.

Convergence data is shown in Fig. 6. Unfortunately, the instrument lines were not installed until the top heading was about 30 m (100 ft) distant, but the data still showed that the tunnel was converging distinctly with time, which accumulated to as much as 50 mm (2.0 in.) during partial bench excavation and shutdown for portal remedial work. The tunnel continued to creep with time until additional ground support was installed. The top heading convergence as measured was ultimately 60 mm (2.4 in.) and the convergence line across the upper bench (which was installed in a timely manner) recorded ultimately 46 mm (1.8 in.) of movement.

The convergence measurements were made where the ground was most probably the weakest for all tunneling in saprolite. In addition compared to the inbound tunnel, the overburden loading was distinctly higher. With all of these contributing factors, it is not surprising that this tunnel deformed about twice that of the inbound tunnel. If the same rationale is used as for the inbound tunnel, the actual convergence would be estimated as twice what was measured, for a total of 120 mm (4.6 in.), or about 0.9% of the tunnel diameter. Had the initial lining not been strengthened, the movement would have been substantially more. For comparison, theoretical elastic amount is on the order of 80 mm (3.1 in.).

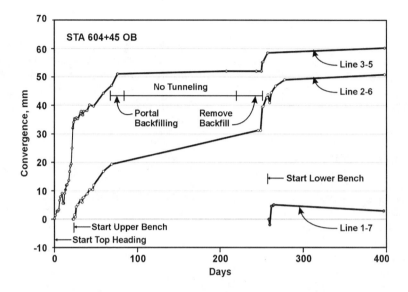

Figure 6. Convergence Measurements in Outbound Tunnel

Clearly, this tunnel needed the additional ground support. The measurements indicated that deformations were continuing and would likely become unacceptable. The cut face 'Q' failure, however, overtook the on-going engineering evaluations of just the tunnel convergence and mandated a positive plan of strengthening the outbound tunnel lining.

Instrumentation for these tunnels, although not perfect, was very successful and quantitatively demonstrated stability of the tunnel with the initial lining. The instrumentation success is attributed to the coordinated engineering efforts of people knowledgeable about the design and expected behavior being on site to monitor the instruments.

Conclusion

Large tunnels were successfully constructed in saprolite, a very weak rock having the character of a hard soil. Several factors combined favorably to make the project successful: thorough site investigation with emphasis on complete geologic and geotechnical characterization, understanding and using concepts of ground reinforcement, analysis techniques that modeled behavior, learning from past experience, proactive engineering including geotechnical instrumentation during construction, and the will and faith to make it happen.

Acknowledgements

The owner is the Hawaii State Department of Transportation. Chief Engineer was Tetsuo Harano, for whom the tunnel was named. Douglas Tanaka was the State's Project Manager and was dedicated to working with the many issues this project encountered. His spirit in making the project a success is gratefully acknowledged. Members of the State's Tunnel Review Board were Ralph B. Peck, A. A. Mathews, Ronald E. Heuer, and Charles Metcalf.

The Authors were all at one time Parsons Brinckerhoff employees and were involved with this project from beginning to end. Bill Hansmire was the Project Manager for the whole tunnel project. Lee Abramson was the Project Engineer for the tunnel design. Glenn Boyce was the Geotechnical Engineer responsible for field testing, analysis, and design, which included personally performing the boundary element analyses. Clayton Mimura was first responsible for site investigation field work by Geolabs-Hawaii, and then provided engineering support with Parsons Brinckerhoff later in construction. Peter Bibbes of Parsons Brinckerhoff was the Design Manger.

References

ASCE (1998). "Paradise Crossed," *Civil Engineering*, ASCE, Vol. 68, July, pp. 42-45.

Abramson, L.W. and Hansmire, W.H. (1989a). "Exploration and Drainage – Combined Usage of Pilot Tunnels," *Proceedings of the International Congress on Progress and Innovation in Tunnelling*, Vol. 1, pp. 9-16.

Abramson, L. W. and Hansmire, W. H. (1989b). "Geotechnical Exploration for the H-3 Highway Trans-Koolau Tunnel," *Proceedings*, 1989. Rapid Excavation and Tunneling Conference, Los Angeles, SME Inc., Vol. 1, pp. 40-64.

Abramson, L. W., Hansmire, W. H., and Boyce, G. M. (1993). "Performance of Tunnel Portals in Weathered Rock," *Int. J. Rock Mech, Min Sci & Geomech. Abstr*, Vol. 30, pp. 1449-1452.

ASCE (1989). Avoiding and Resolving Disputes in Underground Construction, Successful Practices and Guidelines, Prepared by the Technical Committee on Contracting Practices of the Underground Technology Research Council, sponsored by the American Society of Civil Engineers (Construction Division) and the American Institute of Mining Engineers, 24 p. plus appendices.

Boyce, G. M., and Abramson, L. W. (1991a). "Interstate Route H-3 Alternative Dowel Testing Program," *Proceedings*, 1991. Rapid Excavation and Tunneling Conference, Seattle, SME Inc., pp. 121-140.

Boyce, G.M. and L.W. Abramson (1991b), "Plate Load and Pressuremeter Testing in Saprolite," Geotechnical Engineering Congress, 1991, Geotechnical Special Publication No. 27, ASCE, Vol. I, pp. 52-63.

Boyce, G.M. and L.W. Abramson (1991c). "Where Does Rock Begin for Three Highway Projects?", Detection of and Construction at the Soil/Rock Interface, Geotechnical Special Publication No. 28, ASCE, pp. 57-82.

Cording, E. J. (1984). "State of the Art: Rock Tunneling," *Tunnelling in Soil and Rock*, ASCE, pp. 77-106.

Dodds, D. J. and King, E. H. (1972). "Rock Mechanics Instrumentation, Trans-Koolau Pilot Tunnel", *Proceedings*, North American Rapid Excavation and Tunneling Conference, Chicago, SME, AIME, Vol. 1, pp. 683-700.

Hansmire, W. H., Critchfield, J. W., Nicholls, M. H., and Kolell, C. R. (1993). "Construction of Trans-Koolau Tunnel," *Proceedings*, 1993 Rapid Excavation and Tunneling Conference, Boston, SME Inc., 1081-1099.

Heflin, L., Wagner, H., Donde, P. (1991). "U. S. Approach to Soft Ground NATM," *Proceedings*, 1991 Rapid Excavation and Tunneling Conference, SME, pp. 141-155.

Hoek, E. and Brown, E. T. (1980). *Underground Excavations in Rock*, Institution of Mining and Metallurgy, London, 527p.

Hopper, R. C. Lang, T. A., and Mathews, A. A. (1972). "Construction of Straight Creek Tunnel, Colorado," *Proceedings*, North American Rapid Excavation and Tunneling Conference, SME Vol. 1, pp. 501-538.

Mandel, H. M. (1974). "Environmental Considerations in the Design of the Route H-3 Tunnels, Hawaii", *Proceedings of the International Congress on Progress and Innovation in Tunnelling*, Rapid Excavation and Tunneling Conference, Chicago, SME, AIME, Vol. 1, pp. 559-581.

McOllough, P.R. (1981). "Eisenhower Memorial Tunnel – How Colorado Department of Highways Improved Contracting Practices and Management," *Proceedings*, 1981 Rapid Excavation and Tunneling Conference, SME, pp. 1651-1667.

Peck, R. B. (1981). "Weathered-rock portion of the Wilson Tunnel, Honolulu," *Soft-Ground Tunneling, Failures and Displacements*, D. Reséndiz and M. P. Romo, eds, Balkema, pp. 13-22.

Soft Ground Tunneling for Taipei Rapid Transit Systems

Daniel H. Ju[1], Za-Chieh Moh[2] and Richard Hwang[3]

Abstract

This paper discusses a few problems encountered during shield tunneling for constructing the rapid transit systems in the City of Taipei. Also discussed are two cases in which the New Austrian Tunneling Method was used in soft ground. In one of these two cases, compressed air was used as an auxiliary measure for maintaining the face stability, and in the other ground improvement was applied. Although settlements were large, the success well proves the applicability of the method in very poor ground.

Introduction

The laying of the sewerage line along Mingtsu Road in the City of Taipei in 1976 is believed to be the first application of shield tunneling technique in Taiwan. Since then, the number of shield machines used, mainly for laying sewer lines and water mains, increased drastically year by year and reached its peak in 1994 when the construction of the Taipei Rapid Transit Systems (TRTS) was the most active. For constructing the Initial Network of TRTS alone, for example, 30 shield tunneling machines have been used. In addition to shield tunneling, the New Austrian Tunneling Method was used to mine two sections of routes of considerable lengths with success. In one of these cases, compressed air was used, and in the other ground improvement was applied, as auxiliary measures for maintaining face stability.

---

[1] Director, Northern District Project Office, Department of Rapid Transit Systems, Taipei Municipal Government, Taiwan
[2] President, Moh and Associates, Inc., Taipei, Taiwan
[3] Manager, Rapid Transit Projects, Moh and Associates, Inc., Taipei, Taiwan

In view of the soft nature of subsoils, these two cases are indeed milestones in the history of soft ground tunneling in Taiwan.

Taipei Rapid Transit System

The Initial Network, refer to Figure 1, of the Taipei Rapid Transit Systems consists of six lines, namely, the Mucha, Tamshui, Hsintien, Nankang, Panchiao and Chungho Lines, with a total of 79 stations and a total length of route of 86.8 km. The Mucha Line was open to revenue services in March, 1996 and the Tamshui and Chungho Lines were open in December of 1997 and 1998, respectively. The total ridership of all these three lines passed the mark of 100 million on 23 December, 1998. The rest of lines, except the extension of Panchio Line to Tuchen & the extension of Mucha Line to Neihu, will be open in the year of 2000. The Mucha Line is a medium-capacity system while the rest of lines are heavy- capacity systems. Mucha Line is now carrying an average of 40,000 passengers and Tamshui and Chungho

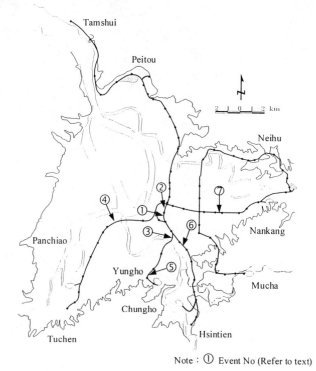

Note : ① Event No (Refer to text)

Figure 1. Initial Network of Taipei Rapid Transit Systems

Lines, together, are carrying an average of 200,000 passengers on each weekday.

Geology

An east-west and a north-south soil profiles across the Taipei Basin are presented in Figure 2. As can be noted that at the surface is a thick layer of the Sungshan Formation. Toward the east and the north, silty clay dominates while in the central city area, where the Taipei Main Station is located, the six-sublayer sequence is evident. Toward the west, the stratigraphy becomes complex with silty sand and silty clay seams interbedded in a rather complicated manner. Toward the south, ground becomes gravelly. A typical CPT profile obtained in the central city area of Taipei is shown in Figure 3 and the soil strengths obtained in laboratory tests are given in Figure 4. The soft nature of subsoils in the Sungshan Formation is readily apparent.

The Sungshan Formation is underlain by the so-called Chingmei Gravels which contains gravels and sands of various sizes and is extremely permeable and rich in water reserve. This gravelly layer was the sole water supply for the entire Taipei City prior to the 70's. It was responsible for several major failures during the underground construction of TRTS.

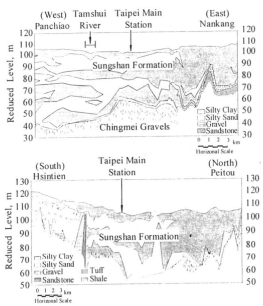

Figure 2. Geological profiles of the Taipei Basin

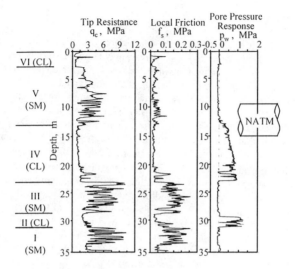

Figure 3. CPT profile in Central Taipei

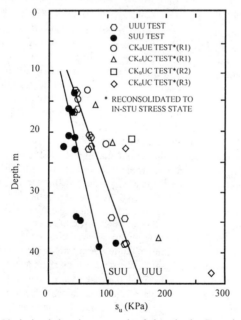

Figure 4. Undrained shearing strength of clays in the Sungshan Formation

Shield tunneling

Of a total length of 86.8 km in the Initial Network, 19 km was completed by shield tunneling. There were 58 tunnel drives, averaging 655m per drive, mined by using 28 earthpressure balancing type and 2 slurry type shield machines. Tunnels are either 5.4m or 5.6m in their inner diameters and reinforced concrete segments are typically 250mm in thickness and 1m in length. Tunnels are generally buried at depths of 10m to 20m below ground surface, with a few exceptions in which the tunnel inverts were as deep as 35m. The progress of tunneling was in general satisfactory except that at a few locations the progress was much hampered by obstacles. The first major obstacle was a 125mm (5 in.) diameter steel casing of a borehole left in place in a previous site investigation encountered in the Down-Track Tunnel in the Hsintien Line (Event 1 in Figure 1). Steel fragments choked the screw conveyor and had to be removed by sending a worker into the earth chamber. This was much more difficult than what one might expect because of the high groundwater table together with high permeability of sands in Sublayer V (refer to Figure 3). Chemical grouting was attempted in vain in front of the face to stop water from entering the chamber. The operation was abandoned because of the fear that further grouting might glue the shield to the ground to the extent that driving would not be able to resume. Finally, pumping was carried out to lower the groundwater table to a level below the tunnel invert for the worker to be able to enter the chamber and to stay there safely for removing the steel fragments and repairing the damaged conveyer.

Steel fragments frequently appeared in spoil removed from tunnels, however, other than the one mentioned above, no serious problems were reported. This was due to the fact that modern shield tunneling machines have sufficient power and the cutters are strong enough to cut steel members as long as they are not too large in size. There were cases in which small RC piles and thin sheet piles were cut through. On the other hand, there were cases in which ground treated by jet grouting was too hard for shield machines to go through. Figure 5 shows a situation encountered during tunneling in constructing the Tamshuei Line (Event 2 in Figure 1). A sinkhole of roughly 3m (10 in.) in depth and 75 m^3 ($1m^3$ = 1.3 cubic yard) in volume was found in front of the shield machine as the specialist subcontractor looked for a missing settlement rod installed for monitoring ground settlement. The cavity was covered by the RC pavement which did not show any signs of subsidence. In this case, jet grouting had been used to treat the ground at the back of the diaphragm wall to prepare for launching of the shield machine. Because the treated ground was too hard, driving of the shield was difficult since the very beginning and chemical had to be injected into the earth chamber as lubricant. Even so, worker had to go into the earth chamber to free the cutter from time to time. It was reported that the temperature of the spoil in the earth chamber was as much as 60°C (140°F). It is postulated that, as the cutter reached the end of the treated zone, a mixed-face situation was encountered. As most of the face was still in the treated ground, the shield advanced rather slowly. On the other hand, portion of the face was already in

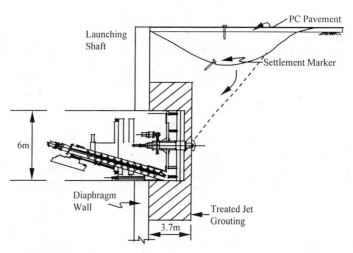

Figure 5. Problem with Ground Treatment during Launching of Shield Machine

the natural ground and soil could easily be excavated and "sucked" into the earth chamber. From what was observed and experience learned elsewhere, it may be concluded that if the treated ground has an unconfined compressive strength of 4 MPa (40 tsf) or above, shield driving is likely to encounter difficulties. This, however, certainly will depend on the capacity of the machine used and the uniformity of ground treatment. To avoid similar events from happening, it may be a good idea to have a transition zone with weaker strength at the end of the ground treated by jet grouting.

There were a few major collapses of ground which occurred either during launching from shafts or during arriving of shield machines at shafts. Although ground treatment is routinely carried out as illustrated in Figure 5, there is no way to ensure that treatment is uniform and the treated ground is perfectly watertight. The Chingmei Gravels is extremely permeable and is very rich in water reserve. Therefore, if the opening made on diaphragm wall is too close to the Chingmei Gravels, soil surrounding the water path quickly liquefied once leakage occurred and the flow usually became uncontrollable in hours. In one case in which leakage occurred when the tunnel portal was enlarged for installing the flexible joint (Event 3 in Figure 1), ground subsided by several meters and a section of tunnel was damaged. A total of 23 rings were seriously distorted and had to be replaced (Hwang, et. al., 1998). In another case in which ground subsided by several meters when the shield machine was making the breakthrough on the diaphragm wall (Event 4 in Figure 4), 39 rings in one tunnel drive and 34 in the other were damaged and two shield machines were submerged (Ju, Duann and Tsai, 1998).

Large tree trunks, up to 1.5m in diameter and 5m in length, were often encountered, usually at depths of 10m to 20m, during deep excavations (Ju, Kung and Duann, 1997). During TRTS constructions, pieces of wood were frequently removed from the spoil during tunneling, however, few problems were reported. A large tree trunk nevertheless did stop the shield machine during excavation for the Up-Track tunnel of the Chungho Line (Event 5 in Figure 1). It was reported that the daily progress rate was reduced from 43 rings (1m per ring) to 4 rings on the day prior to the event. It appears that this tree trunk had been pushed by the shield machine by more than ten meters. As the advancement of the shield machine was obstructed, the earthpressure balancing mechanism was destroyed and soil was "sucked" into the earth chamber in a manner similar to what was observed in Event 2 (refer to Figure 5). A sinkhole of 5m (16 ft) in diameter occurred right above the head of the shield machine. Jet grouting was carried out in front of the shield and compressed air was applied for workers to enter the earth chamber to free the cutter. Two pieces of drift wood, 500mm (20 in.) and 400mm (16 in.) in length, were recovered in the earth chamber.

NATM Tunneling

In this part of the world, the so-called "New Austrian Tunneling Method" appears to have deviated from its original context of being principally an observational method for tunneling and has been adopted to mean nearly all types of tunneling without using shields. However, the essence of the method is missing in the way that contracts are rigid and do not allow for the flexibility of varying the designs during constructions.

The twin tunnels in a 222m section in th Contract CH221 of the Hsintien Line were bored by using the NATM method (Event 6 in Figure 1). They were buried in the Sungshan Formation with their crowns at depths varying from 8m to 11m below the ground surface, refer to Figure 3 for profile. The soft ground called for the use of compressed air to a maximum of 1.35 bar. Construction was carried out in such a way that the two tunnels were inter-connected, as shown in Figure 6, by a cross drift so that both tunnels were able to be pressurized by using a single set of compressed air facility. Excavation was carried out in five stages. Stage 1 excavation was carried out in free air for providing a space to house the compressed air plant. The rest of excavation was carried out in compressed air. Air pressure was not released till both tunnels were fully excavated and primary lining was completed.

The upper heading was kept at a distance of 2m to 4m ahead of the lower heading. As depicted in Figure 7, lattice girders were installed at 1m intervals and the tunnels were protected by shotcrete, 250mm in thickness, and wire mesh as primary lining. For maintaining the stability of the headings, steel lagging sheets, 6mm in thickness, 200mm (8 in.) to 300mm (12 in.) in width and 2m (6 ft) in length, were closely spaced to make a canopy. The tunnels were finally lined by 350mm (14 in.) reinforced concrete as permanent lining.

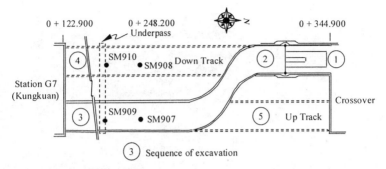

Figure 6. Plan of NATM tunnels in Contract CH221

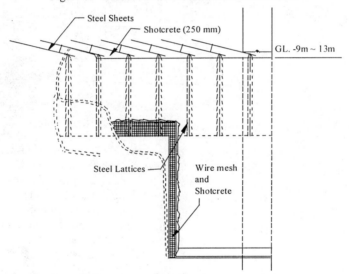

Figure 7. Profile for NATM tunnels in Contract CH221

The consumption of compressed air was about 110 m^3/min (1 cubic meters = 1.3 cubic yard), refer to Figure 8, when tunneling was carried out in the Up-Track tunnel in Stages 2 and 3 excavation before a layer of gravel was first encountered at the face at the halfway of the drive. It increased to 270 m^3/min by the time the heading reached the end of drive of the Up-Track tunnel. It was maintained at 170 to 190 m^3/min during the Stage 4 excavation for mining the Down-Track tunnel. Again, as the gravel layer was encountered, the air consumption increased to a maximum of 280 m^3/min and the four compressors, with a power of 340 kilo-watts each, was fully loaded. As the tunnels were fully lined, the air consumption dropped to 140 m^3/mm.

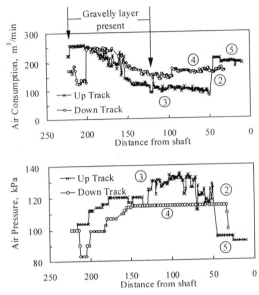

Figure 8. Air pressure and consumption for NATM tunnelling

Figure 9 shows the settlement records and, as can be noted that, gro settlements were significantly affected by two events. In the first event, an explo due to the ignition of gas leaking from a gas line caused much panic of residents. a precaution, the pressure of compressed air was lowered from 1.2 bar to 0.4 bar was maintained at that level for about half a month. In the second event, a malfunction of a transformer disrupted the electrical supply and the pressur compressed air dropped to 0.2 bar in 12 hours before the transformer was repl and the electrical supply was back to normal.

The second application of NATM Method in soft ground was in Con CN256B of the Nankang Line, 54m (177 ft) for the Up-Track and 44m (14 for the Down-Track (Event 7 in Figure 1). Figure 10 is a plan showing the layo the tunnels and the locations of settlement markers. A longitudinal section of the Track tunnel is given in Figure 11. Jet grouting to the west of the underpass carried out previously and the shells of the two shield machines were left in plac the contractor of Contract CN256. Jet grouting underneath the underpass was ca out by the contractor of Contract CN256A when the underpass was constructed. grouting to the east of the underpass was carried out by this contractor, i.e. contractor of Contract CN256B.

The progress of the Up-Track tunnel was quite satisfactory with gr settlements in general less than 50mm (2 in.). Tunnel convergence and settlement of the crown were within 10mm (0.5 in.). Settlements above the Down-Track tunnel were rather large, up to 240mm (9.5 in.). The drastic difference in behavior between the two tunnel drives was due to the fact that ground treatment was difficult for the Down-Track tunnel because of the presence of utilities. The recovery of cores was 50% for the Up-Track tunnel and below 10% for the Down-Track tunnel.

Discussions

Notwithstanding the many difficulties encountered, the tunneling operation in the construction of the Taipei Rapid Transit Systems is deemed successful. Ground loss in general ranges from 1% to 2% which are far less than what were observed previously. The drastic improvement in performance was due to the prompt grouting of tail voids. In quite a few sections of routes, ground heaves were observed. Ground heaves were found to be caused by back grouting of tail voids, rather than by the pressure on the tunnel face. The progress of tunneling was impressive with a peak production rate of 47 rings a day. However, such an extraordinary rate is not advisable as ground settlements tend to become large because back grouting may not be able to catch up with the progress.

Although ground settlements were large in comparison, the successful completion of the two sections of tunnels using the NATM method was a remarkable achievement in consideration of the softness of subsoils. However, it must be admitted that NATM tunneling in soft ground shall be left as the last option because it is a highly risky operation. This is particularly true if compressed air is used as an auxiliary measure for maintaining face stability. Many labors suffered from diver's disease (aeroembolism) due to improper decompression. Ground treatment by grouting is a viable alternative. However, the quality of treatment is difficult to be ascertained, particularly in clays. Furthermore, the presence of utilities is often a major cause for poor treatment.

References

Hwang, R. N., Moh, Z. C., Yang, G. R., Fan, C. B., Chao, C. L. and Wong, R. K. (1998). Ground freezing for repairing a damaged tunnel, Special Lecture, 13th Southeast Asian Geotechnical Conference, 16~20, November, Taipei, Taiwan

Ju, D. H., Duann, S. W. and Tsai, K. H. H. (1998). Ground freezing for restoration of damaged tunnel, Proc., 13th Southeast Asian Geotechnical Conf., Taipei, Taiwan

Ju, D. H., Kung, N. W. and Duann, S. W. (1997). The distribution of driftwoods in the Taipei Basin and the effects to underground works, Proc., 7th Sym. on Current Researches in Geotechnical Engineering, 28~30, August, Chinshan, Taiwan (in Chinese)

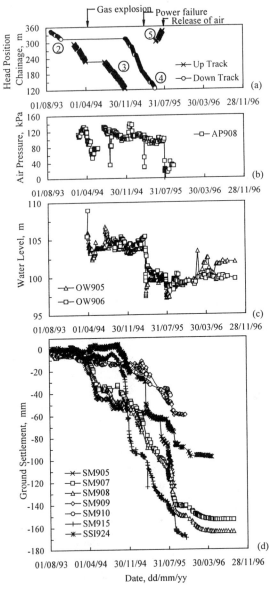

Figure 9. Progress of CH221 NATM tunnelling and instrument readings

GEO-ENGINEERING FOR UNDERGROUND FACILITIES 621

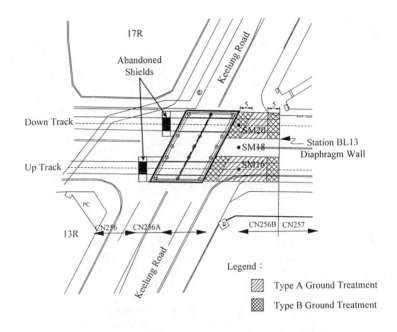

Figure 10. Layout of CN256B NATM tunnelling

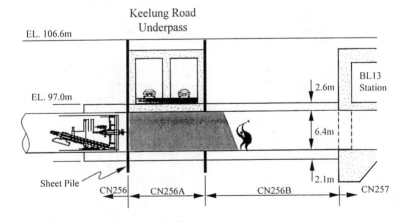

Figure 11. Schematic view of CN256B NATM tunnelling

UNDERGROUND CONCRETE CONSTRUCTION

Daniel P. O'Connor[1]

ABSTRACT

During the past 25 to 35 years, underground rock excavation production rates have increased tremendously. For example in long tunnels, where Tunnel Boring Machines (TBMs) have largely replaced drill and blast techniques, average advance rates in many cases have increased by a factor of 3 to 5 times. Also, other types of underground excavation have been improved by the use of shaft blind hole and raise drilling equipment, road headers, hydraulic drills and improved explosives.

Underground concrete construction has been enhanced by the use of mechanical excavation because overbreak is essentially eliminated, reducing concrete quantities to as much as one-half of what would be needed in a drill and blast operation.

Unfortunately, on many projects the benefits gained by high rates of excavation are diminished by poor production rates during the concrete phase of the project. Other than the gains realized from smaller quantities, concrete placement rates have not kept pace with the faster underground excavation rates.

In this discussion a number of the special problems associated with forming and placement of concrete underground, often in wet conditions, are examined. Suggestions are also offered for methods to overcome impediments to higher production rates without a reduction in overall concrete quality.

---

[1] Senior Project Advisor, Kajima Engineering and Construction, Inc., 901 Corporate Center Drive, Suite 201, Montery Park, CA, 91754

INTRODUCTION

Everyone takes credit for the great improvements in underground excavation. On Early TBM projects the owners and engineers advertised projects with options allowing conventional or mechanical excavation. The contractors and equipment manufacturers took the risks that proved the value of mechanical excavation and caused the quantum leap in excavation rates. Later, after TBMs, raise drills, etc. were proven, owners and engineers modified designs to conform to mechanical excavation and in many cases ruled out conventional excavation.

After the first few early projects proved the capability of TBMs and shaft drills, there have been constant improvements and refinements resulting in the high production rates seen today worldwide.

It is an entirely different story when it comes to underground lining production. There has been no leap in production rates due to the introduction of unique new equipment similar to the development of TBMs. Designs and specified procedures, in most cases, have not emphasized production. Forming and placing systems are not too different than they were 40 - 50 years ago.

Certainly, there have been improvements to concrete placement such as the introduction of new concrete pumps, placing equipment and improved concrete mixes, but overall the resultant increase in advance rates cannot compare with increases in excavation. There have been occasions where contractors have made good concrete production, but in other situations, contractors attempting to increase production rates have been frustrated by lining designs, standard specifications or other restrictions that have limited production to unreasonably low rates.

The purpose of this paper is to offer suggestions, from a contractor's viewpoint, to modify certain common requirements with the goal of increasing underground concrete production rates and decreasing costs.

DESIGN

Tunnel contractors are in a unique position when it comes to the evaluation of the design and procedures imposed on tunnel projects by various owners and engineers. Since the tunnel and underground industry is so small most contractors see the design work of numerous agencies and designers when the projects are in the bid stage.

Thickness and Reinforcing

It is clear by comparing designs, put out to bid by the same or different agencies, that there is something wrong with the principles and procedures being used to design underground linings because the results are so widely different. If, for example, several miles of large tunnel in City "A" is completed with a thin concrete lining with no reinforcing, and then a much smaller diameter tunnel in City "B" is built with a thicker lining and double mat reinforcing steel, something is wrong. Either one (or both) of the designs is incorrect. Or if, after the several miles of tunnel are put into service in City "A" without reinforcing, then several years later ten miles of tunnel are built in the same city requiring heavy reinforcing, something is wrong. Or if a subway agency in City "C" requires a heavily reinforced double mat of steel on early subway tunnels, then essentially eliminates the use of reinforcing on projects of similar depths and geological conditions twenty-five years later, it is easily concluded that one design or the other was incorrect.

Although a contractor almost never gets to see details of a lining design it is evident that the wide discrepancies in the designs result because the lining-ground interaction was ignored or given little value in the case of the heavier linings.

The use of reinforcing steel and thick linings obviously increases cost but also has negative effects on placing procedures and on rates of production.

There are numerous other wide inconsistencies between designs and lining specifications that increase costs and hinder production rates. The differences are large between different designers on such subjects as: lengths between joints, stripping times and strengths, mix designs, use of fly ash and concrete placing restrictions.

Construction Joints

A key factor that affects the opportunity for high lining production rates is the requirement of short lengths between construction joints. The wide difference between agencies in this regard is astounding from a contractor's viewpoint. For example, one U.S. Federal Government agency operating in western states has allowed unlimited lengths between joints, but another unbending U. S. Federal governmental agency limits joints to 9 meters (thirty (30) feet). In just as an astounding incongruity, an eastern city subway agency limits joints to 15 meters (fifty (50) feet), while a western city subway agency allows joints spaced at 91 meters (300 feet). Discrepancies also exists in the type of joints allowed by various agencies or engineers, as will be discussed later.

Concurrent Excavation and Concrete Lining

If concrete lining can be done concurrently with the excavation, project duration can be reduced and costs reduced. The existence of intermediate shafts, as is common on many sewer projects where shafts are needed for manholes or drop facilities, allows the muck hoisting locations to be moved closer to the advancing tunnel face and gives the opportunity to start lining independent of the excavation operation. If possible, contractors should be given the opportunity to use intermediate shafts.

A worthwhile goal is the development of a mucking system that will allow placement of concrete lining during excavation without interference. The use of conveyor belt muck transportation, which is becoming widespread, may be a step in that direction. To enhance further development, designers should simplify lining designs and allow options that foster contractor innovation.

Curves

Designers usually choose minimum curve radiuses based on TBM geometry, and seldom consider concrete lining requirements. To maintain high advance rates in curves, radiuses should be kept in the range of 300 meters (l,000 feet) to 450. meters (1,500 feet), if possible.

Most forms are moved in sections of 7.5, 9, or 12 meters (25, 30 or 40 feet). Designers should select a curve radius using calculations based on cords of these lengths.

Standard Diameters

Sometimes if excavation and forming equipment is available from other projects contractors can save money and increase production if the diameter can be adjusted to match existing equipment. Therefore, many owners or designers provide for an option to change diameters. All designers should make a similar provision.

CONCRETE DELIVERY

Delivery of Concrete in Tunnels

Compared to surface work, it always takes longer to make delivery to underground operations. Furthermore, there are more elements of work in the delivery increasing the risk of delay. If the concrete, for example, is being delivered in a tunnel using rail cars at a speed of 16 km/hr (10 to 12 mph), 20 to 30 minutes travel time is common. To this must be added time to deliver the concrete from the mixer to the shaft bottom and to pump it into the forms.

In situations where concrete is delivered by pumping from surface, times are also longer. For example, 450 meters (1500 feet) of 150 mm (6 in) concrete pump line contains 8 cu. m (11 cy) of concrete. If the rate of pumping is 30 cu. m/hr (40 cy/hr), it takes a minimum of 16 minutes to deliver the concrete from the pump.

Designers of tunnel concrete projects should avoid the insistence on the use of standard concrete mixes and allow options including high slumps and retarding additives because of the long length of time that it takes to deliver concrete to the tunnel forms.

It is a traditional practice to clean pump lines and to assist the pumping of concrete and movement in the forms by the introduction of compressed air into concrete pump line. The intermittent use of air into the line ("air slugging") near the forms propels the concrete into the top of the forms. Surprisingly, some specifications prohibit the use of air slugging even though it has been proven to be an effective way to assist in the placement of tunnel concrete with no detrimental effect to the final product.

Many specifications prohibit the moving of concrete in the forms by vibrators, but it is virtually impossible to adhere to this restriction in most underground situations. The restriction about moving concrete with vibrators is generally believed to reduce segregation, but in general, virtually no segregation occurs.

Delivery of Concrete in Shafts

The major reason why the shaft sinking is so difficult is because gravity must be overcome to hoist muck. During concrete placement, gravity becomes an ally. The simplest way to deliver concrete in shafts is by dropping it in a pipe. Many civil engineer designers are horrified when they hear about dropping concrete free-fall because of the fear of segregation. In fact, thousands of cubic meters of concrete have been effectively dropped in a pipe in mines and on civil projects without reduction in quality.

TYPES OF TUNNEL LINING

The major methods of lining tunnels or other underground structures are:

1. Unlined.
1a. Partially lined with isolated sections covered with shotcrete, lined with concrete or repaired with joint treatment.
2. Shotcrete.
3. Cast-in-place, full round, telescopic single pass.
4. Cast-in-place multiple-pass lining such as "invert first" or "invert last".
5. One-pass precast concrete segments.
6. Precast as a primary lining with a monolithic final lining.
7. Precast pipe.

The choice of which method is used depends on many factors including the length and diameter of the tunnel. A short tunnel cannot justify the purchase of a long set of forms or expensive set up costs, so lower advance rates are acceptable. For long tunnels, clearly a telescopic full round one pass system is the most productive and most economical method. Precast pipe or precast segments are the most productive alternative in some situations and should not be overlooked.

Often there are peculiar differences between projects, so to the extent possible, contractors should be given latitude regarding the selection of the lining method. Innovation cannot occur if a method is rigidly specified.

TUNNEL FORMS

The essential design of forms has not changed for several decades. There have been few innovations and little job to job improvements similar to what has happened in TBM excavation. When a TBM is purchased, a contractor can be reasonably certain that future projects will ultimately share the cost to develop a high production excavation system. In the case of forming equipment, there is no stimulus for contractors to take the risk to develop new ideas because of wide difference between requirements from agency to agency. Innovation has been stifled by restriction to lining procedures.

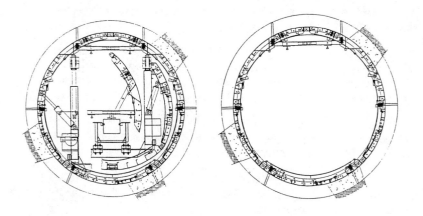

Figure 1. Full Round Telescopic Forms
(Courtesy Everest Equipment Inc.)

CONCRETE MIXES

Concrete mixes for underground work have special requirements that are sometimes ignored by specification writers who choose standard requirements that were developed for surface work. For example, concrete that must be pumped long distances should have a slump of 130 to 180 mm (5 to 7 in.) compared to the 25 mm to 100 mm (1 to 4 in.) slump that is standard for surface construction. The high slump is also required to make the concrete adequately workable behind forms and through reinforcing steel.

Because it takes longer to get underground concrete from the mixer to the forms, initial set times must be delayed. On the other hand, concrete must attain adequate strength to allow stripping of the forms early.

The mix designs for most surface work emphasize the reduction of cement for economic reasons. Accordingly, water and slump are reduced to keep the high water cement ratio needed for strength. The emphasis on reducing cement makes good sense for normal surface work because cement is the most expensive component of the mix and labor costs are relatively low. On the other hand, in underground situations, labor and potential delay costs far outweigh the cost of some additional cement in the mix. The best way to assure that a mix is adequate, is to use one with a cement and fly ash content of at least 350 kg to 400 kg cement per cubic meter (6 1/2 to 7 bags/cy).

If long pumping distances are involved, then the gradation of the aggregates can seriously affect pumpability and workability in the forms. Special adjustments may be required if the combined gradation curve falls outside certain limits. In some situations fine and coarse aggregates have to be changed to make a mix pumpable.

Mix designs for surface work must provide for durability, reversal of stress, change in temperature, freezing and thawing and other protection against salts and other conditions that do not affect underground work. Mix design for underground work must combat a different set of adverse conditions ranging from very mild ones for water tunnels to severe sulfate attack in sewer tunnels. Mixes should be designed to overcome the specific conditions of the project rather than being "cut and pasted" from specification that were developed for surface work.

NEED TO MAINTAIN A CYCLE

In order to obtain high production rates in an underground concrete operation the project must be set up with a sophisticated mechanical forming system usually designed specifically for the project. Also, other standard and specially built equipment must be provided to support the forming operation. In addition, crews on a two or three shift basis must be trained and organized so that all phases of the operation cannot falter, otherwise the 24 hour cycle will be disrupted and the target daily production rates will not be attained. Among the items of work that must be accomplished to complete a cycle are:

1. Washing of tunnel crown and walls.
2. Utilities, (water, air, discharge and concrete pump lines) must be removed, passed through the forms and replaced on the finished walls.
3. Pumping water.
4. Invert cleanup and removal of muck and concrete spillage.
5. Scaling of rock and water control.
6. Placement of reinforcement.
7. Removal of construction joint bulkhead.
8. Stripping of invert and arch forms, movement through existing forms, placement of invert and arch in 7.5, 9 or 12 mm (25, 30 or 40 ft.) sections.
9. Fine adjustment of forms and blocking with spuds to crown roof, walls and invert.
10. Construction of a bulkhead.
11. Pointing and patching of recently stripped concrete.
12. Curing.
13. Set up of concrete pump and pump lines.
14. Placement of 1 to 5 cu m (2 -6 cy) of a lubricating mix.
15. Placement and vibration of concrete up to a sloping joint or vertical bulkhead.
16. Placement of a final truck or car of concrete.
17. Blowing out of and clean up of pump line.
18. Cleanup of spillage.
19. Removal of spuds.
20. Green cutting of the construction joint.

If any element (except for the finishing and curing) is delayed then the whole cycle is disrupted and the daily production is either lost or reduced.

Attainment of efficient cycle times is complicated by the fact that the work must be done around the clock, time must be allowed for changing and transportation of crews and materials and crews working at separate locations must be coordinated. Times of the individual elements must be shrunk to provide a cushion to compensate for problems in other elements which are inevitable. Often owners and engineers new to underground concrete work are shocked by proposals to make the changes needed. Some of these are discussed here in the following paragraphs.

Form Stripping

A key requirement to maintain a high production cycle is the need to strip forms quickly. Most stripping strength and time requirements were developed for rectangular structures and not arches which are essentially self-supporting even at low concrete strengths. A simple structural analysis will show that the strength to support the dead weight of recently placed concrete is very low. If the stripping requirements are 110 kg/sq. cm to 140 kg/sq. cm (1,600 to 2,000 psi) and/or 16 to 20 hours, then there is no possibility of attaining an efficient cycle. Stripping must start toward the end of the second shift. Unlike structures on surface for which stripping requirements were developed, conditions underground are ideal for concrete to attain early strengths because the concrete is confined and not exposed to the elements.

Long before the concrete work begins experimentation and early strength testing of proposed mixes should be started. Adjustments to the mix should be made, even if it requires the addition of cement to assure that stripping times can be reduced and provide an adequate margin of safety. Several trials may be required to provide the confidence needed. If possible, testing after lining starts should be avoided because that will add one more element to the cycle.

Length between Construction Joints and Modified Sloping Joints

As was stated previously, obviously high production is not possible if construction joints are required at 9, 15 or 30 meters (30, 50 or 100 feet.) The length between joints should be unlimited.

The points in the cycle that risks are greatest for serious delay are at the beginning and end of the pour. By allowing the modified sloping construction joints (partial vertical bulkheads to one-quarter to or three-quarters of the diameter) the pour can be stopped quickly and the long pump line can be cleaned out efficiently avoiding the greatest source of delay. Many owners and engineers who have experience with this joint can verify that it provides high quality construction.

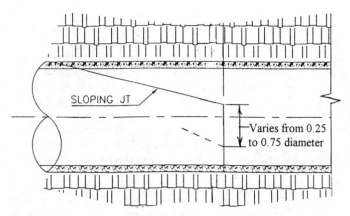

Figure 2. Modified Sloping Joint

Curing

In most tunnels special applications of moisture or curing compounds is not needed because humidity is generally high. ACI 308-92 gives a simple procedure to determine if curing is needed. Curing should not be required unless temperature, humidity and other conditions dictates its use.

Keyways and Water Stops

There is a wide difference of opinion of engineers regarding the use of water stops and keyways on underground projects. The requirement for such items eliminates the possible use of the modified sloping joint and adds to the number of items that eats up the time needed to complete the cycle and adds another source of delay. The application of surface concrete requirements which have marginal or no benefit to underground concrete work but retard production rates are wasteful.

WATER CONTROL AND GROUTING

In the past twenty or so years a system of plastic membrane lining has been introduced which attempts to stop water inflow through the lining. This has worked on some jobs successfully, but it is expensive and adds another item to the cycle with another source of potential delay. Another method being used by some water agencies is to specify a total impervious welded steel liner, but this is also very expensive.

A more practical method of water control, although not as positive as the membrane or steel lining, is the use of the traditional combination of water control grouting in the surrounding ground along with the concrete lining. While not totally impervious, this combination can attain very good results if a few procedures are followed.

For years tunnel men have been using "panning" (usually metal sheets) anchored to the rib and crown to divert water inflow away from the fresh concrete. The panning then leaves a path by which grout can be pumped to seal off the water to a low level. For very light flows the concrete can seal some of the water. For larger flows some sort of drain pipe in addition to the panning should be provided to make sure that the water is adequately drained off. It is extremely important that water is not allowed to flow through the fresh concrete as it will wash away the fine cement, fly ash and sand, leaving the concrete porous. The use of panning in conjunction with pre-grouting prior to and/or after excavation along with grouting after the lining has reached strength can do wonders to reduce water inflow. Surprisingly, panning is seldom specified by designers. The reason for this may be that few references are found in publications.

Contact Grout and Lubricating Mix

Contrary to the apprehensions of many engineers who fear low quality caused by modified sloping joints and other procedures used by tunnel contractors, the concrete at or near the construction joints is usually very good. One reason for the good results is the fact that it is standard practice to pump a few cubic yards of a lubricating mix (same mix less coarse aggregate) at the start of the pour. The mix helps avoid plugged lines and delays, but also provides a bonding coat of mortar on the sloping joints and compensates for any tendency for segregation that occurs in the early part of the pour.

Contact grout is a method to provide testing and insurance that a high quality lining is in place. Some shrinkage of the lining does occur and for various reasons the lining sometimes does not get completely filled. The grout is introduced into the lining through holes near the crown either drilled or formed by the use of pipe. Except in unusual circumstances the grout should be a relatively thick, neat cement mix without sand.

CONCLUSION

Remember that the door was opened by owners and designers which allowed the development of Tunnel Boring Machines by designing projects and writing specifications which provided contractors with wide options of construction methods. This freedom, along with acceptance of risks by the contractors and equipment manufacturers, provided the impetus for the innovations that have resulted in equipment and methods capable of high excavation production rates often seen today.

In contrast, in the lining phase of many projects, production is choked by the imposition of standard specifications which were developed for types of construction with far different requirements than are applicable to underground concrete construction.

Concrete production rates cannot increase to levels appropriate to the high rates of excavation unless agencies and designers discontinue requirements which limit or retard production such as: short lengths between construction joints, unnecessary reinforcing steel, low slump concrete, the lack of options which discourage the use of existing forms and long times for form stripping, full vertical joints, etc.

There have been several projects completed recently and in past years which verify that the anti-productive specifications can be abandoned without reduction in long term quality of underground concrete construction. This lead needs to be followed by many more agencies and designers. When contractors recognized a trend of owners to emphasize high lining rates by rejecting certain requirements, they will accept the risks needed to develop the equipment and procedures required to increase lining production to levels appropriate to high TBM advance rates.

REFERENCES

1. O'Rourke, T. D., Editor, 1984, "Guidelines for Tunnel Lining Design", prepared by the Technical Committee on Tunnel Lining Design of the Underground Technology Research Council of the ASCE Technical Council on Research.
2. Hurd, M. K., 1981, "Formwork for Concrete", prepared under the direction of ACI Committee 347 Formwork for Concrete.
3. U. S. Department of the Interior, Water and Resources Services 1981 "Concrete Manual".
4. Eckardstein, K. E., 1983, Schwing "Pumping Concrete and Concrete Pumps", a concrete placing-manual.
   American Concrete Institute, 1994, "ACI Manual of Concrete Practice:
5. ACI 211.1-91, "Standard Practice for Selecting Proportions for Normal, Heavyweight, and Mass Concrete";
6. ACI 308-92, "Standard Practice for Curing Concrete";
7. ACI 306R-88, "Cold Weather Concreting".

Experience with Steel Seal Membranes for
Liners in Pressure Shafts and Tunnels

Alberto Marulanda P.[1]
Ramiro Gutiérrez R.[2]

Abstract

One of the most important aspects in the design of pressure tunnels is to avoid hydraulic fracturing in the zones where the minimum stress in the rock mass is lower than the operating pressure, which is accomplished by means of an impervious liner that guarantees that the water inside the tunnel will not reach the rock. The most commonly used impervious liner is the concrete encased, heavy steel liner.

In hydroelectric projects with heads up to 250 m, it is possible to reduce the cost of the required impervious liner, using a much thinner steel seal embedded between two concrete rings, the innermost of which provides the required strength against buckling due to external pressure. In recent years several materials such as copper sheets with bituminous coating, rubber membranes, and thin steel membranes have been used in this type of lining. However, there is very little information regarding either design criteria or actual behavior.

This paper presents the design criteria, installation systems, and construction details of two different cases of thin steel membrane liner embedded in concrete rings.

---
[1] Partner and Technical Vicepresident, Ingetec S.A. Cra. 6 No. 30A-30. Santafé de Bogotá. Colombia.
[2] Specialist Engineer. Associated. Chief of Tunnels Department. Ingetec S.A. Cra. 6 No. 30A-30. Santafé de Bogotá. Colombia.

Introduction

Growing needs of electric power have fueled the development of hydroelectric projects with ever increasing operating pressures. The basic design criteria for pressure tunnels and shafts, unlined and lined with concrete or steel liners, applied in the historical cases ahead mentioned, are based on concepts of the minimal level stress in the rock mass, their permeability and long term stability under different work conditions, which were treated in general form in other project by Marulanda et all (1987). In pressure tunnels where the stress level in the rock mass is not enough to withstand the operating pressure, or, where the permeability is so high that excessive leakage could be expected it is necessary to provide an impervious liner. Very often the required imperviousness is obtained by means of a concrete encased heavy steel liner (conventional liner) or a thin steel seal membrane between two concrete rings.

Seal membranes - General criteria

The main purpose of this type of liner is to provide an impervious barrier to avoid hydraulic fracturing of the rock mass using the least amount of steel. This type of liner is not suitable for conduits where large external pressures could be expected, since the thickness of the innermost concrete ring required to withstand external pressure would be too large, thus making the system both impractical and costly.

The liner consists of a thin steel seal membrane or a high-strength plastic membrane between two concrete rings. The membrane bridges the cracks that are created in the concrete due to the internal pressure.

It can be considered that the rock mass takes a part of the load imposed to the liner by the internal pressure, thus reducing the required thickness of the membrane. Therefore, the design of this type of liner requires a thorough knowledge of the deformation characteristics of the surrounding rock mass.

One of the characteristics that needs to be determined is the depth of the disturbed zone around the excavation perimeter. This is a variable heavily dependent on the rock quality, the stress level, and the excavation method (Brekke and Ripley, 1987). Tunnels excavated with a TBM present a thinner disturbed zone than those excavated with drill and blast. The thickness of the disturbed zone in a tunnel excavated with a TBM is generally less than 0,30 m, whereas that in a tunnel excavated with drill and blast varies between 2,0 m and 5,0 m.

Tunnel excavation with a TBM gives a quite uniform circular shape that gives a good fit with the external concrete ring. This is an additional advantage over the drill and blast method when a liner with steel seal is being considered.

Internal and external pressure

As mentioned above, the accurate determination of the deformation characteristics of the rock mass, is very important to adequately assess the percentage of load that will be taken by the rock mass, thus reducing the thickness of the membrane. Nevertheless for steel sheets there is a minimum thickness that can be handled during construction.

The determination of the membrane thickness must consider the full dynamic head. The analysis could be done using the theory of elasticity to determine the distribution of stresses and the compatibility of the radial deformations of the concentric concrete, steel, and rock (both disturbed and sound) rings. Whenever there are considerable variations either in the geometry of the lined cross section or a large anisotropy in the rock mass, it is convenient to use finite element techniques to determine areas of stress concentrations in the membrane, impossible to detect with common analysis techniques. Figure 1 shows the stresses in the steel membrane in the case of an anisotropic rock mass modelled with finite element techniques. Anisotropy of the rock mass was modelled by means of a soft, thick stratum encased in a much harder rock mass, intercepting the cross section. The same figure shows how stress amplification takes place in the membrane, depending on the degree of anisotropy.

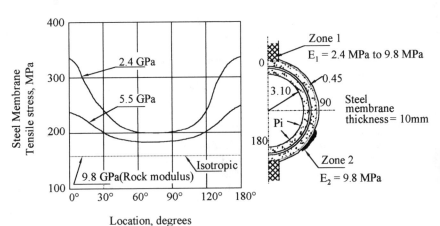

Figure 1. Anisotropic analysis

The design of liners with steel membranes in Colombia, considering load transfer to the rock mass, has been done with maximum allowable stresses in the membrane, due to internal pressure, between 30% and 65% of the yield strength. The membrane thickness is that required to withstand the internal pressure, considering load transfer to the rock mass.

The inner concrete ring is designed to withstand, not only crushing when submitted to the full external pressure, but shear stresses due to concentrated or unbalanced loads.

The design external pressure usually equals the maximum water pressure that could act upon the lining once the tunnel has been dewatered.

The main question lies in what is the maximum water pressure that could act upon the lining. It has been usually considered that the maximum water pressure equals that of a water column 1.3 times higher that the rock cover. In large head projects, the water pressure evaluated as described above is much lower than the operating pressure. Case histories such as that of the Robiei Project (Haldemann, 1987) show that, depending on the geologic conditions, this criterion may be insufficient. In the Robiei Projects the lining failure led to the implementation of a drainage system to control the external pressure. However, the system was not absolutely efficient and the lining failed again.

These experiences have led to the conclusion that, in large head projects, it is safer to consider that the maximum external pressure acting on the lining equals the full hydrostatic pressure in the tunnel. This criterion rules out the use of liners with steel seals for high head projects, since the thickness of the inner concrete ring required to meet the criterion would be too large, and thus, it would be too expensive.

It might be considered that an elaborate system of drainage galleries provided with systematic drainage boreholes is an effective measure to control external pressures, and that the use of such system would make a liner with steel seal suitable for a high head project. However, it should be kept in mind that the effectiveness of such drainage systems can not be guaranteed, increasing the risk of a lining failure due to external load.

When using a steel membrane it is essential to carry out contact grouting at the rock concrete, concrete-membrane, and membrane-concrete interfaces to obtain a uniform contact between the concentric rings. This measure allows not only the load transfer to the rock mass, but also a reduction in the tangential stresses in the inner concrete ring and the membrane due to the internal pressure, and the elimination of bending moments and shear stresses in the inner concrete ring due to external pressure.

To avoid separation of the inner concrete ring from the membrane during construction it is necessary to provide the inner side of the membrane with a number of steel plates or tie bars, evenly distributed along the perimeter, to give the required bond between the two rings.

From the scheduling point of view, the steel membrane liner offers an advantage over the conventional steel liner, since it takes less time to build. Time savings usually come from continous concrete casting with telescopic forms, thus eliminating transitions between concrete lining and steel lining, as well as the fact that no treatment is required to prevent corrosion of the membrane. Good logistics and an adequate coordination are necessary to reduce the erection and welding times, since these jobs for a conventional liner are carried out at the manufactures's shop, whereas for a steel membrane they need to be done in-situ.

Finally, a steel membrane liner can be more readily available than a conventional heavy steel liner in those countries where the latter is not manufactured locally.

Experience in a shaft - General

The Mesitas Hydroelectric Project comprises two surface powerhouses in cascade joined by 18 km of tunnel and 8,5 km of surface penstock. The project was designed for a 70 m^3/s peak flow with an installed capacity of 1 200 Mw, using water from the Bogotá river, which carries the city's common sewage. The main underground excavations is the 4,8 m diameter, 11,1 km long Granada Tunnel, which has a maximum internal pressure of 2,1 MPa. The 4,20 m diameter, 240 m high, vertical surge shaft is located 290 m upstream from the outlet of the Granada tunnel. The shaft was raise bored to a diameter of 1,8 m , and later on expanded to a diameter of 4,20 m . In the shaft a 5 cm thick shotcrete layer was used as temporary support during construction and no further final liner was employed. The tunnel was steel lined almost up to the surge shaft. (See Figure 2).

The Granada tunnel crosses sedimentary rocks of cretaceous age, partially covered by surface Quaternary deposits. The rocks belong to the Guadalupe Formation, which consist mainly of sandstone with siltstone interbeds. The bedding strike is generally normal to the tunnel axis, with dips varying between 10° and 85°.

Events

Large scale hydraulic fracturing phenomenon ocurred during first conveyance filling. Six hours after tunnel filling, movements and strong noises were detected in the mountain and leaks increased rapidly to 15 l/s. Water springs of about 30 l/s were observed in the scarp near the outlet portal and in the vicinity of some creeks. Piezometers installed near the surge shaft showed at all time during the test filling, a pore pressure of about 30% of the internal pressure. When the leakage increment ocurred, this value raised to 50%. During the inspection of the surge shaft and tunnel, it was determined that the cracking generated by the hydraulic fracture on the rock mass was concentrated in the middle third and upwards of the surge shaft, with an orientation semiperpendicular to the tunnel. No cracks were detected at tunnel level.

The height of ground, at the tunnel-shaft intersection was more than twice the operating head as shown in Figure 2. Obviously for hydraulic fracturing to develop, the whole area was heavily unstressed, due to stress relief into the valley, caused by a very complex topographical location. It is an excellent example that conventional empirical rules used for deciding upon steel liner requirements are not adequate if stress field is not well understood.

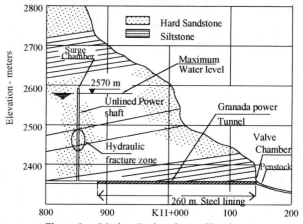

Figure 2. Mesitas Project Surge Chamber Area

After the hydraulic fracturing event an elaborated drainage system was provided to control hydraulic fracturing, thus allowing the project startup. Drainage boreholes drilld from an adjacent opening, in direction to the damaged zone, intercepted the fractures caused by hydraulic fracturing. These boreholes were performed in order to reduce the water pressure in the cracks. The drainage reduced the pore pressure in 40% and with this reduction, the open fractures were closed. The drainage system avoided a general movement of the scarp along the failure plane generated by the hydraulic fracturing of the shaft, proving to be an effective controlling measure from the geotechnical standpoint. However, the drainage system handled large quantities of highly contaminated water, which were discharged into nearby creeks. Since these creeks were the source for the water supply of adjacent townships, the leaks were causing a sanitary problem that called for an impervious liner. It was then decided to use a an impervious liner all along the surge shaft.

Design criteria

Conventional heavy steel liners were not manufactured in Colombia and it was too difficult to import them within the required time frame, and thus, it was decided to use a steel seal. In this alternative it was possible to reduce the required thickness of the membrane to one available locally. It was also decided to install the steel membrane liner only in the shaft stretch where hydraulic fracturing had taken place.

Strains and crack widths were available from the first filling logs . A back analysis of these data allowed the determination of the deformation characteristics of the rock mass in three different stretches of the shaft. A minimum gauge of 5/16" (7,9 mm) was chosen to avoid handling problems during the liner construction.

Figure 3 shows the internal pressure variation along the shaft, the membrane thickness selected for different stretches on the shaft, and the membrane tensile stresses with and without consideration of load transfer to the rock mass.

The analysis showed that, considering load transfer to the rock mass, the maximum working tensile stress is about 65% of the yield strength.

The liner was designed to withstand an external pressure equal to the internal pressure. A 30 cm thick inner concrete ring was provided all along the shaft. A 21 MPa compressive strength concrete was used in the upper 100 m where the maximum external pressure was 1 MPa, whereas a 34,5 MPa compressive strength concrete was used for the lower 120 m where the maximum external pressure was 2,1 MPa. Figure 4 presents a shaft profile indicating thicknesses of the steel membrane, as well as a detailed cross section.

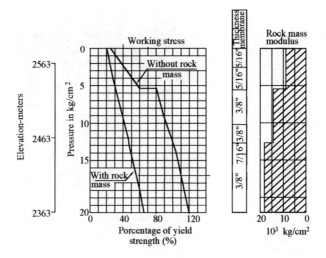

Figure 3.   Steel Stress Levels

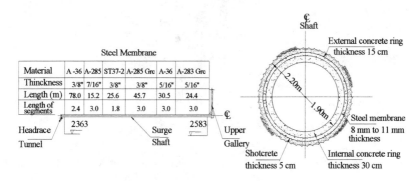

Figure 4.   Steel Lining

Contact grouting at the rock-concrete, concrete-membrane, and membrane-concrete interfaces were not deemed necessary since concrete pouring in a shaft reduces the possibility of voids formation behind the membrane. Besides, a thorough control of concrete mixes and adequate vibration of fresh concrete reduced the possibility of concrete shrinkage.

Fixture system

Steep topography in the area of the surge shaft lead to the use of an aerial tramway in lieu of an access road during the construction stage. The tramway was dismantled prior to the first filling, as soon as the Contractor finished the lining works in the shaft. Therefore, the liner construction with steel seal membrane progressed upward. The contractor installed winches at the upper end of the shaft, carried out the scaling of the shaft to obtain a uniform cross section, and built a reinforced concrete base at the shaft's bottom.

Two winches were installed at the top of the shaft. The first winch allowed the displacement of a sliding platform and the form to cast the inner ring.

The platform served to install the steel membrane, carry out the welding and cast the inner ring. The second winch was used to hoist personnel and supplies through a circular hole in the platform. Figure 5 presents a diagram of the system used in construction.

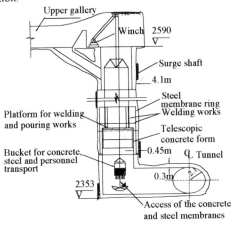

Figure 5. Fixture System

The steel membrane was formed with 90° segments. Each segment was provided with appropriate bracings to avoid buckling and changes in shape. The segments were joined together by means of longitudinal welds with support plates and butt transverse welds. All the welding was carried out from inside the liner. The welds were tested with penetrating ink.

Once the steel membrane was in place, the sliding form moved upward to cast the inner and outer concrete rings. The average lining rate, including erection, welding and pouring of the inner and outer concrete rings, was about 1 m/day. About eight months were required to complete the 219 m long liner.

Experience in a tunnel - General

The case history described below is that of Los Rosales Tunnel, which is a part of the Bogotá water supply system. The 9,0 km long tunnel crosses the mountain range east of Bogotá with a slope of 2 in 1 000. The maximum working pressure will be 0,45 MPa.

The tunnel was excavated with a 3,6 m diameter TBM through sedimentary rocks of Upper Cretaceous age, mainly hard sandstone, siltsone and some shale. Support during excavation was provided by means of a 12 cm thick ring consisting of precast concrete segments.

The initial design considered a 22 cm thick concrete ring all along the tunnel to guarantee the long term stability, and the use of a conventional steel liner (Thickness 11,1 mm) in areas of low cover and near the outlets. See Figure 6.

The initial design called for a 2,30 m diameter, 7/16" (11,1 mm) thick steel liner, provided with 23 cm wide, 3 cm thick, reinforcing rings spaced every 2,49 m . This type of liner shown in Figure 6, was to be used in five tunnel stretches of varying length between 38 and 150 m . The total length of steel liner was 413 m with a weight of 450 ton.

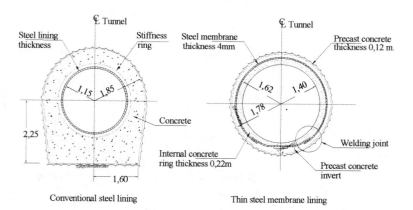

Figure 6. Tunnel Cross Sections

The use of a steel membrane as an alternative to the conventional steel liner was proposed. The alternative was chosen considering that the tunnel was to be excavated with a TBM, that both the internal and external pressures were low, and that this type of liner would eliminate the need of reinforcing rings and the transition between concrete lining and steel lining.

Design criteria

According to the general criteria previously described, a 4 mm thick, A-36 steel membrane was designed to withstand the full internal water pressure without considering load transfer to the rock mass, in which case the maximum tensile stress would be 74% of the yield stress. Since the membrane is embedded between two concrete rings, the outermost of which is supported by the rock mass, load transfer actually takes place, even when the deformation modulus of the rock mass is low (about 1,5 MPa). In this case the maximum tensile stress in the membrane is about 22% of the yield stress.

A 22 cm thick, 21 MPa concrete ring was cast inside the membrane. This can withstand an external pressure of 1,2 MPa, which corresponds to the rock cover and the local water table.

As part of the design criteria a rock-dowel contact grouting layout was considered and executed during the excavation and support works. For the grouting six 30 cm radial holes drilled in the rock with a grouting pressure of 0,2 MPa were used. Aditionally, contact grouting on three staggered holes every 5 m were designed; this contact grouting required a special nozzle. Finally, steel membrane-concrete ring contact grouting for the crown were carried out with holes at 10° to the tunnel axis, staggered every 5 m .

As part of the original steel lining design for the outlet, this lining had to join a surface penstock. Therefore, the steel seal lining was adapted to the conventional lining geometry by means of a transition portion. This transition will overcome the tensile stresses due to closure of the valves in the outlets. The anchoring length for the pipe in the conventional lining was defined as 30 m , long enough to assure a complete transmission of the stresses to the rock. For the extreme of the conventional lining a joint between the conventional and membrane linings was designed. This joint assures a good watertightness.

The steel membrane was transported inside the tunnel using a rail mounted vehicle. This vehicle placed the membrane against the tunnel walls by means of three hydraulic jacks. Once in place, the membrane was fixed on one end to the precast concrete segments using "L" shape tacks while on the other end circunferencial welding between new and previously installed membranes was carried out. The invert steel membrane segment was placed and welded afterwards.

Following the circunferencial weldings, a checking process using penetrating inks along the whole length was done. Once a steel membrane placing work was done the telescopic form was mobilized to cast the inside concrete ring. Following the dowel and ring steel membrane erection contact grouting was performed. This type of lining was done in a continuous manner with no interruptions, with an average rate of 3,1 m/day. Figures 7 and 8 present details about the movilization and the system used for installation and construction of the membrane concrete liner.

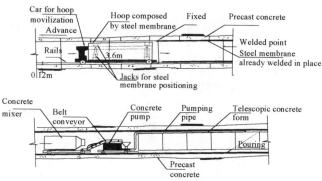

Figure 7. Fixture System

Figure 8. Car for Hoop Movilization

Remarks

To design this impervious lining a knowledge on the characteristics of confinement, permeability, deformation and anisotropy of the rock mass is required. Load transfer to the rock mass should be analysed for a better steel membrane thickness definition. Contact grouting for the rock-concrete ring, and for the membrane-concrete rings must be provided to guarantee an adequate and uniform interaction. The purpose of the contact grouting is to fill void spaces between placed concrete, the steel membrane and the surrounding rock mass. Low pressures are normally used, and the water-cement ratio is often specified, depending on grout takes in individual holes.

Reduction of the membrane thickness implies a lower erection weight and less welding works, which could be carried out from inside.

A good control of the welding and contact grouting works must be scheduled to guarantee a functional liner.

Good logistic planning is required to obtain higher performance rates than those obtainable in the conventional steel lining construction.

References

Brekke, T.L. and Ripley, D.8., 1997. Design Guidelines for pressure tunnels and shafts. EPRI, Document RP-1975-17-17-30.

Haldemann, K., 1987. Robiei Central. Power Conduit Repairs. Tunnels Repairs. Munich University

Dolcini, G. Grandori, R. And Marconi, M. 1990. Water supply revamp for Bogotá. Tunnels & Tunnelling. Pag. 33. September.

Marulanda, A., Eslava, L. F. and Broch, E. 1987. Desing of unlined high pressure tunnels for Guavio hydropower plant in Colombia. Underground power plant. 101 - 110.

Seepage Control Measures for an Underground Powerstation in a Semi-Arid Region

Jason E. Hedien[1], Associate Member, ASCE

Abstract

Potable water is a scarce resource in semi-arid regions such as the Middle East, and conservation of fresh groundwater supplies is imperative to sustained development in these areas. Feasibility level studies for a hydro-electric pumped-storage project located in the region required an estimate of the effects construction of underground caverns and tunnels may have on local groundwater regimes. This paper provides a summary of the various explorations and investigations conducted during the feasibility stage, analyses performed to estimate seepage quantities into the tunnels and caverns during project construction and operation, and the proposed schemes to manage groundwater inflows into the underground structures of the project.

Introduction

Pumped storage projects are designed to store hydroelectric power for use during times of peak electricity demand. This particular project consists of over 1100 meters of access tunnels, 1700 meters of water conducting tunnels and shafts, an underground powerstation complex, and upper and lower water storage reservoirs. The tunnels and shafts vary from 3 to 8 meters in diameter. The underground powerstation consists of three separate chambers: a 9 by 110 by 20 meter spherical valve gallery; an 18 by 150 by 45 meter powerhouse cavern; and a 15 by 150 by 15 meter transformer gallery. A general layout of the project's underground structures is given in Figure 1.

---

[1] Senior Geotechnical Engineer, Harza Engineering Company, 233 South Wacker Drive, Chicago, IL 60606

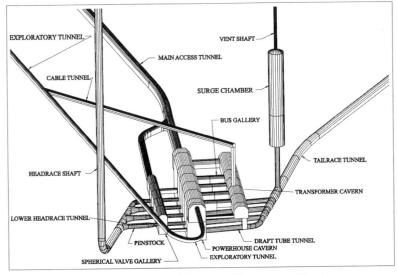

Figure 1. General Layout of Underground Structures

Since there is no reliable source of surface water suitable for use on the project in the area, all water requirements will be met by using groundwater supplies. Water requirements include fresh water needed for project construction, for filling of the reservoirs and power tunnels, for replenishment of evaporated water from the reservoirs and for dilution of dissolved solids in reservoir water that increase in concentration over time. In addition to exploiting local groundwater supplies to meet project water requirements, it is also anticipated that construction of the tunnels, shafts and powerstation complex, located approximately 400 meters below the ground surface, will cause drainage of groundwater into the underground structures. Depletion of local groundwater supplies as a result of pumping from water supply wells and sumps in the underground structures may affect local industries that currently draw water from the aquifers in the vicinity of the project. Due to the project's close proximity to the sea, there is an added concern of increasing groundwater salinity over time as brine from the sea replaces fresh water drawn from the local aquifers. A substantial increase in salinity would render the groundwater in the area unsuitable for use by industries and the project without extensive treatment.

Project Geology

Feasibility level investigations were conducted to evaluate the geologic conditions and material properties affecting project design and construction. Investigations conducted to date include subsurface borings, geological mapping, excavation of test pits, laboratory testing, geophysical surveys and the construction of a 1000-meter long exploratory tunnel. Data obtained from construction of the exploratory tunnel will be discussed in more detail in a later section.

The geologic strata exposed and drilled at the site consist of over 800 meters of sedimentary rock and alluvium ranging in age from Lower Cretaceous (approximately 100 million years ago) to the present. Most of the rock units are shallow marine, carbonate sequences with interbeds of calcareous shale or marl. Sandstone is also present in several of the older and more recent formations. Virtually all of the underground structures will be constructed in dolomite and limestone strata.

Project Hydrogeology

Hydrogeologic investigations conducted to date include analyses of existing well data in the area along with subsurface geologic information to help define the regional groundwater system. A test well program is planned for the next phase of the project to better understand the local aquifer characteristics and the effect of groundwater withdrawl on them.

The feasibility level hydrogeologic investigations have revealed three aquifer systems that affect the project. The deepest aquifer is a regional artesian system confined approximately 40 meters below the powerhouse invert in impervious shale, marl and limestone. Estimates of groundwater potential for this aquifer suggest that approximately 10 million cubic meters per year (MCM/yr) can be safely utilized for project requirements. Deep wells drilled into this aquifer are currently envisioned to supply water for the initial filling of the reservoirs and make-up water during project operation. The structures of the powerstation complex and water conductors for the project will intersect two interconnected aquifers. The primary aquifer of concern consists of dense carbonate rock with groundwater contained in fractures and joints. The transmissivity of this aquifer, estimated using data from groundwater wells operating in the area, is on the order of 280 square meters per day (m^2/day). Using an average aquifer thickness of around 100 meters, the permeability of this aquifer is estimated to be on the order of 3×10^{-3} cm/s. Preliminary estimates of the groundwater exploitation potential for this aquifer indicate that up to 2 MCM/yr may be utilized by the project without adversely affecting local industry.

Most of the project's underground structures will lie below the groundwater table. From the investigations performed to date, water can be expected in the headrace shaft at about 370 meters below the ground surface. Groundwater in the exploratory tunnel was present in fractures and joints with partial confinement by shaley interbeds. The piezometric level of this aquifer lies approximately 50 meters above the crown of the powerhouse chamber. The entire tailrace tunnel alignment will also lie below the groundwater table.

Exploratory Tunnel Program

Construction of the 1000-meter long, 4-meter wide by 4-meter high, exploratory tunnel provided an opportunity to obtain first-hand experience with the actual conditions that will be faced in the underground excavations. A generalized profile of the tunnel below groundwater is shown in Figure 2d. Construction of the first 600 to 700 meters of the exploratory tunnel encountered little to no groundwater inflow. Inflows progressively increased with depth beyond this point.

Criteria set forth in the construction contract documents set 8 cubic meters per hour (m^3/hr) per 100 meters of tunnel as the maximum allowable groundwater inflow into the tunnel. In an attempt to meet this requirement, water control grouting was performed. Since there was an added concern of encountering high groundwater inflows during tunnel excavation, pre-excavation grouting was utilized. The pre-excavation grouting program for the exploratory tunnel consisted of drilling and grouting periphery holes around the circumference of the tunnel 24 meters ahead of the next excavation cycle, with the intent of creating a 5-meter thick grout envelope around the tunnel perimeter. Guidelines developed by the supervising engineer required pre-excavation grouting whenever inflow from feeler holes drilled 24 meters ahead of the face was greater than 60 liters per minute (L/min), or 2.5 L/min per meter of hole. Measured inflows from these feeler holes ranged from 5.3 L/min/m of hole to 19.6 L/min/m of hole in the last 282 meters of tunnel, resulting in pre-excavation grouting of the entire tunnel length below the groundwater table.

Mixes for pre-excavation grouting were made of Type II Portland cement injected with initial water to cement ratios of 5:1 by weight, decreasing progressively to 4:1, 3:1, 2:1 and 1:1 as grouting at each cycle progressed. Bentonite was added to the mixes at a ratio of 2% by weight whenever inflows from grout holes exceeded 150 liters per minute. Injection pressures varied from 1 to 15 bars, depending on groundwater head and rock mass conditions encountered in probe holes drilled ahead of the tunnel face. Injection rates were generally kept at around 50 liters per minute. A summary of the drilling and grouting performed during construction of the exploratory tunnel is provided in Figure 2.

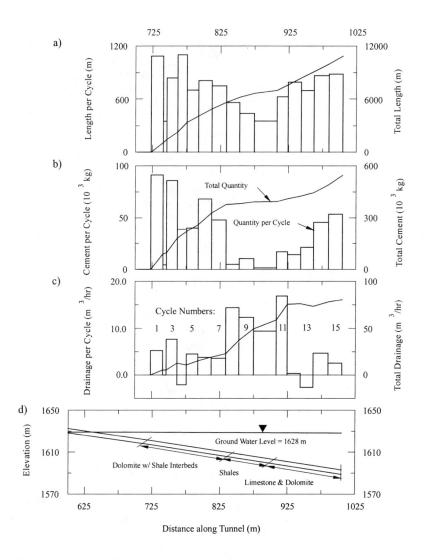

Figure 2. Summary of Exploratory Tunnel Data: a) grout hole drilling quantities; b) grouting quantities; c) inflows quantities; d) tunnel schematic.

Upon completion of excavation, the amount of water being pumped from the tunnel portal amounted to about 80 m^3/hr. Figure 2c provides a summary of the inflow into the last 282-meters of tunnel during construction. About 50 m^3/hr of water (58.3 m^3/hr per 100 m) was attributed to the shale unit, where grout takes were relatively low. In the other units, drainage ranged from 6.1 m^3/hr to 13.8 m^3/hr per 100 m of tunnel. It is postulated that the large inflows encountered in the shaley zone were due to swelling and slaking of the shale upon excavation. This swelling and slaking may have re-opened joints in adjacent rock, causing an increase in the effective permeability of the rock.

Seepage Control Measures

Design of seepage control measures for the underground structures of projects such as this is usually aimed at preventing large uncontrolled inflows during construction, and limiting inflows during project operation to prevent damage to instruments and machinery. Seepage control measures for this particular project must also minimize the impacts of underground construction on groundwater supplies in the area. The quantity of inflow needs to be limited during construction so that excavation and subsequent rock support and lining activities can be efficiently undertaken, and so that the quantity of water to be pumped to the surface and treated is reasonable. During project operation, the quantity of inflow into the powerstation complex and the associated access tunnels needs to be limited to minimize the long-term impacts on the local groundwater regime with respect to aquifer depletion and potential brine intrusion.

Experience gained from excavation of the exploratory tunnel indicates that water inflow control measures will be needed for all excavations below the water table during project construction and operation. Pre-excavation grouting has been proposed as the means to control water inflows during excavation of the underground structures. For the water conveyance and access tunnels, this will involve grouting around the periphery of the excavations ahead of the tunnel face, as was done during construction of the exploratory tunnel. A different approach is envisioned for the powerstation caverns, involving curtain grouting around the complex from a grout tunnel constructed via the exploratory tunnel. The grout curtain will extend upwards to a shaly zone located approximately 10 meters above the powerhouse cavern crown, and downwards below the invert of the powerhouse cavern as shown in Figure 3.

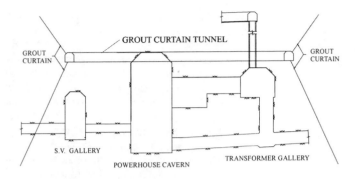

Figure 3. Section of Grout Curtain Perimeter Tunnel

To further reduce the effects of underground construction on the local groundwater regime, all access, construction and water conveyance tunnels will be lined immediately after excavation. For the construction, access and tailrace tunnels, the linings will consist of cast-in-place concrete. All other water conveyance tunnels (the high-head tunnels entering the powerstaion complex) will be steel lined. Consolidation grouting will be performed through the concrete linings where zones of local high inflows are encountered. The powerstation cavern walls will be treated with shotcrete, but no permanent lining system for these caverns is anticipated at this time. Additional grouting via the grout curtain tunnel surrounding the powerstation complex will be performed to further reduce any unacceptably high inflows into these structures.

Estimate of Grouting Effort

To estimate the magnitude of the grouting effort to be undertaken during project construction, it is assumed that the effort of pre-excavation grouting will be similar to that experienced in the exploratory tunnel (in terms of material quantities), and that a similar reduction in inflow rates will be achieved. Various ways of extrapolating and normalizing the exploratory tunnel grouting data have been explored (e.g. relating treatment effort to unit area of excavated surface or to unit volume of rock surrounding an excavation); however, the results are similar no matter which method is used. Parameters used here to estimate material quantities for pre-excavation grouting are based on the unit area of excavated surface in the access and water conveyance tunnels. Estimates of material quantities reported here for construction of the grout curtain around the powerstation complex are based on the surface area of the grout curtain perimeter.

A weighted average of drilling and grouting quantities for grout cycles 5 to 15 in the exploratory tunnel has been used to develop the quantity of drilling per square meter (m/m^2) of tunnel perimeter and the grout take per lineal meter of grout hole (kg/m of cement solids). These cycles are assumed to represent the average pre-excavation grouting effort that will be required during construction of the underground structures of the project. About 2 m/m^2 of excavated surface area and injection of about 37.3 kg/m of cement solids are anticipated for drilling and grouting of pre-excavation grout holes, respectively. Based on the excavated surface area of all water conveyance and access tunnels, as well as the surface area of the grout curtain perimeter, drilling of approximately 185,000 meters of grout holes and injection of approximately 6.9 X 10^6 kg of cement solids will be required.

Table 2 – Post-Excavation Grouting Parameters

Ground Conditions:	Hole Spacing (m):	Hole Length:	Grout Take (kg/m of hole):
Good	2.5	½ D_E	15
Average	2.5	½ D_E	30
Poor	1.8	¾ D_E	70

D_E = fraction of effective tunnel diameter.

Estimates of post-excavation grouting quantities are based on ground conditions anticipated throughout the underground excavations. Table 2 summarizes the parameters assumed for good, average and poor ground conditions expected in the underground structures of the project. These parameters were developed using experience from previous projects in similar geologic environments. Post-excavation grouting for the underground features of the project is expected to require drilling of approximately 75,500 meters of grout holes and injection of approximately 2.9 X 10^6 kg of cement solids. A summary of the estimated pre-excavation and post-excavation grouting quantities is presented in Table 3.

Table 3 – Estimated Water Control Quantities

Feature:	Pre-Excavation Grouting Drilling (m):	Pre-Excavation Grouting Grouting (10^3 kg):	Post-Excavation Grouting Drilling (m):	Post-Excavation Grouting Grouting (10^3 kg):	Peak Estimated Inflow (m³/hr):
Headrace Tunnels	13,900	520	14,800	370	63
Tailrace Tunnels	79,200	2950	27,000	990	204
Powerstation Complex	21,900	820	29,300	1,070	169
Access/ Construction Tunnels	69,500	2590	4,400	480	145

Prediction of Groundwater Inflows

An estimate of groundwater inflow quantities into the project's tunnels and caverns during project construction and operation has been made to determine the potential impacts on the local groundwater regime, and to develop a water treatment program for handling water pumped from the tunnel and cavern excavations. Again, data from construction of the exploratory tunnel was used to develop parameters for the estimate. Drainage measurements from excavation cycles 12 through 15 (see Figure 2c) are considered to be representative of conditions to be encountered in the construction of the remaining underground structures. Using these measurements an average rock mass permeability after pre-excavation treatment was back calculated utilizing the following equation from Goodman et al. (1965):

$$q = \frac{2\pi kH}{\ln(2H/r)} \quad (1)$$

where:
  q = inflow per unit length of tunnel (m³/s/m);
  k = rock mass permeability (m/s);
  H = average head below groundwater (m);
  r = radius of tunnel (m).

For the exploratory tunnel, the average head below groundwater was 35 meters. The average inflow from grout cycles 12 through 15 was 6.1 m³/hr per 100 meter of tunnel, or 17 X 10^{-6} m³/s/m. This results in a back calculated rock mass permeability of 3 X 10^{-7} m/s, or 3 X 10^{-5} cm/s.

Using the back calculated rock mass permeability parameter and the dimensions of the underground structures, an estimate of maximum quantities of water inflow that will need to be pumped from the various underground features can be made utilizing Equation 1. Results for the various water conductors, access tunnels and caverns are summarized in Table 3. The peak inflows in Table 3 will not occur at the same time during the construction period. Groundwater inflows will increase as tunnels and caverns are excavated, and diminish as inflows as tunnels are lined. Using the current project construction schedule and proposed grouting program outlined earlier in this paper, an estimate of how groundwater seepage quantities will vary with time throughout the construction of the project was made. The results are presented in Figure 4. The maximum rate of inflow during construction is on the order of 250 m^3/hr, reducing to approximately 70 m^3/hr at the end of underground construction.

Effects on Local Groundwater

Although additional investigations will be performed in the next phase of the project to perform a more detailed estimate of seepage quantities into the underground structures, predicted inflows using the exploratory tunnel as a model provide a basis for examining potential impacts of project construction on the local groundwater regime. Hydrogeologic studies summarized earlier in this paper indicate that approximately 2 MCM/yr can be drained from the aquifer intersected by the underground structures in the long term without adversely impacting current users of the aquifer in the area. Figure 4 indicates that inflows will average less than 1.0 MCM/yr during underground construction and be reduced to approximately 70 m^3/hr, or 0.6 MCM/yr, in the long term. Thus, although a local draw-down of groundwater levels can be expected, adverse effects to current users of the aquifer in terms of depletion of local groundwater supplies are not anticipated.

Another concern relating to the effects of underground structures on local groundwater is the potential for brine intrusion from sea water as a result of prolonged aquifer depletion. Representative samples of groundwater collected during construction of the extension of the exploratory tunnel indicate that the water in this location is relatively fresh. Over the lifetime of the project, water quality within this formation could change, gradually becoming more saline as a result of brine intrusion. Salinity levels in wells currently operating in the area have increased over the past decade, indicating that this is a possibility. The current plan to line all access and water conveyance tunnels is anticipated to minimize groundwater inflows into the underground structures of the project, thus minimizing the risk for brine intrusion.

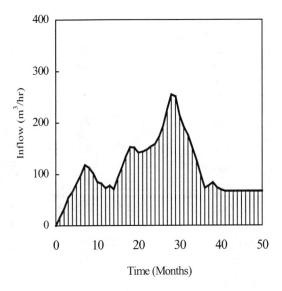

Figure 4. Estimated Inflows During Construction

Conclusions

Feasibility level studies for a hydroelectric pumped storage project in a semi-arid region required an estimate of the effects underground construction may have on the local groundwater regime. Data obtained from various geologic investigations, hydrogeologic investigations and construction of a 1000-meter long exploratory tunnel has been utilized to develop seepage control measures for project construction and operation. These measures, which were designed to accommodate any unforseen conditions encountered during construction and operation, include pre-excavation and lining of access and water conducting tunnels, and the construction of a perimeter tunnel around the powerstation complex. Pre-excavation grouting of project water conductors and access tunnels consists of drilling and grouting holes around the periphery of excavations ahead of the tunnel face. Additional consolidation grouting through the tunnel linings will further reduce groundwater inflows. Curtain grouting will be performed from the powerstation complex perimeter tunnel to limit seepage into the caverns during excavation and during project operation, if it is required.

Parameters for estimating pre-excavation grouting quantities and pumping requirements to evacuate water inflows from all of the underground structures were developed from data obtained during construction of the 1000-meter long exploratory tunnel. It is estimated that pumping will reach a peak rate on the order of 250 m^3/hr during construction, then gradually decrease to a steady-state rate of approximately 70 m^3/hr as tunnels are lined.

It is anticipated that inflows into the powerstation will not adversely affect the groundwater supply for current users in the area in the foreseeable future, and investigations to be conducted in the next phase of the project will be aimed at further defining the effects of underground construction on the local groundwater supplies in this area. These investigations will include extension of the exploratory tunnel into the powerhouse cavern, additional sampling and testing of the rock-mass throughout the underground structures of the project, and construction of test wells and monitoring wells to further establish aquifer characteristics in the project area.

References

Freeze, R.A. and Cherry, J.A. (1979). *Groundwater*. Prentice-Hall, Inc, Englewood Cliffs, New Jersey.

Goodman, R.E., Moye, D.G., Van Schalkwyk, A., Javandel, I. (1965). "Ground water inflows during tunnel driving." *Engineering Geology*, No. 2, pp. 39-56.

Heuer, R.E. (1995). "Estimating Rock Tunnel Water Inflow." *Proceedings, Rapid Excavation and Tunneling Conference* (RETC), Society for Mining, Metallurgy and Exploration, Littleton, CO, pp. 41-60.

Seismic Performance of the Yerba Buena Island Tunnel

Hubert K. Law[1] and Ignatius P. Lam[2]

Abstract

A two-step analysis procedure was adopted to assess the performance of a tunnel liner subjected to a design earthquake. The first step was to compute seismic induced deformations of the tunnel without the presence of the liner due to seismic wave propagation through the island rock. The second step involved imposing the deformations of the tunnel onto the structural liner through rock springs which accounted for interaction between the liner and surrounding rock.

A comprehensive field investigation program was also conducted to study the mechanical behavior of the rock. The findings from the field investigations were incorporated in the wave propagation analyses to define appropriate seismic loading criteria for the tunnel liner. Parametric studies were conducted to consider the variation of rock properties observed in the field.

Introduction

The Yerba Buena Island Tunnel is shown in Figure 1 which has a double deck structure with a reinforced concrete liner. The tunnel was constructed in 1930 on Yerba Buena Island located between the cities of San Francisco and Oakland. This island provides a natural break between east and west spans of the San Francisco-Oakland Bay Bridge. The tunnel length, portal to portal, is 165 m. The tunnel cross section provides a clear width of 20 m at the arch spring and a clear height of 15 m at the arch crown. The thickness of the reinforced concrete liner varies from 1 to 2 m.

The 1989 Loma Prieta Earthquake damaged the east span of the San Francisco-Oakland Bay Bridge resulting a number of injuries and closing the bridge to traffic for

---

[1] Project Engineer, Earth Mechanics Inc., 17660 Newhope St., Suite E, Fountain Valley, CA 92708

[2] Principal, Earth Mechanics Inc., 17660 Newhope St., Suite E, Fountain Valley, CA 92708

one month. The vulnerability of this vital transportation link between the two cities to future earthquakes is critical, and thus the California Department of Transportation has called for a seismic retrofit study of the Yerba Buena Island Tunnel. The seismic performance criteria for the Yerba Buena Island Tunnel has been established to be consistent with that of the San Francisco-Oakland Bay Bridge. As the Bay Bridge is classified as an "important" bridge, the tunnel must be repairable and must not interfere with the ability of the tunnel to support traffic under a Safety Evaluation Earthquake.

Figure 1. Yerba Buena Island Tunnel (Upper Deck Shown)

Site Geology and Seismicity

The rocks of Yerba Buena Island comprise the Franciscan Formation, an assemblage of ancient (Mesozoic, i.e. about 100 to 200 million +/- years old) rocks. The Franciscan Formation on the island, like much of the San Francisco Bay region, is locally covered by non-indurated silty and sandy soil and sediments of Quaternary age (i.e. less than about a million years old). These young sediments are largely slopewash (colluvium) and windblown deposits. Much of these deposits are derived from in-place weathering and decomposition of the Franciscan bedrock formation and later reworked by wind, water, and gravitational slope creep.

There are no known major fault zones at Yerba Buena Island. The nearest known major faults are the San Andreas fault about 19 km to the west, and the Hayward fault about 11 km to the east. The San Andreas fault is capable of generating about a moment magnitude (M_w) 8 earthquake and the Hayward fault an M_w 7.5 event. The recurrence intervals for the maximum earthquakes are estimated to be about 225 to 500 years for the M = 8 earthquake on the San Andreas fault and about 150 to 500 years for a M=7 event on the Hayward fault.

Field Investigations

A comprehensive field investigation program has been conducted to understand material properties of the island rock. The program includes geologic mapping, rock coring and lab testing, down-hole seismic velocity logging, pressure-meter testing, video imaging of boreholes, and specialized geophysical measurement based on Spectral Analysis of Surface Wave (SASW).

Geologic mapping was conducted from the ground surface to collect information on the orientations of discontinuous planes, characteristics of joint surfaces, and weathering conditions. Rock was cored at three locations as shown in Figure 2; and the boring depths range from 32 m to 75 m. Point load tests, unconfined compression tests, and direct shear tests have been conducted on selected core samples.

Seismic wave velocities were measured in the boreholes with down-hole geophysical sounding using OYO Model 170 suspension P-S wave logging system which provided continuous P- and S- wave velocity profiles. Also, down-hole pressure-meter tests were conducted in one of the boreholes to measure Young's modulus of the rock. Upon completion of drilling, the boreholes were photographed continuously by a down-hole digital video camera. The digital video imaging provided important information on bedding and joint orientation, and joint aperture.

The main objective of conducting SASW tests in this tunnel project is to obtain shear wave velocity of the rock behind the tunnel liner along the length of the tunnel wall without drilling through the tunnel liner. This is non-destructive testing which consists of placing geo-phones on the side wall and then impacting the wall with a hammer to generate surface wave signal. The sensors were moved progressively away from the impact source to receive the body wave from a deeper rock. About 20 SASW tests were conducted along the tunnel wall and on the pavement; most tests were performed at about shoulder height above the east bound road way (lower deck).

Based on these site investigations, all of the rocks observed on the island are sedimentary rocks of primarily graywacke sandstone. Weathering in a form of highly fractured rock is found near the surface, and the weather zone ranges from 15 to 30 m. The rocks in proximity to the tunnel are generally sound with very few rock joints. From the SASW measurements, the shear wave velocities of rock surrounding the tunnel range from 760 to 2430 m/sec. The unconfined compressive strengths of the core samples range from about 20-100 MPa in the weathered sandstone and about 100-200 MPa in the less-weathered sandstone at depth.

Analysis Approach

The analysis procedure is a two-step approach involving interaction of rock and

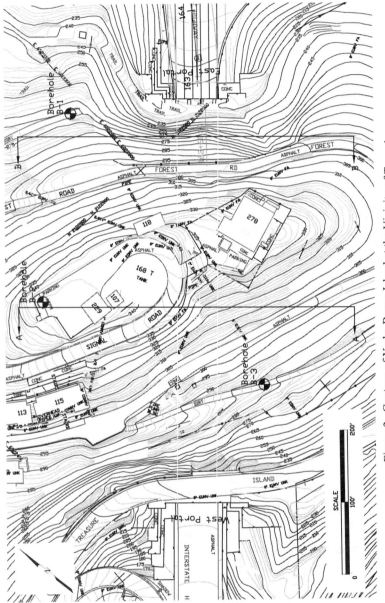

Figure 2. Contour of Yerba Buena Island at the Vicinity of Tunnel

liner. The first step consists of computing ground motions at the perimeter of the tunnel cavity subjected to the design earthquake without presence of the tunnel liner. In the second step, the computed ground deformations are imposed on the structural liner through rock springs. This approach explicitly assumes that the elastic restoring force as well as inertia of the liner are small compared to the surrounding rock, and therefore the feedback from the tunnel liner would not alter the ground deformations computed in the first step. With this approach, more detailed analyses such as implementation of plastic hinges on the structure can easily be conducted in the second step without much computational effort.

The sections follow describe the wave scattering analyses to compute seismic-induced deformations of the tunnel cavity and the development of rock springs to be used in the second step. The second step analysis was conducted by structure engineers, and it will not be reported here.

Wave Scattering

Two transverse cross sections were considered as shown in Figure 2; the section AA passes through peak of the island, while the section BB is near the east portal with a smaller ground cover over the tunnel. The finite element mesh for section AA is shown in Figure 3. In spite of the amount of data from the field investigations, the knowledge of material properties for the island is far from being complete. Given these circumstances, sensitivity analyses were conducted using three different scenarios for rock properties. These three sets of rock properties are denoted as "low", "medium", and "high" shear wave velocity profiles. The "low" shear wave velocity set uses a uniform Vs of 760 m/sec for the entire island except clay. The "medium" shear wave velocity set corresponds to Vs of 760 m/sec from the surface through elevation +30 m and gradually increasing to 1520 m/sec at elevation -150 m. The "high" shear wave velocity set features Vs of 1220 m/sec to 2130 m/sec for the sound rocks and 760 m/sec for the weathered rocks near the surface. A transmitting boundary capable of preventing wave reflection was used at the finite element boundary to model a semi-infinite space.

From the wave scattering analysis, two components ground motions (horizontal and vertical) were computed for all the nodes around the tunnel cavity which are shown in Figure 4 for one of the sensitivity studies. From these computed ground deformations, several modes of differential displacement were considered. As shown in Figure 5, the differential displacement between two nodes at a same elevation causes a potential compression/tension (e.g., second deck), while that between two nodes at different elevations could be attributed to racking. The changes in rotation angle would lead to flexural deformation of the liner. Figure 6 presents time histories of the mentioned differential displacements and changes in rotation angles at some strategic locations, and Figure 7 shows some snap-shots of the tunnel configuration. Tabulations of the peak differential displacements and rotation angles are presented in Table 1 for the six cases. Based on the quantities presented in the table, the section AA with the "medium" shear wave velocity profile was chosen as the benchmark case due

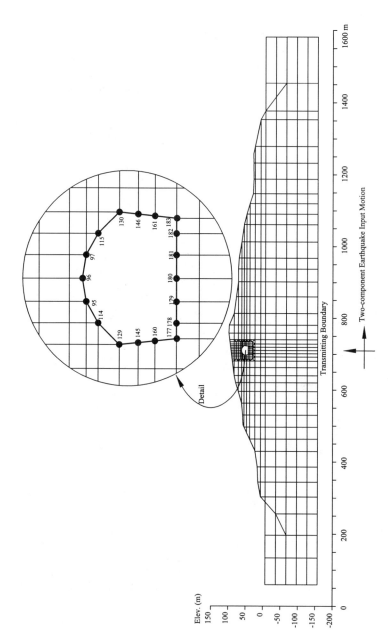

Figure 3. Finite Element Mesh for Section AA

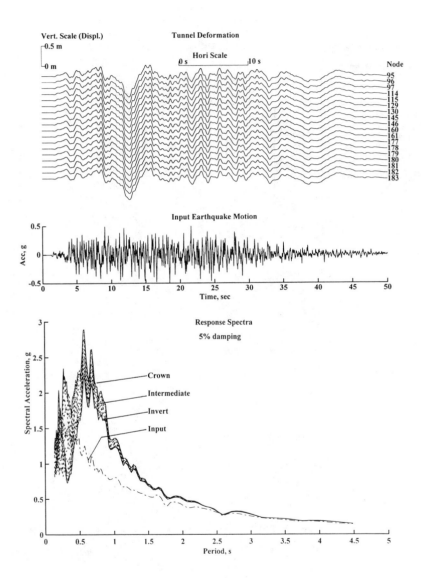

Figure 4. Input Acceleration and Ground Motions of Tunnel Cavity

666 GEO-ENGINEERING FOR UNDERGROUND FACILITIES

to large differential displacements as well as overall shaking intensity for the liner.

Table 1 Tabulation of Peak Differential Displacements and Rotation Angles

Section/ Velocity Profile	Differential Displacement, mm					Rotation Angle, rad		
	Node 145 - Node 146	Node 177 - Node 183	Node 129 - Node 177	Node 130 - Node 183	Node 145 - Node 146	Node 129	Node 96	Node 130
	Horiz.	Horiz.	Horiz.	Horiz.	Vertical			
AA/Low	4.44	2.5	6.3	6.3	7.6	5e-4	1.2e-4	4e-4
AA/Med	5.1	4.3	8.9	8.9	12.7	8e-4	2e-4	7e-4
AA/High	2.5	1.8	3.8	3.8	4.8	2.5e-4	8e-4	2.5e-4
BB/Low	2.3	1.8	3.8	3.8	6.8	2.5e-4	1.5e-4	2.5e-4
BB/Med	2.5	1.8	5.0	5.0	7.6	4e-4	1.8e-4	3.5e-4
BB/High	1.8	0.9	2.5	2.5	4.3	1.5e-4	1e-4	1.5e-4

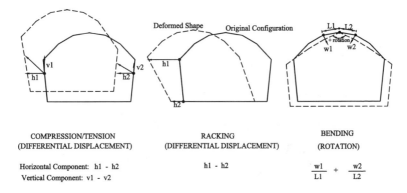

Figure 5. Modes of Differential Displacement

Rock Springs

The procedure to extract the rock springs is based on 3-dimensional finite

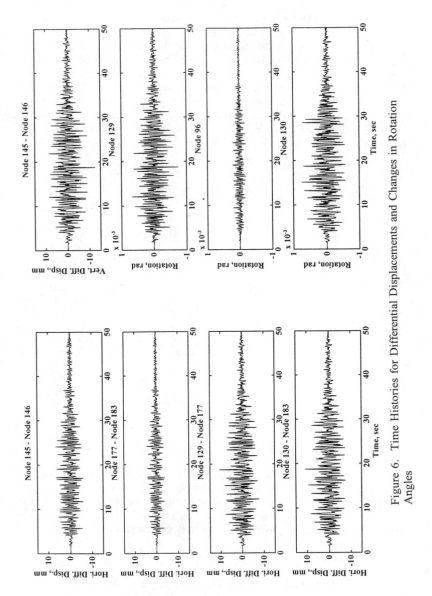

Figure 6. Time Histories for Differential Displacements and Changes in Rotation Angles

element analysis, as shown in Figure 8, where the tunnel cavity is lined with very stiff beam elements while the surrounding rock is assumed elastic. This 3-D section represents a typical cross section of the tunnel. The liner and rock are connected with non-linear elements. The purpose of the non-linear elements is to capture tension gap and slippage of the liner relative to the surrounding rock. It was decided that no tension was allowed in the radial direction while the limit state for slippage was set at a value corresponding to a concrete-rock bond strength of about 550 kPa. The tunnel liner (the system of stiff beams) was then prescribed 1 unit of translational displacement as a rigid body, and the reaction between the liner and the surrounding rock (i.e, force developed at the non-linear element) is taken as stiffness at each of the nodal points. The coupling effects were ignored for simplicity.

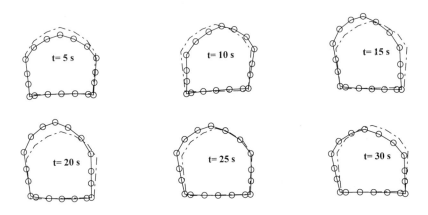

Figure 7. Snap-shots of Tunnel Configuration

Conclusions

Using the seismic induced ground deformations chosen as the benchmark case and the rock springs mentioned above, the structural engineers have conducted vulnerability assessment for the liner. Their parametric studies considered variation of initial stress (geo-static stress) and possible voids behind the liner that can be reasonably expect. Based on these studies, retrofit of the liner was not recommended.

Acknowledgments

The authors fully acknowledge the professional assistance of the Caltrans Contract Management team, the peer review panel members, and all the consultants

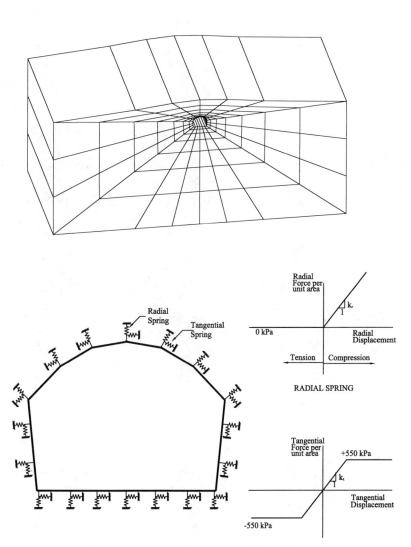

Figure 8. Development of Rock Springs

to this project. HNTB Corporation conducted the studies on seismic vulnerability assessment of the liner.

References

Asakura, T. and Sato, Y. (1996), "Damage to Mountain Tunnels in Hazard Area," Special Issue of Soils and Foundations, Japanese Geotechnical Society, January, 301-310

Earth Mechanics, Inc. (1998), "Report on Field Investigation," Prepared for HNTB and Caltrans, Division of Structures, Sacramento CA, Contract No. 59X855, February

Earth Mechanics, Inc. (1998), "Report on Wave Propagation Analyses," Prepared for HNTB and Caltrans, Division of Structures, Sacramento CA, Contract No. 59X855, March

Goodman, R. E. (1980), Introduction to Rock Mechanics, John Wiley & Sons.

HNTB Co. (1998), "Seismic Vulnerability Assessment and Conceptual Retrofit of Yerba Buena Island Tunnel," Prepared for Caltrans, Division of Structures, Sacramento, CA, Contract No. 59X855-A/1, March

Hudson, M., Idriss, I. M. and Beikae, M. (1994), User's Manual for QUAD4M, A Computer program to Evaluate the Seismic Response of Soil Structures Using Finite Element Procedure and Incorporating a Compliant Base, Department of Civil & Environmental Engineering, University of California, Davis

Hoek, E. and Brown, E. T. (1980), Underground Excavations in Rock, The Institution of Mining and Metallurgy, London

Lysmer, J. M. and Kuhlemeyer, R. L (1969), "Finite Dynamic Model for Infinite Media," J. of the Engineering Mechanics Division, ASCE, Vol 95, No. EM4, August, pp 859-877

Marsh, E. J. (1992), "Two Dimensional Nonlinear Seismic Ground Response Studies," Ph.D. Dissertation, University of Auckland, New Zealand

O'Rourke, T.D. and Shiba, Y. (1997), "Seismic Performance and Design of Tunnel," Technical Paper for National Center for Earthquake Engineering Research, Project Task No. 106-B(F), September

Eagle Ridge Mall Sinkhole
Haines City, Florida

James D. Hussin, Member, ASCE[1]
Joseph A. Eduardo, Member, ASCE[2]
Dean A. Elliott [3]

Abstract

In the summer of 1998, a sinkhole developed overnight in the center of the Eagle Ridge Mall in Haines City, Florida. The surface depression measured approximately 4.3 m (14ft) deep and 12.2 m (40 ft) in diameter. The area was quickly partitioned off with temporary walls. Flowable fill was used to fill the depression and allow access for exploration equipment. Standard Penetration Testing (SPT) encountered raveled soils to a depth of about 30-m (100 ft) beneath the center of the depression. Subsequent SPT and geophysical exploration was performed to identify the lateral extent of the feature. Adjacent mall areas continued to slowly settle as the exploration proceeded. A compaction grouting program was selected to stabilize the feature. During the course of this stabilization, a secondary feature developed in an adjacent department store. The remediation program was extended to treat this area also. This paper discusses the formation of sinkholes in central Florida and the use of compaction grouting to stabilize sinkholes. A case history is presented describing the grouting, monitoring and evaluation of the remedial program at Eagle Ridge Mall.

Introduction

The mall property occupies the eastern half of an intra ridge plain, which lies on the west side of the Lake Wales Ridge. Pre-development land surface elevations range between approximately 33.5m (110 ft) Mean Sea Level (MSL) in

---

[1] Director of Engineering and Business Development, Hayward Baker Inc., 6850 Benjamin Road, Tampa, FL 33634
[2] Senior Project Engineer, Ardaman & Associates, 1525 Centennial Drive, Bartow, FL 33830
[3] Project Manager, Hayward Baker Inc., 6850 Benjamin Road, Tampa, FL 33634

the wetlands of the eastern half of the property to 38 m (135 ft) MSL at the top of a low hill in the western half of the land. To accommodate mall construction, the topography was altered by cutting down the crown of the hill and infilling low-lying areas to create the present landform on which the mall is centered.

In accordance with usual and customary procedures for the region, evaluation of the potential for sinkhole development was included in the site's design phase geotechnical study. Results of the study had indicated that there was no reason to suspect sinkhole conditions below the building footprint itself. However, two years after mall completion, a subsidence crater developed within the mall building below the slab-on-grade concrete floor, at the intersection of the main mall corridor and the side corridor leading to an anchor store. The surface depression measured 4.3-m (14 ft) deep and 12.2 m (40 ft) in diameter. Figure 1 shows a plan view of the mall layout with the sinkhole location, and Figure 2 details the floor slab level survey results in the sinkhole area.

After an initial examination of the crater and the surrounding area, the geotechnical engineer recommended immediate backfilling of the crater with flowable, granular fill to prevent further sloughing of the sides of the crater into the hole. A total of about 150 m^3 (200 yd^3) was required to fill the crater. Following a geotechnical investigation, a remediation program of compaction grouting was recommended to stabilize the problem geologic conditions located below the subsided floor slab. During the course of the stabilization program, a second sinkhole developed in an adjacent department store. Remediation was extended to include this feature.

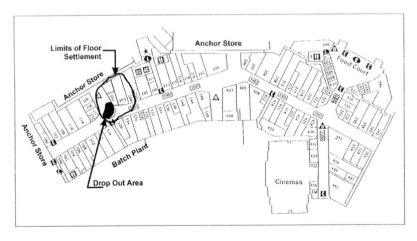

Figure 1: Plan View of Mall Showing Sinkhole Location

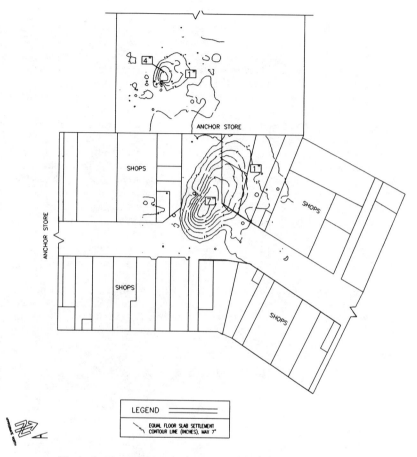

Figure 2: Floor Slab Level Survey Results in the Sinkhole Area

How Florida Sinkholes Form

Parts of Central Florida are susceptible to sinkhole subsidence because of the prevailing hydrogeologic conditions in this locality. A typical subsurface profile consists of up to 27.4 m (90 ft) of marine deposited, quartz-rich sediments overlaying the calcareous limestone formation. Generally, a variably thick, weathered cohesive soil layer (typically clayey sand or clayey silt) separates these two strata. Cavities in the limestone formation, ranging in size from cracks less than one inch thick to cavities large enough to enclose the subject mall building, are formed by dissolution of the limestone mass over time, by the down-dip flow of the mildly acidic precipitation through fissures and joints in the bedrock mass.

Nearly all of Florida is underlain by a stacked aquifer system, consisting of an unconfined surficial aquifer, which contains Holocene and Pleistocene age sediments underlain by confined Intermediate aquifer, whose transmissive zones are composed primarily of limestone and dolomite layers, which lies above the Floridian aquifer, whose transmissive zones are also composed primarily of limestone and dolomite layers. A vertical barrier to the flow of water between the surficial and the first confined (bedrock) aquifer, formed by a variably thick layer of cohesive sediment, creates significant differences between the potentiometric surfaces of these two aquifers. Sinkholes appear at the land surface when infiltrating water from the surficial aquifer sediments drains down through solutionally-enlarged vertical joints in the bedrock surface. This water flow causes a continuous migration of the overlying unconsolidated sediments into the joints. This migration of the overlying sediments relieves the pressure on the skeletal structure of the unconfined aquifer sediments, creating a zone of very loose unconsolidated soil above the solutionally-enlarged joint. This zone of very loose, unconsolidated soil is periodically enlarged by the above described events until the roof of this zone finally collapses and the surface depression (crater) appears (Figure 3).

As central Florida becomes more densely populated, the frequency of sinkholes impacting developed property increases. Sinkhole subsidence can be rapid or slow (Beck, 1984). Fortunately, the process is generally slow enough to avoid loss of life. However, when sinkholes occur near or within existing facilities, structural damage is common.

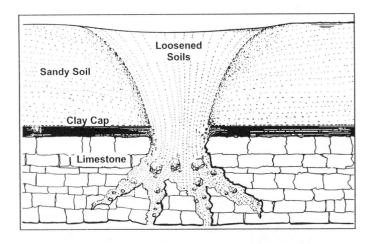

Figure 3. Typical Florida Sinkhole Formation

Exploration of the Eagle Ridge Mall Sinkhole

A Ground Penetrating Radar (GPR) survey was performed to detect anomalous subsurface conditions. A series of eight Standard Penetration Test (SPT) borings were then conducted within the building envelope at locations selected based on the visual examination of the damaged and subsided areas in combination with the GPR survey results. Four Standard Penetration Test (SPT) borings were also conducted at selected locations around the perimeter of the mall to determine the nature of the geologic conditions below suspect depression areas that developed on the parking lots, and to provide information on the possible lateral extent of the discovered sinkhole conditions below the mall building envelope. Borehole depths ranged from a minimum of 23.9-m (78.5 ft) to a maximum of 35.4 m (116 ft) below the existing floor and ground surface. A continuous drilling and sampling procedure was performed within the upper 1.8 to 4.6 m (6 to 15 ft) of the SPT boring, and within depth intervals where the SPT N value was less than 4 blows per 0.33 m (4 blows per foot), in order to detect subtle changes in soil stratigraphy and pertinent engineering properties within these critical depths.

Subsurface Conditions

Two SPT borings log provides are presented in Figure 4. Boring TH-1 was taken near the center of the subsidence crater in the mall common area, and boring TH-8 was performed outside the perimeter of the depression. These borings are representative of the subsurface conditions at their respective locations. Typically, where the bedrock has not been solutioned (corroded) over geologic time, and perceptible voids have not developed, the top of intact rock was encountered at a depth of 18 to 25 m (60 to 80 ft). Where rock had been corroded, the top of intact rock was much lower, ranging from 27.4 to 30.5+ m (90 to 100+ ft).

Soil boring TH-1 was drilled near the center of the crater after it had been filled with flowable fill. The soil deposits below the flowable fill were in an extremely loose, disturbed state. This is evidenced by the frequently easy penetration of the soils by the weight of the drilling rods and the attached sampling spoon. Essentially, no significant resistance to penetration was encountered in soil boring TH-1 below a depth of 12.2-m (40 ft) until the surface of the bedrock was encountered at 30.8 m (101 ft) below the floor. Drilling into the bedrock revealed a hard carbonate material to the maximum drill depth of 35.4-m (116 ft). No rock cavity (void zone) was found in the bedrock. Drilling fluid circulation was lost several times in this boring, both within the overlying soils and within the rock mass at a depth of about 34.4 m (113 ft).

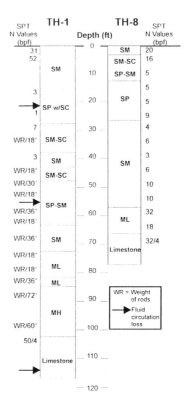

Figure 4: SPT Boring Logs

Soil boring TH-8 was drilled about 18 feet outside the perimeter of the floor surface depression. The soil conditions encountered here were similar to those encountered during the design phase exploration of the mall site. Essentially, the overburden soils were generally more firm and the surface of the rock was found at a depth of 20.4m (67 ft).

At several boring locations away from the sinkhole, where the surface of the rock was found below 27.4 m (90 ft), very weak, raveled soil zones were also encountered, and drilling fluid circulation loss was recorded. The soil above a depth of about 21.3-m (70 ft) had not yet been affected by a loss of skeletal soil structure. These subsurface conditions are considered to be incipient sinkhole conditions. That is, subsidence of the overlying land surface could occur at some time in the future whenever triggering mechanisms occur. The triggering mechanism, in this case, was the sudden change in the head difference between the potentiometric head in the surficial and confined aquifers.

Observation of piezometers installed in the mall common area during the exploration phase of this project in June 1998 indicated that the water level in the near surface soils was approximately 3.4 m (11 ft) below the floor and the water level in the bedrock unit was approximately 7.6 m (25 ft) below the floor, indicating a differential piezometric head of about 4.3 m (14 ft) between the surficial and bedrock aquifers.

Triggering Mechanism

Sinkholes are typically triggered by the development of an internal stress level change in the soil mass located above the rock void. This stress can be created by the application of an external force such as the applied weight of a large amount of new fill soil or by a heavy structure above the rock void location. It may also be created by an increase or decrease in the difference in the piezometric head between the surficial and bedrock aquifers.

The geotechnical engineer concluded that an unusually large amount of rainfall in the 1997-98 winter months (December 1997 to March 1998), followed by an unusually dry period (April 1998 to June 1998), had caused a wide fluctuation in the groundwater level over a short period of time. The consequent stress changes in the soil mass, combined with the natural difference between the piezometric head of the surficial and bedrock aquifer bodies, was sufficient to overcome the arching effect of the soil overlying the rock cavern zones below the mall structure, thus creating a sinkhole crater at the land surface.

Remediation Program

Conventional options to stabilize sinkhole features at unobstructed (open field) sites include:

1. Injection of compaction grout into the soils.

2. Collapse and densification of the soils using the vibro-replacement technique.

3. Collapse and densification of the soils using the high-energy impact of a heavy weight dropped on the ground surface.

4. Installation of piles to bear upon competent bedrock.

Because of the existence of the mall building, and the desire to restore the structure to its original condition, the second and third options were clearly not viable. Pile installation to support undermined portions of the mall was possible, but would require the use of piles capable of being easily spliced and an installation system able to be operated in low head-room. The cost of the large number of piles that would be involved was also deemed to be cost prohibitive. Compaction grouting was therefore recommended to seal off openings to cavernous rock zones, reduce any existing direct hydraulic connection between the surficial and bedrock aquifers in corroded bedrock zones, fill void zones above the rock surface, and compact the loose, raveled soil conditions created by the formation of the sinkhole crater.

Compaction Grouting to Stabilize Sinkholes

The compaction grouting process involves the injection of low slump grout into overburden soils or rock cavities. The grout does not permeate the soil but remains in a relatively spherical ball, displacing and thus densifying adjacent loosened soils as it increases in size (Henry, 1986).

Compaction grouting was developed in the United States and was initially used as a technique for in-situ densification of granular soils to control structural settlement. Subsequently, use of the technique was expanded to include the redensification of soils loosened during soft ground tunneling (Baker, Cording and MacPherson, 1983). Extensive monitoring and instrumentation accompanied the latter application. Data gathered during soft ground tunneling applications served to increase the understanding and knowledge of the capabilities of compaction grouting for sinkholes, since the migration of loosened soils into the void created by a tunneling machine is similar to the way in which a naturally occurring sinkhole is formed (Henry, 1986).

The use of compaction grouting to stabilize sinkholes generally includes several stages. Grout casings are initially installed into the area requiring stabilization until they penetrate a selected distance into the surface of the limestone formation at the bottom of the disturbed (raveled) soil zone. The pipes are typically located in a primary square grid that varies from 3 to 9 m (10 to 30 ft) on center, depending on subsurface conditions and project objectives. The casings are typically installed with a rotary wash drill rig. Compaction grouting then begins. The injected grout consists of a mixture of a sand cement water and additives, which is designed to flow easily through the pipes, and which has a relatively low slump, to prevent the grout from traveling away from the intended location. The initial grout typically has a higher slump to allow the grout to travel along the limestone surface, reestablishing the barrier between the limestone and the overlying sands. Water content is then reduced to produce a lower slump grout, generally less than 76 mm (3 in). The grout is pumped as the pipe is extracted in 0.3 to 0.6-m (1 to 2 ft) increments. This procedure displaces and thus densifies the loose raveled sands.

Grouting at each interval continues until either a significant increase in grout pumping pressure occurs, a predetermined quantity of grout is injected, or ground heave is observed. Based on review of the grouting results at the primary locations, secondary locations at the midpoint of the primary grid might be added, especially if the maximum grout quantity was injected at the adjacent primary locations before significant pressure increase was observed.

Eagle Ridge Mall Sinkhole Stabilization Program

The objective of the compaction grouting program was to stabilize the sinkhole feature by sealing the surface of vertical rock joints, or capping the voids in the bedrock and then densifying the loosened overburden soils.

The compaction grout was batched on site, occasionally supplemented by a ready-mix supplier. The batch plant was set up in the parking lot adjacent to the entrance nearest to the sinkhole. The grout was then pumped from the batch plant through 100 mm (4-inch) steel pipes (slickline) as far as 137 m (450 ft) to the injection location.

To execute the remediation program, two low headroom rotary drill rigs were used to install the grout casing. The rig used in the more confined department store was electric powered to eliminate engine fumes. Grout casing was installed to the base of the raveled soils, which ranged from about 18 to 46 m (60 to 150 ft) in depth. Spacing of the primary injection locations was 6.1 m (20 feet). Figure 5 shows the contours of the depth to the top of bedrock developed with the casing installation information.

Compaction grout was injected as the casing was extracted in 0.6-m (2 ft) intervals. Grouting at each interval continued until one of the following criteria was achieved:

1. Pressure:
    - Maximum grout pressure increase (over the pressure necessary to initiate grout take) varied linearly between 400 and 200 psi for grout slumps ranging between 100 and 178 mm (4 and 7 in).
    - Maximum grout pressure of 200 psi increase for grout with slump between 178 and 228 mm (7 and 9 in).

2. Quantity:
   - Maximum grout quantity of 3.8 m^3 (5 yd^3) per 0.6 m (2 ft) interval.
3. Heave:
   - Maximum cumulative total of 12.7 mm (0.5 in) of surface heave at any building column.
   - Maximum cumulative total of 25.4 mm (1 in) of lift of existing building slab.

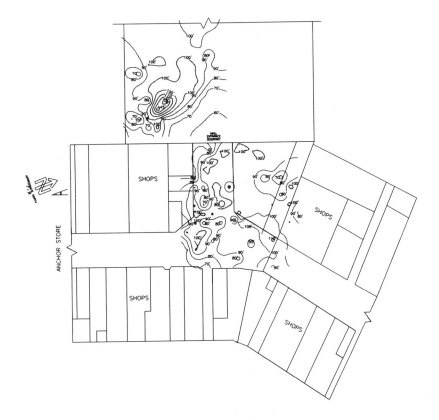

Figure 5: Contours of the Depth to the Top of Bedrock

The grout slump used for this project was greater than what is typically used at sites having unobstructed access to permit pumping over the long distance. The grout slump was generally maintained in the four to seven inch range.

Secondary locations were identified and grouted based on the volume injected and pressure recorded during the primary program. A total of 195 injection locations and approximately 10,300 m^3 (12,320 yd^2) of grout was required to stabilize the site. Figure 6 shows the approximate grout quantities injected at each location. Figure 7 shows grouting underway in the mall area.

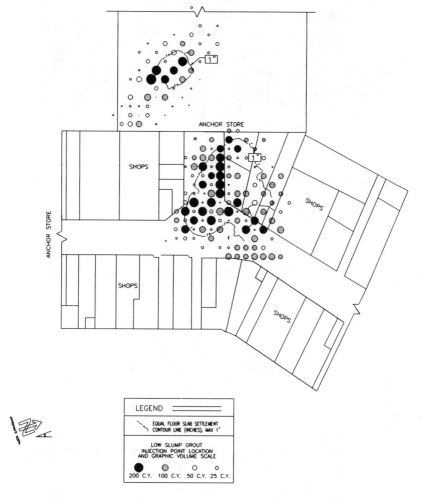

Figure 6: Grout Injection Locations and Approximate Quantities

682 GEO-ENGINEERING FOR UNDERGROUND FACILITIES

Figure 7: Grouting Operations Underway in the Mall Area

Program Monitoring and Evaluation

Grout Injection Monitoring

A technician representing the geotechnical consultant maintained quality control during the grouting operation. This included monitoring of drilling termination depth, grout properties, overall grout quantities and the refusal criteria. The technician's observations were instrumental in selecting secondary injection locations.

Structural Monitoring

During the course of grout injection operations, monitoring was conducted to check the vertical position of building columns located within the area where sinkhole subsidence had occurred. Initially, this monitoring was conducted several times daily to provide advance warning of excessive movements of portions of the structure and thus allow a rapid alteration in the sequence of grout injection points in order to treat a rapidly deteriorating area. Once movement of the columns had slowed to acceptable levels, monitoring frequency was reduced to once-per-day measurements. Column monitoring continued after completion of grouting operations until movement had not been detected for several weeks.

After completion of grouting, elements of the mall structure and floor slab were monitored to determine if post-injection movement of the structure was occurring. This monitoring consisted of measurement of the water level in the surficial aquifer, combined with survey measurement of the elevation of the floor slab above areas where grout had been injected.

To confirm acceptable long-term performance of the compaction grouting program, monitoring of the vertical position of the building columns will be conducted on an annual basis, using benchmarks and column marks established during the remediation program. To avoid disturbing the mall customers, survey points were installed on the roof, above all the mall columns

Summary

Central Florida is underlain by solution-prone, calcereous limestone bedrock. Sinkholes sometimes occur, creating a significant amount of structural damage. Compaction grouting is a proven method for the remediation of active sinkholes and for the pre-treatment of sites where sinkhole conditions are identified. The controlled injection of low slump grout can seal and cap the surface of the bedrock and stabilize raveled soils by displacement and densification. Highly mobile, low-headroom equipment, requiring only a limited work area, makes compaction grouting a particularly viable option where sinkholes occur within structures. At Eagle Ridge Mall, the compaction grouting program successfully stabilized the sinkholes, as evidenced by monitoring results. Stabilization operations were confined to the area immediately adjacent to the affected portion of the mall, allowing the remainder of the mall to maintain normal business operations.

Acknowledgements

The authors extend their appreciation to geotechnical engineers Ardaman & Associates, Inc. and specialty geotechnical contractor Hayward Baker Inc. for their cooperation in the preparation of this paper. Appreciation is also extended to technical editor Chris Herridge for her valuable assistance.

References

Beck, G. B. (1984). "Sinkhole Terminology." *Proceedings of the First Multidisciplinary Conference on Sinkholes.*

Baker, W.H., Cording, E.J., and MacPherson, H.H. (1983). "Compaction Grouting to Control Ground Movement During Tunneling," *Underground Space, Vol. 7*, pp 205-212

Henry, J. F. (1986). "Low Slump Compaction Grouting for Correction of Central Florida Sinkholes." *Proceedings of Environmental Problems in Karst Terranes and Their Solutions,* Bowling Green, KY

Geophysical Surveys of a Known Karst Feature, Oak Ridge Y-12 Plant, Oak Ridge, Tennessee

William E. Doll[1], Jonathan E. Nyquist[2], Philip J. Carpenter[3], Ronald D. Kaufmann[4], and Bradley J. Carr[1]

Abstract
  Geophysical data were acquired at a site on the Oak Ridge Reservation, Tennessee to determine the characteristics of a mud-filled void and to evaluate the effectiveness of a suite of geophysical methods at the site. Methods that were used included microgravity, electrical resistivity, and seismic refraction. Both microgravity and resistivity were able to detect the void as well as overlying structural features. The seismic data provide bedrock depth control for the other two methods, and show other effects that are caused by the void.

Introduction
  In this paper, we describe a suite of geophysical measurements that were used at a site on the Oak Ridge Reservation (ORR), Tennessee where a mud-filled void is known to extend from 18 m to at least 30 m depth. The void was discovered during installation of monitoring wells in Bear Creek Valley, adjacent to the Oak Ridge Y-12 Plant. Because the Y-12 plant is a source of radioactive contaminants, mercury, and solvents, any pathway that might accelerate offsite migration of contaminants is of concern and must be investigated.

  The study site lies on the lower northeast flank of East Chestnut Ridge, about 90m west of Scarboro Rd. (Figure 1). Approximately 6-23 m of alluvium, colluvium and saprolitic residuum overlie bedrock across the site. The bedrock consists of the Copper Ridge Dolomite (Knox Group), Maynardville Limestone and Nolichucky Shale (Conasauga Group) (Shevenell et al., 1992). These units dip approximately 45 degrees to the southeast (Hatcher et al., 1992).

  The subcropping Maynardville Limestone has a stratigraphic thickness of approximately 135 m at the eastern end of Y-12 (Shevenell et al., 1992). It is primarily composed of dolomitic, oolitic, and micritic limestone. The largest cavities and highest hydraulic conductivities of the Conasauga Group have generally been encountered in the Maynardville Limestone, which has also been classified as part of the Knox aquifer on the ORR (Solomon, 1992).

---

[1]Envir. Sciences Div., Oak Ridge National Laboratory, PO Box 2008, Oak Ridge, TN 37831-6038

[2]Department of Geology, Temple University, 1901 N 13th St., Philadelphia, PA 19122-6081

[3] Department of Geology, Northern Illinois University, DeKalb, IL 60115-2854

[4]Technos Inc., 3333 NW 21st St., Miami, FL 33142

Evidence for lateral karst drains (long subsurface channels, parallel to strike) on the ORR includes gravel and cobbles in cavity fill material, secondary travertine, and rapid head responses to storms (during monitoring with pressure transducers) of wells completed in cavities (White, 1993). The Maynardville Limestone along Scarboro Road is believed to be an "exit pathway" for groundwater leaving the ORR (Shevenell et al., 1992). Between 1991 and 1993, a series of monitoring wells were installed across this exit pathway to monitor changes in head and groundwater geochemistry, identify fractures and cavities, and to map lithologies and thickness variations in the Maynardville Limestone. Fractures and cavities were encountered during drilling of all of these wells with the greatest concentration of cavities at a depth of less than 46 m. No cavities were encountered deeper than 72m (Shevenell and Beauchamp, 1994).

Two wells encountered unusually large cavities. GW734 intersects a mud-filled cavity at 18m that extends to at least 30 m. Pressure transducers in this cavity respond within 1-6 hours to storm events, suggesting this cavity is hydraulically connected to the surface. White (1993) suggests this cavity could represent a higher-order "east-flowing lateral drain" in the subsurface karst network. GW722 (46 m south of GW-734 and 4.3m higher at ground surface) also intersected a water-filled cavity between 30-38 m which may be connected to the GW734 cavity (Shevenell et al., 1992). The top of "weathered" bedrock in GW734 is reported in driller's logs at 10.7m, and the top of "fresh" bedrock

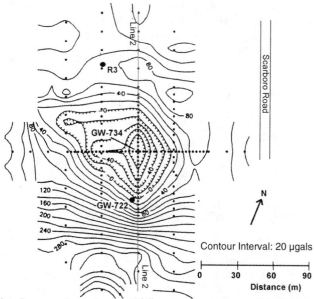

Figure 1. Complete Bouguer Anomaly map for the area surrounding well GW-734. Locations of measurement points, well GW-722 and resistivity sounding R3 are overlain.

is reported at 13.1 m. In GW722, weathered bedrock is reported at 16.5m and fresh bedrock is reported at 22.3m.

Geophysical methods may be used for direct detection of karst cavities, or may be used to infer underyling cavities by imaging "near-surface indicators", shallow features (e.g. sinkholes) that are associated with deeper karst cavities. Carpenter et al. (1998) discussed the evidence for near-surface indicators at the Scarboro Road site with emphasis on ground penetrating radar and shallow electromagnetic measurements. Here we discuss methods that achieve greater penetration, and in some cases direct detection of the cavity.

Microgravity Survey

Microgravity techniques have proven very effective in locating cavities and a large number of case studies have recently been published in which microgravity has been used to identify caves, mine rooms, shafts, and tunnels. Caves and voids typically have anomalies of 10-100 microgals, which require relative elevation accuracies of 0.3-3 cm at measurement points to enable adequate data reduction. Knowledge of near-surface density variations is essential in mapping cavities from microgravity data. Greenfield (1979) notes that a 10 microgal change can be produced by only a 24 cm (10 in) soil thickness change, when the density difference between soil and bedrock is 1 g/cm^3.

A "complete" Bouguer gravity anomaly map (Figure 1), is one that eliminates the effects of topographic features in and adjacent to the map area. In this case, we used a digital elevation database with a 30 m grid spacing to remove effects of topography within 2.4 km of GW-734. On this gravity map, the area which is greater than 90 m north of GW734 is relatively flat, and there is a significant gradient between 45 m north and 90 m south of GW734 (Figure 1).

GPR data for a filled depression north of GW734 (Carpenter et al., 1998) and density values derived from reports and recent measurements were used to constrain two-and-a-half dimensional (2.5D) gravity models for the Scarboro Rd. site. The Hypermag workstation software for gravity modeling (Saltus and Blakely, 1993) was used to perform the 2.5D modeling. Mud densities measured at 18.3, 19.9, and 21.3 m depth in GW734 in the spring of 1993 were 1.25, 1.53, and 1.6 g/cm^3 respectively (Shevenell, 1994). Fresh Maynardville Limestone has measured densities of 2.6-2.8 g/cm^3 and weathered Maynardville limestone has measured densities of 1.8-2.5 g/cm^3 (Bailey and Withington, 1988; Wilson, 1991; King and Haase, 1988). Because the terrain corrected data show a steep gravity gradient to the south, we also include formation contacts in our model. Fresh Copper Ridge Dolomite at GW131 has an average density of 2.8 g/cm^3. Unweathered Nolichucky at GW239 exhibits an average density of 2.3-2.4 g/cm^3 in the Scarboro Road area.

Modeling was conducted along four profile lines to provide a pseudo-3D image of the void. Figure 2 shows the model for Line 2 which is centered on GW734 and has a north-south (plant grid; rotated 34° counter-clockwise from true north) orientation. The remaining gravity models are given in Carpenter et al., 1995. Density contrasts shown on the figures are based on the measurements discussed in the previous paragraph, and are all relative to unweathered Maynardville Limestone. A trial-and error approach was used to assign a density contrast (0.03 g/cm^3) to the filled doline that was identified in GPR data (Carpenter et al., 1998). This value is reasonable if we assume that the doline is filled with a high proportion of chert fragments, or if it has a perched water table or high moisture content. The east-west extent of the void was used to set the in- and out-of-plane distances (for two-and-a-half dimensional modeling) used in the north-south models, including the model for Line 2 (Figure 2). The gravity models indicate that the void has a width (north-south) of 53-91 m, a length (east-west) of 91-122 m, and a height of 7.6-18.3 m. Because its cross-sectional shape changes along strike, the models should be considered representative, and the actual shape may differ significantly from the model shape.

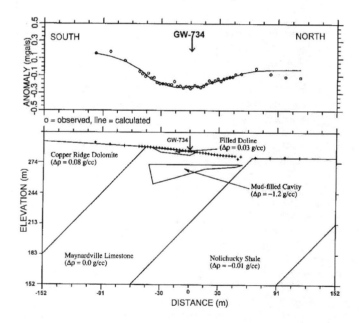

Figure 2. Gravity model for Line 2, which is oriented parallel to grid north (N34°E).

None of the gravity models include effects of changes in density of unconsolidated materials or weathered bedrock because these changes are poorly constrained. It is very possible that these effects contribute to the observed anomalies to the extent that the thickness or lateral extent of the void as shown in the models is somewhat exaggerated. It is encouraging to recognize that if a number of very shallow inhomogeneities were present, the measured gravity values would reflect them, and the gravity field would have much more scatter than observed. Gravity models do not consider the variable thickness of the weathered layer, although we now have estimates of that thickness from seismic refraction data (described later in this paper). It is likely that three-dimensional effects of the void shape play a more important role than the variability of the thickness of the weathered layer. However, it should be noted that the models here assume that fresh bedrock is continuous up to the surface. Furthermore, the models assume a distinct continuous contact between mud and limestone, and it is much more likely that a "halo" surrounds the main void which consists of weathered fractures which may be partially filled with mud or water having densities which differ from that measured. Anomalies may also be due to the cumulative effects of more than one void separated by unweathered rock.

Resistivity Measaurements

Direct-current (DC) resistivity techniques have been very popular in karst studies due to the low equipment cost, relatively simple field procedures and generally rapid interpretation of data. Graphical techniques are commonly used to locate cavities with a variety of array configurations (Kirk and Werner, 1981; Owen, 1983). Milanovic (1981) describes the use of azimuthal resistivity surveys in defining karst flow pathways. These resistivity methods, however, are highly sensitive to interference from near-surface heterogeneity, they employ, at times, empirical rules without a theoretical basis, and they produce a high percentage of false anomalies. Also, most of the examples presented in these papers concern air-filled (vadose) caves; very few, if any, purport to map saturated caves or caves filled with sediment. A more effective use of DC resistivity, as noted by Kirk and Werner (1981) and Milanovic (1981), may be as a reconnaissance tool in which karstified areas or stratigraphic intervals may be identified through bulk resistivity changes.

Electromagnetic (EM) earth conductivity surveys provide essentially the same information as resistivity profiles, but EM surveying is far quicker. Relatively few papers, however, have been published utilizing EM surveys in karst investigations. Kirk and Werner (1981) noted that EM methods generally worked well in identifying shallow clay-filled sinkholes related to deeper karst features. Time-domain electromagnetic profiles have been combined with seismic studies by Nelson and Haigh (1990) to identify sinkhole-prone areas and possible cavities in the lateritic soils of northern Australia. Yuhr et al. (1993) describes a set of vertical dipole EM profiles over a cave system in southwest Texas. Paleo-collapse features exhibited a 10 mS/m decrease in apparent conductivity, whereas a profile over a solution doline in limestone exhibited a 4 mS/m decrease in apparent conductivity. At our site, time domain methods failed, presumably due to power line interference at the site (Carpenter et al., 1995). DC resistivity methods were more effective.

Schlumberger resistivity array soundings were made at the Scarboro Rd. site (Figure 3). Two perpendicular soundings were made over the large Maynardville cavity. The east-west sounding (S36EW) was along strike (Figure 3a) whereas the north-south sounding (S36NS) was along dip (Figure 3b). Resistivities ranged from 300 ohm-m to almost 1000 ohm-m for the upper 16 ft (5 m) which is surprisingly high for a clay-rich saprolitic residuum. This may reflect a high chert content in the near-surface, very well-drained materials, or cavities due to soil piping above bedrock caverns (Panno et al., 1994). All three soundings exhibit low resistivity materials below 66 ft (20 m) depth, probably reflecting the clay-filled cavity.

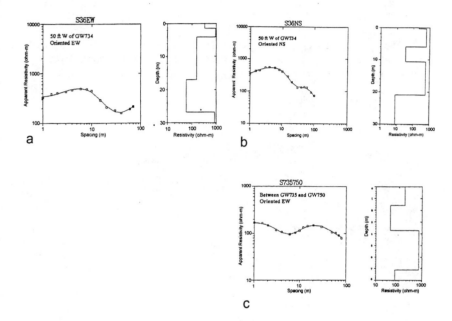

Figure 3. Schlumberger resistivity results for three soundings. Soundings a and b were centered near well GW-734 (E-W and N-S respectively) while sounding C was centered where no void is thought to occur, at point R3, Figure 1.

Soundings north and south of the cavity showed no evidence of a low resistivity layer at depths of 60-100 ft (18-30 m). Sounding S735750 centered at point R3, Figure 1, was acquired approximately along strike over the lowermost Maynardville Limestone. Apparent resistivities were relatively constant (90-200 ohm-m) and the inversion produced the shallow layered model shown in Figure 3c.

Multielectrode Resistivity Surveys

Apart from the conventional resistivity soundings, we had a brief opportunity to use a Sting/Swift 28-electrode resistivity system (Advanced Geosciences Incorporated) to collect multielectrode resistivity at the Scarboro site while the instrument was rented for another project. The Sting/Swift combination consists of a standard Sting DC resistivity instrument used in conjunction with a Swift relay system that allows the user to make a sequence of measurements automatically using a string of electrodes connected by a special cable. The system can make over 100 apparent resistivity measurements in less than an hour. The measurements form the basis for a 2-D data inversion program that interactively adjusts a finite difference forward model to find the smoothest resistivity cross section that matches the observed data (Loke,1996; Loke and Barker, 1996).

We collected data along two perpendicular lines centered on GW-734 using a 28-electrode cable with a 6 m electrode separation and an electrode firing sequence that resembles a combination Wenner-Schlumberger sounding. Figure 4a shows the

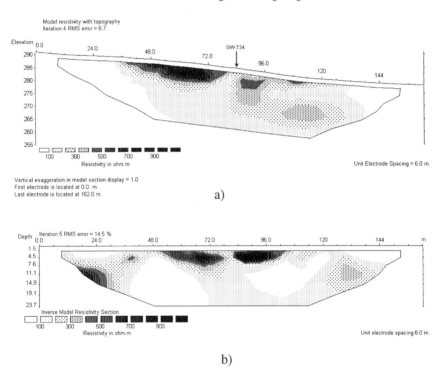

Figure 4. Multielectrode resistivity inversion for a) a profile along Line 2 (Figure 1) and b) an orthogonal profile that interesects Line 2 at well GW-734.

topographically corrected inversion for the north-south line. The general pattern seen in the conventional resistivity data was a resistive near-surface layer associated with the material filling the doline overlying a more conductive layer. The same pattern is seen in the multi-electrode resistivity data, however the 2-D inversion captures the structure of the doline and agrees well with the radar data. The east-west line (Figure 4b) is similar, but only passes over the southernmost edge of the doline.

Some caution must be used in interpreting these data, because a terrain conductivity survey (Carpenter et al., 1995) found two water mains running across the survey area from southeast to northwest. The line in Figure 4a crosses both of these pipes, which are seen as conductive areas at the north and south ends of the line. Near the center of the resistivity line, away from the two pipes, the shallow resistivity data is unaffected by these utilities. However, for the deeper sounding data collected at the largest electrode separations, the current probably spread out laterally far enough that some of it moved through the pipes. This would exaggerate the decrease in apparent resistivity with depth as the inversion software tried to fit the data with a 2-D model. Similarly, the east-west line (Figure 4b) shows low resistivities at the surface at both ends of the line that are associated with a road (at the east end) and a saturated pit and possible pipeline at the west end. Although the deep conductor is compatible with what would be expected from a mud-filled void, it is suspect because of possible effects from pipelines. This profile does, however, show a response to the doline that may be out-of-plane. Consequently, we can confidently conclude that the multielectrode resistivity data mapped the doline fill material, but it is questionable whether or not the deeper part of the resistivity sections are conductive because of the influence of the underlying cavity, the adjacent metal water mains, or a combination of both. The thin, shallow low resistivity zone north of GW-734 and deeper, thicker low resistivity zone south of the well in Figure 4a have a resemblance to the shape of the void in the gravity model (Figure 2) and might provide a basis for an alternative gravity model.

Seismic Refraction Surveys

Seismic refraction surveys have rarely been used to directly image karst features. This is because refraction methods are in principle unable to resolve decreases in velocity as a function of depth (such as a mud-filled void embedded in limestone), and are vulnerable to errors where the geology is highly heterogeneous and three-dimensional. They have been used to determine average rock velocities for various layers, which then can be related to the dynamic moduli and density which reflects the degree of karstification of the rock mass as a whole. Refraction is used in this context most frequently at dam sites, or in foundation studies for large structures such as nuclear power plants (Milanovic, 1981).

Two seismic refraction experiments were conducted at the Scarboro Rd. site in February and June, 1994 (Carpenter et al., 1995). These used hammer and shotgun sources that lacked energy required to image deeper than about 6m. In 1998, seismic refraction data were acquired with a Bison Elastic Wave Generator source and a

Geometrics 48-channel Strataview seismograph with 10 Hz. geophones at 3m spacing.

Figure 5 shows an interpretation of the seismic refraction data that was produced from five shots fired into a 146m spread (48 10Hz geophones at 3m (10-ft) spacing, oriented N-S and centered on GW-734) and processed with the SIPT-2 seismic refraction processing package. SIPT-2 uses a delay-time method to obtain a depth model. The program assumes layers with constant velocities that increase with depth (i.e. v1< v2 < ...).

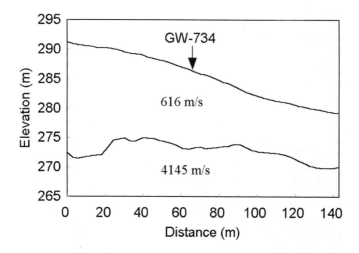

Figure 5. Bedrock depth along Line 2, derived from delay-time analysis of seismic refraction data. Bedrock and surface layer velocities are shown.

As an alternative, first arrival times were picked for 110 shots fired at 1.5m increments into the same 146m geophone spread. These data were entered into a tomographic refraction statics module in the FOCUS seismic reflection processing software package. This module, which is designed to calculate a surface layer correction for seismic reflection data, uses a tomographic method to calculate surface layer thickness and bedrock velocity, assuming a constant surface layer velocity (Figure 6). The delay-time calculation provides a depth to bedrock that is in agreement with the driller's log measurements described earlier, but shows little effect from the void. Although the depth to bedrock that results from the tomographic inversion is not reasonable, it is interesting to note that there are two localities along the profile that are characterized by a depression in the bedrock surface and a low bedrock velocity. These are indications of a localized low velocity zone that causes delays in travel times for rays that pass through (or around) it.

seismic and gravity inversions. Shear velocities might also be included in such an inversion, using SASW methods (Stokoe et al., 1994) or similar methods (e.g. Park et al., 1998).

Acknowledgements

Funding to initiate this effort was provided by the ORNL Groundwater Program Office. Surveying costs were funded by Y-12 Engineering. Most of the funding for this work was provided by the Y-12 Environmental Restoration Program. We would like to thank Les Beard, Dean Kieswetter and Jeff Wilson for their assistance in collection of the resistivity field data for this study. Oak Ridge National Laboratory is managed by Lockheed Martin Energy Research Corp. for the U. S. Department of Energy under contract number DE-AC05-96OR22464.

References

Bailey, Z.C. and D.B. Withington, 1988. Well construction, lithology, and geophysical logs for boreholes in Bear Creek Valley near Oak Ridge, TN: U.S.Geol. Surv. Water-Resources Invest. Rep. 88-4068, Nashville, 21 pp.

Carpenter, P. J., W. E. Doll, and R. D. Kaufmann, 1998, Geophysical character of buried sinkholes on the Oak Ridge Reservation, Tennessee, Journal of Environmental and Engineering Geophysics, v.3, p. 133-146.

Carpenter, P. J., W. E. Doll, and R. D. Kaufmann, 1995, Geophysical surveys over karst features near the Oak Ridge Y-12 Plant, Oak Ridge, TN, report Y/ER-200, Martin Marietta Energy Systems, Inc., 88 pages.

Greenfield, R.J. 1979. Review of geophysical approaches to the detection of karst. Bull. Assoc. Engineering Geol. 16:393-408.

Hatcher, R.D., Jr., P.J. Leminski, R.B. Dreier, R.H. Ketelle, R.R. Lee, D.A. Leitzke, W.M. McMaster, J.L. Foreman and S.Y. Lee. 1992. Status report on the geology of the Oak Ridge Reservation. ORNL/TM-12074. Oak Ridge National Laboratory.

King, H.L. and C.S. Haase. 1988. Summary of results and preliminary interpretation of hydrogeologic packer testing in core holes Well GW-131 through Well GW-135 and CH-157, Oak Ridge Y-12 Plant, Y/TS-495. Martin Marietta Energy Systems, Inc., Oak Ridge Y-12 Plant, Oak Ridge, Tenn.

Kirk, K.G. and E. Werner. 1981. Handbook of Geophysical Cavity-Locating Techniques With Emphasis on Electrical Resistivity, Federal Highway Admin., U.S. Dept. of Trans., Implem. Package FHWA-IP-81-3, (U.S. Govt. Printing Office #1981-341-428:1112)

Loke, M. H., 1996, RES2DINV ver. 2.1: Rapid 2D resistivity inversion using the least-squares method: Penang, Malaysia.

Loke, M. H. and Barker, R. D., 1996, Rapid least-squares inversion of apparent resistivity pseudosections by a quasi-Newton method: Geophysical Prospecting, v. 44, p. 131-152.

Milanovic, P.T. 1981. Karst Hydrogeology. Water Resources Publications, Littleton, CO.

Nelson, R.G. and J.H. Haigh. 1990. Geophysical investigations of sinkholes in lateritic terrains. pp. 133-153. In S.H. Ward (ed.), Geotechnical and Environmental Geophysics, Volume III, Society of Exploration

Geophysicists, Tulsa, OK.

Owen, T.E. 1983. Detection and mapping of tunnels and caves. pp. 161-258. In A.A. Fitch (ed.), Developments in Geophysical Exploration Methods, v. 5. Applied Sci. Publ., London.

Panno, S. V., C. P. Wiebel, P. C. Heigold, and P. C. Reed, 1994. Formation of regolith-collapse sinkholes in southern Illinois: Interpretation and identification of associated buried cavities. Environmental Geology v. 23, pp. 214-220.

Park, C. B. , R. D. Miller, and J. Xia, 1998,Ground roll as a tool to image near-surface anomaly, Expanded Abstracts, 1998 Ann. Meeting. Soc. Expl. Geophy., p. 874-877.

Saltus, R.W. and R.J. Blakely, 1993. HYPERMAG -- An interactive, 2- and 2 1/2-dimensional gravity and magnetic modeling program: Version 3.5, U.S.G.S. Open-file report 93-287.

Shevenell, L.A. 1994. Chemical characteristics of waters in karst formations at the Oak Ridge Y-12 Plant, Y/TS-1001. Martin Marietta Energy Systems, Inc., Oak Ridge Y-12 Plant, Oak Ridge, TN, 83 pp.

Shevenell, L.A. and Beauchamp, J. 1994. Evaluation of cavity occurrence in the Maynardville Limestone and Copper Ridge Dolomite at the Y-12 Plant using logistic and general linear models, Y/TS-1022, Martin Marietta Energy Systems, Inc., Oak Ridge Y-12 Plant, Oak Ridge, TN, 42 pp.

Shevenell, L.A., R.B. Dreier and W.K. Jago. 1992. Summary of fiscal years 1991 and 1992 construction, hydrologic and geologic data from the Maynardville Limestone exit pathway monitoring program. Y/TS-814. Martin Marietta Energy Systems, Inc., Oak Ridge Y-12 Plant, Oak Ridge, TN.

Solomon, D.K., G.K. Moore, L.E. Toran, R.B. Dreier and W.M. McMaster. 1992. A hydrologic framework for the Oak Ridge Reservation. ORNL/TM- . Oak Ridge National Laboratory.

Stokoe II, K. H., G. W. Wright, A. B. James, and M.R. Jose, 1994, Characterization of geotechnical sites by SASW method, in Geophysical characterization of sites, ISSMFE Technical Committee #10, edited by R. D. Woods, Oxford Publishers, New Delhi.

USEPA, 1993. Use of Airborne, Surface and Borehole Geophysical Techniques at Contaminated Sites: A Reference Guide (3 volumes). EPA/625/R-92/007. U.S. Environmental Protection Agency.

White, W. B., 1993. Carbonate terrain hydrogeology at the Oak Ridge Reservation. Internal Correspondence, Feb. 9, 1993. Martin Marietta Energy Systems, Oak Ridge, Tennessee.

Wilson, J.M. 1991. East Chesnut Ridge hydrogeologic characterization -- a geophysical study of two karst features. ECE-91-002. Environ Consult. Engineers, Inc. Prepared for Martin Marietta Energy Systems, Inc., Oak Ridge Y-12 Plant, Oak Ridge, TN.

Yuhr, L., R. Benson and D. Butler. 1993. Characterization of karst features using electromagnetics and microgravity: a strategic approach. pp. 209-228. In R.S. Bell and C.M. Lepper (eds.), 1993 Symp. on the Appl. of Geophys. to Environ. and Engin. Prob., Environmental and Engineering Geophysical Soc., Englewood, CO.

Case Study: Site Characterization Methods in Karst
Mackey, J.R.[1], Roth, M.J.S.[2], and Nyquist, J.E.[3]

Abstract

Most geotechnical investigations in karst areas investigate, among other things, the depth to a highly variable bedrock surface and the possibility of dissolution features (voids or filled fractures) that may cause sinkholes. In areas of thinly mantled clay soils, such as cover much of the northeast United States, geotechnical investigations traditionally use borings and probes to obtain the necessary subsurface information. The reliability of this method of investigation is a function of the variability of the bedrock surface and the size of the voids or fractures of interest. At sites with a highly variable bedrock surface and small dissolution features, the results from borings and probes are often insufficient to characterize the site. Recent developments in multi-electrode earth resistivity tests indicate that this may be a reliable and economical method for obtaining subsurface data at these sites. Results of a few geologic and geoenvironmental applications of this method have been published, but an evaluation of the method as a geotechnical investigation tool in an urban setting has not been reported. In this study, the authors collected 19 resistivity lines at an urban site. The results were compared with boring and probe data as well as with subsurface data obtained during construction at the site. The earth resistivity results were found to correlate well with top of bedrock but correlations with void locations were inconclusive. Additional research is recommended to address the reliability of the method in locating voids and the effects of three-dimensional variability on the results.

---

[1] Student, Department of Civil and Environmental Engineering, Lafayette College
[2] CONTACT AUTHOR: Assistant Professor, Department of Civil and Environmental Engineering, Lafayette College, Easton, PA 18042
[3] Associate Professor, Department of Geology, Temple University, Philadelphia, PA

Introduction

The geotechnical investigation of a site in an urban area challenges even the most experienced engineer. In addition to conducting a subsurface investigation of the soil and rock conditions, the engineer must evaluate the site history to establish locations of past structures, existing and abandoned utilities, and to assess the impact that the proposed construction will have on adjacent structures. In areas of karst geology, the engineer must also investigate subsurface features that may be associated with future sinkhole formation and the nature of the irregular bedrock surface associated with karst geology.

Traditional methods of site investigation in the karst areas of the northeast United States include boring and other probing methods (Thomas and Roth, 1997). Unfortunately, these methods have a low probability of encountering voids or filled fractures and the data obtained may be insufficient to characterize bedrock topography. Conventional geophysical tools, e.g. ground penetrating radar, electromagnetic, gravity and seismic, typically are not used in this area (Thomas and Roth, 1997) due to problems with depth of penetration in clay soils, interpretation, or high associated costs. This paper presents a case study of the geotechnical investigation and construction findings of an urban site in karst geology. This site was investigated using traditional site investigation methods and multi-electrode resistivity testing.

Background on Resistivity Testing

Earth resistivity testing was first used in the early 20th century. The method involves introducing D.C. current into the subsurface through two probes while measuring the voltage drop across two additional probes. The physical arrangement of the probes is systematically changed to obtain information about the resistivity of the subsurface. Earth resistivity testing is described in most texts on geophysical testing (Telford et al., 1976, Sharma, 1978, among others). The process with a four-probe system is very time consuming and the data obtained is difficult to interpret beyond simple changes in resistivity in one-dimensional soundings or profiles.

The advent of multi-electrode earth resistivity systems has greatly changed the normal procedure involved in conducting D.C. resistivity tests. Multi-electrode resistivity systems consist of a line of electrodes connected by cables, an earth resistivity meter and a switching system. The line of electrodes is laid out in the field with equal spacing between electrodes. The switching system runs through a patterned series, inducing current between pairs of electrodes and measuring voltage drops between other pairs. The result is a combined sounding and profile for the ground beneath the line. Rather than moving the equipment to obtain each data point, the multi-electrode earth resistivity system allows for the collection of multiple points without equipment relocation.

A two-dimensional resistivity model of the subsurface may be determined using the data obtained during the test. The apparent resistivity pseudosection is matched using a finite-difference forward modeling subroutine and is adjusted with a least-squares optimization technique (deGroot-Hedlin and Constable, 1990, Loke and Barker, 1996). This technique subdivides the subsurface into rectangular blocks of calculated resistivities to create a model pseudosection that matches the measured pseudosection (Loke, 1996).

Previous Work

Multi-electrode earth resistivity systems have only become commercially available in the last decade and published studies are somewhat limited. The systems are expensive and their reliability has not been well documented. A trained individual is needed for interpretation of results and resistivity profiles must be conducted in conjunction with intrusive methods in order to interpret the results appropriately.

Griffiths and Barker (1993) studied the use of multi-electrode earth resistivity equipment in modeling complex geology. Their paper describes four resistivity studies. Two studies are hydrogeological investigations in Zimbabwe that focus on interpretation of weathered bedrock. The remaining two study sites are in Staffordshire, England. The object of one test was to observe faulting in bedrock and the other was to locate the transition between salt and fresh groundwater. Because the goal of these tests was to determine large-scale features in the subsurface, the tests performed for this study used large electrode spacings.

Dahlin (1996) presents work using multi-electrode tests at four study sites in Sweden and includes boring correlations for some profiles. The resistivity data was used to investigate sediment cover over an aquifer, clay thickness and bedrock depth, dyke material type and quality determination, and sediment analysis.

Earth resistivity methods have also been successfully used in locating karst features. Lambert (1997) reports the use of a multi-electrode system to investigate a site with sinkholes in Missouri. Site soils are listed as clays and the bedrock surface is shown as highly variable. Karst features were identified during the course of testing.

These studies show the viability of the use of earth resistivity systems for exploration in clay soils and karst geology. However, none of these studies investigated the use of multi-electrode testing for evaluating the smaller scale subsurface features and variations that are of concern to geotechnical engineers when designing foundations.

Case Study

The study site is located in the Great Valley region of Pennsylvania and is underlain by the Allentown Formation. The Allentown formation is composed of dolomitic limestone and is characterized by an irregular bedrock surface. This formation also contains subsurface voids and filled fractures associated with the formation of sinkholes.

The footprint of the proposed structure is shown in Figure 1. The proposed structure is a two to three story, 9200 m^2 addition to the existing building to the north. The building footprint is approximately 3600 m^2 and the first floor is at an elevation of 83 m. It is believed that during the 1800's the site was part of or near a quarry, and may have also been used as a teaching facility by the Lafayette College's mining engineering program. The site was developed as part of the

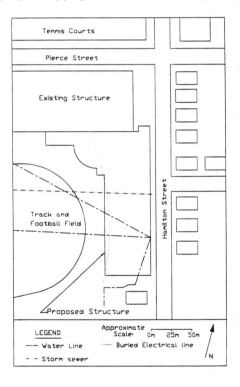

Figure 1. Site plan and utility locations

College's football stadium during the early 1900's. Miscellaneous fill materials were placed on the site at this time. Prior to the current construction activities, the site was a grassy sloped area between Hamilton St. and the football field. Known utilities at the site include a water line, a buried electrical line, and a 0.8 meter diameter masonry storm sewer. These utilities are shown in Figure 1.

Geotechnical Site Investigation

The preliminary investigation at the site included nine test borings (Figure 2). The soil materials were determined to be random fill and unconsolidated clayey silts. The bedrock was found to be a dolomitic limestone and was observed to be

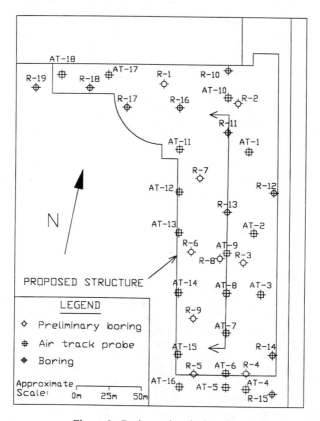

Figure 2. Boring and probe locations

fractured with some open voids. The depth of the bedrock encountered in the borings varied from approximately one meter to 5 meters. Groundwater was not encountered.

Based on the results of the preliminary investigation, additional subsurface exploration was recommended. Ten additional test borings and eighteen percussion rock probes were conducted (Figure 2). A typical cross section of the site based on the preliminary and secondary investigation is given in Figure 3. The site was generally characterized by the geotechnical engineers as having rock at shallow depths in the southern portion of the site and rock at greater depth in the northern portion of the site. The engineers did anticipate some variability in the rock surface and recommended rock probes be conducted at every column location during construction.

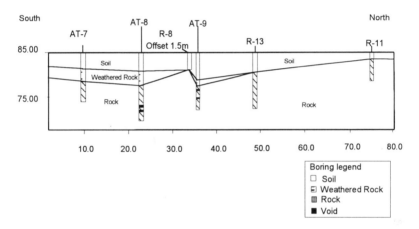

Figure 3. Typical cross section based on boring results

Multi-electrode earth resistivity testing was conducted at the site after the secondary site investigation was completed but before construction. These tests were conducted to evaluate the effectiveness of the earth resistivity method in clayey soils and in an urban setting and were completed after the geotechnical engineers had made their foundation recommendations for the site. Nineteen resistivity lines were conducted at the site using 28 and 56 probe resistivity systems. Figure 4 illustrates the locations of the tests. The tests were conducted using dipole-dipole, Schlumberger, and Wenner arrangements of the electrodes.

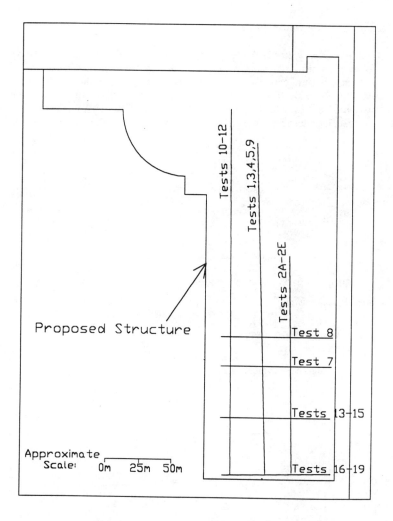

Figure 4. Locations of earth resistivity lines

Testing was not conducted in the northern portion of the site due to limited accessibility. Figure 5 illustrates a typical result from the resistivity tests (Test No. 5). (The results from borings located along this line are also illustrated.) This test was completed using 56 electrodes at 1.5m spacing and the dipole-dipole method.

Effect of Utilities on Resistivity Results

The resistivity profile shown in Figure 5 illustrates the effect of conductive utilities on the resistivity results. The water lines appear as limited areas of low resistivity at a depth of approximately 2.5m. Figure 6 illustrates the results from a resistivity test (Test No. 7) run approximately parallel to a utility. In this case, the resistivity profile is completely controlled by the conductive utility, i.e., essentially all the current introduced into the ground by the test is moving along the utility.

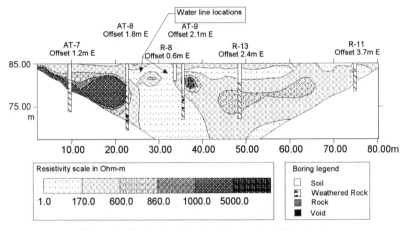

Figure 5. Typical earth resistivity result with borings

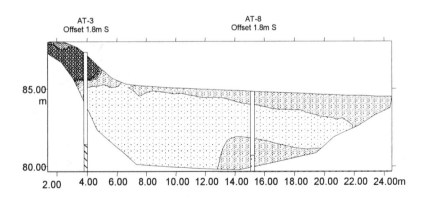

Figure 6. Typical result of resistivity line run parallel to a conductive utility (For resistivity scale and boring legend, refer to Figure 5.)

Comparison of Test Borings and Probes with Resistivity Results

Data from borings completed as part of the geotechnical site investigation were compared to the resistivity profiles. These comparisons indicate that the bedrock is represented by resistivities of approximately 600 Ohm-m or larger. Given this interpretation, the depth to top of bedrock based on the resistivity profiles using a dipole-dipole electrode arrangement was found to correlate reasonable well with the boring results. Actual depths to bedrock as indicated by the borings and probes were typically within one to two meters of the depth indicated by the resistivity results. Other arrangements of the electrodes provided less detail and did not correlate as closely with the boring data.

The resistivity results shown in Figure 5 indicate a drop in the bedrock surface between the AT-8 and AT-9 probes. Because of the presence of the water utilities at the surface in this area, it was initially difficult to interpret the cause of the apparent drop in the bedrock. No borings were located in this area to confirm the finding.

Correlations between the resistivity results and the voids encountered in the bedrock are not readily apparent from the data obtained at the site. A resistivity test will not reveal a void if it is effectively insulated from the applied current by the surrounding bedrock. A void will be revealed only if the void is connected to the surface soil materials by soil- or water-filled fractures in the direction of the applied current. Difficulties in interpreting data concerning the locations of voids are discussed in greater detail in Roth et al. (1999).

Construction Findings and Comparison with Site Investigation Results

During construction of caisson foundations at the site, the contractor obtained detailed data concerning the top of bedrock using percussion rock probes at each column location. Comparisons between the depth to bedrock predicted by the geotechnical engineers using only the preconstruction boring and probe data and the depth determined by the rock probes conducted during construction indicated that the bedrock surface was more variable than expected. The preconstuction data was accurate at the locations tested, but at untested locations, predicted depths varied from actual depths by as much as 11.5 meters. Significant cost overruns were incurred because of the required modifications to foundation depths.

Figure 7 illustrates typical depth to bedrock information obtained from the rock probes and the associated resistivity test (Test 5, the same test result shown in Figure 5). The probes for the column locations were offset approximately 3 meters to the east of the resistivity line. It is apparent from this figure that the area of low resistivity observed and discussed previously agrees with a drop in the bedrock surface in the region near probes J-3 to L-3. The influence of the utilities did not significantly affect the resistivity data obtained at depth.

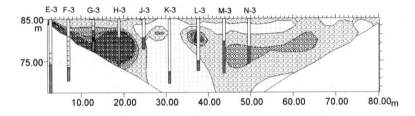

Figure 7. Typical resistivity result with construction subsurface data
(For resistivity scale and boring legend, refer to Figure 5.)

Lessons Learned

From the authors' experiences at the case study site, the following lessons concerning the use of multi-electrode earth resistivity testing were learned:

- because resistivity data at depth is obtained from pairs of electrodes that are spaced further apart, data concerning the subsurface is difficult to obtain in the corners of the site;
- conductive utilities in the subsurface distort all data if a test is conducted approximately parallel to the utility;
- conductive utilities in the subsurface distort a small area of the results if a test is run approximately perpendicular to the utility;
- the dipole-dipole arrangement of electrodes provided results that most closely matched the boring and probe data;
- depth to bedrock can be determined with some accuracy from resistivity results if borings are available for correlation of the results;
- resistivity profiles can provide valuable data concerning bedrock depth between boring locations; and,
- the presence of voids using resistivity results cannot be easily determined.

Conclusions

The authors believe that multi-electrode earth resistivity testing provides significant information in areas with highly variable bedrock surface. However, these tests need to be run in conjunction with borings in order to develop correlations between the results and the bedrock location. Additional study is required to evaluate the reliability of the results and to develop an understanding of the effect of voids on the resistivity results. The three-dimensional effects associated with an irregular bedrock surface also need further study.

Acknowledgements

This study was supported in part by the National Science Foundation under Grant No. CMS-9734899. This support is gratefully acknowledged. The authors would also like to thank John DiGenova of Site Blauvelt Engineers for providing access to the geotechnical investigation results and to Laura Bresette of Bovis Construction for organizing and providing the construction data for the site.

References

Dahlin, T. (1996). "2D resistivity surveying for environmental and engineering applications," First Break, 14, 275-283.
deGroot-Hedlin, C. and Constable, S.C. (1990). "Occam's inversion to generate smooth, two-dimensional models from magnetotelluric data," Geophysics, 55, 1613-1624.
Griffiths and Barker (1993). "Two-dimensional resistivity imaging and modeling in areas of complex geology," Journal of Applied Geophysics, 29, 211-226.
Lambert, D. W. (1997). "Dipole-dipole D.C. resistivity surveying for exploration of karst features," in The Engineering Geology and Hydrogeology of Karst Terranes, B. Beck and B. Stephenson, eds., Balkema, Rotterdam, 413-418.
Loke, M.H. (1996). "RES2DINV ver. 2.1: Rapid 2D resistivity inversion using the least-squares method," Penang, Malaysia.
Loke, M.H., and Barker, R.D. (1996). "Rapid least-squares inversion of apparent resistivity pseudosections by a quasi-Newton method," Geophysical Prospecting, 44, 131-152.
Roth, M.J.S., Mackey, J.R. and Nyquist, J.E. (1999). "A case study of the use of earth resistivity in thinly mantled karst," Proceedings, SAGEEP.
Sharma, P.V. (1986). Geophysical Methods in Geology, 2nd ed., Elsevier, New York.
Telford, W.M., Geldart, L.P., Sheriff, R.E., and Keys, D.A. (1976). Applied Geophysics, Cambridge University Press, Cambridge.
Thomas, B.*, and Roth, M.J.S. (1997). "Site characterization for sinkholes in Pennsylvania and New Jersey", in The Engineering Geology and Hydrogeology of Karst Terranes, B. Beck and B. Stephenson, eds., Balkema, Rotterdam, 281-286.

Geotechnical Engineering
Aspects of Microtunneling in Soft Ground

David C. Mathy[1], Robert A. Kahl[2],
Dru R. Nielson[3]

Abstract

The trenchless installation of pipelines by microtunnel pipe jacking presents new challenges of understanding to the geotechnical engineer. Since the introduction of microtunneling technology to North America in 1984, the number and complexity of microtunnel installations has steadily increased. This paper summarizes some of the key geotechnical engineering aspects of microtunneling through soft ground as experienced in the San Francisco Bay area including: shafts; thrust blocks and reaction walls; jacking force estimates; tunnel stability; oversize materials and obstructions; soil bearing capacity to support the microtunnel machine; spoil removal and solids separation; hydrofracture and slurry loss; shaft backfill and manhole settlement; and ground treatment.

Introduction

Microtunneling is most commonly described as a remotely controlled, guided, pipe jacking process that provides continuous support to the excavation face. The microtunneling process does not require personnel entry into the tunnel (Bennett, 1995). Microtunneling equipment has five independent systems (Iseley, 1997):
- microtunnel boring machine (MTBM),
- jacking or propulsion system,
- spoil removal system,
- laser guidance and remote control system, and
- pipe lubrication system.

---

[1]Principal Engineer, M. ASCE, [2]Senior Engineer, [3]Engineering Geologist, DCM/Joyal Engineering, 484 North Wiget Lane, Walnut Creek, California 94598

All five of these systems are directly impacted by subsurface conditions. Subsurface conditions dictate, in whole or in part, the successful selection of various equipment options and equipment combinations available for microtunneling (e.g., cutter heads, jacking systems, spoil removal and solids separation systems, etc.). Subsurface conditions also influence the design of jacking and reception shafts, tunnel drive lengths, production schedules, pipe materials, and even the feasibility of microtunneling. In fact, there are some subsurface conditions where microtunneling may not be advisable (e.g., small diameter tunnels through cobbles and boulders that are too large in diameter for the microtunnel machine to ingest and crush; tunnels through fills with oversize debris, metal, wood or other obstructions).

In the following sections of this paper, we discuss some of the key geotechnical engineering aspects of microtunneling in soft ground as experienced in the San Francisco Bay area. Each section concludes with general recommendations for geotechnical input into the design and construction of microtunnel pipe jacking projects. The design and construction issues discussed herein are based on direct experience with over 6,000 meters (20,000 feet) of microtunnel pipe jacking installations using slurry MTBM's that provide full earth and hydrostatic pressure balance at the tunnel face. For purposes of this paper, soft ground is defined as estuarine and alluvial soil deposits and very soft sedimentary bedrock that can be excavated using hand tools and methods. Such soils are common throughout the San Francisco Bay area which includes extensive deposits of Bay Mud around the Bay margin, fine to coarse-grained alluvium between the Bay and the coastal mountains which surround the Bay, and very soft sedimentary bedrock within the coastal mountains and their foothills.

Shafts

We begin with shafts in this paper because they often receive the least amount of design attention on a microtunnel pipe jacking project. Typically, a general contractor will build the shafts, turn over the shafts to the microtunnel contractor for pipe jacking and then construct manholes and backfill the shafts. It is very important to develop thorough specifications for shaft construction and to coordinate the shaft specifications with the microtunnel specifications. The very conditions that make microtunneling desirable (i.e., deep excavations, weak soils, high groundwater) make shaft construction difficult and expensive. Even partial failure of a shaft may cause delay of construction and damage to nearby pavements and structures. In addition, most shafts become a part of the permanent project through the construction of manholes. The condition of the shaft bottom will affect manhole stability and settlement and, therefore, the long-term performance of the finished pipeline.

For purposes of shoring design, microtunnel shafts should be considered long-term structures. It is important that no voids or gaps be present immediately behind the shoring. Such gaps will affect the thrust block bearing capacity and may lead to deflection or rotation of the thrust block and jacking frame. Gaps behind shoring also provide an avenue for slurry loss upon launch and retrieval of the MTBM.

In the San Francisco Bay area, shaft shoring systems vary considerably with depth, space constraints, proximity to structures and subsurface conditions and include trench boxes, soldier pile and lagging, sheet piles, cofferdams, caissons or various combinations. Two common examples are discussed herein. In stable soils above groundwater or under dewatered conditions (dry and stable condition) pile and plate systems are common. Steel H-piles are placed in drilled holes and fixed in place with gravel or low strength grout. As the excavation progresses, steel plates are pushed down inside the flanges of the H-piles. Walers are installed as a function of shaft depth. A concrete floor is typically cast in all jacking shafts. In unstable soils (e.g., loose alluvium or Bay Mud) below groundwater, cofferdams consisting of interlocking steel sheet piles are common. The use of sheet piles allows for rectangular shaped pits and longer length jacking frames and pipe segments. Sheet piles are fully interlocked with sufficient toe embedment to ensure base stability. Dewatering is only implemented inside the cofferdam to maintain a relatively dry working base. Due to the unbalanced hydrostatic pressures from groundwater outside the cofferdam, it is very important that this system have no gaps that would allow soil to flow into the excavation and create voids behind the sheet piles. In extreme cases, flow of soil through sheeting gaps may create sinkholes in the ground surface outside the shaft. Intentionally constructed gaps in shoring (e.g., at crossing utilities) require special attention, usually some form of ground treatment, to prevent flowing ground conditions. Walers are installed as a function of shaft depth. A concrete floor is typically cast in all jacking shafts. Where the soils directly underlying shafts consist of loose saturated silts and sands sensitive to vibrations, the sheetpiles are sometimes cut off below grade and left in place rather than risk vibration induced settlements upon vibratory extraction.

It is critical that the geotechnical engineer provide input to the project designers and owner regarding shaft specifications. The specifications should be written to ensure that the types of shafts allowed are consistent with subsurface conditions and anticipated excavation behavior (e.g., no unsupported excavation in flowing ground conditions) and that the shaft construction is coordinated with the microtunnel contractor, particularly with respect to the integration of thrust blocks and reaction walls.

Thrust Blocks and Reaction Walls

Thrust blocks and reaction walls are integral to shaft construction and are a function of shaft type, shaft shape, estimated jacking force and soil conditions around the shaft. Thrust blocks can be as simple as a concrete block or solid steel plate at the rear of the pit. Such simple thrust blocks are common for small estimated jacking forces in dry and stable soil conditions using the pile and plate shaft shoring system described earlier. In a cofferdam style shaft, the thrust block is typically integrated into the rear sheet piles to form a reaction wall. A typical reaction wall evaluation is presented in Stein, et al, 1989. A safety factor of at least 2 needs to be applied to passive earth pressures to limit reaction wall deflection. Where the soils behind the wall are particularly weak and/or where estimated jacking forces are large it may be necessary to structurally tie the shaft together to form a rigid box so that the entire shaft can be used to develop the required jacking reaction. In cases of extremely weak soils, it may be necessary to treat the ground behind the shaft.

Unless the pipe is large enough to allow the installation of intermediate jacking stations (e.g., greater than 900mm [36 inches] inside diameter), all of the jacking capacity of the system is provided by the jacking frame in the shaft. For pipeline installations less than 900mm (36 inches) inside diameter, typical jacking frame capacity is 1,800 to 3,600 kN (200 to 400 tons). The primary design concern with thrust blocks and reaction walls is to provide thrust reaction equal to or greater than estimated jacking forces and limit deflection of the thrust block. Thrust block movement typically takes the form of rotation (rather than straight lateral deflection). Rotation of the thrust block adversely impacts the performance of the jacking frame and can put an eccentric edge load on the pipe. In general, movement of the thrust block should be limited to less than 13 mm (½-inch) (Robinson, 1998).

It is critical that the geotechnical engineer provide input to the project designers and owner both during the design phase and contractor submittal stage to ensure coordination of the thrust block and reaction wall with the shaft shoring design and to ensure that the thrust block and reaction wall capacity is greater than the estimated jacking force.

Jacking Force Estimates

The jacking force is the load required to push the MTBM and the pipeline from the jacking shaft to the receiving shaft. The selection of the microtunnel pipe, drive lengths and shaft spacing, jacking system and jacking thrust block/reaction wall are all a function of the estimated jacking force.

Jacking force is composed of two components: face pressure and friction along the circumference of the MTBM and pipeline. Face pressure is the pressure at the head of the MTBM applied to the face area of the MTBM which develops as the tunnel is advanced. In order to prevent ground surface settlement, the face pressure should exceed active earth pressure plus hydrostatic pressure. In order to prevent ground heave, the face pressure should not exceed passive earth pressure plus hydrostatic pressure. The overall contribution of face pressure on total jacking force is generally relatively low, especially in soft ground conditions, when compared to friction resistance along the circumference of the MTBM and pipeline.

Friction resistance results from friction and/or adhesion forces which develop at the interface of the outside surface of the pipeline and MTBM and the surrounding soil as the MTBM and pipe are advanced. The friction resistance can be estimated using the following formula (Stein et al., 1989):

$R = M \times d_a \times \pi \times L$

where: R = Friction Resistance
M = Skin Friction (Jacking Stress)
d_a = Outside Diameter of Pipe (and MTBM)
L = Length of Pipe (and MTBM)

As seen in the above formula, friction resistance is dependent on the skin friction (jacking stress) parameter. Skin friction is the product of normal stress acting on the pipe wall and a coefficient of friction between the pipe and soil. The geotechnical factors that influence this relationship include soil type, soil stability (ability to hold overcut), swell and squeeze potential, depth of cover, depth of groundwater and type of pipe. These variables are most commonly lumped into a single parameter for skin friction. Methods used to estimate skin friction are summarized in Stein, et. al. (1989). However, jacking stress along the pipeline is not solely a function of geotechnical factors. Extensive case study research (Coller, et. al., 1996) concluded that the following factors also have a significant influence on jacking stress:

- overcut,
- lubrication,
- dewatering,
- steering corrections, and
- initial misalignment.

Design specifications can be used to help minimize jacking force by requiring overcut (consistent with soil type and MTBM steering capabilities), continuous lubrication around the pipeline and limitations on dewatering around the shafts. Steering corrections and initial misalignment are operator controlled and should be addressed in the project specifications. As a rough initial estimate, jacking stress values on the order of 3 to 10 kN/m^2 (0.03 to 0.10 $tons/ft^2$) are often used in soft ground conditions. On a recent microtunnel installation in Martinez, California through 384 feet of San Francisco Bay Mud the jacking stress back calculated at the maximum jacking force was approximately 2 kN/m^2 (0.02 $tons/ft^2$).

The geotechnical engineer should provide the parameters needed by the microtunneling contractor to make jacking force calculations. These geotechnical parameters should include: depth of cover, depth to groundwater, total and buoyant unit weights, standard penetration blow count (N), angle of internal friction, residual angle of internal friction, cohesion, and swell potential. At the design stage, the geotechnical engineer should provide the design team with preliminary estimates of jacking force as well as provide input on methods to control/minimize the jacking force so that economic and feasible drive lengths can be designed. Wherever possible, these estimates should include evaluation of driving records from similar microtunneling projects in the near geographic area under similar geologic conditions. The specifications should then be written to ensure that measures (e.g., overcut, lubrication etc.) are taken during construction to minimize jacking forces and ensure that the thrust block, jacking frame and pipe are adequate to complete each drive.

Tunnel Stability

A careful geotechnical evaluation of tunnel stability is critical to the selection of microtunneling equipment. For purposes of this paper, we are assuming the use of a slurry MTBM which provides full earth and hydrostatic pressure balance at the tunnel face. As such, understanding anticipated tunnel stability is important for the MTBM operator and at the launch and reception shaft portals. The tunnel portals are usually but not always expected to stand unsupported for a short period of time during launch and retrieval of the MTBM. A stable ground condition at the tunnel portal opening is essential to establishing initial line and grade during launching of the MTBM (i.e., avoiding initial misalignment as described above) and establishing an effective launch seal to contain slurry and pipe lubricant and to mitigate against flowing ground conditions and loss of soil from outside the shaft shoring.

The stability of the ground can be initially estimated using the Tunnelman's Ground Classification which is based on soil classification and groundwater conditions (Terzaghi, 1950). Additional analytical methods can be used to further estimate anticipated ground behavior. For cohesive soils, the following formula (Peck, 1969) can be used to evaluate ground behavior where there is no pressure applied to the tunnel excavation (e.g., at unsupported portal openings and overcut annular space around the pipe):

$$N_t = P_z \div S_u$$

where: N_t = Stability Factor
P_z = Overburden Pressure at Tunnel Centerline
S_u = Undrained Shear Strength (defined as one-half of the unconfined compression strength)

Stability factors of less than 1 to 2 indicate the ground is relatively firm, while stability factors of 5 or greater may result in general shear (i.e., sudden loss of ground). Stability factors between 2 and 5 indicate creeping soil which is typically stable in the short-term.

In non-cohesive soils (i.e., silts, sands and gravels without enough clay content to provide cohesion and plasticity) located below the groundwater table, flowing ground conditions are usually assumed. Detailed correlations of tunnel stability for various non-cohesive soils are contained in Bickel (1996).

When unstable ground conditions are anticipated at the tunnel portals, methods should be taken to improve and/or stabilize the ground. Double layered shoring, soil grouting, local dewatering, and/or guide casings can be used to improve and stabilize the ground at tunnel portals. A discussion of the use of grout to stabilize the ground at tunnel portals is presented in the *Ground Treatment* section of this paper. Dewatering outside the shaft (i.e., by wells or well points) is sometimes used to relieve the hydrostatic pressures at the shaft portals and reduce the potential for flowing ground into the portal opening. However, lowering the groundwater may result in increased jacking forces in the vicinity of the shaft and potential ground settlement due to increased effective soil stress. Guide casing has been pushed into Bay Mud deposits at the entry portals in soils which were able to stand for a short period of time and form a temporary "plug" inside the casing.

A thorough geotechnical investigation, particularly at the shaft locations, is critical to the evaluation of portal stability, tunnel face and overcut stability, and the requirements, if any, for stabilization of the soils at the tunnel portals.

Oversize Material and Obstructions

The presence of oversize materials (i.e., cobbles and boulders) and/or obstructions (e.g., fill debris, old foundations, tie-backs) along a project alignment may make microtunneling inadvisable. Identification of potential obstructions is one of the most important challenges in planning and design of a microtunneling project. In the San Francisco Bay Area, boulders, tree roots, railroad spikes and abandoned reinforced concrete foundations have stopped and in some cases damaged MTBMs.

A comprehensive geologic research and reconnaissance combined with a through subsurface investigation is essential in assessing the potential of oversize materials and obstructions along an alignment. For example, large-scale geologic maps can be used to understand the geologic setting and assess the potential for buried cobble and boulder deposits. Reviewing historic air photos can be valuable tool in evaluating the potential for buried obstructions along the alignment by understanding man-made modifications (e.g., past development, fills, etc.) to the landscape. The reader is directed to Mathy (1998) for a more complete discussion of the various resources available to the geotechnical engineer for evaluation of geologic setting.

In areas of suspected oversize material and/or fills a subsurface investigation based solely on traditional small diameter borings (i.e., 150 mm [6-inch] diameter borings) will not be capable of accurate assessment of cobble or boulder size or presence of obstructions. Therefore, in areas of suspected oversize materials and/or fill, subsurface investigations should include large diameter borings (i.e., 0.6 m [2-foot] diameter) and/or test pits. Identification of possible metallic objects, old foundations and other potential obstructions may be possible using geophysical techniques, such as ground penetrating radar.

A comprehensive geotechnical investigation, which includes geologic research and reconnaissance and a rigorous subsurface investigation, is essential to identifying areas of potential oversize material and obstructions. However, even with the most comprehensive geotechnical investigation, there always exists the possibility that oversize materials and/or unanticipated obstructions will be encountered during microtunneling. Project specifications should address oversize materials and obstructions and the course of action to be taken should the MTBM be stopped.

Soil Bearing Capacity to Support the MTBM

In projects involving very soft or very loose soil deposits such as San Francisco Bay Mud, the geotechnical engineer needs to evaluate the ability of these weak soils to support the MTBM during the tunneling process. If the MTBM begins to sink into the soil due to its own weight, it is unlikely that the surrounding soils will have sufficient strength to support steering corrections.

Therefore, once the MTBM begins to sink, it is very difficult, if not impossible, to steer the machine back onto line and grade.

Soil bearing capacity, consolidation properties and liquefaction potential should be considered in evaluating support of the MTBM. The applied load on the soil should be based on the weight of the MTBM, less the soil removed by the MTBM and less the buoyancy of the machine. The net loading should be applied to a 60 to 90 degree arc centered at the base of the MTBM. A full 180 degrees of contract (one diameter) is not appropriate due to the overcut annular space around the MTBM. The weight and length of MTBMs varies from manufacturer to manufacturer and contractor to contractor based on field modifications to individual machines. In general, smaller diameter machines are longer than large diameter machines. For both large and small diameter machines, the weight of the machine is not evenly distributed but is rather front end heavy. For example, the front section of 600 mm (24-inch) to 900 mm (36-inch) O.D. MTBMs can weigh as much as 2,475 kg to 5,400 kg (5,500 to 12,000 lbs) and impose gross contact pressures as high as 20 to 25 kN/m^2 (400 to 500 lb/ft^2). Allowable bearing capacity calculations based on soil strength should include a safety factor of at least 3 to help minimize short-term settlements.

Consolidation settlement should be evaluated based on site-specific laboratory consolidation testing and calculation of time rate of settlement. From these calculations, the percent consolidation and settlement can be estimated for different periods of loading. As long as the MTBM is steadily progressing forward, there should not be enough time for pore pressure drainage under the applied MTBM loading. However, work stoppage for extended periods (e.g., more than a few weeks for mechanical repairs, obstructions, etc.) when the MTBM is supported by normally or underconsolidated soils can be a concern.

Where the MTBM will be supported by loose, saturated sands, the potential for soil liquefaction and loss of bearing capacity due to machine vibrations should be evaluated.

In those cases where the soil strength is too weak to safely support the MTBM or where normally or underconsolidated soils will consolidate under short-term MTBM loading or where long-term areal consolidation is a concern, or where machine vibrations may induce liquefaction, ground treatment may be necessary. The geotechnical engineer must provide this input to the project designers and owner since such ground treatment can be a major expense to the project.

Spoil Removal and Solids Separation

A slurry-type microtunnel system uses pressurized slurry to transport cuttings to the ground surface. In a slurry-type spoil removal system the cuttings are mixed with drilling fluid in a chamber behind the cutting head and the slurry mixture is pumped through slurry hoses up to the ground surface where the cuttings are removed from the slurry and clean drilling fluid is returned to the face of the MTBM. Methods of removing the solids from the slurry vary from simple settlement tanks to complex slurry separation plants.

In congested urban settings, slurry separation plants are typically used to remove cuttings from the slurry. The purpose of the separation plant is to efficiently remove the solids from the slurry so the cleaned drilling fluid can be recycled and the separated solids can be disposed.

The slurry separation plant is a major element of a microtunneling operation with numerous options and combinations of equipment. Separation plants include various components for solids separation each of which is capable of removing a certain range of soil particle sizes. In order for the microtunneling contractor to assemble an efficient slurry separation plant, the soil grain size distribution along the tunnel alignment needs to be determined by the geotechnical engineer. The various separation plant equipment components and combinations available to the contractor include shakers (to remove coarse sands and gravels); desanders - usually 150 mm (6-inch) hydrocyclones (to remove fine to medium sand); desilters - usually 100 mm (4-inch) hydrocyclones (to remove silts); and centrifuges (to remove clays).

The contractor needs grain size distribution data along the tunnel alignment to make appropriate choices of the various separation equipment and potential need for flocculants (e.g., polymers) to add to the process. Geotechnical investigation reports should include full particle size analysis (i.e., sieve analysis and hydrometer) to determine the grain size down to 2 microns on soil samples which are representative of the soils to be encountered within the tunnel alignment.

Hydrofracture and Slurry Loss

Pressurized slurry at the face of the MTBM counterbalances tunnel face pressure and transports cuttings to the ground surface. Hydrofracture can occur when the slurry pressure at the face of the MTBM exceeds the overburden pressure of the surrounding soils. Under conditions of shallow cover pressurized slurry can push its way to the ground surface creating clean-up problems. Loss of slurry pressure due to hydrofracture can also affect cuttings transport through the slurry hoses and therefore may impede the progress of the MTBM.

In congested urban settings, it is not uncommon for the pipeline alignment to pass closely under, over or parallel to existing utilities. Granular backfill used around existing utilities can provide a relatively permeable medium for slurry loss and transport including up and through the pavement surface. Localized heave of asphalt and concrete pavements can occur creating traffic hazards and the need for pavement repair. At creek or culvert undercrossings with shallow cover slurry can find its way into the natural water course. This may lead to fines and clean-up costs imposed by the local storm water discharge authority.

The geotechnical engineer should provide input to the designers and owner regarding the risks of hydrofracture and slurry loss and identify those locations of shallow cover or close proximity to existing utilities, creeks or culverts or other avenues for slurry loss. The project specifications should include contingency plans for containment, clean-up and MTBM operations modifications should hydrofracture and slurry loss occur.

Shaft Backfill and Manhole Settlement

Finished manhole settlement can occur as a result of one of, or a combination of, the following:

- recompression settlement
- consolidation settlement
- vibration induced settlement

Removal of native soils from shafts results in rebound of the shaft bottom due to overburden removal. Recompression settlement is caused by the weight of the shaft backfill around the finished manhole recompressing the underlying soils. Recompression occurs immediately upon backfilling and is a function of shaft depth, soil type, groundwater conditions, and shaft dimensions. Assuming a relatively shallow shaft bottom not disturbed by construction activity, recompression settlement can be on the order of a few centimeters. Disturbance of the shaft bottom by overexcavation, softening, boiling, etc., can produce significantly greater settlement and can only be avoided by careful excavation procedures. The amount of rebound/recompression should be evaluated by elastic theory. Manhole elevations can then be adjusted to account for recompression (i.e., built high) and adjusted from shaft to shaft during construction based on actual field performance.

Consolidation settlement occurs in normally consolidated or underconsolidated soils where the net weight of the finished manhole and shaft backfill exceeds the weight of native soil removed from the shaft excavation. The impact of any increase in overburden pressure on the soils beneath the shaft should be evaluated based on the results of consolidation testing. Normal aggregate backfill may be significantly heavier than the soils removed from the shaft. In shafts excavated in San Francisco Bay Mud, aggregate backfill may outweigh the Bay Mud by as much as 60%. The use of lightweight aggregates approximating the in-situ weight of Bay Mud can be used to mitigate long-term consolidation settlements.

Vibration induced settlement of a finished manhole can occur as a result of vibratory extraction of sheet pile shoring - especially where there is significant toe embedment of the sheet piles. In areas of loose silts and sands localized liquefaction results in settlement of the manhole structure. Manhole settlements of up to 150 mm to 230 mm (6 to 9 inches) have occurred as a result of vibratory extraction of sheet piles (Goodwin, 1998). In these soil conditions, sheet piles should be cut-off below grade and left in place or the ground should be treated to minimize liquefaction potential.

The geotechnical engineer should provide an analysis of each of these forms of manhole settlement through subsurface investigation at each shaft location in order to recommend appropriate mitigation measures.

Ground Treatment

Ground treatment may be necessary to improve and/or stabilize very weak soils to allow construction and ensure long-term performance of the pipeline installed by microtunnel pipe jacking. Ground improvement techniques can increase the in-situ soil strength and decrease the soil permeability and compressibility. Some of the techniques available for ground improvement include compaction grouting and jet grouting.

Compaction grouting is a relatively common method of ground treatment which displaces and compacts soil in-situ by injection of low slump sand-cement grout bulbs at relatively high pressures. Jet grouting improves the soil by injection of a cement slurry radially mixing the in-situ soil and cement at high pressures to produce a soil-cement ("Soilcrete") column. Compaction grouting and/or jet grouting can be used to:

- increase soil strength and decrease groundwater inflow prior to shaft excavation;

- increase soil strength and provide additional support behind reaction walls;
- improve soil stability and stand-up time at tunnel portals;
- in-fill voids between the shaft shoring and soil that may have occurred during shaft excavation and shoring;
- control diving of the MTBM in very weak or underconsolidated soils;
- mitigate long-term settlements in normally consolidated or underconsolidated soils, especially where future loading (e.g., fills) over the pipeline alignment is anticipated; and
- mitigate liquefaction and seismic induced settlement.

Input from the geotechnical engineer is needed to identify areas where ground treatment may be required and to aid the designer in determining the most economical and feasible method of ground treatment.

Appendix - References

1. Bennett, R.D., Guice, L.K., Khan S., Staheli, K. (1995, "Guidelines for Trenchless Technology: Cured-in Place Pipe (CIPP), Fold and Formed Pipe (FFP), Mini-Horizontal Directional Drilling (Mini-HDD), Microtunneling", Trenchless Technology Center Technical Report #400.

2. Bickel, J.O., Kuesel, T.R., King, E.H., (1996), Tunnel Engineering Handbook, Chapman & Hall, New York.

3. Coller, P, Staheli, K., Bennett, D., Post R., "A Review of Jacking Forces by Both Theoretical and Empirical Methods as Compared with 20 Years of Practical Experience", Conference Proceedings, International No-Dig '96, New Orleans, LA, April 1996.

4. Goodwin, J. Gill, J., Mathy, D., (1998), "Recipe of Successful Microtunneling", Proceedings of the 1998 Pipeline Division Conference - Pipelines in the Constructed Environment, ASCE, pp. 124- 132.

5. Iseley, T., Gokhale, S.B., (1997), "Trenchless Installation of Conduits Beneath Roadways", National Cooperative Highway Research Program, Synthesis of Highway Practice 242, Transportation Research Board.

6. Mathy, D.C., Nielson, D.R., (1998), "The Influence of Geologic Setting on Microtunneling", Proceedings of the 1998 Pipeline Division Conference - Pipelines in the Constructed Environment, ASCE, pp. 113-122.

7. Peck, R.B. (1969), "Deep Excavations and Trenching in Soft Ground", Proceedings, 7th International Conference on Soil Mechanics and Foundation Engineering, State of Art Volume, pp. 225-290.

8. Robinson, A.J., Thomson, J.C., (1998), "Shafts for Microtunneling and Pipejacking, Layout, Design & Construction", North American No-Dig 1998, NASTT, pp. 197-217.

9. Stein, D., Mollers, K., Bielecki, R. (1989), Microtunneling, Ernst & Sohn, Berlin.

10. Terzaghi, K. (1950), "Chapter 11 - Geologic Aspects of Soft Ground Tunneling", Applied Sedimentation, P. Trask, ed., John Wiley and Sons, NY.

Folsom East 2 Construction Proving Project: Field Evaluation of
Alternative Tunneling Methods in Cobbles and Boulders

Kimberlie Staheli[1]
David Bennett[2]
Michael A. Maggi[3]
Michael B. Watson[4]
Bruce J. Corwin[5]

Abstract

A Construction Methods Proving Project was conducted to test the feasibility of using trenchless construction methods in cobbles and boulders. The project, conducted for the Sacramento Regional County Sanitation District, evaluated the performance of microtunneling and mechanized open-shield tunneling in Folsom, CA. The performance results and evaluation of each tunneling method were used to specify tunneling techniques for the Folsom East 2 Project, consisting of over 37 km (23 miles) of new interceptor pipeline. This paper describes the testing project and the performance results of each trenchless construction method.

Introduction

The Folsom East 2 (FE-2) project for the Sacramento Regional County Sanitation District (SRCSD) was completed during 1997 and 1998 and involved the design of 4,270 meters (14,000 feet) of 137- to 167-cm (54- to 66-inch) diameter interceptor sewer. The FE-2 project provides a replacement pipeline for the existing 84-cm (33-inch) interceptor in the City of Folsom, California. The FE-2 pipeline will provide relief capacity as well as future growth potential. The FE-2 is part of the Bradshaw-Folsom Interceptor, a major

---
[1] Principal Engineer, Bennett / Staheli Engineers, P.O. Box 821433, Vicksburg, MS 39180.
[2] Principal Engineer, Bennett / Staheli Engineers, P.O. Box 821433, Vicksburg, MS 39180.
[3] Principal Engineer, Sacramento Regional County Sanitation District, 9660 Ecology Lane, Sacramento, CA 95827
[4] Project Engineer, Black & Veatch Engineers, 8801 Folsom Blvd., Suite 120, Sacramento, CA 95826.
[5] Project Manager, Black & Veatch Engineers, 8801 Folsom Blvd., Suite 120, Sacramento, CA 95826.

project being undertaken by the SRCSD to upgrade service within the Sacramento area. The FE-2 is to be installed in varied soil conditions. The most prevalent material consists of cobbles and boulders in silty- and sandy-gravel. The cobbles and boulders range in size up to 61-cm (24-inches) in diameter. The overall project will ultimately consist of 37 km (23 miles) of interceptor extending from FE-2 at the upper end to the SRCSD treatment plant.

Feasibility Study

A Desk-Top Feasibility Study was performed by the Black & Veatch team, consisting of Black and Veatch, Bennett / Staheli Engineers, Kleinfelder, and Montgomery-Watson, to evaluate the technical feasibility of tunneling in the cobble material. The study concluded that it might be possible to successfully use trenchless methods in cobbles; however a field test was considered necessary to validate the feasibility of trenchless methods (Bennett / Staheli, 1997). A Construction Methods Proving Project (CMPP) was planned to field-test the most promising trenchless alternatives. The SRCSD Board believed that the desktop study and subsequent Construction Methods Proving Project would provide information about the feasibility of using trenchless technology on FE-2 and other sections of the Bradshaw/Folsom Interceptor Project. The project, as approved by the Board, included field evaluation of microtunneling and mechanized open shield tunneling construction in cobbles and boulders. The cobbles and boulders were believed to significantly increase the potential for excavation problems and risks of obstructing progress. The lack of clay fines and resulting marginal stability of the ground increased the risk of subsidence and ground losses above the tunnel. The project included instrumentation and monitoring of machine performance, and the evaluation of each tunneling method presented in a final CMPP report.

Construction Methods Proving Project

The SRCSD and Black & Veatch Construction, Inc. (BVCI) entered into a contract in October 1997 for the construction of the test tunnel project. The CMPP project consisted of a proposal to install 152 meters (500 feet) of interceptor using an open shield tunneling technique and 152 meters (500 feet) of interceptor using microtunneling techniques. It was intended that the project would be a partnership between the SRCSD and all parties involved to achieve a successful result. The estimated construction cost was $2,000,000 with a contingency fund of $180,000 for unknown construction challenges.

Subcontract bids were solicited from pre-qualified tunneling sub-contractors. Elmore Pipe Jacking was selected in November 1997 to construct the two tunnels with different tunneling techniques. Elmore bid Akkerman tunneling machines for both mechanized shield tunneling and microtunneling. Elmore elected to use 175 cm (69-inch) nominal diameter Hobas Pipe, a fiberglass reinforced pipe, for the interceptor.

The two trenchless construction methods, microtunneling and mechanized shield tunneling using an open shield machine with a rotating cutter bar, were each set up for the construction of 152 meter (500-foot) tunnels using pipe jacking methods, installing 3 meter (10-foot) sections of Hobas pipe. The tunneling subcontractor, Elmore Pipe Jacking, used a central pipe jacking shaft for both tunnels, excavating in opposite directions. Tunneling parameters were carefully monitored and recorded for evaluation. In addition, ground treatment for reducing subsidence associated with mechanized shield construction was used and evaluated for a segment of the open shield tunnel drive. Instruments, including load cells, vertical borehole extensometers, and embedded settlement points, were installed to provide information on jacking force distribution and ground disturbance above the tunnel. Load cells were incorporated into the pipeline to measure axial load distribution and to isolate frictional loads along the pipeline.

Geotechnical Conditions

Part of the FE-2 alignment traverses areas known to have cobbles. The cobbles may be naturally deposited in a matrix of fines; alternately, the cobbles may have been re-deposited in a relatively loose state as a result of gold dredging operations. The geological conditions are described in detail in Volume II of the CMPP final report (Bennett / Staheli 1998; Kleinfelder, 1997). The cobbles and boulders in a sand matrix dominate the ground conditions and are expected to have significant impacts on interceptor construction, especially for the deeper pipelines.

The site selected for the CMPP was under Folsom Boulevard near Aerojet Road, in Folsom, CA. The site was selected because the geotechnical conditions were believed to be representative of ground conditions along the FE-2 alignment. The geotechnical conditions at the elevation of the pipeline consisted of dense sand and gravel with cobbles and boulders. The majority of the rocks were large gravel and small cobbles, 5- to 10-cm (2- to 4-inches) in diameter; however, it was common to encounter boulders up to 41-cm (16-inches) in diameter. The largest rock removed through the tunneling shield was 56-cm (22-inches) in its longest dimension. The unconfined compressive strength of five tested specimens ranged from 186,140 to 268,900 KPa (27,000 to 39,000 psi). The sand, gravel and cobbles extended 1.2 to 2.4 meters (4 to 8 feet) above the pipe crown, and were overlain by approximately 0.9 to 1.5 meters (3 to 5 feet) of silty sand and 0.3 to 0.9 meters (1 to 3 feet) of sandy silt. Groundwater was not encountered during the tunneling operations.

Mechanized Open-Face Shield Tunneling

An Akkerman mechanized open shield machine with a rotating cutter bar for the field test was chosen and proposed by Elmore Pipe Jacking (EPJ). The proposal was accepted, based on EPJ's persuasive arguments that the machine would work satisfactorily, in lieu of the specified shield with digger boom. The articulated and steerable open shield was operated from inside the tunnel and used in a pipe jacking mode. A short belt conveyer was used to transport the excavated soil (muck) from the face to tracked muck carts at the rear of the shield, which were used to move the muck to the jacking shaft. The machine was launched on Wednesday, February 18, 1998. The average daily progression rate was 4.7 meters (15.4 feet). In the challenging geotechnical conditions, the shield suffered from the high impact loading on the cutter bar due to excavation of the boulders and cobbles. As a result, the machine was repeatedly in need of minor field repairs (tightening of bolts, adjustment of bearings, etc.), that significantly decreased the overall daily progress rate. Figure 1 shows a frontal view of the mechanized shield.

Figure 1. Mechanized Open-Face Shield

The jacking operations during the mechanized tunneling were erratic due to difficulties with stabilization of the excavation face. At times, very little or no material was in contact with the cutter wheel of the shield, and jacking pressures were low.

Jacking forces and torque increased dramatically when the cutter wheel was buried. The machine could not produce enough torque to remove the rocks on the periphery of the machine, and the wheel stalled at regular intervals. Once the obstructing rock was removed and pressure could be re-applied to the thrust jacks, the pipe would "lurch" forward rather than slide forward into place.

Over-excavation was an extensive problem with the mechanized shield tunneling equipment. The machine was not capable of providing any tunnel face stability. Throughout the tunneling operation, stability continued to be a problem as the action of the rotating cutting bar would cause soil above and in front of the machine to run into the machine's open face. In addition, the periphery bits and overcut bars would drag rocks around the outer edge of the machine, loosening and dislodging the material above the machine, causing extensive over-excavation and ground losses above the pipe. The smallest void created during any pipe advance measured 1.5 m^3 (2 cubic yards) over a 3-meter (10-foot) tunneled length or 18% of the tunnel volume. The largest measured over-excavation volume was 8.4 m^3 (11 cubic yards) for a 3-meter (10-foot) tunneled length, or 97% of the tunnel volume. Several vertical holes were drilled to determine the extent of the voids over the pipe, and a "two-sack" sand-cement grout mixture was used to fill the voids. Figure 2 shows the over-excavation during mechanized shield tunneling.

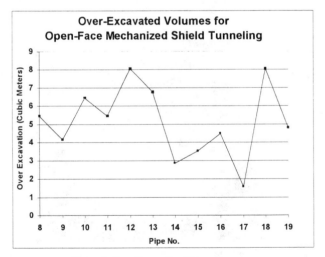

Figure 2. Over-Excavation Volumes

Due to extensive over-excavation and ground losses above the tunnel during the first 37 meters (120 feet) of tunneling, ground stabilization measures were employed in an attempt to stabilize the tunnel face and reduce ground losses. A sodium silicate chemical grout was used for permeation grouting from the ground surface. Following the chemical grouting procedure, the excavation face appeared to be more stable and the volume of over-excavation decreased. However, the void created during tunneling in the grouted zone was still unacceptably large.

The chemically treated ground contributed to significant steering problems that resulted in the breakage of the pipe behind the shield. At 43 meters (142 feet) into the drive, a portion of the invert of the pipe buckled immediately behind the shield. From inspection of the pipe and the measured axial load, it was clear that the pipe did not fail in compression due to the axial load, but rather from a concentrated load from a hard boulder on the outer surface of the pipe. The large boulder became lodged beneath the invert of the shield. As the leading edge of the first pipe section advanced over the boulder, wedging and gouging by the boulder caused the pipe to buckle. Tunneling continued after the pipe broke; however, the mechanized shield tunneling portion of the test was terminated due to the large, uncontrollable settlement events.

The pressure delivered to the hydraulic cylinders propelling the shield was monitored to measure the jacking forces. However, after 24.4 meters (80 feet) of the drive, it was evident that the hydraulic pressure gauge was not functioning properly. Therefore, jacking forces measured by the pressure gauge were highly unreliable. Fortunately, the load cells nested in the pipeline provided accurate measurements of the jacking load during the launch of the Intermediate Jacking Station (IJS) shell. The maximum measured jacking load was 2545 KN (286 tons) over the 61 meters (200-foot) drive.

The normalized frictional loading (skin friction) on the outside of the machine was 10.5 to 15.3 KN/m^2 (0.11 to 0.16 $tons/ft^2$). These relatively large friction loads were due to the high friction angle and normal loads, caused by the soil falling onto the crown of the shield. The normalized frictional load on the pipeline was 4.6 KN/m^2 (0.048 $tons/ft^2$) of pipe surface area. Frictional loading on the pipeline was lower than on the shield, likely due to lubrication with bentonite that reduced the friction factor. However, it was obvious that the bentonite lubrication was not evenly distributed around the pipeline. The pressurized bentonite would flow to the face of the machine through the voids above the crown, instead of filling the annular space.

Microtunneling

Microtunneling, a remotely controlled, guided, pipe-jacking process that is capable of providing continuous support to the excavation face, was also tested. The slurry microtunneling machine was a soft-ground machine with a rotating cutter head and slurry system. Akkerman Trenchless Equipment manufactured the system selected by Elmore Pipe Jacking. The slurry system was used for removal of the excavated material and consisted of slurry charge and discharge lines that extended from the slurry charge tanks to the face of the machine. The charge and discharge lines conveyed slurry to the face and cuttings to the slurry separation unit. A return pump was located in the jacking shaft, and a booster pump was located in the trailing section of the machine. The return pump served as the main pumping unit and pumped the excavated material from the shaft to the separation unit. The booster pump facilitated the movement of the cuttings from the machine's face to the shaft for transportation to the separation unit. Figure 3 shows the microtunneling shield.

Figure 3. Microtunneling Machine

The microtunneling machine was launched on Monday, April 27, 1998. During launch, extensive returns of the slurry (composed only of water and soil cuttings) entered the shaft through small openings and joints in the trench boxes used for shaft support. The high-pressure water eroded the face and transported fines and small rocks into the shaft from the face and crown of the machine. A large void was created on the outside of the jacking shaft, above the buried machine and trailing section. Horizontal probing from the launch face revealed a void that measured 4.9 meters (16 feet) in length from the shaft wall, extending over the crown of the machine. A "two-sack" cement, sand, and grout mixture was used to fill the void.

Three chronic problems during the microtunneling operations were related to the severe ground conditions, inadequate slurry system, and method of operation: 1) clogging of the slurry lines, 2) rapid, excessive wear on the pumps and lines, and 3) extremely high slurry flow rates required to transport the large, dense cuttings in a water-based slurry. The angular rock pieces in the cuttings repeatedly clogged the slurry return lines during tunneling, forcing the tunneling operations to stop while the lines were cleared. The clogging occurred at two main locations: 1) the outlet of the return pump in the shaft, and 2) the outlet of the booster pump in the microtunneling machine. Clogging of the pumps and slurry system was exacerbated by the physical configuration of the pumps, which had 90-degree elbows at the inlet and outlet locations.

On the first day of tunneling, the time spent clearing the slurry lines exceeded the tunneling time by a factor of 3 to 1, with tunnel advance time of only 56 minutes. During the first two days, the longest period of uninterrupted tunneling was 28 minutes. The wear on the slurry transport system from the abrasive crushed cobbles was severe. Excessive wear occurred on the pump housings, slurry lines, couplings, bypass valve, and slurry separation system. High slurry flow rates caused tremendous subsidence of the material in front of the tunnel machine. During excavation of the first 6 p pe sections (18 meter (60 feet)), no stabilizing additives were introduced to the slurry. As a result, it was not possible to lower the slurry flow rates without clogging the slurry system because the excavated material would settle in the slurry lines at lower velocities.

The machine torque readings and advance rates were highly erratic. Torque on the head controlled the advance rate of the machine. Figure 5 shows the erratic torque readings recorded during the first 10 meters of microtunneling. Torque readings throughout the drive displayed a similar trend.

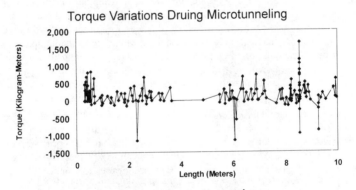

Figure 4. Microtunneling Torque Readings

Because of the boulders, the torque reading would frequently spike, occasionally causing the machine to stall. Consequently, advance rates were extremely variable, cycling with the torque. The operational tunneling cycle was as follows: when the torque increased, the operator markedly lowered the advance rate but continued to rotate the head and circulate slurry, eroding material at the face without advancing the machine. This resulted in the development of a large void at the face, and consequently the torque would drop dramatically as the material was excavated away from the cutter wheel. The operator would then increase the advance rate, moving the machine forward until the cutter wheel contacted the face of the voided excavation, causing the torque to rapidly increase once again. The cycle was repeated throughout the drive, resulting in the creation of massive voids.

The jacking forces developed during microtunneling were relatively high over the short drive length of 26 meters (86 feet). The normalized jacking load was 13.4 KN/m^2 (0.14 tons/ft^2) over the drive length of 26 meters (86 feet). A set of load cells was installed 11.3 meters (37 feet) behind the face of the machine to measure frictional loads. The frictional loading over the 15 meters (49-foot) segment of the pipeline between the load cells and the jacking shaft was 12.4 KN/m^2 (0.13 tons/ft^2). The calculated jacking stress on the shield was 17.2 KN/m^2 (0.18 tons/ft^2). Friction loading on the pipe was lower during microtunneling than during mechanized shield tunneling. This could be explained by improved lubrication techniques during microtunneling, i.e. the use of greater volumes of lubricant injected under pressure into the annular space between the pipe and bore during the microtunneling operations.

Significant subsurface settlements occurred during microtunneling but were "bridged over" by the 23-cm (9-inch) asphalt pavement on Folsom Boulevard. Careful subsurface monitoring was vital to accurately document installation procedures, techniques, and as-built conditions of the pipeline. Prior to the launch of the microtunneling drive, a significant number of subsurface settlement monitoring devices were installed, including multiple point borehole extensometers (MPBX's) and simple embedded rebar single monitoring points. Post-tunneling evaluation revealed that the embedded rebar settlement points gave a much more accurate reading of actual settlement events, since the relatively stiff tubing of the MPBX's tended to resist elongation and bridge over the settlement. While tunneling the 3rd pipe, between 15- to 18-meter (50 and 60 feet) from the launch shaft, large voids were discovered when the rebar settlement point on the centerline of the tunnel dropped to 138 cm (54.5 inches). After documentation of the large settlement event, the void was filled with a weak, two-sack cement and sand grout mixture. Tunneling operations resumed with no change to operational procedures. Not surprisingly, massive over-excavation continued during the installation of the 4th and 5th 3-meter (10-foot) pipe sections (18 to 24 meters (60 to 80 feet) from the tunnel shaft). The rebar settlement point located 20.4 meter (67 feet) from the shaft dropped 104 cm (41 inches). After completely tunneling the 6th pipe section (24 meters (80 feet) into the drive), additional vertical holes were drilled so that the void could be filled with weak cement slurry.

Because of the extensive over-excavation, bentonite was added to the slurry to raise the slurry viscosity to 42 seconds per liter, as measured by the Marsh funnel, to reduce ground losses due to over-excavation. It was expected that increasing the viscosity would allow the excavated material to remain in suspension at lower slurry flow rates, reducing the incident of slurry line clogging. In addition to adding bentonite, the slurry flow rate was lowered to reduce over-excavation. However, even with the higher viscosity slurry, lowering the slurry flow rate caused the slurry system to continually plug. Because the operator could not reduce the flow rate without clogging, the operator tried to increase the advance rate to match the excavation rate. The increased advance rate caused the torque to substantially increase and the machine began to stall regularly. After 62 minutes of tunneling, and virtually no advance, the head stalled and the operator was unable to get the head rotating again. A 0.9 meter (3-foot) diameter vertical auger rig was used to drill to the face of the machine to remove boulders, cobbles, or other material that was preventing the machine from rotating. This operation allowed the microtunneling machine to rotate once again.

After the head was freed, in anticipation of advancing the tunnel further, the slurry was further thickened to a marsh funnel viscosity of 55 seconds per liter. The head was pushed 0.6 meters (2 feet) into the 0.9 meter (3-foot) diameter auger hole, and tunneling was resumed. Tunnel progress was very slow, taking 3 hours and 15 minutes to tunnel through the auger hole. Tunneling resumed in virgin ground for approximately 15 cm (6 inches), at which time the slurry lines again clogged. Two rocks, each approximately 7.6 cm (3-inches) in diameter, were discovered in the slurry lines. Since the rocks were larger than the 4.4 cm (1-3/4-inch) port openings in the chamber, the microtunneling subcontractor believed that the face had sustained significant damage from the excavation process. Additional rocks larger than 4.4 cm (1-3/4-inches) were found in the muck piles on the site. An access shaft was constructed to expose the face of the machine and assess the damage.

Exposure of the machine revealed a substantial amount of wear on the cutting head. The gage cutters were severely worn and the crusher bars showed extensive wear. The most significant damage to the machine was on the slurry ports. At two locations, the wear was severe enough to allow communication between adjacent ports. The large rock discovered in the slurry line likely entered through the damaged ports. Photographs were sent to Akkerman Trenchless Equipment, Inc. with a request for advice on whether the MTBM could be repaired in the field to allow the drive to continue. Akkerman's representative concluded that field repairs were not feasible and the head would have to be shipped to the factory for repair. In addition, the Akkerman letter stated that "there is a high likelihood that a repaired head will end up looking and performing as it does now, upon completion of another drive of similar length," i.e., a repaired machine could not be expected to tunnel any further than 90 feet without needing additional repairs. Based on Akkerman's assessment of the necessary shop repairs, the difficulties experienced with the slurry system, and the microtunneling subcontractor's inability to maintain face stability and prevent massive ground losses, the microtunneling portion of the test was terminated.

Conclusions

Based on the performance information collected during the CMPP, it was clear that the mechanized open-face wheeled excavator was inappropriate for the ground conditions encountered. The machine tested did not provide face stability, resulting in unacceptably large ground losses. Ground stabilization measures were not effective in reducing localized settlement to within acceptable levels. The mechanized open-face wheeled excavator would not be allowed as an excavation technique for installing the pipeline on the FE-2 project.

The microtunneling machine was capable of excavating through the hard, abrasive rocks. However, progress was very slow and face stability was never maintained. Excessive ground loss was observed throughout the drive. Rapid wear of components occurred due to the highly abrasive rock cuttings. Overall performance of the MTBM system, as set up and operated, was unsatisfactory. It was concluded that the slurry system, operating with water-based slurry, and the design of the slurry chamber ports were incompatible with the severe ground conditions. The rapid wear that resulted on all moving parts of the MTBM led to the inescapable conclusion that microtunneling was not economically feasible under these conditions.

Based on the results of this test, tunneled portions of the FE-2 project were specified to be constructed using conventional tunneling methods consisting of an open-face digger shield machine with breasting capability or a closed-face rotating cutter head machine. A two-pass tunnel lining system was specified.

References

Bennett / Staheli Engineers (1997) "Trenchless Construction Methods Desk-Top Feasibility Study Folsom East Interceptor Section 2," Sacramento Regional County Sanitation District, Sacramento, CA, September, 1997.

Bennett / Staheli Engineers (1998) "Construction Methods Proving Project Final Report," Sacramento Regional County Sanitation District, December, 1998.

Kleinfelder (1997) "Geotechnical Investigation Report Folsom East Interceptor Section 2, Sacramento County, California," Sacramento Regional County Sanitation District, Sacramento, CA, 1997.

Jacking Loads Associated with Microtunneling

David Bennett[1], M. ASCE, and Edward J. Cording[2], M. ASCE

Abstract

This paper presents a basis for substantial improvements in the prediction of jacking forces, through the development of a conceptual model and calculation approach that correctly model behavior. In addition, the range of impacts of various operational and geometric factors on jacking forces were evaluated and quantified, using case history data and experimental data.

Jacking forces were measured by testing three microtunneling systems in an instrumented 100-m long test facility with six different soil sections. Five case histories in sands, clay, and silts, totaling over 4,500 m of microtunneling in diameters from 400- to 2,000-mm were also analyzed.

The conceptual model accounts for the effects of ground conditions, construction sequence, operational factors, and geometric factors on the distribution of the face pressure and skin friction components of jacking forces as the shield is advanced.

Introduction

Experiences related to some recent U.S. projects underscore the need for better understanding of the interrelationships among soil characteristics, microtunneling machine characteristics, setup, operation, jacking pipe characteristics, and the prediction and control of jacking loads. Instances of machines becoming stuck as a result of insufficient jacking

---

1 Principal, Bennett/Staheli Engineers, PO Box 821433, Vicksburg, MS 39182
2 Professor of Civil Engineering, University of Illinois at Urbana/Champaign, Urbana, IL

capacity have occurred, and jacking pipes have failed due to overstressing from higher than predicted jacking loads. If the various methods used to calculate jacking forces are examined, one could conclude that different variables are important, depending on the method of calculation used. For example, if the "Rule of Thumb" (Lys 1993) for calculating jacking forces based on force per unit area of pipe is accepted, the only important factors are pipe surface area and force per unit area, or jacking shear stress. The face pressure component is ignored. With this method, the effects of diameter, groundwater levels, overcut, lubrication, soil properties (density, cohesion, friction angle), steering, initial misalignment, and near shaft phenomena are lumped into the selection of a jacking stress value.

Guidance has been developed in Japan and Germany for predicting jacking loads (Stein et al. 1989), based on ground behavior and empirical data summarized from work by Salomo (1979), Scherle (1977), Weber (1981), and Herzog (1985). Milligan and Norris (1993) have developed guidance based on pipe jacking tests in the UK. Others have evaluated jacking forces in clay using undrained shear strength (Haslem, 1986). A statistical/probabilistic approach for predicting jacking forces has also been attempted (International Society for Trenchless Technology, International Working Group No. 3 (ISTT-IWG No. 3 1994 draft). Regardless of the approach used, the calculation of jacking forces is important.

Factors that must be considered in calculating the skin friction component are the effective coefficient of friction for frictional materials, adhesion for cohesive materials and the effective normal stresses and pipe areas over which these stresses act. The determination of the coefficient of friction for cohesionless materials and adhesion for cohesive materials is somewhat less problematic than the determination of the normal stresses and contact areas of the pipe over which these stresses act. Evaluation of all the factors can be troublesome, due to the effects of time and external influences.

Values for coefficient of friction have been published based on test results and back calculation for various soils and pipe materials (Stein et al. 1989). Values of adhesion between various soils and materials have also been published (Winterkorn and Fang 1975). The tabulated values are given as reasonable ranges for adhesion between a given soil type and material. However, these tabulated values are inappropriate for use in microtunneling because they were determined primarily from pile tests and because the effects of overcut and lubrication have not been evaluated.

Significant evidence has been collected from commercial projects and research that indicates that overcut and lubrication can substantially decrease frictional resistance (Milligan and Norris 1993; Bennett and Staheli 1996; Coller et al. 1996; Bennett, 1998). For the selection of normal loads used in prediction of jacking loads, the trap door or silo arching model can be used to estimate vertical stresses on the pipe (Terzaghi 1943). This model is not a precise model for a circular tunnel, but, as shown in Figure 1, it illustrates the important relationships between the movement of the soil mass into the annular space around the pipe and the reduction in stresses on the pipe.

Vertical soil stress, calculated using the trap door arching theory (Terzaghi 1943) is:

$$P_V = \frac{\gamma b - 2c}{2K \tan \delta} + \gamma h_c \left(e^{(-2\frac{Kh_c}{b} \tan \delta)} \right) + \gamma h_c \left(e^{(-2\frac{Kh_c}{b} \tan \delta)} \right)$$

where
    γ = effective unit weight of soil
    h_c = height of soil cover above pipe crown
    b = the equivalent width of the soil column acting on the pipe
    δ = the angle of wall friction
    K = coefficient of earth pressure
    c = soil cohesion

As the trap door displaces downward, shear stresses develop in the yielding column of soil above the trap door, and the vertical stress above the trap door decreases to approximately $\frac{\gamma b - 2c}{2K \tan \delta}$. Thus, the minimum vertical stress acting on the door is proportional to the width of the trap door and soil density, and decreases as soil strength increases. To extend this model to vertical stresses on a pipe, when the overcut is adequate,

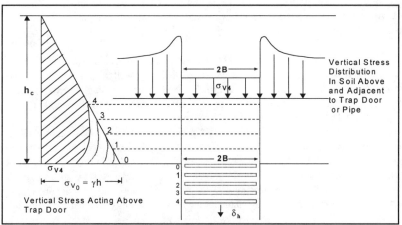

Figure 1. Analogy of Terzaghi's Trap Door Experiment Illustrating Vertical Stresses Acting on Pipe

a sufficient amount of displacement occurs in most soils to reduce stress to a value that is proportional to the diameter of the pipe and unit weight of the soil that settles onto the pipe.

Conceptual Model for Jacking Forces

The conceptual model illustrated in Figure 2 shows the distribution of face pressures and skin friction forces along the length of the shield and pipe as the shield is advanced. The conceptual model graph begins with an ordinate value equivalent to the initial face pressure component. The slope of the graph over the initial portion is usually relatively steep, because of the high skin friction stresses that develop on the unlubricated portion of the shield and pipe and the effects of dewatering near the jacking shaft (Figure 2.a). The dewatering effects are most pronounced for drives in sand. Initial misalignment and subsequent sharp steering corrections heighten the contrast between this initial interval and subsequent intervals. However, these events usually cause more erratic peaky behavior. The slope of the graph flattens out considerably in the lubricated portion of the drive (Figure 2.b). Near the end of the drive, the slope often increases again, indicating the influence of dewatering near the reception shaft (Figures 2.c and 2.d). The change in slope over a given interval of the jacking force graph is often a result of progressive changes in behavior that start before the interval. In addition, the effects of time in specific intervals result in changes in behavior that may be manifested only later in the jacking records. For example, the soil may squeeze into the overcut, if the stability ratio is higher than about 4, or the effects of lubrication in the lubricated zone may diminish with time, as the lubricant and soil mix. The effects of shaft dewatering may increase as the radius of influence increases with time. The interval affected by dewatering is always larger than the short intervals over which the sharp spikes in the jacking force graphs are manifested. The conceptual model was devised to aid in understanding behavior, and the impacts of the various factors that influence jacking forces.

The face pressure component can be calculated, based on Rankine earth pressures, groundwater pressure, and area of the face. Since this component occurs only at the face of the shield, its position is referenced to the shield. The face pressure component of jacking forces, F_p, is usually a small component of overall jacking forces. It can be significant for large pipe diameters, short drives, and high groundwater heads, or when microtunneling through stiff to hard clays and rock. Actual face pressures are usually based on production goals and balanced against MTBM power. That is, the MTBM is typically pushed as fast as possible, while maintaining torque within acceptable limits. Face pressure is typically higher in clays than in sands. Sands tend to run in, unless supported, leading to active earth pressure conditions. Even when penetration rates are at maximum, face pressure and torque are usually low in sands. In clays, available torque controls advance rate and face pressure.

GEO-ENGINEERING FOR UNDERGROUND FACILITIES 735

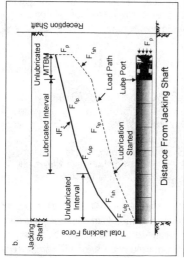
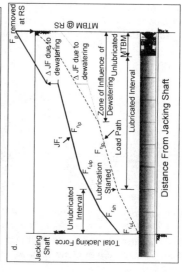
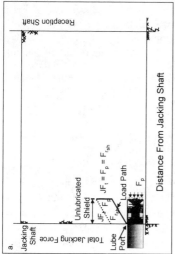
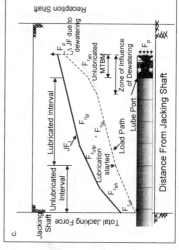

Figure 2. Conceptual Model for Jacking Forces.

The skin friction component is calculated using the general approach:

$$F_r = \sigma'_n \mu A_c L$$

where

σ'_n = average effective normal stress acting on shield and interval of pipe
μ = average friction factor acting on shield and pipe
A_c = circumferential area of pipe, or portion of circumferential area over which normal stress and friction factor act
L = length of drive, or interval of drive considered

The normal stress acting on the shield and pipe varies with position and time. The normal stress is non-uniform around the circumference and along the pipeline. The normal stress increases with time if the soil swells or squeezes in around the pipe. Concentrations occur at undulations caused by steering corrections, or when mixed ground conditions are encountered. Complicating this picture, changes that occur at a given location may be manifested at a later time and position on the jacking record. For example, sudden increases in jacking forces may be manifested over short distances that are caused by dewatering of an interval of the drive that is much larger than the interval over which the increase in jacking forces is observed.

In general, the normal stress acting on the shield and pipe can be related to the effective unit weight of the soil and the pipe diameter, using arching theory, as described earlier (Terzaghi, 1943). The arching factor, C_a, was introduced to quantify this relationship, as shown below:

$$\sigma'_n = C_a \gamma' d_p$$

The interpretation of the jacking force data from the case histories and WES experiments provided evidence that, for sands, the range of C_a was ½ to 1-1/2. For stiff clays, the range was 1/3 to 1. For soft clays and silts, the arching factor ranged from 2/3 to 1-1/2, with the higher values likely due to squeezing behavior. The lower values in stiff clays result from stable annular overcut.

The friction factor also varies with position and time. The friction factor is lowest near lubrication injection points and at the time of injection. The friction factor is highest where no lubricant has been injected, at relatively large distances or elapsed time from the point of injection, and at undulations along the pipeline where high contact stresses may cause the lubricant to be squeezed out into the soil or mixed with the soil. The friction factor may increase during delays, due to soil strength gain with time. The friction factor may be higher in dewatered intervals, as a result of densification and higher particle to particle contact areas. All of these factors have to be considered in evaluating friction factor. However, to make the conceptual model and calculations practical and simple, some averaging or generalization of effects was necessary. Consequently, the intervals over which behavior is considered to be relatively constant were generalized in the model. In most cases, two or three intervals can be used to adequately describe behavior over the entire length of the drive. The first, short interval is characterized by relatively high friction factors and normal stresses, because of dewatering of this interval and because lubrication is typically not injected for the first 6 to 12 m. The second, typically much longer interval is characterized by significantly lower friction factors and normal stresses, because localized shaft dewatering does not affect this interval and because this interval is typically lubricated. The third interval is usually only appropriate if dewatering near the reception shaft is used and has a significant impact on lowering groundwater. When dewatering at the reception shaft is effective, the third interval often is characterized by behavior similar to that observed in the first interval.

In general, the friction factor is related to the effective friction angle of the soil or soil to pipe contact. Because of the relatively large displacements between the soil and pipe, the residual friction angle is appropriate. Residual soil friction angles can be estimated for clays from plasticity and clay size fraction. For sands, grain size distribution and grain shape can be used to provide residual friction angles. Repeated shear tests on remolded disturbed specimens can also be used to provide residual friction angles, as was done during this research (Bennett, 1998). Residual friction angles can be obtained with smaller displacements if precut surfaces are used, or if soil-to-pipe interface tests are conducted. Residual friction angles, however obtained, can be adjusted using the friction reduction factor, C_f, introduced to account for interface shear behavior in the lubricated intervals of microtunneling drives. The values of C_f were determined partly from the evaluation of the results of the direct shear tests, but primarily from the evaluation of the jacking force data from the case histories and WES tests. These analyses indicate that C_f is within the range of 1/2 to 1 for a relatively broad range of soil conditions considered in this research. C_f is applied to the soil friction angle to calculate the interface friction factor as shown below.

$$\mu = \tan C_f \phi_{r\,soil}$$

The majority of the case history jacking force data could be accurately described using a value of $C_f = 1$ for the initial, unlubricated interval of the drive, and $C_f = 1/2$ to $2/3$ for the lubricated interval.

Jacking Forces in Sand.

Jacking forces in sand can be accurately estimated using the conceptual model, by first establishing intervals over which operations and behavior are essentially similar, and then selecting input parameter values that are appropriate for the behavior in that interval. In most cases, a two-interval approach is appropriate, with the first interval approximately 12 m long and the second from 12 m to the end of the drive. For dewatered intervals, or drives above groundwater, the value of unit weight used in the calculations should be γ_{total}.

For intervals below the groundwater, the buoyant unit weight, γ', should be used. Recommendations are summarized in Table 1 for estimates for jacking forces in sands.

Table 1. Calculation of Frictional Jacking Forces in Sands

	Initial Dewatered, Non-Lubricated Interval	Lubricated Non-Dewatered Interval
Upper Bound	$F_r = 1.5 \gamma d_p \tan \phi_r A_p L$ ($C_a = 1.5$; $C_f = 1.0$)	$F_r = \gamma' d_p \tan \frac{2}{3} \phi_r A_p L$ ($C_a = 1.0$; $C_f = 2/3$)
Best Fit	$F_r = \gamma d_p \tan \phi_r A_p L$ ($C_a = 1.0$; $C_f = 1.0$)	$F_r = 2/3 \gamma' d_p \tan \left(\frac{2}{3} \phi_r\right) A_p L$ ($C_a = 2/3$; $C_f = 2/3$)
To account for the impacts of delays, steering corrections and misalignment, the upper bound calculated values of jacking forces may be increased by 1/3.		

Jacking Forces in Clay.

The calculation of jacking forces for projects in clay followed the same approach as used for jacking forces in sands. As shown in Table 2, for the case histories in clay, the lowest jacking forces and stresses were recorded for the drives in stiff to hard clays. Under these conditions, the annulus remains open, lubrication is more effective over longer length and time intervals, and the contact area may be reduced. For projects in very soft to medium silts and clays, the annular space is less stable, and the soil may squeeze in around the pipe. Lubrication is less effective and the pipe and soil are likely in contact over much of the full length and circumferential area. The result is higher jacking stresses.

Table 2. Calculation of Frictional Jacking Forces, F_r, in Clays

Material	Unlubricated Interval	Lubricated Interval
Stiff to Hard Clay: - Upper Bound	$F_r = \gamma d_p \tan \phi_r A_c L$ ($C_a = 1$; $C_f = 1$)	$F_r = 2/3 \gamma d_p \tan 2/3 \phi_r A_c L$ ($C_a = 2/3$; $C_f = 2/3$)
- Best Estimate	$F_r = 2/3 \gamma d_p \tan \phi_r A_c L$ ($C_a = 2/3$; $C_f = 1$)	$F_r = 1/2 \gamma d_p \tan 1/2 \phi_r A_c L$ ($C_a = 1/2$; $C_f = 1/2$)
Soft - Med. Silt, Clay: - Upper Bound	$F_r = \gamma d_p \tan \phi_r A_c L$ ($C_a = 1$; $C_f = 1$)	$F_r = 3\gamma d_p \tan \phi_r A_c L$ ($C_a = 3$; $C_f = 1$)
- Best Estimate	$F_r = 2/3 \gamma d_p \tan \phi_r A_c L$ ($C_a = 2/3$; $C_f = 1$)	$F_r = 1.5 \gamma d_p \tan \phi_r A_c L$ ($C_a = 1.5$; $C_f = 1$)
To account for the impacts of delays, steering corrections, misalignment, etc., the upper bound calculated values of jacking forces may be increased by 1/3.		

The recommendations offered for calculating frictional jacking forces in clays and silts in Table 2 are based on the analyses of the WES experiments and case histories. The case histories include Alki drive 2 in hard clay and mudstone, the large Findlay project in stiff to hard clay, and the Uxbridge and Puyallup projects in soft to medium silts and clays, (Bennett, 1998).

Effects of Operational Factors on Jacking Forces.

As shown in Figure 3, the calculated jacking stresses on the 760-mm-diameter drives of the Findlay project provide significant evidence of the effects of lubrication. No lubrication was used on the first two drives, i.e. drives 30-1 and 30-2. The slope of the jacking force graph and calculated jacking stresses on drives 30-1 and 30-2 were relatively uniform from launch to retrieval. The calculated jacking stress was 0.037 kg/cm^2. This same jacking stress was calculated for the unlubricated portions of drives 30-3 through 30-9. The jacking stresses calculated for the lubricated portions of drives 30-3 through 30-9 were only 0.009 kg/cm^2, a reduction of 76 percent from the jacking stress on the unlubricated portions. Since overcut and all other conditions were essentially unchanged, this difference can be attributed entirely to the effects of lubrication. For the 660-mm-diameter drives on the Findlay project, the calculated jacking stress in the unlubricated interval was 0.007 kg/cm^2. The jacking stresses calculated for the lubricated intervals were 0.005 kg/cm^2, a reduction of 29 percent, again attributable entirely to lubrication, since all other factors were essentially constant. For the 840-mm-diameter Findlay drives, the jacking stress in the unlubricated interval was 0.040 kg/cm^2, compared to 0.0066 to 0.013 kg/cm^2 in the lubricated intervals, a reduction of 68 to 84 percent.

On the Uxbridge and Puyallup projects in soft to medium silts and clays, the jacking stresses were higher, in general, than those calculated for the Findlay project in hard clays. On Uxbridge Drive 1, the jacking stress in the initial unlubricated interval was calculated as 0.014 kg/cm^2, and was 0.032 kg/cm^2 in the latter lubricated interval, or slightly more than double the unlubricated value. Clearly other factors dominated behavior on this drive, most likely squeezing of the soft silt around the pipe that led to higher normal stresses with time and distance. Inspection of the jacking force graphs shows that delays had a significant impact on jacking forces on this project. It is quite plausible that, as a result of these delays and the soft soils, (with stability ratio of 3.6 to 5) the overcut progressively closed in around the pipe, rendering lubrication ineffective and resulting in a progressive steepening of the jacking force graph. On the Puyallup project, lubrication was more effective, reducing jacking stresses from 0.055 to 0.017 kg/cm^2, a reduction of 69 percent. The normal stress used in the calculations was approximately the same value for both the unlubricated and lubricated intervals. The use of lubrication reduced the friction reduction factor from 1 to 1/2, a 50 percent reduction.

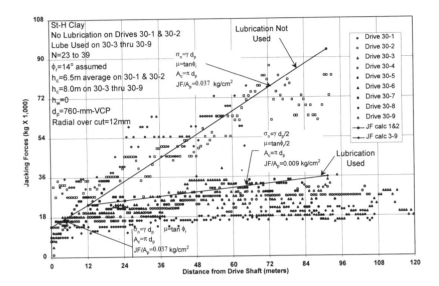

Figure 3. Comparison of Actual and Calculated Jacking Forces vs Distance from Drive Shaft, Findlay Project. All Drives with 760-mm OD Vitrified Clay Pipe. (Effects of Lubrication Illustrated.)

On the WES experiments, it was not possible to completely separate the impacts of lubrication from the impacts of overcut, and delays complicated attempts to isolate effects of lubrication. No lubrication was used on the auger machine test. Since the overcut on this test was 9 mm, or essentially the same value as for the retrievable machine test, it was possible to use these two tests to evaluate effects of lubrication in different soils. The C_f values were 1/2 on the retrievable machine test and were 1 on the auger machine test, a reduction of 50 percent. From the preceding analyses, the benefits of lubrication are estimated to be 30 to 85 percent.

The WES auger and slurry machine tests provided an opportunity to evaluate the influence of overcut on jacking forces. The radial overcut was 9 mm on the auger machine test, and was 2.5 mm on the slurry machine test, while most other factors remained constant. Based on this limited data, the inadequate overcut on the slurry machine test resulted in an average calculated increase of 50% in jacking stresses in clays and sands.

The Findlay project in stiff clay and the Newark project in silty sand each had drives with different diameters, which allowed the effects of diameter on jacking forces to be evaluated. A significant portion of the difference of calculated jacking stresses on the

Findlay project 660-mm- and 760-mm-diameter drives may have been due to lower average normal stresses acting on the 660-mm drives. The smaller unsupported span of the 660-mm bores possibly resulted in longer "stand up" time and lower contact stresses, compared to the 760- and 840-mm drives. This hypothesis is supported by the much lower assumed normal stresses and computed jacking stresses on the 600-mm-diameter drives on the Newark project, compared to the 900-mm-diameter drives. On the Newark project, the overcut was 9 mm for all 600- and 900-mm-diameter drives. The computed jacking stress for the initial unlubricated portions of the 600-mm-diameter Newark drives was only 1/3 to 1/2 of the computed jacking stresses on the same intervals of the 900-mm drives. The differences in jacking stresses between 600- and 900-mm-diameter drives on both Newark and Findlay support the hypothesis that normal stresses acting on the pipe increase proportional to increases in pipe diameter.

The analysis of the case histories and the WES experiments failed to produce a correlation between depth of cover and jacking forces. The case history analyses showed that the normal stress acting on the pipe can be correlated with the product of pipe diameter and soil density multiplied by an arching factor, C_a. The calculation of normal stress using total overburden stress clearly results in over-prediction for all but very shallow projects.

The analysis of the effects of delays indicated transient increases in jacking stresses were approximately 9 to 50 percent. These results were observed on the three WES experiments, some of the Newark drives in sand, the Uxbridge project, and the Puyallup project, both in soft silts and clays (Bennett, 1998). One surprise was that the duration of delays did not correlate with an increase in jacking forces or with an increase in jacking stresses. All delays, even short delays required to add pipe sections, tended to result in transient increases in jacking forces. When the drive resumed, the transient increases tended to dissipate rather quickly and the total jacking forces usually returned to about the same level or only slightly higher levels than before the delay. The impact of delays on jacking forces and stresses increased with distance. The approach that produced the most consistent interpretation was obtained by calculating the increase in jacking stress, as a percentage of jacking stress before the delay. This approach takes into account the circumferential area of pipe in the ground at the time of the delay. This area deserves further study since owners and engineers sometimes prescribe costly operations 24 hours a day, 7 days a week, to minimize increases in jacking forces caused by delays.

Dramatic increases in jacking forces can occur as a result of dewatering near shafts, especially for projects in sands. The significant impacts of dewatering are illustrated using observations from two projects in sands, the Newark and Alki projects. On the Alki project drive 1, jacking forces increased sharply from 250,000 to 470,000 kg over a 3 m interval, as shown in Figure 4, after the dewatering wells near the reception shaft were activated in preparation for retrieval of the MTBM. At the end of the shift on Thursday, with

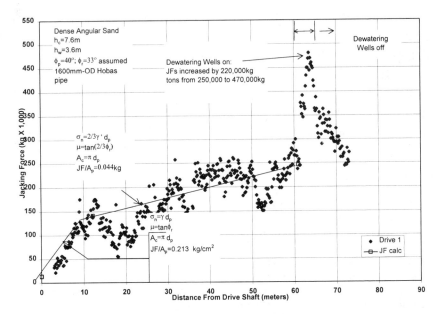

Figure 4. Comparison of Actual and Calculated Jacking Forces vs Distance from Drive Shaft, Alki Transfer Project, Drive 1. Dramatic Effects of Dewatering near Reception Shaft also Illustrated.

the MTBM at 60 m, the wells were activated. The jacking forces at this point were 250,000 kg. On Friday morning, the jacking forces rose to 470,000 kg within 3 m. The drive was stalled because the pipe capacity was exceeded. After some discussion, the wells were deactivated over the weekend. On Monday morning, the drive resumed, and jacking forces quickly fell to approximately the same level as before dewatering. On the Newark project, drive 36-8, the increase in jacking forces was about 40,000 kg, attributable to dewatering when the wells were activated in the silty sands near the reception shaft (Figure 5).

It is difficult to reliably estimate the effects of dewatering on jacking forces using data that are typically available. However, it is clear that the effects in sands can be large, and the interval influenced by dewatering is much larger than the short interval over which the spike in jacking forces may be observed. To minimize the effects, and allow successful completion of the drive, dewatering should only be used to the extent necessary to launch and retrieve the machine.

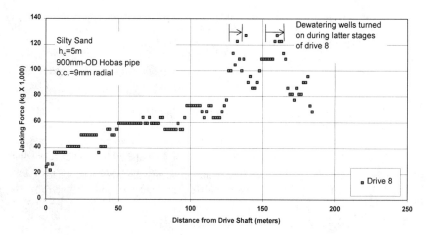

Figure 5. Jacking Forces vs Distance from Drive Shaft, Newark Project, 900-mm Diameter Drive 8, Illustrating Effects of Dewatering.

Conclusions and Recommendations

Jacking forces exert significant impacts on economic and technical feasibility of microtunneling and pipe jacking projects. Failure to control jacking loads within safe limits can lead to pipe failure, excessive movements of the thrust block and accompanying line and grade problems, and inability to complete the drive.

To control jacking forces, the practices listed below are recommended.

- Radial overcuts of 6 to 18 mm should be used.

- Continuous, pressurized lubrication sufficient to fill the annular volume around the pipe is recommended.

- Gradual steering corrections should be used to reduce sharp increases in jacking forces associated with sudden steering corrections.

- Care should be exercised to ensure accurate initial alignment. The jacking forces applied to the pipe are highest at the jacking shaft. Initial misalignments result in eccentric loading and greatly reduced pipe capacity.

- Dewatering should be limited to the locations and time periods essential for launch and retrieval of the MTBM. Dewatering can have substantial impacts on jacking forces in sands.

To predict jacking forces, the conceptual model and calculation approach described in this paper may be used for a wide range of ground conditions, operational, and geometric factors. The calculations should be made by separately evaluating face pressure and skin friction components over the intervals described by the conceptual model. The key factors are selection of appropriate normal stresses and friction factors, and understanding the impacts that operational and geometric factors can exert on jacking forces.

Normal stresses should be estimated based on arching theory. The results of this research indicate a strong relationship exists between normal stresses and soil effective unit weight and pipe diameter. The arching factor, C_a, was introduced to quantify this relationship for projects in sands and clays. The C_a values provided in this paper are recommended for use in calculations. The use of pipe weight to estimate normal stresses was shown to provide unconservative, lower bound estimates for most conditions (even in clay). Pipe weight controlled normal stresses only for hard clays with substantial overcuts.

Friction factors should be selected based on soil residual friction angles. The friction reduction factor, C_f, may be used to relate the interface friction between soil and pipe, or soil to lubricant to pipe to the soil residual friction angle. As testing practice evolves, practical tests may be devised to provide direct measurement of interface friction angles that accurately model field behavior. However, it is considered more practical, for the near term, to use residual soil friction angles determined from correlations with soil index properties, literature, or from repeated shear tests, with an appropriate value of C_f, determined from the case history analyses.

The equations given in Tables 1 and 2 may be used for jacking force calculations in sands, silts, and clays. The conceptual model, calculation approach, and recommended values of C_a and C_f must be applied with judgment, and tempered with experience. The upper bound calculations should be used in design for good workmanship and typical conditions. To account for inexperienced crews, steering corrections, dewatering, misalignment, and delays, the calculated values may be increased by 1/3.

References

Bennett, David and Staheli, Kimberlie, 1996. "Controlled Field Tests of Retrievable Microtunneling System with Reaming Capabilities," U.S. Army Corps of Engineers Technical Report CPAR-GL-96-2.

Bennett, Robert D., 1998. "Jacking Loads and Ground Deformations Associated with Microtunneling", Ph.D. Dissertation, University of Illinois at Urbana-Champaign, Urbana, IL

Coller, Phillip, Staheli, Kimberlie, Bennett, David, and Post, Ray, 1996. "A Review of Jacking Forces by Both Theoretical and Empirical Methods as Compared with 20 Years of Practical Experience," *Proceedings of the International NO-DIG '96*, New Orleans, April 1996.

Haslem, R. F., 1986. "Pipe Jacking Forces: from Theory to Practice," *Proceedings of Infrastructure, Renovation and Waste Control, North West Association Centenary Conference*, pp 173-180, Institution of Civil Engineers.

Herzog, M., 1985. Die Pressenkräfte bei Schildvortrieb und Rohrvorpressung im Lokkergestein. BMT, Issue 6, pp. 236-238. (As summarized by Stein et al. 1989.)

International Society for Trenchless Technology, 1994. Statistical Analysis of Jacking Forces", International Working Group No.3 (draft)

Lys, Robert, Jr., 1993. "Microtunneled VCP: The American Experience," *Proceedings, Water Environment Federation Collection Systems Specialty Conference*, June 1993, Tucson, AZ.

Milligan, G. W., and Norris, P., 1993. "Pipejacking Research Results and Recommendations," U. K. Pipejacking Association, London.

Salomo, K. P., 1979. Experimentelle und theoretische Bestimmung der Pressenkräfte und der Bodenverformung beim Vortrieb eines Vorpreßrohres in rolligen Böden. Dissertation. Technische Universität Berlin. (As summarized by Stein et al. 1989.)

Scherle, M., 1977. Rohrvortrieb Part 2. Bauverlag, Wiesbaden/-Berlin. (As summarized by Stein et al. 1989.)

Stein, D., Möllers, K., and Bielecki, R., 1989. "Microtunneling: Installation and Renewal of Non-man Size Supply and Sewage Lines by the Trenchless Construction Methods," Ernst and Sohn, Berlin, Germany.

Terzaghi, K., 1943. *Theoretical Soil Mechanics* John Wiley and Sons.

Weber, W., 1981. Experimentelle Untersuchungen in rolligen Böden zur Dimensionierung von Preßbohranlagen. Dissertation. Wissenschaftlicher Bericht aus der Arbeit des Institutes für Baumaschinen und Baubetrieb der Rheinisch-Westfälisch Technischen Hochschule Aachen, RWTH Aachen. (As summarized by Stein et al. 1989.)

Winterkorn, H. F., and Fang, H. Y., 1975. *Foundation Engineering Handbook*. Winterkorn and Fang ed., Van Nostrand Reinhold Company, New York, NY.

Tunneled Sewers in Houston, Texas

Michael Hasen, P.E., Member, ASCE[1]

ABSTRACT: This paper presents an overview of the subsurface conditions that can impact tunnel and micro-tunnel construction in Houston, Texas for sewer projects. These conditions include surface faulting, groundwater conditions, overconsolidated soils, environmental conditions and the impact of adjacent facilities.

INTRODUCTION

Tunneling for sewer projects is commonly performed by pipe jacking techniques whereby sections of pipes are jacked directly into the soil behind a cutterhead assembly. On larger tunnels, a primary liner is often constructed as the tunnel is excavated, and pipe is placed and grouted within the primary liner. The following sections address subsurface conditions, tunnel excavation, access shaft construction, and urban environment issues as they relate to typical tunneling in the Houston area.

SUBSURFACE CONDITIONS

Soil conditions in the Houston, Texas typically comprise Pleistocene stiff to very stiff sandy clay and clay with intermittent medium dense to dense sand layers to depths of over 30 meters. The clays are generally overconsolidated, they can be highly plastic, and sometimes have significant secondary structure in the form of slickensides. The sands are generally fine and silty, with most grains usually finer than 0.3 millimeters (No. 50 sieve) and between 10 and 30 percent finer than 0.075 millimeters (No. 200 sieve).

---

[1] Senior Engineer and Operations Manager, HVJ Associates, Inc., 6120 S. Dairy Ashford Road, Houston, Texas 77072, (281) 933-7388.

Groundwater

Geotechnical construction issues are also related to groundwater conditions that are controlled by the location and gradation of sand layers. Groundwater can be encountered as shallow as 3 to 5 meters below ground surface. Since the water bearing sand layers are typically encountered at depths of at least 3 meters, it is often assumed that the clay layer at the surface forms an impermeable cap. This would prevent rapid fluctuations in groundwater levels due to transitory causes such as rainfall events.

Piezometer monitoring results indicate that the water levels in the sand layers can respond relatively quickly to rainfall events. We have measured increases in groundwater levels of 0.6 meters within a sand layer twelve hours after a heavy rainfall event. At this location, the sand layer was located at about 5-meter depth beneath a layer of very stiff clay. We theorize that this observed behavior might be due to seepage through pervious trench backfill. Seepage along secondary structure features in the soil such as slickensides is another potential explanation for this response.

Cemented Sand

In parts of Houston intermittent cementation has been found within the sand layers. This cementation results in very high Standard Penetration Test (SPT) N values, with 50 blows for penetrations of 50 to 100 millimeters not unusual. Recovery of this material is difficult, since it is too dense to sample with a normal SPT sampler, and the cementation is typically too intermittent to allow recovery by coring. We have encountered such conditions at sites generally northwest of downtown Houston. The cemented sand material outcrops along the banks of Buffalo Bayou west of I-610 (West Loop) as shown on Figure 1. Hasen and Chelliah (1998) present a detailed discussion of the cemented sands and their impact on construction.

We have encountered cemented sand conditions at depths ranging from 3 to 16.5 meters. However, this material is more commonly encountered between depths of about 6 and 12 meters. The thickness of these layers is often 75 to 300 millimeters; however, we have encountered layers as thick as 1.2 meters. Multiple layers of cemented sand are often encountered in the same boring. The depth, degree of cementation, and thickness of the cemented layers varies appreciably between borings spaced as closely as 50 meters apart.

Core sampling was performed in borings at several sites to obtain intact cemented sand core in order to perform unconfined compression tests. We have generally not been successful in recovering intact cores of the material. We speculate that the cemented sand is either too friable to core reliably, or that the cementation is typically in relatively thin layers which are destroyed during the coring process. In one soil boring we obtained a 430-millimeter long cemented sand core at a depth of 8.5 meters. We measured a compressive strength of 7.5 MPa on a sample obtained from the core.

Geologic Faulting

In the Gulf Coast region of Texas, over 200 faults are known or suspected to be active. Many of these faults are located in the Greater Houston area as shown on Figure 1. These faults also extend offshore several hundred kilometers and inland north of the Conroe area. Fault movements are relatively slow. Evidence of fault activity includes laterally persistent, abrupt changes in elevation (scarps), in particular an abrupt change in slope along the fault with a return to the slope away from the fault.

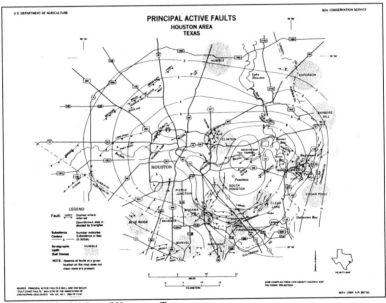

Figure 1 – Fault Map of Houston, Texas

Most of the faults in the Texas Gulf Coast region are considered growth (down-to-the-coast) faults in which the dip angle of the fault near the ground surface averages 75 degrees. Antithetic (complementary) faults occur in conjunction with very active, larger growth faults. These antithetic faults tend to dip toward the active growth fault and terminate into it. Antithetic faults tend to be much shorter in length, and have smaller displacements. Localized faulting is also associated with the presence of subsurface salt diapirs that tend to produce short faults extending radially from the flanks of the dome. Movement rates of these faults range from less than 3 to over 25 millimeters per year. Typical rates are less than 12 millimeters per year.

TUNNEL EXCAVATION

Tunnels for sewers in Houston have commonly been constructed as piped-jacked or primary lined tunnels. Pipe-jacked tunnels are constructed by a one-pass method with or without man entry with diameters commonly ranging from 0.75 to 2 meters. Tunnel construction by this technique involves pipe jacking following a hand-shield excavation or a tunnel-boring machine. The pipe serves as construction lining and tunnel support for stability during construction and as the sewer pipe after construction. Primary lined tunnels are installed using a 2-pass method for tunnel diameters of 1.5 to 4 meters. A primary liner is installed to provide tunnel support commonly comprised of rib and timber lagging or bolted steel plates. The sewer pipe is then placed inside the constructed tunnel, and the open annulus is grouted. Dewatering in advance of construction is common.

Stiff to Very Stiff Clay

For the stiff to very stiff cohesive soil materials, the stability of a tunnel face is determined by its existing undrained shear strength. Stable soil conditions are usually encountered for tunnels driven in these materials. The stability of the unsupported face of the tunnel may be evaluated by a ratio of the overburden pressure divided by the undrained shear strength of the clay soil assuming atmospheric air pressure in the tunnel. This ratio is referred to as the Overload Factor (OF) as described by Peck (1969). OF values of two or less are common on tunnel projects in the Houston area. Generally, a value for the overload factor of 4 or less is desirable. Such a value represents a practical limit below which tunneling may be carried out without difficulties. Higher OF values will frequently lead to large deformations of the soil around and ahead of the tunnel opening, with the associated problem of increased subsidence and possible deformations of the lining. Settlement during tunneling in typical clay soils is usually small and does not cause adverse impacts to facilities commonly located in a typical right-of-way.

Sand and Silty Sand

For silty sand, clayey sand, clayey silt, sandy silt, and silt soil strata, the stability of a tunnel face is essentially governed by the groundwater conditions and by the construction method. Due to the depth as which tunneling becomes an attractive construction method; these soils are generally encountered below the water table during tunneling. Below the groundwater level, or in wet conditions, the stability of the face depends on whether any slight cohesion that may be present is able to withstand the seepage forces from the water flowing into the tunnel. In general, a flowing ground condition will prevail and construction equipment and procedures providing positive support to the face are needed to ensure face stability. Tunnel construction in Houston often occurs at hydrostatic heads of 3 to 8 meters above the crown.

Large ground loss can result from uncontrolled flowing ground. The potential for such ground loss exists wherever water-bearing sands or silts are encountered along the alignment. Careful dewatering of such layers will reduce the potential for development of flowing conditions, but local experience shows that complete dewatering is difficult to achieve. Dewatering is commonly performed using eductors spaced as closely as two to three meters on center. Deep wells are sometimes used, but often the sand layer below the base of the tunnel is too thin for this approach to be practical. Generally, dewatering does not cause significant consolidation settlement of overlying clay layers due to the overconsolidated nature of the clay.

Success of the dewatering program generally depends on having an adequate lead-time in advance of construction and having proper well spacing. Local experience in dewatering for tunnel construction indicates two common problems. First, it is difficult to fully dewater the base of a permeable layer immediately above an impermeable layer. Second, due to the interbedded nature of the soils, all the water-bearing zones may not be intercepted even by closely spaced wells. Either of these conditions can result in the presence of unstable water-bearing soils even though a dewatering system has been installed. Also, seepage due to rainfall, or from adjacent utilities, which the dewatering system cannot accommodate can lead to incomplete dewatering of a target layer.

If large ground loss occurs, grouting can stabilize the formation. Once the formation is stabilized, any facilities that have been damaged must be repaired, and fill is placed to make up the lost ground. Cement stabilized sand fill is often used in order to reduce settlement of the fill.

<u>Cemented Sand</u>

Cemented sands are difficult to excavate and penetrate compared to the more commonly encountered clay and uncemented sand soils. Contractors on projects in the Houston area will not normally make allowance for working with cemented sand material in their bids. Therefore, when it is anticipated that construction will encounter cemented sand it is important to clearly identify it in the construction documents to avoid claims. The presence of cemented sand material on a project does not mean that construction will be difficult. It does mean that "typical" construction procedures and equipment may not be appropriate for the project. Based on its material properties, cemented sand is not a particularly difficult material to excavate. Many tunnels have been excavated through more difficult material.

Equipment that is well suited to good tunnel production in typical Houston soil conditions is often unable to excavate even small amounts of cemented sands. At this point the tunneling equipment becomes stuck because it cannot simply back up to exit the tunnel. Backing up is impossible because the pipe or primary liner constructed behind the tunneling equipment reduces the tunnel diameter behind the equipment. A "rescue shaft" must be excavated from the ground surface to the equipment in order to extract the equipment. A rescue operation involves substantial expense and schedule delay, and typically results in construction claims.

Loss of ground problems during tunneling in Houston are often related to tunneling in sands below the water table. Since most cemented sands are below the water table, loss of ground problems are a real concern during an interruption in tunneling due to cemented sands. Therefore, in addition to the time and expense, the need to perform a rescue may result in damage to adjacent facilities due to loss of ground.

Where cemented sands are encountered within or near the tunnel alignment, an evaluation of the tunnel depth should be made to determine whether the tunnel depth can be changed to avoid the cemented sand material. On wastewater projects there is often little latitude for vertical alignment adjustments due to upstream or downstream control on the flowline. On projects where cemented sands are encountered their presence should be clearly documented on the plans and in the specifications. Tunneling equipment should be selected such that the cutterhead is suited to both soft and hard ground. Tunneling equipment guidance should be developed on a project specific basis, and may be included in the project specifications and baseline report.

Risk should be recognized in structuring the bid. If a single pay rate will be used for tunneling over the entire project, then the contractor must estimate what impact cemented sand conditions will have on overall schedule. If separate pay rates are used for "normal" and "hard ground" tunneling, then the only uncertainty to the contractor is the production rate of the proposed equipment in those conditions. Definition of "hard ground" tunneling in the specifications and baseline report is dependent on the material properties determined during the geotechnical investigation.

Fault Crossings

Where a tunnel crosses a fault the tunnel needs to be protected against damage due to fault movement subsequent to construction. At a fault crossing the pipe is generally installed within a stiffened section through the primary movement zone. The sections of the pipe immediately upstream and downstream of the stiffened section are designed with sufficient flexibility to accommodate the estimated movement over a 50-year design life. Manholes are required on the upstream and downstream sides of the fault crossing in order to facilitate pumping around a pipe break at the fault location during repairs.

Tunneling impacts due to the fault also exist in the form of erratic stratigraphy and groundwater conditions in the vicinity of the fault. The fault zone is typically a nearly vertical thickness of heavily sheared soils about 3 to 6 meters wide. Due to the faulting the soil stratigraphy can change abruptly when crossing from one side to the fault to the other; a tunnel which has been driven in stiff clay on one side of the fault may encounter water-bearing sand after crossing the fault. Also, groundwater can easily travel along the fault zone. Loss of circulation problems are common when penetrating the fault during drilling, and locally high groundwater condition may exist in the fault zone depending on the rainfall conditions during construction.

Liner Deformations

Deformation of the liner in the horizontal and vertical diameters can be expected due to soil-liner interaction. Experience with liner distortion in the Houston area suggest values in the range of 0.75 percent difference in length of the vertical and horizontal diameters; with shortening of the vertical diameter in most cases. To the extent that the tunnel liner reduces the soil deformation due to the rigidity of the liner, bending moments will be developed in the liner. The lining will be adequate with respect to bending if it can be deformed, without overstress, by an amount equal to the expected change in diameter.

Buckling of the liner can be a problem if non-uniform support of the liner occurs. This sometimes happens if a local overcut situation occurs during tunneling which is not properly backfilled. Buckling can also occur if the liner is used as reaction for the tunneling equipment, and the tunneling equipment unevenly applies thrust loads.

ACCESS SHAFT CONSTRUCTION

Access shafts constructed for tunneling operations vary in size depending on whether the shaft is a drive or receive shaft, size of tunneling machine, and size of jacked pipes. Small diameter shafts (less than about 3 meters diameter) are typically constructed by auguring and slipping a corrugated metal pipe culvert or caisson under its own weight into the ground as augering proceeds. Large diameter shafts are formed using steel liner plates, bolted to form a cylindrical shaft as excavation proceeds. Steel ring segments are used to support the interior of the steel lined plates if necessary. Braced excavations constructed of soldier piles and lagging are also used. The space between the soil and the excavation support elements is sometimes grouted to create a watertight excavation. Access shafts for tunneling operations typically are excavated approximately 1.25 meters below pipe invert depth.

Shaft Bottom Stability

Bottom instability results from inadequate shear strength in clay soils to resist stress relief at the base of the excavation, or from piping of water bearing granular soil. These modes of failure result in loss of ground at the ground surface outside the shaft and heave of the excavation base inside the shaft. In the stiff to very stiff clays, the factor of safety against bottom heave is generally 1.75 or greater for typical access shafts. Where access shafts are constructed in sandy soils adequate dewatering to lower groundwater levels to at least 0.6 meters below the base of the excavation is critical to prevent bottom instability.

Loss of Ground

Installations for access shafts sometimes experience loss of ground outside of the excavation due to sloughing of material into the excavation. If proper construction procedures are followed, little or no loss of ground should occur. If loss of ground is excessive, it may cause damage to structures, pavement and services located near the excavation. If loss of ground does occur, soft disturbed soils may develop beneath existing pavement and utilities located close to the excavation location. Large ground loss often results from inadequate dewatering which leads to piping of the fine, sandy soils through the sheeting. Corrective measures to address loss of ground problems often include improved dewatering and/or grouting around the shaft from the ground surface or within the shaft. Repairs associated with loss of ground often include replacement of paving near the top of the shaft, and making up for ground loss through placement of cement stabilized sand fill.

URBAN ENVIRONMENT

Tunneling is commonly used in the Houston area for sewers at depths of about 6 meters or greater. The need for sewers at these depths is driven by two typical causes: (1) the need for a sewer to cross a bayou where an elevated crossing would adversely constrict the floodway, and (2) the need to install new sewers in areas where conflicts with existing utilities or other existing facilities prevents installing the sewer at shallow depth. The presence of existing utilities and the need to prevent floodway restrictions are common in the developed, urban sections of Houston. Impacts on tunneling including environmental conditions, existing utilities, and right-of-way restrictions are discussed below.

Environmental Impacts on Construction

Buried utility projects in the Houston area commonly include a Phase I Environmental Assessment in order to review environmental conditions along the project alignment. Potential impacts to the project include health and safety issues associated with construction in a contaminated area, budget overruns due to unforeseen or changed condition claims, project delays, and disposal of contaminated media. Zlotnik and Hasen (1997) present a detailed discussion of Phase I ESAs for Public Sector Corridor Projects.

The purpose of the Phase I ESA is to identify areas of potential contamination along the project alignment which may impact the project. If suspected contamination is identified during the Phase I ESA, subsurface soil and/or groundwater sampling and testing are performed to confirm the presence and extent of contamination. The majority of the contaminated sites adjacent to public right-of-way are underground petroleum storage tanks. If such a site is close enough to the project alignment, contaminated soil and/or groundwater can be encountered during construction.

Tunneling projects in the Houston area have encountered seriously contaminated conditions, including sites with a meter or more of phase separated product above the water table. In some cases, the project alignment can be altered to avoid the contaminated area by shifting the alignment to the opposite side of the street or by shifting the alignment to a different street. In other cases the alignment must proceed through the contaminated area, and issues including worker health and safety, contaminated soil disposal, and treatment of water produced from the dewatering system are addressed in the project plans and specifications.

We have found that the historical research and site reconnaissance phases of the Phase I ESA are critically important in identifying potentially contaminated sites. Many unregistered sites with underground storage tanks exist in the Houston area, and many of these sites turn out to have contaminated soil and or groundwater in the public right-of-way adjacent to the site. It takes competent research and knowledgeable field personnel to identify unregistered underground storage tank sites along an alignment that may impact a project.

Existing Utilities

As discussed previously, the presence of existing utilities is often times responsible for the need to install a new utility deep enough to require the use of tunneling. However, adjacent utilities can have an impact on the tunneling process itself primarily due to groundwater impacts. Tunneling projects in the Houston area have been adversely impacted due to leakage from existing utilities overwhelming the capacity of dewatering systems installed for the tunneling project.

Also, tunneling projects in the Houston area have been impacted due to leakage of groundwater from trench backfill associated with adjacent utilities. In the Houston area the predominate soil type is very stiff clay, and utility trenches were commonly backfilled with sand or sand and shell many years ago. During rainfall events, the groundwater can collect in the trench and infiltrate into the tunneling zone with adverse effects. In severe cases, the trench backfill itself can migrate into the tunnel bore if significant flows of groundwater between the trench and tunnel develop.

Right-of-Way Restrictions

Right-of-way restrictions can cause a tunnel to be located in unusual places. Tunnels have been driven between the existing pile foundations of a bridge because there was not adequate right-of-way to locate the tunnel outside the foundation. This required careful evaluation of the impact of the tunnel on the foundation capacity, and the impact of the foundation location on the tunnel alignment.

CONCLUSIONS

This paper has described common tunneling conditions and problems in Houston, Texas. The tunneling environment in Houston is generally good, with stiff clays usually encountered. The fine grained and silty nature of the sand layers make dewatering challenging and critical during tunneling in these materials. The presence of conditions such as fault zones, cemented sands, environmental contamination, adjacent utilities, and adjacent facilities are important to recognize during design and address during construction in order to have a successful project.

REFERENCES

Hasen, M. and Chelliah, D. (1998). "Impact of Cemented Sands on Infrastructure Projects in Houston, Texas" Proceedings, ASCE Texas Section Spring 1998 Meeting, South Padre Island, Texas.

Peck, R.B. (1969) "Deep Excavations and Tunneling in Soft Ground," Proceedings, Seventh International Conference on Soil Mechanics and Foundation Engineering, Mexico City, State-of-the Art Volume.

Zlotnik, E. and Hasen, M. (1997) "Phase I ESAs for Public Sector Corridor Projects," Proceedings, ASCE Texas Section Fall 1997 Meeting, Arlington, Texas.

DESIGN AND PLANNING OF URBAN UNDERGROUND CONSTRUCTION USING PIPE BURSTING TECHNIQUES

Samuel T. Ariaratnam,[1] AM ASCE, Jason Ş. Lueke,[2] and Peter Strychowskyj[3]

Abstract

Rehabilitation of our aging underground infrastructure continues to be of great concern resulting in the examination of alternative techniques to address this issue. One such construction technique rapidly gaining acceptance by municipalities across North America is pipe bursting. In comparison to conventional open cut techniques of pipe replacement, pipe bursting has many advantages. Experience has shown that, in almost all circumstances, pipe bursting provides a less expensive alternative to traditional open cut replacement, while increasing overall productivity. This paper describes considerations in designing and planning of pipe bursting operations. Factors including construction methods, pipe materials, and project site characteristics are addressed.

Introduction

Underground utilities have been in place for over a century in many North American municipalities. Increased demand on existing utilities, as a result of urbanization, has resulted in decreased lifespans. Infact, many have functioned far beyond any reasonably anticipated design life. These underground systems deteriorate and require replacement or rehabilitation to maintain their functionality. Maintaining this large network of underground sewer, water, and gas pipelines is difficult and costly. The problem is compounded by the significant impacts that a major repair or replacement project can have on the

---

[1] Asst. Prof., Dept. of Civ. and Envir. Engrg, Univ. of Alberta, Edmonton, Canada T6G 2G7
[2] Grad Res Asst., Dept. of Civ. and Envir. Engrg, Univ. of Alberta, Edmonton, Canada T6G 2G7
[3] General Mngr., Trenchless Replacement Services Ltd., 4015-8th St., Calgary, Canada T2G 3A5

daily life, traffic, and commerce of the area served by and along the pipeline in question. Pipe bursting provides a trenchless option for the replacement of underground infrastructure. The process includes various static, pneumatic, and hydraulic expansion methods of breaking an existing pipe and simultaneously installing, by pulling or pushing, a new pipe of equal or larger diameter. The design and planning of such operations is critical to ensuring success of this emerging trenchless process. As more and more municipalities are faced with the daunting task of repairing or replacing their aging buried infrastructure, there should be greater emphasis placed on tools to assist in design and planning these operations.

Trenchless Pipe Replacement Using the Pipe Bursting Process

The pipe bursting process is defined as the replacement of the host pipe by fragmenting the existing conduit and installing the product pipe in its place (CCET 1991). Generally, bursting is accomplished by pulling a cone shaped bursting head through the host or original pipe. The front end of the bursting head is smaller than the inside diameter of the existing host pipe, while the tail end of the head is slightly larger than the outside diameter of the new product pipe. The leading face of the bursting head is tapered to provide a smooth transition from the small to large diameter on the head. By pulling the head through the host pipe, it is burst while simultaneously expanding the cavity that the original pipe was situated. This void provides the required space for the new pipe as well as reduces friction during installation. The new product pipe is attached to the tail end of the bursting head, while a chain, cable or rod assembly is attached to the front end. As the rods, cable, or chain are pulled, the bursting head fragments the pipe while at the same time pulling the new pipe into place.

Anatomy of a Pipe Bursting Project

One of the principal advantages that pipe bursting has over conventional methods of pipeline replacement is the minimal amount of excavation required to replace existing lines. Pipe bursting is also recognized as the only method of pipeline replacement techniques that is able to replace existing lines with new pipe that is of identical size or larger. For these reasons, pipe bursting is becoming a viable alternative to traditional open cut pipe replacement and, in some cases, pipe relining. Typical replacement pipe sizes range from 50 to 400 mm (2 to 16 inches) in diameter, and lengths between 50 and 100 m (150 and 300 ft). Diameters up to 910 mm (36 inches) have been accomplished in St. Petersburg, Florida (Thomas 1996), and lengths up to 470 m (1550 ft) in Stockbridge, Massachusetts (Saccogna 1997).

In general, the pipe bursting project is divided into sections or lengths that the bursting equipment being used can burst based on the geometry and layout of the total length of pipe being replaced. The length that can be burst is highly dependent on the type of pipe being burst, degree of upsize, soil conditions, geometry of the original installation, and the type of bursting equipment and method used. In addition, the new pipe, whether it is continuous or sectional, will dictate the type of equipment required and the pit setup the greatest.

For the installation of continuous pipe, such as high-density polyethylene, access pits must be excavated at each end of the pipeline to be replaced. On one end of the line, the machine pit is excavated into which the pipe bursting machine that pulls or directs the bursting head is located. Opposite the machine pit is the insertion pit through which the new pipe or product pipe and bursting head are inserted into the existing or host pipe. The setup for a typical burst using static pipe bursting is shown in Figure 1, in this case, the pulling mechanism could consist of rods, chain, or cable. Any services along the pipe route connected to the host pipe must be disconnected prior to the start of the burst with access to the lateral connections achieved through service pits.

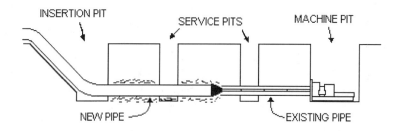

Figure 1. Typical Configuration for Continuous Pipe Installation

If sectional pipe is used as the product pipe for the installation, a slightly different setup is required. Again, access pits are excavated at each end of the line to be replaced, except in this case both pits are considered machine pits. The installation of sectional pipe, usually clay or polyvinyl chloride, requires that constant force be applied to the pipe to keep the joints together during installation. This may be achieved by using a chain or cable run through the product line from the bursting head to a trailing plate on the last pipe section, or alternatively by using a push-pull setup. In the push-pull setup, the bursting head would be pulled by one machine in the pulling pit, while in the opposite pit, the pipe section would be pushed by another machine as illustrated in Figure 2. In this setup, a constant pressure is applied to the new pipe during installation by maintaining the push force slightly higher than the pulling force. This requires the synchronization of the machine forces; however, allows for large diameter installations to be achieved. One such installation occurred in St. Petersburg, Florida, where 230 m (770 ft) of 900 mm (36 inch) diameter vitrified clay pipe was successfully replaced with Hobas pipe (Thomas 1996).

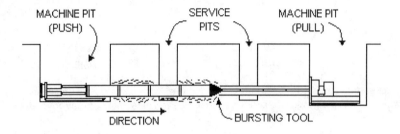

Figure 2. Typical Configuration for Sectional Pipe Installation

The size of the machine pit depends on the size and type of pipe bursting equipment used. Machine pits used in static pipe bursting can range in size from 4050 mm by 2500 mm (13.3 by 8.2 ft) to the size of a manhole. Some types of bursting equipment only require the insertion of a mechanical arm or leg with a pulley into a manhole to direct and pull a cable or chain. Depending on ground conditions and depth of the host pipe, shoring may be required, though sloped walls are also an option. Shoring is generally preferred to keep the footprint of the excavation to a minimum. This assists in minimizing surface disruption and the cost of restoration.

Insertion pits are generally smaller than the machine pits. As a rule of thumb, for static bursting methods using continuous pipe, the length of the insertion pit should be 12 times the diameter of the new product pipe plus a length to account for the slope depending on the depth of the excavation at a ratio of 1.5 to 2.5 run to 1 depth. The slope ratio largely depends on the bend radius of the product pipe. The width of the insertion pit need only be 1200 mm (4 ft). Again, if no shoring is used, the trench walls should be sloped at an appropriate angle to account for ground characteristics.

If sectional pipe is used as the product pipe, the pit length will depend more on the length of one section of pipe with allowance for worker space to aid in the placement of the pipe. Width, like the length, depends more on the space required for the handling of the pipe during the lowering of the pipe.

Service pits may be excavated with a minimal surface footprint. The size of pit depends on the depth of excavation and the maneuverability of the excavation equipment in the confined space of the pit. Generally, a service pit need only be 1200 mm (4 ft) in diameter to provide enough space for a worker to disconnect and reconnect the lateral. These pits may be shored using large diameter steel pipe sections, depending on the pit depth.

Pipe Bursting Systems

There are three bursting systems currently used in the North American pipe bursting industry. These include the static, pneumatic, and hydraulic expansion systems. The main difference between each method is the manner in which force is generated and transferred to the host pipe during bursting operations.

Static Method

Static methods burst the pipe using static forces, or forces that are not generated using potential energy. A large pulling force is applied to the cone shaped bursting head through rods, cable, or chain. The bursting head then is pulled through the pipe causing the pipe to fail in tension by the radial force applied to the pipe wall from the cone within the pipe. As the host pipe is burst, the bursting head pushes the broken pipe pieces into the soil as it displaces the surrounding soil, thus creating a cavity for the new product pipe.

The majority of static pipe busting equipment is modeled after high-powered hydraulic jacks, mounted horizontally rather than vertically. The smaller units usually use two hydraulic cylinders to develop the required pulling force, while the larger units usually use four or more. Mounted in the center of the pistons is a mechanism to grab the chain or rod during the pulling operation. As the rod or chain is pulled by the machine, it is disconnected and the gripping assembly moves forward to grab another section of rod or link of chain. This process is repeated until the installation is complete. If cable is used it is usually pulled by a winch.

Pneumatic Method

The pneumatic method of pipe bursting is designed around a bursting head that displaces the soil using a hammering force developed from a compressed air system. Using compressed air, the bursting head is able to develop a hammering rate of 180 to 580 blows per minute (TT Technologies 1997). The cone shaped bursting head is driven through the soil like a nail being driven into a wall. Each blow impacted by the bursting head into the pipe creates an impact load in the pipe, applying a "hoop" stress into the pipe causing it to burst in tension. In addition the hammering action creates force in the longitudinal orientation, causing failure in shear as the pipe is ripped. The shape of the head, combined with the percussive action push the pipe fragments into the soil providing the space necessary for the installation of the product pipe.

With this method of pipe bursting, the bursting head is guided through the pipe with the use of a tensional cable inserted through the pipe prior to bursting. This cable is attached to the bursting head and provides constant pulling tension, through the use of a winch, to keep the bursting head in contact with the host pipe and aligned with its path, as well as assist in pulling the new host pipe into place. The main driving force that allows the progression of the bursting head through the pipe comes form the percussive hammering action of the head itself. Both the air compressor and the winch are set at constant pressure and tension which allows the operation to proceed with little operator intervention until the pipe section is burst. To power the bursting head compressed air lines must be run through the new product pipe

Hydraulic Expansion Method

This method of pipe bursting is defined by the method in which the host pipe is burst. Rather than the pipe being burst from the transfer of a pulling or hammering force radial into the plane of the pipe diameter, the bursting head expands radially fragmenting the pipe from inside. Using hydraulic cylinders, the head expands to burst the pipe, then contracts to allow the winch to pull the cable and advance the head incrementally forward. The winch or pull on the cable does not assist in the bursting of the pipe, but pulls the head to help displace any residual soil formation as well as pull the product pipe into the expanded cavity.

Like the pneumatic pipe bursting system, the hydraulically expanding bursting head requires a power source to provide energy to burst the pipe. In this case, a portable power unit on the surface provides power for the hydraulic cylinders, with hydraulic hoses run down the entire length of the product pipe.

Pipe Considerations

Pipe bursting has been successfully accomplished on a variety of host pipe material types (Poole et al. 1985; Everett 1997). In general, host pipes can be grouped into two categories based on how the pipe fails in the bursting operation: brittle and plastic. Brittle pipe includes cast iron, clay tile, reinforced and non-reinforced concrete, and asbestos cement materials. Plastic pipe includes polyvinyl chloride and high-density polyethylene pipes. The main difference between these two categories is how failure occurs; brittle pipe fails more in tension while plastic pipe fails more in shear and usually requires larger deformations.

By failing in tension, brittle pipe develops high tensile "hoop" stresses around the circumference of the pipe. When the pipe fails the failure is sudden and the fragments are jagged and sharp. When brittle pipe bursts, a sound similar to that of popcorn popping can be heard on the surface in the vicinity of the bursting head beneath. Alternatively, plastic failures are more characterized as "ripping" the pipe apart. This pipe will tend to fail and expand along a seam. As a result, contractors usually assist the bursting of plastic pipe by cutting a notch in the pipe to provide a stress concentration area for the tear seam to start. In general, pipe that fails in a brittle manner requires less force to burst than plastic pipe. The only exception is heavily reinforced concrete pipe that still bursts in a brittle manner. One such burst of a reinforced concrete pipe occurred in Phoenix, Arizona, where 158 m (521 ft) of 610 mm (24 inch) diameter reinforced concrete pipe was burst using the static pipe bursting method and replaced with sectional vitrified clay pipe 610 mm (24 inch) in diameter (Holstad and Webb 1998; Miller 1998).

For product pipe material, there are a number of choices available ranging from sectional concrete or clay, polyvinyl chloride, steel, or high-density polyethylene. Virtually any type of pipe can be installed using the pipe bursting method. Consideration must be given to whether the product pipe is sectional or continuous. It is generally preferred to use continuous pipe for installations since the operation can proceed without stopping to add a section of pipe.

Typically, high-density polyethylene pipe is used for continuous installations (Howell 1995). It is the preferred material for pipe installation due to its high tensile strength, flexibility and its acceptance by municipalities and industry. HDPE pipe is safe to use for fresh and wastewater transport as well as pressurized gas. It is also easy to handle and transport due to its lightweight, and is easily connected into a continuous string using butt fusing processes.

Soil Movements

One of the primary concerns associated with pipe bursting is the interaction of the displaced soil with surrounding utilities and the surface. These soil movements have the potential of damaging existing buried lines and structures around the line being burst. All methods of pipe bursting displace soil in the process of bursting even if the replacement pipe is the same size as the existing pipe. Soil displacement can be attributed to the larger diameter of the bursting head as well as the thickness of the existing pipe wall. If the host pipe is upsized, the amount of soil displacement is greatly increased. In general, the soil properties and installation characteristics of the original pipe determine the magnitude and effect of ground movements.

To determine the effects of pipe bursting on utilities in close proximity to the pipe being burst, Leach and Reed (1989) conducted a series of field bursting trials at the Water Research Center. They installed cast iron and clay pipes of various diameters at varying depths, then burst these pipes and replaced them with high-density polyethylene pipe with varying upsizing factors. To determine the effects of the soil interaction on adjacent utilities, instrumented ductile iron pipes were buried both perpendicular and parallel to the pipe being burst. Additionally, British Gas conducted monitoring programs on contract work in a similar manner to that undertaken in the controlled field trials. The data collected was used to develop proximity charts to determine safe distances for buried utilities and the effects of surface heave.

Swee and Milligan (1990) conducted laboratory tests on the effect of soil characteristics on the amount of soil movement observed with pipe bursting. Laboratory tests simulating field replacements of scaled down bursting operations were conducted using sand, clay, and a combination of sandy clay backfill in tanks that had transparent walls to observe the ground movements associated with the displacement process. Using positional markers in the cut away sections, time lapse photography was used to determine the vector displacement of the soil particles during bursting. The different soils enabled the measurement of displacements varying with drained and undrained soil conditions as well as air void content and density. These tests provided information related to ground movements for bursts conducted in cohesionless and cohesive soils, and determine a predicted zone of influence. Additional research pertaining to ground movements relating to this work was conducted by Heinz et al. (1992).

The Trenchless Technology Center at Louisiana Tech University conducted a study on the ground movements associated with pipe bursting to obtain a safe distance for utilities from the replacement pipe. Ground movements were measured in terms of vibration and permanent vertical displacements at different depths and offsets from the pipe burst. The study found that the vibrations caused by the bursting process are generally less than the threshold for cosmetic cracks to initiate damage to buried structures (Atalah et al. 1997).

Though numerous studies have been performed in various ground conditions and with varying utility configurations, it may be best to exercise caution when bursting in close proximity to buried utilities. Perhaps the best method to protect utilities that cross the bursting operation is to spot excavate at the planar intersection of the pipe and remove the soil from around the utility. In this manner, localized stress concentrations can be dissipated and the burst more confidently performed.

Upsizing

One of the primary reasons one would favor pipe bursting over other methods of pipe rehabilitation is its ability to upgrade the capacity of the line by increasing its diameter. The degree to which a line can be upsized is related to the soil conditions, geometry of the installation, and the amount of soil that must be displaced and compacted. As the upsize factor increases, the amount of soil that must be compacted increases. Subsequently, this increases the amount of force required to push or pull the bursting head through the host pipe. Additionally, as the length and diameter of the new product line increases, so does the amount of force or energy to pull in the product pipe to overcome the friction of the soil around the pipe (Gokhale et al. 1996). As the length of the installation increases, the factor to which the host pipe can be expanded decreases as shown in Figure 3.

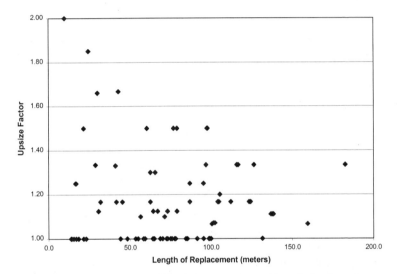

Figure 3. Relationship between Upsize and Installation Length

Data collected from 85 installations performed using the static pipe bursting method demonstrate the trend illustrating that the upsize factor greatly influences the length that can be installed. For example, if one were interested in bursting 100 m (300 ft) of pipe, it may be expected that the upsize be less than 1.5 times that of the original pipe diameter. Subsequently, if only 50 m (150 ft) were to be burst, one may expect to achieve an upsize close to 1.7. Soil conditions, original diameter of the host pipe, and material are major contributing factors in determining the degree to which a pipe may be upsized. Therefore, Figure 3 is only intended to provide pre-design guidance.

Project Planning

Unlike traditional vertical or horizontal construction projects where project planning and management is a constant task through out the project, almost all of the planning for a pipe busting project occurs prior to bursting. The bursting component plays a minor role in the total project.

Typical pipe bursting projects range in installation length between 100 to 200 m (300 to 600 ft). This length is generally the distance between manholes in an urban setting. Additionally, bursts over 300 m (600 ft) have also been accomplished.

Consideration must be given to the arrangement of the machine and insertion pits, since the most time consuming operation in the pipe bursting process is the setup of the machine and the excavation of pits. Therefore, the number of setups and amount of excavation should be reduced. Prior to bringing equipment on site, the planner must consider the arrangement of pipe sections and manholes. It is best to use machine pits more than once to minimize moving the equipment and setting up the power plant. If a section of pipe that is being burst is a continuation or is in line with another section of pipe joined with a manhole, that manhole should be a machine pit. With this arrangement after the first section of pipe is burst, rather than the rods or chain being disconnected as they are pulled, the rods can be shunted (or chain pulled) down the next section of pipe to be burst. This increases the productivity of the operation by shunting rods (or chains) down one section of pipe while bursting another, eliminating removal and reconnection of rods for shunting them down the next line segment.

For example, if the line between manhole (MH) 93 and 98 (Figure 4) were to be replaced by pipe bursting, the machine and insertion pits would need to be located such that they could be used more than once during the burst. For this scenario, the line is considered to be a concrete gravity storm sewer, flowing from west to east. During the installation of the line, it is best to pull the pipe down the grade to keep the proper grade when complete. This project would be best setup by making MH 98 an insertion pit, MH 97 a larger machine and insertion pit, MH 96 a machine pit, MH 95 an insertion pit, manhole 94 a machine pit and MH 93 an insertion pit. Using this setup the machine need only be transported and setup three times, while minimizing the total excavation.

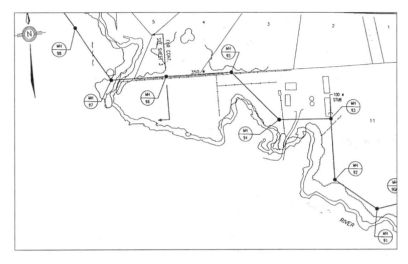

Figure 4. Example Pipe Bursting Project

The installation would begin by inserting the new pipe at MH 98 and pulling the pipe through to MH 97. Manhole 97 would be used as a machine pit for the first section then as an insertion pit to install the second section. Next the pipe bursting equipment would be transferred to MH 96, where it would pull the pipe in through the insertion pit at MH 97, while simultaneously shunting the rods or pulling the chain or cable to MH 95. This is done so that the machine can be turned around and the next section installed without shunting rods or chain again. The product pipe can now be pulled from MH 95 to 96.

For the next installation, the machine would be moved to an excavation at MH 94, and pipe pulled from MH 95 and then MH 93 would complete the installation. In this situation, rods could not be shunted through the section between MH 94 and 93 due to the change in alignment of the line. Alternatively, if chain or cable were used, they could be pulled to MH 93 to facilitate that insertion of pipe. Using this setup procedure the total installation time can be reduced and excavation limited.

Conclusions

Proper design and planning of construction operations can result in significant savings through increased productivity and reduced risk. The unique nature of pipe bursting requires special considerations that are unlike traditional open cut pipe replacement methods. Proper identification of host pipe composition and replacement pipe material greatly influences the method of bursting employed. Site logistics planning, in conjunction with proper equipment selection, should result in improved productivity and successful installations.

References

Atalah, A., Sterling, R., Hadala, P., and Akl, F. (1997). "The effect of pipe bursting on nearby utilities, pavement, and structures." *Trenchless Technology Center Technical Report*, Ruston, Louisiana.

Committee on Construction Equipment and Techniques. (1991). "Trenchless excavation construction methods: classification and evaluation." *J. Contr. Engrg. and Mgmt.*, ASCE, 111(3), 521-536.

Everett, J.G, "Pipebursting." *Construction Business Review*, McLean, Virginia, Vol. 7, No. 1, 10-11.

Gokhale, S.B., Falk, C., and Stein, D. (1996). "A soil-structure interaction model for the dynamic pipe bursting system." *No-Dig Engineering*, Vol.3, No.4, 5-8.

Heinz, H.K., Cassie, J.W. and Evans, P.A. (1992). "Soil displacements associated with pipe bursting." *Proceedings of the 10th Annual Canadian Tunneling Conference*, Banff, Alberta, 317-327.

Holstad, M. and Webb, R. (1998). "Pipe bursting pilot project utilizing 24-inch VCP." *Proceedings of No-Dig '98*, Albuquerque, New Mexico, 542-550.

Howell, N. (1995). "A polyethylene pipe philosophy for pipeline renovation." *Proceedings of No-Dig International '95*, Dresden, Germany.

Leach, G. and Reed, K. (1989). "Observation and assessment of the disturbance caused by displacement methods of trenchless construction." *Proceedings and of No-Dig '89, Fourth International Conference and Exhibition on Trenchless Construction for Utilities,* London, United Kingdom, 67-78.

Miller, P.J. (1998). "First large diameter clay pipe pulled and pushed in Phoenix." Trenchless Technology Magazine, Peninsula, Ohio, July, 30-32.

Poole, A.G., Rosbrook, P.B. and Reynolds, J.H. (1985). "Replacement of small diameter pipes by pipe bursting." *Proceedings of No-Dig International '85*, London, United Kingdom, 147-159.

Saccogna, L.L. (1997). "Pipe bursting saves the day." *Trenchless Technology Magazine*, Peninsula, Ohio, September, 28-29.

Swee, J.L.K, and Milligan, G.W.E. (1990) "Pipebursting: model tests." *Proceedings of No-Dig International '90 Osaka*, Japan, H.3.1 – 8.

Thomas, A. (1996) "'Push-pull' pipebursting restores sewer at thunderdome." Trenchless Technology Magazine, Peninsula, Ohio, September, 36.

TT Technologies, Inc. (1997). "Grundocrack: pneumatic pipe bursting system." *Company Brochure*, Aurora, Illinois.

Columbia Slough Consolidation Conduit Tunnel

Arlan H. Rippe[1], P.E., F.ASCE;
Michael Feroz[2], P.E., M. ASCE; and Frank S. Buehler[3], P.E.

Abstract

Combined sewer overflows (CSOs) have been identified as a significant source of surface water pollution in many American cities. In recent years, the quality of surface water has become a major concern of the City of Portland, Oregon, and regulatory agencies. As a consequence, Portland has undertaken an extensive program to identify and separate the flows from storm and sanitary sewers. The Columbia Slough Consolidation Conduit (CSCC) tunnel is a key element in alleviating this problem in Portland. It is 3.7 m (144 in) diameter and 2,548 m (8,361 ft) long, underground conveyance and storage tunnel. The subsurface conditions disclosed by the explorations were considered to present a challenge for conventional tunneling. As a consequence, the City set up the project to share the construction risk. This paper describes the design and construction of this soft ground tunnel project, with shallow ground cover and close proximity to an existing old concrete sewer. Also described are the means of sharing risks between the owner and the contractor.

Introduction

The CSCC is a key element of the system that is being constructed to alleviate the CSO problem in north Portland. This project is being constructed in three segments. Segment 1 is 827 m (2,712 ft) long and 3.7 m (144 in) inside diameter, and has been constructed as cut and cover reinforced concrete pipe (RCP). Five sections of Segment 1 were installed by boring and jacking under the Union Pacific railroad and Columbia Boulevard. Segment 2, the central part of CSCC, is 2,548 m (8,361 ft) long and 3.7 m (144 in) diameter cast-in-place concrete soft ground tunnel. Segment 3 is 2,118 m (6,950 ft) long and 1.8 m (72 in) inside diameter RCP and is being installed as cut and cover.

---

[1] President, Squier Associates, Inc., 4260 Galewood Street, Lake Oswego, OR 97035
[2] Resident Engineer, CSCC Tunnel, Sverdrup Civil Inc., 5001 N Columbia Blvd, Portland, OR 97203
[3] Construction Manager, CSCC Project, City of Portland Bureau of Environmental Services, 5001 N Columbia Blvd, Portland, OR 97203

Preliminary investigations for the initial selection of possible conduit routes were performed by Woodward-Clyde Consultants (WCC) along a corridor centered on Columbia Boulevard (refer, Vicinity Map, Figure 1), a highly traveled industrial thoroughfare in north Portland. Preliminary and final tunnel design was completed by a team led by KCM, Inc., with geotechnical engineering and environmental site assessments performed by Squier Associates, Inc., and assisted by Fujitani, Hilts and Associates.

Presented in this paper is a brief description of the geological and geotechnical conditions that were considered in the design and construction of the soft ground tunnel. Important interpretive elements, such as stratigraphic relationships, site geology and geologic materials, physical properties and engineering characteristics of soils, and ground water conditions along the alignment for tunnel design and support are described.

Figure 1 - Vicinity Map

Field Explorations

The geotechnical field explorations program for final design consisted of 39 Becker hammer borings, three closed-bit Becker penetration correlation probes, one mud rotary boring, three hollow stem auger (HSA) borings, three large diameter (0.9 m, 3 ft) auger boreholes (LDB), three large test pit excavations, and eight hand-auger borings. The closed-bit Becker holes were drilled adjacent to holes made with the open-bit Becker drill to correlate the penetration resistance of the two drilling techniques along with the standard penetration test results. A summary of the correlations is shown in Figure 2.

Vibrating wire piezometers were installed in ten borings. The large diameter auger borings and test pit excavations were made to supplement the ground performance data obtained from the previous Becker drill borings, especially with respect to representative gradation and maximum particle size, and to observe the ability of the ground to stand unsupported. The hand-auger borings were completed for environmental soil sample collection. Seismic velocity survey profiles along portions of the alignment were completed in order to evaluate and interpret conditions between the bore holes and test excavations.

Subsurface material samples were collected by split-spoon, thin-walled Shelby tubes, and bulk sampling methods. Geotechnical field observation consisted of initial soil classification, Standard Penetration Test (SPT) blow counts (N-values), Becker hammer penetration resistance blow counts, drilling penetration rates, ground water depths, and location of sand layers, cobble deposits, and boulders.

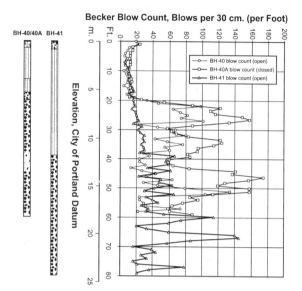

Figure 2 - Becker Blow Count vs. Depth

Geologic Setting

Geologic Units: Topographic elevations within the CSCC project area range from 11.6 m (50 ft) above City of Portland Datum (CPD) at the west end to 27 m (81 ft) along the central portion of Segment 2.

Geology along the CSCC consists of recent river alluvium overlying unconsolidated, fine-grained to coarse-grained catastrophic flood deposits, which in turn, mantle partially lithified Troutdale Formation sediments. A typical geologic profile along the tunnel is depicted in Figure 3. The recent alluvium is composed of variable amounts of fine sand, silt, and clay in continuous to discontinuous units. The thickness of the sediments overlying Columbia River Basalt bedrock exceeds 460 m (1500 ft) in the deepest portions of the basin. The recent deposits range from 6 to 18 m (20 to 59 ft) thick along the Columbia Slough, thinning towards the south. These units are hydraulically connected, to a limited extent, with the Columbia Slough, and consequently, the Columbia River.

Pleistocene-age catastrophic flood deposits are present across the entire CSCC project, overlying the older Troutdale Formation. The deposits were emplaced by multiple high energy turbulent floods (the Missoula Floods) that came down the Columbia River during the glacial periods. In the CSCC project area, they underlie the recent alluvium. The flood deposits range from silt and fine sands, to sandy gravel with discontinuous layers containing cobbles and boulders. The flood deposits can be separated into two main units: an upper fine-grained unit of sandy silt and silty sand (*fine-grained flood deposits*), and a lower coarse-grained unit of sandy gravel with varying amounts of cobbles and boulders (*coarse-grained flood deposits*).

GEO-ENGINEERING FOR UNDERGROUND FACILITIES

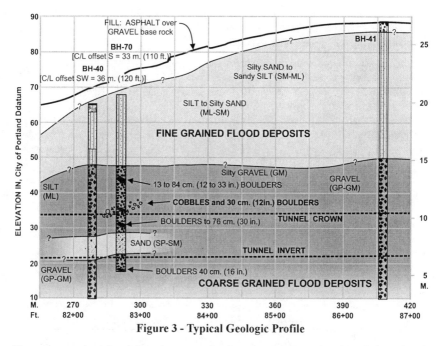

Figure 3 - Typical Geologic Profile

The coarse-grained flood deposits are predominantly gravely sand and sandy gravel with some cobbles (75 to 300 mm, 3 to 12 in, in size) and boulders (greater than 300 mm, 12 in). The cobbles and boulders of the coarse-grained flood deposits are in a sandy gravel matrix. These flood deposits consist of materials from igneous and metamorphic sources. The sands consist, predominately, of minerals derived from basaltic and granitic rocks, with minor components derived from metamorphic quartzite. The gravel consists predominantly of pieces of basalt with minor amounts (10 - 15%) quartzite. Cobbles and boulders consist predominantly of basalt with 5 to 10% quartzite. Quartz minerals and quartzite pieces are the hardest components of the coarse-grained deposits, with a relative hardness of approximately 7 on the Mohs scale. Feldspar and mafic minerals within the sands range in hardness from 5 to 6 on the Mohs Scale. Basalt hardness, and associated strength, is related to the mineral composition and freshness of the rock pieces. Basalt pieces found in the CSCC alignment are slightly weathered, with unconfined compressive strength values ranging from 170 to 490 MPa (25,000 to 71,000 psi).

Ground Water: Ground water hydrogeology of the CSCC consists of an alluvial gravel ground water aquifer, overlain by shallow (perched) ground water. Ground water levels were monitored in the piezometers and monitoring wells installed within the CSCC project area. Some of these were installed during a preliminary phase of investigation. Ground water generally flows from upland areas, toward the major discharge points in the Portland basin, such as the Columbia and Willamette Rivers. Based on about two years of ground water level monitoring, excavations and tunneling in Segment 2 were not expected to encounter the regional ground water table. However, local perched ground water was expected, especially during the seasonally wet periods.

Seismicity: An earthquake hazard evaluation for the CSCC project was based on published regional geologic information and site-specific data. Most of the elements of the CSCC were designed in accordance with the Uniform Building Code (UBC, 1994) Seismic Zone 3. The earthquake hazard evaluation addressed the seismicity of the site and, in particular, the type and characteristics of design earthquakes. Based on this information, the relative risks of liquefaction, lateral spreading, and seismically induced ground settlement were evaluated. Localized areas along the CSCC alignment will be impacted by seismic hazards.

Laboratory Testing

Geotechnical testing included: visual classification, moisture content, Atterberg limits, liquid limit and plastic limit, unit weight, grain-size analysis (both mechanical and hydrometer analysis), shear strength estimates using a torvane shear device, direct shear tests, consolidated-undrained triaxial compression tests, point load tests on cobble samples, and compaction tests. Table 1 presents the geotechnical parameters that were used in the design.

Table 1 - Geotechnical Design Parameters

Geologic Unit	USCS Group Symbol	c - Cohesion	$\phi°$ - Angle of Internal Friction	γ - Unit Weight, Mg/m^3 (lb/CF)
Fill (F)	ML, SM, CL-ML	0	32	2.12 (13.5)
Silty Fine-Grained Flood Deposits	CL-ML, ML, SM	0	32	2.12 (13.5)
Gravely Coarse-Grained Flood Deposits	GP, GP-GM, GM, GW-GM, SP-SM, SW, SP	0	38	2.12 (13.5)

Tunnel Design & Construction Concerns

The coarse-grained facies of the flood deposits are predominantly sandy gravel and gravely sand with layers of packed cobbles and boulders. Subsurface explorations and construction experience, indicate that the particle size characteristics of the deposits can vary abruptly both vertically and horizontally. Boulders ranging in size from 0.3 to 1.2 m (1 to 4 ft) have been reported within these sediments. Large boulders, some greater than 3 m (10 ft), are present at other locations within the Portland area. As a consequence, there was considerable concern that boulders, especially the potential for very large boulders, would cause problems in the advancement of a tunnel. In order to evaluate this concern, a geologic study was completed to quantify the risk of boulders.

Large boulders have been encountered in excavations to the east of the project, toward the mouth of the Columbia River Gorge. These boulders were deposited during the episodic Missoula Flood events, as the floodwater issued from the mouth of the Columbia Gorge. Boulder transport distance and size relationships based on measurements of transported boulders and distance from the boulder source areas have been performed for deposits in the Portland area downstream of the Columbia Gorge mouth (Beeson, 1995). These studies show that the size of basalt boulders decreases to a maximum diameter of about 0.6 to 1.0 m (2 to 3 feet) within about 13 to 16 km (8 to 10 mi) west of the mouth of the Columbia Gorge. The CSCC project area is about 24 km (15 mi) west of the mouth of the Columbia Gorge. Based on relationships of particle size versus downstream distance for boulder deposits, the maximum size range of flood-transported boulders in the CSCC project area was estimated to be 0.6 to 1.2 m (2 to 4 ft) in average size (Baker, 1978).

Ice-Rafted Boulders: Large ice-rafted erratic boulders of granitic or metamorphic lithologies occur in the Portland basin near the 90 to 125 m (295 to 427 ft) CPD elevation. Blocks of ice during the Missoula Floods carried these boulders. Of the known ice-rafted erratics, many have diameters greater than 1 m (3.3 ft). Large ice-rafted boulders could be present in the CSCC project area, but since the project elevation is less than 30 m (100 ft) CPD, the presence of large erratic boulders dropped from melting icebergs was considered highly unlikely.

Boulders Encountered in the Tunnel: The cobbles and boulders observed in the large diameter boreholes and test pit excavations generally had an elongate shape. The minor axes were about one-half of the major axis (dimension). Figure 4 shows that the long dimension slopes downward generally toward the west, in the direction of flow of the transporting water. The cobbles and boulders encountered in the tunneling had the same general elongate shape, and observations at the face of the tunnel boring machine revealed bedding as depicted in the figure. Several boulders were encountered toward the east end of Segment 3 with a long dimension approaching about 1.5 m (5 ft).

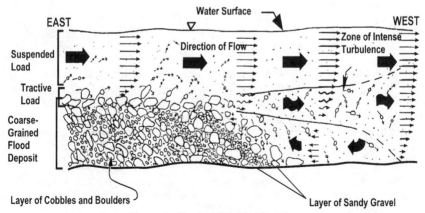

Figure 4 - Typical Boulder Bedding

Geotechnical Design Considerations

Tunnel Liner Design: Segment 2 was constructed as a conventional two-pass tunnel, with an initial support system consisting of steel ribs and wood lagging. The permanent lining is cast-in-place concrete. The subsurface materials disclosed by the borings revealed that mixed-soil face conditions (fine-grained flood deposits overlying the coarse-grained flood deposits) would be encountered in portions of the tunnel. In this regard, the face typically consisted of loose to medium dense sandy silt and silty sand, and medium stiff silt overlying very dense silty to sandy gravel with cobbles and boulders. Layers of sand and silty sand were encountered in the coarse-grained flood deposits. Through parts of the alignment, the tunnel bore was entirely in very dense sandy gravel with cobbles and boulders. Overall, for the design of the initial support system, the subsurface materials were considered granular.

The initial support system of steel ribs and wood lagging is considered sufficiently flexible to develop passive pressure at the springline. It can be analyzed as a simple compression ring subjected to uniform radial pressure. Alternatively, a uniform design soil pressure (expressed in height of soil above the tunnel) for a flexible steel rib and lagging liner can be obtained from references (Proctor and White, 1977). Accordingly, the design loads for temporary supports in tunnels at depths greater than 1.5D for a round tunnel, where D is the tunnel diameter, equal 0.5 Dγ for loose sand ("running ground") above the ground water table. Since much of the tunnel alignment is shallow, and full arching of the soil probably will not develop, a design load equal to Dγ was used, where γ is the design unit weight of soil (γ = 2.12 Mg/m^3, 135 pcf). This design approach did not consider benefits from any stand-up time for "raveling" ground.

The Marston formula (Water Pollution Control Federation Manual of Practice, 1970) was used to estimate the loads on the tunnel for final liner design, as follows:

$P_V = C_t B_t (\gamma B_t - 2c)$ where: P_V = total load on tunnel

C_t = load coefficient

B_t = horizontal width of the tunnel

γ = Unit weight of the overburden soil

c = soil cohesion

The load coefficient, C_t, is a variable corresponding to the soil type and ratio of soil cover to tunnel width and c is a variable for different soil types. C_t varies between 1.0 and 2.0 for the heights of cover along Segment 2. For live loads along Segment 2, considering the minimum depth of cover, adding a surcharge height in the above analysis represented the standard truck load and impact factor. For a cover depth of one tunnel diameter, the live load from a standard truck would be less than 10% of the total dead load, or a surcharge equal to about 0.3 m (1 ft).

Even though an unreinforced concrete lining might be structurally possible, the design of the permanent lining was prudently conservative to allow for future changes and contingencies, such as:

- Changes in loading and activities with time along the busy industrial arterial and collector streets (Columbia Boulevard and Argyle Street);
- Deterioration of the primary support boards;
- The shallow depth of the tunnel; and
- Seismic design issues.

Tunneling Settlements: When a soft ground tunnel is driven, changes occur in the stress regime of the soil. These changes result in settlement at the ground surface above the tunnel, or of structures overlying or adjacent to the tunnel. These changes and the resulting settlements are caused by ground losses. These can occur due to the following:

- Face Losses - soil movement into the face of the tunnel shield from raveling, caving, flowing, running, or squeezing;
- Shield Losses - soil movement into overcutting of the shield as a result of plowing, pitching, or yawing, and from the voids created by overcutters or encountering obstructions such as boulders;
- Tail Losses - soil movement into voids created or left unfilled after the installation of the initial support system.

These various effects are additive, and result in the maximum settlement usually above the crown of the tunnel. Empirical methods are used to estimate potential settlement based upon these causes and related to general soil type. Peck (1969) proposed a relationship for the estimation of volume loss during tunneling. The original proposal has been updated subsequently by more recent tunneling experience (Monsees, 1996). For a single tunnel, the volume of surface settlement is assumed equal to the estimated volume of lost ground. A "settlement trough" is assumed to be created with the bell-shape of a probability curve. In general, the greater the soil cover over the tunnel, the less the anticipated maximum settlement above the crown of the tunnel due to the widening of the settlement trough. The width of the trough is represented by the β-angle, which emanates upward and outward from the springline of the tunnel.

Where tunneling difficulties were encountered, primarily due to encountering cobble and boulder deposits, surficial settlements in soils above the tunnel crown caused damage to underground utility lines. Settlement damage was particularly observed when volume loss exceeded 3% to 5%.

Settlement Limitations: A maximum allowable angular distortion for structures and buildings of 1/600 was used for design. This allowed for an adequate factor of safety, and provides suitable conservatism. This distortion and settlement limitation is for buildings and structures; however, the impact of total and differential settlement on pavements and curbs also requires consideration. To prevent pavement disruption, a less conservative distortion limitation of 1/300 was used for non-critical and low speed pavements. In addition, the impact of settlement trough development on pavement drainage patterns and vehicular ride comfort required consideration. In this regard, for new roadway pavements,

Oregon Department of Transportation limits smoothness to within 7 mm in 4 m (¼-inch in 13 feet), which is approximately equivalent to a 1/600 angular distortion. Where measured settlement was expected or observed to exceed these established predicted limits, then settlement mitigation was required.

Grouting: Various methods of grouting are available to stabilize the ground adjacent to and beneath structures and utilities as a means to protect them from excessive ground movements, including permeation grouting, jet grouting, and compaction grouting. For the CSCC project, both compaction and stabilization (permeation) grouting was considered. Stability analysis indicated that the risk of ground movement and settlement due to loss of face control and caving during tunneling was unacceptable where the tunnel encroached within about 9 m (30 ft) of an existing 2.6 m (102 in) interceptor sewer. Therefore, where the new conduit encroached within this zone, a stabilization grout curtain was specified. A contractor-initiated value engineering proposal was approved that shifted the tunnel alignment laterally away from the existing interceptor, thereby eliminating the requirement for the stabilization grout curtain to protect the existing interceptor. However, the revised alignment moved to the edge of the heavily traveled Columbia Boulevard.

Compaction grouting uses a stiff grout mix injected above the tunnel and below structures to be protected. During grouting, the grout forms a bulb around the injection point and the soil is displaced and compacted. Injection is done through pipes that have been pre-installed. The grout is injected into the zone of loosened ground above the advancing tunnel. By carefully controlling where the grout is injected, injection pressure, volume injected, and when the grout is injected, ground support can be restored before the settlement trough reaches the ground surface or an adjacent structure. Compaction grouting is no substitute, however, for maintaining adequate workmanship, ground support at the tunnel face, and careful installation of the initial support system.

Compaction grouting was specified and accomplished as a means of risk sharing. However, void filling with low pressure grout was also accomplished. This was required where there were nests of cobbles and boulders encountered that caused excessive ground loss.

Geotechnical Considerations For Tunneling

During project predesign and through the 30% design level, a portion of Segment 2 along Columbia Blvd. was considered to be viable for either open cut construction or tunneling. However, tunneling was ultimately selected as the preferred method because of the depth of cover (max. 17 m, 57 ft). Decision factors included: the relative costs of construction, risk to the adjacent existing 2.6 m (102 in) interceptor, and avoidance of significant traffic disruption along Columbia Boulevard. However, the subsurface conditions disclosed in the explorations indicated that tunneling would be difficult. Presented in the following subsections are brief discussions of tunneling methods and equipment used for the CSCC tunnel.

Tunnel Machines: The boulders encountered in the explorations suggested that the tunneling equipment wouldrequire access to the face of the machine to remove boulders that cannot be dislodged or broken into smaller pieces. The fine-grained flood deposit was classified as "slow raveling" and the coarse-grained flood deposit was classified as "fast raveling" to "cohesive running", with the remote possibility of encountering "flowing" sand (The Tunnelman's Ground Classification For Soils System, Heuer, 1974, and Heuer and Virgens, 1987). These classifications suggested that an open-face digger-type machine with hydraulic breasting boards was a viable alternative. However, a wheel-type tunnel machine was also considered to be usable. Due to the extent of cobbles and boulders, this type of tunnel machine was viewed as having a disadvantage, and may be slower, when compared to the open faced shield with breasting boards.

Slurry face tunnel machines and earth pressure balance tunnel machines were considered inappropriate for this project due to the presence of cobbles and boulders, and because access to the face would be difficult. Tunnel machines consisting of shields without the capability to provide at least some face support were considered inappropriate for this project as well because of the considered potential for "flowing" ground.

The final decision for use of equipment type rested with the contractor, who selected the wheel-type Lovat ME181SE machine. The wheel was 4.52 m (184 in) diameter. It had four gage cutters and four flood doors. The TBM was powered by three electric water cooled motors, each 186 kW (250 hp), which drove the cutterhead at variable reversible speeds.

Steering and Propulsion: The presence of the cobbles and boulders in the periphery of the leading edge of the tunnel shield caused some difficulty in maintaining line and grade. In addition, the propulsion system forces needed sufficient stability from the initial support system to provide reaction to push the machine forward. Typically, the thrust was transmitted axially through the wood lagging until dissipated fully by friction between the lagging and the ground. Because of disturbance at the face, the ground was loosened around the initial support system. As a consequence, there was some difficulty in maintaining this stability. On occasion, some longitudinal restraints were required between adjacent ribs in order to provide sufficient reaction to thrust the TBM ahead.

Instrumentation

Instrumentation was installed to monitor the performance of the ground during and after construction of the conduit and other appurtenant facilities. The instrumentation program was designed in order to assure that the work is proceeded as expected, or to identify performance that was different from predicted. The instrumentation program included the following: Fixed Point End Anchors, Free End Drive Point Anchors, Structure Surface Settlement Points, Crack Gages, Slope Inclinometers and Probe Extensometers, and Tilt-meter Plates.

Risk Sharing

As described above, the tunneling was considered to be difficult. Project risks are usually handled several ways, including risk assignment, risk avoidance, and risk sharing. The CSCC tunnel traverses developed areas, to avoid settlement damage to structures, and the resulting claims, approaches of risk sharing were adopted. The contract provided implementation of compaction grouting through the areas that had been identified to be susceptible to damage from "unavoidable" tunneling settlement. In this way, the City was assured that the successful bidder had sufficient budget to accomplish the work to the City's satisfaction without delays and claims. Unfortunately, many projects with disputes can be attributed to construction bids that were "unreasonably optimistic" in order to "win the bid". The successful contractor should be entitled to fair compensation for taking necessary risks. Bids that have too high of a contingency in them penalize the Owner.

Risks are generally assigned to the party that can effect greatest control over them. Attempts are made to avoid risks by thorough investigations and designs. The risk sharing concepts that were adopted for this project included a bid package based on a Geotechnical Baseline Report (GBR). The GBR formed the basis for the bidders to select their construction means and methods compatible with subsurface conditions. Inclusion of geotechnical and environmental data helped the bidders to reduce risks for the unknown. As a result the winning bid was about $22.6 million, some 13% below the Engineer's Cost Estimate. The basic conclusion is that money spent on baseline data during the design phase can save a substantial sum during the construction phase. In addition, a formal partnering workshop was included at project startup. Since all complex projects can result in interpretive disagreements, a disputes resolution board (DRB) was convened. Project insurance is another technique that was used to assign and share risks. The owner controlled insurance program (OCIP) is an excellent method to reduce the tendency for "finger pointing" if problems do arise. This removes another source of uncertainty, and hence, risk.

Conclusions

The detailed instrumentation program implemented during construction phase provided the data and the opportunity to immediately mitigate damages due to ground movements. Lateral and vertical soil displacements measured through the instrumentation program showed soil displacements to be greatest over the crown of the shield which resulted from the void created by dislodging and removal of cobbles and boulders from the periphery of the tunnel bore. Ground loss was also observed behind the shield due to untimely and/or incomplete expansion of the initial support system. Immediate void fill grouting program behind the tunnel shield, after the shove, stabilized the ground movements.

In areas where the ground cover was shallow, the settlement trough was much narrower at the surface when compared to a similar area with deep ground cover. In deep ground cover areas, the settlement was maximum at the centerline of the crown and spread out to about 6 m (20 ft) on either side of the centerline of the crown. A correlation between settlement and ground loss from muck car count showed ground loss to be about 5% of the tunnel volume. This would have resulted in more than 100 mm (4 in) of settlement without implementation of the grouting program

References

Baker, V.R., (1978) "Paleohydraulics and Hydrodynamics of Scabland Floods", in Baker, V.R. and Nummedal, D., 1978, The Channeled Scabland, National Aeronautics and Space Administration, pp. 59-79.

Beeson, Marvin H. *Distribution of Missoula Flood Transported Boulders in the Portland Basin*, letter to Squier Associates, September 23, 1995.

Bickel, J.O., Kuesel, T.R., King, E.H., (1996) *Tunnel Engineering Handbook*, Second Edition, Chapman and Hall, New York, p. 544, 1996.

Heuer, R.E. and Virgens, D.L. (1987) Anticipated behavior of silty sands in tunneling, *Rapid Excavation and Tunneling Conference*, Louisiana, New Orleans, Society of Mining Engineers, Inc., Littleton, Colorado, v.1, p. 221-237.

Heuer, R.E., (1974) Important ground parameters in soft ground tunneling, *Subsurface Exploration for Underground Excavation and Heavy Construction*, New England College, Henniker, New Hampshire, American Society of Civil Engineers, NY, p. 41-55.

Monsees, J.E., (1996) "Soft Ground Tunneling" Chapter 6, *Tunnel Engineering Handbook*, Ed. Bickel, et al, 1996.

Peck, R.B. (1969) Deep excavations and tunneling in soft ground. *State-of-the Art Vol., Seventh International Conference on Soil Mechanics and Foundations*, Mexico City.

Proctor, R.V. and White, T.L., (1977) Earth Tunneling With Steel Supports, Commercial Shearing Company, Youngstown, Ohio, 247p.

Water Pollution Control Federation, (1970) Design and Construction of Sanitary and Storm Sewer, WPCF Manual of Practice No. 9, Washington, D.C., p. 201.

NON-HALITES AND FLUIDS IN SALT FORMATIONS, AND EFFECTS ON CAVERN STORAGE OPERATIONS

Robert L. Thoms[1] and Richard M. Gehle

Abstract

Non-halites occurring in salt formations that may affect cavern storage include both "insolubles" and "hypersolubles", relative to halite. Insolubles, such as shale and anhydrite, may cause falls that damage hanging tubing in caverns. Hypersolubles, such as potash, may cause highly irregular cavern shapes and possible loss of cavern containment. Fluids include both gases and liquids. Intrusion of formation gas into caverns has contaminated or increased the flammability of stored materials. Examples of non-halite and fluid occurrences that have significance for cavern storage operations are described. Currently available techniques for controlling the potentially negative effects of such occurrences are discussed and summarized.

Introduction

Storage in salt caverns has proved to be highly successful wherever suitable host salt formations exist. However, salt formations vary in character, and some are more suitable for storage applications than others. An "ideal" salt formation for the construction and operation of storage caverns would be composed entirely of halite (NaCl). Unfortunately, such formations do not exist.

The configuration of salt formations ranges from near-horizontal bedded deposits to near-vertical domes, with anticlines falling in between these extremes. See Fig. 1. Salt formations vary in composition from relatively pure rock salt (halite) in some domes, to distinct layers of halite and other minerals in most beds. All salt deposits incorporate some "non-halite" minerals. These may be classified into two groups, "insolubles" or "hyper-solubles", according to their

---

[1] AGM, INC., PO Box 10358, College Station, Texas, 77842

relative solubility in halite saturated brine. Fluids are also generally present, to some extent, in both gaseous and liquid phases in salt formations.

The geology of the host salt formation, including the presence of non-halites and fluids, is an important consideration in the construction and operation of storage caverns. The objective of this paper is to describe the occurrence of non-halites and fluids in salt formations, and to cite any associated negative effects on construction and operation of caverns. Current and proposed methods for mitigating such effects will also be noted.

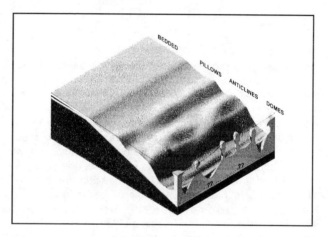

Figure 1. Types of Salt Formations

Non-Halites and Fluids that Occur in Salt Formations

The previously noted two-group classification of non-halites into insolubles and hypersolubles is based on considerations that arise mainly in the construction of storage caverns by solution mining of rock salt. Insolubles frequently occurring in salt formations include anhydrite, gypsum, shale and clay, dolomite, and sandstone; and hypersolubles include potash salts such as sylvite, carnallite, and bischofite.

Fluids that occur in salt formations include both gases and liquids. They often occur together, however considerably more attention has been directed to the occurrence of gases in salt than to liquids. To some extent this is because gas outbursts were a major safety issue in rock salt and potash mines for some time before gas intrusion was noted as a potential concern for storage in salt caverns (Baar, 1977; Martinez and Thoms, 1977). Frequently encountered gases include methane, carbon dioxide, hydrogen sulfide, and nitrogen. Of these, methane generally causes the most concern because it is flammable and not uncommon in salt formations. Hydrogen sulfide has reportedly caused fatalities in German mines (Baar, ibid.), but has not been reported to occur in significant quantities in Gulf Coast salt mines (Belchic, 1960; Hoy, et al., 1962). Liquids include connate (formation) brines and occasional hydrocarbons (Vaughn, 1925). Connate brines tend to be rich in calcium, magnesium, and/or potassium, relative to meteoric waters. A technique for distinguishing between connate and meteoric brines on the basis of their isotopic character has been developed by Knauth (1980).

Forms of Occurrence

Occurrence of Non-halites
The distribution and configuration of non-halites in salt formations depend upon the initial depositional configuration and subsequent tectonic history of the host formation. Remnants of initially continuous beds of non-halites have been used as "markers" to analyze the tectonic and hydrologic history of salt formations (Talbot and Jackson, 1987; Richter-Bernburg, 1987).

Formations that have undergone little movement since deposition tend to be regularly interbedded, and occur in more or less continuous layers. The distribution and continuity of non-halites in formations that have undergone large movements and deformations depends upon their relative ability to flow plastically with rock salt. Members of the insolubles group are generally less plastic than halite, and thus tend to be locally discontinuous where the rock salt in formations has undergone considerable movement, e.g., in the crests of domal salt stocks and relatively thick anticlines.

Kupfer (1967, 1976) has postulated that more pure halite is found near the tops of domal stocks because it is more mobile than insolubles, e.g., anhydrite. The less mobile insolubles, where present in domal and anticlinal crests, occur as discrete blocks and particles. These were broken away from the primary beds, and rafted along with the salt that flowed into the anticlines and domes. The resulting structures can be complicated. For example, anhydrite boudinage can be observed in rock salt mines in Gulf Coast domes (Muehlberger, 1960). The near vertical and folded layering of anhydrite that is visible in rock salt mines in the U.S. Coastal Basin of domes can be used to interpret salt movements within the mature stocks (Kupfer, 1976, ibid.).

In contrast to insolubles, hypersoluble potash deposits in mines in salt domes tend to exhibit more deformation over time than halite. For example, old mine roads through potash-rich zones have completely closed in some German mines, whereas roads through mainly halite have not. (An example of a closed room in potash was pointed out to the senior author during a field trip to the Asse Mine in 1984.) Thus potash deposits may also be strongly deformed where extensive folding and movement of rock salt (halite) layers have occurred.

Potash deposits are present to some extent in a number of U.S. domes. A layer of sylvite was found next to a layer of sandstone in the Cote Blanche Mine in South Louisiana (Golder Associates, Appendix II by Kupfer, 1978). Kupfer suggested the sandstone was a wind blown deposit that occurred with the precipitation of the sylvite in an arid environment. The Palangana Dome in South Texas exhibits relatively abundant potash deposits that have been explored and mapped in some detail by Hofrichter (1968). Apparently the deposits were not considered rich enough to compete with potash production from the bedded salt formations in the western U.S. and Canada.

The "caprock" that is present over many domes in the U.S. Coastal Basin begins as an accumulation of insolubles, e.g., anhydrite, resulting from dissolution of the underlying salt stocks by circulating groundwater (Seni, 1987). The thickness of caprock can be used to estimate the amount of salt dissolution required for its formation. For example, Martinez (1991) has estimated that the dissolution of approximately 4 (vertical) miles (6.4 km) of salt stock wouild be required to form a caprock of 1000 ft (305 m) thickness if the salt contained 5% anhydrite. Mature caprock can incorporate primary anhydrite, secondary gypsum and calcite, and other minerals such as sulfur, gas, and hydrocarbons (Halbouty, 1979). Caprock related issues have impacted cavern storage in a number of Coastal Basin domes, but they are not considered in this paper.

Non-halites and formation fluids tend to occur more frequently within "anomalous zones" (AZs) in the salt stocks of Gulf Coast domes (Kupfer, 1980; Neal, et al., 1993). AZs can occur in linear trends extending over 1000 m, thus qualifying as "megascopic scale" features in the salt stocks (Jackson and Talbot, 1986). With reference to Fig. 2, Kupfer (1976), and Talbot and Jackson (1987) have proposed different geologic models for the occurrence of AZs in salt stocks.

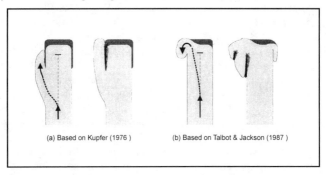

(a) Based on Kupfer (1976)   (b) Based on Talbot & Jackson (1987)

Figure 2. Anomalous Zone Development

The model of Kupfer suggests that AZs exhibit remnants of insolubles and fluids that were trapped against the flanks of a more mature stock by younger salt "spines" flowing up and over them. With additional salt movement, the once exterior insolubles and fluids were incorporated into boundary AZs between spines that comprise the current salt stock. By contrast, the model of Talbot and Jackson implies that AZs include insolubles and fluids that were trapped against dome flanks by toroidal flow of salt coming from above. This type of salt flow has been identified in some German domes, and similar movements may have occurred to some extent in the Cote Blanche Dome (Molinda, 1989.) The model of Talbot and Jackson poses a possible concern for storage caverns that intersect AZs near dome flanks; i.e., it suggests that boundary AZs may be connected at depth to relatively porous formations outside of the salt stock.

Occurrence of Fluids

Encounters with brine at shallow depths in at least two salt mines in Gulf Coast domes have been associated with major AZs that transected the stocks (Kupfer,1980; Jacoby, 1977). Anomalous Zones were initially referred to as "Shear Zones" by Kupfer (1976). Some brines encountered in mines initially appeared to be connate, but with time became increasingly meteoric in character, thus indicating the development of a leak from outside the salt stock.

As noted previously, gas encounters in salt mines can take the form of violent outbursts of large quantities of granular rock salt and flammable gas, i.e., methane. Gas outbursts were extensively studied first in Europe, perhaps because their potentially disastrous effects were initially experienced there (Gimm and Pforr, 1964; and Baar, ibid.). Gas outbursts, and other anomalous features causing problems in U.S. Gulf Coast mines were described by Kupfer (1980) at the Fifth International Salt Symposium held in 1978. At the same meeting, Thoms and Matinez (1980) focussed on gas outbursts ("blowouts") and their cause. A disastrous gas outburst and explosion subsequently occurred the following year in the Belle Isle Mine in southern Louisiana (Plimpton, et al., 1979). This was followed by a series of U.S. Bureau of Mines studies on gas encounters in Gulf Coast salt mines (Iannacchione, et al., 1982; Molinda, ibid.). A comprehensive report on gas releases from salt has recently been produced by Ehgartner, et al., (1998).

Gas outbursts in Gulf Coast mines typically leave near vertical openings, or "chimneys", in the salt "roof" above the mining level. The measured height of an outburst chimney in the Jefferson Island Mine in southern Louisiana was 278 ft (84.7 m) (Mine Safety and Health Administration, 1978). The primary force that drives outbursts has been discussed by a number of investigators (Ehgartner, ibid.; Mahtab, et al., 1983). For example, the weight of the pressurized salt, once exposed by blasting, has been speculated to be the primary driving force for outbursts, since outburst chimneys did not appear to occur in mine floors. On the other hand, Thoms and Martinez (Ibid) proposed that coupled gas pressure and geostatic stress in the salt, along with shock waves from blasting, were the primary factors affecting outbursts.

Figure 3 depicts an outburst chimney and "basement alcove" in the Cote Blanche Mine in South Louisiana that was sketched from memory by the senior author. The chimney was viewed first, about 1977, and then the combination of chimney and alcove was viewed later, about 1981. The chimney resulted from a gas outburst that occurred during the first-pass ("low room") excavation of approximately 8 m height. A photo of this chimney comprises Fig. 4 in Thoms and Martinez (Ibid). The basement alcove resulted from an outburst that occurred some months later during "benching" of an additional 17 m of salt beneath the initial mining level.

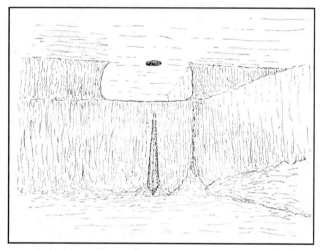

Figure 3. Gas Outbursts in Cote Blanche Salt Mine

The top of the alcove began about 1 to 2 m beneath the low room floor, and enlarged noticeably with depth to the bottom of the bench level. At the time of their respective occurrences, neither of the outbursts appeared to penetrate downward into the floor salt. The shape of the alcove suggests that a "bleed-off" of gas had occurred into the low room prior to benching out the underlying salt floor. However, some gas apparently remained contained in salt within only 1 to 2 m of the open surface of the floor. Gas containment by the salt obviously increased rapidly with confinement below the low-room mine opening, since the diameter of the alcove depicted in Fig. 3 increased with depth from the upper mine floor.

The depth dependent size of the outburst that caused the chimney alcove of Fig. 3 is consistent with observations on a larger scale in Gulf Coast mines, i.e., the severity of outbursts increases with depth of mining. These observations tend to corroborate the proposal by Thoms and Martinez (ibid) that gas pressure is the primary driving force for outbursts in Gulf Coast salt domes. The effect of gravity plays a major role in "clearing the throat" of chimneys so that the outburst can continue to "exhale" into the mine opening (Gimm, ibid.). Different factors may dominate at other sites, and especially in other salt formations.

The previous example of tight gas containment by salt in the Cote Blanche Mine illustrates that the violent effects of outbursts in Gulf Coast rock salt mines may not be effectively mitigated by drilling gas-drain holes into the formation ahead of advancing faces (Baar, ibid; Plimpton, et al., ibid.). That is, drainage of the gas-rich zone was limited to the salt in close proximity to the extensive surface area of the floor. The smaller drainage areas offered by wall surfaces of holes would be even less effective for draining the large volumes of gas-rich salt typically involved in outbursts. On the other hand, this same containment property makes salt caverns satisfactory for gas storage, which is of major interest for this Conference.

In situ fluid pressures in salt have seldom been measured. However, some sparse data exist through measurements made in boreholes drilled and sealed-off from mines in salt formations. For example, gas pressures of 1200 to 1250 psi (8.273 to 8.618 MPa) were measured in boreholes drilled from the now abandoned Belle Isle Mine, which was located at a depth of 1225 to 1253 ft (373.4 to 381.9 m) in southern Louisiana (Plimpton, et al., 1979). This gas pressure was of lithostatic magnitude. (The Belle Isle Mine was intentionally abandoned and flooded on November 15-18, 1985, approximately 5 years after the Jefferson Island Mine was accidentally flooded on November 20, 1980).

With regard to liquid pressures in salt, a value of 600 psi (4.14 MPa) was measured in calcium chloride brine encountered in an exploratory borehole in the Jefferson Island mine at a depth of 1300 ft (396.2 m) from the ground surface (Kumar, 1979). The brine appeared initially to be connate in character, however its pressure was of hydrostatic (or brine head) magnitude, which implied connectivity to permeable formations outside of the salt stock. The borehole was eventually sealed with a packer, indicating that the brine inflow had not stopped with time, and that the intercepted "brine pocket" was connected to sources outside of the salt stock.

Other boreholes in salt have encountered brine that eventually ceased flowing, indicating containment of the source within the salt formation (Hoy, et al., ibid.). Unfortunately, no data were found on measured fluid pressures for such cases. Based on the gas pressure measurements in the Belle Isle Mine, it appears likely that fluid pressures in brines and other liquids confined completely within Gulf Coast salt stocks would be of lithostatic magnitude.

Gas encounters in salt formations are strongly site specific in character. They vary in frequency and magnitude between salt basins, and within individual formations. For example, working rock salt mines were present until late 1980 in each of the "Five Islands" domes in southern Louisiana. The five domes trend linearly from southeast to northwest over a distance of about 72 km (45 mi). The salt stocks of these domes can be visualized as diapiric fingers extending vertically upward from a common extensive salt ridge at depth (Kupfer, 1968). Despite their common source of salt, the mines of the Five Islands have exhibited different characteristics relative to gas encounters. Four of the mines were re-classified as "gassy" by the Mining Safety and Health Administration (MSHA) following the previously noted disastrous gas outburst and explosion in the Belle Isle Mine in 1979. The outburst in Belle Isle reportedly released about 600,000 cu ft (17,000 m^3) of methane (Plimpton, et al., ibid.). By contrast, the Avery Island Mine has never been classified as gassy, and it is located between two other gassy mines in the same Five Islands group of domes. The Weeks Island Mine, located about 13 km (8 mi) to the southeast of Avery Island, is still operating. The Jefferson Island Mine, located about 10 km (6 mi) to the northwest of Avery, was classified as gassy prior to flooding in 1980.

The previous examples provide some background for a few brief comments on why fluids occur in "pressurized pockets" in salt stocks of domes. First, the in situ permeability and shear strength of salt generally varies within the stocks of domes. This is a primary factor affecting the locations and configurations of fluid occurrences. (See Golder Associate, 1977, regarding varying salt permeability in the Cote Blanche salt mine.) A related factor is that fluids are mobile in evolving salt stocks. That is, fluids can migrate over geologic time along relatively permeable zones in deforming salt until stopped by a barrier of less permeable salt. Relatively permeable zones would be more likely to exist in stringers of weaker salts and some non-halites, such as sandstone and potash, interspersed through the salt mass. As noted previously, brine and gas are frequently found together in salt formations, and brine likely plays a role in the mobility of the fluid couple.

Following the previous comments, gas "pockets" in Gulf Coast domes are proposed to occur as zones of gas-rich, and often damp salt that is confined under lithostatic pressure within a sheath of essentially impermeable salt. The "chimneys" formed by outbursts from gas rich zones can vary in configuration from near-spherical shapes, with a diameter about 1 to 10 meters, to the previously noted cylindrical shapes with heights of about 85 m. The latter types of zones contain large volumes of gas, and have been associated with some of the more violent outbursts that have occurred in Gulf Coast mines. Some vertically extensive gas-rich zones may be connected at depth to permeable formations outside of the salt stock, as noted previously.

Long-term migration of fluids through salt stocks cannot be validated through conventional tests because of the time periods involved. However, studies have indicated that gas will flow through salt, provided that the pressure driving the gas is larger than the minimum geostatic stress in the salt (Fokker, et al., 1993). This suggests that gas flow could occur in salt formations that experienced non-isotropic stress fields. Fluid migration over long terms in salt may also occur by means other than flow though effective porosity, e.g., by preferential dissolution (Martinez and Kumar, 1979). Salt creep is enhanced by the presence of brine (Spiers, et al.. 1988), and this too may play a significant role in the long term mobility of fluids in salt. (It should be noted that long-term fluid mobility will not affect the effective containment of salt caverns that have been engineered for storage or disposal operations.)

Effects on Storage Caverns

Effects of Non-Halites

The occurrence of Insolubles in salt formations has a major effect on the solution mining (construction) and operation of storage caverns. Higher percentages of dispersed insolubles in salt increase the amount of energy required to solution mine storage space (Chabannes and Dussaud, 1995). Insoluble particles transported with brine during solution mining will also tend to plug disposal wells (Cole,1992). Insolubles that occur in coherent layers or blocks in salt fomations increase the potential for "falls" during both construction and operation of caverns (Thoms and Neal, 1992). The falls can separate or damage "hanging strings" of tubing placed in the cavern for solution mining or transfer of stored fluids. Layers of insolubles in bedded salts also generally affect the shapes of solution mined caverns because of preferential dissolution of the more soluble salts in the formation. In older brining operations in bedded salts this typically resulted in cavern shapes that resembled inverted Christmas trees, wherein the horizontal portions of "trees" are formed by protruding ledges of insolubles. This effect is reflected in the schematic of a brine cavern incorporated in the letterhead of the Solution Mining Research Institute (SMRI).

The "fall potential" of insolubles that occur in bedded salt caverns depends upon the stratigraphy of the host salt formation. Important site specific factors include spacing, thickness, strength, and tendency of the individual stratum to disaggregate in brine. These factors are typically evaluated during the planning stage of caverns on the basis of logs and core samples from exploratory bore holes. In Gulf Coast salt stocks, insolubles occur in configurations of near vertical "stringers" that may extend over several hundred feet. The stringers are composed mainly of salt, but incorporate relatively high percentages of anhydrite, shale, and/or sandstone. In salt anticlines, the distribution and character of insolubles depend upon the distance traveled from the source beds, thickness of salt in the anticline crest, and plasticity relative to salt.

Falls associated with insolubles have occurred in all types of salt formations, ranging from beds to domes, and are most prevalent during the solution mining of caverns. Once in operation, caverns used for storage of liquids appear to have more operational problems due to falls than do gas storage caverns (compressed natural gas or air). For example, Munson, et al., (1998) have reported on a series of falls that have occurred over a period of years in caverns of the US Strategic Petroleum Reserve (SPR). Evans, et al., (1994) reported that three falls occurred during solution mining, and one during brine removal ("dewatering"), in caverns constructed for storage of natural gas in bedded West Texas salt. However, no falls were reported in a later paper after gas storage was apparently underway in these caverns (Istvan, et al., 1997). The histories of caverns, engineered for gas storage in both bedded and domal salt, would be interesting to compare after operational periods of about 5 to 10 years.

More frequent damage associated with falls in brine "wells" and liquid storage caverns, as compared to gas caverns, may be due in part to the energy transfer properties of the stored media. The energy of pressure waves generated by falls can be transferred through liquids to hanging strings, thereby causing them to swing against cavern walls and be damaged. In gas caverns the energy transfer between pressure waves and hanging strings is less efficient, and direct hits on strings may be necessary to cause damage. The probability of falls causing damage through energy transfer to hanging strings in liquid-filled caverns is thus higher than in gas caverns.

Another factor contributing to falls in liquid caverns is the use of the "brine displacement method", wherein brine is injected into caverns to displace stored fluids. The brine may not be completely saturated, and so dissolution of salt can occur. This increases the potential for falls. The potential for salt falls may increase with time in liquid storage caverns even if no additional brine is injected. Some insoluble ledges (especially shale) will weaken in brine with time and fall. It appears possible that preferential dissolution of cavern walls could also over time due to convection cells within the resident brine in caverns (Lolan, et al., 1998). In theory this could result in an increased potential for falls where salt is being preferentially removed from cavern walls.

The presence of relatively permeable sandstone in salt stocks may contribute to leaks into storage caverns. An example of such a leak in an oil storage cavern of the Louisiana Offshore Oil Port (LOOP), located in the Clovelly Dome in South Louisiana, has been presented by McCauley, et al., (1998). As noted in the previous section, sandstone and/or other anomalous features can be incorporated into boundary Anomalous Zones (AZs) of salt stocks by differential salt flow during dome evolution. One indication of an AZ is a linear depression, or "trough", in the overlying surface of the salt stock (ref). The leaking cavern described by McCauley, et al., appears to be located beneath a trough that runs up the flank of the Clovelly Dome. This suggests that an encounter with a boundary AZ may be affecting this cavern leak.

The presence of hypersolubles in salt formations raises the possibility of loss of control of cavern shape doing solution mining. Some brine caverns in Gulf Coast salt domes have been abandoned because of highly irregular shapes that developed in potash-rich salt (Thoms and Neal, 1992). The main concern was that flow paths might be formed through dissolution of potash stringers that

extended to the boundary of the salt stock, thus causing loss of cavern containment.
Sandstone that is encountered in Gulf Coast salt mines typically exhibits no effective porosity, because potential voids between sand particles are filled with halite crystals. But if sandstone occurs in association with potash, as previously noted in the Cote Blanche Mine, then the voids would be filled with a hypersoluble material. The potash would be readily dissolved out of the sandstone during solution mining of a cavern in halite, leaving behind a sandstone matrix with an induced effective porosity. This may explain why some sandstone stringers in the walls of salt caverns exhibit apparent permeability, as in the previously noted LOOP cavern (McCauley, ibid.).

Effects of Fluids
Formation fluids can produce "bumps" and indications of liquid hydrocarbons during drilling of wells and solution mining of caverns in salt formations. As early as 1925, with reference to the Belle Isle Dome in South Louisiana, Vaughn referred to "the high gas pressure and oil so commonly found within the salt body itself". A well drilled for the purpose of solution mining a cavern in the Sorrento Dome, also in South Louisiana, was abandoned because of high gas pressure encountered in the salt stock that "cut" heavy (dense) drilling mud (Martinez and Thoms, 1977).
Gas in salt formations can pose a long-term concern if it causes persistent gas intrusion into storage caverns. In some cases significant gas flow into caverns from the host salt formation has occurred over a period of several years. Storage of Ethylene in a cavern in the Sorrento Dome was abandoned because of its contamination by abundant intrusion of methane gas. The possible sources for the gas were discussed by Martinez and Thoms (ibid.), but no definitive conclusions were reached. That is, it was not clear if the gas was coming from within the salt stock itself, or from reservoirs outside of the stock.
Ehgartner, et al. (ibid.), and Hinkebein, et al., (1995) have produced comprehensive reports on possible sources and effects of intrusion and accumulation of natural gas in some of the SPR caverns in Gulf Coast domes. The primary concern here is that the intrusion of gas increases the flammability potential for most operations involving handling and storing oil in the caverns. By contrast, effects of gas intrusion into caverns storing natural gas could be small, provided the intruding gas was of "pipeline quality". One effect might be to generate unusual data for the inventory of stored gas.
As stated previously, investigators have speculated about the size of the gas reservoirs associated with persistent and abundant intrusion of gas into some caverns in Gulf Coast domes. As also noted earlier, outburst chimneys associated with gas-rich zones in salt mines have been measured to extend vertically to about 100 m. However, this was in a mine at a depth of less than 500 m, and such zones could be even more extensive at greater depth, where

they would also contain more gas under larger corresponding lithostatic pressure. This would amount to a considerable gas reservoir contained completely within the salt. Alternatively, some caverns might be connected at depth to even larger gas reservoirs outside of the salt, especially if AZs of the type depicted in Fig. 2 are intersected. To summarize, at this point gas intrusions into caverns can be described as site specific phenomenon with essentially unresolved primary sources.

Mitigating Measures

General Measures
The issues raised for storage caverns by occurrences of non-halites and fluids in salt formations are generally site specific in character. This justifies a comprehensive site exploration program early in the planning stage of storage facilities. Findings from such a program are essential for assessing general site feasibility, and for selecting cavern locations that exhibit tractable occurrences of non-halites and formation fluids.

Measures for Non-Halites
Solution mining of caverns can be planned following analysis of the percentage of insolubles in salt cores representative of cavern intervals.. An adequate "sump" should be provided in the cavern design to accommodate insolubles that remain in the cavern following solution mining (Richner, et al., 1992). Appropriate filters can be selected to prevent plugging of brine disposal wells (Cole, ibid.).
As previously noted, a major effect of coherent insolubles in salt caverns is to cause falls during solution mining and storage of liquids. Cavern siting and design are coupled factors that can be varied to reduce the potential for falls. Restrictions on these factors are usually imposed by needs for affordable access to pipelines and specified volume of storage. Cavern configurations in bedded formations are usually dominated by formation stratigraphy, as compared to caverns in domes. Recent cavern designs have featured both arched roofs sited in halite strata, and more nearly spherical shapes in bedded salts, in striving for stability of storage caverns. To reduce the potential for falls in domal salt stocks, it appears prudent to avoid siting caverns under indicators of AZs, e.g., under discernable troughs or trends of depressions on the surface of the salt. This does not guarantee that no falls will occur in caverns outside of such areas.
Horizontal solution mining for brine production in bedded salt formations has been underway for some time (Conti and Schumacher, 1991). However, constructing gas storage caverns by controlled horizontal solution mining is still in an early stage. Associated methodologies have been proposed (Thoms and Gehle, 1993: Gruschow, 1998); computer programs have been developed (Gronefeld and Saalbach, 1998; Kunstman and Urbanczyk, 1995; Russo, 1995;

Saberian, 1995), and at least one field study has been performed (Charnavel and Durup, 1998). It appears likely that a number of competing methods will be developed for solution mining horizontal caverns suitable for storage in bedded salt formations. This should alleviate, but not eliminate, problems associated with occurrences of insolubles, e.g., falls, in bedded salt.

Potash encounters during solution mining in domal salt stocks are indicated by the presence of potassium and/or magnesium in brine returns. They can usually be mitigated by the injection of adequate amounts of blanket material to protect the cavern roof. If containment is threatened by highly irregular and indistinguishable cavern walls, then it may be necessary to re-site caverns to shallower depths or to other locations.

Measures for Fluids

Encounters with gas and liquids during solution mining of caverns can usually be handled with current oilfield technology, e.g., by employing "blowout preventers." Once caverns are constructed, and provided no other salt dissolution occurs, gas intrusion occurs as a relatively steady flow phenomenon. Liquid intrusion into constructed caverns apparently has not been a significant problem.

Gas inflow rates can be reduced by increasing cavern pressure, but will usually increase again once wellhead pressures are reduced. A practical approach for "gassy caverns" is to use them for storage of materials that are not degraded by the intruding gas. Alternatively, stored materials must be treated and "degassed" upon withdrawal from such caverns (Munson, ibid.).

Conclusions

Non-halites and fluids that occur in salt formations raise issues that should be addressed for efficient construction and utilization of storage caverns. A comprehensive exploratory program will provide data that are necessary for identifying potential problems, and for performing realistic appraisals of the suitability of specific sites for storage caverns. Sites that appear appealing from a preliminary survey of surface characteristics may sometimes incorporate subtle geologic features that result in continuing falls, gas intrusion, and leaks in-caverns. Such caverns may turn out to be less than satisfactory for long-term storage operations.

Dedication

This paper is dedicated to Joseph Didier Matinez, mentor and colleague in saline studies. Professionally active to the end, Joe died on February 11, 1998.

References

Baar, C.A., 1977. Applied Salt-Rock Mechanics 1, Elsevier, Amsterdam, New York : 89-90.

Belchic H.C., 1960. The Winnfield Salt Dome, Winn Parish, Louisiana. Guide Book, 1960 Spring Field Trip, Shreveport Geological Society.

Chabannes, C.R., and Dussaud, M., 1995. Solution Mining Caverns in Bedded Salt - The Effect of Insoluble Content, SMRI Spring Meeting, April 30-May 3, New Orleans, 8 pp.

Charnavel, J., and Durup, J.G., 1998. First GDF Horizontal Salt Cavern Experiment, SMRI Spring Meeting, April 19 – 22, New Orleans: 40 – 53.

Conti, P.F., and Schumacher, M.J., 1991. Solution Mining in the 'Nineties – Medium Radius Horizontal Drilling Technology, SMRI Fall Meeting, Oct. 27-30, Las Vegas.

Cole, V.R., Jr., 1992. Brine Disposal Suspended Solids Removal, SMRI Fall Meeting, October 20, Houston, 6 pp.

Ehgartner, B., Neal, J., and Hinkebein, T., 1998. Gas Releases from Salt, SMRI Spring Meeting, April 19 – 22, New Orleans: 68-106.

Evans, L.J., Istvan, J.A., and McDonald, D.T., 1994. Solution Mining of Caverns in the Salado Bedded Salt Formation of West Texas for Storage of Natural Gas, SMRI Spring Meeting, April 24 – 27, Houston, 7pp.

Fokker, P.A., Kenter, C.J., and Rogaar, H.P., 1993. The Effect of Fluid Pressures on the Mechanical Stability of (Rock) Salt, Proc. Seventh Symp. of Salt, V. 1, Kakihana, H., Hardy, H.R., Hoshi, T., Toyokura, K., Eds., Elsevier, Amsterdam: 75-82.

Gimm, W.A.R., and Pforr, H., 1964. Breaking Behavior of Salt Rock Under Rockbursts and Gas Outbursts, 4th Int. Conf. Strata Control Rock Mech., Columbia Univ., New York: 434 – 449.

Golder Associates, 1977. Geotechnical Study of Cote Blanche Mine, for Gulf Interstate Engineering Company.

Gronefeld, P., and Saalbach, B., 1998. Numerical 3D Simulation of Horizontal Leaching Process, SMRI Fall Meeting, Oct. 4 – 7, Rome:141 – 164.

Gruschow, N., 1998. Solution Mining of Thin Inclining Potash Deposits, SMRI Fall Meeting, Oct. 4 – 7, Rome: 165-177.

Halbouty, M.T., 1979. Salt Domes, Gulf Region, United States and Mexico, Second Edition, Gulf Publishing Co., Houston, 561 pp.

Hofrichter, E., 1968. Stratigraphy and Structure of the Palangana Salt Dome, Duval County, Texas, in Saline Deposits, Geological Society of America (GSA) Special Paper 88: 365-379.

Hoy, R.B., Foose, R.M., and O'Neil, Jr., B.J., 1962. Structure of Winnfield Salt Dome, Winn Parish, Louisiana, Am. Assoc. of Pet. Geol. Bull, 46(8): 1444-1459.

Iannacchione, A.G., Finfinger, G.L., Kohler, T.M., and Hyman, D.M., 1982. Investigation of Methane Emissions From an Advancing Face in the Belle Isle Domal Salt Mine, Louisiana, RI 8723, BuMines, 24 pp.
Istvan, J.A., Evans, L.J., Weber, J.H., and Devine, C., 1997. Rock Mechanics for Gas Storage in Bedded Salt Caverns, Int. J. Rock Mech. & Min. Sci. 34(3-4): 295-304.
Jacoby, C.H., 1977. Geology-Hydrology of Avery Island Salt Dome, Y/OWI/SUB-77/16253/1, Office of Waste Isolation, prepared for U.S. Energy Research and Development Administration, 79 pp.
Knauth, L.P., Kumar, M.B., and Mattinez, J.D., 1980. Isotope Geochemistry of Water in Gulf Coast Salt Domes, J. Geophys. Res., 85 : 4863-4871.
Kumar, M.B., 1979. Hydrological Studies of Salt Mines, in An Investigation of the Utility of Gulf Coast Salt Domes for the Storage or Disposal of Radioactive Wastes, J.D. Martinez, R.L. Thoms, C.R. Kolb, M.B. Kumar, R.E. Wilcox, and E.J. Newchurch. Prepared for U.S. Department of Energy, E511-02500-A-1, by the Institute for Environmental Studies, Louisiana State University, Baton Rouge, p. 139-196.
Kunstman, A.S, and Urbanczyk, K.M., 1995. Modelling of Horizontal Cavern Leaching, presented at SMRI Fall Meeting, October 22-25, San Antonio.
Kupfer, D.H., 1967. Mechanism of Intrusion of Gulf Coast Salt, in Proceedings of Symposium on the Geology and Technology of Gulf Coast Salt, Kupfer, Ed., School of Geoscience, Louisiana State University, Baton Rouge: 55-56.
_____1968. Relationship of Internal to External Structure of Salt Domes, in Diapirism and Diapirs – Memoir No. 8, Amer. Assoc. Pet. Geol.
_____ 1976. Shear Zones Inside Gulf Coast Salt Stocks Help to Delineate Spines of Movement, Amer. Assoc. of Pet. Geol. Bull. 60(9): 1434-1447.
_____ 1980. Problems Associated with Anomalous Zones in Louisiana Salt Stocks, USA, Proc. Fifth Int. Symp. on Salt (Hamburg, 1978), N. Ohio Geol. Soc., Cleveland, V. I : 119 – 134.
Lolan, W.E., Valadie, R.J., and Ballou, H.E., 1998. Remote Operated Vehicle (ROV) Design, Cavern Survey and Gel Plugging Agent Application to Repair Louisiana Offshore Oil Port (LOOP) Cavern 14, SMRI Fall Meeting, Oct. 4 – 7, Rome: 327 – 345.
Mahtab, M.A., 1984. Occurrence and Control of Gas Outbursts in Domal Salt, in Proc. First Conf. on Mechanical Behavior of Salt, Hardy, H.R., and Langer, M., (eds.), held at Penn. St. U., Nov. 9-11, 1981, Trans Tech Pubs, Clausthal-Zellerfeld: 775-789.
Martinez, J.D., 1991. Salt Domes, American Scientist, Sep.-Oct., 79 (5), published by Sigma Xi: 420-431.

Martinez, J.D., and Thoms, R.L., 1977. Hydrologic Isolation of Mined Openings in Salt Domes, in An Investigation of the Utility of Gulf Coast Salt Domes for the Storage or Disposal of Radioactive Wastes, J.D. Martinez, R.L. Thoms, C.G. Smith, C.R. Kolb, E.J. Newchurch, and R.E. Wilcox. Prepared for The Office of Waste Isolation, Department of Energy, Y/OWI/SUB-4112/37, by the Institute for Environmental Studies, Louisiana State University, Baton Rouge: 203-206.

Martinez, J.D., and Kumar, M.B., 1979. Mechanisms of Migration of Formation Waters in Salt Domes, in An Investigation of The Utility of Gulf Coast Salt Domes for the Storage or Disposal of Radioactive Wastes, etc: 221-239.

McCauley, T.V., Ratigan, J.L., and Sydansk, R.D., 1998. Characterization of the Brine Loss Zone and Development of a Polymer Gel Plugging Agent to Repair Louisiana Offshore Oil Port (LOOP) Cavern 14, SMRI Fall Meeting, Oct. 4-7, Rome: 391- 406.

Mine Safety and Health Administration (MSHA) Storage Task Force, 1978. Fact Finding Investigation of Southern Louisiana Salt Domes, Weeks Island, Cote Blanche Island, Jefferson Island, Belle Isle, Avery Island, 43 pp.

Molinda, G.M., 1989. Investigation of Methane Occurrence and Outbursts in the Cote Blanche Domal Salt Mine, Louisiana, RI 9186, BuMines, 21 pp.

Muehlberger, W.R., 1960. Internal Structure and Mode of Uplift of the Grand Saline Dome, Van Zandt Co., Texas, U.S.A., Proc. 21st Int'l. Geol. Congr., Copenhagen, Denmark, 18:28-33.

Munson, D.E., Molecke, M.A., and Myers, R.E., 1998. Interior Cavern Conditions and Salt Fall Potential, SMRI Spring Meeting, April 19 – 22, New Orleans: 226 – 239.

Plimpton, H.G., Foster, R.K., Risbeck, J.S., Rutherford, R.P., King, R.G., Buffington, G.L., and Traweek, W.C., 1979. Final Report of Mine Explosion Diaster, Belle Isle Mine, Cargill, Inc., Franklin, St. Mary Parish, Louisiana, June 8, 1979, Mine Safety and Health Administration (MSHA), 156 pp.

Richner, D.R., 1992. In Situ Mining of Soluble Salts, in SME Mining Engineering Handbook, H. Hartman, Senior Ed., Society for Mining, Metallurgy, and Exploration, Inc., Littleton, Colorado: 1493-1512.

Neal, J.T., Magorian, T.R., Thoms, R.L., Autin, W.J., McCulloh, R.P., Denzier, St., and Byrne, K.O., 1993. Anomalous Zones in Gulf Coast Salt Domes with Special Reference to Big Hill, TX, and Weeks Island, LA, SAND92-2283, Sandia Report for U.S. Department of Energy, Albuquerque, 65pp.

Richter-Bernberg, G., 1987. Deformation Within Salt Bodies, in Dynamical Geology of Salt and Related Structures, Lerche, I., and and O'Brien (Eds.), Academic Press, New York: 653 - 677.

Russo, A., 1995. HORSMIC, Horizontal Solution Mining Model, presented at SMRI Fall Meeting, Oct. 22 – 25, San Antonio.

Saberian, A., 1995. A Preliminary Model for Horizontal Well Leaching, presented at SMRI Fall Meeting, Oct. 22-25, San Antonio.

Senni, S.J., 1987. Evolution of Boling Dome Cap Rock with Emphasis on Included Terrigenous Clastics, Fort Bend and Wharton Counties, Texas, in Dynamical Geology of Salt and Related Structures, ibid.).

Spiers, C.J., Urai, J.L., and Lister, G.S., 1988. The Effect of Brine (Inherent or Added) on Rheology and Deformation Mechanism in Salt Rock, in The Mechanical Behavior of Salt, Proc. Second Conf., Hardy, H.R., and Langer, M., Eds., Trans Tech Publications, Clausthal-Zellerfeld: 89-102.

Talbot, CJ., and Jackson, M.P.A., 1987, Internal Kinematics of Salt Diapirs, Amer. Assoc. of Pet. Geol. Bull, 71: 1068-1093.

Thoms, R.L., and Gehle, R.M., 1993. Feasibility of Controlled Solution Mining from Horizontal Wells, SMRI Fall Meeting, Oct. 24 – 27, Lafayette, Louisiana.

Thoms, R.L., and Martinez, J.D., 1978. Blowouts in Domal Salt, Proc. Fifth Inter. Symp. on Salt, Northern Ohio Geological Society, Cleveland: 405-411.

Thoms, R.L., and Neal, J.T., 1992. Effects of Anomalous Features on Solution Mining of Storage Caverns in Domal Salt, SMRI Fall Meeting, October 18 – 22, Houston, 10 pp.

Vaughn, F.E., 1925. The Five Islands, Louisiana, Am. Assoc. Pet. Geol. Bull., 9:756-797.

Risk Assessment and Safety Criteria for Underground Caverns

Massimo Guarascio[1] - Carmela Di Bella[2]

Abstract

Assessment of the safety is based on a reliable knowledge and theoretical modeling of both the geology and geo-structure of the site and of the phenomena causing concern. Major considerations are the integrity of the vessel and the potential short and long term impact on the surrounding environment. Therefore, assessment of the risk include the evaluation of ground movements, subsidence, sinkholes, excessive closure, leakage and ground water contamination. General criteria and recommendations based on real case histories are discussed.

Introduction

Bedded salt layers and salt domes are a reliable environment for underground storage because of the 'self healing' properties of the rock at depth, due to creep. For long time the design of a salt cavern has had the only purpose of maximizing the extraction ratio without considering environmental and safety problems. In the last few decades those concerns have gained a lot of importance and therefore it is not possible to carry out a designing project without considering the impact that the project has on the surrounding environment and without ensuring to it proper safety measurements.

---

[1] University of Rome "La Sapienza", Via Eudossiana, 18, 00184 Rome-Italy- Mining Italiana S.p.A. Piazza Barberini, 52-00187 Rome-Italy

[2] Environmental Engineer – Mining Italiana S.p.A., Piazza Barberini, 52-00187 Rome-Italy

Risk Assessment: Safety Analysis

Safety assessment allows the designing of caverns to be used for production, hydrocarbon storage or waste disposal in such a way that at all times during the construction, operation and post-operative phases (after completion or filling), they present no risk to the surrounding environment. The creation of a cavern within a salt layer for any type of purpose necessarily effects the natural equilibrium of the underground and determines changes. Some effects and expressions of this type of operations may result undesirable for functionality of the structures and for the geo-environmental safety.

The geo-environmental safety verification required during planning and design of the underground openings are to be determined for the various specified cases with appropriate criteria. In general it is necessary to investigate the rock mass and overburden with regard to its structure and its material properties, but also the natural spatial spread of these variables. In this way it is possible to create a model of the rock that simulates as closely as possible the actual conditions of the rock (i.e. the geology, stress conditions and constitutive laws). Afterwards, it is necessary to know the type, the properties, the injection method and the short and long term behavior of the cavern fill. Theoretical models for the various phenomena causing environmental concern are also required. Limiting situations have to be specified, such as:

- Whether or not during and after the construction of the cavity non permitted deformations of the cavern itself or of the ground level can occur (**subsidence**).
- Whether the load-bearing behavior of the rock mass is sufficient to avoid sudden or gradual collapse of the subsurface space (**sinkholes**).
- Whether delays in the backfilling of the cavern can determine an **excessive closure** of the cavern that will compromise the economical convenience of the use for storage.
- Whether the injected materials have a **stabilizing effect** over a longer term.
- Whether the deformations due to the convergence in the rock mass and in the overburden cause unacceptable changes to hydrological conditions (e.g. **groundwater flows, leakage**...).

The most difficult task is to include the essential parameters (i.e. those decisive for the long term stability and the serviceability of the repository) in these models, neglecting unimportant aspects. Primary causes are to be allocated for the limiting situation, and secondary conditions, with average values of their spreads, must be defined.

The analysis can therefore only be carried out for a specific case by combination of engineering geology, geotechnical and rock mechanical investigations, monitoring and mining experience. It is obvious that modeling only reaches a certain level of accuracy, since the actual behavior of a complex

geological site will always remain unknown up to certain extent. It is therefore necessary to take into account a series of uncertainties, i.e. lack of knowledge, such as:
- Geological uncertainties, connected with the formation of the rocks, history and the characteristics of the deposits (i.e. geometry);
- Modeling uncertainties, connected with the choice of a representative model of the rock mass behavior, initial and boundary conditions;
- Uncertainties connected with the geo-mechanical characterization of the deposit, scattered values of geo-mechanical properties, testing of geotechnical data...
- Uncertainties connected with the actual phenomenology stress/strain (i.e. interaction solid-liquid-gaseous phases, seepage...)

The reliability analysis considers systematically the above mentioned uncertainties and allows to study how they spread and what is their influence on the safety of the site.

Risk Assessment: Master of the Uncertainties

There are two main ways to master the uncertainties. One way is to apply a deterministic approach using a safety factor. This can be considered as a measure of the uncertainties because the less precise are the data, the less reliable are the measurements, the less representative is the constitutive model, the greater the safety factor is. For instance the stability of a cavern can be evaluated using the limiting equilibrium and arching capacity theory (see [SMRI, 1998] and [Mining Italiana, 1998]) and the grade of risk can be quantified in a safety factor type criterion.

Another possible solution is to perform a parameter study able to investigate the effects of the variability of input parameters, like mechanical data or design variables, to the performance of the system. Examples of both type of approach are provided as case histories.

Case histories

The first case history to be examined is Belvedere Spinello, a salt mine leached with the solution mining method (see figure 1), located in the southern part of Italy (for more details see [Guarascio, 1994]). For this mine two types of vulnerability analysis have been carried out, one to assess the stability of the pillar between caverns and another to establish the possibility of sinkholes occurrence. In both cases the first step has been the characterization of the geostructure by means of:
- Boreholes

- Laboratory tests (to assess strength parameters and deformability)
- Electric properties (georesistivity and geoconductivity)
- Sonar surveys
- 2D and 3D seismic surveys
- Subsidence measurements

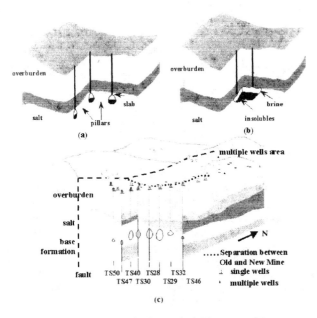

Figure 1: (a) Single well exploitation method. The spans of the cavern roof are generally smaller than those obtained with the multiple well method. Subsidence and sinkholes are not likely to occur. (b) Multiple well exploitation method. For large span of the cavern roof the overburden subsides and collapses into the brine. The collapsed materials move upwards and if the volume of the cavern is large enough the outcrop of a sinkhole occurs. (c) 3D view of the Belvedere Spinello Brinefield with the indication of the borderline between the old mine (exploitation with the multiple wells method) and the new mine (exploitation with the single well method).

Check measurements have been carried out such as:

- Controls in the cavern (echometries and γ-logs)
- Controls on the surface (subsidence measurements)
- Controls inside the wells (γ-logs...)

Numerical calculations have been carried out to investigate the **load bearing behavior and stability of caverns** in the 'New Mine' of the brinefield. For this reason three different calculation models were developed.

- Model I, to consider the configuration of caverns TS28 and TS29 as well as neighboring cavern TS32. This model was used in particular to evaluate the stability of the pillar between the two caverns.
- Model II, to investigate the stability of the pillar at a less favorable height-width-ratios and the stability of the roof area of caverns TS28 and TS29.
- Model III, to take into consideration the configuration of a single fictitious cavern with volume approximating that of the total volume of caverns TS28 and TS29 for both the current as well as for the planned solution mining operations. This model was used in particular to investigate the load bearing behavior of the rock salt in the cavern roof area and in the overlying salt.

All models took into consideration the cover rock, the rock salt and the underlying rock formation. They considered the possibility of occurrence of different types of failure (fracture failure as a result of tensile stresses, of deviatoric stresses and of high creep rates). Different scenarios have been considered, an the obtained results can be summarized as follows:

- With respect to the occurrence of failure as a result of tensile stresses, calculations revealed locally limited minor tensile stresses only at the beginning of the calculated time period and particularly in the zone of the cavern roof. A possible failure of the rock formation load bearing ability as a result of excessive tensile stresses was not concluded from the calculations. The results of time dependent calculation variations showed that the initial high effective stresses occurring in the pillar reduced quickly over a short period of time. This stress reduction, which also made apparent in the low stress intensity factor, was accompanied in some cases with significant effective strain in the middle pillar zone. The long term bearing behavior of the pillar was discovered to deform considerably over time as a result of creep and gradually evaded the initial high load. At the same time the rock zone above and below the caverns become more and more involved in load bearing which leads to the formation of a large scale pressure arch over the whole cavern field. This means that the cover rock and the underlying rock formations become increasingly loaded over a large area.
- Possible failure of load bearing ability of the rock salt formation as a result of excessive deviatoric stresses cannot occur as in general no critical changes in the stress intensity index are revealed. At the beginning of the calculated time period relatively high effective stresses occur in certain places in the rock salt formation, in particular in the strongly loaded pillar between TS28 and TS29, in

the pillar between TS29 and TS32, as well as in the roof zone of the caverns. These relatively high effective stresses result in matching high stress intensity indices (in any of the cases exceeds 100%). In the long term (10, 12 years) the intensity index reduces considerably as a result of creep of the rock salt formation, and always lies below the long term permissible value or within the long term permissible zone of 25 – 35%.

- The assessment of the calculation results with respect to the occurrence of high creep rates and associated creep fracture shows that high values only occur in the pillar zone and in the roof zone of the caverns. However, these creep rates generally lie at least one order of magnitude below the permissible limiting value. A possible failure of the rock formation load bearing ability as a result of high creep rates and associated creep fracture cannot be concluded from the calculation performed.

The assessment showed a generally positive evaluation of the load bearing behavior of TS28 and TS29. Same conclusions can be applied to most of the caverns of the New Mine, with only exception of cavern TS33 because of its unusual shape and position.

In the same mine a vulnerability analysis versus **potential formation of sinkholes** has been carried out. The vulnerability analysis was not carried out during the designing phase, but afterwards as a back analysis to confirm the 'inescapability' of sinkholes formation in a restricted area overmined with multiple wells (Old brinefield). Most probably, if the analysis would have been carried out during the designing phase, the vulnerability of the site would have been proved in advance, and, with an adequate choice of mining method and geometry, the risk could have been reduced to zero.

In 1985 a first evaluation was performed, one year after the occurrence of a major sinkhole and concurrent flooding at the site. The initial evaluation was based on interpretation of the local geology, stratigraphic data obtained from well logging, and data on the subsurface conditions obtained from electro-resistivity surveys on the site. This evaluation was made through the application of the limiting equilibrium and arching capacity theory which was found to be likely the only effective approach given the type of geological and geo-structural information really available. The results have been ranked with respect to the "worst case" (sinkhole C1), and it has been decided to consider as "high vulnerable" the locations having a value in the rank greater than 50% of the value of the worst case.

Based on those data, two areas of different sinkhole potential were identified, one at the north end of the field and one at the central-south of the brinefield (see figure 2). A sinkhole, designated as C3, occurred at the northern area of the brinefield identified as a "high risk area" in the initial 1985 evaluation.

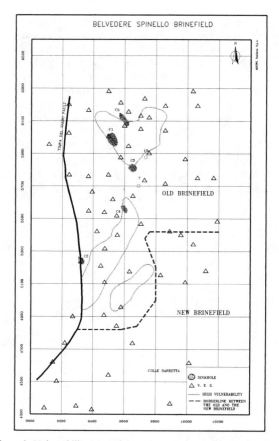

Figure 2: Vulnerability Map of the Belvedere Spinello Brinefield (1985)

A second evaluation was performed by MINING Italiana in 1992 incorporating additional data obtained from salt production, subsidence survey measurements, additional sinkholes occurrences, recorded casing collapse and shearing at various wells, loggings from additional wells drilled in the area and results from a second electro-resistivity and micro-seismic surveys carried out at the site.

The surface of the mine has been divided into a grid with a mesh of 20*20 meters. The basic concept used to interpret the collected data is that for a given location P(x, y) the total vulnerability (W) is given by the product of coefficients, which represent the relative factorized vulnerability (Wi) with respect to each single cause when all the others are kept constant. The final W can be ranked with respect to a reference situation (a location where a sinkhole has occurred and therefore considered the most vulnerable site in the area, in our case sinkholes C1 and C2) or with respect to the worst theoretical situation estimated on the basis of the geomechanical model (Limiting Equilibrium and Arching Capacity Theory). The latter is used in cases where no sinkholes have developed.

Considering the available data, after an appropriate clustering procedure, six coefficients have been identified for the purpose of a factorized vulnerability definition of the Belvedere Spinello Mine. In details the six coefficients are:

1. The coefficient W_1, which takes into account the data collected during the electro-resistivity survey. It is equal to:

$$W_i = 1 + \frac{1}{\rho} \cdot \ln \frac{h_b}{h_t}$$

where ρ is the local resistivity and h_b and h_t are the bottom and top depths of the overall georesistivity horizon (assumption has been made that the conductive surface is limited by a resistivity value lower than 2 Ωm. See figure3).

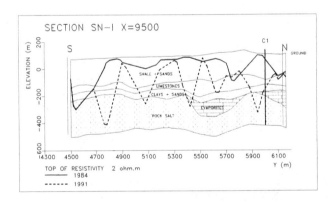

Figure 3: Stratigraphic cross sections of Belvedere Spinello Brinefield with resistivity measurements. The differences between the top of the conductive surface (2 Ωm) measured in 1984 and in 1991 are due to the abandon of the multiple wells exploitation method. Isolated peaks are a consequence of localized collapses.

2. The W_2 coefficient takes into account the influence of the stratigraphy to the occurrence of the sinkhole. The hypotheses made are that the formation of a sinkhole is directly proportional to the thickness of the salt layer (possibility of leaching large caverns), directly proportional to the thickness of the evaporite layer (this stiff layer is responsible for the holding of the overburden materials: the bigger is the evaporite layer, the bigger is the volume of leached cavern) and inversely proportional to the thickness of the overburden (if this layer is thick enough, it is possible that due to the bulking factor of the overburden materials the underground void are completely filled). Therefore the coefficient W_2 is made by the sum of three parameters, which take into account the thickness of the overburden, the thickness of salt layer and the thickness of the evaporite layer each one normalized to the maximum thickness recorded in the mine area ([Thickness(x,y) /maximum thickness]*100). Therefore:

$$W_2 = (W_{21}+W_{22}+W_{23})$$

3. The W_3 coefficient takes into account the size of the cavern. It is equal to 100 in case the site is located exactly on top of a single cavern or in the area of influence of a leached cavern in the old mine; is equal to [100-X] in case the site is located at a distance X from the above mentioned locations and is equal to 0 in case X>100.
4. The W_4 factor takes into account the magnitude and distribution of ground subsidence. It is given by the sum of two coefficients, the first one that gives account for the absolute value of subsidence, and the second one that gives account for the variance of subsidence, and is ranked to the maximum value of the sum recorded in the mine area.

$$W_4 = [W_{41} + W_{42} / \mathbf{max}\ (W_{41} + W_{42})]*100$$

5. The W_5 coefficient takes into account all the phenomenologic evidences connected with anomalous underground situations. Numerically this coefficient is equal to 100 if breaking of well casing and brine spills and lowering of the wellhead have occurred in a location far from the area in which sinkholes outcropped (>50 m) and assumes a decreasing value if these phenomena did not occur all together in the same location.
6. The W_6 coefficient takes into account the micro-seismic data. After a first screening of the enormous amount of recordings, the average montly energy has been considered for each registration point and the factor W has been given by the ratio between the overall energy and the maximum overall energy.

This second evaluation of vulnerability confirmed the potential for sinkhole formation in the northern area of the field and resulted in an extension of the central-south area towards the center of the brinefield.

Further refinements of this vulnerability evaluation have been carried out through the years in the areas where the general analysis has showed possible risk. More parameters, such as the effect of the brine volumes left in place, the results of geo-mechanical checks, tightness tests, geo-electric measurements and 2D and 3D reflection seismic, have been included.

Another case history is provided by the geostatistical analysis for rock mass characterization in the Masua Mine. This is a led and zinc mine located in Sardinia, Italy, where the need for a further deepen of the exploitation level required an evaluation of the stability of the rooms, with consequent more detailed characterization of the geostructure.

The design parameters for numerical modeling of the mine structure were determined by geostatistical analysis of available rock data, including joint mapping, borehole and core logging. The parameters describing rock discontinuites were treated as 'regional variables' and therefore variograms of each variable were constructed. The analysis of rock mass data by geostatistical procedure allowed to infer the spatial variability of the RMR index[3], of the in situ modulus of deformability E_d and of the empirical constants (m and s) entering in the criterion of failure proposed by Hoek and Brown[4]. The results obtained applying the Kriging method (see figure 4) were found to be more realistic than those ones obtained with the Monte Carlo simulation, because the variability was more gradual and without sharp differences from one location to the adjacent.

Having a reliable spatial characterization of the rock mass, it was possible to built with accuracy a 2D and 3D model of the structure in order to apply traditional Finite Element or Finite Difference Methods for an effective evaluation of the stability.

---

[3] The RMR index is a rock mass classification scheme developed by Bieniawski. It takes into account several parameters such as the strength of the intact rock material, the rock quality designation, the spacing of joints and the ground water conditions. For various ranges of each parameter a rating value is assigned. The overall Rock Mass Rating is obtained for individual parameters (see ref. Bieniawsky, 1976).

[4] The failure criterion proposed by Hoek and Brown is given by the equation:
$$\sigma_1 = \sigma_3 + (m\sigma_c\sigma_3 + s\sigma_c^2)^{1/2}$$
σ_1 = maximum strength
σ_c = uniaxial compressive strength of intact rock
σ_3 = minor principal stress
m,s = constants that depend on the properties of the rock. For more details see ref. (Hoek and Brown, 1980).

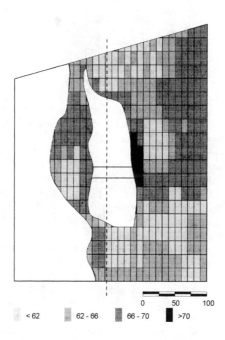

Figure 4: Cross Section of the Masua Mine:
RMR index distribution (Kriging Method)

Recommendations & Conclusions

It is not possible to provide universal procedures valid for every type of storage because of the unique characteristics of each site. It is only possible to outline general recommendations and criteria.

A reliable knowledge and theoretical modeling of both the geology and the geostructure of the site and of the phenomena causing concern is required at first. A vulnerability study can be a really useful tool for a cautious design of the brinefield. Once this mapping is drawn, continuous monitoring of different parameters, directly correlated with the underground situation, is necessary to keep under control the evolution of the dissolution process and to update the risk mapping of the site. These measurements can be considered as warning lights able to give in real time a feeling of the situation. The main type of measurements are:

- Subsidence measurements
- Volume measurements.
- Observation of casing breakage
- Geo-resistivity
- Pressure measurements

Subsidence is probably the effect that is more directly correlated with evolution of the leaching process. A certain subsidence is compulsory in case of salt dissolution, as a consequence of the creation of voids and of salt creeping . When the process develops gradually and safely, the rate of subsidence ranges between precise values that are function of stratigraphy, depth and rate of dissolution, and the shape is the classical bowl-shape. If the rate of subsidence increases or a change in superficial shape (from the wide bowl shape to the sharp inverted sombrero shape) is detected, this mean that anomalies are taking place.

Measurements of the volume of brine extracted from the cavern are used to assess the actual state of dissolution of a cavern: they are input data for computer simulators able to provide the shape and extension of a cavern through a mass balance evaluation.

Practical experience shows that casing breakages have been recorded in the neighborhood of a sinkhole, previously to the occurrence of the sinkhole itself. Therefore a continuos checking of the integrity of casing allows to monitor underground movement before they develop into a sinkhole.

Leakage out the cavern can be revealed through measurements of the electric properties of soils and rocks (geo-resistivity and geo-conductivity) like the so called vertical and horizontal electro-sounding (i.e. V.E.S. and H.E.S.). Brine is a high conductivity medium and when a permeable layer is infiltrated with saturated brine it can show a geo-resistivity much smaller than the 'no-infiltrated' value. Therefore, in a stratified formation (e.g. Belvedere Spinello) where a thick layer of high permeability (sand or sandstone) is interbedded with low permeability units (clay, evaporitic rocks) the V.E.S. can identify infiltrations of saturated brine checking the presence of eventual stratigraphic or georesistivity discontinuities. In many cases the results have shown to be reliable.

Well head pressure measurements, carried out continuously, can give information about the tightness of the cavern: sudden drop of internal cavern pressure Pi can be justified by the opening of cracks that will allow leakage of brine.

References

Bieniawski Z.T., "Rock mass classifications in rock engineering", in *"Exploration for Rock Engineering"*, 1, A.A. Balkema, Cape Town, 1976.

Fernandez G., Hendron A.J.jr., "Underground Storage in Salt Deposit", Journal of Geotechnical Engineering, 1996.

Guarascio M., Barla G., Scavia C., Antonellis M., "Characterization of rock mass by geostatistical analysis at the Masua Mine", CEE Report, 1987

Guarascio M., Fernandez G., Thoms R.L., "Updated Evaluation of Sinkhole Potential at Belvedere Spinello Brinefield", SMRI 1994 Fall Meeting, Hannover, Germany.

Guarascio M., "High Resolution 3D Seismic Applied to Subsidence Evaluation", Wassman's Day, Amsterdam, 1997.

Guarascio M., Fernandez G., "In situ testing for Rock Salt Characterization", SMRI 1998 Fall Meeting, Rome, Italy.

Hoek, E., Brown E.T., "Underground excavations in Rock", Instn. Min. and Metall., London, 1980.

KBB-Mining Italiana, "Rock-mechanical Investigations for Belvedere Spinello Brinefield", Hannover, September 1995

Langer, M., "Use of solution-mined caverns in salt for oil and gas storage and toxic waste disposal in Germany", Engineering Geology, 35 (pg. 183-190), 1993.

Langer M., "Modeling of Site Evaluation and Safety Assessment", Unesco/CoGeoenvironment-Training Workshop on Geoscience For Environmental Planning, Salzmechanik XIV, 1998.

Mining Italiana, "Vulnex Code: User's Manual", Internal Report, 1998.

SANDIA REPORT *"Waste Isolation Pilot Plant Shaft Sealing System Compliance Submittal Design Report"*, Albuquerque (New Mexico), 1996.

Sheorey P.R., "Empirical Rock Failure Criteria", A.A.A. Balkema, Rotterdam, 1997.

SMRI- "Guidelines for Safety Assessment of Salt Caverns", Fall 1998 Meeting, Rome-Italy

Wallner M., "Sensitivity and Uncertainty analysis applied to Cavity Design Computations in Rock salt", Salzmechanick XIV, April 1996

Wallner M., - Modeling the Consequences of future geological Processes by Means of Probabilistic Methods-, Salzmechanik XIV, April 1996

Wassmann, Th., Hans, "Mining subsidence above cavities created by solution mining of rocksalt", 7th International Symposium on Salt (pg. 425-431) 1993

Experience in Underground Storage of Crude Oil in Salt

James K. Linn, PhD[1]
Jon Culbert[2]

Abstract

The U.S. Strategic Petroleum Reserve (SPR) currently stores nearly 88 million cubic metres (550 million barrels) of crude oil in salt underground openings, leached and mined. Experience over 20 years has provided some very important lessons and technology improvements, and pitfalls of underground oil storage in salt. On the very positive side, the ability to monitor the integrity of solution mined caverns has increased greatly, tools used to design and control leaching of salt caverns have become almost routine, and our understanding of the underground or "geo" tank has improved immensely. In the process however, we have learned that all salt "geo" tanks are not suitable for hydrocarbon storage even though they are stable for years, and that underground long term storage of some hydrocarbons has some unanticipated costs.

In this paper we discuss SPR contributions to technology improvements in testing, salt modeling and understanding of underground crude oil storage in salt, using examples of actual cavern operations. We also discuss salt cavern limitations that have been learned through hard experience. Specifically, cavern pressure monitoring and modeling capabilities which can enable loss detection of very small volumes of storage oil are highlighted, along with state of the art salt creep modeling, leaching, and cavern design capabilities. Mitigative and diagnostic techniques for fluid leaks are also discussed.

---

[1] Sandia National Laboratories, Manager, Underground Storage Technology Department MS 0706, P. O. Box 5800, Albuquerque, NM 87185, Ph. 505 844-6813, Fax: 505 844-0240

[2] U. S. Department of Energy, Assistant Project Manager, Strategic Petroleum Reserve Project Office, FE 4431, 900 Commerce Road East, New Orleans, LA 70123, Ph. 504 734-4368, Fax. 504 734-4299

Background

The U.S. Strategic Petroleum Reserve (SPR) was established by Congress in 1975 to store up to 157 million cubic metres (1 billion barrels, 1000 MMB) with an authorization of 118 million cubic metres (750 MMB) of crude oil to be used as a strategic buffer in case of an oil embargo. The decision was made to store the crude oil in caverns (and/or salt mines) in Gulf Coast salt domes. Congress furthermore mandated the fill schedule which dictated that approximately one third of the reserve was "existing" space procured, with the remainder to be developed or leached (solution mined). Existing leached cavern space was purchased at Bryan Mound salt dome in Texas, and West Hackberry, Bayou Choctaw, and Sulphur Mines in Louisiana. Additionally, an existing salt mine at Weeks Island in Louisiana was acquired. New leached caverns were developed at Bryan Mound, Bayou Choctaw, West Hackberry, and Big Hill salt dome in Texas.

Oil fill began in 1978 and continued into the early 1990's reaching a maximum of slightly less than 94 million cubic metres (600 MMB), which is being maintained currently. Small oil withdrawals have been conducted to exercise the total system, conduct small test sales, and in support of the oil shortage due to the Gulf War. Otherwise the oil has been stored in nearly an undisturbed state for up to 20 years.

Many geotechnical or underground problems have been encountered. Some of the problems have been significant and one required oil relocation and site decommissioning, whereas others have been overcome through technology improvements. This paper addresses the major geotechnical problems that have been encountered, and the resulting technology improvements or solutions developed. The general applicability of salt for long term hydrocarbon storage is briefly discussed.

Technology Improvements

Salt Mechanics: Salt is a visco-plastic material; it slowly creeps with time. It is also a fairly strong material in compression, but weak in tension, and highly impermeable and practically insoluble in hydrocarbon liquids. These characteristics greatly impact its use as a hydrocarbon containment material. SPR has been unique in its pioneering applications of salt mechanics, due to the availability of new information being concurrently developed by Sandia and the Department of Energy sponsored salt research at the Waste Isolation Pilot Plant (Munson, 1997), and access to greatly improved computational capabilities.

Salt mechanics finite element calculations have been uniquely responsible for establishing suitable cavern shapes, locations, and operational limitations. Of concern with caverns for hydrocarbon storage is development of suitable shape(s) and operational limits such that salt surrounding the cavern(s) and above the caverns(s) is maintained in compression. Also extremely important is the nearly inevitable slow loss of volume due to salt creep as a result of the lower pressures within the cavern(s) as compared to lithostatic pressures in the salt surrounding the cavern. Initial finite element applications for SPR consisted of a few hundred nodes, and tens of time steps. Currently, finite element creep calculations are performed with hundreds of thousands of nodes and thousand of time steps. As discussion below in understanding the Weeks Island sinkhole, SPR has completed several extremely large simulations of geomechanics problems (possibly the largest geotechnical simulation ever completed?)

Solution Mining Code: Codes to aid in the leaching of caverns existed prior to SPR (Saberian and Podio, 1977). Such codes had been developed by brining companies and others to predict cavern shapes. In the early days of the reserve, SPR tested the existing codes when available, but found that they lacked flexibility in the hydrocarbon (in this case crude oil) fill schedule, and sometimes failed to closely match obtained cavern roof shapes. Sandia was tasked with developing a new code, SANSMIC (Russo, 1983), which had the flexibility of periodic or near continuous oil injection while continuing to leach. Improvements in the fluid plume models within the code were also incorporated which significantly improved the roof modeling, especially for larger caverns. Basic input information included injection water rates, temperature, and salinity, initial well or cavern size, injection and brine removal heights, and non-salt insoluble information. Calculated output information included produced brine salinity, cavern shape, and cavern size. Estimates were made of the insoluble percentages carried out with the brine versus precipitated to the bottom in the cavern. An example comparing calculated with measured data is given in Figure 1.

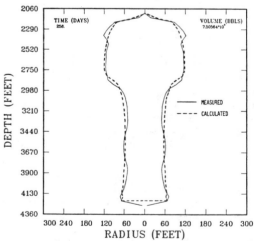

Figure 1. Code Prediction versus Measured Cavern Shape

More recent improvements have been in the packaging of the code for desktop machines, and better interactive displays. Additional features to utilize depth-dependent insoluble content information have also been added.

Nitrogen Testing: During the early cavern acquisition phase, it was recognized that many existing caverns had wells which were many years old. The integrity of such wells was a major concern, as was the ability to test new wells, and to identify new leaks that did occur over time. In 1980 caverns were normally simply pressure tested, which consisted of elevating the cavern pressure by fluid injection, and monitoring any pressure loss over a specific time interval. Typically, brine was injected to raise the cavern pressure to maximum operating pressure (usually .78 to .85 of lithostatic pressure) and the cavern was accepted if the pressure decayed no more than 69 kPa (10 psi) over 24 hours. SPR was a major force in pioneering the development of the nitrogen well test (Goin, 1983). In this test, sufficient nitrogen (an inert gas) was injected into the well to drive the fluid interface slightly below the deepest cemented casing. Through the knowledge of the cross sectional area of the wellbore and the gas/oil or brine interface location obtained from logs, leaks in the wellbore system were detected as changes with time in the interface. The original SPR leak detection or well acceptance criteria was less than 16 cubic metres per year (100 barrels per year). Since this early nitrogen well testing development by SPR (and others), general acceptance of the method by the states has occurred, and several commercial nitrogen well testing services now exist. Improvements have continued including the development by service companies and others of sophisticated computer software that nearly automates the testing of wells.

Pressure Monitoring: Pressure monitoring of hydrocarbon storage caverns has been done by cavern operators, in general, for many years. SPR however, has been uniquely able to utilize pressure monitoring for real time identification of small leaks. This has been accomplished by development of a finite element cavern pressurization model CAVEMAN (Ehgartner, et al., 1995) with cavern specific parameters. The model incorporates all time dependent phenomena within the cavern, i.e., creep, salinity, temperature, and historical pressurization data, to predict future pressurization. New cavern pressurization data is continuously compared with predictions; differences exceeding established limits are then flagged for leak investigations. Leaks in the range of a few tens of cubic metres per year can be identified within weeks. The model is applicable to long term uninterrupted storage of liquid hydrocarbons. Figure 2 shows an actual cavern pressurization history and the model predictions.

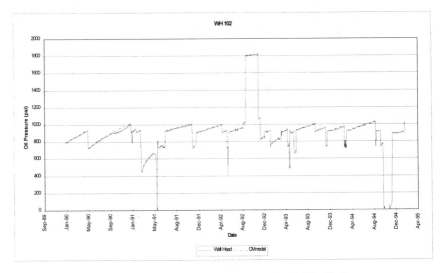

Figure 2. Cavern Pressurization versus Model Prediction

Technical Challenges

The development of caverns and long term storage of over 75 million cubic metres (half billion barrels) of crude oil in salt caverns has not been without technical problems. Perhaps most important is the appreciation of the dynamic behavior of salt storage caverns. Caverns are continuously changing in shape and volume due to creep, salt dissolution, and occasional slabbing. Additionally, the stresses within the salt near the caverns are continuously changing, small amounts of gas are evolving from the salt, and thermal energy is being exchanged between the salt and crude oil. The impact of the cavern changes is dependent upon the stored liquid, in this case crude oil. The changes also effect the surrounding environment, most noticeable by surface subsidence.

The decision to store crude oil in a former room and pillar salt mine was without precedent. The mine potentially offered unlimited drawdowns (withdrawals of oil), at minimal operational costs by operating the mine as a "tank". In contrast, planned drawdowns of solution mined caverns was achieved by displacement of oil with surface waters, resulting in additional solutioning (cavern enlargement), thus withdrawals are limited by the cavern spacing and size (SPR was planned for a maximum of 5 drawdown/refill cycles). Geotechnical problems, however, caused DOE to decide to abandon the mine and relocate the oil. The problems encountered in both caverns and storage mines are discussed below.

Gas Intrusion: Most salt, if not all, contains minute amounts of gas, usually methane but occasionally other hydrocarbons, or carbon dioxide. Historically, encounters with gas in salt was often catastrophic in the form of blowouts, explosions, etc. (Ehgartner, et al., 1998 and Hinkebein, et al., 1995). As in all liquid hydrocarbon storage caverns, the liquid is stored at pressures below lithostatic. Gases contained within the salt (or impurities?) are slowly released. The gases are then absorbed by the crude oil, raising the bubble or boiling point of the oil. Additionally, geothermal heating of stored oil (Bauer and Hinkebein, 1993) also contributes to an increased bubble point. Unfortunately, the commercial pipeline infrastructure cannot transport high bubble point oil due to the safety problems encountered during tank storage. Oil in many of the SPR caverns has had to be degassed (i.e., brought to the surface, heated, allowed to degas at low pressure, and then re-injected). As a result there is an ongoing program of monitoring the gas content and temperature of all stored oil..

Salt Falls or Salt Slabbing: SPR has a documented history of damaged cavern piping, and oil or brine pressure surges, many of which have been attributed to hanging string impact from salt slabs released from the walls or ceilings. Cavern conditions have been extensively studied in an attempt to identify factors contributing to the salt falls (Munson, et al., 1998). The numbers of documented events has been very site specific, even cavern specific within a site, and not dependent on time since leaching, see Figure 3. The exact cause of the events or falls has not been conclusively established although ongoing analyses indicate that salt impurities may be a major contributing factor. The overall impact on the caverns is minimal (in the time frames of interest), but the cost of working over caverns and replacing lost piping is significant. It is of interest to note that at least one other site currently storing crude oil is reporting no salt falls, although their operation has some slight near continuous leaching due to frequent oil withdrawal and refill cycles using surface stored brine (McCauley, 1998).

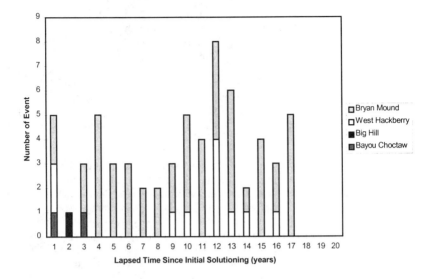

Figure 3. Salt Falls at SPR Sites

Weeks Island: As noted earlier, the Weeks Island SPR site was formerly a two level room and pillar salt mine which converted to oil storage beginning in 1978. The mine levels (11.4 Mm^3 (72 MMB)) were located below a series of mined manways and the oil was accessed via submersible pumps suspended through a bulkhead from the manways. From 1980 through 1992 the only major geotechnical problems were occasional bouts with water leaks in the mined shafts (accesses to the manways) and in one manway drift which was "wet" when first mined. In 1992, a sinkhole was discovered above the mine, Figure 4, raising concerns about a possible leak into the mine. Water already in the mine due to water from the oil (about 1 % Bottom Sediment &Water), etc., made confirmation of the leak extremely difficult. Continued enlargement of the sinkhole (1993 and 1994) prompted drilling around and "below" the sinkhole seeking information on the cause of the sinkhole (Sattler, 1996). A permeable flow meter (Ballard, 1996) was emplaced 21 m (70 ft) below the salt in the solutioned zone found below the top of salt. Additionally, high resolution seismic surveys were conducted near the salt interface below the sinkhole, and isotopic analyses of water samples from below the oil were performed (Bauer, et al., 1996). Data from the flow meter indicated a slow but continuous velocity of water downward toward the mine, and a major salt dissolution zone or "throat" was found directly below the sinkhole. A chemical tracer was injected in the zone that was detected later in the brine below the oil, thus confirming the leak path into the mine. Salt mechanics analyses were used to provide a technical understanding of the conditions potentially causing the leak, i.e., permeability changes associated with the salt creep and possible influence from a shear zone. Detailed salt mechanics analyses indicated that the salt over the aligned edges of the two mine levels was being put in tension by the ongoing creep closure of the mine and resulting subsidence, see Figure 5.

Figure 4. Sinkhole at Weeks Island

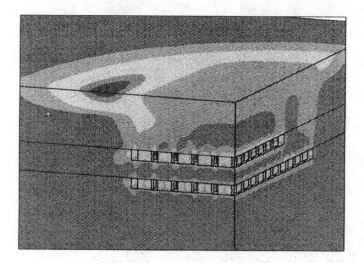

Figure 5. Dilantant Zone Over Mine Edges

Pressurization of the oil storage levels was not feasible due to the repository operational design. Effective grouting of the mine leak was assessed to be improbable, and the prognosis for additional sinkholes (caused by water leaks into the mine) was high. In December 1994 DOE announced its decision to abandon the site, and an oil withdrawal plan was formulated. Of major concern was the possibility of an uncontrolled water influx into the mine during the oil withdrawal operation. Sediment permeability studies (Ostenson, 1994) and leaching analyses indicated that uncontrolled leakage was possible during the oil withdrawal, hence a DOE decision to inject saturated brine into the throat of the leak via drill holes, and form a "freezewall" directly over the leak. Control of a salt mine leak via injection of saturated brine was unique, but has proven effective at least for the short term. The freezewall was accomplished by drilling three rings of holes 20 m in diameter slightly into salt and circulating chilled refrigerant sufficient to freeze the brine, water, and sediments directly over the leak. Brine injection was maintained below the freezewall to balance the hydraulic pressure across the freezewall.

The primary oil withdrawal was completed by late fall 1996, and brine refill of the mine was initiated. Several planned pauses for skimming of oil released from pools and loose salt piles (mining leftovers) have occurred. The difficulties in skimming have been significant. Oil-brine emulsions with greatly increased viscosity, higher than anticipated salt particulate levels, and floating debris (wood, plastics, etc.) have made effective skimming of the oil (about 7.5 cm thick) impossible. Fortunately, although it was unplanned to leave significant oil in the mine when Weeks Island was selected, the remaining oil in the mine is not an environmental hazard. Much of the unrecovered oil (currently estimated at over .15 Mm^3 (1 MMB)) is trapped in the loose salt (estimated at up to 2.25 Mm^3 (15 MMB) equivalent volume). The remainder will be trapped against the rough mined roof surface as the mine is eventually brought up to full hydrostatic pressure. Currently, brine refill of the lower level is near completion and total refill is scheduled for June 1999.

Continued surface subsidence of the mine (a maximum of 0.3 m per yr) following brine refill and hydraulic pressurization will be largely mitigated. Analyses indicate that long term continued subsidence should be reduced to about 3 to 5 percent of the pre-withdrawal rates. However, during the period of time since oil withdrawal occurred and the mine has been at near atmospheric pressure, the observed surface subsidence from the salt creep has nearly tripled. Effects on surface structures located above the mine boundaries have been noticeable but not critical to operations.

The obvious but painful conclusion from the Weeks Island experience is that liquid hydrocarbon storage in shallow salt mines is fundamentally unsound. The geotechnical factors that eventually led to the abandonment of the site will be fairly universal and are now understood. The effectiveness of brine injection to control salt mine leaks has been demonstrated for at least the short term. The usefulness of high-resolution seismic surveys and very sensitive flowmeters as effective diagnostic tools for mine water leaks was demonstrated.

Conclusions

Use of underground salt structures for long term crude oil storage has been occurring in the United States for nearly 20 years. The experience has been largely positive, but some geotechnical problems have been encountered. Most significant of the overall lessons learned has been the fundamental problems associated with using shallow salt mines for crude oil (or other liquid hydrocarbon) storage. Although SPR was the first crude oil storage in the U.S., it was not the first liquid hydrocarbon storage in salt in the U.S. or first crude oil storage (West Germany had already a reserve). Many technological improvements have been made to improve the creation, testing, monitoring, and understanding of underground salt caverns. Fundamental effects due to very small amounts of trapped gas within the salt, salt impurities, and salt creep are now much better understood and quantified. It is anticipated that the use of underground salt structures for hydrocarbon storage will continue, and possible greatly expand as the full need for natural gas storage becomes clear. The use of salt caverns for other purposes such as waste disposal has not been addressed but ongoing research and permitting requests indicate the commercial interest in the area.

References

Ballard, S., "The in situ permeable flow sensor: a groundwater flow velocity meter", Ground Water, v. 34, p. 23-240, 1996.

Bauer, S.J. and Hinkebein, T.E., "An Assessment of Implication of Geothermal Heating of SPR Oil", SAND93-0005, Sandia National Laboratories, Albuquerque, NM, December 1993.

Bauer, S.J., Ehgartner, B.L., Linn, J.K., Lott, S.E., Hinkebein, T.E., Molecke, M.A., Munson, D.E., Neal, J.T., Sattler, A.R., Ballard, S., Bertoldi, M.J., Gump, R.E., Mills, K.E., Lamb, D.W., Thompson, S., Myers, R.E., "Summary of Events and Geotechnical Factors Leading to Decommissioning of the Strategic Petroleum Reserve (SPR) Facility at Weeks, Island, Louisiana, SAND96-2263, Sandia National Laboratories, Albuquerque, NM, October 1996.

Ehgartner, B., Ballard, S., Tavares, M., Yeh, S., Hinkebein, T., Ostensen, R., "A Predictive Model for Pressurization of SPR Caverns", Solution Mining Research Institute, San Antonio, TX, October 23-24, 1995.

Ehgartner, B., Neal, J., Hinkebein, T., "Gas Releases from Salt", SAND98-1354, Sandia National Laboratories, Albuquerque, NM, June 1998

Goin, K.L., "A Plan for Certification and Related Activities for the Department of Energy Strategic Petroleum Reserve Oil Storage Caverns", SAND83-2005, Sandia National Laboratories, Albuquerque, NM, December 1983.

Hinkebein, T.E., Bauer, S.J., Ehgartner, B.L., Linn, J.K., Neal, J.T., Todd, J.L., Kuhlman, P.S., Gniady, C.T., Giles, H.N., "Gas Intrusion into SPR Caverns", SAND 94-0023, Sandia National Laboratories, Albuquerque, NM, December 1995.

McCauley, T, LOOP, private conversation with J. Linn, April 20, 1998, Rome SMRI Spring Meeting.

Munson, D., "Constitutive Model of Creep in Rock Salt Applied to Underground Room Closure", Int. J. Rock Mech. Min. Sci., Vol. 34, No. 2, pp. 233-247, 1997.

Munson, D.E., Molecke, M.A., Neal, J.T., Sattler, A.R., Myers, R.E., "Strategic Petroleum Reserve Caverns Casing Damage Update 1997", SAND98-0090, Sandia National Laboratories, Albuquerque, NM, January 98.

Ostensen, R., Letter to J. K. Linn, "Preliminary Permeability Analysis from Weeks Island Pumping Test", Sandia National Laboratories, Dated 11/16/94.

Russo, A.J., "A User's manual for the Salt Solution Mining Code, SANSMIC", SAND83-1150, Sandia National Laboratories, Albuquerque, NM, September 1983.

Saberian, A. and Podio, A.L., "A Computer Model for Describing the Development of Solution-Mined Cavities, IN SITU, 1(1), p. 1-36, 1977.

Sattler, A. R., "Log Analysis of Six Boreholes in Conjunction with Geologic Characterization Above and on Top of the Weeks Island Salt Dome, SAND 96-0413, Sandia National Laboratories, Albuquerque, NM, June 1966.

Design of a Deep Cavern Intersected by Weak Shale

Bhaskar B. Thapa[1], Associate Member, ASCE and Peter A. Dickson[2]

Abstract

This paper describes the design and construction problems anticipated for the underground powerhouse of a hydroelectric project in the Middle East. The site conditions, with particular reference to a weak shale layer in dolomite, and powerhouse layout are described. Important material properties and the estimation of their values are presented. The ground response to excavation of the large cavern, as seen from numerical analysis, is discussed. Specific construction problems with regard to the shale layer are outlined. Design considerations and measures for control of shale squeezing are given.

Introduction

This paper describes work that was performed during geotechnical studies carried out to demonstrate the technical feasibility of an 800 MW hydroelectric pumped-storage project in the Middle East. This project is being designed to provide peaking capacity and energy storage and will involve pumping water from a lower reservoir to an upper reservoir during off-peak hours when there is surplus low-cost energy from base-load power plants. Electricity is generated by returning the water from the upper reservoir to the lower reservoir. Generation would take place during peak hours to meet a portion of the system demand.

The principal project features consist of upper and lower reservoirs, a water conductor system of tunnels and shafts connecting the two reservoirs, and an underground power station complex. The power station comprises three parallel underground chambers respectively housing the inlet valves, the four pumping-generating units, and the transformers. The rated head of the units is 395 m and the overall length of the water

---

[1] Geotechnical Engineer, Harza Engineering Company, Sears Tower, 233 South Wacker Drive, Chicago, IL 60606-6392
[2] Senior Geologist, Harza Engineering Company, Sears Tower, 233 South Wacker Drive, Chicago, IL 60606-6392

conductors joining the reservoirs is about 1,870 m. Extensive geologic investigations have been performed, including deep core drilling and test adit construction.

Cavern Layout

The main power station cavern is part of a large underground complex of headrace tunnels, draft tubes, equipment caverns, tailrace, and access tunnels. Figure 1 is a schematic illustration of this underground complex that connects the upper and lower reservoirs.

The rock mass is composed of stratified, late Cretaceous-age units of limestone, dolomite, chalky marl, shaley marl, and shale. The main cavern is 44 m high, 18 m wide and 150 m long, and will be located in gently dipping dolomite strata containing two important shale layers as shown in Figure 2. The upper shale is 4-m-thick and found at a minimum of 5 m above the cavern crown. The lower 2-m-thick layer will range between 6 and 9 m below the crown along the length of the cavern. The cavern crown is about 440 m beneath the ground surface.

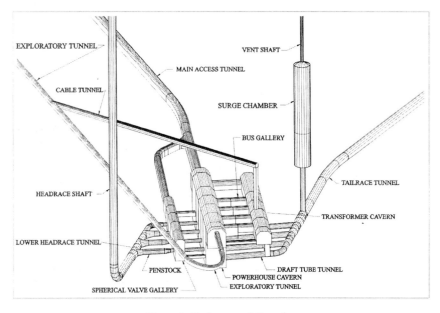

Figure 1: Underground Complex

The subject of this paper is the influence of the shale layers on the stability of the powerhouse excavation. More specifically, the combination of the high stresses due to deep rock cover and sandwiching of weak shale material between competent dolomite layers may be expected to lead to localized shale failure. Large deformations from such an event could initiate undesirable stresses elsewhere and other modes of failure. Additionally, the shale will exhibit time dependent squeezing behavior as observed in the exploratory adit. Design studies were performed to evaluate the feasibility of constructing the cavern, and to assess construction measures to control the influence of the shale on cavern stability.

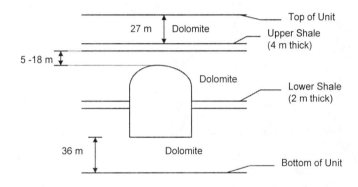

Figure 2: Model of powerhouse cavern

Material Parameters

The material properties used in the analyses are listed in Table 1. The friction angle and cohesion for the dolomite represent the average strength envelope to triaxial test results on drill core samples. The elastic modulus and tensile strength for both the dolomite and shale are taken from project test data reports as is the shale friction angle. The cohesion for the shale is based on a Mohr-Coulomb relation between the unconfined compressive strength, obtained from testing, the friction angle, and cohesion. The *in situ* stress field is assumed to be lithostatic with K=1; additional stress field data were not available at this initial design stage.

Estimates of residual strength of the rocks were also needed, as will be explained later. Since these were not available from laboratory tests at this design stage, they had to be estimated indirectly using the fact that residual strengths are a function of confining pressure. Thus residual strength was estimated by interpolating between the case with residual strength equal to peak strength at brittle-ductile transition pressure and the case with residual strength being zero for an unconfined case[3]. The mean confining pressure of 10 MPa expected around the cavern is used for this interpolation. Goodman (1989) provides brittle-ductile transition pressures for some rocks. For compaction shales the maximum transition pressure of 20 MPa is conservatively used so that a residual strength is estimated as (10/20) x peak strength, which gives a cohesion of 0.85 MPa and a friction angle of 10 degrees. For dolomite, Goodman's reported value for limestone is used as a proxy - using an average transition pressure of 50 MPa, the dolomite residual parameters become (10/50) x peak strength, which gives a cohesion of 1.95 MPa and a friction angle of 25 degrees.

Table 1: Material Properties

Rock Unit	Friction (degrees)	Cohesion (MPa)	Elastic Modulus (MPa)	Tensile Strength (MPa)	Unconfined Compressive (MPa)
Dolomite	56	3.9	5,394	2.94	29
Shale	20	1.7	980	0.10	7.7 (min)

Anticipated Ground Response

The mean confining pressure around the powerhouse cavern is expected to be about 10 MPa, which is greater than the lowest unconfined compressive strength test result of 7.7 MPa for shale. Thus the shale is expected to be overstressed. This was confirmed by an elastic analysis of the powerhouse cavern model performed using the hybrid finite/boundary element stress analysis program *Phases* (Rock Engineering Group, 1992). Primary support consisting of 7-m long rock dowels on 1.2-m centers was used in this initial analysis. The analysis showed that a large zone of overstressed rock develops along the cavern walls as the excavation is taken to grade. This zone is largest at the lower shale layer and tapers off towards the crown and invert. When the entire cavern has been excavated, the overstressed zone extends 16 m beyond the cavern wall at the location of the lower shale layer. In comparison, the overstressed zone is only 2.5 m thick in the crown area and 5 m at the invert. The large overstressed zone along the cavern wall is a potential cause of instability and requires further consideration. This condition was investigated further using an elastic-brittle-plastic material model. This type of analysis allows for material failure and stress redistribution.

The program *Phases* was used to analyze the stresses and deformations for the cavern with the rock modeled as an elastic-brittle-perfectly plastic material. While several sections of the cavern were investigated with the shale layers at varying locations on the

---

[3] It may be recalled that at the brittle-ductile transition pressure, the peak and residual strengths become identical. Below this pressure, the rock will have peak, residual, and intermediate strengths.

cavern wall and crown, the results from only the most critical section are presented. On this critical section, the bottom of the upper shale is 14 m above the crown and the top of the lower shale is 10 m below the crown. A model with the upper shale at its lowest anticipated position of 5 m above the crown shows that the upper shale layer does not become overstressed after cavern excavation and that the crown moves down only about 2.5 cm. Thus the upper shale layer is not anticipated to present a major stability problem for the cavern though it will have an important influence on groundwater movement.

The residual strength parameters estimated above were used to model post-peak strength in the elastic-brittle-perfectly-plastic model. The typical rock support used in this analysis consists of 9-m-long, fully grouted rock dowels (19-mm diameter, 100 kN capacity) on 1.2 m centers. The elastic modulus of the bolts is assumed to be 200,000 MPa. Support from shotcrete is also taken into account in the model. A 2-m-high and 2-m-deep block of concrete is also included in the model, placed in a mined-out section of the cavern wall shale layer. This block is designed to control squeezing as explained later.

Figure 3 illustrates the deformations predicted by *Phases* for this elastic-brittle-perfectly-plastic analysis. Deformations are shown in absolute values (meters) and by contours. Figure 4 depicts the principal stress arching caused by the shale layer. These figures indicate that the cavern can be expected to come to equilibrium under the expected stress field and rock conditions. The analysis reveals that a plastic zone containing weakened rock may be expected to form around the shale layer in the cavern walls. Vertical stresses are expected to arch further back inside the cavern walls as the shale peak strength is exceeded. The shale will carry reduced vertical stresses consistent with its post-peak strength, and the cavern wall comes to equilibrium at values of stress and deformation shown by the *Phases* results.

The *Phases* result indicates that the cavern reaches equilibrium with a maximum inward deformation of about 7.5 cm on the cavern wall at the location of the lower shale layer. This amount of deformation is considered to be acceptable based on observations of squeezing and failure of walls in an exploratory tunnel at the project site. In the exploratory tunnel, two locations were observed where a squeezing shale layer contributed to weakening and failure of the overlying dolomite. At one location, 4.5 cm of shale deformation had occurred and did not cause dolomite instability. At the other location, 11 cm of movement led to collapse of the overlying dolomite wall. While the shale deformation predicted by *Phases* at this design stage is considered acceptable, more or less primary support may actually be required to control deformations once construction exposes actual rock conditions.

Squeeze Control Design Measures

As indicated previously in this paper, the lower shale is expected to be overstressed and exhibit time dependent squeezing if left to itself. The *Phases* analysis shows that elastoplastic equilibrium is achieved with about 7.5 cm of shale movement in the sidewalls of the cavern. The rock mass may be expected to be somewhat weakened by cracks formed during the process of reaching this elastoplastic equilibrium. Uptake of moisture (groundwater) by the loosened shale would further weaken the rock over time. Further weakening will lead to more cracking and the process will continue indefinitely unless support is provided to prevent shale failure after achieving elastoplastic equilibrium. Such behavior was observed in the test adit.

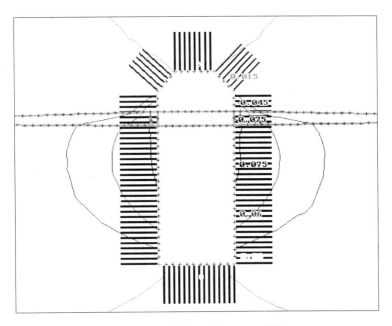

Figure 3: Cavern deformations at equilibrium

The intention of providing the 2-m by 2-m concrete block is to prevent squeeze soon after reaching elastoplastic equilibrium. Construction sequencing would include limiting excavation depths (especially in proximity to the main shale layers), provision of immediate support, and groundwater control. It is suggested to mine out the lower shale unit as it becomes exposed along a construction bench, or even along the full length of the cavern, using temporary supports. The void will then be immediately backfilled with concrete so that the concrete will have gained sufficient strength prior to the start of shale overstressing. Contact grouting is to be performed around the concrete block to ensure quick load transfer. Special drainage features must also be installed at this location where groundwater movement tends to be impeded by the shale aquicludes – a condition readily observed in the test adit. Once the cavern excavation passes this concreted shale stratum, additional anchors may be installed through the concrete block to supply increased block sliding resistance.

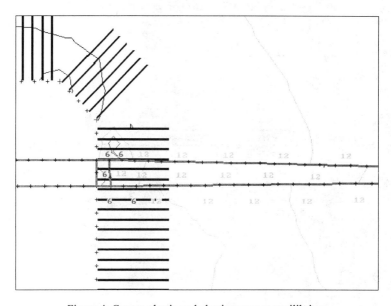

Figure 4: Cavern plastic and elastic zones at equilibrium

The *Phases* model does not address the sliding stability of the concrete block. This was addressed separately using a simple block model shown in Figure 5. After elastoplastic equilibrium is achieved, *Phases* predicts a vertical stress of about 6 MPa between the concrete block and the dolomite and an average vertical stress of about 10 MPa between the shale and the dolomite adjacent to the concrete block. The horizontal stress between the block and the shale predicted by *Phases* is also about 6 MPa at elastoplastic equilibrium. As mentioned above, the shale is expected to weaken further after the elastoplastic equilibrium due to the absorption of water by the loosened shale. The pressure on the block will then increase over time, as the shale tends to fail under the same vertical stress and lowered strengths. To arrest this progressive failure, the block must be able to supply a pressure to prevent active failure of the weakened shale, just as a retaining wall holds back cohesive soil. At this initial design stage where detailed data are not available, it is assumed that the moisture weakens the shale to a state where it has only half of its residual strength, i.e. the friction angle and cohesion are assumed to be 7 degrees and 0.43 MPa. With these lowered strength parameters and an effective vertical stress of 10 MPa acting on the shale, the pressure required to prevent active failure of the shale is estimated to be 7.1 MPa. This is somewhat higher than the elastoplastic equilibrium pressure predicted by *Phases* as would be expected. A sliding stability analysis of the block indicates that the factor of safety against failure is 1.2. Higher safety margins can be achieved by increasing the block size or number of anchor bolts when better material data become available. For example, increasing the block size to 2-m high by 3-m deep results in a sliding safety factor of 1.8.

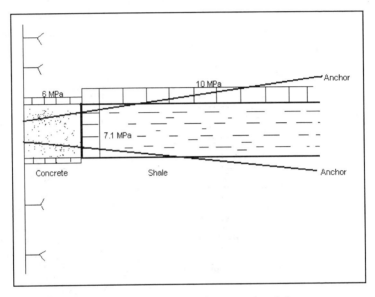

Figure 5: Concrete block supporting squeezing shale

Conclusions

The stability of the deep cavern described in this paper will be strongly influenced by the lower shale layer that intersects the cavern walls. The shale affects both the immediate and long-term response of the cavern to the excavation. It has been shown that both immediate elastoplastic response and delayed squeezing can be controlled by using appropriate support systems and excavation sequencing. It is expected that a stable cavern can be constructed at this site using the methods described in this paper.

References

Goodman, R.E. (1989), Introduction to Rock Mechanics, John-Wiley

Rock Engineering Group (1992-93), *Phases*, University of Toronto

EXPECTED PERFORMANCE OF THERMAL STRESS RELIEF ELEMENTS FOR GROUND SUPPORT AT THE YUCCA MOUNTAIN REPOSITORY

Yiming Sun[1], Richard M. Nolting III[1] and John Cogan[1]

ABSTRACT

Ground support is being designed for emplacement drifts of a potential high-level radioactive waste repository at Yucca Mountain in Nevada. The emplacement drifts, to be emplaced with nuclear waste packages that produce wall rock temperatures approaching 200°C, must provide access for at least 150 years before repository closure. Because excessive thermal stresses will be induced in ground support components, incorporation of stress relief elements is being considered to accommodate the thermally induced deformation of rock mass and ground support materials, thereby lowering stresses to acceptable levels. Numerical models are used to examine the performance of stress relief elements. Results indicate that the stress relief elements can substantially reduce hoop stresses induced by elevated temperatures. The magnitude of stress reduction depends upon rock mass property and location and deformability of the elements. To facilitate movement of the elements the interface of the lining and the rock should be non-bonding.

1. INTRODUCTION

Ground support is being designed for underground excavations for a potential high-level radioactive waste repository at Yucca Mountain, 170 km northwest of Las Vegas, Nevada. Emplacement drifts (5.5 m in diameter), with a total length of about 120 km and arranged in a near planar layout, will be excavated in a volcanic welded tuff, called the TSw2 thermal/mechanical unit, which lies about 200 m to 400 m below the ground surface (Nolting et al 1998). These emplacement drifts will accommodate 70,000 metric tons of initial uranium (MTU) of waste that is packaged in canisters, or waste packages. The waste packages will generate heat up to about 18 kW per package, which produces

---

[1]Morrison Knudsen Corporation, 1261 Town Center Drive, Mail Stop 423, Las Vegas, Nevada 89134

wall rock temperature approaching 200°C (CRWMS M&O 1998). Therefore, thermally-induced stress is considered to be most critical to the performance of the ground support system in the emplacement drifts.

There are two major factors that contribute to the stress and strain in the ground support during the heating process: (1) rock mass deformation due to thermal expansion of the ground and (2) thermal expansion of ground support materials. Therefore, the stress in the ground support may be several times higher than when subjected only to in-situ stress loads (CRWMS M&O 1998). Based on current knowledge of rock properties and loads conventional design of ground support for ambient conditions is not able to fully prevent the ground support from overstressing under thermal loading conditions. As a consequence, special measures, i.e., the use of stress relief elements (deformable joints), have been incorporated in the ground support design.

2. ROCK MASS PROPERTIES AND LOADING CONDITIONS

2.1 ROCK MASS PROPERTIES

The repository host rock, primarily the TSw2 thermal/mechanical unit, is highly jointed with frequent vertical joints and some horizontal joints. Some physical properties of the TSw2 unit, such as thermal conductivity, specific heat and coefficient of thermal expansion, are temperature-dependent (CRWMS M&O 1998). For example, the thermal conductivity value decreases from 2.13 W/m·K for wet rock to 1.50 W/m·K for dry rock, and the coefficient of thermal expansion increases from $7.14 \times 10^{-6}/°C$ at ambient temperature to $13.09 \times 10^{-6}/°C$ at 200°C. The mechanical properties, such as modulus of elasticity, cohesion and friction angle, also have a wide range, depending on the rock mass quality category (RMQ). The RMQ categories, ranging from 1 to 5, represent variations in the rock mass and reflect the influence of rock joints on the rock mass properties. A low RMQ value indicates lower strength rock while a high RMQ value indicates higher strength rock. The modulus of elasticity for the TSw2 unit ranges from 7.76 GPa for RMQ=1 to a nearly intact value of 32.61 GPa for RMQ=5.

2.2 Loading Conditions

The in-situ vertical stress state at the repository horizon ranges from 5 to 10 MPa due to variation in the thickness of overlying rocks. The horizontal-to-vertical stress ratio ranges from 0.3 to 1.0 (CRWMS M&O 1998).

The design thermal load for the repository is related to waste package heat output, waste package spacing, and drift spacing. In this study, a thermal load of 85 metric tons of uranium (MTU) per acre is used (CRWMS M&O 1998). Based on this thermal load with 21 PWR waste packages and a drift spacing of 28 m from center-to-center, the waste package spacing is calculated to be 15.18 m. The service life for the ground support systems in the emplacement drifts to sustain the thermal load and maintain fully functional drifts is at least 150 years following waste emplacement.

3. STRESS RELIEF ELEMENTS FOR THE GROUND SUPPORT SYSTEM

Stress relief elements (deformable joints) are being designed for the ground support in the emplacement drifts to accommodate the rock and lining deformation induced by elevated temperatures. These elements can be incorporated into either precast concrete segments or steel sets. The function of the stress relief elements can be described as follows: During lining installation and under in-situ stress load, the strength and stiffness of the stress relief elements are sufficient to prevent the elements from collapsing or deforming too much to allow load to be transferred through the elements. Under thermal load, when stress in the elements reaches a certain level, the elements deform a prespecified amount. As a result, stresses in the ground support induced by rock deformations are reduced.

Use of stress relief elements in ground support is unconventional but practically achievable. In most documented cases (Grabe 1996), deformable linings are used in tunneling with either high overburden or high water pressure. In either case excessive ground deformation is anticipated. For example, Strobbausl (1996) presents a deformable lining system that includes deformable elements installed in the longitudinal joints of the lining combined with a deformable annular grout between the lining and the rock.

3.1 Materials and Behavior of Stress Relief Elements

Stress relief elements can be made of a variety of materials. In each case the material must either contain voids which can collapse, or must be plastic so that it can be extruded under pressure. Some materials with voids include crushable, granular, prefabricated metal honeycombs, metal springs, foamed plastics, and cellular concrete (Hoff 1964).

Figure 1 illustrates an idealized behavior for a stress relief element material. As the load increases from zero, an initial elastic reaction takes place which is characterized by an elastic modulus of compression, k_e. This modulus is well below that for lining concrete and reflects the amount of air entrapped as small bubbles within the material. No failure takes place in the air/cement composite at this time. With continued loading, a yield point Y, is finally reached, after which the curve flattens considerably. Then the collapse of hardened cement paste between individual air bubbles begins. Subsequent displacements take place with continued collapsing activity, thus avoiding significant load increase. Collapsing material weakens as localized failure progresses but because of confinement between the beam flanges, a general failure cannot occur. This material state is characterized by a very low deformation modulus, k_y, defined as the stiffness of the joint. With continued loading, full collapse is eventually reached at the point C. This is called the locking point of the material. The material is now a crushed mass confined within the beam web. Additional loading can only result in very small displacement which is characterized by a very high modulus, k_c. The value of this modulus will be on the order of the modulus of elasticity of lining concrete.

In practice a precast segmental concrete lining with stress relief elements in place is installed in an emplacement drift at a thrust load below the point Y. As heating of the drift proceeds after waste emplacement, the rock thermal expansion will cause the drift perimeter to slowly close around the lining. With time the drift closure will reach the displacement which corresponds to the yield load of the compressible material, and joint yielding will begin. Further perimeter contraction will then only lead to modest stress increase in the lining due to the collapse of the joint material. Thus, the lining will contract without inducing a high stress. As long as the strength of the lining lies above the full collapse load value, identified by the point C, the lining will not fail during the joint yielding. This means that the critical parameters for the design of a yieldable joint are the elastic modulus, k_e, the plastic modulus, k_y, and the loads and displacements at points Y and C. The material yield load should be high enough to allow for lining installation thrusts, and the total displacement between initial yield and full collapse must allow for the expected shortening of the lining perimeter due to the thermal expansion of rock.

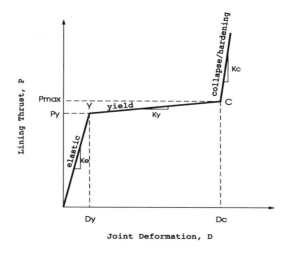

Figure 1. Idealized Behavior for a Compressible Joint Material

Figure 2 illustrates concepts using stress relief elements in lining systems. In Option A, a steel spring is sandwiched between two segments. This system would have a yield point followed by a plastic state as it collapses. In Option B, steel tubes are placed between two plates. As the load increases, the tubes flatten. Plastic collapse occurs when a yield stress is reached. A compressible material can be used to fill the tubes. In Option C, crushable material such as cellular concrete is filled at a joint. The yield stress and deformation depend on the properties of the crushable material and can generally be developed. The concepts demonstrated in Figure 2 can also be applied to the lining system with steel sets.

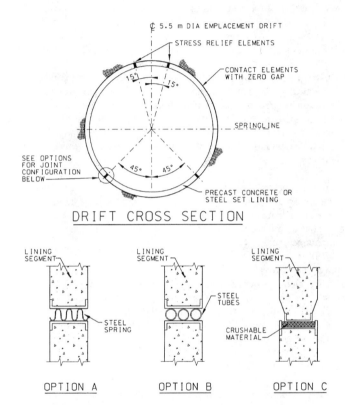

Figure 2. Illustration of Concepts for Stress Relief Element in Lining System

3.2 Functional Requirements for Stress Relief Elements

Effectiveness of the stress relief elements depends on the conditions of the interface at the lining and the rock, location of the elements along a lining ring, and element deformability or stiffness.

To facilitate the action of stress relief elements, an unbonded circumferential interface or discontinuity between the lining and the rock is essential to allow the lining and the rock to slide relative to each other. An unbonded interface is typical of the contact between an expanded precast segment ring and the excavated rock surface, at least in an ideal situation. However, because an expanded precast segment concrete or steel set lining will never fit exactly the bored surface of the opening as the radius of the lining is usually slightly less than that of the bore, areas with gaps or no contact between the lining and the rock will exist. As a consequence, this interface is grouted to provide a more uniform contact area for load transfer between the lining and the rock. Thus, where stress relief elements are used the type of grout must not significantly inhibit shear movement.

Location of stress relief elements on a lining ring is selected to maximize the stress relief effect. Theoretically, the elements should be placed near the crown and invert if the loading condition is predominantly in the horizontal direction, or be located near the springline if the loads act predominantly in the vertical direction. However, practical considerations are taken into account to avoid the complexity and difficulty of lining fabrication and installation, for example, it would be impractical to locate a stress relief joint at the invert.

The minimum number of the stress relief elements on a lining ring can be determined based on the anticipated deformation of the lining and the expected collapsible distance of each element. For example, if the lining is anticipated to deform about 40 mm and each stress relief element can absorb about 10 mm of displacement, the number of the elements should be equal to 4.

In this study, four stress relief elements are embedded in the lining, as illustrated in Figure 2. The elements are located near the crown and invert where the anticipated horizontal stresses are much higher than elsewhere in the lining because the thermally-induced rock deformation is predominantly in the horizontal direction (CRWMS M&O 1998). In addition, the modeled interface between the lining and the rock is not bonded but allows for uniform loading.

4. NUMERICAL MODELING APPROACHES

An unventilated emplacement drift will experience an increase in temperature due to the heat output from the waste packages. Radiation dominates in the heat transfer process from the waste packages to the drift wall. Natural convection by air is considered to have a negligible effect. Within the rock mass, heat flow will occur by conduction due to the thermal gradient between the high-temperature drift wall and the low-temperature rock away from the drift. Results from an on-going in situ heater test indicate that these are the heat transfer mechanisms being observed (SNL 1998).

The ANSYS and FLAC computer codes are used for thermomechanical modeling of the stress relief elements. In the ANSYS models, the circumferential interface is simulated with two-dimensional point-to-surface contact elements. The contact elements represent two surfaces of the lining and the rock, and are capable of supporting only compression in the direction normal to the surfaces and shear in the tangential direction. The shear and compression are proportional at each contact element. No initial "gap" is assigned to the contact elements, representing a full initial contact between the concrete or steel set lining and the rock surface. The only physical property of the contact elements is the coefficient of friction, and its value may depend on the types of materials that form the interface.

The stress relief elements are simulated with two-dimensional point-to-point contact elements. Similar to the point-to-surface contact elements, this type of contact element is also capable of supporting compression in the normal direction. The numerical parameter that controls the amount of compression or "penetration" under loading is the normal stiffness. The value of the normal stiffness can be estimated from the allowable stress for the lining divided by the allowable displacement (compressible distance) of a stress relief element. Note that performance of the stress relief elements under multiple heating and cooling cycles, such as may occur during the repository preclosure period, is not discussed here.

5. NUMERICAL EVALUATION

5.1 Rock Temperature

Temperatures in the rock mass were calculated for a period of 150 years after waste emplacement based on a thermal model with ANSYS. Results show that for a thermal load of 85 MTU/acre an average peak temperature of about 163°C at the drift wall is reached at about 55 years following emplacement (Figure 3). The temperature will stay above boiling many centuries past the period of 150 years.

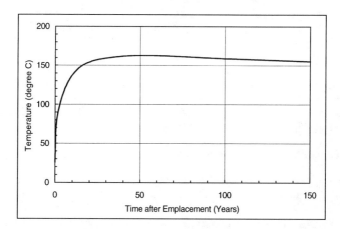

Figure 3. Time Histories of Temperatures on Drift Wall for Thermal Load of 85 MTU/acre

5.2 Rock Mass Stress and Deformation

A thermomechanical model using the FLAC code indicates that compressive tangential stress at the crown of an emplacement drift, as shown in Figure 4, increases with time and rock mass modulus, reaching maximum values of about 23 and 98 MPa at about 60 years for rock mass quality categories of 1 and 5, respectively. Potential rock yield, based on the strength-to-stress ratios, is anticipated to extend to a depth of about 0.3 m to 1.0 m, depending on the rock mass modulus. The results suggest that the rock mass generally behaves elastically, except for local areas in a zone immediately surrounding the opening.

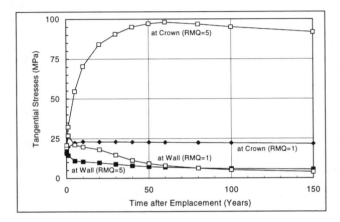

Figure 4. Time Histories of Tangential Stresses at Crown and Springline under In-Situ Stress of 10 MPa and Thermal Load of 85 MTU/acre for Rock Mass Quality Categories 1 and 5

Emplacement drift horizontal and vertical closures (changes in drift diameter) induced by elevated temperatures are presented in Figure 5. These closures also vary with time and rock mass modulus of elasticity, ranging from about 26 mm (inward) in the horizontal direction to about 7 mm (outward) in the vertical direction, in response to the predominantly higher horizontal thermal stresses.

5.3 Effect of Stress Relief Elements

Ground support options considered in this study include 200-mm-thick precast concrete segments and 1.5-meter-spaced W6×20 steel sets (CRWMS M&O 1998). The long-term behavior of the concrete segments and of the steel sets, subjected to combined in-situ stress and thermal loads, is simulated using the ANSYS computer code. Two cases, with a fully-bonded and an unbonded interface between the lining and the rock, representing ground supports with and without stress relief elements, are modeled to evaluate their effectiveness in reducing lining stress.

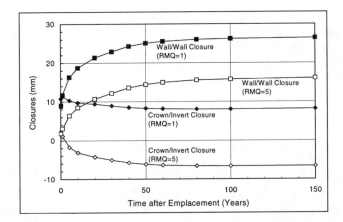

Figure 5. Time Histories of Drift Closure under In-Situ Stress of 10 MPa and Thermal Load of 85 MTU/acre for Rock Mass Quality Categories 1 and 5

5.3.1 Concrete Lining

Figure 6 shows time histories of stresses in a concrete lining modeled with and without stress relief elements. Results indicate that concrete stresses induced by elevated temperatures are very sensitive to the rock mass modulus of elasticity when the concrete is fully bonded to the rock. The stress values are calculated to be greatest for the highest rock mass quality category of 5. Without considering stress relief elements in the lining, the maximum compressive stresses, occurring at about 100 years following waste emplacement, would be about 75 and 96 MPa for rock mass categories of 1 and 5, respectively. These calculated stress values exceed the reference concrete ultimate stress limit of 34.5 MPa.

When stress relief elements are used, stresses in the concrete lining (Figure 6) are shown to be substantially lower, ranging from about 11 MPa to about 19 MPa depending on the location of the stress relief elements in the lining ring and the rock mass modulus of elasticity. It is indicated that the stresses are higher for the lower rock mass modulus of elasticity, which is different from the case where the lining is bonded to the rock. In addition, the stresses at the crown are lower than at the invert because the stress relief elements used in the models for this particular segment configuration are closer to the crown than to the invert, as illustrated in Figure 2. To maximize the stress relief effect, the stress relief elements should be placed as close to the crown and invert as practically possible.

The magnitude of stress reduction in the concrete lining also depends on the stiffness or the collapsible distance of the stress relief elements. Figure 7 presents the time histories of compressive stresses at the crown of a concrete lining with different values of deformation (stiffness) for the stress relief elements. As expected, the greater the stiffness, the higher the stresses. To maintain the stresses in the lining within an allowable level, a sufficient amount of deformation for the stress relief elements should be allowed.

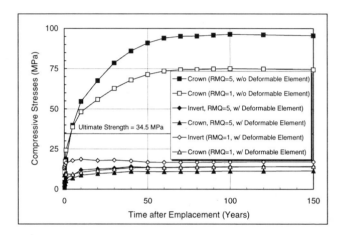

Figure 6. Concrete Lining Compressive Stresses, Showing the Effects of Stress Relief (Deformable) Elements and Rock Mass Category

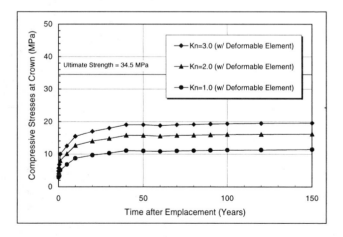

Figure 7. Concrete Lining Compressive Stresses, Showing the Effects of Stiffness of the Stress Relief (Deformable) Element

5.3.2 Steel Sets

Time histories of stresses at the crown and invert of steel sets (W6×20) are illustrated in Figure 8. Results indicate that when the steel sets are modeled, albeit unrealistically, with full bonding to the surrounding rock they will be overstressed. Maximum compressive stresses are approximately 785 and 843 MPa occurring at 100 years after waste emplacement for rock mass quality categories of 1 and 5, respectively. These values are of course well above the reference steel yield limit of 248 MPa. However, steel sets are not bonded to the rock, and assuming a full bonding between the steel and the rock will certainly overestimate the stresses in the lining.

When stress relief elements, combined with an unbonded interface between the steel sets and the rock, are modeled, the calculated maximum stresses in the steel sets, as shown in Figure 8, are about 102 MPa and 152 MPa at the crown and the invert, respectively, for rock mass category of 5. These results indicate that the stresses in the steel sets, as was the case in the precast segment concrete lining, are sensitive to the location and stiffness of the stress relief elements and the rock mass modulus of elasticity.

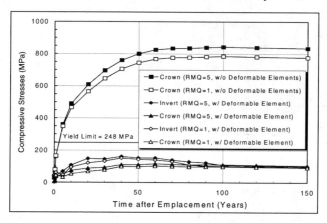

Figure 8. Steel Set Compressive Stresses, Showing the Effects of Stress Relief (Deformable) Elements and Rock Mass Category

6. CONCLUSIONS

Incorporation of stress relief elements in ground support systems composed of precast concrete segments and of steel sets appears to be an effective way to accommodate the thermally-induced deformation of drift rock and ground support materials, thereby lowering stresses to acceptable levels. Results of numerical analysis indicate that the stress relief elements can substantially reduce hoop stresses induced by elevated temperatures. The magnitude of stress reduction depends upon the rock mass modulus of elasticity, location and stiffness of the elements, and condition of the interface between the lining and the rock. To maximize the stress relief effect, the elements should be placed close to the crown and the invert because the thermal loads are predominantly in the horizontal direction. In addition, the interface between the lining and the rock should be non-bonding in order to facilitate movement of the elements.

ACKNOWLEDGEMENTS

The Yucca Mountain Project (YMP) is a U.S. Department of Energy project. TRW Environmental Safety Systems, Inc. (TESS) is DOE's M&O contractor for the YMP. Design, geotechnical analysis, and site investigations are being conducted by several subcontractors to TESS. Repository subsurface design is being performed by Morrison Knudsen Corporation.

REFERENCES

CRWMS M&O (Civilian Radioactive Waste Management System Management and Operating Contractor) 1998. *Repository Ground Support Analysis for Viability Assessment*. BCAA00000-01717-0200-0004 REV 01. Las Vegas, NV.

Grabe, W. 1996. Design of Gaskets for Deformable Tunnel Lining Joint Configuration: New Developments, in *Tunnel Boring Machines: Trends in Design & Construction of Machanized Tunnelling*. Proceeding of the International Lecture Series TBM Tunnelling Trends, Hagenberg, Austria, Edited by H. Wagner & A. Schulter. pp. 239-242. Rotterdam, Netherlands: A. A. Balkema.

Hoff, G. C. 1964. Shock-Isolating Backpacking Materials - A Review of the State of the Art, *Proceedings of the Symposium on Soil-Structure Interaction*. pp. 138-154. Tucson, AZ: University of Arizona.

Nolting, R. M., F. Duan & Y. Sun 1998. Ground Support for Emplacement Drifts at Yucca Mountain. *Proceedings of 8th Congress of the International Association for Engineering Geology and the Environment*, Edited by D. P. Moore & O. Hungr. Vol. 5, pp. 3775-3782. Rotterdam, Netherlands: A. A. Balkema.

SNL (Sandia National Laboratories) 1998. *Drift Scale Test Drift Status Report #1: Evaluation and Comparative Analysis of the Drift Scale Test Thermal and Thermomechanical Data (Results of 12/3/1997 through 5/31/1998)*. Albuquerque, NM.

Strobbausl, S. 1996. TBM Tunnelling under High Overburden with Yielding Segmental Linings: EUREKA Project EU 1079 - "CONTUN", in *Tunnel Boring Machines: Trends in Design & Construction of Mechanized Tunnelling*. Proceedings of the International Lecture Series TBM Tunnelling Trends, Hagenberg, Austria, Edited by H. Wagner & A. Schulter. pp. 61-68. Rotterdam, Netherlands: A. A. Balkema.

WATERPROOFING TUNNELS WITH GEOSYNTHETICS IN EUROPE

Alberto M. Scuero[1]   John A. Wilkes[2]   Gabriella Vaschetti[3]

1.0 ABSTRACT

Geosynthetics can be successfully employed in underground construction. PVC geomembranes have been used as barriers against water and gas infiltration from the surrounding ground, in highway, subway, railway and hydropower tunnels. This paper addresses design, installation and quality control of two case histories: Vaglia/Firenzuola a high-speed railroad project in Italy currently under construction and Spalov, a hydropower tunnel in the Czech Republic completed in 1998.

2.0 INTRODUCTION

The use of geosynthetics in the field of underground construction has been evolving over the last 20 years. Many projects for underground structures such as road, subway and railway tunnels now include these materials as an important element of the design. In particular, a geomembrane system can constitute a dependable and long lasting barrier to infiltration of water and gas present in the surrounding ground. Water and hazardous gas infiltrations can deteriorate the concrete structure and the elements present within the structure itself (vehicles, appurtenances, monitoring equipment, people). For these reasons, infiltration of water and gas must be avoided. The use of a flexible geomembrane system with a drainage system prevents water and gas infiltrations. This paper describes two different applications of geosynthetics to tunnels. The first project describes the Vaglia and Firenzuola railroad tunnels where a geosynthetic system is used to waterproof the tunnel to prevent ground water intrusion. The second project describes the Spalov hydropower tunnel, where a geosynthetic system is installed and left exposed to waterproof the tunnel and increase flow by decreasing the surface roughness.

3.0 VAGLIA AND FIRENZUOLA HIGH-SPEED RAILROAD TUNNELS

The Vaglia and Firenzuola tunnels are part of a new high-speed railroad system in Italy connecting Bologna with Florence. This new section will increase traffic capacity on the most crowded railroad line in Italy. The project, to be accomplished over a period of

[1]Director, CARPI Tech, Via Monte Bianco, 5, 28041, Arona (NO) Italy
[2]President, CARPI USA, 1948 Franklin Road, Suite A2, Roanoke, VA 24014
[3]Consultant, Geosynthetics Hydro, Via Monte Bianco, 5, 28041, Arona (NO) Italy

78 months, involves construction of a new line measuring a length of 78.3 km, 73.1 km of the railroad line will be underground, with a total of 9 tunnels. Vaglia (6,696 m) and Firenzuola (14,339 m) are among the longest tunnels and are the only ones currently under construction.

Design of the tunnels varies according to the nature of the hydrogeological conditions. The conceptual scheme, from the natural ground towards the inside of the tunnel, consists of steel beams to maintain tunnel shape, a shotcrete layer, a waterproofing system, and a final concrete structure. The Vaglia and Firenzuola 1 tunnels are being waterproofed by CARPI with a total of 257,000 m^2 to be installed over a period of 3 years. Firenzuola 2 entails installing the same geomembrane waterproofing system with a total of 250,000 m^2.

According to modern tunneling techniques, excavation is immediately followed by steel beam installation and shotcreting, which limits inward movement. The impermeable geomembrane system is positioned over the shotcrete layer before the final concrete structure is cast. The geomembrane prevents the intrusion of water and gas while the concrete constitutes the ballasting structure for the geomembrane. The effectiveness of the waterproofing system depends not only on the quality of the geomembrane manufactured, but also on the installation and on the quality control procedures. Any damage occurring during installation of the liner, reinforcement, or concrete must be detected and repaired before the final concrete structure is cast.

The stability of the geomembrane relies on the efficiency of its anchorage technique. Typically, the geomembrane is anchored to the shotcrete at point or line attachments, as will be described below. The geomembrane must be secured before the final concrete structure is cast.

Most underground structures are subject to permanent piezometric head, thus the hydrostatic design must consider the stability of the structure itself and the existing equilibrium. A drainage system is usually provided between the geomembrane liner and the shotcrete layer. This system is passively drained so that water infiltrating behind the liner is transmitted, collected and discharged. In this way, hydrostatic pressures are controlled and relieved.

The choice of the geomembrane material must consider the impermeability, the constructability, and the suitability of the material to an underground environment. The material used in underground tunnels in Europe is flexible polyvinylchloride (PVC) due to:

(a) its flexibility and elasticity, which enables it to accommodate the irregularities of excavation,
(b) its resistance to puncture and burst, which guarantees its integrity at protrusions and cavities of an irregular substrate,
(c) its easy handling, suitable to challenging underground installation conditions,
(d) its easy and reliable welding, even at low temperatures,
(e) its being self-extinguishing, and
(f) its resistance to chemical aggression of microorganisms

The type of PVC geomembrane typically used is 2 mm in thickness, having two different colors with the side facing the rock as a dark color, while on the side facing the final concrete layer the membrane is a light color (for a thickness of approximately 0.3 mm). The different colors allow easy and quick visual detection of any damage occurring to the liner during supply, storage or installation, as any tear or hole in the thin light side is signaled by appearance of the darker underlying material. Transparent PVC geomembranes are also used.

The waterproofing system consists of a drainage collection and discharge system, of a drainage layer, and of a waterproofing liner. (Figure 1)

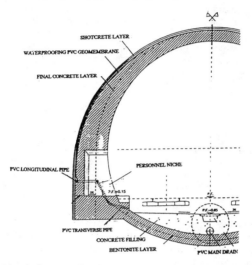

Figure 1. The drainage collection and discharge system at Vaglia and Firenzuola consists of longitudinal microperforated PVC pipes, and of transverse PVC pipes discharging into the main drain every 25 meters, typically at personnel access locations (Fig. 1). When the piezometric head exceeds 50 m, an additional 1.5 meter wide drainage layer is installed over the whole span of the tunnel, in the middle of the 25 m span between discharge points for additional drainage.

Vaglia and Firenzuola tunnels are subject to piezometric heads varying from 40 m to 110 m and from 120 m to 190 m respectively. Design and construction of the waterproofing system was thus a very demanding task.

The drainage layer must facilitate the in-plane flow of water present behind the liner towards the collection and discharge system. A high in-plane transmissivity geotextile was installed for this purpose over the shotcrete layer constituting the primary support. The chosen nonwoven polypropylene (PP) geotextile, area mass 400 g/m^2, provides the required drainage capability and puncture protection for the waterproofing geomembrane against the fairly rough surface of the shotcrete layer. The nonwoven geotextile has excellent mechanical resistance capabilities and is therefore capable to efficiently sustain mechanical stresses such as applied by a rough substrate. Anchorage of the geotextile to the shotcrete was made by steel impact anchors fitted with PVC disks, which provided the surface for subsequent anchorage of the waterproofing liner. The four anchors designed for each square meter were eventually increased during installation for more secure anchorage.

The waterproofing liner is a two-color PVC geomembrane, manufactured by coextrusion to produce a one-layer material, with homogeneous characteristics in its whole mass and a total thickness of 2.0 mm. Temporary anchorage of the liner, necessary before placement of the final concrete inner ring, is provided by the PVC disks attached to the anchors of the drainage geotextile. The PVC geomembrane was heat-welded on these disks, each approximately 75 mm in diameter, and 8 mm thick. Resistance of the PVC disks is inferior to resistance of the PVC geomembrane: when subjected to high stresses, the PVC disks break before failure of the geomembrane can occur, so that the integrity of the geomembrane is not compromised. The PVC liner is supported by the inner concrete ring applied over the shotcrete and PVC layers

The geomembrane was manufactured in long panels, so as to span the entire perimeter of the tunnel. The PVC sheets were installed starting from one side of the invert around to the crown and then over to the other side. Adjacent sheets overlapped for at least 8 cm and were subsequently heat-welded by the hot air, double track automatic method.

The installation of a single roll of geomembrane to cover the entire perimeter of the tunnel was facilitated by the use of a special movable steel structure designed and constructed specifically for these tunnels. The structure consisted of two arches following the shape of the tunnel, and supporting a 4 m long gantry crane. The installing crew worked on the gantry crane, which is hinged to the structure and traveled over the entire span of the arches (see Fig. 2). The crane was equipped with a device for hoisting and unrolling the geomembrane sheets as the crew was securing them to the PVC strips. The structure represents a major improvement in installation technique, providing a quicker and more installation efficient.

The installation plan includes the numbers of rolls including their size and location. The waterproofing installation plan with the general contractor is important because the concreting of the final ring should take place within 20 days after installation of the geomembrane. This precaution is designed to avoid the hydrostatic pressure build-up behind the geomembrane and on the anchorage. It also assures that the waterproofing geomembrane specialists are present during civil works to monitor final concrete operations and insure procedures do not compromise the geomembrane integrity.

Figure 2. Vaglia tunnel, with the movable steel structure in the background

Reliable welds are necessary to construct a continuous liner. The use of PVC membranes allows execution of fast, reliable welds. Welds are made by an automatic welding machine, with manual welding being accomplished only at locations where automatic welding is not feasible. Testing of automatic welds is performed by injecting air at a constant pressure (2 bar) in the 10 mm test channel created between the two welds. If the prescribed pressure is maintained, or does not decrease below 10% of the initial value for the next 10 minutes, the weld is accepted. The Owner's quality control plan initially requires testing 10% of the automatic welds in each concreting section. In case a weld does not comply with the weld specifications, another two welds of the same section are tested. If both welds are accepted, then the unacceptable weld is repaired according to the additional established repair procedures. If either of the other two welds does not comply, all welds of that concreting section are tested. All of the tested manual welds, both those between sheets and those used for cover strips, are tested by the ASTM D4437 Mechanical Point Testing method. In addition, random samples are tested with the vacuum box method.

Before installation of formwork for concreting, the entire surface of the liner is visually inspected. Any visual defects are immediately repaired, in accordance with specified repair procedures.

The Quality Control procedure after installation of the liner system includes verifying the efficiency of the drainage system. After each section is concreted, water in injected into the drainage system and the discharge is monitored to check communication between the transverse pipes and the longitudinal pipe. If water does not flow from the transverse pipe, the system is clogged. It must be cleaned of the concrete that has penetrated the system. Cleaning is done by pressurized water with the inspection hose provided in each section.

Experience acquired at the tunnels described in this paper showed that the primary cause of damage detected after installation of the geomembrane was due to operations related to placement of the reinforcement, movement of the formwork and concreting operations. Clogging of transverse drains caused by additives in the shotcrete was also experienced, indicating a need for increasing the awareness of the workers and of cleaning the drains.

4.0 SPALOV HYDROPOWER TUNNEL

The Spalov Hydro Power Tunnel located in the Czech Republic is a tunnel excavated in granite rock and lined with concrete, with a cement- stones ratio of 1: 7-9. The thickness of the concrete is about 15 cm. The height of the tunnel is 2.7 m and the width at the base is 3.6 m. The radius of the vault is 1.8 m. The maximum velocity of the water is 1.6 m/sec with a maximum flow of 12 m^3/sec. The Spalov tunnel has suffered degradation of the concrete, mainly at the upstream end. The degradation is mainly due to the dynamic action of the flow and to freezing/thawing cycles.

In the initial upstream stretch of about 300m, the main damage entailed complete disappearance of the existing concrete lining in several spots. The bottom of the tunnel was similarly affected by some diffused scouring in the central section. The tunnel vault was generally dry. Some water infiltration along the vertical walls were channelled through small plastic pipes, 10mm diameter.

Prior to the rehabilitation works, the surface of the walls were covered by a 2 to 3 cm thick layer of sediment, which could be easily removed by hand. In some locations of the walls and at the springline of the vault, the cement mortar previously placed as a remediation to the scouring created by the flowing water was detached and pieces up to 20 cm x 30 cm in area could be removed by hand.

Cleaning operations have been conducted at irregular intervals of time, with the last cleaning occurring in 1993. The addition of a waste water treatment plant upstream from the Spalov Tunnel in 1994 by the town of Selimy has greatly reduced the quantity of sediments that accumulate on the internal surfaces of the tunnel. Furthermore, the presence of a screen at the inlet prevents the intrusion of rocks with a diameter greater than 10 cm.

In June 1998, HYDROPOL , acting as a consultant to VCE (Owner of the facility), was planning to restore the integrity of the concrete liner and to replace the stoplogs at the upstream end of the tunnel. At the time, two alternatives methods were considered for the repair of the existing concrete liner:

1. demolishing the damaged parts, transporting the debris to the downstream end, placing anchors and reinforcement steel mesh, and applying shotcrete in layers up to 5 cm of thickness;
2. installing a patented CARPI waterproofing system, consisting of a geocomposite mechanically fastened to the existing surface by a combination of stainless steel profiles, anchored to the existing subgrade.

The project needed to start in early September 1998 and be concluded by October 15, 1998, to coincide with the testing of the new turbines in the power house. During the same period of time, the stoplogs of the upstream portal would also be reconstructed.

The removal of the stoplogs at the downstream portal, to allow their reconstruction, would prevent the diversion of the water from the river in case of flood. These unpredictable conditions meant the tunnel was subject to flooding during the repair work.

VCE, on suggestion of HYDROPOL, decided to use the patented CARPI geomembrane solution because of the following advantages over the shotcrete option:
1) the geomembrane provides a full permanent protection to the concrete of the tunnel
2) maintenance will be greatly reduced
3) cleaning operations will be minimized
4) construction works would be quicker
5) in case of a flood the installation equipment can be easily removed, and
6) the geomembrane solution was price competitive.

The contract between VCE and CARPI was signed on August 14 and the installation works started on September 7, 1998. The project, which required 27 working days, was completed in October 1998 and then successfully impounded.

The proposed solution did not require concrete ballasting of the bottom yet it still allowed for passage of pedestrian traffic or light equipment with smooth rubber wheels. Cleaning of the bottom will be performed by light equipment with a blade covered with rubber. For cleaning operations, it would also be possible to use a special light floating platform on which a high pressure water pump is mounted. This equipment eliminates the use of shovels or other mechanical devices which could damage the geocomposite.

The installation of a waterproofing geomembrane on the internal surface of a free flow or pressure tunnel results in some desirable long term performance results. The geomembrane is installed quickly at a competitive cost with the lifecycle cost being a fraction of other alternatives.

Hydro tunnels can be lined with geomembranes in either an exposed position or covered by a layer of concrete. The geomembrane has to be designed to withstand the different operating conditions: an exposed geomembrane has to withstand the dynamic actions, negative pressure, abrasion, and impact of the flowing water, while remaining watertight, on the other hand, a geomembrane covered by concrete or soil has to provide only watertightness. The advantage of an exposed geomembrane system is that the reduced coefficient of friction of the tunnel lining allows large increases in flow (typically 40 - 90%).

HYDROPOL requested that the geomembrane be left exposed in order to prevent the reduction of the cross section of the tunnel and to take the maximum advantage of the smoothness of the geomembrane, thereby increasing the water velocity and the flow capacity. During the installation, an exposed geomembrane reduces the quantity of material to be transported inside the tunnel because the volume and weight of the geosynthetic liner is much less than a concrete liner system. Additionally, an exposed geomembrane system reduces the preparation works because drainage material smoothes out rough spots without additional surface preparation. Statistically, most damage to the geomembrane occurs while placing the protective layer.

The objective of the lining of a hydro tunnel is to reduce seepage through the structure with the ultimate goal of making it watertight. The use of geomembranes as an impermeable liner enables the designer to achieve an impermeable barrier several orders of magnitude ($k = 1 * 10^{-12}$ cm/sec permeability) more impermeable than traditional systems. A geomembrane allows covering the entire tunnel surface including bridging cracks (even those occurring after installation), fissures, and the construction joints. The elasticity of a PVC geomembrane is approximately 230%, which allows the geomembrane to accommodate significant movements in the structure after installation.

Some of the benefits of an exposed tunnel lining geomembrane system are:

1. Long Service Life
2. Elimination of Leakage
3. Zero Maintenance Costs
4. Low Installation Costs
5. Modular Installation

There are also some significant quality benefits for using a geomembrane system. The face of the structure is sealed covering all fissures, cracks and construction joints. The elasticity of the PVC geomembrane allows the system to accommodate new cracks or fissures as they develop. The drainage system provides definitive objective measurements for system performance: water flow through drainage system and hydrostatic pressure behind geomembrane. By measuring these two quantities, the customer can verify the proper operation of the waterproofing system. Another advantage of having the geocomposite left exposed is the increase in the water flow. There is documented evidence from canals that the increase of the coefficient in the Gauckler Manning formula goes from 53 (the existing surface prior to the rehabilitation works) up to 85, after the installation of an exposed geomembrane system.

The main elements of the geomembrane liner for Spalov Hydro Power Tunnel are:

A drainage layer that will collect infiltrating water. It consists of a geonet, about 8 mm thick, fastened to the inner tunnel surface by means of impact anchors. The geonet covers longitudinally the invert of the tunnel and is covered by the waterproofing geocomposite

The drainage discharge system will discharge the water infiltrating through the existing concrete liner. If not discharged, the infiltration water would build up pressure between the concrete liner and the new waterproofing geocomposite. The drainage system at Spalov consists of discharge valves, made of the same geocomposite material, placed at various locations on the invert level and on the walls and vault (not to interfere with sediments and traffic). The discharge valve will open only when the inside pressure in the tunnel is less than the pressure from the surrounding soil (usually when the tunnel is emptied). The valves will discharge the waters conveyed by the drainage layer inside the tunnel.

The preliminary design assumed the placement of 2 valves on the invert at 6m spacing, 2 valves on the walls at 6m spacing, 2 valves at the spring line of the vault at 6m spacing, and 2 valves on the vault at 6 m spacing. In total, the design allowed for about 2 valves per linear meter of tunnel. However final distribution of the valves was decided onsite after the cleaning of the inside surface of the tunnel identified areas of potential infiltration.

The waterproofing geocomposite consists of a 2.5 mm thick PVC geomembrane (Figure 3), extruded in two layers by flat die, and heat coupled, during extrusion, to a geotextile, with mass/unit area of 500 g/m^2.

The upper layer of the geocomposite, which will remain exposed to the flow of the water, is a light grey color and the inner layer is a black darker color. Any installation damage or scratching are easily detected by the appearance of the darker color of the inner layer.

Figure 3. The Spalov hydropower tunnel during installation of waterproofing system.

The fastening system consists of lines of stainless steel profiles, anchored to the existing concrete. The profiles are installed along the longitudinal axis of the tunnel. There are two types of linear fastening: the pretensioning profiles and flat profiles. When both types are used, they are installed in alternate lines.

The pretensioning profiles are made of two parts, the internal and the external profiles. The internal profile is anchored to the concrete structure by means of chemical anchors. The external profile is connected to the internal profile by means of a special adjustable threaded nut. The external profile will be covered by a strip of geomembrane, heat welded to the geocomposite underneath. The connection of the two profiles creates a clamping effect on the geocomposite. Pretensioning insures the geocomposite will adhere to the existing surface of the concrete. The pretensioning profiles also provide a redundant, secondary drainage system for the water accumulating behind the geocomposite.

The flat profiles are made of stainless steel, about 30 mm x 3 mm in length and fastened by impact anchors. They are covered by the waterproofing geocomposite or geomembrane.

The waterproofing system (Figure 4) is sealed at the upper upstream section (beginning of the waterproofing section) and the lower downstream section (ending of the waterproofing scheme) by means of a stainless steel profile (60 mm in length x 6 mm in width), fastened by chemical anchors, 12 mm in diameter, 100 mm in depth. The stainless steel profile compresses the geocomposite against the existing surface.

The system has been installed on hundreds of hydraulic structures, with excellent performance. It has also been tested to a hydrostatic head of 250 meters.

Figure 4. Inspection of geomembrane attachment prior to welding final cover strip.

Surface preparation prior to the installation of the geomembrane system requires the following:

a) cleaning of all surfaces
b) removal of all loose pieces of the existing concrete
c) removal of debris
d) for the bottom: anchoring and filling with concrete, cement mortar or any other suitable material for those areas where the depression of the cavity is more than 5 cm compared to the original smooth concrete liner
e) for the walls and the vault: placement of an antipuncturing layer of polyethylene geotextile, (1500 g/m2), for those areas where the depression of the cavity is more than 5 cm compared to the concrete liner. Repairs with conventional concrete with polypropylene fibres is required only for those cavities where the fastening profiles will be installed.

5.0 ADDITIONAL HYDROPOWER TUNNEL LINER CASE HISTORIES

In 1997, the lining of the Thissavros Diversion pressure tunnel in Greece was completed. The tunnel is 11 meters in diameter, with a water pressure of 155 meters and maximum water velocity of up to 4 m/sec. The original use of the tunnel was diversion at the time of construction of the dam. The tunnel was originally lined with concrete, which started to crack. After the completion of the dam, the tunnel was modified to become the bottom outlet. The exposed CARPI geomembrane system was installed. Water infiltrating from the reservoir generates a negative uplift pressure on the membrane of about 50 meters. When the tunnel is in operation, the water pressure inside (up to 155 m) the tunnel keeps the waterproofing geocomposite compressed against the supporting concrete, when the tunnel is empty the drainage system is capable to discharge the negative pressure without damaging the geocomposite. Because of the topography of the tunnel, the drainage discharge system was made with one way valves. A few weeks after the membrane was installed and the tunnel put into operation, the tunnel had to be emptied to execute some extra grouting works associated with the gates. This situation allowed the testing of the behaviour of the geocomposite and of its drainage system under the design load. The system worked perfectly. After the grouting, the tunnel was filled again and it will remain full of water, with no plan to dewater it again over its entire life.

One of the oldest exposed tunnel lining projects (1987) is the design and installation for the rehabilitation of the Gandellino pressure tunnel, owned by ENEL (the Italian National Energy Company). A PVC flexible geocomposite was used to retain the water flow and a second rigid PVC sheet was installed on the vault to counteract the negative pressure when the tunnel is empty. After 11 years of service, the performance of the liner has been excellent with no geomembrane repairs required. The tunnel is part of a pump storage scheme, and is dewatered several times per day.

Design and Performance of deep piled circular cofferdam
Alan Powderham[1]

Synopsis

This paper describes design and construction of the large circular cofferdam in the Central Terminal Area of London's Heathrow Airport. The cofferdam is 60 m in diameter and 30 m deep and the principal feature in the recovery solution following the collapse of the station tunnels during construction in October 1994. It utilised 182 large diameter stepped secant/contiguous for the outer wall and 255 large diameter bored piles for the base slab. The design and construction had to deal with disturbed and unstable ground and major subsurface obstructions. There were also severe space limitations and environmental constraints. A key issue was the control of ground movements and the associated effects on the cofferdam itself and protection of adjacent structures. Major cost and time savings were achieved through the use of the Observational Method (Peck, 1969, Powderham 1998) and Value Engineering (VE) (ICE, 1996; Powderham and Rutty, 1994).

Introduction

The Heathrow Express (HEX) Rail Link opened in June 1998 and is a major new connection between Central London and Heathrow Airport. It provides a frequent, direct service between the airport terminals and London's Paddington Station (Fig.1) with a journey time of fifteen minutes.

In October 1994 sections of the station tunnels in the Central Terminal Area (CTA) collapsed during construction. The primary linings for these large diameter platform and concourse tunnels were then being constructed with sprayed concrete linings (SCL), (Fig. 2).

Tunnel invert is 30m below ground surface. Fortunately, there was no loss of life or injuries but substantial damage to the works and certain adjacent structures occurred. Potential delay to the project at this stage resulting from the collapse was estimated to be about eighteen months. An important early decision in the recovery strategy was the formation of the Solutions Team. This followed from the single team approach developed by BAA (NCE, 1998; Lownds, 1998). The Solutions Team members were selected from the main stake-holders of the project: client BAA, main contractor Balfour Beatty (BB), lead designer Mott MacDonald (MM) and the loss adjusters Brocklehurst with their consultant Ove Arup and Partners.

---

[1]Director, Civils and Transportation, Mott MacDonald, Croydon, England

854 GEO-ENGINEERING FOR UNDERGROUND FACILITIES

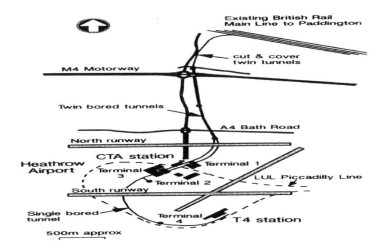

The dominant element in the recovery solution for the CTA was the large circular cofferdam (Figs. 2, 5). The design had to address criteria which included substantially disturbed ground, water filled voids, major obstructions and significant spatial and environmental constraints. Apart from the collapsed tunnels, other obstructions were mass and reinforced concrete and large buried construction plant. The risk management

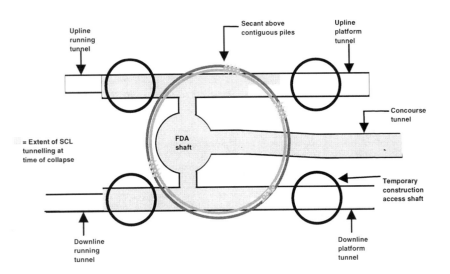

Fig. 2. Circular cofferdam

strategy developed needed to address the new range of weakened ground conditions, particularly the worst credible criteria which required establishing appropriate contingency plans. The Solutions Team members were already very familiar with the project and the tunnelling work required comprehensive review particularly that utilizing sprayed concrete linings. The CTA where the collapse occurred demanded particular focus.

After ground treatment to stabilise the ground, 182 secant piles were installed to form the outer ring. These large bored piles are 40 metres long and reduce in diameter at a depth of 20 metres to continue as discrete "contiguous" piles. Permanent lateral support is provided by reinforced concrete rings cast directly against the piles in sequence with cycles of excavation.

The ground, cofferdam and adjacent structures were comprehensively monitored with a Range of instrumentation including electro-levels, piezometers, extensometers, strain gauges and precise spatial survey. The 255 bored piles for the base slab were installed during July 1996 and construction of the base slab completed by September 1996. The application of the observational method was central to the overall design and construction strategy to manage risk and maximise opportunities to recover time. (Wheeler, 1996; Wallis, 1996).

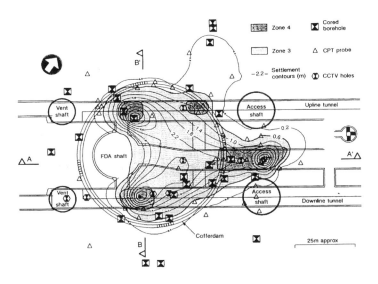

Fig. 3. Settlement contours of London Clay with predicted zones of disturbance Circular cofferdam option

Site Investigation

A site investigation, to evaluate the changed conditions, was initiated very soon after the collapse (Powderham & Rankin, 1997). The ground conditions in the CTA, prior to collapse, were relatively uniform with approximately 6 m of Terrace Gravels overlying the London Clay which has a thickness of around 60 m at this location. The London Clay overlies the Woolwich and Reading Beds which in turn overlie the Chalk which is present at a depth of approximately 90 m below ground level.

The recovery strategy required early establishment of a ground model. This involved an iterative approach and as new information became available it enabled development and refinement of the model. A key element was comparing the original ground horizon levels with those post failure. The focus was the top of the London Clay which had originally been about 6 metres below ground surface and the new levels were mainly assessed from a series of shallow boreholes. A series of deep boreholes was also carried out to assess the condition in and adjacent to the collapsed tunnels. Detailed core logging was used with emphasis placed on visual descriptions. Investigation and design development were proceeding in parallel and it was important not to create unnecessary delay with a prolonged programme of laboratory testing.

Two sets of data for the top of the London Clay, pre and post collapse, were collated. To achieve the best estimate of the contours for these two London Clay horizons, the data was statistically evaluated through a process known as kriging. The difference between the two kriged surfaces was plotted as contours of settlement as a result of the collapse, this is shown in Figure 3.

This process indicated that there were four localised areas of highly disturbed ground. In view of the large collapsed volume, approximately 6,000 cu.m, and the subsequent amount of excavation required, it was considered that there was likely to be significant time dependent softening initiated as a result of the collapse. The excavation of the cofferdam would also create a further reduction in stresses leading to a prediction for soil strengths much lower than for typical conditions in London Clay.

On the basis of the site investigation and predictive numerical analysis four zones were assigned within the London Clay (Figs. 3 and 4). Zone 1 was undisturbed intact

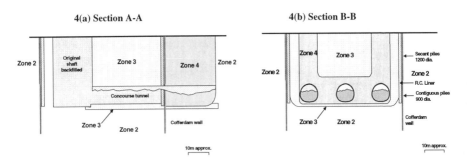

Fig. 4. Cross Sections showing predicted disturbance

London Clay but this was considered to lie beyond the active wedge of soil on the outside of the cofferdam. So the performance of the cofferdam would be principally influenced by Zones 2 to 4. Each of these were assigned two sets of bounding soil properties. The first represented "moderately conservative" (MC) parameters for the mass behaviour of that zone on the cofferdam as a whole. The subsidiary set were "worst credible" (WC) values representing local influences that may occur where pockets of the most severely disturbed soil in that zone could result in adverse loadings on the cofferdam ring. These properties are summarised in Table 1.

	Level	Zone 2		Zone 3		Zone 4	
		MC	WC	MC	WC	MC	WC
c_u (kPa)	118 - 108 mTD	50 + 7 d	30 + 7 d	30 + 7 d	0 + 7 d	0 + 7 d	10 + 1.5 d
	108 - 93 mTD	105 + 3.5 d	85 + 3.5 d	85 + 3.5 d	55 + 3.5 d	55 + 3.5 d	(=0.25 σ'_v)
γ_B (kPa)/m		19.5	19.5	19.5	19	19	16
$\varnothing'$ (degree)		25	25	25	25	25	21
$c/$ (kPa)		10	5	5	0	0	0
Strain (%)		< 0.1	0.2	0.2	0.5	1	N/A
Eu/cu		700	500	500	350	150	150
kh (m/sec)		(1 x 10^{-8} to 1 x 10^{-10})		(1 x 10^{-7} to 1 x 10^{-9})		(1 x 10^{-3} to 1 x 10^{-7})	
kv (m/sec)		Kh x 10^{-1}		Kh x 10^{-1}		kh x 1	
K_o (1)		1.0	0.8	0.8	0.6	0.6	0.6

d = Depth below ground level MC = Moderately Conservative WC = Worst Credible
Zone 4 extends to +95 T.D. Zone 3 extends to +93 T.D. Below +93 T.D. Zone 2 MC should be used.

Table 1 London Clay Soil Parameters

Basis of Solution

Early brainstorming workshops produced a wide range of schemes including micro-tunnelling and jacked tunnels, but the team quickly came to the conclusion that a large cofferdam would be a key feature of the recovery solution. This was seen as a basic requirement to encompass most of the disturbed ground and the majority of the damaged subsurface structures. These structures included the fuel depot access (FDA) shaft (Fig. 2), reinforced concrete piled surface slabs and escalator box, the original 20m diameter 30m deep fuel depot shaft, and the three partially constructed large diameter SCL platform and concourse tunnels. The shaft and tunnels had been filled with around 13,000 cu.m of concrete as an emergency measure for short term stabilisation of the collapse. There was clearly a need for a robust design for the cofferdam which would be built in highly disturbed ground. Initial options favoured a top-down construction sequence based on square or various arrangements of rectangular plan layouts. However, top-down construction had the following major disadvantages:
(i) It required early decisions for the layout of the final internal structure and the associated design to be developed. This applied particularly to floor layout and levels and internal columns and piles. Being on the critical path for design development and approval, these design requirements risked substantial delay.
(ii) Although any permanent slabs may have been able to provide a measure of temporary lateral support during construction, a substantial amount of additional temporary strutting was still likely.

(iii) A prime reason for such a large cofferdam was to enable safe access and removal of major obstructions. Top down construction could severely inhibit such access. The presence of major obstructions throughout the 30m depth would conflict with the construction of the permanent lateral supporting slabs.

For these reasons, although top-down construction is often the preferred method for large excavations in urban environments, it was rejected here in favour of a bottom-up sequence. However, with a rectilinear shape bottom-up construction would maximise the amount of temporary strutting required. For example the amount estimated for a 65m square cofferdam was about 5,500 tonnes of structural steelwork. Apart from the cost and programme implications this would also present significant restrictions to access

Fig. 5. Cofferdam during construction April 1996

and working space. Ongoing concept development therefore concentrated on ways to reduce the need for such strutting. The type of construction for the outer wall of the cofferdam needed to be decided early in the concept development since the whole process for the way forward demanded close integration of design with construction methods.

Circular Cofferdam

Shapes were considered but by December 1994 a circular cofferdam was selected as the preferred option. At 60 m in diameter and 30 m deep it offered a dramatically simple solution. Larger circular cofferdams had been constructed but not in such disturbed and variable ground conditions or utilising a bored piled wall (Fig. 6). The circular cofferdam brought the following major advantages:

a. Complete elimination of temporary cross strutting maximising available space for construction operations
b. It minimised the total volume of excavation. This was because it was possible to arrange the two permanent ventilation shafts to the south close but external to the cofferdam rather than being contained within a rectangular arrangement. Since these two shafts are in the relatively undisturbed Zone 2, the circular cofferdam still encompasses the majority of the disturbed ground including most of the areas of greatest settlement. In comparison with the square cofferdam option there was about 20,000 cu.m less bulk excavation. Apart from major cost savings this afforded important environmental and programme benefits, particularly since construction in

the centre of a busy airport could significantly affect airport operations. The ground was also likely to be contaminated with aviation fuel which had been stored in this location. Bio-remediation was undertaken to mitigate this risk.

c. The symmetry of the solution allowed a uniformly progressive step by step sequence of construction for the cycles of excavation and the casting of the inner reinforced concrete liner supporting the piles. This rhythm and symmetry greatly facilitated the progressive monitoring of ground and structural movements so that the associated trends, and in particular any adverse ones, could be detected at an early stage.

This latter aspect was also highly compatible with the application of the observational method which was part of the overall risk management strategy for construction of the cofferdam and central to the realisation of further potential cost and time savings.

Piled Wall Construction

The site conditions created particular requirements for construction methods. A fundamental aspect of the cofferdam was the design and construction of the outer wall. Diaphragm (or slurry) walls are typically adopted for this type of construction but the combination of heavy obstructions and potentially extensive voids were critical in eliminating this form of wall with its dependence on bentonite slurry support for panel excavation. However, a primary consideration was the need to provide a good cut-off to the groundwater in the gravels particularly in view of its potential contamination. So groundwater inflow had to be controlled along with the need to fully retain any loose, disturbed ground caused by the collapse. The outer wall also needed to be reasonably stiff and robust. Large diameter piles secanted to an appropriate depth offered the potential to satisfy all the criteria. Secant piles were therefore adopted as the basis for development of the wall design. Depth, diameter, spacing and construction tolerances now needed careful consideration in conjunction with construction methods and sequences.

Given the challenges presented by the disturbed ground and obstructions which, apart from reinforced concrete structures, included large items of buried construction plant entombed in mass concrete (Fig. 8), emphasis was placed on keeping the bored piling plant reasonably within its operating range. Thus 1200mm

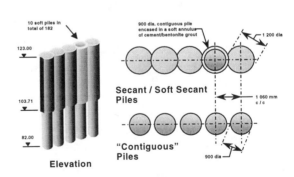

Fig 6. Secant/"Contiguous" piles

diameter secant piles were used for the top 20m which stepped in to continue as 900mm diameter "contiguous" piles for the next 20m (Fig 6). The centres of the piles were set at 1060mm. Both male and female piles were reinforced generally with bar reinforcement except above the tunnels where structural steel sections were used. It is essential to relate design to construction and this was particularly relevant to the piled wall where considerable interactive effort and support was provided to the team by Stent Foundations, the piling subcontractor. The piles were constructed using an oscillator and casing through an accurately constructed reinforced concrete guide wall. The specified minimum vertical tolerance of 1 in 150 was satisfied and was generally achieved close to 1 in 200.

As noted the principal function of the secant piles was to provide a barrier to groundwater and continuous support to any zones of weakened soil. At a depth of twenty metres the secant piles would reach the original level of the crown of the tunnels. The vast majority of the disturbed ground would be encountered over this depth and the secant piles would thus form an effective barrier. It was accepted that some limited zones of ground treatment could be necessary to complete this cut-off. Pre-treatment of the ground prior to piling focussed on the highly disturbed zones above and around the collapsed tunnels. A programme of grouting was initiated to infill any remaining voids in these zones and so minimise risks and potential delays arising from the need for additional contingency ground treatment. In practice the secant piles performed extremely well overall, uniformly controlling ground movements and maintaining a very effective cut off to groundwater.

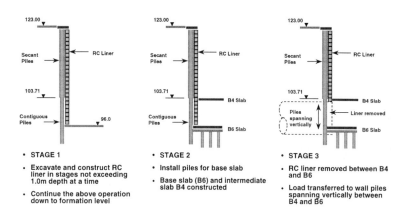

Fig 7. Original Scheme – Construction Sequence

Design Development

Following approval of the design concept for the cofferdam which included the location, diameter and wall type, design effort focused on detailed design development linked to the needs of construction. Given the importance of recovering programme, a fast track design and construction approach was required. This was greatly aided by the formation of an integrated single team including the establishment of site based designers during construction.

As noted previously an advantage of the circular cofferdam was its inherent integrity and simplicity of form. It provided a functional base scheme with the outer ring of piles, the reinforced concrete liner, the piled base slab and an intermediate slab just above tunnel crown. This allowed its fast track design and construction to be independent from the design of whatever internal structure above the intermediate slab was eventually developed. The circular geometry eliminated the need for any permanent cross strutting above the intermediate slab. To this depth the secanted pile wall principally had a temporary function with the inner reinforced concrete liner designed to take all of the ground loading. The "contiguous" piles extending below the secanted section were designed to span between the intermediate and base slabs while the base slab piles were required to control ground heave and provide vertical stability.

Observational Method
Application of the observational method formed an integral part of the risk management strategy for the overall design and construction of the cofferdam. Its use was envisaged right from the start of the design concept. Careful consideration was given throughout the design development to enhance compatibility with construction methods. The

Fig. 8. Piling obstruction from plant buried in concourse tunnel

observational method addressed the performance of the cofferdam during construction. The critical observations were the deflections of the piled wall but viewed in the context of overall ground movements and the potential effects on the cofferdam itself and adjacent structures. In relating design to construction in the context of the observational method an emphasis was placed on simplicity and ease of monitoring. Such aspects as construction rhythm and symmetry have already been noted. The method was implemented on the basis of progressive modification (Powderham 1998). This approach enhances risk management since the starting design on site is acceptably safe to all parties. The particular conditions at Heathrow demanded a demonstrably robust design and one that could sustain, with appropriate pre-planned contingency measures, the worst credible ground conditions. However, given the particular need to safely recover as much of the delay caused by the collapse as possible, the observational method also offered major opportunities. Here progressive modification could generate further time savings by introducing advantageous design changes on the basis of acceptable feedback on measured performance during construction.

The application of the method was characterised by three main aspects. The principal objective was to control the risk associated with such a major excavation. This focused on wall deflections and ground movements and particularly any trends towards adverse conditions. The second aspect related to contingency measures. The method would allow timely implementation of such measures to maintain and control safety. The design was robust and more conservative than one based on predictions of the most probable conditions (cf Peck 1969). There was therefore potential that the method would be able to acceptably demonstrate that contingencies were not necessary or at least minimise them and so mitigate their effect on time and cost. The added benefit of introducing design changes that would save more time, as noted above, was the third factor. Avoidance of contingency measures or sequential introduction of design improvements are inherent benefits of the progressive modification approach.

The primary instrumentation comprised inclinometers in the piles and adjacent ground along with precise levelling. The inclinometers were formed of series of beam mounted electrolevels. Secondary instrumentation involved piezometers, extensometers and spatial survey. (Fig.11).

Contingency Measures
The critical quantities to be measured were deflection of the piled walls and the associated ground movements. These two factors relate to the flexibility of the

Fig. 9. Base slab construction showing Incorporation base of FDA shaft as VE alternative

Fig. 10. Early tunnel breakthrough

structure and the global movements generated by the unloading created by the bulk excavation within the cofferdam. Two principal contingency measures were developed to address trends which indicated the likelihood of unacceptable wall deflection. These were to introduce thicker stiffer reinforced concrete rings in the cofferdam lining and to excavate down the sides only creating a substantial time lag between the main central excavation. Construction of the reinforced concrete liner rings would then progress significantly ahead of the bulk excavation thus providing early support and limiting wall movement. Parametric studies had indicated that under the worst case scenario maximum bending moments could develop in the contiguous piled section with a deflection of around 75 mm. This was set as the limiting condition for acceptable performance of the cofferdam wall. The intention was to avoid exceeding this limit by applying one or both of the above contingency measures at a sufficiently early stage in the excavation process. To successfully implement such a process if necessary would need early and reliable identification of deflection trends. While the performance of the cofferdam was continuously monitored throughout the construction process, a detailed review was set for when the excavation depth reached 7 m to assess trends. Detection of an adverse trend developing would then have led to implementation of contingency measures, but in the event, no adverse trends developed.

Measured Performance and design improvements
The comparison between average predicted and measured deflections of the cofferdam wall are shown in figure 12. The observed values are overall maxima for the various stages. The performance was notably better than the range of predictions generated for moderately conservative soil parameters. Although the cofferdam offered an attractive structural simplicity, the variable soil/structure interaction together with the construction sequence presented great complexity. It was a challenge to produce an appropriate

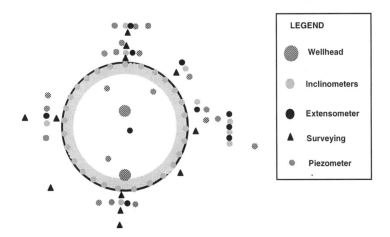

Fig. 11. Cofferdam Instrumentation

model for numerical analysis and the various analyses indicated an average maximum deflection of around 60mm. In practice the average maximum deflection of the piled walls was around 15 mm. The control of the lateral ground movements achieved compares very favourably with other case histories of deep excavations in London Clay (eg Burland & Hancock, 1977 and Marchand, 1993). Average deflections were about 50% less than those predicted for the most probable conditions. These trends were very

evident at the 7m depth review and enabled a variety of advantageous design changes to be implemented. The first change was to increase the depth of excavation and liner ring construction from 1 m to 1.2 m. This change, allowing a faster rate of construction, was undertaken after completion of liner ring 9, all those thereafter being of the increased depth
Another major design change was the introduction of early tunnel breakthroughs, see Figures 9 and 10. The original design plan was to take the lining sequence completely down to base slab level thus maintaining the rhythm of construction and the ease of monitoring (Fig. 7). However, with the performance so demonstrably robust it was decided to break through into the cofferdam from the adjacent shafts at a much earlier stage than originally planned. The effects of the breakthroughs were carefully monitored by implementing each of them progressively in defined stages. The pilot tunnels were sequentially enlarged to full size and temporarily plugged with mass concrete to maintain ring action around the wall of the cofferdam. Early tunnel break-throughs were thus achieved substantially ahead of excavation within the cofferdam. Apart from advancing tunnel construction adjacent to the cofferdam this allowed early progress for track work.

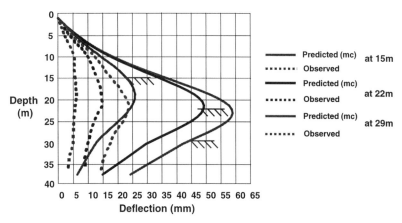

Fig. 12. Cofferdam wall deflections

Conclusions
The cofferdam marked a comprehensive success in an integrated approach to design and construction on a high profile project. BAA merits particular recognition as the initiator for such a positive team environment. The circular shape adopted for the cofferdam created simplicity, symmetry and rhythm. This in turn led to efficiency of function and ease of construction and its monitoring. Recovery of the works was achieved in a demonstrably safe manner with major savings being achieved in time and cost. Overall delay to the project was reduced from eighteen to six months. The success provided a further strong example of the synergy between value engineering and the observational method.[7] The creation of the single team culture made conditions very conducive for

application of these two techniques. Progressive modification[9] brought additional comfort and control in addressing the risks of the variable ground conditions and the uncertainties in soil/structure interaction. Contingencies were avoided and a range of design improvements were introduced during construction delivering substantial time savings.

Acknowledgements

The views and opinions expressed on the Heathrow Cofferdam case history are those of the author and are not necessarily representative of those held by BAA or MM. The author is grateful to BAA for their permission to publish this paper. In addition, the author wishes to recognise the efforts made by all individuals and parties during the investigation, ground treatment, construction and excavation of this recovery solution which has required close and effective teamwork.

References

1. Burland, J. B. & Hancock, R. J. R. (1977) Geotechnical aspects of the design for the underground car park at the Palace of Westminster, London, Structural Engineer, 55, No. 2, 87-100.
2. Institution of Civil Engineers (1996), Creating Value Engineering, ICE design and practice guide, Thomas Telford, London.
3. Marchand, S. P. (1993) A deep basement in Aldersgate Street, London, Proc. ICE, 93 Feb, 19-26.
4. New Civil Engineer (NCE) Supplement, Heathrow Express, London, March 1998.
5. Lownds, S. (1998) Fast track to change on the Heathrow Express, Institute of Personnel and Development, London.
6. Peck, R. B. (1969). Advantages and limitations of the observational method in applied soil mechanics. Geotechnique 19, No. 2, 171-187, ICE, London.
7. Powderham, A. J. & Rutty, P. C. (1994). The observational method in value engineering. Proc. 5th. Int. Conf. Piling Deep Fdns Bruges.
8. Powderham, A. J. & Rankin W. J. (1997). Heathrow collapse recovery solution cofferdam - planning, design and implementation, Proc. Intl. Conf. On Foundation Failures, 251-263, Institution of Engineers, Singapore, 12-13 May 1997.
9. Powderham, A. J. (1998). The observational method – application through progressive modification, BSCES/ASCE, Vol 13, Number 2, 87-110.
10. Wallis S. (1996) Heathrow rises from the mire and shines, Tunnel, International Journal for Subsurface Constructions, STUVA, Cologne, Vol 6/96, 6-22.
11. Wheeler P. (1996) Terminal Condition, Ground Engineering, Vol 29, No. 1, 14-16.
12. Wheeler P. (1996) Arrival Procedures, Ground Engineering Vol. 29, No 7, 18-21

LONG TERM BEHAVIOR OF TUNNELS IN CHICAGO CLAY

Nelson Kawamura[1], Associate Member, ASCE and
Edward J. Cording[2], Member, ASCE

ABSTRACT: A monitoring program was conducted to confirm the adequacy of contractor-proposed design and alignment changes on a flood relief tunnel in Evanston, Illinois. The project provided a unique opportunity to investigate the behavior of a shield-driven tunnel in Chicago Clay and to revisit the pioneering investigations on the Chicago Subway in 1939-1941. This paper focuses on the long-term consolidation and settlement caused by clay disturbance and drainage into the tunnel. Pore pressure decreases and soil volume decreases (consolidation) were measured in a zone extending to three to four radii around the tunnel. An approach has been developed to assess the magnitude and width of the settlement trough that results from consolidation around tunnels in clay deposits.

OBSERVATIONS

Karl Terzaghi and Ralph Peck pioneered systematic field observations in soil tunneling during construction of the Chicago subway in 1939-1941. A total of 12.4 km of twin tunnels were excavated in the Chicago Clay; 9.2 km were excavated without a shield using the liner-plate method, and 3.2 km were excavated in softer clays in the Loop using a semi-blind shield which forced all of the clay to be squeezed through openings occupying 20% of the shield face (Terzaghi, 1942a,b).

The Chicago Clay is lightly overconsolidated and was deposited as two till sheets in glacial Lake Chicago in the waning stages of the last (Wisconsinan) glaciation. The former lake and its clay occupy most of Chicago and extend over 24 km north along the border of Lake Michigan into Evanston, where the 1.3-km-long Evanston sewer tunnel was excavated in 1992 using a Lovat wheeled excavator shield. The shield was articulated in three sections which aided the contractor,

---

[1]Senior Geotechnical Engineer, GeoConsult, P.O. Box 362040, San Juan, PR 00936-2040
[2]Professor of Civil Engineering, University of Illinois, 2230E NCEL, MC-250, 205 N. Mathews Ave., Urbana, IL 61801

McNally Tunneling, in negotiating a very tight (48-m-radius) turn at a street intersection and thereby allowing the shield to be driven from a single shaft rather than jacking in straight segments from multiple shafts through the residential neighborhood.

As in the Chicago subway project, the program of observations in Evanston was focused on measuring movements near the tunnel perimeter as well as at the ground surface so that the source of the ground movements could be evaluated and controlled. Squeeze tests using rods driven into the tunnel face and perimeter provided this information in 1939. In 1992, inclinometers, sondex casings and deep settlement points provided a three-dimensional view of the ground movements around the tunnel and throughout the soil mass. Deep settlement points and surface settlement cross-sections were concentrated at 11 sections, with two of the sections (S-6 and S-9) containing pneumatic pore pressure cells as well as a concentration of ground movement instrumentation.

During passage of the Evanston shield, the clay exhibited undrained behavior, with pore pressure decreasing close to the tunnel as unloading occurred behind the overcutter and tail of the shield and then increasing as the tunnel lining was expanded against the clay. Deep settlement points revealed that ground losses were concentrated at the overcutter at the front of the shield which had a radius 19 mm greater than the body of the shield to aid in negotiating the short radius turn. Ground losses into the tunnel face and behind the tail of the shield prior to lining expansion were negligible. As would be expected for undrained behavior, volume changes in the soil mass, determined from the lateral and vertical measurements, were small. The ground losses at the overcutter resulted in initial surface settlements along the tunnel centerline that typically ranged from 20 to 30 mm, in the range of the 25-mm settlement criterion established prior to construction (Kawamura, 1998).

Over the following two years, pore pressures dropped below their original ambient conditions, confirming that drainage was occurring into the tunnel from the clay surrounding the tunnel. The resulting increase in effective stress produced volume decreases (consolidation) which were measured in a zone extending up to 3 to 4 radii from the tunnel. The consolidation typically added 15 to 30 mm to the centerline settlement. The settlement trough for the long term settlement was wider than the trough for the immediate settlement, a reflection of the fact that the source of the ground movements resulted from volume change in a zone extending to 3 to 4 radii from the tunnel rather than from ground loss occurring at the tunnel perimeter.

TIME DEPENDENT SETTLEMENTS

In Figure 1, surface settlements for the 3.7-m-dia. Evanston tunnel are plotted with time and compared with those obtained in the Chicago subway at one section of the liner plate tunnel and one section of the shield tunnel. Initial settlements during tunnel advance are shown on a linear time scale whereas settlements occurring after ten days are plotted on a logarithmic time scale. The time-rate of surface settlement for the Evanston tunnel is linear with the logarithm of time, and the change in effective stresses is in the recompression range, so that settlements are small.

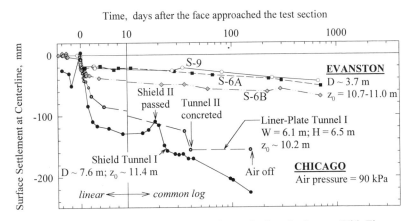

Figure 1. Tunnels in Glacial Chicago Clay - Surface Settlement With Time

The time-dependent surface settlements recorded at the centerline of the 7.62-m Shield Tunnel I of the Chicago subway after the passage of Shield II were approximately an order of magnitude greater per log cycle of time than the rates observed in the 3.67-m Evanston tunnel. The differences are greater than can be attributed solely to the difference in tunnel diameter or the presence of two tunnels in the case of the Chicago subway. As the Chicago subway shields were advanced, significant surface heave developed because of the small opening in the face through which the clay would squeeze as the shield was shoved forward. Subsequent settlements were larger than the heave and continued at a high rate with time, despite the fact that the tunnel remained under an air pressure of approximately 1 atmosphere, which would have minimized drainage into the tunnel.. Terzaghi (1942b) notes that the large settlements were not a result of ground loss behind the tail of the shield because the air pressure and the rapid filling of the tail void with pea gravel would have prevented the clay from invading the tail void. The time-dependent settlement in the Chicago subway shield tunnel is principally attributed to the consolidation of the severely disturbed and remolded clay around the tunnel perimeter that resulted from the use of semi-blind shields with a small opening in the face through which the clay and obstructions ahead of the face had to pass. In contrast, the Evanston tunnel, was excavated with a wheel which cut the clay to the outside perimeter of the shield and did not exert a pressure on the face great enough to heave the clay. Time-dependent surface settlement above the Liner-Plate Tunnel I of the Chicago subway was minimal after Tunnel II was concreted. Inasmuch as the readings were taken before the removal of compressed air, the potential effect of the long term drop in piezometric levels on consolidation settlement is unknown.

Consolidation around the tunnels caused not only an increase in the magnitude of surface settlement, but also an increase in the half width of the surface settlement trough. For instance, in Test Section S-9 of the Evanston tunnel, the surface settlement was 18 mm at the end of tunneling and had increased by an additional 20 mm 244 days after the end of tunneling (Figure 2). The half width of the surface settlement trough, which was 6.5 m at the end of tunneling, increased to 14.5 m at 244 days. Typically, the half width of the surface settlement trough in sections S-6A, S-6B and S-9 after 244 days was approximately twice the half width at the end of tunneling.

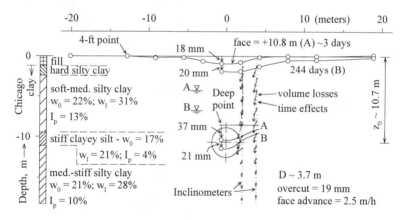

Figure 2. Evanston Tunnel, Section S-9 - Time Effects on Ground Movements

The tunnel at these sections was mined in ground which can be reasonably well represented by the soil profile and index properties presented in Figure 2.

Settlements near the tunnel crown at Test Section S-9 were 37 mm due to ground loss during tunneling and increased an additional 21 mm over the long term (Figure 2). A significant portion of the additional settlement may be attributed to settlement of the tunnel itself as a result of consolidation of the disturbed and undisturbed regions below the invert of the tunnel. The consolidation of the undisturbed region below the invert may be explained by the decrease in piezometric levels because of the dissipation of excess pore water pressure and gravitational drainage into the tunnel, leading to an increase in effective stress. This observation confirms the time-dependent settlement of the shield tunnels of the Chicago subway reported in Terzaghi (1942b). The invert of the Chicago tunnels settled as much as 50 mm, which was primarily ascribed to consolidation of the disturbed region below the tunnel.

TIME-DEPENDENT VOLUMETRIC CHANGE AROUND THE TUNNEL

The field observations from the Chicago subway tunnels and the Evanston tunnel suggest that the principal sources of the time-dependent ground movements are the soil disturbance at the periphery of the tunnel and the long term decrease in piezometric levels around the tunnel (Kawamura, 1998). These sources of ground movements are responsible for the time-dependent volumetric change around the tunnel, which leads to the increase in the surface settlement trough. The effect of time-dependent lining deflection on the surface settlement trough was generally found to be minimal, particularly after the placement of the final concrete liner. Since the readings were terminated during the ongoing primary consolidation, the magnitude of the actual secondary compression in both tunnels is unknown.

Soil Disturbance at the Periphery of the Tunnel

The structural disturbance typically reduces the undrained shear strength and the permeability, and increases the compressibility of soil. The magnitude of the changes in soil properties depends on the extent to which the soil is disturbed, as well as on its sensitivity to the mode of shearing, either active or passive. The structural disturbance around the tunnel depends mainly on the method of tunneling, the shear strength of soil, and the sensitivity of soil structure. Endo and Miyoshi (1978) conducted the only pertinent investigation known on the disturbance-induced changes in soil properties around the tunnel. They found that the strength of soil measured around the tunnel immediately after the passage of a semi-blind shield was about 50% of the initial value. Their investigation was carried out in Subway Line No. 10 in Tokyo, which was mined in soft alluvial silt with undrained shear strength of 30 to 50 kPa.

The thickness of the disturbed region may be estimated on the basis of time-dependent volumetric changes measured around the tunnel using inclinometers and settlement points. This was the case for Section S-9 of the Evanston tunnel, in which the block elements typically underwent volumetric compression, as shown in Figure 3a . The volumetric changes in the fill and hard brown silty clay may not be

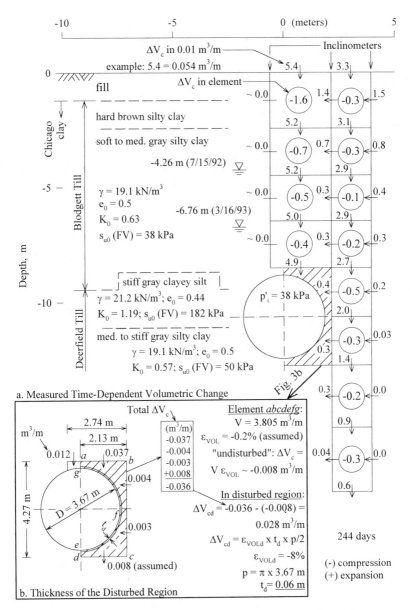

Figure 3. Evanston Tunnel, Test Section S-9 - Consolidation Effect Only

totally ascribed to the effect of consolidation because these layers are typically located above the groundwater level, and therefore, they are not fully saturated. The time-dependent volumetric strains ε_{vol}, which were evaluated by dividing the time-dependent volumetric changes, ΔV_c, by the volumes of respective elements, are generally less than 0.25%. The small time-dependent volumetric strains provide good indication that the clay in the elements was undisturbed. The values of ΔV_c and ε_{vol} were compared as proposed in Cording and Hansmire (1975).

At the periphery of the tunnel, in a soil zone assumed to be completely disturbed, an average volumetric strain of 8% was adopted, on the basis of the available laboratory testing results (Ali, 1993; Romo, 1984; Shogaki, 1996). This led to a computed thickness of an equivalent disturbed zone, t_d, on the order of 0.06 m, as indicated in Figure 3b.

As for the Shield Tunnel I of the Chicago subway, a thickness of the disturbed region on the order of 0.6 m was computed below the invert by assuming a volumetric strain of 8% and minimal lateral displacements. This value will decrease if, contrary to Terzaghi's hypothesis, consolidation is also assumed to occur in the undisturbed region below the tunnel. In any circumstances, the extent of the disturbed region around the Shield Tunnel I is expected to be larger than around the Evanston tunnel. This is because the Shield Tunnel I was excavated by pushing a semi-blind shield (Terzaghi, 1942b), which caused the soil ahead of the face to be displaced rather than excavated, leading to heave and substantial soil disturbance. In the Evanston tunnel, the soil was excavated by a rotating cutter head and brought into the shield, such that soil disturbance was limited to a thin peripheral layer caused mechanically during the shield passage.

Time-Dependent Drop in Piezometric Levels

The piezometer readings in the Evanston tunnel decreased gradually with time during the first 5 months after the end of tunneling. Since the readings were not taken in zones beyond a distance of 2.4 m from the tunnel wall, where consolidation also occurred, the 5-month period would be less than the duration of the primary consolidation. The decrease in piezometric level was caused by the dissipation of the excess pore water pressure generated during tunneling, and by the gravitational drainage into the tunnel which behaved as a horizontal drain (Kawamura, 1998), as illustrated in Figure 4.

Pneumatic piezometer readings in the Evanston tunnel taken about two years after tunneling presented values typically on the order of 14 to 21 kPa below the pore water pressure measured prior to tunneling. Such a decrease in the initial pore water pressure with time endorses the assumption of the gravitational drainage of groundwater into the tunnel. The time-dependent drop in piezometric levels due to the effect of gravitational drainage can be captured in Figure 5 for the shield tunnels of the Chicago subway.

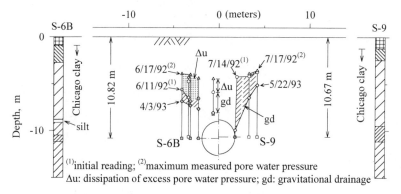

Figure 4. Evanston Tunnel - Changes in Piezometric Levels

It was found that there was enough recharge around the Evanston and Chicago tunnels such that the groundwater flow occurred under constant phreatic surface. A criterion for constant phreatic surface was presented by Fitzpatrick (1980), who claimed that drawdown of the phreatic surface is unlikely to take place around tunnels in soil with coefficient of permeability, k_{soil}, less than 1×10^{-8} m/s. This condition is met by the Chicago clay whose k_{soil} is in the order of 5×10^{-10} m/s (Peck, 1942).

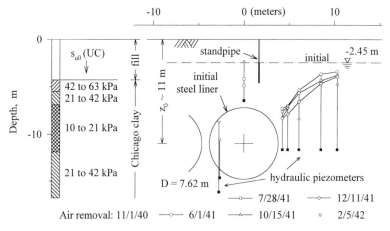

Figure 5. Shield Tunnels of the Chicago Subway, Contract S3, Station 87+70 Long Term Piezometric Levels (data from Terzaghi, 1942b; Peck, 1942)

LONG TERM SURFACE SETTLEMENT TROUGH

The long term decrease in piezometric levels along the springline elevation controls to a great extent the width of the consolidation settlement trough. This is particularly true above tunnels excavated in normally to slightly overconsolidated clays, in which the long term changes in effective stress generally tend toward the compression range. The extent of the drop in piezometric levels at the springline position is a function of the drainage boundary conditions and the permeability of subsoil. In this scenario, the role of soil disturbance and the long term increase in effective stress around the tunnel may decrease to some degree the permeability of soil as a result of the reduction in its void ratio. This may lead to a smaller radius of influence of the flow around the tunnel, i.e., the width of the surface consolidation settlement trough decreases..

The effect of consolidation around the tunnel spreads toward the ground surface in a fashion similar to the upward propagation of the immediate volume losses. The width of the consolidating region at the springline level and the soil profile above the springline level are deemed to be two major controlling factors which affect the width of the surface consolidation settlement trough.

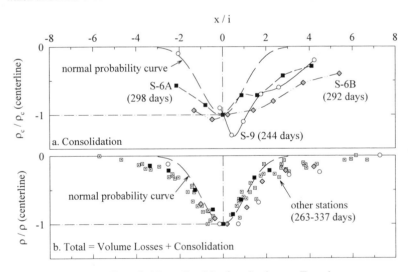

Figure 6. Evanston Tunnel - Normalized Surface Settlement Troughs

The width of the consolidation settlement trough can be examined on a normalized scale, as previously suggested by Shirlaw et al. (1994). In this investigation, the consolidation settlement trough is normalized to the distance to the inflection point, i, of the immediate surface settlement trough, as shown in Figure 6a. The normalized consolidation settlement trough tends to diverge gradually from the normal probability curve as the normalized distance x/i increases, indicating the widening of the surface settlement trough due to the formation of a consolidating region around the tunnel. The normalized width of the consolidation settlement trough is in the range of 5 to 6, as compared to 2.5 for the normalized width of the immediate settlement trough, i.e., the width of the settlement trough in the Evanston tunnel doubled with time.

In terms of the normalized half width, w_{St}/i, of the total settlement trough, values in the range of 3.7 ± 0.4 on the left side, and 3.9 ± 0.7 on the right side of the Evanston tunnel were computed, also demonstrating that it is not a normal probability curve. Figure 6b shows that the settlement ratio $\rho_c/\rho_c(CL)$ at the normalized distance, x/i, of 2.5 averages 0.27 ± 0.03 on the left side and 0.25 ± 0.09 on the right side of the tunnel, as compared to 0.04 for the normal probability curve.

ASSESSMENT OF THE PRIMARY CONSOLIDATION SETTLEMENT

An empirical approach to evaluate the surface settlement above tunnels in clay due to primary consolidation is discussed on the basis of field observations analyzed in foregoing sections.

The general expression for the volume of the surface settlement trough, V_{Spc}, due to primary consolidation is:

$$V_{Spc} = \Delta V_{cd} + \Delta V_{pc} + V_{Lc} \quad (1)$$

where: ΔV_{cd} = time-dependent volumetric change in the disturbed region; ΔV_{pc} = time-dependent volumetric change in the "undisturbed" region; and V_{Lc} = time-dependent volume lost into the tunnel during primary consolidation, which for practical purposes may be assumed equal to zero.

The volumetric change in the disturbed region, ΔV_{cd}, can be estimated, assuming that the volumetric strain, ε_{VOLd}, is equal to 8%, and the disturbed region for a controlled shield tunneling averages a thickness, t_d, of 0.06 m. Therefore, the expression of the volume change, ΔV_{cd}, becomes a simple function of the tunnel radius, R, as shown below:

$$\Delta V_{cd} = \varepsilon_{VOL} \cdot t_d \cdot 2 \cdot \pi \cdot R = 0.08 \cdot 0.06 \cdot 2 \cdot \pi \cdot R \quad \Rightarrow \quad \Delta V_{cd} = 0.03 \cdot R \quad (2)$$

where, R is given in meters, and ΔV_{cd} in m^3/m.

The general expression for the volume change in the undisturbed region, ΔV_{pc}, around the tunnel is:

$$\Delta V_{pc} = \varepsilon_{VOL} \cdot V_0 \qquad (3)$$

in which the volumetric strain, ε_{VOL}, is defined as:

$$\varepsilon_{VOL} = \frac{\Delta e}{1+e_0} = \frac{C_x \cdot \log \frac{\sigma'_{mean0} + \Delta\sigma'_{mean}}{\sigma'_{mean0}}}{1+e_0} \qquad (4)$$

where: Δe = change in void ratio in the undisturbed region due to the increase in average effective mean stress, $\Delta\sigma'_{mean}$; C_x = either compression or recompression index of the undisturbed soil under triaxial consolidation; e_0 = *in situ* void ratio. For effective stresses in the recompression range, the index C_x may be assumed equal to $1.3 \cdot C_r$ on the basis of the results of triaxial consolidation under equal all-round pressure obtained on Boston Blue Clay samples (Ali, 1993).

The shape and size of V_0 are functions of various factors, such as the depth to the groundwater level, depth to the incompressible unit, soil profile, soil properties and soil permeability to liner permeability ratio. For practical purposes, it is more convenient to approximate the actual consolidating region to an equivalent area that can be easily computed, and which results in a V_0 comparable to the actual V_0. For instance, an annular ring surrounding the tunnel may conveniently define the equivalent area. Field observations from several tunnels in clay, including the Evanston tunnel, indicated that the outer radius of the equivalent annular consolidating region may range from 3.1 to 3.9 times the radius of the tunnel. In other words, the average thickness of the equivalent consolidating region is assumed to be on the order of 1 to 1.5 times the diameter of the tunnel (Kawamura, 1998).

Once the volume of the consolidation settlement trough, V_{Spc}, is determined, the primary consolidation settlement, ρ_{pc}, at the centerline of the tunnel can be computed by approximating the area defined by V_{Spc} to a triangle with equivalent area. The volume of the consolidation settlement trough can, therefore, be expressed as:

$$V_{Spc} = \rho_{pc} \cdot w_{Spc} \qquad (5)$$

where: w_{Spc} = half width of the consolidation settlement trough, which for the Evanston tunnel and other tunnels in clay excavated using open face and wheeled shields, was found to be on the order of twice the half width of the immediate settlement trough, w_S, i.e., $w_{Spc}/w_S = 1.99 \pm 0.48$. For many of the tunnels in clay excavated with the use of EPB shield, where the gravitational drainage into the tunnel was presumably small, a w_{Spc}/w_S ratio in the range of 1.10 ± 0.10 was obtained (Kawamura, 1998). The value of w_S can be obtained by using the expression presented in Cording and Hansmire (1975): $w_S = R + z_0 \cdot \tan\beta$. Data from 25 tunnels in clay indicate that the average β is on the order of $38.5° \pm 8.2°$ (Kawamura, 1998).

Using the empirical approach discussed in this section, the values of V_{Spc}, w_{Spc} and ρ_{pc} were computed in test sections S-6B and S-9 of the Evanston tunnel, as summarized in Table 1. The computed and measured consolidation settlements can only be compared qualitatively because a precise value of the duration of primary consolidation, t_p, is not available from the field observations. In this numerical example, the thickness of the disturbed region was assumed to be equal to 0.06 m with the condition that the shield advance is performed under well-controlled tunneling techniques such that only small ground movements are allowed.

Table 1. Primary Consolidation Settlement in the Evanston Tunnel

Station		V_{Spc} (m^3/m)	w_{Spc} (m)	ρ_{pc} (mm)	Time (days)
S-6B	Measured	0.53	24.1	22	292
	Computed	0.65	20.6	32	t_p
S-9	Measured	0.41	20.5	20	244
	Computed	0.44	20.6	21	t_p

CONCLUSIONS

The principal sources of the time-dependent ground movements in the Chicago subway tunnels and the Evanston tunnel were: a) soil disturbance at the tunnel boundary; and b) decrease in piezometric levels due to the dissipation of the excess pore water pressure and gravitational drainage into the tunnel. They were responsible for the volumetric change around the tunnel, which resulted in the widening and deepening of the surface settlement trough during the course of the primary consolidation. The half widths of the consolidation and total surface settlement troughs in the Evanston tunnel increased, on average, by factors of 2 and 1.5, respectively. Both half widths were determined by transforming the areas defined by the respective troughs to equivalent areas of a triangle.

The effect of the primary consolidation on surface settlement trough can be assessed by computing the time-dependent volumetric changes, ΔV_{cd} and ΔV_{pc}, in the disturbed and "undisturbed" regions, respectively. To compute ΔV_{cd}, a volumetric strain, ε_{VOLd}, of 8% and a thickness of the disturbed ring, t_d, on the order of 0.06 m are suggested. The use of a t_d of 0.06 m implies tunneling under well-controlled construction methods, as was the case in the Evanston tunnel. To compute ΔV_{pc} in the recompression range, a triaxial recompression index, C_x, equal to $1.3 \cdot C_r$, and an outer radius of the equivalent consolidating region in the range of 3 to 4 times the radius of tunnel may be used.

ACKNOWLEDGMENTS

McNally Tunneling supported the instrumentation program of the Evanston tunnel. The authors wishes to acknowledge the following persons: Mr Larry Lenahan and the staff of McNally Tunneling for their support and cooperation during the monitoring work; Mr. Robert A. Bauer of Illinois State Geological Survey for the employment during the first author's stay in Illinois; Mr. Alan R. Crumley and Dr. Tirso A. Alvarez of GeoConsult for their insightful comments.

REFERENCES

Ali, S. (1993). Engineering properties of undisturbed samples of Boston Blue Clay. Ph. D. Thesis, University of Illinois, Urbana, Illinois, 399 p.

Cording, E. J., and Hansmire, W. H. (1975). Displacements around soft ground tunnels. Proceedings of the Fifth Pan-American Conference on Soil Mechanics and Foundation Engineering, Buenos Aires, Vol. IV, p. 571-633.

Endo, K., and Miyoshi, M. (1979). Closed-type shield tunnelling through soft silt layer and consequent ground behaviour. Tunnelling Under Difficult Conditions, Proceedings of the International Tunnel Symposium, Tokyo 1978, Kitamura, I., editor, Pergamon Press, New York, p. 329-334.

Fitzpatrick, L. (1980). Lining leakage and consolidation around soft-ground tunnels. Master of Science Thesis, Cornell University, 250 p.

Kawamura, N. (1998). Immediate and long term behavior of the Evanston tunnel in Chicago clay. Ph.D. Thesis, University of Illinois, Urbana, 600 p.

Peck, R. B. (1942). Pendleton Levee failure: discussions. Proceedings of the American Society of Civil Engineers, Vol. 68, No. 7, September, p. 1797-1800.

Romo, M. P. (1984). Settlements induced by soft ground tunneling. Proceedings of the International Conference on Case Histories in Geotechnical Engineering, Rolla, Missouri, Vol. III, p. 1269-1279.

Shirlaw, J. N., Busbridge, J. R., and Yi, X. (1994). Consolidation settlements over tunnels: a review. Canadian Tunnelling Canadien, an annual publication of the Tunneling Association of Canada, 13 p.

Shogaki, T. (1996). A method for correcting consolidation parameters for sample disturbance using volumetric strain. Soils and Foundations, Vol. 36, No. 3, Sept., p. 123-131.

Terzaghi, K. (1942a). Liner-plate tunnels on the Chicago (ILL.) subway. Proceedings of the ASCE, Vol. 68, No. 6, June, p.862-899.

Terzaghi, K. (1942b). Shield tunnels of the Chicago subway. Journal of the Boston Society of Civil Engineers, Vol. XXIX, No. 3, July, p. 163-210.

Field measurements during construction
of a sewage tunnel.

F.M. Abdrabbo [1]   F.M. El-Nahhas [2]   H.A. Abd El-Lateef [3]

Abstract

An in situ monitoring program was carried out during construction of an Earth Pressure Balance (E.P.B) tunnel of Alexandria waste water project. The instrumented area was chosen where the 2.86m shielded tunnel was advanced pipe jacking through layers of silty clay and calcareous sand deposits. Surface soil displacements above the tunnel were measured during the tunnel advance using an array of surface settlement points. Piezometers were also utilized to monitor the ground water response at the advanced shafts. This paper presents field measurements compiled from the instrumentation monitoring program. The general trends of the ground and ground water responses to tunneling under the encountered conditions are identified. The results are compared with the available guide lines of the prediction of soil subsidence above shielded tunnels.

Introduction.

About 3.5 kilometers of Alexandria central zone waste water tunnels as shown in Figure 1 were constructed using closed face shielded tunnels advanced pipe jacking. These tunnels were excavated through layers of silty clay and calcareous medium to fine sand under about 10 meters of ground water head. At the design stage, prediction of ground settlement above a tunnel usually focuses on the subsidence at, or near, ground surface in order to examine the safety of existing buildings founded on shallow foundations. On the other hand, ground control during actual construction requires more detailed evaluation of the actual sources of ground loss into the tunnel as well as the distributions of volume changes in the surrounding soils. This can be adequately achieved by monitoring subsurface ground displacement during construction. The general behavior of the tunnel is discussed on the basis of the measurements and trends of soil and ground water responses to tunnel advance.

---

[1] Professor, Alexandria University, Egypt, 21544
[2] Professor, Ain Shams University, Egypt, 21645
[3] Research student, Alexandria University, Egypt, 21544

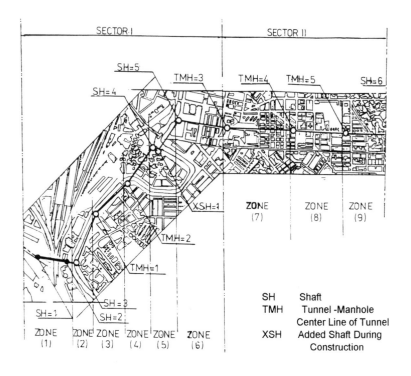

Figure 1. Location of Instrumented Area

As illustrated in Figure 2, The development of the soil subsidence above shielded tunnels exhibits different phases before reaching its final shape. The magnitude and variation of soil displacement within the deposit are greatly influenced by details of construction phases. Furthermore, the ground response to tunneling is dependent on the physical and geotechnical characteristics of the soil mass, (Peck,1969).The relationship between the volume of the surface settlement trough and volume of ground loss into the tunnel was examined by (Cording and Hansmire,1975).Prediction of the ground subsidence associated with urban shallow shielded tunnel based on the shape of final settlement trough of ground surface could be insufficient. Such prediction does not take into account the spatial pattern of soil movement, specifically, the longitudinal gradient of ground subsidence, as shown in Figure 3, (EL-nahhas,1986). The field measurements were compared with the surface settlement reported by (Peck,1969) and longitudinal gradient settlement reported by (EL- nahhas,1986).

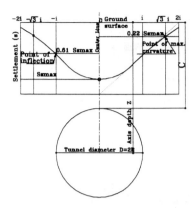

Figure 2. Surface Settlement Trough (after Peck, 1969)

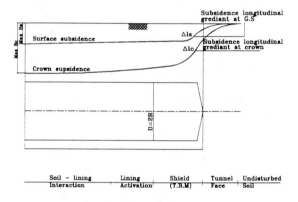

Figure 3. Spatial Mode of Ground Subsidence Above Advancing Shielded Tunnels
(after El-Nahhas, et al, 1986)

Site Condition and Tunnel Construction.

The 2.86m diameter shielded tunnel advanced pipe jacking from Shaft 2 to shaft 1, where the instrumented area was chosen, Figure1, at depth to center line of about 13m below ground surface. The subsurface soil profile at the instrumented area consists of about 3.5m fill, underlain by about 7m of silty clay, intermixed with little broken shells, followed by a layer of graded calcareous slightly silty medium to fine sand, 9m thick and a sandy lime stone layer, Figure 4. The surfacial fill layer is a heterogeneous mixture of silt, clay, pieces of limestone, sand, gravel and brick fragments. The ground water level at the instrumented area is approximately 2.5m below ground surface.

The soil layers encountered at this site extend laterally over most of the central zone of Alexandria city.Geotechnical consideration of these deposits and average properties are given by (Hamza, 1996).

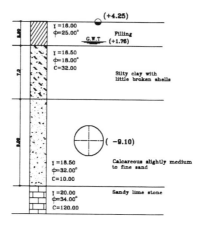

Figure 4. Soil Profile at The Instrumented Area

Construction of the tunnel was carried out by using pipe jacking technique. Precast concrete pipes 2.25m finished I.D where used. The tunneling machine was provided with closed face earth pressure balance shield (E.P.B., T.B.M.). During construction the soil face ahead of the tunnel and along reach of tunnel is treated by using jet grout, whenever required. During the excavation and jacking the soil around the pipes was grouted by high jelling lubricating bentonite, treated with polymers, through grout holes inside the pipes to facilitate the movement of the jacking pipes by reducing the friction along the pipes and to fill the gap of soil around the pipes .Figure 5 illustrates the head of the used tunneling machine .The bore diameter is 2.865 m, while the outer pipe diameter 2.81 m.

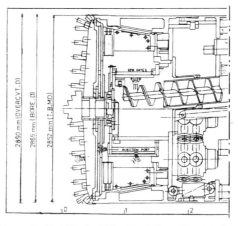

Figure 5. The Head of The Used Tunneling Machine

Instrumented Area.

As shown in Figure 1, the instrumented area was chosen south of Shaft 2. The instruments were distributed on three sections A,B and C perpendicular to the center line of tunnel and at a distance of 42m, 72m and 104m from Shaft 2, Figure 6. An extensive numbers of geotechnical instruments were installed in the ground before the shield advance under the area. The instrumented sections cover an area extending about 25m around the center line of the tunnel. The layout of instrumented area and instruments is shown in Figure 6. The instruments were distributed along sections A,B and C at spacing of 5m. Soil settlement near the ground surface was monitored using surface settlement points (SSP) , also building settlement points (BSP) were used to measure the settlement of nearby buildings. Two stand pipe Piezometers (PZM), Piezometer 1 and Piezometer 2 were constructed nearby Shafts 2 and 1 respectively, with their tips into the sand layer, to measure the change in ground water level.

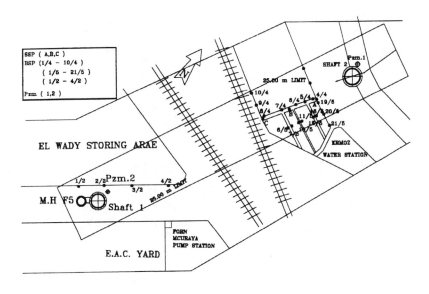

Figure 6-a. Instrumentation Layout.

Field Measurements.

Figure 7 shows representative records of the measured soil settlement at the ground surface using surface settlement points installed above the tunnel and attached to the buildings around the tunnel center line. The development of settlement troughs as related to the distance from center line of the tunnel for the three sections A,B and C are illustrated in Figure 8.

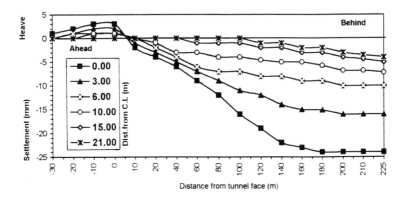

Figure 7-a. Surface Settlement at Section A

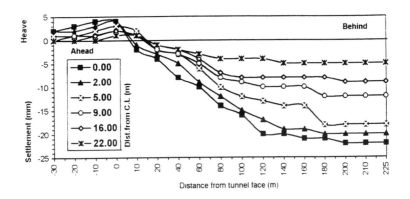

Figure 7-b. Surface Settlement at Section B

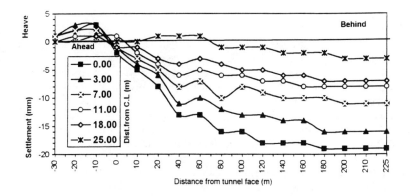

Figure 7-c. Surface Settlement at Section C

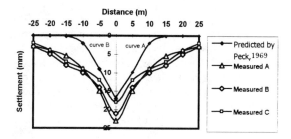

Figure 8. Predicted and Measured Settlement

Figure 9 presents the data compiled from the two stand pipe Piezometers concerning the change in ground water level. The water levels were related to a datum 1.5 m above the average ground surface. Complete records of the measurements of the remaining instruments are given else where by (Abd Lateef1999).

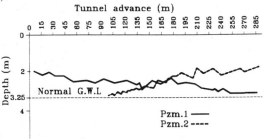

Figure 9. Development of Pore Water Pressure During Tunnel Advance

Discussion of Results.

Based on the measurements complied from the instruments installed at the instrumented area, the following observations were recorded :
Soil near the ground surface directly above the tunnel experienced heave movements of up to 4mm when the tunnel face was approaching. Downward settlement occurred when the shield advanced, reaching a maximum value of about 24mm, which equivalent to 0.9% of bore diameter.

To predict the settlement trough above tunnel using the normal probability curves shown in Figure 2, volume of the measured settlement trough as a percentage of the excavated volume of the tunnel is needed .A bore diameter of 2.865 m was implemented in the calculations, the achieved results are presented in figure 8,curve A. Curve B in the same figure, presents settlement trough above the pipe, but based on the volume of the pipe instead of the excavated volume. The maximum difference between the values of curves A and B is 2.70%. The values of the predicted settlement trough based on the pipe diameter are closer to the measured values than those predicted using bore diameter. Settlement measurements indicated that the volume loss varies from 4 to 6 % of the excavated volume of the tunnel. Figure8 indicated that the width of settlement trough as defined by (Peck,1969) was in the order of 5 to7 times the tunnel excavated radius. Figure 8 illustrated that the maximum slope of the settlement through varied from 1/700 to 1/1000,while Figure 7 indicated that the maximum longitudinal gradient of the settlement which defined by (El-Nahhas, 1986), varied from 1/100 to 1/300.

The large excess water head, ahead of the face of the tunnel, measured by stand pipe Piezometers 1and2 at different stages of construction is about 1.1 m, Figure 9. The pore water pressure at Piezometer 1 is affected by the advancement of the tunneling process along a distance up to 285 m, which equivalents to about 100 times the tunnel diameter, while Piezometer 2 indicates that the excess pore water pressure increases as the tunnel process approach the measuring point The pore water pressure at Piezometer 2 is affected by tunneling process, where the tunnel machine at a rear distance of 170m, which equivalent to about 60 times the tunnel diameter.

Conclusions

The measured soil subsidence and changes in ground water level can be used to predict the response of soil during tunneling process at different areas having similar subsurface conditions. Comparisons of the distribution of the predicted surface settlements with the measured values complied during construction indicated that the maximum surface settlement predicted by Peck is about 65 to 85% of the measured values.

Acknowledgements

The authors wish to thank the Egyptian organization for the execution of the Alexandria waste water project for permission to publish the measured data The support provided by the Alexandria and Ain Shams universities is also acknowledged.

References

1- Abd El-Lateef, H.A.(1999) . "Effect Of Tunneling On Adjacent Structure." Ph.D. Thesis, Alexandria university, Egypt, (In progress).
2 - Cording, E.J. and Hansmire, W.H. (1975) ."Displacements around soft ground tunnels."General report, In Proc. of the 5th. Pan-American conf. on SMFE,Buenos Aires,Vol. 4,pp.571-633.
3 - El -Nahhas, F.(1986). "Spatial mode of ground subsidence above advancing shielded tunnels." Proc. of Int. Congress on large under ground openings, ITA, Italy, Vol.1,pp 720-725.
4 - Hamza Assoc. (1996). "Alexandria central zone flow conveyance system."Geotechnical Investigation and Geotechnical design approach .
5 - Peck R.B. (1969). "Deep excavation and tunneling in soft ground." Proc. of VII IC SMFE, Mexico, stat - of - the - art volume, pp225 – 290.

Performance of Multiple Retention Systems
During Cut and Cover Tunnel Construction

Ted D. Bushell[1] Member ASCE, Daniel W. McCarthy[2], and Elaine McCluskey[3]

Abstract

The 8 GeV tunnel connects the Main Injector and existing Booster Accelerator Rings at Fermilab in Batavia, Illinois. The tunnel is a combination of cast-in-place and precast tunnel sections supported on a cast-in-place slab. To protect adjacent structures from the 9 to 12 meter-deep excavation, a rigid retention system was required. The main retention system consisted of a secant pile wall reinforced with steel beams. On the opposite side of the excavation where movements were of less concern, the retention system consisted of drilled-in-place soldier piles and lagging. The bracing system consisted of an upper level of cross lot struts and a lower level of tiebacks. In less critical areas, conventional driven soldier piles and lagging were used with cross lot braces. Performance monitoring utilized inclinometers, settlement indicators as well as crack monitors and dial gauges. The average lateral movement at the existing foundation level was 1.27 cm with a maximum lateral movement of the existing Booster Accelerator Ring of 0.6 cm. Predicted lateral deflections were in the range of 1.3 to 2.5 cm. Settlement of the Booster Building was 0.1 to 0.2 cm which was less than the established tolerance of 0.63 cm.

Introduction

The Main Injector Project consists of a new 150 GeV accelerator on the Fermilab site in Batavia, Illinois to increase proton delivery by a factor of two or three. The project includes 4572 m of tunnel enclosures, eleven service buildings, and a 345 kV substation. The new accelerator ring connects to the existing Booster Accelerator Ring via the 8 GeV tunnel connection which is described in this paper.

---

[1]Principal Engineer, STS Consultants, Ltd., Vernon Hills, Illinois 60061
[2]Instrumentation Engineer, STS Consultants Ltd., Vernon Hills, Illinois 60061
[3]Structural Engineer, FESS Engineering Group, Fermi National Accelerator Laboratory, Batavia, Illinois 60510

Description of Project and Site

The 8 GeV Booster Connection Project is located adjacent to the existing Booster Accelerator Ring (Booster) and the two-story Southwest Booster Building. The new 8 GeV tunnel has dimensions of 2.4 m wide by 3 m high and is a combination of cast-in-place concrete and precast concrete U-shaped tunnel sections set on a cast-in-place slab. Most of the tunnel floor is located 9.5 m below grade (11 m excavation) but at the Booster, the floor rises to 6.8 m (9 m excavation).

The site was relatively flat prior to construction containing roads, parking lots, and grassy areas adjacent to a two-story laboratory building faced with precast concrete panels. The building was to remain occupied and functional during construction. To achieve this, a maximum settlement tolerance of 0.63 cm was established for the Booster Building.

Subsurface Conditions

The soil conditions at the site varied rather significantly along the length of the structure. At the south end of the site, clay and gravel fill were encountered from the ground surface to a depth of 2 m. However, at the north end of the structure, loose to dense sand and gravel fill was encountered to a depth of 7.5 m. This granular backfill was saturated at a depth of 3 to 5 m (groundwater table) and identified as the backfill surrounding the existing Booster. Underlying the granular fill, natural, gray, stiff to hard silty clay glacial till was encountered.

Earth Retention System

Since the existing Booster Building and Booster Accelerator Ring had to remain in service during and after construction of the new enclosure, the selected retention system needed to minimize ground movements. The existing Booster Building contained sensitive pre-cast concrete elements. Also, the soil borings indicated that the Booster Building foundations were supported on the granular fill overlying the Booster Accelerator Ring. This granular fill was found to extend laterally into the proximity of the proposed excavation. The presence of the granular fill complicated construction of the retention system since this type of soil is sensitive to vibration, erodible, and loses strength with decreased confinement. Three different retention systems were selected to accommodate the varying soil and site conditions. Fig. 1 is a plan view of the upper and lower bracing levels depicting the various wall systems utilized.

The north portion of the excavation, particularly the east side along the Booster Building and Booster Accelerator Ring, was an area of primary concern for which several retention schemes were considered. Initially, a slurry wall was considered, but for such a limited section of wall, this solution was too costly. Conventional soldier pile and lagging systems were not considered appropriate near the existing building since ground loss would have likely occurred while excavating to install lagging in the granular soil. Steel sheet piling was also considered but the vibrations caused by pile driving would have undoubtedly caused excessive settlement and damage to occur within the existing building.

Another option that was considered was grouting to stabilize the sand and gravel fill underlying the Booster Building and driving steel sheeting for the retention system. The grout zone would solidify the granular fill to prevent it from densifying and settling during the driving of sheeting and subsequent excavation. However, problems with this scheme included high fines content in the granular backfill making it difficult to grout and concern over damage to the Booster Accelerator Ring from grouting pressure.

After considering all of the above systems and the constraints imposed on the retention system, a secant pile wall was selected for the east side of the retention system along the Booster Building. A secant wall is essentially a combination of a soldier pile and lagging wall and a slurry wall. The secondary (unreinforced) secant piles act as lagging that is installed prior to excavation so that ground loss does not occur during installation of the lagging as with a conventional lagged wall. Similar to a slurry wall, this wall is relatively stiff and water tight which minimizes movement and ground loss. Most importantly, this type of wall can be installed without driving or vibrating steel members.

Fig. 2 contains sections through the north (Section 1) and south (Section2) ends of the excavation illustrating the secant pile wall. The secant wall consisted of a series of 61 cm diameter, 9 to 11 m long drilled shafts which were backfilled with a one to two bag lean grout mixture. Alternating piles called primary piles were reinforced with full length HP 30.5 x 24 (cm x kg) beams. Unreinforced secant piles were installed between the primary piles. These secondary piles were set back 22.9 cm behind the primary piles and overlapped the adjacent primary piles 15.2 cm as shown in Detail A on Fig. 1.

For the retention system on the west side of the excavation away from the Booster Building, a soldier pile and lagging system was selected. This side of the excavation was not adjacent to any structures and hence, some ground movement could be tolerated. Also, granular backfill was not anticipated along the west wall. However, all soldier piles lying within 23 m of the Booster Building were installed by drilling in lieu of driving to minimize vibrations. Figs. 1 and 2 depict the drilled-in-place soldier piles located across the excavation from the secant piles. The soldier piles consisted of 9 to 11 m long, W 53 x 46 (cm x kg) steel beams placed in 76 cm drilled holes which were backfilled with grout. Soldier piles were spaced at 1.5 to 1.8 m. The lagging consisted of 7.6 cm rough cut mixed hardwood.

At the south end of the excavation, a driven soldier pile and lagging system was used since this area was outside the 23-meter zone discussed above. Soldier beams, consisted of HP 30.5 x 24 (cm x kg) and 35.6 x 40 (cm x kg) beams. The majority of the soldier beams ranged from 10.7 to 12.2 m in length and were spaced at 1.8 to 2.1 m. Lagging consisted of 7.6 cm rough cut mixed hardwood.

Bracing System

To minimize movement of the adjacent structures, two levels of bracing were used in the northern portion of the earth retention system as illustrated in Fig. 2. Since sensitive structures were located along the east wall, the bracing system consisted of an upper cross lot brace and a lower level set of tieback anchors on the east side only. Hence, the west wall had one level of bracing while the east wall had two bracing levels. The only exception to this was at the deep sump location in the center of the excavation where a lower level of tiebacks was also used along the west wall as shown in Fig. 2.

Upper level walers at the north end were situated at elevations of 223 to 224 and consisted of W 61 x 59 (cm x kg) beams on the east wall and W 61 x 66 (cm x kg), W 61 x 47 (cm x kg), and W 61 x 53 (cm x kg) beams on the west wall. The top level of cross lot braces consisted of 30 and 41 cm (0.95 and 1.27 cm wall thickness) diameter steel pipe struts and were typically spaced at 3 to 6 m.

The tiebacks used along the east wall consisted of soil anchors containing three, 1.5 cm strands with a design load of 436 kN. The anchors were spaced at 1.83 m on center and inclined 17.5° from the horizontal to avoid the existing Booster. The anchors were typically 15 m long with a 9 m bonded zone and a 6 m free length. The lower level waler along this tieback level consisted of a double channel C 38 x 15.4 (cm x kg).

At the south end of the excavation, where driven soldier piles were used, the bracing consisted of one level of cross lot braces since movement was not extremely critical in this area. The main waler levels along the east and west walls were situated at elevation 222. This steel waler typically consisted of a W 61 x 59.5 (cm x kg). The cross lot braces in this area were typically spaced at 5.2 m.

Lateral Earth Pressure and Design Loads

Lateral pressures for the secant pile retention system were computed by the retention system contractor using an apparent pressure diagram since the soil profile is a mixture of both granular and cohesive soils. The total Rankine earth pressure, increased by an overload factor of 1.3, was computed and converted into an equivalent trapezoid according to Terzaghi, Peck, and Mesri (1996). Unbalanced water pressure below elevation 222 was added to the earth pressure as was the surcharge pressure from the Booster Building foundations.

Maximum wall moments were calculated to be approximately 148 kN-m resulting in a pile length of 10.7 m. The upper and lower walers were designed for loads of 168 kN/m and 227 kN/m, respectively, for the 7.7 meter excavation. Strut loads for the upper bracing level ranged from 383 to 970 kN.

The lower level bracing on the east (secant pile) wall consisted of soil anchors. The 227 kN/m waler load resulted in an anchor load of 436 kN per anchor. Soil anchors along the west wall had design loads ranging from 276 to 365 kN.

Construction Procedures

An initial excavation down to elevation 225 was performed to create a working bench. This bench would minimize the height of the earth retention walls and enable the contractor to locate existing utilities in preparation for supporting them during the excavation. The earth retention subcontractor first installed the driven soldier piles on each side of the south end of the project. Once this was complete, the secant wall adjacent to the Booster Building was started at the south end, next to the driven soldier pile wall. The unreinforced piles were drilled first followed by reinforced piles. The 61 cm diameter secant pile shafts were augered to water table followed by installation of a temporary steel casing 1 m into the stiff clay. Once the casing sealed off the water, the shaft was advanced to elevation 213 to 216. The reinforced secant (primary) piles containing the HP 30.5 x 24 beams as well as the secondary piles were backfilled with cement grout. Mix design (compressive) strength for the grout in the secant wall was 6890 kPa. The pour and pull method for the temporary casing was used successfully. No water leaks developed after the secant wall was fully exposed and therefore, no additional grouting was required. Approximately two secant piles were drilled per day.

Upon completion of the secant wall, the drilled soldier pile and lagging wall was begun on the west side of the north section of the project. The grout used in the drilled soldier piles had a compressive strength of 2756 kPa. While the drilling operations at the north end continued, the contractor was excavating and installing lagging for the driven pile wall to the south. Lagging operations proceed to a level 0.7 m below the bracing level, and then walers and struts were installed. During wooden lagging installation, the exposed excavation never exceeded 60 cm. When the drilled walls were completed on the north end, excavation proceeded to just below the upper bracing level, and then walers and struts were installed. The struts were preloaded to 50% of their design load and steel shims were used at the waler/pile connections. The excavation then proceeded to the lower bracing level and the tieback anchors were augered in place. Once all of the tiebacks successfully passed the load test requirements, the excavation was continued to the design bottom elevation, the enclosure built and the excavation backfilled.

Instrumentation Program

Data was collected weekly from February 14, 1996 to May 17, 1996, then once a month until August 29, 1996 when the 8 GeV Beam Enclosure Connector structure was completed and backfilled to elevation 224 m. The location of the various instruments is shown on Fig. 1.

Three inclinometers (I-1, I-2 and I-3) were installed between the Booster Building and the secant pile wall to detect possible movement of the building and underlying tunnel. Detailed multiple inclinometer plots are illustrated in Fig. 3. These plots show the change in movement with time as construction proceeds. The inclinometer data indicates some lateral displacement of soils occurred during the installation of the secant piles and as the excavation progressed. The secant pile installation disturbed the sand zone near inclinometer I-1 as indicated by the movement at elevation 216. This movement did not occur at the other inclinometer locations since they were located farther from the secant piles.

The rate of movement decreased after the mud slab placement and leveled off once the backfill material reached elevation 224. The inclinometers reflected the tieback installation at elevation 220 as illustrated by the inflection of the movement curve. All inclinometers also showed slight movements at approximately elevation 219 between the tieback and the bottom of the excavation. The average maximum movement for all three inclinometers at elevation 225 (bottom of Booster Building footing) was 1.27 cm. Movement at the tunnel elevation of 220 was less than 0.6 cm.

Settlement monitoring points were established on Booster Building columns. In addition, soil settlement indicators were installed outside the building adjacent to four footings. The soil settlement indicators are telltale pipes installed in plastic casing installed in the ground to a depth of approximately 1.22 m. The columns as well as the soil settlement indicators were surveyed on a regular basis.

Fig. 4 is a summary plot of the survey readings obtained. The plot indicates some minor upward movement which may be associated with instrument temperature fluctuations. However, this movement is less than 0.2 cm which is within the normal survey accuracy. The settlement plot also shows approximately 0.2 cm of settlement at day 58 on April 24. The inclinometers also show increased movement in the range of 0.2 to 0.4 cm at this same time. This movement of both the settlement indicators and the inclinometers is most likely associated with the excavation to install the tiebacks on April 20 and 21, 1996. The settlement plots show the final settlements in the range of 0.1 cm or less. Considering the fluctuating measurements, the survey readings indicate only minor settlement took place at the Booster Building.

Crack monitors indicated very slight movement during the construction of the 8 GeV Beam Enclosure Connector. Crack monitors were installed directly east of the inclinometers on the exterior wall of the Booster Building. Detailed plots of the crack monitor movement versus time are shown in Fig. 5. The crack monitor data indicates that the west wall of the Booster Building moved horizontally a maximum of 2 mm.

Dial gauges were installed inside the Booster Building on a first floor ceiling I-beam at two locations. Plots of the dial gauge movement versus time shown in Fig. 5 indicate both horizontal and vertical movement in the range of .02 to .03 mm.

The instrumentation indicates that the building movements were minimal during construction and the retention system performed as anticipated. The timing of these movements appeared to correspond with construction activities which consisted primarily of installing secant piles, excavation and tieback installation. Approximately 1.3 to 2.5 cm of movement was predicted prior to construction using empirical methods presented by Terzaghi, Peck, and Mesri (1996). The maximum measured ground movement was 1.99 cm at elevation 226 at Inclinometer location I-1. This compares well to the predicted range of movement.

Conclusions

The earth retention system for this project was unique in that three different types of retention walls and two bracing methods were all combined on the same project. Also, a secant pile type retaining wall was used at this site for the first time.

Instrumentation provided the necessary data to monitor behavior of a critical existing building. Lateral movement and settlement were less than predicted most likely due to the stiffness of the secant pile wall and details such as pre-stressing struts and the use of steel shims between the steel beams and walers.

Acknowledgements

The authors are grateful to Fermi National Accelerator Laboratory, a Department of Energy National Laboratory, for its kind permission to publish this paper. Structural designers for this project were Fluor Daniel. The general contractor was Rausch Construction Company while the retention system subcontractor was Thatcher Engineering Corporation. The authors would also like to thank Keith Bushell for his assistance in preparing illustrations.

Appendix l. References

Terzaghi, K., Peck, R.B., and Mesri, G. (1996). Soil Mechanics in Engineering Practice, Third Edition, New York, John Wiley and Sons, 549 pp.

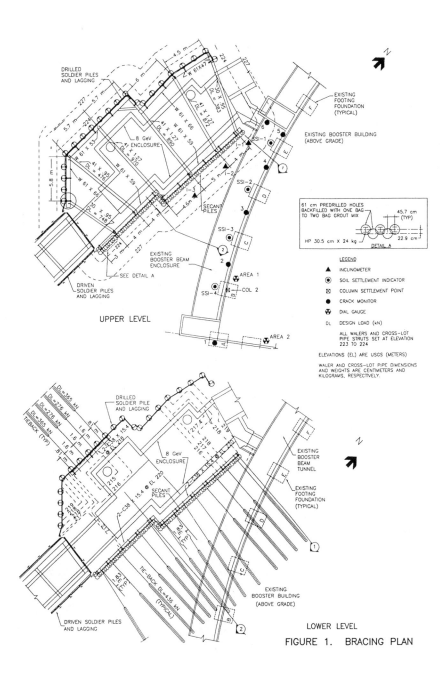

FIGURE 1. BRACING PLAN

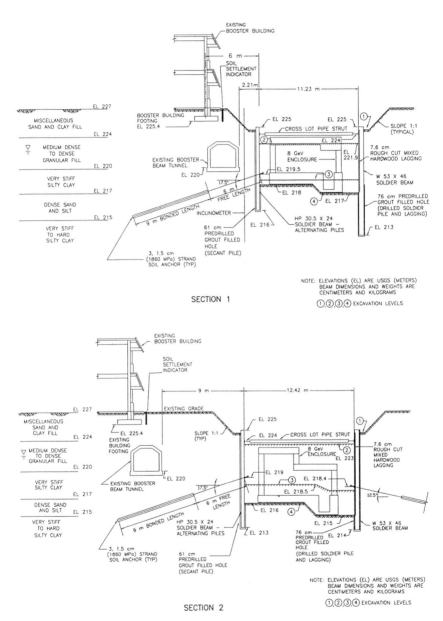

FIGURE 2. RETENTION SYSTEM CROSS SECTIONS

GEO-ENGINEERING FOR UNDERGROUND FACILITIES 897

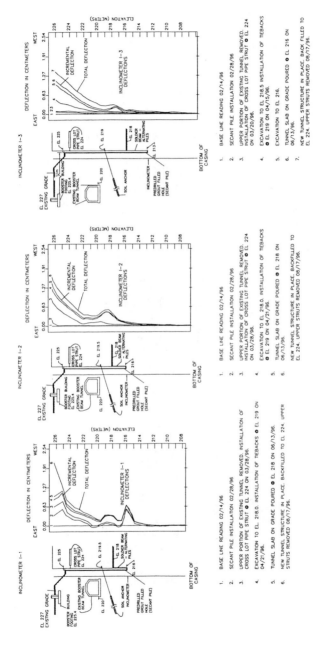

FIGURE 3. INCLINOMETER PLOTS

GEO-ENGINEERING FOR UNDERGROUND FACILITIES

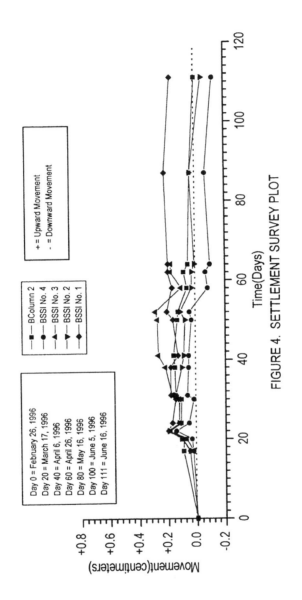

FIGURE 4. SETTLEMENT SURVEY PLOT

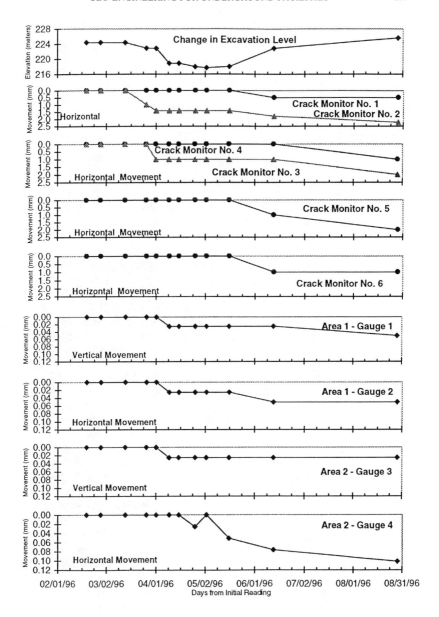

FIGURE 5. CRACK MONITOR AND DIAL GAUGE PLOTS

Performance of Inclinometers in Diaphragm Wall

Donald D. Liou[1], Member, ASCE

Abstract

Trans Tokyo Bay Project is a highway/tunnel/bridge project consisting of two distinct sections. The east section is a 5-km bridge and viaduct section, while the west section is a 10-km undersea tunnel section. The ventilation of the long undersea tunnel is provided through a manmade island, called Kawasaki Island, which is located at the center of the tunnel section. During construction of the tunnel section, this island also served as a work base for the tunneling operation. Kawasaki Island contains a gigantic, cylindrical-shaped diaphragm wall. As of today this diaphragm wall still holds the world record in terms of depth, size and technical difficulties. To monitor the construction and behavior of the diaphragm wall during the long construction period, several inclinometers, along with many other types of instruments, were installed near and on the wall. Although the instrumentation system as a whole provided many valuable construction data, the overall performance of the inclinometers, the most expensive instruments in the system, could only be considered as less than desirable. This paper presents some of the difficulties encountered in the use of these inclinometers. It provides some typical data recorded by these instruments, and discusses the possible reasons for the difficulties encountered. It points out the limitations of inclinometers, and suggests certain measures for preventing similar difficulties in future projects.

Introduction

The Tokyo Aqualine is a direct land link between the eastern and the western sides of Tokyo Bay. It is one of the major infrastructure projects undertaken by the Japanese government in the current decade to relieve problems associated with over-population in the Tokyo-Yokohama metropolitan area. During its construction, this highway project was better known as the Trans Tokyo Bay

---

[1] Associate Professor, Department of Engineering Technology, The University of North Carolina at Charlotte, 9201 University Blvd., Charlotte, NC 28223-0001

GEO-ENGINEERING FOR UNDERGROUND FACILITIES 901

Highway Project. The project is a highway/tunnel/bridge project consisting of two distinct sections. The east section was a 5-km bridge and viaduct section extending westward from Kisarazu City. The west section was a 10-km undersea tunnel section extending eastward from Kawasaki City. The total construction cost of this project is more than 10 billion US dollars.

The ventilation of the long undersea tunnel in the project is provided through a manmade island, called Kawasaki Island, which is located at the center of the tunnel section. The island is located at a site where the original soil is extremely weak and the average seawater level is 28.5 m deep. During construction of the tunnel section, this island also served as a work base for the tunneling operation.

Kawasaki Island contains a gigantic, cylindrical-shaped diaphragm wall. As of today this diaphragm wall, with a 100-m outside diameter, a 2.8-m thickness, and a 119-m depth, still holds the world record in terms of depth, size and technical difficulties. From technical point-of-view, this diaphragm was a showcase project in which the tasks of design, instrumentation, and construction planning were performed side by side. The execution of this wall illustrated how the concepts of constructibility and modular design could be implemented in the design and construction of a real-world project.

To monitor the behavior of the manmade island and to ensure safety during the long construction period, a state-of-art instrumentation system was installed on the island. As a result, many valuable construction data were recorded. Although the performance of the instrumentation system as a whole was quite successful, the overall performance of the inclinometers, the most expensive instruments in the instrumentation system, was less than desirable. Difficulties were encountered during installation of these instruments. When in operation, some of the data recorded by these instruments quickly became a source of puzzle, and presented themselves as a technical challenge to the team responsible for the instrumentation system.

This paper presents part of the experience gained in the use of these inclinometers. It provides some typical data recorded by these instruments, and discusses the possible reasons for the difficulties encountered during the installation and operation of these inclinometers. It points out the limitations of inclinometers, and suggests certain measures for preventing similar difficulties in future projects.

Potential for Heave of Bottom Soil

The diaphragm wall was designed for two primary purposes: ground support and water cutoff. The wall was constructed inside an artificial embankment, as shown in Figure 1. To facilitate the construction, the diaphragm wall was divided into 56 elements, as shown in Figure 2. Each element in the completed cylindrical diaphragm wall was designed to act like a retaining wall by itself. The horizontal curvature in the diaphragm-wall elements also permitted the ring compression to be transferred from one wall element to its neighboring wall elements, even though no cross-lot bracing was actually used.

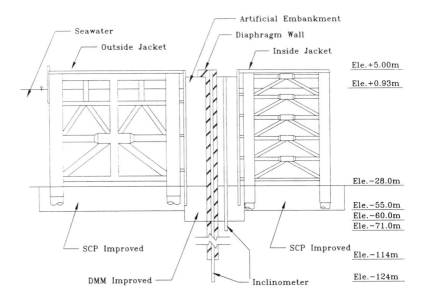

Figure 1. Locations of Inclinometers in Elevation

Following the construction of the diaphragm wall was a soil-excavation phase of construction. In this phase, the seawater, all structural elements and the soil mass above Elevation -69.7m and inside the wall were removed to make room for the construction of the permanent structure. During this phase, the basic deformation of the upper part of the cylindrical wall, the portion that was about 10 to 15 m above the lowest level of interior soil excavation, was expected to be one that corresponded to ring compression. The lower portion of the wall was expected to have predominantly flexural behavior in each vertical plan, due to the requirement of compatibility.

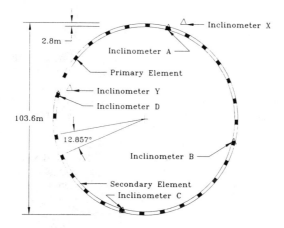

Figure 2. Locations of Inclinometers in Plan

The most serious design and safety concern regarding the diaphragm wall in this construction phase was the heave of the bottom soil inside the cylindrical wall. The danger for heave of bottom soil was mainly caused by the seepage of water through the bottom soil into the excavated space, which clearly had a lower hydraulic head than outside the wall. Therefore, the seepage of water was a result of the unbalanced pressure between the inside and the outside of the cylindrical wall. The potential for the soil mass below the excavation level to heave was expected to reach critical level near the end of the soil excavation phase. Furthermore, between the completion of the soil excavation phase and the sealing of the exposed soil face by a massive concrete mat, the potential for bottom heave was expected to increase with time.

Quality Control Issues

There were two major quality control issues involved in the diaphragm-wall construction phase of the project. They were the quality control of slurry and the quality control of the tremie concrete.

Construction of the Kawasaki Island diaphragm-wall required the process of using a diaphragm-wall machine to drill a vertical shaft into a stable ground. This process destroyed the naturally balanced stress conditions and exposes various soil formations in the ground. Strong soil formations along the shaft might have stable faces, but weak soil formations tended to crumble, collapse, or cave in easily. To mitigate the effects of these undesirable situations, each excavation shaft of diaphragm-wall elements was filled with slurry having suitable density and viscosity.

During the process of excavation, the crumbles of the soil formations continuously contaminated the slurry in the shaft. By circulating and screening the slurry, the soil particles in the slurry were transported to and picked up by a screening system located at the surface. When the shaft excavation was complete, the excavation machine was removed from the shaft. Immediately after the removal, a pan with an area slightly smaller than the cross-section of the shaft was placed into the shaft. Afterwards, there would be a one-day cooling period with no construction activities performed on the shaft. This period allowed some of the soil particles still remaining in the slurry to settle to the pan at the bottom of the shaft. The old slurry in the shaft was later replaced by brand new slurry.

The new slurry in the shaft was in turn replaced by tremie concrete. The quality control of tremie concrete was another important issue during the construction of the diaphragm wall (Liou 1999).

Ground Improvements and Artificial Embankment

To facilitate the diaphragm wall construction, the original soil was improved by both Sand Compaction Pile (SCP) and Deep Mixing methods. SCP improved the properties of a very large and deep---from Elevation -27.0 m to Elevation -55.0 m---area under the man made island. Deep Mixing method improved the area directly under the artificial embankment, as shown in Figure 1.

After the installation of the steel-jacket structures, the annual space between the inner steel-jacket trestle structure and the outside steel-jacket trestle structure was full of sea water. The annual space was filled with sandy fill material resulting an artificial embankment. The composition of the fill material also includes mudstone slurry and cement.

Other Uncertainties

Besides the quality control issues mentioned previously, there were many other uncertainties regarding the behavior of the wall. Two of these uncertainties are discussed below.

One uncertainty was the finished geometry of the diaphragm wall. At the end of the excavation of each diaphragm trench, the geometry of that trench was checked by using ultrasonic echometer. However, even if the geometry of the trench confirmed to the most stringent specifications at the end of excavation, there was no guarantee that the finished diaphragm wall would not have undesirable blemishes, for the faces of some trenches did collapse locally during the concrete placement process.

Another uncertainty was the exact effect of deep wells. As mentioned previously, the heave of soil mass below the excavation level inside the diaphragm wall was a big concern in the interior soil-excavation phase of construction. To reduce the danger of bottom-soil heave, deep wells were installed into the permeable soil strata at several locations immediately outside the diaphragm wall to release the ground-water pressures. As the depth of wells and the difference of water heads became large, the reliability and the dependability of the well system, which was required to operate for a relative long period, would be reduced.

The combination of the quality control issues and uncertainties mentioned above and other factors not mentioned made obvious the need for a better knowledge of the behavior of the diaphragm wall, the manmade island, and its surrounding media. Toward that end, a comprehensive monitoring system involving many different instruments was developed and installed near and in the wall (Liou 1994). It was expected that the conditions be observed by the instrumentation system installed could be used to evaluate and update design judgments made previously, and, if necessary, could guide the remedial measures, making the use of the instrumentation system an integral part of the diaphragm-wall design and construction process.

Locations of Inclinometers

Along with contact earth pressure cells, settlement platforms, piezometers and many other instruments, six inclinometers were installed in and around the diaphragm wall, as illustrated in Figure 2. Two of these instruments were primarily used to monitor the performance of the embankment into which the diaphragm wall was constructed, and they are installed in the embankment near the wall. Besides their main purpose, installing and operating the first two inclinometers also gave the instrumentation team opportunity to gain experience of using inclinometers. The remaining four inclinometers were used to monitor the performance of the finished diaphragm wall, so they were installed in the diaphragm wall itself.

The four inclinometers installed in the diaphragm wall were installed inside the four primary elements approximately facing east, south, west, and north. To facilitate the construction of the diaphragm wall, the wall was divided into two kinds of elements: primary and secondary elements. Primary and secondary diaphragm-wall elements were located alternately, with two primary elements flanked by a secondary element, and vice versa. Primary elements were smaller but more heavily reinforced than secondary elements, and they were constructed before the secondary elements.

Accidents during Installation

The first two inclinometers, the set that was installed in the artificial embankment, had plastic casings. Each cashing had an 83-mm diameter and a total length of 76m. It was made of twenty-five (25) 3-meter-long and one (1) 1-meter-long pieces. The 1-meter piece was located at the very top of the casing. These casing parts were assembled and taped together in the field without using couplings. The casing of inclinometer was inserted into a PVC pipe with a 125 mm diameter. The PVC pipe was also about 76 m long and was made of nineteen (19) 4-meter-long pieces. The casing, the PVC pipe, and another grout pipe were installed into an unsupported borehole pre-drilled into the artificial embankment. The borehole had an inside diameter of 200 mm, and the grout pipe had an outside diameter of 22 mm. The casing was installed according the procedure supplied by the Japanese vendor of the inclinometers. The annular space between the borehole and the PVC pipe and the annular space between the PVC pipe and the casing were backfilled with grout.

Because the casing was relatively long, a mechanical device was used to hang the PVC pipe and the inclinometer casing so that they could be installed near vertically, as shown in Figure 3. This device also served the purpose of overcoming buoyancy before the setting of the grout.

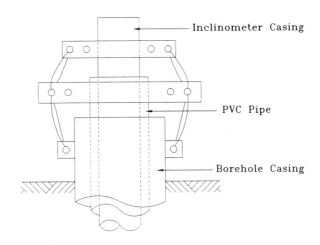

Figure 3. Mechanical Device for Hanging Casing of Inclinometer

The first field test of these inclinometers was conducted the day after installation. During this field test, mishaps occurred on both inclinometers. In the first inclinometer, the probe was stopped at 4 meters from the top of the casing. Because the probe was stopped at a relatively shallow location, the technician who was handling the inclinometer probe tried to pull it out by hand. As a result, the top casing piece and the top PVC pipe piece were pulled out accidentally.

With the mishap at the first inclinometer, technicians were more careful with the second instrument. Initially, things seemed to go smoothly at the second inclinometer. However, when the probe reached a depth of 71 meters from the top of the casing, which was only 5 meters short of the bottom of the casing, it could go no further. As a result, the field test was aborted.

Possible Reasons for the Accidents

After reviewing the available evidence, the reasons for the accidents happened to the first set of inclinometers were attributed to:

1. The bottom of the casing did not properly touch the bottom of the PVC pipe. A gap of a few centimeters might have existed between the bottoms of these pipes, allowing grout to seep into the inclinometer casing.

2. As grout was settling down, it might have developed strong negative skin friction and contributed to the pull out of the first inclinometer casing and PVC pipe. The composition of grout, therefore, needed to be reviewed.

3. The length of overlap between casing pieces was also considered to be inadequate. This might have allowed grout to seep into the casing of the inclinometer and prevented the advance of the probe.

As a result of the debriefing, the casings were washed and reinstalled using a new mix of grout and sand. They were more carefully coupled and taped together. No accidents occurred at these inclinometers again, and they behaved satisfactorily.

Performance of Inclinometers inside the Wall

The casings of the four inclinometers installed in the diaphragm wall were also installed near vertically. Each was 129 meters long. It extended from the top of the diaphragm wall, which was at Elevation 5 m, to Elevation -124 m. The bottom end of the casing was 10-m below the bottom level of the diaphragm wall, which was at Elevation -114 m. The extra length of the casing was required and used as a stable reference for checking purposes. With the extra 10-m length, the bottom end of the casing was not expected to move during the instrument use period.

From the data recorded by other instruments, including the stresses of reinforcements, the stresses of concrete, the recorded wall temperatures, the cylindrical diaphragm wall generally behaved as it had been designed. Once the construction of the wall elements was complete, they formed one integral structure, ready to take the unbalanced pressure between the inside and the outside of the cylindrical wall. The wall started to experience the unbalanced pressure when the seawater and other structural elements inside the wall were removed. As the excavation of the interior soil mass progressed, the pressure exerted on the wall increased correspondingly, and the deformation of the wall also became more and more pronounced. The unbalanced pressure and the deformation of the wall reached their peaks when the excavation of the interior soil mass reached the design level, creating the largest unbalanced pressure on the wall (TTB 1997).

Initially, the data recorded by the four inclinometers located inside the diaphragm wall generally reflected and confirmed the data recorded by other instruments. However, as the seawater, structural elements and some of the soil mass inside the diaphragm wall were removed to make room for the construction of the permanent structure, the whole cylindrical wall shrank slightly. The change of the cylindrical wall was very small. On the average, the radius of the cylinder shrank by about one centimeter only. However, this small change of wall shape had a dramatic impact on the four inclinometers embedded in the wall. The top portion of each inclinometer casing, the portion that was embedded in the diaphragm wall moved with the wall. This created a big change in curvature over a relative short portion of two to three meters of the casing, as shown in the bottom part of Figure 4. The worst among all the situations was that the exact position of the bottom portion, the extra 10 meters below the wall, was uncertain. It was not clear whether the bottom end of the casing remained in its originally installed position or was dragged away from that position.

As mentioned before, the bottom end of the inclinometer casing was used as a stable reference for all the data recorded by the probe. This casing distortion essentially eliminated the stable reference point for these data. Nevertheless, nothing could be done to restore this stable reference, since the bottom end was almost 130 meters below the work platform level.

As in any good quality construction project, the tops of the inclinometers and other control points on the work platform were actually tied by a GPS system. Unfortunately, the data obtained by the GPS system indicated a translational error of about 7-cm over a one-year period.

The impact was dramatic for another not so apparent but fatal reason: the error and the data were of the same order of magnitude. As shown in Figure 4, the largest deformation recorded by the inclinometer at section A of the diaphragm wall was in the order of 10 mm. The deformation of the diaphragm wall was measured by an 884-mm probe. The distance between the wheels of the probe was 500 mm. Therefore, the deformation of the wall was measured at 1-m intervals. The lateral deformation measured by an inclinometer was cumulative from the bottom. At 1-m interval of measurement, it took about 130 measurements for the probe to come up to the top end of the casing. If the error involved in the measurement of an inclinometer is considered as a random walk, after 130 steps the error could easily be in the order of a few mm. Since the error and the data might be of the same order of magnitude, the results of the inclinometer measurement could only be used qualitatively, rather than quantitatively. As a result, at least half of the benefit anticipated from using these inclinometers was lost.

To illustrate the frustration that could be caused by having the error and the data in the same order of magnitude, the lateral deformations recorded at the top of inclinometer A over a period of one month are shown in Figure 5. These deformations were actually recorded every 12 hours. However, only five data points, which were recorded at the beginning of each week, are shown in this figure. In this figure, the lateral deformations as recorded by the inclinometer swayed with time, when in reality it should have been relatively stable.

At the field office, the deformations were originally plotted at their recording intervals, i.e., every 12 hours. This made them a source of puzzle, as many data recorded by other instruments actually swayed at about that interval. For example, the water pressure in a soil stratum recorded by a piezometer swayed near that interval, because of the influence of tide.

Different numerical techniques were applied on these inclinometer data. For example, running average method that is normally used to obtain the average of weekly stock market values was used. However, none of them proved to be of lasting value.

Conclusion and Recommendations

This paper presents part of the experience gained in the use of inclinometers in monitoring the construction and behavior of a large-sized diaphragm wall. It provides some typical data recorded, and discusses the possible reasons for the difficulties encountered during the installation and operation of these inclinometers.

The difficulties encountered during the installation of some of these inclinometers were caused by the three possible reasons: negative skin friction of the grout, the inappropriate coupling between casing parts, and the seepage of grout

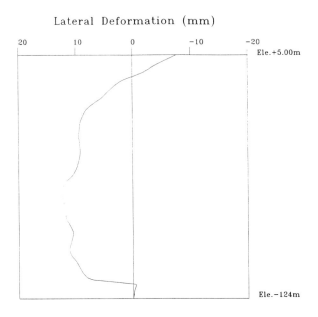

Figure 4. Typical A-Axis Deformation of the Wall Recorded by Inclinometer

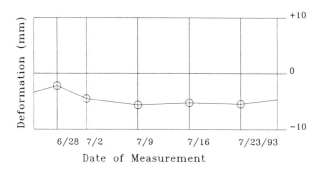

Figure 5. Typical A-Axis Surface Deformations Recorded by Inclinometer

into the inclinometer casing. All of these could be easily avoided in any similar future project, if the installers of inclinometers were adequately aware of past difficulties with inclinometers.

More importantly, this paper points out the limitations of inclinometers: they can only be used in situation where the order of the magnitude of the recorded data is much larger than the error involved. Therefore, inclinometers are appropriate for monitoring the behavior of flexible soil or structures. They are less effective when used in monitoring the behavior of stiff structures, such as the diaphragm wall discussed in this paper. As illustrated in the above, when the error and the data of an inclinometer are of the same order of magnitude, the results of the inclinometer measurement would loss half of their anticipated benefit. They become less useful because they can only be used qualitatively, rather than quantitatively.

Acknowledgment

The writer would like to acknowledge the support of the Taisei-Tobishima-Penta Ocean-Bechtel consortium of the East Kawasaki Island Project for this field study.

References

Liou, D. D. (1994). "Thermal Cracking in the Diaphragm-Wall Concrete of Kawasaki Island." *Proceedings of the International RILEM Symposium*, Munich, Germany, 393-400.

Liou, D. D. (1999). "Thermal Effects in Large-Sized Diaphragm Wall," *ASCE Journal of Performance of Constructed Facilities*, Vol. 13, No. 1, 17-21.

The TTB East Kawasaki Island Project Office, (1997). *The TTB East Kawasaki Island Project Final Work Report,* the Taisei-Tobishima-Penta Ocean-Bechtel consortium.

Tunnel Rehabilitation

Lee W. Abramson, M.ASCE[1] and Marco D. Boscardin, M.ASCE[2]

Abstract

Our aging infrastructure includes underground systems that require repair, rehabilitation, or replacement using methodologies typically associated with tunnel construction. In many cases, the systems have little or no redundancy, so that assessment and rehabilitation construction activities must be performed with short duration interruptions or no interruptions of service. This paper discusses several aspects of tunnel rehabilitation for both the planned design/construction scenario and the emergency scenario. A more detailed discussion of tunnel rehabilitation is being prepared by the American Society of Civil Engineers Underground Technology Research Council Tunnel Rehabilitation.

Introduction

Currently there are no national or industry-wide standards or guidelines for the inspection of tunnels and underground structures. Many guidelines, handbooks, and manuals for inspection of tunnels have been developed by tunnel owners and operating agencies for their own specific situations. However, agency by agency approaches to management of inspections shows considerable variability in both the depth and breadth of inspection procedures. Such procedure variability inevitably raises concerns and questions about the reliability of tunnel structural inspections, the resultant adequacy of tunnel structural maintenance, and the subsequent condition of tunnel structures themselves.

---

[1]Director of Tunnels and Underground Services, Black & Veatch, 720 Third Avenue, Seattle, WA 98104
[2]Branch Manager, GEI Consultants, Inc., 1021 Main Street, Winchester, MA, 01890

As the country's underground tunnel infrastructure grows older, there will be increasing demand for its rehabilitation under both planned and emergency conditions. Owner/agencies, engineers, and contractors will increasingly participate in the inspection, repair, and/or replacement of underground structures to review and maintain the structural integrity of the structures. This paper is intended to serve as an introduction to the components necessary for a successful underground tunnel rehabilitation project. More detailed tunnel rehabilitation guidelines are being developed by the American Society of Civil Engineers Underground Technology Research Council Tunnel Rehabilitation Committee.

For the purposes of this discussions, only tunnels and underground structures with a diameter (lateral/vertical dimensions) of 6 feet (2 meters) or greater are considered. This dimension includes pipelines and many other types of underground construction, such as transit stations and powerhouses, which may not have been constructed utilizing underground tunnel construction techniques, but whose repair and/or rehabilitation now requires that construction take place through an underground opening. The various components of a tunnel rehabilitation design and construction project include data collection, tunnel condition assessment surveys, rehabilitation design, construction considerations and methods, contracting practices, and post-construction monitoring.

Data Collection and Tunnel Condition Surveys

A key element in the rehabilitation of a tunnel is a working knowledge of the original construction and any subsequent modifications. This working knowledge is usually obtained through collection of existing data and performance of a tunnel condition survey. One basic first step in this process is to review existing plans and specifications. It is also necessary to collect and review all available as-built plans, calculations, descriptions, and instrumentation and testing data that was developed during the original construction process and subsequent inspection and modifications. All of the existing historical data collected should be copied and organized for future reference especially during the tunnel condition survey.

Tunnel condition surveys are preferably conducted during periods of inactive use such as partial or complete closure, off-peak times, or maintenance shifts. Transit and highway tunnels can often be closed during specifically defined periods usually at night and on weekends. In some cases, it may not be possible to take the tunnel to be inspected out of service or to enter the tunnel for safety and health concerns, as is often the case for potable water, storm water, outfall, and sewer tunnels. Sometimes, remotely operated vehicles (ROVs) with lights, cameras, and other accessories are required to assess the condition of a tunnel that cannot be entered by inspection teams.

Tunnel inspections are generally best conducted by multi-person teams. The size of the team depends on the size and length of the tunnel as well as schedule and safety restrictions. Because the various measurements need to be made and recorded often involving stooping, reaching, and climbing in poorly lit conditions, inspections can rarely be performed by one person. A two-person team is practically a minimum size inspection team. Often, more than two people are required to increase the production rate, record data, assist with access and egress, maintain safe conditions, etc. When multiple people are involved in an inspection, it is important for each person to have a common understanding of the work plan, objectives, and recording methods so there is not significant variation between the results collected by different members of the inspection team.

Standardized tunnel inspection criteria are necessary for expeditious processing, and evaluation of the tunnel condition data (i.e. the quality and performance of the tunnel system). Coding the information is helpful for consistency of tunnel condition documentation and to assure quality control by providing inspection guidelines to personnel that standardizes their visual observations. This type of coding system can be used for all types of linings such as exposed rock, brick, concrete and steel. Video tape records and photographs are also quite helpful, if not mandatory. Coring and testing using destructive and non-destructive methods may also be part of the inspection program (Russell, 1988).

The inspection of a tunnel is by its very nature a complex and extensive data collection operation. In an effort to systematically catalogue and expand these data and to make them useful over the long-term, one must consider use of efficient computer spreadsheet or data base management programs to sort and manipulate information collected and produce easily-understood information reports (Figure 1). Recently, computer-aided design drafting (CADD), Geographic Information Systems (GIS), and Geographic Positioning Systems (GPS) software programs have been used to record inspection data and prepare rehabilitation plans.

The objective of a tunnel inspection program is to determine the status of the tunnel system and its associated subsystems. This determination is based on review of the original design and the performance of that design over time. The condition survey will also provide valuable information with respect to repairs and other maintenance performed and products used in that particular tunnel environment. It is important therefore to present the results in a clear and concise manner often split up into two volumes containing the factual data in one volume and the interpretation of these data in the other.

Rehabilitation Design

Design of a tunnel rehabilitation project follows the same process as design of a new tunnel:

Figure 1 – Presentation of Tunnel Inspection Data

Tunnel Reach Sect.	Sta. No.	Sam. No.	Feet	pH Readings			Sound Test ·		Gas Detector Readings		Description	pH
				Wtr	Crn	Sides	Hrd	Soft	Ox.	H_2S		
6211 to 6210	144+74	49	0	7	5	5	X		20.3	0	Crown sample	5
	144+50		24								Steel Piping begins	
	144+44		30								Line in General	
	143+86		88								Seam in pipe - no deterioration	
	143+44		130								Steel pipe ends	
	143+28		146		6						Line in General	
6210 to 6209	143+00	50	0	7	6	6	X		20.3	0	Line in General	X
	142+00		100								Line in General	
	141+00		200					X			Line in General	
6209 to 6208	140+14	51	0	7	5	5	X		20.5	0	Collected Crown Sample	X
	140+09	52	5	6	2	2					Crown sample of a 30" incoming pipe	X
	139+14		100					X			Line in General	
	138+44		170								MH 6208 at bend in pipe	

- Assess the need for and goals of the rehabilitation
- Develop operational criteria for the rehabilitation process;
- Evaluate alternative materials, methods, processes, and techniques to accomplish the goals;
- Select one or more design approaches for more detailed evaluation, analysis, specification, and dimensioning;
- Develop contract documents including drawings and specifications for the selected design alternative; and
- Review the progress of rehabilitation in the field and revise rehabilitation methods, as necessary, to accommodate actual conditions encountered during the construction process.

Tunnel rehabilitation can take three general alternative forms:

- Repair that typically consists of superficial treatments to correct small, localized functional or structural deficiencies;
- Restoration that may include general overall repairs to the facility either to return it to an original condition or level of service, or upgrade the structural condition or service capacity of the existing facility to improve its performance; or
- Complete replacement whereby a new or parallel facility is constructed to replace or supplement the existing facility.

The decision to repair, restore, or replace an underground facility should be made after available information on the structural and service conditions have been collected and analyzed, and after those factors prompting the need for rehabilitation have been thoroughly evaluated.

Issues to consider during decisions regarding which of the above alternatives may be most prudent include:

- The nature and degree of deficiencies;
- The age of the existing structure;
- The duration of time available for rehabilitation;
- The presence of alternative or redundant systems;
- The growth in demand for services provided by the facility; and
- The costs associated with various types of rehabilitation.

A balance must be struck between a) the operational need to maintain the existing function of the tunnel and to avoid greater than necessary revenue loss, and b) the need for time-sensitive, or emergency, tunnel rehabilitation due to either the extent or seriousness of tunnel deterioration. Trade-offs between revenue and safety needs are invariably established through investigation and subsequent decision analysis.

Tunnel inspection is critical to determine the types and extent of rehabilitation, but tunnel inspections typically identify the symptoms, and not always the sources of tunnel problems. However, once symptoms have been identified, or the need for tunnel rehabilitation has been established, additional investigation is usually required to determine rehabilitation design parameters and to provide data needed to develop the rehabilitation construction details. The components of the additional tunnel investigations following the inspection are discussed below.

One of the most difficult aspects of tunnel rehabilitation design development is the assessment of the loads acting on the existing tunnel structure, and to determine how those loads will be transferred and act on the rehabilitated structure. These loads generally include soil and rock loads, internal and external water loads, seismic loads, and other mechanical loads imposed by other aspects of the tunnel system (e.g. ceilings, utilities, cranes, etc.). In those cases where a structural failure has occurred, it may be possible to back-calculate the loads that caused the structural failure using estimated properties of the tunnel structural support system. However, the loads estimated to act on an existing tunnel may not always govern rehabilitation design.

In contrast, progressive, piecemeal removal and reconstruction of tunnel supports during rehabilitation construction may result in arching loads to other structural elements, and potentially cause increased loads on some elements. If significant ground movement and softening occurs during reconstruction, loads on the tunnel may increase. In addition, the potential for loads related to swelling, squeezing, and ground creep with time may need to be considered in the load calculations. In most cases, the need to consider these situations will be evident from the existing tunnel performance record. In practice, a range of possible ground loads should be considered for rehabilitation design.

When tunnel rehabilitation is designed with the existing structural support remaining substantially in place, it will be necessary to evaluated the existing and future structural integrity and the capacity of the tunnel lining (if one exists), or other tunnel elements providing structural support. These calculations can be particularly difficult, for example, when the lining consists of concrete, steel, cast iron, or brick and mortar that has deteriorated over many years, especially if exposed to an aggressive environment. The structural capacity of the tunnel support system should be evaluated using methods that consider ground/tunnel interaction, and the adverse impacts of alteration to or corrosion of material properties of the tunnel support elements.

When determining tunnel rehabilitation alternatives, it is critical that the rehabilitation techniques selected are compatible with the existing tunnel structure and the operating conditions under which that structure must function. It is generally desirable to make repairs with materials similar to those used in original tunnel construction.

However, those materials may not be available, or it may be impractical to use them with current construction practices.

One tunnel rehabilitation design task is the selection of appropriate materials for proposed construction. There are many alternative materials and procedures available for use in tunnel rehabilitation. They range from more traditional civil engineering materials, such as steel and concrete, to more modern materials such as polymeric-based membranes and resins and composite material systems. In addition, the techniques to install these materials are varied.

Conventional materials used in modern tunnel construction include wood, steel, and concrete. In general, steel and concrete are the likely options when replacement of deteriorated structural support or enlargement of tunnel space results in major structural rehabilitation (Figure 2). Wood typically assumes the role of temporary support, and may be surrounded by structural concrete, or may be removed upon completion of the structural rehabilitation. Among the advantages of these materials is their familiarity to designers and contractors (properties, behavior, and installation methods). In addition, mixing and/or coating conventional materials with other materials, such as polymers and resins can improve their durability and performance.

Construction materials unique to rehabilitation projects, such as epoxies, plastics, chemical cements, and chemical fillings and coatings can play prominent roles in construction by enhancing the properties and performance of the more conventional construction materials described above. These materials may act as bonding agents or linings when the need for structural rehabilitation is limited or when rehabilitation requires improvement of surface characteristics, protection of underlying structural materials from corrosion, or sealing surface materials to reduce infiltration and exfiltration.

Examples of these newer materials and processes include:

- Resin impregnated fabric that is inflated to conform to the shape of the tunnel and to bond to its surface.
- Polymeric membranes and drainage fabrics can be embedded behind concrete linings to provide impermeable barriers or to reduce hydrostatic pressures on linings.
- Epoxy is used to fill cracks and prevent further deterioration of structural concrete and steel and to coat surfaces to improve hydraulic characteristics and to ease maintenance and cleaning.
- Insulating foams and boards are used to control icing related problems.

Design of rehabilitation programs, whether utilizing conventional or more modern materials and methods, must also consider compatibility of materials and methods, design codes and standards, tunnel operation, construction implications, and contingencies should construction expose conditions not anticipated during design.

GEO-ENGINEERING FOR UNDERGROUND FACILITIES 919

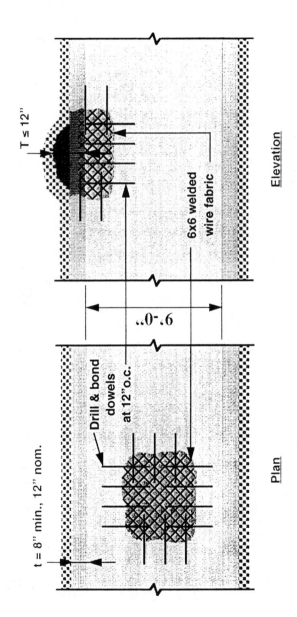

Figure 2 – Typical Tunnel Rehabilitation Repair

Construction Considerations and Methods

During the development of a tunnel rehabilitation program, the conditions in the tunnel and tunnel usage during construction must receive careful attention. The following questions are pertinent:

1. Can the tunnel be taken out of service for the entire construction period? If not, for how long and at what times of day?
2. Will the public or others need use of the tunnel during the construction period? What level of use will be required?
3. What types of operational hazards will workers, equipment, and materials be subjected to during construction (e.g. traffic in transportation tunnels, bad air in sewer tunnels, electrical hazards in utility tunnels, structural instability in unlined or otherwise weakened tunnels, water inflow, adverse temperatures, toxic and other hazardous substances, etc.)?

Lack of sufficient access and staging areas in the tunnel during construction for any reason often limits or totally restricts the ability to perform rehabilitation work that requires taking the tunnel out of service (Figure 3). When severe working restrictions are necessary for operational reasons, the tunnel repair work must be limited to work that can be accomplished with special, limited means and methods. This usually results in longer project schedules and higher costs which may or may not be offset by the alternative losses in revenue. Occasionally, rehabilitation work can be carried out from outside of the tunnel using ground improvement techniques such as grouting.

An increasingly important and frequently occurring issue is control and disposal of construction site waste products such as water, muck, debris, chemicals, coatings, etc. that may or may not fall into the category of contaminated or hazardous materials by regulatory agencies. If hazardous materials are present on the construction site, special procedures and facilities must be in place for proper management and disposal. Such waste-management facilities can require a substantial staging area, well in excess of that needed for the tunnel rehabilitation itself. Disposal of hazardous materials can also add a significant premium to the cost of tunnel rehabilitation. In addition, securing needed permits for discharge/disposal of hazardous materials can significantly impact the construction schedule. Whenever possible, necessary regulatory permits for construction should be obtained as early in the design and construction process as possible, to avoid subsequent project delays.

Working conditions within the tunnel is another important aspect of construction that needs to be addressed. Temperature, humidity, illumination, air quality, water inflow, noxious or toxic substances, structural weaknesses, falling debris, voids, and utilities must be identified prior to construction and addressed during design and in the construction work plan to properly integrate worker needs for comfort, health, and safety. A lack of attention to these matters may result in unanticipated and costly work delays, possible construction claims, and/or interruption of service.

Contracting Tunnel Rehabilitation

The contractual relationship of a tunnel rehabilitation project should consider the nature and immediacy of the work, level of detail and presentation of the work elements, definition of knowns and unknowns, procurement and compensation provisions, degree of interaction between the owner/designer/contractor during construction, and methods for responding to unforseen conditions. If the need for tunnel rehabilitation is anticipated over a substantial period of time based on routine inspections and maintenance, the owner has the flexibility to select from among several contracting alternatives ranging from competitive bidding based on detailed plans and specifications, to sole source negotiation of contract terms based on conceptual sketches and guidelines, to design/build. Often on public sector projects, the only possible type of contract and procurement is through competitive bidding except in emergencies.

Emergency projects are typically necessitated by major disruptions in service or threats to public safety. Given the urgency of these conditions, there will probably not be sufficient time to thoroughly investigate conditions producing the emergency, nor to develop detailed contract specifications prior to commencement of the rehabilitation work. The initial scope of work may not be well-defined because the full nature of the emergency may not be known until causal conditions are subsequently uncovered. As a result, the owner may prefer a more flexible approach to the scope of work, and adaptable payment provisions that can be changed as the scope of work changes.

Project Examples

The following is a brief list of rehabilitation projects and some of the aspects discussed above.

BART Transbay Tube in San Francisco, California: The seismic joint of this immersed binocular tube transit tunnel was found to be at its full extension following the Loma Prieta earthquake. Methods for "resetting" the joint back to its null position were evaluated.

Chambers Creek Tunnel in Tacoma, Washington: Routine maintenance disclosed an unusually high rate of hydrogen sulfide corrosion in one mile of this reinforced concrete pipe lined sewer tunnel. Rehabilitation methods currently being considered include slip lining a new pipeline in or complete replacement.

Chapline Hill Tunnel in Pittsburgh, Pennsylvania: This single track, brick lined railroad tunnel required upgrading before returning to service after years of deterioration due to the seepage of acidic groundwater from nearby coal mining. New drainage systems and a new shotcrete lining were installed.

EBMUD Claremont Tunnel in Oakland, California: The condition of this and other raw water tunnels was evaluated and emergency repair plans were drawn up in the event that an earthquake along the Hayward Fault occurs and causes severe damage that threatens water supply service to much of Oakland.

Interstate 40 Sterling Mountain Tunnel near Asheville, North Carolina: A major rock fall occurred adjacent to this two-lane, one-way highway tunnel destroying part of the tunnel and access portal. Emergency repairs were required including rock portal stabilization, reconstruction of part of the tunnel, and construction of a wider rock fall zone.

Morenci Tunnel in Morenci, Arizona: This two-lane, two way unlined rock highway tunnel was evaluated with respect to rock fall hazard and safety. The tunnel could not be shut down for repairs.

Needles Eye Tunnel near Boulder, Colorado: This historic, unlined, rock tunnel provided access for hikers and bikers along an abandoned railroad. The recreation path was closed as a result of a rock fall that caused a major personal injury.

Oakland-Alameda Tubes in California: Twin, two-lane highway immersed tube tunnels were found to be susceptible to damage due to an earthquake in the Bay Area. Recommended seismic retrofits included making the tube joints more flexible and the surrounding soils less prone to liquefaction using ground improvement methods.

Path Tunnels E&F in New York and New Jersey: These twin transit tunnels under the Hudson River are lined with cast-iron segments and concrete. A construction accident caused major water leakage into the tunnel necessitating an inspection and rehabilitation program. Severe settlement of the tunnels was also disclosed.

Conclusions

1. A key element in the rehabilitation of a tunnel is a working knowledge of the original construction and any subsequent modifications. This working knowledge is usually obtained through collection of existing data and performance of a tunnel condition survey.

2. Tunnel rehabilitation follows a process that includes identifying goals, developing operational criteria, evaluating alternative materials and methods, selecting design approaches, developing contract documents, and reviewing the rehabilitation to accommodate actual conditions encountered during construction.

3. Tunnel rehabilitation may include superficial repair, major restoration, or complete replacement depending on the nature and degree of deficiencies, age of the existing structure, duration of time available for rehabilitation, presence of alternative or redundant systems, growth in demand for services provided by the facility, and rehabilitation costs.

4. Rehabilitation materials and methods must be compatible with the existing tunnel structure and operating conditions. Construction materials unique to rehabilitation projects, such as epoxies, plastics, chemical cements, and chemical fillings and coatings can play prominent roles in construction by enhancing the properties and performance of the more conventional construction materials such as wood, steel, and concrete. Construction planning should be part of the design process with consideration of access, staging, spoil handling/disposal, and working conditions in the tunnel.

5. Often on public sector projects, the only possible type of contract and procurement is through competitive bidding except in emergencies. Emergency projects are typically necessitated by major disruptions in service or threats to public safety. There will probably not be sufficient time to develop detailed contract specifications prior to commencement of the rehabilitation work. As a result, the owner may prefer a more flexible approach to the scope of work, and adaptable payment provisions that can be changed as the scope of work changes.

Acknowledgements

Much of this paper is based on the American Society of Civil Engineers Underground Technology Research Council Tunnel Rehabilitation Committee Guidelines on Tunnel Rehabilitation being prepared under the leadership of Henry Russell and Bruce Beverly with input from Michael Gilbert, Joseph Griffith, Joseph MacElroy, Henry Russell, James Smith, Antoni Zelecheski, Lee Abramson, and Marco Boscardin. The authors gratefully acknowledge the advice and knowledge about tunnels and rehabilitation shared freely by these and other colleagues who are too many to name herein.

References

Russell, H. R. (1988), "The Inspection and Rehabilitation of Transit Tunnels," 1987 William Barclay Parsons Fellowship, Parsons Brinckerhoff Inc., Monograph 3, May.

Figure 3 – Restricted Access on Tunnel Rehabilitation Projects

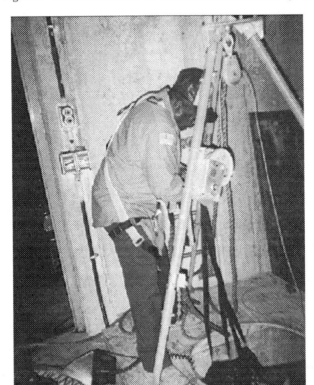

Rehabilitation of Tunnel Liners with Shotcrete

Henry A. Russell, P.E.[1]

Abstract

Parsons Brinckerhoff has been a leader in the area of rehabilitation of underground structures for many years. Recently on numerous projects shotcrete has been used for the rehabilitation of the structural elements of highway, rail and fluid transmission tunnels. Most rehabilitation projects require that the facility is kept in service during the rehabilitation process, and materials cannot be left on site during the progress of the work. As a result of these requirements, shotcrete has become a major process in the structural restoration of the underground structures. Over the last 10 years, Parsons Brinckerhoff has completed numerous projects with the use of polymer and modified shotcrete. Extensive in-house research has been performed and with the cooperation of the facility owners and material manufacturer's modifications to shotcrete mixes and methods of application have led to numerous successful projects. This paper will discuss the use of shotcrete on a recent highway project, which illustrates the advantages of the use of shotcrete for use in tunnel rehabilitation.

Introduction

Shotcrete, is the pneumatic application of cementitious products for the restoration of concrete structures. This process has been in use for over 100 years for the construction and repair of concrete structures both above and below ground. Shotcrete is defined by the American Concrete Institute as a "Mortar or concrete pneumatically projected at a high velocity onto a surface." [2] Since the 1970's the use of low-pressure application of cementitious mortar has been commonplace in Europe and is known as Plastering. Parsons Brinckerhoff has with the cooperation of the Massachusetts Turnpike Authority used both types of shotcrete application to apply polymer cementitious products for the repair of spalls and delaminations in tunnel liners in active highway tunnels. The selection of the process type, and the material to be applied is dependent on the specific conditions for tunnel access and available time for the installation of the repair. This paper will discuss the procedures utilized to identify the extent to the repairs to the liner to be performed, and the work required to implement the reinstatement of the tunnel liner.

The repair of concrete delaminations and spalls in tunnels has traditionally been performed by a form-and-pour method for the placement of concrete, or by the hand application of cementitious mortars that have been modified by the addition of polymers. Both of these

---

[1] Principal Professional Associate, Parsons Brinckerhoff Quade &Douglas, Inc, 120 Boylston St., Boston MA, 02116
[2] American Concrete Institute, ACI 318 Farmington Hills, MI.

methods are not well suited for rail and highway tunnels that are in continuous daily operation. This daily operation usually required that the tunnel could be out of service for very short periods of time and the repair process must be rapid, not infringe on the operating envelope of the daily traffic and be a durable long-term monolithic repair. The use of shotcrete is a logical process that allows for rapid setup, application and ease of transport into and out of the tunnel on a daily basis.

Shotcrete Methods

There are two processes for the application of shotcrete; Dry Process and Wet Process. Both processes have been in use for many years and are equally applicable to all types of work. The Dry process is a method where the mix of cement and additives is blown dry through the hose and at the nozzle water is added to hydrate the mix. The material is blown onto the prepared surface and built up to the original lines of the structure. This process often uses a premoistening system at the hopper to reduce dust, and in the case of the use of polymers, to start the polymer reaction. The dry process provides for low cement/water ratios and in combination with the polymers allows for a shrinkage compensated mix. Dry process is also used without any additives for general repair work in areas where rapid sure and shrinkage is not of major concern. The impact of both dry process and high velocity wet process produces a certain percentage of rebound of material. Rebound material is the harder and larger components of the mix and is dependent on numerous factors, thickness of the repair, shooting velocity, trajectory, and the skill of the nozzleman. This rebound in dry process can range between 20 – 50 % of the material sprayed. The addition of polymers to the mix reduces the rebound to a lower percentage in the range of 20-30%. The rebound material should not be reused and is therefore waste.

The wet process is a method where all the materials are mixed prior to the blowing though the hose and application to surface to be repaired. The application of wet process shotcrete in the repair of tunnel liners is usually limited to the application of polymer/flyash modified repair mortars. The use of low pressure wet process shotcreting as used in tunnel liner repairs is identical to the European repair process referred to as "Plastering". This process is the application of a repair mortar in one continuous full-depth repair by the use of rotary stator or modified swing tube pumps. This process replaces the hand application of the premixed repair mortar. The use of low pressure wet process with polymer/flyash mortar produces essentially no waste or rebound. The only waste form this process is the material that is over shot or wasted in the cleaning of the hose and tools. In general this process allows for less than 5% waste.

The successful application is shotcrete, regardless of the process chosen is the skill of the nozzleman. In the case of the wet process both the nozzleman and the laborer mixing the mortar. A successful repair program requires the nozzleman and the other members of the crew to be skilled and tested on site using mock-ups of the types of areas to be repaired. These mock-ups should closely duplicate the shape and surfaces to be repaired. This testing program is often used to certify the skill of the shotcreting crew and provides for better quality control during the progress of the work by developing an understanding between the Engineer, Owner and Contractor what is an acceptable product for the work.

Surface Preparation

The surface preparation for both processes is performed in the same manner. The removal of all of the unsound concrete is performed by either the use of chipping hammers or the use of hydro-demolition. The unsound concrete is removed to the full depth of the unsound concrete. In cases where chipping hammers are used it has been found that limiting the size of the hammers by weight is the best way to control over excavation. The limiting the weight of the chipping hammers with bit, to be less than 30 lbs. (13.6Kg) reduces the risk of over excavation of concrete. These hammers are too weak to excavate concrete in excess of 4000 psi. (27580 Kpa).

The use of Hydro-demolition requires testing on site, at the beginning of the project to determine what pressures are required to excavate the unsound concrete without removing the sound substrate.

Hydro-demolition should not be used in areas of electrical equipment, cables, or other mechanical equipment that may be effected by the excavation process. The area to be repaired must not have feather edges, and must have a vertical edge of approximately 1/8 inch. This vertical shoulder is necessary to prevent spalling at the edge of the new repair. (Figure 1)

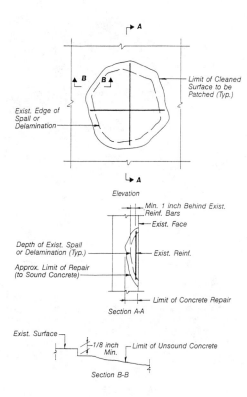

Figure 1. Typical Section at Concrete Repair

After the demolition of the unsound concrete any leaking cracks or construction joints must be sealed to allow the application of the reinforcing steel coatings and the shotcrete. This sealing should be performed using a chemical grout suitable for the type and magnitude of the leakage. In general single component polyurethane grouts are the most successful in effectively sealing most tunnel leaks.

Once the unsound concrete has been removed, reinforcing steel must be cleaned and if a loss of section is evident the damaged reinforcing steel must be removed and replaced. All rust and scale must be removed form the reinforcing steel and any steel liner sections or other structural steel elements. The cleaning is generally to a white metal commercial grade cleaning. Once cleaned the reinforcing steel is to be evaluated for loss of section and if the loss of section is greater than 30% an analysis must be performed, and if required, the steel is to be replaced. In general, mechanical couplers are to be used to reduce the amount of the tunnel liner. (Figure 2)

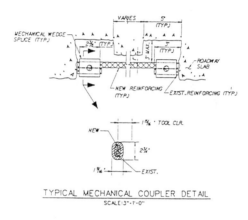

Figure 2. Mechanical Coupler for Reinforcing Steel

After the steel has been cleaned a coating must be placed on the steel to protect the steel from accelerated corrosion due to the formation of an electrolytic cell. Numerous products exist for this purpose, including epoxy and zinc rich coatings. Zinc rich coatings are better suited for this application due to the fact that they do not form a bond-breaker as do many epoxies. This is important since these materials are applied by the use of a paint brush and it is difficult to prevent the concrete surface from being accidentally coated. The application of the zinc rich coating is to be performed within 48 hours of the cleaning and not more than 30 days prior to the application of the shotcrete.

Once the reinforcing and structural steel elements have been cleaned and coated, welded wire mesh is to be placed over the area to be shotcreted. The mesh is placed to within 2 inches of the edge of the repair. The wire mesh is attached to the existing reinforcing and to the substrate by the use of J hooks. The purpose of the wire mesh is to assist in the buildup of the shotcrete and to provide for a monolithic repair that becomes part of the host structure. The wire mesh should be hot dipped galvanized after fabrication and is best if provided to the site in sheets rather than on a roll. If epoxy coated mesh is used it must be in sheets in order to eliminate field touch-up of the cut ends of the mesh. The mesh size for dry process is 2X2 inch mesh (50.8 X50.8 mm) and for wet process 4X4 inch mesh (10.6X101.6 mm). The large mesh is required for the wet process to prevent clogging of the mesh by the shotcrete and therefore creating voids behind the mesh surface.

Shotcrete Application

The application for the shotcrete is performed by a skilled nozzleman and is shot with the nozzle placed perpendicular to the surface to be shot. Prior to the placement of any material the nozzle sprays the surface of the substrate with water to premoisten the interface. The material is placed in a circular motion in a small radius circle. The mix is usually placed form the bottom of the vertical repair and from left to right on horizontal or overhead repairs. In dry process applications that are polymer modified, the polymer acts as a bonding agent and assists in the rapid build up of the material. In wet process in addition to the polymers fibers are often added to assist n the buildup. In both cases the application is rapid. The nozzle in dry applications is kept approximately 3 feet (1 m) for the surface to be repaired, however wet process requires the nozzle to be kept approximately 18 inches (0.5 M) from the surface to be shotcreted. The application of the shotcrete is continuous and is full depth with stainless steel piano wires stretched across the repair area to control the maximum build. After the entire area to be patched is filled with shotcrete the material is allowed to cure for 20-30 minutes, at which time the mix is screeded and troweled to the desired finish. Trying to work the shotcrete prior to this time will result in tearing of the surface and make finishing very difficult. Caution must be exercised to monitor the drying rate of the shotcrete since the times stated here will vary depending on wind conditions and relative humidity.

After the repair has been troweled to the desired finish a curing compound must be sprayed on the surface of the new shotcrete to prevent rapid drying. The manufacturer of the premixed shotcrete will recommend a curing compound best suited for the job site conditions.

Case Study

The following case study illustrates the process for the identification of the structural defects, preparation of the substrate and the application of the cementitious products.

Project Description

The Sumner/ Callahan Tunnels are vehicular tunnels located in Boston, Massachusetts, and connect downtown Boston with Logan International Airport. Each tunnel has over 100,000 trips a day of mixed traffic. They were built in 1934 and 1960 respectively. The tunnels were constructed by the use of compressed air and an open shield, and have steel segmental liners with an interior concrete lining of varying thickness ranging from 12 inches (304 cm.) to approximately 48 inches (1.22 m.). The tunnels are 28 feet (8.51m) in diameter, and are 5280 feet in length (1.60 km). Both tunnels have a supply air duct below the roadway and an exhaust duct above the roadway providing fully transverse ventilation to the roadway. (Figure 3).

The structural liner of the tunnel's exhaust and supply air duct were severely deteriorated. The exhaust duct deterioration was due to the presence of high concentrations of sulfates and sulfites from the vehicle exhaust. The supply air duct was degrading as a result of the presence of road salt laden drainage water laying on the invert of the tunnel liner. This water impounding is common to many vehicular tunnels where the invert of the supply duct is used at the conduit for the tunnel drainage system. The deterioration of the liner was compounded by shallow concrete cover over the reinforcing steel and freeze thaw conditions which occur as a result of the outside atmosphere being pumped into the tunnel for ventilation purposes. In addition to the liner problems, the roadway deck had undergone severe delamination as a result of the aforementioned conditions.

930 GEO-ENGINEERING FOR UNDERGROUND FACILITIES

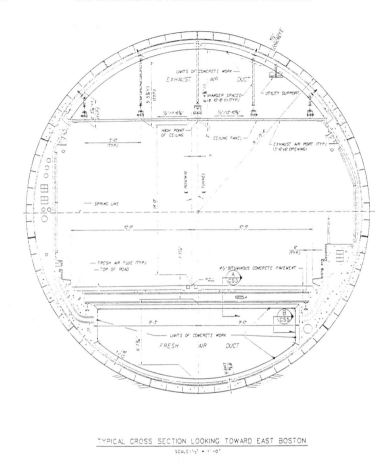

Figure 3. Typical Tunnel Cross Section – Sumner/Callahan Tunnel

Based on our preliminary site investigation it was evident that the repair work was extensive. The High traffic counts dictated only a limited time for the repairs and the requirements of maintaining traffic allowed for only a five (5) hour work window between the hours of midnight and 05:00 AM to perform the work. In addition, the placement of concrete forms was prohibited, and storage of materials and equipment in the tunnel was not allowed. All of these factors required careful attention to detail in the development of the repair methods, and products for the concrete liner restoration. A polymer modified dry process shotcrete was chosen as the repair product for the following reasons:

- High Strength concrete repair (6000 psi, 41370 Kpa)
- Concrete forms not required

GEO-ENGINEERING FOR UNDERGROUND FACILITIES 931

- Shrinkage compensated
- Monolithic repair equal to the original liner design strength
- Easy mobilization and demobilization on a daily basis
- Resistant to chloride and exhaust gases attack
- Rapid cure
- Polymer reduces the rebound (waste)

Site Investigation and documentation of Tunnel Defects

The site investigation was performed by the use of traditional observational methods, which included the sounding of the concrete surfaces and the documentation the defects as described in the Inspection and Rehabilitation of Transit Tunnels[3]. The documentation of the concrete defects is a standardized format with standardized codes used to simplify the cataloguing and presentation of the data (Figure 4).

CALLAHAN TUNNEL ROADWAY DEFICIENCIES
UNDERSIDE CONCRETE

STATION (1+00+99)	DELAMINATION (SF)	SPALL <2" (SF)	SPALL TO 2"-3" REINF BAR(SF)	SPALL BEHIND 4" REINF BAR (SF)	CONSTR. JOINT >1/2" (LF)	CRACK <1/8" (LF)
8	40	-	-	-	-	350
9	400	10	54	-	-	450
10	1350	6	6	1	-	200
11	900	3	-	-	-	54
12	600	-	-	2	-	158
13	1100	-	-	-	-	218
14	900	2	-	-	-	20
15	1100	-	-	-	-	100
16	1400	-	-	1	-	50
17	1100	-	10	10	-	50
18	1050	20	-	-	-	50
19	1550	-	-	-	-	50
20	1250	4	-	30	-	50
21	1350	8	-	-	-	50
22	1125	60	25	-	-	50
23	1400	16	10	-	-	50
24	1450	5	1	-	-	50
25	1130	20	4	-	-	50
26	1140	13	4	-	-	50
27	1175	16	2	-	-	50
28	1275	-	20	-	-	50
29	1175	2	15	10	-	50
30	980	-	5	-	-	50
31	550	4	10	25	-	50
32	525	16	50	200	-	50
33	400	-	-	30	-	50
34	1100	-	-	100	3	60
35	710	-	12	30	-	78
36	600	100	4	-	-	40
37	1200	-	-	30	-	-
38	700	-	30	60	-	-
39	900	-	-	15	-	-
40	1000	2	25	-	-	-
41	1000	-	4	-	-	-
42	1150	2	-	-	-	18
43	1320	-	-	20	-	-
44	1400	-	-	30	-	-
45	1500	20	30	-	-	150
46	1200	-	-	-	-	-
47	1300	15	-	20	-	-
48	1200	-	-	-	-	-
49	1300	16	-	50	-	-
50	800	-	-	-	-	-
51	1000	20	-	100	-	-
52	1200	-	30	-	-	-
53	800	-	5	5	-	138
54	1600	-	-	400	-	130
55	1400	-	-	25	-	36
56	1600	-	-	5	-	18
57	1700	-	20	-	-	-
58	100	-	-	-	-	18
TOTAL	54195	560	372	1054	3	3086

Figure 4. Typical documentation of Tunnel Defects

---

[3] Russell, Henry A. *The Inspection and Rehabilitation of Transit Tunnels*, William Barclay Parsons Fellowship Parsons Brinckerhoff Inc. 1988.

Contract Documents

During the early stages of the design a test section was chosen to test the shotcrete on the invert of he tunnel. The test section required excavation of unsound concrete cleaning of the reinforcing and the installation of new shotcrete. A polymer modified dry process shotcrete was chosen for the project. The product chosen was Renderoc SP as manufactured by Fosroc Inc., of Georgetown Kentucky. This product was chosen due to its excellent track record in Europe. Th product has the following physical properties:[4]

Wet Density	150 lbs/ft^3 (2.4Kg/liter)
Compressive Strength	4800 psi (33.1Mpa) @ 24 hours
	6500 psi (44.8Mpa) @ 7 days
	8000 psi (55.2Mpa @28 days
Chloride Permeability	220 coulombs (very low)
Slant Shear (bond)	1650 (11.3 Mpa) @ 7 days
	2500 (17.2 Mpa @ 28 days
Drying Shrinkage	300 microstrains @ 28 days

The test section revealed that the curing of the product was slower than anticipated. Based on the installation performance during the tests the mix was modified by the manufacturer to accelerate the curing to about 45 minutes while maintaining the shrinkage compensation. This modification in conjunction with use of wire mesh accelerated the installation time (Figure 5). These modifications allowed the product to be placed on tunnel surfaces, particularly on the underside of the roadway slab and be able to return the roadway into service within the time frame required by the contract.

In addition to the improving of the mix for the site, premoistening was of the mix was chosen to reduce the presence of dust from the mixing and shooting operation. This was necessary to keep the cementitious dust from damaging the ventilation fans located within the service buildings.

The repairs to he invert section were performed by the use of an Allentown AG15 rotary gun with a predampner. The machine was dismantled to place it within the supply duct and reassembled to work within the confined space. All of the pre-bagged material for use in the supply duct was hand carried through manholes into the duct for application. This posed a logistical problem since the material is provided in 55 pound (25 Kg) bags and required over 36000 bags. The manufacturer provided special packaging referred to as "supersacks" which held 2970 pounds (1350Kg). This special packaging allowed the use of additional shotcrete machines to be operated from the roadway surface, pumping material through the manholes to accelerate the work. Details of the invert repairs are shown on Figures 5, 6 and 7.

---

[4] Fosroc, Inc, Renderoc SP Technical Data Sheet, Fosroc Inc, Georgetown, KY, 1991.

GEO-ENGINEERING FOR UNDERGROUND FACILITIES 933

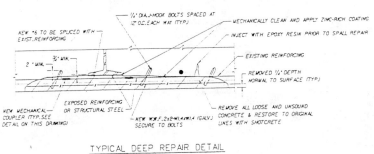

Figure 5. Typical Roadway Slab Repairs

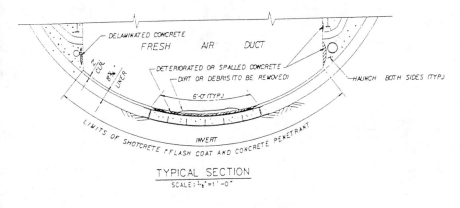

Figure 6. Typical Invert Repair Locations

GEO-ENGINEERING FOR UNDERGROUND FACILITIES

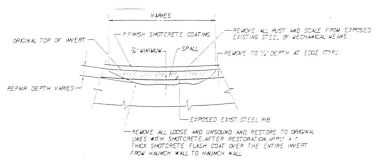

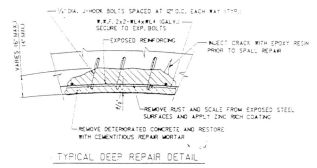

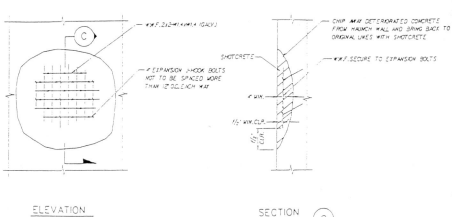

Figure 7. Typical Invert Repairs

Conclusions

The use of shotcrete for the repair and reinstatement of tunnel liners is a viable repair method for the rapid economical repair of these structural elements. The Sumner/Callahan tunnel project was performed over five years and allowed for the tunnel system to be completely rehabilitated without closing down this important urban transportation facility. The project resulted in over 33,000 CF (934.5 m^3) to be placed with no observable effect on the commuters of Boston. In addition to the aforementioned project over 4000 CF (113 m^3) for the reinstatement of the Prudential tunnel in Boston. Currently, Parsons Brinckerhoff has numerous project that are ongoing in New York, Boston, and Oregon where the use of specialty shotcrete products are being used to reduce costs and maintain schedule on new and rehabilitation projects. As result of the success on this project and other similar underground rehabilitation projects the use of polymer modified dry shotcrete is expanding to bridge and garage rehabilitation in New England, largely as a result of the work described here.

References

American Concrete Institute, *ACI 318, Design Manual for Concrete Structures*, January 1998

Fosroc Inc. *Renderoc SP Technical Data Sheet*, Fosroc Inc, Georgetown KY, 1991

Parsons Brinckerhoff Quade & Douglas, Inc, *MTA Contract 361-0072*, January 1992

Russell, Henry A. *The Inspection and Rehabilitation of Transit Tunnels*, William Barclay Parsons Fellowship Parsons Brinckerhoff Inc. 1987.

Rock Grouting: Contemporary Concepts in
Materials, Methods, and Verification

D. A. Bruce[1], M. ASCE, A. Naudts[2], and C. Gause[3]

Abstract

The paper provides a broad overview of contemporary aspects of rock grouting practice. Generic classifications for the various materials and drilling and grouting methodologies are provided, and observations on qa/qc and verification procedures are given. Details are provided of recent seepage prevention works at two major U.S. dams to further illustrate the general principles.

Introduction

The use of drilling and grouting methods to locate and seal all manner of fissures and voids in rock masses has been common throughout the world for over a century. While the goals of such programs have largely remained unchanged, the materials and methods have undergone remarkable change in response to technological advances and increasingly onerous site specific demands. These changes, however, have not been constant in their rate of evolution in any given part of the world. For example, little advance seems to have been made during the 50 year period of intense activity on U.S. Federal dams from the 1920's onwards. One may cite extremely restrictive, prescriptive specifications as the main reason for the languid rate of innovation.

During the last few years, however, the art of rock grouting has entered a new phase of progress, rapidly drawing it towards the status of an engineering science. This paper reviews contemporary methods, highlighting those areas where the most significant advances have occurred, and indeed are still occurring. These developments are then illustrated with

[1]Principal, ECO Geosystems, L.P., Pittsburgh, PA
[2]Manager, ECO Grouting Specialists, Ltd., Toronto, ON
[3]Sales Manager, Master Builders Inc., Cleveland, OH

reference to two recent U.S. projects - one an existing dam seepage remediation, the other a new dam curtain.

Materials

General Classification

There are four categories of materials (Bruce et al., 1997) which can be listed in order of increasing rheological performance and cost:

1. Particulate (suspension or cementitious) grouts, having a Binghamian performance.
2. Colloidal solutions, which are evolutive Newtonian fluids in which viscosity increases with time.
3. Pure solutions, being non evolutive Newtonian solutions in which viscosity is essentially constant until setting, within an adjustable period.
4. "Miscellaneous" materials.

Category 1 comprises mixtures of water and one or several particulate solids such as cement, flyash, clay, or sand. Such mixes, depending on their composition, may prove to be stable (i.e., having minimal bleeding) or unstable, when left at rest. Stable, thixotropic grouts have both cohesion and plastic viscosity increasing with time at a rate that may be considerably accelerated under pressure. Category 1 grouts are most common in rock grouting and are undergoing rapid development as a result of a markedly increased understanding of basic rheological and hydration principles.

Category 2 and 3 grouts are now commonly referred to as solution or chemical grouts and are typically subdivided on the basis of their component chemistries, for example, silicate based (Category 2), or resins (Category 3). They are rarely used in rock grouting, having application largely in "fast flow" sealing operations.

Category 4 comprises a wide range of relatively exotic grout materials, which have been used relatively infrequently, and only in certain industries and markets. Nevertheless, their importance and significance is growing due to the high performance standards which can be achieved when they are correctly used. The current renaissance in the use of hot bitumen grouts is a good example, in cases of extreme seepage conditions.

Developments in Particulate Grouts

Due to their basic properties and relative economy, particulate grouts remain the most commonly used for both routine waterproofing and ground strengthening. The water to solids ratio is the prime determinant of their basic characteristics such as stability, fluidity, rheology, strength, and durability. Five broad subcategories can be identified:

1. Neat cement grouts.
2. Clay/bentonite-cement grouts
3. Grouts with fillers. (Including low mobility or "compaction" grouts)
4. Grouts for special applications (Such as for antiwashout conditions)
5. Grouts for special applications.
6. Grouts with enhanced penetrability.

It should be borne in mind that many particulate grouts alone are unsuited for sealing high flow, high head conditions: they will be diluted or washed away prior to setting in the desired location. However, the recent developments in rheology, stability, and hydration control technologies, and the major advances made in antiwashout additives have offered new opportunities to exploit the many economic, logistical, and long term performance benefits of cementitious compounds (Gause and Bruce, 1997). Water cement ratios are now typically in the range of 2 or 3 as a maximum, many times lower than the "traditional" mixes of the 1930's. These developments have drawn largely from experience with the wide range of additives developed primarily for the concrete industry. It is now common for a routine fissure grouting operation to feature a suite of grout mixes containing several components (in addition to cement and water), to satisfy site specific fluid and set property requirements, (Table 1), while the use of finer grind materials (e.g. DePaoli et al., 1992) has further enhanced penetrability efficiency. At the other end of the aperture spectrum, economic bulk infill mixes (e.g. for karsts, old mineral workings) are being refined using large volumes of relatively inexpensive materials such as flyash, and naturally occurring soils from gravels to clays. Admixture technology is again valuable in such mixes, providing stability, rheology and anti washout properties.

Developments in Other Grout Families

Given that the sodium silicate based grouts are never used in rock grouting, and that cost and environmental concerns rule out the regular use of most solution grouts in rock grouting (with the exception of certain acrylates), major developments have revolved around two groups of materials:

- Polyurethane
  – Water-reactive polyurethane: Liquid resin, often in solution with a solvent or in a elasticizing agent, possibly with added accelerator, reacts with groundwater to provide either a flexible (elastomeric) or rigid foam. Viscosities range from 50 to 100 cP. There are two subdivisions:

Additive	Beneficial Effects	Adverse Effects	Other Comments
Flyash Type C or Type F	Improves grain size distribution of cured grout, Cheap filler with pozzolanic properties. Can be used as a replacement for some of the cement and reacts with the free lime resulting from the cement hydration process. Increases durability and resistance to pressure filtration.	Increases viscosity and cohesion.	Concentrations of Type C flyash in excess of 20% by weight of cement should be avoided.
Bentonite	Reduces bleed and increases resistance to pressure filtration. Slight lubrication and penetrability benefits.	Increases viscosity and cohesion. Weakens grout.	Should be added as pre-hydrated suspension
Silica Fume	Fine grained powder which improves pressure filtration resistance and reduces bleed. Improves water repellency and enhances penetrability. Improves grain size distribution of cured grout.	Increases viscosity and cohesion.	Difficult to handle due to fineness.
Viscosity Modifiers (Welan Gum)	Makes the grout suspension more water repellant, provides resistance to pressure filtration, and reduces bleed.	Increases viscosity and cohesion.	At higher doses, provides some thixotropy to the grout which is helpful for artesian conditions.
Dispersants or Water Reducers (Superplasticizer)	Overprints solid particles with a negative charge causing them to repel one another. Reduces agglomeration of particles thereby reducing grain size by inhibiting the development of macro-flocs. Also reduces viscosity and cohesion.	Depending on chemistry chosen, may accelerate or retard hydration process. This is not necessarily negative.	Dispersants have a distinct life span. Working life depends on dispersant chemistry chosen.

Table 1. Common grout additives (Wilson and Dreese, 1998).

1) Hydrophobic - react with water but repel it after the final (cured) product has been formed
2) Hydrophillic - react with water but continue to physically absorb it after the chemical reaction has been completed.

- Two component polyurethanes: Two compounds in liquid form react to provide either a rigid foam or an elastic gel due to multiple supplementing with a polyisocyanate and a polyol. Such resins have viscosities from 100 to 1,000 cP and strengths as high as 2 MPA. A thorough description of these grouts was provided by Naudts (1996).

- Hot Melts
  - For certain cases seepage cut off applications, hot melts can be a particularly attractive option. Bitumens are composed of hydrocarbons of very high molecular weights, usually obtained from the residues of petroleum distillation. Bitumen may be viscous to hard at room temperature, and have relatively low viscosity (15 to 100 cP) when hot (say 200 degrees C plus). They are used in particularly challenging water-stopping applications (Bruce et al., 1998), remain stable with time, and have good chemical resistance. Contemporary optimization principles require simultaneous penetration by stable particulate grouts to ensure good long-term performance. Although the concept is decades old, it is only in the last three years that the process has been completely "reinvented" to provide a tool of extraordinary value.

Methods

Drilling

There are three generic methods of rock drilling which have been used routinely in rock drilling (the rotasonic method has not yet met wide application for grouting):

- High Rotation Speed/Low Torque Rotary: relatively light drill rigs can be used to extract core samples, when using a core barrel system, or can also be used simply to drill holes, using "blind" or "plug" diamond impregnated bits. Typically used for holes up to 100mm diameter.

- Low Rotational Speed/High Torque Rotary: used with heavier and more powerful rigs to drill holes of greater diameter to considerable depths. The penetration rate also depends on the thrust applied to the bit. Uses a variety of drag, roller, or finger bits depending on the rock, and relates closely to water well or oil field drilling technology.

- Rotary Percussive: the drill bit (cross- or button-) is both percussed and rotated. In general the percussive energy is the determinant of penetration rate. There are two different options:
  - <u>Top drive</u>: where the drill rods are rotated and percussed by the drill head on the rig.
  - <u>Down-the-hole hammer</u>: where the (larger diameter) drill rods are only rotated by the drill head, and compressed air is fed down the rods to activate the percussive hammer mounted directly above the bit.

In principle, the prime controls over choice of drilling method should ideally be related to the geology, the hole depth, and diameter, bearing in mind always the question of lineal cost. Hole linearity and drill access restraints may also have significant impact.

Overall in the United States, rock drilling is largely and traditionally conducted by rotary methods although the insistence on diamond drilling is no longer so prevalent. Top drive rotary percussion is growing in acceptance in certain quarters - with the increasing availability of higher powered diesel hydraulic drill rigs - as long as water or foam flush is used. Holes up to 100mm in diameter to depths of 50m can be drilled economically. Somewhat perversely, certain specialists are beginning to allow air flushed rotary-percussive drilling for routine grout holes. Even when the air is "misted" with some inducted water, most specialists agree that this medium has a detrimental effect on the ability of the fissures to subsequently accept grout (Houlsby, 1990; Weaver, 1991). Such methods are still, of course, wholly applicable for drilling grout holes to locate and fill large voids such as karstic features. It is common to have drilling rigs instrumented to provide real time accurate data on those drilling parameters which in some way reflect directly the geology and ground water conditions.

Grouting

Rock grouting practice largely follows traditional lines (Ewert, 1985), although it would appear that more recent publications by specialists such as Houlsby (1990) and Weaver (1991) have had a refreshing and stimulating impact. There are three basic methods used for grouting stable rock masses:

- Downstage (Descending stage) with top hole packer;
- Downstage with down hole packer; and
- Upstage (Ascending stage).

Circuit grouting is now only very infrequently used.

The competent rock available on most dam sites is well suited for upstage grouting and this has historically been the most common method. Downstage methods have recently had more demand in the U.S. reflecting the challenges and difficulties posed more difficult site and geological conditions in the remedial and hazardous waste markets.

In some cases of extremely weathered and/or collapsing ground conditions, even descending stage methods can prove impractical, and the MPSP (Multiple Packer Sleeve Pipe) Method is now the method of choice. (Bruce and Gallavresi, 1988). This has particular application in remedial rock grouting operations.

The MPSP system is similar to the sleeved tube (tube à manchette) principle in common use for grouting soils and the softest rocks. The sleeve grout in the conventional system is replaced by concentric polypropylene fabric collars, slipped around sleeve ports at specific points along the tube (Figure 1). After placing the tube in the hole, the collars are inflated with cement grout, via a double packer and so the grout pipe is centered in the hole, and divides the hole into stages. Each stage can then be grouted with whatever material is judged appropriate, through the intermediate sleeved ports. Considerable use has been made of MPSP in loose, incompetent, or voided rock masses, especially karstic limestones in recent projects involving the authors in the Philippines, Canada, and the U.S. Such systems permit the use of a wide range of grouting materials, including the hot melts.

Regarding equipment, contemporary practice features the use of highly automated grout preparation and pumping stations. Mixers are high speed, high shear, high output and are capable of batching wide ranges of multicomponent particulate grouts with accuracy and consistency. Electricity is the power source of choice. Pumps must be capable of infinite stepping of injection rate and volume within their operating parameters and are usually electrically and/or hydraulically powered. Higher pressure operations (say above 2 MPa), require piston pumps, while progressive cavity pumps remain common for low pressure work. Other families of grouts require their own batching and delivery systems, usually provided by, or in conjunction with, the materials suppliers.

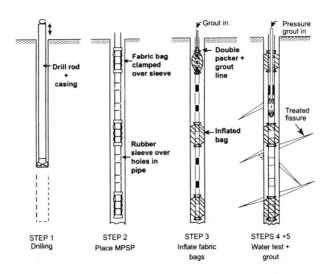

Figure 1. Multiple packer sleeve pipe (MPSP) system.

QA/QC & Verification

General

The fundamental approach to a correctly engineered grout curtain remains

- Investigate site and determine causes/paths of leakage;
- Execute grouting program; and
- Verify performance.

The traditional tools for investigation and verification such as coring, permeability testing, ground water characterization, dye testing, piezometric levels, and outflow monitoring have been supplemented by a range of geophysical tests in certain applications, and by sophisticated data collection, analysis and presentation instrumentation.

However, it is in the qa/qc programs now exercised during the execution of grouting works that the most significant progress is being made. As reported by Wilson and Dreese (1998), the potential of electronic measurement devices mated with computers was recognized almost as soon as widespread use of computers came into being in the early 1980s. The first trials were conducted at Ridgeway Dam by the U.S. Bureau of Reclamation (USBR). The problems with the first system were numerous, but it led to the USBR embarking on development of a comprehensive hardware and software system that would provide, generate, and record all the information that was needed for monitoring, control and analysis of grouting (Demming et al., 1985). That system was written in Basic programming language by a USBR software subcontractor, who retained the proprietary rights to the software, and the USBR implemented its use at Stillwater Dam in 1985. Since that time, there have been dramatic improvements in both the number and type of electronic measurement devices, computers and data management software.

At the simplest level, readings from flow meters and pressure transducers are transmitted to an X-Y recorder and manual calculations are then conducted. However, potentially significant head losses and gains from the system and the environment are ignored. The manual manipulation can be erroneous and is usually cumbersome when head difference allowances must be made. The next level allows for computer display of readings and spreadsheet calculations. Although head losses and gains are more easily accounted for, data entry from display to spreadsheet is still required.

The highest level is represented by CAGES - Computer Aided Grouting and Engineering System. The displayed data are automatically adjusted for all necessary correction factors to reflect actual parameters within the stage being grouted. The displayed data include real time plots of the pressure and flow values, and a time plot showing Apparent Lugeon value. This is a calculated Lugeon value adjusted for the viscosity of the grout and which allows evaluation of the geologic formation response during grouting. The software also generates final hole records comprised of actual and adjusted measurements and scaled time plots of all parameters throughout the entire grouting operation. A final level of sophistication, which is not in general use, includes remotely activated control valves to allow adjustment of flows and pressures during grouting. Computer assisted grouting combined with the application of the Apparent Lugeon Theory and Amenability Theory (Naudts, 1995) provides the knowledgeable grouting practitioner with real time data acquisition and a sound, scientific basis for decision making. As a consequence, every stage in every hole can be correctly brought to a natural refusal by informed manipulation of grout pressure, injection rate, rheology and grain size.

Case Histories

Tims Ford Dam, TN (Bruce et al., 1998a)

Background. Tims Ford Dam is an embankment structure on the Elk River approximately 14 km west of Winchester, TN. This water regulating Tennessee Valley Authority (TVA) structure is about 460m long with the crest at Elevation 227.4m. The right (west) abutment of the dam intersects orthogonally a natural ridge running nearly north-south, and consisting of clay and weathered chert overburden overlying a karstic foundation of various limestones. The crest of this right rim abutment varies in elevation from 287m to about 292m with the top of rock generally around Elevation 274m. The maximum pool elevation is at Elevation 270.7m.

The Problem. Seepage through the right rim was recorded from first impoundment in 1971, prompting some local grouting. However, a major seepage at Elevation 260m, about 290m upstream of the dam center line persisted. It grew steadily each year until 1994 to about 15,000 l/min, but increased dramatically in 1995 to over 29,000 l/min following a record reservoir drawdown. TVA determined that a remedial grouting program be effected to reduce this flow to less than 4,000 l/min at maximum pool by sealing major karstic features thought to be present at that location.

An exploratory drilling and water testing program defined the geographic extent and depth of the remediation.

The Solution. A multirow grout curtain was designed, approximately 240m long. The holes were inclined at 30 degrees to the vertical to encourage intersection of (sub) vertical features and were oriented in opposite directions in the two outside rows. Primary holes in each row were foreseen at 12-m centers, with conventional split spacing methods to be employed (to 3-m centers or closer). The central, tightening, row was vertical. The grouting was to be executed between Elevations 270.7 and 256m - locally deeper if dictated by the stage permeability tests conducted prior to the grouting of each stage.

Because of the suspected high flow conditions, the downstream curtain row holes that encountered voids and active flow conditions were designated to be grouted with fast-setting (1 to 3 minute set time) hydrophillic polyurethane resin to provide an initial semi-permanent flow barrier. Holes that did not encounter voids or active flow were to be grouted with cementitious grouts. Upon completion of the downstream row, it was anticipated that the active flow conditions would be mitigated, thus allowing the entire upstream row followed by the third, central, closure row to be grouted with cementitious grouts to form a permanent and durable grout curtain. The grouting was designed to be performed using upstage methods, although it was anticipated that poor foundation conditions could locally require utilization of downstage methods. The grout holes were to be cased through the overburden from the surface to the top of the curtain.

The Specifications contained provisions that required monitoring and limitations to outflow pH and turbidity to protect the downstream environment. TVA agreed to draw down the reservoir to Elevations 260.6m (3m below minimum normal pool) to minimize hydraulic gradient and flow through the rim. The curtain was to be constructed by first grouting the far ends, so conceptually channeling the flow through a middle zone which would then be treated.

Highlights of Construction.
- When drawdown of the reservoir reached Elevation 261.8m, the outflow from the leak completely and naturally stopped. As a consequence, much of the grouting work could be done in "no flow" conditions, therefore, largely eliminating the need for the polyurethane grouts, and extending the applicability of cement based formulations.
- Larger than anticipated open or clay-filled features were encountered by the down-the-hole drilling methods, especially in the upper 6m or so of the curtain. For technical, commercial, environmental and scheduling reasons, such features were treated with a low mobility "compaction grout" (slump 50 to 150mm; containing also water reducing and antiwashout agents).
- A suite of cement-based grouts (Table 2) was developed to permit the appropriate match of mix design and "thickening sequence" to the particular stage conditions as revealed by drilling and permeability testing (both multi- and single-pressure tests).
- In response to conditions revealed during the treatment, observations of the seepage and further dye testing, extra groups of holes were added the north end of the curtain, including 11 orthogonal to the original curtain, to allow specific treatment of key features.

- About 1,550m^3 of compaction grout, 1,530 liters of polyurethane, and 605m^3 cement based grouts were injected into a total of 250 holes (comprising 3,400 lin.m of rock drilling).

Effect of Treatment. Throughout the work, closest attention was paid in real time to data from the drilling, water testing, and grouting activities in addition to information from leak monitoring, piezometers and dye testing. The curtain was thus brought to an engineered refusal. During refilling of the reservoir, the leak had been totally eliminated with the level at Elevation 265m, when, for financial reasons, the work was terminated. The most recent reading, with the lake at Elevation 269m, indicates a seepage of around 950 l/min (net of surface runoff contributions) - about 5% of the flow at the equivalent lake elevation prior to grouting. Data from piezometers and dye testing support the existence of an efficient and durable curtain.

Ingredient	Unit	Mix A	Mix B	Mix C	Mix D
Water	lb	141	141	94	94
Bentonite	lb	4.7	9.4	4.7	4.7
Cement	lb	94	94	94	94
Rheobuild 2000	oz	15	30	20	30
Rheomac UW450	oz	0	0	0	5
Volume of batch	gal	20.8	21.0	15.1	15.1
Specific gravity		1.39	1.4	1.53	1.53
Bleed	%	<5	<1	<1	0
Kpf	min$^{-1/2}$	<0.104	<0.042	<0.042	<0.042
28-Day Compress.	psi	500	500	800	800
Marsh time	sec	35	50	60+	100+
Stiffening time	hh:mm	4:30	4:30	4:00	4:00
Hardening time	hh:mm	10:30	8:30	8:00	8:00
Water and slurry volumes					
Bentonite slurry volume	gal	8.0	16.1	8.0	8.0
Additional water volume	gal	9.9	2.8	4.2	4.2

Table 2. Compositions and properties of cement grout mixes, Tims Ford Dam, TN.

Penn Forest Dam, PA (Wilson and Dreese, 1998)

Background. The new Penn Forest Dam is being constructed to replace the old Penn Forest Dam, which was a severely ailing earthen embankment dam. The new dam is being constructed with roller compacted concrete just upstream from the old dam, and is approximately 54m high and 600m long. It includes a three-line grout curtain, designed to have a maximum residual permeability of 3 Lugeons on a 4.5m width. The lines were 1.5m apart to a depth of 42m.

Construction Concepts. An accelerated construction schedule resulted in the grouting being split into two separate consecutive contracts, the first for one line (A); the second for the other two lines (B and C). Due to the short design period duration and other factors, the A-Line grouting contract was issued specifying conventional methods (for example, neat cement grouts, agitator tank dipstick measurements, and pressure gages). However, sufficient time was available to design the second contract using "advanced" methods, such as balanced, stable cement based grouts and computer assisted grouting.

Mix Design. Whereas Line A used neat cement mixes of w/c from 3 to 0.7, Lines B and C used a suite of multicomponent mixes comprising Type III cement, flyash, bentonite, welan gum and dispersant, as determined during extensive preconstruction field testing. Particular attention was paid to minimizing the pressure filtration coefficient (below 40×10^{-3} min$^{-1/2}$) to promote efficient penetration and long term durability.

Injection Monitoring and Control. Lines B and C were injected using the CAGES software, which according to Wilson and Dreese, provided many advantages over the traditional manual methods:

- Real time data are obtained at much smaller time intervals (5 to 15 sec. frequency vs. 5 to 15 min. frequency).
- Eliminates potential for missing critical events such as pressure spikes.
- Data obtained are more accurate.
- Higher grouting pressures can be used with confidence.
- Formation response to procedure changes (mix or pressure) is shown instantly.
- Damage to formation due to over-pressuring can be easily detected and mitigated.
- Significant acceleration of pressure testing and grouting operations.
- More consistent grouting procedures due to central control location.
- Reduction in inspection manpower requirements.
- Provides detailed, permanent graphic records showing the entire time history for each operation on each stage.

The authors also found that the advanced system required less grout to reach the target permeability, largely as a result of the enhanced penetrability of these stable grouts. Financially, the construction cost savings were about 10%, the inspection cost savings 25%; and the construction schedule savings 25%, relative to those incurred during the previous, traditional grouting phase.

Final Remarks

These two case histories illustrate how recent advances in rock grouting technology can provide treatment of the highest quality in a controlled, engineered fashion. The use of multicomponent, balanced particulate grouts, allied with computer aided control and evaluation systems, has also proven to be highly cost effective when compared to traditional approaches. The engineer therefore has increased confidence in his ability to achieve stringent performance goals for the many applications of rock grouting.

References

Bruce, D.A. and F. Gallavresi (1988). "The MPSP System: A new method of grouting difficult rock formations." ASCE Geotechnical Special Publication No. 14, "Geotechnical Aspects of Karst Terrains," pp. 97-114. Presented at ASCE National Convention, Nashville, TN. May 10-11.

Bruce, D.A., J.A. Hamby, and J.F. Henry. (1998a) - "Tims Ford Dam, Tennessee: Remedial Grouting of Right Rim", Proceedings of the 1998 Annual Conference Association of State Dam Safety Officials, October 11-14, Las Vegas, Nevada, Paper No. 70.

Bruce, D.A., A. Naudts, and W.G. Smoak. (1998b). "High flow seepage reduction in major structures: materials, principles, and case histories." Proceedings of ASCE Conference, Boston, MA, October 18-22, Geotechnical Specialty Publication No. 80, pp. 156-176.

Bruce, D.A., Littlejohn, G.S. and A. Naudts (1997) "Grouting Materials for Ground Treatment: A Practitioner's Guide, *Grouting - Compaction, Remediation, Testing*, ASCE, Geotechnical Special Publication No. 66, Ed. By C. Vipulanandan, pp. 306-334.

Demming, M., J.L. Rogers, and A. Tula. (1985) "Computer Applications in Grouting." *Issues in Dam Grouting.* ASCE, Denver, pp. 123-131.

DePaoli, B., B. Bosco, R. Granata, and D.A. Bruce. (1992). "Fundamental Observations on Cement Based Grouts (2): "Microfine Grouts and the Cemill® Process", *Grouting, Soil Improvement and Geosynthetics.* ASCE, New Orleans, pp. 486-499.

Ewert (1985) "Rock grouting with emphasis on dam sites." Springer-Verlag, Berlin

Gause, C. and D.A. Bruce. (1997). "Control of Fluid Properties of Particulate Grouts (1) General Concepts". *Grouting - Compaction, Remediation, Testing,* ASCE, Geotechnical Special Publication No. 66, Ed. By C. Vipulanandan, pp. 212-229.

Houlsby, A.C. (1990). "Construction and design of cement grouting". John Wiley & Sons, 442 p.

Naudts, A. (1995) "Grouting to Improve Foundation Soil". *Practical Foundation Engineering Handbook.* McGraw-Hill, New York, Section 5B.

Weaver, K.D. (1991). "Dam foundation grouting." ASCE Publication, 178 p.

Wilson, D. and T. Dreese. (1998) "Grouting Technologies for Dam Foundations", Proceedings of the 1998 Annual Conference Association of State Dam Safety Officials, October 11-14, Las Vegas, Nevada, Paper No. 68.

POLYURETHANE GROUTING FOR SEALING LEAKAGES IN TUNNELS

BERT P. KRIEKEMANS

Abstract

Waterleakages through rock existed before mankind. And leakages through crack and joints in concrete have been a problem for a long time. When concrete moves, it often cracks, The size of the crack depends on the amount of movement; waterleakages can cause freeze-thaw damage, corrosion, structural weakness, and even failure. Cold weather may bring hazardous ice formations underfoot or overhead. Polyurethane Resins (PUR) have been successfully resolving water-control problems in concrete repair for over 20 years. A wide variety of materials and application procedures are currently being used, depending on the specific results required.

Types of Polyurethane Grouts

Polyurethane Resins must be broken into two (2) chemical base classification; either Toluene Diisocyanate (TDI) or Diphenylmethane Diisocyanate (MDI). Through this chemical base classification it is found that the Resins can be broken down further into the nature of the chemical reaction; either hydrophilic or hydrophobic. Finally, the cured end products after full reaction must be classified into 4 basic types: rigid foams, solids, gels and flexible foams.

Hydrophobic - Hydrophilic

First a review of the chemical base difference between hydrophilic and hydrophobic will be made to assist in understanding how the Resins will react [1]when injected into flowing or weeping cracks in any substrate.

---

[1] President, DeNeef Construction Chemicals, Inc. Waller, TX USA

Leaking Joint - right half already Injected & Sealed

Hydrophilic resins love water by their definition. A hydrophilic resin will incorporate or use the complete water molecule in its chemical reaction. This type of resin has a TDI chemical base and tends to absorb as much water as is available while the chemical reaction takes place. The cured-in-place end product can vary greatly, from a foam to a flaccid gel depending on the amount of water introduced. To achieve the cured-in-place end product desired, a hydrophilic resin should be applied using a multi-ratio pumping system so the applicator can control the final results.

Hydrophilic foams are the most flexible materials having high elongation as well as high tensile and bond strengths to either wet or dry concrete substrate. They tend to have a higher shrinkage rate when tested by ASTM-D-1042 standards (the use of this standard should be reviewed because shrinkage is measured in a non-adhesive state). This type of foam is an excellent repair material in moving non-structural cracks.

A *hydrophilic gel* on the other hand has little or no bonding capabilities, elongation, or compressive strengths. This type of end-product should only be injected into the soils outside of a substrate where they will fill voids between the soil particles and build a matrix with soils to become a mass or grout curtain. Typically, a gel is pumped at ratios of 5:1 to 12:1 (water to resin) depending on the void ratio in the soil. In the past, most hydrophilic resins were not 100% solids and used acetone as a diluent to lower the viscosity. Today, however, the newest technology can offer an improved product by using non-flammable and non-toxic diluents in place of acetone; thereby lowering the threshold limit value (TLV) and permissible exposure limits (PEL) for worker safety.

Hydrophobic resins, on the other hand, fear water by definition. A hydrophobic resin does not incorporate the complete water molecule in its chemical reaction. This type of resin has an MDI chemical base. During its chemical reaction with water, it will only use the portions of the water molecule necessary for the curing process and will displace all remaining water. With variations in the types of polymers used in manufacturing, a wide variety of cured end-products may result. Hydrophobic resins are available in solids, rigid foams and flexible foams. These foams have high tensile strengths and have good bond strengths to dry concrete. Most often, these foams are injected with a single component pumping system with which you can/may obtain almost infinite control of the chemical reaction by using the appropriate accelerators. Foaming can be adjusted to as low as five (5) seconds in high flowing conditions or as long as three (3) hours where deep penetration is required.

When evaluating the cured-in-place products a hydrophobic resin produces, we can classify them into three (3) basic categories: solids, rigid foams and flexible foams. Each cured resin type is best suited and intended for a specific application.

Solid resins have multiple applications. They can be used as low viscosity soil injection resins which are injected into soils to control water migration or they may be used to control hazardous waste from infiltrating aquifers. Soil type **hydrophobic** resins can build extremely high compressive strengths in soils and are widely used in curtain grouting. When being used in containment areas with an extreme pH, these resins will harden with only the moisture available and are generally not affected by the pH of the waste. Other solid type resins used are similar to that of plural component epoxies. They can be injected into voids or concrete cracks where structural integrity or structural rebonding is needed. They bond extremely well to concrete surfaces and protect the rebar from further corrosion.

Rigid hydrophobic resins are highly expansive and can be used for void filling, soil consolidation and non-moving cracks. Extreme caution must be taken when these resins are used in soils because slabs or walls can be lifted due to its highly expansive nature. Rigid foams were the original polyurethane resins used in water remedy projects and have been highly successful in stopping large volumes of water in a wide variety of applications. As with all technological advances, many new polymers have been found and now a wide variety of flexible hydrophobic resins have evolved and, in many applications, have replaced the rigid type materials in crack or joint repairs.

Flexible hydrophobic resins have risen to the forefront of the industry and are specified for a variety of repairs in older construction as well as new construction. This type of resin is used in moving cracks and construction joints, above or below grade, to seal and protect the substrate from deterioration. With the injection of these resins, reinforcing steel can now be re-encapsulated with a flexible material that will move with the natural thermal cycles and maintain a flexible gasket to alleviate further deterioration of the concrete substrate.

Application techniques and placement methods have been developed to assist the applicator in resolving water-control projects both permanently and successfully.

Tunnel Application: Chicago flood in 1992

Background

Early January 1992 a section of the nearly century-old freight tunnel under the north branch of the Chicago River at Kinzie Street was discovered to have been damaged. It was theorized that damage had been caused by a recent pile driving operation. Before the damage could be repaired, a breach occurred, resulting in river water flooding significant sections of the freight tunnel system. Resulting in the closure of the Chicago Stock Exchange for almost a week. The Consulting Engineers were called in by the Chicago Transit Authority to assess the conditions (numerous inflows) and recommend a course of action to mitigate any possible damage the adjacent flooded freight tunnels posed to the subway tunnels and to the CTA's elevated rapid transit system above.

Inspection Program

Two types of subway tunnel inspections were conducted. One type, a detailed mapping of all cracks, would be used to monitor any movement of the structure. The second type, a "wet" crack inspection, would be used to evaluate the magnitude and extent of the water intrusion caused by the adjacent flooded freight tunnels.

Within 48 hours a soil boring program was under way to assess the impact of the exfiltration of the freight tunnel flood waters into the soils adjacent to the subway tunnels and the elevated rapid transit system's supporting structure (columns on shallow spread footings). Within 72 hours instrumentation was installed within the subway tunnel structure and on the elevated structure to provide more accurate measurements of any movements.

During the first inspection tour it was determined that the inflows into the subway tunnel structure were widespread and substantial enough that they needed to be stopped as soon as possible to prevent further damage to the subway electrical systems..

Determining What And Where To Grout

The subway tunnel inspections very quickly identified the areas of highest inflow into the Dearborn Street and State Street Subway tunnels. From the "wet" crack inspection data location plans for the chemical grouting plan were developed. The locations were prioritized in the field (based upon further visual observation of the magnitude of the inflow) and marked for polyurethane grouting.

The boroscope inspections identified three significant voids, which were actually abandoned sections of the freight tunnels, near the crown of the Dearborn Street Subway. Since there was not any floodwater within these sections of the freight tunnel, it was presumed the bulkheads shown on the subway "as builts" did exist and apparently were not leaking, at least not yet. Data obtained from subway "as builts" and boroscope observations pinpointed the exact areas for cementitious grouting.

Leaking Joint-right half already Injected & Sealed

Installation Methods

Site Constraints and Set-Up

Access to the work sites was limited. The primary grouting location in the Dearborn Street Subway was midway (nearly 300 meters) from either of the two available access points, the Clark/Lake Station and the emergency exit for the southbound tube at Lake and Canal. All equipment and material had to be hand carried down a circular staircase, or down through the station and then down 1000' of track. As the 600 volt traction power had to be shut off, the contractor had to provide light plants and mobile generators mounted outside the emergency exit.

The Contractor and a subcontractor, were hired by the CTA to perform the recommended grouting programs in the Dearborn Street and State Street Subways. A major contractor hired by the City of Chicago was on site to seal the breach in the freight tunnel.

The Contractor forces began erecting scaffolding for the chemical grouting operation in the Dearborn Street Subway. Scaffolding was erected initially at the six (6) highest priority locations of inflow.

Equipment

The chemical grouting equipment consisted of the following:

- A 5-gallon mixing and holding vat
- A GRACO EM 490 2.5-horsepower pump
- A delivery hose with a valved 3/8"-nozzle
- Heavy duty concrete drills and bits

Two sets of equipment were used to work on two locations simultaneously. Holes were drilled into the concrete tunnel wall at a point adjacent to the crack beginning at the tunnel crown and working downward. The grout was injected into the drilled holes until the inflow was stopped. The number of holes drilled and subsequently injected with Hydro Active grout at any one inflow location depended upon how quickly the inflow as stopped.

The cementitious grouting program began in Dearborn Street Subway on April 30, 1992, following completion of the Hydro Active Polyurethane grouting program. The cementitious program was undertaken to fill voids behind the tunnel lining and restore an even pressure distribution on the tunnel. Cementitious grout was chosen because it is more rigid and less expensive than chemical grouts.

A diesel-powered CTA maintenance train was used to deliver materials and equipment to the site. The scaffolding used to reach the crown of the subway tunnel was affixed to the bed of one of the cars of the CTA maintenance train, thus the contractors operation could easily move between sites. The train was also used to deliver water to the grouting sites. The cementitious grouting equipment included the following:

- A ChemGrout CG550P System (pump, mixer and vat)
- A delivery hose and valved nozzle with pressure gauge

A series of holes were drilled into the void areas marked earlier by the consultant and the cementitious grout was pumped in. The grout used consisted of a 1:1 mixture of Portland cement and silica sand.

Crews and Support Personnel

The Polyurethane grouting operation was conducted using two 12-hour shifts. Each shift consisted of two chemical grout crews and one support crew. The chemical grout crews consisted of three staff (a foreman and two grout technicians). The support crew consisted of a Construction Foreman and 4 laborers. The entire construction operation was under the control of a Construction Superintendent.
As with the chemical grout operation, the cementitious grout operation was conducted round-the-clock using two 12-hour shifts. The cementitious grouting operation was performed by contractor forces consisting of a superintendent, a foreman, an operator, and 4 laborers. Consultants provided one full-time observer per shift for each of the grouting operation.

Results

Short Term

Both the Polyurethane and cementitious grouting programs proved to be successful in aiding the speedy return of the CTA's rapid transit subway facilities to revenue services.

Subsequent inspections did not uncover any further leakage in the repaired areas.

Long Term

The results of the emergency non-structural crack repair program points to the conclusion that a proper Polyurethane grout injection program is a viable solution to eliminating unwanted water intrusion into tunnels and other underground structures.

ACKNOWLEDGEMENTS

Chicago Flood Report; Frederic H. Parkinson III, Parsons Brinckerhoff, Quade and Douglas, Inc.,
Kenny Construction
Hayward Baker Company

Groundwater Control at Difficult Geological Interfaces

Arthur B. Corwin, P. E., ASCE member[1]
Thomas J. Tuozzolo, ASCE member[2]
Paul C. Schmall, P. E., ASCE member[3]

Abstract

Often in the construction of underground structures, we are not so fortunate as to be able to excavate and construct within the depth of a single homogeneous geological stratum. Where groundwater control is required and water bearing soils are underlain by an aquiclude or impermeable layer within the depth of excavation, ground water control is typically handled with the use of closely spaced wellpoints to lower groundwater levels as close as practical to the geological interface, generally requiring the addition of trench drains to intercept residual seepage. Where absolute water cut-off is required, the combined use of predrainage dewatering methods and cut-offs methods has been successful. The implementation of three cut-off methods: thin diaphragm jet grout wall, slurry trench, and ground freezing, is discussed herein in conjunction with the excavation and construction of several underground structures.

Introduction

A common challenge encountered with site dewatering is the behavior of groundwater at pronounced changes in geology. The condition that must always be approached with caution is where high permeability, coarse grained soil overlies a clay layer or low permeability soil or rock within the depth of dewatering, or when that transition from high to low permeability occurs immediately or several feet beneath subgrade elevation. At such locations, complete drainage of the permeable soil is physically not possible and some quantity of water will remain perched above the interface and necessitate the use of open pumping techniques such as trench drains and sumps to handle residual seepage into the excavation. These difficult draining conditions are commonly referred to in the trade as interface problems.

---
[1]Arthur B. Corwin, P. E., Executive Vice President, Moretrench American Corporation, 100 Stickle Avenue, Rockaway, NJ 07866
[2]Thomas J. Tuozzolo, Project Executive, Moretrench American Corporation, 100 Stickle Avenue, Rockaway, NJ 07866
[3]Paul C. Schmall, P. E., Chief Engineer, Moretrench American Corporation, 100 Stickle Avenue, Rockaway, NJ 07866

Wellpoints[3] are typically used at such geological interfaces where water must be drawn down as close as possible to an underlying clay layer. The spacing of the wellpoints may depend on several factors: the total amount of groundwater lowering required below static levels, soil permeability characteristics, the uniformity and cohesiveness of the soil and subsequent susceptibility of soil movement, and the ease and practicality of handling seepage that passes between the wellpoints.

Wellpoints[3] are generally considered the best practice for controlling interface seepage water that is within practical depth ranges of 15 to 20 feet per wellpoint level; however, there are often requirements for absolute water cut-off dictated by specific project demands. Where seepage is unacceptable, it is common to combine cut-off and predrainage dewatering methods to achieve a cost effective complete water cut-off. Cut-off methods commonly utilized are slurry trenches, diaphragm walls, tight steel sheet piling, chemical grouting, jet grouting, soil mixing, and ground freezing. Several specific examples of combined efforts of dewatering and cut-off are cited herein to address seepage at difficult geological interfaces.

Use of a Thin Jet Grout Diaphragm Wall in Conjunction with Wellpoint Dewatering

Since its construction in 1907, The New York City Metropolitan Transit Authority Lenox Avenue subway line, which services the borough of Manhattan, has suffered significant water infiltration resulting in the movement of the underlying soils, voids beneath the slab and deterioration of the unreinforced concrete invert. An early 1907 article in a local newspaper called the problem "a stream flowing in the subway".

This constant flow of water and movement of sediment, combined with the constant weight of an average 360 metric ton (400-ton) train traveling on the tracks every three minutes during rush hour, has caused progressive undermining and deterioration of the structure between 110th and 117th Streets resulting in subsidence of the track beds. The subsidence became so severe that trains were limited to speeds of only 16 km/hr (10 mph). Past efforts have been made to stop the problem by injecting grout beneath the subway invert. Although these efforts have remedied the situation for a short period of time, the Transit Authority concluded that a permanent repair had to be made.

The solution to the water seepage problem consisted of a $138 million project to replace approximately 730m (2400 linear feet) of the tunnel invert. Hence, the Metropolitan Transportation Authority/New York City Transit underwent contract C-33204, "The Lenox Avenue Line "A" Rehabilitation". With earlier rehabilitation contracts, adjacent sections of the invert were replaced and a permanent wellpoint dewatering system was installed. On one of these earlier contracts, specifically in the area of 117th Street (at the northern end of 116th Street Station), the contractor experienced high permeability coarse sand immediately beneath the structure underlain only a few feet below by silt. The thickness of the coarse sands was not great enough to allow complete drainage of the sands to below invert with wellpoints. Although the groundwater pressures had been significantly lowered by the operation of the wellpoint system in this area, significant water flows were experienced as the groundwater ran across the top of the silt and into the excavation area. Tight steel sheeting installed from within the close confines of the subway tunnel (in addition to sumps and trench drains) was ultimately utilized to partially cut off the inflow of water from the coarse sands to permit work to proceed. This past experience forced the Metropolitan Transportation Authority to undergo extensive preconstruction studies for this contract C-33204.

A Geotechnical and Hydrogeological Data study was performed to evaluate subsurface conditions for this adjacent contract section which would be executed between 110th and 117th Streets. Similar to an adjacent contract section which experienced groundwater control difficulties, the existing groundwater table, was approximately 3.8 to 4.2 m (12 to 14 feet) above the base of the existing subway, and had to be lowered below the proposed elevation of the new concrete invert so that the construction could be performed in the dry. Several pump tests were performed in the area of the 116th Street station (115th to 117th Streets), where previous invert reconstruction efforts experienced dewatering difficulties, and where the geotechnical information indicated that the low conductivity layer (silt) rose to intercept the base of the invert structure. The pump tests confirmed very high hydraulic conductivity of the formation near the invert and limited radial response of the formation to pumping stress. Yields of 1.89 to 3.15 l/s (30 to 50 gpm) were experienced from small diameter wells installed to the top of the silt, confirming the existence of the suspected difficult dewatering conditions.

The results of the soil samples taken during the geotechnical study determined that the site geology was made up of a stratum of fill which extended to a depth of approximately 3 m (10 feet), followed by a stratum consisting of a medium to coarse sand to a maximum depth of 14m (20 feet). The next stratum consisted of a fine sand to a maximum depth of 14m (45 feet) below grade followed by a low permeable silty & clay layer which terminated the soil study at a depth of approximately 16m (52 feet) below grade. The existing groundwater table was determined to be located 3.8 to 4.2 m (12 to 14 feet) below grade. (See Figure 1)

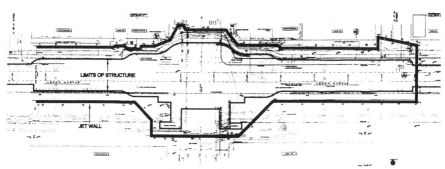

LENOX AVENUE SUBWAY RECONSTRUCTION PROJECT
JET GROUT WALL LAYOUT
FIGURE 1

In the area of the 116th Street station, it was noted that the low conductive silt layer rose, as previously recorded, just below the elevation for the new invert. Based on past history and the hydrologic information gathered, this presented an immense problem. The low permeable silt layer was high and close to the subway invert; it interfaced with the higher permeable sand formation. Since the interface was approximately at the bottom of the proposed invert it would have been impractical to install a dewatering system to handle the high volume of flow from the highly conductive sands. With this in mind, the contract construction dewatering specifications recommended a temporary wellpoint system in conjunction with a groundwater flow barrier in the area of 115th to 117th Streets. The recommended groundwater flow barriers were interlocking sheeting, slurry walls and jet grouting.

Following a formal bid letting, the contract to replace the deteriorated invert was awarded to A. J. Pegno Construction Corp. (AJP). The project had a very tight schedule and imposed a $30,000 liquidated damages clause for every day the project went over schedule. Realizing that a major part of the project was to design and install a dewatering system that would control the high groundwater flows, especially in the area of the 116th Street Station, AJP hired Moretrench American Corporation (MTA) to design and install a dewatering system. Besides providing construction groundwater control, flows ranging from 94.63 to 126.18 L/S (1,500 to 2,000 gallons per minute) for the entire project, the system had to solve the problem of groundwater control at the silt and sand interface between 115th and 117th Streets so that construction of the new invert could take place.

With that in mind, MTA installed a wellpoint dewatering system that operated in conjunction with a hydraulic barrier which encircled the 116th Street Station between 115th and 117th Streets. (See Figure 2)

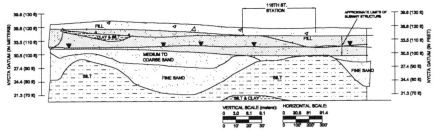

LENOX AVENUE SUBWAY RECONSTRUCTION PROJECT
GEOLOGIC PROFILE

FIGURE 2

The hydraulic barrier, selected to cut-off the interface between the sand and silt, was a thin diaphragm jet grout wall. Although frequently used in Europe, to the writers' knowledge, this jet grouting technique had not been performed on a full-scale project in the United States prior to this installation. This procedure was chosen after evaluating several other methods. The project was on highly traveled New York City streets which were underlain by a spider web of existing underground utilities including a 100 year old brick sewer. Due to these constraints other methods such as slurry walls and steel sheet piling were not practical and could not provide adequate closure to control seepage. Jet grout columns was an option, but would have required approximately 4 times the number of injection points, thus making it less economical, and resulting in more spoils to contend with. In addition, the large number of grout holes required would have created the potential for more disturbance to existing underground structures and utilities. Since the cut-off wall did not have to provide structural resistance and was only required to act as a hydraulic barrier, the thin diaphragm wall was the best choice.

The thin diaphragm wall, measuring 5 to 15 cm (2 to 6 inches), was constructed by utilizing the double fluid jet grout method. A double fluid system involves introducing two components during injection, both grout and air. The system works by injecting grout at high pressures through two center cement nozzles. The grout is used to erode and mix the soil. The air is supplied at a lower pressure rate through a nozzle which surrounds the cement nozzle. The air displaces the soil around the cement nozzle and increases the erosion length an efficiency. By introducing air into this system, the grouting becomes more effective in cohesive soils. For this particular project the cement nozzles were located 150 degrees apart, because this was the desired wall orientation required to provide a continuous interlocking wall. (See Figure 3)

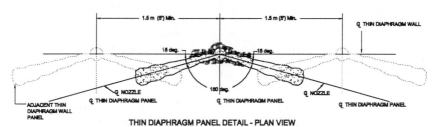

THIN DIAPHRAGM PANEL DETAIL - PLAN VIEW
LENOX AVENUE SUBWAY RECONSTRUCTION PROJECT

FIGURE 3

Prior to the installation of the thin diaphragm wall, a 1.2 m (4 foot) deep trench was excavated at the proposed location of the diaphragm wall to collect the spoils generated from the installation. The thin diaphragm wall was then constructed by first drilling a 75 to 100 mm (3 to 4 inch) diameter hole with a standard roller bit. The holes (injections) were spaced 1.5 m (5 feet) apart. The depth of the drill hole was directly related to the required depth of the thin diaphragm wall. The hole was drilled 6m (20 feet) through the coarse grain sand followed by an additional 3m (10 feet) into the silt/clay stratum, to provide the necessary cut-off barrier. Once the bottom of the hole was reached, fluid (grout and air) was introduced through the drill hole under pressures in the range of 380 Bar (5500 psi). The fluid eroded the soil thus creating a mixture of soil and grout. The soil and grout mixture, "soilcrete", had a compressive strength on the order of 3.45 to 6.9 MPA (500 to 1,000 psi). This continued as the jet grout rods were lifted at a uniform rate. The top of the wall terminated approximately 6m (20 feet) from the bottom of the hole, approximately 1.5 m (5 feet) above the water table. Unlike, jet grout columns, the rods are not rotated during extraction, thus providing the geometric shape of a 150 degree wall consisting of interlocking baffles. The resulting barrier of interlocking barriers has a plan width of .76 m (2.5 feet). It was very consequential that prior to jet grouting the orientation of the nozzles remain at a constant alignment so that the wall was constructed in the correct plane. As with all jet grouting, quality control played an important part in the final result of the wall and also eliminated the possible damage that could have been caused to the underground utilities while grouting under such high pressures. The jet grouting was conducted on both the east and west sides of the tunnel between 115th and 117th streets and across 117th Street, thus forming a U-shape. (See Figure 4)

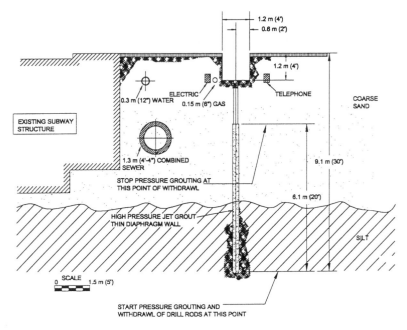

SECTIONAL VIEW OF JET GROUT WALL

FIGURE 4

The design intent of the cut-off wall was not to eliminate the need for dewatering, but to cut-off direct recharge to the high permeability sands immediately below the structure which could not be adequately handled with the use of wellpoints otherwise. Economically, it was not practical to attempt to completely cut off all groundwater flow to the construction work with a very deep jet grout wall penetrating all permeable soil strata. The jet grout wall was laid out to fully penetrate the highly permeable coarse sands and key into the underlying silt. Past testing showed that the underlying silt was somewhat permeable and would require dewatering to provide a stable subgrade. In addition to the jet grout cut-off wall, wellpoints were installed along both sides of the station inside of the hydraulic barrier on 3 m (10 foot) center-to-center spacing to dewater the underlying silt material and prevent undermining of the shallow jet grout wall due to the higher groundwater pressures outside of the jet grout wall.

The effectiveness of the jet grout wall was immediately apparent in observing the dramatic change in water levels inside and outside of the jet grout wall. Water levels inside the confines of the jet grout wall were observed to be below subgrade elevation, in the underlying silts, and water levels outside of the jet grout wall indicated an appreciable saturated thickness of coarse sands. Several feet of groundwater lowering was observed in the coarse sands external to the jet wall, presumably due to massive pumping efforts associated with dewatering of the balance of the project in very permeable soil. Dewatering of the entire 116th Street Station area was completed with standard wellpoint installation, and excavation occurred without any high volume inrush of groundwater. The project parties concluded that the thin jet grout diaphragm wall was a great success in controlling the interface problem at the Lenox Avenue Line "A" Reconstruction Project. This system played an important role in the project and assisted in the project being completed ahead of the tight schedule.

Use of a Slurry Trench with Deep Well Dewatering

A combination of a slurry trench with dewatering successfully controlled groundwater for a large sloped excavation immediately adjacent to the Arkansas River[1]. The Murray Hydro-Electric Project required excavation through a very permeable sand and gravel aquifer extending directly to bedrock, 26 m (85 feet) below high water in the River. The excavation was carried through 22 m (72 feet) of alluvium from original ground surface to bedrock. Rock excavation was subsequently carried to a depth of 11 m (36 feet). A perimeter system of deep wells was installed outside of the deepest excavation cut, and pumped at combined flows of up to 1577.29 L/S (25,000 gpm) to lower the water level to 4.6 m (15 feet) above the bedrock. A short soil/bentonite slurry trench near the toe of the slope, inside the perimeter system of wells, cut-off and held back the residual head at the soil-rock interface to permit excavation to final subgrade in the dry. (See Figure 5)

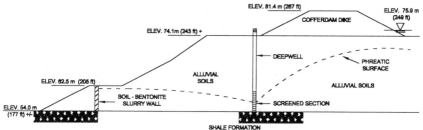

SECTION THROUGH COFFERDAM AND EXCAVATION
MURRAY HYDRO PROJECT SECTIONAL VIEWS
FIGURE 5

The project faced some severe constraints which favored the complete residual water cut-off provided by the slurry trench. The safety of the immediately adjacent existing lock and dam was a major concern. The possibility of water bleeding into a toe drain at the soil raised concerns about movement of material at the base of the slope. If a slope failure occurred, the river might bypass the dam through the excavation. If it did, control of the river for navigation would be lost.

Test borings drilled at the proposed excavation site, together with records of United States Army Corps of Engineers (COE) borings taken during construction of the existing dam, and a COE Design Memorandum, were used to define the geology at the site. The alluvium was typically stratified into three distinct layers: An upper stratum of sandy silt and silty sand (ML, SM); an intermediate layer of medium to coarse sand (SW, SP); and, a lower layer consisting of about 4.6 m (15 feet) of coarse sand, gravel and cobbles. Drilling indicated the likelihood of boulders in the lower layer. The bedrock was soft to moderately hard, unweathered, fissile, carbonaceous silty clay shale. Very few joints were found in the test borings. Most joints in the cores were quartz healed. Water pressure testing of the shale during the investigation indicated minimal water take at applied heads of 11 kPa/m (1/2 psi/foot) of boring.

A field pumping test confirmed the high permeability of the middle and lower zones of the alluvium, deduced from the soil descriptions and sieve analyses. In the latter part of the test the lower pool dropped more than a meter (several feet) and the observation wells reacted quickly, indicating the probability of close connection between the river water and the groundwater.

An earthen cofferdam 1200 meters (4000 feet) in perimeter was designed to enclose the excavation, with its base set at existing ground surface. The cofferdam was constructed of compacted silty sand borrowed from the upper alluvial layer within the excavation boundary. Slopes of the dike were two to one vertical. A berm 3 m (10 feet) in width was provided inside the cofferdam at ground surface for dewatering wells. (See Figure 6)

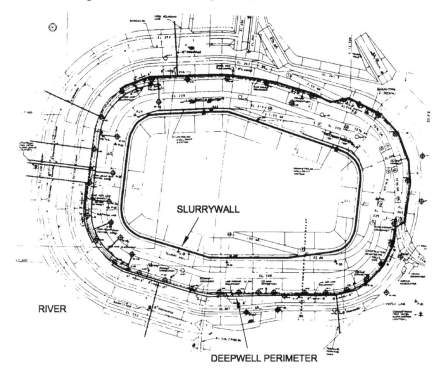

MURRAY HYDRO PROJECT PLAN

FIGURE 6

A full depth slurry trench, from the top of the protective cofferdam to bedrock was considered. At a depth of 25 m (85 feet), removal of the boulders in the lower alluvium would have been difficult. Since the slurry trench cutoff would have to support up to 24 m (80 feet) of differential head, an effective keyway into the weathered upper shale would be necessary, which would be difficult at that depth. Schedule constraints were also a vital consideration. The entire slurry trench would have to be completed before excavation began, adding more than six weeks to an already tight critical path.

A deep well dewatering system was designed to dewater the very permeable aquifer to within 4.6 m (15 feet) of the top of rock. When dewatering a deep excavation such as this to impermeable bedrock, the conventional method has been to lower the water level most of the required distance with wells, and the last 4.6 m (15 feet) is controlled with wellpoints. At Murray, this method presented difficulties. Given the high permeability of the alluvium, and the proximity of the river, much of the water being pumped by wells along the river would transfer to the wellpoints. The required capacity of the wellpoint system was estimated to be very high. An innovative solution for these conditions was developed: a shallow slurry trench 0.61 m (2 feet) wide, keyed into the shale. With this cutoff, sufficient depth of water remained at the deep wells so that they continued functioning as required, after the excavation had been carried to rock.

As soon as a portion of the excavation reached elevation 63 m (208 feet), construction of the slurry trench began using a backhoe. With a trench depth of only 9 m (30 feet), the backhoe had ample reserve power for handling boulders, which ranged up to 1.4 m (4.5 feet) in size. The reserve power also made it possible to rip a keyway in the weathered upper zone of the shale.

Four shallow wells and two sumps were constructed to remove the water stored in the alluvium inside the slurry trench. Excavation to rock proceeded with very little difficulty. Slopes were stable. When equilibrium was reached, leakage through the slurry trench was less than 4 l/s (60 gpm) around a perimeter at toe of slope of 575 m (1900 feet).

Rock excavation below the alluvium proceeded with near vertical slopes. Seepage from the exposed rock joints in the cut resulted in a minimal amount of water, easily handled by localized sumps. The shale was found to be sound, with tight, discontinuous joints, as indicated by the cores. The maximum depth of the rock cut was 11 m (36 feet). Subgrade was reached close to the planned schedule. The excavation was successfully kept dry until construction and backfill was completed up to upstream pool.

Use of Ground Freezing

The technique of ground freezing has been used successfully at difficult geological interfaces where absolute water cut-off is required. It is a process where soil pore water is frozen in-situ to provide strength and impermeability to the frozen soil mass. Ground freezing is most effectively used in difficult ground conditions where the combination of earth support and absolute groundwater cut-off is required and can not be economically provided by other means.

Ground freezing is typically applied to the ground with the installation of a series of closely spaced freeze pipes through which chilled brine is recirculated to draw heat from the ground and freeze the groundwater in place. Frozen soil grows around each freeze pipe like columns of frozen soil. With adequate time and the proper spacing of freeze pipes, the frozen columns will interlock and form either frozen soil walls or frozen soil masses. Reasonable thicknesses of frozen soil walls, generally on the order of 1 to 2 m (4 to 6 feet), have sufficient compressive strength to support a circular shaft excavation to considerable depths.

The effectiveness of ground freezing at difficult geological interfaces comes from the ability to propagate the growth of the freeze without actually disturbing or intruding the virgin ground with equipment or substances. Considerably sized frozen structures or contiguous frozen soil masses can be formed with the installation of relatively small diameter pipes generally spaced 0.9 to 2.5 m (3 to 8 feet) apart. The effectiveness of the technique remains unhindered by difficult ground conditions such as cobbles and boulders which are often found at overburden - rock interfaces, dramatic variances in permeability, rubble fill, and the interference of existing structures, that often prevent the installation of other cut-offs.

Probably the most common application of ground freezing is for the construction of deep shafts that penetrate through water bearing overburden and into rock. Freezing provides both support of the excavation and complete water cutoff through the overburden, weathered rock interface and down into competent rock. With the use of a lesser cut-off method which cannot conform perfectly to the top of rock or to adequate depth within the weathered rock, grouting of the interface is generally performed from within the shaft and often at high water pressures. This exercise becomes riskier with increased depth to the top of rock and increased external water pressure.

In Kansas City[2], histories of previous shaft sinking problems were examined when a new shaft was proposed close to the Missouri River . Records indicated that relatively heavy inflows of water had occurred near the soil and rock contact. After considering various options for groundwater control at the new shaft, the owners and advisors came to the conclusion that freezing presented the least risk and uncertainty. The interface was well defined, with gravelly sand resting directly on unweathered shale rock. Freeze pipes were installed to 145 ft total depth, and a 30 ft penetration into rock. After freeze wall formation, the shaft was excavated and lined into competent rock without incident.

A good example of a geologic terrain ideally suited to ground freezing occurs in Milwaukee[2]. During the 1980's and early 1990's, the Milwaukee Metropolitan Sewerage District constructed a major series of waste water conveyance and storage tunnels through diverse Silurian bedrock formations buried beneath up to 60 m (200 feet) of extremely variable glacial deposits. Many shafts were required to access and service the deep tunnel system, and all of them traversed the geologic interface. The Sewerage District commissioned a series of detailed geotechnical reports revealing potentially hazardous transitions from saturated soils to rock in the shaft excavations. Ground freezing quickly established itself as the preferred technique: the City of Milwaukee has seen more urban ground freezing than any other city in North America. It is interesting to note that freezing was in some cases necessary to repair defective slurry wall and cased drilled shaft installations which were defeated by the combination of poor practices, difficult geology and abundant groundwater. (See Figure 7)

GEO-ENGINEERING FOR UNDERGROUND FACILITIES   967

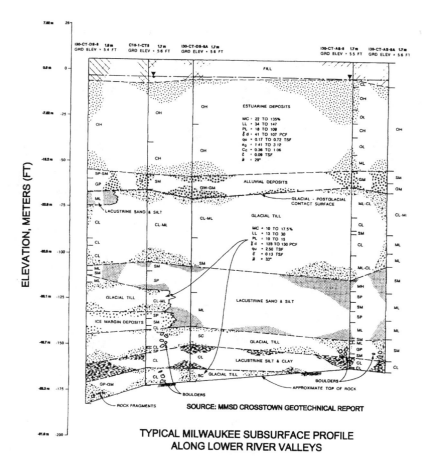

**TYPICAL MILWAUKEE SUBSURFACE PROFILE
ALONG LOWER RIVER VALLEYS**

FIGURE 7

Conclusion

One of the most difficult ground conditions to deal with when it is necessary to excavate below the water table is a glacial interface. This is especially true when the interface consists of a highly permeably stratum such as sand and gravel which is underlain by a much less permeable or impermeable layer such as silts, clays or bedrock. This paper discusses several methods that can be employed to deal with this condition by pumping or eliminating the situation by using various cut-off techniques. These included jet grouting at the diaphragm wall, slurry trenching and ground freezing . The several case histories presented illustrate where these methods have been used successfully.

References

[1]Corwin, Miller and Powers, "Combining Slurry Trench and Dewatering of a Large Deep Excavtion", RETC Proceedings, Chapter 8, 1989

[2]Maishman and Corwin, "Shafts for Water Tunnels; Ground Freesing in Kansas City, MO and New York, NY", RETC, 1993

[3]Schmall, Maishman and Corwin, "Groundwater Control Methods Used in Construction", Construction Congress, ASCE, 511-519, 1995

Characteristics of Urethane-Grouted Sand for In Situ Seepage Control

M. A. Gabr[1], M. ASCE

Abstract

The use of urethane (Scotch-Seal^{Tm}) Chemical Grout 5610 and its constituents in the formation of in-situ seepage barrier by means of permeation grouting is investigated. Seepage and strength properties of urethane-grouted sand are assessed through a laboratory testing program. The optimum grout mix was determined to be 50% grout time extender (GTE), 5% grout reinforcer (GR), and 50% water. Permeation grouting using this mix reduced the hydraulic conductivity of the test sand by 6 orders magnitude from 2×10^{-2} mm/s to 2×10^{-8} mm/s. The hydraulic conductivity was not affected by shrinkage due to drying. After 36 days of air drying, the steady state hydraulic conductivity of grouted sand was measured to be 2×10^{-8} mm/s. Swelling was however significant as the total swelling magnitude of grouted sand specimens was 43% under a normal pressure of 14.7 kN/m^2 and 10% under 147 kN/m^2. On the other hand, the rate of shrinking was 2.8×10^{-4} mm/min with residual increase in volume of approximately 33% (residual increase in volume is defined as percent volume gained during swelling but not recovered during shrinking) under normal pressure of 14.7 kN/m^2. The residual increase in volume after the completion of the dry cycle was approximately 6% under 147 kN/m^2 normal pressure. Furthermore, the shear strength of the grouted sand was significantly reduced as the grouted sand specimens were immersed in water. The total cohesion measured in UU triaxial testing was reduced from 12.4 kPa after 7 days to 5.5 kPa after 36 days immersion in water. In comparison, UU triaxial tests on dry specimens yielded a cohesion of 35 kPa and a friction angle of 24°. The urethane grout seems to be adequate for the formation of a seepage barrier. Special attention however should be paid to the loss of strength as a function of the saturation conditions. It is recommended that, as part of the seepage ingress design, field saturation conditions be simulated in a laboratory testing program. Such testing should be site-specific and the data on the degradation in shear strength as well as volume changes associated with swell/shrink behavior should be incorporated in the design.

---

[1] Associate Professor, Department of Civil Engineering, CB 7908, North Carolina State University, Raleigh, NC 27695-7908

Introduction

The ingress of water into tunnels is undesirable during construction and operation. Control measures utilized to limit water infiltration include grouting to seal fissures and reduce the hydraulic conductivity as presented by Clarke (1982), installation of liners as presented by Schneidkraut (1993), and ground freezing as presented by Suzuki et al (1994). Grouting involves grout mix formulation, injection/delivery methods, and monitoring the injected material to ensure adequate placement. Lining installation includes the construction of a temporary or permanent lining system using cementitious and/or polymeric materials. Ground freezing provides for temporary support as well as excluding groundwater in water-bearing stratum during construction.

Permeation grouting is a viable method of forming in situ barrier for seepage control. McLaren et al. (1986) defined permeation grouting as the process of filling voids and fissures in rocks or soil. The injection pressures must exceed the pressure of water in order to displace it, but must not be high enough to expand or create fissures in the soil mass. Permeation grouting common applications include: seepage control in soils, rocks and excavations, air and fire control in underground mining, increasing bearing capacity of granular soils and shattered rock, slope stability improvements, sewer and water line rehabilitation, and strengthening brick and masonry structures. Sherer (1986) conducted a laboratory study on sands grouted with various chemical grouts, at pressures of approximately 69 kPa, to investigate the formation of a barrier for containment of hazardous waste. Results indicated the ability of urethane grouts to withstand the effect of various chemicals including base, and acid solutions. Lowther and Gabr (1997) presented results on the seepage characteristics of urethane-grouted sand and the associate swelling behavior under saturation conditions in waste environment. Vipulanandan et al (1997) addressed the shrink/swell behavior of urethane-grouted sand specimens and indicated that Calcium Chloride ($CaCl_2$) was effective in reducing the shrinkage of the grout and grouted sand by 40%. At the same time, ($CaCl_2$) however resulted in increasing swelling by 10% to 20%.

Permeation grouting using urethane is investigated as a means of barrier formation for seepage control. In addition to acting as barrier to limit water ingress during tunneling, the grout will lend a cohesion component to the soil to facilitate the tunneling process. Seepage and strength properties of urethane-grouted sand are assessed through a laboratory testing program. The experimental program is comprised of viscosity testing on the grout material, and immersion, hydraulic conductivity, shrink/swell, and strength testing on grouted sand. Results are presented and discussed in relation to the proposed usage for seepage control in tunnels.

Experimental Program

The experimental program presented herein was used to determine the fluid properties of various grout mix ratios and, upon selection of an optimum mix, the seepage and strength characteristics of grouted-sand specimens.

Grout Material

The urethane grout used in this research was Scotch-Seal^{Tm} Brand Chemical Grout 5610 by 3M. Chemical Grout 5610 is most commonly used for water line rehabilitation and sealing leaks and infiltration problems in sewer systems. The grout was comprised of five main ingredients: a urethane prepolymer, acetone, toluene 2,4 - diisocyanate (2,4 TDI), toluene 2,6 - diisocyanate (2,6 TDI), and benzoyl chloride. The latter four compounds are subject to reporting requirements of the Emergency Planning and Community Right-to-Know Act (EPCRA) of 1986 and 40 CFR 372. However, the grout is supplied by 3M in premixed containers. Water (pH between 5 and 9) alone can be used as the activator, however, there are additives available to customize the grout formulation for a specific need. Additives to the urethane grout include :

1. Gel time extender (GTE) which extends the time period before gelation of the grout occurs. The material is relatively inert and is comprised entirely of water and sodium lauryl sulfate.

2. Accelerator additive to decrease the gelation time and is comprised of N,N-dimethyl ethanolamine.

3. Gel reinforcing (GR) agent to increase the strength of the Chemical Grout 5610 and reduce the possibility for gel shrinkage under dry conditions. Additionally, it is used to increase the grouts resistance to freeze/thaw cycling, and chemical and biological degradation.

Test Soil and Sample Preparation

The soil used for the testing program was air dried uniformly graded Ottawa sand. Table 1 shows the geotechnical properties of the sand. The sand was placed, in lifts, into a 305 mm diameter column (see Figure 1) and compacted using a vibrating plate following the procedure described by Lowther and Gabr (1997). The prepared samples had in-place dry unit weight of 16 kN/m^3 which corresponded to a relative density of approximately 50%. The grain size distribution of the test sand is presented in Figure 2 and is shown in relation to the criteria presented by Baker (1982) for groutable soils.

A 19 mm in diameter and 0.4 m long CPVC injection well was used for the injection of the urethane grout into the sand sample. A surcharge of approximately 7 kN/m^2

was applied on the surface of the sand sample using a steel plate. The grout was injected to the soil at the rate of 2 liters/min using a Masterflex peristaltic pump. Approximately 4.5 liters of grout were required to form a grout bulb with 90% grout saturation. After the grout was cured for 24 hours, the grout bulb was removed from the column and specimens were retrieved from the bulb for subsequent testing.

Table 1. Geotechnical Properties of Ungrouted Sand

Property	Value
As-Placed Unit Weight (γd)	16 kN/m^3
Moisture Content (ASTM 2216-92)	0.1%
Specific Gravity (ASTM 854-92)	2.65
Minimum Dry Unit Weigh (γd_{min}) (ASTM 4254-92)	14.9 kN/m^3
Maximum Dry Unit Weigh (γd_{max}) (ASTM 4253-92)	17.3 kN/m^3
Angle of Internal Friction (degree) (ASTM 2850-95)	35
Coefficient of Hydraulic Conductivity (mm/s) (ASTM 5084-92)	2x10^{-2} mm/s

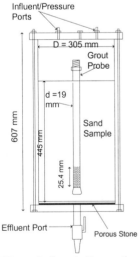

Figure 1. Sample Preparation Column and Injection Well

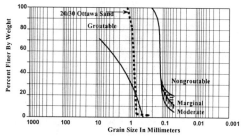

Figure 2. Grain Size Distribution of Test Soil in Relation to Criteria Set by Baker (1982)

Viscosity Results

Viscosity affects the ability to inject the grout into the soil and can be a limiting factor depending on its value. Viscosity testing was conducted in accordance with ASTM D 4016-93 *Standard Test Method for Viscosity of Chemical Grouts by Brookfield Viscometer*. Tests were performed using 550 ml samples prepared in a 600 ml beakers at temperature of 24° C. It should be noted that GTE and 5610 gel have freezing points of approximately 7°C and 15°C, respectively. This observation is of importance since the freezing points may prohibit the use of this grout in certain weather conditions.

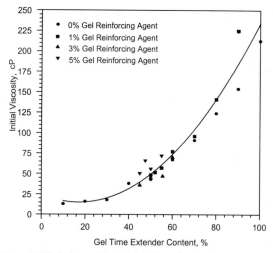

Figure 3. Variation of the Initial Viscosity as a Function of the % Gel Time Extender and % Gel Reinforcing Agent

The increase in the initial viscosity as a function of the gel time extender is shown in Figure 3. No dependancy trend in the viscosity values as a function of the percent gel reinforcing agent was observed despite an initial viscosity value of 150 (centi poise) cP for the gel reinforcing agent (GR). The initial viscosity increased from 12 cP to 50 cP as the gel time extender increase from 10% to 50%. Furthermore, and as shown in Figure 4, the gelation time decreased from 19 minutes to 12 minutes as the percent reinforcing agent was increase from 0% to 5% for a mix containing 50% gel time extender. Gel time requirement should be taken into account when an injection pattern is devised to avoid clogging of the injection lines.

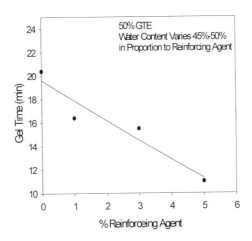

Figure 4. Gelation Time as a Function of % Reinforcing Agent for 50% Gel Time Extender

Immersion Testing

Immersion testing of sand specimens grouted with the various mix ratios was performed in order to discern the extent of degradation and the needed amount of gel reinforcing agent in the activator solution. Specimens grouted using the following three mix ratios were immersed in containers filled with distilled water for a period of five days:

a) 70% GTE: 30% H20: 0% GR
b) 50% GTE: 48% H20: 2% GR
c) 50% GTE: 45% H20: 5% GR

As shown in Figure 5, specimens formed using mix a and mix b experienced extensive degradation while specimens formed using mix c remained relatively intact.

Mix c was designated as the optimum mix for subsequent performance testing and throughout this paper, the 50% GTE: 45% H20: 5% GR grout mix will be referred to as mix c.

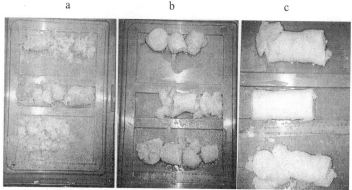

Figure 5. Water Immersion for 5 days -Triplicate Samples: a) 70%GTE, 30% H_2O, 0% Reinforcing Agent b) 50%GTE, 48% H_2O, 2% Reinforcing Agent and c) 50%GTE, 45% H_2O, 5% Reinforcing Agent

Hydraulic Conductivity

The results from the hydraulic conductivity testing performed in accordance to ASTM 5084 on candidate grout mixes were presented by Lowther and Gabr (1997) and indicated that the introduction of grout mix c into the sand resulted in a 6 order of magnitude reduction (2×10^{-2} mm/s to 1×10^{-8} mm/s) in hydraulic conductivity. In addition, data indicated that the hydraulic conductivity of the grouted-sand was not dependent on the composition of the activator solution as the hydraulic conductivity of the grouted sand remained on the order of 10^{-8} mm/s with increasing the gel reinforcing agent (GR) from 0% to 5%.

Further results illustrating the effect of drying on the hydraulic conductivity are presented in Figure 6. Grouted specimens (with mix c) initially yielded a hydraulic conductivity of 1×10^{-8} mm/s. After 36 days of air drying, the steady

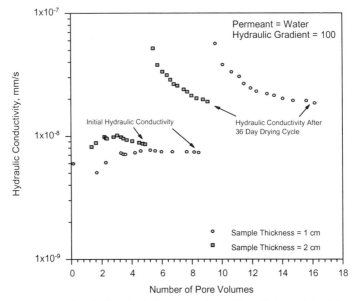

Figure 6. Hydraulic Conductivity as Function of Pore Volume After 36-days Drying Cycle

state hydraulic conductivity was 2×10^{-8} mm/s. These results were consistent for specimen thicknesses of 10 mm and 20 mm. These two thicknesses were used in order to discern the effect of any tension cracks formed during drying (the hypothesis was if tension cracks were formed, an effect on the 10 mm specimens would have been apparent but not necessarily so for the 20 mm specimen).

The absence of visible tension cracks coupled with the consistent k values for the 10 mm and 20 mm indicated no adverse effect of drying on the hydraulic conductivity of the grouted sand and therefore on its ability to function as a seepage barrier. A discussion related to shrink/swell volume change is presented in the next section.

Swelling and Shrinkage

The swell/shrink deformation characteristics of sand specimens, grouted with mix c, were measured under vertical normal pressures of 14.6 kN/m² and 146 kN/m². The tests were performed in conformance with ASTM 4546-90 *Standard Test Method for One-Dimensional Swell or Settlement Potential of Cohesive Soils using Method B*. The specimens used for testing were 63 mm in diameter and 19 mm in thickness. The vertical deformation was monitored during a hydration cycle of 13 days that was followed by a drying cycle of approximately 13 days. Figure 7 shows the swell/shrinkage deformation as a function of time.

After the completion of the swell cycle, the total swelling magnitude under a normal pressure 14.7 kN/m^2 was 43% versus 10% under 147 kN/m^2 of normal pressure. The rate of swelling was on the order of 1.1×10^{-3} mm/min during the first 10,000 minutes and under a normal pressure of 14.7 kN/m^2. Toward the end of the swell cycle, this rate was on the order of 2×10^{-4} mm/min. Under the same normal pressure, the rate of shrinking was 2.8×10^{-4} mm/min with residual increase in volume of approximately 33% (residual increase in volume is defined as

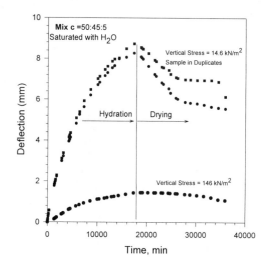

Figure 7. Swell/Shrink Deformation Cycles

percent volume gained during swelling but not recovered during shrinking).

In the case of 147 kN/m^2 normal pressure, residual increase in volume after the completion of the dry cycle was approximately 6%. Vipulanandan et al (1997) presented results on the weight variation of grouted sand specimens as a function of shrink/swell cycles. Results indicated that Calcium Chloride ($CaCl_2$) was effective in reducing the shrinkage of the grout and grouted sand by 40%. At the same time, however, ($CaCl_2$) resulted in increasing swelling by 10% to 20%.

Strength Characteristics

Strength testing included unconsolidated undrained (UU) tests on grouted sand specimens (grouted with mix c) immersed in waster for 1, 7, and 36 days as well as unconfined compression tests on dry specimens. The UU tests were conducted confining pressures of 13.8, 27.6, and 41.3 kPa, respectively, for each time duration. Figure 8 shows the Mohr-Coulomb failure envelopes as a function of immersion duration (recall the ungrouted sand has $\phi = 35°$.) After one day immersion, ϕ was measured equal to $16°$ with c = 16.5 kPa. The test specimens (3 specimens) were partially saturated in this case as volume change took place and therefore the friction angle of $16°$.

After 7 days, the total shear strength parameter was cohesion = 12.4 kPa and zero friction angle. After 36 days of immersion in water, the test specimens exhibited a cohesion of 5.5 kPa and zero friction angle. In comparison, UU triaxial tests on three grouted sand specimens, tested dry, yielded a cohesion of 35 kPa and a friction angle of $24°$.

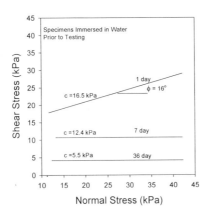

Figure 8. Failure Envelopes for Saturated Specimens: UU Testing

Unconfined compression testing was conducted on dry grouted specimens in accordance with ASTM 4219-89 *Standard Test Method for Unconfined Compressive Strength Index of Chemical Grouted Soils*. The test were conducted on grouted specimens having water content of 16%. The results are shown in Figure 9 and indicated unconfined compressive strength (Su) of 25 kPa after one day and approximately 50 kPa after 7 days. Approximately, 95% of the strength gain was achieved in 7 days. The test specimens failed at an average strain of 8% with a defined failure plane having an angle of $45°$ with the horizontal.

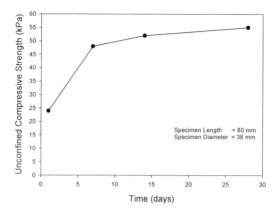

Figure 9. Unconfined Compressive Strength as a Function of Time

Variation in Strength

Eight grouted sand specimens (using mix c grout) were retrieved from various locations within the grout bulb in order to examine the variation in the unconfined strength within the grouted mass. Figure 11 shows the grout bulb before and after specimens retrieval. Specimens were retrieved from the top, center, side, and bottom of the bulb. Figure 12 shows Su distribution as a function of location. The mean Su value for all specimens tested was 35 kPa. Two specimens from the bottom showed an average Su of 42 kPa while two specimens from the middle showed an average value of 28 kPa. The variation in strength as a function of location can be considered insignificant given the statistical variability of the testing technique.

Summary and Conclusions

Permeation grouting using urethane was investigated as a means of forming a barrier for seepage control in tunnels. Seepage and strength properties of urethane-grouted sand were assessed through a laboratory testing program. The experimental program included viscosity testing on the grout material and immersion, hydraulic conductivity, shrink/swell, and strength testing on grouted sand. The optimum grout mix was determined to be 50% GTE, 5%GR, and 50% water. Based on the results presented in this study, the following conclusions are advanced:

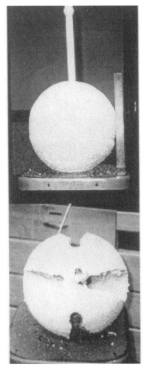

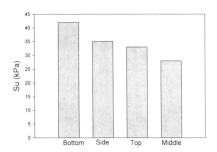

Figure 11. Variation of Su with Specimen Location

Figure 10. Grout Bulb Before and after Specimen Retrieval

1. Permeation grouting using urethane reduced the hydraulic conductivity of the test sand by 6 orders magnitude; from 2×10^{-2} mm/s to 1×10^{-8} mm/s.

2. The hydraulic conductivity was not affected by shrinkage due to drying. After 36 days of air drying, the steady state hydraulic conductivity of grouted sand was measured to be 2×10^{-8} mm/s.

3. Total swelling magnitude of grouted sand specimens was 43% under a normal pressure of 14.7 kN/m^2 and 10% under 147 kN/m^2 normal pressure.

4. The rate of shrinkage was 2.8×10^{-4} mm/min with residual increase in volume of approximately 33% (after the completion of 13 day of dry cycle) under normal pressure of14.7 kN/m^2. In the case of 147 kN/m^2 normal pressure, residual increase in volume after the completion of the dry cycle was approximately 6%.

5. The shear strength of the grouted sand was significantly reduced as the grouted

sand specimens were immersed in water. The total cohesion measured in UU triaxial testing was reduced from 12.4 kPa after 7 days immersion in water to 5.5 kPa after 36 days immersion in water. In comparison, UU triaxial tests on three specimens, tested dry, yielded a cohesion of 35 kPa and a friction angle of 24°

The urethane grout seems to be adequate for the formation of a seepage barrier as the hydraulic conductivity of the grouted sand was on the order of 10^{-8} mm/s. Special attention however should be paid to the loss of strength as a function of the saturation conditions. It is recommended that field saturation conditions be simulated in the laboratory for testing the degradation in shear strength as well as evaluate volume changes associate with swell/shrink behavior. Such testing should be site-specific and can be viewed as a part of the design process.

Acknowledgment

This study was sponsored by the Federal Energy Technology Center. Jason Lowther performed the lab experiments on the grout material and the grouted sand. His assistance is greatly appreciated. David Turner provided assistance with the test equipment and injection probes.

References

American Society of Testing and Materials (1998). "Test Methods for Soil and Rocks", Vol. 4.08, ASTM, Philadelphia, PA.

Baker, Hayward W. (1982) "Planning and Performing Structural Chemical Grouting," Proceedings of the Conference on Grouting in Geotechnical Engineering, ASCE, New Orleans, LA pp. 515-539.

Clarke W. J. (1982) " Performance Characteristics of Acrylate Polymer Grout" Grouting in Geotechnical Engineering, Proceedings, ASCE, New Orleans, pp. 418-432.

Lowther J. and M. A. Gabr (1997) "Permeability and Strength Characteristics of Urethane-Grouted Sand" American Society of Civil Engineers, Geotechnical Special Publication GSP No. 66 , Logan, pp 258-270.

Mclaren R.J., and Balsamo, N.J. (1986). Fly Ash Design Manual for Road and Site Applications. Vol. I and 2. Electric Power Research Institute. GAI Consultants Inc., Monroville, Pittsburgh

Schneidkraut, Peter S. (1993) "Final Tunnel Linings on Oakwood Beach West Branch Interceptor, Staten Island, New York," Proceedings - Rapid Excavation and Tunneling Conference Proceedings of the 1993 Rapid Excavation and

Tunneling Conference Jun 13-17 Boston, p 295-304.

Sherer, R. (1986). "Permeability Behavior of Grouted Soils Subjected to Hazardous Wastes." Master Thesis, University of Cincinnati, Cincinnati, OH.

Suzuki, Shuichi; Mizuno, Ryoichi; Kimura, Katsuhiko (1994) "Construction of Extension Shield Tunnel Using Ground Freezing Method" Proceedings of the 7th International Symposium on Ground Freezing, October 24-28, Nancy, France, A.A. Balkema pp 289-293.

Vipulanandan, C., Jasti V., Magill D., and Mack D. (1997)" Control of Shrinkage and Swelling in Polymeric Grouts and Grouted Sand," American Society of Civil Engineers, Geotechnical Special Publication GSP No. 66 , Logan, pp 271-288.

Successful Urban Rock Tunneling
in Spite of Encountering Gasoline, Cleaning Solvents, and Methane Gas

Raymond W. Henn, M.ASCE[1] and David Rogstad[2]

Abstract

A 4.8 km (3 mile) section of the 32.2 km (20 mile) Dallas Area Rapid Transit's Light Rail Starter System is located underground in Austin Chalk. The tunnel alignments are within the City of Dallas, Texas, and are almost entirely under the North Central Expressway as well as several commercial and residential structures. There were two portals and various shafts along the alignment. The twin bore 6.6 meter (21.5 foot) diameter tunnels were excavated with a "free air" Tunnel Boring Machine (TBM). At two different times and locations along the southbound tunnel alignment all tunneling operations had to be stopped when first, gasoline and cleaning solvents and later methane and other gases were encountered. The first occurrence happened approximately 91.5 meters (300 feet) from the North Portal at the start of the first tunnel drive, when explosive levels of a combination of gasoline and cleaning solvents entered the tunnel face via a system of open faults. In this case, a 2.4 meter (8 foot) diameter exploratory shaft was excavated just in front of the TBM cutterhead to help determine the source of the problem. Additionally, a 3.7 meter by 3.7 meter (12 foot-by-12 foot) exploratory tunnel was driven parallel and then ahead of the TBM tunnel using a permissible (explosion proof) roadheader. The second occurrence was when explosive levels of methane and other gases were encountered approximately 3 km (10,000 feet) into the southbound tunnel drive. The first step was to determine the source of the gases using geologic mapping and geochemical methods. In this case, the TBM's and the tunnel's electric and ventilation systems were upgraded. Both of these occurrences were

---

[1] Raymond W. Henn, M. ASCE, Vice President & Director of Construction Services, Haley & Aldrich, Inc., 110 16th Street, Suite 900, Denver, Colorado 80202.

[2] David Rogstad, Regional Manager, Frontier-Kemper Construction, Inc., 4634 East Marginal Way South, Suite C-140, Seattle, Washington 98134.

clarified and handled as differing site conditions. Tunneling operations were shutdown for a total of 166 days as a result of these two differing site conditions. This paper reviews the general geologic conditions, the construction methods and equipment used, and the TBM and tunnel ventilation and electrical systems modifications made as a result of the unanticipated conditions encountered.

Introduction

The Dallas Area Rapid Transit (DART) began construction of its 32.2 km (20-mile) Light Rail Starter System in 1992. The Starter System consisted of a 4.8 km (3 mile) section of twin 6.6 meter (21.5 foot) diameter tunnels. The $86 million NC-1B Tunnel Contract was awarded to S.A. Healy Company. The work included 10.4 km (34,750 feet) of running tunnel, 21 cross passageways, two underground and one open-cut station excavations, five ventilation shafts with associated chambers and connecting tunnels, two emergency exit shafts with connecting tunnels, and three inclined escalator shafts, as well as cut-and-cover and above-ground structures.

The running tunnels were excavated with a TBM which was assembled and launched from the North Portal. The other underground excavations were completed using five roadheaders of various types and sizes. The construction plan called for the TBM to first excavate the southbound running tunnel, to be disassembled at the South Portal, trucked back to the North Portal, reassembled, and then excavate the northbound running tunnel. After excavation of the North Portal, which would later be the site of the Mockingbird open-cut station, two 18.3 meter (60-foot) long starter tunnels, required to launch the TBM, were excavated by roadheader. Two additional 72 meter (236 foot) long starter (TBM exit) tunnels were excavated at the South Portal.

At two different times and locations along the southbound running tunnel alignment, all tunneling operations had to be stopped when first gasoline and cleaning solvents and later methane and other gases were encountered. The first occurrence happened approximately 91.5 meters (300 feet) from the North Portal at the start of the first tunnel drive, when explosive levels of a combination of gasoline and cleaning solvents entered the tunnel face via a system of open faults. In this case, a 2.4 meter (8 foot) diameter exploratory shaft was excavated just in front of the TBM cutterhead to help determine the source of the problem. Additionally, a 3.7 meter by 3.7 meter (12 foot-by-12 foot) exploratory tunnel was driven parallel (along the northbound running tunnel alignment) and then ahead of the TBM tunnel using a permissible (explosion proof) roadheader. The second stoppage was when explosive levels of methane and other gases were encountered approximately 3 km (10,000 feet) into the southbound tunnel drive. The first step after tunneling was stopped was to determine the source of the gases using geologic mapping and geochemical methods. As a result of the presence of the gases, the TBM's and tunnel's electric and ventilation systems were upgraded. Both of these occurrences were clarified and handled as differing site conditions.

Tunneling operations were shutdown for a total of 166 days as a result of these two differing site conditions.This paper reviews the general geologic conditions, the construction methods and equipment used, and the TBM and tunnel ventilation and electrical systems modifications made as a result of the unanticipated conditions encountered.

Geology

The surface geology consists of fills made up mainly of moderate to high plasticity clays and silty or sandy clays, with varying amounts of construction debris such as rock, brick, concrete, and asphalt fragments, as well as occasional roots and organic materials. At the South Portal, the fill thicknesses ranged up to 1.5 meters (5 feet). The fill materials at the North Portal ranged from 0.3 to 1 meter (1 to 3 feet) in depth. Underlying the fills are terrace deposits along the southern half of the tunnel alignment and residual clay along the northern half of the tunnel alignment.

The terrace deposits were formed by the archaic Trinity River floodplain. The terrace soils are subdivided in two strata, the upper and the lower. Near the South Portal location, the upper stratum extends to a depth of approximately 5.5 to 6.7 meters (18 to 22 feet) below ground surface and consists of moderate to highly plastic soils. The lower stratum which extends down to approximately 12.5 to 14.0 meters (41 to 46 feet) below ground surface consisted of mainly granular soil. The fines in the lower stratum can be classified as low plasticity to non-plastic soils.

The residual soils found at the North Portal are derived from the complete decomposition of the Austin Chalk formation. These residual soils are generally thin, less than 3 meters (10 feet).

The lower terrace and residual soils overlay the Cretaceous Age-sedimentary limestones of the Austin Chalk formation. Generally, a layer of soft weathered limestone overlies the fresh unweathered limestone. The weathered limestone thickness varies from approximately 1 meter (3 feet) at the South Portal to approximately 2 meters (6.5 feet) at the North Portal.

Except for the first approximately 72 meters (236 feet) in each of the two starter (TBM exit) tunnels just north of the South Portal (which was excavated in a mixed-face condition [lower terrace, weathered Austin Chalk, and unweathered Austin Chalk]), the entire 10.6 km (34,750 feet) of running tunnel was excavated in unweathered Austin Chalk.

The unweathered Austin chalk is generally moderately hard and slightly to moderately argillaceous, asphaltic, with traces of fossils and occasional layers of calcareous shale, shaly limestone, bentonitic material, and chalky layers. Traces of pyrite were found in the limestone and in the bentonite. The unconfined compressive strength of the Austin

Chalk ranges from 14 MPa to 21 MPa (from 2,000 psi to 3,000 psi). The Austin Chalk is nearly flat lying, dipping to the east-southeast at about 0.5 degrees. The total thickness of the Austin Chalk in the Dallas area is about 168 meters (550 feet). At the North Portal, the truncated thickness of the Austin Chalk is about 48 meters (150 feet).

Faulting is frequent, with small offsets, typically less than 1.5 meters (5 feet). Joint systems are noticeable in the near subsurface and prominent in the North Portal area. During tunnel excavation over 200 faults were mapped. At only one fault zone were offsets greater than 5 meters (20 feet) observed. Underlying the Austin Chalk to the west is the Eagle Ford Shale and to the east is the Taylor Marl.

Equipment

The 10.6 km (34,750 foot) of running tunnel was excavated using a rebuilt 6.6 meter (21.5 foot) diameter Robbins open-face, mainbeam TBM. The cutterhead was fitted with 68 disc cutters, 425 mm (17-inch) in diameter. The cutterhead rotated at 6.2 rev/min powered by eight, 250 HP electric motors. The four forward thrust cylinders exerted a force of 10,387kN (2,335,000 lbs), resulting in 200kN (45,000 lbs) of thrust per cutter. The TBM was supplied to the job as a "free air" machine since the tunnel was classified as nongassy. Tunnel ventilation was supplied by a 200 HP portal fan which delivered approximately 24 cms (50,000 cfm) of air to a discharge point on the upper deck of the TBM's trailing floor approximately 30 meters (100 feet) to the rear of the cutterhead. Air was supplied via a 1,500 mm (60-inch) diameter flexible duct fan line. A fan line storage canister or cassette, located on the upper deck of the trailing floor, was capable of storing 365 meters (1,200 feet) of flexible fan line. A 175 HP fan was installed at the exhaust side of the storage canister. During normal "blow" ventilation, this fan was shut off and allowed to "free wheel" as air was blown from the portal towards the working face. If it became necessary to reverse the air flow, the portal fan would be shut off and the fan located at the canister would blow from the face towards the portal. This two-fan configuration allows for reversing the air flow when utilizing a flexible fan line.

The TBM trailing floor consisted of 9 decks and was approximately 61 meters (200 feet) long. The system was supplied by Boretech.

Muck removal was by a continuous conveyor system utilizing a 750 mm (30-inch) wide belt. The conveyor capacity was 800 tons per hour, the conveyor speed was 9 meter/sec (800 ft/min). The system was supplied by Continental Conveyor.

Differing Site Condition No. 1 - Liquid Hydrocarbons

On November 17, 1992, TBM tunnel excavation of the southbound running tunnel was stopped when explosive levels of liquid hydrocarbons were encountered flowing into the tunnel from the rock in the area of the TBM cutterhead. Subsequent analysis of the liquid showed it to be a combination of gasoline and cleaning solvents.

The TBM excavation had only advanced 91.5 meters (300 feet) from the face of the North Portal when excavation was stopped. At this point, the crown of the tunnel was approximately seven meters (23 feet) below the ground surface. The exact flow path by which the liquid was entering the tunnel (for example, via a joint, bedding plane, fault, or other geologic feature) could not immediately be determined.

Within a day and a half of the work stoppage, a series of soil borings were drilled from the surface on a grid pattern in front of and to the left and right of the TBM cutterhead. No hydrocarbons were detected in the borings. On the third day of the work stoppage, a 2.4 meter (8 foot) diameter auger drill was used to excavate an exploratory shaft immediately in front of the TBM cutterhead (Figure 1). The bottom of the shaft was excavated to approximately the same elevation as the tunnel invert. The shaft excavation revealed an open fault approximately 200 mm wide by 550 mm high (8 inches by 22 inches) located approximate at spring line on the right rib of the tunnel (Figure 2). The open fault extended into the rock mass as far as the eye could see. Some hand excavation was performed to fully expose the open fault. The strike of the fault was approximately perpendicular to the tunnel center line.

After the discovery of the open fault, a permissible (explosion-proof) roadheader was rented and shipped from Pennsylvania to Dallas for the excavation of a 3.7 meter by 3.7 meter (12 foot-by-12 foot) exploratory tunnel (Figures 1 and 3). The exploratory tunnel was first driven parallel to the southbound running tunnel along the centerline of the northbound running tunnel. Once beyond the TBM cutterhead, the exploratory tunnel was turned and excavation continued along the centerline of the southbound running tunnel in front of the TBM (Figure 1). The decision to excavate an exploratory tunnel was based on the belief that the fault previously discovered in the exploratory shaft could be one in a series of faults. The exploratory tunnel first crossed the previously discovered fault which was exposed in both the right and left sidewalls (ribs) of the exploratory tunnel. After making the turn and excavating along the centerline of the southbound running tunnel, a second open fault was found approximately 15.2 meters (50 feet) from the first fault. The exploratory tunnel was excavated an additional 106.7 meters (350 feet) beyond the second fault, then excavation stopped. The decision to terminate excavation of the exploratory tunnel was based on a detailed geologic investigation which included geologic mapping, analyzing the structural geology, and a review of the geomorphology of the area based on early (1930s) aerial photographs.

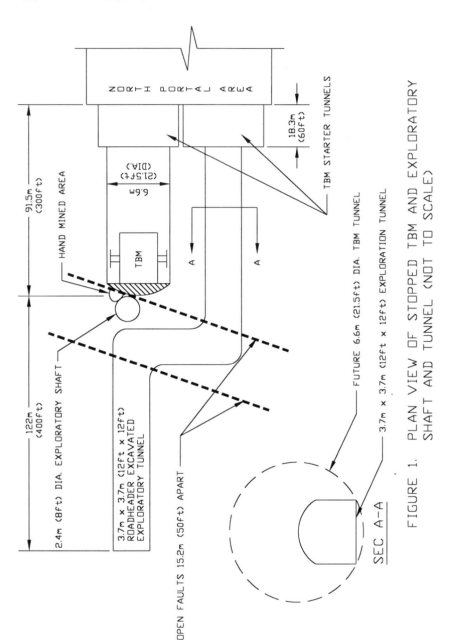

FIGURE 1. PLAN VIEW OF STOPPED TBM AND EXPLORATORY SHAFT AND TUNNEL (NOT TO SCALE)

GEO-ENGINEERING FOR UNDERGROUND FACILITIES 989

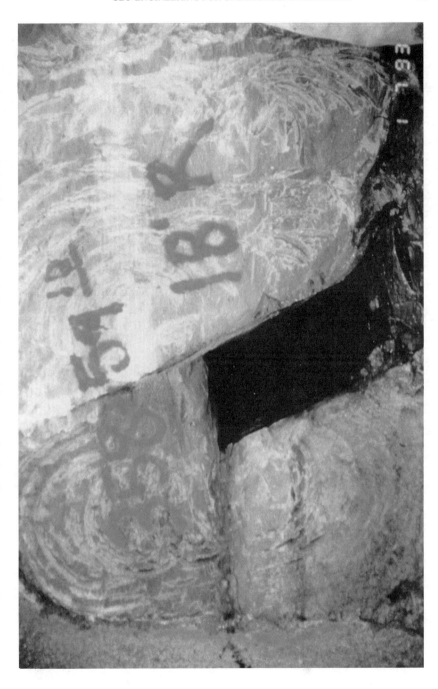

FIGURE 2. Open Fault 200 mm Wide-by-550 mm High (8 inches-by-22 inches)

FIGURE 3. 3.7-by-3.7 (12 ft-by-12 ft) Exploratory Roadheader Tunnel

It is believed the gasoline and cleaning solvent had seeped through the soil and fractured rock from surface industrial facilities over a period of many years. It appeared the gasoline and cleaning solvent were floating on top of a groundwater pool which was located within the open fault. The groundwater pool was completely drained when the TBM excavated into the fault. Once drained, no additional hydrocarbons entered the tunnel via the fault. The fault was monitored for approximately six months. No additional gasoline nor cleaning solvents were detected. The roadheader-driven crossover (connection tunnel) between the northbound and southbound running tunnels was formed and concreted full, after which TBM tunneling of the southbound tunnel was resumed. Shortly after tunneling had been restarted, it was again stopped for 13 days to allow for the installation of steel ring sets and the shotcreting of the 91.5 meters (300 feet) of TBM tunnel that had been excavated just prior to the shutdown. This additional 13-day work stoppage was necessary because the untreated Austin Chalk between the portal and the TBM had "air slaked" over the 74-day shut down, leaving this section of the tunnel unstable. After lengthy negotiations, the Contractor was paid approximate $2.5M for this work stoppage and the related extra work. Tunnel excavation was stopped a total of 87 days as a result of this differing site condition.

Differing Site Condition No. 2 - Gaseous Hydrocarbons (Methane and Other Gases)

On April 13, 1993, tunnel excavation of the southbound running tunnel was again stopped when explosive levels of the methane and other gases were encountered. This stoppage occurred approximately 3 km (10,000 feet) from the North Portal. The gases were first detected by the gas monitoring equipment located on the TBM. The equipment sounded the alarm and the TBM was shutdown. During the shutdown, which lasted several hours, the tunnel air quality was checked manually. The air quality was found to be acceptable and the TBM was restarted. After advancing approximately 2 meters (6 feet), the gas monitoring equipment once again detected methane gas and the TBM was shutdown. After manually testing the air quality, the TBM was again restarted. This cycle of alarms, shutdowns, air testing, and restarts was repeated two more times before it was decided to completely shutdown tunneling and conduct a detailed geologic investigation of the area surrounding the TBM's location.

Within three days of the shutdown, geologic data collection was completed. Based on the data, a geologic map was drawn which showed joints, faults, and discontinuities in the rock. The map depicted geologic conditions from just behind the TBM cutterhead to an area at the rear of the trailing floor, a distance of approximately 61 meters (200 feet). Using the map, joint and fault locations were selected for geochemical testing. At these selected locations, a 13 mm by 200 mm (1/2 inch by 8 inch) hole was drilled. Upon completion of drilling, the hole was corked and left undisturbed for approximately 24 hours. Then, using a hypodermic needle inserted through the cork, an air sample from the drill hole was collected. The air samples were sent to a laboratory for analysis using a gas chromatograph.

The results of the testing showed the presence of methane, ethane, ethene, propane, propene, isobutane, and N-butane.

As a result of the suite of gases detected, two conclusions were reached; one about the depth at which the gases were formed and the second about the relative age of the gases. The first conclusion was that the gases were formed at a depth of between 1.5 km to 2.1 km (5,000 to 7,000 feet). This was based on the presence of the higher order gases which form at higher temperatures and pressures than are required for the formation of methane. The second conclusion was that the gases entering the tunnel were old and weathered (for example, the presence of ethene and propene, which are weathered forms of ethane and propane, respectively). It was further speculated that the gases, originating at depth, migrated to the tunnel horizon through the many joints and faults sometime in the geologic past. The gases were trapped in the joints and faults where they weathered in place. The trapped gas was then released into the tunnel environment, when the TBM excavated through the joints and faults.

Because of the presence of gas, the tunnel was voluntarily reclassified from "free air" to potentially gassy. This reclassification required a modification of the TBM's electrical systems, the tunnel lighting and power supply systems, and the tunnel/TBM ventilation systems. The potentially gassy classification requires that once 20 percent of the lower explosive level (LEL) of methane concentration in the tunnel atmosphere is reached (equates to one percent methane by volume of the atmosphere) all Class I Division 2 rated electrical equipment be switched off (see Figure 4). The remaining essential electrically powered equipment must be rated for operation in a Class I Division I environment (explosive). It was determined that the ventilation, groundwater discharge pumping and lighting systems were essential. All of these systems were upgraded to meet OSHA Class I Division 1 standards for the gassy environment. All other systems were modified and configured to automatically shutdown when 20 percent of the LEL was reached. The ventilation system was the most essential system to maintain because lowering the concentration of gas in the atmosphere is accomplished by dilution with fresh air. The original design of the TBM's ventilation system consisted of a 200 HP portal fan which delivered approximately 24 cms (50,000 cfm) of air to a point on the upper deck of the TBM's trailing floor, approximately 30 meters (100 feet) to the rear of the cutterhead. The air was supplied via a 1,500 mm (60-inch) diameter flexible duct fan line. As part of the modifications a custom-built, elliptical-shaped fiberglass fan line, approximately 19,350 sq cm (3,000 sq in) in cross-section was installed. The fiberglass fan line connected to the discharge end of the existing ventilation system and delivered the air to a discharge point about 10 meters (33 feet) from the TBM cutterhead.

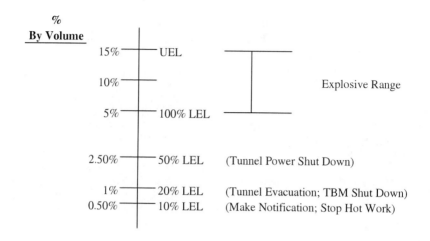

% LEL	% Volume
10%	0.50% (Make Notification; Stop Hot Work)
20%	1.00% (TBM Shut Down; Evacuate)
50%	2.50% (Tunnel Power Shut Down)
100%	5.00% ⎤
	Explosive Range
	15.00% ⎦

Notes:

UEL = Upper Explosive Limit
LEL = Lower Explosive Limit

FIGURE 4. Effects of Methane Levels on Performing Tunnel Excavation

After lengthy negotiation, the Contractor was paid approximately $3M for this work stoppage and related extra work. Tunnel excavation was stopped a total of 79 days as a result of the differing site conditions.

Schedule Recovery

The combined critical path delay caused by the two differing site conditions shutdowns was 166 days. As a result, an extensive acceleration and schedule recovery plan was developed. The project had four 7.6 meter (25-foot) diameter, 42 meter (138-foot) deep ventilation shafts with associated chambers and connecting tunnels. The Contractor had planned to drill the shafts to the full diameter, removing the muck to the surface. All four shafts were excavated in this manner. However, the chambers and connecting tunnels were planned to be excavated by roadheader, with muck removal via the previously excavated TBM running tunnels. The original plan showed the chamber and connecting tunnels on the critical path of the construction schedule. As a result of several meetings between the Contractor, Construction Manager, and the Owner, a change order was issued to the Contractor to remove the chamber and connecting tunnel muck to the surface via the ventilation shafts, thus removing these activities from the critical path. The muck was hoisted to the surface using a crane and muck bucket arrangement. Through this resequencing of work, 105 days of critical path schedule were recovered.

Conclusion

The DART tunnel project contract documents required the use of both Partnering and a Dispute Review Board (DRB). In spite of these two large and complex differing site conditions and various other changes, the project was completed without a single dispute being brought to the DRB. The authors, who both worked at the job site on the project, believe this was due almost entirely to fair and open communications, empowerment of the field staffs, and a team approach of wanting to get the job done on the part of the Contractor, Construction Manager, and the Owner. This team approach was a critical element in successfully completing the project under these less than ideal conditions.

Design Considerations Due to Groundwater Contamination Minnesota Library Access Center

Stephen T. Jansen[1], Bruce D. Wagener[2], and Gordon J. Girtz[3]

Abstract

The University of Minnesota - Minnesota Library Access Center (MLAC) is a unique underground facility designed for long-term storage of paper-based library and archival materials. Extensive groundwater contamination from off-site sources was identified in the project area prior to the start of construction. Key modifications and additions to the initial design have been necessary to control contaminated groundwater during construction and to ensure proper management as part of facility operation. Environmental monitoring and specialized construction techniques have also been required to ensure proper identification, management and disposition of contaminated bedrock and groundwater encountered during construction. The actions being implemented are protective of the facility and its contents, and have been approved by state and local regulatory agencies.

---

[1]M.S., P.G., Vice President, Peer Environmental & Engineering Resources, Inc., 7710 Computer Avenue, Suite 101, Minneapolis, MN 55435; 612-831-3341, 612-831-4552 (fax), sjansen@peerenvironmental.com.

[2]M.C.E., P.E., P.G., Project Engineer, CNA Consulting Engineers, 2800 University Avenue Southeast, Minneapolis, MN 55414; 612-379-8805, 612-370-8160 (fax), bdw@cnaengineers.com.

[3]M.S., M.P.H., M.B.A, Institutional Environmental Affairs Program Director, University of Minnesota, Boynton Health Services, Room W140, 410 Church Street, Minneapolis, MN 55455; 612-626-3595, 612-624-1949 (fax), girtz001@maroon.tc.umn.edu.

Introduction

The Minnesota Library Access Center (MLAC) is being constructed by the University of Minnesota on the West Bank campus, along the western bluff of the Mississippi River, in Minneapolis, Minnesota (see Figure 1). The MLAC is a unique underground facility designed for long-term storage of paper-based library and archival materials. Extensive groundwater contamination from off-site sources was identified in the project area prior to the start of construction. This paper describes the key modifications and additions to the initial design, which have been necessary to control and to properly manage contaminated groundwater during construction as part of facility operation. Also discussed are environmental monitoring and specialized construction techniques, which have been required to ensure proper identification, management and disposition of contaminated bedrock and groundwater encountered during construction.

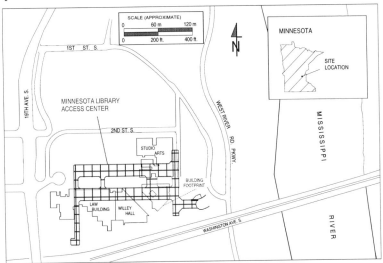

Figure 1 – MLAC project location.

Project Description

Figure 2 shows the layout of the facility components. The finished facility will consist of two caverns which measure 183 meters (m) long by 20 m wide by 7 m high (600 feet [ft] by 65 ft by 23 ft) with associated tunnels, two vertical access shafts, a drive-in portal, and a 7,340 square-meter (79,000 square-foot) 4-story building with two sub-levels. The caverns are constructed 15 to 29 m (50 to 95 ft) below grade within bedrock. Two pre-cast concrete interliners (buildings) will be constructed within the caverns to house library and archival materials. The constant temperature and humidity provided by the mined space will provide a superior and stable condition for storage of paper-based collections.

Construction of the MLAC started in June 1997 and is currently in progress. The planned completion date is November 1999. The project team includes CNA Consulting Engineers (civil engineer), Peer Environmental & Engineering Resources, Inc. (environmental consultant), Stageberg Beyer Sachs, Inc. (architect), the University Environmental Affairs (environmental project management), and McLauchlin Armlin North & Associates (the owner's representative). C.S. McCrossan Construction, Inc. completed the first phase of construction, which included excavation of the underground mined space and access shafts. M.A. Mortenson Company is completing the second phase of construction, which includes construction of the 4-story building, the pre-cast buildings, permanent groundwater drainage systems, and finish work.

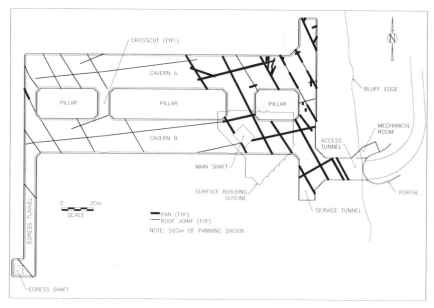

Figure 2 – Layout of the MLAC.

Subsurface Conditions

The geology in the project area consists of approximately 9 m (30 ft) of unconsolidated soil (overburden) underlain by Ordovician-age sedimentary bedrock including the Platteville Limestone (9 m or 30 ft thick), the Glenwood Shale (1.2 m or 4 ft thick), and the St. Peter Sandstone (49 m or 160 ft thick). The underground space for the MLAC is mined within the shale and sandstone. The two vertical access shafts extend from ground surface through the overburden, limestone, and shale, into the sandstone. The foundation for the 4-story building is founded in the uppermost portion of the limestone. Figure 3 shows a subsurface cross-section of the project area. Photograph 1 shows the northern cavern under construction.

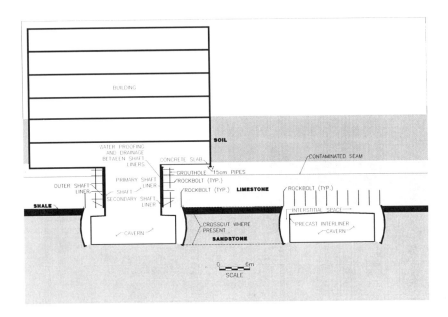

Figure 3 - Subsurface cross-section of project area.

The base of the limestone unit forms the roof of the underground mined space. The upper 4.6 m (15 ft) of the limestone unit is characterized by several thin horizontal bedding planes, at least two of which are believed to be thin volcanic ash (clay) beds. A prominent, laterally continuous bedding plane occurs approximately 2.4 to 3 m (8 to 10 ft) below the top of the limestone unit. Vertical fracture (joint) sets also occur within the limestone. The primary vertical joints trend northwest to southeast, with a secondary set trending roughly northeast-southwest.

Groundwater Conditions

Groundwater immediately below the MLAC project area occurs in the overburden, the limestone, and the sandstone. The shale acts as a confining layer due to its low permeability. Groundwater within the overburden and limestone is perched above the shale. In the limestone unit, groundwater occurs within the horizontal bedding planes and vertical joints. Groundwater within the limestone flows east-southeast, and discharges from outcrops at the river bluff southeast of the project area. The uppermost portion of the sandstone is unsaturated, with groundwater occurring at a depth of 7.6 to 9 m (25 to 30 ft) below the top of the unit. This upper unsaturated interval of the sandstone unit makes construction of the underground space possible. Groundwater flow within the St. Peter Aquifer is eastward, and ultimately discharges into the Mississippi River. Collection of groundwater from the overburden, limestone, and sandstone is required as part of construction and future operation of the MLAC.

Photograph 1 - View of northern cavern looking west from service tunnel.

Environmental Monitoring

Based on the initial geotechnical and environmental investigations, it was anticipated that MLAC construction activities would encounter contaminated groundwater and bedrock. An environmental contingency plan was developed prior to the start of construction. The contingency plan established the necessary procedures for monitoring, sampling and testing, as well as the proper handling and management (storage, treatment and disposal) of contaminated media encountered during construction. The Minnesota Pollution Control Agency (the State regulatory agency) approved the plan, prior to its implementation.

Site investigations as well as ongoing construction monitoring, sampling and testing have identified and better characterized groundwater contamination within the limestone, and to a lesser extent the sandstone. The contamination sources have been determined to originate upgradient of the project area. Contaminants identified in the limestone include semi-volatile organic compounds (primarily polynuclear aromatic hydrocarbons[PAHs]) and volatile organic compounds (VOCs) consistent with coal tar and petroleum materials. In addition, light non-aqueous free phase liquid (LNAPL) has been observed within bedding planes and vertical fractures in the limestone. Contaminants identified within the sandstone include VOCs (primarily chlorinated solvents) and semi-volatile organics.

To date, approximately 1,450 metric tons (1,600 short tons) of contaminated limestone and shale bedrock have been removed and have been or are targeted for off-site thermal treatment. A total of 356 million liters (9.4 million gallons) of contaminated groundwater have been collected by temporary drainage systems and construction dewatering. The majority of this water has been routed through a separator tank, then discharged to the on-site sanitary sewer system under a permit granted by the local Publicly-Owned Treatment Works. A small volume of the water was pretreated using a portable carbon filtration system prior to discharge to the sanitary sewer.

Design/Construction Considerations

The project design includes collection/drainage systems to control groundwater currently affecting the project area. These include a cavern roof panning system, access shaft drain systems, and a foundation drainage system for the 4-story building. Petroleum and PAH compounds present in groundwater may chemically react with polyvinyl chloride (PVC) materials and solvent-welded connections (included in the original design). The nature of the ground water contamination impacting the project dictates, from a design prospective, that primary components of the drainage systems be constructed of high-density polyethylene (HDPE) materials, which are more resistant to PAH and petroleum compounds. Modifications to the cavern roof panning system and associated piping to address biological fouling have also been a significant design issue. The drainage piping from all systems will be routed to sumps, which will allow routing of the water to the on-site sanitary sewer, storm sewer or pretreatment equipment depending upon contaminant concentrations.

The current design also includes hydraulic barrier features that will reduce the flow/volume of contaminated groundwater entering the facility, which in turn will reduce operation and maintenance costs and potential indoor air quality issues resulting from the contamination. These features include vertical cement grout curtains around the building foundation and access shafts, cement grouted rockbolts in the cavern roof, and waterproofing/vapor systems constructed of air/vapor retarding membrane around the building foundation and pre-cast storage buildings.

Collection/Drainage Systems

Cavern Roof Panning System. During design of the project a joint mapping program was conducted to characterize joints in the limestone. The river bluffs along the

Mississippi provided an excellent location to conduct the mapping. Two nearly vertical joint sets were identified as described below:

	Strike Azimuth	Average Joint Spacing
Joint Set 1	140°/320°	12 m (39 ft)
Joint Set 2	60°/240°	7 m (23 ft)

Based on this information and the orientation and square footage of the proposed facility, an estimated 760 m (2,500 ft) of joint trace was expected in the roof of the facility. During construction, 1,060 m (3,480 ft) of joints were actually encountered. Approximately 560 m (1,840 ft) of these joints leaked groundwater and required water control. The groundwater was collected using custom fabricated panning system as described below. Figure 2 shows the joints encountered and where panning was installed underneath the joints.

The panning consisted of 33 centimeter (cm) (13 inch[in]) wide by 12.7 cm (5 in) deep by 5 millimeter (mm) (3/16 in) thick HDPE troughs covered with 2.5 to 7.6 cm (1 to 3 in) of shotcrete. The pans have 5.1 cm (2 in) diameter dropouts spaced at 2.4 m (8 ft) intervals. The dropouts are currently connected to 5.1 cm (2 in) temporary PVC piping. The permanent carrier piping will be 5.1 cm (2 in) HDPE. Access and inspection of the pan interior is provided by a 7.6 cm (3 in) diameter cleanout port installed at a 30 degree angle to the pan. Photograph 3 shows a typical pan installation.

Shotcrete was reinforced with #4 diameter rebar welded to the nuts of 46 cm (18 in) long x 2.5 cm (1 in) diameter rock anchors installed on 1.5 m (5 ft) centers on each side of the pan. The rebar provided additional support to the pan before and during shotcrete placement. Wire ties were also required to provide additional stability of the pan against the impact of shotcrete placement. The gap between the cavern roof and the top edge of the pan was sealed with mortar to prevent migration of shotcrete to the interior of the pan.

During construction, the panning system discharge is being routed to a temporary holding tank, then into the on-site sanitary sewer system. Ultimately, the panning discharge will be routed to a mechanical/sump room within the southern cavern. Valves on the piping will allow for directing the water to the sanitary sewer, storm sewer, or pretreatment equipment, depending upon contaminant concentrations.

The panning system was a necessary part of the original design, regardless of the presence of groundwater contamination. The specially fabricated HDPE panning materials and piping were chosen over lower cost materials such as PVC, due to the material's chemical resistivity. Longitudinally-cut HDPE pipe was also considered for panning, but the pipe sections had a tendency to curl when cut. Cleanouts on the panning and carrier piping were added to allow access to control biological fouling.

Access Shaft Systems. Two access shafts connect the underground mined space to the surface as shown in Figures 2 and 3. Groundwater within the overburden and limestone necessitated the installation of shaft drainage systems.

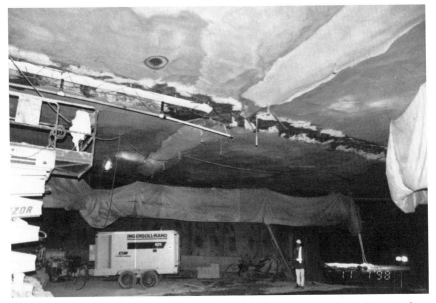

Photograph 2 - Installation of HDPE panning. Note one pan has been shotcreted.

Each shaft structure consists of a primary concrete liner placed directly against the limestone and a secondary concrete liner core that extends from the cavern floor to the top of the 4-story building. Between the two liners, waterproofing and drainage board will be placed to control water that leaks through the primary liner. The waterproofing systems are discussed in a following section of this paper. The drainage board selected was 2.5 cm (1 in) thick molded sheet HDPE drainage panel. The secondary concrete liner is placed directly against the waterproofing/drainage system.

At the base of the shaft liner, water is collected in a gutter, then directed via piping to the mechanical/sump room, then routed to the sanitary sewer, storm sewer or pretreatment equipment depending upon contaminant concentrations.

The primary concrete liners and limestone grouting (discussed in a following section of in this paper) have significantly reduced the leakage of groundwater into the shafts. Currently, approximately 2 to 8 liters/minute (0.5 to 2 gallons per/minute) of water leaks through cracks in the primary liner in both access shafts.

Building Foundation System. The foundation for the 4-story building was excavated to 1 m (3.3 ft) below the top of the limestone unit. A drain tile system was required, due to the occurrence of groundwater near the top of the limestone. The system includes drain tile lines around the building's perimeter along the inside and outside of the footings. A network of drain tiles was also installed under the floor slab inside the foundation walls to control potential upwelling of groundwater along vertical joints, and from the underlying deeper bedding planes.

The drain tile consisted of 15 cm (6 in) diameter perforated PVC pipe with a polypropylene geotextile fabric surrounded by a drainage aggregate. The aggregate was also wrapped with the geotextile fabric. A cross-section of the perimeter drain tile is shown in Figure 4.

The presence of LNAPL (free product) and contaminated groundwater within the building excavation area necessitated several design modifications. These included: 1) recessing the interior drain tile lines within trenches chipped into the limestone substrate, and 2) placement of a mud slab in areas where the limestone was over-excavated to bring the excavation up to the finished elevation, and to eliminate void space under the floor slab where product may accumulate. Additional modifications are also being considered to allow monitoring and future management of organic vapors under the floor slab if necessary.

Operation and Maintenance (O & M). Shortly after installation of temporary panning, biological growth consisting of water mold (tentatively identified as genus Saprolegnia) developed on several wet joints in the underground mined space. The water mold is aerobic and is believed to be utilizing certain groundwater contaminants as a food source. The water mold has been observed to grow on the limestone substrate in patches and stringers along the joints (see Photograph 3). The mold has caused periodic blockage of the temporary panning dropouts and carrier piping. Modifications have been made to the permanent panning system to manage the mold. An O & M program is being developed, which includes panning and piping video logging and flushing with chlorinated water or biocide to control mold growth.

Hydraulic Barrier Features

Vertical Grout Curtains. Limestone was excavated in the building subbasement, the main shaft, and egress shaft. Contaminated groundwater in these excavations could create construction and long term operation and maintenance difficulties. Vertical grout curtains were installed to reduce the flows in these excavations.

After overburden excavation was completed but before rock excavation began, the grout curtains were constructed by pumping cement grout into vertical drill holes in the interior of the excavations and around the excavation perimeter. The grout mixture consisted of four bags of cement per 95 liters (25 gallons) of water. Grouting was specified to be done in stages as follows. In the first stage, the 7.6 cm (3-in) diameter

1004 GEO-ENGINEERING FOR UNDERGROUND FACILITIES

Photograph 3 - View of water mold growing along vertical joint at base of the limestone (after temporary pan has been removed).

holes were drilled in the interior of the excavations to the depth of the contaminated bedding plane, approximately 2.7 m (9 ft) below the top of the formation. Grout was pumped into each hole while adjacent holes were sealed with packers. After all holes were pumped, they were re-drilled to a total depth of approximately 5.2 m (17 ft) and then the grouting procedure was repeated. In the second stage, holes were drilled around the perimeter of the excavations at a spacing of 3 m (10 ft) and grouted in the same manner as the first stage holes. Third stage grout holes were drilled and grouted at a 3 m (10 ft) spacing between the stage 2 holes.

The purpose of grouting in two stages was to seal off the top of the formation, which packer tests performed during the design phase indicated it to be more permeable than the lower portion of the limestone. Once the shallower rock was grouted, grouting of deeper, tighter seams could be ensured because grout would not be lost in the shallower rock.

Cemented Grouted Rockbolts. Roof support in the caverns and tunnels is provided by cement-grouted rockbolts manufactured by Williams Form Engineering, London, Ontario, Canada. The rockbolts are 3.2 m (10.5 ft) long, hollow core #11 bars with a mechanical anchor on the up-the-hole end and a square plate, washer, and nut on the roof surface. One rockbolt was installed for every 4.6 m^2 (50 square ft) of roof area. To limit the area of unsupported roof, the contractor was required to install rockbolts within 4.3 m (14 ft) of the excavation face as the excavations progressed. Each rockbolt was stressed to 307 kilo-Newton (kN) (69,000 pounds) immediately after installation and mechanical anchor expansion. Rockbolts were grouted by pumping slightly expansive cement grout through the rockbolt annulus. Complete grout encapsulation was proven by communication of grout through a short vent tube installed through the rockbolt plate. The vent tube was then pinched off and a pressure of 103 kN/m^2 (15 psi) was maintained for 2 minutes to grout any rock voids.

Hollow cement-grouted rockbolts were chosen over less expensive rockbolt types for two reasons. First, prestressing the rockbolts allowed the rockbolts to take on load before deformation of the rock occurred. This was important to prevent the opening of bedding planes in the rock, which could increase groundwater flow. Second, pressure grouting had the potential to seal open seams in the rock. Out of the 2,200 rockbolts installed, roughly 30 leaked groundwater before grouting. No rockbolts leaked groundwater after grouting.

Building Foundation Water Control System. Following grout curtain installation, groundwater continued to enter the subbasement excavation because the upper 1 m (3 ft) of weathered limestone could not be grouted. Grouting at pressures required to seal the open seams would have lifted the bedrock. The water control system was installed around the lower perimeter of the subbasement excavation to reduce inflow of groundwater (see Figure 4). The system consists of a shotcrete and bentonite barrier placed on the bedrock ledge and on the lower portion of the soil retention system. A shotcrete-smoothing layer was placed against the 0.75 m (2.5 ft) high bedrock ledge, followed by a cant of granular or spray-on bentonite at the toe of the ledge. A bentonite sheet-waterproofing layer was installed at the toe of the ledge and to the base of the soil retention system.

Waterproofing/Vapor Systems. Waterproofing/vapor systems were installed in the following areas: 1) subbasement walls, footings, and floor slab; 2) main shaft and egress shaft secondary walls; 3) cavern pre-cast building roof, walls, footings, and floor slab; and 4) portal structure. Waterproofing installation is continuous from the subbasement walls and floor to the main shaft secondary walls to the cavern pre-cast interliner roof, walls, and floor. Specified design provides a continuous barrier to groundwater and vapors on all below grade construction. The waterproofing membrane chosen was butyl rubber manufactured by Carlisle Coatings, Inc., Sapulpa, Oklahoma, because of its low vapor permeability rating and resistance to the groundwater contaminants. On walls, below floor slabs, and on pre-cast interliner roofs, a 0.15 cm (60-mil) thick membrane is being installed. Below footings and on portal structure horizontal surfaces, a 0.30 cm (120-mil) thick membrane is being installed.

Additional Water Control Measures

The state regulatory agency has requested that an up-gradient groundwater control system be installed within the contaminated limestone unit to supplement the original and modified MLAC project design features including water collection and hydraulic barriers. A number of different water control alternatives have been evaluated. An additional hydrogeologic investigation is currently in progress to determine the need for and type of additional water control measures that may be warranted.

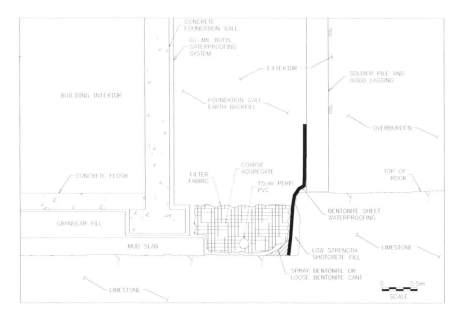

Figure 4 - Building foundation water control system.

Conclusions

A major underground library storage facility is being successfully completed in a project area affected by groundwater contamination. This facility will allow the University to fully utilize its surface and subsurface real estate, and will create a more compact campus with considerable growth potential. A unique combination of collection/drainage and hydraulic barrier features has been implemented as part of the MLAC project design. These features have been modifications to the original design, or have been added during construction to address the groundwater contamination. The State regulatory agency has deemed the actions described herein as protective of the facility and its contents. The State has requested additional water control measures upgradient of the MLAC project, which the project team is currently evaluating.

An Update on Deep Mixing Technology Worldwide

D. A. Bruce[1], M. ASCE and M. E. C. Bruce[2], M. ASCE

Abstract

The authors have recently completed the first part of a two-part report on the Deep Mixing Method. The first volume deals with the chronology of evolution, applications, classification of the various methods, the international market, and possible links to expansion in the U.S. The second volume, due later in 1999, is to deal with treated soil properties and QA/QC issues. This paper provides a synopsis of the first volume.

Background

The Federal Highway Administration (FHWA) commissioned a global state of practice review from the authors following the Tokyo Conference on Deep Mixing and Jet Grouting in May 1996. That conference may be judged by our profession in retrospect to be one of the more significant expressions of technical knowledge on a narrow range of subjects to have impacted current and future U.S. specialty geotechnical construction practice. Not only were the historical leaders of technology from Japan and Scandinavia present, but there was a significant proportion of attendees from North America and Europe to ensure that the rich volume of data openly presented would have a global impact in specialty geotechnical engineering circles. For the first time, these specialists communicated freely and openly in the English language about retrospective, introspective and prospective aspects of the industry. This was particularly welcome from the Japanese and Scandinavian practitioners, whose fundamental and excellent research and development findings had hitherto been available largely in their native language. For example, Terashi (1997) reported over 200 technical papers on Deep Mixing Method (DMM) were published each year in the Japanese language alone, so rendering their contents beyond the scope of occidental readers.

[1]ECO Geosystems, L.P., Pittsburgh, PA, USA
[2]geotechnica, s.a., Pittsburgh, PA, USA

In the United States, there is a rapidly growing demand in a variety of markets for the benefits that DMM can provide. Mass ground treatment schemes in Boston, MA for tunneling, can be compared with earth retention projects in Milwaukee, WI, seismic retrofits for dams in the Rockies, and deep foundation systems in the San Francisco Bay Area with respect to variety, intensity and technological ingenuity. Yang (1997), provided an informed and comprehensive review.

Most recently, the University of Wisconsin, Milwaukee has started to offer a short course on DMM, at which the leading users and practitioners in the country share their knowledge - and compare their skills - with their peers, while deep mixing specialists are increasingly represented at ASCE GeoInstitute conferences (e.g., Logan, 1997; and Boston, 1998) both as exhibitors and authors. The Deep Foundations Institute has, as of October 1998, established a Deep Mixing Subcommittee, one of whose goals is to establish technical links with the major national and trade organizations worldwide. National groups are hosting specialty conferences, such as the Swedes in October, 1999 in Stockholm.

In the face of this exponential growth of knowledge, technology, and applications, it is timely that the first part of the FHWA state of practice should emerge. It attempts to provide a summary of the technology as it enters the next century. This paper provides a brief synopsis, which should supplement and update the authors' previous overview on the subject (Bruce et al., 1998a and b). The paper follows the same structure as the FHWA Report but focuses only on the more recent developments. A second FHWA-sponsored volume, dealing with treated soil properties and QA/QC issues, is scheduled for later in 1999.

Scope and Definition

DMM remains an in situ soil treatment technology whereby the soil is blended with cementitious and or other materials, either in dry or wet (slurry grout) form. The greatest amount of the work conducted globally involves vertical penetration by one or a number of mixing shafts to create discrete columns or panels. Depending on the application, these elements may be constructed to overlap to provide a variety of geometries of treated soil. The FHWA study addresses only these vertical, rotary methods.

However, there are an increasing number of methods under development which create either mass treatment by using inclined auger or conveyor technology or by using vertical beams with lateral jetting capabilities to provide thin, but continuous in situ membranes. Such applications mainly serve the environmental market - containment fixation, and retention, respectively - and are typically viable to relatively shallow depths (10m). Nevertheless, future studies of DMM should entertain these methods alongside our conventional groups of methodologies.

Historical Evolution

The FHWA study listed some 82 events considered significant in the growth of DMM since the original U.S. concept in 1954, and the independent Japanese and Scandinavian exploitations in 1967. Most of these key events have occurred in the last decade, emphasizing the ever increasing rate of development by contractors, consultants, and owners - including federal agencies in the case of Japan, China, France, Sweden, and Finland. This theme is revisited in later sections.

Another mark of the significance of deep mixing as an engineering tool worthy of retrospective study is the series of reviews by Porbaha and coworkers (1998 a and b), sponsored by the Science and Technology Agency of Japan which closely detail both commercial and research progressions in these last 25 years.

Applications

The main groups of applications remain:

- Hydraulic cutoff walls
- Excavation support walls
- Ground treatment
- Liquefaction mitigation
- In situ reinforcement, piles and gravity walls
- Environmental remediation.

Globally, the novelty now arises when local methods are used for new applications, or when established methods are used in new geographic areas, often by contractors who are seeking to develop their own variant of the method in response to a particular project's challenges. Thus we may anticipate in the next decade's technical press a plethora of case histories dealing with environmental and liquefaction mitigation, and in situ earth reinforcement from practitioners in countries as diverse as the U.K., Indonesia, and Australia, based on the authors' current project awareness.

The viability, both technically and commercially of DMM in its various potential applications and settings will continue to be challenged by solutions based on other technologies and cultural preferences, and rightly so: deep mixing is not the panacea for all specialty geotechnical problems. However, when the goal is ground treatment, improvement or retention, the ground and site are relatively unobstructed, and the depth is limited to about 40m, then deep mixing will most probably be a viable option in countries with easy commercial access to the technology.

Classification of Methods

A total of 24 different methods - mostly fully operational and patented - were identified by the FHWA survey (Figure 1). The classification adopted is based on the nature of the "binder" (grout, or dry); the method of soil blending (rotary alone, or rotary with jet assistance); and the location at which most of the soil/binder blending occurs (along the shaft of a long auger, or only at the mixing tool located at the end of a rod). This classification, of course, only applies to those deep mixing systems employing vertical mixing principles (as discussed above). A new "arm" to this classification will be necessary to accommodate the "mass", or "lateral jetting" variants.

The authors have received peer reviews of this proposed classification from specialists worldwide, and have monitored global practice for three years to date. The generic classification of Figure 1 has in patent terms, "satisfied" these challenges, and so is considered appropriate.

Regarding the future, the constructional developmental trends are towards improving the quality of the mixing process (e.g., Systems 11 and 12); using less expensive binder components (e.g., System 6-FGC); obtaining larger diameter of treatment via jet assistance (e.g., Systems 18 and 21); and improving the level of computer assisted control (most systems, but especially in the U.S., Systems 3, 20, 23, and 24).

International Perspective

In the United States, the authors found 11 contractors who execute, or have the resources to execute, DMM. One group, comprising Condon-Johnson Associates, GeoCon, Hayward Baker, Inquip Associates, Millgard Corporation, and Terra Constructors, are essentially U.S. owned, offer other geotechnical and/or environmental services, and appear to have no involvement in, or dependency upon, foreign resources or licenses. The second group, including Raito, SCC Technology, SMW Seiko, Stabilator USA and Trevi ICOS, are either wholly owned U.S. subsidiaries of foreign companies, or operate only and exclusively under foreign license in the U.S. These companies offer the variants shown against their names in Figure 1.

Following first use in 1986, it has only been since 1992 that DMM has achieved national prominence, with most contractors offering services only since the mid 1990s. It is estimated that the annual volume of DMM conducted for geotechnical applications is currently running between $50 and 80 million with the bulk of the work being conducted in Boston, and the San Francisco Bay Area. The annual market for environmental applications is around $20 million, and unlike the faster growing geotechnical market potential, is growing at only 5 to 10 percent per annum. These estimates are based on interviews with practitioners country-wide.

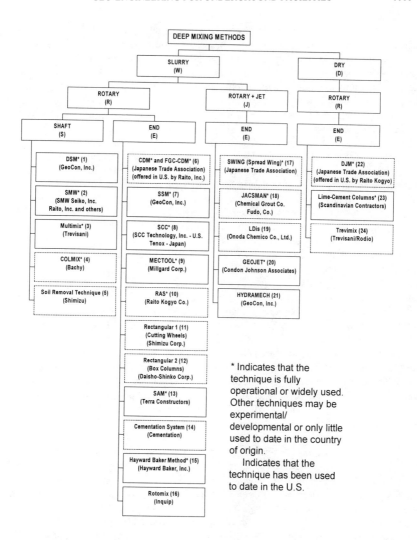

Figure 1. Classification of Deep Mixing Methods based on "binder" (Wet/Dry); penetration/mixing principle (Rotary/Jet); and location of mixing action (Shaft/End).

The level of DMM activity in Japan remains by far the highest in the world. Building upon the Government-sponsored research work in 1967, full scale DMM systems have been used commercially since 1974, and appear to have grown especially quickly in annual volume since the early 1980s. The Japanese contractors, in close cooperation with the Federal Government, manufacturers, suppliers and consultants have continued to develop and enhance DMM technology in response to technical and commercial challenges. Trade associations, often comprising dozens of members, serve the technologies of CDM, DJM, SWING, and Mixed Walls, for example. These associations organize annual conferences and collect and publish data on market volume: a service not yet available in the U.S. Data on annual volumes of ground treated were published by Bruce et al. (1998) from which it may be inferred that the annual DMM volume in Japan is valued at $250 to 500 million, most of it related directly to seismic mitigation.

Activity is increasing in China, especially for harbor and port development at estuarine cities, and has traditionally involved Japanese input. DMM has also been used elsewhere in S.E. Asia, including Taiwan (Liao et al., 1992) and Hong Kong. The total regional market outside of Japan is smaller than in Japan, but exact figures are not readily available.

Like the Japanese, the Swedes began researching in 1967 via a series of laboratory and field tests. The original coworkers included the Swedish Geotechnical Institute, private consultants, and piling companies. This cooperative model has endured, and a wealth of information has been generated about the technical and commercial aspects of the Lime Cement column method in Sweden, and more recently by similar groups in Finland. Their focus remains on ground improvement and pile/soil interaction solutions for very soft, highly compressible clayey and/or organic soils. Therefore, and again in contrast with typical Japanese and U.S. practice, relatively light and mobile equipment has been developed producing single columns only up to 0.8m in diameter, to relatively shallow depths (typically not more than 25m) and with low unconfined compressive strengths (0.15-0.20 MPa).

Market growth has been particularly rapid since 1989 in both countries where the combined volume is around $30 to 40 million annually. Contractors from Sweden and Finland are also active in other countries including Norway, the Baltic States, Holland, U.K., and Hong Kong (in addition to one company in the U.S.) The region's commitment to DMM development is clearly underlined by the formation, in Sweden, of the Deep Stabilization Research Center, and in Finland, a National Structures Research Programme, both in 1995. In each case, national resources have been assembled - similar to the Japanese model - and the findings are to be published in 2000 and 2001, respectively.

The Future of DMM in the United States

Notwithstanding the benefits and advantages which contemporary DMM techniques can offer, there remain a number of factors, often interrelated, which act as potential barriers to market entry for prospective contractors, and/or controls over market growth. These include:

- Demand for the product: given the national trends towards urban construction and redevelopment, seismic retrofit and environmental clear up - all challenges to be solved in situ - then demand for DMM will continue to increase.
- Awareness of the product: a wider range of active specialty contractors and consultants, more prolific technical publications, short courses and the coincidence of several high profile DMM projects nationwide have combined to elevate awareness of DMM in general engineering circles, and will so continue to increase demand.
- Bidding methods/responsibility for performance: the authors believe that the interests of a rapidly developing and complex technology like DMM in the U.S. are best served by "design-build" concepts. Thus, the rate of growth of DMM will be influenced strongly by the rate at which innovative contract procurement and administration vehicles are adopted nationwide.
- Geotechnical limitations: DMM has been developed to treat relatively soft, unobstructed soils and fills in sites with good access. There are other practical limitations as to depth, strength, and durability of the treatment. Extreme care should be taken not to overextend the limits of DMM capability without due regard to a true appreciation of the fundamentals of its evolution. Otherwise, inappropriately applied, designed, and constructed work will lead to owner disappointment, or worse.
- Technology protection: most of the 24 methods show on Figure 1 are protected in their technology by Patent or similar. Thus new potential contractors must either invent their own system, or acquire a foreign license. The latter seems more realistic, given the timetables and costs involved in conducting basic research and development.
- Capital cost of startup: given the high levels of technical sophistication, and large physical scale of most systems, startup costs are high. In addition, the larger projects may require several machines and so committed capital expenditures may easily rise to several million dollars. The equipment must also be regularly maintained and upgraded leading to the general conclusion that DMM is a "cash hungry" technology for the contractors who offer it - although the potential return on investment is high. Thus, the field of potential contractors is practically limited by the levels of their own financial resources.

Final Remarks

The many different types of DMM continue to undergo major developments and have experienced notable international success for over two decades. These techniques have rightly become trusted, valuable, and competitive engineering tools for treating, improving, and retaining soils in a wide range of applications.

However, to maintain the rapid growth of the last ten years, DMM must continue to be applied appropriately, designed correctly, constructed efficiently, and restricted sensibly to the natural restraints of soil conditions and mechanical capability. Despite its market potential, it remains relatively costly for contractors to acquire, and so the number of potential competitors will probably remain relatively small. Following this logic, the authors conclude that DMM may well become a commodity product - but a product which can be provided only by a relatively small number of producers. The comparison with the circumstances of the petroleum industry is clear, but with the positive observation that the reserves of the DMM producers are not, in the fundamental sense, finite.

Acknowledgments

The work described herein was funded by the FHWA under the director of Dr. Al DiMillio. The FHWA report was reviewed by and contributed to by a large number of specialists worldwide, too numerous to mention. Special thanks, however, to Pete, Dan, Osamu, George, Maasaki, James, Goran, Johan, and David.

References

Bruce, D.A. M.E.C. Bruce, and A.F. DiMillio. (1998a). "Deep Mixing Method: A Global Perspective", *Civil Engineering*, Vol. 68, No. 12, December, pp. 38-41.

Bruce, D.A. M.E.C. Bruce, and A.F. DiMillio. (1998a). "Deep Mixing Method: A Global Perspective", *Soil Improvement for Big Digs*. American Society of Civil Engineers, Annual Conference, Boston, MA, October 18-22, Geotechnical Specialty Publication No. 81, pp. 1-26.

Federal Highway Administration. (1998). "An Introduction to the Deep Mixing Methods as used in Geotechnical Applications", Prepared by ECO Geosystems, Inc. Vol. 1., November.

Porbaha, A. (1998). "State of the art in deep mixing technology: part I: basic concepts and overview." *Ground Improvement*, Journal of the International Society of Soil Mechanics and Geotechnical Engineering (TC-17), Thomas Telford, Vol. 2, No.2, pp. 81-92.

Porbaha, A., H. Tanaka, and M. Kobayashi. (1998). "State of the art in deep mixing technology: part II: applications." *Ground Improvement*, Journal of the International Society of Soil Mechanics and Geotechnical Engineering (TC-17), Thomas Telford, Vol. 2, No.3, pp. 125-139.

Terashi, M. (1997). Personal communication.

Jet Grouting in Cohesive Soils

Lawrence B. Gruner, M. ASCE[1]

Abstract

The specialty practice of jet grouting has been used successfully for two decades in the United States, primarily for the stabilization of loose sands and gravels for a wide variety of tunneling, excavation support, and underpinning applications. Recently, jet grouting has been used in clays and cohesive soils, providing similar results as those found in unconsolidated and cohesionless soils, due primarily to recent advances in pumping and grout delivery systems.

Two case studies of jetting in cohesive soils are explored herein including analyses of the soil conditions, jetting parameters, and grout mixtures. The first project involved the underpinning of the Argentine Naval Commission building in Washington, DC. More than 230 jet grout columns were installed to depths of up to 18m (60') with nominal column diameters of 76.2 cm (30"). The majority of these columns were installed in low-headroom conditions.

The second case study involves the use of jet grouting for pre-excavation support and water control prior to a microtunneling operation in a heavy industrial site in New Jersey. Here, the jetting was used to provide underpinning support for critical foundations in and around the path of the proposed microtunnel, and to create waterproof jacking and receiving pits for the tunneling machine. The latter system was used in lieu of conventional sheeting and dewatering methods due to the fact that a known contaminant plume was identified near the construction site. The overlapping sidewalls and horizontal bottom plug created a dry "bathtub" effect in these pits, and the plume was contained.

Lawrence Gruner serves as the Technical Director of the Geotechnical Construction Group of Layne Christensen Company, headquartered in Milwaukee, WI.

The Argentine Naval Commission Building

Halfway between the White House and the Capitol Building, on Pennsylvania Avenue in downtown Washington, DC, lies another limestone-clad building, belonging to the Argentine Naval Commission. The ANC Building was built in two segments and these components are referred to as the East Wing and the West Wing. The original building, now the West Wing, was built 100 years ago in 1898. The East Wing was added in 1921.

The foundation of the West Wing consists of brick masonry load-bearing walls, masonry arch floors, and steel framing. The East Wing has one-way concrete floor joists, beams, and columns, bearing on brick load-bearing masonry walls. The combined building stands four stories high with a steeply pitched wood framed roof with red slate shingles.

The subsurface soils typically consist of a soft to medium stiff gray and brown clayey silt and silty clay with little to fine medium sand, ranging in thickness from 8.4-10.5m (28-35'). There is also a thinner layer of silty sand mixed with fill directly beneath the existing foundation. Below the stiff clay is a lense of medium dense to very dense brown fine to coarse gravelly sand (Haley & Aldrich, 1992).

Two episodes of subterranean construction occurred in adjacent properties in 1984-85 (Phase I) and in 1988-89 (Phase II). During these construction episodes, the ANC Building experienced horizontal and vertical movements. It has been determined that an inadequate temporary support system was used to construct the building closest to the ANC foundation during Phase I activities (Haley & Aldrich, 1992).

During Phase II construction of the adjacent building, tiebacks extending beneath the foundations of the East Wing of the ANC Building caused the loss of soil during their installation, and an accelerated settlement was noticed. Other factors during the Phase II construction that caused settlement were the lateral movement of the excavation support system and a loss of ground behind the lagging.

Ultimately, the East Wing rotated away from the original West Wing with the largest gaps exhibited in the attic between the two wings of the ANC Building. The dilemma that now faced the owner's structural and geotechnical engineers was how to arrest the settlement of this complex foundation. A variety of options were explored over a number of years and jet grouting was finally determined to provide the best engineering solution to this dilemma. Foundation repairs were not performed immediately due to litigation that extended over a period of several years.

Haley & Aldrich's Silver Spring, Maryland office prepared a specification for the jet grouting procedure. The intent of the jet grouting was to stabilize the bearing walls and foundations of the structure by the creation of in-situ soilcrete columns extending from the bottom of the existing foundations to depths of approximately sixty feet from grade.

The columns were specified to be a nominal 76.2cm (30") diameter and the in situ soilcrete was to be a minimum of 3450 kPa (500 PSI) in a 28 day unconfined compressive grout strength. These columns were to achieve an expanded diameter in the upper 1.5m (5') of the column. This expanded cross section would ensure the intimate contact between the top of the soilcrete column and the bottom of the existing limestone block foundation. Although the design loads of these soilcrete columns was not specified in the bidding documents, it was assumed that if the columns were installed meeting the above parameters, that future settlement of the ANC Building would be negligible.

Fonditek International, a division of the Layne Christensen Company, approached this jetting project by proposing the use of the single fluid method. This method was chosen based on the project requirements combined with the complexity of the existing foundation. It was felt that more control could be implemented in the underpinning of the existing foundation, composed of individual blocks with poorly mortared joints, by using the single fluid method. Other advantages of using the single fluid system include higher soilcrete strengths and a reduction in the amount of spoils. The amount of grout spoils would prove to be a critical factor in the low-headroom jetting performed in the basement areas.

The work was performed with a Puntel PX-1200 drill for the exterior jetting and the Puntel PX-609 for the interior work. The grout plant consisted of a Puntel PTMA-1000 mixing plant, a Tecniwell TW351 jetting pump, and a 42 ton capacity cement silo. The exterior drilling was performed in one stroke, using a 18.6m (62') drill steel and grout monitor, while the interior work had to be performed using .9m (3') drill steels due to limited headroom conditions.

The following jet grouting parameters were utilized in the test section and were continued in the production grouting work:

Grout Mix Design

- Portland Cement, ASTM C-150, Type II; no additives
- Compressive strength of soilcrete at 28 days (fc') = 3459 kPa (500 PSI) (minimum)

Quality Control

- Grout sample cylinders collected daily and sent for independent testing at independent laboratories
- Specific gravity of grout for each batch checked by the use of mud balance.
- Jet grouting parameters were monitored continuously during column construction. Coring of production columns was performed at selected locations as directed by the engineer. Core samples were sent for independent laboratory testing and results presented as for the CCT program.

- Test pits were excavated after the installation of the initial production columns to evaluate the effect of the jetting on the East Wall and to assess the need for further foundation wall stabilization.

Building Monitoring Plan

Survey monitoring points and crack gauges were installed at strategic locations on the interior and the exterior of the building. Fonditek collected daily movement readings at monitoring points adjacent to column locations by optical survey. These readings were recorded and submitted to the client and to the owner's engineer on a daily basis. All drilling and grouting was done with a laser level in place, in the area of the jetting activities (Micciche et al., 1998).

Movement of the existing building was minimal during the jet grouting procedures. Monitoring of the building has been continued, by others, subsequent to the completion of the jet grouting with no significant movement observed.

Enlarged Column Diameter Plan

Enlarged column diameters were required for a vertical distance of approximately 1.5m (5') extending from the bottom of each existing foundation downward. This was accomplished by the drilling and jetting of several short "stub" columns, oriented on a variety of shallow angles beneath the footing, in an effort to create a soilcrete pile cap for each vertical jet grouted column (see Figures 1 & 2).

Jet Grouting Production Columns

Upon completion of a test grouting section, production columns were installed on the interior and along the exterior walls of the ANC building. Since approximately one half of the ground floor was occupied by a local banking firm and the building was located in a busy downtown district, all work had to be performed at night.

Outside of the building, pavers surrounded the building and extended on all sides to the neighboring street curbs. Special care had to be taken to provide sufficient trenches to handle the drilling and jet grouting spoils. In the basement, spoils were pumped from trenches cut in the floor slab into 55 gallon drums, and transported to the street by elevator. The drums were then disposed of off site. The single fluid method of jet grouting yields the least amount of spoils (vs. the double and triple fluid methods) due to the fact that a majority of the cement grout injected mixes with the in-situ soils rather than returning to the surface.

Figure 1
The underpinning of the ANC Building by Jet Grouting

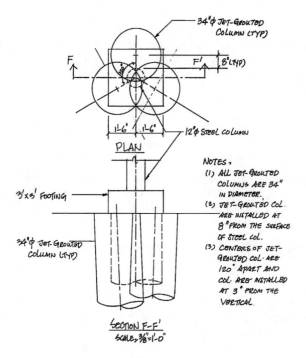

Figure 2
Typical jet grouting pattern beneath existing column detail - ANC Bldg

A comparison of the three main jet grouting systems can be summarized as follows:
- Single fluid system (F1) - The fluid is a cement-based grout where the jet simultaneously erodes and injects through horizontal ports in the monitor (drill steel). It involves only a partial replacement of the soil.
- Two fluid system (F2) - This method similarly injects a jet of high pressure cement - based grout, but includes a cone of compressed air around the grout stream, accelerating the cement particles. F2 typically creates a larger column diameter than F1 and gives a higher degree of soil replacement (and therefore spoils, as well).
- Three fluid system (F3) - Here a minimum of two separate jets are used to create the column. The upper jet combines the forces of water and and an envelope of compressed air used to erode and excavate the soil around the drill rod. A lower jet injects a cement-based grout, under lower pressure, filling the void created by the upper jet. Theoretically, this method creates the largest diameter column, but generates the greatest volume of spoils of the three methods (Bruce et al., 1994).

The results of the jet grouted in-situ soilcrete testing was also a success. Random core samples were taken in the completed jet grouted columns and 28 day unconfined compressive soilcrete strengths exceeded the 3450 kPa (500 PSI) minimum required by the specifications. Many of the cores tested exceeded 6900 kPa (1000 PSI) in just 14 days. Net strengths ranged from 4250 to 16740 kPa (616 to 2426 PSI) (corrected strength) in 12 to 14 day breaks (see Figure 3).

Sample ID/Depth (m/ft)	Age (days)	Area mm²/(sq.in.)	Load Kg(lbs) kPa/(psi)	Net Strength	Correction Factor kPa/(psi)	Corrected Strength
TC1-5/ (16.5m/55')	12	2567/3.98	4050/9000	15600/2261	0.99	15442/2238
TC1-2A/ (5.1m/17')	12	2709/4.2	4680/1040	17084/2476	0.98	16739/2426
TC1-1B/ (3m/10')	12	2709/4.2	2380/5290	8625/1250	0.98	8452/1225
TC2-4A/ (8.1m/27')	13	2857/4.43	1260/2800	4360/632	0.975	4250/616
TC2-3B/ (6.6m/22')	13	2857/4.43	2362/5250	8176/1185	0.975	7969/1155
TC2-2A/ (3.9m/13')	13	2857/4.43	4252/9450	14717/2133	0.975	14352/2080
TC2-1A/ (2.7m/9')	13	2857/4.43	4140/9200	14331/2077	0.975	13972/2025

Notes:
- Design strength is 3450 kPa (500 PSI).
- Sampling was performed by others.
- Compressive strength was performed in accordance with ASTM C-42.
- Type of specimen: Core sample (approx. 5.08cm [2"] diameter).
- Laboratory testing performed by Geotech Engineers, Inc., Beltsville, MD

Figure 3
Representative in-situ soilcrete strengths created by jet grouting in clay soils, ANC Bldg, Washington, DC (Hon et al., 1997)

Jet Grouting and Microtunneling in NJ

In an aged pharmaceutical plant located in the center of New Jersey's industrial "alleyway", midway between New York City and Philadelphia, a microtunneling system was installed. These microtunnels wove their way, through the underground maze of building foundations, from jacking pit to receiving pit, interconnecting the conduits for a future sewer system. In most instances, several 1.52m (60") diameter laterals branched off from a pit, extending to the far reaches of the facility grounds. Tunneling runs of 150m to 300m (500' to 1000') were typical, necessitating pinpoint accuracy for the orientation and advancement of the microtunneling machine.

Jet grouting was chosen as the desired method to provide pre-excavation support, to be installed prior to any tunneling or excavation for the jacking and receiving pits. Jet grouted columns were installed beneath all existing foundations that might be influenced by the tunneling operation. This involved the drilling and jetting of a variety of angle holes on the exterior and from within the interior of buildings along the microtunnel path. Most of the interior injection work required the use of low headroom equipment necessitating the use of 1.12m (3') drill rods for the drilling and jetting work. The accuracy and placement of the jet grouted soilcrete columns was equally as important as the orientation of the microtunneling operations.

The soils conditions throughout the site were a "mixed bag" but the predominant soils types were cohesive soils with clays and silty clays. Random layers of fill were encountered, as might be expected in an old industrial site in the Northeast. In these layers, clay and silty clay was found, binding the various rubble fill. These layers were usually encountered at shallower depths, indicating that they may have been the remains of former foundations, retrofitted during the metamorphosis of this ever-expanding industrial facility.

The jet grouting pattern and grout strength requirements were developed by the owner's structural and geotechnical engineer, PS&S, Warren, NJ. The grout mixes, pressures, orientation of grout columns, and all of the requirements necessary to provide a successful underpinning operation were the responsibility of the specialty jet grouting contractor, Fonditek International. Since this situation also involved structural underpinning, the single fluid method of jet grouting was chosen by the specialty contractor.

The same drilling equipment utilized for the ANC building was employed on this project. The Puntel 1200 was used for the exterior work, and the Puntel 609 for the low headroom work (see Figure 4). A Tecniwell 400 fully automated jet grout plant was used for the majority of the jetting on this project. Due to the magnitude of the project, the Tecniwell 351 was also used with a separate drilling and jetting crew to meet the owner's compressed schedule of construction.

Shortly after entering into the project, it was realized that jet grouting would play yet another role in the construction of this microtunnel complex. It was determined that a plume of potentially contaminated groundwater existed within the confines of the construction site. Rather than use standard sheeting and shoring to construct the jacking and receiving pits, jet grouting was chosen in lieu of these methods to seal out groundwater seepage into the pits. The problem with standard sheeting and shoring methods was that they could not provide a dry pit without the need for sump pumps or shallow well points. If pumping was required, the migration of this contaminated plume was a risk, and the contaminated water would have to be contained, treated as a hazardous material, and be disposed of accordingly.

Jet grouting technology could provide a "bathtub" effect by creating vertical walls of soilcrete as well as a horizontal bottom plug tying into the vertical walls. The vertical

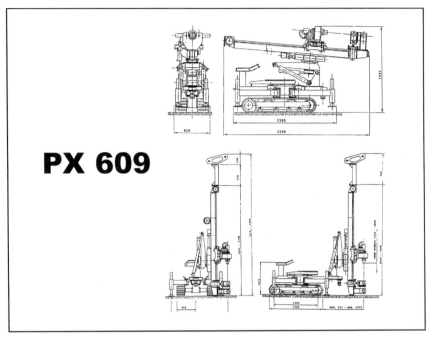
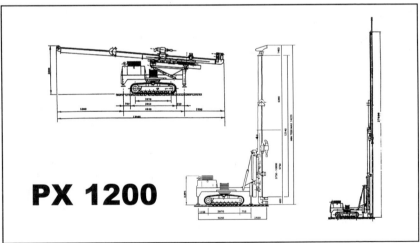

Figure 4
The interior and exterior drilling equipment for Jet Grouting

walls were created by placing a pattern of overlapping columns around the perimeter of the proposed pit. This created a continuous secant wall of grout restricting the migration of soils and groundwater into the proposed pit area. The horizontal bottom plug was created in a similar manner, but here, only the bottom 1.2 -1.5m (4-5') of each element was jetted with the standard cement-based grout. The method was a success, with most of the pits leaking at a rate of less than 5 gpm.

At this juncture in the project, the specialty contractor introduced another innovation in jetting technology. This method combines jet grouting with the technology of drilled micro piles. On the vertical sidewalls of the pits, the jet grouted columns were advanced simultaneously with a 13.75cm (5.5") OD micro pile. By this methodology, the micropile was centralized within each soilcrete column. Again, the single fluid method was chosen for this procedure. This internal reinforcement in each of the jet grouted columns was used to resist the lateral soil loads at each of the pit locations. This eliminated the need for traditional lagging, and only two strategically placed whalers were needed to complete the pit construction.

It should be mentioned that the two fluid method was utilized in a test section for the pit construction, but proved too disruptive to the surrounding structures. It became apparent the layer of fill material near the surface became the path of least resistance for the high pressure air (the second tube in the two fluid process). The percolation of air, introduced by the two fluid system, into the soils, caused a disruption at the surface extending up to 15m (50') in diameter, from the point of injection. The triple fluid system was also discussed as an alternative, but was discounted as an option based on the results of the two fluid test. It is important to note that each project should be carefully analyzed, weighing the pros and cons of all three jet grouting systems on a case-by-case basis. Subsequent to this analysis, it is equally important to perform a test section before committing to a method for the production jetting.

Conclusion

The jetting system with internal reinforcement offered many attractive features for these applications:
- Jet grouting could be used as the Support of Excavation (SOE) by utilizing internal reinforcement.
- Jet grouting secant columns eliminated lagging and potential for loss of ground.
- Jet grouting drilling systems could more easily advance through obstructions.
- Injected grout volume can be predicted with better accuracy than other grouting methods.
- Duration of the work can be predicted with better accuracy due to the finite nature of the jetting process (e.g.-standard rates of withdrawn, rotation, grout pressures, and pump rates)

- Single-fluid (F1) jet grouting system allows for better control of ground movements vs. previous experience with Double-fluid (F2) and Triple-fluid (F3).
- Jet grouting could underpin existing foundations adjacent to the excavation, as required (Micciche et al., 1998).

Core testing of the soilcrete columns revealed that unconfined compressive strengths of the in-situ material on both projects were typically in the range of 6900 to 17250 kPa (1000 to 2500 PSI). The column diameters were in the range of 762 to 915cm (30 to 36") utilizing the single fluid method. The "waterproofing" effect of the pit excavations on the NJ project was so effective that sump pumps were virtually unnecessary except to handle some run-off from episodes of heavy rain. The pits again acted like "bathtubs", as rainwater did not drain from the pits after heavy rainfall.

These successful results in cohesive soils, primarily clays, can be attributed to several factors. These factors include: automated quality control; state-of-the-art drilling and pumping equipment; innovative jetting nozzle design; and highly skilled practitioners of jetting technologies continuing to improve their methodologies.

Acknowledgements

I wish to thank Dave Schoenwolf and Brian Zalenko with Haley & Aldrich's Silver Spring, MD office for their engineering expertise and support on the ANC Bldg project,and likewise, "Ding" Carbonelle with PS&S's Warren, NJ office, on the NJ industrial project. I am also grateful to the Layne Christensen's Brockton, MA Fontditek International office, especially Peter Iovino, Romano Micciche, and Rupert Hon, without whom these projects would not have been realized or constructed.

References

Bruce, Donald A., Petros P. Xanthakos, Lee W. Abramson, 1994, "Ground Control and Improvement," Chapter 8, "Jet Grouting"

Haley & Aldrich,Silver Springs, MD, Spiegel Zamenecnik & Shah, Inc., Washington, DC, 1992, Report on Probable Cause of Settlement Damage to the Argentine Naval Commission Building, Washington, DC

Hon, Rupert and Romano Micciche, 1997, Technical Submittal - Argentine Naval Commission Building, Washington, DC

Micciche, Romano,P.E., Thomas Hurley, and Giovanni Aurilio, 1998, ASCE 1998 Conf. Case History, "Controlled Jet Grouting for Viaduct Pier Excavation Support, Initial Leverett Circle Connectors, Central Artery (I-93)/ Tunnel (I-90), Boston, MA"

DESIGN AND CONSTRUCTION OF A JET-GROUTED BARRIER WALL

by D.M. White[1], M. Koelling[2], D.W. Ashcom[1], and J. Kurrus[3]

ABSTRACT

An innovative application of jet-grouting technology has been successfully used to construct a below-grade barrier wall to contain petroleum hydrocarbon contamination in Spokane, Washington. The wall was constructed to intercept the flow of groundwater and Bunker C fuel. Bunker C, a heavy petroleum hydrocarbon, flowed slowly downgradient over a period of years from underground storage tanks at a steam plant operated by a Spokane utility. A 540-foot-long hydrocarbon plume was delineated; hydrocarbons and groundwater flow through a permeable sand and gravel formation within troughs in relatively impermeable bedrock and fine silt. A 280-foot-long wall was constructed as a hydraulic barrier to intercept hydrocarbons and control groundwater flow. The wall consisted of overlapping columns of jet grouted Soilcrete, which reduced the permeability of the sand and gravel formation by approximately four orders of magnitude. The wall was anchored into the bedrock and silt, with a base elevation at depths ranging from 26 to 52 feet below grade and with a top elevation at 25 feet below grade. The height of the wall was varied to provide a level top, with the exception of an area left 2 feet lower to serve as a below-grade weir. A subsurface pumping system (hydraulic control system) allows the removal and treatment of the groundwater on the upgradient side of the barrier and disposal of the treated water into the municipal wastewater-collection system. Instrumentation associated with the pumping system allowed documentation of the effectiveness of the wall at maintaining the negative head differential. A SCADA system allows monitoring and control of this system from remote locations. Jet grouting allowed the wall to be constructed with a minimum of disruption to traffic on a busy downtown street, while working within an area of multiple utilities. The paper documents construction of the barrier wall and early operation of the hydraulic control system to verify the functionality of the wall. Additional data concerning the time-related strength gain characteristics of the soil-cement-flyash-bentonite mix that forms the Soilcrete also is presented.

[1] M. ASCE, AGI Technologies, Bellevue, WA
[2] M. ASCE, Hayward Baker, Inc., Seattle, WA
[3] M. ASCE, Washington Water Power, Spokane, WA

INTRODUCTION

The Central Steam Plant was built in 1915 and was once the source of steam heat for many of the buildings in downtown Spokane, Washington. The plant was operated using various fuels, including coal, wood, and eventually oil. Bunker C oil, a thick, slow-flowing type of oil, was used prior to the plants closure in 1986. The fuel was stored in seven underground concrete storage tanks that were constructed between 1966 and 1975 and ranged in capacity from 72,000 to 140,000 gallons each. The steam plant ceased operations in 1986 and all tanks were removed from service and cleaned by 1987. A structural investigation indicated that control and construction joints in the tanks were the most likely causes of oil leakages (Construction Technologies Laboratory, 1993).

A remedial investigation determined that the Bunker C oil had seeped deep into the ground to the north of the plant. The limits of the spill were mapped during an extensive investigation by Washington Water Power (WWP) and its consultants (see Figure 1) and the oil was found to have spread about one block north of the spill site. A Cleanup Action Plan was developed for the Site (defined as the area affected by petroleum hydrocarbons at concentrations greater than cleanup levels, as shown on Figure 1) that included the underground barrier, hydraulic control system, selected soil removal, oil recovery, construction of bioventing wells, and other features. This paper focuses on the design and construction of the underground barrier wall, built to ensure that oil cannot migrate further.

GEOLOGY

The geology of the site consists of three primary units: unconsolidated sand and gravel (Missoula Flood Deposits), silt (Latah Formation), and bedrock (Columbia River Basalt). As shown on Figure 2, the upper sand/gravel unit is encountered at the surface across the entire site. The sand/gravel unit increases in thickness from 10 to 20 feet along the southern boundary of the site to about 50 feet near the northern boundary. The sand/gravel unit is a high hydraulic conductivity (estimated as 6×10^{-1} cm/s, AGI, 1997) sand, gravel and cobble material deposited by the Missoula Flood at the close of the most recent ice age. Elsewhere (upgradient) in the Spokane Valley, the formation provides groundwater for high capacity irrigation and drinking-water wells as the Spokane-Rathdrum Prairie Aquifer. It is considered a sole-source aquifer in these locations, but has no relation to groundwater at the site.

The silt is part of the Latah Formation and underlies the sand throughout most of the western part of the site. The silt ranges in thickness from 0 to about 15 feet and generally fills in depressions in the bedrock surface. The hydraulic conductivity of the silt is at least three orders of magnitude less than the overlying gravels and the silt is considered a confining bed.

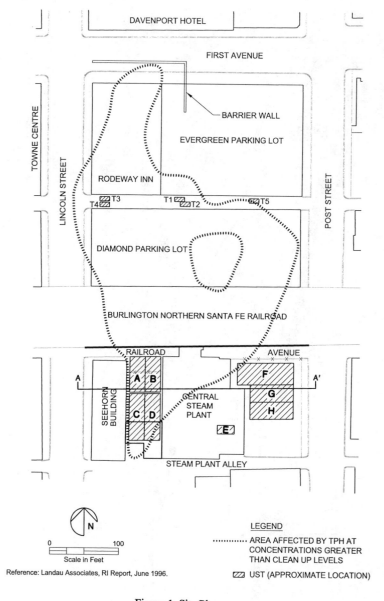

Figure 1. Site Plan.

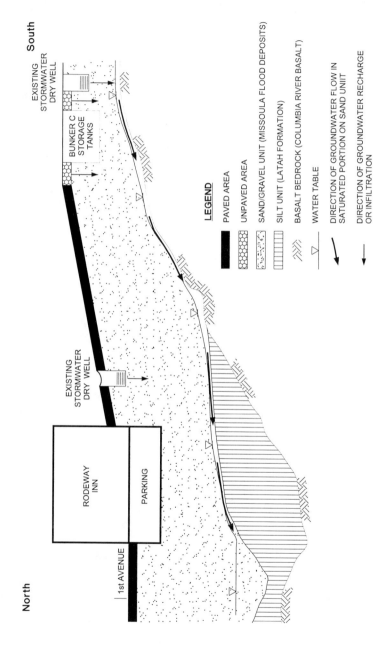

Figure 2. Conceptual Hydrogeologic Model.

Bedrock underlies the sand/gravel or silt (where present) across the entire site and consists of slightly fractured basalts that confines groundwater in the upper gravels. These basalts are present in outcrops at the Spokane River about 0.5 mile north of the site, where they form the Spokane Falls.

Groundwater generally flows from south to north at the site. The sand/gravel unit hydraulic gradient ranges from about 0.04 to 0.08 ft/ft across most of the site but decreases to approximately 0.0005 ft/ft near the north site boundary. Site groundwater is not hydraulically connected with the Spokane-Rathdrum Prairie Aquifer and does not constitute a practical source of drinking water.

PETROLEUM PRODUCT RELEASE

An estimated 75,000 gallons of petroleum product leaked from the underground storage tanks at the Central Steam Plant. This petroleum is now located in soils as shown on Figure 1. The petroleum generally is limited to a zone extending from the confining bed upwards to a maximum thickness of 4.5 feet, the top of which generally is located at least 25 feet below ground surface. Free phase petroleum is present in monitoring wells in the area of contaminated soil. Free phase petroleum product is recoverable only in the northern part of the site.

REMEDIATION GOALS

The goals of the remedial action were to protect human health and the environment. Site cleanup levels were established for the site considering the characteristics of the contaminated medium (soil, groundwater, etc.), the characteristics of the hazardous substances present, migration and exposure pathways, and potential receptor points in accordance with the Washington Model Toxics Control Act (MTCA, WAC 173-340). Site conditions minimize risk to human health and the environment due to human contact with soil contaminants. Soil cleanup levels protective of groundwater are more conservative than those based on contact; consequently, soil cleanup levels were developed to protect groundwater. Although site groundwater discharges to a larger saturated zone north of the site that is not likely to be used as a source of drinking water, protection of site groundwater is important to limit potential exposure of humans and to protect other beneficial uses of the water. Groundwater cleanup levels were established accordingly.

REMEDIAL ALTERNATIVES

Alternatives for addressing site contamination and meeting the remedial action goals were evaluated in a Feasibility Study (Landau, 1995). Six cleanup alternatives were presented and included various elements from the following list of individual cleanup actions:

- institutional controls
- monitoring
- tank closure
- surface-water diversion
- paving
- subsurface barrier wall construction
- product recovery
- bioventing
- liquid-phase bioremediation
- shallow soil excavation
- deep soil excavation

No single alternative could remove all contaminated media from the site, especially beneath the railroad viaduct and buildings, so all alternatives considered included institutional controls and groundwater monitoring. These are required by MTCA whenever contaminated materials are left in place.

The Ecology selected Cleanup Action included the institutional controls and monitoring, closure of all tanks on site, free-phase product recovery from the affected soil, paving to reduce the effects of infiltrating water (surface-water diversion), construction of a subsurface barrier wall, excavation of shallow contaminated soils, and bioventing to stimulate *in situ* biodegradation and ultimate destruction by bacteria of petroleum. The subsurface barrier wall was an important element of the overall remediation plan. The barrier wall was constructed at the downgradient edge of the free-phase petroleum product plume. The wall prevents any further northward migration of the free-phase petroleum product and enhances the ability of product recovery systems to perform by increasing overall product thickness and groundwater thickness upgradient of the wall. AGI Technologies was selected as the design-build contractor for the remedial action and utilized the services of Hayward Baker, Inc. to construct the barrier wall. The wall is shown in plan and section on Figures 1 and 3, respectively.

BARRIER WALL TECHNOLOGY SELECTION AND DESIGN

The Feasibility Study provided a recommended alignment for the barrier wall. Several technologies were considered and evaluated and on a preliminary basis, jet grouting was identified as the preferred technology for construction of the wall. The design-build contract for the wall required validation of the preferred technology, further evaluation of the physical dimensions of the wall and development of construction specifications for the barrier material.

Burke (1992) reported that *in situ* barriers have been utilized for many years in geotechnical applications, primarily to reduce dewatering costs on civil projects. Conventional seepage barrier construction for this work has typically included injection (permeation) grouting and slurry trenches. The extension of barrier technology and construction has more recently been applied to environmental applications where control of contaminant migration is required. Method innovations used in North America in the past decade to supplement the more conventional systems have included jet grouted walls, soil mixed walls, injected pile walls, and liner panel walls.

The technical and construction considerations to be addressed and met by any barrier wall installation for this project included:

1. Limitation of horizontal permeability;
2. Continuity of the barrier;
3. Barrier contact and adjoinment with bedrock;
4. Capability to place barrier at depth and not disrupt general hydrology regime;
5. Minimization of disruption to utility pipes;
6. Minimization of disruption to downtown city street.

Several barrier methods were capable of meeting one or more of the foregoing issues. During preliminary phases of the project, soil mixing, slurry trench construction, and injection grouting were discussed as alternative barrier methods. A brief discussion of these methods is available in the aforementioned Burke paper and their use is well documented in the literature. However, the jet grouting method was determined to best satisfy the technical and construction considerations.

JET GROUTING METHOD

Jet grouting is a Ground Modification system used to create *in situ*, cemented formations of soil (Soilcrete). Applications of the system fall into three broad categories: (1) under-pinning and/or excavation support; (2) stabilization of soft soils; and (3) groundwater or pollution control. The procedure includes drilling (typ. 6 inch) boreholes using grout or drilling mud to stabilize the hole. Erosion of the soil is initiated at the design depth with high velocity injection of cutting and replacement fluids in a bottom-up procedure. This continues with consistent, uniform rotation and lifting to create column geometry, while expelling eroded spoil out of the top of the borehole. Designed integration of adjacent columns creates a Soilcrete mass. Geometries and physical properties of the Soilcrete are engineered to provide a predictable degree of improvement.

METHOD JUSTIFICATION

A number of factors form the basis for utilizing a jet grout barrier for this project. The severely sloping basalt contour along the wall alignment, with significant elevation change within short horizontal distances, presented difficulty in creating a tight barrier contact to the bedrock surface with most barrier alternatives. Jet grouting provides an excellent sealing capability at any sloping bedrock surface by grouting into the bedrock and installing Soilcrete in the overlying soil and/or fractured rock in the same process. Secondly, the barrier did not have to be installed up to the street grade. The "buried barrier," with top elevation shown in the section (Figure 3), was best installed by jet grouting to specifically target the cutoff zone at depth. Other methods would have incorporated the entire soil profile in mixing/trenching procedure that would have constructed a barrier up to the street grade. With regard to the buried top of barrier, an added design advantage for the jet grout method, not initially considered, was the capability of the method to install a weir (as shown in the section) that would allow for control of subsurface flow (in the event of hydraulic control system failure) and simplify groundwater monitoring. Thirdly, existing near-surface utility pipes are best bypassed by jet grout drill holes, allowing construction of the barrier at deeper depths without disruption to the utilities. Lastly, street disruption is minimized by utilizing a portable drilling unit and a portable batching/pumping plant, allowing traffic lanes(s) to remain open.

As reported by Welsh and Burke (1995), jet grouting began to be utilized in the United States in the late 1980s as a vertical and horizontal barrier to prevent groundwater migration. This paper cites 16 case histories where jet grouting was successfully used to control contaminated and uncontaminated groundwater flow.

WALL DESIGN

The barrier wall design incorporated the soil boring information from additional holes drilled near to the proposed final alignment. A review of this information allowed for refinement of the plan limits of the alignment and depths of the barrier. This refinement was made to best locate the barrier to maximize efficiency of the remediation and subsequent treatment of the extracted groundwater.

The requirements of barrier wall continuity and limitation of permeability were determined to best be met within the subsurface profile for this project by utilizing the triple rod (fluid) system of jet grouting installed as overlapping circular columns. The triple rod system pumps grout, air, and water through different lines to the jetting tool. High velocity coaxial air and water (6,000 psi) form the erosion medium. Grout is injected at a lower velocity from a separate jetting nozzle below the erosion jet. This separation of the erosion process from the grouting process yields a higher quality grouted soil mass (Soilcrete). During preliminary design, consideration was given to different geometries of Soilcrete to include single and double panel walls. Individual panels of Soilcrete can be installed to overlap from each drill location, thereby creating the wall. However, the redundancy and conservatism of the circular column overlap was determined to best serve the requirements of the barrier for this project. The circular columns were designed with a diameter of 3.5 feet, spaced at 2.5 feet, to create an overlap and barrier width of 2.5 feet. The design required drilling and grouting a minimum of 5 feet into the Latah Silt and/or Basalt Bedrock below the Spokane Sand/Gravel to insure tight barrier contact at each drill location.

The grout mix for the Soilcrete design included cement, fly ash, bentonite, and water. These materials were proportioned accordingly to balance permeability and strength requirements. The design intent included meeting the Consent Decree requirement of a barrier with a permeability of less than 1×10^{-5} cm/s. An average

GEO-ENGINEERING FOR UNDERGROUND FACILITIES 1035

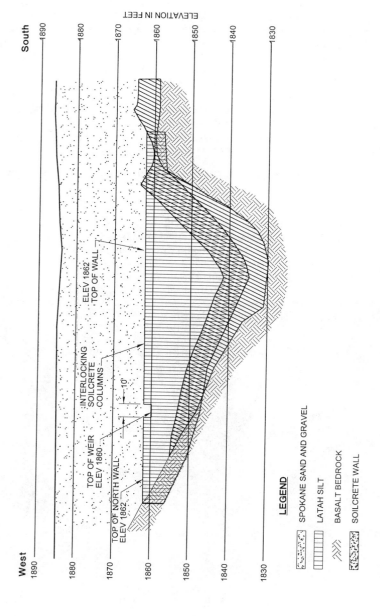

Figure 3. Barrier Wall.

permeability of 1×10^{-6} cm/s was targeted for the work. The strength requirement included creating a Soilcrete barrier product that was tested to be at least as strong as the surrounding soil. This was equated to approximately 25 psi compressive strength for purposes of this project.

BARRIER WALL CONSTRUCTION

Installation of the barrier utilized an hydraulic rotary drill rig (as shown in the Figure 4 with special jet grout modifications to control the fluid volumes, fluid pressures, and jetting tool rotation and lift speeds. This drilling and jetting unit was supported by a portable batching/pumping plant that was located in a parking lot within a few hundred feet of the drill and off the street.

Initially, the pavement section was removed along the proposed barrier alignment for a width of 18 inches to allow access of the jetting tool to the soil profile and to provide a waste trench to contain the spoil. During the barrier installation, the drill rig was in the street to allow at least one lane of traffic flow. The individual Soilcrete columns were then installed along the approximate 280-foot length of L-shaped alignment. The columns were installed in a predetermined "hop-skotch" sequence to allow for jetting closure between two previously installed columns, as shown on the plan. An approximate total of 120 Soilcrete columns were installed. The barrier itself ranged in height from about 0 to 30 feet, following the bedrock valley contour. These barrier depths correspond to approximate depths of 30 to 55 feet below the street elevation. Construction of the test columns and production columns was performed in August and September 1997.

Figure 4. Hydraulic rotary drill rig.

SOILCRETE TESTING PROGRAM

Construction quality-control testing of the barrier wall included unconfined compression and triaxial shear strength testing and triaxial chamber permeability testing on Soilcrete samples retrieved from the wall. The samples were wet-cast after retrieval from the freshly mixed Soilcrete using the construction drilling tools. They were transported to the testing laboratory and stored in a water bath at room temperature until needed for testing. Larger numbers of unconfined compression and smaller numbers of triaxial shear strength tests were performed to assist in verification that the design strength of 25 psi was achieved. Permeability tests were conducted under appropriate confining stresses to verify that the design permeability was achieved.

Permeability test results indicated that the average permeability of the Soilcrete samples was 6×10^{-6} cm/s, meeting the requirements of the Consent Decree and slightly greater than the target permeability. More interestingly, the permeability of the 16 samples tested varied over less than an order of magnitude, between 2.2×10^{-6} and 10×10^{-6} cm/s. This consistency was unexpected but was validated by an interlaboratory testing program between the AGI Technologies and University of Akron geotechnical laboratories.

Strength test results were surprisingly more variable than permeability and the subject of some early concern as the mix did not gain strength with time as quickly as originally anticipated (see Figure 5). The variability in strength test results seems particularly high when it is compared to the relatively low variability of the permeability test results. The triaxial strength tests showed the same variability and similar average as the unconfined compression tests plotted on the figure. The high proportion of bentonite added to control the permeability of the Soilcrete was

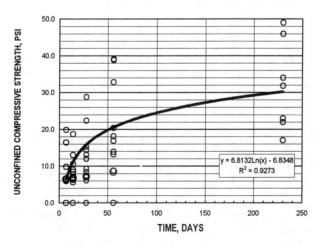

Figure 5. Soilcrete unconfined compressive strength versus time.

believed to be responsible for the delay in strength gain and the variability of strength. A conclusive explanation of the variability in the strength test results has not been made, despite a detailed statistical analysis and close observation of the tested samples. Ultimately, the wall achieved satisfactory strength.

OPERATIONAL TESTING

The operational test of the wall occurred in late September 1997 as the final Soilcrete column was constructed and the barrier closed. With the real-time monitoring capability of the project instrumentation and control system, engineers were able to observe water levels in the already installed monitoring/treatment wells immediately upgradient of the barrier wall start to rise within an hour of the installation of the final Soilcrete column. Later testing showed that the pumping system was able to control water levels behind the barrier. This allowed operation of the treatment system and demonstrated the effectiveness of the barrier wall as a hydraulic control structure. AGI Technologies continues to operate the remediation system for Washington Water Power. The SCADA system installed allows monitoring and control of the full remediation system from AGI's Bellevue, Washington office more than 200 miles away from the site or from any other location.

CONCLUSIONS

The project demonstrated that a functional underground hydraulic structure could be constructed using overlapping Soilcrete columns. Jet grouting technology has advanced to a stage that allows engineers to build and operate critical structures with confidence.

REFERENCES

AGI Technologies. (1997). "Final Engineering Design Report, Central Steam Plant Oil Spill Remediation, Spokane, Washington." Prepared for Washington Water Power Company, Bellevue, Washington.
AGI Technologies. (1998). "Final Cleanup Action Report, Central Steam Plant Oil Spill Remediation, Spokane, Washington." Prepared for Washington Water Power Company, Bellevue, Washington.
Burke, G.K. (1992). *"In Situ* Containment Barriers and Collection Systems for Environmental Applications." Canadian Geotechnical Society, Southern Ontario Section, Symposium on Slurry Walls, Toronto, Ontario, Canada.
Construction Technologies Laboratory. (1993). "Final Report, Inspection and Evaluation of Fuels Oil Storage Structures." Prepared for Washington Water Power Company, Spokane, Washington.
Landau Associates. (1995). "Final Draft Feasibility Study--The Washington Water Power Company Steam Plant." Prepared for Washington Water Power Company, Spokane, Washington.
Welsh, J.P., and Burke, G.K. (1995). "Vertical Cutoffs and Bottom Sealing by Jet Grouting." *Proceedings of a Speciality Conference sponsored by the Geotechnical Engineering and Environmental Engineering Divisions.* ASCE. February 24-26, 1995, New Orleans, Louisiana.

Short Aggregate Piers Reinforce Soils Near Tunnels

Richard L. Handy[1], Member, ASCE, Nathaniel S. Fox[2], Member, ASCE, and Kord J. Wissman[3], Member, ASCE

Abstract

Short aggregate piers (*Geopier*™ foundation elements; *Rammed Aggregate Piers*™) were used to reinforce loose, collapsing sand and clay soils over and around an existing railroad tunnel. During construction, an unbraced corrugated steel plate arch tunnel lining had deformed laterally under pressure from sand backfill, leaving voids along the lower tunnel walls. In time sand slowly infiltrated into the voids from above, causing the road and sidewalk slabs to settle and tilt approximately 150-200 mm (6-8 inches) downward from the crest of the tunnel. The damaged slabs were removed, and soil near and over the tunnel was laterally compacted and reinforced with short aggregate piers prior to slab replacement. Based in part on these results, an option was suggested to use compacted short aggregate piers to help reduce stresses induced by foundations for a new building overlying a 90-year-old sewer tunnel.

A supplemental influence from compaction of short aggregate piers may be to increase resistance to soil liquefaction during earthquakes. Total energies are comparable to those from deep dynamic compaction, and have the advantage of being distributed throughout the length of the pier instead of being concentrated at the ground surface. The maximum depth of influence is about 4.5m (15 ft).

Introduction

Short aggregate piers (Geopier foundation elements) and the associated matrix soil

---
[1]Distinguished Professor Emeritus, Dept. of Civil and Construction Engineering, Iowa State University, Ames, IA 50011
[2]President, and [3]Chief Engineer, Geopier™ Foundation Co., Inc., 11421 East Aster Dr., Scottsdale, AZ 85259

normally are used to support buildings and other structures. This is the first reported use as soil reinforcement around tunnels and for pavement support.

The Problem

It was not the most momentous construction project in Madrid, Iowa (pop. 2526), and not as exciting as when the school burned down earlier in the week, but the new railroad tunnel underpass did contribute an element of surprise after the street on either side settled so much that an unsuspecting driver easily could go airborne and land on the other side in a series of spine-jarring bounces. As the street barricades went up, a local wryly observed, "I don't mean to pry, but is that really the way this is supposed to be?"

Investigation

The corrugated steel plate arch tunnel replaced a turn-of-the-century railroad overpass bridge. Unfortunately, as the unbraced tunnel was being backfilled, lateral soil pressures pushed the sidewalls inward (Figure 1). After construction, soil on either side of the tunnel crest gradually settled, creating a pavement geometry like a shallow roof with a peak in the middle. Faced with an untenable situation, the contractor took bankruptcy and left others to work out the deficiencies.

A post-construction geotechnical report described loose sand fill to the depth of the borings, 3 m (10 ft), on both sides of the tunnel, and attributed the excessive settlement to inadequate compaction. However, this did not explain the narrowed and lopsided shape of the tunnel. The construction specifications had called for temporary bracing to be installed to hold the tunnel shape while the fill was being placed, and apparently through ignorance of the magnitude of the potential damage from lateral soil pressures, the bracing was omitted.

Figure 1. Installing compacted short aggregate pier reinforcing members along a deformed railroad tunnel.

The distortion shown in Figure 1 suggests that backfilling may have proceeded first on the south (right) side, pushing the top of the tunnel to the north. Then as would be expected, filling on the north side then did not overcome passive pressures and push the tunnel lining back, but only caused the sides to deform inward. That in turn would tend to create voids next to the lower walls of the tunnel lining as they were pushed away from the previously compacted backfill. Thus, because of the lack of temporary tunnel bracing, soil compaction may have aggravated instead of preventing the future settlement problem.

The hypothesis concerning distortion of the tunnel lining during backfilling was supported by observations of a high groundwater table in sand backfill close to the tunnel, perhaps attributable to upward displacement by sand falling down into the lower, water-filled voids. Groundwater mounding was contained on one side by the tunnel lining and on the other by clay soil, and at the ends by a concrete facing that was added to prevent slope erosion (Figure 1).

Repair options

An early clue that the problem might be serious and ongoing was when lower walls flanking the ends of the tunnel pushed in, in part because of the lack of a resisting top arch. A repair was made by tying back the end walls with horizontal ground anchors. While these appeared to stop further inward movements of tunnel walls, they did not stop pavement settlements.

The most obvious repair option for the tunnel itself was to start over, removing the pavement slabs and approach fills, re-shaping and bracing the tunnel lining, and replacing and compacting the backfill. Disadvantages of this procedure were high cost, a long construction time during which rail traffic would be interrupted, and sensitivity to weather conditions. Another option, structural slabs supported on piles and grade beams, would be an expensive overkill.

The thickness of soil cover over the tunnel was about 1.2 m (4 ft), too thin for geofabric reinforcement to have much effect. Furthermore, the depth and extent of collapsible soil next to the tunnel indicated that a deeper method of soil improvement was needed. Pressure grouting would be difficult because of ready escape routes through sides of the embankment.

It therefore was decided to reinforce the existing loose backfill with specially compacted short aggregate piers (Geopier foundation elements) that densify the soils adjacent to the piers and increase lateral stresses in the soil matrix, aiding the transfer of surface pressures away from the tunnel walls. One possible disadvantage would be a tendency to further warp the tunnel lining during installation and lateral compaction of the piers. This disadvantage was mitigated by the restraining weight of the soil in place over the tunnel, a situation that was unlike that of the initial backfilling. Additionally, observations of the tunnel were made during pier installation to allow procedures to be changed if problems should arise.

Horizontal stresses

Horizontal in-situ stress measurements at other sites with the K_o Stepped Blade (Handy et al. 1982) indicate that Geopier densification can impose limiting passive conditions in soils close to the aggregate piers, with lateral pressures dissipating linearly with radial distance outward from each pier. The lateral stress regime extends approximately one diameter below the drilled depths of the piers prior to compaction.

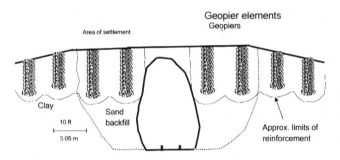

Figure 2. Pier positions relative to the tunnel. Compaction expands the aggregate piers downward and laterally, compressing and stabilizing the encompassing soil.

Design

The distorted tunnel was approximately 7.3 m (24 ft) high, so two lines of 760 mm (30 inch) diameter by 3.6 m (12 ft) long piers were constructed in the sand backfill on either side of the tunnel to extend down to near the spring line, in order to densify the soil and transfer load to the tunnel arch (Figure 2). These two primary lines were flanked by two additional lines on each side extending to 3 m (10 ft) depth, thereby creating a reinforcing layer in the shape of shallow flanking arches pointing slightly downward towards the tunnel lining. Each line is comprised of five piers, giving a total of 40 compacted short aggregate piers spaced 2.4m (8 ft) and 1.5m (5 ft) apart parallel and transverse to the roadway, respectively.

Figure 3. Dumping a pre-measured amount of aggregate that when compacted will comprise a single lift in the pier.

A bulb of free-draining, coarse aggregate was rammed at the bottom of each pier. Previous studies have shown that the additional depth of densification approximately equals the diameter of the pier, but in loose sand susceptible to vibration the influence may extend deeper. An open-graded aggregate was selected for the basal bulb to reduce the potential for excess pore pressure and even possible liquefaction of the sand.

A special circular ram beveled at 45° around the perimeter is used to induce lateral compaction and high lateral stresses in soil between and adjacent to the piers. The ram used for 760 mm (30-inch) piers is 660 mm (26 in.) in diameter, with the bevel occupying over 50 percent of the surface contact area.

The normal compaction procedure is to ram the basal bulb for 1 minute or to refusal, after which the pier is built up in nominal 300 mm (1 ft) thick layers of well-graded aggregate, each layer being compacted for a minimum of 20 seconds and to refusal with a 400-600 cycle per minute hydraulically driven ram. A discussion of the total energy involved is presented later in this paper.

Construction

Temporary casing was required for pier borings in the sand backfill to prevent collapse into the open borings. The casing was incrementally pulled upward prior to dumping and ramming each aggregate layer (Figures 3 and 4). Casing was not required for borings in the clay.

Figure 4. Special beveled ram being lowered through the temporary casing to compact the next layer of the aggregate pier.

Results

Future deflections of the pavement slab are predicted to be less than 25 mm (1 inch), and the reinforcement appears to be performing satisfactorily.

A tentative tunnel application with a different focus

New construction in an old city often impinges on older—in some cases much older—buried structures that still are functional and in use. Such is the case in Memphis, Tennessee, where a 1.8x1.5 m (5x6 ft) inside-dimension concrete sewer tunnel constructed prior to 1914 underlies the site of a planned multistory commercial building.

Test borings revealed 2 to 8 m (6 to 27 ft) or more of random fill over and around the 90+ year-old tunnel that still is in service. Historical maps indicate that the tunnel was constructed along an old bayou that was filled to match adjacent grade elevations. Construction options included moving the tunnel, or leaving it in place and defending it from the anticipated foundation loads. For reasons of economy and scheduling the latter procedure was selected.

The tunnel will be structurally strengthened and is being lined internally. One option is to strengthen and reinforce the overlying fill with compacted short aggregate piers. In contrast to piling, these would not extend down to the level of the tunnel, but as in the

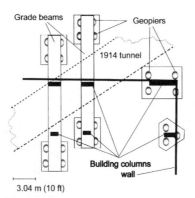

Figure 5. Support members and grade beams to protect an existing tunnel from anticipated foundation loads.

case of piling they would not be installed directly over the tunnel, but off to the sides, where groups of three or four, nominal 760 mm (30 inch) diameter, 2.1 m (7 ft)-long piers and the associated soils are planned to support concrete pads that in turn will support grade beams and footings for the building (Figure 5).

Calculated transfer of pier compaction stresses to the tunnel

A potential drawback to the use of Rammed Aggregate Piers over an existing structure is the dynamic tamping force. A force of 260 kN (60 kips) has been measured at the base of the 0.45 sq m (4.91 sq ft) tamping foot, giving a tamping stress of 580 kPa (12,000 psf) at the contact with the soil (Lawton, 1998). This of course will vary depending on dynamic properties of the soil. From elastic theory and based on an influence factor of 0.02, the maximum dynamic stress transmitted to the tunnel during compaction will be approximately 12 kPa (250 psf). This is over twice the static load pressure that will occur from the aggregate pier-supported building. Thus the pier compaction may be regarded as a proof test of the tunnel structure, in which case the tunnel will be inspected before, during, and after pier installation. Based on elastic analysis of the stress distribution and on experience, settlement of the completed building structure is anticipated to be less than 25 mm (1 inch). If another option such as the use of drilled shafts is selected to support the grade beams, a small amount of differential settlement therefore may be expected.

Rammed aggregate piers to improve earthquake resistance

Memphis is near the New Madrid, Missouri epicenter of some of the largest earthquakes recorded in North America. Four earthquakes of about magnitude 8 occurred within this seismic zone during the winter of 1811-1812. The New Madrid region constitutes the highest level of seismicity in the central and eastern parts of the US. A peak ground acceleration of 0.3-0.35 g is associated with an earthquake that has a 5 percent probability of being exceeded in 50 years (USGS, 1998).

Compaction of short aggregate piers is somewhat analogous to the deep dynamic compaction (DDC) ground improvement method. The DDC method consists of densifying the soil be repeatedly dropping a heavy weight from a crane, and has been shown to be effective for densifying potentially liquefiable soils to depths of about 10-12 m (30-35 ft) (Broms, 1991). A simplistic view is that dynamic compaction creates localized ground vibrations that do the same job as earthquakes, prior to building on potentially unstable ground.

A disadvantage of the DDC method is that high levels of energy must be applied at the ground surface in order to densify soils at depth. The maximum compaction depth varies with the square root of the impact energy, so doubling the impact energy increases nominal penetration depth only by $\sqrt{2}$ (Broms, 1991). High ground surface energy levels induce relatively large peak particle velocities that may be damaging to brittle structures, which of course includes old concrete tunnels.

The energy level per stroke during construction of 760 mm (30 inch) diameter aggregate piers is about 5 kNm (3700 ft-lb), roughly a thousand times less than typical energy per blow with DDC. On the other hand, whereas 3 to 8 blows are typical at each location with DDC, over 3500 blows will be delivered to construct a 3.7 m (12 ft) long aggregate pier. Thus the total energy inputs are of the same order to magnitude. However, compaction energy for the piers is applied almost uniformly throughout the length of the pier. Of particular importance is that the distributed, lower impact compaction should significantly reduce the risk of damage to nearby structures.

In contrast to the positive depth control from the use of the rammed aggregate method, the penetration depth of DDC (in meters) varies from 0.3 to 0.8 times $\sqrt{WH}$, where W is in tonnes and H is in meters (Broms, 1991). Thus, to attain a compaction depth of 3.7 m (12 ft), a 10 tonne weight would be dropped from 2 to 15 m, depending on the soil conditions.

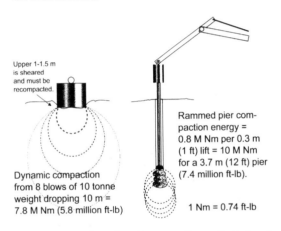

Figure 6. Compaction energies are comparable to reach a comparable depth, but pier compaction is more uniform and gives better depth control.

Additional earthquake protection will derive from the removal of compressible materials and replacement by densified aggregate. Recent research by Lawton (1998) with instrumented full-size short aggregate piers on an Interstate Highway 15 bridge section in Salt Lake City, Utah, indicates that Geopier-supported footings moved only a few millimeters (fractions of an inch) after being subjected to over 60 push-pull cycles simulating loadings from a magnitude 7.5 Richter scale earthquake. Static load tests indicate that aggregate piers are substantially stiffer than stone columns, attributable to the higher degree of compaction. Hydraulic conductivity of the piers varies depending on gradation of the aggregate, an open gradation being preferred where there is a possibility for either positive or negative pore pressures developing during or after compaction.

Conclusions

Deep soil reinforcement with compacted aggregate piers is an economical method for protection or repair of underground tunnels, including tunnels that will be subjected to additional loading. The application to support pavement approach slabs also suggests another potential use, to remedy or prevent the troublesome road and highway "bump at the end of the bridge."

Short aggregate piers constructed according to the Geopier™ method involve compaction energies comparable to those from deep dynamic compaction (DDC), even though individual impact energies are lower by orders of magnitude. Compaction energies are distributed evenly and efficiently throughout each pier length, affording a positive control of treatment depths and areas. The piers and their method of installation also increases soil stiffness. These several influences suggest that short aggregate piers should be an efficient and effective means for improving earthquake resistance.

Acknowledgments

The authors thank Rick Howe, PE, Stanley D. Lindsey and Associates, Ltd., for furnishing design details for the Memphis tunnel protection system using short aggregate piers, and Evert Lawton, Ph.D., PE, for furnishing information on the Salt Lake City tests.

References

Broms, Bengt (1991). Deep compaction of granular soils. In *Foundation Engineering Handbook*, H. Y. Fang, ed., pp. 814-832. New York; Van Nostrand Reinhold.

Handy, R. L., Remmes, B., Moldt, S., Lutenegger, A. J., and Trott, G. (1982). "In situ stress determination by Iowa Stepped Blade." *J. Geotech. Engrg.* 108, 1405-1422.

Lawton, Evert (1998). Geotechnical testing at the I-15 Bridge over South Temple, Salt Lake City, Utah. Preliminary report prepared by the Department of Civil Engineering, University of Utah, Salt Lake City, Utah.

USGS (1998). National Seismic Hazard Mapping Project. Geologic Hazards Team, Golden, Colorado. Data are available from the Web Site: http://geohazards.cr.usgs.gov/eq/

DEEP MIXING TECHNOLOGY FOR DEEP EXCAVATION

Ali Porbaha[1], Hideyuki Asada[2], and M. Javad Fatemi[3]

ABSTRACT

In light of recent acceptance of deep mixing (DM) technology in the United States to overcome technical problems associated with infrastructure development in the urban areas with soft ground condition, there is a need to transfer the experiences gained in Asia, where DM technology has been applied for more than two decades. Accordingly, the aim of this paper is to examine the application of DM for deep excavation of very soft ground with an emphasis on case histories from Japan, China, Singapore and Taiwan where soft ground is a major problem for geo-engineers. Guidelines are provided for design of braced and tieback excavations with DM stabilized bases to prevent heaving, coupled with a collective evaluation of relevant technical issues.

INTRODUCTION

Underground construction of facilities is inevitable in many metropolitan areas due to lack of space and the need to expand the existing transportation networks. Therefore, it is of great importance to identify the relevant technologies that have been practiced in different countries.

Deep mixing technology, which has been steadily progressed in Japan (see next section) and Sweden, has been applied for solution of various problems in soft ground engineering for more than two decades. It has been applied, for instance, for

---

[1] **Professional Associate, Technical Research Institute, TOA Corporation, Yokohama, JAPAN**
   (formerly; Fellow of Science and Technology Agency, Port and Harbour Research Institute)

[2] **Research Engineer, Technical Research Institute, TOA Corporation, Yokohama, JAPAN**

[3] **Assistant Professor, Department of Civil Engineering, Shiraz University, Shiraz, IRAN**

prevention of sliding failure, reduction of settlement, excavation support, controlling seepage, preventing shear deformation to mitigate liquefaction, and for remediation of contaminated ground.

Acknowledging the need to increase the awareness of the practitioners in a global perspective on how relevant technologies are being applied, the aim of this article is to examine the application of deep mixing for excavation control. Although several projects using DM for excavation have been reported in North America (see, for example: ASCE, 1997; and ASCE, 1998), the emphasis here is to present excavation related projects in several Asian countries including; Japan, China, Singapore, and Taiwan. Presentation of these projects is aimed to assist engineers in identifying and assessing the situations in which the technology has been successfully applied. This article is in conjunction with those presented by Porbaha (1998a, and 1998b), and Porbaha et al. (1998, and 1999).

DEEP MIXING

Deep mixing is a soil stabilization technique that mixes in situ soil with a stabilizing reagent, such as cement, lime or some chemical admixtures, in the form of slurry or powder, to improve engineering properties of the soft ground. Specially designed machines with several shafts equipped with mixing blades and stabilizer injection nozzles are used to construct in-situ treated soil columns in various patterns and configurations (Porbaha, 1998b).

In the last three decades the Japanese construction industry has contributed extensively to develop relevant techniques for characterization of the stabilized geomaterial, and development of machinery for installation and quality control of both marine and land-based projects (see, for example, Okumura et al., 1972; Kawasaki et al., 1981; Saitoh et al., 1982; Hirama et al., 1983; Chida, 1982; Terashi et al., 1983; Inatomi et al., 1985; Nishibayashi, 1985; Suzuki et al., 1988; Tateyama et al., 1992; Tanaka, 1993; Matsushita et al., 1993; Mitsuhashi et al., 1996).

DEEP EXCAVATION

Soft soil deposits are widely found in many regions around the world. On the other hand, the construction of new underground facilities in the urban areas which require excavation at depth are inevitable.

Deep excavation of soft ground, particularly in the vicinity of exiting structures, has always been a challenge for geotechnical engineers. In the case of deep excavation of soft ground, stability against heaving is a major problem during construction. Accordingly, due to stability problems coupled with the stringent requirement to avoid settlement, ground stabilization in the following circumstances are unavoidable to restrain heaving and excessive damage to surroundings of the construction site:

1. Deep excavation in soft soil ground condition where passive resistance of a diaphragm wall is insufficient;
2. For excavation sites in the urban areas surrounded with high rise buildings or structures which are sensitive to settlement; and insufficient space does not allow construction of deep diaphragm wall penetration for stability requirements.

Figure 1 shows various applications of the deep mixing for excavation control in terms of a measure to increase resistance against uplift and heaving, in addition to restraining earth pressure. Thus, the functions of DM for base stabilization are:

- Heave reduction during construction;
- Improving stability by increasing passive resistance; and
- Increasing bearing capacity for the foundation of the planned underground facility.

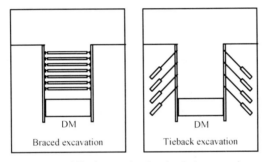

(a) Base stabilization to reduce heaving during excavation

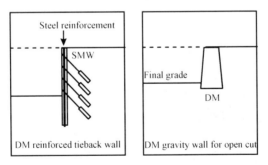

(b) Retaining systems to restrain active earth pressure during excavation

Figure 1. Various applications of DM for excavation control

Several projects of deep excavation of soft ground and remedial measures to counteract base heave using deep mixing are examined in this section. The case histories related to base stabilization (Case a in Fig. 1) are presented here, and those projects in which DM was used as the retaining system to restrain active earth pressure are not presented due to allowable page limits. A short summary of the case histories are provided, therefore, the readers may refer to the original document for further details of the projects.

(a) Excavation for the entrance of an undersea tunnel: To complete the entrances of undersea tunnels on a reclaimed land in Tokyo bay a cut-and-cover method was adopted. The excavation is 13 to 35 m deep, 37 to 40 m wide, and about 500 m long. The reclaimed ground is very soft comprising 5 to 8 m loose sand and silt with high water content, and soft silt to clayey soil in the lower layers up to depth of 38 m. The N values from SPT (Standard penetration test) is between 0 to 4 for clay; and 2 to 20 for top sand; and 10 for bottom sand. Since the ground is extremely soft, steel sheet piles were installed all around the excavation along with dewatering. To prevent heaving and to increase passive earth pressure, bottom of the excavation was improved from 16.6 to 31.6 m using deep mixing in contact column patterns. Jet grouting was injected subsequently to fill the gap between the sheet piles and the DM columns. After ground improvement excavation was carried out by installing six stages of struts to the depth of 20.9 m from the ground surface. Field monitoring confirmed that DM was effective in preventing heaving, however, deformation of sheet piles and axial stress on the struts exceeded the design values in many locations (Inoguchi et al., 1995). Based on the analysis of the field data, it was found that back calculated moduli were much lower than those applied in the initial stage of the construction. Details of the project along with the deformation of the steel pipe pile are shown in Figure 2.

(b) Excavation adjacent to a railway in operation: The bottom of a 13 m deep excavation at the construction site of Kawasaki in Kanagawa is comprised of soft soil with N values ranging between 3 to 7, and high groundwater level of about 1 m below the surface. The site was located in the urban area surrounded by railways and a steel tower. Therefore, the effect of excavation on the surrounding structures was a serious problem. DM method using low displacement jet columns was adopted to stabilize bottom of the excavation. DM columns were 550 mm in diameter and the thickness of the stabilized zone was 3 m. The maximum displacement was 6 mm measured by an inclinometer installed in the column walls (Ueki et al., 1996). Details of the project are presented in Figure 3.

(c) Braced excavation for an immersed tunnel using DJM: The dry jet mixing (DJM), is a deep mixing technique that mixes dry powdered hardening reagents such as cement or quicklime powder into the soft ground. Nakagawa et al. (1996) reported

the use of DJM, jet grouting and steel-concrete composite piles to support an excavation 19 to 21 m in depth, 48 m in width and 66.2 m in length in soft ground for an immersed tunnel. 372 precast concrete piles of 800 mm in diameter were installed to a depth of 64m. Figure 4 shows the ground condition of the site, and other details of the project.

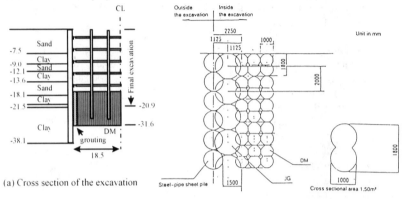

(a) Cross section of the excavation

(b) Configuration of treated ground

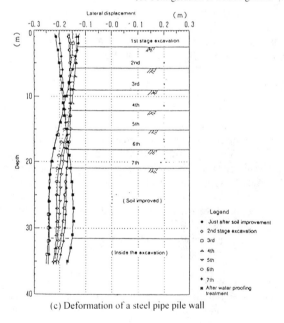

(c) Deformation of a steel pipe pile wall

Figure 2: Excavation for the entrance of an undersea tunnel (Inoguchi et al., 1995)

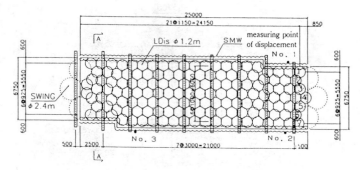

(a) Plan view of the treated ground

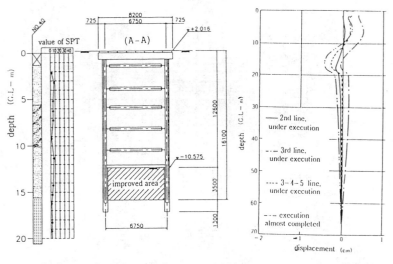

(b) Cross section of the excavation   (c) Displacement measured at point No.1

Figure 3: Excavation adjacent to a railway in operation (Ueki et al., 1996)

d) Excavation for Tokyo International Airport: For Tokyo International airport the excavation work (19.5 deep and 35 m in width) came to a halt at several sites due to large heave (300 mm) that occurred at the base. DM method was applied to a depth of 29 m to increase the passive earth pressure and the resistance against base failure. Tanaka (1993) described the details of the project.

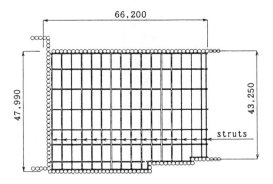

(a) Plan view of the treated ground

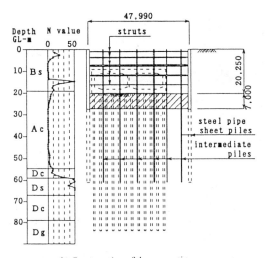

(b) Cross section of the excavation

Figure 4: Excavation for an immersed tunnel using DJM (Nakagawa et al., 1996)

(e) Excavation for the subway station in Singapore: Hulme and Krishnan (1998) described some technical challenges for construction of mass rapid transit in Singapore. The excavation for a central station was 18 m deep, 225m long and 23 m wide. DM using lime columns of 1 to 1.2 m in diameter were applied for base stabilization to prevent damage to the surrounding buildings due to deep excavation. The diaphragm walls were constructed around the perimeter of the station as

retaining wall with seven stage struts. Chew et al. (1993) described the details of the project.

(f) Excavation for high rise building in Taiwan: For the 11.9 m deep excavation adjacent to seven-story buildings with limited space, the DM columns of 0.8 m in diameter were installed at 2 m spacing, leading to an improved area ratio of 12.5%. The improved depth was 11 to 17 m below ground surface. The design unconfined compressive strength was 2 MPa. The deep mixing columns were installed after the completion of diaphragm wall (0.6 m thick), and thereby no ground heave occurred within the construction site. Excavation was successfully carried out without any sign of heaving. The maximum measured wall displacement was 20 mm. Hsieh et al. (1995) described the details of the project.

(g) Excavation adjacent to a river in China: For a 9 m deep excavation, 62.3 m by 44 m in area adjacent to a river in China, the initial design was to install concrete piles of 1.3 m in diameter up to the depth of 15 m. However, the construction was halted due to severe quicksand that occurred at the site. As a remedial measure DJM columns of 6 to 11 m in length in two rows were constructed in parallel to the piles. Zhang and Xu (1996) described the details of the project.

DESIGN APPROACHES

The methods for analysis of heaving for braced excavation is carried out based on bearing capacity concept or by examining the moment equilibrium, as illustrated schematically in Figure 5 (JSSMFE, 1993). In the former approach the vertical pressure at the excavated level (P_v) is compared with the ultimate bearing capacity of the base (q_d); whereas in the latter approach the driving (M_d) and the resisting (M_r) moments are compared to examine the factor of safety (SF) against heave action.

For the case of braced excavation of Tokyo International Airport, the stability analysis was carried out based on moment equilibrium approach with the assumption that center of the slip circle acting at the lowest strut (Tanaka, 1993). Stability factor N_t was estimated using the equation proposed by Peck (1969) i.e.,

$$N_t = \gamma \cdot H / S_u, \qquad (1)$$

where γ is the unit weight of the untreated soil, H is the excavated depth, and S_u is the undrained shear strength of the untreated soil at the bottom of excavation.

For the case of tieback excavation, the dimensionless stability charts based on the Janbu method (1954) are available (O'Rourk and O'Donnell, 1997) to evaluate the deep rotational slip of tie-back excavation. This method was applied for the analysis of excavation using DM for the Central Artery/Tunnel project in Boston.

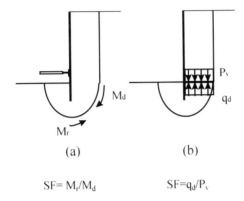

Figure 5: Analysis of heaving based on (a) moment equilibrium (b) bearing capacity concept

SUMMARY AND FINAL REMARKS

This paper examines the case histories of deep mixing technology in several Asian countries, as applied for base stabilization of deep excavation in soft ground. The following conclusions are drawn by comparing the case histories related to excavation:

- In terms of stabilizing reagent, cement and lime columns have been applied in both slurry and powder injection techniques.
- In terms of improvement pattern the overlapped, touched, and/or discrete columns were employed.
- The maximum reported depth of improved ground for these projects is 31.6 m.
- The thickness of the improved ground varies depending on soil and stress conditions at the base, and it reached to a maximum of 10.7 m for these projects.
- In terms of analysis, the stability number for braced excavation ranges between 4 to 5. For the CA/T project, the reported stability number for the tieback excavation is 5.57.

Drawing further collective conclusions from these case histories are not possible due to insufficient reported data including:

- Complexity of the site condition;
- Quality of the soil at the bottom of the excavation;
- Scale of the excavation in terms of depth and area;
- Sensitivity of the structures adjacent to the excavation to settlement;

- Groundwater condition and dewatering of the site;
- Issues related to stabilization (such as: strength, rigidity of the columns, configuration and pattern of the improved ground, replacement ratio, type of the stabilizer, and so on);
- Time lag between ground improvement and excavation;

Apart from being environmentally friendly in terms of noise and vibration, DM has the advantage of rapid solidification that reduce the construction time and thus the construction cost (Porbaha et al., 1999). Therefore, the overall construction cost may be offset by reducing the construction period. However, one major problem in using DM for underground construction in urban areas is the scale of the machinery to be employed for slurry preparation, injection and in situ mixing. Therefore, attempts should be made in downsizing scale of the machinery to increase the scope of applications in the populated urban areas.

REFERENCES

ASCE (1997) In situ ground improvement, reinforcement and treatment: A twenty year update and a vision for the 21st century, Ground reinforcement committee, Geo-Institute Conference, Logan, UT, July 16-17.

ASCE (1998) Geotechnical Special Publications No. 81, 83, 84, and 86, presented in Geo-Congress-98, Boston, MA, October 18-21.

CHEW, H.H., TAKEDA, T., ICHIKAWA, K. and HOSOI, T. (1993) Chemico-lime pile soil improvement used for soft clay ground, Proceedings of 11th Southeast Asian Geotechnical conference, Singapore, 319-324.

CHIDA, S. (1982) Dry Jet Mixing method, state-of-the-art on improvement methods for soft ground, JSSMFE 69-76.

HIRAMA K. and TORIIHARA, M. (1983) Dynamic properties of cement-mixed soils,-Influences of cement content and duration of curing, Report of Ohbayashi Corp, Technical Institute, No.26, 92-96.

HSIEH, H., S., WU, L.H., LU, F.C. and LIN, Y. K. (1995) Application of JG and DMP to reduce excavation induced diaphragm wall deflection, Proceedings of 10th Asian Regional Conf. on SMFE, China, 403-406.

HULME, T.W., and KRISHNAN, R. (1998) Singapore MRT-A construction challenge., Big Digs around the world, ASCE Geotechnical Special Publication Number 86, 79-102.

INATOMI, T., KAZAMA, M. and IMAMURA, T. (1985) Model tests on vibrational characteristics of improved ground by deep mixing method: Technical Note of the Port and Harbour Research Institute, N0.520: 1-38.

INOGUCHI, H., KUBOTA, N., UEKI, H., YAOYAMA, T., and SAKAJO, S. (1995) Symposium on underground construction in soft ground, Tokyo, 147-150.

JANBU, N. (1954) Stability analysis of slopes with dimensionless parameters, Ph.D. Thesis, Harvard University, Cambridge, Mass.

JSSMFE (1993) Underground construction in soft ground, Japanese Society of Soil Mechanics and Foundation Engineering, Tokyo, 76.

KAWASAKI, T., SUZUKI, Y., and SUZUKI, Y. (1981) On the deep mixing chemical mixing method using cement hardening agent, Takenaka Technical Research Report No. 26, Nov., 13-42.

MATSUSHITA, M., IHARA, S., HOSOYA, Y., MATSUO, T. and KUWAJIMA, H. (1993) Soil improvement method controlling at low strength for excavation of ground treated by the deep mixing method, Proceedings of the 28th JSSMFE, Vol.2, 2565-2568.

MITSUHASHI, H., TSUKADA, Y., CHIDA, S. and TERUI, N. (1996) Quality estimation of stabilized soil by deep mixing method applying Rotary Penetration test, Japanese Geotechnical Society, Tokyo, 121-126.

NAKAGAWA, S., KAMEGAYA, I., KUREHA, K., and YOSHIDA, T. (1996) Case history and behavioural analysis of braced large scale open excavation in very soft reclaimed land in coastal area, Geotechnical aspects of underground construction on soft ground, Tokyo, 179-184.

NISHIBAYASHI, K. (1985) Experimental research into mixing devices used for the deep mixing method of soil improvement, Proc. of 20th annual meeting, JSSMFE, 1747-1750

OKUMURA, T., MITSUMOTO, T., TERASHI, M., SAKAI, M. and YOSHIDA, T. (1972) Deep lime mixing for soil stabilization, Report of Port and Harbour Research Institute, Vol. 11, No.1, 67-106.

O'ROURKE, T. D., and O'DONNELL, C. J. (1997) Field behavior of excavation stabilized by deep mixing, J. of Geotech. and Geoenv. Eng., ASCE, Vol. 123, No.6, 516-524.

PECK, R.B. (1969) Deep excavations and tunneling in soft ground, Proceedings of VII ICSMFE, state of the art volume, 225-290.

PORBAHA, A. (1998a) Ground improvement engineering - the state of the US practice, Discussion to the paper No. GI-030, Ground Improvement, Journal of ISSMGE, Vol. 2, No. 4, 189-190.

PORBAHA, A. (1998b) State of the art in deep mixing technology, Part I: Basic concepts, Ground Improvement, Journal of ISSMGE, Vol. 2, No.2, 81-92.

PORBAHA, A., TANAKA, H., and KOBAYASHI, M. (1998) State of the art in deep mixing technology, Part II: Applications, Ground Improvement, Journal of ISSMGE, Vol. 2, No.3, 125-139.

PORBAHA, A., ZEN, K., and KOBAYASHI, M.(1999) Deep mixing technology for liquefaction mitigation, Journal of Infrastructure Systems, ASCE, Vol. 5, No.1, 21-34.

SAITOH, S., BABASKI, R., HIKITA, S., HARUKI, T., IOROI, Y. and KIYOTOH, T. (1982) Studies on improved ground by deep mixing method- sampling and unconfined compression test of large-sized specimen - Takenaka Technical Research Report No. 28, 1-11.

SUZUKI, K., BABASAKI, R. SUZUKI, Y. and FUJII, N. (1988) Centrifuge model studies on stability of wall type improved ground by DMM, Proc. of the 43th JSCE, 148-149.

TANAKA, H. (1993) Behavior of braced excavations stabilized by deep mixing method, Soils and Foundation, Vol. 33, No. 2, 105-115.

TERASHI, M., TANAKA, H. and KITAZUME, M. (1983) Extrusion failure of ground improved by the deep mixing method, Proc. of the 7th ARCSMFE, Vol.1, 313-318.

UEKI, H., HASEGAWA, K., SUZUKI, K., and BESSHO, M. (1996) Development of a high pressure jet mixing method for displacement reduction, International symposium for ground improvement geosystems, Tokyo, 767-772.

ZHANG, Z. and XU, Z. (1996) Application of DJM method to soft soil improvement, Proceedings of International Conference on soft soil Eng., Ninjing, 980-985.

CASE STUDY OF SUPPORT OF CRITICAL UTILITIES DURING EXCAVATION

Sharp, Julie[1], Splitter, Ted[2], and Sturman, John[3]

Abstract

A project involving excavation and removal of leaking underground storage tanks was conducted on a site with a number of physical limitations. A high-pressure water supply line, which was constructed in 1907, ran within approximately 18cm (7in.) of one of the tanks. Additionally, a parallel high-voltage electrical line ran approximately 2m (5 ft) away from the tanks. During the construction phase, neither of the lines could have service interrupted. The tolerance for displacement of the water line was essentially zero. Due to the presence of free-phase gasoline on the groundwater and associated explosion hazards, support methods were limited to non-sparking methods. The soil conditions consisted of beach sand with a shallow water table, approximately 3 m (10 ft.) below grade. Thus, the potential for the tanks to pop out of the excavation due to buoyancy was addressed in the approach. The site was located in a pedestrian area with a large volume of tourism. Care was taken to minimize the gasoline vapors emanating from the former tanks.

A number of alternative methods and combinations of methods were considered to provide excavation support to remove the tanks, associated piping, and to place piping for groundwater remediation. Based on the modeling of lateral load/deflection (p-y) curves, the selected support combination consisted of a pier and haunch system to hold the water line and a steel sheet pile wall to protect the electrical line from being undermined due to the excavation. The field installation required a series of drilled concrete piers due to 3.7 m (12 ft) spacing between joints in the water line.

---

[1] Senior Engineer, LFR Inc., 1900 Powell St., 12th Floor, Emeryville, CA 94608

[2] Principal Engineer, LFR Inc., 1900 Powell St., 12th Floor, Emeryville, CA 94608

[3] Senior Engineer, LFR Inc., 407 Center St., Bethlehem, PA 18018

Field surveying was conducted during the project to verify that the water line did not experience displacement greater than design criteria during the excavation and to correlate field measurements with calculated analyses. Water supply service and electrical service were provided continuously during the project. The tanks and appurtenant piping were successfully removed from the site. Groundwater remediation was later completed, the site was transferred, and the fuel leak case was closed.

Introduction

Leaking underground fuel storage tanks (USTs) near Fisherman's Wharf in Monterey, California prompted overseeing regulatory agencies to require the property owner to remove the USTs and associated piping. The Site layout is shown on Figure 1. Engineering analysis focused on the removal of two 38,000-liter (10,000-gallon) USTs, as explained below, however, five other tanks and approximately 60m (200 ft.) of underground product piping were removed during this project.

A number of constraints complicated this project. Most significantly, a high-pressure (690 kPa [100 psi]) water supply line which served over 250,000 people in the area was only about 18 cm (7 in) north of the northern UST. The water company told us that the line could not be shut down or relocated, and that it could not tolerate any horizontal or vertical deflection. The line was constructed in 1907 and consisted of ductile iron pipe, 41 cm (16 in.) in inside diameter, with wall thickness of 2 cm (3/4 in.) and joints spaced every 3.7 m (12 ft.). The joints were bell-and-spigot type with lead sealant. The top of the pipe was at a depth of about 1m (3 ft.) below the ground surface. When exposed, the pipe appeared to be in fair condition.

If a break in the water line was to occur, we expected that much of the site, including petroleum-affected soils, would have been washed into Monterey Bay. The water company informed us that after a break was reported, it would take them ½ to ¾ hour to close the two valves necessary to isolate the break.

A second complication was that a high-voltage (21,000-volt) electrical line was situated about 1.5 m (5 feet) south of the southern UST. To add to the complexity, this electrical line ran under a well used, asphalt/concrete (A/C) paved bicycle path. The City required that this path remain open during our tank removal project, or, if this was not possible, that an alternate paved path be constructed nearby for bicyclist's use during our project. The owner selected the latter option, thereby funding construction of a 230 m (750 feet) long by 2.5 m (8 ft) wide A/C paved path, which was removed after UST removal activities were completed and the original path was restored.

Additionally, due to the leakage from the USTs, hydrocarbon vapors were a concern both for potential exposure to the public (the site was near a beach and public wharf) and for explosion hazards associated with sparking equipment. The critical features around the USTs are shown on Figure 2.

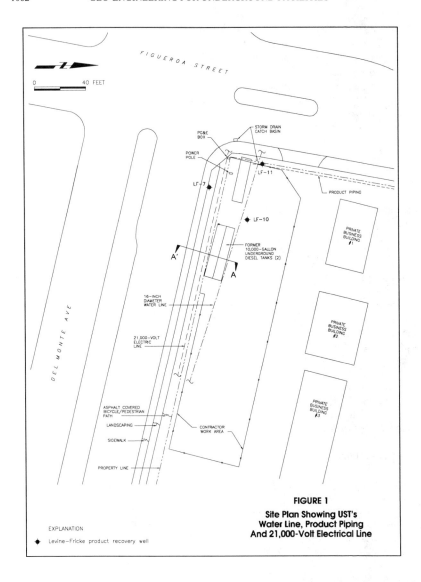

FIGURE 1
Site Plan Showing UST's
Water Line, Product Piping
And 21,000-Volt Electrical Line

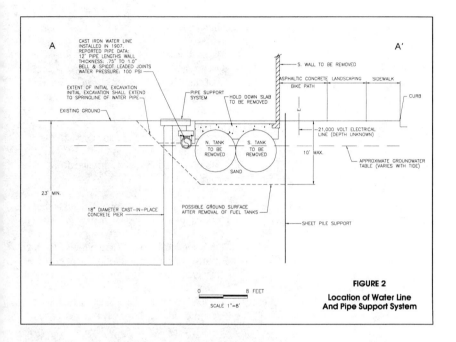

FIGURE 2
Location of Water Line
And Pipe Support System

The USTs were situated directly below a ½ m (1.5 ft) thick concrete hold-down slab. The tops of the USTs were approximately ½ m (1.5 ft) below the ground surface and the USTs were 2 m (6.7 ft) in diameter. The excavation necessary to remove the USTs and some petroleum-affected soil would undermine the water line and electrical line if support mechanisms were not constructed. To conduct this project safely and effectively, we needed to develop ground support systems which could allow removal of the USTs and associated piping without compromising the integrity of the surrounding lines or allowing excess exposure to petroleum vapors.

Site Soil Conditions

Soils consisted of poorly-graded beach sand, SP (Unified Soil Classification System). Using Standard Penetration Test (SPT) empirical correlations and a Factor of Safety of 2, we estimated a skin friction value of 12 kPa (250 psf) and an internal angle of friction of 35 degrees. We estimated a moist soil density of 18.9 kN/m^3 (120 pcf). The groundwater ranged from 2 to 3m (6 to 10 ft) below ground surface based on tidal and storm influence. After encountering the water table, the soil conditions did not significantly change with depth. The shallowest bedrock unit was estimated to be greater than 15m (50 ft) below the ground surface.

Selection of Water Line and Electrical Line Support Methods

The primary question of the approach to this project was how to best support the water line and allow removal of the USTs. The support system needed to support the water line from only one side in order to permit removal of the USTs. As noted above, no sparking could be produced during installation. In general, the system needed to have very high reliability due to the risks involved with the project.

After considering a variety of alternatives, we selected a pier and haunch system to support the water line, and steel sheetpiling to provide earth support for the electrical line. The sheetpiling consisted of 10 m (30 ft) long, z-section sheets, driven and/or vibrated approximately 8 m (25 ft) below ground surface.

The pier and haunch system consisted of eight reinforced concrete piers drilled and installed 1 m (3 ft) north of the water line. The concrete was a 7-sack mix with a strength requirement of 3,000 psi in 3 days. Each pier contained four #7 reinforcing bars. The piers were drilled to 7 m (20 ft) depth and were extended 1 m (3 ft) at the top so the top of the pier was above the water line. The water line was connected to the piers via three steel beams as shown in Photograph 1. U-bolts were placed under the bottom of the water line and were bolted to the bottom beam. The connection is also shown in Photograph 1. The question remained as to whether the piers could carry the vertical and lateral load with little or no deflection once the USTs were removed and the excavation was open.

Photograph 1-Water Line Support System

Analysis of Potential Water Line Deflection

Modeling the field conditions consisted of selecting design loading conditions and soil loading criteria.

Loading Conditions

To start the modeling of the load-deflection (p-y) relationship, we considered the following loads on each pier: the weight of the water line itself, the weight of the water within the line, the haunch and steel beams, and the lateral soil load based on the assumption that the sand on the tank side would fall away. The lateral soil load on each pier was calculated to be 1.7 kN (3,800 lbs). Additionally, the moment due to the offset of the loading was calculated to be 12kN-m (9,000 ft-lbs).

Design Criteria

Given the selection of the support system and the soil conditions, it was our opinion that some amount of vertical and lateral deflection could be expected. Since the water company would not assist us in selecting a design deflection, based on our experience, we selected a vertical and lateral deflection of 0.25 cm (0.1 in) as the design criteria. This was based on our level of confidence in the accuracy of our modeling, our estimate of the realistic amount of displacement the pipe could experience over a section without producing a leak or rupture, and the accuracy of our survey measurements during field monitoring. The sizing of support members was based on these design criteria. However, we made all reasonable efforts to reduce the potential for deflection. In the vertical direction, the piers were loaded using the pipe and water loads as a reaction through the bolting process thus pre-loading and pre-settling the piers. In the lateral direction, lateral earth forces (active pressure) were reduced by removal of soil in the upper portion of the pier against the pier and support system members.

Vertical Deflection

The design settlement for the piers of 0.25 cm (0.1 in), was used to obtain proposed pier diameters, through the use of Meyerhof's equations of area of influence versus allowable pressures versus blow counts corresponding to 2.5 cm (1 in) of settlement. An equivalent area larger than the cross-sectional area of the pier was obtained by including the area around the base of the pier created by extending a line from 1/3 of the way up the pier at a 30 degree angle down to the elevation of the bottom of the pier. The load applied to this equivalent area was used to calculate an equivalent pressure. This pressure was compared to pressures used in Meyerhof's equations corresponding to 2.5 cm (1 in) of settlement, to backcalculate 0.25 cm (0.1 in) settlement. Results of this analysis indicated that a pier of 0.5 m (1.5 ft) diameter would settle less than 0.25 cm (0.1 in). Using this diameter, analysis of lateral deflections was then conducted.

Lateral Defection

Using the computer program "PILE D/G" the pier diameter and depth used in the settlement analysis were input to compute the lateral deflection. PILE D/G is based on Reese's solution of the differential equation involving a beam on an elastic foundation (1977). We calculated that, based on using 8 piers located between waterline joints (and offset 1 m [3 ft] from the water line) along the anticipated sidewall of the excavation, the piers would have to be 0.5 m (1.5 ft) in diameter and 7 m (23 ft.) deep to obtain a lateral deflection of approximately 0.25 cm (0.1 in). Figure 3 shows the p-y curve generated.

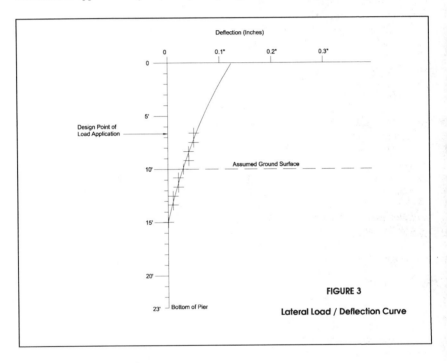

FIGURE 3
Lateral Load / Deflection Curve

Field Installation

The field installation of the pier and haunches was labor-intensive. Heaving sands were encountered on two of the borings. Steel casings were used in the boreholes as the augers were advanced below the groundwater table. The sands above the groundwater table produced relatively little caving. We think this was mainly due to the capillary forces created by a tidal groundwater condition. Piers were formed above the initial excavation using sonatubes.

When the piers had reached design strength, the steel beams were added. The upper beams (W12X65) were bolted to four threaded rods that extended out of the top of each pier. These beams were placed perpendicular (in plan view) to the water line and extended from the top of each pier to the water line. The continuous beams (W10X22) were then bolted to the bottom of the upper beams so that they were situated directly above the water line. The lower beams (W10X22) were then bolted to the bottom of the continuous beams on either side of the exposed water line joints. U-bolts were positioned under the bottom of the water line and bolted to each of the bottom beams. Photograph 1 shows the pier and haunch configuration.

When the pier and haunch system was complete, we were ready to begin UST removal. The 0.5 m (1.5 ft) thick reinforced concrete hold-down slab situated above the USTs was broken up and removed. Removal activities were completed as quickly as possible, to reduce the time the excavation would be open. The USTs were removed, leaving an excavation of about 12 m (40 ft) by 3 m (10 ft) and 3 m (10 ft) deep (see Photograph 2). Although some sloughing occurred, the excavation walls held remarkably well. The electrical line on the south side was protected by the steel sheetpiling. The east and west walls of the excavation experienced little sloughing. Both water and free-phase petroleum hydrocarbons entered the excavation while it was open, and we pumped out and removed as much product as time allowed. Both USTs were observed to have suffered serious corrosion damage. Many holes were noted in the USTs, and significant pitting and pocking were apparent.

Photograph 2 – Support System After Tank Removal

Decommissioning

The excavation was backfilled in such a way that subsequent vapor extraction and/or bioventing could be conducted. The bottom 2 m (7 ft) was backfilled with 4 by 2 cm (1.5 by ¾ in) drain rock encapsulated within a Mirafi 140NS geotextile. Approximately one m (3 ft) of Lapis #110 sand was placed and compacted above that, and the upper 20 cm (8 in) was backfilled/compacted with 2 cm (3/4 in) Class 2 aggregate base. Slotted PVC piping was placed in the excavation for possible future use as bioventing and vapor extraction system components. A product recovery system that had temporarily been dismantled, was re-installed. The steel sheetpiling was removed. The void beneath the water line was filled by water jetting. The U-bolts and steel beams of the water line support system were dismantled and removed. The concrete piers were left in place, and metallic tape was placed on the top of each pier at the request of the water company.

The UST and piping removal was successfully completed without damage to the adjacent utility lines, interruption of service, or significant release of gasoline vapors to the area in the vicinity of the project.

Monitoring Results

A surveyor was retained to monitor lateral and vertical locations of the water line before and after the excavation. Elevations of seven exposed bell joints were surveyed. No movement was detected in either direction to an accuracy of 0.25 cm (0.01 ft). This was in agreement with our modeling which calculated approximately 0.21 cm (0.1 in) in both the vertical and lateral directions.

Summary and Conclusions

Difficult access conditions existed at the site requiring an innovative support system that would allow the planned excavation without endangering the safety of the in-service water line. Soil design criteria were selected based on relative density measurements in the beach sand and engineering experience. Allowable vertical and lateral movement criteria were selected by the design team based on experience and attainable performance. An engineered support system was selected which consisted of drilled reinforced concrete piers offset from the waterline, a steel haunch and beam system, and U bolts on either side of the pipe joints. The system was designed and installed.

The system met all of the design criteria and supported the water pipeline during the tank removal and backfill process. Measurements of the pipeline before and after excavation and partial backfilling were performed to verify the proper functioning of the support system.

The calculations of vertical and lateral deflection based on published and experienced based soil strength values and standard computational methods were sufficiently accurate to predict the measured performance. Prudent construction practices to reduce pipeline and support system deflections aided in the successful outcome of this project.

Acknowledgements

The authors with to acknowledge Mr. Norm Brudigam P.E. who assisted on the project with structural calculations and support system design.

References

Resse (1977) Laterally Loaded Piles: Program Documentation, Journal of the Geotechnical Engineering Division of ASCE, v. 103, No. GT4.

Geosoft (1987) PILED/G Computer program and manual: Lateral Load Analysis of Drilled Piers and Piles,

Prediction of Settlement in Bangkok Ground Due To Tunnel Excavation

Tiew Wannipa [1]
Teachavorasinskun Supot [2]

Abstract

A non-linear elastic soil model was formulated based on the results obtained from the isotropically consolidated undrained triaxial extension tests. Tests were carried out on the undisturbed samples of soft to medium clays collected from a site nearby a location where previous information of ground movement due to tunnel excavation was available. The proposed model related the reduction of the normalized stiffness; i.e., ratio of the tangent Young's modulus to its initial value, to the effective stress ratio; i.e., ratio of the deviator stress and the effective mean stress. The two parameters required by the model were empirically dependent only on the initial confining stress. While the effect of soil types used in the study was negligible. The soil model was used in the finite element analysis program 'CRISP' to calibrate its applicability in comparison to other models; e.g., Modified Cam Clay and Drucker-Prager models. In the program, the initial stiffness was also a function of the initial effective mean stress. It was found that the computed vertical and lateral movements obtained from the proposed model using a stress relaxation of 3% were similar to the observed ones. Note that during construction of the tunnel, the measured amount of ground loss was also about 2%. The other two soil models tested required very high percentages of stress relaxation to match observed field deformations; especially the Drucker-Prager model.

Introduction

The first tunnel for supporting the mass rapid transit system in Bangkok will be bored in the beginning of 1999 and it is very important that the response of Bangkok ground to tunnel excavation is better understood. Although there have been several small water supply and sewerage tunnels constructed in the past ten

---

[1]Graduate student, [2]Lecturer, Department of Civil Engineering, Faculty of Engineering, Chulalongkorn University, Phayathai Rd., Pathumwan, Bangkok 10330, THAILAND

years, the knowledge and experience developed from those past projects were very limited due to the lack of good monitoring and managing systems (Thongyot, 1996 and Teachavorasinskun, 1997).

The changes of stresses around tunnel opening can be schematically shown in Table 1. The paths of stresses may be summarized as following;
1. Above the *crown*, the overburden stresses, σ_v, is almost kept constant with increasing in the horizontal shear stress, τ_h and τ_v.
2. At the *spring line*, the lateral stress, σ_h, is decreases with almost constant overburden stress, σ_v.
3. Beneath the *invert*, the overburden stress, σ_v, decreases with nearly constant lateral stress, σ_h.

It can be seen that the changes of stresses of the ground during tunnel excavation are mostly unloading, except for the soil mass above the crown. In order to predict the ground movement, the constitutive equation should be properly formulated according to these actual stress paths as recommending in Table 1.

Table 1. Schematically Changes of Stresses at Various Locations Around the Tunnel Opening Before and After Excavation

Location	Before tunnel excavation	After tunnel excavation
Crown (Direct shear test)	σ_v ↓, ← σ_h	σ_v ↓, τ_h, ← σ_h, τ_v
Spring line (Lateral unloading using triaxial)	σ_v ↓, ← σ_h	σ_v ↓, ← $\sigma_h - \Delta\sigma_h$
Invert (Triaxial extension)	σ_v ↓, ← σ_h	$\sigma_v - \Delta\sigma_v$ ↓, ← σ_h

In the present study, a series of triaxial extension tests were performed out and a simple stiffness variation model expressing the reduction of the Young's modulus with shear stress was proposed. The performance of the model was, then, calibrated by using the FEM analysis. Triaxial extension tests on Bangkok clay (Nong Ngoo Hao clay) had been carried out previously as reported in Li (1975) for isotropically consolidated samples and in Santichaianant (1995) for K_0-consolidated samples. However, those results were not accurate enough for using in modelling the behavior of soil.

Samples and Equipment

The undisturbed samples were collected from a borehole located on the campus of Chulalongkorn University. The site was chosen because it is located close to the existed water supply line constructed by tunneling using the earth pressure balance (EPB) method and ground movement data were collected during construction. The soil profile at the borehole could be divided into two zones; i.e., soft clay (-3.0 ~ -9.0 m) and medium clay (-9.0 ~ -14.0 m). The physical properties of the soft and medium clays at levels of -4.0, -8.0 and -12.0 m from the ground surface are summarized in Table 2.

Figure 1 shows pictures of the triaxial apparatus employed in the study. The hydraulic servo-valve used for controlling the vertical load is installed on the top platen of the triaxial cell and, therefore, no loading frame is required. Load cell and LVDT are placed inside the triaxial cell in order to increase the accuracy load and displacement measurements. All samples were trimmed to have a diameter and height of 35 and 70 mm, respectively. After the sample was pressure saturated to attain a B-value greater than 95%, the sample was left to consolidate under the prescribed isotropic stress for 12 hours and, then, sheared to failure under the extension mode. The analog output from all transducers was converted and digitally stored for further processing. A conventional odeometer test was also performed to obtain some other necessary parameters for the Modified Cam Clay model.

Table 2 Physical properties of the soils used in the experiment

Location	w (%)	LL (%)	PL (%)	PI (%)
Soft clay (-4.0 m)	55.6	62.2	19.6	42.6
Soft clay (-8.0 m)	58.4	77.6	23.8	53.8
Medium clay (-12.0 m)	56.3	83.8	26.5	57.3

Figure 1 Picture of the Triaxial Apparatus Used in the Study

Triaxial extension test results

Figure 2 shows the relationships between the deviator stress, $q = \sigma_v' - \sigma_c'$, and the axial strain, ε_a, obtained from a series of triaxial extension tests on the undisturbed samples collected from three different depths from the ground surface. Note that σ_v' and σ_c' are the effective vertical and confining stresses, respectively. The undrained shear strength ratios, $S_u/(\sigma_v')_{ini}$ derived from each test are summarized in Table 3. It can be seen that the samples collected at -8.0 and -12.0 m. are very similar. The corresponding stress paths obtained from those tests are shown in Figure 3. These are used for the determination of the critical state parameter used in the Modified Cam Clay model.

Table 3 Summary of the Measured Undrained Shear Strength Ratios and Initial Young's Modulus (from Triaxial Extension Test)

Sample	Initial overburden stress, $(\sigma_v')_{ini}$ (kPa)	Undrained shear strength ratio, $S_u/(\sigma_v')_{ini}$	Initial Young's modulus, E_{ini} (Mpa)
- 4.0 m	30	0.30	6
-8.0 m	50	0.49	13
	250*	0.41	32
-12.0 m	80	0.46	14
	250*	0.50	40

* Addition test performed at higher stresses

The initial tangent Young's modulus, E_{ini}, was computed at the axial strain level of 0.1%. Values of E_{ini} calulrate obtained from each test are shown in Table 3. The effect of the initial confining stress on the measured initial Young's modulus can be expressed using the conventional power law, where the power factor, n, is about 0.65. The normalized Young's modulus, E_{tan}/E_{ini}, is plotted against the total stress ratio, q/p, and shown in Figure 4. It was found that a simple exponential function could be used to fit the test data. The proposed function can be written as:

$$\frac{E_{tan}}{E_{ini}} = m \exp\left(n \frac{q}{p}\right) \quad (1)$$

where m and n are constants. The parameter m should have a value of 1.0. It was observed from the test results that parameter n is solely a function of the initial confining stress as can be seen in Figure 5.

The increase in the initial confining stress results in the decrease in the computed value of n. For the sake of easiness in implementation in the FEM program, this set of data is fitted by the simple expression as:

$$n = -0.0261\sigma'_c + 12.945 \qquad (2)$$

The variations of soil types used in the present study do not seem to affect the value of parameter n.

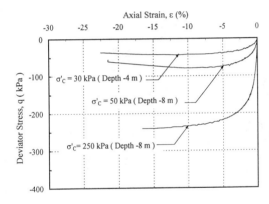

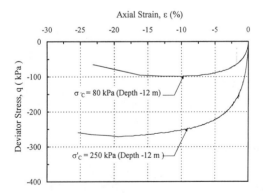

Figure 2 Stress-Strain Relationship Obtained From Undrained Triaxial Extension Tests

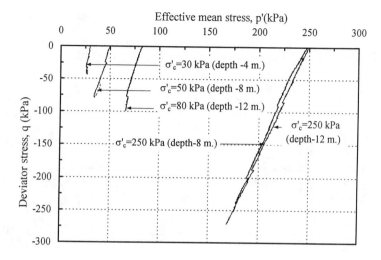

Figure 3 Stress Paths Obtained From Undrained Triaxial Extension Test

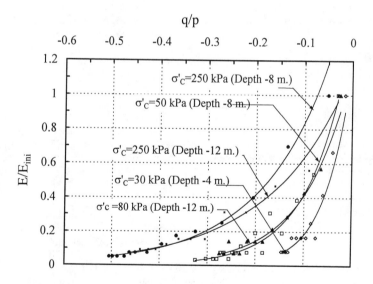

Figure 4 Relationship between the normalized Young's modulus and stress ratio.

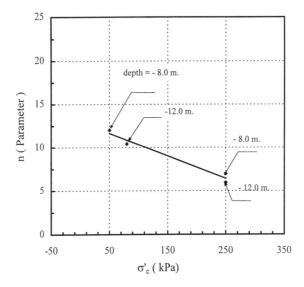

Figure 5 Variation of Parameter *n* with the Initial Confining Stress.

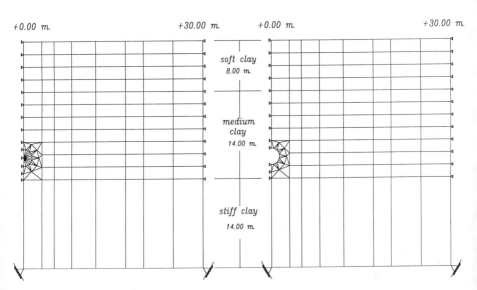

Figure 6 FEM Domain Used in the Analysis (tunnel diameter = 2.66 m).

Analytical result

The proposed stiffness variation model (eqn.(1) and (2)) was used in a FEM program called 'CRISP' (Britto, 1987) to calibrate its applicability. To be able to accurately set up the initial stiffness of the soil element, eqn.(3) which expresses the ratio of initial Young's modulus at the value p of interest to initial Young's modulus at a reference value of p was used.

$$\frac{(E_{ini})_p}{(E_{ini})_{p_{ref}}} = \sqrt{\frac{p}{p_{ref}}} \qquad (3)$$

where $(E_{ini})_p$ and $(E_{ini})_{p_{ref}}$ = the initial Young's modulus when the initial mean stress are p and p_{ref}, respectively,

The FEM domain used in the analysis is shown in Figure 6. The tunnel, with an inner diameter of 2.66 m, was constructed in the medium clay layer using the Earth Pressure Balance (EPB) method. The thickness of the segmental reinforced concrete lining is about 0.20 m. During construction, the ground movements at a section nearby Chulalongkorn University were monitored (Thongyot, 1996). The properties of soils tested in the present study may, therefore, be assumed to be representative for the analysis.

Figure 7 shows the computed and observed ground surface settlements. The analyses run using three different material models; i.e., the Modified Cam Clay model, Drucker-Prager model and the model proposed by this study. The stress relaxation adopted in the analysis was randomly varied to fit the observed result. For the proposed model, a stress relaxation of 3% is found to be the most appropriate. Note that if the gap between liner and soil was not developed and stiffness of liner been much higher than that of the surrounded soil, liner will carry all load and very small deformation can be taken place. This is not the actual situation taken place in the field. It is finally decided that the liner element was removed and the total excavated forces transformed from the excavated elements were applied step by step. The ratio of the amount of applied excavated forces to the total one was formed stress relaxation. With this very low amount of stress relaxation, the Drucker-Prager model greatly underestimates the settlement of the ground surface,while the Modified Cam Clay model overestmated the maximum settlement by nearly two times . Note that with the elastic-perfectly plastic model (Drucker-Prager model), the stress states will be brought to failure only when the total amount stress relaxation is assumed. It should also be noted that the amount of ground loss estimated during construction was about 2-3%, which is very similar to the stress relaxation requires for the proposed model. In a tunnel of different dimensions, this may not be true. The distribution of the vertical ground movement from the ground surface to the crown level is shown in Figure 8. The result computed from the proposed model is very similar to that observed from the field. For the lateral movement, the result from the inclinometer installed near the tunnel is shown in Figure 9. All computed values overestimate the observed data. However, it may be seen that the proposed model gives the closest prediction of the computed values.

Conclusion

The very simple stiffness variation model was found to be successfully applied estimating the ground movement during tunnel excavation in Bangkok clay. The model was formulated based on the results obtained from a series of undrained triaxial extension tests performed on undisturbed soft to medium clay samples. The samples were collected from a nearby a location where previous ground movement records due to tunnel construction were available. The ground movements computed by the FEM analysis using the proposed model were fairly very similar to the monitored results especially for the vertical ground movement. The proposed model was considered to be to other models because it is simple and required only one model parameter for the analysis.

Reference

Britto,A.M. and Gunn,M.I. (1987):"Critical State of Soil mechanics Via Finite Elements,Ellis Horwood Limited, London.

Li, Y. G.(1975):"Stress-Strain Behavior and Strength Characteristics of Soft Nong Ngoo Hao Clay Under Extension Condition", Master Thesis, Asian Institute of Technology, Bangkok, Thailand.

Santichaianant, K., (1995):"K_0-Consolidated Triaxial Extension Behavior of Soft Nong Ngoo Hao Clay," Master Thesis, Asian Institute of Technology, Bangkok, Thailand.

Teachavorasinskun, S. (1997)"Tunnel Excavation in Bangkok: Analytical and Field Measurement",Proc. of 9th Int. Conf. on Computer Methods and Advances in Geomechanics,Wuhan,China,November 1997.

Thongyot, T. (1996):"Ground Movement Associated with 11-km Water Transmission Bored Tunnels in Bangkok Subsoils," Master Thesis, Asian Institute of Technology, Bangkok, Thailand.

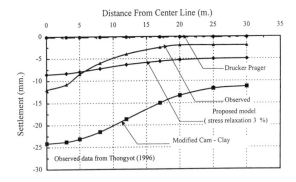

Figure 7 Computed and Observed Ground Surface Settlement

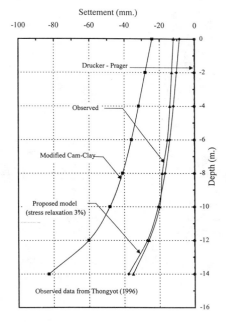

Figure 8 Computed and Observed Distribution of Vertical Ground Movement

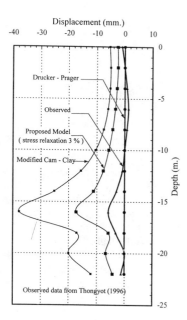

Figure 9 Computed and Observed Lateral Ground Movement

A Field Observation and Prediction in Soft Ground Tunnel
By Santiago F. Zapata[1], Jeffry Budiman[2], Theodore Maynard, Members, ASCE

Abstract

An instrumentation program was carried out during the construction of a 2.4 meter diameter tunnel to install water supply line. The objective of the instrumentation was to monitor the ground movement during construction of the tunnel and the field data was used to compare with the prediction from analytical solutions. The elevation of the center of the monitored tunnel was 6.7 m below the existing ground surface; the ground water table was -4.3 m., the instrumentation consists of three extensometers, two piezometers, and four inclinometers. Analytical analysis was performed using the ABAQUS finite element program where modified Cam-Clay and Drucker-Prager soil constitutive models were utilized.

Inroduction

An instrumentation program was carried out during the construction of a 2.4 m diameter tunnel to install water supply line. The City of Chicago supervised the construction of the tunnel.

Durting the Summer 1992, the City of Chicago supervised a segment of tunnel starting at West Lawrence Avenue and stretching Northwest along Higgins Avenue approximately two miles to North Austin Avenue. The tunnel is constructed into hard

---

[1] Formerly Doctoral Graduate Student, Dept. of Civil ánd Arch. Eng., Illinois Institute of Technology, 3201 S. Dearborn, Chicago, IL 60616.
[2] Assoc. Prof., Dept of Civil and Arch. Eng., Illinois Institute of Technology, 3201 S. Dearborn , Chicago, IL 60616.
[3] Formerly Chief Soil Mechanics Section, Department of Transportation, City of Chicago, 1224 West Van Buren Street, Chicago, IL 60607.

cohesive soils, the crown to a depth of 5.50 m (18ft.) below ground surface. The instrumented section is located at North Avondale Avenue, the tunnel section with a length of 9.6 m (32).

Regional Geology

Almost all of the subsoil in the Chicago Area was deposited during the glacial epoch. The glacial deposits consist largely of material picked up in the vicinity which is composed primarily of clay and silt. The deposits consist of large masses of clay that were laid down by the last great continental ice sheet.

Local Geology

Some areas in the Chicago area were lower than the level of Lake Michigan, on account of the raising of the grade, the upper eight to twenty feet of soil consist of miscellaneous fill (crushed stone, cinder, gravel, sand, broken bricks, topsoil, etc.). Beneath the fill is commonly found a gray, lake-bottom silt of variable quality, but usually loose and compressible. It has a thickness of about six feet. Approximately at datum is a crust of partially oxidized blue and yellow clay which acquired its stiffness by desiccation during a temporary low-water stage of the geological forerunner of Lake Michigan, (Peck,1948). Although the strength and thickness of this crust vary , the layer is one of the most important features of Chicago subsoil, because is the only material of high quality at shallow depth. Most of the early sky-scrapers were founded upon it, and the crust undoubtedly served to spread the loads from their footing before the pressure was transmitted to the soft clay beneath. The soft clay deposit beneath the oxidized crust was the principal seat of settlement of early buildings.

Tunnel Construction

Tunnel operation was accomplished using a Lovat Model M-90 Tunnel Boring Machine (TBM). The TBM was specifically modified for this project to provide the required 2.44 m (8 ft) diameter bore and a length of 6.1 m (20 ft). The skin plate of the shield was 3.2 cm (1.25 in) thick. This TBM was fully shielded including a 360^0 tail shield and was equipped with an articulated cutting head. Six jacks with a theoretical capacity of 100 tons were distributed uniformly around the shield just inside the cylindrical wall. Actual hydraulic pressure limited each jack capacity to 80 tons. The jacks were designed to react on steel ribs of the temporary lining installed behind the shield. These jacks had a stroke of 1.68 m (5.5 ft) permitting erection of a maximum of five 1.5 m (5) ft. long sets within the tail shield. The average rate boring was 2.0 m/h.

Instrumentation

To measure the three-dimensional pattern of soil displacements during tunneling, a 9.75 m (32 ft) tunnel section was instrumented. The instruments were installed in a staggered grid arrangement. One inclinometer was constructed through the center line of the tunnel, and the others on the side with various distances from the tunnel center. The extensometers were also installed in the similar configuration. Piezometers were installed on the side of the tunnel.

Four longitudinal alignments designated as A, B, C, and D were selected for instrumentation as shown in Figure 1. Four (4) inclinometers designated as I1, I2, I3 and I4, were installed to measure the lateral movements. Three (3) multiple-position extensometers, designated as E1, E2 and E3 were installed at location as shown in Figure 2 to measure vertical displacements above and next to the tunnel. Two (2) piezometers, designated as P1 and P2, were installed to measure of pore water pressure in the soil. Instruments were regularly monitored as the tunnel advanced through the test section. Field measurements were correlated to the position of the shield relative to the particular instrument. In addition, standard surveying techniques measured ground surface settlements. For these ground surface settlement a survey reference point was installed 30.5 m (100 ft) east of the test section.

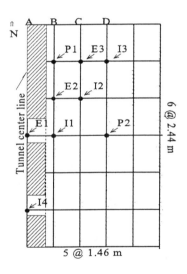
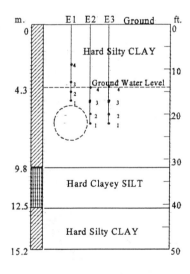

Figure 1. Location of the Instrumentation   Figure 2. Anchors of Extensometers E1, E2 and E3

The field instrumentation was designed to measure the three dimensional patterns of displacements during tunneling.

Finite Element Analysis

The ABAQUS (1996) program was used for FEM analysis. Time dependent responses of ground movement are not considered in this work.

Finite Element Mesh.

In this paper the mesh consists of 276 three-dimensional elements and 476 nodes. The finite element mesh consists of 8-node linear brick, reduced integration and hourglass control, C3D8R. In addition to the above case, the results obtained for a three-dimensional mesh with hybrid elements are also compared, these elements are primarily intended for use with incompressible and almost incompressible material behavior. C3D8R elements are considered to be superior than C3D8H elements, 8-node linear hybrid, because the former element in the ABAQUS' input file the program is asked to invert a singular matrix. The displacements from this analysis are designated as: U1, transversal displacement, U2, vertical displacement and U3, longitudinal displacement.

Soil Parameters.

Two types of elastoplastic models were considered: modified Cam-Clay and Drucker-Prager. The modified Cam-Clay model is rate independent, isotropic, strain hardening, elasto-plastic, non linear, and associated flow rule. Yielding does not imply failure in this work-hardening theory.

The Drucker-Prager failure surface is perfectly plastic (no hardening), but plastic flow on this surface produces inelastic volume increase, which causes the cap to soften.

The parameters for the Cam-Clay and Drucker Prager models are presented in Table 1 and 2 respectively.

Pressure at the face of the shield.

A uniform pressure is applied to the exposed face elements. The average pressure levels for slurry shield analysis is 114 kPa (2400 psf), Kasali (1982). Finno (1983) applied uniform induced pressure for EPB shield of 71.9 kPa (1200 psf) for the longitudinal analysis, and for the transverse analysis the pressure was 57.5 kPa (1200 psf). An average heaving pressure of 60.7 kPa (1260 psf) was used by for Shirasuna (1985) in his research to induce displacements in his model and compare to those observed in the field.

Information obtained from contractor's engineers who worked on in this project shows the average induced pressure was about 48.4 kPa (1000 psf) for the TBM shield.

The analysis presented in this section were made using the same induced pressure reported by the contractor.

Table 1. Cam-Clay Material Parameters for Chicago Clay

Soil parameter		Silty Clay	Clayey Silt
Slope of normal compression line	λ	0.03913	0.03696
Slope of unloading-reloading line	κ^*	0.00478	0.00348
Initial void ratio	e_0	0.3686	0.4050
Slope of critical state line	M	1.26	1.38
Effective angle of shearing resistance	ϕ'	31.5^0	34.5^0
Poisson's ratio	ν	0.30	0.30

$^*\kappa = C_r / 2.3$

Table 2. Drucker-Prager Model Material Parameters for Chicago Clay

Soil parameter		Silty Clay	Clayey Silt
Slope of unloading-reloading line	κ^*	0.010	0.0080
Initial void ratio	e_0	0.3686	0.4050
Angle of shearing resistance	β	37.7^0	45.9^0
Uniaxial Compression Yield Stress	σ_c^0	285 kPa	320 kPa
Poisson's ratio	ν	0.30	0.30

$^*\kappa = C_r$

Displacement Patterns

The first stage is to examine the effect of the induced pressure of the TBM in the surrounding soil of the tunnel, and the second stage the surface settlements developed after the mining of the tunnel. Over this face the advancing shield induces a heave pressure, which is transmitted through the model. This analysis focuses in the behavior of three vertical slices S1, S2 and S3 as shown in Figure 3 where the inclinometers I1, I2 and I3 are located. Vertical slice S1 is nearest to the tunnel line and vertical slice S3 is the furthest. Second stage, this analysis focuses on the longitudinal slice AB which is located directly above the tunnel and should be the most affected by the mining as shown in Figure 3.

Contours of computed values in meters over the slices S1, is shown in Figure 4. This lateral or transversal displacement (direction U1) is positive if the horizontal movement of the soil is towards to the side of the tunnel otherwise is negative. The longitudinal displacement (direction U3) is positive if the horizontal movement of the soil is away from the front of the shield otherwise is negative.

Transversal movements during shield passage

The transversal movements are shown for the situations, where the face of the shield is approximately 3.6 m (12 ft) behind the reference line, and approximately 6.1 m (20 ft). ahead of the reference line of S1. As the TBM shield approaches the reference line, the lateral movements are away from the tunnel. After the shield passes by, the lateral movements are toward to the tunnel. The predicted and observed movements for the advancing shield are shown in Figure 5.

Longitudinal movements during shield passage.

These measurements or observations show the amount of displacement which occur towards or away from the front of the shield. The predicted and observed movements for the advancing shield are shown in Figure 6.

Vertical movements.

After shield passage, the ground surface settle. Data from the ground surface survey shows data the maximum settlement was 1.2 cm (0.04 ft) over the center line of the tunnel. Readings was taken before and after the shield pass by the reference line, the maximum settlement was recorded at the fifth day after the shield passe by the reference line.

Figure 7 shows two finite element models modified Cam-Clay and Drucker-Prager which predict the vertical ground movement of the longitudinal slice AB after the shield passage. The Cam-Clay model predicted a vertical movement from 2.2 cm to 2.7 cm (0.87 in to 1.06 in) and the Drucker-Prager model predicted a vertical movement from 0.6 cm to 0.7 cm (0.23 in to 0.27 in). This analysis show that the ground surface does not settle uniformly.

Piezometer Measurements

The results of piezometers P1 and P2 reading are summarized in Figure 8. The figure provides the pore water pressure (pwp) development in the soil in a chronological sequence. First response shows as the shield approaches to the piezometer was measured, then pwp response as the shield passes the piezometers locations The readings were approximately constant and then reach the equilibrium, 35 kPa (5.20 psi) and 30.8 kPa (4.40 psi) for P1 and P2 respectively. The piezometer P1 is located at a radial distance of 15 cm (0.5 ft) from the tunnel (at the tunnel springline) increased

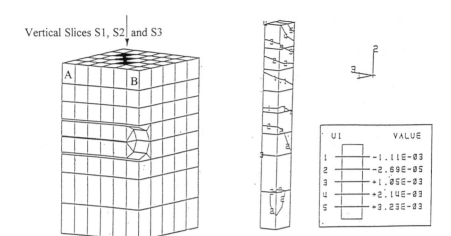

Figure 3. F.E.M. for 3-D Analysis of Tunnel Showing Top of Vertical Slices S1, S2, and S3, and the Longitudinal Slice AB on the Top

Figure 4. Modified Cam-Clay Model: Lateral Movement on Slice S1 in Meter

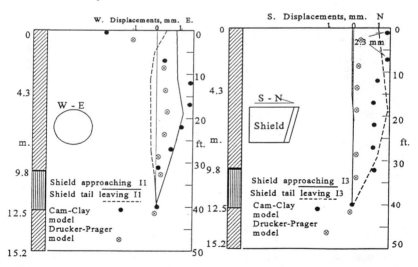

Figure 5. Observed and Predicted Lateral Displacements, Inclinometer I1

Figure 6. Observed and Predicted Longitudinal Displacements, Inclinometer I3

approximately 10 percent in porewater pressure (pwp) as the TBM approached to reference line and then decreased sharply when the shield continues beyond this location. The pwp in P2, which is located at a radial distance of 3.9 m (12.8 f) from the shield (near at the tunnel invert) increased approximately 7 percent and held constant for one day and then decreased as shield passed this location.

Palmer and Belshaw (1978) reported pwp increases about 10 percent as the tunnel approaches and decreases of as much 40 percent as the shield passes. In this analytical solution, negative pwp was obtained close to the tunnel while small positive values were observed in the elastic region further away from the tunnel.

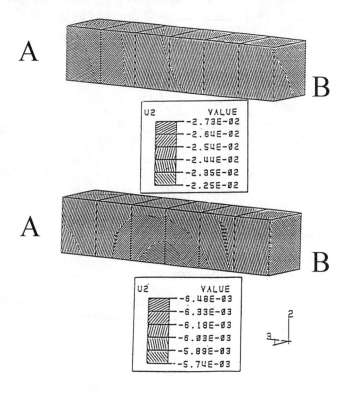

Figure 7. Predicted Vertical Displacements (U2) on Longitudinal Slice AB, Top: Prediction with Modified Cam-Clay Model. Bottom: Prediction with Drucker-Prager Model

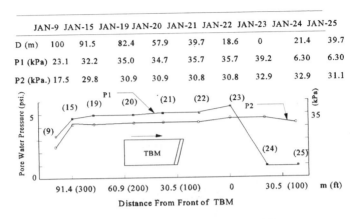

Figure 8. Pore Water Pressure Along Tunnel Line

Extensometer Measurements

Cross sections through each of the three instrument lines in this section are shown in Figure 2. Anchor points for each extensometer are shown by small solid circles. Also included in Figure 2 are the surface geologic strata that was logged from the soil borings for instruments installations, Figure 9 shows the vertical displacements of extensometers E1, E2 and E3.

Field data from E1 reveals that the vertical deformations above the crown of tunnel varies with the elevations. Three anchor points show downward movements, the magnitudes of displacements increases with the distances from the surface. However, anchor 1 from extensometer E1 shows reverse trend for unknown reason.

Six extensometers readings on E1, E2 and E3 were taken from January 20 to January 31 in two days intervals. These readings were taken manually with the help of micrometer.

Displacement Field

It is useful to examine the development of the displacements during construction operation. Figure 10 shows displacement field on a vertical plane through the enter line. It shows that with 74.2 kPa (1000 psf) pressure at the face, the soil undergoes heaving at the face, and, as the machine progressed, the movement into the tail void become apparent. These movement lead to the settlement of the ground surface as the shield passes a given reference point.

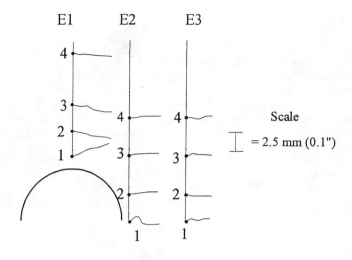

Figure 9. Extensometers Readings: E1, E2 and E3

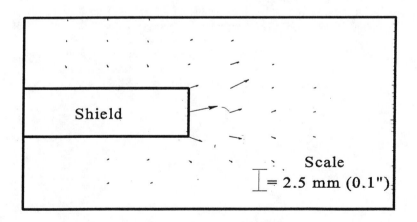

Figure 10. Displacement Field on Vertical Plane Along the Center Line of the Tunnel

Conclusions

Control of ground movement during tunneling is a major concern related to construction in urban environments. The new shield tunneling equipment which are radical departures from conventional methods to minimizing ground movements during construction. In spite of the heightened interest in this new construction technology, however, the actual mechanism of ground support generated by these machines is not fully understood, moreover, the variability of the ground conditions and the material properties significantes affect the overall behavior.

References

ABAQUS User's Manual, Version, 5.4, Hibbit, Karlsson and Sorenson. Hibbit, Karlson and Sorenson, Inc., Providence, R.I., 1996.

Bathe, K., "Finite Element Procedures in Engineering Analysis", Prentice-Hall Inc., Englewood Cliffs, New Jersey, 1982.

Finno, R. J., "Response of Cohesive Soil to Advanced Shield Tunneling", Ph. D. Thesis, Stanford University, Standorf, CA, 1983.

Kasali, J. G.,"Three-Dimensional Finite Element Analysis of Advanced Shield Tunneling", Ph. D. Thesis, Standorf University, Standorf, CA, 1981.

Palmer, R. B., and D. J. Belshaw, "Deformations and Pore Pressure in the Vicinity of a Precast, Segmented, Concrete-Lined in Clay", Canadian Geotechnical Journal, Vol. 17, pp. 174-184, 1980.

Peck, R. B., "History of Building Foundations in Chicago", Bulletin of University of Illinois, Published by University of Illinois, Urbana, Vol. 45, No. 29, pp. 10-13, 1948.

Shirasuna, K., "Finite Element Analysis on Cohesive Soil Behavior Due to Advanced Shield Tunneling", Ph. D., Thesis, Department of Civil Engineering, Virginia, Polytechnic Institute and State University, Blaksburg, Virginia, 1985.

Adding Realism to the Planning of Rock Tunnel Boring Machine Projects
Christopher Laughton[1], Priscilla Nelson[2], Yousof Abd Al-Jalil[3]

Abstract

Experienced tunnel planners will openly acknowledge that the idea of a Tunnel Boring Machine (TBM) drive proceeding according to any preconceived plan is illusory. However, despite this fact the planner will commonly develop a cost and schedule assuming that the tunnel will be excavated at a constant advance rate for the full length of the tunnel. This is a naïve assumption that fails to convey to management the very real possibility that the advance rate will vary and that the time and cost to complete the tunnel are liable to deviate from those defined in the project plan. Such plans rapidly loose their credibility once the tunneling begins and, in hindsight, they often prove to have been both inaccurate and over-optimistic. On site, operational inefficiency, breakdowns, high water inflow and adverse rock conditions all contrive to make the TBM advance rate far less predictable and less rapid than a study of the project plan would lead one to believe.

This paper summarizes how sets of case history TBM performance records have been collected and are used as input to a suite of models that predict both the time and cost to complete the excavation of a TBM-bored tunnel. The case history data faithfully report conditions and TBM performance over the full length of the drive. The models use these data sets to simulate machine performance on new sites. These predictions specifically take account of the impact of rock mass and machine factors on performance. The output from these models is presented in a probabilistic format that provides the manager with an objective definition of risk, expressed in terms of cost and schedule.

The data base format and the output from one of the models are described. The model was used to support the development of a conceptual design for the Very Large Hadron Collider (VLHC) project, a potential "next generation" high energy particle accelerator. The model output has also been used to identify the major cost drivers of the tunneling operation, and research is currently underway to identify cost-reduction measures that could be introduced in to the TBM-system.

Introduction

Owners and contractors who select the TBM option for tunneling may obtain cheap, rapidly constructed tunnel, but they also subject themselves to a high level of tunneling risk. In selecting the TBM, the success of the whole project becomes heavily dependent upon the performance of one equipment system. The performance of this system is, in turn, dependent on the reliability of its subsystems and the cooperation of the rock mass, which is both the material excavated and an integral part of the tunnel structure. In considering tunneling options, it thus behooves the planner of a tunnel project to develop a good understanding of both the strengths and weaknesses of a TBM operating in a given rock mass. In some instances, the planner may be able to develop such an understanding by drawing upon local tunneling experience in the same rock units. However, in most instances there will be a need to research and identify relevant experience from other remote sites. To service this need, TBM performance data bases have been compiled and published, including those of Bruce and Morrell (1970), Krause (1976), Ikeda and Nishimatsu (1980), Innaurato et al. (1980), Marin (1980, 1986), Muirhead (1982), Fourmaintreaux (1983), Sinha (1986), Parkes (1988), Giafferi (1989), Kuwahara (1990), Touran and Martinez (1991), Kovari et al. (1993), Howarth (1994), and the Association

---
[1] Fermi National Accelerator Laboratory
[2] National Science Foundation
[3] Terramar Inc..

Française des Travaux en Souterrain (AFTES) (1994). Proprietary data bases have also been compiled by individual consultants and manufacturers. However, their usefulness to the uninitiated planner may be less than that of the published data. Gertsch (1994) notes that proprietary data bases often include reference to non-standard tests and classifications which may be difficult to interpret or apply, in a predictive context with confidence.

One shortcoming of most of the data bases cited above is the absence of utilization (U) reporting. Data bases have tended to focus on Penetration Rate (PR) and cutter wear (CW) reporting. This focus was prompted by the marginal viability of earlier TBMs working in stronger, more abrasive rocks. The data base developed at the University of Texas at Austin provides summaries of TBM performance over units of time and distance, from start-up to hole-through, using the mean and standard deviation for all the performance parameters needed to develop meaningful predictions of TBM performance, namely PR, U, CW and advance rate (AR) (Nelson et al., 1994, Laughton, 1998).

Describing the TBM Case History Data Base

The data base uses layout, rock mass, and equipment parameters to describe each case history site. These parameters allow the planner to find case history sites with similar characteristics to those of the site for which a performance prediction is required. The performance parameters stored in the data base can then be used, either directly or through the application of predictive models, to support the development of realistic performance expectations for TBMs operating under "like-site" conditions.

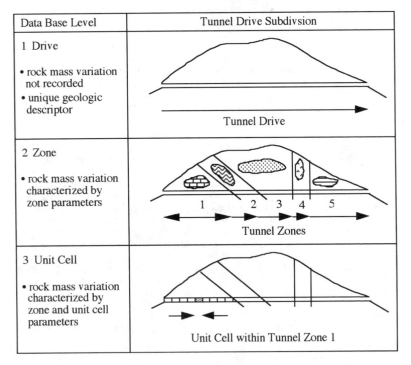

Fig. 1 : Data Base Levels Relative to the Degree of Rock Mass Definition

Four storage levels of data are recorded in the data base. The storage levels are used to describe the average performance of the TBM over smaller increments of time and/or distance. From Levels 1 through 3, a finer correspondence is established between rock mass conditions, rock mass behavior and TBM performance; this equates to a reduction in the time or distance of unit integration. Level 4 stores operational research parameters related to modeling the frequency and severity of TBM sub-systems' breakdown (Time Between Failures, TBF) and repair (Time To Repair, TTR), as recorded using reliability functions. A schematic representation of the physical scales of the reporting units for the Levels 1 through 3 is shown in Fig. 1.

In the data base, rock mass conditions, behavior and TBM performance are defined by reference to the drive, zone and unit cell. These terms are used to reflect the level of local geotechnical definition afforded by the site documentation. The tunnel drive is the full length of tunnel excavated by the TBM. The drive is sub-divided into zones using the basic steps outlined by Dearman (1988):

1) determination of the general or regional geologic structure,
2) identification of quasi-homogeneous rock mass zones, and
3) delineation of major faults and discontinuity features at tunnel depth.

A unit cell is a short length of tunnel (10 m) within a zone for which average TBM performance parameters, rock mass condition and behavior classes are recorded. These parameters are derived from the back-analysis of detailed contractor shift reports and geologic maps.

The data base storage levels are maintained in an Excel spreadsheet. The reporting format was selected to facilitate the entry and sharing of new data. Only standard and commonly reported site layout, geotechnical parameters, and machine specifications are used in the data base, with focus placed on identifying and registering the parameters that have been demonstrated to impact TBM performance.

Describing the Geotechnical Conditions in the Tunnel

Geotechnical parameters recorded in the data base provide for a Basic Geotechnical Description (BGD) of each zone. Where scaled geologic maps of the tunnel were compiled on site, the zone-based BGD was complemented by a pragmatic cell-by-cell classification of both rock mass conditions and behavior. A ranking of block size and alteration was recorded for each cell to define a rock mass condition or state. A complimentary ranking or classification of rock mass behavior was made based on the presence of fracturing and overbreak, as indicated schematically in Fig. 2.

Behavior-Classes	Evidence of Instability at the Heading	
I: Field-Massive	• bedding only • no mapped joints	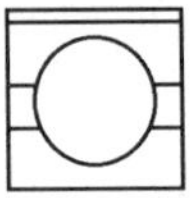
II: Fractured But Stable	• mapped joints present • no overbreak	
III: Local Fracture-Bound Overbreak	• local overbreak • limits of overbreak defined by planar fracture surfaces	
IV: Local Soil-Like Overbreak "Pockets"	• local overbreak • limits of overbreak defined by uneven weathered/altered surfaces	
V: Extreme Mining Areas	• extended overbreak length • heavily fractured and/or altered/weathered materials	

Fig. 2: Rock Mass Behavior Classification

These behavior classes were developed from the back-analysis of tunnels that had been excavated under relatively low stress, and low water inflow environments. On the tunnel maps, instability was attributed to gravity-driven fall-out of material, either discontinuity-bound wedges/blocks or soil pockets. This simple classification is not intended to provide a comprehensive classification of rock mass behavior. However, past experience would suggest that this pragmatic ranking can promote a higher level of uniformity in describing rock mass behavior for many tunnels where adverse behavior is caused by discontinuity features (Brannsfors and Nord, 1979, and Inokuma et al., 1994).

Target Applications of the Data Base
The TBM data base is designed to support the development of performance estimates during all stages of project development. At the earliest stage of the project, when only a modicum of site-specific data is available, the Level 1 data set can be used. As the site characterization progresses the Level 2, 3 and 4 data base levels can be used to provide for more accurate matching of the new and case history sites and more precise modeling of the excavation process, as shown in Table 1.

Table 1: Site Investigation Stages, after Dumbleton and West (1976)

Investigative Program Stage	Site Characterization Objectives	Tunnel Engineering and TBM Design Objectives	Data Base Levels
Desk Studies	define rock units, and general geologic structure	establish feasibility and select viable project option(s)	1
Investigation of Individual Rock Zones	define local structure, fluid and stress regimes, evaluate intact strength and abrasion	predict typical tunnel conditions, select construction method(s) and design	1-2
Investigation of "Suspect" Zones of Weakness	define extent and likely behavior of "suspect" stretches" of tunnel fault, fold or altered-zones	predict and mitigate any problematic tunnel conditions	2-3
Excavation	record rock mass structure, test intact strengths	record in-tunnel conditions, evaluate and optimize performance	3-4

The Level 1 data set is designed primarily to support desk studies, as conducted to evaluate project feasibility. The data set is general and descriptive in nature, but provides quantitative input to support alignment decisions. The Level 2 data set is designed to support selection of final alignments, shortlisting of methods and means, and identification of compatible design options. The Level 3 data set is designed to support the final selection of methods and means. The Level 4 data base is designed to support operational analyses of the integrated TBM excavation system based on a knowledge of the interdependence between activities, the activity durations, and the reliability of the TBM subsystems.

Describing Case History TBM Performance

TBM performance parameters, PR, U, AR and cutter wear data are taken directly from the source documents and summarized using basic statistical parameters (mean and standard deviation). The use of the standard deviation value provides the end-user with an indication of the degree of variation in TBM performance experienced on site. For simulation purposes, this performance data is fitted to parametric probability density functions (PDFs) such as truncated normal distributions. These performance PDFs can, in turn, be used in simulation to provide for a description of the excavation outcome, defined in probabilistic terms. The fitting process and use of PDFs in simulation are discussed in detail in Nelson et al. (1994).

In the Level 2 data base, the impact of major breakdowns and the traversal of Extreme Mining Areas (EMAs) on TBM performance was noted. Major breakdowns were defined as TBM breakdowns greater than one week in duration, these were typically associated with main bearing repair or replacement. EMAs were identified when the routine operation of the TBM was effectively suspended for a duration of more than one week. EMAs were typically associated with major instability and/or high water inflow at the heading.

The Level 3 TBM Performance Model

TBM performance can be simulated using a variety of modeling techniques that are specifically tailored to the limitations of the input. When site characteristics are not well defined, performance input is most appropriately modeled using the multiple case histories data sets that can be accessed in Level 1 and 2. Here, AR, PR or U data sets can be sampled and PDFs fitted and progress through a rock zone simulated on a daily, weekly or monthly basis.

If there is adequate similarity between the new site and a site within the Level 3 data base, simulation can be made on a cell-by-cell basis using the process shown schematically in Fig. 3.

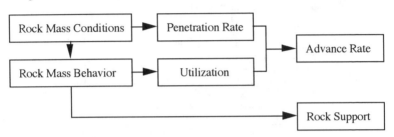

Fig 3: Prediction of TBM Performance using a Level 3 Data Set

To support a simulation of TBM performance using Level 3 data sets, the rock mass conditions are first generated within the zone. This is done using sets of probability data derived directly from the back-analysis of tunnel maps. Firstly, a Markov Transition matrix is used to model the cell-by-cell variability in representative block size along the tunnel. Once the sequence of cell block sizes has been generated a corresponding alteration state is predicted and a Rock Mass Condition Classification (RMC) of the cell defined.

Once the rock mass condition has been defined for each cell, case history-derived probabilities are again used to predict the class of rock mass behavior (RMB). PR and U PDFs are associated with each RMC and RMB class and these are sampled to simulate the variation in PR and U as a function of the rock mass conditions. Simulation of performance in each zone can be repeated until a histogram is adequately populated or until a PDFs can be confidently fitted for the time and cost to complete tunnel. The rock support levels of each cell is predicted probabilistically based on the RMB class.

Application of the Model to a Tunnel Project

The data base modeling has been applied to a number of projects. The simulation summarized here describes the input and results of a simulation undertaken to develop an estimate of a basic set of TBM-related excavation costs required to construct a particle physics accelerator. The accelerator is to be housed in a 34 km long tunnel, sited in dolomitic limestone. The simulation model supports the establishment of a baseline cost estimate and will help identify potential areas where TBM tunneling costs can be reduced. Construction planning and packaging of the project is based on the development of four construction units of one large access shaft and 8.5 km of tunnel, as shown in Fig. 4. The excavated diameter of the accelerator tunnel is 3.7 m.

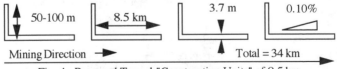

Fig. 4: Proposed Tunnel "Construction Units" of 8.5 km

For prediction of TBM performance, the rock mass is modeled using the data base Limestone 1 model. This geotechnical model uses probability data obtained from the back-analysis of a dolomitic limestone tunnel driven in Chicago that forms part of the Tunnel and Reservoir Project. Performance PDFs were selected from case history data sets to simulate PR and U for each classified cell along the alignment.

TBM and Rock Support Selection

For the tunneling simulation two 3.7 meter diameter open, high power TBMs and conveyor units were selected. The cutterheads were mounted with 430 mm cutters rotating at a fixed rate of 13.4 RPM. Two TBMs operating in parallel were selected to reduce the overall risk in the boring venture; potentially allowing one TBM to compensate for any under performance of the other.

Rock support was selected for three behavior classes as shown in Fig. 5, the steel support elements were assumed to be placed on a shove-by-shove basis.

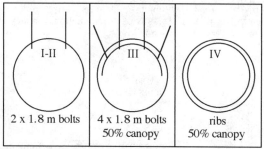

Fig. 5: Support Installed in the Behavior-Classified Rock Mass

The three classes of rock mass shown in Fig. 5, correspond to the four types of rock mass behavior observed in the limestone model. Classes I & II correspond to stable rock mass behavior under massive and fractured states respectively. Classes III and IV are associated with fracture- or alteration-controlled instability, respectively. Class V conditions were not observed in the mapped case history but were simulated probabilistically assuming an average distance between EMAs of 50 kilometers. The class V conditions correspond to the rock support and treatment required to negotiate an EMA. The EMA spacing corresponds to the experience of the TARP project in the mining of the layer-cake limestones, typical of the region.

Cost Input Data

The costs of large pieces of equipment, including TBM, back-up systems and conveyors, are new costs fully amortized over the length of the tunnel. The prices of equipment were based on budget quotes obtained from manufacturers.

Table 2: Unit Pricing of the Main Consumables

Items	Cost, $	Unit	Comment
Cutter Wear & Spares	21.2	cub.meter	normal wear of parts
Support Costs, I, II, III, IV	50-250	lin.meter	50, 100, 130, and 250 $
TBM Electricity	126.0	TBM h	1260 kWh @ 0.1 $
Site Electricity	100.0	Shift h	1000 kWh @ 0.1 $
Maintenance Allowance	100.0	per shift	for weekend maintenance

An allowance was added to capital costs of new equipment to account for in-tunnel utilities, rails, rolling stock and surface equipment (these items are assumed to have been largely amortized on previous work.). On this basis the total equipment cost to the project was estimated as being $532 per linear meter.

Table 2 shows the main consumable cost items associated with the TBM process. Cutter wear in the limestone was budgeted at $2 per cubic meter with spares budgeted at an equivalent rate. Power cost was taken as a nominal 10 cents per kWh. Machine power was accumulated on a TBM operating hour basis and site power accumulated on a shift hour basis. An additional maintenance allowance was added to account for weekend work conducted on the TBM and conveyor systems.

A shift crew size of 18 persons was used for estimating purposes and an average hourly rate of $35 assumed. The size of the crew and average hourly rate were considered to be a fair representation of local practices based on reviews of recent project estimates.

Simulation Results

To model TBM performance within this rock mass, forty simulations of a 4.25 km section of tunnel were performed. The results of the simulation, reported in terms of shift hours and unit excavation price, are shown in Fig. 6.

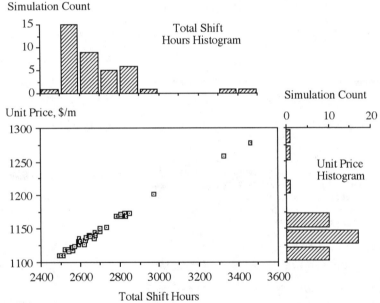

Fig. 6: Results for 40 Simulations of a 4.25 Length of TBM Driven Tunnel

The average cost for the construction of a meter of tunnel is 1147 $/m and the average advance rate (AR) 1.59 m/hr. The case history data on which the model input was based indicates that under field massive conditions the TBM would achieve average PR and U values of 6 meters/hour and 34 % respectively, resulting in an AR of 2.04 meters/hr. The simulation results in an average performance value substantially below that predicted for field massive tunneling (from 2.04 to 1.59 meters per hour). The reduction in performance noted is caused by the imperfect nature of the rock mass as varying rock mass conditions cause variation in TBM performance.

The outliers shown at the higher end of the scatter and histogram plots underscore the significant impact that the traversal of EMA conditions can have on overall performance. For the limestone rock mass such occurrences are rare and their impact on overall machine performance reduced. However, in less uniform geologic regimes (for example, basalt flows and highly deformed/faulted metamorphic or granitic rock masses) EMA delay can be expected to have a major negative impact on the bottom line of the excavation activity and the planner would be wise to focus on the detection and mitigation of likely EMA features before committing the TBM to the underground.

The break-out of costs for a single simulation are shown in the pie-chart in Fig. 7. In this limestone rock mass it becomes clear that issues of stability, abrasion and general maintenance are only secondary cost contributors. The most important cost drivers are clearly the costs of labor and equipment.

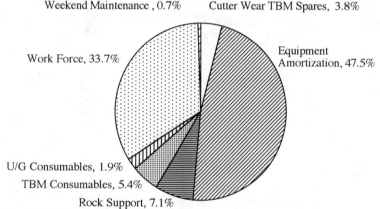

Fig. 7: Pi-Chart Sowing Distribution of Costs from Simulation 40

The results of these analysis have prompted proponents of the accelerator tunnel, Fermi National Accelerator Laboratory, to start studying new ways in which tunneling costs can be reduced. This work is being undertaken in partnership with The Robbins Company and Kenny Construction Company.

A detailed cost breakdown of the costs associated with building the accelerator tunnel, including shaft work, grouting and finish-out work that was not included in the scope of the TBM cost model has now been made. This cost breakdown is being used as a basis for identifying potential avenues for reducing the cost of the TBM construction process. The results of this initial study will likely be a list of recommendations to improve the utilization of the machine, increase the amount of automation of the underground activities and reduce the amortization of equipment costs.

Conclusions

The data base and modeling work described here is not promoted as a universal solution to the recording and simulation of rock masses and prediction of TBM performance. The amount of data back-analyzed has been too small to support such a claim. The simplicity of the simulation model is largely attributed to the relatively narrow range of rock mass conditions observed in the mapped tunnels. However, more common types of adverse rock mass behavior were encountered in the case history tunnels and a model established that can be further developed to incorporate the impact of other factors such as water and stress.

The model is simple to implement and the resulting rock mass classifications are compatible with the limitations the underground environment necessarily imposes on the practitioner. The parallel classification of the tunneled rock mass relative to both rock mass conditions and behavior provides for the establishment of a pragmatic method to record and model the effects that tunnel conditions have on TBM performance.

In predicting TBM performance, the planner must strive not to underestimate the geotechnical challenges and/or overestimate the machine's ability to counter them. It is hoped that the data base and modeling work will provide an objective framework within which TBM operations can be consistently logged and future performance more realistically projected.

References

Association des Travaux en Souterrain (AFTES), Working group No. 4, Mécanisation de l'Excavation (1986), "Choix d'un type de tunnelier ou de bouclier mécanisé," Tunnels et Ouvrages Souterrain, No. 76, July-August, pp. 161-167.

Brännfors, S. and G, Nord (1979) "Design of Underground Structures with Respect to Modern Construction Methods from a Contractor's Viewpoint." Proceedings, 4th International Symposium of the International Congress on Rock Mechanics, Montreux, Switzerland, Vol. 3, pp. 303-310.

Bruce, W.E., and R.J. Morrell, (1970), "Rapid Excavation in Hard Rock: a State-of-the Art Report," Proceedings, Symposium on Deep Tunnels in Hard Rock, Milwaukee, Wisconsin, November, pp. 187-219.

Dearman, W.R. (1988) "Classification Systems, Design of Underground Structures Based on Classification Systems," Proceedings, International Symposium on Engineering Geology and Underground Construction," Lisbon, Portugal, pp. II-5-30.

Dumbleton E. G. and G. West (1976), "A Guide to Site Investigation Procedures for Tunnels," Transport and Road Research Laboratory Report, Department of the Environment, Report No. 740, 23 p.

Fourmaintreaux D. (1983), "Performances des Tunneliers au Rocher: Resultats Actuels, Previsions et Controle," Proceedings, 5th International Congress on Rock Mechanics, Melbourne, Australia, pp. D211-215.

Gertsch, R.E., (1994), "Mechanical Mining: Challenges and Directions," Mining Engineering, Vol. 46, No. 11, pp. 1250-1253.

Giafferi, J-L. (1989) "L'expérience d'Electricité de France en Matière de Creusement de Galeries au Tunnelier," Revue de l'Energie, Vol. 40, No. 410, March-April, pp. 302-309.

Howarth, D. F. (1994), "Database of TBM Projects Undertaken between 1950 and 1990 and an Assessment of Associated Ground-Strength Limitations," Underground Space Technology, Vol. 9, No. 2, pp. 209-213.

Ikeda, K and Y. Nishimatsu (1980), "The Effects of Geotechnical Properties of Rock Formation on the Productivity of Tunnel Boring Machines," Interim Report from the Working Group 'Research, presented at the International Tunnelling Association General Assembly, Brussels, 10 p.

Innaurato, N., R. Mancini and A. Sampaolo (1980), "Italian Experience with Tunnel Boring Machines and General Considerations on Boring Machine-Rock Interaction," Proceedings, Eurotunnel '80 Conference, Basle, Switzerland, September 1980, pp. 67-77.

Inokuma, A, S. Suzuki, I. Maruyama, J. Mitsuo, T. Sakai and U Tokunaga (1994), "Studies on the Present State and the Mechanism of Trouble Occurrence in Tunnel Construction in Japan," Proceedings, International Tunnel Association Conference, Cairo, Egypt, 1994, pp. 239-246.

Kovari K., R. Fechtig and C. Amstad (1993), "Experience with Large Diameter Tunnel Boring Machines in Switzerland," Proceedings of the Symposium of the International Tunneling Association, Amsterdam, pp. 485-496.

Kuwahara, S. (1990), "Overview of the International Tunnelling Association Tunneling Project Summary Reports," Tunnelling and Underground Space Technology, Vol. 5, No. 3, page 191.

Laughton, C. (1998), "Evaluation and Prediction of Tunnel Boring Machine Performance in Variable Rock Masses", University of Texas at Austin, 327 p.

Marin, G. (1986), "60 kilomètres de Creusement au Tunnelier. Le point sur l'Expériences d'Electricité de France," Tunnels et Ouvrages Souterrain, No 75, May-June, pp. 110-125.

Nelson, P.P., Y. Abd Al-Jalil, C.Laughton (1994), "Tunnel Boring Machine Project Data and Construction Simulation," Geotechnical Engineering Report GR94-4, 323 p.

Parkes, D. B. (1988), "The Performance of Tunnel-Boring Machines in Rock," Construction Industry Research and Information Association, Special Publication No. 62, 56 p.

Sinha, R. S. (1986), "Tunnels: Machine Excavation - Rate of Progress - Machine Data," Bureau of Reclamation Report No. REC-ERC-86-8, July, 37 p.

Shoring for Leslie Station: Design Assessment and Construction Performance

J Westland[1], S.J. Boone[1], P. Branco[2], D. MacDonald[3] and M. Meschino[4]

Abstract

This paper describes the design and construction of an excavation made for a new subway station in Toronto. Overburden soils at the station site consist of nearly normally consolidated silty clay, glacial till, and dense sand. A wing of a college building that abuts the excavation edge is supported by shallow spread footings. Design assessments predicted that porewater pressure losses in the clay soils, both through the excavation wall and through the underlying fissured till, could result in damaging building settlement. The design assessments included empirical and numerical modelling of various shoring systems. Underpinning of the structure was judged to be too costly and disruptive to the tenants compared to other means of protection. As a result of the design-stage work, the retaining system type, a minimum wall stiffness, and a porewater pressure replenishment system were made part of the minimum contract requirements. A detailed monitoring program was undertaken on behalf of the transit authority; it included multiple piezometer arrays, inclinometers, and measurements of both ground and building settlement. Construction data are presented in relation to the design predictions and the system performance is discussed in this paper.

Introduction

Leslie Station is being constructed as part of the 6.4 km long Sheppard Subway that will run east-west between Don Mills Road and Yonge Street in Toronto, Ontario. The Sheppard Subway includes 3.8 km of twin bored tunnels and five underground stations, all of which are to be constructed using cut-and-cover techniques. The subway alignment traverses the broad gently sloping Don River valley (about 3 km wide and 50 m deep). Leslie Station is located at the low point of the alignment, just west of the Don River.

[1] Golder Associates Ltd., 2180 Meadowvale Boulevard, Mississauga, Ontario L5N 5S3
[2] Thurber Engineering Ltd., 170 Evans Avenue, Suite #101, Etobicoke, Ontario M8Z 5Y6
[3] Hatch Mott MacDonald, 2800 Speakman Drive, Mississauga, Ontario L5K 2R7
[4] Yolles Partnership, 163 Queen Street East, Toronto, Ontario M5A 1S1

The station is being constructed on a narrow site which slopes upward from Sheppard Avenue to an existing community college complex. The corner of a four-storey residence building is within 1 m of the station, resulting in the building foundation touching the back of the shoring wall. At the residence location, the station excavation will be 14 m deep. The station design features a lower platform level and an upper concourse level, with the concourse level being below grade adjacent to the college residence. However, on account of the sloping ground profile, the concourse is at grade adjacent to Sheppard Avenue, providing level access from the street.

The project design was undertaken by a team that included the Section Designers (Moriyama and Teschima Architects, Yolles Partnership Engineers and Thurber Engineering), the Program Managers (Delcan–Hatch Joint Venture, Golder Associates) and the Toronto Transit Commission (TTC). The approach to managing subsurface risk and assessing impacts on adjacent facilities followed the strategy outlined in Westland et al (1998) and Boone et al (1998).

Subsurface Conditions

Ground conditions at the site were investigated by a phased borehole investigation program. Emphasis was placed in the later phases on obtaining strength and deformation properties of the weak soils that underlie the college residence and on confirming the founding elevation of the college residence by use of test pit excavations.

The site stratigraphy is illustrated by Figure 1, which also indicates the founding level of the college residence and the outline of the station excavation. The lower deposits (Upper Till, Middle Sand / Silt and Middle Clay) are heavily over-consolidated and are interpreted to have been deposited during the last glaciation and the preceding inter-glacial period. Standard Penetration Test (SPT) "N" values in these deposits were all above 30 blows / 0.3 m penetration, with many greater than 50 blows. The Upper Till has an undrained shear strength of greater than 300 kPa and the angle of internal friction for the Middle Sand / Silt Deposit was inferred to be at least 36°.

The Recent Clay and Recent Sand / Silt Deposits are part of the "Peel Ponds" deposits that are believed to have been deposited in post-glacial lakes, after the last glaciation. The Recent Sand / Silt deposit had SPT "N" values ranging from 10 to 30 blows / 0.3 m penetration, while the undrained shear strength profile of the Recent Clay in the vicinity of the Seneca College Building is illustrated by Figure 2.

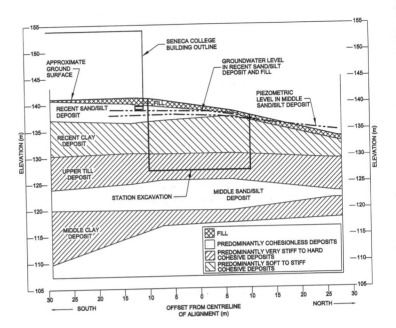

Figure 1. Stratigraphic section through Leslie Station excavation, adjacent to college residence.

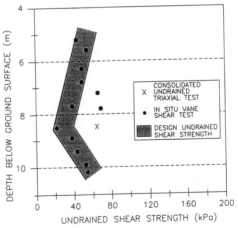

Figure 2. Profile of undrained shear strength in Recent Clay Deposit.

Emphasis was placed on measuring strength and deformation properties of the Recent Clay Deposit. The field and laboratory test program included in-situ vane shear strength tests, pressuremeter tests, undrained triaxial tests (with pore-pressure measurement), triaxial permeability tests and oedometer tests on horizontally and vertically trimmed samples. The physical properties used for analyses of ground movement are provided in Table 1, later in this paper.

The data from the Recent Clay Deposit show a fairly wide scatter. The deposit is not homogenous and displays a rhythmic, varved-like structure at some locations and is more massive, with a sand and gravel content of up to 40 percent at other locations. This structure accounts for the variability of strength and deformation measurements.

Building Assessments and Protection

The TTC design manual required that shoring systems be selected and designed to limit damage to adjacent structures to slight or less (refer to Burland et al 1977). Initially a "Level 1" (i.e. preliminary) assessment of the impact of the excavation on the residence was carried out. For the Level 1 assessment, empirical data on expected ground movements from braced excavations in various soil types was used to estimate the pattern and magnitude of ground movement (Rumsey and Cooper 1984). From the pattern and magnitude of ground movement, key parameters affecting building damage were calculated (horizontal strain and angular distortion) and used to assess the potential for building damage (Boscardin and Cording 1989). Not surprisingly, the Level 1 assessment at the Seneca College building suggested severe damage would result if the station excavation was supported by a soldier-pile and lagging wall, with the maximum settlement estimated to be about 65 mm. A more detailed "Level 2" assessment of building damage was then undertaken and included refinements to movement predictions for building protection alternatives and refinements to the assessment of the structure's tolerance to movement.

The primary mechanisms of settlement of the residence building were considered to be deformation of the shoring structure as the excavation progressed and settlement due to consolidation of the Recent Clay from porewater pressure losses. Porewater pressures were likely to be reduced by dewatering of the Middle Sand / Silt Deposit and seepage through the shoring wall. Underpinning the structure to isolate it from the effects of the ground movement was considered, but was not adopted. Ground settlement was expected to occur for a significant distance back from the shoring line and underpinning works would have had to extend over a similar distance. Furthermore, the structure is supported by both exterior and interior strip footings. Underpinning of the interior footings would have caused significant disruption to the college residents and was considered too expensive. Further, underpinning carried with it risks of ground loss in the Recent Sand / Silt which could itself have induced damaging ground movement.

It was considered feasible to limit shoring-induced ground movements through the use of a contiguous caisson wall – the local standard approach for limiting excavation induced ground movements. A concern with this approach was the higher permeability (as compared to the Recent Clay) of the lean concrete which forms the infill piles of such walls and the small gaps that are inevitable with deep installations of this kind. The relatively high permeability of the wall and its gaps would allow the porewater pressure in the Recent Clay to dissipate, inducing consolidation settlement of the clay. A less permeable wall, such as a concrete diaphragm wall, was considered for the site but was not specified because it was considered that mobilizing a non-local contractor for a relatively short section of wall would not be cost-effective. A compensation grouting program was considered to maintain the level of the residence as consolidation and shoring wall deformation occurred; however, after an initial cost comparison the alternative of providing a stiff contiguous caisson wall, together with a system of infiltration wells to maintain the porewater pressure in the Recent Clay was considered to be the lowest cost alternative available to meet the settlement tolerance of the structure.

The settlement tolerance of the structure was further considered in the "Level 2" assessment process. An inspection of the existing structure was carried out and found that:
- settlement effects were concentrated at control joints within the structure, with a maximum differential settlement being about 10 mm;
- existing hairline cracking was visible on some internal and external walls; and
- environmental damage (spalling) of the brick cladding had occurred.

The tolerance to additional settlement was established by application of the maximum crack width criteria of 5 mm, consistent with the "slight" damage category. By assuming a parabolic settlement pattern, and amplification of crack widths at the top of the structure due to bending, it was determined that a maximum settlement of 10 mm could occur at the edge of the building closest to the excavation. It was recognized, however, that the existing system of construction joints and expansion joints could accommodate some of the building movement without inducing damaging cracks. The actual tolerance of the structure was considered to be an additional settlement of about 15 mm and performance criteria for the contractor were established based on a maximum tolerable settlement of 13 mm (½ inch).

Analysis of Ground Movements

The analysis of ground movements for the selected building protection scheme was carried out in the following stages:
- evaluation of short-term displacements due to the sequential excavation and erection of the support system; and
- evaluation of time dependent settlements due to consolidation of the Recent Clay from dewatering the Middle Sand / Silt and seepage through the wall.

The ground movement analysis was an iterative process, in which various feasible shoring designs were modelled to determine a system that could meet the settlement tolerance of the adjacent structure. Key system design parameters were then extracted from the analysis and imposed as minimum and maximum design criteria for the shoring wall. This approach was adopted since it provided the contractor flexibility with respect to means and methods while ensuring that key design aspects, necessary for the successful execution of the project, were incorporated into the designs of all bidders.

A schematic of the modelled shoring system that was found to meet the settlement criteria and was used as a basis for developing the shoring design criteria is provided by Figure 3.

The evaluation of short-term displacements was carried out using the finite element method. Table 1 presents details of the soil properties used in the analysis. Soil properties and anchor loads were selected to provide lower and upper bounds for soil displacements at the residence building as indicated by the results provided in Figure 4.

Modelling of construction was carried out by sequentially activating structural elements (retaining wall, struts and soil anchors) and eliminating excavated soil elements. Soil anchors were modelled as a combination of constant loads and springs to simulate the pre-loading of the anchors and the variation of anchor loads due to movements of the retaining wall. Initial trial modelling indicated that short-term pore-pressure changes in the Recent Clay and Upper Till, as a result of excavation and application of anchor and strut loads, were very small and were neglected in the subsequent analyses. The results of the analyses are presented in Figure 4.

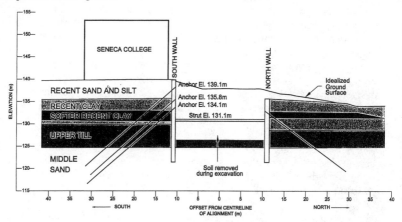

Figure 3.   Schematic of shoring system and ground conditions used in numerical model.

Table 1 Soil Properties used for Numerical Analysis

SOIL PROPERTIES Material	γ (kN/m3)	ϕ' (deg)	Cu (kPa)	E (MPa)	μ	k_r (m/sec)	Cc	Cr	e0	Cv (m2/s)
Recent Sand and Silt	20	30	-	15 to 25	0.35	-	-	-	-	-
Recent Clay										
4.5 to 7.5 m and 9.5 to 10.5 depth	20.5	-	40	8 to 14	0.45	3.00E-10	0.4	0.1	0.7	1.00E-07
7.5 to 9.5 m depth	20.5	-	25	5 to 14	0.45	3.00E-10	0.4	0.1	0.7	1.00E-07
Upper Clay Till	23	-	300	40 to 60	0.45	6.00E-11	-		-	5.00E-07
Middle Sand	21	36	-	150 to 200	0.25	6.00E-07	-		-	-

Time dependent settlements due to consolidation of the Recent Clay were assessed in two separate stages:
- analysis of transient pore-pressure changes due to dewatering of the Middle Sand / Silt deposit; and
- evaluation of settlements due to one dimensional consolidation of the Recent Clay deposit.

The transient seepage analysis was carried out using the finite element method. Surface settlements due to consolidation were assessed assuming one-dimensional consolidation of the Recent Clay with the change in effective stress calculated as the difference between the in-situ and transient pore-pressures. Constant in-situ hydrostatic piezometric head boundary conditions were assigned along the replenishment wells, located 1 m behind the retaining wall and at the top surface of the Recent Clay. A constant piezometric head was assigned to the dewatering wells, 3 m behind the retaining wall in the Middle Sand, in order to obtain a phreatic surface just below the base of excavation. Porewater pressure conditions were "captured" at several time intervals after the start of construction to allow assessment of the progress of consolidation settlement with time. A construction duration of 19 months was assumed for design purposes. Without maintaining porewater pressures in the clay, it was estimated that consolidation of the clay could result in about 2 mm of settlement per month in addition to excavation-induced settlements. For the 19 month duration, the analysis suggested that the porewater pressure replenishment system would be successful in maintaining near hydrostatic pressure conditions in the Recent Clay, "upgradient" of the line of replenishment wells. The time dependent surface settlements behind the retaining wall are shown on Figure 4. The one-dimensional consolidation model used in the time-dependent settlement analysis does not reflect the rotation of principal stresses in the soil and the development of shear stresses between adjacent columns of soil that take place close to the retaining wall.

The analysis of the ground displacement indicated that the short-term settlements of the college building were expected to be small and were not significantly sensitive to variations of soil properties within the anticipated range of values provided in Table 1. The analysis indicated that time dependent settlements were expected to be larger than the short-term settlements. During the 19 month period of operation of the dewatering system, for average soil properties and anchor loads, the analysis predicted a maximum surface settlement of 16 mm.

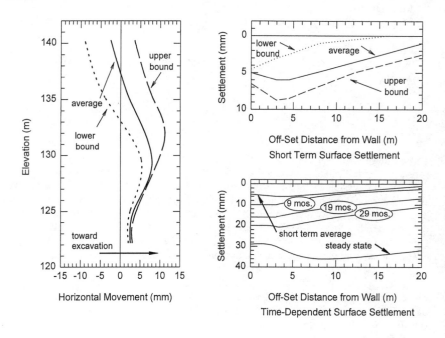

Figure 4. Horizontal and vertical ground movement behind shoring wall from numerical modelling. Time-dependent settlements include effect of porewater replenishment system.

Contractual Strategy

The TTC's approach to temporary works is to assign design responsibility to the contractor, who is required to meet certain minimum and maximum design criteria (such as resistance of a minimum earth pressure) and to meet maximum ground and structure movement performance criteria that are verified by a field monitoring program. For the shoring at the college residence, the type of wall (contiguous caisson wall) and the use of a porewater replenishment system were specified, in addition to various design criteria.

The design criteria were derived from the analyses and modelling that was undertaken. Key specified criteria for the wall included:

Minimum Length of Contiguous Caisson Wall	- 47 m
Minimum Flexural Stiffness of Wall (EI)	- 230,000 kNm^2/m of wall
Minimum Concrete Strength of Wall	- 25 MPa for piles with steel
	- 4 MPa for infill piles
Elevations of Horizontal Restraints	- 4 levels at specified elevations
A Minimum Horizontal Working Load/m length of Wall at each Restraint Level	
Minimum Toe Penetration Below Excavation Base	- 5 m for piles with steel
	- 1 m for infill piles
Minimum Restraint Pre-Load	- 50% Working Load for Struts
	- 110% Working Load for Tie-Backs

These criteria provided some assurance that a wall system would be designed that was consistent with the analysis that had demonstrated that satisfactory protection of the building was feasible. Further, with the prescriptive criteria, review of the contractor's submittals would be based on numerical criteria, minimizing differences in engineering opinion between owner and contractor. The specified criteria still allowed the contractor design flexibility to suit his equipment and preferred methods. For example, the number, type, spacing and installation method for the horizontal restraints was determined by the contractor, as were the design details of the wall to meet flexure criteria.

A similar contract approach was adopted for the porewater pressure replenishment system. The contract specified that two independent systems were to be provided: one in the Recent Sand / Silt Deposit and one in the Recent Clay Deposit. The maximum well spacing was specified, as was the length over which the system was to operate. Monitoring of flow rates into each independent gravity fed system was also specified.

Construction Performance

The Contractor designed and constructed a contiguous caisson wall and a porewater replenishment system adjacent to the college residence that complied with the minimum and maximum criteria set forth in the Contract Documents. While the design phase analysis assumed that the lowest level of support would be provided by a strut, the contractor opted for a fully tied-back wall. The constructed wall included the following features:
- contiguous piles (914 mm diameter) where the "king" piles included two wide-flange beams welded side-by-side to meet the specified stiffness;
- four levels of tie-backs inclined at 20° to 30°, with maximum working loads of up 1,360 kN, and
- all tie-backs were about 150 mm diameter and were post-grouted.

Performance of the retaining, dewatering, porewater pressure replenishment systems, and building response were all monitored by the TTC during construction. Settlement of the building was monitored by precise electronic leveling surveys (accuracy of about ±1 mm) of bolts set into the concrete foundation wall and sleeved steel rods set 1.8 m into the ground. Crack gauges were installed on the outside wall joints. Inclinometers installed within the wall and in the ground behind the wall were utilized for measuring lateral movements. Porewater pressure in the clay was observed using arrays of vibrating wire piezometers isolated at various depths and at various distances from the excavation. In general, groups of instruments (arrays) were arranged in lines running perpendicular to the excavation line to observe the lateral influence of construction activities. Arrays were placed at a number of critical locations including the college residence and also including areas of soldier-pile and lagging wall where the consequences of larger ground movements were judged to be less severe, but where control of excavation activities was still necessary to minimize damage to utilities.

Figure 5 illustrates the performance of the two wall system types in the vicinity of the college building. Typically, the caisson wall deformed in cantilever mode outward by about 5 mm prior to installation of the top row of tie-backs. During tie-back stressing, the caisson wall moved back into the soil mass a few millimeters. This behaviour repeated for successive tie-back levels. Performance of the soldier-pile and lagging wall, where two of the support levels consisted of struts, exhibited significantly more lateral ground deformation.

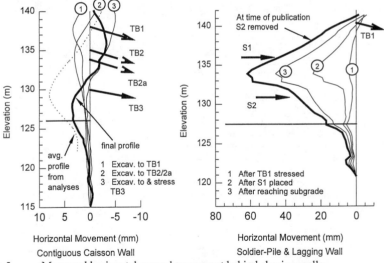

Figure 5.   Measured horizontal ground movement behind shoring walls.

Porewater pressure responses from two example arrays are illustrated in Figure 6. In general, the arrays behind the soldier-pile and lagging walls showed an initial response to recharge prior to excavation. During and following excavation, the porewater pressures dropped to a steady-state, with pressure-head losses of about 2.5 m (25 kPa) below the pre-construction levels. Porewater pressure losses relative to off-set distances from the excavation are also illustrated in Figure 6. At the end of excavation, the bottom row of piezometers located near the interface between the fissured till and clay deposit (at Elevation 131.0 m) indicated a nearly stable porewater pressure elevation head drop of 1 m to 1.5 m at a distance of 10 m to 12 m from the excavation. The piezometers within arrays closer to the building and directly influenced by the maintenance systems exhibited a similar response during the excavation period. Problems arose because tie-back installations broke a number of replenishment wells and provided pathways for porewater pressure losses through the tie-back heads. Concern about ground losses through tie-back holes during their construction and subsequent leakage prompted a reduction in feed-water flows to the porewater pressure maintenance system. After excavation was completed, the contractor made additional efforts to seal the tie-backs, resumed appropriate flow rates, and porewater pressures responded accordingly. Porewater pressures within the clay mass beneath the building were then maintained within about 10 kPa (1 m) of pre-construction levels, regardless of distance from the excavation.

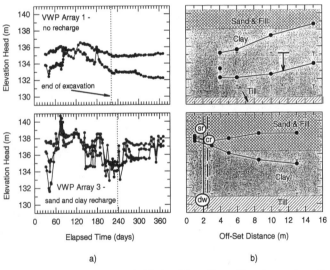

Figure 6. Porewater pressure responses behind solider-pile and lagging and contiguous caisson walls: (a) head vs. time for wells 3 m to 4 m behind excavation; b) head vs. distance behind excavation at time of publication.

As illustrated on Figure 7, the majority of settlement occurred during excavation. Given the porewater pressure losses during excavation, separation of consolidation and excavation-related settlements is difficult. Based on the settlement patterns of ground monitoring points in areas where the excavation was supported by a fully tied-back soldier-pile and lagging wall system (where little lateral movement occurred) it has been estimated that settlement due to consolidation was on the order of 1 mm to 2 mm per month during excavation. Very little time-dependent settlement was observed behind the contiguous caisson wall and porewater pressure replenishment system, once it was fully functional.

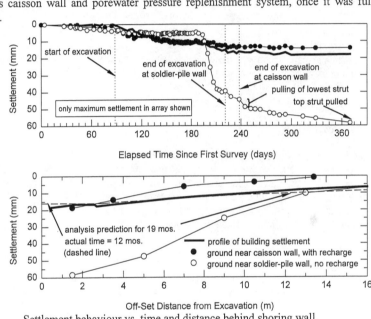

Figure 7. Settlement behaviour vs. time and distance behind shoring wall.

The performance data suggest that the time-dependent consolidation effects brought on by porewater pressure changes were reasonably modelled for areas without the replenishment system, or when the replenishment system was not fully functional. However, when fully functional, the replenishment system has proven to be more effective than modelled, likely due to the varved nature of the clay.

The pattern of building settlement is more linear than the pattern of ground settlement, suggesting the building stiffness relative to that of the founding soils has attenuated the characteristic parabolic settlement profile. To date the damage to the residence structure has been negligible.

Conclusions

The Leslie Station shoring project provides a valuable case history that demonstrates that:
- reasonable prediction of shoring performance and ground movement can be made with present analytical tools that are supported by thorough field and laboratory data;
- porewater pressure replenishment in clay deposits is an effective method to reduce consolidation settlement, where water-tightness of the shoring wall cannot be assured;
- building damage assessments based on the assumption that the structure's deformation will match the ground movement pattern are somewhat conservative;
- tied-back walls with relatively high pre-load are an effective means of minimizing horizontal strains; and,
- an owner can effectively translate the results of design stage analytical modelling of temporary works into enforceable temporary works design criteria without unnecessarily constraining the contractor's means and methods and assuming full responsibility for the design.

Acknowledgements

The writers thank the TTC for permission to publish this paper and acknowledge the contributions made by S. Poot, who managed the monitoring database for this project, M. Pando, who undertook the design-stage numerical modelling, the staff at Morrison Hershfield who undertook surveys of monitoring points, the staff at Peto MacCallum who read the geotechnical instruments and the staff of Acres International who undertook the site investigation and laboratory testing.

References

Boone, S.J., Branco, P., and Garrod B. (1998). *Building and utility damage assessments, risk, and construction settlement control*. Tunnels and Metropolises, Proceedings of World Tunnel Congress 1998, Sao Paulo, Brazil, A. Negro and A. Ferreira, eds., Balkema, Rotterdam, pp. 243 – 248.

Burland, J.B., Broms, B.B., and deMello, V.F.B. (1977), *Behaviour of Foundations and Structures, State of the Art Report*, Proc., 9th Int'l. Conf. On Soil Mechanical Found. Eng. II, Tokyo, Japan, pp. 495 – 546.

Boscardin, M.D. and Cording, E.J. (1989). *Building Response to Excavation - Induced Settlement*, Journal of Geotechnical Engineering, ASCE 115(1), pp. 1 – 21.

Rumsey, P.B., and Cooper, I. (1984). *Groundwater Movement and their Effect on Structures – Chapter 5. Trenches in Soil*, Edited by Attwell, P.B., and Taylor, R.K. Surrey University Press, Glasgow, England.

Westland, J., Busbridge, J.R., and Ball, J.G. (1998). *Managing Subsurface Risk for Toronto's Rapid Transit Expansion Program*. Proceedings of North American Tunnelling 1998, L. Ozdemir, ed., Balkema, Rotterdam, pp. 37 – 45.

List of Symbols

γ	- Unit Weight of Soil
ϕ'	- Angle of Internal Friction
μ	- Poisson's Ratio
Cc	- Coefficient of Compression (Initial Loading)
Cr	- Coefficient of Compression (Reloading)
Cu	- Undrained Shear Strength
Cv	- Coefficient of Consolidation (Vertical)
E	- Deformation Modulus
Kv	- Vertical Permeability

True Behavior of High Pressure Tunnel for
The Guavio Hydroelectric Project

Alberto Marulanda P. [1]
Ramiro Gutiérrez R. [2]

Abstract

Good engineering practice involves not only the development of functional and safe projects, but also has to provide an economic solution. The tendency when there is lack of knowledge is to over design the possible solutions. In high pressure tunnel projects where improved design methods are necessary, there is an increasing possibility of overdesigns that can make a project financially unfeasible. This paper describes a Colombian case treating the design, adjustments and behavior of the tunnels and shafts at the Guavio Hydroelectric project, which has some of the highest head pressure tunnels in the world, and actually have more than six years of operation without any problem.

Introduction

The Guavio Hydroelectric proyect is located 150 km east of Colombia's Capital Santa Fé de Bogotá. This project at the moment represents the country largest electric plant with an installed capacity of 1600 MW, the project also includes some of the largest an deepest power tunnels built, a rockfill Dam with central clay core 250 m high, an underground power house 220 m long, 18 m wide and 37 m high located at a depth of 550 m below the surface.

---
[1] Partner and Technical Vicepresident, Ingetec S.A. Cra. 6 No. 30A-30. Santafé de Bogotá. Colombia.
[2] Specialist Engineer. Associated. Chief of Tunnels Department. Ingetec S.A. Cra. 6 No. 30A-30.Santafé de Bogotá. Colombia.

The power tunnels include an upper tunnel 13,3 km long, two lower 1,2 km power tunnels, a vertical power shaft 550 m high, a surge shaft 480 m high, and a 5,2 km long tailrace tunnel that discharges into the Guavio River. The maximum hydrostatic pressure head is 1100 m (10,8 MPa), reached in the lower power tunnel near the underground power house.

The project begun its construction in the year of 1981 and was finished towards December 1992. A general profile view can be seen in Figure No. 1

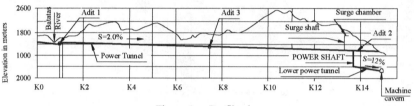

Figure 1.   Profile view

The NATM philosophy was applied as the construction technique, and was the contractual base for the further development of the project tunnel and shaft's excavations.

General Geology

The power tunnels and shafts crossed good quality sedimentary rocks from the paleozoic and juratriassic ages, with some localized fault and shear zones in cretaceous rocks. The regional structure corresponded to the eastern flank of the Esmeralda anticline. The tectonic activity was caracterized by some regional faults and inverted geological secuences.

The Cretaceous was represented by Cáqueza formation, where this formation overlays discordantly the paleozoic rocks and was lithologically formed by dark gray shales and siltstones.   The Juratriassic rocks located in the eastern part of the project, are composed by two units; JR_3 with black siliceous shale, of medium to thick stratification, and intercalated sandstone beds, and gray quartzitic sandstones, thick stratification, and JR_2 composed by argillite and gray quartzitic sandstone, and finally red sandstones.   The paleozoic rocks are the oldest rocks included in the Farallones group of carbonian and devonian age.   The carbonian rocks were represented by interbedded argillites and quartzites.   The devonian metasediments were composed by hard siliceous rocks, green phyllites, quartzites and gray slates. The power tunnel crossed several major faults: La Vega de San Juan, Colonias, Garabato, Montegato, and Esmeralda faults.

The rocks were classified between low and high strength, with medium to low moduli. The average strength values ranging between 25 MPa and 195 MPa and a Young's modulus between 9 GPa and 20 GPa (see Figure 2), being the paleozoic rocks the ones with the highest strength and modulus, and the cretaceous rock the ones with the lowest strength and modulus.

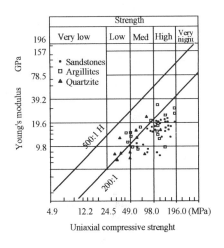

Figure 2. Strength and modulus of the rocks

Lining design criteria

The basic design criteria used at the tunnels and shafts at the Guavio Project was reported by Marulanda et all (1987); being the most important aspects those involving deformability and stability conditions, leakage due to high permeability of the rock mass and hydraulic fracture of the surrounding rock mass.

- Stability conditions

Unlined tunnels and shafts were normally accepted when the rock was self supporting and no expectations for rock loosening due to internal pressure or to action of water was foreseen. For shotcrete lining the same criteria was used while the rock mass deformations would not induce cracking of the shotcrete layers.

With this criteria, shotcrete was used as a lining in sectors of very laminated rock to avoid progressive erosion of rock in zones of fractured hard rock for controlling the falling of small rock blocks, in order to prevent a continuos loosening of the mass.

Concrete liners were considered when the elastic rock mass parameters were such that the rock could take a great deal of the induced internal pressure without excessive strain. Cracking in the concrete lining was accepted since, when the internal pressure is removed, tension cracks will tend to close, without a loss of the structural capacity of the lining for external pressure.

Strain compability of the rock-concrete boundary was performed by sensibility analysis, relating tunnel pressure with rock mass modulus and their associated crack widths, keeping an elastic deformation response. When several cracks could occur, reinforced concrete was used, with a limit deformation of 1,0 cm. Figure No. 3 shows a sensibility analysis example in the upper power tunnel.

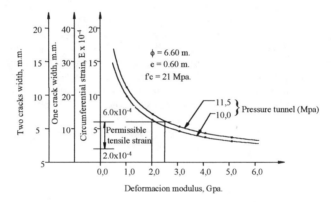

Figure 3. Sensibility analysis of the concrete lining

- Water leakage

Design criteria considered was that in pressure tunnels and shafts, the rock mass itself is the basic element to guarantee imperviousness. Unless, an impervious lining is placed in areas of high permeability or the permeability of the rock mass is reduced by means of grouting, there will always be the possibility to have water leakage, that may reach the ground surface or near underground excavations.

As the permeability of rock is normally negligible, it is the jointing and the faulting of the rock mass and the type and amount of joint infilling material, that are of importance when an area is evaluated for leakage.

When assessing the expected water leakage conditions of an unlined tunnel, it is essential to consider the location of the water table with respect to the tunnel internal water pressure head. If the original water table is higher than the hydrostratic tunnel pressure head, after the filling of the tunnel, the area around the tunnel would be recharged up to the hydrostatic pressure level, becoming lower than the original water table, and then tunnel leakage would be minor and diminishing with time, even if the rock mass is highly pervious; under those circumstances the rock mass permeability is of minor importance. If the water table is lower than the hydrostatic pressure level and overburden to hydrostatic pressure head ratio were less than one, it would be possible to have water losses to the surface.

Neither a plain concrete lining nor a reinforced concrete lining are considered impervious elements when placed in a high pressure tunnel. If water losses occur, the mountain would be recharged and, upon dewatering of the tunnel, the concrete would be subjected to external water pressures, that will dissipate through the cracks due to the high pressure levels. If there were large fractures in a rock mass which were able to store large volumes of water, it is the common practice to use drainage holes for quick pressure release.

The basic concept for the use of drainage holes is that in zones for very low permeability is acceptable that the concrete lining or the shotcrete be under external pressure, because, under small movements of the concrete lining, the external pressure, will be relieved. Since the water bulk modulus is higher than the lining modulus, the movement of the lining will be enough to relieve the pressure. The result will be different when the rock storage capacity is higher, since a permanent piston effect will always be acting upon the lining. Considering the aforementioned concepts, s steel lining or extensive consolidation grouting works are mandatory when the possibilities of great leakage are foreseen.

- Hydraulic fracture

Hydraulic fracture of a rock mass surrounding a pressure tunnel occurs when the tensile stress, induced by water pressure, exceeds the confining stresses within the rock and causes rupturing of the mass, either along pre-existing discontinuities or through intact rock, creating water channels that can cause uncontrolled loss of water, which can have adverse effects in near underground stability works or in adyacent hillsides. The only alternative to guarentee that the phenomenon of hydraulic fracturing does not occur is an impervious liner.

Power shaft location and lining

Initial design considered a concrete lined shaft, plus consolidation grouting, located in such way that the conventional rules of thumb that intended to compensate the internal pressure with the overburden weight was adecuate, and no steel lining was required.

However, past experience encountered during test filling in the near Chivor II Project, showed that high leakage could occur, still fullfilling those rules, because there exists topographycal protuberances that allowed stress relief. Then, a stress profile was revised according to Broch criteria (1982, 1984), being the new location 200 m upstream of the original one. Later, due to adverse geological conditions found during the upper power tunnel excavation, the shaft was moved about 100 m upstream of the aforementioned site.

Two dimensional finite element models and design charts, Broch (1984), were used to verify total field stress in the areas of complex topography. However due to model limitations and uncertainties, it was decided to carry out in-situ hydraulic fracture tests in areas near to the upper and lower elbows, to have a more realistic picture of the minimum principal stress values. (See Figure No. 4).

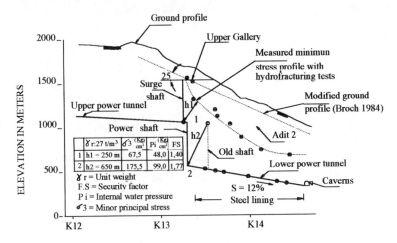

Figure No. 4. Hydraulic fracturing test and steel lining - lower tunnel and power shaft

The tests indicated that the real stresses in the rock mass were lower than the ones that could be estimated by means of the revised profile or by design charts and the finite element model results.

Final lining for tunnels and shafts

As a result of the aforementioned concepts in the upper and lower tunnels, there were three basic liner sections: a concrete liner (non reinforced and reinforced), a steel liner and an unlined section; these cross sections are shown in Figure No. 5. Additionaly, grouting works consisted in sistematic rings with 8 holes per ring, one diameter long, spaced each

5,0 m, located where overburden to hydroestatic pressure head ratio was minor of 1,0. In the surge shaft only shotcrete was placed, and grouting and drainage holes were implemented. In the power shaft and lower tunnel, concrete lining and steel lining was adopted following the aforementioned criteria.

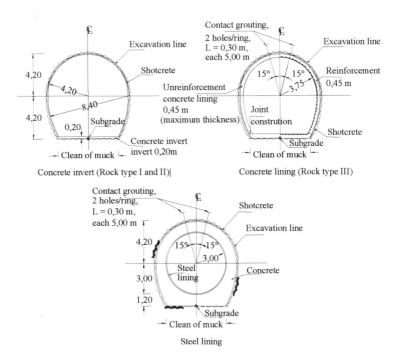

Figure 5. Upper power tunnel. Lining. Cross sections

The final lining applied in the power tunnels and shafts is shown in the following table.

Tunnel or shaft	Concrete Liner		Steel Lining	Unlined	Remarks
	unreinforced	reinforced			
Upper power tunnel	46%	5%	4%	45%	
Surge Shaft		6%		94%	With radial grouting and drainage along the shaft.
Power shaft	72%	9%	19%		With radial grouting along the shaft.
Lower tunnel	11%	3%	86%		With radial grouting.

Power shaft construction incidents

The power shaft excavation was initially drilled as a 30 cm pilot hole. It was reamed to a 2,1 m diameter hole and a depth of 200 m ; rockfalls occurred plugging the shaft. Aproximatley 2300 m^3 of fall material, compossed by rock fragments ranging between 0,2 m and 0,3 m and water flows of 30 l/s , were recorded at the shaft's bottom. Alimak method was abandoned and the construction procedure adopted was the full-face sinking to 7,1 m diameter by drilling and blasting from the upper elbow to the plugged zone. An exploratory program with 50 drill holes and a gallery between elevations 890 and 873 was developed in order to establish the cavern's geometry and the characteristics of the rockfall material. Figure No. 6 shows the general geometry of the cavern evaluated and the shaft zone involved.

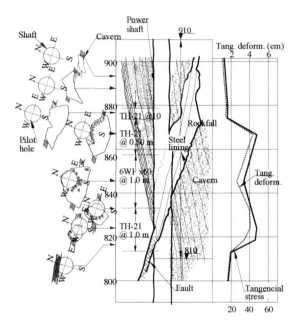

Figure No.6.   Power shaft plugging

In order to evaluate the behavior of the lining in the power shaft's cavern zone, a two-dimensional finite element model was carried out in order to simulate the anisotropic conditions encountered and the geometric variations caused by the cavern. This model showed that the maximum tensile stresses and deformations under internal pressure could be produced between elevations 810 m and 875 m, where part of the shaft was in contact with the rockfall material. (See Figure No. 6).

Different solutions with concrete and steel lining in the cavern zone where analyzed in order to stabilize the rockfall, but also to prevent high leakage.

Finally, only a 100 m long steel lining was used in the critical part, between elevations 810 and 910 covering all the rockfall and the cavern zone. At both ends of the steel lining, reinforced concrete and radial grout curtains (20 m long) were installed.

Other alternatives which involved grouting activities and concrete lining were not used because they could present variations in the drilling and grouting time; also the steel lining was more reliable and could reduce future uncertainities.

Upper elbow of the power shaft

Where the power tunnel intercepts the power shaft and the adit 2, horizontal and vertical flow direction changes occur, generating forces that induced high anisotropic stresses and strains. (See Figure No. 7).

A FEM analysis was also used, to take into account the disturbed rock zones, the special geometry problem involved and the water pressures induced. Different rock conditions and fracture and elasticity modulus were used. The final models included a 45 cm concrete thickness, 2,0 m of fractured rock zone (E = 12,86 GPa),1,0 m to 4,0 m of sound rock (18,3 GPa), internal pressures up to 4,6 MPa equal to the maximum static pressure, and a flow rate of 200 m^3/s for the water induced forces.

As a result of parameter variation in the model, a sensibility analysis was done that showed that for any typical deformation of the rocks, the steel amount did not reduce the working stress levels, mainly caused by a high rock modulus in the elbow area. The stress distributions were calculated, helping to optimize the geometry, and particularly the transition from a horse shoe shape to a circular lining, improving the hydraulic work and lowering the stresses. Finally, a typical reinforced liner (working at 60 % of the yield point) with steel area of 26 cm^2/m transversally and 14,0 cm^2/m longitudinally was placed.

Plugs Design

The upper power tunnel was constructed through three access adits, where concrete plugs were located.These structures correspond to critical points of the high pressure conveyance.

Plugs were designed to satisfy an acceptable shear strength and hydraulic gradient along the plugs, according to local rock conditions, geological features, geometry and access requirements. Design criteria was developed from behavior of other plugs in projects currently in operation in Colombia. Figures No. 8 and No. 9, show the design criteria related with maximum shear stresses and gradients adopted in Guavio.

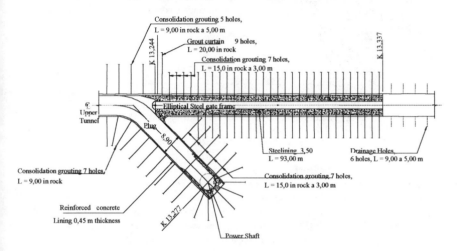

Figure No.7. Adit 2. Plug and grouting works

The three plugs had rectangular galleries which were designed structurally to guarantee under external pressure the working capacity of the section. In this cases of very high external hydraulic pressures, the practice was to provide a steel lining design to absorb full external pressure located near to the steel gate frames. Concrete's shear strength was checked in cases where the slenderness ratio of the invert slab was greater than 3,0. In these cases flexure induced stresses were also checked. The plugs were also analyzed as a massive structure where temperature and shrinkage cracking must be controlled. Another consideration taken in account was stress concentration induced by the steel gate frame. The frame and reinforcement were dimensioned to maintain shear and tension stresses within the working range in accordance with ACI recommendations.

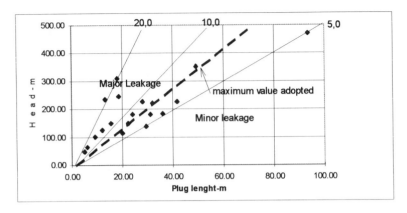

Figure No. 8. Plug design criteria. Hydraulic gradients

The permeability of the rock mass around the plugs was controlled with contact and consolidation grouting. The contact grouting consisted in two holes drilled every 3,0 m alternately 15 degrees from the crown axis and penetrating a minimum of 30 cm into the rock. The consolidation grouting consisted in a radial curtain located upstream of the plug with (12 holes ring with a length of at least 2,0 diameters), and consolidation radial grouting spaced every 3,0 m to 5,0 m with 8 holes per ring and one tunnel diameter in depth. Finally, in the downstream zone of the plugs drainage rings with 5 holes per ring with half diameter length was implemented. Each plugs had piezometers and vibrating wire strain and concrete stress cells. Figure No. 7 shows the plant view of adit 2 and plug with grouting and drainage curtain details

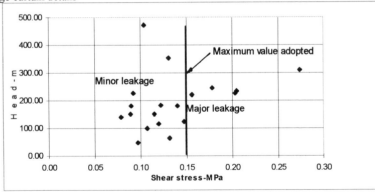

Figure No. 9. Plug design criteria. Shear stress.

Guavio filling test

Filling test was basically done in stages to allow hydrological reestablishment and to limit deformation of the rock mass and cracks in the shotcrete and concrete liners. This process took at least two months, with a controlled watering rate about 1 m/h to 5 m/h.

The filling test begun in October, 1992, and was temporaly stopped for some minor adjustments in the gate of the adit 1. Second stage filling was done and stopped by middle November, again because some adjustments were to be made in the adit 1 and to allow hydrological equilibrium. Finally in November the third filling stage was completed succesfully.

In order to verify the behavior of tunnels and shafts, in each plug adits and outlet zones, piezometric stations and measurements of leakage were anticipated and observed. Similarly surface monitoring of the water flow in creeks, springs, and phreatic levels were done.

In Figure No. 10 and Figure No. 11, total internal pressure and water leakage from three adits are shown. From those figures, adit 1 and 2 show a slight increase of leakage and once total pressure was reached, the inflows were stabilized. In adit 3 a major increase of leakage is related with the existance of very permeable zones encountered during excavation that produced large inflows that reduced with time during tunnel construction. Once the tunnel was filled the stabilized adit inflow is very similar to the peak encountered during construction.

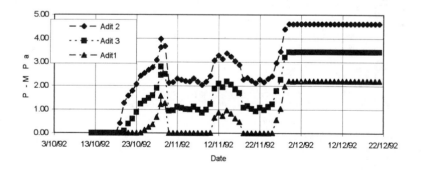

Figure No. 10. Internal pressure in the adits.

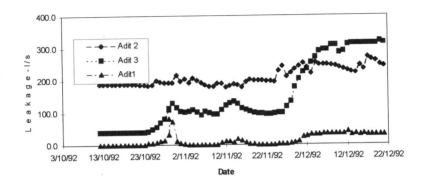

Figure No.11. Total leakage in the adits.

Figure No. 12 shows the water flows behavior in two main creeks located near to the adit 3. Those measurements showed a complete stabilization of the water flows, that recovered the normal conditions. (Previous to tunnel excavation).

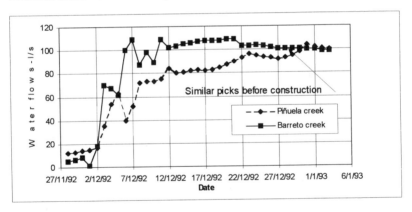

Figure No. 12 . Water flows in surface creeks.Adit 3 Zone

Figure No. 13, shows a special case of the measurements observed in the addit No 2 and its plug. In this figure it's observed that for a final internal pressures up to 4,6 MPa, the piezometric ratios were stable and had a very good behavior with ratios near 40% for the piezometer nearest the pressure tunnel and a ratio near 7%, behind the plug.

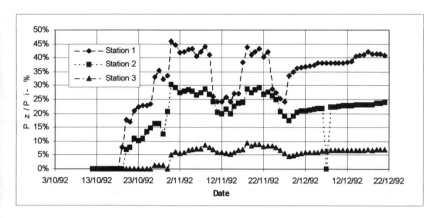

Figure No.13. Total piezometric ratios in the adit 2.

In general the ground water monitoring proved a good behavior along the powe conduits, with no major changes with respect to the normal conditions. After six years c current operation the Guavio Hydroelectric Project has not detected leakage or stabilit problems in the tunnels or shafts.

Conclusions

Modern engineering practice along with a good design criteria, can provide short steel liners, and long-low cost-unlined or unreinforced concrete liners in tunnels and shaft: even when they involve high water pressure. In special cases an appropriate model ca explain the behavior of particular geotechnical conditions, allowing better designs engineers

References

Broch, E., 1982. Development of Unlined Pressure Shafts and Tunnels in Norway. Roc: Mechanics - Caverns and pressure shafts (Wittke, editor). Rotterdam, the Netherlands, A. A Balkema. 545-554.

Broch, E., 1984 - Unlined High Pressure Tunnels in Areas of Complex Topography. Wate Power and Dam Construction. Volume 36, 11:21-23.

Marulanda, A, Eslava, L.F. and Broch, E., 1987. Design of Unlined High Pressure Tunnel for Guavio Hydroelectric Project in Colombia. Underground Hydropower Plants.Volume 1 Oslo; Norway. 101-111.

Geotechnical Lessons Learned for San Diego's South Bay Ocean Outfall

Jon Y. Kaneshiro, Luciano Meiorin, Stephen J. Navin, and Svante Hjertberg*

Parsons Engineering Science, La Jolla, CA

*Consultant, Coronado, CA

Abstract

For more than 65 years, raw sewage has flowed down the hills of Tijuana, Mexico into the Tijuana River, crossing the U.S.-Mexican border. The sewage contaminates the low-lying river valley in San Diego and then flows into the Pacific Ocean, creating a health hazard for beaches at Border Field State Park, Silver Strand Beach in Coronado, and the most severely affected, Imperial Beach. To alleviate these problems, the South Bay Ocean Outfall (SBOO) was constructed using state-of-the-art marine and geotechnical investigations, liner design, and tunnel boring machine design. The South Bay Ocean Outfall was commissioned on November 30, 1998 and effluent began flowing on January 13, 1999. This paper describes some of the geotechnical lessons learned.

Project Description

The South Bay Ocean Outfall (SBOO) will convey up to a total of 14.6 m^3/sec of treated wastewater effluent originating in both Tijuana (8.8 m^3/sec) and the City of San Diego (5.8 m^3/sec) (Figure 1). Design flows are 7.6 m^3/sec (average dry weather flow) and 14.6 m^3/sec (peak flow). A peak flow of 11.3 m^3/sec will be conveyed through the SBOO by gravity, and the capacity can be increased to 14.6 m^3/sec in the future by addition of a pump station. The construction of the South Bay International Water Treatment Plant (SBIWTP) is proceeding in parallel with the SBOO project (Figures 1 and 2). The South Bay Land Outfall (SBLO) is a 3.66-m diameter by 3,750-m long pipeline constructed in 1991-1994, which connects the SBIWTP with the SBOO. The SBOO is designed for a service life of 75 years. The total cost including design of the SBIWTP, SBLO, and SBOO when completed is $ 380 million. The SBOO construction cost is $140 million. The owners are the International Boundary and Water Commission (60%) and the City of San Diego (40%). The City administered the contract packages through Sverdrup/ICF Kaiser Engineers who performed the construction management. Parsons Engineering Science is the designer and provided shop drawing review and design services during construction. Project oversight provided by the EPA, IBWC, U.S. Army Corps of Engineers and the State of California.

The SBOO construction schedule is 39 months. The construction of the SBOO was divided into three contract segments (Figure 2). Contract Package 1 and Contract Package 3 are completed and are discussed in McBain et al. (1998). Contract Package 2 is the tunnel and underground structure segments, and was chosen over a conventional sea floor pipeline to avoid environmental and seismic concerns, as well as to provide an estimated 25% savings of $40 million, even considering tunneling risks (McBain et al., 1998).

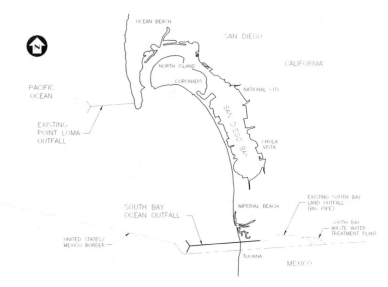

Figure 1. Site Location and General Project Arrangement

Contract Package 2 carries effluent from the end of the SBLO through an anti-intrusion structure (to prevent seawater from backing up) down an 11-m diameter drop shaft to an elevation of about -50m below sea level. At that point, a tunnel is being constructed with an internal diameter of 3.35 m and an overall length of approximately 5.8km. The tunnel will terminate about 4.27 km offshore where it will connect to a 2.8-m diameter by a 47-m long riser, where effluent will be conveyed vertically to a seabed pipeline of Contract Package 3. Notice to proceed was given in September of 1995 to Traylor Brothers/Obayashi, a J.V. (Evansville, IN / San Francisco, CA) who were low bidder at $88,285,000. Major subcontractors include Case Foundation (Chicago, IL) for the Riser and Layne Northwest (Pewaukee, WI) for the drop shaft freeze wall and dewatering. Major suppliers include Boretec/Mitsubishi J.V. (Solon, OH / Kobe, Japan) for the earth pressure balance tunnel boring machine (EPBM), Sehulster Tunnels/Pre-Con (Rancho Cucomunga, CA) for the precast concrete segmented one-pass liner system, and NESCO Fabricators (Mare Island, CA) for the Riser.

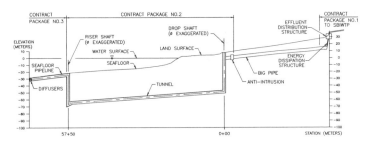

Figure 2. Generalized Project Profile

Geotechnical and Design Considerations

The SBOO had unique technical challenges pushing the envelope of the state-of-the-art for difficult tunneling conditions. The main tunneling challenge was the difficult geology under a maximum of 7 bars for unprecedented distances, as shown in Figure 3. No other tunnel in the world to date has this combination of high pressure and long distance (Navin et al., 1995). The tunnel is entirely within the San Diego Formation consisting of very stiff to hard clays; very dense silts; very dense, fine silty sands; clean sands; clayey, silty and sandy gravels/cobbles/boulders, (with boulders up to 1 m diameter); and well-cemented concretions. Consequently, the use of a universal tunnel boring machine (TBM) (either of the slurry or EPBM type) was required.

The design featured a one pass, five piece precast concrete segmented liner, designed for the full overburden and to withstand an internal pressure (over hydrostatic) of 1.5 bars under normal operating conditions and 2.7 bars during pumping. The segments are connected to form a monolithic pipe by using hoop steel linked by heavy bearing plates across the segment joints (Kaneshiro et al., 1996). In addition, 15 active faults and one dormant fault (Schug et al., 1995) (Figure 3) cross the SBOO alignment requiring special design considerations (Kaneshiro et al., 1996a).

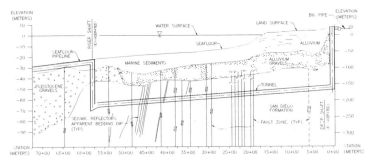

Figure 3. Generalized Geological Profile of Tunnel Alignment

The key seismic sources are the Rose Canyon fault and the Coronado Banks fault, which is about 10 km west of the Riser. These faults are capable of producing a Magnitude of 7-1/4 and 7-1/2 earthquake, respectively. A probabilistic seismic hazard analysis and a probabilistic fault displacement analysis were performed, which gave a peak ground acceleration of 0.63 g and a nominal fault displacement of 76 mm with a probability of exceedance of 10 percent in 75 years.

Geotechnical Lessons Learned

As indicated above the investigation, design, and construction faced many geotechnical and design challenges. Meeting these challenges contributed to a successful project. Some of the major issues the project faced are described under the following topics and provides an illuminating perspective to geotechnical lessons learned.
- Contract documents and bid environment
- Drop shaft differing site condition
- TBM startup and dewatering
- Tunneling conditions and TBM modifications
- Offshore riser and design deficiency claims

Contract Documents and Bid Environment

The contractual setting and bid environment can set the framework for a successful project as they relate to tunnel and geotechnical engineering issues. Some examples for SBOO follow.

The Contract Documents followed the "innovative" U.S. tunneling contracting practices (USNCTT, 1974; O'Rourke, 1984; Matayas, 1996), including:
- full disclosure of geotechincal information via a Geotechnical Design Summary Report (GDSR or Geotechnical Baseline Report);
- a differing site condition clause;
- a mobilization item for the TBM;
- placing the Contractor's bid in escrow;
- a three party disputes review board; and
- partnering (Kaneshuro et al., 1998) and risk evaluations during construction.

Recognizing that very specialized experience was required, at the 50% and 95% design submittals, contractors and machine manufacturers worldwide were solicited for comments and interest in the project. In particular, the only true EPBM completed projects in the United States at the time of design were the N2 tunnel in San Francisco (1981) and the Anacosta River Crossing (1984) in Washington, D.C. Other numerous EPBM or slurry tunneling projects under difficult conditions were primarily in Europe and Japan. From the designer's point of view it was desirable to have a joint venture of a U.S. firm with a more experienced European or Japanese contractor. The City of San Diego did not allow pre-qualifications. Instead, qualifications on minimum contractor and personnel experience were specified. This also applied for the TBM manufacturer, segment supplier, and freeze wall contractor.

In slight contrast to what is documented by Riley (1997) on TBM procurement for the SBOO and other projects, the SBOO designers viewed the TBM specifications as performance based. This followed U.S. practice of providing minimum requirements of an EPBM or slurry machine as dictated by ground conditions (Navin et al., 1995). The specifications required changing the main bearing from within the TBM, and the ability to use compressed air and freeze requirements from the TBM if needed for major repair of the cutterhead. These specifications were implemented to avoid the kind of machine problems on the St. Clair and Storebaelt projects, and more recently, the Sheppard Ave. Project.

Minimum contractor qualifications, TBM manufacturer experience, and TBM technical proposals, segment manufacturer experience, and gasket manufacturer experience were required as part of the bid. The contract also required performance bonds for the value of the supply work for the TBM, the segments, and the gaskets. With such qualifications and experience requirements, two bids were received, both being teamed with Japanese contractors. One contractor, not teamed with a foreign contractor, dropped out at bid opening. They objected to the fact that the Contract shed too much risk on the contractor. The only disappointment was there were no European joint ventures that bid.

An important aspect was to recognize that, and educate the Owner that, tunneling projects can have large cost overruns due to differing site conditions, and consequently adequate funding must be made available. With over half of contract package 2 by tunnel construction, and over one-third by tunnel construction for all three contracts, the SBOO project, albeit with state-of-the-art technologies, had risks. The "Greenbook" (APWA, 1998) uses 25% variation in quantities before a contractor may seek compensation. Sperry (1988) suggests 15% related to geotechnical conditions for underground projects and as much as 25% for the entire project. To see that adequate funds were available, the City contracts allowed 10% contingencies for their capital improvement projects. In addition, the contract bid was set up so as to provide about 5% of the entire contract amount in contract allowances as authorized by the Owner.

Other contractual and reasonable means for providing a framework for a successful project were implemented as follows.
- For SBOO, the differing site condition clause was modified to allow for a 15% variation in ground conditions, with a debit or credit adjustment if the conditions were worse or better than expected. The 15% variation in quantities was based judgement and 8 offshore and 5 onshore borings and the projected geology based on the formational dip. It should be noted that there was much debate over better than expected conditions, conservative versus realistic baselines, and that the definition of worse or better conditions were on the opposite end of soil gradations depending on the type of TBM the contractor selected.
- With respect to equipment rental rates, the Supplementary General Conditions noted that with the exception of demonstrated financing and insurance, costs shall not be included for plant and equipment, specifically acquired for the project, such as, the TBM and its trailing gear and the fixed platform for the riser construction.

Drop Shaft Differing Site Conditions (DSC)

Installation of over 40 freeze holes (<1m spacing) to a depth of 90 m around the perimeter of the drop shaft began in December of 1995. The freeze subcontractor monitored the freeze wall with temperature profiling of the freeze holes and temperature probes. In May of 1996, however, during excavation at about 24 m below the ground surface, leakage on the order of 450 liters/minute was observed in the bottom of the shaft. With the realization that there was a hole in the freeze wall, the freeze subcontractor turned off the sump pumps in the shaft. Before backfilling of the shaft could begin, a large sinkhole developed adjacent to the shaft and a large inflow of water filled the shaft to

Figure 4. Sinkhole Adjacent to Shaft

within 7 m of the surface, approximately at the level of the surrounding groundwater table. The sinkhole was backfilled with shaft muck and the shaft itself was backfilled with imported sand. The freeze plant was shut off to reduce the risk of introducing brine into the surrounding area. Fortunately, no loss of freeze pipes occurred, and the two freeze plants were put on line to begin refreezing. Continuous temperature profiling and monitoring ensued. Closure of the freeze wall, however, was not occurring in the predicted time. A warm spot manifested as a high groundwater velocity zone on the northwest side of the shaft adjacent to the Tijuana River Valley was suspect. Four holes were converted to Nitrogen freeze holes in the suspect zone. Within three weeks, the closure was achieved as indicated by a continuous rise of cold water within the monitoring well inside the shaft. The concrete lining for the shaft was completed in November of 1996. There was a five-month delay in construction of the drop shaft, due to the hole in the freeze wall during excavation. While the GDSR warned *"that localized zones may be present in the basal alluvial gravels where groundwater velocities may be relatively high,"* all team members (partners) convened immediately over numerous meetings to solve the problem, without assigning culpability. An expert with freeze wall construction experience was hired to provide an independent assessment of the freeze subcontractor's means and methods. On the one hand, the expert thought the freeze subcontractor proceeded to excavate too early before the telltale sign of a continuous rise of water in the monitoring well inside the shaft. On the other hand, the expert indicated given other circumstantial factors, however, he would have made the same decision as the freeze subcontractor had done to proceed with excavation.

After escalation of this issue to the project manager level of partnering, it was accepted that the warm spot on the freeze wall was due to moving groundwater possibly related to underground stream flows in the adjacent Tijuana River Valley. This was later confirmed by geologic mapping of the drop shaft wall, which showed a thicker sequence of cobbles not identified by the boreholes. The Contractor and Subcontractor were equitably compensated for $2.7 million and 4 months time extension for the high velocity groundwater zone. Also, they requested compensation of $1.3 million for difficult drilling in the northwest quadrant. Through project level partnering an amount of $0.5 million was agreed upon.

TBM Start-up and Dewatering

Initial start up operations lasted from January to May 1997. Startup activities included scheduled stops for tunnel eye construction; installation of trailing gear, screw conveyor, umbilical support (hydraulic lines approaching 200 m in length), shaft elevator, train switches, muck car shaft guides, and other miscellaneous structures.

In order to turn the EPBM under, the Contractor chose to dewater the shaft eye with an array of ten wells spaced from 9 to 17 m along the tunnel length and 4.5 to 6 m on either side of the tunnel centerline. The contractor identified an aquitard, which would prevent the TBM from experiencing the full hydrostatic head of 5.5 bars before the complete screw auger could be installed.

On breakout, about 75 m^3 of soil was lost, estimated from the amount of muck removed from the shaft. About half of the soil was attributed to loss through the seals at the shaft and the tunnel eye, and the other half was attributed to imbalance of TBM advance versus material excavation. Concern arose over the possibility of large voids under the batch plant and crane pad as well as loss of passive support for the tunnel liner. About a week after the loss of ground, these concerns were alleviated when the loss of ground manifested itself as a 75 m^3 sinkhole at the surface. The sinkhole was promptly backfilled with concrete. The sinkhole caused the aquitard to be breached, making the dewatering wells ineffective in reducing the pressure to a manageable level.

The Contractor expressed concern that an early intervention using freezing or grouting would be required to inspect and perform maintenance on the cutterhead. Several reasons for poor start-up were considered, including a DSC's and improper TBM cutterhead design. Since the initial tunnel reach contained abrasive gravels, cobbles, and boulders, the use of compressed air alone would not prevent flowing ground. The Geotechnical Design Summary Report indicated that *"If access to the cutterhead is required in these areas for an unplanned inspection/repair, the only likely method by which this can be effected is through the use of ground freezing."* The TBM specifications required that the TBM have the capabilities to install freeze pipes ahead of the cutterhead. Interpretation of who should pay for an unplanned event was dependent upon the circumstances, and was challenged by the Contractor that it should be paid for regardless of the situation. Fortunately, this was never challenged to its fullest extent because the start-up problems were overcome. The mining difficulties and resulting sinkholes were attributed to Contractor learning curve, hydraulic, and operational related issues.

Tunneling Conditions and TBM Modifications

The geologic conditions at SBOO stirred the debate of earth pressure balance machine (EPBM) versus slurry machine TBM technology. An EPBM maintains muck in a pressure chamber (earth plenum) and in front of the cutterhead via a screw conveyor. The screw conveyor is controlled to permit pressure in the plenum to be matched (or balanced) to the insitu pressure at the excavated face, preventing uncontrolled soil displacements. A slurry machine maintains pressure at the excavated face by injection of slurry within the excavated soils. The spoils are mixed in the plenum behind the cutterhead, where the spoils are conveyed via slurry pipes and pumps.

The EPBM is advantageous for its higher excavation rates in clay, silt, and silty sand while the slurry machine is advantageous for its (perceived) security and faster advance in cohesionless soils, including gravels, cobbles, and boulders (GCB). Slurry machine technology, however requires costly pumping and a slurry processing plant. On some projects, the two technologies had merged with control of the tunnel face via a screw conveyor, and conveyance of spoils via slurry technology. The distribution of the geologic materials (Table 1) favored EPBM technology; however, EPBM as viewed by some was at risk with respect to the distribution of cohesionless materials. As noted by Nishitake (1987), however, EPBM's fitted with disc cutters with the mixing of spoils and breaking up of GCB in front of the cutterhead had been achieved on several projects in Japan. The contract allowed the contractor's to choose between the two technologies. The Contractor selected an EPBM, affectionately christened Molita.

The GDSR divided up the tunnel profile into four Reaches. Within each Reach, the ground was classified according to the four soil types as shown in Table 1. Table 1 also shows the predicted versus the actual conditions. As can be seen the variation was less than 15% total, and conditions were more favorable for an EPBM. The performance of Molita is demonstrated by the monthly progress shown in Table 2.

Table 1 Summary of Predicted versus Actual Geologic Conditions

Reach/ Length	GCB		Silty Sand		Sandy Silt		Clay/Cohesive Silt	
	Predict	Actual	Predict	Actual	Predict	Actual	Predict	Actual
I/927m	29%	5.3%	44%	26.4%	18%	51.4%	9%	16.9%
II/2728m	5%	5.8%	12%	4.5%	8%	13.5%	75%	76.2%
III/1112m	41	41.3%	24%	33%	35%	16.1%	0%	9.6%
IV/1023m	9	2.9%	5%	5.6%	23%	3.8%	63%	87%
Total	16.5%	12.1%	18.1%	13.6%	17.4%	18.3%	48%	56%

Table 2 Monthly Tunnel Production

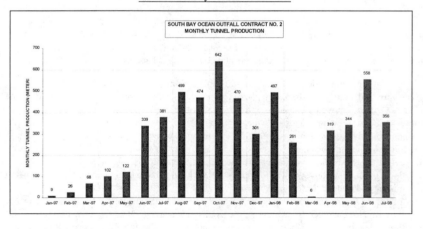

Table 3 shows the Contractor's anticipated average advance (meters/month, meters/week, and meters/day), the Engineer's anticipated advance, and the actual advance from Table 1. Table 3 shows the overall average advance, overall average not including the initial 185 rings (222 m) for startup, and the actual overall advance not including startup and or time for TBM modifications. The Engineer's anticipated advance rate is normalized for a 6-day work week. The Engineer also allowed for 2.5 months for contingencies, which is not included.

Table 3 Tunnel Overall Average Advance Comparisons

	Period	Overall	Overall w/o Startup	Overall w/o Startup & Mods.
Engineer's Anticipated Advance Rate	Month	367 m	386 m	-
	Week	85 m	89 m	-
	Day	14 m	14.8 m	-
Contractor's Anticipated Advance Rate	Month	325 m	346 m	-
	Week	75 m	80 m	-
	Day	12.5 m	13.3 m	-
Actual Advance Rate (Table 2)	Month	312 m	388 m	436 m
	Week	73 m	91 m	100 m
	Day	12 m	15 m	17 m

Beginning the week of January 12, 1997 mining began. It was completed on July 22, 1998, after the contractor had installed 4,731 rings or 5,775m. Best monthly production was 642 m. For a 6 day a week mining cycle, the best weekly advance was 173m. Best daily production rate was 30 rings (36.6 m) for three 8 hour shifts. Best instantaneous penetration was about 80 mm per minute, corresponding to a best single shove of 1 ring (1.2 m) in under 15 minutes. Best ring erection time was under 25 minutes.

Details of the unique EPBM are discussed in Navin et al. (1996) as originally ordered and in Robinson and Jatczak (1999) as modified in the field. In preparation for tackling the abrasive GCB, Kinnametal™ hard facing was added to Molita's cutterhead face in addition to 27 tungsten carbide wear bars on the gauge (Figure 5). Also, "grizzly" bars were added to restrict the size of boulders allowed in the earth plenum and the helical and axial screw conveyor.

The non-production in March 1998 (Table 2) was due to incompatibility of the screw conveyors and mixing in the earth plenum when GCB was encountered in Reach III. The Contractor had removed some of the grizzly bars in the three cutterhead openings to allow for more efficient flow of materials when mining clays, silts, and sands. When GCB was encountered, the larger openings compromised the spoil conveyance. To rectify the problems, the following action was taken.
- two additional boulder gates were installed in addition to the existing four guillotine gates,
- the helical screw conveyor was moved farther forward into the earth plenum to provide better mixing and remove the "dead zone",
- a boulder stop, consisting of bars/restrictions welded to the beginning of the axial screw to prevent larger boulders from entering the axial screw conveyor was installed, and
- Pressurization of the screw conveyors to provide for additional pressure balance, not compensated by the normal pressure drop (per flight of screw) when in GCB.

The specifications required the use of foam for spoils conditioning and to reduce cutterhead wear for the long drives required between cutterhead inspection and maintenance. Two types of foam were employed, Type A being a surfactant and Type B being polymer and cellulose for gravelly soils. Typical foam ratio by volume were 30% in clays, silts, and silty sands with a target range of 45% in gravelly soils, and when in GCB was as high as 80 to 100%. The shield also used bentonite under difficult conditions and for maintenance stops.

The tunnel face stability was a critical factor for access to the cutterhead of the TBM for maintenance. The GDSR dictated a minimum of three interventions in Reach II and one intervention in Reach IV for cutterhead maintenance. The EPBM was required to have a compressed air lock, with a minimum design pressure of 3 bar, to increase the standup time of the tunnel face. As indicated in the GDSR and by parametric studies, for many reaches of the tunnel, the ground was expected to be unstable or flow, particularly when high external pressures were present.

Figure 5. EPBM (note armor plating on face)

The contractor had about 10 interventions. In some cases, the ground was unstable and the shield required advancing to better ground. In two cases, the shield's plenum was pressurized to about 1.5 bar. On two occasions, the author inspected the ground conditions (under free air), revealing a very stable face. At one station the conditions revealed geologic structure, soil joints.

Offshore Riser DSC and Design Deficiency Claims

The offshore riser is a pipe encased by tremie concrete within a vertical shaft (Figure on right). The riser connects to the tunnel with a steel-lined elbow and connects to the seabed pipe with a horizontal outlet in the riser head. Fabrication of the riser structure began in October 1995 and was completed in the spring of 1996. Offshore riser shaft construction began in October 1996 and included excavating the upper 7.6 m of marine sediments and driving a two-stage casing. The 3.96 m ID upper stage casing was driven through Pleistocene Gravels into approximately 1.5 m of the San Diego Formation. A full length 3.65 m ID lower stage casing was driven through the upper casing to within 3 m above the planned tunnel crown. The void between the casings was grouted.

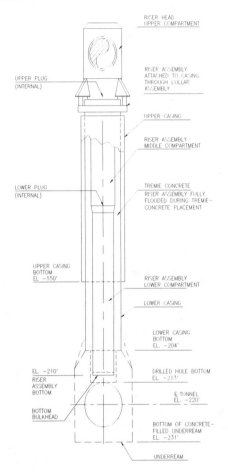

The underreaming of the area below the lower shaft casing to allow replacement of soil with a low heat of hydration 10 to 13 MPa concrete in the tunnel-to-riser connection zone was performed in November of 1995. The concrete plug provides the target for the TBM to pass through. The narrow range of concrete strength was specified to provide a stable medium through which the connection is mined, and be easily excavated by the TBM.

Unfortunately, the concrete never attained the proper strength, only being on the order of 1 MPa. In addition, voids on the order of 45 m^3 developed between the lower casing and the soil. Subsequently, the voids were grouted underwater from inside the casing. The concrete plug was redrilled and replaced with the proper mix in July 1997.

The riser Subcontractor claimed a substantial sum and a 34 day contract extension due to excess clays and cobbles in the Marine Sediments and excess cobbles in the Pleistocene Gravels. The Designer acknowledged that the clay layer in the marine sediments was thicker than what was expected, but noted the construction difficulties were attributable to improper methods for installing sheet piles. The Designer also contended that the numerous cobbles in the Pleistocene

Gravels was as expected. These differing site condition issues went before the Disputes Review Board (DRB). The DRB concluded that a differing site condition (DSC) did exist and excess cobbles were encountered for about 25 percent of shaft excavation in the Pleistocene Gravels, and that Contract Documents did not adequately address this. However, the DRB concluded excess cobbles were not encountered in the Marine Sediments.

The Subcontractor also claimed a substantial sum and a 203 day contract extension over lack of strength gain of the concrete plug attributable to the Designer's direction, and overly restrictive specification and impossible-to-meet specifications. They also claimed that the design relied upon embedment of the concrete plug at the bottom edge of the lower casing (el. -204 ft). The Construction Manager (CM) had evidence of improper concrete batching and addition of too much water. Also, the Contractor had filed a lien on the laboratory that performed the trial batch test of the concrete mixes.

The Designer performed analysis and field tests to show that the concrete plug was properly designed and that the drill was improperly stabilized, undermining the weak concrete, and causing the subsequent inflow of soils. Rather than putting this issue before the DRB, it was agreed to escalate the issue to the executive level of partnering and work towards a global settlement for the DSC and design deficiency issues. Both issues were settled for less than 10% of the claim.

Conclusions

Effluent flows, ranging between 0.5 to 1 m^3/sec, began on January 12, 1999, thus fulfilling a decade long grass-root effort to restore San Diego's South Bay beaches. SBOO has received domestic and international attention for the unique requirements and state-of-the-art TBM specifications and tunnel liner design, including numerous awards and the California State Council ASCE 1998 Project of the Year Award. The illuminating construction events discussed provide geotechnical lessons learned for this most technically challenging project. The lessons learned are summarized as follows.

- On tunnel projects, successful completion depends as much on the art of tunnel design and prediction of tunneling conditions as the contractual setting and bid environment. Adequate contingencies, although not used for SBOO, laid the ground work for a successful project.
- Freeze wall shafts require careful consideration before proceeding to excavate. Special attention/scrutiny during design specification and construction inspection is recommended.
- Molita including field modifications to handle GCB was the correct choice for the ground conditions. SBOO thus provides a useful case history for the debate between EPBM vs. slurry machine.
- Geotechnical DSC's issues, no matter how illogical, may be blamed for problems that are associated with contractor's means and methods.
- Partnering, through all the PEOPLE who formed the SBOO design and construction team, proved to be a successful contract mechanism to sort out the geotechnical issues from the others, whether or not the issues worked in favor or against the contractor or the predicted baseline conditions.

References

APWA [American Public Works Association], 1998, Standard specifications for public works construction, *Building News, Inc* (pub.).

Kaneshiro, J.Y., G.W. McBain and S.J. Navin, G. Colzani, F.B. Estep, R.H. Lee and J.M. Mueller, 1998, Partnering Under Pressure - San Diego's South Bay Ocean Outfall Construction Experience, in K.Y. Lo, ed., Proceedings of the 15th Candian Tunnelling 1998, Vancouver, September 24-26.

Kaneshiro, J.Y., S.J. Navin, and G.E. Korbin, 1996, Unique precast concrete segmented liner for the South Bay Ocean Outfall project, in L. Ozdemir, ed., Proceedings of the International Conference on North American Tunneling '96 and the 22nd General Assembly of the International Tunneling Association, Washington, DC, April 21-24.

Kaneshiro, J.Y., G.E. Korbin, and J.D. Hart, 1996a, Fault crossing design and seismic considerations for the South Bay Ocean Outfall, in L.F. Catalano, ed., Pipeline Crossings 1996, ASCE Proceedings of the Specialty Conference, June 16-19.

Matayas, R.M., A.A. Mathews, R. J. Smith, and P.E. Sperry, 1996, Costruction Dispute Review Board manual, McGraw-Hill.

McBain, G.W., J.Y. Kaneshiro, S.J. Navin, F.X. Collins, and L. Meiroin, 1998, To tunnel or not to tunnel, that is the question, WEFTEC'98, October.

Navin, S.J., J.Y. Kaneshiro, L.J. Stout, and G.E. Korbin, 1995, The South Bay Tunnel Outfall project, San Diego, California, in G.E. Williamson and I.M. Gowring, eds., Proceedings of the Rapid Excavation and Tunneling Conference, June 18-22.

Navin, S., J. Kaneshiro, G. Korbin, L. Home, and G. Williamson, 1996, Unique tunnel boring machine for the South Bay Ocean Outfall project, Proceedings of the 14th Canadian Tunneling Conference, October 9-11.

Nishitake, S., 1987, Earth pressure balanced shield machine to cope with boulders, in Jacobs, J.M. and R.S. Hendricks, Proceedings of the Rapid Excavation and Tunneling Conference, New Orleans, LA, June-14-17.

O'Rourke, T.D. (ed), 1984, Underground Technology Research Council's Technical Committee on Tunnel Lining Design, Guidelines for tunnel lining design, ASCE, NY.

Riley, J.J., 1997, Owner responsibilities in the selection of tunnel boring machines, with reference to contractual requirements and construction conditions, Proceedings World Tunneling Congress, 23rd General Assembly of the International Tunneling Association, Austria, April.

Robinson, B. and M. Jatczak, 1999, Construction of the South Bay Ocean Outfall, Proceedings of the Rapid Excavation and Tunneling Conference, Orlando FL, June 21-23.

Schug, D.L., J.D. Reitman, T.K. Rockwell, Y. Moriwaki, and J. Kaneshiro, Characterization of offshore fault hazards South Bay Tunnel Outfall, in Annual Meeting Abstracts of the Association of Engineering Geologists, Sudbury, MA.

Sperry, P.E., 1988, Costing contingencies, Civil Engineering (ASCE), April.

USNCTT [U.S. National Committee on Tunneling Technology, Subcommittee on Contracting Practices], 1974, Better contracting for underground construction, National Academy of Sciences, Washington D.C.

POSITIONING FOR LEAST COST: MANAGING RISKS IN UNDERGROUND CONSTRUCTION

Lawrence H. Roth, P.E., G.E., F. ASCE

INTRODUCTION

Underground construction is unfortunately often an adversarial process. Unforeseen problems, such as differing site conditions, can lead to projects of lower quality, or can cause cost escalation and schedule creep. When these problems occur, disputes can arise and stakeholders frequently find themselves at odds, each looking to the other for relief. *Positioning for least cost*, as described in this paper, is systemic in construction. While many stakeholders participate in underground construction, this paper focuses on how the owner, the contractor, and the designer position for least cost, and how this can affect project outcomes.

As part of positioning for least cost, stakeholders attempt to avoid risk by transferring it to others, creating an emotionally charged atmosphere that can lead to disputes. This paper describes several risk management tools that emphasize risk sharing rather than risk transfer. These tools can help avoid disputes, or facilitate timely and cost-effective resolution when disputes occur. The paper focuses on the use of Geotechnical Baseline Reports (GBRs) as a way to allocate risks and reduce uncertainty, and illustrates the use of GBRs in a case history. The paper also describes alternative dispute resolution (ADR) techniques aimed at avoiding the time and high cost of litigation, and discusses several risk management tools that have evolved from ADR.

POSITIONING FOR LEAST COST

What does *positioning for least cost* mean? The answer is found in the strategies and tactics that stakeholders use to achieve their project objectives.

Owners. Owners are motivated to reduce project costs to improve return-on-investment (ROI). In general, owners fall into two broad categories. Owners in the first category -- for example many in the private sector -- position to reduce the *total* cost of the project,

---

Principal Engineer, Subsurface Consultants, Inc., Lafayette, California

including the costs associated with site investigations, with project design, and with construction management and quality control. Spending less on these activities inevitably results in increased uncertainty, which increases risk. Uncertainty -- for example in ground conditions to be encountered -- increases the likelihood of claims, which increases costs, or leads to disputes, and in some cases, litigation.

Owners in the second category -- for example many in the public sector -- are motivated to reduce the *capital cost* of construction, since that is the cost that generally receives the most scrutiny from project stakeholders such as taxpayers. In this case, the owner positions to control construction cost, and may therefore be resistant to pay for change, even change that may be legitimate. This can lead to schedule delay and increased cost, either through payment of claims for extras, or for dispute resolution.

In the latter case, there is cause for optimism if the owner seeks to control construction cost by reducing uncertainty. This type of owner is often willing to pay more for site investigations, more for project design, or more for construction management and quality control. By paying for more thorough site investigations, the owner aims to reduce uncertainty in ground conditions, traditionally the aspect of construction with the greatest uncertainty and risk. Increased investment in site investigations reduces the likelihood for cost over-runs, for payment of claims, or for disputes associated with differing site conditions leading to a better chance of controlling construction cost. The owner may, however, perceive an *increase* in total project costs because of increased investment in site investigations and engineering.

Contractors. Contractors are motivated to be competitive and to build sales volume by reducing contingencies in their bid prices. After a project is awarded and construction begins, the contractor positions for least cost to increase its profits. Reduced contingencies, and the tendency to cut corners, increases the contractor's risk and increases the contractor's desire to transfer risks to others. This increases the likelihood of claims for extras, again leading to increased construction cost.

Designers. Increasingly, designers are being selected for projects based on fee rather than qualifications. This causes designers to position for least cost by limiting scope to reduce cost and increase profits. Limiting costs for site investigations, for thoroughness of design, or for construction management and quality control increases uncertainty. In this case, increased uncertainty can lead to increased bid prices, to cost over-runs resulting from claims for change, and to increased construction cost.

Result. Positioning for least cost is expensive. To achieve their objectives, stakeholders focus on risk avoidance and risk transfer, rather than risk sharing. In this atmosphere of uncertainty, we should not be surprised that project quality suffers, or that costs increase because of claims and dispute resolution. The inevitable result of stakeholder positioning for least cost is not *least cost*, but rather *increased cost* for some or for all.

RISK MANAGEMENT

When disputes occur, positioning for least cost manifests itself making resolution difficult, often escalating to litigation. Unfortunately, settling disputes through the court system is a time-consuming and costly process. Not surprisingly, in the past two decades the construction industry has devised and implemented many new tools to reduce the cost and time of dispute resolution. For the most part, dispute resolution tools, such as the ADR techniques described below, are invoked after a dispute occurs. However, ADR tools such as Dispute Review Boards (DRBs) and Escrow Bid Documents (EBD) are increasingly being used before disputes occur to help manage risks.

Other tools that evolved from ADR -- for example partnering -- help stakeholders do a better job identifying and allocating risks. Still other tools -- for example GBRs -- are aimed at reducing project uncertainty to reduce risk. A few of the more common dispute resolution and risk management tools in use today are described below.

Alternative Dispute Resolution. ADR comprises methods used to resolve disputes without resorting to litigation. Disruptions to business, and the frustrating expenditure of money and time to fight lawsuits, has led to an increasing use of ADR. In 1992 more than 40,000 cases were handled by private mediation and arbitration firms (Pollock, 1993). Tracking the legal experiences of 406 companies between 1990 and 1992, the Center for Public Resources estimates a savings of $150 million in fees for legal services through the use of alternatives to conventional litigation (Pollock, 1993).

ADR is not new to the construction industry; for example, arbitration has been used for nearly a century. Traditional arbitration, however, is often adversarial (i.e., *win-lose*). In 1974, ASFE/Professional Firms Practicing in the Geosciences introduced the concept of mediation/arbitration as a conciliatory ADR technique (i.e., *win-win*) to help relieve the professional liability problems confronting geotechnical engineers.

In *Alternative Dispute Resolution for the Construction Industry*, ASFE describes 20 different ADR techniques, such as mediation, mediation/arbitration, mini-trials, and summary jury trials (ASFE, 1988). In all cases, successful ADR techniques eliminate the need for formal civil trials, and reduce the cost and time for discovery that is a major part of the litigation process. On the project site, rapid resolution of disputes means that attention can be focused on construction quality and on timely completion. ASFE (1988) cites several benefits of conciliatory ADR:

1. ADR is flexible, offering many options for dispute resolution. Stakeholders in a dispute can select a technique suitable for a specific circumstance, and the selected method can be tailored to meet project needs.

2. ADR saves time and money. It can be implemented quickly following a dispute, often in a matter of days after the incident giving rise to the claim. Most ADR

techniques are designed for expedited resolution, in sharp contrast to the months or years required for a dispute to wend its way through the labyrinthine court system.

3. ADR permits disputants to select the neutral party who will arbitrate, mediate, or otherwise facilitate resolution. (Note that some ADR techniques, such as DRBs, recognize that the best time to select a neutral party is at the start of a project before bias is created by the situation in dispute.)

4. ADR processes are open, accessible, and user-friendly, permitting disputants to be intimately involved and in control from the start.

5. ADR permits flexible and creative solutions, not limited to cash settlements resulting from litigation.

6. ADR processes can be confidential; proceedings need not be public as in court.

7. By emphasizing fairness, ADR encourages non-adversarial processes and business-like solutions aimed at preserving relationships, rather than polarizing them.

Dispute Review Boards. The DRB is an ADR technique that has been used extensively in underground construction, and has also been used on highway, bridge, and dam projects (ASCE, 1991). The DRB is often stipulated in the bid documents, although it can be created at any time during a project.

Most DRBs comprise three professionals, one selected by the owner, one selected by the contractor, and a third selected by the first two. DRB members should be experienced with the project-specific design and construction issues, with cost and scheduling, with interpreting contract documents, and with claims analysis. To be successful, project stakeholders must acknowledge the expertise of Board members, respect them as project participants, and trust their objectivity and impartiality (ASCE, 1991).

The DRB generally participates throughout the project, meeting regularly at the construction site so that the Board can see work in progress. If a dispute arises, the DRB is convened to evaluate relevant background material, and to hear the positions of the disputants. In general, DRB hearings are informal, problem-solving sessions. Attorneys are discouraged, as is excessive reliance on technical experts. The goal of the DRB is to determine merit of position, and to resolve the dispute at the project level. DRBs offer many benefits:

1. The likelihood of disputes influences contractors' bid pricing strategies. By including a DRB in the contract, the owner expresses its willingness to be reasonable in the event of a dispute. The presence of the DRB may encourage the bidder to shave contingencies in its bid, resulting in savings.

2. DRBs can move quickly, making timely decisions to keep projects moving without significant delay. When a dispute occurs, it can be resolved while relevant technical

information is readily available, and before stakeholders have become entrenched in their positions.

3. Although DRB decisions are generally non-binding, the contractual commitment of project stakeholders to the DRB means that decisions are usually accepted.

4. DRBs can often prevent disputes from escalating by fostering trust and respect between project stakeholders who rely on the DRB as impartial and knowledgeable. With experts close at hand, the tendency for posturing and game-playing is reduced since stakeholders know that disputes will be quickly and effectively addressed.

DRBs resolve disputes more cost-effectively than litigation. In one study, the use of DRBs was tracked on 166 projects worth $10.5 billion. DRBs heard 225 disputes and successfully resolved 208. Only one dispute went to litigation and was resolved before adjudication. DRBs save cost and time, and generally add only 0.1-0.3 percent to total project costs (*Engineering News-Record,* July 11, 1994).

Escrow Bid Documents. Both ASCE (1991) and Hatem (1998) recommend the use of EBD for complex underground projects. The EBD technique independently preserves the contractor's bid documents (considered to be *trade secrets* that remain the contractor's property at all times) for use in avoiding or resolving disputes, and for negotiating equitable adjustments to the contract price.

ASCE (1991) describes typical procedures for EBD. Normally, the three lowest bidders submit sealed EBDs shortly after bid opening. Prior to award, the EBD of the apparent low bidder is examined to verify its authenticity, completeness, and legibility. It is not reviewed to evaluate the contractor's qualifications, estimating techniques, or proposed construction methods; nothing in the EBD changes the terms of the contract. The EBDs of unsuccessful bidders are returned following award of the contract.

The EBD should clearly document the estimated costs for performing the work in each bid item, and should include quantity takeoffs, calculations, quotes, consultants' reports, notes, assumptions, and other information that the bidder used to arrive at its bid price. Hatem (1998) also recommends that the EBD include a detailed description of the anticipated means and methods of construction, as well as how they might be affected by differing site conditions.

When not in use, the EBD is stored in third-party escrow. If needed to help resolve a claim or dispute, the EBD is reviewed by representatives of both the owner and the contractor, and if there is a DRB, by members of the Board. On completion of the contract, the EBD is returned to the contractor. EBDs benefit both the contractor and the owner by leading to a more honest and cooperative relationship on the job (ASCE, 1991).

Partnering. Since construction can be adversarial, stakeholders may be suspicious of one another and protective of their own financial and legal positions (Work Systems, 1994). Partnering is a structured process designed to coalesce project stakeholders into a team by identifying goals held in common, and by causing them to recognize that their individual goals may be in conflict. By expressing and understanding their individual and shared goals, participants in a partnering process seek to improve quality, avoid litigation, and identify cost-effective ways of doing business.

The agreement to partner does not replace the contract which remains the legal agreement. With the contract as the starting point, partnering generally results in additional, jointly-developed documents that describe:

1. The organizational structure within which the partners will work.
2. The project goals held in common.
3. The strategies to meet those goals.
4. The measures of success in meeting the goals.
5. The methods to be used for communication and conflict resolution.

Although a partnering agreement is not a formal contract, agreements made in the partnering process can be legally binding in the event of dispute (ACEC & AIA, 1993).

Partnering is accomplished in working sessions throughout the life of the project. Normally, an initial session focuses on identifying and overcoming the barriers to partnership formation, on establishing common ground between partners, and on improving communication. With common ground established, partnering leads to a charter that defines the nature and responsibilities of the team, the limits of its authority, and the methods of conflict resolution (Work Systems, 1994). The charter also defines goals, measures of success, and incentives (ACEC & AIA, 1993).

Successful partnering is dependent on understanding the process, committing to the partnership, creating a structure in which to make the process work, and persevering through the process (ACEC & AIA, 1993). All key individuals in the project must understand partnering and be involved in the process of defining goals and the methods for attaining them on the project. Organizational leaders must be committed to partnering; must demonstrate openness, trust, and cooperation; and must be willing to change long-held, potentially negative attitudes.

Partnering can offer significant benefits. In a study of projects administered by the U.S. Army Corps of Engineers, Weston (1992) found that there was nine percent less cost growth from claims and change orders on projects with partnering compared to projects without partnering. Between 1988 and 1994, the Corps used partnering on over 200 projects, none of which were litigated (*Engineering News-Record,* July 11, 1994). Partnering can help reduce costs, minimize disputes, and increase project satisfaction at a cost that might add about 0.2 percent to the contract price (Weston, 1992).

Geotechnical Baseline Reports. As described by Essex (1997) and Edgerton (1998), the primary purpose of a GBR is to establish a contractual statement (i.e., a baseline) of the geotechnical conditions anticipated during construction of projects that have significant risk of claims for differing site conditions. GBRs define the geotechnical conditions used to form the basis of design, and that should be used as the basis for bids. The goal of the GBR is to reduce uncertainty by defining the baseline and by allocating risk. The contractor accepts risks for conditions that are consistent with or more favorable than those indicated by the baseline, while the owner accepts risks for conditions that are less favorable than those indicated by the baseline.

On most projects, the geotechnical information gathered for design is made available to bidders for their interpretation. Contract documents typically disclaim responsibility for interpreting the geotechnical data. Since bidders generally have neither the time, the resources, nor the expertise to undertake their own pre-construction investigations, they rely on their own interpretation of the data presented. In some cases, bidders fail to recognize likely problems with geotechnical conditions. In other cases, bidders make overly-optimistic interpretations of anticipated conditions to sharpen their bid prices. These factors often contribute to claims for additional compensation or time when the actual conditions encountered are more adverse than those assumed.

These claims are usually made under the Differing Site Conditions (DSC) clause, almost always a standard clause in contract documents. The DSC clause is intended to relieve the contractor of the risk of encountering conditions materially different from those indicated, or normally expected. The DSC clause provides a remedy to contractors under the construction contract, and assigns the risk for differing site conditions to the owner. In contrast, by establishing a geotechnical baseline, and making it a part of the contract documents, the risks of encountering differing site conditions are allocated between the owner and the contractor. In return for accepting the risks associated with conditions less favorable than the baseline, the owner should receive more competitive bids since bidders can reduce their contingencies for risk. Essex (1997) states that clear and precise baselines enhance the benefits of the DSC clause.

In the past several years, geotechnical baseline reports have gone by many names. Early on, the baseline was defined simply by making the geotechnical report *a part of* the contract documents. Later, the geotechnical report that was made part of the contract documents was called a Geotechnical Design Summary Report (GDSR). Essex (1997) recommends calling the interpretive report included in the contract documents a Geotechnical Baseline Report (GBR) to be used by:

1. The designer, as a basis for design and the anticipated cost of construction.
2. Bidders, as a contractual indication of anticipated conditions, and the risks to be allocated to the contractor.
3. The contractor, as the basis for selecting construction means and methods.
4. The contractor and the owner's construction manager for identifying potential differing site conditions during construction.

5. The owner and the contractor to resolve disputes related to conditions that vary from the baseline.

Essex (1997) states that the geotechnical baseline is a *contractual baseline*, not necessarily *geotechnical fact*. Establishment of the baseline requires interpretation of geotechnical conditions by interpolating between or extrapolating beyond locations where geotechnical conditions are known (for example, at boreholes). Baselines directly influence bid prices, as well as adjustments to the contract price and time. The owner's tolerance for risk must influence the setting of the baseline. By setting a conservative, or pessimistic baseline, the risk for differing site conditions is allocated to the contractor and the cost is built into the bid price. Setting an un-conservative, or optimistic baseline, means that the cost for differing site conditions will be defined by change order.

RISK ALLOCATION USING GBRs

The baseline allocates risk between owner and contractor. A potential pitfall to effective use of GBRs can occur if an unscrupulous or unwitting owner views the baseline as the designer's warranty of conditions to be encountered. In this case, the risks borne by the owner can be transferred to the designer without fair and equitable compensation. This and other potential pitfall are described below.

Potential Pitfalls of GBRs

The contractor may assume that the baseline is conservative, and base its bid on an overly-optimistic assessment of conditions. In this case, the contractor may prove unwilling to accept the consequences for its optimism, and may try to re-allocate risk by making meritless claims.

The designer may fail to clearly communicate the purpose, use, and risk-transfer ramifications of GBRs to the owner. Or, the designer may fail to propose an appropriate baseline because:

1. There were inadequate site investigations to characterize geotechnical conditions.
2. The geotechnical conditions were misinterpreted.
3. The anticipated means and methods of construction were misunderstood.
4. The owner's tolerance for risk was misjudged.
5. The state-of-the-art for site investigations was not sufficiently advanced to accurately quantify important items in the baseline.

An owner who sets *construction* cost containment, rather than *total* cost containment as its goal will be resistant to pay for change, even legitimate change. Or, the owner may see GBRs as a means to transfer risk. This can occur if the owner sees the baseline as a warranty of conditions to be encountered, and as a consequence looks to the designer for satisfaction of DSC claims.

Successful Use of GBRs

For GBRs to be successful, the contractor must prepare a responsible bid, and must realize that it has accepted the risk of conditions that are consistent with, or are more favorable than the baseline. In addition, the contractor must provide means and methods that are appropriate for baseline conditions.

Designers must use cost-effective site investigations to propose realistic baselines. Designers must also understand the range of means and methods that might be employed, and must write clear and definitive baselines compatible with other parts of the contract documents. Finally, designers must ensure that owners understand and accept the purpose of the GBR, as well as the baseline and its consequences.

Owners must provide appropriate funding for site investigations to properly characterize baseline conditions. While the owner may look to the designer for advice, it is critical that the owner select the baseline consistent with its appetite for risk. The owner must also provide adequate funding to document conditions exposed by construction, and maintain adequate reserves to quickly pay for valid DSC claims. Finally, the owner must accept the fact that it has set the baseline, and that the baseline does not constitute the designer's warranty of subsurface conditions.

Can risks be fairly allocated using GBRs? It depends on project stakeholders:

1. Understanding and accepting the ramifications of GBRs as a contract mechanism.
2. Understanding and accepting their respective roles and responsibilities.
3. Accepting responsibility for the risks as allocated.
4. Recognizing that there is no guarantee against claims for differing site conditions.
5. Understanding that a differing site condition is not by itself proof of defective design.

GBR Case History

The Project. In the 1980s, the author was involved in what is thought to be one of the first uses of the GBR concept on a project other than tunnel construction. In this project, a large metropolitan sewerage district contracted for design and construction of a major new pump station requiring an excavation 12-15 m (40-50 feet) deep over an area of 46 by 61 m (150 by 200 feet). The site was constricted, and nearby structures were settlement-sensitive, water-holding masonry tanks supported on shallow timber piles. The owner's goals included keeping the treatment plant in operation, minimizing damage to existing facilities, and containing total project cost.

The designer, an international firm with in-house geotechnical design capability, recognized that the site soil conditions comprising 11 m (35 feet) of soft clay overlying dense glacial till, and the proximity of settlement-sensitive structures required a sophisticated excavation bracing system for temporary earth support. With the owner's concurrence, the designer made the geotechnical report a part of the contract documents (well before the term GBR had come into common use).

The geotechnical report included data and interpretations that established the baseline for, among other things, the design and performance of the earth support system. The contract documents instructed the contractor to design and install the excavation bracing system using its own choice of earth support systems to maintain safety and prevent ground movement. The selected system was required to prevent soil movement that could damage adjacent structures, or damage or delay the work. The documents also required the earth support system to conform to regulatory requirements, and to incorporate the sequence of excavation and installation of lateral supports in the loading diagram.

The contract documents required the contractor to select soil, hydrostatic, and surcharge loading for design of the earth support system, but in no case was the selected loading diagram to be less than the one shown in the geotechnical report. As a further requirement, the earth support system was to be extended to the glacial till stratum defined in the report. Deformations were limited to less than 50 mm (2 inches), and instrumentation was to be installed and monitored so that action could be taken if deformations exceeded those specified.

The Problem. During construction, the contractor, local and well-experienced, submitted a design for an earth support system based on a loading diagram less than that specified in the geotechnical report. In addition, the contractor's proposed earth support system did not extend to the glacial till as shown in the report and as specified. Concerned for excessive deflection, and for the possibility of base failure, the designer rejected the submittal. The contractor challenged citing its responsibility for means and methods, and for design of temporary works.

After negotiation, the contractor acquiesced and constructed an earth support system that conformed to the contract requirements. The contractor, an experienced building contractor with little underground construction experience, was not trying to "cut corners," but was clearly not familiar with the requirements of a geotechnical report made part of the contract documents (i.e., a GBR). In this case, it is likely that the GBR concept helped prevent the problem from escalating into a costly and time-consuming dispute.

Lessons Learned. In hindsight, the problem may have been avoided if the bid process had clarified and emphasized the GBR concept. And, although the geotechnical report was bound with the contract documents, the geotechnical report and the contract documents could have been better coordinated to reduce confusion. To its credit, the contractor accepted its responsibility under the contract, and constructed an earth support system that met the owner's objectives and the designer's performance goals. In this case, the GBR concept ensured construction of a satisfactory earth support system.

CONCLUSIONS

With competing goals to increase ROI and profits, owners, contractors, and designers position for least cost often leading to disputes and to higher costs for some or for all. When disputes occur, ADR techniques can reduce the cost and time to resolve them as compared to conventional litigation. ADR can be a fair, non-adversarial process that aims to preserve relationships. Risk management tools, such as dispute review boards, escrow bid documents, and partnering have proven effective in helping project participants realize their goals by doing better jobs of identifying and managing risks.

GBRs can be used to reduce risk by reducing uncertainty in construction. Successful use of GBRs requires that contractors prepare responsible bids, provide appropriate means and methods, and accept the risk for conditions that are consistent with, or are more favorable than the baseline. Designers must propose realistic, clear, and definitive baselines, and ensure that owners understand the purpose of the GBR, as well as the baseline and its consequences. Owners must provide appropriate funding to characterize baseline conditions, to document conditions exposed by construction, and to pay for valid DSC claims. The owner must select the baseline consistent with its appetite for risk, and must accept the fact that the baseline does not constitute the designer's warranty of subsurface conditions.

REFERENCES

ACEC & AIA. *A Project Partnering Guide for Design Professionals.* American Consulting Engineers Council and The American Institute of Architects. Washington, D.C. 1993.

ASCE. *Avoiding and Resolving Disputes During Construction -- Successful Practices and Guidelines.* The Technical Committee on Contracting Practices of the Underground Technology Research Council. American Society of Civil Engineers. New York, New York. 1991.

ASFE/Professional Firms Practicing in the Geosciences. *Alternative Dispute Resolution for the Construction Industry.* Silver Spring, Maryland. 1988.

Engineering News-Record. "Industry Pounds Away at Disputes." McGraw Hill. New York, New York. Pages 24-27. July 11, 1994.

Edgerton, William W. "Site Investigations: A Guide." *Civil Engineering.* American Society of Civil Engineers. Reston, Virginia. June 1998.

Essex, Randall J., Editor. *Geotechnical Baseline Reports for Underground Construction -- Guidelines and Practices.* The Technical Committee on Geotechnical Reports of the

Underground Technology Research Council. American Society of Civil Engineers. New York, New York. 1997.

Hatem, Esq., David J. *Subsurface Conditions -- Risk Management for Design and Construction Management Professionals.* John Wiley & Sons. New York, New York. 1998.

Pollock, Ellen J. "Mediation Firms Alter the legal Landscape." *The Wall Street Journal.* New York, New York. March 22, 1993.

Weston, D. C. *An Analysis of Project Performance for Partnering Projects in the U.S. Army Corps of Engineers.* The University of Texas at Austin. Reproduced by the National Technical Information Service, U.S. Department of Commerce. Springfield, Virginia. 1992.

Work System Associates, Inc. *The Evolution of Strategic Partnering.* Marlborough, Massachusetts. Volume 5, May 1994.

ACKNOWLEDGMENT

The author wishes to thank Bill Rudolph, John Bachner, Jim Schneider and Mark Stanley for their thoughtful review and comment.

PROFESSIONAL STANDARDS FOR THE GEOTECHNICAL ENGINEER
A Legal Perspective

Karen P. Layng
Justin L. Weisberg[1]

Introduction

The services of the Geotechnical Engineer have a significant impact on the success of almost every construction project. For example, geotechnical engineering services such as the design of foundations for nuclear power plants and the determination of the soil characteristics beneath a landfill demonstrate the gravity of geotechnical engineering decisions on the safety and welfare of the general public. To protect the public, statutes have been enacted to regulate the practice of Geotechnical Engineers and ensure that those practicing within this field have demonstrated a minimum level of knowledge and experience. Such statutes are also intended to ensure that those who provide geotechnical engineering services follow certain professional practices. In addition to governmental regulation of Geotechnical Engineers, minimum standards of care have been established through the opinions and practices within the professional engineering community. When Geotechnical Engineers fail to comply with such standards or regulations, they can be found liable for the damages which result. The following paper will explore some of the practices that have resulted in discipline of, or liability to, Professional Engineers with a focus upon

---

[1]Karen P. Layng is a Partner in the litigation practice area of Vedder Price, Kaufman & Kammholz and heads the firm's construction law group. Justin L. Weisberg, P.E., M.ASCE is an attorney with the law firm Vedder, Price, Kaufman & Kammholz and is a member of the firm's construction law group. Ms. Layng and Mr. Weisberg counsel engineers on professional liability, construction, environmental and insurance coverage issues. Vedder, Price, Kaufman & Kammholz is a full service law firm with offices in Illinois, New York and New Jersey. The views expressed herein are their own and should not be attributed to their firm or their clients. If you have any questions concerning the issues raised in this paper, please feel free to contact either Ms. Layng or Mr. Weisberg at (312) 609-7500.

Professional Regulations

Most states restrict the practice of professional engineering to those persons who are registered pursuant to the regulations determined by the governing Board of Engineers. "Professional Engineering Practice" has been defined as:[1]

> ... the consultation on, conception, investigation, evaluation, planning, and design of, and selection of materials and methods to be used in, administration of construction contracts for, or site observation of, an engineering system or facility, where such consultation, conception, investigation, planning, design, selection, administration, or observation requires extensive knowledge of engineering laws, formulae, materials, practice and construction methods. A person shall be construed to practice or offer to practice professional engineering, within the meaning and intent of the Act, who practices, or who, by verbal claim, sign, advertisement, letterhead, card, or any other way, is represented to be a professional engineer, or through the use of the initials "P.E." or the title "engineer" or any of its derivations or some other title implies licensure as a professional engineer, or holds himself out as able to perform any service which is recognized as professional engineering practice.
>
> Examples of the practice of professional engineering include, but need not be limited to, . . . Geotechnical engineering, including, subsurface investigations; soil classification, geology and geohydrology, incidental to the practice of professional engineering...

Generally, an individual must be registered as a Professional Engineer before he or she can offer to provide engineering services. A cover letter by an individual not registered but listing experience in engineering design, product development, failure analysis, and investigative procedure, has been found to constitute the unauthorized practice of professional engineering. In that decision, the Court determined that even holding oneself out as able to perform a service which constituted professional engineering, without first holding such a license, violated the Illinois statute.[2]

Many states require an individual practicing within the field of geotechnical engineering to be registered as a Professional Engineer or, as required in some states, to be specifically registered within the sub-discipline of civil engineering. A few states, such as California, have even gone so far as to recognize registration in the specialized field of geotechnical engineering and require Civil Engineers to obtain additional experience and testing to obtain such certification.[3]

Violations Which Commonly Lead To Disciplinary Proceedings

The following paragraphs will list some of the violations which commonly result in disciplinary proceedings against the Professional Engineer. The forms of discipline applied within most states include fines and/or suspension or revocation of the engineer's registration. It is noteworthy that professional registration has been considered a property right which entitles the registrant to due process, including a hearing prior to revocation.[4]

Permitting Seal To Be Affixed To Plans Or Specifications Not Prepared By Engineer Or Under His Personal Supervision

The seal of the Professional Engineer designates that the subject plans and specifications have been prepared either by or under the supervision of the engineer who affixed her seal upon the plans. Most states consider it a disciplinary offense for an engineer to affix his seal upon plans or specifications which have not been prepared by him or under his immediate supervision.

At least one Court has questioned a reviewing board's strict interpretation of the word "prepared." The owner of plans originally prepared for one building retained an engineer not responsible for the original plan preparation to review the plans for use in the construction of an identical building, in a different location. The engineering review board suspended the engineer's license for affixing his seal to plans not prepared by him or prepared by a subordinate under his direct supervision. The Court, in overturning the suspension, determined that the review of the plans by the engineer and the engineer's recommendation for further investigation of the soil conditions at the proposed location constituted "preparation" under the intent of the statute.[5] However, other states have determined that the act of one Professional Engineer affixing his seal to plans originally prepared by an engineer at another firm constitutes misconduct.[6]

Misconduct

Misconduct relates to some deviation from a fixed duty or definite rule of conduct.[7] In one case, the engineer was determined to be guilty of misconduct for performing welding on a construction project without proper certification. The Court determined that while welding did not fall within the definition of professional engineering, responsible supervision of construction did fall within the practice of professional engineering. The Court further found that welding done by the supervising engineer constituted misconduct where the state administrative code required any person engaging in welding to be certified.[8]

In a separate decision, a misrepresentation by the design engineer stating that the engineer had reviewed the structural integrity of a design change request by the fabricator when no engineering calculations were indeed performed, was deemed supportive of a finding of misconduct.[9]

However, where an engineer billed amounts in excess of those originally agreed to for the engineer's fees and services, was not held sufficient to support a finding of misconduct.[10]

Incompetence

Incompetence refers to some demonstrated lack of competence or ability to perform professional functions. Generally, a single error in itself will not support a finding of gross negligence. However, several errors demonstrating a lack of ability to perform professional engineering have been found to support a determination of incompetence. For example, the design of a building which was deficient with respect to the foundation, structural beams, steel roof joists, certain steel columns, the masonry bearing walls and the anchorage of the roof to the building, supported a finding of incompetence.[11]

Gross Negligence

A finding of gross negligence requires a higher degree of failure to exercise ordinary care in a given situation.[12] Whether admitted negligence is of such a degree as to constitute gross negligence is a determination which requires experience, technical competence and specialized knowledge of the profession and its standards. Accordingly, such determinations are typically considered within the province of the state-appointed engineering boards. However, the judiciary will often review the basis of the Board's decision, and will overturn or remand a board ruling if there is an obvious flaw within the Board's published reasoning in making its decision. In one case, the Appellate Court remanded a finding of gross negligence back to the Board because the Board's decision was based on the engineer's failure to rectify a mistake after notice. There, the Court found that there was insufficient evidence to indicate that the engineer actually received notice of the defective open web truss design. The Board, therefore, remanded the case back to the Examining Board for further consideration of whether the Engineer was guilty of gross negligence.[13]

Other courts have defined gross negligence as a reckless act which demonstrates a wilful and wanton abrogation of professional responsibility.[14] In *Duncan v. Board of Architects*,[15] the Court considered whether the findings of the Board supported its determination that Duncan was grossly negligent in the structural design of the Kansas City Hyatt House. A structural failure in the Kansas City Hyatt House resulted in the collapse of two walkways, killing 114 people. The Court articulated several factors supporting the Board's determination of gross negligence. The structural drawings provided by Duncan to the steel fabricator contained several serious errors. The connections appeared to be fully designed, although Duncan testified that his intent was to have the connections designed by the fabricator. The drawings did not indicate that the connections were to be designed by the fabricator and omitted important load calculations necessary to enable the fabricator to design the connections.

The box beam hanger connection drawings provided by Duncan did not meet the design requirements of the Kansas City Building Code. The drawings failed to properly identify the type of welds required, the need for bearing plates and/or stiffeners, and erroneously identified the hanger rods as standard rather than high-strength steel. The drawings were determined to be so deficient as to demonstrate a conscious indifference to the required duty of a structural engineer.[16] It was also noted by the Court that Duncan failed to determine the structural integrity of a proposed change to the hanger rod configuration by the fabricator, which was the main cause of the collapse. The failure of Duncan to recheck the steel design in the atrium as requested by the owner and architect, even though Duncan commented that such calculations had been performed, further supported the Board's finding of gross negligence.[17] *Duncan* illustrates the heightened responsibility cast upon engineers when the safety of the public is at issue.

Civil Liability

The following paragraphs analyze the issues and defenses involved in a claim for damages against a professional engineer which are based upon negligence or malpractice.

Economic Loss

Generally, the economic loss doctrine prevents parties not in privity with the Engineer from recovering economic losses resulting from defects within the Engineer's plans.[18] For example, in the prevalent situation where the owner contracts separately with the Engineer and the Contractor, and Contractor suffers economic losses on a project resulting from defective plans, the Contractor cannot bring a claim against the Engineer who prepared the plans.

There are three often recognized exceptions to the economic loss doctrine: 1) where the plaintiff sustained personal injury or property damage from a tortious event, i.e., a sudden and dangerous occurrence; 2) where the plaintiff's damages were proximately caused by defendant's false representation, i.e., fraud; and 3) where the plaintiff's damages were proximately caused by a negligent representation by a defendant in the business of supplying information for the guidance of others in their business transactions.[19]

A majority of states, including Illinois, have found that engineering is not considered a business for the purpose of supplying information to others as referenced in exception 3 noted above.[20] A minority of cases have held that engineering is a type of business which supplies information to others upon which it can be anticipated that third parties would rely, thereby, excepting engineering from the economic loss rule.[21]

A party may recover pecuniary losses resulting from the negligent representations of an engineer, where there is actual privity of contract between the parties or a relationship so close as to approach that of privity.[22] Courts have held that regardless of privity, a

Contractor may recover economic losses attributable to the architect's negligence against the architect administrating the subject construction project in situations where the architect had final authority to determine compliance with the construction project.[23]

Professional Standard of Care

Normally, a finding of liability against a Geotechnical Engineer requires expert testimony supporting the contention that the services provided did not meet the accepted standards of engineering practice.[24] Sound engineering practice for residential foundation design, according to the testimony offered in one case, required consideration of the soil conditions at the site. The required soil investigation included a determination of the types of soil at various depths, whether the soil would be classified as a stiff clay, a medium clay or a sandy silt, and the plasticity of each layer of soil.[25] Although the Engineer's conduct is measured against the accepted standard of care for a similarly situated professional within the "same area," Courts have held that the standards of care applicable to Geotechnical Engineers are nationally similar, therefore, eliminating the need for testimony of a local expert.[26]

It has been opined that the accepted practice for the determination of soil conditions at a particular site requires soil samples to be taken and scientifically tested. However, accepted practice would allow an engineer to rely on data from soil samples taken from previous projects in close proximity to the subject project. The characteristics from both the soil underlying the subject project and the soil from the nearby location must seem consistent with the native soil described in local survey manuals. Courts have upheld dismissals of suits against engineers, because plaintiffs failed to meet their burden of proving that the engineers breached the prevailing standard of care for Professional Engineers, or that any breach of the professional standard of care caused the plaintiffs' damages.

Limitation of Liability

Generally, the engineer's liability is limited to those areas of responsibility which are assumed by contract, or are assumed by the engineer during the performance of his services. For example, the engineer is rarely responsible for liability arising out the means and methods of construction. In a case where a worker was injured due to the collapse of a trench, the Court upheld the dismissal of the claims against the engineering firm, holding that proper design of shoring for the excavation was part of the means and methods of the Contractor's work and not the responsibility of the engineer. The Court further noted that, although the Engineer made geotechnical information available for the engineer's review, the Project Manual placed the responsibility on the Contractor to make its own determination of the soil conditions. The Court determined that the engineer, therefore, had no responsibility for the collapse of the trench.[27] However, courts have found the Engineer liable where he had actual knowledge that the Contractor was proceeding unsafely.[28] It is important in the contract to specifically limit those services for which the Geotechnical

Engineer is not accepting responsibility. One such limitation should be environmental services. For example, in one case, the Court determined that the failure to detect contaminants during a geotechnical study could support a finding of negligence against the engineer.[29]

Conclusion

Geotechnical Engineers are subject to scrutiny by both the government and the general public. Governmental proceedings are generally in front of a reviewing Board of Engineers, and are based on a violation of statute or regulation. Administrative proceedings can lead to fines and/or revocation or suspension of registration. Administrative discipline is generally based on a demonstrated inability to practice as an engineer or a disregard of proper engineering practices. Civil proceedings, on the other hand, can be based on a single act which failed to meet the accepted standard for geotechnical engineering practice, and which resulted in some loss by the plaintiff. The civil remedy is generally limited to monetary damages to account for the loss suffered by the plaintiff.

Given the increased level of regulations, governing, and liability arising out of the actions of geotechnical engineers, these professional must familiarize themselves with applicable standards and regulations, related to every project on which they work.

In addition, the insurance industry is quickly responding to the expanding number of lawsuits and are offering new products to cover potential errors and/or omissions by such professionals. Geotechnical engineers would be well-served by discussing their existing coverages and/or need for further insurance with their risk managers or insurance agent/brokers. Through education and potential insurance, geotechnical engineers can reduce their potential legal damages, losses, and risks.

References

1. 735 ILCS 325/4 (o).

2. *Van Breeman v. Department of Professional Regulation*, 296 Ill.App. 3d 363, 365, 694 N.E.2d 688 (2nd Dist. 1998).

3. See *Horeczko v. State Board of Registration and Land Surveyors*, 232 Cal. 3d 1352, 284 Cal. Rptr. 149 (2nd Dist. 1991).

4. *Leland v. Mississippi State Board of Registration For Professional Engineers and Land Surveyors, et al.*, 841 F. Supp. 192, 198 (S.D. Miss. 1993).

5. *Indiana State Board of Registration for Professional Engineers and Land Surveyors v. Nord*, 600 N.E.2d 124, 129 (Ind. App. 3rd Dist. 1992).

6. *Caito v. Board of Registration for Professional Engineers*, 1998 WL 426028, 4 (R.I. Superior Ct. 1998).

7. *Vivian v. Board of Architects*, 61 Wis. 2d 627, 635, 213 N.W.2d 359 (1974).

8. *Vivian*, 61 Wis. 2d 639.

9. *Duncan v. Missouri Board of Architects*, 744 S.W.2d 524, 541 (Missouri Eastern Dist. App. 1988).

10. *Webb v. Department of Professional Regulation Board of Professional Engineers*, 595 So.2d 1103, 1104 (Fla. 5th Dist. Ct. App. 1992)[Overturning suspension by the Board].

11. *In The Matter of William E. Bruce*, 97 N.C. App. 138, 387 S.E.2d 82, 83 (1990).

12. *Vivian*, 61 Wis. 2d at 635.

13. *Vivian*, 61 Wis. 2d at 638.

14. *Duncan*, 744 S.W.2d at 524.

15. *Duncan v. Missouri Board of Architects*, 744 S.W.2d 524, 541 (Missouri Eastern Dist. App. 1988).

16. *Duncan*, 744 S.W.2d at 540.

17. *Id.*

18. *Firemans Fund Insurance Company v. SEC Donohue*, 176 Ill. 2d 160, 679 N.E.2d 1197, 1200 (1997).

19. *Firemans Fund*, 679 N.E.2d at 1199.

20. *Id.*

21. *Southland Construction, Inc. v. Richeson Corp.*, 642 S.E.2d 5, 9 (Fla. App 1994).

22. *Plantation House and Garden Products, Inc., v. R-Three Investors, et al.*, 670 N.Y.S.2d 505, 506 (1998).

23. *HBS Contractors, Inc. v. National Fire Insurance of Hartford*, 501 S.E.2d 372, 376 (N.C. App. 1998).

24. *Harris v. Williams*, 679 So.2d 990, 991 (La. App. 2d Cir 1996) [Finding the Geotechnical engineer not liable for damage to home from differential settlement, because no expert testimony was offered to suggest a defect in the design or construction of the foundation].

25. *Harris*, 679 So.2d at 995.

26. *Martin v. Barge, Waggoner, Summer and Cannon, et al.*, 894 S.W.2d 750, 752 (Tenn. App. 1994).

27. *Burns v. Black & Veatch*, 854 S.W.2d 450, 454 (Mo. App. 1993).

28. *Balagna v. Shawnee County*, 233 Kan. 1068, 668 P.2d 157, 164 (1983).

29. *Gordon et al. v. National Railroad Passenger Corporation, et al.*, 1997 WL 298320, 13 (Del. Ch. 1997).

Removal of Subsurface Obstructions - An Alternative Contractual Approach

D.J. Mason III, PE[1]; Robert S. J. Berry[2]; David J. Hatem, Esq.[3]

Introduction

Subsurface obstructions (defined as objects, materials, or structures which stop the progress of excavation or other subsurface operations) are commonly encountered during all types of underground construction projects. Encountering subsurface obstructions typically leads to contractor claims for additional cost and time. Such claims, when legitimate, obviously lead to project cost and schedule growth. Such claims, when not substantiated, can also indirectly lead to additional costs and other negative project impacts which are manifested by prolonged disputes, unnecessary expenditure of management effort and resources, and a general deterioration of the relationship between the owner or management consultant and the construction contractor. These outcomes have historically been accepted as the industry norm and inherent in the process of subsurface construction.

A heightened scrutiny of cost growth on public projects has stimulated the need for the development of alternative contractual approaches for controlling generally predictable areas of common cost growth such as subsurface obstructions. The somewhat recent advent of the use of geotechnical baselines in construction contracts, as described in the ASCE publication titled "Geotechnical Baseline Reports for Underground Construction - Guidelines and Practices" and edited by Randall J. Essex is one such example of alternative approaches being implemented within the industry. This paper describes one contractual approach to controlling

---

[1] Senior Manager, Keville Enterprises, Inc., 120 Boylston St., Suite 505, Boston, MA 02116.

[2] Construction Contract Administration Manager, Bechtel/Parsons Brinckerhoff, 1 South Station, Boston, MA 02110.

[3] Burns & Levinson LLP, 125 Summer St., Boston, MA 02110-1624

cost and schedule issues associated with subsurface obstruction removal, which is founded to some degree in the geotechnical baselining concept. This paper also addresses the issues of allocation of risk between owners and construction contractors, and how to allocate that risk for subsurface obstructions to construction contractors in an equitable manner. Finally, the paper summarizes the advantages of using this particular approach from the authors' perspectives, and highlights contractual considerations that need to be addressed when developing such alternative approaches.

Background

Many heavy/civil construction contracts require extensive excavation and underground work such as slurry wall construction, load bearing element construction, drilled shaft foundation construction, support of excavation installation, pipe jacking, and micro tunneling. All of these operations face the risk of encountering subsurface obstructions (as previously defined). This risk varies depending on the characteristics of the subsurface geologic, local, and geographic considerations, and is heightened greatly in areas with subsurface strata comprised of historic fill.

A prime example of a project with extensive excavation and underground work through historic fill layers is the Central Artery/Tunnel (CA/T) Project in Boston, Massachusetts. The Project is building a series of slurry wall and cut-and-cover tunnels and associated underground utilities in downtown Boston, through historic fill layers that are typically 30 to 60 feet thick. These fill layers are strewn with a multitude of subsurface obstructions such as buried foundations, piles, sea walls, wharf structures, rip rap, miscellaneous construction debris, granite blocks, abandoned railroad structures, abandoned utility structures, and many other objects, materials, and structures. Through historic reviews, subsurface investigations, general knowledge, and information from prior construction projects, it was known that these and other types of obstructions exist in the Boston subsurface throughout the CA/T Project area, and that they would be encountered during all types of excavation and subsurface operations.

Although the CA/T Project performed fairly extensive subsurface investigations which confirmed the general presence of these obstructions, it was virtually impossible and certainly impracticable to pinpoint the exact locations and quantities of all of the various types of obstructions. In the execution of the earliest CA/T construction contracts, approximately 43% of all CA/T Project change orders were due to differing site conditions, and the majority of these were as a result of subsurface obstructions. The high incidence of subsurface obstructions made it challenging to manage and minimize Project cost and schedule growth. With several billion dollars worth of construction not yet completed, it was realized that the majority of the exposure to cost and schedule growth due to subsurface obstructions still existed. In order to improve the management and mitigation of this exposure to uncontrolled cost and schedule growth, the CA/T Project took a pro-active position in developing and implementing the contract provisions for subsurface obstruction removal that are described in this paper.

Contract Provisions

This section describes the contract provisions that were developed by the CA/T Project for risk allocation and compensation for the removal of subsurface obstructions. All of the topics contained in these contract provisions should be considered when developing other subsurface obstruction removal specifications.

1. List of Potential Obstructions

In order to clearly indicate the various types of *potential* obstructions that were anticipated in the subsurface, an extensive list of the anticipated objects, materials, and structures that would in all probability be encountered during excavation and subsurface operations was included in the obstruction removal provisions. The list was developed from data included in the geotechnical reports, information from historic reports, and from prior experience working in the Boston subsurface. The contract provisions also clearly and specifically stated that the potential obstructions listed, regardless of location, were hereby indicated for the contract. The items were specifically designated as *potential* obstructions, and were not to be deemed as obstructions unless and until they met the contractual definition of an obstruction. This distinction between *potential* obstructions and obstructions was made because it was realized that not all items encountered in the subsurface would actually obstruct the excavation and subsurface work.

2. Basic Contractual Definitions

The term obstruction does not have a single common meaning which is clear and unambiguous enough to use the term absent contractual definition. In order to ensure that the contract specifications for the removal of subsurface obstructions were clear, unambiguous, and not subject to different interpretations, it was essential to establish clear contractual definitions for obstructions. The following basic definitions which contractually establish what obstructions are were included in the contract specifications. Two basic definitions of obstructions were established, one for obstructions during mass excavation and utility trench excavation, and one for obstructions during the more complicated subsurface operations that are performed essentially in the blind such as slurry wall excavation and drilled shaft drilling. Note that obstructions are defined based on their affect on subsurface operations, and are not based on the physical attributes (except for the 1 cubic yard criteria) of the obstructions themselves. This approach prevents contractors from claiming that every single item encountered in the subsurface is an obstruction, keeps the focus of obstruction removal tied directly to the impact of obstructions on the excavation and subsurface operations, and uses the definitive operative phrase "unavoidably and completely stops the progress" as the basis for determining whether or not an object meets the definition of an obstruction.

During mass excavation and utility trench excavation, obstructions were defined as man-made or man-placed objects, materials, or structures, greater than 1 cubic yard in size, which unavoidably and completely stop the progress of the excavation in the volume occupied by the obstruction, despite the contractor's diligent efforts. Unit price pay items were established for removal of items meeting this definition. During these operations, bedrock, boulders and other naturally occurring hard deposits, utilities and utility structures indicated on the contract drawings, and timber piles were specifically excluded from being compensated as obstructions. The contract provisions also specifically stated that all costs and time associated with removing these specifically excluded items, and all other objects, materials, and structures encountered during these operations that did not meet this definition of obstructions were to be compensated via the applicable excavation payment item.

During slurry wall and load bearing element excavation, drilled shaft foundation drilling, pipe jacking, excavation support wall installation, and subsurface drilling operations, obstructions were defined as man-made or man-placed objects, materials, structures, or boulders, which unavoidably and completely stop the progress of the excavation or subsurface work for more than 1 hour, despite the contractor's diligent efforts. Unit price pay items were established for removal of items meeting this definition. The two basic differences between this definition and the definition of obstructions during mass and trench excavation are that this definition includes boulders, and this definition requires the progress of the work to be stopped for more than 1 hour. Boulders were included because their presence was inherent in the subsurface of the CA/T Project, and it was felt that the type of object would not always be readily discernable during these blind subsurface operations. The criteria that an obstruction had to stop the progress of the work for more than one hour was included because it was felt that some temporary stoppage of the work due to items in the subsurface was inherent in these operations, and that diligent efforts by the contractor would in some cases break through subsurface items within an hour. During these operations, bedrock and other naturally occurring hard deposits (except for boulders), and utilities and utility structures indicated on the contract drawings were specifically excluded from being compensated as obstructions. All costs and time associated with removing these specifically excluded items, and all other objects, materials, and structures encountered during these operations that did not meet this definition of obstructions were to be included in the payment items associated with the particular structure that is being excavated or drilled (e.g., slurry wall, load bearing element, drilled shaft, etc.).

These definitions and exceptions can be tailored to meet the particular subsurface operations, requirements, and conditions of any project involving any type of subsurface work, but it is essential that any such definitions are stated as clearly and unambiguously as possible, with no room for varied and subjective interpretations.

3. Notification and Determination

The contract specifications were developed to include a clear description of the procedures that must be followed to invoke measurement and payment under the obstruction removal unit pay items. Measurement and payment were not invoked unless an item encountered during a subsurface operation met the applicable definition of obstruction during that type of operation. The obstruction removal specifications required the contractor to notify the owner immediately upon encountering any potential obstruction which unavoidably and completely stopped the progress of the work despite the contractor's diligent efforts. Note that *unavoidable and complete* stoppage of the work is a key component in each of the definitions of obstruction. Upon receiving such notification, the owner was required to make a determination as to whether the item was an obstruction (in accordance with the definitions and exceptions), and under which pay item it fell. If the contractor disagreed with the owner's determination, it was specifically required to proceed diligently in accordance with the owner's direction, and to file a notice of claim within 3 days. This notice did not relieve the contractor of its obligation to diligently proceed with the work. In order to ensure that this notification and determination process occurred as quickly and efficiently as possible, the owner had to ensure that its field personnel (inspectors, engineers, etc.) were familiar with these procedures, and that the contractor's personnel were vigilant in their monitoring of subsurface operations and providing timely notification.

4. Means and Methods

The obstruction removal provisions did not detail or specify the means and methods for removing obstructions. Means and methods were left to the discretion of the contractor. The specifications did, however, contain numerous specific statements which required the contractor to diligently remove obstructions by the most practical means and methods, using all the appropriate equipment and personnel necessary for the effective removal of obstructions. The contractor was also required to have on hand at all times readily available equipment, tools, materials, and personnel appropriate for the effective removal of all types of potential obstructions delineated in the obstruction removal provisions. Additionally, the obstruction removal provisions required the contractor to submit its planned means and methods for removing all types of potential obstructions listed in the obstruction removal provisions during all types of excavation and subsurface operations. This submittal requirement amplified that the contractor was responsible for choosing its own means and methods for obstruction removal, and also documented the contractor's planned approach.

Cost Considerations

Several cost considerations were factored into the development of the obstruction removal provisions. This section describes these cost considerations.

1. Change Order Costs Versus Competitively Bid Costs

In the authors' experience, it is generally accepted in the industry that construction work procured via change orders has a higher price than construction work procured via competitive bidding. The higher price of change order work is primarily due to the fact that the costs of post-award change orders may include standby and delay costs, resequencing costs, premium time costs, acceleration costs, inefficiency costs, and other impact costs, all of which are typically sought by contractors when pursuing change orders. If the same work is properly incorporated into a competitively bid contract prior to bid opening, then all such costs are most likely either included in the competitively bid price at a reduced rate or not included at all, since overloading costs in a bid price increases the contractor's risk of not submitting the lowest bid and thus not being awarded the contract. In developing the contractual mechanism for the removal of subsurface obstructions, the CA/T Project took this into consideration and decided to incorporate the costs of subsurface obstruction removals into the base bid contract, rather than pay for them as change orders. This was accomplished by including a variety of unit price payment items for removal of subsurface obstructions in the contract provisions.

2. Unit Price Pay Items

Because of the uncertainty in quantities of obstruction removals, the CA/T Project decided to use unit price pay items for obstruction removals. This approach enabled the CA/T Project to pay for obstruction removals only if they were encountered rather than as a lump sum, and allowed the CA/T Project to use a variation in estimated quantities clause to cover (within a reasonable range) overruns and underruns of estimated quantities. Separate unit price pay items were established for the removal of obstructions during all of the major excavation and subsurface operations such as mass excavation, utility trench excavation, slurry wall excavation, drilled shaft drilling, pipe jacking, and support of excavation installation. During mass excavation and utility trench excavation, obstructions would be readily measurable and therefore these obstruction removal pay items were assigned units of cubic yards. During the blind subsurface operations such as slurry wall excavation obstructions would not be readily measurable, and the length of time to remove an obstruction could vary considerably depending on the type of obstruction, the location of the obstruction, and the type of slurry wall excavator being used (hydraulic clam, hydro mill, etc.). These obstruction removal pay items were therefore assigned units of crew hours. The payment items for the removal of obstructions specifically included the removal of obstructions that were indicated on the Contract Drawings, as well as obstructions that were not indicated on the Contract Drawings but whose general presence or nature was indicated in the contract documents or reference documents.

3. Minimum Unit Prices

There are several different reasons why contractors may choose to submit penny or dollar bids. One such reason may be that a contractor may feel that the estimated quantities for a unit price pay item are too great, and thus money included in that pay item may not be collected. The contractor would therefore choose to include all of its money for that particular unit pay item in a lump sum bid item and bid one penny or one dollar for the unit item, thereby guaranteeing itself payment for its own estimated quantity of the unit pay item. Penny or dollar bidding would have defeated the CA/T Project's objective of paying for obstruction removals only if they were encountered. In order to prevent contractors from submitting penny or dollar bids for the obstruction removal pay items with units of crew hours, the CA/T Project established minimum unit prices for the obstruction removal pay items with units of crew hours.

The provisions for the minimum unit prices required that for those obstruction removal pay items with units of crew hours, bidders had to bid an amount that was either equal to or greater than the stipulated minimum unit price, and clearly stated that any bid price submitted that was lower than the minimum unit price would be deemed a bid at the minimum unit price. These provisions also clearly stated that bidders were solely responsible for the adequacy and completeness of their unit price bids based on their own independent review and consideration of all aspects of the work, and their anticipated means, methods, techniques, procedures, and sequences to be utilized in the performance of the work. Additionally, the provisions clearly stated that the stipulated minimum unit prices did not expressly or by implication constitute any representation or warranty that the minimum unit prices were adequate for the performance of the work covered by the unit prices.

4. Costs Included in Obstruction Removal Unit Pay Items

The provisions for the obstruction removal unit pay items were written to include all costs associated with removing obstructions in the pay items, including all direct and indirect costs to remove the obstructions, all inefficiency costs, and all idle time, resequencing, delay, and other time-related costs.

5. Variations in Estimated Quantities

The CA/T Project realized that the estimated quantities provided for the obstruction removal pay items were subject to variation, due to the fact that it was virtually impossible to estimate quantities of obstructions contained in the subsurface. Accordingly, the provisions were written such that the obstruction removal pay items were subject to a variation in estimated quantities clause, which allowed either party to demand renegotiation of any obstruction removal unit price when the actual quantity either overran or underran the estimated quantity by more than 25%. The contracts also contained provisions which clearly stated that the estimated quantities were approximate only, and that the CA/T Project did not represent or warrant that actual quantities would equal the estimated quantities.

Schedule Considerations

The CA/T Project is comprised of a series of individual construction contracts that are building segments of tunnel which together will comprise the overall Interstate 93 and Interstate 90 tunnel system in downtown Boston. All of these individual construction contracts have multiple interfaces with other contracts. Accordingly, there is a high risk that schedule delays in one of these contracts will ripple into one or more of the other interfacing contracts. Such delays could ultimately delay the entire CA/T Project, which would result in high cost overruns. Furthermore, change orders due to subsurface obstructions were very likely to create schedule delays since the majority of the underground work typically fell on the critical paths of contract schedules. The CA/T Project considered these points when developing the obstruction removal provisions, and clearly stated in the provisions that all time associated with performing 125% of the estimated quantities of the obstruction removal pay items was included in the contract time, and that no time extension would be considered for obstruction removals unless the actual quantities exceeded 125% of the estimated quantities. In this manner the CA/T Project minimized the potential for schedule delays and the associated cost growth due to subsurface obstructions. The provisions also reserved the owner's right to seek time credits if the actual quantities underran the estimated quantities of the obstruction removal pay items.

Features Which Make Subsurface Obstruction Removal Provisions Equitable for Bidders and Contractors

In developing the obstruction removal provisions, the CA/T Project endeavored to ensure that the provisions were equitable to bidders and contractors and thus included several features in the provisions which met this objective. First, the provisions were structured with separate obstruction removal pay items for each type of excavation and subsurface operation. This allowed bidders to submit separate prices for removing obstructions during each of the various operations based on their own chosen means and methods. The provisions also contained estimated quantities for each of the obstruction removal payment items, which eliminated the typical requirement for the contractors/bidders to assume what quantities of obstructions would be encountered during each excavation and subsurface operation based on their own interpretation of the indications of the contract documents.

Contractors were allowed to set their own prices for the obstruction removal pay items by virtue of their bid prices. This eliminated the need for the contractors to spend management effort and resources in developing and negotiating change orders for obstructions. Contractors were also given the contractual right to renegotiate the prices for the actual quantities of obstruction removal pay items that were in excess of 125% of the estimated quantities. Additionally, contractors were able to seek time extensions for these overrun quantities. In the case where the actual quantities of the obstruction removal pay items were less than 75% of the estimated quantities, contractors were again allowed to renegotiate the unit prices, thereby allowing them to seek price adjustments and compensation for unabsorbed indirect costs.

The obstruction removal payment items for the more difficult and more costly blind subsurface operations such as slurry wall excavation and drilled shaft foundation drilling were assigned units of crew hours. It was felt that this was the most equitable and biddable approach since the contractors' costs for these pay items would be driven primarily by the time that their crews spent removing the obstructions, rather than the size of the obstruction.

The unit price pay items provided a mechanism through which the contractors could requisition payment for obstruction removal as soon as the obstructions were removed. This approach provided better cash flow for the contractors than the slower change order process would have.

The CA/T Project also solicited input and feedback from the Construction Industries of Massachusetts which is a consortium of local construction contractors during the development of the obstruction removal provisions. This group provided some constructive suggestions and ideas through this process.

Sharing of Risk

When instituting contract provisions like those described in this paper, owners must consciously determine how the risk associated with subsurface conditions is to be shared between the contractor and the owner. Some owners may wish to allocate all of the subsurface risk to the contractor. In this case, the contract would likely be written with no differing site conditions clause, and with clear statements that the contractor is solely responsible for all site and subsurface conditions. This could be mistakenly construed as having excellent cost and schedule control since there would be no risk of cost or schedule growth due to subsurface conditions, however, this would be offset by the higher bid prices that would result from contractors covering their exposure in their bids. Prudent contractors would also be likely to include the costs of performing their own subsurface investigations in their bid prices. Also, this contracting approach may not be legally permissible in public sector contracting.

At the opposite end of the spectrum, an owner could assume the majority of the risk associated with subsurface conditions itself by including a differing site conditions clause, and issuing change orders for subsurface conditions which vary materially from the indications of the contract documents. This approach would provide the owner with the least amount of cost and schedule control, but would also result in lower bid prices.

In the case of the obstruction removal provisions described in this paper, the risk of subsurface obstructions is distributed more evenly between the owner and the contractor. All time and costs associated with removing 125% of the estimated quantities of obstructions are included in the base bid contract price and contract time, thereby providing the owner with good cost and schedule control. The contractor's sole compensation for obstruction removals, whether differing site conditions or otherwise, is its own bid prices for the unit price pay items. However, the contractor is entitled to seek relief from these unit prices if the estimated quantities are either overrun or underrun by more than 25%. The net result is relatively low bid prices for obstruction removals and no cost or schedule growth for quantities up to 125% of the estimated quantities.

During the development of these provisions consideration was given as to whether paying for obstruction removals by the crew hour (as in the case of blind subsurface operations such as slurry wall excavation and drilled shaft drilling) was exposing the CA/T Project to too much risk because of the chance that contractors may be inclined to take their time removing obstructions. The Project felt that this concern was outweighed by the aggressive overall schedules of each contract that the contractors were required to meet, the contract provisions which clearly stated in several places that at all times the contractor shall diligently remove obstructions and shall use the most appropriate and practical means and methods to remove the obstructions, and the contract provisions which clearly stated that all time associated with removing the estimated quantities of obstructions was included in the contract time. It was also felt that including unit prices with units of crew hours was the most equitable and biddable approach since the contractors' costs would be driven by the time that their crews spent removing the obstruction rather than the size of the obstruction.

Modified Application of Geotechnical Baselining

The typical application of geotechnical baselines in construction contracts occurs in the form of definitive statements within the contract documents which establish objective, measurable contractual quantities or expectations for geotechnical conditions. For example, a contract may state that within a particular soil layer, the contractor shall anticipate encountering 100 boulders greater than 1 cubic yard in size per 1,000 cubic yards of excavated soil. The net effect of this approach is to contractually definitize what is anticipated in the subsurface, rather than leaving what should be anticipated open to a broad range of reasonable yet different interpretations. Geotechnical baseline statements like this therefore establish contractual thresholds which, if materially exceeded, constitute a mutual understanding between the parties which define grounds for an equitable adjustment in cost and/or time.

The contract provisions for the removal of subsurface obstructions that are described in this paper are an example of geotechnical baselining taken one step further than normal. These provisions do not stop at providing a mere statement of the quantities of subsurface obstructions that are anticipated. These provisions go one step beyond in that they provide unit price pay items to compensate the contractor for removing subsurface obstructions as they are encountered, rather than waiting to see if the estimated quantities are exceeded and then granting a change order. As described above, the CA/T Project felt that this approach was more equitable to the contractors. Furthermore, if the provisions had stopped at providing mere statements of the anticipated quantities of subsurface obstructions, the CA/T Project's and contractors' field personnel would still have been required to monitor and track quantities actually encountered in order to establish when baseline quantities were exceeded and thus when change orders were due. Thus, the incorporation of the unit pay items added virtually no additional monitoring requirements in the field than would have been required in the typical geotechnical baseline application.

Benefits of This Approach

There are several benefits to using the contractual approach for the removal of subsurface obstructions that is described in this paper. One of the primary benefits is that it provides a streamlined contractual payment mechanism via the unit price pay items which greatly diminishes claims and disputes over subsurface obstructions. This streamlined payment mechanism also provides contractors with better cash flow throughout the excavation and subsurface operation phases of their contracts, and avoids the prolonged change order process.

The contractor's prices for removing obstructions are clearly and contractually established. This eliminates any conjecture about the contractor's price at the time of bid should any disputes arise.

The owner receives the benefit of better cost control since all costs for removing 125% of the estimated quantities of obstructions are included in the competitively bid contract award price, including all impact and all time related costs. Cost growth due to subsurface obstructions is therefore limited only to those quantities of obstructions that are encountered in excess of 125% of the estimated quantities. The owner also receives the benefit of better overall prices for removal of obstructions due to the fact that obstruction removals are competitively bid rather than negotiated as change orders.

The owner also receives the benefit of better schedule control since the time for removing 125% of the estimated quantities of subsurface obstructions is included in the contract time. The potential for schedule growth is also limited only to those quantities of obstructions that are encountered in excess of 125% of the estimated quantities. On large, complex projects with numerous adjacent and interfacing contracts such as the CA/T Project, avoiding schedule growth on individual contracts in this manner results in the prevention of schedule growth due to obstructions from rippling into adjacent and interfacing contracts.

This approach establishes a condition wherein the owner only pays for subsurface obstructions when they are actually encountered and removed, via unit price pay items, rather than having the costs of all subsurface obstruction removals buried in a lump sum contract pay item. The use of unit price pay items for obstruction removals rather than lump sum pay items also eliminates the need for owners to expend management effort and resources pursuing credits when anticipated lump sum baseline quantities of obstructions are not actually encountered.

Summary

Cost and schedule control is typically the primary focus of any construction project. It is difficult to control either if subsurface obstructions are handled as change orders as has historically been the case in the industry. The contractual approach to subsurface obstruction removal that is described in this paper provides one possible solution to this problem. Owners should consider the use of this approach to achieve a higher level of control of schedule and costs than is typically possible when treating subsurface obstructions as change orders. This approach also provides a great deal of flexibility to owners in allocating the risk associated with subsurface obstructions between themselves and contractors. The basic concepts described in this paper can be tailored to accommodate a variety of different situations and cost, schedule, and risk requirements, by adjusting contract definitions, estimated quantities, units of measure, and other aspects of the contract provisions. The provisions can also be expanded upon to include the removal of obstructions during other subsurface operations such as soil mixing, pile driving, mined tunneling, etc.

It is certainly not easy to overcome the common industry mind set that subsurface obstructions inherently result in change orders and their associated cost growth and schedule growth. In order to effectively implement the contractual approach to subsurface obstruction removal described herein a certain comfort level must be developed by both the owner and the contractor. Both contractor and owner representatives must work together to ensure that they each have a clear understanding of how risk for subsurface conditions is allocated between them and how the contract provisions are to be administered, and both must be constantly aware of the objective of using the provisions as a fair, equitable, and timely means of providing compensation for subsurface obstruction removal. Some CA/T Project contractors have adapted rather readily to the concept, while others are still somewhat skeptically developing a comfort level. Those contractors that have adapted most readily are the contractors that submitted realistic and reasonable bid prices for the unit price pay items.

When accepted and properly administered by both the owner and the contractor, this alternative contractual approach to subsurface obstruction removal can serve as an overall benefit to both parties.

Temporary Support of Buildings
Who Really Assumes the Responsibility?

By Andrew J. Ciancia[1], Member, ASCE and Kevin J. O'Neill[2]

Abstract

The New York City Building Code imposes absolute liability upon an excavator for any damage to an adjacent building caused by an excavation that exceeds 3.1 m (10 ft), even if the building arguably has a pre-existing defect. Legal problems can arise for the engineer, in particular, the geotechnical engineer, in providing recommendations for a temporary support system at the request of the client or contractor, even if the engineer is not the "engineer of record." This paper discusses both the technical and legal aspects of several case studies involving temporary support systems constructed in New York City. The paper also provides recommendations for limiting the legal exposure of the engineer. A thorough understanding of both the technical and legal aspects of a building code, in conjunction with a comprehensive study of the site, is important when the engineer participates in the design or construction of any inherently high-risk temporary support system.

Introduction

During the past three years, there has been a surge in new building construction in New York City. Many projects involve deep excavations located directly adjacent to older buildings and busy sidewalks and streets. The need to occupy buildings as soon as possible, coupled with the very high cost of construction, has put pressure not only on the foundation contractor but also on the engineer, most likely the geotechnical engineer, to complete difficult excavation projects on an accelerated schedule.

Many technical and legal challenges face the engineer in designing temporary retaining systems, especially in New York City. The general practice is to have the foundation contractor assume all responsibility for designing, constructing, and inspecting

---
[1] Vice President, Langan Engineering and Environmental Services, P.C., New York, New York 10001.
[2] Partner, Gogick & Seiden, LLP, Attorneys at Law, New York, New York 10004.

temporary retaining systems such as sheeting and shoring or rock bolting. The design team, comprising the architect, the engineer, and others, is generally responsible for the permanent design aspects of the new structure. However, in many cases, the client or contractor may request the engineer to inspect a field condition or to provide design recommendations for revising a temporary support system. If the engineer participates in a design decision, the engineer may become liable, possibly without being fully aware of the potential legal consequences.

This paper begins with an overview of the special legal requirements imposed by the New York City Building Code (NYC Building Code) (1995), followed by discussions of two projects involving deep excavations. The paper finishes with suggestions as to how the engineer can provide recommendations for a temporary support system and the prudent steps that can be taken to limit, if not eliminate, legal exposure.

Building Code Requirements

The NYC Building Code imposes the responsibility to protect structures that are adjacent to excavations upon one of two parties: the owner of the adjacent structure, or the person causing the excavation to be made. For this paper, the person, firm, or group responsible for the excavation is defined as the "excavator". If the excavation is carried to a depth of 3.1 m (10 ft) or less below curb level, the NYC Building Code places the responsibility for protection on the owner of the adjacent building. For excavations that exceed a depth of 3.1 m (10 ft), the NYC Building Code places the responsibility to protect the adjacent structure on the excavator.

The 3.1-m (10-ft) standard is considered a "bright-line" standard in the eyes of the law. In other words, the code section imposes absolute liability upon the excavator for any damage caused by the excavation even before it exceeds 3.1 m (10 ft) below the legally established curb level. The owner of the adjacent structure needs only to show that the excavator failed to make adequate precautions to protect the adjoining structure, and that the excavator's activities were the proximate cause of any damage. The owner of the adjacent structure only has the obligation to provide the excavator permission to enter and inspect the adjoining structure prior to the excavation.

If the excavator intends to dig more than 3.1 m (10 ft) below curb level, the absolute duty to protect adjacent structures arises at the commencement of the excavation. The New York Courts have been clear with regard to the 3.1 m (10 ft) test and its application even to situations where the excavation has not reached 3.1 m (10 ft), as explained by one New York Court as follows:

> "If a man intends to make an excavation more than 3.1 m (10 ft) in depth below the curb, he is liable for injury to the adjoining property which results from the excavation before reaching the intended depth. That is, if he intended to go 4.5 m (15 ft) below curb line, and the injury occurred when the excavation had only

reached 2.1 m (7 ft), the excavator would be liable, because the intention to go more than 3.1 m (10 ft) in depth relieved the neighbor of any duty to protect his own property" (Foster v. Zampieri 1912).

An expansive of interpretation has been given by the courts in construing the NYC Building Code section with regard to deep excavations as demonstrated by the Court in a case (Harder Realty 1946) where the excavator was required to protect two buildings affected by the excavation, despite the fact that the buildings were separated from the excavation site by a 15.4 m (50 ft) wide public thoroughfare. The court held that any attempt to restrict the application of the NYC Building Code section to premises actually touching the excavated area would be "inconsistent with the obvious intention of the Code to protect all property in the vicinity of deep excavations from injury". The Court expressly characterized deep excavations (greater than 3.1 m depth) as "extra hazardous activities" requiring a liberal interpretation of the applicable NYC Building Code section to avoid frustration of the important public policy embodied in the NYC Building Code.

The NYC Building Code, and the cases that have interpreted it, expands an excavator's duty to protect adjacent buildings from the responsibility imposed by common law. In common law, an excavator's obligations are merely to use reasonable care in the performance of the work. Liability only extends to damage resulting from the excavator's negligence. The absence of absolute liability does not relieve excavators of the obligation to avoid damage to adjacent structures. In common law, excavators have an obligation to maintain the lateral support of adjacent structures to avoid damage to those structures. The common law rule is that the excavator is liable for injury to any land that is within the "natural zone of support" (Gordon 1917).

The NYC Building Code, however, makes an excavator absolutely responsible for protecting the adjacent building, at the excavator's own expense, wherever the safety of that adjacent building "may be affected" by an adjacent excavation. The excavator's responsibility to provide protection applies "at all times" when undertaking an excavation, and is not affected by any claimed pre-existing defect in the adjacent structure, even if the adjacent structure was arguably not constructed in accordance with the requirements of the NYC Building Code.

It should be noted that absolute liability imposed by the NYC Building Code for the protection of adjoining structures does not make the excavators liable for damages to adjacent structures that the excavation did not cause. In other words, an adjacent property owner claiming damage to its structure must demonstrate that the excavator caused the damages for which the adjacent property owner seeks a recovery. The case law indicates that the adjacent property owner must show that the duty to protect a structure was violated by the excavator, causing the adjacent property owner damage.

The requirement that the adjacent owner demonstrate damage to its structure caused by an excavation, as a condition of the lawsuit, has tempted some excavators to defend such claims by alleging that the damage was caused by a "pre-existing condition" in the adjacent structure. However, the requirement that the adjacent property owner establish "cause" as to actual damages, does not in any way diminish the duty imposed by the Code upon deep excavators to protect adjacent structures. While an adjacent property owner must always prove the extent of its damage, the property owner does not have to show that the excavator was negligent, nor that the adjacent property owner's building was free from defects which exacerbated the damage that would otherwise have been caused by the excavation.

In a 100-year-old case (Cohen v. Simmons 1894), the lessee of a building adjoining a deep excavation site sued the owner of that site. The lessee claimed that the absence of proper shoring and protection by the excavator damaged the walls of the lessee's premises. The excavator responded that he should not be held liable for "enlarging" defects that were present before the excavation ever began unless it was established that he had performed his excavation negligently. The trial level Court rejected the excavator's position, and a jury verdict was rendered against the excavator. The excavator appealed, and the Appellate Division affirmed that finding of the Court below, noting that the statute imposed an absolute duty to protect adjoining structures from injury without regard to whether the excavator was negligent, and without regard to the existence of pre-existing defects in the adjacent structure which were "enlarged" by the excavation.

Case Studies

Two case studies of temporary support systems dramatically draw into focus the consequences of the NYC Building Code requirements. The first case involved a 21.5-m (70-ft) deep cut that was excavated adjacent to a high-rise apartment building. The apartment building was originally constructed without the perimeter basement foundation wall, unbeknownst to the excavation contractor. In the second case, the geotechnical engineer retained by the owner was responsible for the design of the entire temporary support system of a 23.1-m (75-ft) deep excavation. As part of the bid, the excavation contractor was responsible for installing rock anchors and instrumentation in accordance with the contract documents.

Case 1: Mixed-Use Tower

The first case study involves a 42-story mixed-use tower located on the upper east side of Manhattan. In order to facilitate below-grade parking, four levels of basements were designed that required a 21.5-m (70-ft) deep excavation, mostly in hard mica schist rock. The foundation contractor was provided with boring logs of the site and was given the responsibility of designing and installing the temporary retaining systems, including sheeting, shoring, and rock anchors. On this project, the design team (including the geotechnical engineer) was not given the responsibility for designing the temporary support. Unbeknownst to the owner, the design team, and the contractor, an adjacent apartment building had apparently been constructed, in part, on a carriage house foundation in the mid-1960's. As a result, the apartment building did not have a perimeter basement foundation wall. Refer to Figure 1a for a sectional view of the excavation and the adjacent building.

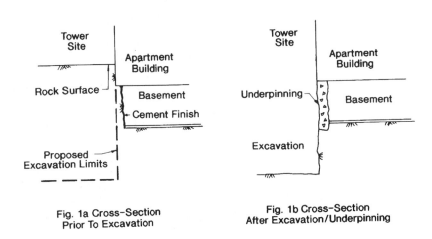

Fig. 1a Cross-Section
Prior To Excavation

Fig. 1b Cross-Section
After Excavation/Underpinning

Approximately two months into the excavation program, the "defective" condition was discovered by the contractor when, while chipping the rock, he inadvertently made a horizontal hole through the basement "wall" of the apartment building. The excavation was suspended for several months until the adjacent building could be studied and repaired. The solution was to construct a new foundation wall under a portion of the apartment building by the underpinning method. The rock had to be removed in 1-m (3-ft) wide sections under the wall of the apartment building and replaced with reinforced concrete, as shown in Figure 1b.

The tower owner needed to hire a specialized underpinning contractor to construct a new foundation wall under the apartment building because the contractor onsite was not sufficiently qualified to perform this specialized work. The tower owner subsequently claimed that he was forced to replace the portion of the adjacent apartment building that rested on the old carriage house foundation. The cost was substantial in terms of the expense to repair the adjacent foundation, and in terms of the delay to the project. The tower owner brought suit against the apartment building owner, as well as the architect and engineer associated with the construction of the apartment building. The tower owner claimed that the existing carriage house foundation constituted a defect not caused by the deep excavation, which he should not have been responsible to repair. The apartment building owner, architect, and engineer responded that it was the duty of the tower owner and contractor to protect adjacent structures that "may be" damaged by deep excavations and that the NYC Building Code imposed an absolute, non-delegable duty to perform all work necessary to protect the apartment building from damage. The apartment building owner also pointed out that the claimed "defective condition" had not caused any damaged to the structure in the 20 to 25 years since its construction.

The Court was never required to rule on the merits of the claim and did not decide the parties' respective rights and obligations under the Code, because the Court dismissed the action as barred by the applicable statue of limitations. However, it appears that the New York City Administrative Code would have required the tower owner to shoulder the burden of performing whatever work was necessary to avoid any damage to the apartment building which "may have been affected" by the tower deep excavation.

Case 2: High-Rise College Building

This project involved a public agency that was responsible for the construction of a high-rise college building in mid-town Manhattan. The site measured approximately 61 m (200 ft) by 45.7 m (150 ft) in plan dimensions and was surrounded by three busy city streets on the north, east and south sides and two 50-year old, 8- to 11- story apartment buildings on the west side. Included in the design of the college building was a two-level gymnasium located entirely below grade that necessitated an excavation up to 23.1 m (75 ft) deep, mainly in mica schist rock. Due to the depth and location of the excavation, the public agency retained a geotechnical engineer to be responsible for not only the subsurface study at the site but also for the design of the temporary support system. The two old, rather fragile, occupied apartment buildings were of particular concern.

The geotechnical engineer agreed to be responsible for the design of the temporary retaining system provided specific requirements were made part of the construction documents. The following requirements were agreed to and implemented:

1. The foundation contractor had to be pre-qualified. During the pre-bid process, the contractors were required to show their experience and procedures on similar projects (at least 15 m (45 ft) deep) with respect to blasting operations, line drilling, rock anchor installation, and underpinning/shoring. Also as part of the pre-qualifications submittal, the contractor would have to describe sequencing used on similar projects in relation to blasting and tieback installation adjacent to occupied structures. The owner and design team reviewed the written responses from contractors, interviewed the short-listed firms, and selected the foundation construction firm based on both qualifications and price.

2. The geotechnical engineer would have a full-time resident at the site to supervise the installation and testing of rock anchors, to geologically map the exposed mica schist rock, and to install and monitor instrumentation (extensometers, tell-tales, crack monitors and seismographs). The engineer would monitor the work to verify that the contractor was performing the

excavation activities in accordance with the contract documents and also to revise, as needed, the design of the temporary system as subsurface conditions were exposed in the excavation.

3. The excavation specifications had to be prepared to limit the unsupported rock bench heights adjacent to the two apartment buildings (west excavation vertical face) to no more than 1.8 m (6 ft). In this manner, there was careful co-ordination between blasting and rock anchor installation such that the anchors were installed relatively quickly after newly excavated and unsupported rock faces were produced by the blasting operations. The typical rock anchor spacing on the west excavation rock face was 2.1 m (7 ft) on center. (refer to Fig. 2).

4. Criteria had to be established for allowable vibration levels, as measured inside adjacent structures. Vibration levels were recorded and analyzed with respect to maximum charge per delay versus scaled distance. Revisions to the blasting program were made as needed to account for field conditions and the responses of the buildings to the blast vibrations.

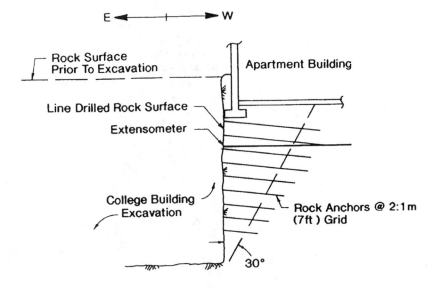

Fig. 2 Excavation Cross-Section

The 23.1-m (75-ft) deep excavation was successfully completed. Based on extensometer and optical survey readings, the west excavation face moved horizontally and vertically less than 1.3 cm (0.5 in). However, the apartment buildings did experience some cracking of the exterior brick facades and interior plaster walls and ceilings as a result of the adjacent chipping and blasting operations. It was subsequently discovered that the foundation walls of the apartment buildings were poured directly against the rock surface (Figure 2). This coupling effect of the existing walls of the apartment buildings attached to the rock face appeared to be a conduit for transmitting rock chipping and blasting vibrations directly into the apartment buildings.

The owners of the apartment buildings brought suit against the public agency, the foundation contractor, and the structural engineer for the damage caused by the blasting operations but not by the temporary support system. The damages were settled out of court.

The geotechnical engineer was not included in the suit. Reportedly, during the discovery period, the lawyers for the owners of the apartment building reviewed the comprehensive project data developed by the geotechnical engineer and determined that there was no case in relation to the temporary support system.

Conclusions

Local building codes, such as the New York City Building Code, may place tremendous liability on the engineer and contractor for temporary support systems. Engineers who provide design recommendations in the field to a contractor for a temporary retaining system (rock anchors, underpinning, sheeting, shoring, etc.) may be taking on considerable risk without being fully aware of code requirements and consequences. Engineers are advised to make sure that their contracts clearly demonstrate their lack of involvement with an excavation, if they are not involved, or, alternatively, the engineer should be furnished with ample authority to require the implementation of a support system which is commensurate with the risk involved.

An engineer taking on any responsibility for designing a temporary retaining system should be fully aware of the legal requirements of the building code of the locality. To minimize liability, responsibilities of the engineer and contractor should be clearly established during the early design stages of the project. Construction documents should be comprehensive, and field monitoring is critical. Project engineers should train their field personnel on the legal ramifications of local building codes. Liability may extend beyond damage resulting from negligence, and the engineer may be absolutely responsible for protecting an adjacent building, no matter what condition it is in.

References

Cohen v. Simmons, 21 N.Y.S. 385 (1st Dept. 1892), Aff'd, 142 N.Y. 671, 37 N.E. 571 (1894),

Foster v. Zampieri (140 A.D., 471, 473, 125 N.Y.S. 422, 424 (1st Dept.) aff'd 206 N.Y. 704 (1912); quoting from dissent in Blanchard v. Savarese, 97 A.D. 58, 89 N.Y.S. 664,667.

Gordon v. Automobile Club of America, 180 A.D. 927, 167 N.Y.S. 585, 587 (1st Dept. 1917).

Victor A. Harder Realty and Construction Company v. City of New York, 64 N.Y.S. 2d, 310, 319 (Sup. Ct. N.Y. City 1946).

New York City, Department of Buildings, Building Code of the City of New York, amended 1995.

Perspectives On Risk Assessment In Tunneling

By: Donald P. Richards[1], Fellow, ASCE

Abstract

Underground construction, whether in the civil or mining industry, has long been accepted as a "risky business", from both a safety and a financial perspective. These risks can and should be identified, assessed, and reduced as much as possible in the planning, design, and construction phases of any major underground construction project. Geotechnical factors are often a major contributor to these risks. The author offers his perspective, based somewhat upon his international experience, on simplified methods to accomplish these tasks, including a case history application where the proposed techniques were used.

Introduction

It has been generally accepted in the heavy construction industry in the 20th century, that underground construction is relatively risky in terms of safety, but also in terms of financial risk, whether due to variable ground and/or groundwater conditions, or other factors that may influence the construction process and/or progress and therefore the construction schedule, which directly relates to cost. Although there are currently available numerous sophisticated "risk models" to evaluate construction risk, these usually require professional staff with significant previous similar training and experience in the use of statistics to make these models useful and meaningful. It is the intent of this discussion to present a simplified risk model that can be used by the non specialized construction professional to make a preliminary evaluation of risk, whether by the owner, designer, or contractor; and in the planning, design, or construction phase of a project. The implications of this risk model may be different for the owner, designer and contractor, since their perceptions of risk may vary as well as their exposure to these risks.

---

[1] Senior Tunnel Advisor, LACHEL & Associates, Inc., Golden, CO 80401.

For tunneling as a specific construction industry entity, Brierly and Cavan (1987) and Brierly (1988) have expressed the opinion that not only the perceptions of risk can vary, but also the control of risk can vary, with most risks distributed and/or controlled by either the owner or the contractor, with little risk to the designer/consultant. However, with the legal and accounting professionals currently taking an active role in underground construction contracts, the designer/consultant liability and exposure seems to be increasing, and therefore they too are not only a party exposed to their own types of risk, but they are also in a position to either increase or reduce the risk for the owner and contractor, by the way in which the tunnel project has been investigated, designed, and presented in the contract documents, and how the construction is managed. With respect to this, Huse (1992, 1993) has offered several recommendations on management of underground construction contracts, and these have been discussed in detail in the following paragraphs. Since the highest risks in underground construction generally seem to be related to variations in subsurface geological conditions, these types of considerations will be highlighted, and methods of risk evaluation presented.

The Risk Issue

In underground construction, whether it be in the civil or mining industry; whether it is in soft ground or hard rock or somewhere in between; whether it is in tunneling, shaft sinking, cavern construction, or in open cut; and whether it is at home in the USA or abroad somewhere, there will always be some degree of risk involved. This risk can be identified at any stage of a project but the earlier the better in order to properly account for it in the planning, design and construction. Anderson (1998) has appropriately noted that the cost of prevention is many times less than the cost of rectifying the consequences of any "disaster". In addition, both Anderson, and Thompson and Perry (1992) have strongly recommended that risk control is best started early in a project. The probability and magnitude of the risk may vary as a function of the complexity of the project, the complexity of the geology and the ground water regime, and the adequacy of the design for the actual site conditions. Other contributing factors include the experience of the owner, designer, and contractor on similar projects and in similar ground conditions, and the appropriateness of the contractor's selection of equipment and construction methods and sequences.

As a concept, risk is no different in underground construction than in any other discipline, although the specific risks and the key factors that contribute to them may be different. Syme (1995) notes several generic questions that must be considered in any risk assessment, a few of which have been adapted below as they may apply to underground construction:

- How good are the knowledge base and methods for estimating the risks?
- How are risk estimates incorporated into the decision making process?
- What factors influence an individual's perceptions of risk?

- How is it determined "how safe is safe?"
- How are perceptions of risk incorporated into public and/or project policy?

As noted by Alhalaby and Whyte (1994), Haddad and Mullins (1995), and Abbott (1998), the systematic management of risk can usually be developed in three stages: (1) Risk or hazard identification; (2) Risk analysis (quantification of their effects); and (3) Risk management (planning a response to specific risks to mitigate or reduce their potential impacts). The discussions presented herein will concentrate mainly on the second stage, with the first stage as a secondary issue. The last stage will not be addressed since it is too project specific. A recent example of an owner incorporating risk management into overall project management has been outlined by Erwin and Sinha (1996), in which the owner controlled the insurance program for the project. If this approach were adopted by other owners, the "construction risk assessment" could be part of the "project risk assessment", which would have already included a risk assessment during the planning and design phases of the project.

In engineering and construction in general, the most serious effects of risks are (1) failure to keep within the cost estimate (or for the contractor, within the bid price), (2) failure to achieve the required completion date (with consequent cost implications for the owner or penalties for a contractor), and (3) failure to meet the required quality and operational requirements. Factors contributing to the development of these types of effects may include, but not necessarily be limited to issues such as labor problems (strikes etc.), weather delays (excessive rain, snow, ice, wind etc.), acts of God (floods, hurricanes, tornadoes etc.), material shortages (shortage of supply or shipping delays), political problems (riots, war, revolution etc.), and financial issues such as reductions in funding or currency devaluation on an overseas project (Arkell, 1980; Selsdon and Palmer, 1988). These types of issues are not unique to underground construction. Of a more critical nature to underground construction are the subsurface ground and ground water conditions: i.e. have they been defined adequately, are they expected to vary, have the construction equipment, procedures, and sequences been properly selected for the expected conditions, are the construction schedule and budget sufficient to account for the anticipated conditions and any reasonable variations?

Proposed Risk Model

As noted previously, it is not the intent of this discussion to address risk models that require rigorous evaluations by geostatistics and probability (Giles, 1994; Cole et. al., 1994). Rather, the intent is to present simplified techniques that can be used by non specialized professional construction engineers to identify appropriate project risks, and to quantify their potential project impact on a preliminary and relative basis so that those that appear to be the most serious can be addressed in more detail by specialized staff as required.

In a simplified fashion as noted by Roylett (1995), Barnes (1995), Haddad and Mullins (1995), and Nicholson et al. (1995), risk can be considered as being a function of both **frequency** (probability or frequency with which any particular hazard will manifest itself), and **consequence** (the impact or outcome of the hazard manifestation). Thus for any particular hazard identified on a particular project (regardless of the nature of the hazard), a risk assessment can be made using the methods suggested below. Although not discussed herein, once a key risk is identified (based upon either frequency or consequence or both), it can then be evaluated further to address methods to eliminate, reduce, or otherwise manage the risk. This step should be done in the planning and design stages, so that by the time the project gets to construction, the key issues have already been addressed and hopefully minimized accordingly. **Risk (R)**, can therefore be considered to be the product of **frequency (F)** and **consequence (C)**, or **R = F x C**. For each specific risk identified for a project, regardless of the source of the risk or the stage of the project, such an evaluation can and should be made as a minimum effort in risk evaluation.

Risk Assessment

In evaluating the likelihood of occurrence, several categorization methods have been suggested by other authors, for example Nicholson et al. (1995), or Cole et al. (1994). The categorization with the widest range of possibilities offers a more definitive, yet subjective, evaluation, and might imply a false level of precision in the evaluation. It should be noted however, as pointed out by Rohrmann (1995), that risk judgments are often prone to biases, with probabilities for rare but prominent hazards tending to be overestimated, and probabilities for commonplace hazards underestimated. These tendencies must be considered when assigning the numerical values in making this type of evaluation. In addition, as noted by Haddad and Mullins (1995), data used to assess probability should relate, as far as possible, directly to the process under study. For purposes of this discussion, a categorization as modified from Nicholson et. al. (1996) and Burchell (1998) is suggested as a basic guideline, including numerical ratings as noted below.

1. Improbable (not considered probable, extremely unlikely)
2. Remote (expected to occur once every 1000 projects, unlikely to occur)
3. Occasional (expected to occur once every 100 projects, likely to occur)
4. Probable (expected to occur once every 10 projects, likely multiple occurrences)
5. Frequent (expected to occur during every project, frequent occurrences likely)

Similarly, the consequences of any particular risk can be categorized either on a generic relative numerical scale, or related to money, health, or project impact as modified from Nicholson et. al.(1966) and Burchell (1998). As noted by Kampmann et. al. (1998), the significance of any particular risk is often related to the owner's or project's risk acceptance criteria. For illustrative purposes in this discussion, a generic relative scale is offered for consideration as follows.

1. Insignificant (no anticipated delays, or significant injuries or costs)
2. Minor (potential for serious injury, delay (hrs) and extra costs can be accommodated)
3. Significant or serious (potential for fatality, delay (< 1 day) costs significant)
4. Major (single fatality, delay (< 1 week), possible re-design, substantial cost impacts)
5. Catastrophic (multiple fatalities, delay of weeks or months, project close-down possible, litigation likely, cost impacts excessive)

With these risk factors now having been identified and assigned numerical values, they can then be plotted on an X-Y graph to develop a "picture" of the results of the risk model not only to illustrate which are the critical risks, but also to show the relative distribution of all risks considered. A typical risk plot is shown conceptually in Figure 1, as modeled after Nicholson et al. (1996). Abbott (1998) proposed a similar type of presentation, but only in a qualitative manner. The plot of numerical values as suggested herein offers a quantitative (but admittedly subjective) assessment of project risks for underground construction, as a "first estimate" to identify the most serious risks requiring further evaluation by more sophisticated techniques as suggested by Haddad and Mullins (1995). A very similar system is being used for risk assessment on underground contracts in Singapore for the Metro on the North East Line Project, except that the numerical scales are reversed; i.e. increased numerical value was indicative of lower consequences or probability, instead of higher as suggested herein, and either will work equally well (Burchell, 1998).

As noted in Figure 1, the identified risks can then be grouped in terms of their seriousness, with the proposed categories of "tolerable", "as low as reasonably possible", and "intolerable", as suggested by Nicholson et. al. (1996). Furthermore, if the more serious risks are located on the lower half of the graph toward the "consequences" axis, they might be considered to be more serious than if the probability is high, but the consequences low, even though the overall risk may still be categorized as "intolerable". It might even be considered, or advantageous, to make a risk assessment in several steps, i.e. make an initial assessment, and then if some risks appear to be unacceptable, then apply mitigation measures to the planning, design, or construction procedures as appropriate, and then repeat the process to see how much the perceived risk has been reduced in probability. It should also be noted that regardless of the effectiveness of an applied mitigation measure, the consequence of a hazard would not change unless the project boundary conditions were changed also.

These categories of both frequency and consequences are normally assigned by owners, designers, or contractors based upon their experience on similar projects and/or in similar ground and/or ground water conditions. This categorization can be taken one step further as required, by relating it to life, property or money independently as

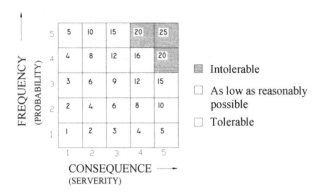

Figure 1: Conceptual Plot - Risk Frequency vs. Consequence

suggested by Cole et. al. (1994). It should also be recognized as noted by South (1985), that the risks as seen by a contractor may differ significantly from the self-same risks as seen by an owner. In addition, a specific contractor may be perceived as a risk by an owner, but this can normally be resolved by a bidder pre-qualification process. The converse is also possible, but a contractor then has the option either not to bid or to put in a higher price. Kampmann and Summers (1998) even report on the use of a risk assessment process used prior to and during contract negotiations to choose a successful contractor for the Copenhagen Metro System.

In addition, the detailed assessment of a specific risk as perceived by a private owner may be different than the assessment of the same risk by a governmental agency. It should also be remembered that that risks may vary continually throughout the life of a project and risk perceptions may change as the project progresses. Above all, risk allocation is part of the continual process of communication, and poor communication is usually caused by the involvement of many different disciplines, lack of attention to detail, insufficient planning, and fear. Last but not least, the attitudes and experience of the legal system under whose jurisdiction the project will be accomplished can also have a definite impact upon the perceptions of a particular risk.

The recommendations of the USNCTT (1974), Kuesel (1979), and the ITA (1988) for better contracting practices and the principles of allocation and the sharing of risk in underground construction should have helped to alleviate or distribute some of the perceived risks. Varangaokar and Dharap (1997) have noted some of the major "hazards" of "one-sided" contracts with little or no provision for risk sharing, including lack of provision for different excavation rates in different ground conditions, lack of provisions for variable payment depending upon the amount of ground water encountered, penalties for delays when the excavation equipment is provided by the owner, technicalities in the specifications that may preclude payment for materials actually used on site, failure of owner to provide contractual support - allowing a time extension but not extra payment, and adoption of new taxes and/or regulations after contract award.

However, it is evident that many owners are reluctant to adopt the "risk sharing" attitude. There seems to exist in some cases the perception that if the contractor has all the risk and accepts this condition when preparing his bid, then a "fixed price" contract can be negotiated with the potential for fewer claims, thereby supposedly "saving" money in the long term. This concept is somewhat of a fallacy however, especially on larger value contracts, since the contractor will almost always put a higher contingency in his bid if he knows that he has all the risk, resulting in a higher contract price than if the owner shared the risk. Then if the uncertainties never materialize into actual problems, he makes a handsome additional profit at the owner's expense, but by the owner's own choice.

Although the above discussion was intended to present a "simple" risk analysis model, it may be desirable, or at times necessary to modify the technique slightly in order to better represent particular conditions where a certain risk may be perceived as much more important than other risks. In such a case, the "weighting" technique as outlined by Cancelli and Crosta (1994) might be adopted. Also, it may occasionally be of interest to evaluate a "what if" scenario in which the originally assumed project boundary conditions are varied in order to perform a sensitivity analysis in terms of risk as suggested by Conway (1994).

Geotechnical Factors

As noted above, geotechnical factors are often a major contributor to risks in underground construction. It is vital to the "success" of a project to evaluate risk in the planning, design stages of a project, as a function of the project complexity, geological complexity, and the extent of the subsurface investigation, as outlined by Whyte and Tonks (1994). In this fashion, if project or geological complexity indicates the potential for excessive (more than is considered acceptable) risk if the subsurface ground and/or ground water conditions are not more thoroughly defined, then the risk mitigation process might include expanding the subsurface geotechnical investigation to better define or clarify the "uncertainties", thereby reducing the level of risk. This should be strongly considered by the owner and the designer, and maybe even by the contractor, depending upon the contractual conditions. Some geotechnical factors which commonly contribute to problems (and risk) in tunneling are noted below, but this is intended only as a general guideline, not as a comprehensive listing.

Soft Ground Tunnels:

- Unanticipated presence and/or quantity of ground water;

- Adverse or aggressive ground water chemistry;
- Toxic or explosive gases;
- Unexpected variations in soil gradation, stiffness, strength, density etc.;
- Unanticipated silt lenses, e.g. "varved" clay;
- Unanticipated presence, size or frequency of cobbles and boulders;
- Presence of "buried" objects, piling, boats, well casings, old utilities etc.;
- Preferential weaknesses such as slickensides;
- Liquefaction potential;
- Presence of sensitive clays;
- Presence of unweathered blocks or protrusions of bedrock.

Rock Tunnels:

- Presence of unanticipated fault or shear zones;
- Presence of hard inclusions (e.g. chert, ironstone etc.) or intrusive dikes;
- Buried valleys with soil like conditions, and possibly unexpected ground water;
- Excessive or unanticipated abrasive mineralogy;
- Unanticipated excessive weathering or residual soil development;
- Toxic or explosive gases;
- Adverse or aggressive ground water chemistry;
- Adversely oriented discontinuities;
- Squeezing or swelling ground conditions;
- Change in orientation, frequency, or condition of discontinuities;
- Presence of "blocky" ground and/or zones of crushed material;
- Presence of unanticipated karstic features.
- Unanticipated presence and/or quantity of ground water;

Even if the above types of geotechnical features are considered, the perceived level of risk may vary, depending upon the extent of experience of the owner, designer, and/or contractor on similar projects and in similar ground and ground water conditions. As an example, a contractor with extensive experience in the glacial tills of the Great Lakes area in the US would likely have a different perspective than a contractor with no local experience in the area, or a contractor with extensive hard rock tunnel experience may not know how to effectively and efficiently drive a tunnel in saturated sediments using a slurry or EPB tunnel boring machine.

If these types of geotechnical risks are identified and assessed early in the initial stages of project development it is less likely that they will cause problems later. If the owner, designer, and contractor can function as a "team" with the common goal to succeed, rather than to be continually in adversarial positions, then the odds of success are greatly improved. The full development and disclosure of subsurface conditions and identification of "uncertainties" will contribute to the proper selection of appropriate construction methods and sequences, thereby reducing the risk.

It may not always be possible to easily expand the subsurface geotechnical investigation. As an example, on a metro or sewer tunnel project in a major city, site access and depth limitations for additional borings are usually much less restrictive than for a railroad tunnel in rugged mountainous terrain. In such a case, it is often easier and more critical to do a detailed study of any available geological maps, supplemented by a comprehensive field mapping program of the project area, and projecting this to tunnel depth and interpreted along with the few available borings. Similarly, the geotechnical data base in a large urban area with extensive previous underground experience could be expected to be much larger and better defined than for a rural project with little or no local related subsurface investigation or underground construction history.

It should also be noted that the type, extent and reliability of the geotechnical information presented in the contract documents, as well as the method of presentation, and their relationship to other contractual documents (specifications clauses regarding construction claims, unforeseen ground conditions, differing site conditions etc.) can influence the risk assessment perspective.

The development and use of the Geotechnical Design Summary Report (GDSR) in underground construction in the US was intended to improve contractual relations, and to reduce and/or share the risk (Bohlke,1996). However, despite twenty or more years of use, numerous "improvements", and several changes (Brierly, 1998) in name and presentation format (Geotechnical Data Report - GDR, Geotechnical Interpretative Report - GIR, Geotechnical Baseline Report - GBR), contract litigation in the underground construction industry is still extensive, as can be verified by our lawyer "friends" working in the tunneling industry. In recognition of the relative ineffectiveness of these documents at times, the Underground Technology Research Council (1996) held a forum to present the perspectives of owners, designers and contractors on the subject of how to present geotechnical data in contract documents, and the opinions presented in that forum should be strongly considered by all parties working in the tunnel construction industry. Perhaps those in the US tunneling industry should reassess the benefits of the GDSR (and its sister reports) - are they working to serve their intended purpose? Has their purpose changed? Are there lessons to be learned from international tunnel construction contracting practices where the GDSR is not used very often if at all? These types of considerations have recently been discussed independently by Hatem (1998), and by Samuels (1998).

Regardless of how the risk sharing is distributed between the contractor and the owner, what type or types of geotechnical documentation were included in the contract documents, the level of detail in the documentation, and the specific details of the specifications regarding "differing site conditions", it is still recommended that a

Disputes Review Board (DRB) be a contractual provision. The contractual aspects of resolving geotechnical related disputes have been addressed in detail by Thomas et. al. (1992a, 1992b) for conditions in which the contract documents do not and do contain a "differing site conditions" clause respectively. This has been recommended by the USNCTT (1974), ITA (1988), and UTRC (1991) as a beneficial step toward better contracting practices. Experience in the US in the last ten years or so has demonstrated that many contractual disputes in tunnel contracts can be resolved in this manner rather than going to the long and costly step of litigation where the only "winners" are the attorneys. The effectiveness of the DRB approach to reducing litigation has recently been discussed by Smith (1994) and by Mathews et. al. (1995). However, only time will tell if this trend increases or even if it continues.

Another contractual provision to be considered would be a requirement for the bidders to include a risk assessment as part of their tender package. This would force them to evaluate the risks in preparing their bid, and would allow the owners to see how the potential contractors perceive the project in terms of risk. It may also be enlightening in developing a better project approach to risk sharing as discussed above.

Other Contributing Factors

As noted previously, there are many factors contributing to risk in underground construction besides those of a geotechnical nature as noted above. Although it is beyond the scope and intent of this discussion to address these "other issues" in detail, Casey (1979), Abramson (1998), Neff (1998), Anderson (1998), and Abbott (1998) have presented fairly comprehensive listings, a few of which are noted below (as well as a few added by the author) as a reminder or checklist.

- Environmental regulations regarding disposal of dewatering discharge and excavation muck;
- Environmental restrictions regarding noise, dust, vibrations etc.;
- Transportation load and clearance limits, route and/or time restrictions;
- Local, state, and federal regulation restrictions and limitations (including safety and health);
- Complications and inefficiencies from local union regulations;
- Availability, qualifications, and experience of local labor resources;
- Risk of damage to adjacent properties (or utilities) from construction work; acceptable settlements and/or distortions, baseline measurements of initial conditions, measurement methods, and associated third party "risks";
- Construction delays (weather, labor, permits, material supply, other uncertainties etc.) and associated liquidated damages penalties;
- Scope expansion or creep by owner resulting in exceeding budget and/or schedule;
- Nature of waterproofing details; constructability, leakage criteria and warranties;

- Nature of details for construction and/or expansion joints; constructability, leakage criteria, compatibility with interior finish requirements;
- Nature, constructability, and suitability of temporary ground support requirements;
- Inappropriate selection of construction equipment, methods, and sequences to accommodate the anticipated ground and ground water conditions and risks;
- Breakdown in communications or sharing of critical project information;
- Inappropriate, misleading or misunderstood contractual arrangements;
- The potential for "multi-causal" events with a complex interaction of contributing factors;
- Underground fires;
- Underground floods, natural or caused by construction;
- Major equipment breakdowns or malfunction.

In addition, from a contractor's perspective, Desai and Kurup (1997) have provided the following list of issues that are frequently known to cause construction delays and increase contract costs, both major sources of risk as noted above. While these were noted specifically for an "overseas" environment, they are valid anywhere, but perhaps to varying degrees, depending upon the knowledge and experience of the owner, time and budget available for preparation of contract documents, and the specific details of the project.

- Inadequate pre-bid geotechnical investigation of ground and/or ground water conditions;
- Lack of proper planning and inadequate or incomplete bid package;
- Unrealistic cost and schedule estimates from the initial stages;
- Inadequate infrastructure facilities available at the job site;
- Unrealistic or unfair contractual conditions leading to major contractual disputes ;
- Lack of motivation and commitment on the part of the owner (insufficient or ineffective decision making process);
- Pressure to produce the low bid for contract award (without a pre-qualification process to "normalize" the bidders as equally qualified);
- Lack of the contract documents to allow for the latest in construction technology.

Of further note, is that perceptions of risk for a domestic underground construction contract, may vary significantly if the same project were to be built in similar ground conditions in Europe or in a developing country. As an example, Gould (1992) has outlined some of the legal concerns for contracting in Europe VS in the USA, and even notes differences between different European countries. These differences may decrease somewhat with the implementation of the European Common Market. However, in developing countries, the differences from US practice may be dramatic.

Example Case

The author had the opportunity to use the suggested risk assessment method on a major soft ground bored tunnel crossing beneath the Nile River for Line 2 of the Metro in Cairo, Egypt. The tunnel was excavated 9.4m in diameter through saturated sands included a single permanent pre-cast concrete segmented tunnel lining. Several categories of risks were considered, including design risks, construction risks, and long term risks during the service period associated with outside influences both inside the tunnel and external to the tunnel (details are given in Richards et. al., 1998).

Design risks considered included tunnel structural integrity, buoyancy, corrosion, seismic (shaking), seismic (liquefaction), leakage, geologic data reliability, river bed protection design, and overall structural stability. Construction risks considered included major mechanical failure, loss of face control, segment gasket failure, unexpected geological variations, tail seal leakage, articulation joint leakage, cutter breakage, scour of the river bed over the tunnel, mechanical damage to the completed tunnel lining, unintentional over excavation, and inadequate placement of tail void grout. Long term risks internal to the tunnel included water leakage, fire, train accidents, sabotage, excessive corrosion, and loss of structural stability. Long term risks external to the tunnel over the design life included unanticipated construction activity over the tunnel, unanticipated scour of the river bed above the tunnel, inadequate or poor long term monitoring of the tunnel performance and condition, inadequate or poor quality remedial measures, river traffic accidents over or near the tunnel, exceptional earthquake exceeding design levels, exceptional river flows exceeding design levels, degradation of river bed protection over the tunnel, and some unpredictable combination of the above events occurring simultaneously.

After the first round of risk identification and assessment, major apparent risks were evaluated further to develop methods to reduce the level of risk, and then the risk assessment repeated to reflect the "refinements". The resulting risk plot (frequency vs. consequences) is shown in Figure 2. One of the most interesting risks identified, and that which required the most amount of effort to mitigate, was the long term potential for excessive scour of the river bed over the top of the tunnel during "flood" conditions. As a result of the risk studies, the owner undertook extensive hydraulic studies of the river in the Metro crossing area in order to determine if supplemental river bed protection measures were required to reduce the long term risk to the Metro tunnel.

Summary

Underground construction always has been and always will be a "risky business". However, the risks associated with tunneling can be identified and evaluated using the

GEO-ENGINEERING FOR UNDERGROUND FACILITIES   1197

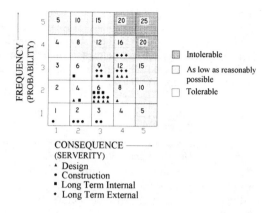

Fig. 2: Risk Frequency vs. Consequence for Nile Crossing

simplified techniques suggested. For those risks assessed to be serious, more sophisticated risk models based upon statistics and probability can be used to quantify the risk in order to develop appropriate risk reduction measures. Since geotechnical conditions often have a major influence upon risks in tunneling, these have been addressed specifically, including critical factors introduce risk and that should be systematically considered and evaluated since they frequently cause problems during construction. In addition, discussions have been included regarding the presentation of the results of the subsurface geotechnical investigation as it relates to risks in design and construction of underground facilities.

Conclusions

Risk is an unavoidable and inevitable aspect of the underground construction process. However, the risks can usually be minimized, but never totally eliminated, by proper risk management. This should be assessed independently by all concerned parties, including the owner, designer, and contractor. As noted by Anderson, 1998b, project risk assessment should be initiated in the planning process and continue through design and construction. During the process of planning, design, and construction, the risks should be periodically re-evaluated, as knowledge and experience improve and perceptions of risk change. In this fashion, both "probability of occurrence" and "magnitude of the consequences" can be managed in an effort to reduce risk to an acceptable level for all parties involved.

The following suggestions are offered as further ideas for consideration in the ongoing efforts to share risk between the Contractor and the Owner, and hopefully to revitalize the tunneling industry in the US as a reputable professional endeavor led by **Engineers**, not by unions and attorneys. There is no doubt that attorneys play a useful role in the contracting process, but this author questions if it should be a dominant role. Some of these ideas are already in use but sometimes forgotten, some are probably new to and possibly radically different from current practices in the industry, and some are possibly controversial.

- Concentrate on professionalism in the industry

    - Allow sound engineering principles to govern designs and critical decisions, rather than fear of legal actions;
    - Maintain high ethical and moral standards, even during dispute negotiation;
    - Maintain teamwork approach to resolve key issues - problems must be properly and thoroughly understood before an appropriate engineering solution can be developed;
    - Plan profit margin based upon high quality and quantity of workmanship, not upon the number of attorneys on the staff.

- Modify the Contractor selection and bidding/contract award process

    - Pre-qualify and form a short list as with engineering services;
    - Allow the selected Contractor to do his own limited site investigation to develop further information on key geotechnical issues from his perspective;
    - Require the selected Contractor to perform his own risk assessment for the project, and compare it to the risk assessment by the Owner/Designer;
    - Develop the final contractual Geotechnical Baseline Report as a team effort, with input from and mutual agreement by the Contractor, considering his own investigation and risk assessment;
    - Require the selected Contractor to develop his final bid price after completion of the above steps;
    - **Negotiate** the final contract price after the Contractor has completed the above tasks, as is done with engineering services.

- Modify the contract execution process

    - Always (not optional) include a provision for a Disputes Review Board;
    - Always include a Value Engineering program, with sharing of cost savings;
    - Systematically require the submittal of detailed method statements (for review and comment, not necessarily for approval) for critical or sensitive tasks such as excavation methods and sequences, TBM specifications before it is ordered, ground treatment methods and sequences, and ground water control methods and sequences;

- Modify disputes resolution process
    - Allow innovation in the construction process, and work together to make it work, rather than fear legal action if it is not perfect upon initial application in the field.
    - Provide contractual mechanism to share cost where the actual site conditions are demonstrated to be worse than those agreed upon in the GBR;
    - Provide contractual mechanism to share savings where the actual site conditions are demonstrated to be better than those agreed upon in the GBR;
    - Include unit rates in the contract for items such as different ground support methods, dewatering as a function of quantity, stand-by time, etc., so that either costs or savings can be easily identified;
    - Involve the DRB only when the contractual parties cannot agree upon the impact of the differing site condition (either better or worse) upon the cost adjustment (either downward or upward respectively).

The above noted types of considerations should improve the contractual working relationships in underground construction, eliminate frivolous claims, and allow the Owners and Contractors to benefit from one another's experience rather than to be adversaries. In the scenario proposed above, failure by an owner to make use of the selected Contractor's experience would be the loss of a valuable resource. If we as engineers in the tunneling industry can achieve these goals, then we perhaps still have a chance to control the industry ourselves, rather than to continue to have it controlled by our attorney colleagues.

Acknowledgments

The author is grateful to his professional colleagues; Dennis Lachel, Eric Shanholtz, Tony Burchell, Dr. Jim Monsees, and Dr. Gary Brierly, for their interest, their willingness to take the time to review the early draft versions of this paper, and their constructive criticism, which hopefully allowed the author to make improvements. The author also expresses his thanks to the National Authority for Tunnels, Ministry of Transport, and Parsons Brinckerhoff International, both of Cairo, Egypt, for permission to use the Nile crossing case history data.

References

Abbott, E. L., 1998, "Preparation of Contract Documents for Subsurface Projects," in *Subsurface Conditions: Risk Management for Design and Construction Professionals*, D. J. Hatem, ed., pp. 95-128, John Wiley and Sons, New York.

Abramson, L.W., 1998, "Root Causes of Disputes on Tunnel Construction Contracts", *Proc. North American Tunneling '98*, A.A. Balkema, Rotterdam.

Alhaby, N. M. H. and I. L. Whyte, 1994, "The Impact of Ground Risks in Construction on Project Finance", in *Risk and Reliability in Ground Engineering*, pp. 54-67, ICE, London.

Anderson, J. M., 1998, "An 'ERASE' Approach to Eliminate and Control Risks in Tunneling Before Work Starts," *Proceedings Conference Tunnels and Metropolises*, Vol. 1, pp. 135-139, Sao Paulo, Brazil, A. A. Balkema, Rotterdam.

Anderson, J., 1998b, "Minimizing Underground Construction Risks Requires Maximum Engineering Effort", Tunnelling and Underground Space Technology, Vol. 13, No. 4, pp. 365-368, Pergamon, Oxford.

Arkell. J., 1980, *Overseas Contracts: Conception to Completion*, "Paper 8: Finance, ECDG, Tax and Law", Proceedings Conference, ICE, London, pp. 79-84.

Barnes, P. H., 1995, "Conflicting Notions of Risk: The Chasm of Conflict Between Institutional Regulators and the Public," *Proceedings Conference on Integrated Risk Assessment*, pp. 15-21, Newcastle, N. S. W., Australia, A. A. Balkema, Rotterdam.

Bohlke, B., ed., 1996, *Geotechnical Reports in Underground Construction*, ASCE, Underground Technology Research Council, Washington, D. C.

Brierly, G. S. and B. Cavan, 1987, "The Risks Associated with Tunneling Projects", *Tunneling Technology Newsletter No. 58*, U. S. National Committee on Tunneling Technology.

Brierly, G. S., 1988, "Discussion on 'The Risks Associated with Tunneling Projects' ", *Tunneling Technology Newsletter No. 64*, U. S. National Committee on Tunneling Technology.

Brierly, G. S., 1998, "Subsurface Investigations and Geotechnical Report Preparation," in *Subsurface Conditions: Risk Management for Design and Construction Professionals*, D. J. Hatem, ed., pp. 49-94, John Wiley and Sons, New York.

Burchell, A.J., 1998, Personal communication to the author.

Cancelli, A. and G. Crosta, 1994,"Hazard and Risk Assessment in Rockfall Prone Areas", in *Risk and Reliability in Ground Engineering*, pp.177-190, ICE, London.

Casey, J.J., 1979, "Identification and Nature of Risks in Construction Projects - A Contractor's Perspective", *Proc. RETC*, Vol. 2, pp. 1705-1712, AIME, Littleton.

Cole, K. W., S. T. Jarvis, and A.J. Turner, 1994, "To Treat or Not to Treat Abandoned Mine Workings: Towards Achieving a Dialog Over Risk and Reliability", in *Risk and Reliability in Ground Engineering*, pp.1-28. ICE, London.

Conway, J. J., 1994, "Application of Risk Management in Tunnel Feasibility Studies", in *Risk and Reliability in Ground Engineering*, pp.227-240, ICE, London.

Desai, S. K. and T. S. M. Kurup, 1997, "Contract Management for Accelerated Completion of Underground Works in Hydro-Electric Projects - Contractors' Perspectives," *Proceedings Intl. Conference Tunnelling Asia '97*, pp. 383 - 390, A. A. Balkema, Rotterdam.

Erwin, P. and S. Sinha, 1996, "Coordinating Risk Management with Project Management in OCIP Projects", *Proc. North American Tunneling '96*, A.A. Balkema, Rotterdam.

Giles, D. 1994, "Geostatistical Interpolation of Techniques for Geotechnical Data Modeling and Ground Condition Risk and Reliability Assessment," in *Risk and Reliability in Ground Engineering*, pp. 202-214, ICE, London.

Gould, R.N.H., 1992, "Contract Negotiations: Managing Risks of Foreign Contracts - Tender Rules and Regional Differences, The European Perspective", *Proc. Seminar on International Construction Law - Opportunities and Risks in the '90s*, American Bar Association, Chicago.

Haddad, S., and D. P. Mullins, 1995, "The Role of Risk Classification and Prioritization in an Integrated Assessment Regime," *Proceedings Conference on Integrated Risk Assessment*, pp. 23-35, Newcastle, N. S. W., Australia, A. A. Balkema, Rotterdam.

Hatem, D.J., 1998, Geotechnical Baselines; Professional Liability Implications", Tunnelling and Underground Space Technology, Vol. 13, No. 2, pp.143-150, Pergamon, Oxford.

Huse, J. A., 1992, "Control of Tunnelling Contracts", Tunnels and Tunnelling, September, British Tunnelling Society.

Huse, J. A., 1993, "Preparation and Presentation of Construction Claims", Tunnels and Tunnelling, January, British Tunnelling Society.

International Tunneling Association (ITA), 1988, "Recommendations on Contractual Sharing of Risk", Tunnelling and Underground Space Technology, Vol. 3, No. 2.

Kampmann, J., S. D. Eskesen, and J. W. Summers, 1998, "Risk Assessment Helps Select the Contractor for the Copenhagen Metro System," *Proceedings Conference Tunnels and Metropolises*, Vol. 1, pp. 123-128, Sao Paulo, Brazil, A. A. Balkema, Rotterdam.

Kuesel, T.R., 1979, "Allocation of Risks", *Proc. RETC*, Vol. 2, pp. 1713-1724, AIME, Littleton.

Mathews, A.A., R.M. Matyas, P.E. Sperry, and R.J. Smith, 1995, "Disputes Review Boards - Current Status and Recommendations", *Proc. RETC*, pp. 157-184, SME, Littleton.

Molak, V., 1997, *Fundamentals of Risk Analysis and Risk Management*, Lewis Publishers.

Neff, T. L., 1998, "Risk Management Considerations for Complex Subsurface Projects," in *Subsurface Conditions: Risk Management for Design and Construction Professionals*, D. J. Hatem, ed., pp. 129-161, John Wiley and Sons, New York.

Nicholson, D., C-M. Tse, and C. Penny,1996, "Risk Management", Section 3.11 in The Observational Method in Ground Engineering: Principles and Applications, Draft Funders Report CIRIA Research Project RP 527, London.

Richards, D. P., P. Ramond, and O. Ezzeldine, 1998, " Design and Construction of the First Bored Tunnel Across the Nile River", *Proceedings Conference Tunnels and Metropolises*, Vol. 1, pp. 173-179, Sao Paulo, Brazil, A. A. Balkema, Rotterdam.

Rohrmann, B., 1995, "Technological Risks - Perception, Evaluation, and Communication," *Proceedings Conference on Integrated Risk Assessment*, pp. 7-13, Newcastle, N. S. W., Australia A. A. Balkema, Rotterdam.

Roylett, B. G., 1995, "Risk Financing in Planning, Design and Operation of Major Infrastructure," *Proceedings Seminar on Acceptable Risks for Extreme Events in The Planning and Design of Major Infrastructure*, pp.61-69, Sydney, N. S. W., Australia, A. A. Balkema, Rotterdam.

Samuels, R., 1998, "Geotechnical Baselines - Part of the Problem, or a Solution", *Proc. North American Tunneling '98*, A.A. Balkema, Rotterdam.

Selsdon, L. and R. H. Palmer, 1988, *Overseas Projects*: Crucial Problems, "Paper 6: Financial Problems - A Bankers View", Proceedings Conference, pp.45-48, ICE, London.

Smith, R.J., 1994, "Disputes Review Boards for Tunneling Projects Using the ASCE Model", Tunnelling and Underground Space Technology, Vol. 9, No. 1, pp. 37-45, Pergamon, Oxford.

South, A. J., 1985, *Management of International Construction Projects*, "Paper 3: Risk Analysis and Allocation", Proceedings Conference, pp. 15-21, ICE, London.

Syme, G. J., 1995, "Community Acceptance of Risk: Trust, Reliability and Consent," *Proceedings Seminar on Acceptable Risks for Extreme Events in The Planning and Design of Major Infrastructure*, pp. 31-39, Sydney, N. S. W., Australia, A. A. Balkema, Rotterdam.

Thomas, H . R., G. R. Smith and R. M. Ponderlick, 1992a, "Resolving Contract Disputes Based on Misrepresentations", ASCE Journal of Construction Engineering, Vol. 118, No. 3, September.

Thomas, H . R., G. R. Smith and R. M. Ponderlick, 1992b, "Resolving Contract Disputes Based on Differing-Site-Condition Clause", ASCE Journal of Construction Engineering, Vol. 118, No. 4, December.

Thompson, P. A. and J. G. Perry (eds.), 1992, *Engineering Construction Risks*, Thomas Telford, London.

Underground Technology Research Council,1991, Avoiding and Resolving Disputes During Construction - Successful Practices and Guidelines, ASCE, New York.

Underground Technology Research Council,1996, Opinion Papers on Geotechnical Reports in Underground Construction, Forum held in Washington, D. C., April.

U.S. National Committee on Tunneling Technology (USNCTT), 1974, "Better Contracting for Underground Construction," prepared by USNCTT Standing Subcommittee No. 4, Contracting Practices, National Research Council, National Academy of Science - National Academy of Engineering, Washington, D. C.

Varangaonkar, R. D. and V. M. Dharap, 1997, "Tunnelling Contracts With Controversial and One-sided Conditions," *Proceedings Intl. Conference Tunnelling Asia '97*,pp. 391 - 397, A. A. Balkema, Rotterdam.

Whyte, I. L., and D. M. Tonks, 1994, "Project Risks and Site Investigation Strategy", in *Risk and Reliability in Ground Engineering*, pp.100-111, ICE, London.

Subject Index

Page number refers to the first page of paper

Airport terminals, 853
Anchors, 200
Aquifers, 647
Arching, 495
Arterial highways, 570
Asia, 1048
Australia, 553
Axial loads, 417

Barrier design, 140
Barriers, 1027
Bays, 377, 706, 1130
Bending moments, 417
Bentonite, 121
Blasting, 24
Bolts, 526
Boreholes, 471
Boring, 255, 622, 1091, 1130
Boston, 570
Boulders, 255, 720
Braced excavation, 393
Bridge construction, 327
Building codes, 1176
Buildings, 121, 1102, 1176
Buildings, office, 212
Buried pipes, 483

California, 82, 121
Case reports, 24, 162, 188, 231, 405, 483, 507, 634, 695, 936, 1048, 1060, 1185
Caverns, 647, 780, 797, 810, 821
Cements, 1048
Centrifuge, 495
Chemical grouting, 950, 969
Classification, 188, 1007
Clays, 231, 866, 1070

Coal mines, 339, 353, 365
Cofferdams, 853
Combined sewers, 768
Compaction, 1039
Compaction grouting, 353, 671
Compressed air, 610
Compressibility, 41
Concrete, 900, 1116
Concrete construction, 622
Concrete deterioration, 912
Constitutive models, 231, 1080
Construction, 821, 853
Construction methods, 756
Contaminants, 105, 121, 140, 684
Contamination, 995, 1027
Contracts, 582, 912, 1164, 1176
Control, 82, 105, 377, 957
Core walls, 121
Cost control, 1164
Cost minimization, 1143
Cracks, 925
Creep, 315
Crude oil, 810
Culverts, 243
Curved profiles, 526

Dams, 936
Data collection, 271
Deep foundations, 24, 377
Deep soil mixing, 1007, 1048
Deformation, 231, 365, 377, 393, 514, 659, 900
Delaminating, 925
Densification, 671
Design, 66, 162, 176, 282, 327, 553, 570, 582, 594, 756, 821, 853, 912, 1027

Design criteria, 442, 634, 995, 1116
Diaphragm wall, 900
Differential settlement, 1048
Displacement, 339, 1080
Drainage, 866, 995
Drilling, 93, 936, 1016

Earth pressure, 405, 417
Earth reinforcement, 162
Earth structures, 327
Earthquake engineering, 483
Embankments, 327, 900
Environmental factors, 442
Europe, 841
Excavation, 41, 162, 176, 200, 231, 255, 303, 377, 471, 507, 514, 957, 1016, 1048, 1060, 1070, 1102, 1176

Failures, 471
Field tests, 507, 879
Finite element method, 243, 514, 526, 538, 659
Flooding, 315
Floods, 768
Foundations, 327
Fractures, 459
Frozen soils, 176, 957
Fuzzy sets, 282

Gasoline, 1060
Geological surveys, 255, 282
Geology, 442, 459
Geophysical surveys, 16, 271, 294, 353, 684
Geosynthetics, 841
Geotechnical engineering, 1, 16, 29, 93, 706, 1007, 1130, 1143, 1155, 1164, 1185
Geotextiles, 841
Gravity sewers, 866

Ground motion, 29, 162, 393, 514, 1070
Ground support equipment, 829
Ground water, 66, 176, 553, 582, 647, 746, 957
Ground-water flow, 82
Ground-water pollution, 140, 797, 995
Grouting, 82, 188, 647, 936, 950, 1007

H beams, 405
Hawaii, 594
Heaving, 1048
Highways, 553, 594
History, 24
Hydraulic conductivity, 140, 768
Hydraulic fracturing, 634
Hydraulic performance, 66
Hydraulic structures, 995
Hydraulics, 1027
Hydroelectric powerplants, 459, 538, 821, 841, 1116
Hydrogen sulfide, 429

Imaging techniques, 294
In situ tests, 140
Inflatable structures, 243
Inspection, 66, 912, 925
Instrumentation, 176, 243, 365, 594, 888, 900, 1080
Intercepting sewers, 768

Jacking, 570, 706, 731
Japan, 900
Jet grouting, 1016, 1027

Karst, 684, 695

Laboratory tests, 140
Lateral forces, 339
Lateral stress, 1039

Leakage, 66, 442, 780, 797, 950, 1060, 1116
Legal factors, 1155
Liability, 1155, 1176
Light rail transit, 429, 582, 983
Limestone, 671, 684
Liquefaction, 1039
Loads, 731
Logging, 93

Machinery, 622, 1091, 1130
Mapping, 271
Market research, 1007
Measurement, 303
Membranes, 634, 841
Methane, 429, 983
Microtunneling, 53, 706, 720, 731, 746, 1016
Migration, 105, 684
Mines, 315, 339, 538
Mining, 271, 303, 780, 810
Mixing, 1007
Models, 731, 1070, 1091
Monitoring, 162, 176, 294, 339, 365, 879
Mud, 706

Nevada, 829
New York City, 24, 1176
Nitrogen, 810
Numerical analysis, 538, 1102
Numerical models, 231, 514

Outfall sewers, 1130
Overburden, 365
Overflow, 53

Parking facilities, 212
Partnering, 983, 1143
Performance, 594, 829, 853, 866, 888
Performance evaluation, 659

Permeability, 82
Petroleum, 1027
Piers, 1039, 1060
Piezometers, 1080
Pile foundations, 377
Pile tests, 188
Piles, 188, 417, 853, 888, 1102
Pipe jacking, 720, 879
Pipe laying, 756
Pipelines, 720
Planning, 756, 1091
Plasticity, 231
Polyurethane, 950, 969
Polyvinyl chloride, 841
Pore pressure, 866
Pore water, 1102
Predictions, 731, 879, 1080
Preloading, 393
Pressure tunnels, 66, 442, 459, 634, 1116
Prestressing, 200
Probability, 282
Professional practice, 1155
Pumped storage, 538

Quality control, 936

Radioactive waste disposal, 829
Rapid transit railways, 610
Rehabilitation, 912, 925
Reinforcement, 671, 1039
Repairing, 327, 912, 925, 950
Reports, 1143
Retaining walls, 405
Retention, 888
Risk allocation, 1164
Risk analysis, 797, 1185
Risk management, 1143
Rock excavation, 1091
Rock joints, 459, 526
Rock masses, 936
Rock properties, 365, 471

Rock strength, 66, 303
Rocks, 829
Roofs, 507

Salt domes, 780, 797, 810
Salt removal, 315
San Francisco, 377, 706
Sand, 417, 746, 969
Sandstone, 553
Sea walls, 53
Sealants, 969
Sealing, 969
Seepage, 647, 957
Seismic design, 429
Seismic response, 659
Seismic studies, 684
Seismic surveys, 303
Seismic tests, 294
Settlement analysis, 1
Sewage, 879
Sewer pipes, 53
Sewers, 29, 53, 720, 746, 756
Shafts, 634, 706, 1116, 1130
Shafts, excavations, 353
Shale, 821
Shear strength, 969
Shear stress, 483
Shoring, 200, 1102
Shotcrete, 200, 243, 925
Singapore, 29
Sinkholes, 315, 327, 671, 684, 695, 810
Site evaluation, 271
Site investigation, 93, 695
Slurries, 720
Slurry walls, 105, 176, 900
Soft soils, 41, 231, 610, 706, 1080
Soil components, 1016
Soil deformation, 121
Soil mechanics, 495
Soil nailing, 212
Soil pressure, 243

Soil properties, 121
Soil sampling, 93
Soil strength, 1039
Soils, 294, 514
Soil-structure interaction, 243, 294, 483
Solutions, 780
Solvents, 983
Stability, 459
Stability analysis, 471
Stabilization, 353, 671, 1016
Standards, 1155
State-of-the-art reviews, 1, 1007
Stiffness, 162, 365
Storage facilities, 780
Storage tanks, 1060
Stress analysis, 483, 526
Stress relaxation, 829, 1070
Struts, 393, 888
Subsidence, 315, 327, 339, 353, 365, 797
Subsurface investigations, 16, 93, 255, 353, 594, 695, 746, 1164
Subway tunnels, 1
Subways, 1, 24, 29, 82, 1102
Supports, 162, 176, 200, 393, 507, 1176
Surface waves, 303

Taiwan, 610
Technology, 16
Temperature effects, 829
Temporary structures, 1176
Tennessee, 684
Test equipment, 731
Test procedures, 140
Tests, 495
Texas, 746
Thailand, 1070
Theories, 1, 24
Tieback restraint systems, 41, 405, 417, 888

Toe aprons, 417
Transportation, 16
Trenchless technology, 93
Trusses, 507
Tunnel construction, 514, 553, 570, 582, 594, 888, 1080, 1091, 1185
Tunnel linings, 429, 634, 659, 925, 1116
Tunnel supports, 526
Tunneling, 16, 53, 255, 271, 429, 495, 610, 622, 1185
Tunnels, 29, 243, 294, 634, 647, 746, 768, 841, 866, 879, 912, 950, 969, 983, 1039, 1070, 1130

Uncertainty analysis, 282
Underground construction, 1, 16, 41, 255, 303, 471, 507, 622, 756, 1143, 1164, 1185
Underground storage, 780, 797, 810, 1060

Underground structures, 212, 282, 429, 483, 495, 538, 582, 647, 821, 957, 983, 995
Underpinning, 188, 1016
United Kingdom, 105
United States, 1, 105
Universities, 995
Urban areas, 695, 756
Utilities, 1060

Ventilation, 983
Vertical loads, 495

Walls, 41, 140, 200, 405, 1027
Washington, 53
Wastewater disposal, 29, 768, 879
Wastewater treatment, 53
Water supply, 442, 1080
Wave propagation, 659
Wisconsin, 212

Yield, 315

Author Index

Page number refers to the first page of paper

Abd Al-Jalil, Yousof, 1091
Abd El-Lateef, H. A., 879
Abdrabbo, Fathi M., 879
Abramson, Lee, 53
Abramson, Lee W., 594, 912
Abu-Farsakh, Murad Y., 514
Adams, Daniel N., 553
Allouche, Erez N., 93
Alvarez, Tirso A., 459
Anderson, Thomas C., 405
Angulo, Mauricio, 255
Ariaratnam, Samuel T., 93, 756
Armour, Tom A., P.E., 176
Asada, Hideyuki, 1048
Ashcom, D. W., 1027
Aubertin, Michel, 471

Bahner, Eric W., 212
Baxter, Diane Y., 121
Bay, J. A., 303
Bazán-Arias, Cathy, 243
Bennett, David, 720, 731
Bentler, David J., 121
Berry, Robert S. J., 1164
Bidhendi, H., 393
Biggar, Kevin W., 93
Boirum, Ralph, 53
Boone, S. J., 393, 1102
Boscardin, Marco D., 912
Boyce, Glenn M., 594
Branco, P., 1102
Bruce, D. A., 188
Bruce, Donald A., 936, 1007
Bruce, M. E. C., 188
Bruce, Mary Ellen C., 1007
Budiman, Jeffry, 1080
Buehler, Frank S., P.E., 768

Bushell, Ted D., 888
Byrne, R. John, P.E., 200

Carpenter, Philip J., 684
Carr, Bradley J., 684
Choi, Hangseok, 140
Ciancia, Andrew J., 1176
Clementino, Renato V., 93
Cogan, John, 829
Cook, Roy F., 82
Cording, E. J., 417
Cording, Edward J., 459, 731, 866
Corwin, Arthur B., P.E., 957
Corwin, Bruce J., 720
Cotton, David M., P.E., 200
Culbert, Jon, 810

Daniel, David E., 140
Davidson, Richard R., 121
Davis, C. A., 483
Dawson, Andrew R., 105
Descour, Jozef M., 271
Di Bella, Carmela, 797
Dickson, Peter A., 821
Divito, Randall C., 327
Doll, William E., 684

Eduardo, Joseph A., 671
Einstein, Herbert H., 495
Elioff, Amanda, P.E., 429
Elliott, Dean A., 671
El-Nahhas, F. M., 879
Eskilsson, Jeppe N., 442
Espinosa-Silva, Augusto, 41
Evans, Jeffrey C., 105

Fatemi, M. Javad, 1048
Fernandez, Americo L., 294
Fernández, Gabriel G., 459
Feroz, Michael, P.E., 768
Filz, George M., 121
Flanagan, Richard, 29
Fox, Nathaniel S., 1039

Gabr, M. A., 969
Gause, Chris, 936
Gehle, Richard M., 780
Girtz, Gordon J., 995
Grabinsky, M., 393
Gruner, Lawrence B., 1016
Guarascio, Massimo, 797
Gutiérrez, Ramiro, 634, 1116

Handy, Richard L., 1039
Hansmire, William H., 594
Hanson, David H., 271
Haramy, Khamis Y., 271
Hasen, Michael, P.E., 746
Hatem, David J., 1164
Hedien, Jason E., 647
Henderson, Carl, 243
Henn, Raymond W., 983
Hjertberg, Svante, 1130
Hoffmann, Arthur G., P.E., 353
Hood, Charles, 582
Hunt, Steven W., P.E., 255
Hussin, James D., 671
Hutchison, Bruce J., 553
Hwang, Richard, 610

Iglesia, Geraldo R., 495

Jansen, Stephen T., 995
Ju, Daniel H., 610

Kahl, Robert A., 706
Kaneshiro, Jon Y., 1130
Kaufmann, Ronald D., 684

Kawamura, Nelson, 866
Koelling, M., 1027
Kousoftas, Demetrious C., 231
Koutsoftas, Demetrious C., 377
Kriekemans, Bert P., 950
Kurrus, J., 1027

Lam, Ignatius P., 659
Laughton, Christopher, 1091
Law, Hubert K., 659
Layng, Karen P., 1155
Lee, N. J., 303
Lemley, Jack K., 16
Lewis, Karl H., 243
Li, Li, 471
Liang, R., 282
Linn, James K., 810
Liou, Donald D., 900
Long, J. H., 417
Louhenapessy, W. G., 526
Luccioni, Laurent X., 231
Lueke, Jason S., 756
Lukas, Robert G., 162
Luke, B. A., 303
Lundin, Tracy K., 582

MacDonald, D., 1102
Mackey, Jennifer R., 695
Maggi, Michael A., 720
Mahar, James W., 339
Mainville, Peter A., 570
Marino, Gennaro G., P.E., 339
Marulanda, Alberto, 634, 1116
Mason, D. J., III, P.E., 1164
Mathy, David C., 706
Maynard, Theodore, 1080
McCarthy, Daniel W., 888
McCluskey, Elaine, 888
McRae, Michael T., 553
Meier, Len, 353
Meiorin, Luciano, 1130
Merritt, Andrew H., 66

Meschino, M., 1102
Mimura, Clayton S., 594
Moh, Za-Chieh, 610
Monsees, James E., P.E., 429
Mueller, C. G., 417

Naudts, Alex, 936
Navin, Stephen J., 1130
Nawari, N. O., 282
Neil, David M., 271
Nelson, P. P., 303
Nelson, Priscilla, 1091
Nielson, Dru R., 706
Nolting, Richard M., III, 829
Nyquist, Jon E., 695
Nyquist, Jonathan E., 684

O'Connor, Daniel P., 622
O'Connor, Kevin M., 365
O'Neill, Kevin J., 1176

Parola, Jerry F., 162
Peck, Ralph B., 1
Pestana, Juan M., 231
Porbaha, Ali, 1048
Porter, Stephen J., 553
Powderham, Alan, 853
Powderham, Alan J., 570

Rice, Philip M., 570
Richards, Donald P., 1185
Richters, Gunars, 582
Rippe, Arlan H., 768
Robertson, Peter K., 93
Rogstad, David, 983
Roth, Lawrence H., P.E., 1143
Roth, Mary J. S., 695
Russell, Henry A., 925

Samuels, Reuben, 24
Santamarina, J. Carlos, 294
Schmall, Paul C., 957

Schmidt, Birger, 29
Scuero, Alberto M., 841
Sharp, Julie, 1060
Siekmeier, John A., 365
Simon, Richard, 471
Sirvinskis, Joe, 212
Splitter, Ted, 1060
Staheli, Kimberlie, 720
Stokoe, K. H., II, 303
Strychowskyj, Peter, 756
Sturman, John, 1060
Sun, Yiming, 829
Supot, Teachavorasinskun, 1070

Taylor, Stephen, 570
Tee, Tan Boon, 29
Thapa, Bhaskar B., 821
Thoms, Robert L., 780
Traylor, R. P., 188
Tumay, Mehmet T., 514
Tuozzolo, Thomas J., 957

Uddin, Nasim, P.E., 538

Valentine, Maureen S., P.E., 327
Van Sambeek, Leo L., 315
Vaschetti, Gabriella, 841

Wagener, Bruce D., 995
Walker, Stanley E., P.E., 327
Wannipa, Tiew, 1070
Warren, Stuart, 82
Watson, Michael B., 720
Weatherby, David E., 405
Weisberg, Justin L., 1155
Westland, J., 393
Westland, John, 1102
White, Dean M., 1027
Whitman, Robert V., 495
Wilkes, John A., 841
Wissman, Kord J., 1039
Wolschlag, Chris J., P.E., 200

Young, Dae S., 507

Zapata, Santiago F., 1080
Zhu, Fusheng, 507